Chemotherapeutics and Disease Control

Chemotherapeutics and Disease Control

ENCYCLOPEDIA REPRINT SERIES

Editor: Mary Howe-Grant

A WILEY-INTERSCIENCE PUBLICATION

JOHN WILEY & SONS

NEW YORK • CHICHESTER • BRISBANE • TORONTO • SINGAPORE

This text is printed on acid-free paper.

Library of Congress Catalog Number: 93-2966
ISBN 0-471-59633-7

Printed in the United States of America

10 9 8 7 6 5 4 3 2 1

CONTENTS

EDITORIAL STAFF

Executive Editor: **Jacqueline I. Kroschwitz**
Editor: **Mary Howe-Grant**
Assistant Editor: **Cathleen A. Treacy**
Editorial Supervisor: **Lindy J. Humphreys**
Copy Editor: **Linda M. Kim**
Indexing Assistants: **Juliet A. Nelson**
Monica Stipanov

CONTRIBUTORS

Frederick J. Antosz, *The Upjohn Company, Kalamazoo, Michigan,* Ansamacrolides (under Antibiotics)

Brian Bannister, *The Upjohn Company, Kalamazoo, Michigan,* Lincosaminides (under Antibiotics)

John Becher, *CDC Drug Service, Atlanta, Georgia,* Anthelmintics (under Antiparasitic agents)

Peter R. Bernstein, *ICI Americas Corporation, Wilmington, Delaware,* Antiasthmatic agents

Daniel P. Bonner, *Bristol-Myers Squibb Pharmaceutical Research Institute, Princeton, New Jersey,* Monobactams (under Antibiotics)

Donald Borders, *American Cyanamid Company, Pearl River, New York,* Survey (under Antibiotics)

Geert Cauwenbergh, *Janssen Research Foundation, Beerse, Belgium,* Antimycotics (under Antiparasitic agents)

B. Cavalleri, *Merrell-Dow Research Institute, Gerenzano, Italy,* Glycopeptides (under Antibiotics)

Peter Cervoni, *Lederle Laboratories, Pearl River, New York,* Cardiovascular agents

Peter S. Chan, *Lederle Laboratories, Pearl River, New York,* Cardiovascular agents

Ian Chopra, *American Cyanamid Company, Pearl River, New York,* Tetracyclines (under Antibiotics)

David L. Crandall, *Lederle Laboratories, Pearl River, New York,* Cardiovascular agents

Louise W. Crandall, *Lilly Research Laboratories, Indianapolis, Indiana,* Polyethers (under Antibiotics)

Terrence Doyle, *Bristol-Myers Squibb Company, Wallingford, Connecticut,* Chemotherapeutics, anticancer

Frank H. Ebetino, *Norwich Eaton Pharmaceuticals Corporation, Norwich, New York,* Nitrofurans (under Antibacterial agents)

George A. Ellestad, *American Cyanamid Company, Pearl River, New York,* Tetracyclines (under Antibiotics)

Michael H. Fisher, *Merck, Sharp, and Dohme Research Laboratories, Rahway, New Jersey,* Avermectins (under Antiparasitic agents)

William O. Foye, *Massachusetts College of Pharmacy and Allied Health Sciences, Boston, Massachusetts,* Sulfonamides (under Antibacterial agents)

Ashit K. Ganguly, *Schering-Plough Corporation, Bloomfield, New Jersey,* Oligosaccharides (under Antibiotics)

Roque Edward Garcia, *Technicon Instruments Corporation, Tarrytown, New York,* Hematology (under Automated instrumentation)

Paul Gherson, *Miles, Inc., Tarrytown, New York,* Clinical chemistry (under Automated instrumentation)

V.M. Girijavallabhan, *Schering-Plough Corporation, Bloomfield New Jersey,* Oligosaccharides (under Antibiotics)

Charles M. Hall, *The Upjohn Company, Kalamazoo, Michigan,* Analgesics, antipyretics, and antiinflammatory agents

Robert L. Hamill, *Lily Research Laboratories, Indianapolis, Indiana,* Polyethers (under Antibiotics)

Joseph J. Hlavka, *American Cyanamid Company, Pearl River, New York,* Tetracyclines (under Antibiotics)

Subhash Karkare, *Amgen, Inc., Thousand Oaks, California,* Cell culture technology

Herbert Kirst, *Eli Lily and Company, Indianapolis, Indiana,* Macrolides (under Antibiotics)

Daniel L. Klayman, *Walter Reed Army Institute of Research, Washington, D.C.,* Antiprotozoals (under Antiparasitic agents)

William H. Koster, *Bristol-Myers Squibb Pharmaceutical Research Institute, Princeton, New Jersey,* Monobactams (under Antibiotics)

Howard Lanza, *Miles Inc., Tarrytown, New York,* Clinical chemistry (under Automated instrumentation)

Klaus R. Linder, *Bristol-Myers Squibb Pharmaceutical Research Institute, Princeton, New Jersey,* Monobactams (under Antibiotics)

George R. Lenz, *The BOC Group Technical Center, Murray Hill, New Jersey,* Anesthetics

George Y. Lesher, *Sterling-Winthrop Research Institute, Rensselaer, New York,* Quinolones (under Antibacterial agents)

Hubert Maehr, *Hoffmann-LaRoche, Inc., Nutley, New Jersey,* Elfamycins (under Antibiotics)

Donald McGregor, *Bristol-Myers Squibb Company, Wallingford, Connecticut,* Aminoglycosides (under Antibiotics)

George H. Miller, *Schering-Plough Research, Bloomfield, New Jersey,* Chloramphenicol and analogues (under Antibiotics)

Helmut Mrozik, *Merck, Sharp, and Dohme Research Laboratories, Rahway, New Jersey,* Avermectins (under Antiparasitic agents)

Tattanahalli Nagabhushan, *Schering-Plough Research, Bloomfield, New Jersey,* Chloramphenicols and analogues (under Antibiotics)

Richard A. Nugent, *The Upjohn Company, Kalamazoo, Michigan,* Analgesics, antipyretics, and antiinflammatory agents

Neal F. Osborne, *SmithKline Beecham Pharmaceuticals, Surrey, United Kingdom,* Carbapenems and penems (under Antibiotics, β-Lactams)

Milton Pelavin, *Miles, Inc., Tarrytown, New York,* Clinical chemistry (under Automated instrumentation)

William Pietro, *York University, North York, Ontario, Canada,* Biosenors

R.J. Ponsford, *SmithKline Beecham Pharmaceuticals, Surrey, United Kingdom,* Penicillins and others (under Antibiotics, β-Lactams)

Nancy L. Reichenback, *Temple University School of Medicine, Philadelphia, Pennsylvania,* Nucleosides (under Antibiotics)

Michael Reuman, *Sterling-Winthrop Research Institute, Rensselaer, New York,* Quinolones (under Antibacterial agents)

Ganapathi R. Revankar, *Triplex Pharmaceutical Corporation, The Woodlands, Texas,* Antiviral agents

John Roberts, *Hoffmann-La Roche, Inc., Nutley, New Jersey,* Cephalosporins (under Antibiotics, β-Lactams)

Roland K. Robins, *ICN Nucleic Acid Research Institute, Costa Mesa, California,* Antiviral agents

Hollis G. Schoepke, *Anaquest, Murray Hill, New Jersey,* Anesthetics

Thomas J. Schwan, *Norwich Eaton Pharmaceuticals Corporation, Norwich, New York,* Nitrofurans (under Antibacterial agents)

Robert Southgate, *SmithKline Beecham Pharmaceuticals, Surrey, United Kingdom,* Carbapenems and penems (under Antibiotics, β-Lactams)

Theodore C. Spaulding, *Anaquest, Murray Hill, New Jersey,* Anesthetics

J.G. Stam, *Pfizer Central Research, Groton, Connecticut,* β-Lactamase inhibitors (under Antibiotics, β-Lactams)

Robert J. Suhadolnik, *Temple University School of Medicine, Philadelphia, Pennsylvania,* Nucleosides (under Antibiotics)

Jefferson W. Tilley, *Hoffmann-LaRoche Inc., Nutley, New Jersey,* Antiobesity drugs

John Turrell, *Technicon Instruments Corporation, Tarrytown, New York,* Hematology (under Automated instrumentation)

Kanwal J. Varma, *Schering-Plough Research, Bloomfield, New Jersey,* Chloramphenicol and analogues (under Antibiotics)

Dan Vlastelica, *Miles, Inc., Tarrytown, New York,* Clinical chemistry (under Automated instrumentation)

Dolatrai Vyas, *Bristol-Myers Squibb Company, Wallingford, Connecticut,* Chemotherapeutics, Anticancer

Timothy Ward, *Millsap College, Jackson, Mississippi,* Analytical techniques (under Biopolymers)

Edmund M. Wise Jr., *The Wellcome Research Laboratories, Research Triangle Park, North Carolina,* Peptides (under Antibiotics)

David Zelmanovic, *Technicon Instruments Corporation, Tarrytown, New York,* Hematology (under Automated instrumentation)

PREFACE

This volume is one of a series of carefully selected reprints from the world-renowed *Kirk-Othmer Encyclopedia of Chemical Technology* designed to provide specific audiences with articles grouped by a central theme. Although the 27-volume Kirk-Othmer Encyclopedia is widely available, many readers and users of this key reference tool have expressed interest in having selected articles in their specialty collected for handy desk reference or teaching purposes. In response to this need, we have chosen original, complete articles related to chemotherapeutics and disease control to make up this new volume. The full texts, tables, figures, and reference materials from the original work have been reproduced here unchanged. All articles are by industrial or academic experts in the field of pharmaceutical and medicinal applications of chemical technology and the final work represents the result of careful review by competent specialists and the thorough editorial processing of the professional Wiley staff. Introductory information from the Encyclopedia concerning nomenclature, SI units and conversion factors, and related information has been provided as a further guide to the contents.

This volume includes comprehensive coverage of the chemical agents and techniques used in the control and detection of the principal disease classes, eg asthma, cancer, cardiovascular disease, and diseases of bacterial, parasitic, and viral origin. Articles on bioanalytical techniques and cell culture technology may also be found herein. Chemical Abstracts Service (CAS) Registry Numbers and chemical structures of the chemotherapeutic agents, alphabetical organization, extensive cross-referencing, and a complete index further enhance the utility of this Encyclopedia. The 37 articles in this Encyclopedia, prepared by 62 leading authorities from pharmaceutical houses, universities, and research institutes, vary in length from six to 93 pages depending on the commercial importance and current scientific interest in the topic. The contents should be of interest to all those engaged in the design, synthesis, analysis, and manufacture of chemotherapeutic agents as well as to those specifically engaged in the application of these materials. The book should be an important research reference tool, desk-top information resource, and supplementary reading asset for teaching professionals and their students.

M. Howe-Grant

NOTE ON CHEMICAL ABSTRACTS SERVICE REGISTRY NUMBERS AND NOMENCLATURE

Chemical Abstracts Service (CAS) Registry Numbers are unique numerical identifiers assigned to substances recorded in the CAS Registry System. They appear in brackets in the *Chemical Abstracts* (CA) substance and formula indexes following the names of compounds. A single compound may have synonyms in the chemical literature. A simple compound like phenethylamine can be named β-phenylethylamine or, as in *Chemical Abstracts*, benzeneethanamine. The usefulness of the *Encyclopedia* depends on accessibility through the most common correct name of a substance. Because of this diversity in nomenclature careful attention has been given to the problem in order to assist the reader as much as possible, especially in locating the systematic CA index name by means of the Registry Number. For this purpose, the reader may refer to the CAS Registry Handbook—Number Section which lists in numerical order the Registry Number with the *Chemical Abstracts* index name and the molecular formula; eg, **458-88-8**, Piperidine, 2-propyl-, (*S*)-, $C_8H_{17}N$; in the *Encyclopedia* this compound would be found under its common name, coniine [*458-88-8*]. Alternatively, this information can be retrieved electronically from CAS Online. In many cases molecular formulas have also been provided in the *Encyclopedia* text to facilitate electronic searching. The Registry Number is a valuable link for the reader in retrieving additional published information on substances and also as a point of access for on-line data bases.

In all cases, the CAS Registry Numbers have been given for title compounds in articles and for all compounds in the index. All specific substances indexed in *Chemical Abstracts* since 1965 are included in the CAS Registry System as are a large number of substances derived from a variety of reference works. The CAS Registry System identifies a substance on the basis of an unambiguous computer-language description of its molecular structure including stereochemical detail. The Registry Number is a machine-checkable number (like a Social Security number) assigned in sequential order to each substance as it enters the registry system. The value of the number lies in the fact that it is a concise and unique means of substance identification, which is independent of, and therefore bridges, many systems of chemical nomenclature. For polymers, one Registry Number may be used for the entire family; eg, polyoxyethylene (20) sorbitan monolaurate has the same number as all of its polyoxyethylene homologues.

Cross-references are inserted in the index for many common names and for some systematic names. Trademark names appear in the index. Names that are incorrect, misleading, or ambiguous are avoided. Formulas are given very frequently in the text to help in identifying compounds. The spelling and form used, even for industrial names, follow American chemical usage, but not always the usage of *Chemical Abstracts* (eg, *coniine* is used instead of *(S)-2-propylpiperidine*, *aniline* instead of *benzenamine*, and *acrylic acid* instead of *2-propenoic acid*).

There are variations in representation of rings in different disciplines. The dye industry does not designate aromaticity or double bonds in rings. All double bonds and aromaticity are shown in the *Encyclopedia* as a matter of course. For example, tetralin has an aromatic ring and a saturated ring and its structure

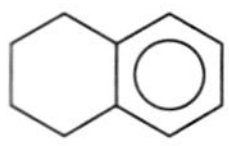

appears in the *Encyclopedia* with its common name, Registry Number enclosed in brackets, and parenthetical CA index name, ie, tetralin [*119-64-2*] (1,2,3,4-tetrahydronaphthalene). With names and structural formulas, and especially with CAS Registry Numbers, the aim is to help the reader have a concise means of substance identification.

CONVERSION FACTORS, ABBREVIATIONS, AND UNIT SYMBOLS

SI Units (Adopted 1960)

The International System of Units (abbreviated SI), is being implemented throughout the world. This measurement system is a modernized version of the MKSA (meter, kilogram, second, ampere) system, and its details are published and controlled by an international treaty organization (The International Bureau of Weights and Measures) (1).

SI units are divided into three classes:

BASE UNITS

length	meter[†] (m)
mass	kilogram (kg)
time	second (s)
electric current	ampere (A)
thermodynamic temperature[‡]	kelvin (K)
amount of substance	mole (mol)
luminous intensity	candela (cd)

SUPPLEMENTARY UNITS

plane angle	radian (rad)
solid angle	steradian (sr)

[†]The spellings “metre” and “litre” are preferred by ASTM; however, “-er” is used in the *Encyclopedia*.

[‡]Wide use is made of Celsius temperature (t) defined by

$$t = T - T_0$$

where T is the thermodynamic temperature, expressed in kelvin, and $T_0 = 273.15$ K by definition. A temperature interval may be expressed in degrees Celsius as well as in kelvin.

DERIVED UNITS AND OTHER ACCEPTABLE UNITS

These units are formed by combining base units, supplementary units, and other derived units (2–4). Those derived units having special names and symbols are marked with an asterisk in the list below.

Quantity	Unit	Symbol	Acceptable equivalent
*absorbed dose	gray	Gy	J/kg
acceleration	meter per second squared	m/s^2	
*activity (of a radionuclide)	becquerel	Bq	1/s
area	square kilometer	km^2	
	square hectometer	hm^2	ha (hectare)
	square meter	m^2	
concentration (of amount of substance)	mole per cubic meter	mol/m^3	
current density	ampere per square meter	$A//m^2$	
density, mass density	kilogram per cubic meter	kg/m^3	g/L; mg/cm^3
dipole moment (quantity)	coulomb meter	C·m	
*dose equivalent	sievert	Sv	J/kg
*electric capacitance	farad	F	C/V
*electric charge, quantity of electricity	coulomb	C	A·s
electric charge density	coulomb per cubic meter	C/m^3	
*electric conductance	siemens	S	A/V
electric field strength	volt per meter	V/m	
electric flux density	coulomb per square meter	C/m^2	
*electric potential, potential difference, electromotive force	volt	V	W/A
*electric resistance	ohm	Ω	V/A
*energy, work, quantity of heat	megajoule	MJ	
	kilojoule	kJ	
	joule	J	N·m
	electronvolt†	eV†	
	kilowatt-hour†	kW·h†	
energy density	joule per cubic meter	J/m^3	
*force	kilonewton	kN	
	newton	N	$kg·m/s^2$

†This non-SI unit is recognized by the CIPM as having to be retained because of practical importance or use in specialized fields (1).

Quantity	Unit	Symbol	Acceptable equivalent
*frequency	megahertz	MHz	
	hertz	Hz	1/s
heat capacity, entropy	joule per kelvin	J/K	
heat capacity (specific), specific entropy	joule per kilogram kelvin	J/(kg·K)	
heat transfer coefficient	watt per square meter kelvin	$W/(m^2{\cdot}K)$	
*illuminance	lux	lx	lm/m^2
*inductance	henry	H	Wb/A
linear density	kilogram per meter	kg/m	
luminance	candela per square meter	cd/m^2	
*luminous flux	lumen	lm	cd·sr
magnetic field strength	ampere per meter	A/m	
*magnetic flux	weber	Wb	V·s
*magnetic flux density	tesla	T	Wb/m^2
molar energy	joule per mole	J/mol	
molar entropy, molar heat capacity	joule per mole kelvin	J/(mol·K)	
moment of force, torque	newton meter	N·m	
momentum	kilogram meter per second	kg·m/s	
permeability	henry per meter	H/m	
permittivity	farad per meter	F/m	
*power, heat flow rate, radiant flux	kilowatt	kW	
	watt	W	J/s
power density, heat flux density, irradiance	watt per square meter	W/m^2	
*pressure, stress	megapascal	MPa	
	kilopascal	kPa	
	pascal	Pa	N/m^2
sound level	decibel	dB	
specific energy	joule per kilogram	J/kg	
specific volume	cubic meter per kilogram	m^3/kg	
surface tension	newton per meter	N/m	
thermal conductivity	watt per meter kelvin	W/(m·K)	
velocity	meter per second	m/s	
	kilometer per hour	km/h	
viscosity, dynamic	pascal second	Pa·s	
	millipascal second	mPa·s	
viscosity, kinematic	square meter per second	m^2/s	
	square millimeter per second	mm^2/s	

Quantity	Unit	Symbol	Acceptable equivalent
volume	cubic meter	m^3	
	cubic decimeter	dm^3	L (liter) (5)
	cubic centimeter	cm^3	mL
wave number	1 per meter	m^{-1}	
	1 per centimeter	cm^{-1}	

In addition, there are 16 prefixes used to indicate order of magnitude, as follows:

Multiplication factor	Prefix	Symbol	Note
10^{18}	exa	E	
10^{15}	peta	P	
10^{12}	tera	T	
10^{9}	giga	G	
10^{6}	mega	M	
10^{3}	kilo	k	
10^{2}	hecto	h[a]	[a]Although hecto, deka, deci, and centi are SI prefixes, their use should be avoided except for SI unit-multiples for area and volume and nontechnical use of centimeter, as for body and clothing measurement.
10	deka	da[a]	
10^{-1}	deci	d[a]	
10^{-2}	centi	c[a]	
10^{-3}	milli	m	
10^{-6}	micro	μ	
10^{-9}	nano	n	
10^{-12}	pico	p	
10^{-15}	femto	f	
10^{-18}	atto	a	

For a complete description of SI and its use the reader is referred to ASTM E 380 (4) and the article UNITS AND CONVERSION FACTORS which appears in Vol. 24.

A representative list of conversion factors from non-SI to SI units is presented herewith. Factors are given to four significant figures. Exact relationships are followed by a dagger. A more complete list is given in the latest editions of ASTM E 380 (4) and ANSI Z210.1 (6).

Conversion Factors to SI Units

To convert from	To	Multiply by
acre	square meter (m^2)	4.047×10^{3}
angstrom	meter (m)	1.0×10^{-10}†
are	square meter (m^2)	1.0×10^{2}†
astronomical unit	meter (m)	1.496×10^{11}

†Exact.

To convert from	To	Multiply by
atmosphere, standard	pascal (Pa)	1.013×10^{5}
bar	pascal (Pa)	$1.0 \times 10^{5\dagger}$
barn	square meter (m^2)	$1.0 \times 10^{-28\dagger}$
barrel (42 U.S. liquid gallons)	cubic meter (m^3)	0.1590
Bohr magneton (μ_B)	J/T	9.274×10^{-24}
Btu (International Table)	joule (J)	1.055×10^{3}
Btu (mean)	joule (J)	1.056×10^{3}
Btu (thermochemical)	joule (J)	1.054×10^{3}
bushel	cubic meter (m^3)	3.524×10^{-2}
calorie (International Table)	joule (J)	4.187
calorie (mean)	joule (J)	4.190
calorie (thermochemical)	joule (J)	$4.184^{\dagger}$
centipoise	pascal second (Pa·s)	$1.0 \times 10^{-3\dagger}$
centistokes	square millimeter per second (mm^2/s)	$1.0^{\dagger}$
cfm (cubic foot per minute)	cubic meter per second (m^3/s)	4.72×10^{-4}
cubic inch	cubic meter (m^3)	1.639×10^{-5}
cubic foot	cubic meter (m^3)	2.832×10^{-2}
cubic yard	cubic meter (m^3)	0.7646
curie	becquerel (Bq)	$3.70 \times 10^{10\dagger}$
debye	coulomb meter (C·m)	3.336×10^{-30}
degree (angle)	radian (rad)	1.745×10^{-2}
denier (international)	kilogram per meter (kg/m)	1.111×10^{-7}
	$tex^{\ddagger}$	0.1111
dram (apothecaries')	kilogram (kg)	3.888×10^{-3}
dram (avoirdupois)	kilogram (kg)	1.772×10^{-3}
dram (U.S. fluid)	cubic meter (m^3)	3.697×10^{-6}
dyne	newton (N)	$1.0 \times 10^{-5\dagger}$
dyne/cm	newton per meter (N/m)	$1.0 \times 10^{-3\dagger}$
electronvolt	joule (J)	1.602×10^{-19}
erg	joule (J)	$1.0 \times 10^{-7\dagger}$
fathom	meter (m)	1.829
fluid ounce (U.S.)	cubic meter (m^3)	2.957×10^{-5}
foot	meter (m)	$0.3048^{\dagger}$
footcandle	lux (lx)	10.76
furlong	meter (m)	2.012×10^{-2}
gal	meter per second squared (m/s^2)	$1.0 \times 10^{-2\dagger}$
gallon (U.S. dry)	cubic meter (m^3)	4.405×10^{-3}
gallon (U.S. liquid)	cubic meter (m^3)	3.785×10^{-3}
gallon per minute (gpm)	cubic meter per second (m^3/s)	6.309×10^{-5}
	cubic meter per hour (m^3/h)	0.2271

†Exact.
‡See footnote on p. xvi.

To convert from	To	Multiply by
gauss	tesla (T)	1.0×10^{-4}
gilbert	ampere (A)	0.7958
gill (U.S.)	cubic meter (m^3)	1.183×10^{-4}
grade	radian	1.571×10^{-2}
grain	kilogram (kg)	6.480×10^{-5}
gram force per denier	newton per tex (N/tex)	8.826×10^{-2}
hectare	square meter (m^2)	$1.0 \times 10^{4\dagger}$
horsepower (550 ft·lbf/s)	watt (W)	7.457×10^{2}
horespower (boiler)	watt (W)	9.810×10^{3}
horsepower (electric)	watt (W)	$7.46 \times 10^{2\dagger}$
hundredweight (long)	kilogram (kg)	50.80
hundredweight (short)	kilogram (kg)	45.36
inch	meter (m)	$2.54 \times 10^{-2\dagger}$
inch of mercury (32°F)	pascal (Pa)	3.386×10^{3}
inch of water (39.2°F)	pascal (Pa)	2.491×10^{2}
kilogram-force	newton (N)	9.807
kilowatt hour	megajoule (MJ)	$3.6^{\dagger}$
kip	newton(N)	4.448×10^{3}
knot (international)	meter per second (m/S)	0.5144
lambert	candela per square meter (cd/m^3)	3.183×10^{3}
league (British nautical)	meter (m)	5.559×10^{3}
league (statute)	meter (m)	4.828×10^{3}
light year	meter (m)	9.461×10^{15}
liter (for fluids only)	cubic meter (m^3)	$1.0 \times 10^{-3\dagger}$
maxwell	weber (Wb)	$1.0 \times 10^{-8\dagger}$
micron	meter (m)	$1.0 \times 10^{-6\dagger}$
mil	meter (m)	$2.54 \times 10^{-5\dagger}$
mile (statute)	meter (m)	1.609×10^{3}
mile (U.S. nautical)	meter (m)	$1.852 \times 10^{3\dagger}$
mile per hour	meter per second (m/s)	0.4470
millibar	pascal (Pa)	1.0×10^{2}
millimeter of mercury (0°C)	pascal (Pa)	$1.333 \times 10^{2\dagger}$
minute (angular)	radian	2.909×10^{-4}
myriagram	kilogram (kg)	10
myriameter	kilometer (km)	10
oersted	ampere per meter (A/m)	79.58
ounce (avoirdupois)	kilogram (kg)	2.835×10^{-2}
ounce (troy)	kilogram (kg)	3.110×10^{-2}
ounce (U.S. fluid)	cubic meter (m^3)	2.957×10^{-5}
ounce-force	newton (N)	0.2780
peck (U.S.)	cubic meter (m^3)	8.810×10^{-3}
pennyweight	kilogram (kg)	1.555×10^{-3}
pint (U.S. dry)	cubic meter (m^3)	5.506×10^{-4}

†Exact.

To convert from	To	Multiply by
pint (U.S. liquid)	cubic meter (m^3)	4.732×10^{-4}
poise (absolute viscosity)	pascal second (Pa·s)	$0.10^{\dagger}$
pound (avoirdupois)	kilogram (kg)	0.4536
pound (troy)	kilogram (kg)	0.3732
poundal	newton (N)	0.1383
pound-force	newton (N)	4.448
pound force per square inch (psi)	pascal (Pa)	6.895×10^3
quart (U.S. dry)	cubic meter (m^3)	1.101×10^{-3}
quart (U.S. liquid)	cubic meter (m^3)	9.464×10^{-4}
quintal	kilogram (kg)	$1.0 \times 10^{2\dagger}$
rad	gray (Gy)	$1.0 \times 10^{-2\dagger}$
rod	meter (m)	5.029
roentgen	coulomb per kilogram (C/kg)	2.58×10^{-4}
second (angle)	radian (rad)	$4.848 \times 10^{-6\dagger}$
section	square meter (m^2)	2.590×10^6
slug	kilogram (kg)	14.59
spherical candle power	lumen (lm)	12.57
square inch	square meter (m^2)	6.452×10^{-4}
square foot	square meter (m^2)	9.290×10^{-2}
square mile	square meter (m^2)	2.590×10^6
square yard	square meter (m^2)	0.8361
stere	cubic meter (m^3)	$1.0^{\dagger}$
stokes (kinematic viscosity)	square meter per second (m^2/s)	$1.0 \times 10^{-4\dagger}$
tex	kilogram per meter (kg/m)	$1.0 \times 10^{-6\dagger}$
ton (long, 2240 pounds)	kilogram (kg)	1.016×10^3
ton (metric) (tonne)	kilogram (kg)	$1.0 \times 10^{3\dagger}$
ton (short, 2000 pounds)	kilogram (kg)	9.072×10^2
torr	pascal (Pa)	1.333×10^2
unit pole	weber (Wb)	1.257×10^{-7}
yard	meter (m)	$0.9144^{\dagger}$

†Exact.

Abbreviations and Unit Symbols

Following is a list of common abbreviations and unit symbols used in the *Encyclopedia*. In general they agree with those listed in *American National Standard Abbreviations for Use on Drawings and in Text (ANSI Y1.1)* (6) and *American National Standard Letter Symbols for Units in Science and Technology (ANSI Y10)* (6). Also included is a list of acronyms for a number of private and government organizations as well as common industrial solvents, polymers, and other chemicals.

Rules for Writing Unit Symbols (4):

1. Unit symbols are printed in upright letters (roman) regardless of the type style used in the surrounding text.
2. Unit symbols are unaltered in the plural.
3. Unit symbols are not followed by a period except when used at the end of a sentence.
4. Letter unit symbols are generally printed lower-case (for example, cd for candela) unless the unit name has been derived from a proper name, in which case the first letter of the symbol is capitalized (W, Pa). Prefixes and unit symbols retain their prescribed form regardless of the surrounding typography.
5. In the complete expression for a quantity, a space should be left between the numerical value and the unit symbol. For example, write 2.37 lm, *not* 2.37lm, and 35 mm, *not* 35mm. When the quantity is used in an adjectival sense, a hyphen is often used, for example, 35-mm film. *Exception:* No space is left between the numerical value and the symbols for degree, minute, and second of plane angle, degree Celsius, and the percent sign.
6. No space is used between the prefix and unit symbol (for example, kg).
7. Symbols, not abbreviations, should be used for units. For example, use "A," not "amp," for ampere.
8. When multiplying unit symbols, use a raised dot:

   N·m for newton meter

   In the case of W·h, the dot may be omitted, thus:

   Wh

   An exception to this practice is made for computer printouts, automatic typewriter work, etc, where the raised dot is not possible, and a dot on the line may be used.
9. When dividing unit symbols, use one of the following forms:

   $$\mathrm{m/s} \quad \textit{or} \quad \mathrm{m{\cdot}s^{-1}} \quad \textit{or} \quad \frac{\mathrm{m}}{\mathrm{s}}$$

   In no case should more than one slash be used in the same expression unless parentheses are inserted to avoid ambiguity. For example: write:

   $$\mathrm{J/(mol{\cdot}K)} \quad \textit{or} \quad \mathrm{J{\cdot}mol^{-1}\cdot K^{-1}} \quad \textit{or} \quad \mathrm{(J/mol)/K}$$

   but *not*

   $$\mathrm{J/mol/K}$$

10. Do not mix symbols and unit names in the same expression. Write:

joules per kilogram *or* J/kg *or* $J \cdot kg^{-1}$

but *not*

joules/kilogram *nor* joules/kg *nor* $joules \cdot kg^{-1}$

ABBREVIATIONS AND UNITS

A	ampere
A	anion (eg, HA)
A	mass number
a	atto (prefix for 10^{-18})
AATCC	American Association of Textile Chemists and Colorists
ABS	acrylonitrile–butadiene–styrene
abs	absolute
ac	alternating current, *n.*
a-c	alternating current, *adj.*
ac-	alicyclic
acac	acetylacetonate
ACGIH	American Conference of Governmental Industrial Hygienists
ACS	American Chemical Society
AGA	American Gas Association
Ah	ampere hour
AIChE	American Institute of Chemical Engineers
AIME	American Institute of Mining, Metallurgical, and Petroleum Engineers
AIP	American Institute of Physics
AISI	American Iron and Steel Institute
alc	alcohol(ic)
Alk	alkyl
alk	alkaline (not alkali)
amt	amount
amu	atomic mass unit
ANSI	American National Standards Institute
AO	atomic orbital
AOAC	Association of Official Analytical Chemists
AOCS	Americal Oil Chemists' Society
APHA	American Public Health Association
API	American Petroleum Institute
aq	aqueous
Ar	aryl
ar-	aromatic
as-	asymmetric(al)
ASHRAE	American Society of Heating, Refrigerating, and Air Conditioning Engineers
ASM	American Society for Metals
ASME	American Society of Mechanical Engineers
ASTM	American Society for Testing and Materials
at no.	atomic number
at wt	atomic weight
av(g)	average
AWS	American Welding Society
b	bonding orbital
bbl	barrel
bcc	body-centered cubic
BCT	body-centered tetragonal
Bé	Baumé
BET	Brunauer-Emmett-Teller (adsorption equation)
bid	twice daily
Boc	*t*-butyloxycarbonyl
BOD	biochemical (biological) oxygen demand
bp	boiling point
Bq	becquerel

C	coulomb
°C	degree Celsius
C-	denoting attachment to carbon
c	centi (prefix for 10^{-2})
c	critical
ca	circa (approximately)
cd	candela; current density; circular dichroism
CFR	Code of Federal Regulations
cgs	centimeter-gram-second
CI	Color Index
cis-	isomer in which substituted groups are on same side of double bond between C atoms
cl	carload
cm	centimeter
cmil	circular mil
cmpd	compound
CNS	central nervous system
CoA	coenzyme A
COD	chemical oxygen demand
coml	commercial(ly)
cp	chemically pure
cph	close-packed hexagonal
CPSC	Consumer Product Safety Commission
cryst	crystalline
cub	cubic
D	Debye
D-	denoting configurational relationship
d	differential operator
d	day; deci (prefix for 10^{-1})
d-	*dextro*-, dextrorotatory
da	deka (prefix for 10^{1})
dB	decibel
dc	direct current, *n.*
d-c	direct current, *adj.*
dec	decompose
detd	determined
detn	determination
Di	didymium, a mixture of all lanthanons
dia	diameter
dil	dilute
DIN	Deutsche Industrie Normen
dl-; DL-	racemic
DMA	dimethylacetamide
DMF	dimethylformamide
DMG	dimethyl glyoxime
DMSO	dimethyl sulfoxide
DOD	Department of Defense
DOE	Department of Energy
DOT	Department of Transportation
DP	degree of polymerization
dp	dew point
DPH	diamond pyramid hardness
dstl(d)	distill(ed)
dta	differential thermal analysis
(*E*)-	entgegen; opposed
ϵ	dielectric constant (unitless number)
e	electron
ECU	electrochemical unit
ed.	edited, edition, editor
ED	effective dose
EDTA	ethylenediaminetetraacetic acid
emf	electromotive force
emu	electromagnetic unit
en	ethylene diamine
eng	engineering
EPA	Environmental Protection Agency
epr	electron paramagnetic resonance
eq.	equation
esca	electron spectroscopy for chemical analysis
esp	especially
esr	electron-spin resonance
est(d)	estimate(d)
estn	estimation
esu	electrostatic unit
exp	experiment, experimental
ext(d)	extract(ed)
F	farad (capacitance)
F	faraday (96,487 C)
f	femto (prefix for 10^{-15})

FAO	Food and Agriculture Organization (United Nations)
fcc	face-centered cubic
FDA	Food and Drug Administration
FEA	Federal Energy Administration
FHSA	Federal Hazardous Substances Act
fob	free on board
fp	freezing point
FPC	Federal Power Commission
FRB	Federal Reserve Board
frz	freezing
G	giga (prefix for 10^9)
G	gravitational constant = 6.67×10^{11} N·m²/kg²
g	gram
(g)	gas, only as in $H_2O(g)$
g	gravitational acceleration
gc	gas chromatography
gem-	geminal
glc	gas–liquid chromatography
g-mol wt; gmw	gram-molecular weight
GNP	gross national product
gpc	gel-permeation chromatography
GRAS	Generally Recognized as Safe
grd	ground
Gy	gray
H	henry
h	hour; hecto (prefix for 10^2)
ha	hectare
HB	Brinell hardness number
Hb	hemoglobin
hcp	hexagonal close-packed
hex	hexagonal
HK	Knoop hardness number
hplc	high performance liquid chromatography
HRC	Rockwell hardness (C scale)
HV	Vickers hardness number
hyd	hydrated, hydrous
hyg	hygroscopic
Hz	hertz
^{i} (eg, Pr^i)	iso (eg, isopropyl)
i-	inactive (eg, *i*-methionine)
IACS	International Annealed Copper Standard
ibp	initial boiling point
IC	integrated circuit
ICC	Interstate Commerce Commission
ICT	International Critical Table
ID	inside diameter; infective dose
ip	intraperitoneal
IPS	iron pipe size
ir	infrared
IRLG	Interagency Regulatory Liaison Group
ISO	International Organization Standardization
ITS-90	International Temperature Scale (NIST)
IU	International Unit
IUPAC	International Union of Pure and Applied Chemistry
IV	iodine value
iv	intravenous
J	joule
K	kelvin
k	kilo (prefix for 10^3)
kg	kilogram
L	denoting configurational relationship
L	liter (for fluids only) (5)
l-	*levo-*, levorotatory
(l)	liquid, only as in $NH_3(l)$
LC_{50}	conc lethal to 50% of the animals tests
LCAO	linear combination of atomic orbitals
lc	liquid chromatography
LCD	liquid crystal display
lcl	less than carload lots
LD_{50}	dose lethal to 50% of the animals tested

LED	light-emitting diode
liq	liquid
lm	lumen
ln	logarithm (natural)
LNG	liquefied natural gas
log	logarithm (common)
LPG	liquefied petroleum gas
ltl	less than truckload lots
lx	lux
M	mega (prefix for 10^6); metal (as in MA)
M	molar; actual mass
$\overline{M}_w$	weight-average mol wt
$\overline{M}_n$	number-average mol wt
m	meter; milli (prefix for 10^{-3})
m	molal
m-	meta
max	maximum
MCA	Chemical Manufacturers' Association (was Manufacturing Chemists Association)
MEK	methyl ethyl ketone
meq	milliequivalent
mfd	manufactured
mfg	manufacturing
mfr	manufacturer
MIBC	methyl isobutyl carbinol
MIBK	methyl isobutyl ketone
MIC	minimum inhibiting concentration
min	minute; minimum
mL	milliliter
MLD	minimum lethal dose
MO	molecular orbital
mo	month
mol	mole
mol wt	molecular weight
mp	melting point
MR	molar refraction
ms	mass spectrometry
MSDS	material safety data sheet
mxt	mixture
μ	micro (prefix for 10^{-6})
N	newton (force)
N	normal (concentration); neutron number
N-	denoting attachment to nitrogen
n (as n_D^{20})	index of refraction (for 20°C and sodium light)
^{n} (as Bu^n), *n*-	normal (straight-chain structure)
n	neutron
n	nano (prefix for 10^9)
na	not available
NAS	National Academy of Sciences
NASA	National Aeronautics and Space Administration
nat	natural
ndt	nondestructive testing
neg	negative
NF	*National Formulary*
NIH	National Institutes of Health
NIOSH	National Institute of Occupational Safety and Health
NIST	National Institute of Standards and Technology (formerly National Bureau of Standards)
nmr	nuclear magnetic resonance
NND	New and Nonofficial Drugs (AMA)
no.	number
NOI-(BN)	not otherwise indexed (by name)
NOS	not otherwise specified
nqr	nuclear quadruple resonance
NRC	Nuclear Regulatory Commission; National Research Council
NRI	New Ring Index
NSF	National Science Foundation
NTA	nitrilotriacetic acid
NTP	normal temperature and pressure (25°C and 101.3 kPa or 1 atm)

NTSB — National Transportation Safety Board
O- — denoting attachment to oxygen
o- — ortho
OD — outside diameter
OPEC — Organization of Petroleum Exporting Countries
o-phen — *o*-phenanthridine
OSHA — Occupational Safety and Health Administration
owf — on weight of fiber
Ω — ohm
P — peta (prefix for 10^{15})
p — pico (prefix for 10^{-12})
p- — para
p — proton
p. — page
Pa — pascal (pressure)
PEL — personal exposure limit based on an 8-h exposure
pd — potential difference
pH — negative logarithm of the effective hydrogen ion concentration
phr — parts per hundred of resin (rubber)
p-i-n — positive-intrinsic-negative
pmr — proton magnetic resonance
p-n — positive-negative
po — per os (oral)
POP — polyoxypropylene
pos — positive
pp. — pages
ppb — parts per billion (10^9)
ppm — parts per million (10^6)
ppmv — parts per million by volume
ppmwt — parts per million by weight
PPO — poly(phenyl oxide)
ppt(d) — precipitate(d)
pptn — precipitation
Pr (no.) — foreign prototype (number)
pt — point; part
PVC — poly(vinyl chloride)
pwd — powder
py — pyridine
qv — quod vide (which see)
R — univalent hydrocarbon radical
(*R*)- — rectus (clockwise configuration)
r — precision of data
rad — radian; radius
RCRA — Resource Conservation and Recovery Act
rds — rate-determining step
ref. — reference
rf — radio frequency, *n.*
r-f — radio frequency, *adj.*
rh — relative humidity
RI — Ring Index
rms — root-mean square
rpm — rotations per minute
rps — revolutions per second
RT — room temperature
RTECS — Registry of Toxic Effects of Chemical Substances
^{s} (eg, Bu^s); *sec*- — secondary (eg, secondary butyl)
S — siemens
(*S*)- — sinister (counterclockwise configuration)
S- — denoting attachment to sulfur
s- — symmetric(al)
s — second
(s) — solid, only as in $H_2O(s)$
SAE — Society of Automotive Engineers
SAN — styrene-acrylonitrile
sat(d) — saturate(d)
satn — saturation
SBS — styrene–butadiene–styrene
sc — subcutaneous
SCF — self-consistent field; standard cubic feet
Sch — Schultz number
SFs — Saybolt Furol seconds
SI — Le Système International d'Unités (International System of Units)
sl sol — slightly soluble
sol — soluble

soln	solution
soly	solubility
sp	specific; species
sp gr	specific gravity
sr	steradian
std	standard
STP	standard temperature and pressure (0°C and 101.3 kPa)
sub	sublime(s)
SUs	Saybolt Universal seconds
syn	synthetic
ᵗ (eg, Buᵗ), *t*-, *tert*-	tertiary (eg, tertiary butyl)
T	tera (prefix for 10^{12}); tesla (magnetic flux density)
t	metric ton (tonne)
t	temperature
TAPPI	Technical Association of the Pulp and Paper Industry
TCC	Tagliabue closed up
tex	tex (linear density)
T_g	glass-transition temperature
tga	thermogravimetric analysis
THF	tetrahydrofuran
tlc	thin layer chromatography
TLV	threshold limit value
trans-	isomer in which substituted groups are on opposite sides of double bond between C atoms
TSCA	Toxic Substances Control Act
TWA	time-weighted average
Twad	Twaddell
UL	Underwriters' Laboratory
USDA	United States Department of Agriculture
USP	*United States Pharmacopeia*
uv	ultraviolet
V	volt (emf)
var	variable
vic-	vicinal
vol	volume (not volatile)
vs	versus
v sol	very soluble
W	watt
Wb	weber
Wh	watt hour
WHO	World Health Organization (United Nations)
wk	week
yr	year
(*Z*)-	zusammen; together; atomic number

Non-SI (Unacceptable and Obsolete) Units		Use
Å	angstrom	nm
at	atmosphere, technical	Pa
atm	atmosphere, standard	Pa
b	barn	cm^2
bar†	bar	Pa
bbl	barrel	m^3
bhp	brake horsepower	W
Btu	British thermal unit	J
bu	bushel	m^3; L
cal	calorie	J
cfm	cubic foot per minute	m^3/s
Ci	curie	Bq
cSt	centistokes	mm^2/s
c/s	cycle per second	Hz
cu	cubic	exponential form

†Do not use bar (10^5Pa) or millibar (10^2Pa) because they are not SI units, and are accepted internationally only for a limited time in special fields because of existing usage.

Non-SI (Unacceptable and Obsolete) Units		Use
D	debye	C·m
den	denier	tex
dr	dram	kg
dyn	dyne	N
dyn/cm	dyne per centimeter	mN/m
erg	erg	J
eu	entropy unit	J/K
°F	degree Fahrenheit	°C; K
fc	footcandle	lx
fl	footlambert	lx
fl oz	fluid ounce	m^3; L
ft	foot	m
ft·lbf	foot pound-force	J
gf den	gram-force per denier	N/tex
G	gauss	T
Gal	gal	m/s^2
gal	gallon	m^3; L
Gb	gilbert	A
gpm	gallon per minute	(m^3/s); (m^3/h)
gr	grain	kg
hp	horsepower	W
ihp	indicated horsepower	W
in.	inch	m
in. Hg	inch of mercury	Pa
in. H_2O	inch of water	Pa
in.-lbf	inch pound-force	J
kcal	kilo-calorie	J
kgf	kilogram-force	N
kilo	for kilogram	kg
L	lambert	lx
lb	pound	kg
lbf	pound-force	N
mho	mho	S
mi	mile	m
MM	million	M
mm Hg	millimeter of mercury	Pa
mμ	millimicron	nm
mph	miles per hour	km/h
μ	micron	μm
Oe	oersted	A/m
oz	ounce	kg
ozf	ounce-force	N
η	poise	Pa·s
P	poise	Pa·s
ph	phot	lx
psi	pounds-force per square inch	Pa
psia	pounds-force per square inch absolute	Pa
psig	pounds-force per square inch gage	Pa
qt	quart	m^3; L
°R	degree Rankine	K
rd	rad	Gy
sb	stilb	lx
SCF	standard cubic foot	m^3
sq	square	exponential form
thm	therm	J
yd	yard	m

BIBLIOGRAPHY

1. The International Bureau of Weights and Measures, BIPM (Parc de Saint-Cloud, France) is described in Appendix X2 of Ref. 4. This bureau operates under the exclusive supervision of the International Committee for Weights and Measures (CIPM).
2. *Metric Editorial Guide (ANMC-78-1)*, latest ed., American National Metric Council, 5410 Grosvenor Lane, Bethesda, Md. 20814, 1981.
3. *SI Units and Recommendations for the Use of Their Multiples and of Certain Other Units (ISO 1000-1981)*, American National Standards Institute, 1430 Broadway, New York, N.Y. 10018, 1981.
4. Based on *ASTM E 380-89a (Standard Practice for Use of the International System of Units (SI))*, American Society for Testing and Materials, 1916 Race Street, Philadelphia, Pa. 19103, 1989.
5. *Fed. Regist.*, Dec. 10, 1976 (41 FR 36414).
6. For ANSI address, see Ref. 3.

R. P. LUKENS
ASTM Committee E-43 on SI Practice

ANALGESICS, ANTIPYRETICS, AND ANTIINFLAMMATORY AGENTS

Pain, pyresis, and inflammation are distinct physiological responses which can occur independently; they are often associated as the body mounts a response to an injury or insult. Each is an important signal from injured tissue and the signal's continued presence can guide a physician in diagnosing and treating the condition which led to the occurrence. Literature and folklore are replete with remedies which claim to alleviate aches and pains of all kinds. Whereas many treatments are baseless, some, especially those based on herbal medicines, serve as the basis for the rational discovery of the drugs that are used in Western medicine for treating pain and inflammation.

The twentieth century has seen considerable progress in the understanding of pain and inflammation and the relationship of one to the other. It is increasingly apparent that the central and peripheral nervous systems are capable of causing the production of mediators which can attract and activate inflammatory cells thereby initiating or amplifying an inflammatory response. At the same time, inflammatory cells are also capable of releasing substances which not only respond to the original insult, but also stimulate the nervous systems.

A number of interdependent physiological mechanisms, each providing new targets for therapeutic intervention, and allowing for the development of new treatments for pain, pyresis, and inflammation, have been discovered. At least eight distinct types of opioid receptors have been identified and the corresponding individual functions are beginning to be understood. Moreover, a great deal has been learned about the role of lipid mediators in the inflammatory response and how antiinflammatory agents control their production.

Endogenous Receptors and Ligands

In 1973 evidence was provided for the existence of receptors which recognize opioid molecules (1). As many as eight opioid receptors are known to exist, but only four, designated as μ (mu), κ (kappa), σ (sigma), and δ (delta), are believed to be important to the central nervous system (CNS). Several portions of the brain, including the hypothalamus and amygdala, the spinal cord, and nuclei concerned with vagal reflexes contain varying concentrations of the different types of opioid receptors.

The development of selective agonists and antagonists has aided in formulating a profile of pharmacological effects for each receptor type (2). The best understood is the μ receptor which plays a role in euphoria, respiratory depression, tolerance, and physical dependence as well as analgesic effects. Studies using highly selective agonists and antagonists imply that there may be μ receptor subtypes (3) suggesting that not all of the ascribed effects result from the same receptor. Antinociceptive activities and possible modulation of selected μ receptor activities can be identified. In addition to analgesia, the κ receptor effects include diuresis, sedation, and physical dependence (4). The effects of the δ receptor are less well understood as there is a lack of highly selective δ receptor ligands.

Shortly following the discovery of the opioid receptors, peptidic substances having opioid activity were detected in the pituitary gland (5). There are three known groups of endogenous opioid peptides: endorphins (6), enkephalins, and dynorphins (see OPIODS, ENDOGENOUS). Examples are given in Table 1. Every member of the three families is the progeny of a distinct polypeptide precursor. The endorphins are fragments of proopiomelanocortin [*66796-54-1*] (POMC), a polypeptide also containing β-lipotropic hormone, adrenocorticotropic hormone (ACTH), and melanocyte stimulatory hormone (α-MSH) (8). β-Endorphin, the largest endogenous opioid peptide, is principally synthesized, stored, and ultimately released from cells in the pituitary gland as well as from neurons in the

Table 1. Opioid Peptides

Compound	CAS Registry Number	Structural formula
β-endorphin	[*60617-12-1*]	H–TyrGlyGlyPheMetThrSerGluLysSerGlnThrProLeu-ValThrLeuPheLysAsnAlaIleValLysAsnAlaHisLys-LysGlyGln–OH
leu-enkephalin	[*58822-25-6*]	H–TryGlyGlyPheLeu–OH
met-enkephalin	[*58569-55-4*]	H–TyrGlyGlyPheMet–OH
dynorphin A(1–17)	[*80448-90-4*]	H–TyrGlyGlyPheLeuArgArgIleArgProLysLeuLysTrp-AspAsnGln–OH

CNS. Although β-endorphin contains the sequence for met-enkephalin, $C_{27}H_{35}N_5O_7S$, the endorphin is not processed to this enkephalin. Instead met-enkephalin is derived from proenkephalin, a polypeptide containing 6 met-enkephalins and 1 leu-enkephalin, $C_{28}H_{37}N_5O_7$ (9). These highly conserved, five amino acid peptides are located in the adrenal medulla, the CNS, and the peripheral nervous systems. Another polypeptide, prodynorphin also contains a leu-enkephalin fragment, as well as other peptidic opioids, including α- and β-neoendorphin and dynorphin A(1–17), $C_{99}H_{155}N_{31}O_{23}$. The latter is further cleaved to dynorphin A(1–8) and dynorphin B(1–3).

When released, endogenous opioids act upon specific receptors, displaying effects like the quintessential opioid, morphine [*57-27-2*], $C_{17}H_{19}NO_3$. These endogenous peptides appear to induce analgesia and depress respiratory function, gastrointestinal transit, and several other physiological functions. Leu-enkephalin binds with relatively high selectivity to the δ receptor, dynorphin A(1–17) to the κ receptor. Like morphine, β-endorphin, and met-enkephalin are μ agonists. When administered intravenously, β-endorphin was three times more potent than morphine on a molar basis even though the large size and charged nature of β-endorphin should preclude its passage across the blood-brain barrier. The smaller enkephalins and dynorphins are potential substrates for peptidase degradation when administered systemically. Synthetic enkephalin analogues, designed to resist degradation, demonstrated analgesia in humans. Unfortunately, like morphine, these compounds do not cause analgesia without tolerance and physical dependence.

Opioid Agonists

Opium is the dried, powdered sap of the unripe seed pod of *Papaver somniferum;* a poppy plant indigenous to Asia minor. Theophrastus described its medical properties in the third century BC, but the Sumerians, ca BC 4000, probably perceived its utility. Arab physicians knew of the drug, and Arab traders carried it to the Orient where it was used as a treatment for dysentery. Paracelsus is credited with repopularizing the drug in western Europe in the early sixteenth century by formulating opium into "laudanum", which is still in use. More than 20 different alkaloids (qv) of two different classes comprise 25% of the weight of dry opium. The benzylisoquinolines, characterized by papaverine [*58-74-2*] (1.0%), a smooth muscle relaxant, and noscapine [*128-62-1*] (6.0%), an antitussive agent, do not have any analgesic effects. The phenanthrenes, the second group, are the more common and include 10% morphine (**1**, R′ = R = H), 0.5% codeine [*76-57-3*], $C_{18}H_{21}NO_3$, (**1**, R′ = H, R = CH_3), and 0.2% thebaine [*115-37-7*], $C_{19}H_{21}NO_3$, (**2**).

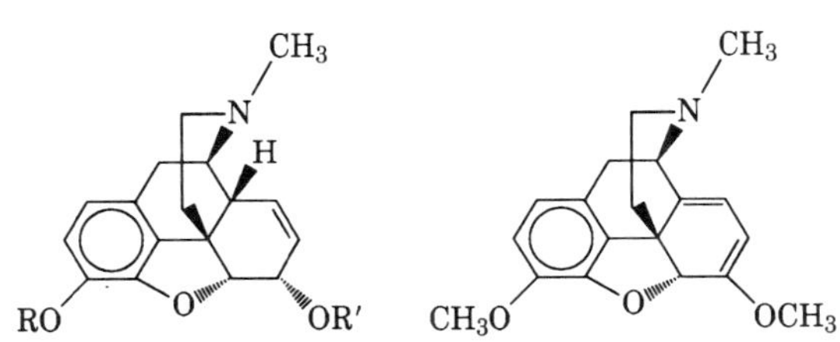

(**1**) (**2**)

The correct structure of morphine, named after Morpheus, the Greek god of sleep, was first proposed in 1925 (10) and was confirmed in 1955 via total synthesis (11). Substitution at one or both of the two hydroxyl groups, one phenolic and the other allylic, give codeine, methyl morphine, where the substitution is on the phenolic hydroxyl, and heroin [*561-27-3*], $C_{21}H_{23}NO_5$, (**1**, R′ = R = $COCH_3$) also known as 3,6-diacetoxy morphine, which serves as a prodrug and converts to morphine *in vivo*. Methylating both hydroxyl groups and introducing an additional double bond into the allylic ring leads to thebaine which has little analgesic action.

Morphine, mol wt 285.3, effectively relieves pain and increases an individual's ability to tolerate a painful experience. It also produces a remarkably broad range of other effects including drowsiness, mood changes, respiratory depression, nausea, decreased gastrointestinal motility, and vomiting. Morphine behaves as a receptor agonist, acting preferentially at the μ receptor, but also exhibiting appreciable affinity for other opioid receptors. A standard therapeutic dose is 10–15 mg, usually administered subcutaneously. Analgesia peaks in about one hour and lasts for four to five hours. Morphine can also be administered orally, but it is only one sixth as potent. When delivered intravenously, morphine-induced respiratory depression is observed at below analgesic dose levels, as early as seven minutes after administration and for as long as 4–5 hours. An important feature of morphine and related drugs is the development of physical dependence on, and tolerance to, some of the effects. Increasingly large doses of drug must be administered to maintain the analgesic effects and the possibility of psychological dependence is a primary limitation to clinical use. However there are studies which suggest that when patients take morphine to combat pain, it is rare to see addiction, and thus morphine usage may be reasonable for the treatment of chronic pain (12). The cessation of drug uptake does produce withdrawl symptoms, such as diarrhea, vomiting, chills, fever, abdominal cramps, and abdominal pain, which are different from those observed during withdrawl from other CNS depressants. The onset and duration of symptoms depend upon the pharmokinetic profile of the drug.

Codeine, mol wt 299.3, is a significantly less potent analgesic than morphine, requiring 60 mg (0.20 mmol) to equal the effectiveness of 10 mg (0.04 mmol) of morphine. However, codeine is orally effective, and it is less addictive and associated with less nausea than morphine. Codeine is used as an antitussive agent, although newer, nonaddictive agents are preferred (see EXPECTORANTS, ANTITUSSIVES, AND RELATED AGENTS).

Introduced in 1898, heroin was heralded as a nonaddictive alternative to morphine. Subsequent clinical experience showed it to be highly addictive and preferred by addicts over morphine (13). Heroin is approximately ten times more potent than morphine, with quicker onset and shorter duration of action. It is orally inactive, but is effective when administered intravenously. Although heroin has not been available in the United States for therapeutic use since 1906, it is still used clinically in other countries for its analgesic properties.

Synthetic and Semisynthetic Agonists

In attempts to discover drugs demonstrating fewer undesirable side effects than morphine, many synthetic analogues have been prepared. Some of these are shown in Table 2. The orally effective levorphanol (**3**) retains the complete morphine carbon skeleton, is more potent than morphine, and produces a correspondingly greater degree of respiratory depression (14). Oxymorphone (**4**, R = OH) and hydromorphone (**4**, R = H), both of which contain a C-6 carbonyl group, are more potent than morphine, but also produce greater side effects. Removal of one ring led to phenazocine (**5**), which is more potent than morphine and has the same duration of action, but is equally addictive.

Table 2. Synthetic Morphine Analogues

Name	CAS Registry Number	Molecular formula	Structure number	Structure
levorphanol	*[77-07-6]*	$C_{17}H_{23}NO$	(**3**)	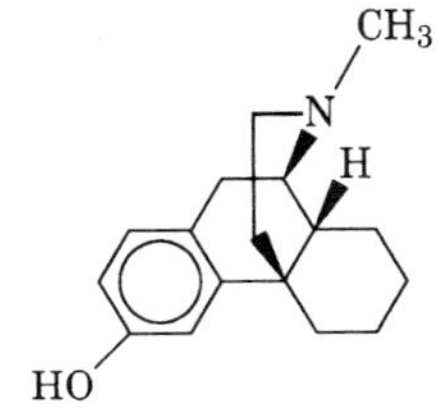
hydromorphone oxymorphone	*[466-99-9]* *[76-41-5]*	$C_{17}H_{19}NO_3$ $C_{17}H_{19}NO_4$	(**4**, R = H) (**4**, R = OH)	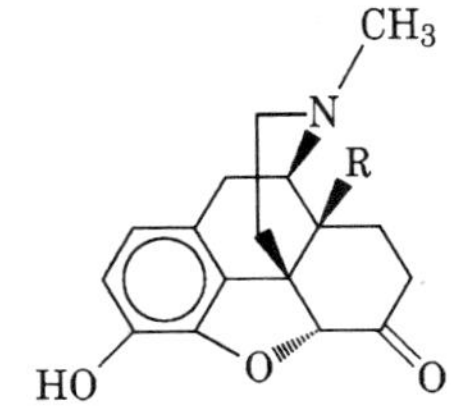
phenazocine	*[127-35-5]*	$C_{22}H_{27}NO$	(**5**)	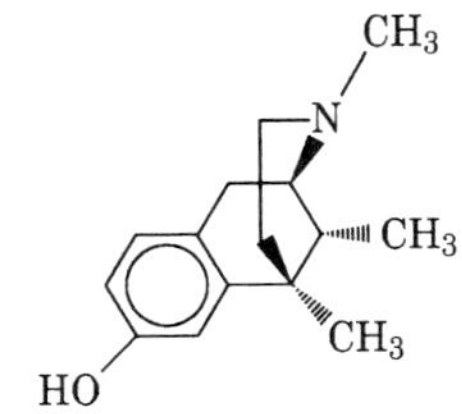
fentanyl	*[437-38-7]*	$C_{22}H_{29}NO_2$	(**6**)	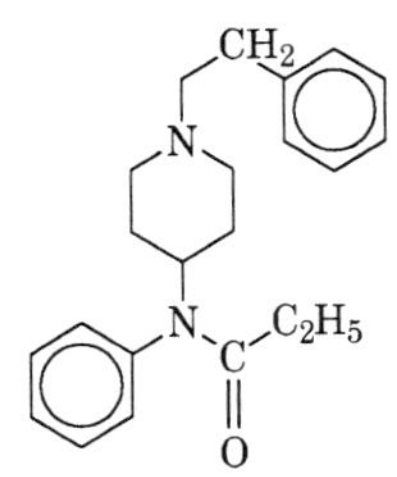

Several common structural features necessary for opioid, analgesic activity have been identified from the action of the analogues. Systematic simplification demonstrated that much of the morphine ring structure could be modified or even eliminated. The piperidine ring, in a chair conformation, (Fig. 1) and in particular the nitrogen atom of that ring, appears essential to pharmacological activity. The nitrogen is believed to attach to an anionic center in the receptor. Also crucial is the presence of the phenyl ring which, through van der Waals forces, binds to the

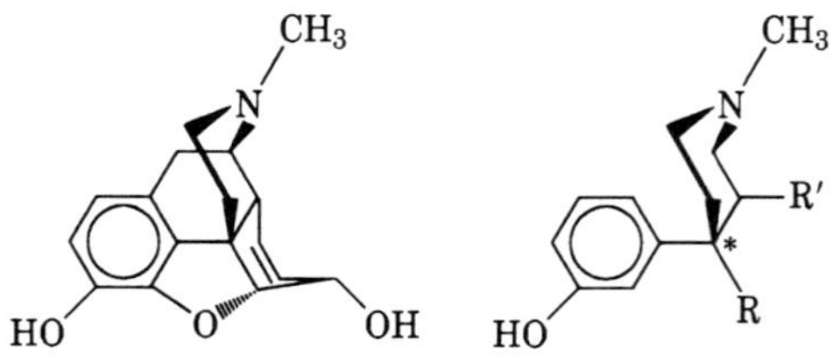

(1) (7)

Fig. 1. Conformational representation of the piperidine ring of morphine (**1**) and analogues meperidine (**7**, R′ = H, R = $COOC_2H_5$) and alphaprodine (**7**, R′ = CH_3, R = OCC_2H_5). The chiral center of interest in structure (**7**) is starred (see text).

hydrophobic portion of the receptor. As in any receptor-ligand interaction, the sterochemistry of the chiral center shown in Figure 1 must be maintained.

In the late 1930s, meperidine [*57-42-1*], $C_{15}H_{21}NO_2$, (Fig. 1) was identified as a potent μ receptor agonist (15). The later analogues, fentanyl (**6**) (Table 2) and alphaprodine [*77-20-3*], $C_{16}H_{23}NO_2$, (Fig. 1), also retain the key 4-phenyl piperidine ring structure. These agents appear to interact more strongly with the κ receptor than morphine, although they are still strong μ agonists. Fentanyl is more potent than morphine, has a much shorter duration of action, and is used to aid the induction and maintenance of inhaled anesthesia. Meperidine and alphaprodine are less potent than morphine, requiring 75 (0.30 mmol) and 40 (0.15 mmol) mg, respectively, to equal the effectiveness of 10 mg (0.04 mmol) of morphine. However, they are more effective orally than morphine, without producing constipating or antitussive effects. Meperidine causes the same degree of respiratory depression as morphine at equivalent doses and is definitely addictive (16).

Phenylpropylamines are another structural class of receptor agonist. The two most important members are methadone [*76-99-3*], $C_{21}H_{27}NO$, (**8**) (17), discovered in Germany during World War II, and propoxyphene [*469-62-5*], $C_{22}H_{29}NO_2$, (**9**).

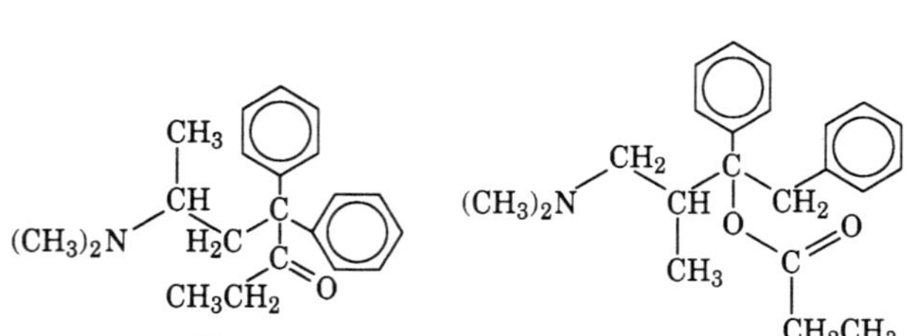

(8) (9)

These compounds can adopt a conformation that mimics a disubstituted pseudopiperidine ring.

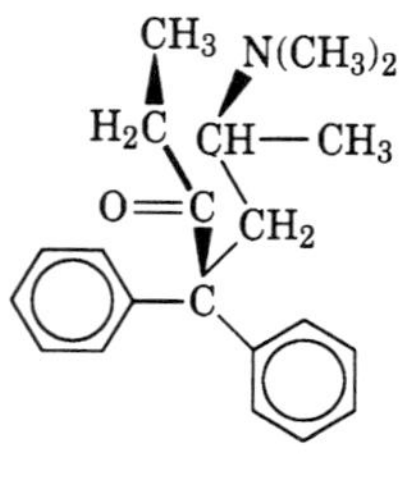

(8)

Methadone is orally active, having analgesic potency as great as morphine and longer duration of action. Although as potentially addictive as morphine, methadone causes less severe but more prolonged withdrawal symptoms and is therefore primarily used for the treatment of compulsive heroin users (18). In contrast, propoxyphene is a weak analgesic, only about half as potent as codeine and is often formulated with aspirin or acetaminophen, resulting in combinations which are more effective than either agent alone. Although propoxyphene has less potential for abuse than codeine, the actual incidence of abuse is equivalent, and hundreds of people die each year in the United States from overdoses.

Opioid Antagonists and Partial Agonists

The replacement of the *N*-methyl group on the nitrogen atom of the piperidine ring of morphine and analogues by allyl, isopropyl, or methyl cyclopropyl, an isopropyl isostere, results in compounds which antagonize opioid responses, especially respiratory depression. Naloxone [*465-65-6*], $C_{19}H_{21}NO_4$ (**10**, R = CH_2CHCH_2), and naltrexone [*16590-41-3*], $C_{20}H_{23}NO_4$ (**10**, R = CH_2–◁), are both derived from oxymorphone (Table 2) and exhibit agonist activity only at doses that are of little clinical significance (19). In the absence of opioid drugs, naloxone does not cause analgesia, respiratory depression, or sedation. However, when administered with an opioid analgesic, the effects produced by the opioid agonist are promptly reversed. The ability to antagonize opioids at all of the different opioid receptors makes naloxone useful for the treatment of opioid overdose. As is typical for opioid antagonists, naloxone induces withdrawal symptoms in patients addicted to opiatelike drugs. Dosage is generally 0.4 mg (0.001 mmol) given intravenously in several doses because of the drug's short half life. Naltrexone has a similar profile, but it is orally active and has a significantly longer half life.

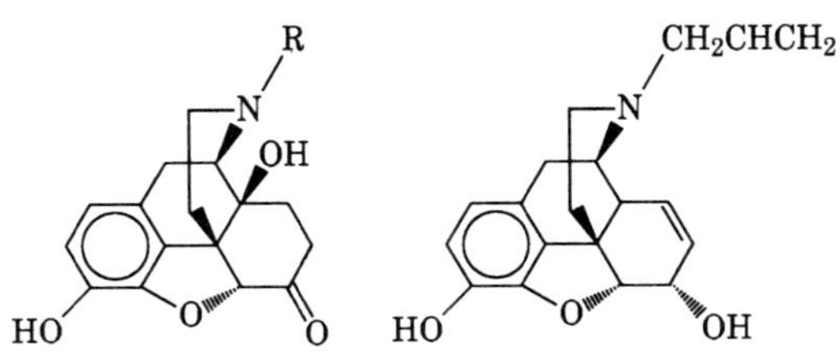

(**10**) (**11**)

Before the introduction of naloxone, nalorphine [*62-67-9*], $C_{19}H_{21}NO_3$ (**11**) was the drug most often used for opioid overdose (20). Originally developed in the early 1940s, nalorphine is a morphine derivative and displays both agonist and antagonist properties. By itself, nalorphine causes typical agonist effects, analgesia and respiratory depression (21), but after administration of an opioid, such as morphine, it acts as an antagonist and negates the opioid's agonist effects (22). Its usefulness as an agonist is limited by the dysphoria it causes, and its use as an antagonist has been supplanted by the more effective naloxone.

The quest for compounds that combined the analgesic properties of morphine, were nonaddictive, and lacked the side effects of nalorphine, led to the development of the drugs shown in Table 3. These compounds have both agonist

Table 3. Drugs Exhibiting Both Opioid Agonist and Antagonist Activities

Compound name	CAS Registry Number	Molecular formula	Structure number	Structure
buprenorphine	[*52485-79-7*]	$C_{29}H_{41}NO_4$	(**12**)	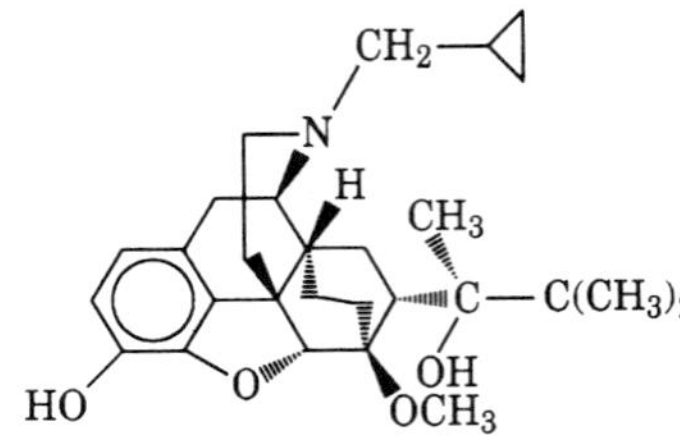
butorphanol	[*42408-82-2*]	$C_{21}H_{29}NO_2$	(**13**)	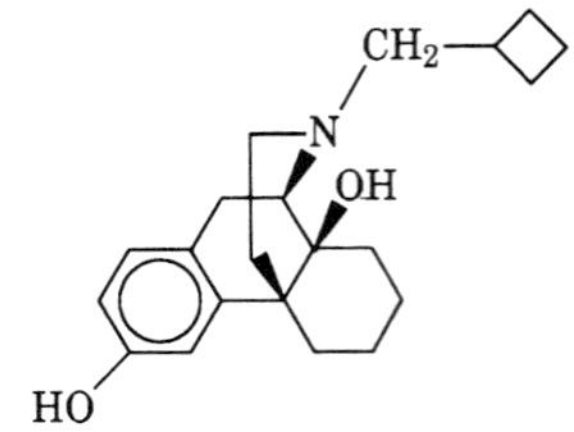
nalbuphine	[*20594-83-6*]	$C_{21}H_{27}NO_4$	(**14**)	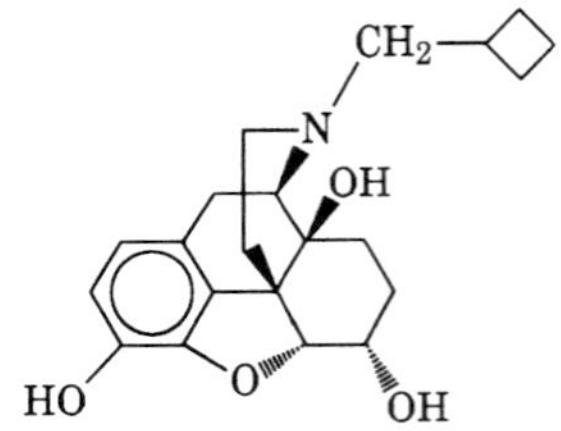
pentazocine	[*359-83-11*]	$C_{19}H_{27}NO$	(**15**)	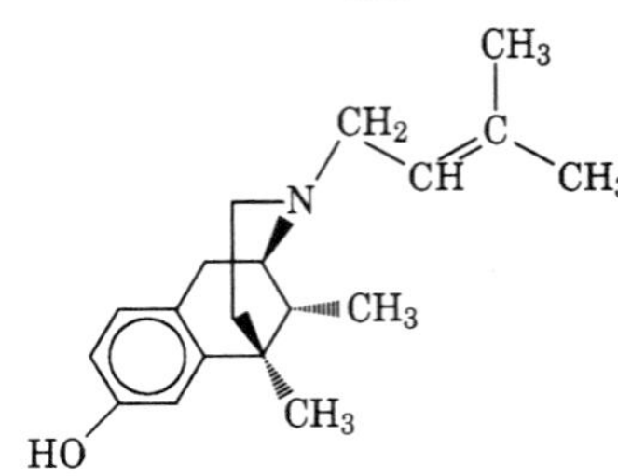

and antagonist activities. Nalbuphine (**14**) (23) and buprenorphine (**12**) (24) are semisynthetic materials derived from oxymorphone (**4**, R = OH) and thebaine (**2**) respectively, whereas pentazocine (**15**) (25) and butorphanol (**13**) (26) are benzomorphan and morphan derivatives. Although structurally similar, they display different receptor affinities: pentazocine is a weak μ-antagonist, but a strong agonist of the κ-receptor; nalbuphine is a competitive antagonist for the μ-receptor, blocking the effects of the morphinelike drugs, but is a partial agonist for the κ and σ receptors; and buprenorphine is a partial agonist for the μ-receptor.

Pharmacologically, the effects of the drugs in Table 3 resemble those of opioid agonists. All four have analgesic potency equal to or greater than morphine and like morphine, they cause respiratory depression. A ceiling effect is reached, however, above which increased doses do not increase respiratory depression or do not produce proportionally greater depression. Buprenorphine causes a slow onset of the depression, and once the effects have begun, naloxone does not readily reverse the symptoms, suggesting that buprenorphine dissociates very slowly from opioid receptors. In all of these drugs except buprenorphine, high doses induce withdrawal symptoms in opioid dependent patients.

Other Analgesic Agents

Most analgesic agents rely on agonism of the μ receptor for their activity, however the ability of the κ and δ receptors to induce analgesia is also well documented (27). For example, the known μ receptor antagonist nalorphine is a κ agonist and still retains analgesic properties. Some nonmorphine analgesics which may preferentially bind to κ and δ receptors are found in Table 4. Spiradoline (**19**, X = H, Y = Cl), a κ agonist identified in preclinical studies, has good analgesic activity and little respiratory or gastrointestinal side effects (28,29). However spiradoline also produced sedative and psychomimetic effects in humans (30) preventing its development. No κ agonist has undergone rigorous clinical investigation, and the concept of a κ agonist analgesic as an alternative to morphine is not yet clinically validated.

Unselective and μ selective antagonists are well known, but there are few known selective κ receptor antagonists. The first of these, norbinaltorphimine [*105618-26-6*], (nor BNI), $C_{40}H_{43}N_3O_6$ has 400-fold selectivity for κ over μ receptors and 250-fold for κ over δ (31). Two derivatives of spiradoline (**19**, X = H, Y = Cl) which lack the saturated furan ring, [*125132-66-3*], $C_{20}H_{27}N_3OS$, where X = NCS, Y = H, and [*122407-13-0*], $C_{20}H_{25}Cl_2N_3OS$, where X = NCS, Y = Cl, are site-directed, irreversible inhibitors of κ receptors, designed through modification of the spiradoline class of compounds (32). Although these drugs inhibit binding to κ receptors by 90%, they fail to bind all of the previously identified κ receptors, which suggests heterogeneity of the κ receptors.

It has also been suggested that the δ-receptor mediates analgesia. A synthetic peptide, D-Pen (2), D-Pen (5) enkephalin (DPDPE), a specific δ agonist (33), has produced analgesia in test animals, implying that agents of this type could be useful (34). Further studies demonstrated that mice deficient in μ, but not δ, receptors still perceive analgesia when given morphine or DPDPE thus confirming δ mediated analgesia (35). Administration of a δ selective enkephalin to morphine tolerant patients produced clinically significant pain relief, demonstrating that problems of tolerance may be avoided by using drugs of different pharmacological specificity (36). Although active *in vivo,* these drugs are rarely used systemically, because of very poor penetration of the blood brain barrier. The ability of an exogenous δ-receptor agonist to elicit meaningful analgesia in humans has not been demonstrated.

An alternative to the administration of exogenous analgesic agents is to encourage the stability of the endogenous agonists. The endogenous ligand, enkephalin, is enzymatically inactivated by enkephalinases (37) and a strategy to inhibit this event may increase its concentration and duration of action (see ENZYMES, INHIBITORS AND ANTAGONISTS). When administered intravenously,

Table 4. Nonmorphine Analgesics

Name	CAS Registry Number	Molecular formula	Structure number	Structure
acetaminophen	[*103-90-2*]	$C_8H_9NO_2$	(**16**, R = OH)	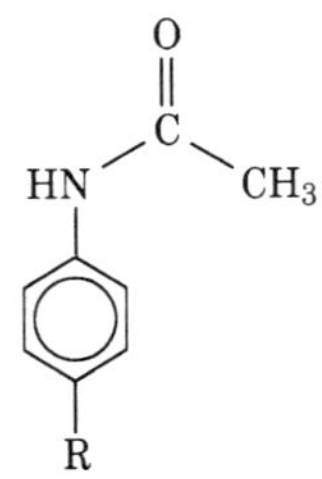
acetanilide	[*103-84-4*]	C_8H_9NO	(**16**, R = H)	
phenacetin	[*62-44-2*]	$C_{10}H_{13}NO_2$	(**16**, R = OC_2H_5)	
acetorphan	[*81110-73-8*]	$C_{21}H_{23}NO_4S$	(**17**, R = $COCH_3$, R′ = $CH_2C_6H_5$)	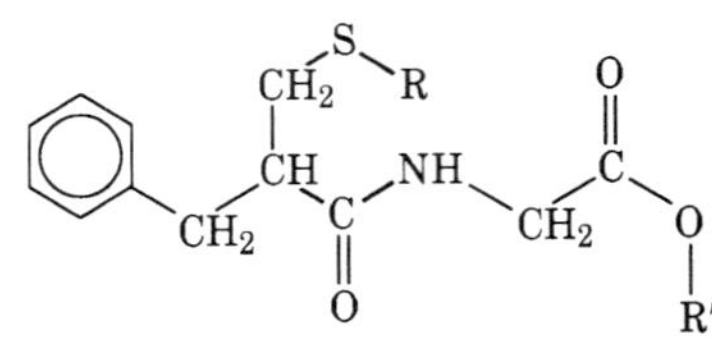
thiorphan	[*76721-89-6*]	$C_{12}H_{15}NO_3S$	(**17**, R = H, R′ = H)	
aspirin	[*50-78-2*]	$C_9H_8O_4$	(**18**)	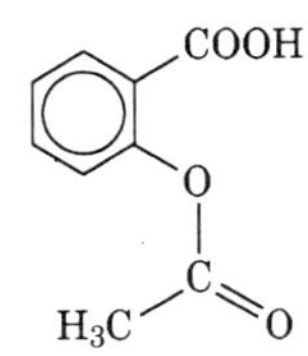
spiradoline	[*87173-97-5*][a]	$C_{22}H_{30}Cl_2N_2O_2$	(**19**, X = H, Y = Cl)	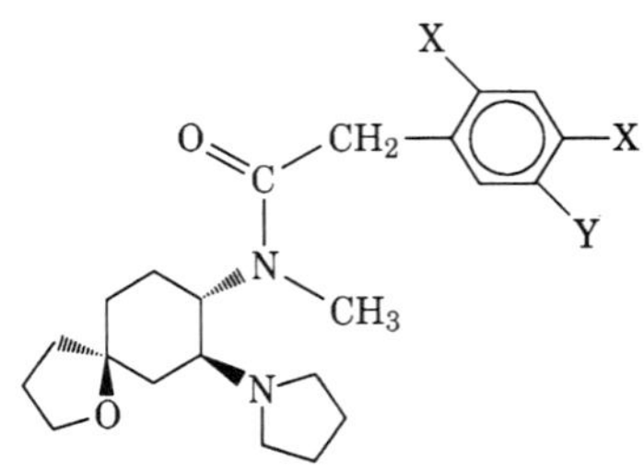

[a] The Registry Number is for the monomethanesulfonate, CH_3SO_3H, salt.

acetorphan, (**17**, R = $COCH_3$, R′ = $CH_2C_6H_5$), a prodrug of thiorphan (**17**, R = R′ = H), the first reported enkephalinase inhibitor, produced analgesia in post-myelography headache pain, but not in shock-induced pain (38). There was no tolerance or dependence observed.

Numerous neuropeptides are believed to be involved with the transmission or inhibition of pain, and the hope is to utilize this approach as a strategy to induce analgesia. Substance P is reported to be a transmitter of nociceptive impulses (39), and therefore antagonists should be analgesic. Capsaicin [*404-86-4*], $C_{18}H_{27}NO_3$, is known to deplete substance P and cause analgesia (40), but its side effects are intolerable. Antagonists to bradykinin [*58-82-2*], $C_{50}H_{73}N_{15}O_{11}$, a substance known to induce pain (41), have shown some success in preclinical trials.

Antiinflammatories and Antipyretics

Most of the time, the powerful analgesia supplied by morphine and the other opioid analgesics is not needed. Rather, a mild analgesic, such as aspirin, the most commonly employed analgesic agent, can be used for the treatment of simple pain associated with headaches, minor muscle pain, mild trauma, arthritis, cold and flu symptoms, and fever.

Aspirin, the oldest of the nonsteroidal antiinflammatory drugs (NSAIDs), is a member of the salicylate group. Salicyclic acid, isolated in 1838 from willow bark, was commonly used throughout the Middle Ages for the treatment of headaches. Its therapeutic benefits were well known even to Hippocrates. Many derivatives have been prepared to reduce the irritation associated with the pure acid. The most common form is acetylsalicyclic acid [*50-78-2*] (**18**). Aspirin is an effective treatment for pain, fever, and for symptoms of acute inflammation. It is often the first treatment for the pain and swelling associated with rheumatoid arthritis. Because of its frequent usage, there are over 10,000 cases of serious salicylate intoxication in the United States every year. Many of these cases involve children, and some are fatal. High doses or chronic use of aspirin may aggravate peptic ulcer symptoms, generate gastric ulceration and bleeding, and produce acid-base and electrolyte imbalance.

Aspirin's pain relief results, not through direct action on the central nervous system, but rather through peripheral action. It has been proposed (42) that aspirin and aspirin-like drugs inhibit the enzymatic production of prostaglandins, a group of endogenous agents which are well known to cause erythema, edema, pain, and fever (43). Aspirin does not act as a prostaglandin receptor antagonist, rather it blocks the cyclooxygenase enzyme catalyzed conversion of arachidonic acid [*506-32-1*], $C_{20}H_{32}O_2$ to cyclic endoperoxides, a prostaglandin precursor as shown in Figure 2. Aspirin controls the synthesis and ultimately the release of prostaglandins from mammalian cells, producing an analgesic effect by decreasing the availability of prostaglandins which typically potentiate the effect of bradykinin on receptors which mediate pain.

The action of endogenous pyrogens on the hypothalmus produces fever, because of a readjustment in the central set point controlling the body's internal temperature. Salicylates and other NSAIDs achieve their antipyretic effect by controlling the prostaglandin-induced release of pyrogens (44). However, hyperthermia induced by exercise, heat stroke, heat exhaustion, or drugs is not affected (45). In these cases the increased temperature is a result of insufficient heat loss. Salicylates block the synthesis of thromboxane A_2, which, in concert with prostacyclin [*35121-78-9*], controls platelet aggregation (46). The result is a decrease in the clotting rate, a condition which lasts long after the analgesic effects of aspirin and other NSAIDs have worn off and which can have serious consequences for those about to undergo surgery, but may be beneficial for those with a tendency towards thrombosis.

A second class of NSAIDs, the so-called coal tar analgesics, are derived from acetanilide (**16**, R = H). Although it is no longer used therapeutically, its analogues, phenacetin (**16**, R = OC_2H_5) and the active metabolite, acetaminophen (**16**, R = OH) are effective alternatives to aspirin (47). They have analgesic and

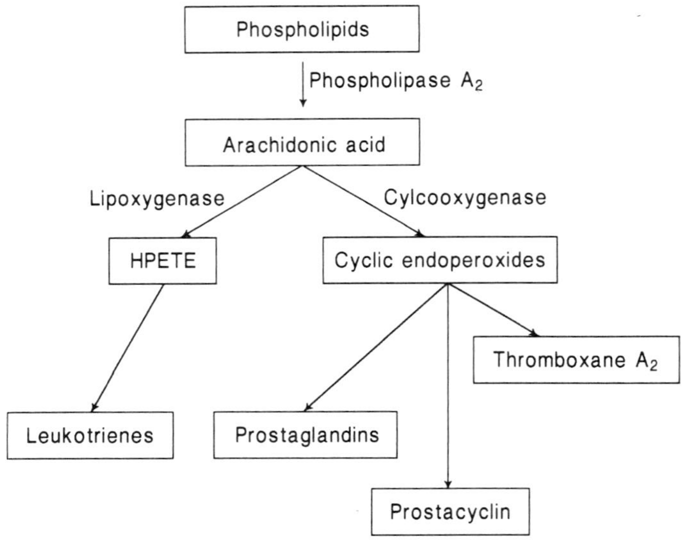

Fig. 2. Arachidonic acid cascade. HPETE = hydroperoxyeicosa-tetraenoic acids.

antipyretic effects that do not differ significantly from aspirin, but they do not cause the gastric irritation, erosion, and bleeding that may occur after salicylate treatment. In contrast to aspirin, however, they are not cyclooxygenase inhibitors and have no antiinflammatory properties. Clinically acetaminophen is preferred over phenacetin, because it has less overall toxicity.

A more recently introduced, nonprescription analgesic is the aryl propionic acid, ibuprofen (**20**) (48), shown in Table 5, which offers significant advantages over aspirin. Ibuprofen, a cyclooxygenase inhibitor, displays good antiinflammatory activity. It is more potent than aspirin and has a lower incidence of gastrointestinal irritation, although at high doses or chronic exposure, gastric irritation, as well as some renal toxicity, has been observed. Ibuprofen is more effective than propoxyphene (**9**) in relieving episiotomy pain, pain following dental extractions, and menstrual pain.

Ibuprofen only recently emerged from an ever growing collection of prescription NSAIDs (49) many of which are shown in Table 5. All of these drugs act by inhibiting the cyclooxygenase enzyme (Fig. 2) and there is little to distinguish in terms of analgesic effectiveness. When prescribed for the treatment of arthritis, patients require large doses for extended periods, exacerbating the side effects. Differences among drugs are found in dose levels, duration of action, and the incidence of gastric ulceration.

The adrenal cortex produces steroidal hormones that are associated with carbohydrate, fat, and protein metabolism, electrolyte balance, and gonadal functions (58). One of these, cortisone [*53-06-5*], $C_{21}H_{28}O_5$ (**31**), demonstrated a remarkable ability to relieve the symptoms of inflammatory conditions (59). Other glucocorticoid steroids, such as dexamethasone [*50-02-2*], $C_{22}H_{29}FO_5$ (**32**, R = F, R′ =

Table 5. Nonsteroidal Antiinflammatory Drugs

Name	CAS Registry Number	Molecular formula	Structure number	Reference	Structure
		Aryl propionic acids			
ibuprofen	[*15687-27-1*]	$C_{13}H_{18}O_2$	**(20)**		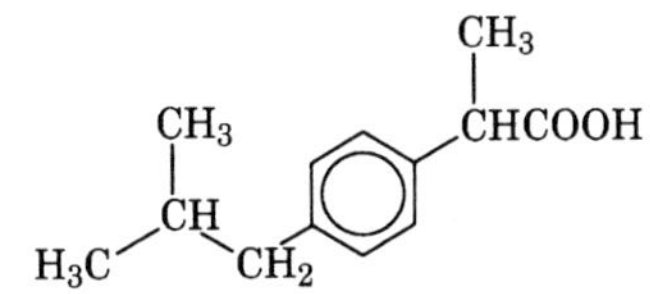
naproxen	[*22204-53-1*]	$C_{14}H_{14}O_3$	**(21)**	50	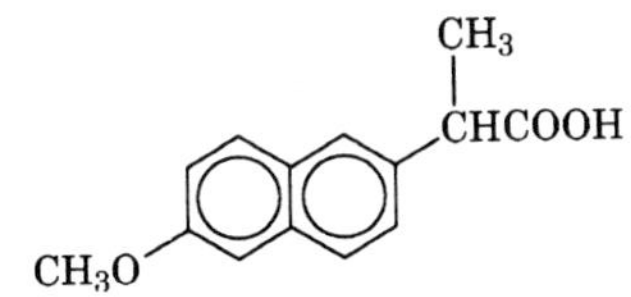
ketoprofen	[*22071-15-4*]	$C_{16}H_{14}O_3$	**(22)**	51	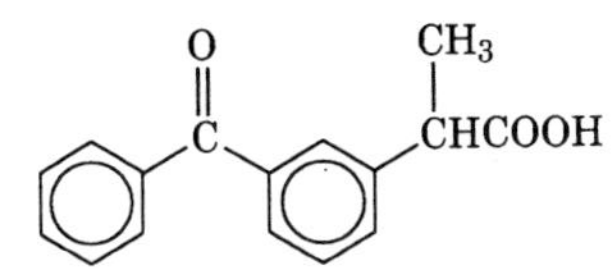
flurbiprofen	[*5104-49-4*]	$C_{15}H_{13}FO_2$	**(23)**	52	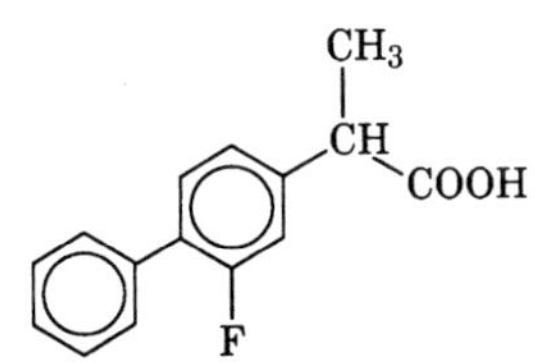

Table 5. (*Continued*)

Name	CAS Registry Number	Molecular formula	Structure number	Reference	Structure
		Indoles			
indomethacin	[*53-86-1*]	$C_{19}H_{16}ClNO_4$	(**24**)	53	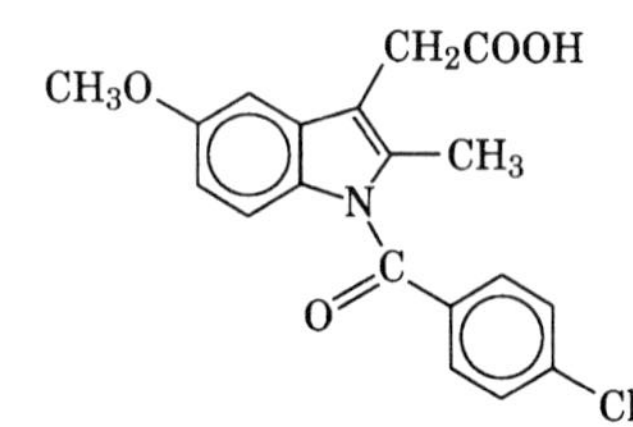
sulindac	[*38194-50-2*]	$C_{20}H_{17}FO_2S$	(**25**)	53	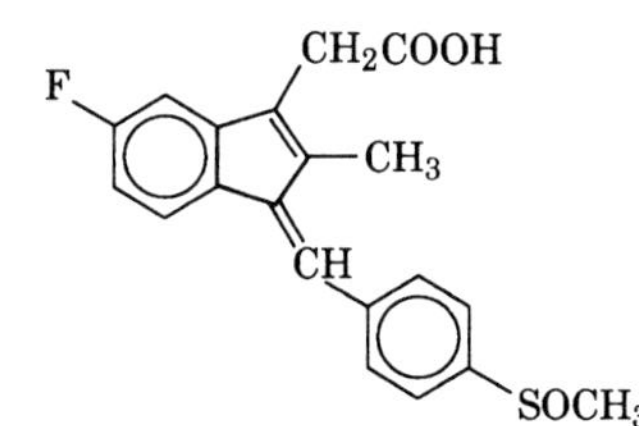
		Fenamates			
diclofenac	[*15307-79-6*]	$C_{14}H_{11}Cl_{12}NO_2Na$	(**26**)	54	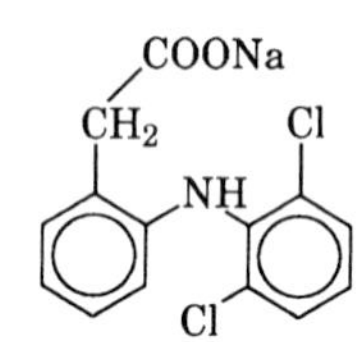

meclofenamate	[*644-62-2*]	$C_{14}H_{11}Cl_2NO_2$	**(27)**	55	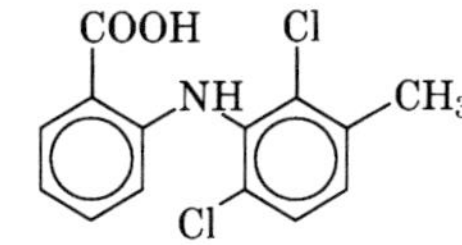
		Others			
piroxicam	[*36322-90-4*]	$C_{15}H_{13}N_3O_4S$	**(28)**	56	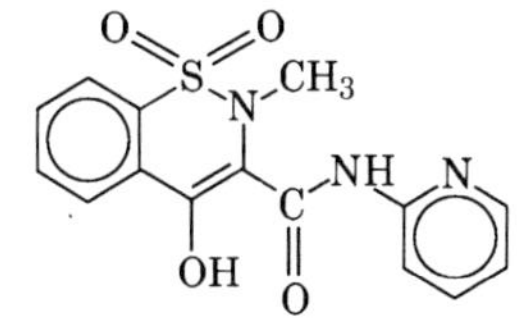
phenylbutazone	[*50-33-9*]	$C_{16}H_{20}N_2O_2$	**(29)**		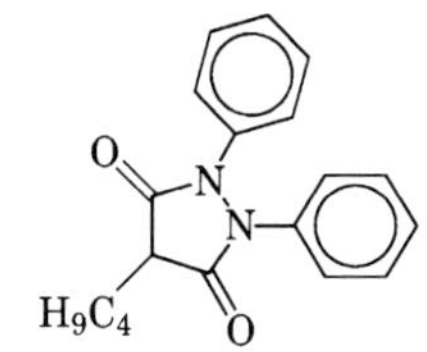
diflunisal	[*22494-42-4*]	$C_{13}H_8F_2O_3$	**(30)**	57	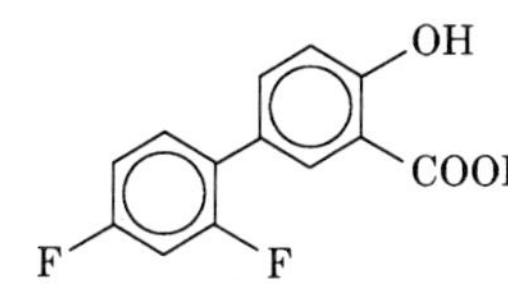

CH_3), and prednisolone [*53-03-2*], $C_{21}H_{28}O_5$ (**32**, R = R′ = H), also have antiinflammatory properties.

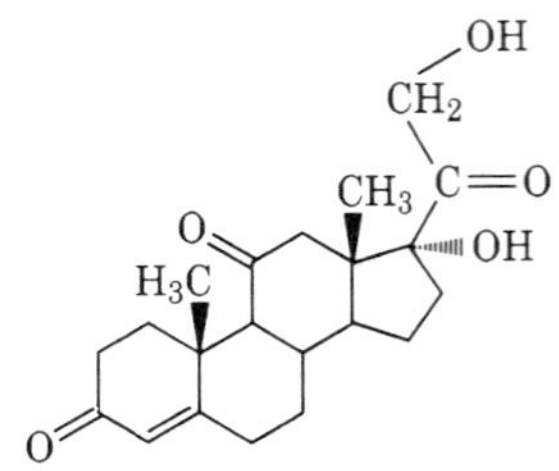

(**31**)

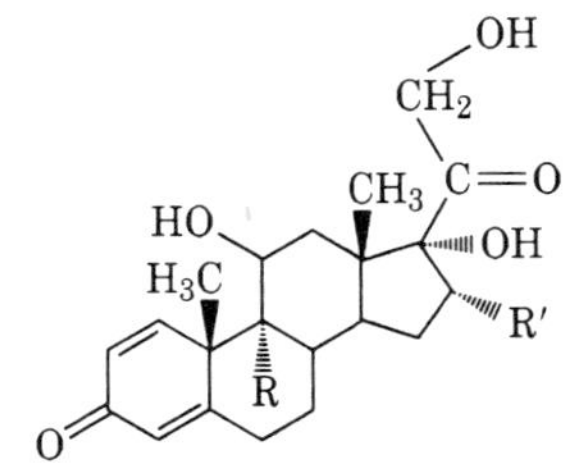

(**32**)

These steroids are capable of preventing or suppressing the development of the swelling, redness, local heat, and tenderness which characterize inflammation. They inhibit not only the acute symptoms of the inflammatory process, such as edema, fibrin deposition, and capillary dilatation, but also the chronic manifestations. There is evidence that glucocorticoids induce the synthesis of a protein that inhibits phospholipase A_2 (60), diminishing the release of arachidonic acid from phospholipids (Fig. 2), thereby reducing chemotaxis and inflammation.

Unfortunately steroids merely suppress the inflammation while the underlying cause of the disease remains. Another serious concern about steroids is that of toxicity. The abrupt withdrawal of glucocorticoid steroids results in acute adrenal insufficiency. Long term use may induce osteoporosis, peptidic ulcers, the retention of fluid, or an increased susceptibility to infections. Because of these problems, steroids are rarely the first line of treatment for any inflammatory condition, and their use in rheumatoid arthritis begins after more conservative therapies have failed.

Analgesic antiinflammatory drugs having alternative mechanisms of action are under study. Whereas conventional therapies inhibit prostaglandin synthesis, other products of arachidonic acid metabolism (Fig. 2), such as the leukotrienes (LTs), have yet to be investigated. The 5-lipoxygenase enzyme begins a cascade of events which converts arachidonic acid to LTB4, LTC4, LTD4, and LTE4 (61). In particular LTB4 plays a significant role in the recruitment of cells to the inflammatory site and is a potent mediator of immune responses. An antagonist of such a mediator might be efficacious in the treatment of inflammation.

The inhibition of both the lipoxygenase and cyclooxygenase pathways of arachidonic acid metabolism could offer improved therapy over conventional NSAIDs. Benoxaprofen [*51234-28-7*], $C_{16}H_{12}ClNO_3$ (**33**), a potent cyclooxygenase and lipoxygenase inhibitor, was introduced in 1982 (62). In clinical trials benoxaprofen showed good analgesic and antiinflammatory activity, but it was removed from the market within months of its release because of unacceptable levels of renal toxicity. The toxicity may not, however, have been related to the mechanism of action (63). Timegadine [*71079-19-1*], $C_{20}H_{23}N_5S$ (**34**), a drug having a similar mechanistic profile, is undergoing clinical trials (64).

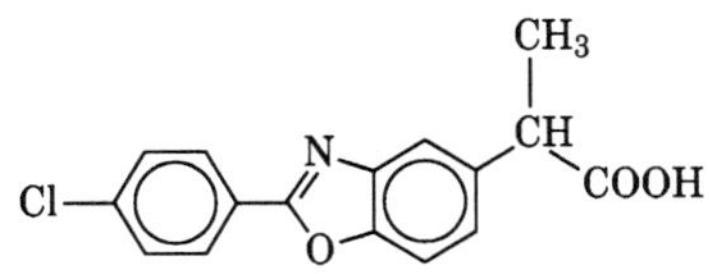

(33)

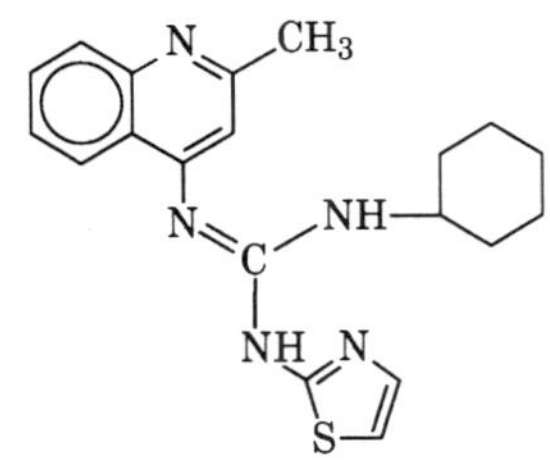

(34)

Advances in the isolation and purification of lymphokine and monokine growth and differentiation factors offer the opportunity to interfer with the inflammatory process at the cellular level (65). The best understood factors are the interleukins (1–9), tumor necrosis factor, and the interferons. Produced by activated macrophages, and other cells, interleukin 1 (IL-1) is a local and hormonal mediator, which induces acute phase responses, local inflammation, and pyrogenicity. It is linked to rheumatoid arthritis, psoriasis, gingivitis, gout, islet destruction in diabetes, and wound healing (66). One of the principal functions of IL-1 is to induce the synthesis and release of interleukin 2 (IL-2) (67). IL-2 is a T-cell growth factor, which stimulates the proliferation of T-cells and the release of numerous

Table 6. Trade Names and Producers of Analgesics, Antipyretics, and Antiinflammatory Drugs

Compound name	Structure number	Trade name	Producer
levorphanol	(**3**)	Dromoran	Hoffmann-LaRoche
phenazocine	(**5**)	Prinadol	Smith Kline & Beecham
meperidine	(**7**, R′ = H, R = $COOC_2H_5$)	Demerol	Winthrop
propoxyphene	(**9**)	Darvon	Eli Lilly
naloxone	(**10**, R = CH_2CHCH_2)	Narcan	Du Pont
naltrexone	(**10**, R = CH_3–cyclopropyl)	Trexan	Du Pont
nalbuphine	(**14**)	Nubain	Du Pont
buprenorphine	(**12**)	Buprenex	Norwich Eaton
pentazocine	(**15**)	Talwin	Winthrop
butorphanol	(**13**)	Stadol	Bristol
acetaminophen	(**16**, R = OH)	Tylenol	McNeil
ibuprofen	(**20**)	Motrin	Upjohn
naproxen	(**21**)	Naprosyn	Syntex
ketoprofen	(**22**)	Orudis	Wyeth
flurbiprofen	(**23**)	Ansaid	Upjohn
indomethacin	(**24**)	Indocin	Merck Sharpe & Dohme
sulindac	(**25**)	Clinoril	Merck Sharpe & Dohme
diclofenac	(**26**)	Voltaren	CIBA-GEIGY
meclofenamate	(**27**)	Meclomen	Parke-Davis
piroxicam	(**28**)	Feldene	Pfizer
diflunisal	(**30**)	Dolobid	Merck Sharpe & Dohme
benoxaprofen	(**33**)	Oraflex	Eli Lilly

T-cell cytokines (68), thus amplifying the immune response. Interleukin 4 is produced by T-cells, but acts principally on B-cells and enhances the B-cell's ability to present antigen to T-cells (69). Interleukin 6 (IL-6) is produced by various cell lines upon stimulation with IL-1 or tumor necrosis factor. It acts as a B-cell growth and differentiation factor and acts in concert with interleukin 3 to induce proliferation of pre B-cells in culture (70).

Economic Aspects

Analgesics and antiarthritics represent significant worldwide pharmaceutical markets. Table 6 lists trade names and producers of some of the more commercially important agents. The 1989 analgesic market was estimated at approximately \$4.0 billion. Of that figure, \$792 million is represented by opioid analgesics, whereas much of the nonnarcotic portion of the market is represented by NSAIDs. The 1989 antiarthritic market approached \$5 billion. Sales of prescription and nonprescription NSAIDs within this market were approximately \$4.5 billion. This figure includes both antiinflammatory and analgesic uses of these agents. The antiarthritic market is expected to continue to grow at a significant rate as the worldwide population ages.

BIBLIOGRAPHY

"Analgesics and Antipyretics" in *ECT* 1st ed., Vol. 1, pp. 851–861, by A. W. Ruddy, Sterling-Winthrop Research Institute, Division, Sterling Drug Inc.; *ECT* 2nd ed., Vol. 2, pp. 379–393, by George deStevens, Ciba Pharmaceutical Company; "Analgesics, Antipyretics, and Antiinflammatory Agents" in *ECT* 3rd ed., Vol. 2, pp. 574–586, by W. F. Michne, Sterling-Winthrop Research Institute.

1. C. B. Pert and S. H. Snyder, *Science* **179,** 1011 (1973); L. Terenius, *Acta Pharmacol. Toxicol.* **32,** 317 (1973); E. J. Miller, J. M. Hiller, and I. Edelman, *Proc. Natl. Acad. Sci. U.S.A.* **70,** 1947 (1973).
2. J. J. Galligan, H. I. Mosberg, R. Hurst, V. J. Hruby, and T. F. Birls, *J. Pharmacol. Exp. Ther.* **229,** 641 (1984).
3. J. S. Heyman, S. A. Mulvaney, H. I. Mosberg, and F. Porreca, *Brain Res.* **420,** 100 (1987); R. J. Sheldon, L. Nunan, and F. Porreca, *J. Pharmacol. Exp. Ther.* **249,** 462 (1989); K.-J. Chang, E. T. Wei, A. Killian, and J.-K. Chang, *J. Pharmacol. Exp. Ther.* **242,** 1 (1987).
4. J. D. Leander, *J. Pharmacol. Exp. Ther.* **224,** 89 (1983); L. A. Dykstra, D. E. Gmerek, G. Winger, and J. H. Woods, *J. Pharmacol. Exp. Ther.* **242,** 413 (1987).
5. J. Hughes, T. W. Smith, H. Kosterlitz, L. A. Fothergill, B. A. Morgan, and H. R. Morris, *Nature (London),* **258,** 577 (1975).
6. C. H. Li and D. Chung, *Proc. Natl. Acad. Sci. U.S.A.* **73,** 1145 (1976).
7. A. Goldstein, S. Tachibana, L. I. Lowney, M. Hunkapiller, and L. Hood, *Proc. Natl. Acad. Sci. U.S.A.* **76,** 6666 (1979).
8. B. A. Eipper and R. E. Mains, *Endocr. Rev.* **1,** 1 (1980).
9. R. V. Lewis, A. S. Stern, S. Kimura, J. Rossier, S. Stein, and S. Udenfriend, *Science* **208,** 1459 (1980).
10. J. M. Gulland and R. Robinson, *Mem. Proc. Manchester Lit. Philos. Soc.* **69,** 79 (1925).
11. M. Gates and G. Tschudi, *J. Am. Chem. Soc.* **78,** 1380 (1956).

12. R. K. Portenoy and K. M. Foley, *Pain* **25,** 171 (1986).
13. W. R. Martin and H. F. Fraser, *J. Pharmacol. Exp. Ther.* **133,** 388 (1961).
14. R. Dixon, T. Crews, C. Inturrisi, and K. Foley, *Res. Commun. Chem. Pathol. Pharmacol.* **41,** 3 (1983).
15. O. Eisleb and O. Schaumann, *Dtsch. Med. Wochenschr.* **65,** 967 (1939).
16. C. J. Glynn and L. E. Mather, *Pain* **13,** 237 (1982).
17. W. R. Martin, D. R. Jasinski, C. A. Haertzen, D. C. Kay, B. E. Jones, P. A. Mansky, and R. W. Carpenter, *Arch. Gen. Psychiatry,* **28,** 286 (1973).
18. V. P. Dole and M. A. Nysander, *J. Am. Med. Assoc.* **193,** 646 (1965).
19. W. R. Martin, *Pharmacol. Rev.* **35,** 283 (1983); K. Verebey, J. Volavka, S. Mule, and R. Resnik, *Clin. Pharmacol. Ther.* **20,** 315 (1976); J. P. Gonzalez and R. N. Brogden, *Drugs* **35,** 192 (1988).
20. F. Konno, C. Kobayashi, T. Morimoto, I. Takayanagi, N. Cho, and M. Hirobe, *Arch. Int. Pharmacodyn. Ther.* **282,** 219 (1986).
21. A. S. Keats and J. Telford, *J. Pharmacol. Exp. Ther.* **112,** 356 (1956).
22. H. Isbell, *Fed. Proc.* **15,** 442 (1956).
23. N. B. Fahmy, *Clin. Anaesthesiol.* **1,** 164 (1983).
24. R. D. Oullette, *J. Clin. Pharmacol.* **22,** 165 (1982).
25. R. N. Brogden, T. M. Speight, and G. S. Avery, *Drugs* **5,** 6 (1973).
26. M. S. Gilbert, R. M. Hanover, B. S. Moylan, and F. S. Caruso, *Clin. Pharmacol. Ther.* **20,** 359 (1976).
27. D. M. Zimmerman and J. D. Leander, *J. Med. Chem.* **83,** 895 (1990).
28. M. F. Piercey and F. J. Einspahr, *J. Pharmacol. Exp. Ther.* **251,** 267 (1989).
29. P. F. Von Voigtlander and R. A. Lewis, *J. Pharmacol. Exp. Ther.* **246,** 259 (1988).
30. G. Peters and S. Gaylor, *Clin. Pharmacol. Ther.* **45,** 130 (1989).
31. P. J. Birch, A. G. Hayes, M. J. Sheehan, and M. B. Tyers, *Eur. J. Pharmacol.* **144,** 405 (1987).
32. B. R. de Costa, R. B. Rothman, V. Bykov, A. E. Jacobson, and K. C. Rice, *J. Med. Chem.* **32,** 281 (1989); B. R. de Costa, L. Band, R. B. Rothman, A. E. Jacobson, V. Bykov, A. Pert, and K. C. Rice, *FEBS Lett.* **249,** 178 (1989).
33. K. Akiyam, K. W. Gee, H. I. Mosberg, V. J. Hruby, and H. I. Yamamura, *Proc. Natl. Acad. Sci. U.S.A.* **82,** 2543 (1985); A. D. Corbett, M. G. Gillan, H. W. Kosterlitz, A. T. McKnight, S. J. Paterson, and L. E. Robson, *Br. J. Pharmacol.* **83,** 271 (1984).
34. F. Porreca, J. S. Heyman, H. I. Mosberg, J. R. Omnaas, and J. L. Vaught, *J. Pharmacol. Exp. Ther.* **241,** 393 (1987).
35. J. L. Vaught, J. R. Mathiasen, and R. B. Raffa, *Pain Suppl.* **4,** S250 (1987).
36. E. S. Krames, D. J. Wilkie, and J. Gershow, *Pain* **24,** 205 (1986).
37. J.-C. Schwartz, B. Malfroy, and S. De La Baume, *Life Sci.* **29,** 1715 (1981).
38. J.-M. Lecomte, J. Constentin, A. Vlaiculesu, P. Chaillet, J. Marcais-Collado, C. Llorens-Cortes, M. LeBoyer, and J.-C. Schwartz, *J. Pharmacol. Exp. Ther.* **237,** 937 (1986); P. Floras, A.-M. Bidabe, J.-M. Caille, G. Simonet, J. M. Lecomte, and M. Sabathie, *Am. J. Neurorad.* **4,** 653 (1983).
39. P. R. Wilson and T. L. Yaksh, *Anaesth. Intens. Care* **8,** 248 (1980).
40. T. Otsuki, H. Nakahama, H. Niizuma, and J. Suzuki, *Br. Res.* **365,** 235 (1986).
41. D. C. Manning, A. A. Khan, J. N. Campbell, A. Meyers, and S. N. Raja, *Pain Suppl.* **4,** S16 (1987).
42. J. R. Vane, *Nature (London)* **231,** 232 (1971); J. B. Smith and A. L. Willis, *Nature (London)* **231,** 235 (1971).
43. R. J. Flower, *Pharmacol. Rev.* **26,** 33 (1974).
44. G. W. Gander, J. Chafee, and F. Goodale, *Proc. Soc. Exp. Biol. Med.* **126,** 205 (1967).
45. J. D. Commer in A. Goth, ed., *Medical Pharmacology,* C. V. Moseby Co., St. Louis, Mo., 1984, p. 366.

46. R. R. Gorman, *Fed. Proc.* **38,** 83 (1979).
47. B. Ameer and D. J. Greenblatt, *Ann. Intern. Med.* **87,** 202 (1977).
48. G. L. Royer, C. E. Seckman, and I. R. Welshman, *Am. J. Med.* **77(1A),** 25 (1984).
49. F. D. Hart and E. C. Huskisson, *Drugs* **27,** 232 (1984).
50. R. N. Brogden, R. C. Heel, T. M. Speight, and G. S. Avery, *Drugs* **18,** 241 (1979).
51. T. G. Kantor, *Pharmacotherapy* (Carlisle, Mass.) **6,** 93 (1986).
52. T. G. Kantor, *Am. J. Med.* **80**(3A), 3 (1986).
53. T. Y. Shen and C. A. Winter, *Adv. Drug Res.* **12,** 90 (1977).
54. P. A. Todd and E. M. Sorkin, *Drugs* **35,** 244 (1988).
55. H. O. J. Collier and W. J. F. Sweatman, *Nature (London),* **219,** 864 (1968).
56. R. N. Brogden, R. C. Heel, T. M. Speight, and G. S. Avery, *Drugs* **28,** 292 (1984).
57. R. O. Davies, *Pharmacotherapy* **3,** 95 (1983).
58. R. C. Haynes and F. Murad in A. Gillman, L. S. Goodman, T. W. Rall, F. Murad, eds., Goodman and Gilman's *The Pharmacological Basis of Therapeutics,* 7th ed., Macmillan Publishing Co., Inc., New York, 1985, p. 1459.
59. S. P. Hensch, E. C. Kendal, C. H. Slocumb, and H. F. Polly, *Mayo Clin. Proc.* **24,** 181 (1949).
60. F. Hirata, E. Schiffmann, K. Venkatacubmanian, D. Salomon, and J. Axelrod, *Proc. Natl. Acad. Sci. U.S.A.* **77,** 2533 (1980).
61. S. Moncada, R. J. Flower, and J. R. Vane in A. Gillman, L. S. Goodman, T. W. Rall, and F. Murad, eds., Goodman and Gilman's *The Pharmacological Basis of Therapeutics,* 7th ed., Macmillan Publishing Co., Inc., New York, 1985, p. 660.
62. H. Engler and I. Szelenyi, *Int. J. Tissue React.* **7,** 133 (1985).
63. C. Weny *Nature (London),* **298,** 597 (1982).
64. R. F. Meyers, J. C. Anthes, D. J. Casmer, and M. I. Siegel, *Inflammation* **9,** 91 (1985).
65. J. Boger and J. A. Schmidt, *Ann. Rep. Med. Chem.* **23,** 171 (1988).
66. W. R. Benjamin, P. T. Lomedico, and P. L. Kilian, *Ann. Rep. Med. Chem.* **20,** 173 (1985); S. B. Mizel, *Immunol. Today* **8,** 330 (1987).
67. J. J. Farrar, W. R. Benjamin, L. Chang, and B. A. Pawson, *Ann. Rep. Med. Chem.* **19,** 191 (1984).
68. M. Fletcher and A. L. Goldstein, *Lymphokine Res.* **6,** 45 (1987); K. A. Smith, *Immunol. Today* **8,** 11 (1987).
69. W. E. Paul and J. Ohara, *Ann. Rev. Immunol.* **5,** 429 (1987).
70. K. Ikebuchi, G. G. Wong, S. C. Clark, J. N. Ihle, Y. Hirai, and M. Ogawa, *Proc. Natl. Acad. Sci. U.S.A.* **84,** 9035 (1987).

RICHARD A. NUGENT
CHARLES M. HALL
Upjohn Company

ANESTHETICS

The term anesthesia comes from the Greek "anaisthai" or insensibility and constitutes a state in which perception of noxious events such as surgical procedures are imperceptible. This state may or may not be accompanied by loss of consciousness. A complete or general anesthetic given by the inhalation or intravenous route produces hypnosis (profound sleep), analgesia, muscle relaxation, and protection against the increase in blood pressure and heart rate resulting from surgical stress (maintains homeostasis). An anesthetic which blocks the neural transmission of painful stimuli through afferent nerves and does not affect the level of consciousness can be classified as a local anesthetic.

The induction of general anesthesia produces a progressive deepening of the anesthetic state and represents, in the anatomical sense, a descending desensitization of the central nervous system (CNS). A progression of clinical signs are useful for estimating the depth of anesthesia. These signs vary somewhat for each anesthetic, but in general four stages can be defined (1): (*1*) State of altered consciousness and analgesia indicative of action on cerebral canticle areas begins. (*2*) Loss of consciousness, often accompanied by delirium and excitement, occurs. Irregular respiration and motor movement result from depression of higher motor inhibitory centers with the release of lower motor mechanisms. (*3*) Stage of surgical anesthesia is reached in which spinal cord and spinal reflexes are abolished providing relaxation of skeletal musculature. Within the four planes in this state are loss of corneal, conjunctival, pharyngeal, and laryngeal reflexes as the depth of anesthesia progresses. (*4*) Onset of respiratory paralysis occurs resulting from significant depression of the medullary respiratory center. Subsequent cardiovascular collapse ensues.

Most signs which require a skeletal muscle reflex would not be apparent after treatment with a neuromuscular blocking drug or would be altered significantly by preoperative drugs. In these cases central nervous system monitoring via an electroencephalogram (EEG), and hemodynamic and blood gas monitoring, help assess the depth of anesthesia. The potency of inhaled agents is expressed as the minimum alveolar concentration (MAC) that is required to prevent spontaneous movement in response to a surgical or equivalent stimulus in 50% of patients (2).

The onset of action is fast (within 60 seconds) for the intravenous anesthetic agents and somewhat slower for inhalation and local anesthetics. The induction time for inhalation agents is a function of the equilibrium established between the alveolar concentration relative to the inspired concentration of the gas. Onset of anesthesia can be enhanced by increasing the inspired concentration to approximately twice the desired alveolar concentration, then reducing the concentration once induction is achieved (3). The onset of local anesthetic action is influenced by the site, route, dosage (volume and concentration), and pH at the injection site.

Theories of General Anesthesia

Although the modern practice of anesthesia is exceedingly sophisticated, the basic molecular mechanism underlying it is still lacking. Even whether the target sites of the inhalational anesthetics are lipid or protein remains unknown, although it is likely that proteins are intimately involved. The inhalational anesthetics are very diverse chemically and extrapolation from the effect of a particular agent to the physiological state of anesthesia is problematic. However, anesthesia theories may be roughly divided into two categories: lipid theories and protein theories (4–6).

Lipid Theories. Although there are many varieties of the lipid theory, all postulate that inhalational anesthetics exert their primary effects by dissolving in the lipid portions of nerves thereby altering the conductivity. This theory is based on observations made at the turn of the twentieth century on the relationship between anesthetic potency and the partition coefficient of the agent between olive oil and gas (7–9). The anesthetic potency of a wide variety of anesthetics, covering four orders of magnitude, correlates exceedingly well with the agent's oil–gas partition coefficient.

The primary site of action is postulated to be the lipid matrix of cell membranes. The lipid properties which are said to be altered vary from theory to theory and include: enhancing membrane fluidity; volume expansion; melting of gel phases; increasing membrane thickness, surface tension, and lateral surface pressure; and encouraging the formation of polar dislocations (10,11). Most theories postulate that changes in the lipids influence the activities of crucial membrane proteins such as ion channels. The lipid theories suffer from an important drawback: at clinically used concentrations, the effects of inhalational anesthetics on lipid bilayers are very small and essentially undetectable (6, 12, 13).

Protein Theories. The direct interaction of inhalation anesthetics and proteins (qv) has been proposed as the cause of anesthesia (12,14). An inhalation agent, whether a noble gas or a fluorinated ether, could dissolve asymmetrically in a protein (6,15). Resultant conformational changes in the protein, if these changes occur, could then cause changes in activity, for example, the chloride ion channel which is modulated by many of the intravenous agents. Quantitative data on the binding of anesthetics to proteins are limited but are related to anesthetic partition coefficients (16,17). Although binding to a soluble protein and modulation of its function are not necessarily coupled (18,19), the proteins of interest are those involved in nerve impulse transmission and are embedded in lipid membranes. Thus dissociation of lipid and protein effects is not possible.

The membrane enzyme luciferase, responsible for light emission in fireflies, is sensitive to anesthetics (20,21), and the concentrations of inhalational agents which inhibit luciferase are the same as those which cause general anesthesia. Studies of various classes of inhalational agents and luciferase demonstrated that above a certain chain length in a homologous series, a point is reached where higher members are not anesthetic. The same cut-off effect in efficacy is observed in anesthesia (22). This effect is not explainable by lipid theory.

Anesthetic Agents

INHALATION AGENTS

An ideal inhalation anesthetic would exhibit physical, chemical, and pharmacological properties allowing safe usage in a variety of surgical interventions. The agent should be odorless, nonflammable at concentrations which are likely to be used in the operating room, and stable both on storage and to soda lime, which is used as the CO_2 absorber in the anesthetic circuit. Induction of, and recovery from, anesthesia should be rapid, and minimal side effects should be observed on the cardiovascular (depression and epinephrine compatibility) or central nervous systems (EEG activation) (see EPINEPHRINE AND NOREPINEPHRINE; NEUROREGULATORS). The drug should not be metabolized.

A number of inhalation anesthetics have been introduced to clinical practice, some of which are listed in Table 1. All agents introduced after 1950, except ethyl vinyl ether, contain fluorine. Agents such as ether, chloroform, trichloroethylene (Trilene), cyclopropane, and fluoroxene (Fluoromar), which were once used, have been displaced by the newer fluorinated anesthetics.

Historical Inhalation Agents. Diethyl ether produces excellent surgical anesthesia, but it is flammable (see ETHERS). Chloroform is a nonflammable, sweet smelling, colorless liquid which provides analgesia at nonanesthetic doses and can provide potent anesthesia at 1% (see CHLOROCARBONS AND CHLOROHYDROCARBONS). However, a metabolite causes hepatic cell necrosis. Trilene, a nonflammable colorless liquid, has a slower onset and recovery and a higher toxicity and chemical reactivity than desirable. Cyclopropane is a colorless gas

Table 1. Properties and Partition Coefficients of Inhalation Anesthetics

Agent	CAS Registry Number	Molecular formula	Partition coefficient[a] Oil/gas	Blood/gas	Boiling point, °C	Year introduced
ethyl ether	*[60-29-7]*	$C_4H_{10}O$	65	12.1	35	1842
chloroform	*[67-66-3]*	$CHCl_3$	394	8.4	61	1847
trichloroethylene	*[79-01-6]*	C_2HCl_3	200–250	9.0	87	1934
fluoroxene	*[406-90-6]*	$C_4H_5F_3O$	47	1.37	43.1	1960
halothane	*[151-67-7]*	$C_2HBrClF_3$	224	2.3	50.2	1956
isoflurane	*[26675-46-7]*	$C_3H_2ClF_5O$	90.8	1.4	48.5	1980
enflurane	*[13838-16-9]*	$C_3H_2ClF_5O$	96.5	1.9	56.5	1972
methoxyflurane	*[76-38-0]*	$C_3H_4Cl_2F_2O$	970	12.0	105	1962
sevoflurane	*[28523-86-6]*	$C_4H_3F_7O$	42	0.6	58	1989
desflurane	*[57041-67-5]*	$C_3H_2F_6O$	18.7	0.42	23.5	
nitrous oxide	*[10024-97-2]*	N_2O	1.4	0.46	−89.5	1850s
cyclopropane	*[75-19-4]*	C_3H_6			−33	1934

[a]Ref. 23.

which has rapid induction (2–3 min) and recovery characteristics and analgesia is obtained in the range of 3–5% with adequate skeletal muscle relaxation (see HYDROCARBONS). The use of cyclopropane has ceased, however, because of its flammability and marked predisposition to cause arrhythmias.

Because of the flammable nature of diethyl ether and cyclopropane, a number of hybrid molecules were prepared in an attempt to improve on the useful characteristics of each, but none were marketed (24,25). Structure activity relationships for halogenated hydrocarbons and ethers (26) indicated that fluorine (qv) was the most suitable halogen to use in order to decrease the potential for cardiac arrhythmias. In addition, the introduction of fluorine decreases flammability and boiling point and increases stability (see FLUORINE COMPOUNDS, ORGANIC). It also decreases anesthetic potency. Fluoroxene, 2,2,2,-trifluoroethyl vinyl ether *[406-90-6]*, the first fluorinated ether prepared and commercialized (27), has oil/gas and blood/gas partition coefficients (Table 1) lower than ether, chloroform, or Trilene, contributing to a rapid onset and recovery. This agent is less potent (MAC = 3.4%) than other fluorinated hydrocarbons introduced later and its flammability is close to MAC. Whereas the agent is stable to soda lime, it is readily oxidized and hydrolyzed to trifluoroethanol and acetaldehyde and is unstable to light (28). Fluoroxene is also metabolized to a significant extent (29,30). Thus it is no longer used.

Clinical Inhalation Agents. *Nitrous Oxide.* Nitrous oxide, described by Priestly in 1772, was first used to relieve severe dental pain in the latter part of the 18th century. Sometime in the mid-1800s N_2O was successfully used as an anesthetic, and its widespread usage coincided with the development of anesthesia machines. Nitrous oxide is a nonflammable, colorless, odorless, and tasteless gas that can exist as a liquid under pressure at room temperature. It is normally stored in cylinders. However, it supports combustion.

A mixture of 20% nitrous oxide and 80% oxygen produces analgesia equivalent to 15 mg of morphine in humans (see ANALGESICS, ANTIPYRETICS, AND ANTIINFLAMMATORY AGENTS) (31,32). Effects of nitrous oxide can be partially reversed using opiate antagonists and it has been suggested that the agent interacts with endogenous opioids (23,33,34) (see OPIOIDS, ENDOGENOUS). Because of analgesic properties, N_2O is added to other inhaled anesthetics and to opioids at concentrations usually between 30 and 70% in oxygen. Nitrous oxide is a very weak anesthetic, MAC = 104%. However, the blood/gas partition coefficient has been established to be about 0.46 (35,36) allowing the agent to come into rapid equilibrium with the inspired and alveolar concentrations (37). Because of poor solubility in blood and other tissues, induction and recovery from nitrous oxide administration is rapid.

Nitrous oxide produces respiratory depression (38,39). It has been shown to produce a direct myocardial depressant effect in dogs (40) and in humans breathing a 40% N_2O/60% oxygen mixture (41); however, this may be offset by the activation of the sympathetic nervous system (42). The combination of nitrous oxide and opioids can produce decreases in myocardial contractility, heart rate, and blood pressure (43).

There appear also to be toxic effects. In animals, nitrous oxide has been shown to inactivate methionine synthetase which prevents the conversion of deoxyuridine to thymidine and thus has the potential for inducing megaloblastic anemia, leukopenia, and teratogenicity (44–46). A variety of epidemiologic surveys suggest positive correlations between exposure to nitrous oxide and spontaneous abortion in dental assistants (47).

Methoxyflurane. Methoxyflurane (2,2-dichloro-1,1-difluoroethyl methyl ether) [*76-38-0*], a clear colorless liquid having a vapor pressure of 2.70 kPa (20.3 mm Hg) at 20°C, is not irritating to the respiratory tract. Its odor has been described as sweet or fruity. Whereas the odor may make this agent suitable for inhalation induction, the blood/gas and oil/gas partition coefficients (Table 1) are very high and thus onset and recovery are the slowest of all anesthetics (48). Methoxyflurane is stable to light and soda lime, but not to oxygen; it is nonflammable and nonexplosive at anesthetic concentrations (28). Methoxyflurane is the most potent of the inhaled anesthetics having MAC = 0.6%. The agent is both a respiratory and cardiovascular depressant (49). It also produces muscle relaxation as the depth of anesthesia increases.

Approximately 50% of an absorbed dose is transformed to fluoride ion, dichloroacetic acid, methoxydifluoroacetic acid, and oxalic acid (50). Because of fluoride ion associated renal impairment, the duration of anesthesia using methoxyflurane must be limited (51,52).

Halothane. Halothane or Fluothane, 2-bromo-2-chloro-1,1,1-trifluoroethane [*151-67-7*], is a colorless liquid with a pleasant odor. Its lower flammability limit, 4.8% in 70% N_2O/30% O_2, renders it essentially nonflammable. It has a vapor pressure of 32.5 kPa (244 mm Hg) at 20°C and is stable to soda lime. However, it is photochemically reactive.

Halothane is a potent anesthetic having a MAC value of 0.77%. MAC requirements, as in other inhaled anesthetics, increase in younger patients and decrease in the elderly (53). Halothane is a cardio–respiratory depressant which produces changes in cardiac output and arterial pressure comparable to the other inhalation agents, isoflurane or enflurane. Halothane does not decrease peripheral vascular resistance as do these others. It is also the least likely of the inhalants to cause an increase in heart rate. The ability of halothane to sensitize the heart to arrhythmogenic effects of epinephrine is the greatest of the inhalants (54), although halothane produces less muscle relaxation than isoflurane (55). Approximately 20% of an absorbed dose of halothane is oxidatively metabolized to trifluoroacetic acid (56,57) and halothane has been implicated in producing liver dysfunction in a very small number of patients (58). Fluoride produced from the biodegradation of halothane or the other agents has little effect on normal kidney function (59). Halothane usage has been declining because of the potential liver effects, although the agent is used where inhalation induction is desired, especially in pediatrics.

Enflurane. Enflurane or Ethrane, 2-chloro-1,1,2-trifluoroethyl difluoromethyl ether [*13838-16-9*], was synthesized in 1963. This agent is a mildly pungent colorless liquid having a vapor pressure of 22.85 kPa (171.8 mm Hg) at 20°C. Enflurane is stable to soda lime and light (28). The minimum flammable concentration is 5.8% in 70% N_2O/30% O_2, placing its flammability between that of halothane and isoflurane. Enflurane is the least potent of the inhaled anesthetics marketed in 1990. It has a MAC value of 1.7% (60), although the solubility characteristics of enflurane are similar to those of isoflurane and halothane (Table 1) and recovery is both rapid and less complicated by nausea than in the case of halothane (61–63).

When compared with isoflurane and halothane at equi-anesthetic concentrations, enflurane produces a greater degree of respiratory depression (64). Hemodynamic parameters such as cardiac output, stroke volume, and blood pressure decrease where heart rate is only slightly changed. Surgical stimulation tends to reverse depressed cardiac output and arterial blood pressure (65). In the presence of epinephrine, enflurane does not potentiate arryhthmias to the same extent as halothane (66). Enflurane provides muscle relaxation and when combined with neuromuscular blocking drugs, the block is enhanced, allowing lower doses of the relaxant to be used (67,68). Unlike other inhalational anesthetics, enflurane at high concentrations causes EEG changes which are seizurelike. Enflurane is oxidatively metabolized in the liver. In humans, about 2.4% of the enflurane is metabolized (69), but clinically significant hepatic dysfunction is rare.

Isoflurane. Isoflurane or Forane, 1-chloro-2,2,2-trifluoroethyl difluoromethyl ether [*26675-46-7*], was prepared in 1965 (28). It is a clear, colorless liquid having a vapor pressure of 31.85 kPa (239.5 mm Hg) at 20°C. It has an ethereal odor which may limit the rate at which the inspired concentration may be increased for a rapid induction. The agent is stable to soda lime and light, and does not react with metals. The minimum flammable concentration is 7% in 70% N_2O/30% O_2. Isoflurane is a potent anesthetic having a MAC value of 1.15% in O_2 (70). MAC requirements are decreased by increasing age, decreasing body temperature, pregnancy, and a variety of depressant drugs. This agent is the least soluble of the modern volatile agents (Table 1) allowing for a rapid onset and recovery.

Isoflurane is a respiratory depressant (71). At concentrations which are associated with surgical levels of anesthesia, there is little or no depression of myocardial function. In experimental animals, isoflurane is the safest of the oral clinical agents (72). Cardiac output is maintained despite a decrease in stroke volume. This is usually because of an increase in heart rate. The decrease in blood pressure can be used to produce "deliberate hypotension" necessary for some intracranial procedures (73). This agent produces less sensitization of the human heart to epinephrine relative to the other inhaled anesthetics. Isoflurane potentiates the action of neuromuscular blockers and when used alone can produce sufficient muscle relaxation (74). Of all the inhaled agents currently in use, isoflurane is metabolized to the least extent (75). Unlike halothane, isoflurane does not appear to produce liver injury and unlike methoxyflurane, isoflurane is not associated with renal toxicity.

Isoflurane is the most widely used inhalational anesthetic and more closely approaches the ideal than other marketed drugs. It has found application in the anesthetic management of all types of surgical procedures.

Developmental Inhalational Agents. *Sevoflurane.* Sevoflurane, 1,1,1,3,3,3-hexafluoro-2-propyl fluoromethyl ether [*28523-86-6*], is nonpungent, suggesting use in induction of anesthesia. The blood/gas partition coefficient is less than other marketed products (Table 1) yet similar to nitrous oxide, suggesting fast onset and recovery. In animal studies, recovery was faster for sevoflurane than for isoflurane, enflurane, or halothane (76). Sevoflurane is stable to light, oxygen, and metals (28). However, the agent does degrade in soda lime (77).

Sevoflurane is less potent than clinically used inhalation agents having MAC of 1.71% and 2.05% in humans (78,79). Sevoflurane is also a respiratory depressant similar to other agents (78,80) decreasing both blood pressure and heart rate (80). Sevoflurane decreases myocardial oxygen blood flow (81) and the potential to sensitize the mycocardium to epinephrine induced arrhythmias is similar to isoflurane (82). Other positive characteristics include sufficient muscle relaxation and an EEG pattern which is devoid of convulsant activity (78,79). Sevoflurane appears to have advantages over current products, especially in regard to onset, recovery, and no increase in heart rate. The primary disadvantage appears to be its instability to soda lime. The drug was launched in Japan in late 1989.

Desflurane. Suprane or desflurane, 1,2,2,2-tetrafluoroethyl difluoromethyl ether [*57041-67-5*], has a structure similar to isoflurane except for the substitution of a fluorine for the chlorine on the alpha methyl carbon. The agent has a vapor pressure of 88.3 kPa (664 mm Hg) at 20°C. The partition coefficients for blood/gas and tissue/gas (Table 1) suggest that this agent should have a faster onset and recovery than current agents including sevoflurane (76,83). Its odor is milder than isoflurane when used for induction. In unpremedicated patients upper airway tolerance was good: only transient coughing and easily managed secretions were observed. Desflurane is stable to O_2, light, and metals, and also to soda lime in systems where sevoflurane, halothane, and isoflurane are forced to degrade (84).

Desflurane is less potent than the other fluorinated anesthetics having MAC values of 5.7 to 8.9% in animals (76,85), and 6% to 7.25% in surgical patients. The respiratory effects are similar to isoflurane. Heart rate is somewhat increased and blood pressure decreased with increasing concentrations. Cardiac output remains fairly stable. Desflurane does not sensitize the myocardium to epinephine relative to isoflurane (86). EEG effects are similar to isoflurane and muscle relaxation is satisfactory (87). Desflurane is not metabolized to any significant extent (88,89) as levels of fluoride ion in the serum and urine are not increased even after prolonged exposure. Desflurane appears to offer advantages over sevoflurane and other inhaled anesthetics because of its limited solubility in blood and other tissues. It is the least metabolized of current agents.

INTRAVENOUS ANESTHETIC AGENTS

The intravenous (iv) anesthetic agents are of two types: those which are used to induce, but not maintain, anesthesia, and those which are useful not only for induction, but also for maintenance. The period of induction is perhaps the most crucial part of the anesthesia. The need is for an anesthetic agent having an extremely fast rate of onset and limited side effects. Fast onset minimizes the stress and agitation which could arise during a more lengthy induction. An ideal maintenance iv anesthetic has been a goal for over a century. Various classes of compounds have been used. Among these are: barbiturates, opioids, and hindered phenols. But the ideal agent for both induction and maintenance has not yet been found. For this reason, balanced anesthesia is often used: a potent opioid is combined with an inhalational agent.

A major difference between the inhalational agents and the iv anesthetics is that the former probably exert their biological activity through physical effects while the latter, in most cases, function through a biological receptor mediated pathway. There are two primary biochemical receptors used for iv anesthesia in the CNS, the gabaergic (90) and the opiate (91). The γ-aminobutyric acid [*56-12-2*] (GABA) receptor controls a chloride ion channel and the GABA receptor itself is modulated by various drug receptors for the barbiturates, anesthetic steroids, and benzodiazepines. An increase in GABA binding results in a variety of effects: anesthesia, sleep, amnesia, muscle relaxation, and an antianxiety effect. The opiate receptor, of which there are various types, controls analgesia (92). Its primary endogenous ligands are the enkephalins and endorphins. However, the more familiar plant alkaloid morphine [*57-27-2*] also binds strongly to this receptor (93). Because anesthesia using iv agents is receptor based, antagonists have been developed which can reverse the activity of these anesthetics.

Barbiturates. The first class of readily accepted iv anesthetics were barbiturates. This group's biological activity results from interaction with a drug receptor that modulates GABA binding (94,95). The first of these agents was thiopental [*76-75-5*] (Pentothal), $C_{11}H_{18}N_2O_2S$, (**1**) which is still the most widely used drug for induction. Thiopentone is an ultra short-acting depressant for the CNS which induces hypnosis and amnesia, but not analgesia, within one arm-brain circulation time (30–40 s) (96). Its action is predictable, induction is smooth, and recovery after a small dose is rapid, accompanied by some somnolence and retrograde amnesia. The agent does, however, cause repiratory and cardiovascular depression and it accumulates as a result of redistribution into the fatty tissues, so repeated doses are not advisable (97). A general problem with thiobarbiturates is a poor therapeutic index. There is continued research into safer, more effective iv anesthetics (98).

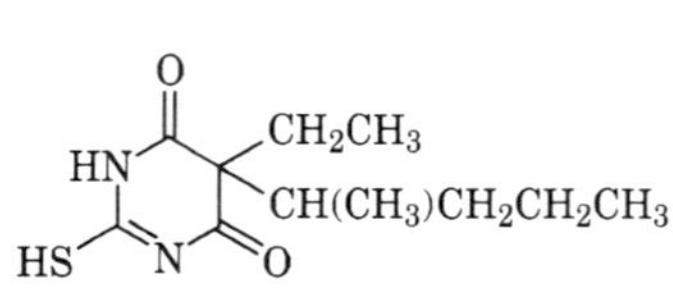

(**1**)

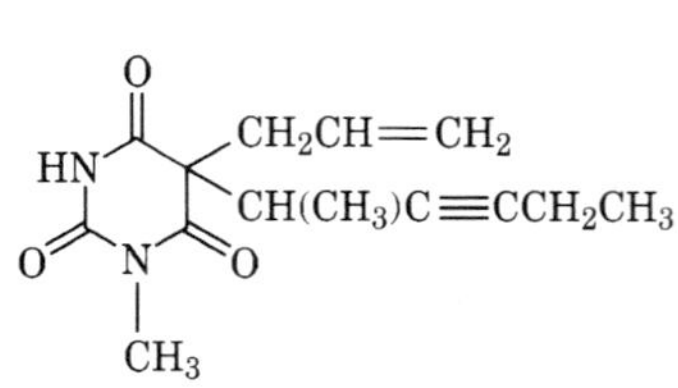

(**2**)

Methohexital [*18652-93-2*] (Brevital), $C_{14}H_{18}N_2O_3$, (**2**) is a barbiturate iv anesthetic induction agent that has a slightly faster onset than thiopentone and less accumulation. The recovery from anesthesia is also slightly faster and better. However, induction is associated with an increased incidence of excitatory phenomena. Methohexital also causes respiratory and cardiovascular depression and is unstable in solution, necessitating reconstitution before use (99).

Benzodiazepines. Librium and Valium were first introduced for their muscle relaxant activity, but became widely used for their antianxiety (minor tranquilizer) activity. The benzodiazepines bind to a drug receptor which modulates the binding of GABA to its receptor (100). Diazepam [*439-14-5*] (Valium), $C_{16}H_{13}ClN_2O$, (**3**) is used as a premedicant to anesthesia to lessen preoperative anxiety and to prevent recall of the operative procedure. Midazolam [*59467-70-8*] (Versed) $C_{18}H_{13}ClFN_3$, (**4**) a water soluble benzodiazepine having a shorter duration of action than diazepam (101), has specific applications in anesthesia. Its primary intramuscular use is as a premedicant similar to diazepam. As an iv anesthetic it is comparable to thiopentone for induction but has a slower onset, approximately 1.5 min after opioid premedication and 2–2.5 min without. Recovery is slower using midazolam than it is using thiopentone.

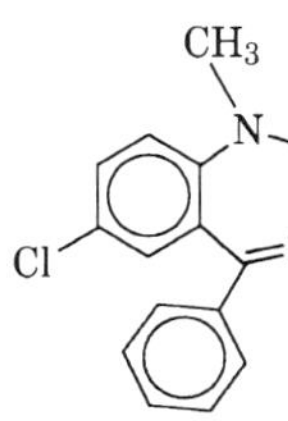

(**3**)

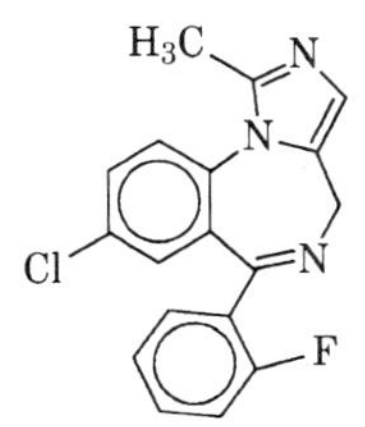

(**4**)

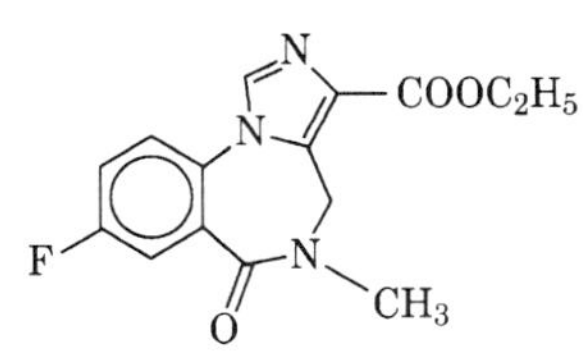

(**5**)

Midazolam and diazepam decrease arterial pressure without a change in heart rate. Like thiopentone, midazolam is a respiratory depressant. Advantages of midazolam are its amnestic effect, coupled with less postoperative depression (102). A reversal agent for the benzodiazepines has also become available. Flumazenil [*78755-81-4*], $C_{15}H_{14}FN_3O_3$, (**5**) displaces the benzodiazepines from their receptor but has little demonstrable activity of its own (103,104).

Etomidate. Etomidate [*33125-97-2*] (Hypnomidate, Amidate), $C_{14}H_{16}N_2O_2$, (**6**) is a relatively simple imidazole carboxylic ester developed in the 1970s as an iv anesthetic that became available in the United States in the mid-1980s (105). Etomidate has almost as rapid an onset and short duration of action as thiopentone, but it is approximately twelve times as potent. In contrast to other agents, it is rapidly and extensively metabolized by liver and plasma esterases. It causes less respiratory and cardiovascular depression than thiopentone and does not release histamine. In addition to pain on injection, there is frequent occurrence of extraneous muscle movement (myoclonus), controllable by benzodiazepine or opiate premedication (106). There is also a relatively high frequency of nausea and vomiting, particularly when opiate premedication is used. Despite these disadvantages, etomidate is extensively used in neuro and cardiovascular surgery (107).

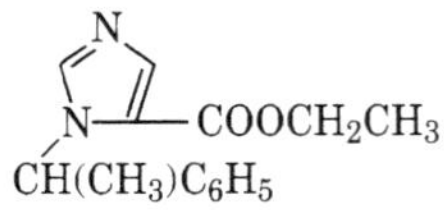

(6)

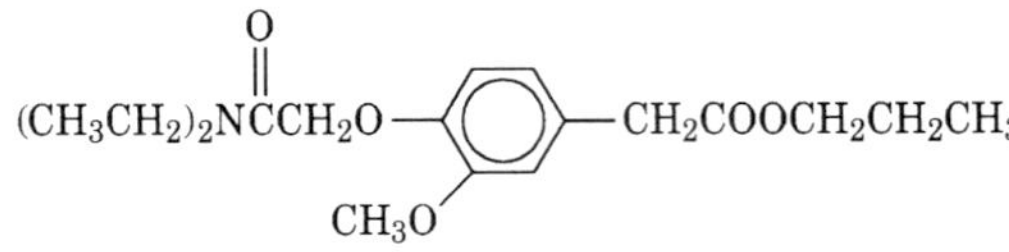

(7)

Propanidid. Propanidid [*1421-14-3*] (Epontol), $C_{18}H_{27}NO_5$, (**7**) a derivative of the propyl ester of homovanillic acid, has been in clinical use in Europe for a number of years. Its main advantage is rapid onset of action and a fast recovery which, like etomidate, is because of rapid metabolism by esterases rather than redistribution (108). Excretion is rapid: 75 to 90% of the drug is eliminated as metabolites within two hours. Propanidid side effects include hypotension, tachycardia, and hyperventilation followed by apnea, as well as excitatory side effects such as tremor and involuntary muscle movement (109).

Propofol. 2,6-Diisopropylphenol [*2078-54-8*], propofol, $C_{12}H_{18}O$, is a newer intravenous anesthetic used for both induction and maintenance (110). This drug, also known as Diprivan, is chemically unlike any of the previously described iv agents and was developed following the discovery of the anesthetic activity of the 2,6-diethyl analogue (111). Propofol itself is insoluble in water and is usually formulated in a soya bean oil emulsion. Propofol induction, with or without an opioid, is smooth and similar to that of other agents, although pain at the injection site and apnea have been reported (112,113). Recovery after induction and maintenance is faster, with fewer side effects, than with thiopentone (114), although sexual disinhibition has been anecdotally reported (115). Propofol is rapidly distributed, metabolized, and eliminated (116).

Opioids. Morphine [*57-27-2*], $C_{17}H_{19}NO_3$, (**8**) the most prevalent and analgesically potent of the naturally occurring opium alkaloids (qv), has been used as an anesthetic premedication for over one hundred years (93). It has also been used as an iv analgesic for the last four decades, and, since 1969, in high doses as an anesthetic agent (117).

Morphine has certain undesirable side effects. Among these are respiratory depression, nausea, and vomiting, depression of the cough reflex, cardiovascular depression and hypotension, smooth muscle contraction (constipation), and histamine release (93). Morphine's onset of action, duration, and low therapeutic indices have prompted a search for a more effective opiate iv anesthetic. Extreme simplification of the complex morphine molecule has resulted in anilido–piperidines, the fentanyl class of extremely potent opiate iv anesthetics (118,119).

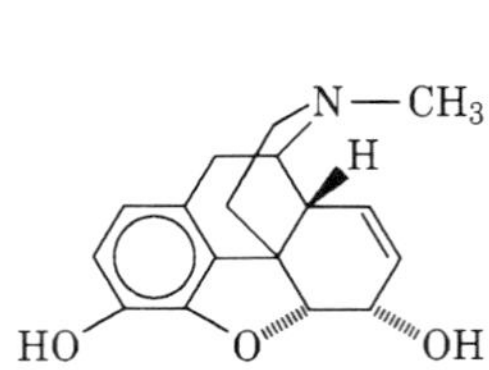

(8)

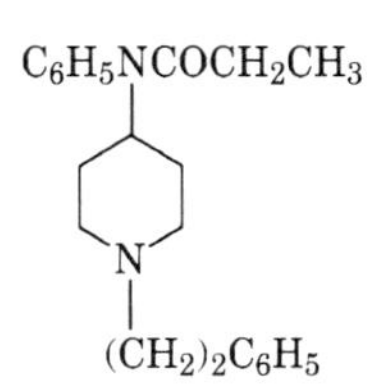

(9)

Fentanyl [*437-38-7*] (Sublimaze, Leptanal), $C_{22}H_{28}N_2O$, (**9**) has been extensively used since its introduction into clinical practice in the 1960s (119). Because of its potency, which is 50–100 times that of morphine, a rapid onset of action and a short duration, its use as an iv anesthetic is widespread. The short duration results from redistribution from the brain to other tissues, rather than elimination. It does, however, have the usual opiate disadvantages; respiratory depression, chest wall rigidity, nausea, and bradycardia. Fentanyl has an extremely wide therapeutic ratio. The size of the dose influences its duration of action which, after iv administration, may last from approximately 30 min to 2 to 3 h (120,121). In cardiac surgery fentanyl is administered in very large doses to produce profound analgesia and suppress cardiovascular reflex responses. This technique is particularly useful for patients with compromised circulation where any increase in cardiac demand could precipitate myocardial ischemia (122).

Approved for clinical use in 1986, alfentanil [*71195-58-9*] (Alfenta, Rapifen), $C_{21}H_{32}N_6O_3$, (**10**, R = 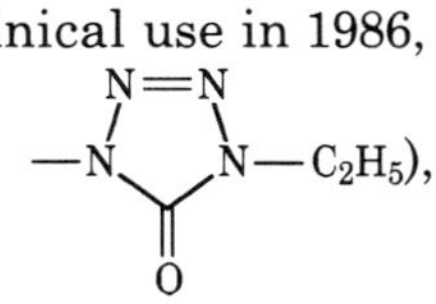), is about one fourth as potent as fentanyl and has a rapid onset. Alfentanil has an extremely short duration of action, from 0.2–0.3 times that of fentanyl (123,124), paralleling its rapid elimination. Thus alfentanil has been given by infusion pumps. Although alfentanil has the usual opiate side effects, its rapid elimination leads to a more rapid recovery and less frequent side effects and complications (125,126).

Sufentanil [*56030-54-7*] (Sufenta), $C_{22}H_{30}N_2O_2S$, (**10**, R = 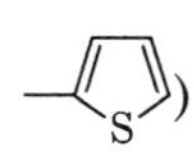) is five to ten times as potent as fentanyl. Sufentanil also has a short duration of action, but has an extremely high therapeutic ratio (127). Sufentanil, with less respiratory depression than either alfentanil or fentanyl, is used in high dose opioid anesthesia for cardiac surgery, and as a low dose supplement in balanced anesthesia for general and outpatient surgery (128,129).

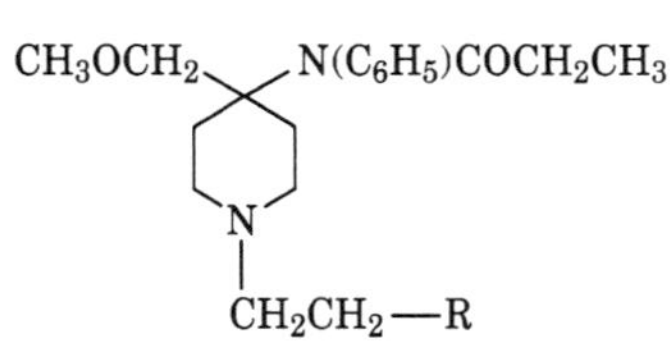

(**10**)

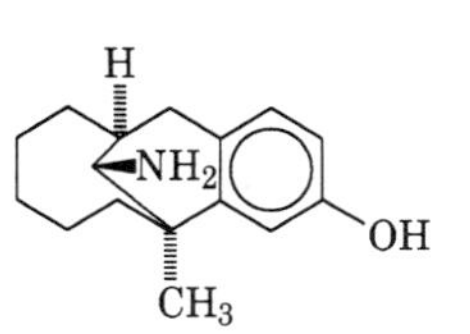

(**11**)

Dezocine [*53648-55-8*] (Dalgan), $C_{16}H_{23}NO$, (**11**) is an opiate having both agonist and antagonist properties which lowers MAC requirements for cyclopropane with a ceiling effect on respiratory depression in rats (130) and reduces the MAC requirement for enflurane in dogs (131). Dezocine was effective against post operative pain, producing sedation as a primary side effect (132). The compound has also produced respiratory depression in volunteers but was reported to have a ceiling effect (133,134), and there has been a suggestion of psychotomimetic effects (134).

Opioid Reversal Agents. At the end of a surgery, when the decision is made to reverse opioid induced activity, the primary concern is to eliminate the respiratory depressant effects inherent in the potent opiates used as anesthetics. This reversal, brought about by opioid reversal agents (Table 2), can also diminish

Table 2. Opioid Reversal Agents

Compound name	CAS Registry Number	Trade name	Molecular formula	Structure number	Structure	Reference
				Antagonists		
naloxone naltrexone	[*465-65-6*] [*16590-41-3*]	Narcan Trexan	$C_{19}H_{21}NO_4$ $C_{20}H_{23}NO_4$	(**12**, R = $CH_2CH{=}CH_2$) (**12**, R = CH_2—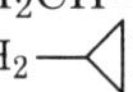)	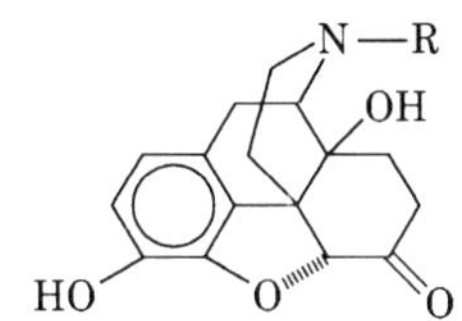	135 136–138
nalmefene	[*55096-26-9*]		$C_{21}H_{25}NO_3$	(**13**)	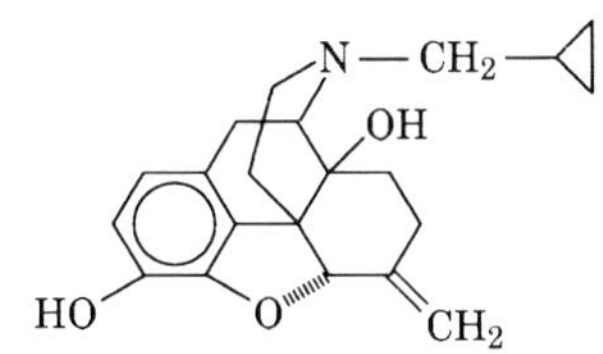	144
				Agonist–antagonists		
levallorphan	[*152-02-3*]	Lorfan	$C_{19}H_{25}NO$	(**14**)	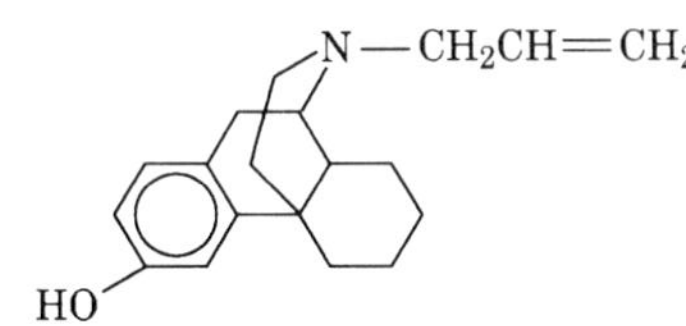	139
nalorphine nalbuphine	[*62-67-9*] [*20594-83-6*]	Lethidrone Nubain	$C_{19}H_{21}NO_3$ $C_{21}H_{27}NO_4$	(**15**, R = $CH{=}CH_2$) (**15**, R = 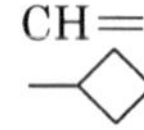)	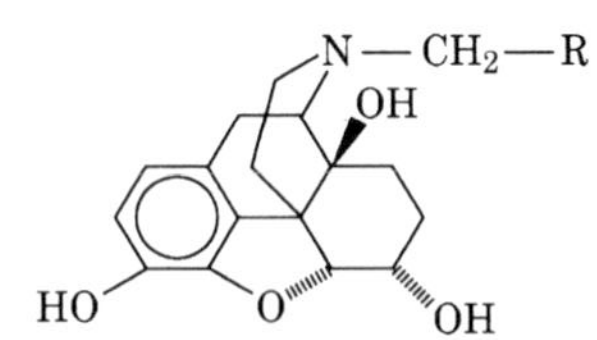	140,141 142,143

the analgesic effects of the opiates. There are two types of reversal agents: pure antagonists and mixed agonist–antagonists. Antagonism of the opiate agonist effect may be viewed as lowering agonist activity below a series of thresholds. The dose can be titrated so that a significant degree of analgesia can remain even though clinically significant respiratory depression is reversed. The antagonists naloxone (**12,** R = C_3H_5) and naltrexone (**12,** R = C_4H_7) are the most commonly used, although the use of the agonist–antagonist nalbuphine (**15,** R = C_4H_7) is increasing because it maintains a moderate level of analgesia while reversing the side effects (143). Nalmefene (**13**), a long acting reversal agent, is in clinical trials (144). The agonist–antagonist nalorphine (Lethidrone) (**15**, R = C_2H_3) is not available in the United States. Levallorphan (**14**) is little used because of nalorphinelike psychotomimetic effects.

Steroid Anesthetics. The anesthetic properties of a number of steroids were discovered in 1941 (145) and Althesin, a mixture of alfaxalone [*23930-19-0*], $C_{21}H_{32}O_3$, (**16,** R = H), and alfadone acetate [*23930-37-2*], $C_{23}H_{34}O_5$, (**16,** R = $OOCH_3$), was introduced in 1971 (146). Althesin appeared to have many of the ideal iv anesthetic properties. However, it was associated with a high incidence of allergic reactions and implicated in the development of renal failure (147); it was withdrawn in 1984. Like the barbiturates and benzodiazepines, the anesthetic steroids bind to a drug receptor which modulates GABA control of a chloride ion channel (148,149).

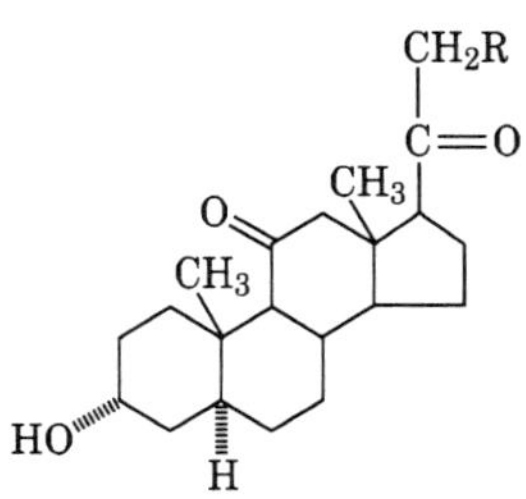

(16)

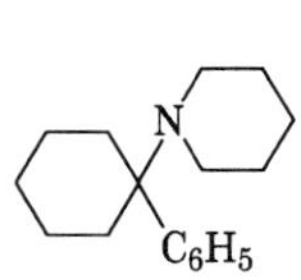

(17)

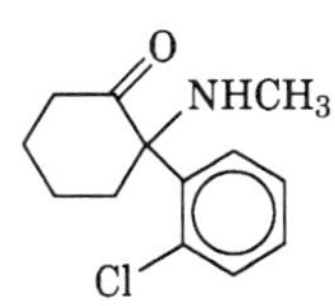

(18)

Other Anesthetic Agents. Dissociative anesthesia is so called because of the feeling of dissociation engendered by administration of drugs related to phencyclidine [*77-10-1*] (PCP), $C_{17}H_{25}N$, (**17**). The anesthesia, different from that produced by the inhalational and other iv agents, is characterized by a trancelike cataleptic state. However, the eyes can remain open and light reflexes are retained. Ketamine [*6740-88-1*] (Ketalar), $C_{13}H_{16}ClNO$, (**18**) is chemically related to phencyclidine (150). Ketamine anesthesia is characterized by rapid onset but slow recovery, a profound analgesia, and a lack of clinical cardiovascular depression. Unlike other iv agents, it can also be administered intramuscularly. Its greatest disadvantage is that, like PCP, it is a potent psychotomimetic. For this reason, it is only used for trauma patients, asthmatics, children, and patients at high risk (151).

Neuroleptic analgesia is so called because the combination of a major tranquilizer, a neuroleptic drug, and a potent opiate produces an anesthetic state characterized by sedation, apathy, and mental detachment (see PSYCHOPHARMACOLOGICAL AGENTS) (152). Innovar [*8067-59-2*], a combination of droperidol [*548-72-*

2], $C_{22}H_{22}FN_3O_2$, (**19**) and fentanyl (**9**) citrate, is used for procedures that do not require muscle relaxation. However, the onset of action is slow.

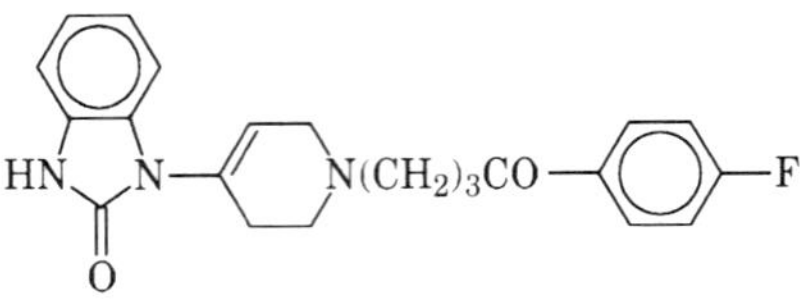

(**19**)

LOCAL ANESTHETICS

Local anesthetics produce anesthesia by blocking nerve impulse conduction in sensory, as well as motor nerve, fibers. Nerve impulses are initiated by membrane depolarization, effected by the opening of a sodium ion channel and an influx of sodium ions. Local anesthetics act by inhibiting the channel's opening; they bind to a receptor located in the channel's interior. The degree of blockage on an isolated nerve depends not only on the amount of drug, but also on the rate of nerve stimulation (153–156).

Local anesthetic activity is usually demonstrated by compounds which possess both an aromatic and an amine moiety separated by a lipophilic hydrocarbon chain and a polar group (153,157). A selection of local anesthetic agents is given in Table 3. In the clinically useful agents, the polar group is an ester (**20**) or an amide (**21**). Activity may be maintained, however, when the polar function is an ether, thioether, ketone, or thioester. The amino esters, more susceptible to hydrolysis than the amino amides, are metabolized in the plasma by cholinesterases. The metabolites of the amino esters are derivatives of 4-aminobenzoic acid (PABA) and these substances are capable of inducing allergic reactions in a small percentage of patients. The amino amides undergo enzymatic degradation in the liver and allergic reactions are rare.

Pharmacological Profile. The profile of the ideal local anesthetic agent depends largely on the type and length of the surgical procedure for which it is applied (157). Procedures could include neuraxial (spinal and epidural) anesthesia, nerve and plexus blocks, or field blocks (local infiltration). In general, the ideal agent should have a short onset of anesthesia and be useful for multiple indications such as infiltration, nerve blocks, intravenous, extradural, spinal, and topical administration. The therapeutic indexes for systemic CNS and cardiovascular toxicity should be high and the agent should be compatible with the vasoconstrictor epinephrine.

The pharmacological profile of local anesthetic agents depends largely on the physicochemical properties given in Table 4 (153). Lipophilicity appears to be the primary determinant of *in vitro* (isolated nerve) potency. Agents which are lipophilic penetrate nerve membranes more easily than more hydrophilic compounds resulting in a greater concentration of agent at the interior sodium ion channel receptor. The *in vivo* potency, however, is also dependent on such other factors as vasodilator and tissue redistribution properties. The duration of anesthesia is primarily dependent on the agent's protein binding capability. Long duration agents, eg, bupivacaine (**22,** R = n-C_4H_9) and tetracaine (**20,** X = H, R =

Table 3. Local Anesthetic Agents

Name	CAS Registry Number	Molecular formula	Structure number	Structure
procaine	[*59-46-1*]	$C_{13}H_{20}N_2O_2$	(**20**, X = R = H, R′ = C_2H_5)	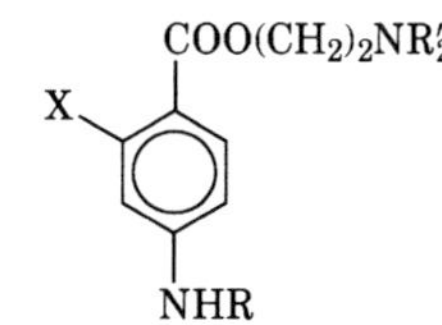
chloroprocaine	[*133-16-4*]	$C_{13}H_{19}ClN_2O_2$	(**20**, X = Cl, R = H, R′ = C_2H_5)	
tetracaine	[*94-24-6*]	$C_{15}H_{24}N_2O_2$	(**20**, X = H, R = *n*-C_4H_9, R′ = CH_3)	
lidocaine	[*137-58-6*]	$C_{14}H_{22}N_2O$	(**21**, X = CH_3, R = H, R′ = R″ = C_2H_5)	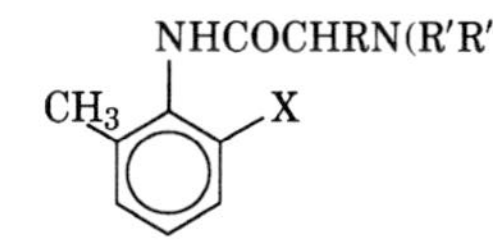
prilocaine	[*721-50-6*]	$C_{13}H_{20}N_2O$	(**21**, X = H, R = CH_3, R′ = H, R″ = *n*-C_3H_7)	
etidocaine	[*36637-18-0*]	$C_{17}H_{28}N_2O$	(**21**, X = CH_3, R = R′ = C_2H_5, R″ = *n*-C_3H_7)	
mepivacaine	[*96-88-8*]	$C_{15}H_{22}N_2O$	(**22**, R = CH_3)	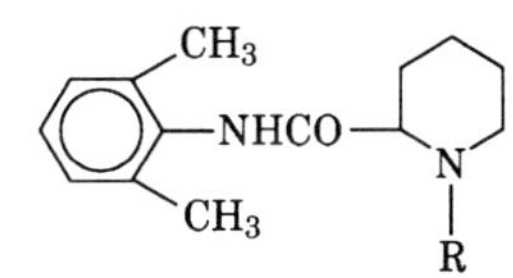
bupivacaine	[*2180-92-9*]	$C_{18}H_{28}N_2O$	(**22**, R = *n*-C_4H_9)	
ropivacaine	[*84057-95-4*]	$C_{17}H_{26}N_2O$	(**22**, R = *n*-C_3H_7)	

Table 4. ***In Vitro*** **Conduction Blocking and Physiochemical Properties of Local Anesthetic Agents**[a]

Agent	Conduction blocking properties[b]			Physiochemical properties		
	Potency	Onset	Duration	pK_a	Lipid solubility	Protein binding, %
			Low potency			
procaine	1	1	1	8.9	0.6	5.8[c]
			Intermediate potency			
mepivacaine	2	1	1.5	7.6	1.0	77[c]
prilocaine	3	1	1.5	7.7	0.8	55[d]
chloroprocaine	4	0.8	0.75	8.7		
lidocaine	4	0.8	1.5	7.7	2.9	64[d]
			High potency			
tetracaine	16	2	8	8.5	80	76[d]
bupivacaine	16	0.6	8	8.1	28	95[d]
etidocaine	16	0.4	8	7.7	141	94[d]

[a]Ref. 153.
[b]Relative to procaine. Data are derived from isolated frog sciatic nerve.
[c]Nerve homogenate binding.
[d]Plasma protein binding.

n-C_4H_9, R′ = CH_3), appear to be more tightly bound to proteins than those possessing short durations such as procaine (**20,** X = R = H, R′ = C_2H_5) and lidocaine (**21,** X = CH_3, R = H, R′ = R″ = C_2H_5) hydrochloride.

The rate of onset of local anesthetic activity *in vitro* is primarily determined by the pK_a of the local anesthetic agent. The protonated species of the agent is the active form whereas the free base possesses little or no activity. However, it is the basic form of the drug which diffuses through the nerve sheath. Consequently the higher the concentration of the free base, ie, the lower the pK_a, the more rapid is the onset of anesthesia. The *in vivo* onset is also determined by the nonnervous tissue diffusability of the compound. Many long duration local anesthetics exhibit long anesthetic onsets. Mixtures of these long duration agents and a short duration drug possessing a rapid onset have been used.

The rate of removal of the local anesthetic from the site of injection also affects its profile. All local anesthetic agents possess some vasodilatory activity at clinically useful concentrations. Agents which are more potent in this regard tend to be absorbed more rapidly by the vasculature. They are less potent anesthetics and have shorter durations than those having lower vasodilatory activity. A comparison of potency, onset, and duration as a function of physiochemical properties is presented in Table 4.

Another clinical consideration is the ability of local anesthetic agents to effect differential blockade of sensory and motor fibers. In surgical procedures such as obstetrics or postoperative pain relief, an agent which produces profound sensory block accompanied by minimal motor block is desirable. On the other hand some procedures such as limb surgery require both deep sensory and motor blockade. In clinical practice, bupivacaine (**22,** R = n-C_4H_9), in low doses, exhibits the former behavior and is used primarily as an extradural agent in obstetrics. The lowest effective extradural concentration of etidocaine (**21,** X = CH_3, R = R′ = C_2H_5, R″ = n-C_3H_7), however, shows both adequate sensory and profound motor blockade so that it is useful in surgical situations where maximum neuromuscular blockade is necessary. In an isolated nerve preparation, bupivacaine blocks unmyelinated C fibers which are mainly responsible for pain perception at a much greater extent than the myelinated A fibers which carry motor impulses. It is postulated that absorption of bupivacaine by the vasculature at the site of injection, combined with the slow diffusion of this agent, results in an insufficient amount of the drug penetrating the large A fibers to cause motor conduction blockade. Clinically, motor block can be observed in some procedures.

Specific Local Anesthetic Agents. Clinically used local anesthetics and the methods of application are summarized in Table 5. Procaine hydrochloride [*51-05-8*] (Novocain), introduced in 1905, is a relatively weak anesthetic having a long onset and short duration of action. Its primary use is in infiltration anesthesia and differential spinal blocks. The low potency and low systemic toxicity

Table 5. Clinically Used Local Anesthetic Agents

Agent	Method of application	Comment
	Amino esters	
procaine	infiltration, spinal	slow onset, short duration
chloroprocaine	peripheral nerve and obstetric extradural	fast onset, short duration, low systemic toxicity
tetracaine	spinal	slow onset, long duration, high systemic toxicity
	Amino amides	
lidocaine	infiltration, iv regional, peripheral nerve block, extradural block, spinal and topical	fast onset, moderate duration, low systemic toxicity, most versatile agent
mepivacaine	infiltration, peripheral nerve block, extradural	similar to lidocaine
prilocaine	similar to lidocaine	safest amino amide, methemoglobinemia at high doses
bupivacaine	infiltration, peripheral nerve block, extradural	moderate onset, long duration, potential for cardiovascular side effects, sensory/motor separation
etidocaine	infiltration, peripheral nerve block, extradural	fast onset, long duration, profound motor block

result from rapid hydrolysis. The 4-aminobenzoic acid formed is, however, thought to be responsible for allergic reactions caused by the drug.

Chloroprocaine hydrochloride [*3858-89-7*] is characterized by low potency, rapid onset, short duration of action, and low systemic toxicity. It is indicated for infiltration anesthesia at 1–2% and for extradural anesthesia at 2–3% when short surgical procedures are performed under regional anesthesia. Chloroprocaine may be mixed with long duration agents such as bupivacaine (**22,** R = n-C_4H_9) to afford a more rapid onset and shorter duration of action than bupivacaine alone.

Tetracaine (**20,** X = H, R = n-C_4H_9, R′ = CH_3) is primarily used in spinal anesthesia providing a slow onset, high potency, and a long duration of action. Its potential for systemic toxicity limits its use in other forms of regional anesthesia. Tetracaine possesses excellent topical anesthetic activity and has been used in endotracheal surface and opthalmic anesthesia.

Lidocaine hydrochloride [*73-78-9*] (Xylocaine), is the most versatile local anesthetic agent because of its moderate potency and duration of action, rapid onset, topical activity, and low toxicity. Its main indications are for infiltration, peripheral nerve blocks, extradural anesthesia, and in spinal anesthesia where a duration of 30 to 60 min is desirable. Because of its vasodilator activity, addition of the vasoconstrictor, epinephrine, increases the duration of action of lidocaine markedly. It is also available in ointment or aerosol preparations for a variety of topical applications.

Mepivacaine hydrochloride [*1722-62-9*], similar in profile to lidocaine, is used for infiltration, peripheral nerve blocks, and extradural anesthesia. It appears to be less toxic than lidocaine in adults but more toxic in newborns. The duration of action is longer than that of lidocaine because of its lower vasodilator activity. Mepivacaine has little topical activity.

Prilocaine hydrochloride [*1786-81-8*] is also similar in profile to lidocaine, although prilocaine has significantly less vasodilator activity. Prilocaine is the least toxic of the amino amide local anesthetics. However, its tendency to cause methemoglobinemia, especially in newborns, has eliminated its use in obstetric surgery.

Bupivacaine hydrochloride [*14252-80-3*] (Marcaine), one of the most potent and longest duration agents in clinical use, is also one of the most toxic. The onset of anesthesia is moderate to long, but it is generally acceptable. Applications include infiltration, peripheral nerve blocks, extradural, and spinal anesthesia at 0.25 to 0.75%, although use at the high concentration is limited by the tendency to produce profound cardiovascular toxicity upon overdose. Advantages are in extradural obstetric procedures and postoperative pain relief because bupivacaine produces profound sensory but little motor block. Additionally, its high efficacy and long duration of action decreases the need for repeated injections.

Etidocaine (**21,** X = CH_3, R = R′ = C_2H_5, R″ = n-C_3H_7), characterized by high efficacy, short onset, and long duration of action, may be used for infiltration, peripheral nerve block, and extradural anesthesia at 1.5 percent. Both sensory and profound motor block result. Thus this agent is useful only in procedures such as lower limb orthopedic and abdominal surgeries where neuromuscular blockade is necessary.

Anesthetic Agents Under Development. Ropivacaine (AL-381) (**22,** R = n-C_3H_7), similar in structure to mepivacaine and bupivacaine, has potency, duration, and CNS and cardiovascular toxicity somewhat lower than those of bupivacaine. Thus the advantage of ropivacaine is the significantly higher cardiovascular and CNS therapeutic indexes (158). Carbizocaine [*76629-87-3*] (BK-95), $C_{21}H_{36}N_2O_3$, (**23**) is an amino carbamate possessing high potency as a topical, infiltration, and peripheral nerve blocking agent. Its potency in the rat sciatic nerve block test is ten times higher than that of bupivacaine (159). Heptacaine [*55792-21-7*], $C_{21}H_{36}N_2O_3$, (**24**), also an amino carbamate, is somewhat more potent and has a slightly longer duration of action than lidocaine. The acute toxicity of heptacaine is comparable to that of lidocaine (160). Carticaine [*23964-58-1*] $C_{13}H_{20}N_2O_3S$, (**25**) may be useful as an infiltration anesthetic with a somewhat longer duration of action than lidocaine and similar toxicity (161).

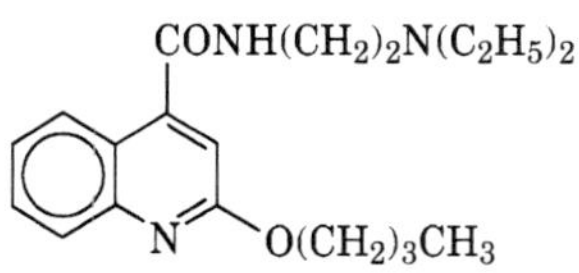

(**23**)

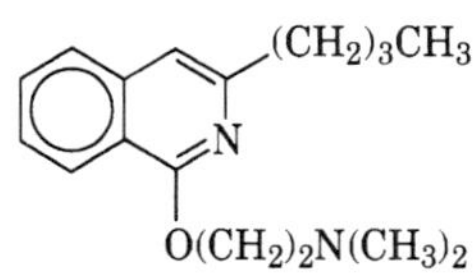

(**24**)

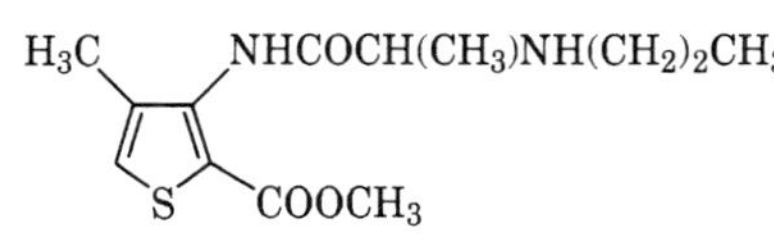

(**25**)

Economic Aspects

The total U.S. market value for the anesthetic agents listed was \$299.9 million in 1990 (162). General inhalation agents, valued at \$154.5 million, comprised over half (51.5%) of the 1990 market. General iv anesthetics were valued at \$111.5 million (37.2%). Local injectable agents, at \$33.9 million, represented the smallest portion of the market (11.3%). U.S. sales for selected anesthesia pharmaceuticals are given in Table 6.

Table 6. 1990 U.S. Sales for Anesthetic Drugs[a]

Drug	Market value, $\$ \times 10^6$
Inhalational agents	
fluothane (halothane)	2.5
ethrane (enflurane)	18.1
forane (isoflurane)	133.9
Injectable (iv) agents	
pentothal (thiopental)	26.0
diprivan (propofol)[b]	29.0
sublimaze (fentanyl)	19.9
alfenta (alfentanyl)	8.6
sufenta (sufentanyl)	28.0
Local injectable	
novocain (procaine)	0.7
xylocaine (lidocaine)	22.5
marcaine (bupivacaine)	10.7

[a]Ref. 162. Actual 1990 sales figures, generic and name brand products combined.
[b]Launched November 1989.

BIBLIOGRAPHY

"Anesthetics" in *ECT* 1st ed., Vol. 1, pp. 902–913, by J. C. Krantz, Jr., University of Maryland; *ECT* 2nd ed., Vol. 2, pp. 393–410, by John C. Krantz, Jr., University of Maryland; in *ECT* 3rd ed., Vol. 2, pp. 684–699, by A. R. Patel, National Cancer Institute.

1. T. C. Smith and H. Wollman, in A. G. Gilman, L. S. Goodman, T. W. Rall, and F. Murad, eds., *The Pharmacological Basis of Therapeutics,* Macmillan Publishing Company, New York, 1985, p. 269.
2. T. F. Hornbein and co-workers, *Anesth. Analg.* **61,** 553 (1982).
3. E. I. Eger, II, *Anesthesiology* **24,** 153 (1963).
4. K. W. Miller and M. E. Wolff, eds., *Burger's Medicinal Chemistry,* 4th ed., John Wiley & Sons, Inc., New York, 1981, Vol. 3, p. 623.
5. I. Ueda and H. Kamaya, *Anesth. Analg.* **63,** 929 (1984).
6. N. P. Franks and W. R. Lieb, *Trends. Pharm. Sci.* **8,** 169 (1987).
7. H. Meyer, *Naunyn-Schmiedebergs Arch. Exp. Pathol. Pharmakol.* **42,** 109 (1899).
8. H. Meyer, *Naunyn-Schmiedebergs Arch. Exp. Pathol. Pharmakol.* **46,** 338 (1901).
9. E. Overton, *Studien uber Narkose,* Fisher, Jena, Germany, 1901.
10. D. D. Koblin and E. I. Eger, II, *N. Engl. J. Med.* **301,** 1222 (1979).
11. K. W. Miller, *Int. Rev. Neurobiol.* **27,** 1 (1985).
12. N. P. Franks and W. R. Lieb, *Nature* (London) **274,** 339 (1978).
13. N. P. Franks and W. R. Lieb, *J. Mol. Biol.* **133,** 469 (1979).
14. J. M. Boggs, T. Yoong, and J. C. Hsia, *Mol. Pharmacol.* **12,** 127 (1976).
15. N. P. Franks and W. R. Lieb, *Nature* (London) **300,** 407 (1982).
16. F. Helmer, K. Kiebs, and C. Hansch, *Biochemistry* **7,** 2058 (1968).
17. K. Kiehs, C. Hansch, and L. Moore, *Biochemistry* **5,** 2602 (1966).
18. B. P. Schoenborn, *Nature* (London) **208,** 760 (1965).
19. F. F. Brown, M. J. Halsey, and R. E. Richards, *Proc. R. Soc. London, Ser. B,* **193,** 387 (1976).
20. N. P. Franks and W. R. Lieb, *Nature* (London) **310,** 599 (1984).
21. N. P. Franks and W. R. Lieb, *Nature* (London) **316,** 349 (1985).
22. N. P. Franks and W. R. Lieb, *Proc. Natl. Acad. Sci. (USA)* **83,** 5116 (1986).
23. M. A. Gillman and F. J. Lichtigfeld, *J. Neural. Sci.* **49,** 41 (1981).
24. J. Adriani, *The Pharmacology of Anesthetic Drugs,* 5th ed., Thomas Company, Springfield, Ill., 1970.
25. E. R. Larsen in P. Tarrant, ed., *Fluorine Chemistry Reviews,* Vol. 3, Marcel Dekker, New York, 1969.
26. F. G. Rudo and J. C. Krantz, *Br. J. Anaesth.* **46,** 181 (1974).
27. J. C. Krantz and co-workers, *J. Pharmacol. Exp. Ther.* **108,** 488 (1953).
28. R. C. Terrell, *Br. J. Anaesth.* **56,** 3S (1984).
29. H. Gion and co-workers, *Anesthesiology* **40,** 553 (1974).
30. D. A. Blake, R. S. Rozman, H. F. Cascorbi, and J. C. Krantz, *Biochem. Pharmacol.* **16,** 1237 (1967).
31. A. D. Finck in E. I. Eger, II, ed., *Nitrous Oxide,* Elsevier, New York, 1985, p. 41.
32. W. P. Chapman, J. G. Arrowood, and H. K. Beecher, *J. Clin. Invest.* **22,** 871 (1943).
33. J. C. Yang, W. C. Clark, and S. H. Ngai, *Anesthesiology* **52,** 414 (1980).
34. M. A. Gillman, L. Kok, and F. J. Lichtigfeld, *Eur. J. Pharmacol.* **61,** 175 (1980).
35. R. Siebeck, *Skand. Arch. Physiol.* **21,** 368 (1909).
36. C. P. Gibbs, E. S. Munson, and M. K. Than, *Anesthesiology* **43,** 100 (1974).
37. E. S. Munson, E. I. Eger, II, M. K. Than, and W. J. Embro, *Anesth. Analg.* **47,** 224 (1978).
38. T. F. Hornbein and co-workers, *Anesth. Analg.* **61,** 553 (1982).
39. P. M. Winter and co-workers, *Am. Soc. Anesthesiol., (Abstracts),* 103 (1972).
40. N. W. B. Craythorne and T. D. Darby, *Br. J. Anaesth.* **37,** 560 (1965).

41. S. Ericksen, G. Johannsen, and N. Frost, *Acta. Anesthesiol. Scand.* **24,** 74 (1980).
42. A. F. Fakunaga and R. M. Epstein, *Anesthesiology* **39,** 23 (1973).
43. J. H. Eisele in E. I. Eger, II, ed., *Nitrous Oxide,* Elsevier, New York, 1985, p. 125.
44. E. Vierra and co-workers, *Anesth. Analg.* **59,** 175 (1980).
45. B. R. Fink, T. H. Shepar, and R. J. Blandau, *Nature* (London) **214,** 146 (1967).
46. G. A. Lane and co-workers, *Science* **210,** 899 (1980).
47. E. N. Cohen and co-workers, *J. Am. Dent. Assoc.* **101,** 21 (1980).
48. E. I. Eger, II, *Anesthetic Uptake Action,* Williams and Wilkins, Baltimore, Md., 1974.
49. B. E. Marshall and H. Wollman in A. G. Gilman, L. S. Goodman, T. W. Rall, and F. Murad, eds., *Goodman and Gilman's Pharmacological Basis of Therapeutics,* Macmillan Publishing Co., New York, 1985, p. 276.
50. R. I. Mazze, J. R. Trudell, and M. J. Cousins, *Anesthesiology* **35,** 247 (1971).
51. R. I. Mazze, M. J. Cousins, and J. C. Kosek, *Anesthesiology* **36,** 571 (1972).
52. W. B. Crandell, S. G. Pappas, and A. Macdonald, *Anesthesiology* **27,** 591 (1966).
53. G. A. Gregory, E. I. Eger, II, and E. S. Munson, *Anesthesiology* **30,** 488 (1969).
54. T. A. Joas and W. C. Stevens, *Anesthesiology* **35,** 48 (1971).
55. R. D. Miller and co-workers, *Anesthesiology* **35,** 38 (1971).
56. R. I. Mazze, *Br. J. Anesth.* **56,** 27S (1984).
57. L. A. Widger, A. J. Gandolfi, and R. A. Van Dyke, *Anesthesiology* **44,** 197 (1976).
58. Subcommittee on the National Halothane Study of the Committee on Anesthesia, National Academy of Sciences—National Research Council, *J. Am. Med. Assoc.* **197,** 775 (1976).
59. M. J. Cousins, L. R. Greinstein, B. A. Hitt, and R. I. Mazze, *Anesthesiology* **44,** 44 (1976).
60. H. Gion and L. Saidman, *Anesthesiology* **35,** 361 (1971).
61. L. H. Storms, A. H. Stark, R. K. Calverley, and N. T. Smith, *Anesthesiology Analg.* **59,** 245 (1980).
62. S. H. Davidson, *Acta. Anesthesiol. Scand.* **22,** 58 (1978).
63. B. J. Stanford, O. M. Plantevin, and J. R. Gilbert, *Br. J. Anaesth.* **51,** 1143 (1979).
64. R. K. Calverley and co-workers, *Anesth. Analg.* **57,** 610 (1978).
65. J. Sanlesson, L. Anestedt, P. O. Jarnberg, and O. Norlander, *Acta. Anesthesiol. Scand.* **22,** 381 (1978).
66. R. A. Johnston, E. I. Eger, II, and C. Wilson, *Anesth. Analg.* **55,** 709 (1976).
67. F. F. Foldes, A. Bencini, and D. Newton, *Br. J. Anaesth.* **52**(Supp. 1), 64S (1980).
68. F. M. Ramsey and co-workers, *Anesthesiology* **57,** A255 (1982).
69. R. E. Chase and co-workers, *Anesthesiology* **35,** 262 (1971).
70. W. C. Stevens, W. M. Dolan, and R. T. Gibbons, *Anesthesiology* **42,** 197 (1975).
71. T. H. Cromwell and co-workers, *Anesthesiology* **35,** 17 (1971).
72. B. Wolfson, W. D. Hetrick, C. L. Lake, and E. S. Siker, *Anesthesiology* **48,** 187 (1978).
73. A. M. Lam and A. W. Gelb, *Anesth. Analg.* **62,** 742 (1983).
74. P. P. Raj, M. J. Tod, and M. T. Jenkins, *South. Med. J.* **69,** 1128 (1976).
75. D. A. Holaday, V. Fiserova-Bergerova, I. P. Latto, and M. A. Zumbiel, *Anesthesiology* **43,** 325 (1975).
76. E. I. Eger, II and B. J. Johnson, *Anesth. Analg.* **66,** 974 (1987).
77. E. I. Eger, II, *Anesth. Analg.* **66,** 971 (1987).
78. T. Katoh and K. Ikeda, *J. Anesthesiol.* **2,** 63 (1988).
79. M. S. Scheller, A. Tateishe, J. C. Drumond, and M. H. Zorrow, *Anesthesiology* **68,** 548 (1988).
80. D. A. Holaday and F. R. Smith, *Anesthesiology* **54,** 100 (1981).
81. S. Akazawa and co-workers, *J. Anesth.* **2,** 227 (1988).
82. Y. Hayashi and co-workers, *Anesthesiology* **69,** 145 (1988).
83. N. Yasuda, A. G. Targ, and E. I. Eger, II, *Anesthesiology* **69,** 3A, A625 (1988).

84. E. I. Eger, II, *Anesth. Analg.* **66,** 983 (1987).
85. B. M. Doorley, S. J. Waters, R. C. Terrell, and J. L. Robinson, *Anesthesiology* **69,** 89 (1988).
86. R. B. Weiskopf and co-workers, *Anesthesiology* **70,** 293 (1989).
87. I. J. Rampil and co-workers, *Anesthesiology* **69,** 298 (1988).
88. D. D. Koblin and co-workers, *Anesth. Analg.* **67,** 534 (1988).
89. D. D. Koblin and co-workers, *Anesth. Analg.* **68,** 147 (1989).
90. P. E. Keane and K. Biziere, *Life Sci.* **41,** 1437 (1987).
91. A. Mansour and co-workers, *Trends Neurosci.* **11,** 303 (1988).
92. T. L. Yaksh, N. R. F. Al-Rodhan, and T. S. Jensen, *Prog. Brain Res.* **77,** 371 (1988).
93. G. R. Lenz, S. M. Evans, D. E. Walters, and A. J. Hopfinger, *Opiates,* Academic Press, Orlando, Fla., 1986.
94. S. K. Kulkarni and M. K. Ticku, *Drugs of Today* **25,** 501 (1989).
95. P. Skolnick, K. C. Rice, J. L. Barker, and S. M. Paul, *Brain Res.* **233,** 143 (1982).
96. J. W. Dundee, *Postgrad. Med. J.* **61**(Supp. 3), 3 (1985).
97. B. E. Marshall and H. Wollman in A. G. Gilman, L. S. Goodman, T. W. Rall, and F. Murad, eds., *The Pharmacological Basis of Therapeutics,* MacMillan Publishing Co., New York, 1985, p. 292.
98. H. Schneiden, *Drugs of Today* **23,** 209 (1987).
99. D. J. Coleman, *Anesthesiol. Widerbelebung* **57,** 74 (1972).
100. C. R. Gardner, *Drug Dev. Res.* **12,** 1 (1988).
101. J. G. Reves, R. J. Fragen, H. R. Vinik, and D. J. Greenblatt, *Anesthesiology* **62,** 310 (1985).
102. M. Gerecke, *Br. J. Clin. Pharmacol.* **16,** 115 (1983).
103. B. Ricou and co-workers, *Br. J. Anaesth.* **58,** 1005 (1986).
104. K. Hillier, *Drugs Future* **6,** 402 (1982).
105. M. Morgan, J. Lumley, and J. G. Whitwam, *Lancet* **1,** 955 (1975).
106. A. Holdcroft, M. Morgan, J. G. Whitwam, and J. Lumley, *Br. J. Anaesth.* **48,** 199 (1976).
107. P. J. Roberts, *Drugs Future* **1,** 461, (1976); **2,** 703 (1977); **3,** 7708 (1978).
108. G. W. Black and R. S. J. Clarke, *Int. Anesthesiol. Clin.* **9,** 171 (1971).
109. R. S. J. Clarke and J. W. Dundee, *Curr. Res. Anesth. Analg.* **45,** 250 (1966).
110. J. W. Flacke, *Semin. Anesth.* **7,** 178 (1988).
111. R. James and J. B. Glen, *J. Med. Chem.* **23,** 1350 (1980).
112. J. L. Apfelbaum, *Semin. Anesth.* **7**(Supp. 1), 21 (1988).
113. R. J. Fragen, *Semin. Anesth.* **7**(Supp. 1), 131 (1988).
114. V. A. Doze, A. Schafer, and P. F. White, *Anesthesiology* **69,** 63 (1988).
115. S. R. W. Bricker, *Anesthesia* **43,** 171 (1988).
116. P. F. White, *Semin. Anesth.* **7**(Supp. 1), 4 (1988).
117. J. D. Borel, *Contemp. Anesth. Pract.* **7,** 1 (1983).
118. *Drugs of Today* **1,** 3 (1965).
119. P. A. J. Janssen, *Br. J. Anaesth.* **34,** 260 (1962).
120. C. J. H. Andrews and C. Prys-Roberts, *Clin. Anesthesiol.* **1,** 97 (1983).
121. E. A. Welchew and J. A. Thornton, *Anesthesia* **37,** 309 (1982).
122. J. G. Bovill, P. S. Sebel, and T. H. Stanley, *Anesthesiology* **61,** 731 (1984).
123. P. A. J. Janssen in F. G. Estafanous, ed., *Opioids in Anesthesia,* Butterworth Publishers, Stoneham, Mass., 1984, p. 37.
124. G. E. Larijani and M. E. Goldberg, *Clin. Pharmacol.* **6,** 275 (1987).
125. C. Rosow, *Semin. Anesth.* **7,** 107 (1988).
126. S. J. Hopkins, *Drugs Future* **6,** 335 (1981); **9,** 460 (1984).
127. T. J. Sandford, Jr., *Semin. Anesth.* **7,** 127 (1988).
128. J. P. Monk, R. Beresford, and A. Ward, *Drugs* **36,** 286 (1988).

129. E. Arigoin-Martelli, *Drugs Future* **2,** 334 (1977); **9,** 386 (1984).
130. J. C. Rowlingson, J. C. Mosceki, and C. A. DiFazio, *Anesth. Analg.* **62,** 899 (1983).
131. R. I. Hall, M. R. Murphy, F. Szlam, and C. C. Hug, Jr., *Anesth. Analg.* **66,** 1169 (1987).
132. U. A. Pandit, S. P. Kothary, and S. K. Pandit, *J. Clin. Pharmacol.* **26,** 275 (1986).
133. T. J. Gal and C. A. DiFazio, *Anesthesiology* **61,** 716 (1984).
134. A. Romagnoli and A. S. Keats, *Clin. Pharmacol. Ther.* **35,** 367 (1984).
135. L. F. McNicholas and W. R. Martin, *Drugs* **29,** 81 (1984).
136. T. M. Purkert and H. M. Ginzburg, *Clin. Anesthesiol.* **1,** 168 (1983).
137. J. P. Gonzalez and R. N. Brogden, *Drugs* **35,** 192 (1988).
138. P. J. Roberts, *Drugs Future* **2,** 45 (1977); **11,** 68 (1986).
139. J. E. Eckenhoff and S. R. Oech, *Clin. Pharmacol. Ther.* **1,** 483 (1960).
140. L. Lasagna and H. K. Beecher, *J. Pharmacol. Exp. Ther.* **112,** 356 (1954).
141. W. R. Martin and C. W. Gorodetzky, *J. Pharmacol. Exp. Ther.* **150,** 437 (1965).
142. P. J. Roberts, *Drugs Future* **2,** 613 (1977).
143. R. Thesen and co-workers, *Pharm. Ztg.* **16,** 31 (1988).
144. *Drugs Future* **9,** 518 (1984).
145. H. Selye, *Endocrinology* **30,** 437 (1942).
146. G. H. Phillipps in J. J. Halsey, R. A. Millar, and J. A. Sutton, eds., *Molecular Mechanisms in General Anesthesia,* Churchill Livingstone, New York, 1974, p. 32.
147. C. Prys-Roberts and J. Sear, *Br. J. Anaesthesia* **52,** 363 (1980).
148. N. L. Harrison and M. A. Simmonds, *Brain Res.* **323,** 287 (1984).
149. B. Wardley-Smith and H. J. Little, *Br. J. Anaesthesia* **57,** 629 (1985).
150. P. F. White, *Semin. Anesth.* **7,** 113 (1988).
151. D. L. Reich and G. Silvay, *Can. J. Anaesth.* **36,** 186 (1989).
152. G. G. Arnault, *J. Am. Assoc. Nurse Anesth.* **37,** 291 (1969).
153. B. G. Covino and H. G. Vasallo, *Local Anesthetics: Mechanism of Action and Clinical Use,* Grune and Stratton, New York, 1976.
154. R. H. deJong, *Local Anesthetics,* Thomas Co., Springfield, Ill., 1977.
155. E. Ericksson, *Illustrated Handbook of Local Anesthesia,* W. B. Saunders Philadelphia, Pa., 1979.
156. J. M. Ritchie and N. M. Greene in A. G. Gilman, L. S. Goodman, and A. Gilman, eds., *The Pharmacological Basis of Therapeutics,* 6th ed., Macmillan Publishing Co., New York, 1980, p. 300.
157. B. G. Covino, *Br. J. Anaesth.* **58,** 701 (1986).
158. B. G. Covino, *Regional Anesthesia Update,* Harvard Medical School, Cambridge, Mass., Nov., 1986.
159. L. Benes, P. Svec, J. Kozlovsky, and A. Borovansky, *Cesk. Farm.* **27,** 167 (1978); S. Radl, *Drugs Future.* **11,** 555 (1986).
160. *Drugs Future* **4,** 491 (1979).
161. *Drugs Today* **12,** 131 (1976).
162. *U.S. Pharmaceutical Market: Hospitals,* Vol. 1, Dec. 1990, IMS Pharmaceuticals Database Division.

George R. Lenz
BOC Group Technical Center

Hollis G. Schoepke
Theodore C. Spaulding
Anaquest

ANTIASTHMATIC AGENTS

Asthma is an extremely complex condition characterized by variable and reversible airways obstruction combined with nonspecific bronchial hypersensitivity (1–3). The cause of asthma, which is not always readily diagnosed (4), remains unknown. Days, if not weeks, are needed to document the spontaneous reversal of the airways obstruction in some patients. Asthmatics experience both an immediate hypersensitivity response and a delayed late-phase reaction, each mediated by a different pathway. Chronic asthma has come to be viewed as an inflammatory disease (5). The late-phase reaction plays a key role in inducing and maintaining the inflammatory state which in turn is thought to induce the bronchial hyperresponsiveness (6). The airways obstruction results from both contraction of airways smooth muscle and excessive bronchial edema. Edema, a characteristic of inflammatory states, is accompanied, in this case, by the formation of a viscous mucus which can completely block the small airways.

Asthma affects 3–5% of the population and is one of the most common chronic illnesses (7–9). Both the frequency and severity of asthma appear to be increasing (10–13). Acute, severe asthma has the potential to be fatal. The disease may first appear in childhood and individuals so affected can suffer recurrent episodes throughout their lives or they may "outgrow" the condition at puberty.

On the other hand, there is also adult-onset asthma. These people show no symptoms as children or as young adults, but suddenly develop symptoms later in life. There have been many reports of bronchial infections preceding the appearance of asthma. However it is not known whether these infections contributed to the development of the disease or whether individuals who are already predisposed to asthma are more likely to experience bronchospasms as a result of a bronchial infection (14).

Clinically, there are several ways of classifying asthma and treatment varies depending upon the classification. Extrinsic asthma, also called allergic asthma, is experienced by adults and, most commonly, by children. In extrinsic asthma it is possible to demonstrate specific causal agents, usually antigens, eg, animal danders, foods, drugs, house dust, pollens, or mold spores. Atopic extrinsic asthmatics usually have elevated levels of circulating immunoglobulin E (IgE), have a family history of inherited allergies (atopy), and commonly show other allergic symptoms such as rhinitis, or have a history of eczema (15). However, especially in occupation-induced asthma, irritants may act as the allergenic agent and IgE does not appear to be involved in some nonatopic extrinsic asthmatics.

Intrinsic asthma, also called idiopathic asthma, usually develops in adulthood. In intrinsic asthma allergic factors are not demonstrable. Episodes of intrinsic asthma may be triggered by a variety of stimuli, eg, emotional state, exposure to cold air, or inert dusts. Both intrinsic and extrinsic asthmatics can be prone to exercise-induced attacks. Individuals who experience a combination of extrinsic and instrinsic asthmatic reactions have mixed asthma. Status asthmaticus refers to an especially acute life-threatening asthma attack which is resistant to normal treatments and which may require hospitalization in order to stabilize the patient.

Current asthma treatments are not curative and historically have relied on pharmacologic intervention with bronchodilators (Fig. 1) (16) to prevent or relieve the symptoms of asthma. More recently the focal points of both treatment and research efforts have shifted from bronchodilators to agents which reduce the underlying inflammatory state (5,6,17). Management of extrinsic asthma usually includes manipulation of the patient's environment to minimize or completely eliminate the causal agent. This can be especially beneficial in occupationally induced asthma. Treatment of extrinsic asthmatics may also include hyposensitization, or desensitization, by exposure to small and increasing amounts of the known antigen. Although hyposensitization is widely used by allergists for the treatment of allergic rhinitis, its use and potential benefit in the treatment of strictly defined asthma remains controversial (18–20). The cascade of events involved in the asthmatic response and potential points for pharmacolgic intervention are shown in Figure 1.

The order of preference of drug treatment varies from country to country (21,22) (Table 1). In part this difference reflects the lack of a "magic bullet" for the treatment of asthma. However, this difference may also be explained by differences in marketing approval. Some agents are not yet available in the United States and the exposure of U.S. physicians to some of the newer inhalation formulations has been limited. However, the use of inhaled, rather than oral, agents has begun to increase in the United States (23,24) as indicated by the changes in the market for prescription antiasthmatic drugs. In the period from

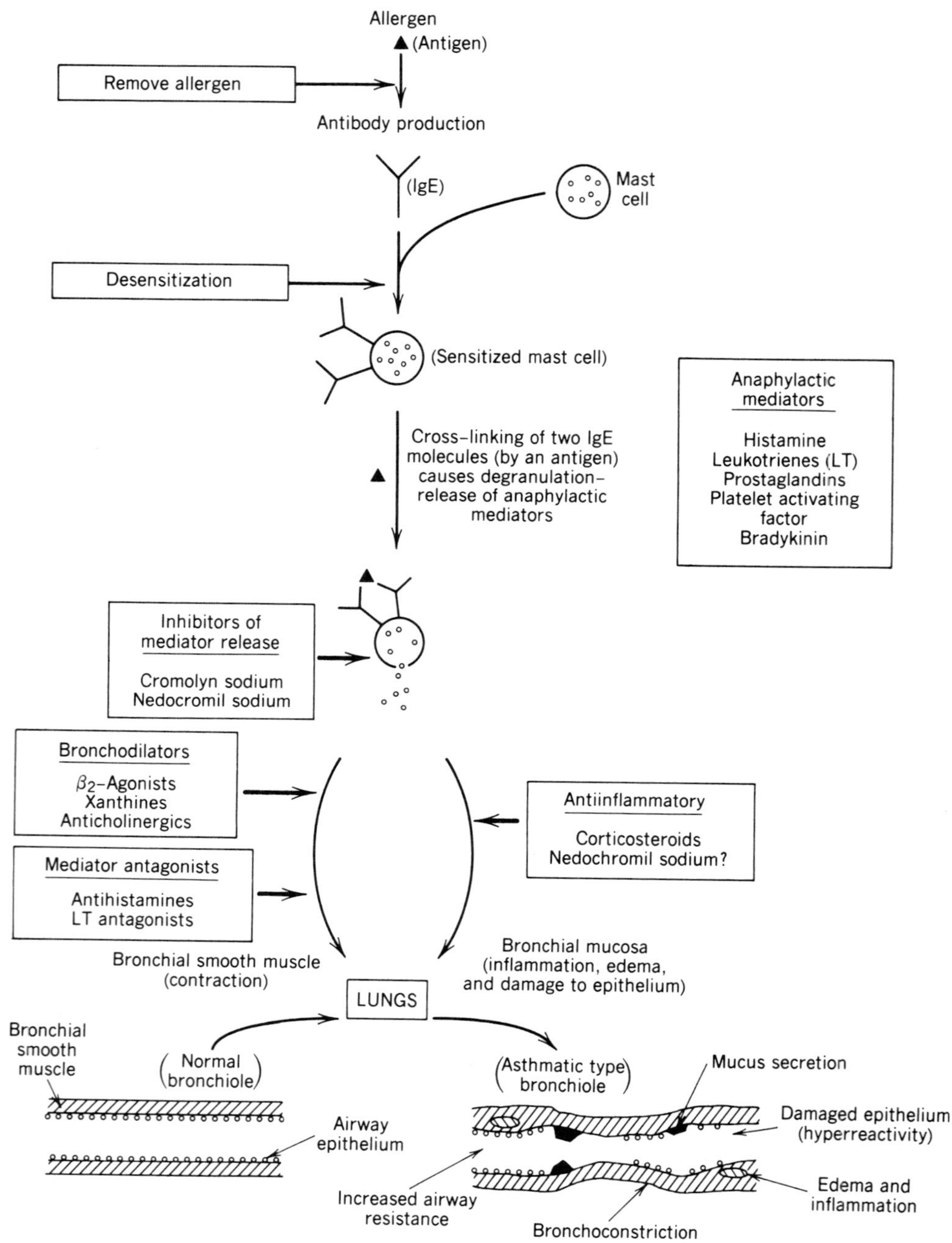

Fig. 1. Schematic diagram showing possible sites of action of antiasthmatic drugs.

1984 to 1989 the share of new prescriptions for inhaled β_2-agonists increased from 18% to 26%, whereas that for oral xanthines dropped from 39% to 27% (25).

Improvements in asthma treatment include the development of more effective, safer formulations of known drugs. The aerosol administration of β_2-agonists or corticosteroids results in a decrease in side effects. Also, the use of reliable

Table 1. Order of Preference for First Choice Antiasthmatic Agent Maintenance Therapy

	Order of preference		1989 U.S. market share, %[c]
	U.S.[a]	UK[b]	
oral xanthines	1	4	~27
inhaled β_2-agonist	2	1	~26
oral β_2-agonist	3		~18
inhaled steroid	4	2	~18
inhaled inhibitor of mediator release	5	3	~5

[a]Ref. 21.
[b]Ref. 22.
[c]Ref. 25. Courtesy of IMS International, a company of Dun & Bradstreet Corporation.

sustained release formulations has revolutionized the use of oral xanthines which have a very narrow therapeutic index (see CONTROLLED RELEASE TECHNOLOGY). For many individuals, asthma symptoms tend to worsen at night and the inhaled bronchodilators do not usually last through an entire night's sleep (26,27).

β-Adrenergic Stimulants

β_2-Agonists are widely used in the symptomatic treatment of asthma. Although both oral and aerosol formulations of these bronchodilators have been available for many years, advances have occurred in delivery technology with the development of dry powder aerosols (qv) (see DRUG DELIVERY SYSTEMS) (28). The ease of usage of these breath-activated systems has improved patient compliance and therapeutic response. There are several detailed reviews on β_2-agonist therapy of bronchial asthma (29–31), and on the structure-activity relationships of this class of drugs (32).

The modern usage of β_2-agonists for the treatment of asthma dates to 1903 when the effect of injected epinephrine [*51-43-4*] (adrenaline), $C_9H_{13}NO_3$, (**1**, R = CH_3) was investigated (see EPINEPHRINE AND NOREPINEPHRINE) (33). As in some other modern treatments, eg, xanthines and anticholinergics, the roots of β_2-agonist therapy for asthma can be found in historical records which document the use of herbal extracts containing ephedrine [*299-42-3*], $C_{10}H_{15}NO$, (**2**) as bronchodilators. Epinephrine and ephedrine are structurally related to the catecholamine norepinephrine [*51-41-2*], $C_8H_{11}NO_3$, (**1**, R = H), a neurotransmitter of the adrenergic nervous system (see NEUROREGULATORS).

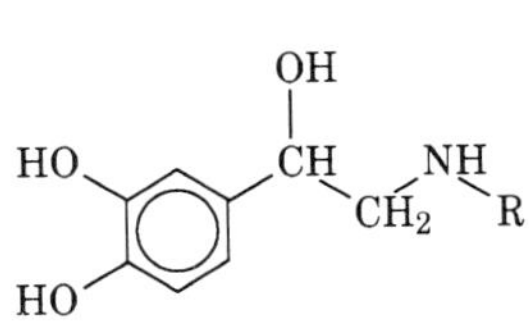

(**1**)

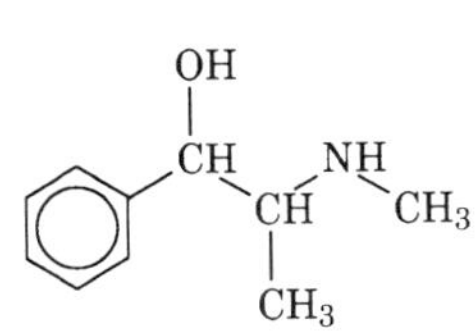

(**2**)

Endogenous catecholamines have an extremely short half-life and reuptake into sympathetic nerve storage vesicules is the main mechanism for removing norepinephrine from circulation. To some degree reuptake also occurs with epinephrine. This process is relatively selective, however, and does not readily take place with nonnaturally occurring catecholamines. Catecholamines that escape reuptake are rapidly catabolized by two widely distributed enzymes, catechol-*O*-methyltransferase (COMT) and monoamine oxidase (MAO). The former transfers a methyl group to the 3-hydroxy of catecholamines. The resultant 3-methoxy compounds have no adrenergic agonist activity and in some cases may be adrenergic antagonists. MAO is less discriminatory, cleaving the carbon–nitrogen bond of primary and secondary amines with concommitant oxidation of the carbon atom.

Ephedrine, which is not a catecholamine, has weak oral activity as a bronchodilator and although it has some direct action at adrenergic receptors, its predominant mode of action is by displacing norepinephrine from storage vesicules. β_2-Agonists which are in use or are under investigation are the result of quests for improved selectivity, retention of potency, oral activity, and longer duration of action.

The adrenergic nervous system is one of the two main branches of the autonomic nervous system (34), which regulates the so-called automatic functions of the body, eg, the actions of various organs such as the lungs and heart. The other main branch of the autonomic nervous system is called the cholinergic nervous system and it has acetylcholine [*51-84-3*], $CH_3COO(CH_2)_2N(CH_3)_3^+$, (see CHOLINE; ENZYME INHIBITION) as a neurotransmitter. More recently, a third branch to the autonomic nervous system has been identified in human (35) and animal airways (36,37). In this nonadrenergic, noncholinergic nervous system both the neurotransmitter and the importance in lung function are in dispute (38). The adrenergic and cholinergic nervous systems often act in opposition to one another (34).

Division of the receptors in the adrenergic nervous system into two classes (α and β) was proposed in 1948 (39) when a difference in the rank order of potency of epinephrine (**1**, R = CH_3), norepinephrine (**1**, R = H), and isoproterenol [*7683-59-2*], $C_{11}H_{17}NO_3$, (**1**, R = $CH(CH_3)_2$) was noted to depend on the organ examined. Further subdivision into groups β_1 and β_2 was proposed in 1967 (40). Both types of β-adrenoceptors are found throughout the body. One of the more significant roles for β_1-receptors is in cardiac tissues where they mediate contraction. Although the effect of triggering β_2-receptors is tissue-dependent, stimulating those found in bronchial and vascular smooth muscle induces relaxation (Fig. 2). Stimulating the adrenergic receptor results in activation of adenylate cyclase which in turn catalyzes production of cyclic-3′,5′-adenosine monophosphate (cyclic AMP). The cyclic AMP then acts as a second messenger activating a protein kinase which selectively phosphorylates myosin kinase resulting in inhibition of calcium-dependent smooth muscle contraction. Also, activation of β_2-adrenoceptors results in the inhibition of cholinergic neurotransmission in human airways (41) and this too should cause smooth muscle relaxation.

Elevation of cyclic AMP levels is also known to inhibit the release of inflammatory and contractile mediators from mast cells (42). The good clinical efficacy of β_2-agonists may be related to this action because some members of this

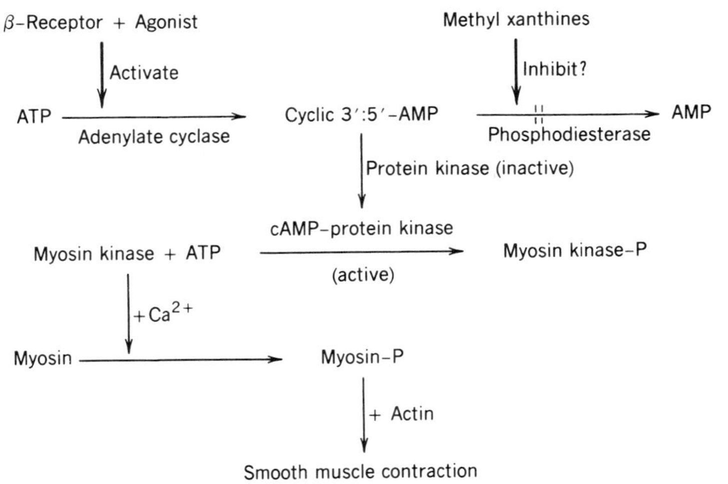

Fig. 2. Proposed mechanism of inhibition of smooth muscle contraction by β_2-agonists, where AMP is adenosine monophosphate, cAMP is cyclic-3′:5′ adenosine monophosphate, ATP is adenosine triphosphate, and -P is an attached phosphate.

class of drugs inhibit mediator release at the same concentrations at which they relax smooth muscle (43). In contrast to their effectiveness against immediate bronchoconstriction, β_2-agonists do not inhibit the late asthmatic response or reverse bronchial hyperreactivity (44,45).

Because of the widespread nature of adrenoceptors, nonselective β-agonists can produce many undesirable side effects. Therefore, before adrenergic agonists could become widely used in the treatment of asthma, some selectivity in action was needed. Whereas epinephrine and ephedrine have significant agonist activity at both α and β adrenoceptors, isoproterenol is a selective agonist at the β-receptor (39). However, isoproterenol does not distinguish between the β_1 and β_2 receptors and it is not active orally.

Aerosol administration of isoproterenol produces a prompt (2–5 minutes) intense bronchodilatation of relatively short (1 h) duration. The lack of β_2-selectivity leads, in many cases, to tachycardia and blood pressure elevation. Also, use of isoproterenol, like all other known β-agonists, results in a down-regulation, or desensitization, of β-adrenergic receptors. This desensitization is only partial, and after time (depending on dose, patient, and agent), a stable, less responsive state is achieved in which β-agonists remain effective. Isoproterenol has been widely used for many years.

A significant advance in β-agonist therapy occurred with the discovery of metaproterenol [*586-06-1*], $C_{11}H_{17}NO_3$, (**3**, R = $CH(CH_3)_2$). Replacing the catechol subgroup with a resorcinol unit results in a compound which is no longer susceptible to metabolism by COMT and therefore has a longer (4 h) duration of action. Metaproterenol has a selectivity profile that is similar to that of isoproterenol, but it is 10–40 times less potent *in vitro* (46). However, metaproterenol is active if given orally and, therapeutically, it is administered either orally or by aerosol. Better selectivity of action is achieved by the aerosol route, although large or frequently repeated aerosol doses may also cause side effects.

Changing the *N*-substituent of metaproterenol to a *tert*-butyl group gives rise to terbutaline [*23031-25-6*], $C_{12}H_{19}NO_3$, (**3**, R = $C(CH_3)_3$) (47), an orally active agent having a duration of action of from 6–7 h. The presence of a tertiary carbon alpha to the amine protects against the action of MAO. Also, it had been shown earlier for catecholamines that increasing the steric bulk around the nitrogen results in improved β_2-selectivity (32). This finding carries over to the resorcinol series: terbutaline stimulates β_2-receptors to a significantly greater extent than β_1-receptors.

Terbutaline has about six times the selectivity of isoproterenol and is claimed to produce fewer side effects. Terbutaline inhibits the release of mediators from sensitized human lungs at clinically relevant concentrations (31), and has been shown to have a prophylactic effect in bronchial challenge studies in addition to being a bronchodilator. Also, nasal administration of fenoterol [*13392-18-2*], $C_{17}H_{21}NO_4$, (**3**, R = $CH(CH_3)CH_2C_6H_4OH$), a closely related β_2-agonist, in clinical studies on the treatment of allergic rhinitis has clearly shown inhibition of release of allergic mediators (48,49). Both aerosol and oral formulations of terbutaline are in use, although examination of an oral slow-release formulation for use in nocturnal asthma revealed that patients preferred long-acting theophylline preparations because of the incidence of undesired side effects (50).

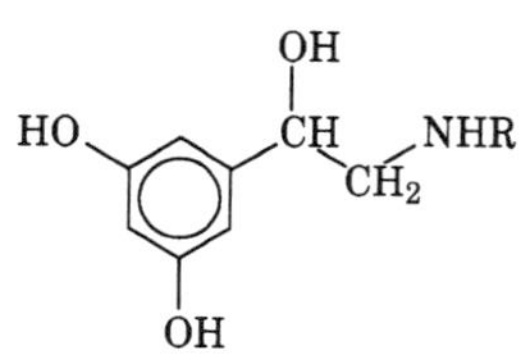

(**3**)

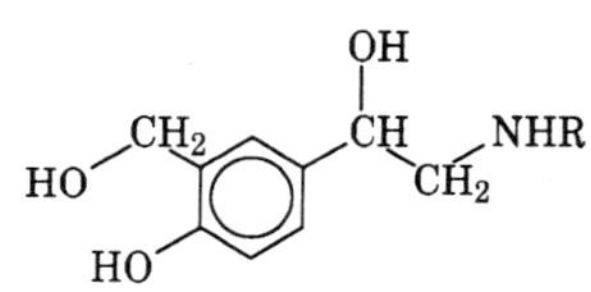

(**4**)

Variation of the substituents on the aromatic ring led to the discovery that the 3-hydroxy could be replaced with a variety of groups, eg, CH_2OH, $NHCH_3$, $NHCONH_2$, $NHS(O)_2CH_3$, that successfully mimic the electron donating and hydrogen bonding properties of the hydroxide moiety at the adrenergic receptor. The 3-CH_2OH analogue has been the most studied in this series of compounds. Albuterol [*18559-94-9*], $C_{13}H_{21}NO_3$, (**4**, R = $C(CH_3)_3$) also called salbutamol (51), is the most widely prescribed β_2-agonist for either aerosol or oral administration both in the United States and worldwide.

Replacement of the hydroxy with CH_2OH results in resistance to both COMT and conjugation. Also, because the *tert*-butyl group confers resistance to MAO, albuterol is metabolically quite stable and animal studies have shown that it is excreted unchanged and as the phenolic glucuronide (31). *In vitro,* albuterol has been shown to inhibit mediator release and to have about 59 times the β_2/β_1-selectivity of isproterenol (32). In contrast, clinical studies using po albuterol reveal only a seven- to tenfold improvement in selectivity as compared to isoproterenol (52). Studies using aerosol administration show that albuterol has fewer side effects, more rapid onset, and at least the same duration of action (5 h) as the immediate release oral formulations. It is believed that the cardiac responses following oral administration result from an indirect reflex stimulation of cardiac output. This reflex results from the lowering in blood pressure that the β_2-agonist induced relaxation of vascular smooth muscle produces. Because of this finding the preferred mode of administration of β_2-agonists is by aerosol even with newer, more selective agents.

More recent research efforts have focused on the development of longer acting β-agonists which could be administered less frequently and be more efficacious in controlling nocturnal asthma. There are several agents currently under clinical evaluation. Bitolterol [*30392-40-6*], $C_{28}H_{31}NO_5$, (**5**) represents one approach to extending duration of action. It is a prodrug having no direct effect at adrenergic receptors; it must be hydrolized by esterases to the active phenolic agent. Bitolterol, presumably because of differential rates of hydrolysis, gives unpredictable bronchodilation if given orally (53). However it is hydrolyzed efficiently in the lung and an aerosol formulation of it has been compared, albeit unfavorably, with sustained release theophylline for control of nocturnal asthma (27).

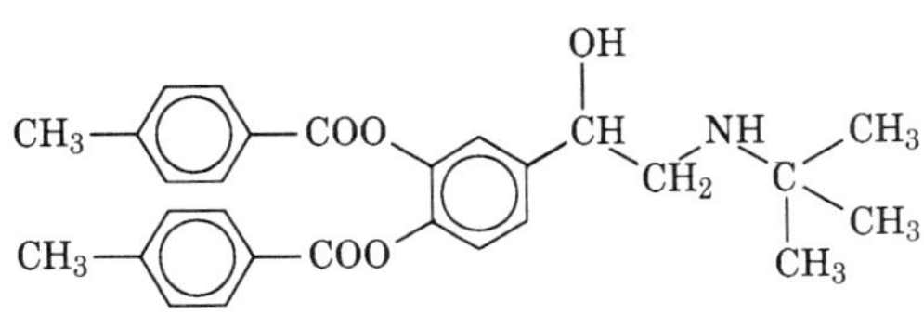

(**5**)

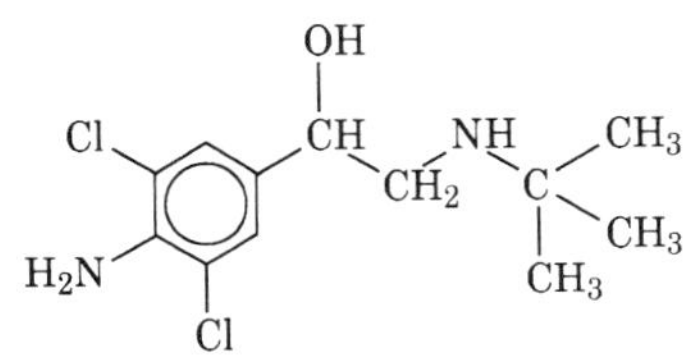

(**6**)

Clenbuterol [*37148-27-9*], $C_{12}H_{18}Cl_2N_2O$, (**6**) (54), a nonphenolic analogue of terbutaline, is not susceptible to COMT or to conjugation. It has a long plasma half-life (20 h), but does not seem to differ in pharmacological half-life from albuterol (55).

Salmeterol [*89365-50-4*], $C_{25}H_{37}NO_4$, (**4**, R = $(CH_2)_6O(CH_2)_4C_6H_5$) (56), is a long-acting analogue of albuterol in which the amine substituent is a long lipophilic chain. Although the plasma half-life of salmeterol is the same as that of albuterol, salmeterol has a much longer (12 h) duration of action following aerosol administration and tests to determine its efficacy in nocturnal asthma are underway (55). *In vitro,* salmeterol is about five times as potent as salbutamol and also shows an unusually extended duration of action (57) which might result from exo-receptor binding (56).

Although β_2-agonists are useful in the treatment of asthma, the profiles would be improved if the bronchodilating effects could be further separated from the side effects. It is not clear how this could be accomplished, but it has been shown that a β_3-receptor exists in humans and other uncharacterized subclasses of receptors have also been postulated (58). The investigation of these other β-receptors could lead to more selective agents.

Xanthine Derivatives

For many years oral xanthines, shown in Table 2, were the preferred first-line treatment for asthma in the United States, and if the aerosol and oral formulations of β_2-agonists are considered separately, as they are in Table 1, this was still

Table 2. Xanthine and Xanthine Derivatives Used as Oral Antiasthmatic Agents

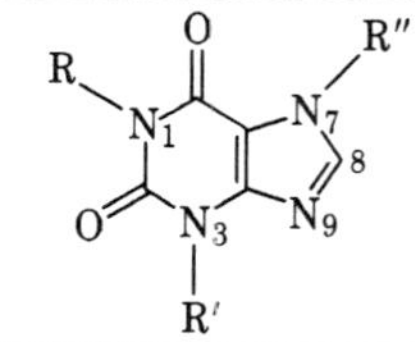

Compound	CAS Registry Number	Molecular formula	Structure number	R	R′	R″
xanthine	*[69-89-6]*	$C_5H_4N_4O_2$	(**7**)	H	H	H
theophylline	*[58-55-9]*	$C_7H_8N_4O_2$	(**8**)	CH_3	CH_3	H
theobromine	*[83-67-0]*	$C_7H_8N_4O_2$	(**9**)	H	CH_3	CH_3
caffeine	*[58-08-2]*	$C_8H_{10}N_4O_2$	(**10**)	CH_3	CH_3	CH_3
enprofylline	*[41078-02-8]*	$C_8H_{10}N_4O_2$	(**11**)	H	n-C_3H_7	H

the case in 1989. Within this class of compounds theophylline (**8**), or one of its various salt forms, such as aminophylline *[317-34-0]* (theophylline: ethylenediamine::2:1), have been the predominant agents. Theophylline, 1,3-dimethylxanthine *[58-55-9]*, is but one member of a class of naturally occuring alkaloids. Two more common alkaloids are theobromine (**9**), isomeric with theophylline and the principal alkaloid in cacao beans, and caffeine, (**10**), 1,3,7-trimethylxanthine *[58-08-2]*, found in coffee and tea.

The bronchodilating effect of caffeine has been recognized for hundreds of years. In the western world the first description of a caffeine preparation for asthma was made in 1859 (59) by a Scottish physician who recommended strong black coffee as a bronchodilator. In many parts of the world, however, use of xanthines is less frequent than in the United States.

Historically, the use of xanthines has been hampered by poor aqueous solubility, rapid but highly variable metabolism, and the existance of a low therapeutic index. Solubility problems were partially solved by the preparation of various salt forms, eg, aminophylline. However, it was since recognized that the added base in aminophylline only increases solubility by increasing pH and thus does not affect the rate of absorption from the gut (65). Thus, in more recent medical practice, theophylline is commonly dispensed in anhydrous form and aminophylline is only recommended for iv administration.

The development of easy-to-use assays for determining theophylline blood levels afforded a handle on maintenance of effective but nontoxic levels. The relatively good availability of such assays in the United States probably contributed to the historical preference for theophylline treatment by U.S. physicians. Careful titration of the dose must be done on a patient-by-patient basis because individual rates of metabolism vary widely. Most (~85%) of an oral dose of theophylline is metabolized by liver microsomal enzymes. As a result many drugs, eg, cimetidine *[51481-61-9]*, anticonvulsants, or conditions, eg, fever, cigarette smoking, liver disease, which affect liver function alter theophylline blood levels.

Common side effects of theophylline therapy include headache, dyspepsia, and nausea. More serious side effects such as lethal seizures or cardiac arrythmias can occur if blood levels are too high. Many derivatives of theophylline have been prepared in an effort to discover an analogue without these limitations (60,61). However, the most universal solution has resulted from the development of reliable sustained release formulations. This technology limits the peaks and valleys in serum blood levels that occur with frequent dosing of immediate release formulations. Controlled release addresses the problems inherent in a drug which is rapidly metabolized but which is toxic at levels ($>20\ \mu g/mL$) that are only slightly higher than the therapeutically efficacious ones (10–20 $\mu g/mL$). Furthermore, such once-a-day formulations taken just before bedtime have proven especially beneficial in the control of nocturnal asthma (27,50,62).

The effectiveness of theophylline in the treatment of asthma seems to result from a combination of biological properties which are not clearly understood (63). Detailed discussions of the possible role of xanthines in asthma may be found in references 64–66.

The effects of xanthine alkyl substitution on bronchodilation have been summarized as follows (60,61): N_1 alkylation is essential for adenosine antagonism, bronchodilator and toxic potency may increase; N_3 alkylation is essential for increased bronchodilator potency, toxicity may increase; N_7 alkylation results in decreased bronchodilator and toxic potency; C_8 alkylation has no effect on bronchodilator potency, but toxic potency and adenosine antagonism may increase; and N_9 alkylation results in general loss of potency.

Theophylline's predominant mode of action appears to be bronchodilation. However, it has also been shown that prophylactic administration of theophylline provides some protection from asthma attacks and suppresses the late-phase response (67,68). Some researchers believe that at therapeutic serum concentrations theophylline may inhibit the development of airway inflammation (69). There are conflicting reports on the effect of theophylline on allergen-induced bronchial hyperresponsiveness: some clinical studies report a reduction in hyperresponsiveness, others do not (69,70). Theophylline clearly does not reverse the general bronchial hyperresponsiveness over the course of long-term therapy (71). Because of the relationship between hyperresponsiveness and inflammation (Fig. 1), these findings argue against theophylline having a significant antiinflammatory component.

Initially, it was believed that the ability of xanthines to inhibit phosphodiesterase (PDE) led to bronchodilation (Fig. 2). One significant flaw in this proposal is that the concentration of theophylline needed to significantly inhibit PDE *in vitro* is higher than the therapeutically useful serum values (72). It is possible that concentration of theophylline in airways smooth muscle occurs, but there is no support for this idea from tissue distribution studies. Furthermore, other potent PDE inhibitors such as dipyridamole [*58-32-2*] are not bronchodilators (73). Finally, although clinical studies have shown that neither po nor continuous iv theophylline has a direct effect on circulating cyclic AMP levels (74,75), one study has shown that iv theophylline significant potentiates the increase in cyclic AMP levels induced by isoproterenol (74).

An alternative mechanism which has more recently been suggested is antagonism by theophylline of adenosine receptors. This is a well-documented effect which has been blamed for causing many of the theophylline side effects. Adenosine has a bronchoconstrictor effect when given by inhalation to asthmatics but not when given to controls (76). Also, asthmatics release adenosine into the circulation following antigen-induced bronchoconstriction (77). Therefore, it may be relevant that theophylline is an antagonist of adenosine at therapeutically useful concentrations (78). Arguing against this theory is the fact that enprofylline (3-propylxanthine [*41078-02-8*]), (**11**) is simultaneously claimed as a more potent bronchodilator than theophylline and not being an adenosine antagonist (67,79,80). Furthermore, theophylline relaxes airways smooth muscle *in vitro* probably independently of adenosine antagonism because significant concentrations of adenosine should not be present under *in vitro* conditions.

Although the benefit of theophylline treatment in asthma is without question, its use is decreasing as a result of the introduction of other effective drugs which do not have the same potential for serious side effects. It is as yet unclear whether enprofylline is a better agent or whether better drug design will require a better understanding of theophylline's mechanism of action. A theophyllinelike agent, lacking the side effects profile, would clearly be advantageous in the treatment of asthma.

Steroids

Steroids are widely used for the preventative treatment of asthma. Especially useful in the management of severe cases, their usage appears to be increasing. However, they have had a checkered history in asthma treatment. Shortly after the first synthetic corticosteroids became available, po cortisone (**12**) (Table 3) was tried as an antiasthmatic agent and showed remarkable success (81). However, within a few years the number and severity of side effects reported from the systemic administration of nonselective corticosteroids was deemed unacceptable (82–84). In an effort to reduce systemic side effects, treatment of asthmatics with inhaled cortisone was explored (85). Although this therapy was successful, steroid-induced side effects such as adrenal suppression and retention of both sodium and water still occurred, resulting in the long held view that whereas corticosteroids are effective antiasthmatic agents, they should only be used when all other therapy fails to control the disease.

Resurgent interest in the use of steroids for the treatment of asthma has been prompted by several developments. First, came the discovery of steroids such as prednisone (**13**), beclomethasone dipropionate (**14**), and budesonide (**15**). These newer, more selective, compounds successfully separate glucocorticoid (antiinflammatory) and mineralocorticoid (electrolyte regulation) activities and do not suppress serum hydrocortisone. Both beclomethasone dipropionate and budesonide were initially developed as topical agents for dermatological indications. Later they were reformulated using aerosol delivery systems to produce antiasthmatic agents having good efficacy which do not produce most of the side effects found with oral steroids (86,87,88). In part, the selectivity of both of these agents derives from the ability to get into the circulatory system where they are rapidly converted into less active metabolites that do not behave as systemic glucocorticoids. Steroid usage has increased because of the identification of asthma as an inflammatory disease. Thus aerosolized antiinflammatory steroids are often used for first-line asthma treatment (5,6).

Although the mechanism of glucocorticosteroid action in bronchial asthma is not fully understood, various possibilities have been discussed in depth (89,90).

Table 3. Steroids Used as Antiasthmatic Agents

Name	CAS Registry Number	Molecular formula	Structure number	Structure
cortisone	[*53-06-5*]	$C_{21}H_{28}O_5$	**(12)**	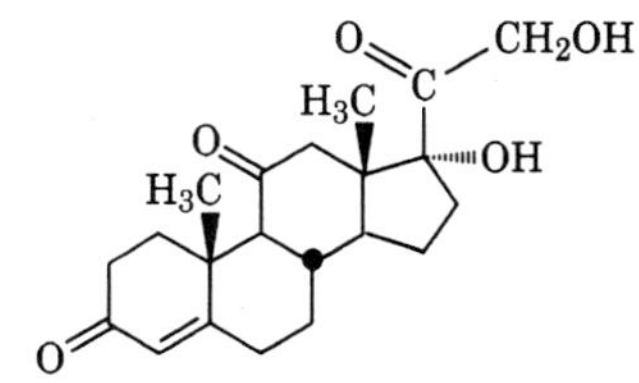
prednisone	[*53-03-2*]	$C_{21}H_{26}O_5$	**(13)**	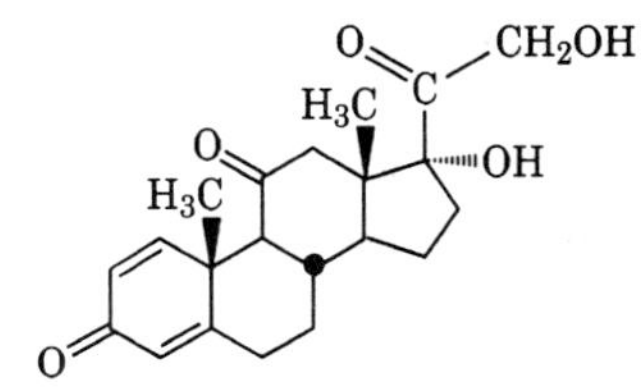
beclomethasone dipropionate	[*5534-09-8*]	$C_{22}H_{29}ClO_5$	**(14)**	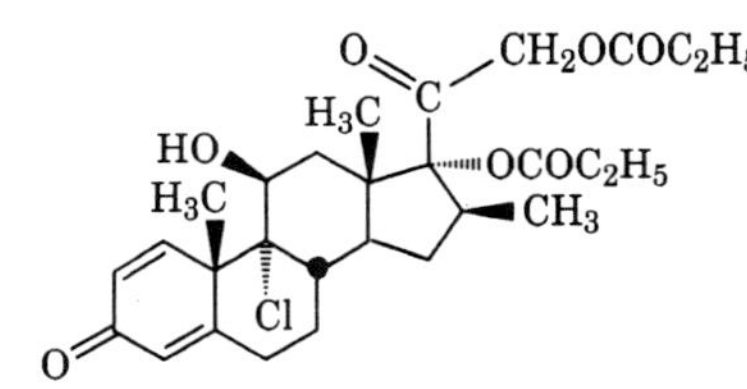
budesonide	[*51333-22-3*]	$C_{25}H_{34}O_6$	**(15)**	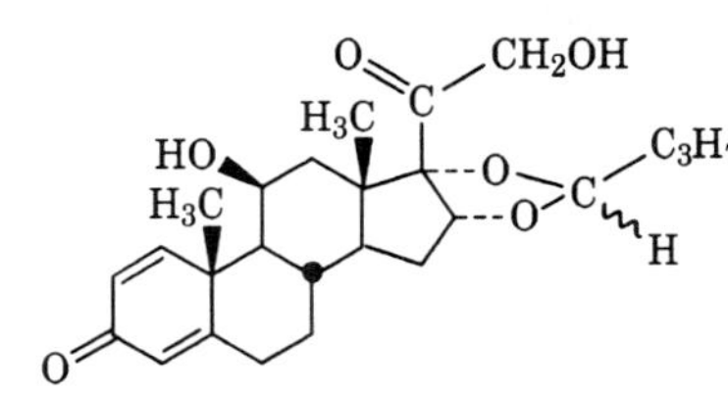

The time course of action of steroids is slower than that of β_2-agonists or theophylline and therefore steroids are not considered to be bronchodilators. It is known that steroids bind to cytosolic receptors. The steroid-receptor complex then enters the cell nucleus where the complex acts at specific sites and affects protein synthesis. The effect is to reduce the inflammatory response as well as the concentration of bronchoconstricting mediators. In addition, glucocorticoid treatment is known to reverse β_2-agonist induced adrenergic subsensitivity and to increase the number of β_2-adrenergic receptors in lung cells (91). The resultant increase in sensitivity to the natural circulating levels of norepinephrine could help induce bronchorelaxation.

Single dose or short-term treatment with aerosolized steroids inhibits both the late asthmatic response and allergen-induced bronchial hyperresponsiveness (45,92). However it does not affect the early asthmatic response nor does it induce bronchodilation (45,92). Long-term treatment with steroids protects against both the early and late asthmatic responses and also reduces bronchial hyperresponsiveness (44,71,86,93). Over time, the airways relax (dilate) and measures of airway function, such as forced expiratory volume in one second (FEV_1), gradually return to almost normal levels.

Aerosolized steroids clearly play an important role in the present-day management of asthma (87). They are reasonably safe and work best when taken prophylactically. Patient compliance, however, remains a significant problem. In part this problem is typical of any aerosolized agent. But in the case of steroids, the problem is exacerbated because a patient needs to take the steroids for several weeks before the full benefit is expressed. Oral or injected steroids (especially prednisone) are the antiasthmatic agents of last resort and are widely used to treat status asthmaticus. An agent that could mimic the actions of steroids but which would work faster and/or without side effects might be the ideal antiasthmatic agent.

Inhibitors of Mediator Release

Whereas disodium chromoglycate [*15826-37-6*] (DSCG), $C_{23}H_{14}Na_2O_{11}$, (**16**) enjoys some modest success as a preventative antiasthmatic agent, it has never achieved the same level of popularity in the United States as it has in other markets, such as in the United Kingdom. Its properties have been covered in detail in several reviews (94,95). DSCG was discovered by testing by an allergic man without the support of animal models (94,96). This uncommon approach is now viewed as unacceptable. DSCG has been unique in its class for many years, partly because early tests designed to determine the mode of action showed that DSCG is not an antagonist of any of the known mediators of asthma nor does it induce bronchodilation (95).

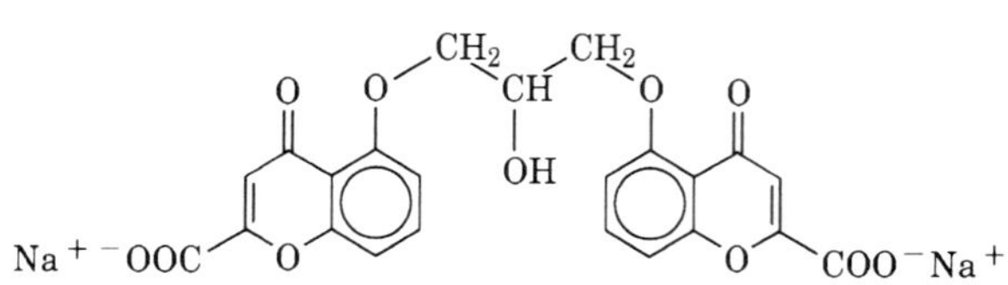

(**16**)

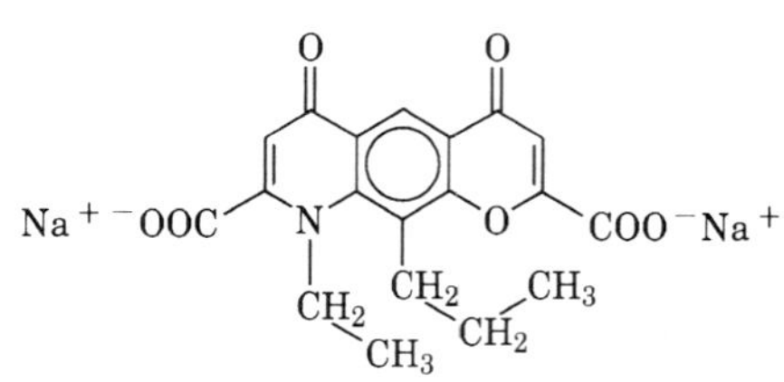

(**17**)

To aid in both the understanding of DSCG's effective mechanism and the discovery of an active analogue, animal models were developed (97). It was believed that DSCG's mode of action is by stabilization of mast cells and inhibition of mediators' release (42,98). The tremendous effort expended by the pharmaceutical industry initially resulted in failure. These investigations have been reviewed in depth (97), and it is now believed that improper disease models were employed (99,100). These early models measured the stabilizing effect of the test agents on connective tissue mast cells. Newer models have focused on mucosal (lung) mast cells and a more recent proposal is that DSCG and a related compound, nedocromil sodium [*69049-74-7*], $C_{19}H_{15}NNa_2O_7$, (**17**), both inhibit the effect of sensory nerve activation, thereby interfering with bronchoconstriction (101).

DSCG is very poorly absorbed following oral dosing and is therefore administered by aerosol inhalation, four times a day (94,95). Metabolism is minimal:

DSCG is excreted mostly as unchanged drug. Thus, although it is considered one of the safest available antiasthmatic agents, it is not a cure-all and seems to work best as an adjunct to other therapy, allowing a reduction in the dosage of the other agents (94). Clinical studies show that DSCG offers no protection if administered following antigen challenge (94,95), but if administered prophylactically, it protects both extrinsic and intrinsic asthmatics (94,102,103). DSCG is uniquely effective against both the early and late-phase asthmatic responses; it also protects against exercise-induced asthma (45,104,105). The effect of DSCG on nonspecific bronchial hyperreactivity is not at all clear; clinical studies have produced conflicting reports (45,104–106).

With the shift in preference to aerosolized agents, nedocromil sodium (**17**) has been introduced as a follow-up agent to DSCG (**16**) (107–109). Nedocromil sodium, structurally distinct from DSCG, is significantly more potent in humans, and can be administered twice daily. Like DSCG, it is administered only via an aerosol, is not significantly metabolized, and is excreted as unchanged drug. The rate limiting step for the duration of action seems to be absorption. A clinical trial has shown that, if given four times a day at 4 mg per dose, nedocromil sodium is comparable with, and equivalent to, inhaled beclomethasone dipropionate (**14**) in nearly all parameters (110).

Anticholinergic Agents

At this writing anticholinergic agents are not widely used for the symptomatic treatment of asthma, although compounds such as atropine [*51-55-8*], $C_{17}H_{23}NO_3$, (**18**) have been used for centuries (111). Inhalation of the smoke produced by burning herbal mixtures, such as Datura Stramonium, provided bronchodilation and relief from some of the symptoms of asthma. The major active component in these preparations was atropine or other closely related alkaloids (qv).

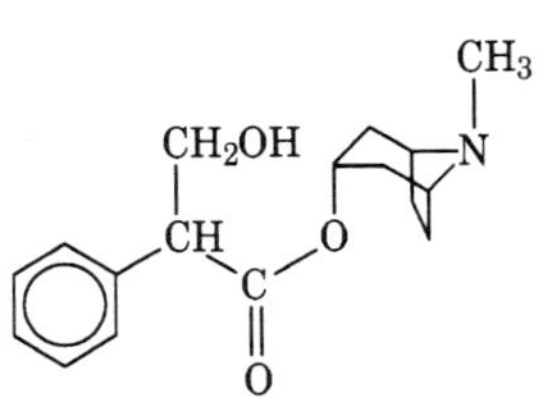

(**18**)

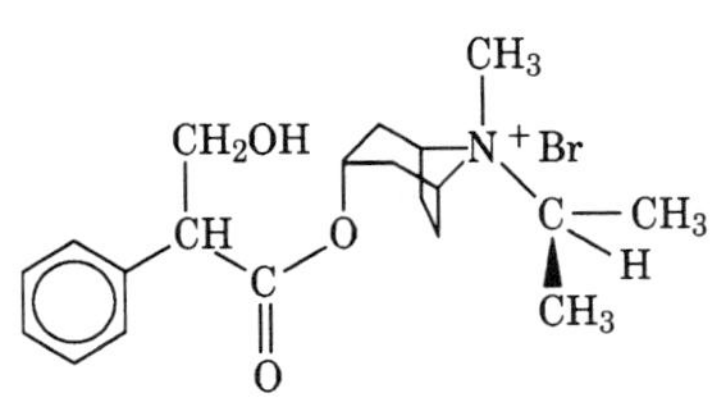

(**19**)

The beneficial effect of anticholinergics in asthma relies upon bronchial smooth muscle exhibiting a cholinergically mediated tone (resting state tension) (112,113). Receptors in the cholinergic nervous system are divided into two main classes, muscarinic (M) and nicotinic (113). Anticholinergic agents exert their bronchodilating effect by blocking the muscarinic-receptors found in bronchial smooth muscle. This blockade inhibits the normal cholinergic-induced tone. In addition, it has been shown that cholinergic receptor stimulation results in inhibition of adenylate cyclase, thereby reducing cyclic AMP levels (114). Blockade of this effect should result in indirect bronchodilatation (Fig. 2). Although atropine is effective in preventing exercise-induced asthma (103), it and other anticholinergic agents have no effect on bronchial hyperesponsiveness, the release of other mediators, or on the inflammatory process (115,116). Significant dose related side-effects such as blurred vision, dry mouth, and inhibition of gastric motility occur. These side effects result from systematic distribution of the drugs, including penetration into the central nervous system and their widespread antagonism of other muscarinic receptors.

Ipratropium bromide [*22254-24-6*], $C_{20}H_{30}BrNO_3$, (**19**) (115) is an example of a newer anticholinergic agent. This isopropyl quarternary salt derivative of atropine is not lipid soluble or well-absorbed through the gut, nor does it readily cross the blood-brain barrier. Using aerosol administration, this agent has a much lower incidence of side effects. Ipratropium bromide has gained limited acceptance as an antiasthmatic agent and seems to be more useful in patient populations that show limited response to β_2-adrenergic agonists (117). Clinical studies suggest that a synergistic effect may result from the co-administration of a β_2-agonist and an anticholinergic agent (117,118).

Research efforts on anticholinergic agents have been influenced by the finding that muscarinic-receptors can be divided into subtypes: M_1, M_2, and M_3 (119–121). It has been suggested that all three muscarinic-receptor subtypes may play a role in asthma etiology (119). Neither atropine nor ipratropium bromide successfully differentiates between subtypes (120). Because of the possibility that indiscriminate blockade of muscarinic receptors may increase release of acetylcholine via a feedback mechanism, it seems possible that agents which selectively block the smooth muscle receptors (M_3) may have better profiles than present drugs (120). However, more than half the cholinergic receptors in human lung are of the M_1 subtype, so this view may represent an oversimplification (119).

Agents Undergoing Clinical Evaluation

Antihistamines. Antihistamines are not recommended for the symptomatic treatment of asthma, although recent studies have shown that histamine (**20**) (Table 4) released from mast cells may account for 50% of the immediate asthmatic response (see HISTAMINE AND HISTAMINE ANTAGONISTS) (122). Reviews on antihistamines in asthma therapy may be found in references 124–127. The possible role of histamine in the anaphylactic response was first recognized in the early 1900s (123). Almost 40 years passed before compounds became available which were sufficiently nontoxic to be tried in clinical studies. These early agents, tested as therapeutic agents for allergic rhinitis and asthma (128), were only weakly active as antihistamines. They were not specific for histamine receptors and activity at muscarinic (cholinergic) and adrenergic receptors contributed both to bronchodilating actions and to a wide variety of side effects which limited use.

The discovery that histamine receptors, just like adrenergic and cholinergic receptors, may be divided into subtypes H_1 and H_2, combined with the recognition that the bronchial receptors were of the H_1 type, set the stage for a newer generation of antihistamines. Clinical investigation in the late 1970s of the H_1-antihistamines clemastine (**21**) (129) and chlorpheniramine (**22**) (130,131), shown in Table 4, revealed that these agents had a bronchodilating effect and provided some protection against exercise-induced asthma. But the side effects, induced at

Table 4. Histamine and Antihistamines

Name	CAS Registry Number	Molecular formula	Structure number	Structure
histamine	[*51-45-6*]	$C_5H_9N_3$	**(20)**	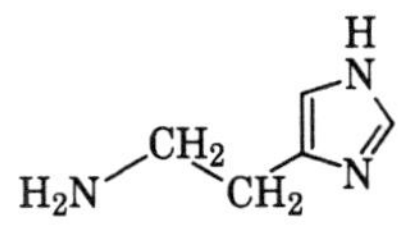
clemastine	[*15686-51-8*]	$C_{21}H_{26}ClNO$	**(21)**	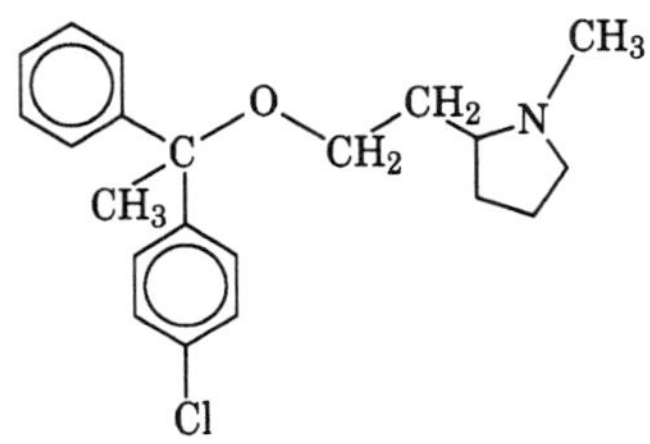
chloro-pheniramine	[*132-22-9*]	$C_{16}H_{19}N_2Cl$	**(22)**	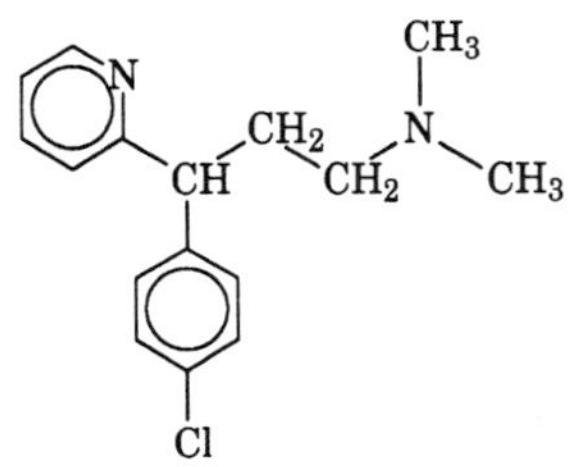
terfenadine	[*50679-08-8*]	$C_{20}H_{41}NO_2$	**(23)**	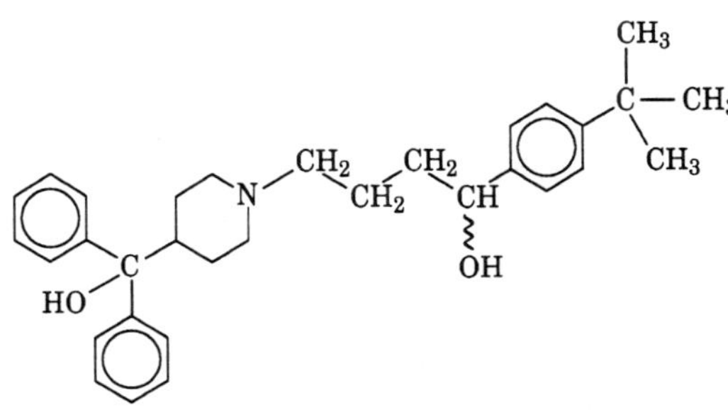
astemizole	[*68844-77-9*]	$C_{28}H_{31}N_4FO$	**(24)**	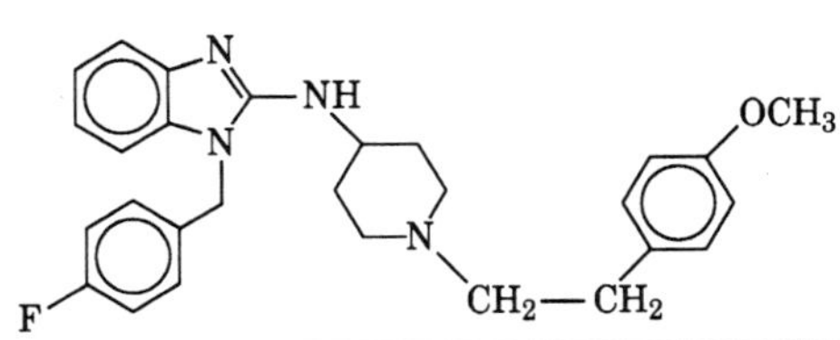

H_1-receptors in the central nervous system which produced sedation and at cholinergic receptors, limited the maximal doses that could be given. These drugs were not useful as antiasthmatic agents.

Two newer potent selective H_1-antagonists, terfenadine (**23**) (132) and astemizole (**24**) (133), have been developed which have neither the sedative nor the anticholinergic liabilities of the earlier agents. Both of these compounds have proven efficacious in the treatment of hay fever and produce very few side effects, prompting a re-evaluation of the role of antihistamines in asthma treatment.

Astemizole has very unusual pharmacokinetic parameters. It shows delayed onset both *in vitro* and *in vivo* (126) and also has a very long duration of action. *In vivo* formation of the desmethyl metabolite, equiactive to astemizole as an H_1-antagonist, occurs. This metabolite is cleared at only 1/10th the rate of astemizole. Clinically, it may take two days following a dose of astemizole for symptom alleviation to occur in the treatment of hay fever (134). However, the effect is very long lasting: after a single 10-mg tablet, significant antihistaminic activity can be demonstrated in volunteers for about 20 days (135). Astemizole seems to delay the onset of exercise-induced asthma but not affect the severity of the bronchoconstriction (136,137). Also, prolonged astemizole treatment offers some protection from the symptoms of seasonal asthma with significant attenuation of the early component of bronchoconstriction but little effect on the late-phase response (138).

Oral terfenadine has a relatively rapid onset (134), and a sufficiently long duration of action that it needs to be dosed only once or twice a day. Clinical studies show: it effectively blocks histamine-induced bronchospasm (124); it causes modest bronchodilation (139,140); and it provides limited protection against either exercise-induced asthma (140,141) or challenge using nebullized water (139). The bronchodilatory response to 120-mg oral terfenadine has been shown to approach that of 200-μg of inhaled albuterol (**4,** R = $C(CH_3)_3$) (142). Terfenadine, however, has no effect on the dose of methacholine needed to cause a 20% fall in FEV_1 (143). Also, it has only a very modest protective effect against antigen-induced asthma, that effect being partial blockade of the immediate bronchoconstrictor response (144). Studies on its use in the treatment of seasonal atopic asthmatics have shown minor, but statistically significant, effects which have correlated with a reduction in the need to use other bronchodilators (125,145).

Leukotriene Antagonists. Over 50 years ago a second component, besides histamine, of the immediate type hypersensitivity reaction was identified (146). This discovery was named the slow reacting substance of anaphylaxis (SRS-A). Because it was believed that SRS-A played a primary role in the etiology of asthma and other diseases, the discovery generated a great deal of interest. However, determination of SRS-A structure and development of either antagonists or biosynthesis inhibitors was severely hampered by the limited availability of SRS-A from biological sources and chemical and biological instability (147). In 1980 it was shown that SRS-A was made up of a mixture of leukotrienes (148), which were products of the 5-lipoxygenase pathway of arachidonic acid metabolism (see PROSTAGLANDINS). Tests showed that the pure leukotrienes are 100 to 1000 times as potent as histamine in terms of bronchial spasmogens both *in vitro* and in clinical studies, prompting a resurgence of interest in the development of leukotriene antagonists and biosynthesis inhibitors. At this writing, only work in receptor antagonists has progressed far enough to show promise in the clinic. The leukotrienes are listed in Table 5 as are some antagonists.

The first SRS-A antagonist, FPL-55712 (**26**) (149), was discovered before the structures of the leukotrienes were determined. Although this compound is relatively weak as an antagonist and suffers from a very short half-life *in vivo,* it played an important role both in leukotriene structure elucidation and as a model for later antagonists. In work structurally related to FPL-55712, LY-171883 was

Table 5. Leukotrienes and Leukotriene Antagonists

Name	CAS Registry Number	Molecular formula	Structure number	Structure
			Leukotrienes	
leukotriene C_4	[*72025-60-6*]	$C_{30}H_{47}N_3O_9S$	(**25**, RSH = glutathione)	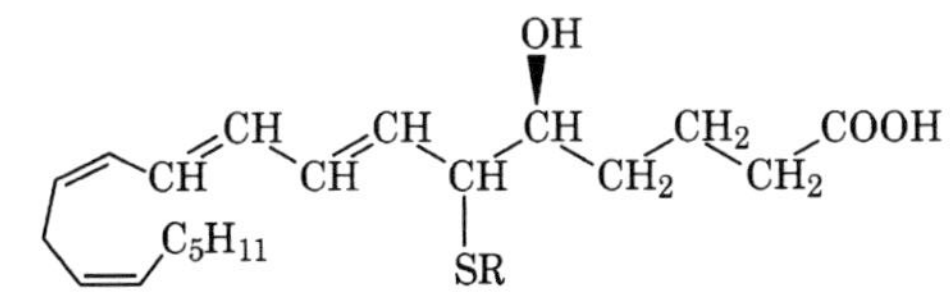
leukotriene D_4	[*73836-78-9*]	$C_{25}H_{40}N_2O_6S$	(**25**, RSH = cysteinylglycine)	
leukotriene E_4	[*75715-89-8*]	$C_{23}H_{37}NO_5S$	(**25**, RSH = cysteine)	
			Leukotriene antagonists	
FPL-55712	[*40786-08-1*]	$C_{27}H_{30}O_9$	(**26**)	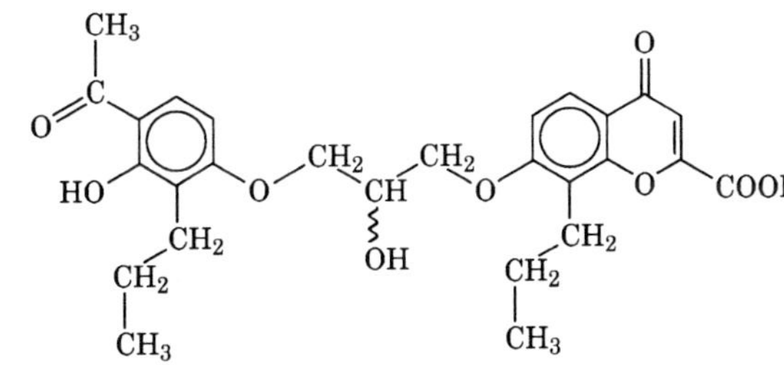
LY-171883	[*88107-10-2*]	$C_{16}H_{22}N_4O_3$	(**27**)	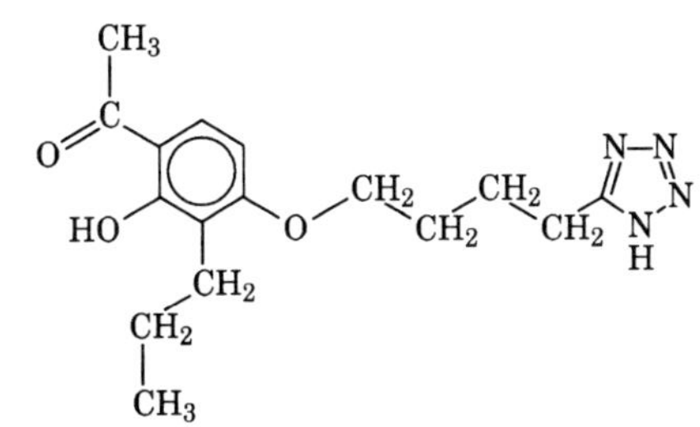

Table 5. *(Continued)*

Name	CAS Registry Number	Molecular formula	Structure number	Structure
MK-571 (formerly L–660711)	[*115104-28-4*]	$C_{24}H_{23}N_2O_3ClS_2$	(**28**)	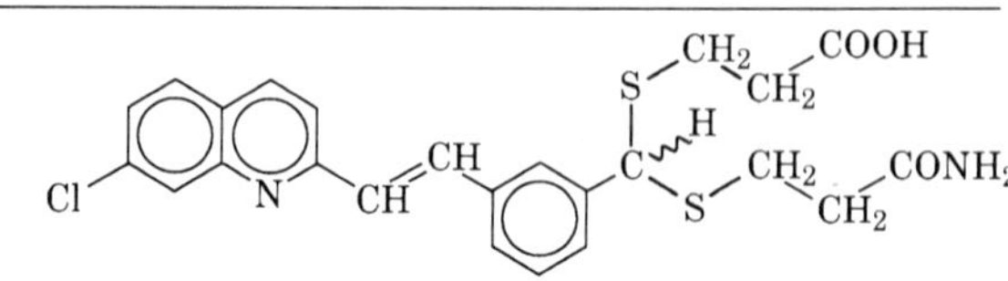
ICI-204219	[*107753-78-6*]	$C_{31}H_{33}N_3O_6S$	(**29**)	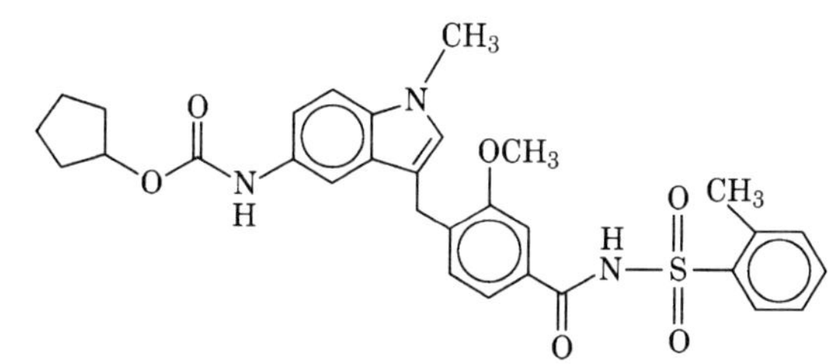

developed (**27**) (150). LY-171883 was evaluated in several clinical trials before development was stopped. Orally administered, LY-171883 blocked slightly the response to aerosol LTD_4, improved pulmonary function (FEV_1) in mild asthmatics (151), decreased the sensitivity of asthmatics to cold air-induced bronchoconstriction (152), and significantly reduced the bronchoconstrictor response to inhaled antigen (153). However, in all these studies the beneficial effects were minimal.

Many structurally diverse newer leukotriene antagonists that are at varying stages of development have since been discovered (154,155). Two compounds which are especially potent and which seemed most promising as potential antiasthmatic agents are MK-571 (formerly L-660711) (**28**) (156), and ICI–204219 (**29**) (157). Both of these agents advanced to the clinic but the development of MK-571 has since been discontinued. However, studies show that two hours after a single 40-mg po dose, ICI–204219 caused a greater than 100-fold shift in the dose-response curve to aerosol LTD_4 (158), and caused a greater than tenfold shift in the dose-response to inhaled antigen (159).

Other Drugs. In addition to the drugs already discussed, a wide variety of other agents have been investigated for antiasthmatic potential. For example, a large but unsuccessful effort went into the search for a prostaglandin-type bronchodilator in the 1970s. Within this general area the uncertain role of thromboxane as a mediator in asthma has prompted limited interest in the exploration of thromboxane A_2 synthesis inhibitors and/or antagonists (160,161).

In 1990 5-lipoxygenase inhibitors, ie, leukotriene biosynthesis inhibitors, began undergoing clinical evaluation (161,162). Also, rapidly approaching clinical evaluation are a structurally diverse group of platelet-activating factor (PAF) antagonists (162). In humans, PAF is a potent bronchospastic agent that also induces bronchial hyperreactivity. There is significant effort being expended to discover and develop potent PAF antagonists as antiasthmatic agents (163,164).

The key to a cure for asthma is presumably to be found in a better understanding of its etiology. Until then, there are many approaches being utilized (165). As an example, there is interest in discovering a drug which would specifically affect calcium ions in airway smooth muscle and cause bronchodilation. Similarly, if selective potassium channel openers can be found, they should act as bronchodilators. Alternatively, a selective inhibitor of the correct phosphodiesterase isozyme should mimic the bronchodilating effects of the β_2-agonists without inducing the side effects. Formation of both the prostanoid and leukotriene bronchoconstrictors as well as platelet activating factor should theoretically be inhibited, if a selective phospholipase A_2 inhibitor can be discovered. Finally, the roles of the nonadrenergic, noncholinergic nervous system (38) and the neuropeptides, eg, tachykinins, in asthma have only begun to be explored.

Economic Aspects

Total sales of prescription bronchodilators and antiasthma products in 1989 were approximately \$1.2 and \$3.3 billion in the North American and world markets, respectively (166). The three largest shares of the world market were held by: Glaxo Holdings plc, Schering-Plough Corporation, and Boehringer Ingelheim Corporation.

In 1989 β_2-agonists were the largest class of prescription bronchodilators and antiasthma agents. Total sales reached approximately $1.2 billion. Albuterol (**4**, R = $C(CH_3)_3$) was the best-selling (ranked by dollar value) antiasthmatic agent (167). Most of the remaining β_2-agonist sales were for terbutaline (**3**, R = $C(CH_3)_3$) and metaproterenol (**3**, R = $CH(CH_3)_2$). Inhaled corticoids were the next largest class. Corticoid sales, which reached approximately $400 million, were predominantly accounted for by beclomethasone diproprionate (**14**) and budenoside (**15**). Sales of nonprescription, over-the-counter, antiasthma products are only a small fraction of the prescription market.

BIBLIOGRAPHY

"Antiasthmatic Agents" in *ECT,* 3rd ed., pp. 758–782, by D. D. Miller, Ohio State University.

1. M. E. Gershwin, ed., *Bronchial Asthma,* Grune & Stratton, Orlando, Fla., 1986.
2. D. R. Buckle and H. Smith, eds., *Development of Anti-Asthma Drugs,* Butterworths, London, UK, 1984.
3. E. B. Weiss, M. S. Segal, and M. Stein, eds., *Bronchial Asthma: Mechanism and Therapeutics,* 2nd ed., Little, Brown and Co., Toronto, Canada, 1985.
4. Committee on Diagnostic Standards, American Thoracic Society, *Am. Rev. Respir. Dis.* **85,** 762 (1962).
5. Editorial, *Ann. Allergy* **62,** 481 (1989).
6. F. E. Hargreave, *J. Allergy Clin. Immunol.* **83,** 525 (1989).
7. D. M. Fleming and D. L. Crombie, *Br. Med. J.* **294,** 279 (1987).
8. A. J. Woolcock, *Chest* (suppl.) **90,** 40S (1986).
9. R. R. Dodge and B. Burrows, *Am. Rev. Resp. Dis.* **122,** 567 (1980).
10. S. R. Benatar, *N. Engl. J. Med.* **314,** 423 (1986).
11. P. G. J. Burney, *Lancet* **2,** 323 (1986).
12. E. D. Robin, *Chest* **93,** 614 (1988).
13. R. M. Sly, *Ann. Allergy* **53,** 20 (1984).
14. Ref. 3, p. 18.
15. J. W. Gerrard, C. G. Ko, P. Vickers, and C. D. Gerrard, *Ann. Allergy* **36,** 10 (1976).
16. C. G. A. Persson, *Trends in Pharmacol. Sci.* **3,** 312 (1982).
17. P. J. Barnes, *J. Allergy Clin. Immunol.* **83,** 1013 (1989).
18. D. W. Moran and A. W. Wheeler in Ref. 2, pp. 349–390.
19. R. W. Fox and R. F. Lockey in Ref. 1, pp. 371–392.
20. P. S. Creticos, *J. Allergy Clin. Immunol.* **83,** 554 (1989).
21. J. E. Hodgkin, *Chest* (Suppl.) **90,** 62S (1986).
22. T. J. H. Clark, *Chest* (Suppl.) **90,** 67S (1986).
23. A. F. Barker, *West. J. Med.* **150,** 303 (1989).
24. M. T. Newhouse and M. B. Dolovich, *N. Engl. J. Med.* **315,** 870 (1986).
25. *National Prescription Audits 1984–1989.* IMS International, a Company of the Dun & Bradstreet Corporation, 1989.
26. Editorial, "Asthma at Night," *Lancet* **1,** 220 (1983).
27. C. W. Zwillich and co-workers, *Am. Rev. Respir. Dis.* **139,** 470 (1989).
28. S. P. Newman, in J. A. Bristal, ed., *Annual Reports in Medicinal Chemistry,* Vol. 25, Academic Press, New York, 1990, pp. 178–189.
29. H. S. Nelson, *J. Allergy Clin. Immunol.* **77,** 771 (1986).
30. C. E. Reed, *J. Allergy Clin. Immunol.* **76,** 335 (1985).

31. L. E. Martin, J. C. Hobson, J. A. Page, and C. Harrison, *Eur. J. Pharmacol.* **14,** 183 (1971).
32. C. Kaiser in D. L. Temple Jr., ed., *Drugs Affecting the Respiratory System, ACS Symp. Ser.* **118,** 251 (1980).
33. J. G. M. Bullowa and D. M. Kaplan, *Med. News* **83,** 787 (1903).
34. S. E. Meyer in A. G. Goodman, L. S. Goodman, and A. Gilman, eds., *The Pharmacological Basis of Therapeutics,* 6th ed., MacMillan, New York, 1980, p. 56.
35. J. B. Richardson and J. Beland, *J. Appl. Physiol.* **41,** 764 (1976).
36. J. B. Richardson, *J. Allergy Clin. Immunol.* **56,** 473 (1975).
37. J. F. Souhrada, E. Melzer, and P. Grantham, *Respir. Physiol.* **40,** 199 (1980).
38. J. B. Richardson in Ref. 3, pp. 123–126.
39. R. P. Alquist, *Am. J. Physiol.* **153,** 586 (1948).
40. A. M. Lands and co-workers, *Nature (London)* **214,** 597 (1967).
41. K. J. Rhoden, L. A. Meldrum, and P. J. Barnes, *J. Appl. Physiol.* **65,** 700 (1988).
42. F. L. Pearce and J. Clements, *Biochem. Pharmacol.* **31,** 2247 (1982).
43. I. F. Skidmore in A. B. Kay, ed., *Asthma: Clinical Pharmacology and Therapeutic Progress,* Blackwell Scientific, Oxford, UK, 1986, p. 171.
44. K. F. Kerrebijn, E. E. M. van Essen-Zandvliet, and H. J. Neijens, *J. Allergy Clin. Immunol.* **79,** 653 (1987).
45. D. W. Cockcroft and K. Y. Murdock, *J. Allergy Clin. Immunol.* **79,** 734 (1987).
46. J. D. S. McEvoy, A. Vall-Spinosa, and J. W. Paterson, *Am. Rev. Respir. Dis.* **108,** 490 (1973).
47. J. J. McPhillips in M. E. Goldberg, ed., *Pharmacological and Biochemical Properties of Drug Substances,* Vol. 1, American Pharmaceutical Association, Washington, D.C., 1977, p. 311. A review on terbutaline.
48. M. J. Schumacher, *J. Allergy Clin. Immunol.* **66,** 33 (1980).
49. P. Borum and N. Mygind, *J. Allergy Clin. Immunol.* **66,** 25 (1980).
50. M. Heins and co-workers, *Eur. Respir. J.* **1,** 306 (1988).
51. R. T. Brittain and D. M. Harris in Ref. 45, p. 257. A review on albuterol.
52. J. W. Paterson, R. J. Courtney-Evans, and F. J. Prime, *Br. J. Dis. Chest* **65,** 21 (1971).
53. I. Kass and T. S. Mingo, *Chest* **78,** 283 (1980).
54. *Arzneim.-Forsch.* **26,** 1403 (1976). A series of articles on the properties of clenbuterol.
55. A. Ullman and N. Svedmyr, *Thorax* **43,** 674 (1988).
56. J. Bradshaw and co-workers, *Br. J. Pharmacol.* **92,** 590P (1987).
57. D. I. Ball and co-workers, *Br. J. Pharmacol.* **92,** 591P (1987).
58. L. J. Emorine and co-workers, *Science,* **245,** 1118 (1989).
59. H. Salter, *Edinburgh Med. J.,* 1109 (1859).
60. C. G. A. Persson, *Agents and Actions* (Suppl.) **13,** 115 (1983). A review on Theophylline structure-activity relationships.
61. C. G. A. Persson, *Trends in Pharmacol. Sci.* **3,** 312 (1982). A review on Theophylline structure-activity relationships.
62. W. W. Arkinstall, M. E. Atkins, D. Harrison, and J. H. Stewart, *Am. Rev. Respir. Dis.* **135,** 316 (1987).
63. C. G. A. Persson, *J. Allergy Clin. Immunol.* **78,** 780 (1986).
64. N. Cushley and S. T. Holgate in Ref. 2, pp. 205–223.
65. M. Weinberger and L. Hendeles in Ref. 3, pp. 646–674.
66. For a review on "Recent Innovations in Theophylline-Like Bronchodilator Drugs" see R. J. Seidehamel and D. L. Temple, Jr., in D. L. Temple, Jr., ed., *Drugs Affecting the Respiratory System, ACS Symp. Ser.* **118,** 285 (1980).
67. R. Pauwels and co-workers, *J. Allergy Clin. Immunol.* **76,** 583 (1985).
68. S. Godfrey, *J. Allergy Clin. Immunol.* **65,** 97 (1980).
69. R. A. Pauwels, *J. Allergy Clin. Immunol.* **83,** 548 (1989).

70. D. W. Cockcroft and co-workers, *J. Allergy Clin. Immunol.* **83,** 913 (1989).
71. J. I. Dutoit, C. M. Salome, and A. J. Woolcock, *Am. Rev. Respir. Dis.* **136,** 1174 (1987).
72. J. B. Polson, J. J. Krzanowski, A. L. Goodman, and A. Szentivanyi, *J. Clin. Exp. Pharmacol. Physiol.* **5,** 535 (1978).
73. R. E. Ruffin and M. T. Newhouse, *Eur. J. Respir. Dis.* **62,** 123 (1981).
74. P. W. Trembath and J. Shaw, *Br. J. Clin. Pharmacol.* **6,** 499 (1978).
75. F. H. Parrott and co-workers, *Am. Rev. Respir. Dis.* **113,** 156 (1976).
76. M. J. Cushley, A. E. Tattersfield, and S. T. Holgate, *Br. J. Clin. Pharmacol.* **15,** 161 (1983).
77. J. S. Mann, A. G. Renwick, and S. T. Holgate, *Clin. Sci.* **65,** 22P (1983).
78. B. B. Fredholm, *Eur. J. Respir. Dis.* **61** (suppl. **109**), 29 (1980).
79. K. R. Chapman and co-workers, *Am. Rev. Respir. Dis.* **139,** 688 (1989).
80. C. G. A. Persson, I. Erjefalt, and J. A. Karlsson, *Acta Pharmacol. et Toxicol.* **49,** 317 (1981).
81. H. M. Carryer and co-workers, *J. of Allergy* **21,** 282 (1950).
82. H. Arnoldsson, *Acta Allergol.* **6,** 1 (1958).
83. M. McAllen, *Respiration* (suppl.) **27,** 250 (1970).
84. H. C. Smyllie and C. K. Connolly, *Thorax* **23,** 571 (1968).
85. M. L. Gelfand, *N. Eng. J. Med.* **245,** 293 (1951).
86. J. M. Henriksen, *Br. Med. J.* **291,** 248 (1985).
87. J. H. Toogood, B. Jennings, and J. C. Baskerville in Ref. 3, pp. 698–713.
88. E. O. Meltzer, J. P. Kemp, M. J. Welch, and H. A. Orgel, *Am. Rev. Respir. Dis.* **131,** 732 (1985).
89. M. Kaliner, *J. Allergy Clin. Immunol.* **76,** 321 (1985).
90. Supplement: "Corticosteroids: Their Biological Mechanisms and Application to the Treatment of Asthma", *Am. Rev. Respir. Dis.* **141,** S1 (1990).
91. C. M. Fraser and J. C. Venter, *Biochem. Biophys. Res. Commun.* **94,** 390 (1989).
92. R. Dahl, J. M. Henriksen, S-A. Johansson, and P. Venge in A. B. Kay, ed., *Asthma: Clinical Pharmacology and Therapeutic Progress,* Blackwell Scientific, Oxford, UK, 1986, p. 331.
93. A. J. Woolcock, K. Yan, and C. M. Salome, *Clin. Allergy* **18,** 165 (1988).
94. R. N. Brogden, T. M. Speight, and G. S. Avery, *Drugs* **7,** 164 (1974).
95. J. S. G. Cox and co-workers, *Adv. in Drug Res.* **5,** 115 (1970).
96. R. E. C. Altounyan, *Acta Allergol.* **22,** 487 (1967).
97. J. L. Suschitzky and P. Sheard in G. P. Ellis and G. B. West, eds., *Progress in Medicinal Chemistry,* Elsevier Science, B.V., **21,** 1 (1984).
98. J. S. G. Cox, *Nature (London)* **216,** 1328 (1967).
99. T. C. Stokes and J. Morley, *Br. J. Dis. Chest* **75,** 1 (1981).
100. T. S. C. Orr, *Drugs* **37** (Suppl. 1), 113 (1989).
101. C. M. S. Dixon and P. J. Barnes, *Br. J. Clin. Pharmacol.* **27,** 831 (1989).
102. Brompton Hospital/Medical Research Council Collaborative Trial, *Br. Med. J.* **4,** 383 (1972).
103. S. Godfrey and P. Koenig, *Pediatrics* **56,** 930 (1975).
104. U. G. Svendsen, *J. Allergy Clin. Immunol.* **80,** 68 (1987).
105. I. L. Bernstein, *J. Allergy Clin. Immunol.* **76,** 381 (1987).
106. P. Haber, G. Keyer, and C. O. Burghuber, *Respiration* **55,** 44 (1989).
107. T. S. C. Orr and co-workers, in A. B. Kay, ed., *Asthma: Clinical Pharmacology and Therapeutic Progress,* Blackwell Scientific, Oxford, UK, 1986, p. 265.
108. J. P. Gonzalez and R. N. Brogden, *Drugs* **34,** 560 (1987).
109. S. Lal, S. Malhotra, D. Gribben, and D. Hodder, *Thorax* **39,** 809 (1984).
110. K. C. Bergmann, C. P. Bauer, and A. Overlack, *Curr. Med. Res. Opin.* **11,** 533 (1989).
111. B. Gandevia, *Postgrad. Med. J.* **51** (suppl. 7), 12 (1985).

112. G. E. Pakes in Ref. 2, p. 159.
113. P. Taylor in A. G. Goodman, L. S. Goodman, and A. Gilman, eds., *The Pharmacological Basis of Therapeutics,* 6th ed., MacMillan, New York, 1980, p. 211.
114. C. A. Jones, J. M. Madison, M. Tom-moy, and J. K. Brown, *Am. J. Physiol.* **253,** C97 (1987).
115. R. Bauer and co-workers in M. E. Goldberg, ed., *Pharmacological and Biochemical Properties of Drug Substances,* Vol. 2, American Pharmaceutical Association, Washington, D.C., 1979, p. 489. A general review on ipratropium bromide.
116. P. H. Howarth and co-workers, *Am. Rev. Respir. Dis.* **132,** 986 (1985).
117. F. E. R. Simons, *J. Allergy Clin. Immunol.* **80,** 239 (1987).
118. A. S. Rebuck and co-workers, *Am. J. Med.* **82,** 59 (1987).
119. J. Maclagan and P. J. Barnes, *Trends Pharmacol. Sci.* (Suppl. 1) **10,** 88 (1989).
120. P. J. Barnes, P. A. Minette, and J. Macalagan, *Trends Pharmacol. Sci.* **9,** 412 (1988).
121. R. B. Barlow, K. N. Burston, and A. Vis, *Proceed. Br. Pharm. Soc.* 141P (1979).
122. P. Rafferty, R. Beasley, and S. T. Holgate, *Am. Rev. Respir. Dis.* **136,** 369 (1987).
123. H. H. Dale and P. P. Laidlaw, *J. Physiol.* **41,** 318 (1910).
124. P. Rafferty, *Ann. Allergy* **63,** 389 (1989).
125. S. T. Holgate and J. P. Finnerty, *J. Allergy Clin. Immunol.* **83,** 537 (1989).
126. P. Rafferty and S. T. Holgate, *J. Allergy Clin. Immunol.* **84,** 144 (1989).
127. W. E. Pierson and F. S. Virant, *Ann. Allergy* **63,** 601 (1989).
128. C. H. Eyermann, *J. Allergy* **17,** 210 (1946).
129. J. P. R. Hartley and S. G. Nogrady, *Aust. and N.Z. J. Med.* **11,** 106 (1981).
130. V. T. Popa, *J. Allergy Clin. Immunol.* **59,** 54 (1977).
131. R. C. Groggins, A. D. Milner, and G. M. Stokes, *Br. J. Dis. Chest* **73,** 297 (1979).
132. B. L. Martz, J. P. Mumford, J. T. Burke, and J. L. R. Barlow, *Arzneim. Forsch.* **32,** 1153 (1982).
133. P. M. Laduron, P. F. M. Janssen, W. Gommeren, and J. E. Leysen, *Mol. Pharmacol.* **21,** 294 (1982).
134. J. P. Girard, D. Sommacal-Schopf, P. Bigliardi, and S. A. Henauer, *J. Int. Med. Res.* **13,** 102 (1985).
135. J. Callier and co-workers, *Curr. Ther. Res.* **29,** 24 (1981).
136. V. Backer and co-workers, *Allergy* **44,** 209 (1989).
137. M. D. Clee, C. G. Ingram, P. C. Reid, and A. S. Robertson, *Br. J. Dis. Chest* **78,** 180 (1984).
138. S. T. Holgate, M. B. Emanuel, and P. H. Howarth, *J. Allergy Clin. Immunol.* **76,** 375 (1985).
139. R. G. Townley, R. J. Hopp, A. K. Bewtra, and M. Nabe, *Ann. Allergy* **63,** 455 (1989).
140. W. E. Pierson and co-workers, *Ann. Allergy* **63,** 461 (1989).
141. K. R. Patel, *Br. Med. J.* **288,** 1496 (1984).
142. W. O. C. M. Cookson, *Br. J. Clin. Pharmacol.* **24,** 120 (1987).
143. K. R. Patel, *J. Allergy Clin. Immunol.* **79,** 355 (1987).
144. T. B. Chan, D. M. Shelton, and N. M. Eiser, *Br. J. Dis. Chest* **80,** 375 (1986).
145. A. Taytard and co-workers, *Br. J. Clin. Pharmacol.* **24,** 743 (1987).
146. C. H. Kellaway and W. R. Trethewie, *Q. J. Exp. Physiol. Cog. Med. Sci.* **30,** 121 (1940).
147. L. W. Chakrin and D. M. Bailey, eds., *The Leukotrienes, Chemistry and Biology,* Academic Press, New York, 1984. A review on the leukotrienes.
148. S. Hammarstrom and co-workers, *Biochem. Biophys. Res. Commun.* **91,** 1266 (1979).
149. J. Augstein and co-workers, *Nat. New Biol.* **245,** 215 (1973).
150. J. H. Fleisch, *J. Pharmacol. Exp. Ther.* **233,** 148 (1985).
151. M. L. Cloud and co-workers, *Am. Rev. Respir. Dis.* **140,** 1336 (1989).
152. E. Israel and co-workers, *Am. Rev. Respir. Dis.* **140,** 1348 (1989).
153. R. W. Fuller, P. N. Black, and C. T. Dollery, *J. Allergy Clin. Immunol.* **83,** 939 (1989).

154. D. W. Snyder and J. H. Fleisch, *Annu. Rev. Pharmacol. Toxicol.* **29,** 123 (1989).
155. A. Shaw and R. D. Krell, *J. Med. Chem.* in press (1990).
156. J. C. Kips and co-workers, *Am. Rev. Resp. Dis.* **139,** A63 (1989).
157. R. D. Krell and co-workers, *Am. Acad. Allergy Immunol.* **81,** 276 (1988).
158. L. J. Smith and co-workers, *Am. Rev. Respir. Dis.* **141,** 987 (1990).
159. S. R. Findlay, C. Easley, J. M. Barden, and M. Glass, *J. Allergy Clin. Immunol.* **85,** 197 (1990).
160. E. W. Collington and H. Finch in J. A. Bristol, ed., *Annual Reports in Medicinal Chemistry,* Vol. 25, Academic Press, New York, 1990, pp. 99–108.
161. A. Shaw in J. A. Bristol, ed., in Ref. 160, pp. 61–70.
162. J. H. Musser, *J. Med. Chem.* **34** (1991).
163. C. P. Page, D. N. Robertson, and A. J. Coyle "New Anti-Asthma Drugs", in S. O. Donnell and C. Persson, eds., *Agents and Actions Suppl. 23,* Birkhauser Verlag, Basel, Switzerland, 1988, pp. 173–185.
164. C. Page and A. Abbott, *Trends in Pharm. Sci.* **10**(7) (1989).
165. P. J. Barnes, ed., *New Drugs for Asthma,* IBC Technical Services, London, 1989.
166. *The Pharmaceutical Market, World Review 1989,* IMS International, a Company of the Dun & Bradstreet Corporation.
167. *Annual Report and Accounts 1990* Glaxo Holdings plc, London, 1990, p. 13.

PETER R. BERNSTEIN
ICI Americas Corporation

ANTIBACTERIAL AGENTS, SYNTHETIC

QUINOLONES

Quinolone carboxylic acids are a class of totally synthetic antibacterial agents which have the general structure (**1**).

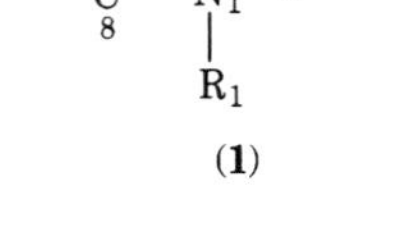

(**1**)

This representation is intended to encompass 4-oxo-3-quinolinecarboxylic acids as well as the corresponding 1,8-naphthyridines, cinnolines, and pyrido[2,3-*d*]-pyrimidines. These classes are illustrated by ciprofloxacin [*85721-33-1*] (**2**) (1), nalidixic acid [*389-08-2*] (**3**) (2), cinoxacin [*28657-80-9*] (**4**) (3), and piromidic acid [*19562-30-2*] (**5**) (4), respectively.

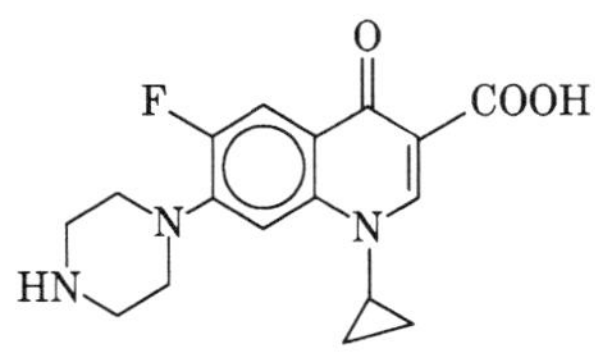

(**2**)

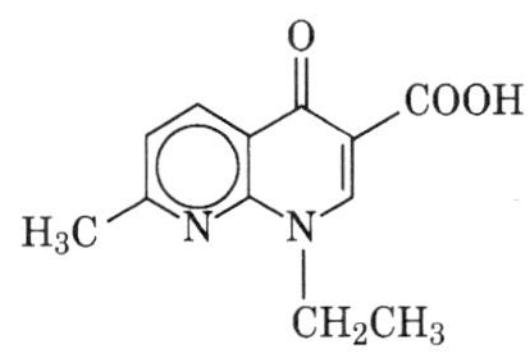

(**3**)

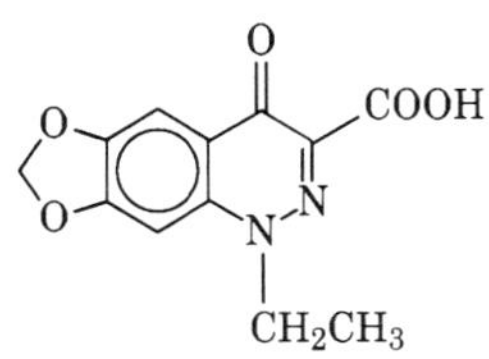

(**4**)

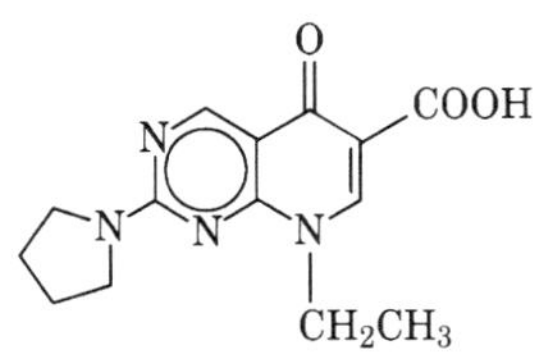

(**5**)

Nalidixic acid was the first quinolone introduced to clinical practice, and it is still in use today for the treatment of urinary tract infections caused by gram-negative pathogens (5,6). Since the introduction of nalidixic acid in the early 1960s, substantial improvements have been made in the antibacterial spectrum and potency of quinolones. Additionally, newer quinolones are well absorbed and distributed throughout the body making them useful for the treatment of a variety of systemic infections including those of the respiratory tract, gastrointestinal tract, the ear, nose, and throat, sexually transmitted diseases, and bone infections (7–9). Because of rapid advances in the development of quinolone antibacterial agents since the 1980s, they are now important tools available to physicians. Established quinolone antibacterial agents are presented in Table 1.

Structure–Activity Relationships

The most important structural features necessary for meaningful antibacterial activity of quinolones include a carboxylic acid attached to the 3-position of the quinolone nucleus and a small alkyl or aryl group in the 1-position. In addition to these more or less rudimentary requirements for activity (10), all of the newer generation quinolones have a fluorine attached to the 6-position and a nitrogen heterocycle attached to the 7-position. This heterocycle is often a piperazine or pyrrolidine derivative. Comparison of the data in Table 2 for nalidixic acid (**3**) and cinoxacin (**4**), nonfluorinated quinolones that lack a nitrogen heterocycle at position 7, with that for norfloxacin (**8**) (11), enoxacin (**7**) (9), and pefloxacin (**9**) (12) underscore the advantage of these modifications in newer generation quinolones. Nalidixic acid and cinoxacin have modest activity against *E. Coli* whereas the three later compounds are much more active against a wider spectrum of bacteria.

Table 1. Commercially Significant Quinolone Antibacterial Agents

Quinolone carboxylic acid[a]	CAS Registry Number	Structure	Company	Indications
ciprofloxacin (**2**)	[*85721-33-1*]	[b]	Bayer	lower RTI[c], skin and soft tissue, bone and joint, UTI[d], and infectious diarrhea
ofloxacin (**6a**)	[*83380-47-6*]	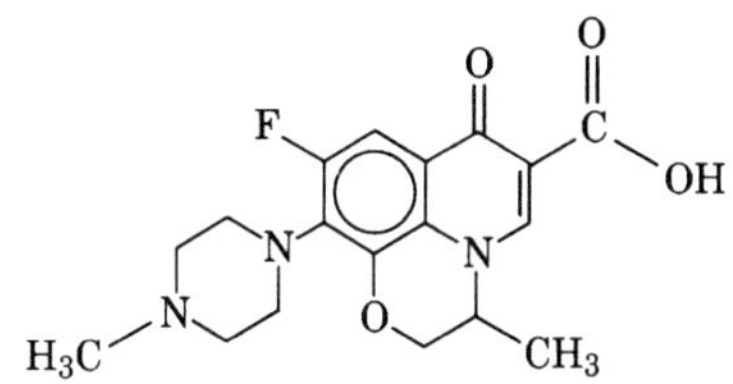	Daiich Seiyku	UTI, GUT[e], GI tract, RTI, skin and soft tissue
enoxacin (**7**)	[*74011-58-8*]	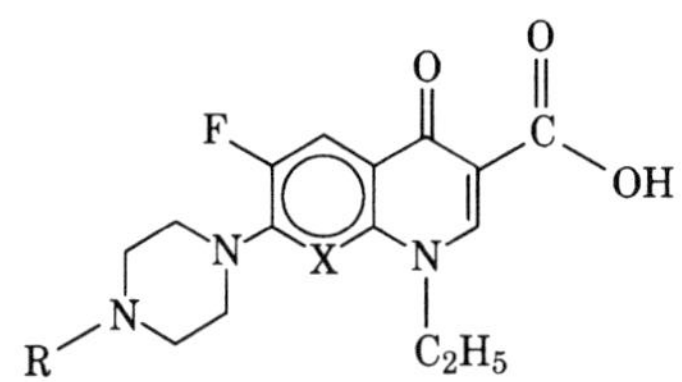 where R = H, X = N	Dainippon	UTI, GUT, skin infections, upper and lower RTI, and GI tract infection
norfloxacin (**8**)	[*70458-96-7*]	**7**, where R = H, X = CH	Kyorin	UTI
pefloxacin (**9**)	[*70458-92-3*]	**7**, where R = CH_3, X = CH	Roger Bellon (Rhone-Poulenc)	UTI, RTI, bone and joint infections, meningitis, septicemia, and endocaditis

[a]The systematic names are as follows: (**2**) 1-Cyclopropyl-6-fluoro-1,4-dihydro-4-oxo-7-(1-piperazinyl)-3-quinolinecarboxylic acid; (**7**) 1-ethyl-6-fluoro-1,4-dihydro-4-oxo-7-(1-piperazinyl)-1,8-naphthyridine-3-carboxylic acid; (**8**) 1-ethyl-6-fluoro-1,4-dihydro-4-oxo-7-(1-piperazinyl)-3-quinolinecarboxylic acid; (**6a**) 9-fluoro-2,3-dihydro-3-methyl-10-(4-methyl-1-piperazinyl)-7-oxo-7*H*-pyrido[1,2,3-*de*][1,4]benzoxazine-6-carboxylic acid; (**9**) 1-ethyl-6-fluoro-1,4-dihydro-7-(4-methyl-1-piperazinyl)-4-oxo-3-quinolinecarboxylic acid.

[b]In text.

[c]Respiratory tract infection.

[d]Urinary tract infection.

[e]Genito–urinary tract infection.

Table 2. Antibacterial Activity of Representative Quinolones[a]

Quinolone	Minimum Inhibitory Concentration, μg/mL						Refs.
	EC	PA	BF	SA	SF	SP	
(3)	2–8	64–256	32	32	>32	128–256	3,13
(4)	2–4	>256	>32	>32	32	>256	3,13
(8)	0.03–1	0.25–4	16–128	0.5–8	2–8	2–8	14–16
(17)	0.06–0.5	0.5–8	8	1	8	1	13,17
(7)	<0.12	2	4	0.5	2	2	13
(9)	0.06–2	0.05–4	8–16	0.25	4	2	13,16
(19)	0.004–0.5	0.5–2	1–4	0.06–0.25	1	0.25–1	13.18
(18)	0.008–0.25	0.03–2	0.5–2	0.008–0.5	0.25	0.06	13,19
(2)	0.008–0.25	0.06–0.5	4–32	0.25–2	0.5–2	0.25–2	14–16
(6a)	0.03–0.25	0.5–8	2–16	0.25–1	2–4	0.5–4	14,16,17
(13)	0.004	2	0.12	0.015	0.03	0.015	13
(16)	0.03–1	1–32	0.06–5	0.008	0.06	0.008–0.03	13,14,17

[a]These determinations were performed at Sterling Drug, Inc. The organisms used for the measurements were *E. Coli* ATCC 1-25922 (EC), *P. Aeruginosa* ATCC 27853 (PA), *B. Fragilis* ATCC 25285 (BF), *S. Aureus* ATCC 29213 (SA), *S. Faecalis* ATCC 29212 (SF), and *S. Pyogenes* ATCC 6301 (SP). Other values are taken from the literature for a large number of organisms of each strain.

The 6-fluorine provides a significant enhancement in antibacterial activity for many quinolones presumably by increasing cellular penetration (11,20) and the inhibitory activity against the enzyme. In the case of norfloxacin, a fluorine at the 6-position provides a compound which is 16-fold more potent against *E. Coli* than the nonfluorinated derivative (21–24). Substituents other than fluorine at the 6-position have provided some enhancement in activity over the 6-H compound, but this has been less than that provided by fluorine (11). In some cases the effect of the 6-fluorine is much smaller, depending on the group at the 7-position (25,26).

In earlier quinolones a number of groups were placed at the 7-position of the quinolone nucleus, including heterocycles attached to this position as well as a number of 6,7- and 7,8-ring fused derivatives. Many of these earlier variations have been extensively reviewed (10). In newer clinically important quinolones, the 7-position usually contains a piperazine derivative (see Fig. 1 and Table 3). Piperazine mimics for quinolones have been introduced in the form of 3-(aminomethyl)pyrrolidines (27). Pyrrolidinyl quinolones of this type, represented by CI-934 [*91188-00-0*] (**12**) (28), PD-117,558 [*99734-97-1*] (**13**) (29), and DS-4524 (**10**) (30) offer improvements over their piperazinyl counterparts in the form of increased gram-positive activity. When the 7-piperazine is replaced by an azole substituent (31), the resulting antibacterial activity is similar to that of the piperazinyl

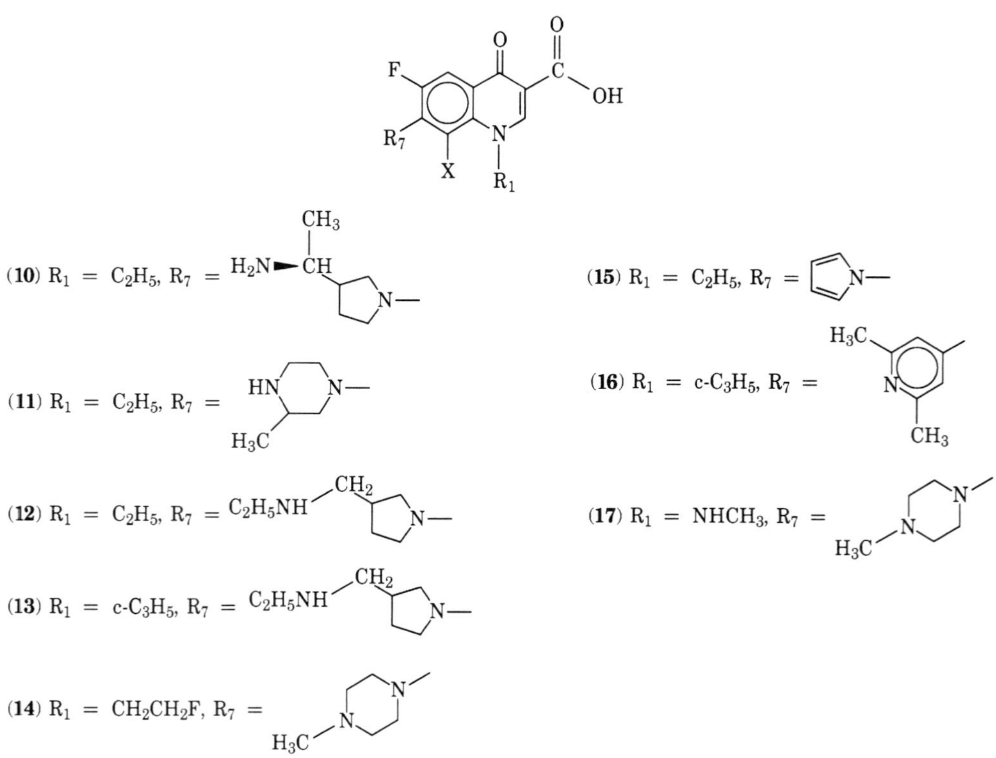

Fig. 1. Piperazine and pyrrolidine substitution at the 7-position of quinolones. For (**10**)–(**14**) X = F; in (**15**)–(**17**), X = H.

relatives. Irloxacin (**15**), an example of the azole replacement for the piperazine, may offer some advantages over the piperazinyl with respect to gram-positive activity, particularly *Staphylococcus* strains (32). Win 57,273 [*123942-02-1*] (**16**), a quinolone with a 2,6-dimethyl-4-pyridinyl group at the 7-position, shows good gram-positive activity particularly against *S. Aureus* and additionally, this agent exhibits excellent potency against anaerobes (14,17). Good activity against anaerobic organisms is still uncommon for quinolones (33).

An ethyl group at the 1-position was found to be optimum in earlier quinolones when compared with a methyl or propyl group at this position, and thus, based on these steric considerations, retained in many subsequent quinolone structures. Other groups at the 1-position include the 2-fluoroethyl substituent in fleroxacin (**14**) (23,34,35), and an aminomethyl group in amifloxacin (**17**) (36,37); these groups all have similar steric requirements. Along with norfloxacin, these agents offer good gram-negative activity, particularly against *E. Coli.*

Steric factors, however, are not the only important considerations for groups at the 1-position. Ciprofloxacin (**2**) (1) with a cyclopropyl group in the 1-position, when compared with its ethyl analogue norfloxacin in Table 2, offers increased antibacterial potency by 4–8 times against many organisms and approximately 30-fold against *E. Coli.* Additionally, ciprofloxacin has improved *in vivo* potency over norfloxacin (38). The increase in activity afforded by the cyclopropyl substituent over that of the ethyl cannot be explained simply on steric grounds and it was suggested that through space, electronic interactions may be important (39). The 2,4-difluorophenyl group in tosufloxacin (**18**) (18) and temafloxacin (**19**) (19) provides these agents with excellent antibacterial activity comparable with that found for ciprofloxacin, and also gives these agents *in vivo* activity superior to that offered by ciprofloxacin. Quinolones with a *tert*-butyl group at the 1-position also compare favorably with ciprofloxacin (40).

Another important development in the structure–activity relationships of quinolone antibacterials came with the introduction of the 1,8-bridged quinolone ofloxacin (**6a**). In this quinolone, the movement of the ethyl group at the 1-position is restricted by "tying" it to the 8-position in the form of a 1,4-benzoxazine ring. *In vitro* activity improvements are found that are more or less comparable to the improvements noted with ciprofloxacin (35,41–43). A variation on ofloxacin is rufloxacin (**20**); this compound lacks the methyl group on the 1,8-bridge and contains a sulfur in place of the oxygen attached to the 8-position. Rufloxacin, although less potent than ofloxacin, is well absorbed and has a longer half life than does ofloxacin (44–46).

When ofloxacin was first introduced it was made available as the racemate. Later the optical isomers were prepared and it was found that the (*S*)-enantiomer, DR-3355 (**6b**), was substantially more active (8–128-fold) than the (*R*)-isomer against a broad range of bacteria (47–50). This chiral preference is not unique to ofloxacin and has been demonstrated in other quinolones as well (51,52). This significant finding has already had an impact on the design of new quinolone antibacterials still in development (53).

The 5-position of quinolones can be substituted by small groups such as halogens, hydroxyl, or amino (54–56). The amino group at this position can be advantageous, particularly when applied to 6,8-difluoro-7-piperazinyl or 6,8-diflu-

Table 3. Quinolones in Development

Quinolone carboxylic acid[a]	CAS Registry Number	Structure	Company
DS-4524 (**10**)	*[107333-96-0]*	*b*	Daiichi
lomefloxacin (**11**)	*[98079-51-7]*	*b*	Hokuriku Seiyaku/Monsanto
fleroxacin (**14**)	*[79660-72-3]*	*b*	Kyorin/Roche
tosufloxacin (**18**)[c]	*[100490-36-6]*	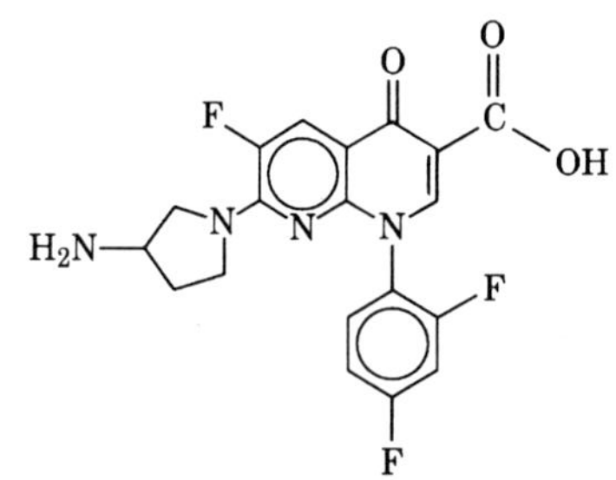	Toyama/Abbott
sparfloxacin (**21**)	*[110871-86-8]*	*d*	Dianippon/Warner-Lambert
DR-3355 (**6b**)	*[10986-85-4]*	*e*	Daiichi Seiyaku
rufloxacin (**20**)	*[101363-10-4]*	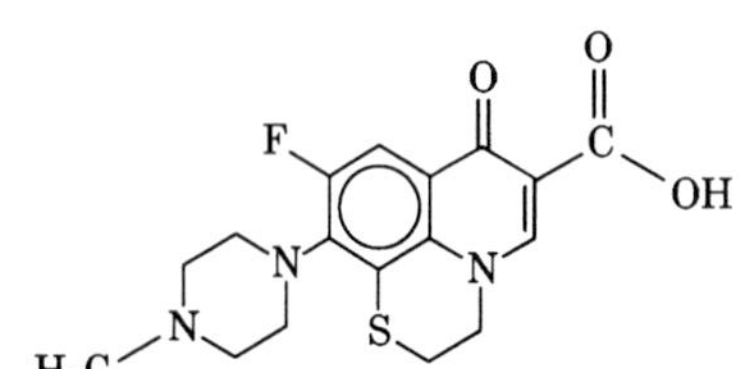	Mediolanum

temafloxacin (**19**)	[*105784-61-0*]	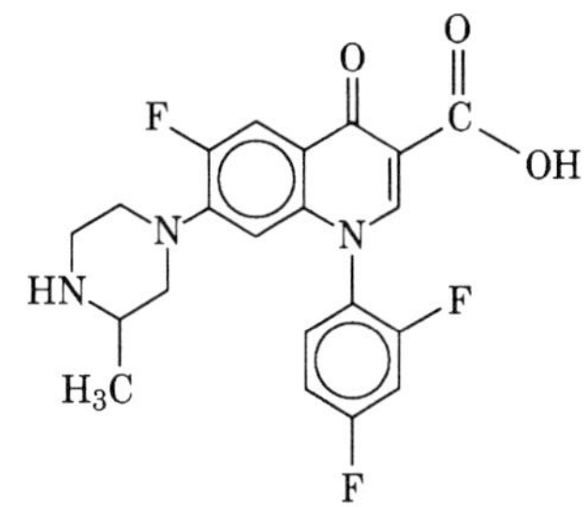	Abbott/Tanabe Seiyaku
amifloxacin (**17**)	[*86393-37-5*]	*b*	Eastman Kodak/Yamanouchi
irloxacin (**15**)	[*91524-15-1*]	*b*	Esteve

[a]The systematic names are as follows: (**10**) (*R**,*S**)-7-(3-(1-aminoethyl)-1-pyrrolidinyl)1-ethyl-6,8-difluoro-1,4-dihydro-4-oxo-3-quinolinecarboxylic acid; (**11**) 1-ethyl-6,8-difluoro-1,4-dihydro-7-(3-methyl-1-piperazinyl)-4-oxo-3-quinolinecarboxylic acid; (**14**) 6,8-difluoro-1-(2-fluoroethyl)-1,4-dihydro-7-(4-methyl-1-piperazinyl)-4-oxo-3-quinolinecarboxylic acid; (**18**) D,L-7-(3-amino-1-pyrrolidinyl)-1-(2,4-difluorophenyl)-6-fluoro-1,4-dihydro-4-oxo-1,8-napthyridine-3-carboxylic acid; (**21**) 5-amino-1-cyclopropyl-6,8-difluoro-7-(*cis*-3,5-dimethyl-1-piperazinyl)-4(1*H*)-oxo-3-quinolinecarboxylic acid; (**6b**) (*S*)-9-fluoro-2,3-dihydro-3-methyl-10-(4-methyl-1-piperazinyl)-7-oxo-7*H*-pyrido[1,2,3-de][1,4]benzoxazine-6-carboxylic acid; (**20**) 9-fluoro-2,3-dihydro-10-(4-methyl-1-piperazinyl)-7-oxo-7*H*-pyrido[123-de]-1,4-benzothiazine-6-carboxylic acid; (**19**) 1-(2,4-difluorophenyl)-6-fluoro-1,4-dihydro-7-(3-methyl-1-piperazinyl)-4-oxo-3-quinolinecarboxylic acid; (**17**) 6-fluoro-1,4-dihydro-1-(methylamino)-7-(4-methyl-1-piperazinyl)-4-oxo-3-quinolinecarboxylic acid; (**15**) 6-fluoro-7-(pyrrol-1-yl)-1-ethyl-1,4-dihydro-4-oxo-3-quinoline carboxylic acid.

[b]See Figure 1.

[c]This agent was launched in Japan during April, 1990.

[d]In text.

[e]See Table 1; (**6a**) represents the racemate. (**6b**) is the (*S*)-enantiomer.

oro-7-pyrrolidinyl quinolones. In contrast to 6,8-difluoro quinolones, when this replacement is applied to ofloxacin, the resulting derivative has reduced antibacterial activity (57). Replacement of the 5-amino group with methylamine or dimethylamine causes activity to drop substantially. Sparfloxacin [*110871-86-8*] (**21**), a representative of 5-amino-6,8-difluoro quinolones, affords modest improvements in gram-positive activity as well as increased *in vivo* potency when compared with both ciprofloxacin and ofloxacin (54).

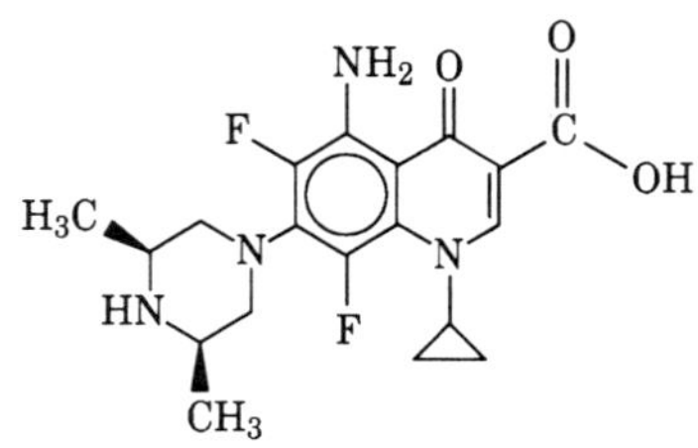

(**21**)

The 8-position of the quinolone nucleus can often be advantageously substituted by fluorine (58) or chlorine (59) to give compounds with improved antibacterial potency over hydrogen in this position. With 1,8-naphthyridines, activity is reported to be approximately equivalent to the quinolone bearing a hydrogen in this position (60). As an example of this, see the data for norfloxacin (**8**) and enoxacin (**7**) in Table 2.

A surprising development involving the 3-position has been presented (61,62). This position traditionally has had a carboxylic acid group and attempts to replace the carboxylate have resulted in lower antibacterial activity. However, it was demonstrated that the carboxylate could be replaced with an isothiazolo ring fused between the 2- and 3-position of the quinolone nucleus. A-62824 [*111279-87-9*] (**22**), illustrates this structural modification as applied to ciprofloxacin.

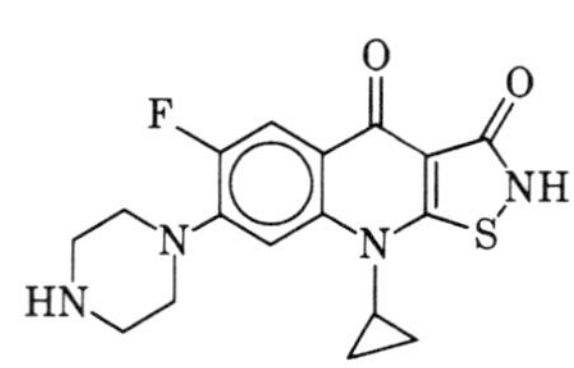

(**22**)

This replacement results in compounds which are several times more active than their 3-carboxylate counterparts and they show this activity against a broad range of bacteria as well as against the target enzyme. No *in vivo* antibacterial activity has been described.

Quinolone–Cephalosporin Codrugs. Quinolones have been covalently linked to cephalosporins in order to generate a codrug containing one molecule of each type of antibacterial agent. An example is the fleroxacin–cefotaxime [*63527-52-6*] combination, Ro 23-9424 [*115622-58-7*] (**23**).

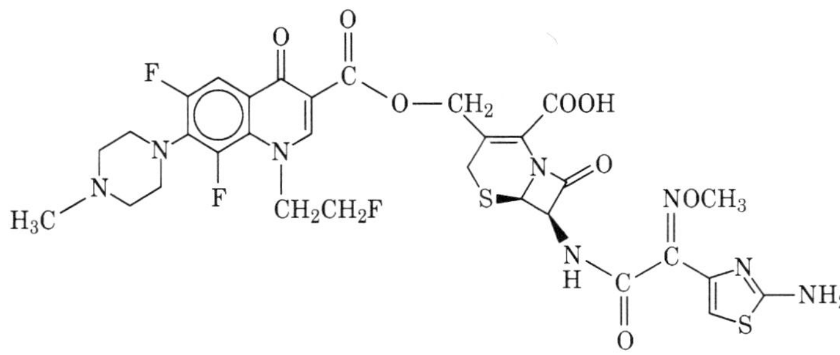

(**23**)

In essence, the cephalosporin acts as a carrier (63) for the quinolone. The quinolone is replaced in the bacterial cell after the action of β-lactamase on the cephalosporin portion of the molecule. This codrug combination represents a relatively new class of antibacterial agents which appear to offer advantages over the separated components (64). A good introductory discussion of these exciting agents can be found (65) (see also ANTIBIOTICS; β-LACTAMS; CEPHALOSPORINS).

Mechanism of Action

Quinolones exert their antibacterial activity by interfering with the replication of bacterial DNA (66,67) by inhibition of the enzyme DNA gyrase through the formation of a quinolone-DNA-enzyme complex (68–70). The critical reaction, catalyzed by the enzyme, is the negative supercoiling of bacterial DNA, a process involving the breaking and resealing of double-stranded circular DNA.

This enzyme has been shown to have two subunits referred to as A and B (71). The B subunit is responsible for supplying the energy needed for the supercoiling reaction (72); this process is inhibited by a class of antibiotics called coumermycins. The coumermycins competitively bind to the B subunit thereby blocking the binding of ATP. Quinolones do not interact with the B subunit, but exert their effect on the processes of the A subunit. One model proposed that quinolones, unlike coumermycins, do not bind to the enzyme, but have DNA as their molecular target (73). This model was later refined indicating that a gyrase-DNA complex is inhibited by quinolones (74). The essence of this refined model (74,75) is that gyrase interacts with DNA causing the double-strand break needed for strand passage in the supercoiling process. Once the DNA strands are cut, the protruding ends provide a site to which quinolones can bind. When this occurs, the supercoiling process is stopped and, along with it, gyrase turnover.

Preparation of Quinolones

The general method by which most newer fluoro quinolones are prepared involves a ring closure reaction to form the quinolone nucleus by forming one of the highlighted bonds, *a* or *b,* shown in general structure (**24**).

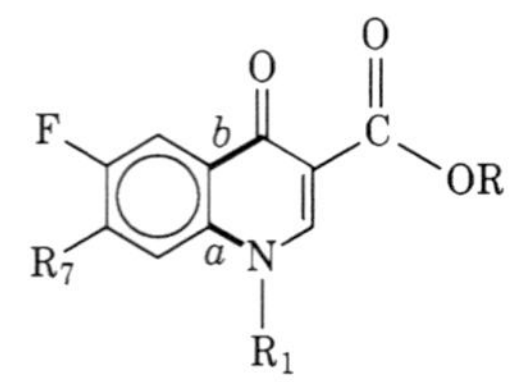

(**24**)

In the general preparation of quinolones by forming the nitrogen aryl bond *a* in the ring closure, typical precursors are prepared as shown in Figure 2. The ring closure involves nucleophilic displacement of a halogen, usually a chlorine or fluorine (76); eg, (**29**) and (**30**) lead to (**31**) [*86483-54-7*] and (**32**) [*123942-15-4*], respectively.

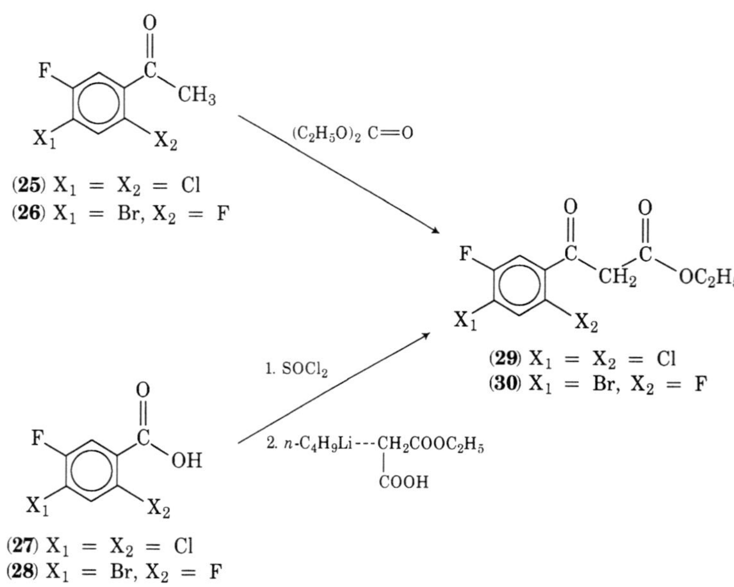

Fig. 2. Preparation of precursors for closure to structure (**24**) at the *a* bond. (**25**) = 2,4-dichloro-5-fluoroacetophenone [*704-10-9*]; (**26**) = 4-bromo-2,5-difluoroacetophenone [*123942-11-0*]; (**27**) = 2,4-dichloro-5-fluorobenzoic acid [*86522-89-6*]; (**28**) = 4-bromo-2,5-difluorobenzoic acid [*28314-82-1*]; (**29**) = ethyl 3-(2,4-dichloro-5-fluorophenyl)-3-oxopropanoate [*86483-51-4*]; (**30**) = ethyl 3-(2,5-difluoro-4-bromophenyl)-3-oxopropanoate [*123942-12-1*].

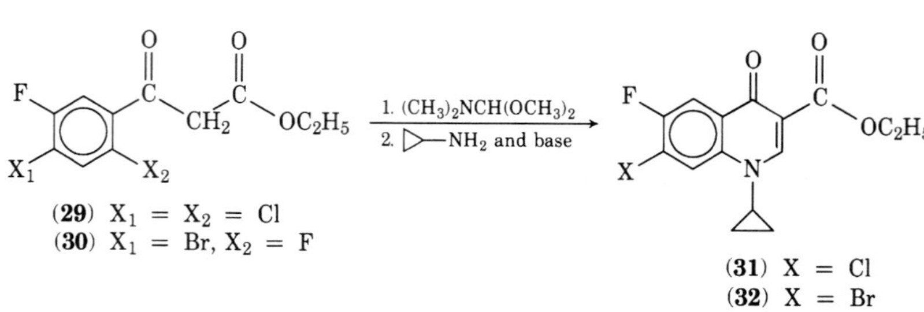

Alternatively, $HC(OC_2H_5)_3$ may be used in step 1. The appropriate R_7 group is attached by a second nucleophilic displacement to give the desired quinolone.

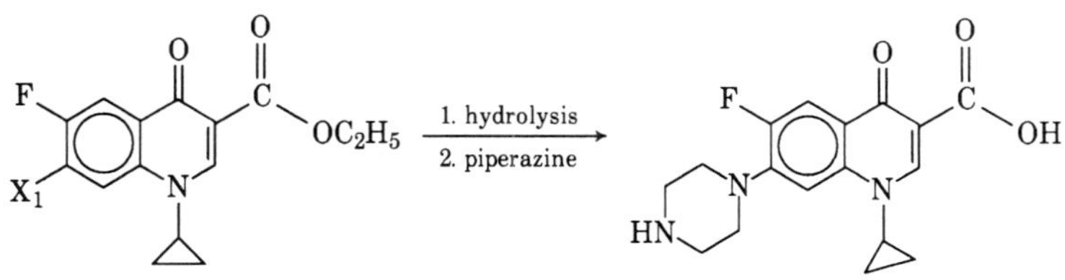

(**31**), (**32**) (**2**)

This is a general reaction as the heterocyclic group of most newer quinolones is derived from a nucleophilic piperazine or pyrrolidine derivative (11,47,53,54).

In the synthesis of Win 57,273 the attachment of the R_7 group, a 2,6-dimethylpyridinyl group, involves formation of a carbon–carbon bond rather than a carbon–nitrogen bond. The method for the attachment of this group is a palladium mediated coupling reaction (77,78) of 4-tributylstannyl-2,6-dimethylpyridine [*122033-61-8*] with a 7-haloquinolone (26).

(**31**), (**32**)

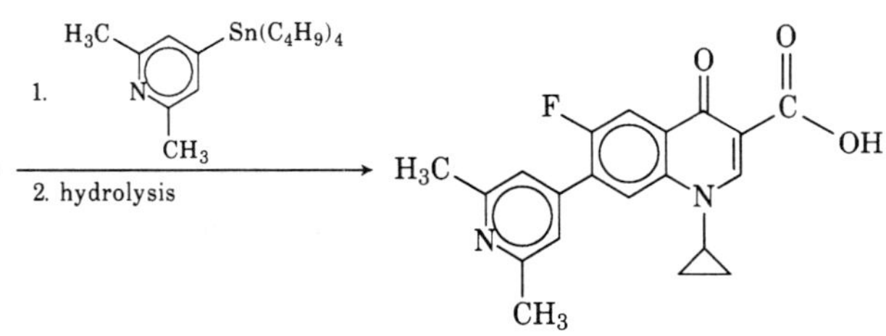

(**16**)

This coupling works best when the halogen at the 7-position is bromine rather than chlorine or fluorine. This represents the first application of this coupling reaction to the intact quinoline nucleus and thus represents an important advance in quinolone chemistry.

The alternative cyclization route is one which forms bond *b* in the ring closure reaction. This process is illustrated for norfloxacin (11) in Figure 3. In this sequence, 3-chloro-4-fluoro-aniline [*367-21-5*] and diethyl ethoxymethylenemalonate [*87-13-8*] are heated to displace ethanol giving the intermediate propanedioate [*70032-30-3*]. Thermal cyclization is achieved in diphenyl ether [*101-84-8*] to form the quinolone nucleus in (**33**) [*75073-15-3*]. Subsequent alkylation and hydrolysis gives (**34**) [*70458-94-5*]; displacement of the 7-chlorine by piperazine [*110-85-0*] gives norfloxacin (**8**). The Gould-Jacobs method described (79) is limited to quinolone targets in which the group on the 1-position is derived from a reasonably reactive alkyl halide.

Safety of Quinolone Antibacterial Agents

For the most part quinolones are well tolerated with few reports of adverse reactions (80,81), but there are some concerns, one of which is that they cause

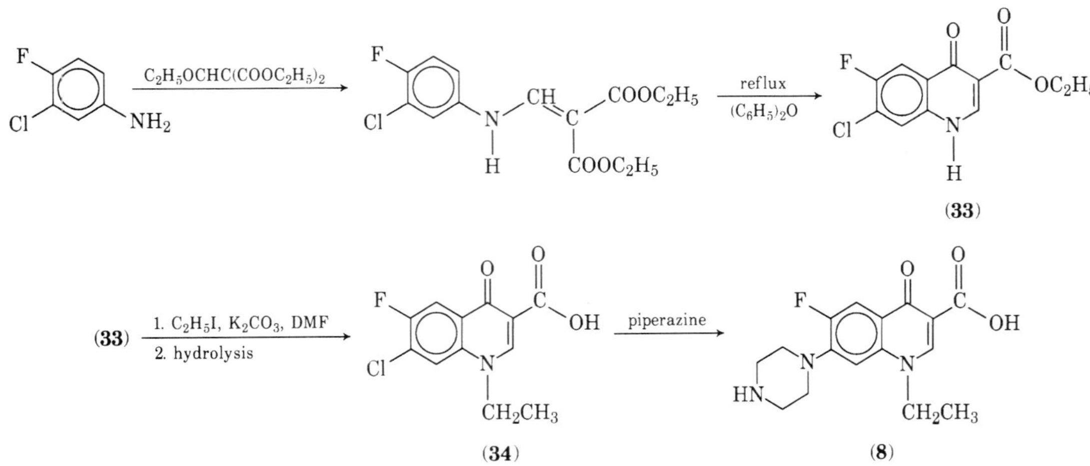

Fig. 3. Cyclization to form ring bond *b* in (**24**).

arthropathy in juvenile animals. This is a condition in which the cartilage of weight-bearing joints is damaged. Because of this concern, quinolones are contraindicated for children and during pregnancy. A second problem is that of adverse side effects including those of the central nervous system (CNS) and also adverse reactions when quinolones are taken with other drugs (82). The CNS side effects may, in part, be related to the affinity of quinolones to γ-aminobutyric acid receptors thus inhibiting the binding of this important neurotransmitter (83). When quinolones are taken along with other drugs, adverse side reactions can stem from a variety of causes. In the case of theophylline [*5967-84-0*], an antiasthmatic, the side effects may be related to the inhibition of the metabolism of this drug (84,85). Adverse reactions should decrease with more recent quinolones such as lomefloxacin, sparfloxacin, and temafloxacin (86–90). A useful discussion of quinolone toxicity and safety has appeared (91).

Resistance. Resistance to quinolones, particularly to newer agents (6), is not a significant problem at this time (92). The resistance that has been found is the result of mutation of the gene for the A subunit of the gyrase enzyme. A second mode of resistance results in decreased drug permeation resulting from a change in the outer membrane protein of the bacterial cell (93).

Economic Aspects

The domestic antiinfective market is in excess of \$3 billion and the global market surpasses \$8 billion (94,95). Earlier quinolones have had little impact on this market as their utility was limited primarily to urinary tract infections. Sales of cinoxacin, for example, are on the order of \$4 million. Newer quinolones, because of their safety, seemingly low propensity toward bacterial resistance, and vastly improved therapeutic utility, now enjoy a much greater share of this market. This is especially true for ciprofloxacin, the most important quinolone antibacterial agent. Sales of ciprofloxacin exceeded \$135 million in the United States in 1988, \$260 million in 1989, and were reported at \$385 million in 1991. Additionally, ciprofloxacin will soon be available as an intravenous formulation allowing it to be used to treat a greater number of patients and thus assuming a larger portion of this market (96–98). The wholesale price of ciprofloxacin (99) for 50 tablets is $84.03, $90.04, and $174.07 for 250 mg, 500 mg, and 750 mg sizes, respectively. This translates to a cost of $3.36 per day for twice a day dosing. Although these costs are higher than other antibacterial agents, overall savings should be realized (95) compared to other types of therapy necessitated by nonorally active agents, including hospitalization (81).

BIBLIOGRAPHY

"Antibacterial Agents, Synthetic, Nalidixic Acid and Other Quinolone Carboxylic Acids" in *ECT* 3rd ed., Vol. 2, pp. 782–789, by G. Y. Lesher, Sterling-Winthrop Research Institute.

1. R. Wise, J. M. Andrews, and L. J. Edwards, *Antimicrob. Agents Chemother.* **23,** 559 (1983).
2. G. Y. Lesher, E. J. Foelich, M. D. Gruett, J. H. Bailey, and P. R. Brundage, *J. Med. Chem.* **5,** 1063 (1962).
3. W. E. Wick, D. A. Preston, W. A. White, and R. S. Gordee, *Antimicrob. Agents Chemother.* **4,** 415 (1973).
4. S. Minami, T. Shono, and J. Matsumoto, *Chem. Pharm. Bull.* **19,** 1426 (1971).
5. D. F. McDonald and H. B. Short, *Antimicrob. Agents Chemother.* **5,** 628 (1964).
6. T. Doi and A. Takeda, *Chemotherapy* **35,** 175 (1987).
7. J. H. Paton and D. S. Reeves, *Drugs* **36,** 193 (1988).
8. P. M. Shah, *Prog. Drug Res.* **31,** 243 (1987).
9. C. Siporin, *Annu. Rev. Microbiol.* **43,** 601 (1989).
10. R. Albrecht, *Prog. Drug. Res.* **21,** 9 (1977).
11. H. Koga, A. Itoh, S. Murayama, S. Suzue, and A. G. Pernet, *J. Med. Chem.* **23,** 1358 (1980).
12. J. P. Gonzalez and J. M. Henwood, *Drugs* **37,** 628 (1989).
13. A. L. Barry, *Antimicrobic Newslett.* **5,** 69 (1988); G. M. Eliopoulos and C. T. Eliopoulos, in J. S. Wolfson and D. C. Hooper, eds., *Quinolone Antimicrobial Agents,* American Society for Microbiology, Washington, D.C., 1989, p. 35.
14. R. N. Jones and A. L. Barry, *Antimicrob. Agents Chemother.* **34,** 306 (1990).
15. G. M. Eliopoulos, A. Gardella, and R. C. Moellering, Jr., *Antimicrob. Agents Chemother.* **25,** 319 (1988).
16. D. L. Van Caekenberghe and S. R. Pattyn, *Antimicrob. Agents Chemother.* **25,** 518 (1984).
17. D. M. Sedlock, R. A. Dobson, D. M. Deuel, G. Y. Lesher, and J. B. Rake, *Antimicrob. Agents Chemother.* **34,** 568 (1990).
18. D. J. Hardy and co-workers, *Antimicrob. Agents Chemother.* **31,** 1768 (1987).
19. P. B. Fernandes and co-workers, *Antimicrob. Agents Chemother.* **32,** 27 (1988).
20. J. M. Domagala and co-workers, *J. Med. Chem.* **29,** 394 (1986).
21. A. King, C. Warren, K. Shannon, and I. Phillips, *Antimicrob. Agents Chemother.* **21,** 604 (1982).
22. A. Ito and co-workers, *Antimicrob. Agents Chemother.* **17,** 103 (1980).
23. R. Wise, *J. Antimicrob. Chemother.* **13** (Suppl. B), 59 (1984).
24. B. Holmes, R. N. Brogden, and D. M. Richards, *Drugs* **30,** 482 (1985).
25. Y. Nishimura and J. Matsumoto, *J. Med. Chem.* **30,** 1622 (1987).
26. M. Reuman and co-workers, *Abstracts, 29th Interscience Conference on Antimicrobial Agents and Chemotherapy,* 1989, Houston, Tex., Abstract 1193, p. 304.

27. J. M. Domagala, C. L. Heifetz, T. F. Mich, and J. B. Nichols, *J. Med. Chem.* **29,** 445 (1986).
28. W. Mandell and H. C. Neu, *Antimicrob. Agents Chemother.* **29,** 852 (1986).
29. J. M. Domagala, C. L. Heifetz, T. F. Mich, and J. B. Nichols, *Abstracts, 26th ICAAC,* New Orleans, La., 1986, Abstract 422, p. 174.
30. Y. Kimura and co-workers, *Abstracts, 29th ICAAC,* Houston, Tex., 1989, Abstract 1192, p. 304.
31. T. Uno and co-workers, *J. Med. Chem.* **30,** 2163 (1987).
32. M. A. Xicota, R. Coli, M. Esteve, M. Moros, and J. Paresi, *Drugs Exp. Clin. Res.* **13,** 133 (1987).
33. E. J. C. Goldstein and D. M. Citron, *Antimicrob. Agents Chemother.* **27,** 657 (1985).
34. L. Verbist, *J. Antimicrob. Chemother.* **20,** 363 (1987).
35. T. Kumada and H. C. Neu, *J. Antimicrob. Chemother.* **16,** 563 (1985).
36. J. B. Cornett, R. B. Wagner, R. A. Dobson, and M. P. Wentland, *Antimicrob. Agents Chemother.* **27,** 4 (1985).
37. M. P. Wentland and co-workers, *J. Med. Chem.* **27,** 1103 (1984).
38. O. Paulsen, *Drugs of Today* **24,** 361 (1988).
39. J. M. Domagala and co-workers, *J. Med. Chem.* **31,** 911 (1988).
40. D. Bouzard and co-workers, *J. Med. Chem.* **32,** 537 (1989).
41. J. F. Chantot and A. Bryskier, *J. Antimicrob. Chemother.* **16,** 475 (1985).
42. K. Sato and co-workers, *Antimicrob. Agents Chemother.* **30,** 777 (1984).
43. J. P. Monk and D. M. Campoli-Richards, *Drugs* **33,** 346 (1987).
44. V. Cecchetti and co-workers, *J. Med. Chem.* **30,** 465 (1987).
45. G. Ravizzola and co-workers, *Drugs Exp. Clin. Res.* **15,** 11 (1989).
46. G. Segre, D. Cerretani, D. Cerri, and L. Moltoni, *Drugs Exp. Clin. Res.* **14,** 747 (1988).
47. S. Atarashi and co-workers, *Chem. Pharm. Bull.* **35,** 1896 (1987).
48. I. Hayakawa and co-workers, *Antimicrob. Agents Chemother.* **29,** 163 (1986).
49. M. Imamura, S. Shibamura, I. Hayakawa, and Y. Osada, *Antimicrob. Agents Chemother.* **31,** 325 (1987).
50. L. A. Mitscher, P. N. Sharma, D. T. W. Chu, L. L. Shen, and A. G. Pernet, *J. Med. Chem.* **30,** 2283 (1987).
51. J. F. Gerster and co-workers, *J. Med. Chem.* **30,** 839 (1987).
52. T. P. Culbertson, J. M. Domagala, J. B. Nichols, S. Priebe, and R. W. Skeean, *J. Med. Chem.* **30,** 1715 (1987).
53. T. Rosen and co-workers, *J. Med. Chem.* **31,** 1586 (1988).
54. T. Miyamoto and co-workers, *J. Med. Chem.* **33,** 1645 (1990).
55. D. B. Moran, C. B. Ziegler, T. S. Dunne, N. A. Kuck, and Y.-I. Lin, *J. Med. Chem.* **32,** 1313 (1989).
56. J. M. Domagala and co-workers, *J. Med. Chem.* **31,** 503 (1988).
57. J. M. Domagala and co-workers, *Drugs Exp. Clin. Res.* **14,** 453 (1988).
58. Belg. Pat. BE 899,399 (July 31, 1984), (to Kyorin Pharmaceutical Co., Ltd.).
59. R. Wise, R. Lockley, M. Webberly, and Z. N. Adhami, *J. Antimicrob. Chemother.* **14** (Suppl. C), 75 (1984).
60. J. P. Sanchez and co-workers, *J. Med. Chem.* **31,** 983 (1988).
61. D. T. W. Chu, P. B. Fernandes, A. K. Clairborne, L. Shen, and A. Pernet, *Abstracts, 27th ICAAC,* New York, N.Y., 1987, Abstract 250, p. 140.
62. D. T. W. Chu, P. B. Fernandes, A. K. Clairborne, L. Shen, A. G. Pernet, *Drugs Exp. Clin. Res.* **14,** 379 (1988).
63. S. Mobashery, S. A. Lerner, and M. Johnston, *J. Am. Chem. Soc.* **108,** 1685 (1986).
64. R. N. Jones, A. L. Barry, and C. Thornsberry, *Antimicrob. Agents Chemother.* **33,** 944 (1989).
65. R. Cleeland and co-workers, *Antimicrobic Newslett.* **6,** 6 (1989).

66. W. A. Goss, W. H. Dietz, and T. M. Cook, *J. Bacteriol.* **89,** 1068 (1965).
67. A. Bauernfeind and G. Grummer, *Chemotherapia* **10,** 95 (1965).
68. G. C. Crumplin and J. T. Smith, *Antimicrob. Agents Chemother.* **8,** 251 (1975).
69. M. Gellert, K. Mizuuchi, M. H. O'Dea, and H. A. Nash, *Proc. Natl. Acad. Sci. USA* **73,** 3872 (1976).
70. M. Gellert, K. Mizuuchi, M. H. O'Dea, T. Itoh, and J. Tomizawa, *Proc. Natl. Acad. Sci. USA* **74,** 4772 (1977).
71. E. C. Engle, S. H. Manes, and K. Drlica, *J. Bacteriol.* **149,** 92 (1982).
72. A. Sugino, N. P. Higgins, P. O. Brown, C. L. Peebles, and N. R. Cozzarelli, *Proc. Natl. Acad. Sci. USA* **75,** 4838 (1978).
73. L. L. Shen and A. G. Pernet, *Proc. Natl. Acad. Sci. USA* **82,** 307 (1985).
74. L. L. Shen and co-workers, *Biochemistry* **28,** 3886 (1989).
75. L. L. Shen, J. Baranowski, and A. G. Pernet, *Biochemistry* **28,** 3879 (1989).
76. D. T. W. Chu, *J. Heteterocycl. Chem.* **22,** 1033 (1985).
77. J. K. Stille, *Angew. Chem. Int. Ed. Engl.* **25,** 508 (1986).
78. T. R. Bailey, *Tetrahedron Lett.* **27,** 4407 (1986).
79. R. G. Gould and W. A. Jacobs, *J. Am. Chem. Soc.* **61,** 2890 (1939).
80. G. Schluler, in H. C. Neu and H. Weuta, eds., *First International Ciprofloxacin Workshop,* Excerpta Medica, Amsterdam, 1985, p. 61.
81. D. C. Hooper and J. S. Wolfson, eds., *Quinolone Antimicrobial Agents,* American Society for Microbiology, Washington, D.C., 1989, p. 249.
82. R. A. Prince, *Antimicrobic Newslett.* **6,** 93 (1989).
83. A. Tsuji and co-workers, *Antimicrob. Agents Chemother.* **32,** 190 (1988).
84. Y. Mizuki, M. Kamaura, T. Yamaguchi, Y. Sekine, and M. Hashimoto, *Arzneim. Forsch.* **39,** 593 (1989).
85. M. Parent, M. St-Laurent, and M. LeBel, *Antimicrob. Agents Chemother.* **34,** 1249 (1990).
86. R. A. Robertson and E. J. Begg, *Abstracts, 29th ICAAC,* Houston, Tex., 1989, Abstract 217, p. 136.
87. A. H. Staib, S. Harder, U. Fuhr, C. Wack, and G. Medert, *Abstracts, 29th ICAAC,* Houston, Tex., 1989, Abstract 218, p. 137.
88. D. E. Nix, A. Norman, and J. J. Schentag, *Abstracts, 29th ICAAC,* Houston, Tex., 1989, Abstract 219, p. 137.
89. P. J. Carpentier, J. Afarian, J. Cerasoli, P. Hsu, and A. G. Pernet, *Abstracts, 29th ICAAC,* Houston, Tex., 1989, Abstract 548, p. 193.
90. J. C. Craft, J. R. Cerasoli, P. Hsu, and A. G. Pernet, *Abstracts, 29th ICAAC,* Houston, Tex., 1989, Abstract 990, p. 270.
91. H. Stahlmann and H. Lode, in V. T. Andriole, ed., *The Quinolones,* Academic Press, New York, 1988, p. 201.
92. J. S. Wolfson and D. C. Hooper, *Rev. Inf. Dis.* **11** (Suppl. 5), S960 (1989).
93. H. Aoyama, K. Sato, K. Kato, K. Hirai, and S. Mitsuhashi, *Antimicrob. Agents Chemother.* **31,** 1640 (1987).
94. G. Weiss, *Barrons,* 35 (Mar. 10, 1986).
95. R. N. Jones, *Antimicrobic Newslett.* **6,** 53 (1989).
96. *FDC Reports: The Pink Sheet* **51**(12), 18 (1989).
97. *FDC Reports: The Pink Sheet* **52**(28), T&G 5 (1990).
98. *FDC Reports: The Pink Sheet* **53**(5), T&G 1 (1991).
99. *New Drug Comment.* **16**(2), 78 (1989).

M. Reuman
G. Y. Lesher
Sterling-Winthrop Inc.

NITROFURANS

The nitrofurans encompass a class of synthetic antibacterials characterized by the 5-nitro-2-furanyl group:

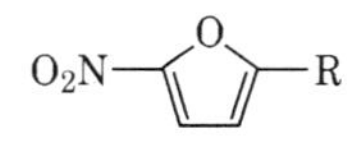

With relatively few exceptions, the R function contains the azomethine (—CH=N—) moiety. The most prominent exceptions contain the olefinic (—C=C—) moiety (Table 1).

The antimicrobial activity of this class of compounds was first reported in 1943 (1) and this finding was described independently four years later (2). After investigation of several derivatives, it was concluded that the 5-nitro function was responsible for the observed antimicrobial activity (3). These preliminary findings focused attention on this area of research and as a result, several thousand nitrofurans have been prepared, and thousands of articles have been published describing the extensive efforts in these areas.

Several nitrofurans have been marketed either regionally or worldwide in the human and veterinary areas because of their broad spectrum of activity, relatively mild toxicity, and low tendency for resistance development. However, accurate total volumes or sales of these products on a global basis are not generally available. In the United States, nearly four million prescriptions were written in 1989 for products containing nitrofurazone or nitrofurantoin. U.S. sales during this period for these products approached $70 million; thus therapy with this class of compounds remains a significant therapeutic alternative.

Chemical Properties

Nitrofurans are generally stable, water-insoluble, crystalline solids, which decompose (darken) upon prolonged exposure to light or alkali. Their strong ultraviolet absorption provides the basis for analytical procedures for these materials. Chemical reduction of these nitro compounds gives rise to amino derivatives (4) as well as products of ring-opening (5,6). Nucleophiles which have been shown to displace the nitro function include methoxy (7,8), halogen (9), azido (10), alkylmercapto (10), and phenylsulfonyl (10). Photochemical hydroxylation of 5-nitro-2-furancarboxaldehyde gives rise to the 4-hydroxy derivative (11).

Biological Mechanism of Action

The mechanism of action of selected members of this class has been investigated. Nitrofurantoin is bactericidal at the concentrations achieved in the urinary tract and highly effective against the corresponding pathogens (12). Since its introduction for use in uncomplicated urinary tract infections, essentially no resistance development has occurred, unlike nearly all other antimicrobials. This is believed to be because of the ability of the nitrofurans to affect multiple cytoplasmic

Table 1. Structures of Representative Antibacterial Nitrofurans

Name	CAS Registry Number	Molecular formula	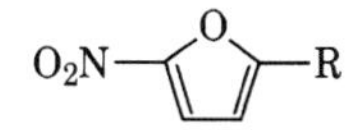 R
furium	[*531-82-8*]	$C_9H_7N_3O_4S$	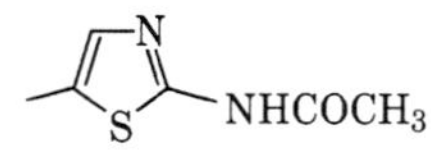
furazolidone	[*67-45-8*]	$C_8H_7N_3O_5$	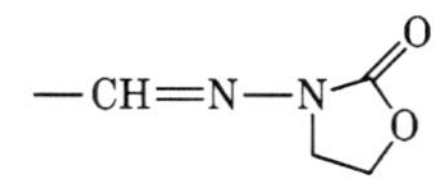
Z-furan	[*710-25-8*]	$C_7H_6N_2O_4$	—CH=CH—CONH₂
furylfuramide	[*3688-53-7*]	$C_{11}H_8N_2O_5$	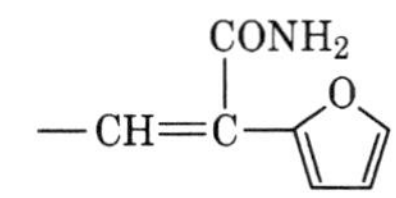
nitrovin	[*2315-20-0*]	$C_{14}H_{12}N_6O_6{\cdot}HCl$	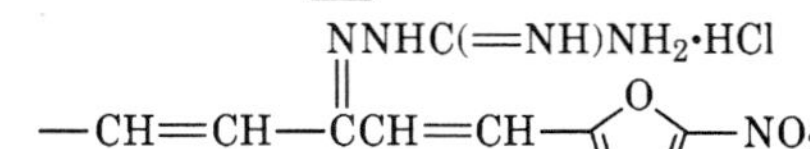
furalazine[a]	[*566-12-7*]	$C_9H_7N_5O_3$	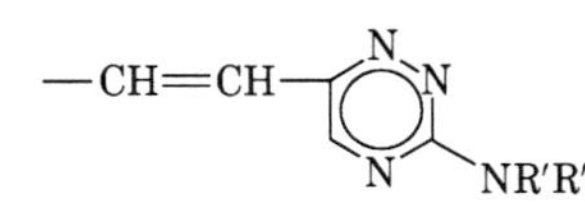
acetylfuratrizine[b]	[*1789-26-0*]	$C_{11}H_9N_5O_4$	
panfuran-S[c]	[*794-93-4*]	$C_{11}H_{11}N_5O_5$	
nifuroxime	[*6236-05-1*]	$C_5H_4N_2O_4$	—CH=NOH (anti)
nitrofurazone	[*59-87-0*]	$C_6H_6N_4O_4$	—CH=N—NHCONH₂
nifuraldezone	[*3270-71-1*]	$C_7H_6N_4O_5$	—CH=N—NHCOCONH₂
nihydrazone	[*67-28-7*]	$C_7H_7N_3O_4$	—CH=N—NHCOCH₃
nitrofurantoin	[*67-20-9*]	$C_8H_6N_4O_5$	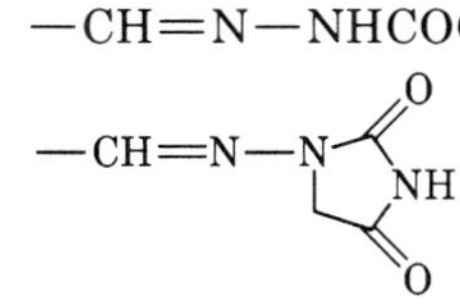
nifuratel	[*4936-47-4*]	$C_{10}H_{11}N_3O_5S$	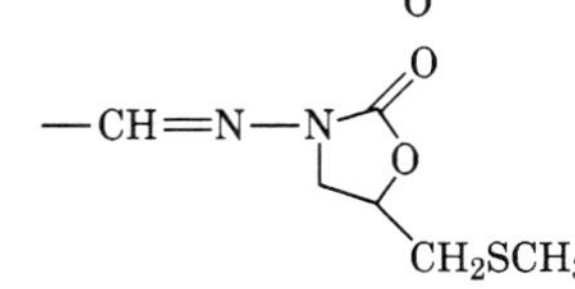
nitrofurathiazide	[*61143-06-4*]	$C_{13}H_{12}N_4O_6S$	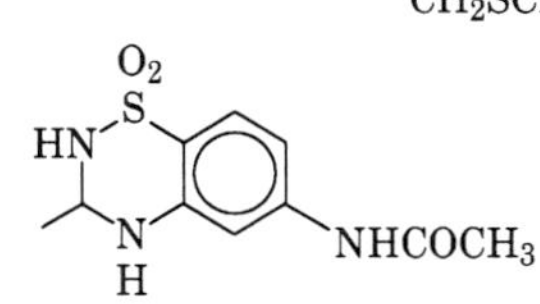
nifurtoinol	[*1088-92-2*]	$C_9H_8N_4O_6$	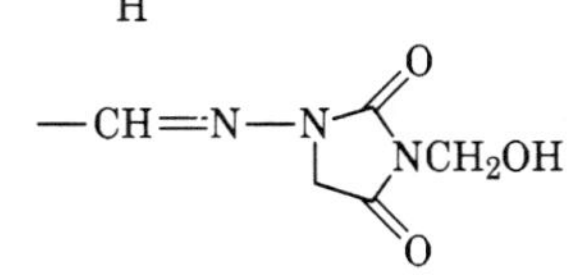
nifurprinol	[*13411-16-0*]	$C_{12}H_{10}N_2O_4$	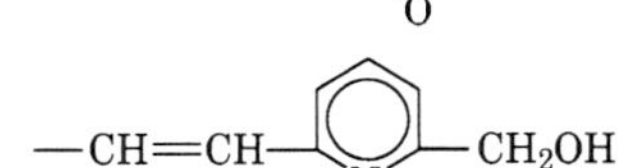

[a] R′ = R″ = H.
[b] R′ = H; R″ = CH_3CO—.
[c] R′ = R″ = —CH_2OH.

targets such as citric acid cycloenzyme, bacterial DNA and RNA, and ribosomes essential for bacterial function. Both nitrofurazone (13) and nitrofurantoin (12) are metabolized to reactive intermediates that facilitate inhibitory processes at multiple sites such that a low probability of simultaneous resistance development exists. Potential intermediates have been postulated to be hydroxylamines, nitrosamines, and free radicals.

Mutagenic and Carcinogenic Activity

Although *in vitro* mutagenicity tests suggest that some nitrofurans in general provoke a positive response, use of *in vivo* mammalian systems has produced equivocal or negative results.

It is not appropriate to generalize the carcinogenicity of this class of compounds. Nitrofurazone appears to increase the incidence of benign mammary tumors in rats. The tumorigenic activity of furazolidone is expressed by an increase in the incidence of spontaneous tumors in both mice and rats. Bioassays of nitrofurantoin in several species of mice and rats failed to reveal any evidence of direct tumorigenic activity. Ovarian tumors have been reported in $B_6C_3F_1$ mice, but these are believed due to an indirect expression of toxicity (14,15).

These three nitrofurans have been used therapeutically for over 30 years with no reports of human neoplasia and the relevance of the animal findings to short term therapy in humans has not been established.

Preparation

Most commercialized nitrofurans are derived from 5-nitro-2-furancarboxaldehyde [*698-63-5*] or the corresponding diacetate.

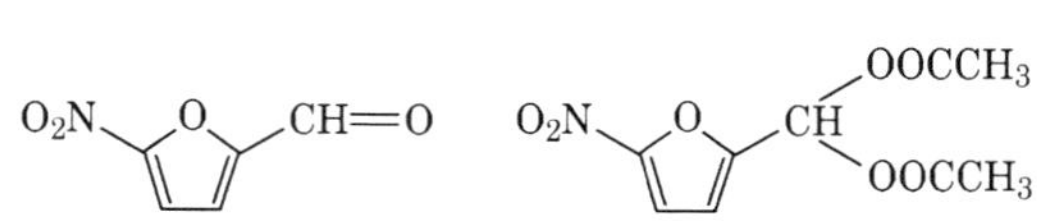

These precursors are prepared by reaction of fuming nitric acid in excess acetic anhydride at low temperatures with 2-furancarboxaldehyde [*98-01-1*] (furfural) or its diacetate (16) followed by treatment of an intermediate 2-acetoxy-2,5-dihydrofuran [*63848-92-0*] with pyridine (17). This process has been improved by the use of concentrated nitric acid (18,19), as well as catalytic amounts of phosphorus pentoxide, trichloride, and oxychloride (20), and sulfuric acid (21). Orthophosphoric acid, *p*-toluenesulfonic acid, arsenic acid, boric acid, and stibonic acid, among others are useful additives for the nitration of furfural with acetyl nitrate. Hydrolysis of 5-nitro-2-furancarboxyaldehyde diacetate [*92-55-7*] with aqueous mineral acids provides the aldehyde which is suitable for use without additional purification.

An alternative route to 5-nitro-2-furancarboxaldehyde requires nitration of 2-furancarboxaldehyde oxime [*1121-47-7*] with mixed acid to give the nitrated oxime [*555-15-7*], and concomitant hydrolysis (22). Furthermore, 2-furancarboxaldehyde derivatives with the R-substituent in place have been nitrated to the desired product (23).

Furium. *N*-[4-(5-Nitro-2-furanyl)-2-thiazolyl]acetamide, has demonstrated activity against bacilli and pathogenic enterobacteria (24). The product, prepared from thiourea and 2-bromo-1-(5-nitro-2-furanyl)ethanone followed by acetylation of the intermediate aminothiazole with acetic anhydride in pyridine (25), is marketed in several countries for both human and veterinary use.

Furazolidone. 3-[(5-Nitro-2-furanyl)methyleneamino]-2-oxazolidinone, is manufactured by the condensation of 5-nitro-2-furancarboxaldehyde with 3-amino-2-oxazolidinone (19,26). The product is employed clinically for gastrointestinal and vaginal infections as well as an antibacterial and antiprotozoal agent in poultry and swine. Furazolidone has been shown to be metabolized to its 4-hydroxyfuryl derivative in rats (27) and to 3-(4-cyano-2-oxobutylideneamino)-2-oxazolidinone in several species (28).

Z-Furan. 3-(5-Nitro-2-furanyl)-2-propenamide, is prepared by condensation of 5-nitro-2-furancarboxaldehyde diacetate with malonic ester followed by PCl_5 chlorination and amination (29). The product was marketed in Japan as a food preservative.

Furylfuramide. α-[(5-Nitro-2-furanyl)-2-methylene]-2-furanacetamide, withdrawn from the market in Japan in 1974 because of mutagenicity, is prepared by condensation of 5-nitro-2-furancarboxaldehyde with 2-furanacetic acid followed by chlorination and amination (30). The isomerization of cis to trans form of furlyfuramide has been shown to occur in the presence of a variety of biological reducing agents (31).

Nitrovin. 2-{3-(5-Nitro-2-furanyl)-1-[2-(5-nitro-2-furanyl)ethenyl]-2-propenylidene}hydrazinecarboximidamide hydrochloride has been marketed for both human and veterinary use as an antibacterial agent. The product, which has also seen use as a veterinary food additive (32), is prepared from 5-nitro-2-furancarboxaldehyde and acetone followed by treatment of the resulting dione with aminoguanidine (33).

Furalazine, Acetylfuratrizine, Panfuran-S. Heating nitrovin in butanol or dimethylformamide at 100–130°C affords furalazine, 6-[2-(5-nitro-2-furanyl)ethenyl]-1,2,4-triazine-3-amine (34). An improved synthesis originates with 5-nitro-2-furancarboxaldehyde and acetone, proceeds through 4-(5-nitro-2-furanyl)-3-buten-2-one followed by a selenium dioxide oxidation to the pyruvaldehyde hydrate, and subsequent reaction with aminoguanidine (35). Furalazine, acetylfuratrizine (36), and the *N*-*N*-bis(hydroxymethyl) derivative, Panfuran-S, formed from the parent compound and formaldehyde (37), are systemic antibacterial agents.

Nifuroxime. The anti form of the oxime prepared from 5-nitro-2-furancarboxaldehyde and hydroxylamine, has been marketed for topical use as an antibacterial or an antifungal agent (1,17).

Nitrofurazone. 2-[(5-Nitro-2-furanyl)methylene]hydrazinecarboximide, the first nitrofuran to be employed clinically, is prepared from 5-nitro-2-furancarboxaldehyde and semicarbazide (19). This product has seen clinical use topically as an antibacterial, for systemic application for bacterial infections in poultry and swine, and also has been employed as a food additive. In rats, nitrofurazone is hydroxylated at the 4 position of the furan moiety (27). The involvement of nitrenium ions has also been postulated in the mechanism of action of nitrofurazone (38).

Nifuraldezone. Aminooxoacetic acid [(5-nitro-2-furanyl)methylene]-hydrazide, nifuraldezone, is prepared by the reaction of 5-nitro-2-furancarboxaldehyde with semioxamazide. The product is useful in the treatment of dysentery in calves (39).

Nihydrazone. Acetic acid [5-nitro-2-furanyl)methylene]-hydrazide has seen use in veterinary medicine both as a coccidiostat and an antibacterial (39).

Nitrofurantoin. The urinary tract antibacterial, 1-[(5-nitro-2-furanyl)-methyleneamino]-2,4-imidazolidinedione, is prepared by reaction of 5-nitro-2-furancarboxaldehyde with 1-amino-2,4-imidazolidinedione (19,40). Crystallization of the material from nitromethane provides a macrocrystalline form (41) which continues to be employed clinically.

In rats, the oxidative and reductive metabolism products have been identified as the 4-hydroxylated furan and [(3-cyano-1-oxopropyl)methyleneamino]-2,4-imidazolidinedione, respectively (27,42). In addition, the ease of electron transfer as a mechanism of activity with nitrofurantoin and nitrofurazone has been studied (43).

Nifuratel. Nifuratel, 5-[(methylthio)methyl]-3-[5-nitro-2-furanyl)-methyleneamino]-2-oxazolidinone, is prepared by treatment of the appropriate methylthiohydrazinopropanol with diethylcarbonate followed by condensation with 5-nitro-2-furancarboxaldehyde diacetate (44). The product has been used in antibacterial, antifungal, and antiprotozoal applications.

Nitrofurathiazide. 6-Acetylamino-3,-4-dihydro-3-(5-nitro-2-furanyl)-2*H*-1,2,4-benzothiadiazine 1,1-dioxide, nitrofurathiazide, is synthesized from 4-acetylamino-2-aminobenzenesulfonamide with 5-nitro-2-furancarboxaldehyde (45). The product has been employed as an antibacterial agent for humans and, in combination with other drugs, has been used for acute or chronic otitis externa of dogs and cats.

Nifurtoinol. Treatment of nitrofurantoin with formaldehyde gives nifurtoinol, 3-hydroxymethyl-1-[5-nitro-2-furanyl)methyleneamino]-2,4-imidazolidinedione, a urinary tract antibacterial (46). The kinetics of decomposition of this drug and its potential as a prodrug of nitrofurantoin have been reported (47).

Nifurprinol. 6-[2-(5-Nitro-2-furanyl)ethenyl]-2-pyridine methanol, is prepared from 5-nitro-2-furancarboxaldehyde and 6-methyl-2-pyridine methanol (48). The product has been used as an antibacterial in fish diseases.

Additional discussions are available in the General References concerning the properties of several nitrofurans. This includes further coverage of the chemotherapeutic and physical properties; antimicrobial activity; bacterial resistance; absorption, distribution, and excretion; clinical use; and safety studies, of this interesting class of antiinfective compounds.

BIBLIOGRAPHY

"Nitrofurans" in *ECT* 2nd ed., Vol. 13, pp. 853–857, K. J. Hayes, Norwich Pharmacal Company; "Antibacterial Agents, Synthetic—Nitrofurans" in *ECT* 3nd ed., Vol. 2, pp. 790–794, F. F. Ebetino, Norwich Eaton Pharmaceuticals Corporation.

1. U.S. Pat. 2,319,481 (May 18, 1943), W. B. Stillman, A. B. Scott, and J. M. Clampit (to Norwich Pharmacal Co.)
2. O. Dann and E. F. Möller, *Berichte* **80,** 23 (1947).
3. M. C. Dodd and W. B. Stillman, *J. Pharmacol. Exp. Ther.* **82,** 11 (1944).
4. F. F. Ebetino, J. J. Carroll, and G. Gever, *J. Med. Pharm. Chem.* **5,** 513 (1962).
5. F. L. Austin, *Chem. Ind.* 523 (1957).
6. J. J. Gavin, F. F. Ebetino, R. Freedman and W. E. Waterbury, *Arch. Biochem. Biophys.* **113,** 399 (1966).
7. J. Olivard and J. P. Heotis, *J. Org. Chem.* **33,** 2552 (1968).
8. L. J. Powers, *J. Pharm. Sci.* **60,** 1425 (1971).
9. H. R. Snyder, Jr., and P. H. Seehausen, *J. Heterocycl. Chem.* **10,** 385 (1973).
10. F. Lieb and K. Eiter, *Justus Liebigs Ann. Chem.* **761,** 130 (1972).
11. J. Olivard, G. M. Rose, G. M. Klein, and J. P. Heotis, *J. Med. Chem.* **19,** 729 (1976).
12. C. C. McOsker, J. R. Pollack, and J. A. Andersen, in L. H. Harrison, ed., *Management of Urinary Infections,* Royal Society of Medicine International Congress and Symposium Series No. 154, Royal Society of Medicine Service Limited, London, 1989.
13. D. R. McCalla, "Nitrofurans" in F. E. Hahn, ed., *Mechanism of Action of Antimicrobial Agents,* Vol. 9, Springer-Verlag, New York, 1979, p. 176.
14. K. A. Stitzel, R. F. McConnell, and T. A. Dierckman, *Toxic. Pathol.* **17,** 774 (1989).
15. W. H. Butler, T. C. Graham, and M. L. Sutton, *Food Chem. Toxicol.* **28,** 49 (1990).
16. H. Gilman and G. F. Wright, *J. Am. Chem. Soc.* **52,** 2550, 4165 (1930).
17. A. P. Dunlop and F. N. Peters, "The Furans," *ACS Monograph Series No. 119,* Reinhold Publishing Corp., New York, 1953, pp. 141–169, 725, 727.
18. U.S. Pat. 2,490,006 (Nov. 29, 1949), W. Kimel, J. H. Coleman, and W. B. Stillman (to Eaton Laboratories, Inc.).
19. H. J. Sanders, R. T. Edmunds, and W. B. Stillman, *Ind. Eng. Chem.* **47,** 358 (1955).
20. U.S. Pat. 2,502,114 (Mar. 28, 1950), M. Witte and C. J. Lind (to Allied Chemical & Dye Corp.).
21. Jpn. Kokai 4377 (Dec. 16, 1950), J. Yoshida and co-workers.
22. U.S. Pat. 2,927,110 (Mar. 1, 1960), G. Gever and C. J. O'Keefe (to Norwich Pharmacal Co.).
23. U.S. Pat. 2,898,335 (Aug. 4, 1959), J. G. Michels (to Norwich Pharmacal Co.).
24. K. Miura and H. K. Reckendorf, *Progress in Medicinal Chemistry,* Vol. 5, Plenum Press, New York, 1967, p. 320.
25. U.S. Pat. 2,992,225 (July 11, 1961), D. E. Dickson (to Abbott Laboratories).
26. U.S. Pat. 2,742,462 (Apr. 17, 1956), G. Gever (to Norwich Pharmacal Co.).
27. A. J. Streeter, T. R. Krueger, and B. A. Hoener, *Pharmacology* **36,** 283 (1988).
28. L. J. Vrooman and co-workers, *Xenobiotica* **17,** 1427 (1987).
29. T. Takahashi and co-workers, *J. Pharm. Soc. Jpn.* **69,** 286 (1949).
30. H. Saikachi and A. Tanak, *J. Pharm. Soc. Jpn.* **83,** 147 (1963).
31. E. D. Clark, P. Wardman, and I. Wilson, *Biochem. Pharmacol.* **33,** 87 (1984).
32. U.S. Pat. 3,264,112 (Aug. 2, 1966), Y. Kodama, T. Fujiwara, and T. Inagaki (to Toyama Chemical Industries Co., Ltd.).
33. Jpn. Kokai 2673 (July 18, 1952), H. Uota and co-workers (to Toyama Chemical Industries Co., Ltd.); Jpn. Kokai 2974 (Aug. 7, 1952), H. Uota and A. Kuriyama (to Toyama Chemical Industries Co., Ltd.).
34. U.S. Pat. 3,159,625 (Dec. 1, 1964), Y. Kodama, T. Maeda, A. Takai, and I. Takamichi (to Toyama Chemical Industries Co., Ltd.).
35. U.S. Pat, 3,349,086 (Oct. 24, 1967) and Fr. Pat. 1,377,650 (Nov. 6, 1964), R. R. G. Haber (to ABIC Chemical Laboratories, Ltd.).
36. A. Takai and I. Saikawa, *J. Pharm. Soc. Jpn.* **84,** 9 (1964).
37. Ref. 36, p. 16.
38. A. J. Streeter and B. Hoener, *Pharm. Res.* **5**(7), 434 (1958).

39. U.S. Pat. 2,416,234 (Feb. 18, 1947), W. B. Stillman and A. B. Scott (to Eaton Laboratories, Inc.).
40. U.S. Pat. 2,610,181 (Sept. 9, 1952), K. J. Hayes (to Eaton Laboratories, Inc.): U.S. Pat. 2,779,786 (Jan. 29, 1957), J. H. Coleman, W. Hayden, Jr., and C. J. O'Keefe (to Norwich Pharmacal Co.).
41. U.S. Pat. 3,401,221 (Sept. 10, 1968), A. R. Borgmann, K. J. Hayes, H. E. Paul, and M. F. Paul (to Norwich Pharmacal Co.).
42. M. B. Aufiere, B. A. Hoener, and M. Voce, *Drug Metab. Dispos.* **6,** 403, (1978).
43. J. R. Ames, M. D. Ryan, and P. Kovacic, *J. Free Radicals Bio. & Med.* **2**(5–6), 377, 1986.
44. U.S. Pat. 3,288,787 (Nov. 29, 1966), G. Massaroli (to Inphar S.A.).
45. U.S. Pat. 3,155,654 (Nov. 3, 1964) and Fr. Pat. M1831 (June 24, 1963), M. H. Sherlock (to Schering Corp.).
46. U.S. Pat. 3,446,802 (May 27, 1969), J. G. Michels (to Norwich Pharmacal Co.).
47. H. Bundgaard and M. Johansen, *Int. J. Pharm.* **5,** 67 (1980).
48. Brit. Pat. 1,053,730 (Jan. 4, 1967) (to Dainippon Pharmaceutical Co., Ltd.).

General References

H. E. Paul and M. F. Paul in R. J. Schnitzer and F. Hawking, eds., *Experimental Chemotherapy,* Vol. 2, Academic Press, Inc., New York, 1964, pp. 207–370; Vol. 4, 1966, pp. 521–536.
K. Miura and H. F. Reckendorf in G. P. Ellis and G. B. West, eds., *Progress in Medicinal Chemistry,* Vol. 5, Plenum Press, New York, 1967, pp. 320–381.
H. Kala and D. Ausborn, *Pharmazie* **26,** 121–135, 193–207 (1971).
E. Grunberg and E. H. Titsworth, *Annu. Rev. Microbiol.* **27,** 317–346 (1973).
R. E. Chamberlain, *J. Antimicrob. Chemother.* **2,** 325 (1976).
G. T. Bryan in G. T. Bryan, ed., *Carcinogenesis,* Vol. 4 (Nitrofurans), Raven Press, New York, 1978, pp. 1–11.
M. Ichikawa in G. T. Bryan, ed., *Carcinogenesis,* Vol. 4 (Nitrofurans), Raven Press, New York, 1978, pp. 13–57.

THOMAS J. SCHWAN
FRANK H. EBETINO
Norwich Eaton Pharmaceuticals, Inc.

SULFONAMIDES

Sulfonamides derived from sulfanilamide (*p*-aminobenzenesulfonamide) are commonly referred to as sulfa drugs. Although several drug classes are characterized by the presence of a sulfonamide function, eg, hypoglycemics, carbonic anhydrase inhibitors, saluretics, and tubular transport inhibitors, the antibacterial sulfonamides have become classified as the sulfa drugs. Therapeutically active derivatives are usually substituted on the N^1 nitrogen; the N^4 position is generally unsubstituted. These features are illustrated by the structures of sulfanilamide (**1**) and sulfadiazine (**2**).

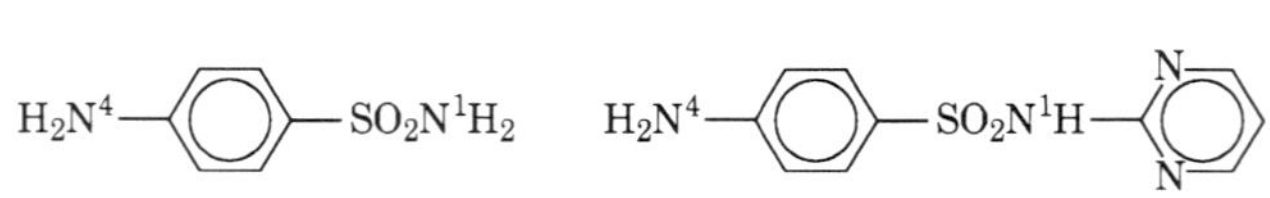

(**1**) (**2**)

Some related antibacterials are also included with the sulfonamides. The azo dye, Prontosil (**3**) is metabolized to sulfanilamide *in vivo,* and was the progenitor of the sulfa drugs. Also, the antibacterial sulfones, eg, dapsone (**4**), are believed to act in a similar fashion on enzymes involved with synthesis of folic acid, leading to bacterial growth inhibition.

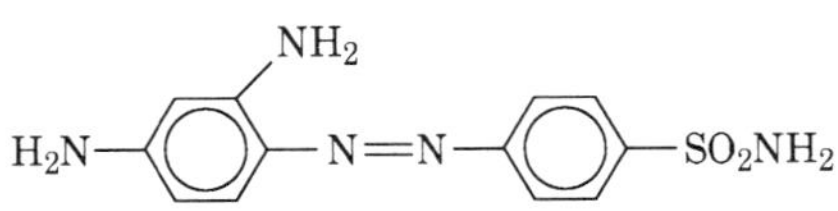

(**3**)

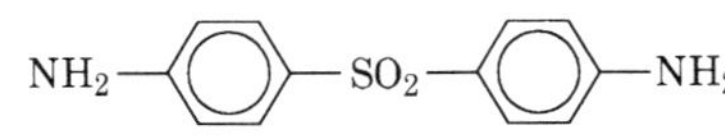

(**4**)

A monograph (1) covers the pioneering period of sulfa drug development and describes over 5000 sulfanilamide derivatives, their preparation, properties, trade names, and biological testing. This review is remarkably complete through 1944. Several thousand additional derivatives have been made since, but no comparable coverage is available. A definitive account of medical applications up to 1960 has been published (2), and a review of experimental antibacterial aspects has been made (3). Chapters on general aspects of sulfonamides and sulfones have appeared (4,5). A review of the clinical efficacy of trimethoprim–sulfamethoxazole has been published (6).

The sulfa drugs are still important as antimicrobials, although they have been replaced in many systemic infections by the natural and semisynthetic antibiotics. They are of great value in third world countries where problems of storage and lack of medical personnel make appropriate use of antibiotics difficult. They are especially useful in urinary tract infections, particularly the combination of sulfamethoxazole with trimethoprim. Their effectiveness has been enhanced by co-administration with dihydrofolate reductase inhibitors, and the combination of sulfamethoxazole with trimethoprim is of value in treatment of a number of specific microbial infections. The introduction of this combination (cotrimoxazole) in the late 1960s (1973 in the United States) resulted in increased use of sulfonamides.

The sulfas also remain clinically useful in the treatment of chancroid, lymphogranuloma venereum, trachoma, inclusion conjunctivitis, and the fungus-related nocardiosis (7). In combination with pyrimethamine, they are recommended for toxoplasmosis (8) and have been used for chloroquine-resistant falciparium malaria (4,9). There has also been some use of sulfas for the prophylaxis of rheumatic fever. The sulfone, dapsone, remains an accepted treatment for all forms of leprosy (4).

The clinical usefulness of the sulfonamides depends not only on antimicrobial effectiveness, but on other factors such as aqueous and liposolubility, protein-binding, half-life, and metabolism. Currently used sulfas vary widely in their absorption, distribution, and excretion patterns. Some of those in clinical practice, past or present, are listed in Tables 1 and 2. These can be grouped according to their rate of absorption and half-life. One group remains largely unabsorbed after oral administration, and is useful for gastrointestinal infections. A second group is characterized by high solubility, quick absorption, and rapid excretion, mainly unaltered, and is widely used for urinary tract infections. Another group is absorbed rapidly but excreted slowly, maintaining adequate blood levels for

Table 1. Sulfonamides of Structure 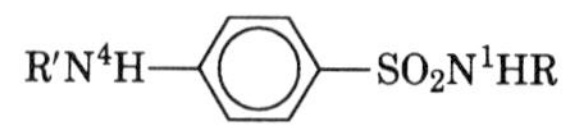

Generic name	R	CAS Registry Number	pK_a	Solubility in water, 25°C, mg/100 mL	Liposolubility, %[a]	Human plasma half-life[a], h	Protein binding[b] % bound
R′ = H							
sulfanilamide	H	*[16-74-1]*	10.5	750	71	9	9
sulfacetamide	$-COCH_3$	*[144-80-9]*	5.4	>670	2.0	7	9.5
sulfadiazine	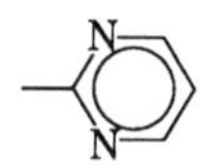	*[68-35-9]*	6.5	8	26.4	17	37.8
sulfadimethoxine	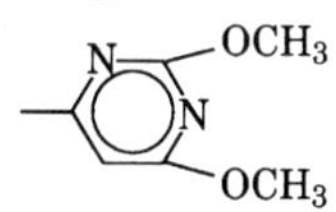	*[122-11-2]*	6.1	<4.6	78.7	40	92.3
sulfaguanidine	$-C(NH)NH_2$	*[57-67-0]*	12.05	100		5	
sulfisomidine	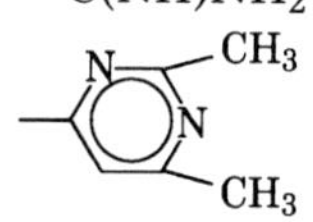	*[515-64-0]*	7.4	~200	19	7.5	67
sulfisoxazole	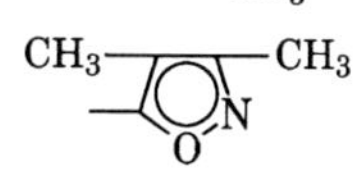	*[127-69-5]*	5.0	350[c]	4.8	6	76.5
sulfamethazine	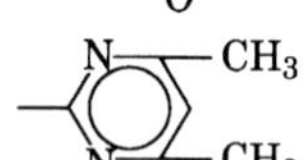	*[57-68-1]*	7.4	~100	82.6	7	66
sulfamethizole	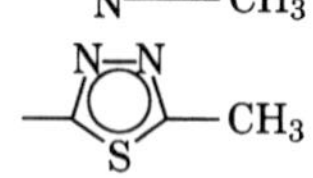	*[144-82-1]*	5.5	25		2.5	22
sulfamethoxazole	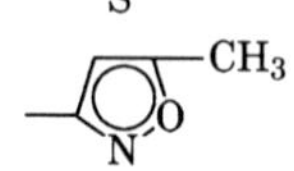	*[723-46-6]*	6.0		20.5	11	60

sulfamethoxypyridazine	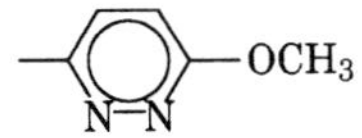	[*80-35-3*]	7.2	110 (pH5)	70.4	37	77
sulfamoxole	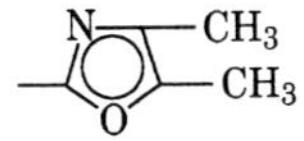	[*729-99-7*]	7.4	85	41.4	11	76.5
sulfaphenazole	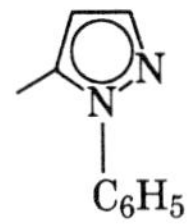	[*526-08-9*]	6.1	150	69	10	87.5
sulfapyridine	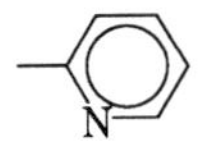	[*144-83-2*]	8.4	30	14	9	70
sulfapyrazine	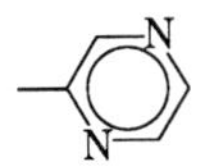	[*116-44-9*]	6.0	5			
sulfathiazole	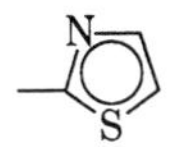	[*72-14-0*]	7.25	60	15.3	4	68
R′ = phthalyl							
N^4-phthalysulfathiazole	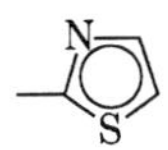	[*85-73-4*]	acid	sol (pH7)			
R′ = succinyl							
N^4-succinylsulfathiazole	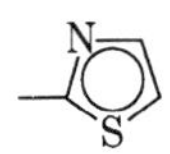	[*116-43-8*]	acid	20			

[a]Ref. 4, pp. 28–31.
[b]At 1.0 μmol/mL.
[c]*N′*-acetylsulfisoxazole [*80-74-0*] water solubility = 7 mg/100 mL.

Table 2. Sulfones and Other Structures Related to Sulfonamides

Generic name	Structure	CAS Registry Number	Solubility in water, 25°C mg/100mL	Human plasma half-life[a]
mafenide	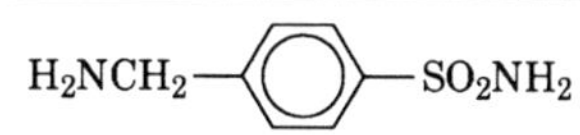	[*138-39-6*]	sol (salt)	
sulfasalazine	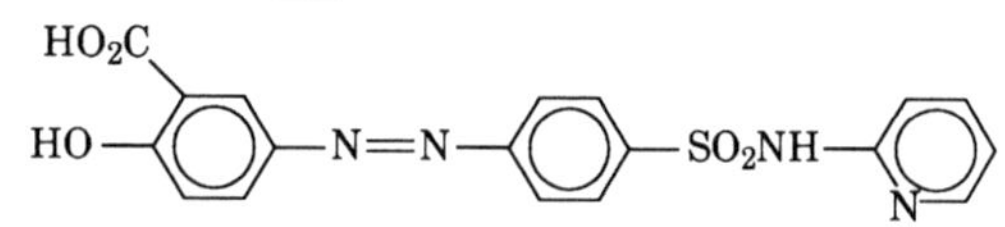	[*599-79-1*]	sol (pH7)	
sulfamidochrysoidine (Prontosil)	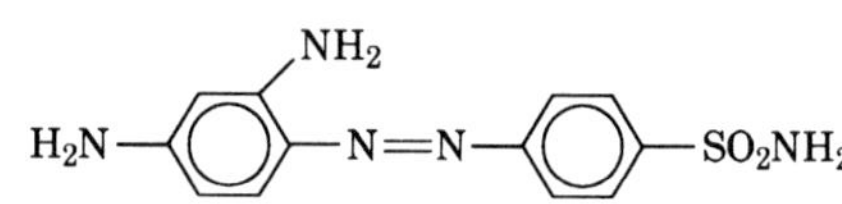	[*103-12-8*]	sol (salt)	
dapsone[b]	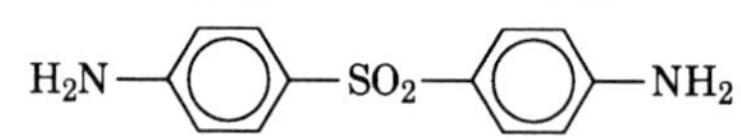	[*80-08-0*]	insol	20 h[c]
acedapsone	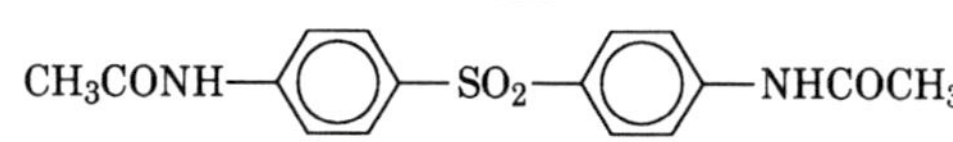	[*77-46-3*]	0.3	43 d[d]

[a]Ref. 4, pp. 28–31.
[b]pK_a of protonated form = 1.3.
[c]Protein binding at 1.0 μmol/mL = 50% bound.
[d]Intramuscularly.

long periods; these drugs are useful for chronic infections and for prophylaxis. Sulfonamides with half-lives up to 10 h are considered short-acting, those with half-lives of 10–24 h are termed medium-acting, and those with half-lives longer than 24 h are long-acting. This wide range of pharmacokinetic properties, along with their ease of administration, broad spectrum antimicrobial activity, and noninterference with host-defense mechanisms is responsible for their widespread use five decades after their discovery.

Their antimicrobial activity includes species requiring a folic acid synthetic pathway, comprising many gram-positive and gram-negative cocci and bacilli, mycobacteria, some large viruses, protozoa, and fungi (5). The action of the sulfa drugs and related sulfones is bacteriostatic rather than bactericidal. It is believed to be due to interference with the assembly of folic acid at the step that adds *p*-aminobenzoic acid (PABA). For a thorough discussion of the mechanisms of action postulated for the sulfonamides, see reference 10.

The discovery of the sulfa drugs, the first to control systemic bacterial infections, had its origin in the investigation of organic dyes for chemotherapeutic purposes initiated by Paul Ehrlich and others in the first decade of the 1900s. The azo compound, Prontosil [*103-12-8*] (**3**), was discovered at the I. G. Farbenindustrie in Germany in 1935 (11) to cure bacterial septicemias. It was fortunate that animal tests were used in the screening procedure, because Prontosil (sulfamidochrysoidine) is inactive *in vitro*. A group at the Pasteur Institute in Paris found that sulfanilamide, a metabolic product, was responsible for the activity (12), and was active against susceptible organisms both *in vitro* and *in vivo* (12,13). This marked the beginning of a worldwide effort to prepare and test derivatives and analogues of sulfanilamide. In less than 10 years, over 5000 compounds of this type were synthesized, and they appear in the 1948 review (1). The effort to find compounds having a broader antimicrobial spectrum and improved therapeutic ratio resulted in the development and clinical use of the N^1-heterocyclic substituted sulfanilamides, exemplified by sulfapyridine, sulfathiazole, and sulfadiazine (all 2-substituted heterocycles). Later research has not produced sulfonamides having greater activity, but compounds having improved pharmacokinetic properties for specific uses have largely replaced the first used sulfonamides.

During the most active period of investigation of sulfanilamide derivatives, 1935–1944, for systemic bacterial infections, the antimycobacterial activity of 4,4′-diaminodiphenylsulfone [*80-08-8*] (DDS, dapsone) was discovered (14). Although neither this compound nor its derivatives proved to be clinically useful for human tuberculosis, it did evolve into the most important type of compound for leprosy (15). The *N,N′*-diacetyl derivative has also found use against certain resistant strains of falciparum malaria.

Therapeutic Aspects

Systemic Infections. The sulfonamides were the first drugs effective against the bacterial septicemias (blood stream infections), but they have also been effective in tissue infections due to streptococci, and fungus-related nocardia (2). However, penicillin and other natural and semisynthetic antibiotics have displaced the sulfas for many uses. The sulfas continue to be used in such conditions as rheumatic fever prophylaxis and in nocardiosis (16). Sodium sulfacetamide solution or ointment is also used in infections of the conjuctiva, and oral administration of the insoluble sulfonamides, phthalyl or succinyl sulfathiazole, phthalyl sulfacetamide, and sulfaguanidine, provide temporary inhibition of the intestinal microbial flora. This is of value in preparing the bowel for surgery (17). Topical use of sulfonamides is generally unsatisfactory and is accompanied by a high risk of allergic sensitization.

Sulfonamides in combination with dihydrofolate reductase inhibitors are of continuing value. Pyrimethamine [*58-14-0*] (**5**) in combination with sulfonamides is employed for toxoplasmosis (7), and a trimethoprim (**6**)-sulfamethoxazole preparation is used not only for urinary tract infections but also for brucellosis, cholera, and malaria.

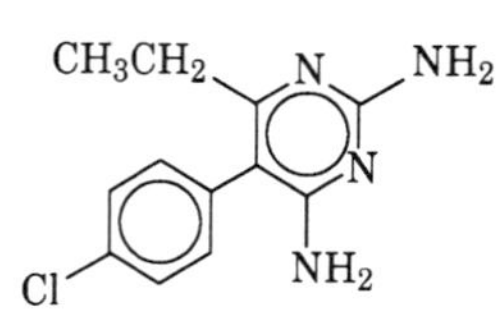

(**5**)

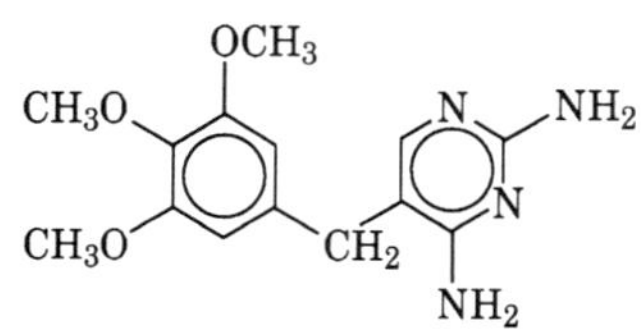

(**6**)

Synergistic effects are accomplished by the use of these combinations. Resistance to this combined therapy, either by bacteria or protozoa, emerges very slowly (18).

Urinary Tract Infections. The sulfa drugs have maintained an important place in treatment of urinary tract infections, where short acting, highly soluble drugs are required (19). Those most used in the United States are sulfisoxazole, sulfamethizole, sulfisomidine, and sulfacytine. For chronic urinary tract infection, the sulfamethoxazole–trimethoprim combination is frequently employed.

Sulfamethoxazole inhibits bacterial synthesis of dihydrofolic acid, and trimethoprim blocks the production of tetrahydrofolic acid by inhibiting the enzyme dihydrofolate reductase. Thus two consecutive steps are blocked in the biosynthesis of nucleic acids and proteins essential to many bacteria. *In vitro* serial dilution tests have shown that the combination of sulfamethoxazole and trimethoprim [*738-70-5*] inhibits the growth of common urinary tract pathogens with the exception of *Pseudomonas aeruginosa*. Table 3 illustrates the enhanced effect of the combination over that of either agent alone.

Table 3. Representative Minimum Inhibitory Concentrations[a] for Urinary Tract Organisms by Sulfamethoxazole (SMX), Trimethoprim (TMP), and Their Combination

Bacteria	TMP alone	SMX alone	Combination[b] of TMP and SMX: TMP	Combination[b] of TMP and SMX: SMX
Escherichia coli	0.05–1.5	1.0–245	0.05–0.5	0.95–9.5
Proteus spp., indole+	0.5–5.0	7.35–300	0.05–1.5	0.95–28.5
Proteus mirabilis	0.5–1.5	7.35–30	0.05–0.15	0.95–2.85
Klebsiella-Enterobacter	0.15–5.0	2.45–245	0.05–1.5	0.95–28.5

[a]MIC in μg/mL.
[b]TMP: SMX = 1:20.

Mixtures of sulfas (eg, the tri-sulfapyrimidines) have also been used for this purpose. Resistant organisms frequently result after a urinary tract infection has been treated with sulfonamides, however.

Other Infections. The slowly excreted sulfonamides (eg, sulfamethoxypyridazine, sulfadimethoxine) are used for treatment of minor infections such as sinusitis or otitis, or for prolonged maintenance therapy. Soluble sulfonamides are sometimes used for protozoal infections in combination with other agents. Pyrimethamine, combined with sulfonamides, has been used for toxoplasmosis or leishmaniasis, and trimethoprim with sulfonamides has been used in some types of malaria. In nocardiosis, sulfonamides have been used with cycloserine [*68-41-7*] (17).

Resistance to sulfonamides is now common for *N. meningitidis,* as well as in cases of bacillary dysentery. Antibiotics have generally replaced the sulfonamides for these purposes. Sulfonamides, particularly sulfisoxazole and sulfadiazine, are of value in treatment of infections due to *Nocardia* species, and sulfonamides are effective for trachoma. Inclusion conjunctivitis is also treated with sulfacetamide ointment. Oral administration of a sulfonamide, eg, sulfisoxazole, has been successful for treatment of lymphogranuloma venereum and chancroid. Dapsone and sulfonamides have also been used for treatment of the skin disorder dermatitis herpetiformis. Sulfonamides have been used for long term prophylaxis of rheumatic fever, but are being replaced by penicillin for this purpose, except in cases of hypersensitivity to penicillin (19).

Because of several unique properties, the sulfonamides continue to occupy an important place in antimicrobial therapy. Development of derivatives that remain largely unabsorbed in the intestinal tract makes them of particular advantage in causing local changes in the bacterial flora. Development of sulfas with high solubility in the urine and low renal toxicity has provided agents with effectiveness in chronic urinary tract infections. Also, the establishment of the value of sulfonamide combinations with the dihydrofolate reductase inhibitors trimethoprin and pyrimethamine has given the sulfas a preferred place in management of some specific microbial infections.

Physiochemical Properties

The sulfa drugs are weak acids, the more important ones generally having a pK_a in the range of 5–8. This acidity, due to the sulfonamide function, makes the sulfas soluble in basic aqueous solution. The pK_a is modified by the presence of the N^1-substituent, but the clinically useful sulfas generally have pK_a values which give the compounds good solubility at physiologic pH. The N^4-acyl derivatives, such as succinylsulfathiazole, phthalylsulfathiazole, and salicylazosulfapyridine, contain a free carboxyl group and are soluble in the physiologic pH range. As pH is lowered, the solubility of the N^1-substituted sulfonamides reaches a minimum in the range of pH 3–5, which corresponds to the solubility of the molecular species. At this pH, either the sulfa molecule or its N^4-acetyl metabolic product may crystallize in the kidneys, giving rise to blockage and irritation. Because most of the sulfas have a free aromatic amino group which can be protonated, the sulfas are soluble as cations with pK_as of ~ 2. Both their aqueous and liposolubilities have been thoroughly studied (20–22).

The amino group is readily diazotized in aqueous solution, and this reaction forms a basis for the assay of sulfas. Aldehydes also react to form anils, and the yellow product formed with 4-(dimethylamino)benzaldehyde can be used for detection in thin-layer and paper chromatography. Chromatographic retention values have been determined in a number of thin layer systems, and have been used as an expression of the lipophilic character of sulfonamides (23). These values have corresponded well with Hansch lipophilic parameters determined in an isobutyl alcohol–water system.

Lipid solubilities of the sulfonamides vary over a wide range. Oil–water partition coefficients using an aqueous ethylene chloride system have been determined (20). The differences among individual members unquestionably influence their pharmacokinetics as well as antimicrobial activity. The longer acting sulfonamides with high tubular reabsorption generally have a high degree of liposolubility (19). Both antimicrobial activity and half-life of the drug are influenced by liposolubility, but these effects cannot be separated from that of the degree of ionization at physiologic pH, so no precise relation between activity and liposolubility has been established.

Numerous studies have been made to find a correlation between the physicochemical properties of the sulfonamides and their bacteriostatic activity. Degree of ionization, lipid–water solubility, and protein binding have all been observed. The presence of a primary aromatic amino group has been found essential for activity, and the presence of N^1-substituents is a significant influence in the degree of acid dissociation of the sulfonamide group. As early as 1942, a study of the relationship between the pK_a and antibacterial activity of an extensive series of sulfonamides showed that a plot of log 1/MIC against pK_a was a parabolic curve, the maximum lying between pK_a 6.0 and 7.4 (24). The maximal activity was thus observed in compounds with a pK_a that fell in the physiologic pH range. The pK_a of most of the active sulfonamides found since then has been in this range as well. These findings (24) were related to the antimetabolite theory (25,26), regarding the structural similarity between the sulfonamide and the natural substrate, PABA, that it was replacing. The theory emphasized the need for polarization of the sulfonyl group, so that it should resemble as closely as possible the geometrical and electronic characteristics of the *p*-aminobenzoate ion.

In another attempt to relate degree of ionization with antibacterial activity, the effect of pH of the medium on the antibacterial activity was studied (27,28). Activity increased with increase in pH only up to the point at which the drug was 50% ionized, and then decreased. The interpretation of this was that sulfonamides penetrate the bacterial cell in the unionized form, but once inside the cell, the equilibrium between ionized and unionized forms is reestablished, and the activity is due to the ionized form. For optimum activity, a sulfonamide should have a pK_a that provides half-dissociation at the physiologic pH in the area where it is absorbed. This observation also provided an explanation of the parabolic relationship between pK_a and MIC (24).

In subsequent studies attempting to find a correlation of physicochemical properties and antimicrobial activity, other parameters have been employed, such as Hammett σ values, electronic distribution calculated by molecular orbital methods, spectral characteristics, and hydrophobicity constants. No new insight on the role of physiochemical properties of the sulfonamides has resulted. Acid dissociation appears to play a predominant role, since it affects aqueous solubility, partition coefficient and transport across membranes, protein binding, tubular secretion, and reabsorption in the kidneys. An exhaustive discussion of these studies has been provided (10).

Biological Mechanism of Action

The sulfa drugs act on microorganisms by limiting or halting growth rather than by a bactericidal action. They inhibit growth of bacteria *in vitro* only if the medium is free of inactivating substances, mainly peptones and *p*-aminobenzoic acid. If *in vitro* activity had been a prerequisite for testing as a drug, neither Prontosil nor sulfanilamide itself would have been selected.

The Woods-Fildes theory postulated that *p*-aminobenzoic acid (PABA) [*150-13-0*] is an essential metabolite for the bacteria that are affected. Woods obtained evidence that PABA antagonizes the activity of sulfonamides and showed that PABA could completely reverse the bacteriostatic activity of sulfanilamide against various bacteria *in vitro*. PABA was later isolated from the various tissues and fluids, and yeast extract, where antagonism to sulfanilamide was found. Further studies showed that inhibition of bacterial growth by sulfonamides in simple media can be reversed not only competitively by PABA, but also noncompetitively by compounds not structurally related to PABA, such as 1-methionine, 1-serine, glycine, adenine, guanine, and thymine (29). The finding that sulfanilamide-inhibited cultures accumulated 4-amino-5-imidazolecarboxamide ribotide (30), a precursor to purine biosynthesis (31), indicated that purine biosynthesis was affected by the sulfonamides.

The primary mode of action of the sulfonamides is, however, the interference with the uptake of PABA for the formation of folic acid [*59-30-3*]. The sulfonamides impede this synthesis and are therefore toxic to those bacteria that synthesize their own folic acid. Mammals cannot synthesize this and related vitamins and depend on food sources for them; the sulfas are therefore not toxic to mammals in this regard.

Subsequent knowledge of the structure, function, and biosynthesis of the folic acid coenzyme gradually allowed a picture to be formed regarding the step in this pathway that is inhibited by sulfonamides. The biosynthetic scheme for folic acid is shown in Figure 1. Sulfonamides compete in the step where condensation of PABA with pteridine pyrophosphate takes place to form dihydropteroate (32). The amino acids, purines, and pyrimidines that are able to replace or spare PABA are those with a formation that requires one-carbon transfer catalyzed by folic acid coenzymes (5).

Direct evidence of the inhibition of dihydropteroate and dihydrofolate synthesis was obtained (33) from work with enzymes from *E. coli*. The synthesis of dihydropteroate from PABA is inhibited by a number of sulfonamides and, in general, the more potent inhibitors of folate biosynthesis are the better bacterial growth inhibitors. Subsequently, evidence was obtained of incorporation of sulfonamides (34,35) in the pteridine moiety, but this incorporation is probably not of physiologic significance (36). Two binding sites on the enzyme involved, dihydropteroate synthetase, have been proposed (37) as being involved in the competi-

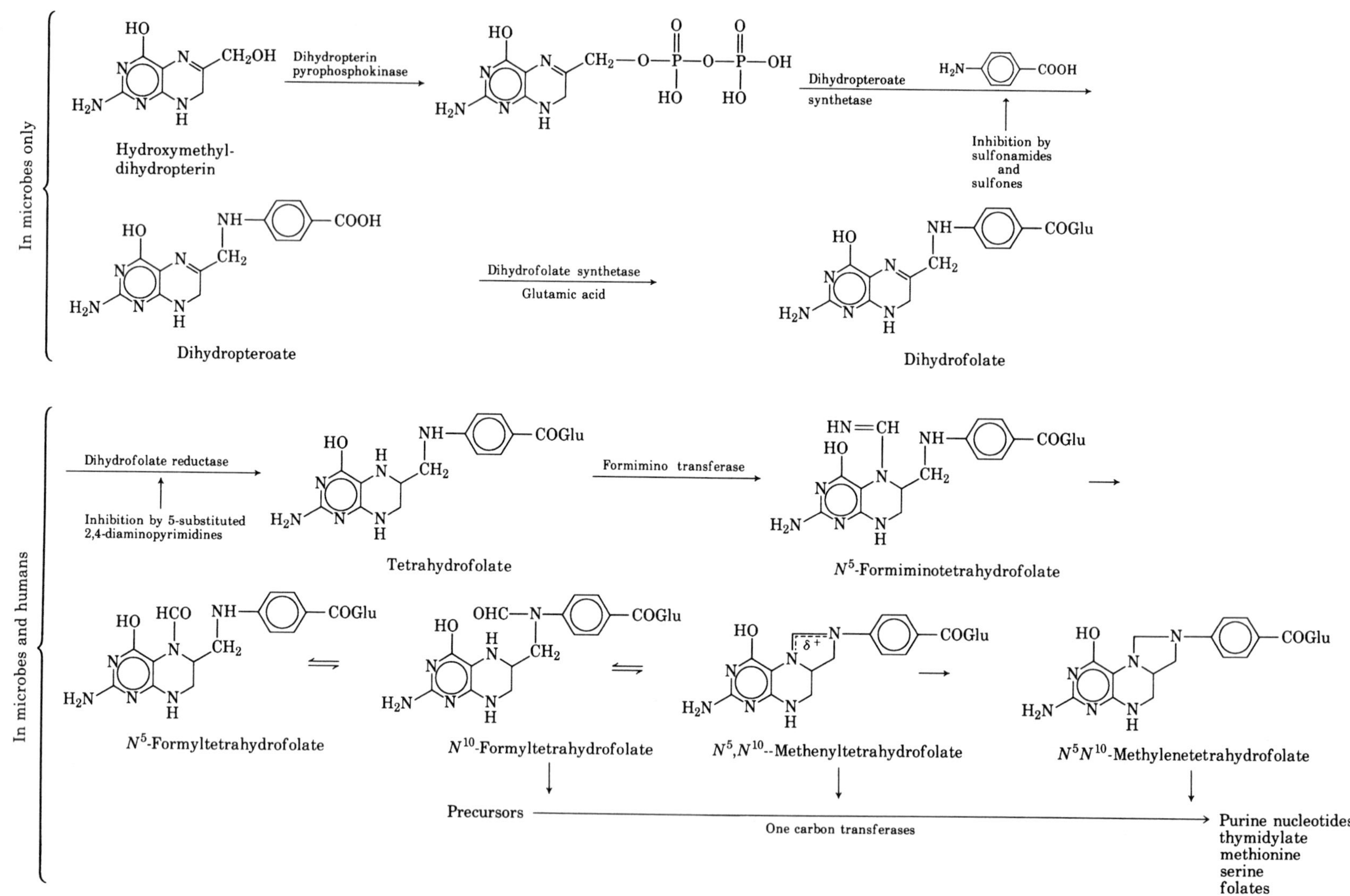

Fig. 1. Folate metabolism and sites of action of antifolates. Courtesy of Lea & Febiger (5).

tion between PABA and the sulfonamides. One is specific for the NH_2 group, and the other is where the acidic group of PABA, or the SO_2 function of the sulfonamide, binds. This view of the inhibition accomodates both sulfonamides and sulfones, since the sulfonamide NH is not a binding function. The present knowledge of the topography of the enzyme surface is not sufficient to explain how the additional phenyl group of the sulfones is accomodated, but it is likely that there is sufficient bulk tolerance for this group as well as the N^1-heterocycles of the sulfonamides. Competition with PABA for these binding sites is most likely the primary mode of action of the sulfonamides and sulfones (see reference 10 for a more complete discussion of this binding).

Development of Resistance. One of the principal disadvantages of sulfonamide therapy is the emergence of drug-resistant strains of bacteria. Resistance develops by several mechanisms: overproduction of PABA (38); altered permeability of the organisms to sulfonamides (39); and reduced affinity of dihydropteroate synthetase for sulfonamides while the affinity for PABA is retained (40). Sulfonamides also show cross-resistance to other sulfonamides but not to other antibacterials. In plasmodia, resistance may occur by means of a bypass mechanism in which the organisms can use preformed folic acid (41).

Structure-Activity Relationships

The following generalizations arose from a review of more than 5000 sulfonamides (1).

1. The amino and sulfonyl groups on the benzene ring should be in the 1,4 positions; the amino group should be unsubstituted or converted to a free amino *in vivo*.
2. Replacement of the benzene ring by other ring systems, or the introduction of additional substituents, decreases or abolishes activity.
3. Exchange of the SO_2NH by SOC_6H_4-*p*-NH_2, $CONH_2$, CONHR, or COC_6H_4R generally reduces activity.
4. N^1-Monosubstitution may result in greater activity, and will increase activity with a number of heterocycles; N^1-disubstitution in general leads to inactive compounds.

A later review (42) confirmed these generalizations. Replacement of the SO_2NH by $SO_2C_6H_4$-*p*-NH_2 gives the sulfones that show reduced activity for some bacterial species, but show good activity against leprosy and some strains of malarial organisms. General structures summarizing the essential features for activity for the sulfonamides (7) and sulfones (8) follow.

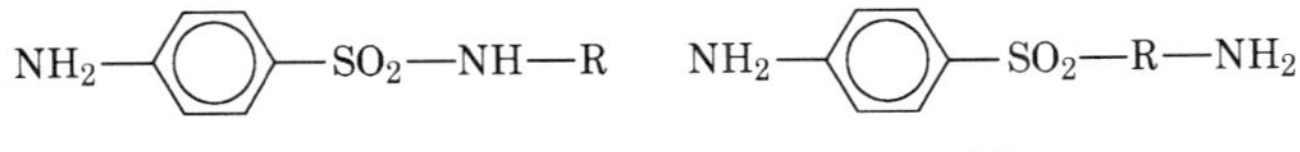

(7) (8)

For the sulfonamides, the best activity is found where R is heterocyclic, but it can also be isocyclic or acyl. For the sulfones, R can be phenylene or a heterocycle; the parent dapsone, where R is phenyl, is the most active.

Other related structures may be active, but through a different mechanism. Separation of the amino group from the ring by CH_2 has provided an active sulfonamide, mafenide. Replacement of the amino by amidino has also given an active sulfone, methyl *p*-amidinophenyl sulfone [*17574-50-4*], for which no metabolite antagonist is known (1). Although not a sulfonamide, *p*-aminosalicylic acid [*65-49-6*] also has its action reversed by PABA. It has bacteriostatic action, but is most specific against mycobacteria (43).

Preparation and Manufacture

The most common method for the preparation of sulfonamides is by the action of *N*-acetylsulfanilyl chloride with the appropriate amine (1). Excess amine or suitable base is used to neutralize the hydrochloric acid formed.

$$CH_3CONHC_6H_4SO_2\,Cl + RNH_2 \xrightarrow{\text{base}} CH_3CONHC_6H_4SO_2NHR + HCl$$

The resulting acetyl compound is usually hydrolyzed with aqueous alkali to give the free amine. Other *N*-acyl derivatives may be used, particularly for the less soluble succinyl and phthaloyl products. The use of *p*-nitrobenzenesulfonyl chloride, followed by reduction of the nitro to an amino function, is much more expensive and is rarely used. *N*-Acetylsulfanilyl chloride [*121-60-8*] is obtained by the chlorosulfonation of acetanilide [*103-84-4*], which is the basic material for most of the sulfonamides.

N^1-Heterocyclic Sulfanilamides. The parent sulfanilamide is manufactured by the reaction of *N*-acetylsulfanilyl chloride with excess concentrated aqueous ammonia, and hydrolysis of the product. Most heterocyclic amines are less reactive, and the condensation with the sulfonyl chloride is usually done in anhydrous media in the presence of an acid-binding agent. Use of anhydrous conditions avoids hydrolytic destruction of the sulfonyl chloride. The solvent and acid-binding functions are commonly filled by pyridine, or by mixtures of pyridine and acetone. Tertiary amines, such as triethylamine, may be substituted for pyridine. The majority of N^1-heterocyclic sulfanilamides are made by simple condensation with *N*-acetylsulfanilyl chloride and hydrolysis.

A variation of this procedure is used for sulfisomidine because of the different character of the amino group in the 4-position of a pyrimidine ring. Two moles of the sulfonyl chloride are condensed with one mole of 4-amino-2,6-dimethylpyrimidine in the presence of triethylamine. The resulting bis(acetylsulfanilyl) derivative is readily hydrolyzed to the product. The formation of the bis(acetylsulfanilyl) derivative has also been employed for other heterocyclic amines, eg, for synthesis of sulfathiazole and sulfamoxole (44), but the 1:1 reaction is probably preferable.

In a few cases, N^1-heterocyclic sulfanilamides have been prepared by the condensation of an active heterocyclic halide with the sulfonamide nitrogen of sulfanilamide or its *N*-acetyl derivative in the presence of an acid-binding agent. Sulfapyridine, sulfadiazine, and sulfapyrazine have been made by this method (1), but the most important application is probably for the synthesis of sulfachlorapyridazine (**9**) and sulfamethoxypyridazine (**10**) (45).

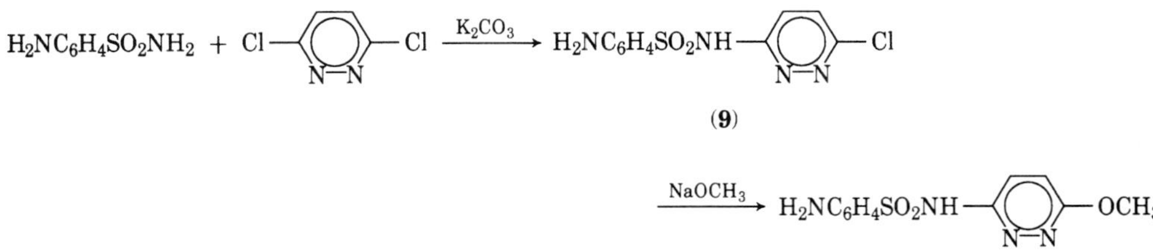

(**9**)

(**10**)

N^1-Heterocyclic derivatives can be formed in some cases by a ring closure to give the heterocycle. Sulfadiazine, sulfamethazine, sulfamerazine, and sulfathiazole have been prepared in this fashion, but also by the usual procedure from the sulfonyl chloride and heterocyclic amine. The synthesis of sulfamethazine from sulfaguanidine is an example of the ring closure method.

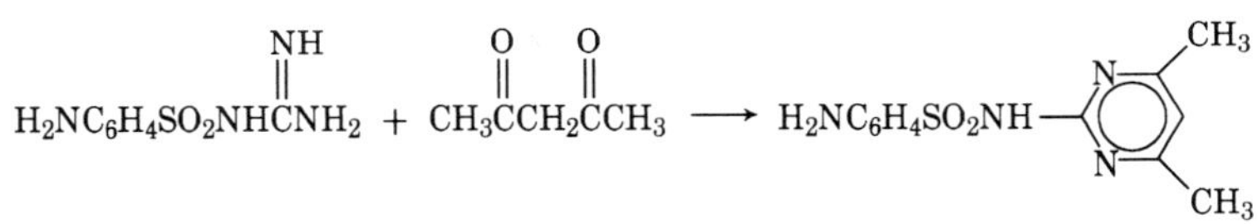

N^1-Acylsulfanilamides. Only two examples of this class still have much use in the United States. Sulfacetamide is prepared by acetylation of N^4-acetylsulfanilamide in pyridine. The resulting N^1, N^4-diacetylsulfanilamide can be hydrolyzed selectively with alkali to give the monacetyl product. Its separation from regenerated sulfanilamide is easily achieved because of the greater acidity of sulfacetamide. Sulfabenzamide may be prepared in similar fashion.

Sulfaguanidine is prepared by condensation of *N*-acetylsulfanilyl chloride with guanidine in presence of alkali. The N^4-acetyl group is removed by acid or alkaline hydrolysis.

N^1-Heterocyclic-N^4-acylsulfanilamides. Two examples of this class still in use are succinylsulfathiazole and phthalysulfathiazole. They are prepared by fusion of sulfathiazole with appropriate anhydrides.

N^1-Heterocyclic-N^1-acetylsulfanilamides. These derivatives may be prepared by condensation of the heterocyclic amine with *p*-nitrobenzenesulfonyl chloride followed by acetylation of the nitro compound. The product may be reduced under mild conditions to give the 4-amino-N^1-heterocyclic-N^1-acetyl derivative. Other approaches, however, have been developed (46,47).

Other Derivatives. Salicylazosulfapyridine [*599-79-1*] (sulfasalazine) can be prepared by diazotization of sulfapyridine and coupling to salicylic acid. Mafenide is prepared by chlorosulfonating *N*-benzylacetamide, reaction of the resulting α-acetamido-*p*-toluenesulfonyl chloride with ammonia, and hydrolyzing the acetyl group.

Dapsone has been prepared by a number of procedures (1,48). One procedure employs the reaction of 1-chloro-4-nitrobenzene with excess sodium sulfide to give the 4-amino-4′-nitrodiphenyl sulfide. This compound, after acetylation of the amino group, is oxidized with H_2O_2 to the sulfone. The nitro group is then reduced to amino, and hydrolysis of the acetyl gives the product.

$$O_2N{-}C_6H_4{-}Cl \xrightarrow{Na_2S} O_2N{-}C_6H_4{-}S{-}C_6H_4{-}NH_2 \xrightarrow{(CH_3CO)_2O} O_2N{-}C_6H_4SC_6H_4NHCOCH_3$$

$$\xrightarrow{H_2O_2} O_2NC_6H_4SO_2C_6H_4NHCOCH_3 \xrightarrow[H^+]{H_2} H_2NC_6H_4SO_2C_6H_4NH_2$$

Analysis

Sulfonamides having a free *p*-amino group are readily assayed by titration with nitrous acid. The sulfonamide function may also be titrated with base, such as lithium methoxide. The majority of the sulfas listed in the U.S. Pharmacopeia XXII, however, are assayed by chromatographic methods, particularly high performance liquid chromatography (49). Sulfonamides for which assays are listed in the U.S. Pharmacopeia XXII-National Formulary XVII include the following: sulfacetamide, sulfabenzamide, sulfadiazine, sulfadoxine, sulfamerazine, sulfamethazine, sulfamethizole, sulfamethoxazole, sulfapyridine, sulfasalazine, sulfathiazole, sulfinpyrazone, sulfisoxazole, sulfisoxazole acetyl, sulfisoxazole diolamine, sulfoxone, triple sulfa, dapsone, and various combinations with prednisolone, pyrimethamine, and trimethoprim.

Toxicity

A small percentage of patients treated with sulfa drugs have shown toxic effects, such as drug fever, rashes, mild peripheral neuritis, and mental disturbance. In general, these effects are more prevalent with higher blood levels, and may accompany poor excretion or overdosing (21). In 1966 the FDA required that two long-acting sulfas, sulfamethoxypyridazine and sulfadimethoxine carry a label warning of the possibility of death due to Stevens-Johnson syndrome, an extremely severe dermatologic reaction (50).

Crystalluria, due to formation of insoluble *N*-acetyl metabolic products, was one of the more serious toxic effects observed with sulfonamide therapy. This is much less of a problem than with the early sulfas, mainly because of the discovery of agents highly soluble at urinary pH, the development of long-acting sulfas that build up adequate blood levels at doses low enough to avoid crystallization, and the discovery of compounds that are excreted chiefly as highly soluble glucuronides. The danger of kidney blockage is also reduced by increasing fluid intake and by administering sodium bicarbonate to alkalinize the urine. The use of mixtures, such as triple sulfa, also reduces the possibility of crystallization in the kidneys.

Blood dyscrasias are quite uncommon, but if they occur may be serious enough to cause discontinuance of the therapy. Both topical and systemic administration of sulfas can cause hypersensitivity reactions, such as urticaria, exfoliative dermatitis, photosensitization, erythema nodosum, and in its most severe form, erythema multiformexudativum (Stevens-Johnson syndrome). In general, however, use of sulfonamide therapy is considered relatively safe.

Economic Aspects

The production of sulfa drugs increased rapidly after their introduction in the 1930s, and reached a maximum of 9000 t in 1943. An abrupt drop to less than half this amount occurred in 1944, with the commercial availability of penicillin. Contrary to some expectations that the sulfas would be totally replaced by antibiotics, their output has remained close to the 1944 level. Increased veterinary use, along with low cost and effectiveness of the sulfas has maintained a market for them. United States production figures of sulfas for selected years are shown in Table 4, in comparison with those for total antibiotics. The most recent figures released for sulfa drugs also include those for other agents, such as antiprotozoan agents and other urinary tract antiinfective agents.

Table 4. United States Production of Sulfa Drugs and Antibiotics, t[a]

Drug	1943	1946	1956	1966	1975	1987
total sulfa drugs	9077	4630	3462	4944	2122	5557[b]
total antibiotics	0	34.5	1784	8756	8312	16,099

[a]From Ref. 51.
[b]Also includes antiprotozoan agents and other urinary tract antiinfectives, but does not include the antileprotic sulfones.

BIBLIOGRAPHY

"Sulfonamides," in *ECT* 1st ed., Vol. 13, pp. 312–316; "Sulfa Drugs," in *ECT* 1st ed., Vol. 13, pp. 271–285, by M. E. Hultquist, American Cyanamid Company; in *ECT* 2nd ed., Vol. 19, pp. 255–261; pp. 261–279; in *ECT* 3rd ed. Vol. 2, pp. 795–808, by L. Doub, Parke-Davis & Co.

1. E. H. Northey, *The Sulfonamides and Allied Compounds,* Reinhold Publishing Corp., New York, 1948, pp. 517–577.
2. L. Weinstein, M. A. Madoff, and C. M. Samet, *N. Engl. J. Med.* **263,** 793 (1960).
3. N. Anand, in J. W. Corcoran and F. E. Hahn, eds., *Antibiotics,* Vol. 3, Springer-Verlag, New York, 1975, pp. 668–698.
4. N. Anand, in M. E. Wolff, ed., *Burger's Medical Chemistry,* 4th ed. Part 2, John Wiley & Sons, Inc., New York, 1979, pp. 1–40.
5. N. Anand, in W. O. Fove, ed., *Principles of Medicinal Chemistry,* 3rd ed., Lea & Febiger, Philadelphia, 1988, pp. 637–659.
6. G. P. Wormser, G. T. Keusch, and C. H. Rennie, *Drugs* **24,** 459–518 (1982).
7. Ref. 2, pp. 900–907.
8. H. A. Feldman, *N. Engl. J. Med.* **279,** 1370, 1431–1437 (1968).
9. W. Modell, *Science* **162,** 1346 (1968).

10. P. G. De Benedetti, in B. Testa, ed., *Advances in Drug Research* Vol. 16, Academic Press, London, 1987, pp. 228–279.
11. G. Domagk, *Dtsche. Med. Wochenschr.* **61,** 250 (1935).
12. J. Trefouel, Mme J. Trefouel, F. Nitti, and D. Bovet, *C.R. Soc. Biol.* **120,** 756 (1935).
13. G. A. H. Buttle, W. H. Grey, and D. Stephenson, *Lancet* **1,** 194 (1937).
14. N. Rist, *Nature* **146,** 838 (1940).
15. S. G. Browne, *Adv. Pharmacol. Chemother.* **7,** 211 (1969).
16. S. G. Browne, *Br. Med. J.* **III,** 725 (1968).
17. F. H. Meyers, E. Jawetz, and A. Goldfien, *Review of Medical Pharmacology,* 3rd ed., Lange Medical Publications, Los Altos, Calif., 1972, pp. 523–527.
18. G. H. Hitchings, ed., *Inhibition of Folate Metabolism in Chemotherapy. The Origins and Uses of Co-trimoxazole,* Springer-Verlag, Berlin, 1983.
19. G. L. Mandell and M. E. Sande, in A. G. Gilman, L. S. Goodman, T. W. Rall, and F. Murad, eds., *Goodman and Gilman's The Pharmacological Basis of Therapeutics,* 7th ed., MacMillan Pub. Co., New York, 1985, pp. 1095–1109.
20. J. Rieder, *Arzniem. Forsch.* **13,** 81, 89, 95 (1963).
21. D. Lehr, *Ann. N.Y. Acad. Sci.* **69,** 417 (1957).
22. E. R. Garrett, J. B. Mielck, J. K. Seydel, and H. J. Kessler, *J. Med. Chem.* **12,** 740 (1969).
23. G. L. Biagi, A. M. Barbaro, M. C. Guerra, G. C. Forti, and M. E. Fracasso, *J. Med. Chem.* **17,** 28 (1974).
24. P. H. Bell and R. O. Roblin, Jr., *J. Am. Chem. Soc.* **64,** 2905 (1942).
25. D. D. Woods, *Br. J. Exp. Pathol.* **21,** 74 (1940).
26. P. Fildes, *Lancet* **1,** 955 (1940).
27. P. B. Cowles, *Yale J. Biol. Med.* **14,** 599 (1942).
28. A. H. Brueckner, *Yale J. Biol. Med.* **15,** 813 (1943).
29. E. E. Snell and H. K. Mitchell, *Arch. Biochem.* **1,** 93 (1943).
30. M. R. Stetten and C. L. Fox, *J. Biol. Chem.* **161,** 333 (1945).
31. W. Shive and co-workers, *J. Am. Chem. Soc.* **69,** 725 (1947).
32. R. Tschesche, *Z. Naturforschung* (C) **2b,** 10 (1947).
33. C. M. Brown, R. A. Weisman, and D. A. Molnar, *J. Biol. Chem.* **236,** 2534 (1961).
34. R. A. Weisman and C. M. Brown, *J. Biol. Chem.* **239,** 326 (1964).
35. L. Bock, G. H. Miller, K. J. Schaper, and J. K. Seydel, *J. Med. Chem.* **17,** 23 (1974).
36. S. Roland and co-workers, *J. Biol. Chem.* **254,** 10,337 (1960).
37. I. Moriguchi and S. Wada, *Chem. Pharm. Bull. (Tokyo)* **16,** 754 (1968).
38. P. J. White and D. D. Woods, *J. Gen. Microbiol.* **40,** 243 (1965).
39. N. Nagate, I. Matsuhisa, K. Inoue, and S. Mitsukashi, *Microbiol. Immunol.* **22,** 367 (1978).
40. O. Skold, *Antimicrob. Agents Chemother.* **9,** 49 (1976).
41. A. Bishop, *Parasitology* **53,** 10 (1963).
42. R. G. Shepherd, in A. Burger, ed., *Medicinal Chemistry,* 3rd. ed., Wiley-Interscience, New York, 1970, p. 255.
43. G. R. Youmans and A. S. Youmans, in R. J. Schnitzer and F. Hawking, eds., *Experimental Chemotherapy,* Vol. 2, Academic Press, Inc., New York, 1964, pp. 445–447.
44. Brit. Pat. 819,019 (Aug. 26, 1959) (to Nordmark-Werke Gesellschaft, Hamburg).
45. J. H. Clark and co-workers, *J. Am. Chem. Soc.* **80,** 890 (1958).
46. U.S. Pat. 2,974,137 (Mar. 7, 1961), G. T. Fitchett, J. E. Gordon, and R. G. Shepherd (to American Cyanamid Co., New York).
47. U.S. Pat. 2,721,200 (Apr. 18, 1955), M. Hoffer (to Hoffmann-LaRoche).
48. L. Doub, in W. H. Hartung, ed., *Medicinal Chemistry,* Vol. 5, John Wiley & Sons, Inc., New York, 1961, pp. 352–354.
49. *United States Pharmacopeia XXII-National Formulary XVII,* 1990.
50. Washington News Section, *J. Am. Med. Assoc.* **195,** 27 (1966).

51. United States Tariff Commission, *Synthetic Organic Chemicals, United States Production and Sales,* U.S. Government Printing Office, Washington, D.C., 1941–1943, 1946, 1956, 1966, 1975; United States International Trade Commission. *Synthetic Organic Chemicals, United States Production and Sales,* Publication 2118, 1987.

WILLIAM O. FOYE
Massachusetts College of Pharmacy
and Allied Health Sciences

ANTIBIOTICS

SURVEY

Antibiotics are chemical substances produced by microorganisms and other living systems that are capable in low concentrations of inhibiting the growth of bacteria or other microorganisms. This inhibitory effect can be *in vitro* or *in vivo*. Antibiotics having both *in vivo* activity and low mammalian toxicity have been extremely valuable in treating infectious diseases. There are over 10,000 antibiotics produced by microorganisms that have been reported in the scientific literature, approximately 6100 of which have been characterized and have had molecular structures assigned (1,2). These substances range from the very simple to extremely complex, but most antibiotics are in the 300–800 mol wt range. Many thousands of semisynthetic variations of the naturally occurring antibiotics have been prepared. Only relatively few (~200) have become commercial products for human and veterinary uses (see also ANTIBACTERIAL AGENTS).

Microorganisms producing antibiotics include antinomyces, fungi, bacteria, and algae. Marine organisms such as sponges and soft corals have also been found to produce antibiotics (3). In some cases symbionic microorganisms such as algae may be responsible for the antibiotic production noted in the higher life forms. Terrestrial plants have also been found to produce some antibiotic substances (4).

The mechanism of action of a number of antibiotics with regard to the inhibition of bacteria, fungi, or other organisms has been established. The more common mechanisms include inhibition of bacteria cell wall biosynthesis, inhibition of protein, RNA, or DNA synthesis, and damaging of membranes (5). Cell wall biosynthesis is a target present in bacteria but not in mammalian cells. Thus the β-lactams, which are very effective against bacteria, are relatively nontoxic to humans. In contrast, antibiotics that damage DNA, like adriamycin, are relatively toxic to both types of cells. However, adriamycin, which was found to be more toxic to rapidly proliferating tumor cells than to most normal cells with slower turnover rates, shows significant selectivity against tumor cells to find clinical application as an antitumor agent (5).

History

Antibiotics were used in folk medicine at least as early as 2500 years ago when the Chinese reported the medicinally beneficial effects of moldy bean curd. Evidence for some type of tetracycline antibiotic usage by the Sudanese-Nubian civilization (350 AD) was reported in 1980 (6). Fluorescent areas in human bones from this era were observed that were identical in location and characteristics to modern bone from patients treated with tetracyclines. Identification of tetracycline in the ancient bones was further substantiated by fluorescence spectrum measurements and microbiological inhibition studies (7).

One of the most profound developments in the history of modern medicine has been the discovery of antibiotics to control infections. The realization that microbial products could cure infectious diseases spanned approximately sixty-five years of discoveries (8). One of the first modern scientific demonstrations was the observation by Louis Pasteur in 1877 that common bacteria inhibited the growth of a pure anthrax culture. Other observations followed. Then in 1928 it was noted by Alexander Fleming that a culture of a green *Penicillium* inhibited the growth of bacteria on an agar plate and in 1932 a paper on the use of the penicillin preparation as a topical agent for treating infected wounds was published. But it was not until 1940 that a systematic investigation of antibacterial substances was made along with a reinvestigation of the properties of penicillin. A stable dry powder was prepared from fermentations of *Penicillium* that possessed strong antibacterial properties and the clinical utility of penicillin, which was shown to be highly effective against bacterial infections yet nontoxic to animals, was clearly visible.

In 1939 the isolation of a mixture of microbial products named tyrothricin from a soil bacillus was described. Further investigation showed this material to be a mixture of gramicidin and tyrocidine. In rapid succession the isolation of actinomycin (1940), streptothricin (1942), streptomycin (1943), and neomycin (1949), produced by *Streptomyces,* were reported and in 1942 the word antibiotic was introduced. Chloramphenicol, the first of the so-called broad spectrum antibiotics having a wide range of antimicrobial activity, was discovered in 1947.

Aureomycin, the first member of the commercially important tetracycline antibiotics, was discovered in 1948.

The number of naturally occurring antibiotics increased from about 30 known in 1945, to 150 in 1949, 450 in 1953, 1200 in 1960, and to 10,000 by 1990 (1,9). Table 1 lists the years of historical importance to the development of antibiotics

Table 1. Year of Discovery or Market Introduction of Some of the More Important Antibiotics

Antibiotic	CAS Registry Number	Year	
		Discovery	Introduction
penicillin		1929	
tyrothricin	*[1404-88-2]*	1939	
griseofulvin	*[1403-70-9]*	1939	
streptomycin	*[57-92-1]*	1944	
bacitracin	*[1405-87-4]*	1945	
chloramphenicol	*[56-75-7]*		1947
polymyxin	*[1406-11-7]*	1948	
chlortetracycline	*[57-62-5]*		1948
cephalosporin C,N,P		1948	
neomycin	*[1404-04-2]*	1949	
oxytetracycline	*[79-57-2]*		1950
nystatin	*[1400-61-9]*	1950	
erythromycin	*[114-07-8]*		1952
novobiocin	*[303-81-1]*	1955	
kanamycin	*[8063-07-8]*		1957
ampicillin[a]	*[69-53-4]*		1962
fusidic acid	*[6990-06-3]*		1961
cephalothin[a]	*[153-61-7]*	1962	
lincomycin	*[154-21-2]*		1963
gentamicin	*[1403-66-3]*		1963
carbenicillin[a]	*[4697-36-3]*	1964	
cephalexin[a]	*[15686-71-2]*	1966	
clindamycin[a]	*[18323-44-9]*		1967
cephaloxidine and cephalothin[a,b]			1969
minocycline[a]	*[10118-90-8]*		1971
amoxycillin[a]	*[26787-78-0]*		1972
cefoxitin[a,c]	*[35607-66-0]*		1978
tricarcillin[a]	*[34787-01-4]*		1979
mezlocillin[a]	*[51481-65-3]*		1980
piperacillin[a]	*[61477-96-1]*		1980
cefotaxime[a]	*[63527-52-6]*		1980
moxalactam[a]	*[64952-97-2]*		1981
augmentin[d]	*[74469-00-4]*		1984
aztreonam[e]	*[78110-38-0]*		1984
imipenem[a,f]	*[74431-23-5]*		1985

[a]Semisynthetic products.
[b]First oral cephalosporins.
[c]First commercial cephamycin.
[d]First β-lactamase inhibitor combination.
[e]First monobactam and a synthetic product.
[f]First carbapenem.

used for treatment in humans. Most of the antibiotics introduced since the 1970s have been derived from synthetic modifications of the β-lactam antibiotics (qv).

Nomenclature

Antibiotics, and all other marketed drug products, are identified by three different types of names: (*1*) a chemical name based on standard rules of chemical nomenclature, (*2*) a generic (nonproprietary) name that is relatively short and follows a systematic procedure of describing a compound structure-type and utility, and (*3*) a trade name that is given by the manufacturer to distinguish it from competitive products. Some of the antibiotics discovered prior to 1961, when the U.S. adopted names (USAN) program started, had received many different names in the chemical literature. The USAN program is a specifically organized effort in the United States directed to producing simple and useful nonproprietary names for drugs while the drug is still in its investigational stage (10). Most of the antibiotics reported in literature are not of commercial interest and are not given generic names. If the chemical name is too complex, a trivial (nonsystematic or semisytematic) name is often assigned by the discoverer. For industrial organizations these names usually appear as alphanumeric coded identifiers such as LL-F28249α, given the generic name of nemadectin when the commercial potential became apparent (11).

A marketed antibiotic's generic name is usually preferred because it is simpler and easier to remember. Tetracycline [*60-54-8*], is an example:

chemical name	4-(dimethylamino)-1,4,4a,5,5a,6,11,12a-octahydro-3,6,10,12,12a-pentahydroxy-6-methyl-1,11-dioxo-2-naphthacenecarboxamide
generic name	tetracycline
trade names	Achromycin (Lederle), Panmycin (Upjohn) Sumycin (Squibb), Tetracyn (Pfizer), Tetrex (Bristol), etc

Classification of Antibiotics

Antibiotics have a wide diversity of chemical structures and range in molecular weight from near 100 to over 13,000. Most of the antibiotics fall into broad structure families. Because of the wide diversity and complexity of chemical structures, a chemical classification scheme for all antibiotics has been difficult. The most comprehensive scheme may be found in reference 12. Another method of classifying antibiotics is by mechanism of action (5). However, the modes of action of many antibiotics are still unknown and some have mixed modes of action. Usually within a structure family, the general mechanism of action is the same. For example, of the β-lactams having antibacterial activity, all appear to inhibit bacterial cell wall biosynthesis.

A chemical classification of some of the commercially more important antibiotic families that is generally consistent with the scheme of reference 12, is given here (2).

Aminoglycosides. Antibiotics in the aminoglycoside group characteristically contain amino sugars and deoxystreptamine or streptamine. This family of antibiotics has frequently been referred to as aminocyclitol aminoglycosides. Representative members are streptomycin, neomycin, kanamycin, gentamicin, tobramycin, and amikacin. These antibiotics all inhibit protein biosynthesis.

Ansamacrolides. Antibiotics in the ansamacrolide family are also referred to as ansamycins. They are benzenoid or naphthalenoid aromatic compounds in which nonadjacent positions are bridged by an aliphatic chain to form a cyclic structure. One of the aliphatic–aromatic junctions is always an amide bond. Rifampin is a semisynthetically derived member of this family and has clinical importance. It has selective antibacterial activity and inhibits RNA polymerase.

β-Lactams. All β-lactams are chemically characterized by having a β-lactam ring. Substructure groups are the penicillins, cephalosporins, carbapenems, monobactams, nocardicins, and clavulanic acid. Commercially this family is the most important group of antibiotics used to control bacterial infections. The β-lactams act by inhibition of bacterial cell wall biosynthesis.

Chloramphenicol. Only chloramphenicol and a few closely related analogues fall into this group. Chloramphenicol, a nitro benzene derivative of dichloroacetic acid, inhibits protein biosynthesis.

Glycopeptides. Vancomycin, avoparcin, and teicoplanin are examples of glycopeptide antibiotics. This family has cyclic peptide structures and biphenyl containing amino acids. Sugars are attached to the peptide unit resulting in compounds frequently in the molecular weight range of 1400–2000. These antibiotics inhibit bacterial cell wall biosynthesis by binding to D-alanyl-D-alanine units found in the cell walls.

Lincomycin. The lincomycins and celesticetins are a small family of antibiotics that have carbohydrate-type structures. Clindamycin, a chemical modification of lincomycin, is clinically superior. Antibiotics in this family inhibit gram-positive aerobic and anaerobic bacteria by interfering with protein biosynthesis.

Macrolides. Antibiotics in the macrolide group are macrocyclic lactones that can be further classified into two main subgroups: (*1*) polyene macrolides that are antifungal agents and include compounds like nystatin and amphotericin B; and (*2*) antibacterial antibiotics represented by erythromycin and tylosin. A number of other subfamilies of antibacterial and antifungal antibiotics fall into the broad category of macrolides.

Polyethers. Antibiotics within this family contain a number of cyclic ether and ketal units and have a carboxylic acid group. They form complexes with mono- and divalent cations that are soluble in nonpolar organic solvents. They interact with bacterial cell membranes and allow cations to pass through the membranes causing cell death. Because of this property they have been classified as ionophores. Monensin, lasalocid, and maduramicin are examples of polyethers that are used commercially as anticoccidial agents in poultry and as growth promotants in ruminants.

Tetracyclines. The tetracyclines are a small group of antibiotics characterized as containing a polyhydronaphthacene nucleus. Commercially the tetracyclines are very important. They have been used clinically against gram-positive and gram-negative bacteria, spirochete, mycoplasmas, and rickettsiae and have veterinary applications in promoting growth and feed efficiency. The mode of action is inhibition of protein synthesis. Some of the more important members of this family are tetracycline, minocycline, and doxycycline.

Production

Most of the microorganisms used to produce antibiotics were isolated from soil samples. These microorganisms occur as heterogenous populations and generally inhabit the top few centimeters of soil. Families demonstrated to produce antibiotics include actinomycetes, bacteria, and fungi. Actinomycetes are in numbers the most productive for antibiotics. Until 1974 antibiotics produced by actinomycetes were almost exclusively from the genus *Streptomyces,* accounting for 95% of the total of about 2000 antibiotics known at that time. Thereafter, the role of "rare" actinomycetes as an antibiotic source became apparent as these latter organisms provided about 25% of the approximately 1100 antibiotics from actinomycete origin discovered in the next six years. The rare actinomycetes include genera such as *Actinoplanes, Micromonospora,* and *Nocardia.* Their capacity to produce diverse antibiotic structures is comparable to *Streptomyces*; however, the isolation of these organisms is less frequent by conventional isolation techniques (13).

To obtain reproducible antibiotic production by fermentation, it is necessary to obtain a pure culture of the producing organism. Pure cultures are isolated from mixed soil sample populations by various streaking and isolation techniques on nutrient media. Once a pure culture has been found that produces a new antibiotic typically on a mg/L scale, improvement in antibiotic yield is accomplished by modification of the fermentation medium or strain selection and mutation of the producing organism. Production of g/L quantities may take years to accomplish.

The vast majority of new antibiotics result from screening soil microorganisms or by semisynthetic modification of naturally occurring antibiotics. One alternative approach, termed directed biosynthesis, involves feeding potential biosynthetic precursors into a fermentation of a known antibiotic in order to produce new components (13,14). This aproach was used in the early development stages of the penicillins to obtain penicillin G. Phenylacetic acid was fed to fermentations of *Penicillium chrysogenum.* Another approach involves bioconversions. A known antibiotic is added to the fermentation of a microorganism that is capable of carrying out a process such as hydroxylation or amination. Modifications of these types can frequently be accomplished with a selectivity that is difficult to obtain by synthetic methods. A procedure called mutasynthesis, first reported with regard to new antibiotics related to neomycin (15,16), is sometimes used.

Genetic engineering (qv) technology has evolved that allows modifications of the microorganisms' DNA to produce new material. The structure of indolizomycin, the first new antibiotic obtained by a procedure known as protoplast fusion (17), was totally different from the antibiotics produced by the grandparental strains of *Streptomyces griseus* and *S. tenjimariensis* used in the fusion experiment. Cloning of genes involved in antibiotic biosynthesis from actinomycetes has become possible. The generation of the mederrhodin A producing *Streptomyces* clone was the first example of antibiotic production using *in vitro* recombinations (18). Gene segments from *S. coelicolor,* the producer of actinorhodin, were introduced into *Streptomyces* sp. AM7161, the producer of a related antibiotic called medermycin. Through this manipulation a clone producing a new medermycin derivative was obtained.

Commercial fermentations are conducted in large bioreactors which are usually referred to as fermentors and are designed for operation in batch, fed-batch, or continuous fermentation (qv) modes (19). The organism is grown at constant temperature in a sterile nutrient medium, containing a carbon source, a nitrogen source, and a small amount of inorganic salts, under controlled conditions of aeration (qv) and agitation. The media for commercial fermentations are usually complex mixtures containing relatively inexpensive nutrients such as molasses, corn steep liquor, or soy meal. A batch fermentation is a closed system containing sterilized medium inoculated with the desired microorganism. Incubation is allowed to proceed under optimal physiological conditions and nothing is added during the course of the fermentation except oxygen, an antifoam agent (see DEFOAMERS), and acid or base to control the pH. In the fed-batch process, nutrients are added in increments as the fermentation progresses. In the continuous fermentation, nutrient solution is added to the bioreactor continuously as an equivalent volume of converted nutrient solution with microorganisms is simultaneously removed from the system for processing.

The batch and fed-batch procedures are used for most commercial antibiotic fermentations. A typical batch fermentor may hold over 150,000 liters. When a maximum yield of antibiotic is obtained, the fermentation broth is processed by purification procedures tailored for the specific antibiotic being produced. Nonpolar antibiotics are usually purified by solvent extraction procedures; water-soluble compounds are commonly purified by ion-exchange methods. Chromatography procedures can readily provide high quality material, but for economic reasons chromatography steps are avoided if possible.

Most of the new commercial antibiotics have resulted from semisynthetic studies. New cephalosporins, a number of which are synthesized by acylation of fermentation-derived 7-aminocephalosporanic acid, are an example. Two orally active cephalosporins called cefroxadine and cephalexin are produced by a synthetic ring-expansion of penicillin V.

Economic Aspects

The world antibiotic market for 1989, estimated at 12.3 billion dollars (20), was the second biggest pharmaceutical category accounting for approximately 9% of the total world pharmaceutical market. Cardiovascular agents (qv) were first. In 1988 antibiotic production in the United States was 13.1×10^6 kg compared to central nervous system depressants and stimulants, 22.3×10^6 kg, gastrointestinal agents and therapeutic nutrients, 45.2×10^6 kg, and vitamins, 17.5×10^6 kg (21). The leading pharmaceutical companies for the manufacture and sales of antibiotics in 1989 are listed in Table 2 (2).

Table 2. Antibiotic Pharmaceutical Manufacturers for 1989

Company	Market share, %[a]
Eli Lilly	11.5
SmithKline Beecham	9.3
Bristol-Myers Squibb	8.6
Shionogi	7.0
Hoffmann-La Roche	6.9
Pfizer	6.4
Merck	6.4
Hoechst	6.3
Takeda	5.1
Fujisawa	4.7
others	27.8

[a]1989 world market estimated at 12.3×10^9 (20).

Japan held 37.5% of the world antibiotic market in 1988, the USA 23.2%, Italy 8.0%, the United Kingdom 5.4%, Germany 3.6%, and other countries 22.3% (20). The disproportionate size of the Japanese market is in part a consequence of the inherent strengths of Japanese industry which include expertise in fermentation technology and intensive chemical manipulation of known structures. In addition, antibiotic prescribing in Japan is extremely popular among doctors as a result of the Japanese reimbursement system.

Uses

Antibacterial Agents. There is a continuous need for new antibiotics primarily as a result of bacterial resistance. There are two aspects to this phenomenon. First, as the more common pathogens are controlled by antibiotics, less common, highly resistant organisms present more prominent health problems. Examples are infections resulting from *Pseudomonas* and *Enterobacter*. Second, and this has been the major driving force in the antibiotic market, particularly in the hospital sector, there is emerging resistance. This type of bacterial resistance can take place through one or more mechanisms including reduced target affinity, target bypass, drug inactivation, reduced permeability of bacterial cells, and drug efflux from the organism. Resistance can be passed from one organism to another through plasmids which are highly mobile segments of bacterial DNA. This transfer of resistance can even occur between different species and genera of bacteria.

The development of new antibiotics to combat resistance, and to provide easier oral administration and improved pharmacokinetics has been successful through synthetic modifications. This approach has been particulary rewarding in the area of β-lactams. The commercial importance of the β-lactams is evident from Table 3 which gives the market share of antibacterials. Fully 62% of the 1989 world antibacterial market belonged to the cephalosporin and penicillin β-lactams (20).

Table 3. 1989 World Antibacterial and Antibiotic Market

Structural classification	Market share, %[a]
cephalosporins	48
penicillins	14
quinolones (synthetic)	11
macrolides	8
aminoglycosides	6
tetracyclines	6
others	7

[a]1989 world market estimated at 12.3×10^9 (20).

Anticancer Agents. In the latter 1980s antibiotics became more important in the treatment of cancer (see CHEMOTHERAPEUTICS, ANTICANCER). In 1987 the antitumor antibiotic adriamycin [*23214-92-84*] was one of the top 50 drugs worldwide. Its revenues were estimated at $231 million (22). Other important antibiotics clinically used for treatment of cancer include bleomycin [*11056-06-7*] and mitomycin C [*50-07-7*]. These compounds are all toxic both to bacteria and to mammalian cells because they interact with DNA. Although antibiotics of this type are not useful as antibacterial agents because of their toxicity and resulting low safety margins, they are useful as antitumor agents. They do show some selective toxicity to rapidly proliferating cancer cells.

Antituberculin Agents. Rifampin [*13292-46-1*], a semisynthetic derivative of rifamycin SV, is a most valuable drug for treatment of tuberculosis, an infection caused by mycobacteria, leprosy, and an expanding range of other infections (23). Cycloserine [*64-41-7*] has been used to a limited extent for treatment of tuberculosis as a reserve drug. Although cycloserine inhibits bacteria by interfering with their cell wall biosynthesis, it has toxic side effects in humans in the form of neurotoxicity. Capreomycin [*11003-38-6*] and to a much lesser extent viomycin [*32988-50-4*], both of which are peptides, have also been used for treatment of this disease.

Antifungal Agents. Antibiotics have been effective in the treatment of fungal diseases (see ANTIPARASITIC AGENTS, ANTIMYCOTICS). Because these types of infections are less prevalent than bacterial infections, the market for antifungal antibiotics has been relatively small. In 1989 the total world market for systemic antifungal agents was $290 million and antifungal antibiotics accounted for approximately 4% (24). However, as the number of immunocompromised patients increases as a result of therapy following organ transplants, acquired immunodeficiency syndrome (AIDS), and anticancer chemotherapy, the need for new and improved antifungal agents may increase significantly. Amphotericin B [*1397-89-3*], a polyene antibiotic, remains the most useful drug for systemic treatment of serious fungal infections despite its inherent toxicity (23). Other clinically used antifungal polyene antibiotics include nystatin and pimaricin [*7681-93-8*], however, applications are essentially limited to superficial infections. They are too toxic for parenteral administration. Griseofulvin [*126-07-8*], which has a structure containing a spiro ring system, is a very good drug for treatment of dermatophyte or ringworm infections.

Antiviral Agents. Although a number of antibiotics have been shown to have some sort of antiviral activity, only vidarabine [*5536-17-4*] (adenine arabinoside) is used clinically against viral infections at this time. As the need for new antiviral agents (qv) increases and new screening procedures are developed, one would expect the discovery of other new effective antiviral antibiotics that could be used safely in human therapy.

Veterinary Applications. Another use for antibiotics is for veterinary applications and for animal feed supplements to promote growth in livestock (see FEEDS AND FEED ADDITIVES). Feed antibiotics used in the United States far surpass all other agricultural applications in terms of kilogram quantities used and approach quantities used in human medicines (25). In 1980 the USA feed antibiotic usage was estimated to be between five and six million kg. The U.S. Council of Agricultural Science and Technology estimates that feed additives save the U.S. consumer approximately $3500 million per year in meat prices, and antibiotic use accounts for most of this.

Feed additive antibiotics achieved sales estimated at $710 million in 1989. This market was dominated by tetracyclines and tylosin [*1401-69-0*] which are believed to account for over 50% of antibiotic sales for this use (26). Other important feed antibiotics include bacitracin, virginiamycin [*11006-76-1*], bambermycin [*11015-37-5*], lincomycin, and spiramycin [*8025-81-8*]. The growth promoting effect of most of these antibiotics is believed to result from subtherapeutic treatment of diseases. In addition, there is another group of antibiotics that effect growth rate and feed efficiency in livestock in the absence of disease. Key products in this area include the polyether ionophores monensin [*17090-79-8*] and lasalocid [*25999-31-9*] and the glycopeptide avoparcin [*37332-99-3*]. In 1989 the total estimated world market sales for these types of growth promoters was $132 million (see GROWTH REGULATORS).

Coccidiosis is a protozoal disease of the intestinal tract of animals that leads to severe loss of productivity and death. The development and widespread use of anticoccidials has revolutionized the poultry industry. The estimated world market for anticoccidial agents in 1989 was $425 million and this was dominated by the polyether ionophore antibiotics monensin, salinomycin [*53003-10-4*], narasin [*55134-13-9*], lasalocid, and maduramicin [*84878-61-5*] (26).

Antimicrobials, a diverse group of products including antibiotics for therapeutic treatment, had estimated worldwide sales of $1500 million in 1989, excluding feed applications, and represent the largest sector in the animal pharmaceutical area. The tetracycline antibiotics represent the largest part of this group having an estimated $475 million in sales. The penicillins with $110 million and macrolides with $250 million in sales in 1989 also represent large segments of this market (26).

Another very important agricultural application for antibiotics is in the area of anthelmintics (see ANTIPARASITIC AGENTS, ANTHELMINTICS). The anthelmintic area is the second largest veterinary pharmaceutical market worldwide. The worldwide anthelmintic market in 1988 was estimated to be $850 million. The largest selling single anthelmintic agent was ivermectin [*30288-86-7*], which had estimated sales of $245 million (26) (see ANTIPARASITIC AGENTS, AVERMECTINS).

Ivermectin is the catalytic reduction product of avermectin, a macrolide containing a spiroketal ring system. Two other related antibiotics having significantly different structural features and biological properties, moxidectin and milbemycin oxime, were more recently introduced into the market. Although these compounds have no antimicrobial activity, they are sometimes referred to as antibiotics because they are derived from fermentation products and have very selective toxicities. They have potent activity against worms or helminths and certain ectoparasites such as mites and ticks.

BIBLIOGRAPHY

"Antibiotics,"in *ECT* 2nd ed., Vol. 2, pp. 533–540, by W. E. Brown, The Squibb Institute for Medical Research; "Antibiotics, Survey," in *ECT* 3rd ed., Vol. 2, pp. 809–819, by H. G. Klein, Klein Medical International.

1. S. E. DeVoe, M. Storms, and K. Haraki, *Antibiotics Properties and Structures Data Base,* maintained at the Medical Research Division, American Cyanamid Co., Pearl River, N.Y., Aug. 1990.
2. A. A. Higton and A. D. Roberts, in B. W. Bycroft, ed., *Dictionary of Antibiotics and Related Substances,* Chapman and Hall, New York, 1988.
3. D. J. Faulkner, *Nat. Prod. Rep.* **5,** 613 (1988); L. S. Shield and K. L. Rinehart, Jr., in M. J. Weinstein and G. H. Wagman, eds., *Journal of Chromatograph Library,* Vol. 15, Elsevier, New York, 1978, p. 309.
4. L. A. Mitscher, *Recent Adv. Phytochem.* **9,** 243 (1975).
5. E. F. Gale, E. Cundliffe, P. E. Reynolds, M. H. Richmond, and M. J. Waring, *The Molecular Basis of Antibiotic Action,* 2nd ed., John Wiley & Sons, Inc., New York, 1981; C. M. Haskell and co-workers, *Cancer Treatment,* 3d ed., W. B. Saunders Co., Philadelphia, 1990, Chapt. 4.
6. E. J. Bassett, M. S. Keith, G. J. Armelagos, D. L. Martin, and A. R. Villanueva, *Science* **209,** 1532 (1980).
7. J. J. Hlavka and J. H. Boothe, eds., *The Tetracyclines,* Springer-Verlag, New York, 1985, Chapt. 1.
8. S. A. Waksman, *The Antibiotic Era,* The Waksman Foundation of Japan, Inc., University of Tokyo Press, Tokyo, Japan, 1975, Chapt. 1 and 2.
9. J. Bérdy and K. Magyar, *Process Biochem.* **3,** 45 (1968).
10. W. M. Heller and C. A. Fleeger, *USAN and USP Dictionary of Drug Names,* 24th ed., United States Pharmacopeial Convention, Inc., Rockville, Md., 1989, pp. 5–12.
11. G. T. Carter and co-workers, *J. Antibiot.* **41,** 519 (1988).
12. J. Bérdy, A. Aszalos, and K. L. McNitt, *CRC Handbook of Antibiotic Compounds,* Vol. 14, CRC Press, Inc., Boca Raton, Fla., 1987.
13. Y. Okami and K. Hotta, in M. Goodfellow, S. T. Williams, and M. Mordarski, eds., *Actinomycetes in Biotechnology,* Academic Press, 1988, Chapt. 2; H. Umezawa, ed., *Index of Antibiotics from Actinomycetes,* Vol. 1 and 2, University Park Press, State College, Pa., 1967 and 1978.
14. A. L. Demain, in L. Ninet, P. E. Bost, D. H. Bouanchaud, and J. Florent, eds., *The Future of Antibiotherapy and Antibiotic Research,* Academic Press, London, 1981, p. 417.
15. K. L. Rinehart, Jr., *Pure Appl. Chem.* **49,** 1361 (1977).
16. W. T. Shier, K. L. Rinehart, Jr., and D. Gottlieb, *Proc. Natl. Acad. Sci. USA* **63,** 198 (1969).
17. S. Gomi and co-workers, *J. Antibiot.* **37,** 1491 (1984).
18. D. A. Hopwood and co-workers, *Nature* **314,** 642 (1985).

19. W. Crueger and A. Crueger, *Biotechnology: A Textbook of Industrial Microbiology,* 2nd ed., Sinauer Associates, Inc., Sunderland, Mass., 1989, Chapt. 5.
20. *County NatWest WoodMac, International Pharmaceutical Service,* County NatWest Securities Limited, London, 1989, Part 4; 1990.
21. *Synthetic Organic Chemicals, United States Production and Sales, 1988, United States International Trade Commission,* U.S. Government Printing Office, Washington, D.C., 1989.
22. *Scrip Yearbook 1989,* PJB Publications Ltd., Richmond, Surrey, UK, 1989.
23. A. Kucers, N. Mck. Bennett, and R. J. Kemp, *The Use of Antibiotics,* 4th ed., William Heinemann Medical Books, London, 1987, pp. 914, 1418–1528.
24. *IMS Data Base,* Intercontinental Medical Statistics Limited, Middlesex HA55HQ, UK.
25. *Animal Pharm, Feed Additives, The Market,* V & O Publications Ltd., Richmond, Surrey, UK, 1987.
26. *County NatWest WoodMac, Animal Health & Nutrition Service,* County NatWest Securities Limited, London, 1990.

DONALD BORDERS
American Cyanamid Company

AMINOGLYCOSIDES

The term *aminoglycoside* is commonly used to refer to members of the class of antibacterial antibiotics, the structures of which are derived from D-streptamine [*488-52-8*] $C_6H_{14}N_2O_4$ (**1**, R = OH), D-2-deoxystreptamine [*2037-48-1*] $C_6H_{14}N_2O_3$ (**1**, R = H), or closely related compounds. The terms *aminocyclitol* and *aminoglycoside-aminocyclitol* are also sometimes used to identify this group of compounds.

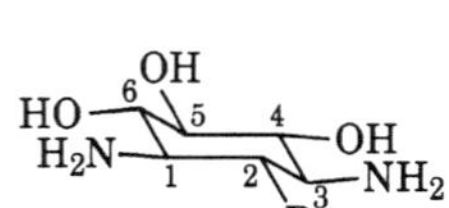

(**1**)

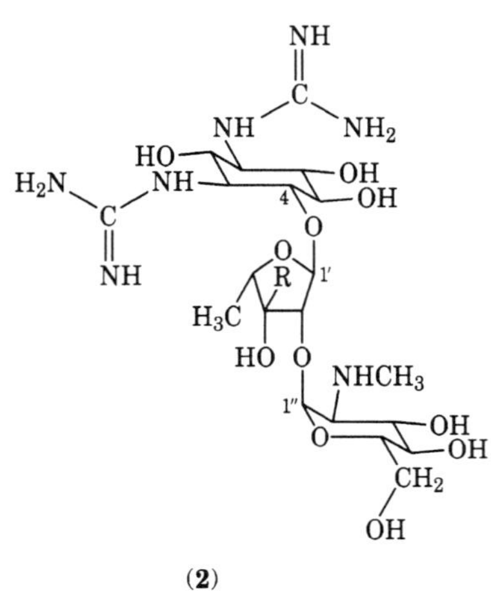

(**2**)

Historically, the first example of an aminoglycoside antibiotic is generally considered to be streptomycin [*57-92-1*] $C_2H_{39}N_7O_{12}$ (**2**, R = CHO), or streptomycin sulfate [*3810-74-0*] (chemical name: *O*-2-deoxy-2-(methylamino)-α-L-glucopyranosyl-(1→2)-*O*-5-deoxy-3-*C*-formyl-α-L-lyxofuranosyl-(1→4)-*N,N′*-bis(aminoiminomethyl)-L-streptamine) isolated in 1944 from a strain of *Streptomyces griseus* (1–4). This discovery, a milestone in the history of antibacterial chemotherapy (5), was the starting point for research and development activities that continue to be important in the management of bacterial infectious disease.

Aminoglycosides in Medical Usage

In 1991 the following aminoglycosides were most important in medical practice. Clinical experience is most extensive for the first four.

1. Gentamicin [*1403-66-3*] or gentamicin sulfate [*1405-41-0*], is a mixture of gentamicin C_1 [*25876-10-2*] $C_{21}H_{43}N_5O_7$ (**3**, R = R′ = CH_3), gentamicin C_2 [*25876-11-3*] $C_{20}H_{41}N_5O_7$ (**3**, R = CH_3, R′ = H), and gentamicin C_{1a} [*26098-04-4*] $C_{19}H_{39}N_5O_7$ (**3**, R = R′ = H) isolated from *Micromonospora purpurea* and related species (6–9). Note that compounds obtained from *Micromonospora* species are given names ending in -micin; those isolated from *Streptomyces* end in -mycin. The chemical name for gentamicin C_{1a} is *O*-3-deoxy-4-*C*-methyl-3-(methylamino)-β-L-arabinopyranosyl-(1→6)-*O*-[2,6-diamino-2,3,4,6-tetradeoxy-α-D-*erthryo*-hexopyranosyl-(1→4)]-2-deoxy-D-streptamine.

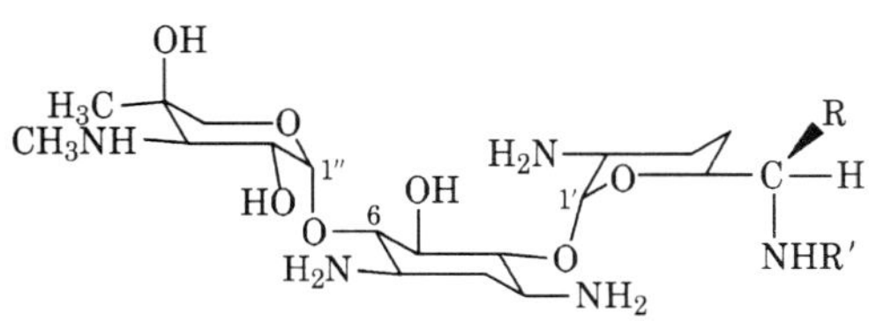

(**3**)

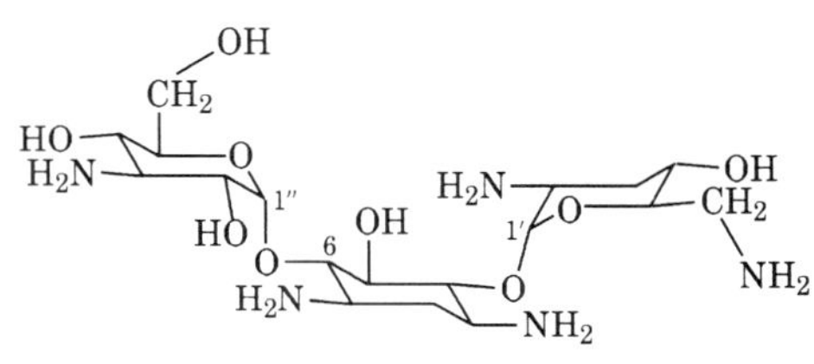

(**4**)

2. Tobramycin [*32986-56-4*] $C_{18}H_{37}N_5O_9$, or tobramycin sulfate [*79645-27-5*], also called nebramycin factor 6, (**4**) (10–16), was isolated from *Streptomyces tenebrarius* and has the chemical name *O*-3-amino-3-deoxy-α-D-glucopyranosyl-(1→6)-*O*-[2,6-diamino-2,3,6-trideoxy-α-D-*ribo*-hexopyranosyl-(1→4)]-2-deoxy-D-streptamine.
3. Amikacin [*37517-28-5*] $C_{22}H_{43}N_5O_{13}$, or amikacin sulfate [*39831-55-5*] (**5**, R = (*S*)-$COCHOHCH_2CH_2NH_2$) is a semisynthetic derivative (17) of kanamycin A [*59-01-8*] $C_{18}H_{36}N_4O_{11}$ (**5**, R = H) which is produced by *Streptomyces kanamyceticus* (18–22). Amikacin has the chemical name (*S*)-*O*-3-amino-3-deoxy-α-D-glucopyranosyl-(1→6)-*O*-[6-amino-6-deoxy-α-D-glucopyranosyl-(1→4)-*N*[1]-(4-amino-2-hydroxy-1-oxobutyl)-2-deoxy-D-streptamine.

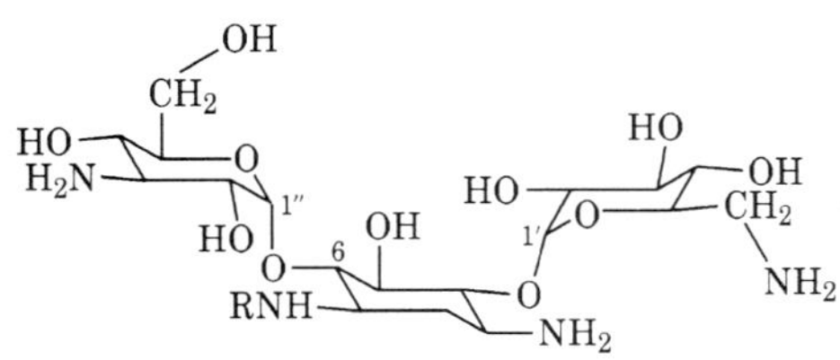

(**5**)

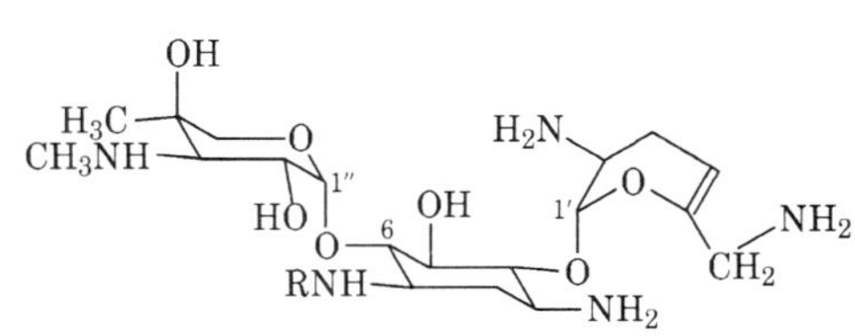

(**6**)

4. Netilmicin [*56391-56-1*] $C_{21}H_{41}N_5O_7$, or netilimicin sulfate [*56391-57-2*], (**6**, R = CH_2CH_3) is a semisynthetic derivative (23) of sisomicin [*32385-11-8*] $C_{19}H_{37}N_5O_7$, (**6**, R = H). Sisomicin is produced by *Micromonospora inyoesis* (24–26). Netilmicin has the chemical name *O*-3-deoxy-4-*C*-methyl-3-(methylamino)-β-L-arabinopyranosyl-(1→6)-*O*-[2,6-diamino-2,3,4,6-tetradeoxy-α-D-*glycero*-hex-4-enopyranosyl-(1→4)]-2-deoxy-N^1-ethyl-D-streptamine.
5. Dibekacin [*34493-98-6*], $C_{18}H_{37}N_5O_8$, (**7**, R = R′ = H), chemical name *O*-3-amino-3-deoxy-α-D-glucopyranosyl-(1→6)-*O*-[2,6-diamino-2,3,4,6-tetradeoxy-α-D-*erythro*-hexopyranosyl-(1→4)]-2-deoxy-D-streptamine, is a semisynthetic derivative (27,28) of kanamycin B [*4696-78-8*] $C_{18}H_{37}N_5O_{10}$ (**7**, R = R′ = OH) (*O*-3-amino-3-deoxy-α-D-glucopyranosyl-(1→6)-*O*-[2,6-diamino-2,6-dideoxy-α-D-glucopyranosyl-(1→4)]-2-deoxy-D-streptamine) (29,30).

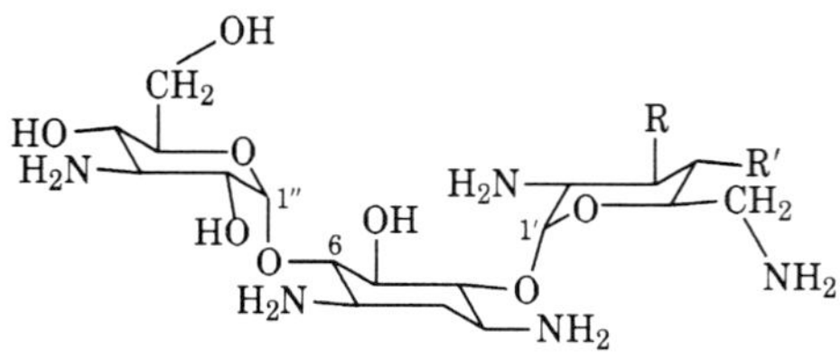

(**7**)

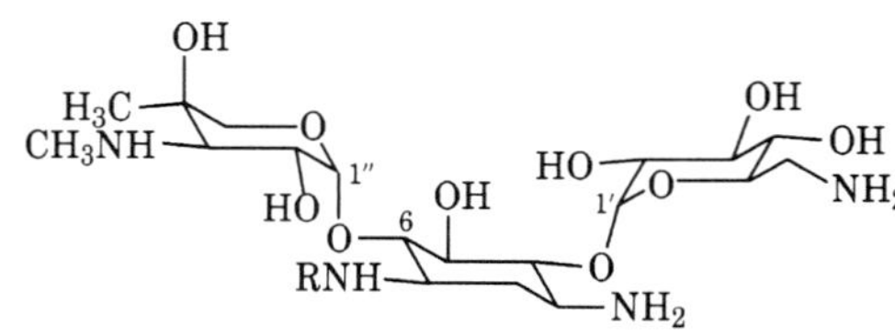

(**8**)

6. Isepamicin [*58152-03-7*], $C_{22}H_{43}N_5O_{12}$, (**8**, R = (*S*)-COCHOHCH$_2$NH$_2$) ((*S*)-*O*-6-amino-6-deoxy-α-D-glucopyranosyl-(1→4)-*O*-[3-deoxy-4-*C*-methyl-3-(methylamino)-β-L-arabinopyranosyl-(1→6)]-N^1-(3-amino-2-hydroxy-1-oxopropyl)-2-deoxy-D-streptamine), which was introduced in 1988, is a semisynthetic derivative (31) of gentamicin B [*36889-15-3*] $C_{19}H_{38}N_4O_{10}$ (**8**, R = H) (32).
7. Neomycin [*1404-04-1*] (33) or neomycin sulfate [*1405-10-3*], isolated from *Streptomyces fradiae* (34–36), is sometimes used for gut sterilization and other nonsystemic applications. Neomycin is a mixture of neomycin B [*119-04-0*] $C_{23}H_{46}N_6O_{13}$ (**9**) (*O*-2,6-diamino-2,6-dideoxy-α-D-glucopyranosyl-(1→4)-*O*-[*O*-2,6-diamino-2,6-dideoxy-β-L-idopyranosyl-(1→3)-β-D-ribofuranosyl-(1→5)]-2-deoxy-D-streptamine) and neomycin C [*66-86-4*] $C_{23}H_{46}N_6O_{13}$. Neomycin C is 5‴-epi-neomycin B. Neomycin B is the main component.

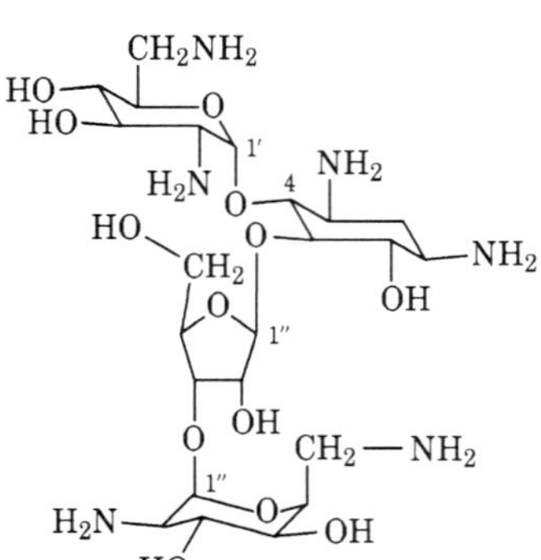

(**9**)

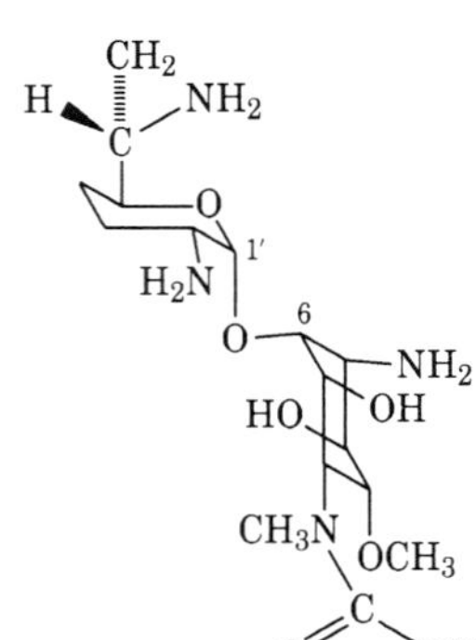

(**10**)

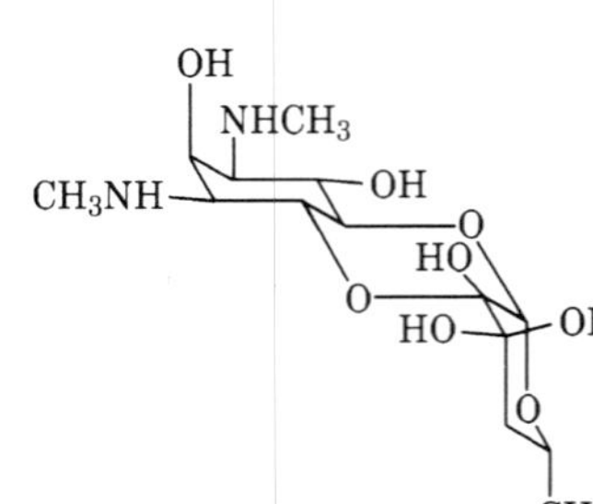

(**11**)

8. Astromicin [*55779-06-1*], $C_{17}H_{35}N_5O_6$, (**10**) (4-amino-1-[(aminoacetyl)methylamino]-1,4-dideoxy-3-*O*-(2,6-diamino-2,3,4,6,7-pentadeoxy-β-L-*lyxo*-heptopyranosyl)-6-*O*-methyl-L-*chiro*-inositol), also known as fortimicin A, is a member of a complex isolated from *Micromonospora olivoasterospora* (37–40). Although astromicin has a structure that is considerably different from those of the other aminoglycosides listed in this section, it is still an aminocyclitol joined to an aminosugar via a glycosidic linkage. Astromicin was introduced in limited markets in 1987.
9. Spectinomycin [*21736-83-4*], $C_{14}H_{24}N_2O_7$, or spectinomycin pentahydrate [*22189-32-8*] (**11**) (decahydro-4a,7,9-trihydroxy-2-methyl-6,8-bis(methylamino)-4*H*-pyrano[2,3-b][1,4]benzodioxin-4-one) is not strictly an aminoglycoside, but its structure does contain a modified streptamine moiety. The ketone group on position 4 exists as the hydrate. Spectinomycin was isolated from *Streptomyces spectabilis* (41–43). Although it has a fairly broad antibacterial spectrum, medical usage is confined to the treatment of *Neisseria gonorrhoea* infections and it is the agent of choice for resistant strains of that species.

Among the older aminoglycoside derivatives, kanamycin A and sisomicin were, at one time, a significant part of medical practice, but have now been largely replaced by the compounds listed in Table 1. Streptomycin is still used in a few restricted situations.

Table 1. Annual Worldwide Estimated Sales of Aminoglycosides

Aminoglycoside			Annual worldwide estimated sales, U.S.$ $\times 10^6$		
Generic name	Trade name	Company	1988	1989	1990
gentamicin	Garamycin and others[a]	Schering-Plough and others[a]	120	130	140
tobramycin	Nebcin	Lilly	170	150	150
amikacin	Amikin	Bristol-Myers Squibb	190	190	200
netilmicin	Netromycin	Schering-Plough	130	130	130
dibekacin	Panimycin	Meiji Seika	60	50	50
isepamicin	Isepacin	Schering-Plough	20	45	45
	Exacin	Toyo Jozo			
Other aminoglycosides			120	95	85
Total			*810*	*790*	*800*

[a]Gentamicin is off patent, and is currently marketed by several companies.

Medical and Biological Properties

General Antibacterial Properties. In the clinical control of bacterial infectious disease, the aminoglycosides gentamicin, tobramycin, amikacin, netilmicin, and to a lesser extent, dibekacin and isepamicin are most commonly used for the treatment of serious infections involving aerobic or facultative gram-negative bacilli, especially in the compromised host. This usage is discussed in the literature (44–51).

Overall, these aminoglycosides are approximately equally effective for the clinical treatment of infections involving susceptible strains of *Eschericia coli, Klebsiella* sp., *Enterobacter* sp., *Citrobacter* sp., *Serratia* sp., *Proteus mirabilis, Morganella morganii, Proteus vulgaris, Proteus rettgeri, Providencia stuartii, Pseudomonas aeruginosa,* and the gram-positive species *Staphylococcus aureus.* Other gram-positive bacteria, such as *Streptococcus* sp., and anaerobes are not generally susceptible to aminoglycosides (52–56). There are some differences in the *in vitro* activities of the various aminoglycosides against these species. Tobramycin (**4**) tends to have better *in vitro* activity against *P. aeruginosa* than gentamicin (**3**), but it has been difficult to show that these differences are reflected in clinical performance. There are also differences in potency. Amikacin (**5**, R = (*S*)-$COCHOHCH_2CH_2NH_2$), for example, is generally less potent than the others, but a decreased toxicity allows the dose to be adjusted for comparable efficacy. Thus, especially in the case of the four aminoglycosides for which there is the greatest amount of clinical experience, ie, gentamicin, tobramycin, amikacin, and netilmicin (**6**, R = CH_2CH_3), there does not appear to be a significant difference in the outcome of treatment of infections (57) due to sensitive strains of these bacteria.

Particular advantages of the aminoglycosides are that, in general, the bactericidal concentration is not significantly different from the growth inhibitory concentration and, furthermore, the bactericidal effect is rapid and concentration dependent (58). In very serious infections such as occur in febrile neutropenic patients or in enterococcal endocarditis, a combination of an aminoglycoside with a β-lactam antibiotic (see ANTIBIOTICS, β-LACTAMS) or vancomycin (see ANTIBIOTICS, GLYCOPEPTIDES) is often used. In addition to extending the spectrum of coverage, such combinations are often found to be synergistic. They have a greater inhibitory effect than would be predicted from additivity alone (59–61). However, under sufficiently concentrated conditions, β-lactams and aminoglycosides can inactivate each other by acylation of one of the amino groups (62).

Bacterial Resistance Mechanisms. The occurrence of antibiotic resistant bacterial strains is an important medical problem. This is especially true for nosocomial infections (63–65). The incidence and type of resistance are highly variable phenomena that can change with time and location. In some cases, these strains can arise in response to the environmental pressure of aminoglycoside usage. However, the incidence of aminoglycoside resistance overall has remained relatively stable throughout the 1980s and into the 1990s (51). Of the aminoglycosides in current use, amikacin and isepamicin (**8**, R = (*S*)-$COCHOHCH_2NH_2$) are least susceptible to the bacterial resistance mechanisms, netilmicin and dibekacin (**7**, R = R′ = H) are intermediate, and gentamicin and tobramycin are most susceptible (66). Resistance to streptomycin is widespread, and its use is currently confined primarily to infections caused by *Mycobacterium tuberculosis, Yersinia pestis,* and *Francisella tularensis.*

Resistance to the antibacterial action of aminoglycosides is mediated primarily by three mechanisms: (*1*) Changes in the bacterial ribosome such that the affinity for the antibiotic is significantly decreased. This mechanism appears to be relatively rare except for streptomycin (67). (*2*) Significant reduction in the rate at which the aminoglycoside passes through the cell wall and cell membrane into the cytoplasm. This phenomenon is presumably responsible for the intrinsic resistance of *Streptococcus* sp. and anaerobic bacteria to the action of aminoglycosides. It may also be an important component of acquired resistance (68). (*3*) Inactivation of the aminoglycoside by plasmid-borne enzymes which catalyze acetylation, phosphorylation, or adenylylation reactions. This is the most common of the resistance mechanisms, and is a significant clinical problem. A large number of aminoglycoside acetyltransferase (AAC), aminoglycoside phosphotransferase (APH), and aminoglycoside nucleotidyltransferase (ANT), also referred to as aminoglycoside adenylyltransferase (AAD), enzymes and related isozymes have been found. These enzymes differ in the substrates on which they operate, their catalytic efficiency, and in the product produced (68–73).

The most clinically significant of the aminoglycoside inactivating enzymes seem to be (74): ANT(2″), an enzyme that catalyzes the transfer of an adenylyl group to the 2″-hydroxy of gentamicin, tobramycin, dibekacin, and kanamycins A and B; AAC(6′)-(various isozymes), enzymes that catalyze the transfer of an acetyl group to the 6′-amino of gentamicin, tobramycin, amikacin, isepamicin, netilmicin, kanamycins A and B, sisomicin, dibekacin, and neomycin; and AAC(3)-(various isozymes), enzymes that catalyze the acetylation of the 3-amino moiety of gentamicin, tobramycin, netilmicin, kanamycins A and B, dibekacin, neomycin, and astromicin. If all three of these enzymes were to operate on tobramycin (**4**), the result would be structure (**12**) (20).

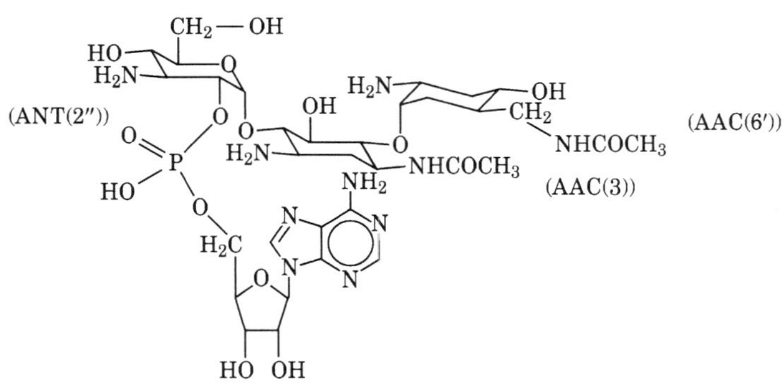

(**12**)

Other enzymes that may play a role clinically are AAC(2′), important in some *Providencia* and *Proteus* species, APH(3′), and APH(3″) (71). Others that have been observed include APH(2″), APH(5″), APH(6), ANT(4′), ANT(3″), and ANT(6). It has been suggesed that, for effective resistance, an organism must have both impaired aminoglycoside transport and levels of aminoglycoside inactivating enzymes such that the rate of inactivation in the cytoplasm is greater than the rate of influx (68).

Pharmacokinetics. The aminoglycosides are not reliably absorbed following oral dosing, so they are administered primarily by intravenous infusion or intramuscular injection (75–77). Distribution throughout the vascular and interstitial space occurs fairly rapidly. Because these antibiotics are polycationic and relatively large, biological membranes are not readily crossed, unless transport mechanisms exist as in the kidney proximal tubule cells, and intracellular levels are generally low. Levels in the cerebrospinal fluid (CSF), bronchial secretions, and saliva are also low. Aminoglycoside excretion takes place almost entirely via the kidneys. Most of the compound is eliminated unmetabolized in a relatively rapid beta phase having a half-life of 2–3 h. A small portion, however, is excreted in a slow gamma phase, half-life 35–200 h. Because of the relatively short beta-phase half-life, dosing is usually two or three times per day. Efficient clearance is dependent on glomerular function and, when glomerular filtration rate is reduced because of age or kidney disease, particular care in dosing must be exercised to prevent accumulation of toxic aminoglycoside levels (78). Some studies have shown that, for certain clinical applications, an aminoglycoside can be administered once a day and efficacy is unchanged while toxicity is decreased (79–82). A possible contributor to the success of this strategy is the observation of a post-antibiotic effect (83), in which growth inhibition of some bacterial species continues for a period of time after the serum concentration of the aminoglycoside drops below the presumed inhibitory level.

A novel approach to the modification of aminoglycoside pharmacokinetics is under investigation (84). Administration of gentamicin encapsulated in egg phosphatidylcholine liposomes has been found to lead to a longer half-life and much higher spleen and liver levels for the gentamicin component. This formulation is undergoing clinical study (85).

Toxicology. A primary limiting factor in the clinical use of the aminoglycosides is the potential toxicity. This toxicity prevents the use of significantly increased doses to cover difficult infections. Also, serum aminoglycoside concentrations can reach toxic levels even when normal doses are used because of variation in glomerular filtration efficiency. All the aminoglycosides in medical practice are capable of causing these adverse reactions (51,57,77,86).

The principal aminoglycoside toxicities are neuromuscular paralysis, ototoxicity, and nephrotoxicity. Neuromuscular paralysis is a relatively rare complication resulting from high aminoglycoside concentrations at the neuromuscular junctions following, for example, rapid bolus intravenous injection or peritoneal instillation, rather than the normal intravenous infusion. The mechanism apparently involves an inhibition of both the presynaptic release of acetylcholine and the acetylcholine postsynaptic receptors (51).

Both auditory (cochlear) (high frequency hearing loss), and vestibular (nausea, vertigo) dysfunctions are experienced (87). Whereas the ototoxic responses are not life-threatening, these effects are especially significant because they are often irreversible and cumulative. The incidence is estimated at from 0.5 to 25%. The wide range is partially a reflection of the difficulty in measuring this kind of toxic effect (51). The mechanisms involved in these reactions are not fully understood, but they apparently involve damage to hair cells in the organ of Corti and ampullar cristae, neither of which can regenerate. All of the commonly used aminoglycosides can cause both cochlear and vestibular effects, but amikacin is more likely to cause the cochlear type of toxicity (88,89) and particularly severe auditory dysfunction has led to the restriction of neomycin use to topical applications. Overall, however, the incidence of clinical ototoxicity seems to be comparable for each of the different aminoglycosides (51,57), with the possible exception of a reduced incidence for netilmicin (86,89,90). In a comparative study using a guinea pig model, the increasing order of cochlear toxicity was found to be netilmicin, dibekacin, amikacin, tobramycin, gentamicin; the order of vestibular toxicity was amikacin, netilmicin, tobramycin, dibekacin, gentamicin (90).

Estimates of the incidence of nephrotoxicity range from 5 to 25%. In most cases, the observed effects are a mild-to-moderate decrease in glomerular filtration rate accompanied by the urinary excretion of renal tubular enzymes. The effects are usually reversible, and their likelihood is reduced by monitoring serum levels and glomerular filtration rate and adjusting the dosing accordingly. However, the possibility of a serious reaction, particularly in the compromised patient, represents a significant deterrent to aminoglycoside usage. It has been estimated that nephrotoxic reactions add an average of $183 in hospital costs for each patient who receives an aminoglycoside (91). Overall, gentamicin and tobramycin may be somewhat more likely to be nephrotoxic than amikacin and netilmicin (51,89), but the differences are generally not significant (92). Preliminary experimental evidence in animal models indicate that isepamicin may be less nephrotoxic (93,94), but clinical evidence is not yet available. Spectinomycin does not have the nephrotoxic and ototoxic liabilities of the aminoglycosides nor does it have the breadth of antibacterial spectrum typical of the aminoglycosides.

Significant progress has been made in recent years toward understanding the mechanisms involved in the nephrotoxic effects of the aminoglycosides (52,77). The scheme which has evolved can be outlined as follows: (*1*) aminoglycoside in the serum is filtered through the glomerulus into the lumen of the kidney tubule; (*2*) an initial charge–mediated binding of the aminoglycoside to the brush border membrane of the kidney proximal tubule cells is followed by a pinocytotic uptake into those cells, possibly utilizing the basic amino acid transport system (95); (*3*) after uptake, the pinocytotic vesicles fuse with preexisting lysosomes, leading to a rapid and substantial accumulation of aminoglycoside in the lysosomes (96); (*4*) in the lysosome, the aminoglycoside appears to inhibit phospholipases, resulting in the accumulation of phospholipids in the form of myeloid bodies (phosphlipidosis) (97,98). Extralysosomal effects are also observed, particularly with respect to mitochondrial function (99–101); (*5*) in a step the mechanism of which is unclear, phospholipidosis and the other effects are followed by cell necrosis, then regeneration (102). This may involve destabilization of the lysosomal membrane or interference with membrane recycling processes. The extent of kidney damage can be viewed as a result of competitive rates of cell necrosis and regeneration (97).

A novel approach to the problem of aminoglycoside nephrotoxicity has been to search for compounds that can inhibit toxicity without compromising efficacy. A number of agents have been reported to reduce aminoglycoside toxicity in animal models; the most extensively studied of these is sodium polyaspartate (103–107).

Mechanism of Antibacterial Action. In spite of the fact that the antibacterial activity of the aminoglycosides has been known since the 1940s, the mechanisms involved are still incompletely understood. Numerous reviews have appeared (eg, 108–113) and the sequence of events seems to be as outlined below.

First, the cationic aminoglycoside binds to anionic groups on the bacterial cell surface. This step is both energy-independent and nonspecific. Then, the molecule passes through the cell wall via porins or self-promoted defects in the wall, traverses the periplasmic space, and arrives at the cell membrane. The next step has been termed the energy-dependent phase I and can be inhibited by compounds that block electron transport. According to one model, the aminoglycoside binds to a transporter molecule and, once a threshold electrical potential is reached, is translocated into the cytoplasm with the involvement of the electron-transport chain. Uptake, to the extent that it occurs during this phase, is small. In the energy-dependent phase II, there is an accelerated and irreversible aminoglycoside uptake into the cytoplasm. This step can be inhibited by inhibitors of both electron transport and protein synthesis, and is dependent on an interaction of the aminoglycoside with the ribosomes. One of the several suggested models for this step envisages a cell membrane that becomes increasing leaky because of the incorporation of faulty proteins resulting from aminoglycoside-induced errors in translation (114,115). In the cytoplasm, the aminoglycoside binds to the ribosome leading to a misreading of the genetic code (116), and the consequent production of abnormal proteins, and, at higher concentrations, an inhibition of protein synthesis. Cell death, the cause of which is uncertain, follows.

Aminoglycoside Biosynthesis. The biosynthesis of the aminoglycosides has been extensively studied and reviewed (117–119). Perhaps the most interesting aspect is the biosynthesis of 2-deoxystreptamine (**1**, R = H), in which the C-1 and C-6 of a D-glucose molecule become the C-1 and C-2 of 2-deoxystreptamine by way of the intermediate 2-deoxy-*scyllo*-inosose. The details of this conversion are still unclear.

Structure-Activity Relationships Among Aminoglycoside Derivatives

The aminoglycosides possess properties that make them valuable for the control of bacterial infectious disease, but they also have distinct limitations, especially in the areas of toxicity and susceptibility to bacterial resistance mechanisms. Thus there has been a large amount of research aimed at reducing the limitations while maintaining the advantages generally by introducing modifications into the basic aminoglycoside molecular structure. In general, three approaches have been used to generate novel aminoglycoside structures. (*1*) Search for microorganisms that produce novel structures directly. This approach has resulted in gentamicin and tobramycin. (*2*) Chemically modify structures of molecules available from microorganisms ie, undertake semisynthesis. Netilmicin, amikacin, dibekacin, and isepamicin have resulted from this approach. (*3*) Generate microorganism mutants that require a modifiable exogenous substrate for the biosynthesis of the aminoglycoside. Because of the difficulty of the chemistry, total synthesis (120,121) has not yet been a major contributor of modified aminoglycosides.

The number of aminoglycoside derivatives reported over the years is very large. Much of the work has been summarized in reviews (120,122–125). Only the structural subclasses and representative derivatives are discussed here. Some of the structures are given in Table 2.

Table 2. Structures Relating to Aminoglycoside Derivatives

Name	CAS Registry Number	Molecular formula	Structure number	Structure
streptidine	[*85-17-6*]	$C_8H_{18}N_6O_4$	(**13**)	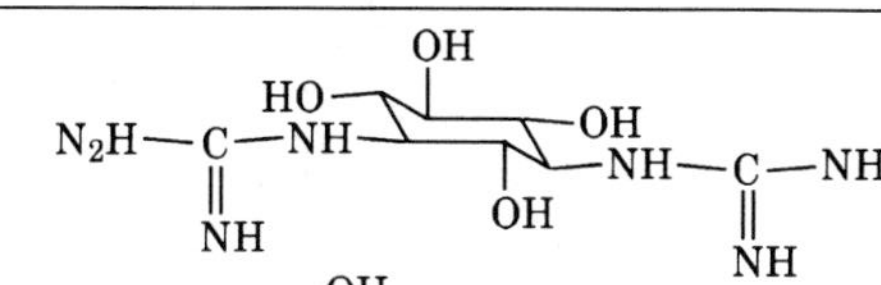
apramycin[a]	[*37321-09-8*]	$C_{21}H_{41}N_5O_{11}$	(**14**)	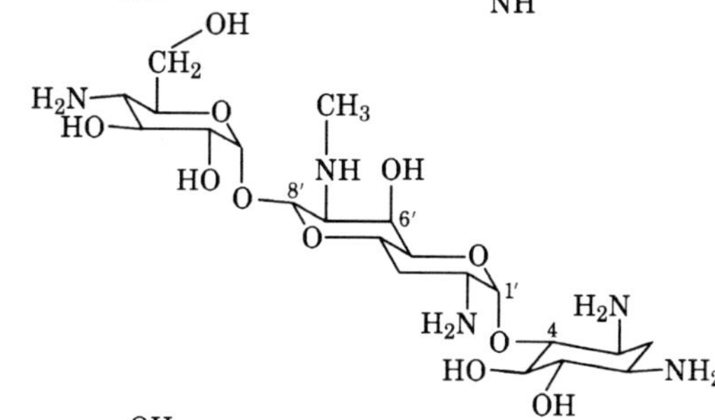
hygromicin B	[*31282-04-9*]	$C_{20}H_{37}N_3O_{13}$	(**15**)	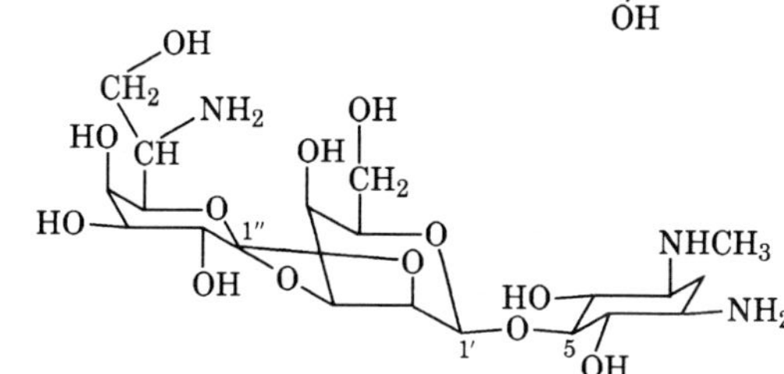
ribostamycin butirosin B	[*25546-65-0*] [*34291-03-7*]	$C_{17}H_{34}N_4O_{10}$ $C_{21}H_{41}N_5O_{12}$	(**16**, R = H) (**16**, R = (*S*)-COCHOHCH$_2$CH$_2$NH$_2$)	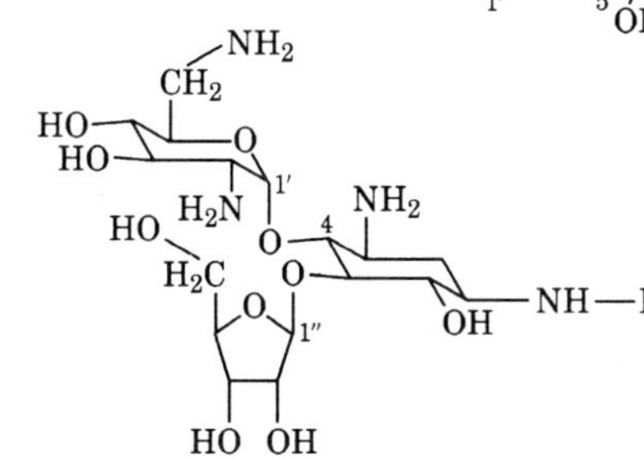

[a]Also known as nebramycin factor 2.

Derivatives of Streptidine. The first aminoglycoside to be isolated was streptomycin (**2**, R = CHO), which is a derivative of streptidine (**13**) shown in Table 2. A number of derivatives have been prepared (126), the most interesting of which was dihydrostreptomycin [*128-46-1*] $C_{21}H_{41}N_7O_{12}$ (**2**, R = CH_2OH), which retained the antimicrobial spectrum of its precursor. This derivative was removed from clinical usage because of a high incidence of ototoxicity. Even minimal modification of the guanidine groups in dihydrostreptomycin led to loss of activity, although a single methyl on the 1-guanidino was tolerated (127). The 3″-deoxy- and 3″-epidihydrostreptomycin derivatives were quite active and resistant to some inactivating reactions (128).

Derivatives Containing a 4-Monosubstituted-2-deoxystreptamine Moiety. Many of these compounds can be viewed as fragments derived from the removal of the *X*-substituent from a 4,*X*-disubstituted-2-deoxystreptamine aminoglycoside, eg, neamine [*3947-65-7*] $C_{12}H_{26}N_4O_6$ which is derived from neomycin B (**9**) by hydrolysis at the 5-position, paromamine [*534-47-4*], $C_{12}H_{25}N_3O_7$ derived from paromomycin I or II (discussed below) by hydrolysis at the 5-position, tobramine [*34051-04-2*] $C_{12}H_{26}N_4O_5$ (nebramine) derived from tobramycin (**4**) by hydrolysis at the 6-position, and gentamine C_{1a} [*35025-95-7*], $C_{12}H_{26}N_4O_4$, derived from gentamicin C_{1a} (**3**, R = R′ = H) by hydrolysis at the 6-position. Some of these compounds are found as fermentation products; alternatively, they can be obtained by chemical hydrolysis of the parent antibiotic. In general, however, the antibacterial activity is poor. An exception is apramycin (**14**), isolated from *Streptomyces tenebrarius,* which has respectable antibacterial activity and is resistant to the actions of some aminoglycoside-inactivating enzymes (129,130). For more details concerning this subclass of derivatives, see references 122 and 125.

Derivatives Containing a 5-Monosubstituted-2-deoxystreptamine Moiety. Few derivatives of this type are known. One is hygromycin B (**15**) isolated from *Streptomyces hygroscopicus* and *Streptomyces eurocidicus* (131–133). The antibacterial activity is poor, but hygromycin B is reported to have anthelmintic activity (see ANTIPARASITIC AGENTS, ANTHELMINTICS).

Derivatives Containing a 4,5-Disubstituted-2-deoxystreptamine Moiety. The first of this subclass to be isolated were the neomycins (B (**9**) and C), followed by paromomycin I [*7542-37-2*] $C_{23}H_{45}N_5O_{14}$, and paromomycin II [*51795-47-2*] $C_{23}H_{45}N_5O_{14}$. The paromomycins are produced by a *Streptomyces rimosus* strain and have the same structures as neomycins B and C, respectively, except that the 6′-substituent is a hydroxy instead of an amino group (33,134–136). A closely related compound is lividomycin B [*37636-51-4*] $C_{23}H_{45}N_5O_{13}$, which has the structure of neomycin B but the 3′-position is unsubstituted, instead of having a hydroxy group, and the 6′-substituent is a hydroxy (137–139). All these derivatives have significant antibacterial activity, but potency is lower than that of gentamicin and all are inactivated by AAC(3). The paromomycins are inactive against *P. aeruginosa* strains.

A related series of aminoglycosides have a structure similar to neomycin B, except that the sugar residue attached to the ribofuranosyl ring at 3″ is missing. The parent compound ribostamycin (**16**, R = H) was isolated from *Streptomyces ribosidificus* (140,141). A modified version of this molecule, butirosin B (**16**, R = (*S*)-$COCHOHCH_2CH_2NH_2$) was isolated from *Bacillus circulans*. Butirosin B was the first aminoglycoside isolated from a bacterium (142–145) and it was found that, by virtue of the (*S*)-4-amino-2-hydroxybutyryl substituent on the 1-amino group, an expanded antibacterial spectrum and resistance to the inactivating enzymes AAC(3) and APH(3′) had been obtained. This observation indicated that a modification at the 1-position on the molecule could influence enzyme-catalyzed events occurring at the 3- and 3′-positions and ultimately led to the preparation of amikacin. The (*S*)-3-amino-2-hydroxypropionyl substituent was found to have a similar effect, resulting in isepamicin. This work showed that structure manipulation could affect the biological properties of an aminoglycoside in a significantly favorable way, and thus was an important impetus to further structural modifications within the 4,5-disubstituted-2-deoxystreptamine class (122,125).

Derivatives Containing a 4,6-Disubstituted-2-deoxystreptamine Moiety. Examples of this class are the kanamycin complex, consisting primarily of A (**5**, R = H), B (**7**, R = R′ = OH), and C. Kanamycin C has the same structure as B except that hydroxy replaces the 6′-amino group (146). The members of the nebramycin complex are closely related structurally to the kanamycins (125). The most important are nebramycin factor 6, which is tobramycin (**4**), and nebramycin factor 5, which is kanamycin B (**7**, R = R′ = OH). Kanamycin A was introduced into medical practice and still sees limited use, even though resistance to it is widespread and activity against *P. aeruginosa* strains is poor. Structural modifications of the kanamycins have been extensively investigated (120) and the most notable outcome of this work were the discoveries of amikacin (**5**, R = (*S*)-$COCHOHCH_2CH_2NH_2$) and dibekacin (**7**, R = R′ = H). Many qualitative structure–activity correlations were found from these studies (120,122,123).

In general, acylation of any of the 3-, 2′-, 6′-, or 3″-amino groups greatly reduces antibacterial activity (147). However, acylation of the 1-amino group with one of a very limited number (148,149) of acyl groups, most notably, (*S*)-4-amino-2-hydroxybutyryl and (*S*)-3-amino-2-hydroxypropionyl, confers greatly improved resistance to bacterial inactivation reactions catalyzed by APH(3′), ANT(2″), and AAC(3), but not to others such as those of some isozymes of AAC(6′) and ANT(4′). The 1-*N*-(2-aminoethanesulfonyl) derivatives have also been reported to have similar properties (150). Resistance to AAC(6′) was improved somewhat in the formimidoyl and acetimidoyl derivatives of amikacin, but at a cost in potency (151). Interestingly, the 1-*N*-palmitoyl-3″-trifluoroacetyl derivatives of several aminoglycosides have been reported to have antiviral activity (152).

Monoalkylation using small alkyl groups of the 1-, 2′- and 3″-amines tends to reduce potency somewhat, whereas alkylation of the 3-amino group reduces activity considerably (147). 1-*N*-Alkylation, as in the case of 1-*N*-acylation, can lead to resistance to some bacterial resistance reactions, eg, 1-*N*-ethyl kanamycin A (153) has some resistance to APH(3′), ANT(2″), and AAC(3). Butakacin [*59733-86-7*] $C_{22}H_{45}N_5O_{12}$, the 1-*N*-(*S*)-2-hydroxy-4-aminobutyl derivative of kanamycin A (154), has antimicrobial properties similar to amikacin. A similar effect was seen with the 1-*N*-(1,3-dihydroxy-2-propyl) derivative of kanamycin B, propikacin [*66887-96-5*] $C_{21}H_{43}N_5O_{12}$ (155). Methylation of the 6′-amine reduces susceptibility to AAC(6′) (156).

Removal of the hydroxyl groups from the 3′-, 4′-, 5-, or 6″-positions does not severely compromise antibacterial activity (157) and, in the case of 3′ and 4′, can protect from the action of APH(3′) and ANT(4′). This type of modification is incorporated naturally in tobramycin (**4**) and semisynthetically in dibekacin (**7**, R = R′ = H), resulting in significantly improved activity against *P. aeruginosa*. Perhaps surprisingly, 5,2′,3′,4′,4″,6″-hexadeoxykanamycin A is fully active, and even the 5,2′,3′,4′,2″,4″,6″-heptadeoxy derivative retains some activity (158). The 2″-hydroxy appears to be the most important for activity (159,160). Removal of the 6′-hydroxy from kanamycin C, however, is detrimental (161) and the introduction of a 3′-fluoro in place of the 3′-hydroxy increases activity somewhat (162). Removal or epimerization of the 5-hydroxy of kanamycin B reduces antibacterial potency somewhat (163). When the 6″-hydroxy of kanamycin B has a carbamoyl group, the resulting derivative is the naturally occurring (*S. tenebrarius*) nebramycin factor 4 which has significant activity (164). In amikacin, replacement of the 6″-hydroxy with a fluorine decreases activity somewhat (165).

The 1-epi analogue of tobramycin (**4**) was slightly less active than the parent, but the 3-epi was almost inactive (166). Epimerization at the 1-position had a more detrimental effect with kanamycin A (166,167).

Four members of the gentamicin complex have been mentioned: B (**8**, R = H), C_1 (**3**, R = R′ = CH_3), C_{1a} (**3**, R = R′ = H), and C_2 (**3**, R = CH_3, R′ = H). Gentamicin A [*13291-74-2*] $C_{18}H_{36}N_4O_{10}$ (**17**, R = R″ = H, R′ = NH_2, R‴ = OH), gentamicin X_2 [*36889-17-5*] (**17**, R = CH_3, R′ = NH_2, R″ = H, R‴ = OH), and gentamicin B_1 [*36889-16-4*] $C_{20}H_{40}N_4O_{10}$ (**17**, R = R″ = CH_3, R′ = OH, R‴ = NH_2) are representative of other members that have been isolated (168).

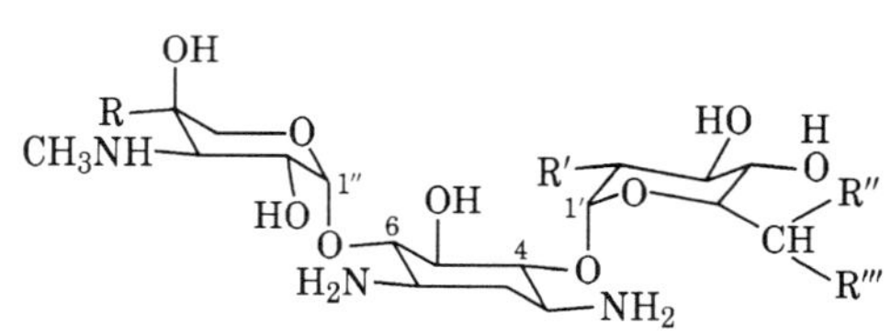

(**17**)

In addition to being antibacterial agents, although not as good as the C_1 + C_{1a} + C_2 combination, these compounds are reported to have antiprotozoal and anthelmintic activity. Many structure–activity correlations can be made within the gentamicin series (120,122,123).

1-*N*-Monoalkylation of members of the gentamicin–sisomicin group has been productive. This modification provides resistance to ANT(2″) and AAC(3) (123). Netilmicin (**6**, R = CH_2CH_3), which is 1-*N*-ethyl-sisomicin, is the most noteworthy example. Methylation of the 6′-amine confers resistance to some AAC(6′) isozymes, eg, micronomicin [*52093-21-7*] $C_{20}H_{41}N_5O_7$ (sagamicin, gentamicin C_{2b}), the naturally occurring 6′-*N*-methyl derivative of gentamicin C_{1a}. The addition of a methyl group to the 6′-carbon, as in verdamicin, has a similar effect. Also, alkylation of the 2′-amino group can provide resistance to AAC(6′) (123).

As in the kanamycin series, an α-hydroxy-ω-aminoacyl group on the 1-amino moiety of a number of gentamicin–sisomicin analogues confers activity against resistant bacterial strains (169). Isepamicin (**8**, R = (*S*)-$COCHOHCH_2NH_2$) is the primary example of this type of modification. In contrast to the kanamycins, 1-*N*-acetyl sisomicin was resistant to inactivation by AAC(3) and ANT(2″) (170). Other active 1-*N*-acylated derivatives have also been described (171,172). The naturally occurring 2′-*N*-formylsisomicin is considerably less active overall than sisomicin itself (173). Guanylation of the 1-, 3-, 2′-, or 6′-amines of gentamicin C components generally reduced potency (174); 2′-*N*-guanylgentamicin C_1 was less nephrotoxic in rats than gentamicin (175).

Considerable variation at the 2′, 3′, 4′, 5′ and 6′ positions can be tolerated. For example, sisomicin (**6**, R = H) can be considered to be a 4′,5′-dehydro derivative of gentamicin C_{1a}, and the related verdamicin [*49863-48-1*] $C_{20}H_{39}N_5O_7$, isolated from *Micromonospora grisea* (176), is the 4′,5′-dehydro derivative of gentamicin C_2 (**3**, R = CH_3, R′ = H). The 2′-position can be H, OH, or NH_2, as indicated by gentamicins C_2 and B_1 (169), the 3′-position can be H or OH, the 4′-position can be H or OH, and the 6′-position can carry an OH, NH_2, or $NHCH_3$, and may or may not have a C-methyl.

The 3″-methylamino in gentamicin C_2 can be changed to amino, ethylamino, *n*-propylamino, or *n*-butylamino with little change in antibacterial activity (177).

1-Deamino- and 1-deamino-1-hydroxy- analogues in the gentamicin series have been obtained by isolation and synthesis (178–180). These derivatives have weak antibacterial activity, although the 1-epi analogues of gentamicin C_1, netilmicin, and sisomicin are highly active (180).

The seldomycin complex, represented by seldomycin factor 1 [*56276-04-1*] $C_{17}H_{34}N_4O_{10}$ (**18**, R = R′ = R″ = R‴ = OH), seldomycin factor 3 [*56276-05-2*] $C_{17}H_{35}N_5O_9$ (**18**, R = R′ = R″ = OH, R‴ = NH_2), and seldomycin factor 5 [*56276-26-7*] $C_{18}H_{38}N_6O_7$ (**18**, R = OCH_3, R′ = R‴ = NH_2, R″ = H), were isolated from *Streptomyces hofunensis* (181,182). Of these, factor 5 has the best antibacterial activity, and is comparable in potency to kanamycin A (**5**, R = H). 3′-Deoxyseldomycin factor 5 (183) was found to be more active than the parent compound, while the 3′-epi analogue was less active (184).

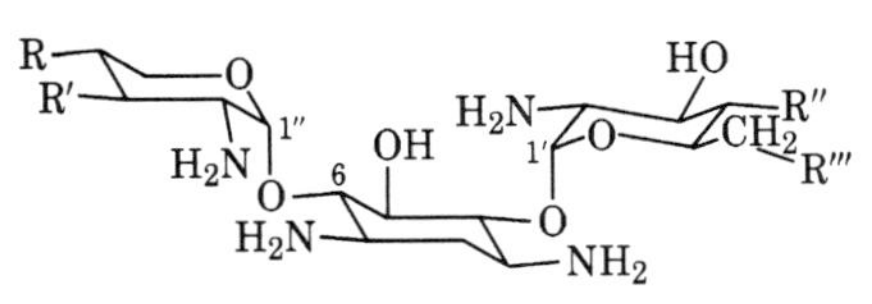

(**18**)

Derivatives Containing a Modified 4,6-Disubstituted-2-deoxystreptamine Moiety. Mutasynthesis is the term used to refer to the process of employing mutated intermediate-requiring organisms to prepare derivatives that would be difficult to obtain by synthetic or semisynthetic means (117,185–188). Whereas this approach has been used to obtain aminoglycoside analogues, the primary contribution has been as a means to vary the 2-deoxystreptamine portion of the molecule. The most significant compound found by this approach is 5-episisomicin, which is produced by a 2-deoxystreptamine-requiring mutant of *M. inyoensis* (189) when the medium is supplemented by 2-deoxy-5-epistreptamine. Compared to the parent sisomicin (**6**, R = H), this analogue has comparable potency, and is resistant to the effects of ANT(2″), some AAC(3), AAC(2′), and some AAC(6′) (190–193). Other mutasynthetically-produced sisomicin analogues include the 2-hydroxy analogue (from streptamine) and the 5-deoxy analogue (from 2,5-dideoxystreptamine). Similar analogues were obtained in the gentamicin series from a mutated *M. purpurea* strain.

Structural Modifications in the Fortimicin Series. The most recent class of aminoglycosides to undergo significant structural modification studies are the fortimicins and related compounds (120,125,185). Fortimicin A, astromicin (**10**), has been introduced into medical practice. Its overall antibacterial activity against sensitive strains of susceptible bacteria is similar to that of kanamycin A (5, R = H), ie, active against most Enterobacteriaceae and *S. aureus,* but not *P. aeruginosa*. However, fortimicin is resistant to most of the bacterial inactivating enzymes with the exception of some AAC(3) and AAC(6′) isozymes. The 1-amino group is acetylated by AAC(3) (194), and the 6′-amino by AAC(6′) (195). A number of fortimicin analogues have been isolated (196), and some of these are shown in Table 3. The presence of a glycyl residue or a proton on the 4-amino group apparently causes a change in the ring conformation (197).

Table 3. Fortimicin Analogues

Name	CAS Registry Number	Molecular formula	Structure number	R	R′	Structure
fortimicin A	*[55779-06-1]*	$C_{17}H_{35}N_5O_6$	**(19)**	H	CH_3	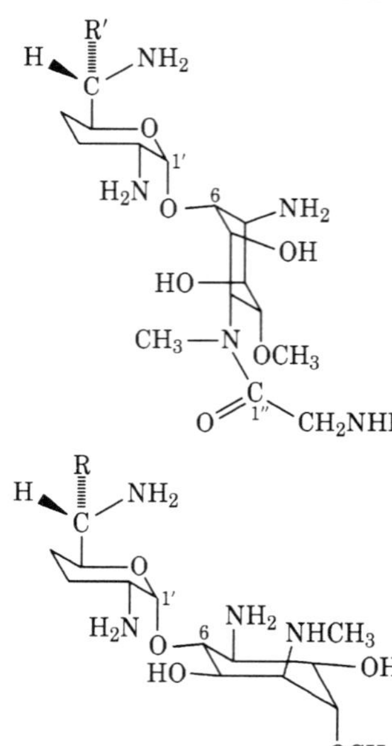
fortimicin C	*[62874-51-5]*	$C_{18}H_{36}N_6O_7$	**(19)**	$CONH_2$	CH_3	
fortimicin D	*[67330-20-5]*	$C_{16}H_{33}N_5O_6$	**(19)**	H	H	
dactimicin	*[73196-97-1]*	$C_{18}H_{36}N_6O_6$	**(19)**	CHNH	CH_3	
fortimicin B	*[54783-95-8]*	$C_{15}H_{32}N_4O_5$	**(20)**	CH_3		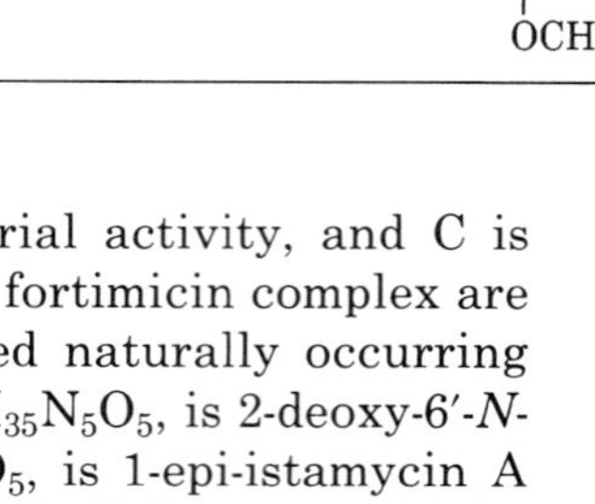
fortimicin KE	*[67330-21-6]*	$C_{14}H_{30}N_4O_5$	**(20)**	H		

Fortimicins B and KE show only weak antibacterial activity, and C is somewhat less active than A or D. Other members of the fortimicin complex are described in the literature (198,199). Two closely related naturally occurring complexes, the istamycins, istamycin A *[72503-79-8]*, $C_{17}H_{35}N_5O_5$, is 2-deoxy-6′-*N*-methylfortimicin D, istamycin B *[72523-64-9]*, $C_{17}H_{35}N_5O_5$, is 1-epi-istamycin A (200,201), and sporaricins, sporaricin A *[68743-79-3]*, $C_{17}H_{35}N_5O_5$, is 1-epi-2-deoxyfortimicin A (202), indicate that the 1-epi, 2-deoxy (203–205), and 6′-$NHCH_3$ (206) modifications are consistent with significant antibacterial activity.

Dactimicin (4-amino-1,4-dideoxy-3-*O*-(2,6-diamino-2,3,4,6,7-pentadeoxy-β-L-*lyxo*-heptopyranosyl)-1-[[[(iminomethyl)amino]acetyl]methylamino]-6-*O*-methyl-L-*chiro*-inositol) and dactimicin sulfate [*73245-91-7*] were isolated from *Dactylosporangium matsuzakiense* (207–209), and have received additional study. The overall antibacterial activity is similar to that of astromicin (**10**) except that dactimicin is less susceptible to the effects of AAC(3) (210–215). Compared to amikacin (**5**, R = (*S*)-$COCHOHCH_2CH_2NH_2$), it is similar except for being more active against *Serratia* species and less active versus *P. aeruginosa*. There is experimental evidence for a relatively low degree of ototoxicity (211) and nephrotoxicity (216–219).

The biosynthesis of the fortimicins has received some initial study (220–222) and a significant amount of semisynthetic modification has been carried out in this series. 3-*O*-Demethylfortimicin A (223,224) was found to be significantly more potent than the parent fortmicin A, especially against *P. aeruginosa*. 3-*O*-Demethyl-2-epi-, 2,3-di-epi-, and 3-*O*-demethyl-2,3-di-epi-fortimicin A are somewhat less active than fortimicin A, but the 3-*O*-demethyl-3-epi- analogue is devoid of activity (225,226). 3-Epi-fortimicin D is poorly active (227), but 3-*O*-demethyl-3-epi-istamycin B is reported to be fully active (228). The 3-amino-3-demethoxy derivative was comparable to the parent (229). In the sporaricin A series, the 3-demethoxy-3-fluoro- and 3-demethoxy-3,3-difluoro- analogues have similar activity to that of the parent (230), as does the 3-*O*-demethyl-3-*O*-(3-amino-2-hydroxypropyl) derivative (231).

The glycyl, or formimidoylglycyl, residue on the 4-methylamino group is important for good antibacterial potency in the fortimicin series. In a series of acyl and alkyl variations of this residue, only the sarcosyl and β-alanyl acyl groups (232) and the (*S*)-$CH_2CHOHCH_2CH_2NH_2$ and (*S*)-$CH_2CHOHCH_2CH_2NHCH_3$ alkyl groups were able to come close to the activity of fortimicin A (233). Replacement of the glycyl with a 2-aminoethanesulfonyl residue led to loss of antibacterial activity (234). Furthermore, the methyl group appears to be critical to antibacterial activity; replacement by a hydrogen or an ethyl group eliminated activity (235).

Whereas 2-deoxy-fortimicin A has full antibacterial activity (205), and the 2-deoxy-3-demethoxy- derivative is even more active than fortimicin A (236), the 2,5-dideoxy derivative has poor activity, as do the 2-deoxy-2-chloro and the 2,5-dideoxy-4,5-dehydro derivatives. The 2-amino-3-*O*-demethyl-2-deoxy analogue has quite good activity (229).

The addition of a 3-hydroxypropyl group to the 7′-position of fortimicin A caused a significant reduction in antibacterial activity (237).

Economic Aspects

Estimated worldwide sales are given in Table 1 for the aminoglycoside derivatives most commonly used in medicine. Although amikacin has somewhat higher sales, gentamicin is by far the most widely used in terms of number of courses of treatment.

Members of the aminoglycoside class of antibacterial antibiotics retain an important role in the control of bacterial infectious disease, especially in the treatment of gram-negative infections in the compromised patient. Semisynthetic modifications of naturally occurring compounds have yielded derivatives with greatly enhanced resistance to bacterial inactivation mechanisms without altering the basic antimicrobial spectrum. The principal factors limiting aminoglycoside usage are nephrotoxicity and ototoxicity. To date, structural modification has not been able to completely dissociate toxicity from antibacterial activity. Current research in the aminoglycoside field is focused on the reduction of toxicity through less frequent dosing, liposomal formulations, and coadministration of toxicity inhibitors.

BIBLIOGRAPHY

"Streptomycin Antibiotics (Survey; Streptomycin; Neomycin)" in *ECT* 1st ed., Vol. 13, pp. 57–81 and 90–94, by S. A. Waksman and H. A. Lechevalier, Institute for Microbiology, Rutgers University; "Streptomycin and Related Antibiotics" in *ECT* 2nd ed., Vol. 19, pp. 33–48, by D. Perlman, University of Wisconsin; "Antibiotics, Aminoglycoside" in *ECT* 3rd ed., Vol. 2, pp. 819–852, by Peter J. L. Daniels, Schering-Plough Corp.

1. A. Schatz, E. Bugie, and S. A. Waksman, *Proc. Soc. Exp. Biol. Med.* **55,** 66 (1944).
2. F. A. Keuhl, R. L. Peck, C. E. Hoffhine, Jr., and K. Folkers, *J. Am. Chem. Soc.* **70,** 2325 (1948).
3. S. Neidle, D. Rogers, and M. B. Hursthouse, *Tetrahedron Lett.* 4725 (1968).
4. S. Umezawa, Y. Takahashi, T. Usui, and T. Tsuchiya, *J. Antibiot.* **27,** 997 (1974).
5. S. A. Waksman, *Science* **118,** 259 (1953).
6. M. J. Weinstein and co-workers, *J. Med. Chem.* **6,** 463 (1963).
7. M. J. Weinstein, G. M. Luedemann, E. M. Oden, and G. H. Wagman, *Antimicrob. Agents Chemother.* 1 (1963).
8. J. P. Rosselet and co-workers, *Antimicrob. Agents Chemother.* 14 (1963).
9. D. J. Cooper, P. J. L. Daniels, M. D. Yudis, H. M. Marigliano, R. D. Guthrie, and S. T. K. Bukhari, *J. Chem. Soc. (C)* 3126 (1971).
10. W. M. Stark, M. M. Hoehn, and N. G. Knox, *Antimicrob. Agents Chemother.* 314 (1967).
11. C. E. Higgins and R. E. Kastner, *Antimicrob. Agents Chemother.* 324 (1967).
12. R. Q. Thompson and E. A. Presti, *Antimicrob. Agents Chemother.* 332 (1967).
13. W. E. Wick and J. S. Welles, *Antimicrob. Agents Chemother.* 341 (1967).
14. K. F. Koch and J. A. Rhoades, *Antimicrob. Agents Chemother.* 309 (1970).
15. Y. Takagi, T. Miyeke, T. Tsuchiya, S. Umezawa, and H. Umezawa, *Bull. Chem. Soc. Jpn.* **49,** 3649 (1976).
16. M. Tanabe, D. M. Yasuda, and G. Detre, *Tetrahedron Lett.* 3607 (1977).
17. H. Kawaguchi, T. Naito, S. Nakagawa, and K. Fujisawa, *J. Antibiot.* **25,** 695 (1972).
18. H. Umezawa and co-workers, *J. Antibiot. Ser. A* **10,** 181 (1957).
19. M. J. Cron and co-workers, *J. Am. Chem. Soc.* **80,** 4741 (1958).
20. K. Maeda, M. Murase, H. Mawatari, and H. Umezawa, *J. Antibiot. Ser. A* **11,** 163 (1958).
21. M. Hichens and K. L. Rinehart, Jr., *J. Am. Chem. Soc.* **85,** 1547 (1963).
22. S. Umezawa, K. Tatsuta, and S. Koto, *Bull. Chem. Soc. Jpn.* **42,** 533 (1969).
23. J. J. Wright, *J. Chem. Soc. Chem. Commun.* 206 (1976).
24. M. J. Weinstein, J. A. Marquez, R. T. Testa, G. H. Wagman, E. M. Oden, and J. A. Waitz, *J. Antibiot.* **23,** 551 (1970).

25. H. Reimann and co-workers, *J. Org. Chem.* **39,** 1451 (1974).
26. D. H. Davies, A. K. Mallams, M. Connelis, D. Loebenberg, E. L. Moss, Jr., and J. A. Waitz, *J. Med. Chem.* **21,** 189 (1978).
27. H. Umezawa, S. Umezawa, T. Tsuchiya, and Y. Okazaki, *J. Antibiot.* **24,** 485 (1971).
28. S. Umezawa, H. Umezawa, Y. Okazaki, and T. Tsuchiya, *Bull. Chem. Soc. Jpn.* **45,** 3624 (1972).
29. T. Ito, M. Nishio, and H. Ogawa, *J. Antibiot. Ser. A* **17,** 189 (1964).
30. S. Umezawa, S. Koto, K. Tatsuta, H. Hineno, Y. Nishimura, and T. Tsumura, *Bull. Chem. Soc. Jpn.* **42,** 537 (1969).
31. T. L. Nagabhushan, A. B. Cooper, H. Tsai, P. J. L. Daniels, and G. H. Miller, *J. Antibiot.* **31,** 681 (1978).
32. J. Weinstein, D. J. Cooper, and P. J. L. Daniels, *Abstracts, 12th Intersci. Conf. Antimicrob. Agents Chemother.* 9 (1972).
33. K. L. Rinehart, Jr., *The Neomycins and Related Antibiotics,* John Wiley & Sons, Inc., New York, 1961.
34. S. A. Waksman and H. A. Lechevalier, *Science* **109,** 305 (1949).
35. K. L. Rinehart and co-workers, *J. Am. Chem. Soc.* **84,** 3218 (1962).
36. T. Usui and S. Umezawa, *J. Antibiot.* **40,** 1464 (1987).
37. T. Nara and co-workers, *J. Antibiot.* **30,** 533 (1977).
38. R. S. Egan and co-workers, *J. Antibiot.* **30,** 552 (1977).
39. R. L. Girolami and J. M. Stamm, *J. Antibiot.* **30,** 564 (1977).
40. Y. Honda and T. Suami, *Bull. Chem. Soc. Jpn.* **55,** 1156 (1982).
41. D. J. Mason, A. Dietz, and R. Smith, *Antibiot. Chemother.* **11,** 118 (1961).
42. P. F. Wiley, A. D. Argoudelis, and H. Hoeksema, *J. Am. Chem. Soc.* **85,** 2652 (1961).
43. T. G. Cochran and D. J. Abraham, *J. Chem. Soc. Chem. Commun.* 4944 (1972).
44. G. L. Mandell, R. G. Douglas, Jr., and J. E. Bennett, eds., *Principles and Practice of Infectious Diseases,* 3rd ed., Churchill Livingstone Inc., New York, 1990.
45. A. Whelton and H. C. Neu, eds., *The Aminoglycosides,* Marcel Dekker, Inc., New York, 1982.
46. *Physicians' Desk Reference,* 43rd ed., Medical Economics Co., Inc., Oradell, N.J., 1989, pp. 715, 738, 787, 1097, 1193, 1794, 1924, 1942.
47. *Aminoglycoside Antibiotics,* H. Umezawa and I. R. Hooper, eds., Springer-Verlag, New York, 1982.
48. R. C. Moellering, Jr. and W. E. Siegenthaler, *Amer. J. Med.* **80,** suppl. 6B, 1–233 (1986).
49. *Rev. Infect. Dis.* **5,** suppl. 2, 210–316 (1983).
50. *J. Antimicrob. Chemother.* **8,** suppl. A, 1–156 (1981).
51. B. A. Cunha, *Pharmacotherapy* **8,** 334 (1988).
52. P. S. Lietman, in ref. 44, p. 269.
53. R. C. Moellering, Jr., in ref. 45, p. 65.
54. C. R. Smith and P. S. Lietman, in ref. 45, p. 497.
55. H. C. Neu, in ref. 45, p. 611.
56. W. E. Siegenthaler, A. Bonetti, and R. Luthy, in ref. 48, p. 2.
57. D. A. Evans, J. Buring, S. Mayrent, B. Rosner, T. Colton, and C. Hennekens, in ref. 48, p. 39.
58. R. D. Moore, P. S. Leitman, and C. R. Smith, *J. Infect. Dis.* **155,** 93 (1987).
59. G. P. Bodey, in ref. 45, p. 557.
60. J. Klastersky and S. H. Zinner, *Rev. Infect. Dis.* **4,** 294 (1982).
61. H. Gaya, in ref. 48, p. 149.
62. J. L. Henderson, R. E. Polk, and B. J. Kline, *Am. J. Hosp. Pharm.* **38,** 1167 (1981).
63. R. C. Moellering, Jr., in ref. 49, p. 212.
64. L. S. Young, in ref. 49, p. 250.

65. H. Umezawa and S. Kondo, in ref. 47, p. 26.
66. K. E. Price, *Antimicrob. Agents Chemother.* **29,** 543 (1986).
67. G. Funatsu and H. G. Wittman, *J. Mol. Biol.* **68,** 547 (1972).
68. L. E. Bryan, *J. Antimicrob. Chemother.* **22,** suppl. A, 1 (1988).
69. J. E. Davies and D. I. Smith, *Ann. Rev. Microbiol.* **32,** 469 (1978).
70. J. E. Davies, *Rev. Infect. Dis.* **5,** suppl. 2, S261 (1983).
71. K. H. Mayer, in ref. 48, p. 56.
72. S. Mitsuhashi and H. Kawabe, in ref. 45, p. 97.
73. H. Umezawa and S. Kondo, in ref. 47, p. 267.
74. *Rev. Infect. Dis.* **5,** suppl. 2, S314 (1983).
75. H. C. Neu, in ref. 45, p. 125.
76. J. J. Schentag, in ref. 45, p. 143.
77. T. Koeda, K. Umemura, and M. Yokota, in ref. 47, p. 293.
78. A. Whelton, in ref. 45, p. 585.
79. S. H. Powell and co-workers, *J. Infect. Dis.* **147,** 918 (1983).
80. F. Clerckx-Braun and co-workers, *Abstracts, 27th Intersci. Conf. Antimicrob. Agents Chemother.* abstr. no. **25** (1987).
81. S. T. Fan, W. Y. Lau, C. H. Teoh-Chan, K. F. Lau, and E. H. Mauracher, *J. Antimicrob. Chemother.* **22,** 69 (1988).
82. R. Maller, B. Isaksson, L. Nilsson, and L. Soren, *J. Antimicrob. Chemother.* **22,** 75 (1988).
83. W. A. Craig and B. Vogelman, *Ann. Intern. Med.* **106,** 900 (1987).
84. C. E. Swenson, K. A. Stewart, J. L. Hammett, W. E. Fitzsommons, and R. S. Ginsberg, *Antimicrob. Agents Chemother.* **34,** 235 (1990).
85. *Drug News and Perspectives* **2,** 365 (1989).
86. H. Mattie, W. A. Craig, and J. C. Pechere, *J. Antimicrob. Chemother.* **24,** 281 (1989).
87. R. E. Brummett and K. E. Fox, *Antimicrob. Agents Chemother.* **33,** 797 (1989).
88. C. L. Bendush, in ref. 40, p. 453.
89. G. Kahlmeter and J. I. Dahlager, *J. Antimicrob. Chemother.* **12,** suppl. A, 9 (1984).
90. A. M. Lerner and co-workers, *Lancet* **1,** 1123 (1983); I. Kitasato, M. Yokota, S. Inouye, and M. Igarashi, *Chemother.* **36,** 155 (1990).
91. J. M. Eisenberg and co-workers, *Ann. Int. Med.* **107,** 900 (1987).
92. R. D. Meyer, in ref. 48, p. 119.
93. G. H. Miller, P. J. S. Chiu, and J. A. Waitz, *J. Antibiot.* **31,** 688 (1978).
94. L. I. Rankin, F. C. Luft, M. N. Yum, R. S. Sloan, C. B. Dinwiddie, Jr., and L. L. Isaacs, *Antimicrob. Agents Chemother.* **16,** 491 (1979).
95. A. Whelton, in ref. 45, p. 191.
96. F. J. Silverblatt, in ref. 45, p. 223.
97. P. M. Tulkens, in ref. 48, p. 105.
98. P. M. Tulkens, *Contrib. Nephrol.* **42,** 168 (1984).
99. C. F. Simmons, R. T. Bogusky, and R. D. Humes, *J. Pharmacol. Exp. Ther.* **214,** 709 (1980).
100. G. J. Kaloyanides, *Contrib. Nephrol.* **42,** 148 (1984).
101. G. J. Kaloyanides, *Fundam. Appl. Toxicol.* **4,** 930 (1984).
102. G. Toubeau and co-workers, *Lab. Invest.* **54,** 385 (1986).
103. P. D. Williams, G. H. Hottendorf, and D. B. Bennett, *J. Pharmacol. Exp. Ther.* **237,** 919 (1986).
104. D. N. Gilbert and co-workers, *J. Infect. Dis.* **159,** 945 (1989).
105. L. S. Ramsammy, C. Josepovitz, B. B. Lane, and G. J. Kaloyanides, *J. Pharmacol. Exp. Ther.* **250,** 149 (1989).
106. D. Beauchamp, G. Laurent, P. Maldague, and P. M. Tulkens, *Arch. Toxicol.* **9,** suppl., 306 (1986).

107. B. K. Kishore, S. Ibrahim, P. Lambricht, G. Laurent, P. Maldague, and P. M. Tulkens, *Abstracts, 29th Intersci. Conf. on Antimicrob. Agents Chemother.* abstr. no. **296** (1989).
108. B. D. Davis, *Microbiol. Rev.* **51,** 341 (1987).
109. H. W. Taber, J. P. Mueller, P. F. Miller, and A. S. Arrow, *Microbiol. Rev.* **51,** 439 (1987).
110. R. E. W. Hancock, *J. Antimicrob. Chemother.* **8,** 249 (1981).
111. Ref. 110, p. 429.
112. S. Mitsuhashi, in ref. 47, p. 205.
113. N. Tanaka, in ref. 47, p. 221.
114. B. D. Davis, *J. Antimicrob. Chemother.* **22,** 1 (1988).
115. M. A. Wyka and A. C. St. John, *Antimicrob. Agents Chemother.* **34,** 534 (1990).
116. T.-P. Hausner, U. Geigenmuller, and K. H. Nierhaus, *J. Biol. Chem.* **263,** 13103 (1988).
117. T. Okuda and Y. Ito, in ref. 47, p. 111.
118. K. Suzukake, K. Tokunaga, H. Hayashi, M. Hori, Y. Uehara, D. Ikeda, and H. Umezawa, *J. Antibiot.* **38,** 1211 (1985).
119. K. Kakinuma, Y. Ogawa, T. Sasaki, H. Seto, and N. Otake, *J. Antibiot.* **42,** 926 (1989).
120. S. Umezawa and T. Tsuchiya, in ref. 47, p. 37.
121. D. A. Cox, K. Richardson, and B. C. Ross, in *Topics in Antibiotic Chemistry,* Vol. 1, P. G. Sammes, ed., John Wiley & Sons, Inc., New York, 1977, p. 1.
122. K. E. Price, J. C. Godfrey, and H. Kawaguchi, *Adv. Appl. Microbiol.* **18,** 191 (1974).
123. T. L. Nagabhushan, G. H. Miller, and M. J. Weinstein, in ref. 40, p. 3.
124. *Aminocyclitol Antibiotics,* K. L. Rinehart, Jr. and T. Suami, eds., ACS Symposium Series 125, American Chemical Society, Washington, D.C., 1980.
125. I. R. Hooper, in ref. 47, p. 1.
126. S. Umezawa, *Adv. Carbohydr. Chem.* **30,** 111 (1974).
127. D. L. Delaware, M. S. Sharma, B. S. Iyengar, W. A. Remers, and T. A. Pursiano, *J. Antibiot.* **39,** 251 (1986).
128. T. Tsuchiya, S. Sakamoto, T. Yamasaki, and S. Umezawa, *J. Antibiot.* **35,** 639 (1982).
129. W. M. Stark, M. M. Hoehn, and N. G. Knox, *Antimicrob. Agents Chemother.* 314 (1968).
130. S. O'Connor, L. K. T. Lam, N. D. Jones, and M. O. Chaney, *J. Org. Chem.* **41,** 2087 (1976).
131. R. L. Mann and W. W. Bromer, *J. Amer. Chem. Soc.* **80,** 2714 (1958).
132. N. Neuss and co-workers, *Helv. Chim. Acta* **53,** 2314 (1970).
133. J. Shoji and Y. Nakagawa, *J. Antibiot.* **23,** 569 (1970).
134. G. L. Coffey and co-workers, *J. Antibiot. Chemother.* **9,** 730 (1959).
135. T. H. Haskell, J. C. French, and Q. R. Bartz, *J. Amer. Chem. Soc.* **81,** 3480, 3482 (1959).
136. R. T. Schillings and C. P. Schaffner, *Antimicrob. Agents Chemother.* 274 (1961).
137. T. Mori, T. Ichiyanaga, H. Kondo, K. Tokunaga, T. Oda, and K. Munakata, *J. Antibiot.* **24,** 339 (1971).
138. T. Oda, T. Mori, Y. Kyotani, and M. Nakayama, *J. Antibiot.* **24,** 511 (1971).
139. T. Mori, Y. Kyotani, I. Watanabe, and T. Oda, *J. Antibiot.* **25,** 149 (1972).
140. T. Shomura, N. Ezaki, T. Tsuruoka, T. Niwa, E. Akita, and T. Niida, *J. Antibiot.* **23,** 155 (1970).
141. E. Akita, T. Tsuruoka, N. Ezaki, and T. Niida, *J. Antibiot.* **23,** 173 (1970).
142. J. D. Howells and co-workers, *Antimicrob. Agents Chemother.* **2,** 79 (1972).
143. H. W. Dion, P. W. K. Woo, N. E. Willmer, D. L. Kern, J. Onaga, and S. A. Fusari, *Antimicrob. Agents Chemother.* **2,** 84 (1972).
144. C. L. Heifitz, M. W. Fisher, J. A. Chodubski, and M. O. DeCarlo, *Antimicrob. Agents Chemother.* **2,** 89 (1972).
145. H. Tsukiura, K. Saito, S. Kobaru, M. Konishi, and H. Kawaguchi, *J. Antibiot.* **26,** 386 (1973).

146. M. Murase, *J. Antibiot. Ser. A* **14,** 136 (1961).
147. S. Sicsic, J. F. Le Bigot, C. Vincent, C. Cerceau, and F. Le Gofic, *J. Antibiot.* **35,** 574 (1982).
148. T. Naito and co-workers, *J. Antibiot.* **27,** 251 (1974).
149. E. Umemura, T. Tsuchiya, and S. Umezawa, *J. Antibiot.* **41,** 530 (1988).
150. E. Akita, Y. Horiuchi, T. Miyazawa, and H. Unezawa, *J. Antibiot.* **36,** 745 (1983).
151. H. Iwasawa, D. Ikeda, S. Kondo, and H. Umezawa, *J. Antibiot.* **37,** 428 (1984).
152. K. Matsuda, N. Yasuda, H. Tsutsumi, and T. Takaya, *J. Antibiot.* **40,** 843 (1987).
153. S. Nakagawa, S. Toda, Y. Abe, H. Yamashita, K. Fujisawa, T. Naito, and H. Kawaguchi, *J. Antibiot.* **31,** 675 (1978).
154. K. Richardson, S. Jevons, J. W. Moore, B. C. Ross, and J. R. Wright, *J. Antibiot.* **30,** 843 (1977).
155. K. Richardson, K. W. Brammer, S. Jevons, R. M. Plews, and J. R. Wright, *J. Antibiot.* **32,** 973 (1979).
156. H. Umezawa, K. Iinuma, S. Kondo, M. Hamada, and K. Maeda, *J. Antibiot.* **28,** 483 (1975).
157. H. Iwasawa, D. Ikeda, S. Kondo, and H. Umezawa, *J. Antibiot.* **35,** 1715 (1982).
158. H. Umezawa, H. Iwasawa, D. Ikeda, and S. Kondo, *J. Antibiot.* **36,** 1087 (1983).
159. P. J. L. Daniels, J. Weinstein, R. W. Tkach, and J. Morton, *J. Antibiot.* **27,** 150 (1974).
160. M. N. Sharma, V. Kumar, and W. A. Remers, *J. Antibiot.* **35,** 905 (1982).
161. S. Toda, S. Nakagawa, and T. Naito, *J. Antibiot.* **30,** 1002 (1977).
162. T. Tsuchiya, Y. Takahashi, Y. Kobayashi, S. Umezawa, and H. Umezawa, *J. Antibiot.* **38,** 1287 (1985).
163. T. Suami, in ref. 86, p. 43.
164. C. E. Higgens and R. E. Kastner, *Antimicrob. Agents Chemother.* 304 (1968).
165. R. Albert, K. Dax, A. E. Stutz, and J. Hildebrandt, *J. Antibiot.* **38,** 275 (1985).
166. K. Igarashi and co-workers, *Carbohydr. Res.* **109,** 73 (1982).
167. Y. Takahashi, T. Tsuchiya, Y. Suzuki, S. Umezawa, H. Umezawa, and S. Fukatsu, *Bull. Chem. Soc. Jpn.* **56,** 1807 (1983).
168. P. J. L. Daniels and co-workers, *J. Chem. Soc. Chem. Commun.* 1629 (1971).
169. P. J. L. Daniels, A. B. Cooper, S. W. McCombie, T. L. Nagabhushan, D. F. Rane, and J. J. Wright, *J. Antibiot.* **32,** suppl., S195 (1979).
170. J. J. Wright and co-workers, *J. Antibiot.* **29,** 714 (1976).
171. M. Phillipe, A. M. Sepulchre, S. D. Gero, H. Loibner, W. Streicher, and P. Stutz, *J. Antibiot.* **35,** 1507 (1982).
172. A. K. Mallams, J. B. Morton, and P. Reichert, *J. Chem. Soc. Perkin I,* 2186 (1981).
173. S. Satoi, M. Awata, N. Muto, M. Hayashi, H. Sagai, and M. Otani, *J. Antibiot.* **36,** 1 (1983).
174. W. Streicher, H. Loibner, J. Hildebrandt, and F. Turnowsky, *Drugs Exp. Clin. Res.* **9,** 591 (1983).
175. E. Wilmotte and co-workers, *Drugs Exp. Clin. Res.* **9,** 467 (1983).
176. M. J. Weinstein, G. H. Wagman, J. A. Marquez, R. T. Testa, and J. A. Waitz, *Antimicrob. Agents Chemother.* **7,** 246 (1975).
177. T. L. Nagabhushan, J. J. Wright, A. B. Cooper, W. N. Turner, and G. H. Miller, *J. Antibiot.* **31,** 43 (1978).
178. K. Shirahata, H. Kase, S. Kitamura, and T. Iida, *J. Antibiot.* **35,** 520 (1982).
179. M. Phillipe and co-workers, *J. Antibiot.* **36,** 250 (1983).
180. D. L. Boxler and co-workers, *J. Chem. Soc. Perkin I,* 2168 (1981).
181. T. Nara and co-workers, *J. Antibiot.* **30,** 17 (1977).
182. R. S. Egan and co-workers, *J. Antibiot.* **30,** 31 (1977).
183. H. Matsushima, Y. Mori, and K. Kitaura, *J. Antibiot.* **30,** 890 (1977).
184. H. Matsushima, K. Kitaura, and Y. Mori, *Bull. Chem. Soc. Jpn.* **50,** 3039 (1977).

185. F. Leitner and K. E. Price, in ref. 45, p. 29.
186. S. J. Daum and J. R. Lemke, *Annu. Rev. Microbiol.* **33,** 241 (1979).
187. K. L. Rinehart, in ref. 86, p. 335.
188. P. J. L. Daniels, D. F. Rane, S. W. McCombie, R. T. Testa, J. J. Wright, and T. L. Nagabhushan, in ref. 86, p. 371.
189. P. J. L. Daniels and D. F. Rane, in D. Schessinger, ed., *Microbiology,* ASM Publications, Washington, D.C., 1979, p. 314.
190. J. A. Waitz, G. H. Miller, E. Moss, Jr., and P. J. S. Chiu, *Antimicrob. Agents Chemother.* **13,** 41 (1978).
191. K. P. Fu and H. C. Neu, *Antimicrob. Agents Chemother.* **14,** 194 (1978).
192. S. A. Kabins and C. Nathan, *Antimicrob. Agents Chemother.* **14,** 391 (1978).
193. R. V. Goering, C. C. Sanders, and W. E. Sanders, Jr., *Antimicrob. Agents Chemother.* **14,** 824 (1978).
194. S. Sato, T. Iida, K. Okachi, K. Shirahata, and T. Nara, *J. Antibiot.* **30,** 1025 (1977).
195. T. Okubo, M. Inoue, M. Nagashima, M. Okii, T. Iida, and S. Mitsuhashi, *J. Antibiot.* **38,** 122 (1985).
196. T. Iida, M. Sato, I. Matsubara, Y. Mori, and K. Shirahata, *J. Antibiot.* **32,** 1273 (1979).
197. N. Hirayama, K. Shirahata, Y. Ohashi, and Y. Sasada, *Mol. Pharmacol.* **23,** 127 (1983).
198. J. B. McAlpine and co-workers, in ref. 124, p. 295.
199. K. Shirahata, G. Shimura, S. Takasawa, T. Iida, and K. Takahashi, in ref. 124, p. 309.
200. Y. Okami, K. Hotta, M. Yoshida, D. Ikeda, S. Konda, and H. Umezawa, *J. Antibiot.* **32,** 964 (1979).
201. D. Ikeda, Y. Horiuchi, M. Yoshida, T. Miyasaka, S. Kondo, and H. Umezawa, *Carbohydr. Res.* **109,** 33 (1982).
202. T. Deushi, M. Nakayama, I. Watanabe, T. Mori, H. Naganawa, and H. Umezawa, *J. Antibiot.* **32,** 187 (1979).
203. J. R. Martin, P. Johnson, J. Tadanier, and A. W. Goldstein, *J. Antibiot.* **33,** 810 (1980).
204. T. Yamaguchi and co-workers, *J. Antibiot.* **32,** 1137 (1979).
205. T. Suami, K. Tadano, and K. Matsuzawa, *J. Antibiot.* **33,** 1289 (1980).
206. J. Tadanier, D. A. Dunnigan, J. R. Martin, L. Freiberg, and M. Cirovic, *J. Antibiot.* **34,** 193 (1981).
207. S. Inouye and co-workers, *J. Antibiot.* **32,** 1354 (1979).
208. K. Ohba and co-workers, *J. Antibiot.* **34,** 1090 (1981).
209. K. Atsumi, E. Akita, and T. Niida, *J. Antibiot.* **35,** 90 (1982).
210. Y. Matsuhashi, T. Yoshida, T. Hara, Y. Kazuno, and S. Inouye, *Antimicrob. Agents Chemother.* **27,** 589 (1985).
211. S. Omoto, T. Yoshida, M. Kurebe, and S. Inouye, *Drugs Exp. Clin. Res.* **13,** 719 (1987).
212. S. Stefani, G. Nicoletti, G. Russo, and M. A. Toscano, *Drugs Exp. Clin. Res.* **13,** 727 (1987).
213. R. Gomez-Lus, M. A. Marco, and M. L. Gomez-Lus, *Drugs Exp. Clin. Res.* **13,** 731 (1987).
214. D. Felmingham and K. Jones, *Abstracts, 28th Intersci. Conf. Antimicrob. Agents Chemother.* abstr. no. **1506** (1988).
215. P. Paglia, G. Molinari, A. Pesce, and E. A. Debbia, *Eur. J. Clin. Microbiol.* **8,** 639 (1989).
216. I. Viano, A. A. E. Bertelli, and C. Dianzani, *Drugs Exp. Clin. Res.* **13,** 737 (1987).
217. L. Giovannini and co-workers, *Drugs Exp. Clin. Res.* **13,** 741 (1987).
218. M. Bonadio and co-workers, *Drugs Exp. Clin. Res.* **13,** 747 (1987).
219. R. Palla and co-workers, *Drugs Exp. Clin. Res.* **13,** 751 (1987).
220. S. Itoh and co-workers, *J. Antibiot.* **37,** 1664 (1984).
221. Y. Okadura and co-workers, *J. Antibiot.* **37,** 1670 (1984).
222. T. Dairi and M. Hasegawa, *J. Antibiot.* **42,** 934 (1989).

223. J. R. Martin, P. Johnson, J. Tadanier, and A. Goldstein, *Antimicrob. Agents Chemother.* **18,** 761 (1980).
224. R. N. Jones, C. Thornsberry, A. L. Barry, R. R. Packer, C. N. Baker, and R. E. Badal, *Antimicrob. Agents Chemother.* **18,** 773 (1980).
225. J. Tadanier and R. Hallas, *J. Antibiot.* **36,** 256 (1983).
226. J. R. Martin, P. Johnson, J. Tadanier, M. Cirovic, and R. S. Stanaszek, *J. Antibiot.* **35,** 46 (1982).
227. P. Kurath, R. S. Stanaszek, and M. Cirovic, *J. Antibiot.* **35,** 1338 (1982).
228. D. Ikeda, S. Gomi, S. Kondo, and H. Umezawa, *J. Antibiot.* **36,** 331 (1983).
229. J. Tadanier, R. Hallas, J. Holms, L. A. Freiberg, and D. Bacino, *J. Antibiot.* **36,** 267 (1983).
230. T. Tsuchiya, T. Torii, S. Umezawa, and H. Umezawa, *J. Antibiot.* **35,** 1245 (1982).
231. I. Watanabe, K. Kamiya, T. Yamaguchi, T. Mori, and T. Twuchiya, *Carbohydr. Res.* **109,** 47 (1982).
232. J. Tadanier, J. R. Martin, P. Kurath, A. W. Goldstein, and P. Johnson, *Carbohydr. Res.* **79,** 91 (1980).
233. M. Sato and Y. Mori, *J. Antibiot.* **32,** 371 (1979).
234. J. Tadanier and R. Hallas, *J. Antibiot.* **34,** 403 (1981).
235. P. Kurath and co-workers, *J. Antibiot.* **34,** 691 (1981).
236. J. Tadanier and R. Hallas, *J. Antibiot.* **35,** 688 (1982).
237. K. Kanei and co-workers, *Carbohydr. Res.* **170,** 47 (1987).

DONALD MCGREGOR
Bristol-Myers Squibb Company

ANTIBIOTICS, ANTHRACYCLINES. See CHEMOTHERAPEUTICS, ANTICANCER.

ANSAMACROLIDES

The ansamacrolides or ansamycins are a family of antibiotics characterized by an aliphatic ansa-bridge that connects two nonadjacent positions of an the aromatic nucleus (1,2). Ansamacrolides can be divided into two groups based on the nature of the aromatic nucleus. One group contains a naphthoquinoid nucleus and includes the streptovaricins, the rifamycins, tolypomycin, the halomycins, the naphthomycins, actamycin, the diastovaricins, kanglemycin, awamycin, and ansathiazin. The other group contains a benzoquinoid nucleus and includes geldanamycin, the maytansinoids, the herbimycins, the macbecins, the mycotrienins, the trienomycins, the ansatrienins, and the ansamitocins. Table 1 lists the naturally occurring ansamacrolides and some of their physical properties. Table 2 summarizes the biological activity of these antibiotics.

Naphthoquinoids

Streptovaricins. The streptovaricins are produced by *Streptomyces spectablis* n. sp. (60,61) and are isolated as a crude complex (62,63). Substance B 44 P, isolated from *Streptomyces* B 44-P1, was shown to be identical with the streptovaricins (64).

Chemical degradation studies carried out on streptovaricins A and C, which

Table 1. Properties of the Ansamacrolides

Ansamacrolide	CAS Registry Number	Producing organism	Molecular formula	MP, °C	$[\alpha]_D$ [a]	Structure no.	Reference
Naphthoquinoid							
streptovaricin A	*[23344-16-3]*	*Streptomyces spectabilis*	$C_{42}H_{53}NO_{16}$	233–243	+610	**(4)**	3
streptovaricin B	*[11031-82-6]*	*Streptomyces spectabilis*	$C_{42}H_{53}NO_{15}$	187–189	+576	**(5)**	3
streptovaricin C	*[23344-17-4]*	*Streptomyces spectabilis*	$C_{40}H_{51}NO_{14}$	189–191	+602	**(6)**	3
streptovaricin D	*[32164-26-4]*	*Streptomyces spectabilis*	$C_{40}H_{51}NO_{13}$	172–175	+590.3	**(7)**	3
streptovaricin E	*[35413-63-9]*	*Streptomyces spectabilis*	$C_{40}H_{49}NO_{14}$	198–202	+412.3	**(8)**	3
streptovaricin F_G [b]	*[35512-37-9]*	*Streptomyces spectabilis*	$C_{39}H_{47}NO_{14}$	222–224	+488[c]	**(12)**	3
streptovaricin G	*[11031-85-9]*	*Streptomyces spectabilis*	$C_{40}H_{51}NO_{15}$	190–192	+473	**(9)**	3
streptovaricin J	*[52275-61-3]*	*Streptomyces spectabilis*	$C_{42}H_{53}NO_{15}$	177–180	+436	**(10)**	3
streptovaricin K	*[23344-16-3]*	*Streptomyces spectabilis*	$C_{42}H_{53}NO_{16}$	162–165		**(11)**	4
streptovaricin U	*[74200-47-8]*	*Streptomyces spectabilis*	$C_{36}H_{49}NO_{10}$	115–140	−110[c]	**(13)**	5
rifamycin B	*[13929-35-6]*	*Nocardia mediterranei* n. sp. *Streptomyces tolypophorus*	$C_{39}H_{49}NO_{14}$	160–164[d]	−11[c]	**(22)**	6,7
rifamycin O	*[14487-05-9]*	*Streptomyces mediterranei* No. 4107 A_2	$C_{39}H_{47}NO_{14}$	160[d]	+71.5[e]	**(23)**	6,8
rifamycin S	*[13553-79-2]*	*Nocardia mediterranei* n. sp.	$C_{37}H_{45}NO_{12}$	146–147	+462	**(24)**	6,9
rifamycin SV	*[6998-60-3]*	*Nocardia mediterranei* mutant ATCC 21271	$C_{37}H_{47}NO_{12}$	140[d]	−4[c]	**(25)**	6,10
rifamycin Y	*[15271-73-5]*	*Nocardia mediterranei*	$C_{39}H_{47}NO_{15}$		+325[c]	**(29)**	11
rifamycin L	*[26117-02-2]*	*Nocardia mediterranei*	$C_{39}H_{49}NO_{14}$	152–153[d]		**(28)**	12
rifamycin W	*[53904-81-7]*	*Nocardia mediterranei* mutant 126	$C_{35}H_{45}NO_{11}$			**(30)**	13,14
rifamycin R	*[59264-04-9]*	*Nocardia mediterranei* mutant ATCC 31066	$C_{37}H_{45}NO_{13}$			**(27)**	15
rifamycin P	*[59232-87-0]*	*Nocardia mediterranei* mutant ATCC 13685	$C_{38}H_{46}N_2O_{11}S$			**(31)**	16
rifamycin Q	*[59232-88-1]*	*Nocardia mediterranei* mutant ATCC 13685	$C_{39}H_{48}N_2O_{12}S$			**(32)**	16
rifamycin verde	*[72428-95-6]*	*Nocardia mediterranei* mutant ATCC 13685	$C_{39}H_{46}N_2O_{12}S$			**(33)**	16
rifamycin G	*[59996-09-7]*	*Nocardia mediterranei*	$C_{36}H_{47}NO_{12}$	250–252		**(26)**	17
tolypomycin Y	*[23412-26-2]*	*Streptomyces tolypophorus*	$C_{43}H_{54}NO_{14}$	120	+326[f]	**(48)**	7
halomycin A	*[56411-51-9]*	*Micromonospora halophytica*	$C_{43}H_{58}N_2O_{13}$	192–194	+100.5	**(54)**	18
halomycin B	*[54356-09-1]*	*Micromonospora halophytica*	$C_{43}H_{58}N_2O_{12}$	178–182	+73.1[c]	**(55)**	19,20
halomycin C	*[64419-06-3]*	*Micromonospora halophytica*	$C_{43}H_{58}N_2O_{13}$		+153	**(56)**	18
naphthomycin A	*[55557-40-9]*	*Streptomyces collinu* Lindenbein Tü 105	$C_{40}H_{46}ClNO_9$	ca 200	+432	**(59)**	21,22
naphthomycin B	*[86825-40-9]*	*Streptomyces galbus* Tü 353	$C_{39}H_{44}ClNO_9$	156–165	+412	**(60)**	23

Table 1. (*Continued*)

Ansamacrolide	CAS Registry Number	Producing organism	Molecular formula	MP, °C	$[\alpha]_D$ [a]	Structure no.	Reference
naphthomycin C	*[86825-87-8]*	*Streptomyces diastatochromogenes* Tü 1892	$C_{39}H_{45}NO_9$		+117.7	**(61)**	23
naphthomycin D	*[105225-04-5]*	*Streptomyces* Tü 2357	$C_{40}H_{47}NO_{10}$		+323	**(62)**	22
naphthomycin E	*[105225-03-4]*	*Streptomyces* Tü 2357	$C_{40}H_{47}NO_9$		+133.5	**(63)**	22
naphthomycin F	*[105225-01-2]*	*Streptomyces* Tü 2357	$C_{46}H_{56}N_2O_{12}S$		+333.4	**(64)**	22
naphthomycin G	*[105225-02-3]*	*Streptomyces* Tü 2357	$C_{45}H_{54}N_2O_{12}S$	188	+254[c]	**(65)**	22
naphthomycin H[g]	*[98525-20-3]*	*Streptomyces* Y-83,40369	$C_{39}H_{44}ClNO_9$	150	+218	**(66)**	24,25
naphthoquinomycin A	*[101190-62-9]*	*Streptomyces* S-1998	$C_{40}H_{47}NO_{10}$	173–182	+212	**(67)**	25
naphthoquinomycin B	*[101190-61-8]*	*Streptomyces* S-1998	$C_{40}H_{47}NO_9S$	171–180	+541	**(68)**	25
actamycin	*[76045-67-5]*	*Streptomyces* E/784	$C_{39}H_{45}NO_{10}$	190–192		**(69)**	26
diastovaricin I	*[102281-52-7]*	*Streptomyces diastochromogenes* (variabilicolor)	$C_{39}H_{45}NO_{10}$	237–238	+351[c]	**(70)**	27
diastovaricin II	*[102228-99-9]*	*Streptomyces diastochromogenes* (variabilicolor)	$C_{44}H_{52}N_2O_{12}S$	160–164	+281[c]	**(71)**	27
kanglemycin A	*[114153-91-2]*	*Nocardia mediterranei* var. *kanglensis* 1747-64	$C_{50}H_{63}NO_{19}$	156[d]	+315.4[c]	**(74)**	28
awamycin	*[87913-35-7]*	*Streptomyces albolongus* C-46466	$C_{38}H_{49}NO_{12}S$	162–165	+949	**(75)**	29
ansathiazin	*[105645-37-2]*	*Streptomyces albolongus* C-46366	$C_{37}H_{47}NO_{13}S$	170–175	−32	**(76)**	29
Benzoquinoid							
geldanamycin	*[30562-34-6]*	*Streptomyces hygroscopicus* var. *geldanus* var. *nova* (UC-5208)	$C_{29}H_{40}N_2O_9$	252–255	+55	**(77)**	30
herbimycin A	*[70563-58-5]*	*Streptomyces hygroscopicus* AM-3672	$C_{30}H_{42}N_2O_9$	230	+137	**(82)**	31,32
herbimycin B	*[76207-83-5]*	*Streptomyces hygroscopicus* AM-3672	$C_{28}H_{38}N_2O_8$	229[d]	+109	**(83)**	33
herbimycin C	*[91700-92-7]*	*Streptomyces hygroscopicus* AM-3672	$C_{29}H_{40}N_2O_9$	203[d]	+210	**(84)**	34
macbecin I	*[73341-72-7]*	*Nocardia* sp. C-14919 (N-2001)	$C_{30}H_{42}N_2O_8$	187–188	+351	**(90)**	35,36
macbecin II	*[73341-73-8]*	*Nocardia* sp. C-14919 (N-2001)	$C_{30}H_{44}N_2O_8$	148[d]	+62	**(91)**	35,36
mycotrienin I	*[82189-03-5]*	*Streptomyces rishirensis* T-23	$C_{36}H_{48}N_2O_8$	117	+92[c]	**(92)**	37,38
mycotrienin II	*[82189-04-6]*	*Streptomyces rishirensis* T-23	$C_{36}H_{50}N_2O_8$	151	+288[c]	**(93)**	37–39
mycotrienol I	*[81904-15-6]*	*Streptomyces rishirensis* T-23	$C_{26}H_{33}NO_6$	94–95	+4.3[c]	**(95)**	40
mycotrienol II	*[84906-73-0]*	*Streptomyces rishirensis* T-23C	$C_{26}H_{35}NO_6$	130–132	+273[c]	**(96)**	40
20-*O*-methylmycotrienin II	*[98873-82-6]*	*Streptomyces rishirensis* T-23	$C_{37}H_{52}N_2O_8$	128	+373[c]	**(94)**	41

trienomycin A[h]	*[97955-33-4]*	*Streptomyces rishirensis* T-23 *Streptomyces* sp. No. 83-16	$C_{36}H_{50}N_2O_7$	135	+174[c]	**(97)**	41–43
trienomycin B	*[100662-01-9]*	*Streptomyces* sp. No. 83-16	$C_{34}H_{48}N_2O_7$	124–126	+170[c]	**(98)**	44
trienomycin C	*[100662-00-8]*	*Streptomyces* sp. No. 83-16	$C_{34}H_{48}N_2O_7$	119.5–121.5	+186[c]	**(99)**	44
ansatrienin A	*[80111-47-3]*	*Streptomyces collinus* strain Tü 1982	$C_{36}H_{48}N_2O_8$			**(100)**	45
ansatrienin A_2	*[85819-32-4]*	*Streptomyces collinus* strain Tü 1982	$C_{34}H_{46}N_2O_8$	115[d]	+115.7	**(101)**	46
ansatrienin A_3	*[85819-31-4]*	*Streptomyces collinus* strain Tü 1982	$C_{34}H_{46}N_2O_8$	117[d]	+119.4	**(102)**	46
ansatrienin B	*[80111-48-4]*	*Streptomyces collinus* strain Tü 1982	$C_{34}H_{46}N_2O_8$			**(103)**	45
maytansine	*[35486-53-8]*	*Maytenus ovatus* Loes. *Maytenus serrata* (Celastraceae)	$C_{34}H_{46}ClN_3O_{10}$	171–172	−145	**(104)**	47,48
maytanprine	*[38997-09-0]*	*Maytenus buchananii* (Loes.) R. Wilczek *Maytenus serrata* (Celastraceae)	$C_{35}H_{48}ClN_3O_{10}$	169–170	−125	**(106)**	48,49
maytanbutine	*[38997-10-3]*	*Maytenus buchananii* (Loes.) R. Wilczek *Maytenus serrata* (Celastraceae)	$C_{36}H_{50}ClN_3O_{10}$	170–171	−122	**(107)**	48,49
maytanvaline	*[52978-27-5]*	*Maytenus buchananii* (Loes.) R. Wilczek	$C_{37}H_{52}ClN_3O_{10}$	175–176.5	−135	**(105)**	48,50
maysine	*[52978-28-6]*	*Maytenus buchananii* (Loes.) R. Wilczek	$C_{28}H_{35}ClN_2O_7$	137–141	−173[f]	**(110)**	48,50
normaysine	*[52978-29-7]*	*Maytenus buchananii* (Loes.) R. Wilczek	$C_{27}H_{33}ClN_2O_7$	187–188	−217[f]	**(111)**	48,50
maysenine	*[52978-30-0]*	*Maytenus buchananii* (Loes.) R. Wilczek	$C_{27}H_{33}ClN_2O_6$	184–185	−57[f]	**(112)**	48,50
colubrinol	*[50657-35-5]*	*Colubrina texensis* Gray (Rhamnaceae) *Trewia nudiflora* L. (Euphorbiaceae)	$C_{36}H_{50}ClN_2O_{11}$	194–196	−94	**(114)**	51,52
colubrinol acetate	*[50499-79-1]*	*Colubrina texensis* Gray (Rhamnaceae)	$C_{36}H_{50}ClN_2O_{11}$	179–182	−127	**(115)**	51
maytanacine	*[57103-69-2]*	*Putterlickia verrucosa* Szysyl. (Celastraceae)	$C_{30}H_{39}ClN_2O_9$	234–237	−119	**(109)**	48,53
maytansinol	*[57103-68-1]*	*Putterlickia verrucosa* Szysyl. (Celastraceae)	$C_{28}H_{37}ClN_2O_9$	173–174.5	−309	**(113)**	48,53
maytanbutacine	*[62414-95-3]*	*Maytenus serrata* (Celastraceae)	$C_{34}H_{45}ClN_2O_{11}$	253–255	−90[f]	**(108)**	48
normaytansine	*[75983-74-3]*	*Maytenus buchananii*	$C_{32}H_{41}ClN_2O_8$	166–167		**(116)**	54
trewiasine	*[78987-26-5]*	*Trewia nudiflora* L. (Euphorbiaceae)	$C_{37}H_{52}ClN_3O_{11}$	182–185	−94	**(117)**	55
dehydrotrewiasine	*[78987-27-6]*	*Trewia nudiflora* L. (Euphorbiaceae)	$C_{37}H_{50}ClN_3O_{11}$	165–170	−90	**(118)**	55
demethyltrewiasine	*[78987-28-7]*	*Trewia nudiflora* L. (Euphorbiaceae)	$C_{36}H_{50}ClN_3O_{11}$	129–142	−126	**(119)**	55
trenudine	*[82390-94-1]*	*Trewia nudiflora* L. (Euphorbiaceae)	$C_{36}H_{48}ClN_3O_{13}$	200–205[d]	−114	**(121)**	55,56
treflorine	*[82390-93-0]*	*Trewia nudiflora* L. (Euphorbiaceae)	$C_{36}H_{48}ClN_3O_{12}$	205–208[d]	−138	**(120)**	55,56
normaytancyprine	*[84123-43-3]*	*Putterlickia verucosa* Szyszyl. (Celastraceae)	$C_{37}H_{50}ClN_3O_{10}$	143–145		**(125)**	57
N-methyltrenudone	*[82400-19-9]*	*Trewia nudiflora* L. (Euphorbiaceae)	$C_{37}H_{48}ClN_3O_{13}$	192–197[d]	−110	**(122)**	56
10-epitrewiasine	*[88198-82-7]*	*Trewia nudiflora* L. (Euphorbiaceae)	$C_{37}H_{53}ClN_3O_{11}$	159–162	−48	**(123)**	52
nortrewiasine	*[88147-94-8]*	*Trewia nudiflora* L. (Euphorbiaceae)	$C_{36}H_{50}ClN_3O_{11}$	155–158	−58	**(124)**	52
ansamitocin P-3	*[66584-72-3]*	*Nocardia* sp. C-15003	$C_{32}H_{43}ClN_2O_9$	190–191	−136	**(127)**	58
ansamitocin P-3′	*[66547-09-9]*	*Nocardia* sp. C-15003	$C_{32}H_{43}ClN_2O_9$	182–185	−134	**(128)**	58
ansamitocin P-4	*[66547-10-2]*	*Nocardia* sp. C-15003	$C_{33}H_{45}ClN_2O_9$	177–180	−142	**(129)**	58

Table 1. (*Continued*)

Ansamacrolide	CAS Registry Number	Producing organism	Molecular formula	MP, °C	$[\alpha]_D$ [a]	Structure no.	Reference
ansamitocin PND-0	[*77353-66-3*]	*Nocardia* sp. C-14482 mutant N-1231	$C_{27}H_{35}ClN_2O_8$	189–191[d]		**(131)**	59
ansamitocin PND-1	[*77353-67-4*]	*Nocardia* sp. C-14482 mutant N-1231	$C_{29}H_{37}ClN_2O_9$		−55.8[f]	**(132)**	59
ansamitocin PND-2	[*77353-68-5*]	*Nocardia* sp. C-14482 mutant N-1231	$C_{30}H_{39}ClN_2O_9$		−56.3[f]	**(133)**	59
ansamitocin PND-3	[*77353-69-6*]	*Nocardia* sp. C-14482 mutant N-1231	$C_{31}H_{41}ClN_2O_9$	226–228[d]	−57.1[f]	**(134)**	59
ansamitocin PND-4	[*77353-70-9*]	*Nocardia* sp. C-14482 mutant N-1231	$C_{32}H_{43}ClN_2O_9$		−56.6[f]	**(135)**	59
ansamitocin PHM-1	[*78619-40-6*]	*Nocardia* sp. C-14482 mutant N-1231	$C_{30}H_{39}ClN_2O_{10}$			**(136)**	59
ansamitocin PHM-2	[*78619-39-3*]	*Nocardia* sp. C-14482 mutant N-1231	$C_{31}H_{41}ClN_2O_{10}$			**(137)**	59
ansamitocin PHM-3	[*78619-38-2*]	*Nocardia* sp. C-14482 mutant N-1231	$C_{32}H_{43}ClN_2O_{10}$	192–194[d]	−148[f]	**(138)**	59
ansamitocin PHM-4	[*78630-36-1*]	*Nocardia* sp. C-14482 mutant N-1231	$C_{33}H_{45}ClN_2O_{10}$		−148[f]	**(139)**	59
ansamitocin P-4-βHY	[*78619-41-7*]	*Nocardia* sp. C-14482 mutant N-1231	$C_{33}H_{45}ClN_2O_{10}$	201–203[d]		**(140)**	59
ansamitocin P-4-γHY	[*78709-93-0*]	*Nocardia* sp. C-14482 mutant N-1231	$C_{33}H_{45}ClN_2O_{10}$	205–207[d]		**(141)**	59
ansamitocin PND-4-βHY	[*78619-43-9*]	*Nocardia* sp. C-14482 mutant N-1231	$C_{32}H_{43}ClN_2O_{10}$			**(142)**	59
ansamitocin deClQ-O	[*78630-38-3*]	*Nocardia* sp. C-14482 mutant N-1231	$C_{28}H_{38}N_2O_7$			**(145)**	59
ansamitocin QND-O	[*78619-44-0*]	*Nocardia* sp. C-14482 mutant N-1231	$C_{27}H_{35}ClN_2O_7$			**(144)**	59
ansamitocin deClQND-O	[*78630-37-2*]	*Nocardia* sp. C-14482 mutant N-1231	$C_{27}H_{36}N_2O_7$			**(143)**	59
ansamitocin PHO-3	[*62414-96-4*]	*Nocardia* sp. C-14482 mutant N-1231	$C_{32}H_{43}ClN_2O_{10}$	227–229[d]	−95.9[f]	**(130)**	59
ansamitocin P-2	[*57103-70-5*]	*Nordardia* sp. C-15003	$C_{31}H_{41}ClN_2O_9$	188–190[d]	−127	**(126)**	58

[a]Determined in $CHCl_3$ unless otherwise noted.
[b]Originally called streptovaricin F.
[c]Determined in methanol.
[d]Decomposes at this temperature.
[e]Determined in dioxane.
[f]Determined in ethanol.
[g]Same as naphthoquinomycin C.
[h]Same as 23-deoxymycotrienin II.

Table 2. Biological Activity of the Ansamacrolides

Ansamacrolide	Biological activity
streptovaricins	antibacterial (gram-positive and mycobacteria), antiviral, inhibitors of reverse transcriptase
rifamycins	antibacterial (gram-positive, gram-negative, and mycobacteria), antiviral, inhibitors of reverse transcriptase
tolypomycins	antibacterial (gram-positive)
halomycins	antibacterial (gram-positive)
naphthomycins	antibacterial (gram-positive), vitamin K antagonist
actamycins	inhibitors of fatty acid synthesis
diastovaricins	antileukemic
kanglemycins	antibacterial (gram-positive)
awamycins	antibacterial (gram-positive), antitumor
ansathiazins	antibacterial (gram-positive)
geldanamycins	antiprotozoal, herbicidal, inhibitors of reverse transcriptase
herbimycins	herbicidal, antitumor, antiviral, inhibitors of tyrosine kinase
macbecins	antibacterial (gram-positive), antiprotozoal, antifungal, antitumor
mycotrienins	antifungal
trienomycins	antitumor
ansatrienins	antifungal
maytansinoids	antileukemic, antitumor
ansamitocins	antiprotozoal, antifungal, antitumor

are the primary components of the crude complex, yielded substances shown in Figure 1. Streptovaricin A (**4**) consumes two moles of sodium periodate to yield varicinal A [*21913-68-8*] (**1**), $C_{13}H_{20}O_7$, which accounts for the aliphatic portion of the molecule, and prestreptovarone [*58074-37-6*] (**2**), $C_{29}H_{29}NO_9$, which accounts for the aromatic chromophore of the streptovaricins (Fig. 2). Streptovaricin G (**9**) is the only other streptovaricin that yields prestreptovarone upon treatment with sodium periodate. Treatment of streptovaricins A (**4**), B (**5**), C (**6**), E (**8**), and G (**9**) with sodium periodate and osmium tetroxide yields streptovarone [*36108-44-8*] (**3**), $C_{24}H_{23}NO_9$, which is also produced by the reaction of prestreptovarone with sodium periodate and osmium tetroxide (4,65). A number of aliphatic products were isolated from the oxidation of streptovaricin C and its derivatives (66).

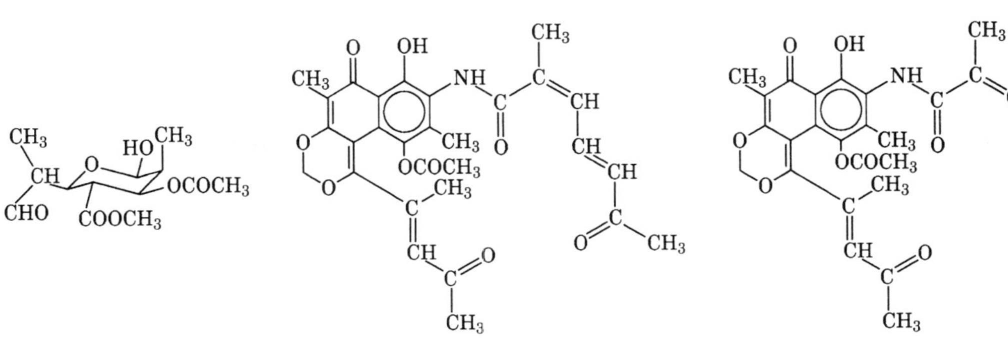

(**1**) (**2**) (**3**)

Fig. 1. Products from the chemical degradation of streptovaricins A and C.

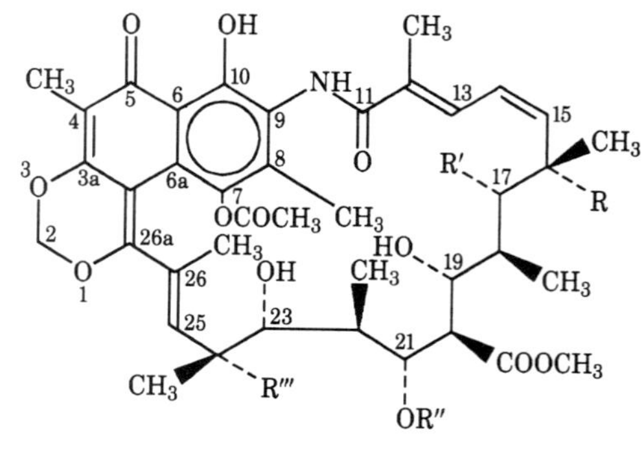

streptovaricin		R	R′	R″	R‴
A	(**4**)	OH	OH	$COCH_3$	OH
B	(**5**)	H	OH	$COCH_3$	OH
C	(**6**)	H	OH	H	OH
D	(**7**)	H	OH	H	H
E	(**8**)	H	O=	H	OH
G	(**9**)	OH	OH	H	OH
J	(**10**)	H	$OCOCH_3$	H	OH
K	(**11**)	OH	$OCOCH_3$	H	OH

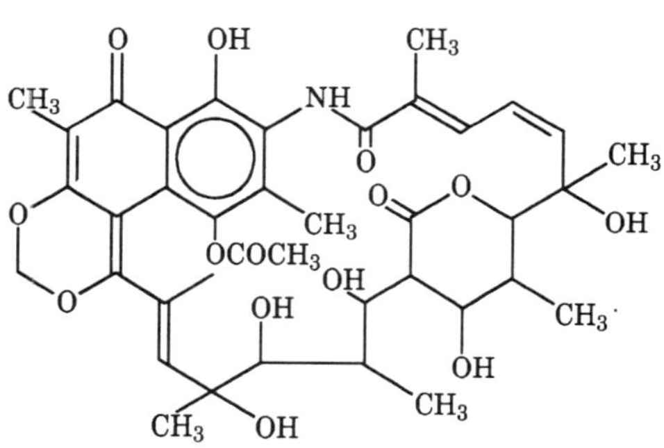

streptovaricin F_G (**12**)

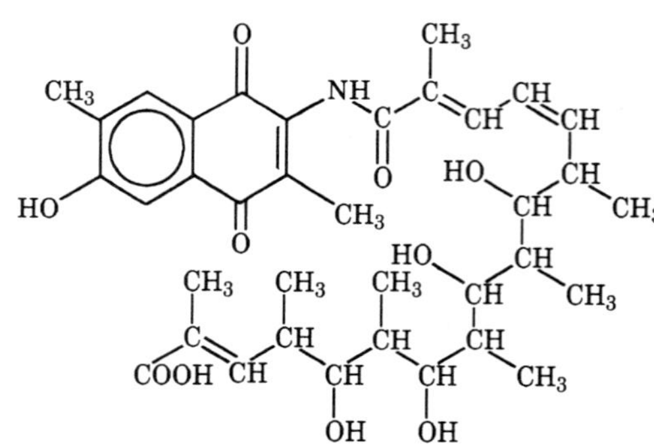

streptovaricin U (**13**)

Fig. 2. Structures of the streptovaricins.

The relationships among the various streptovaricins were shown by the following reactions: streptovaricin E (**8**) converts to streptovaricin C (**6**) upon treatment with sodium borohydride; streptovaricins A (**4**), G (**9**), and K (**11**) yield the same triacetate derivative upon acetylation; streptovaricins B (**5**), C (**6**), and J (**10**) yield the same tri- and tetraacetate derivatives upon acetylation; streptovaricin G (**9**) converts to streptovaricin F_G (**12**) upon treatment with base (4). The open-chain streptovaricin U (**13**) has also been isolated from the streptovaricin complex (5).

The structures of the streptovaricins were confirmed by x-ray crystallographic studies (67,68). The absolute configuration of the ansa-bridge is 6(*R*), 7(*R*), 8(*R*), 9(*R*), 10(*S*), 11(*S*), 12(*R*), 13(*S*), and 14(*R*) with a helicity of *P* (69). Complete mass spectral (4,70,71) and ^{13}C nmr studies have been reported (72).

Chemical Properties and Derivatives. All of the streptovaricins except streptovaricin D (**7**) react with one mole of sodium periodate to yield the corresponding streptovals (4), shown in Table 3. The streptovaricins undergo thermal isomerization to the corresponding atropisostreptovaricins (73). In the natural streptovaricins the ansa-bridge lies above the aromatic nucleus but in the atropisostreptovaricins this bridge lies below the aromatic nucleus. Most spectral properties of the isomers are nearly identical, but the optical rotations, although of approximately equal magnitude, are of opposite sign. If pyridine is used as the solvent, lactonized streptovaricins are obtained similar to streptovaricin F_G (**12**) as well as the atropisostreptovaricins. Treatment of streptovaricin C or D with oxygenated concentrated ammonia–methanol solution yields damavaricin C (**20**) or D (**21**) through loss of the enol acetate and methylenedioxy groups (74) (Table

Table 3. Derivatives of the Streptovaricins

Compound	CAS Registry Number	Molecular formula	Structure number	R	R′	R″
Streptovals						
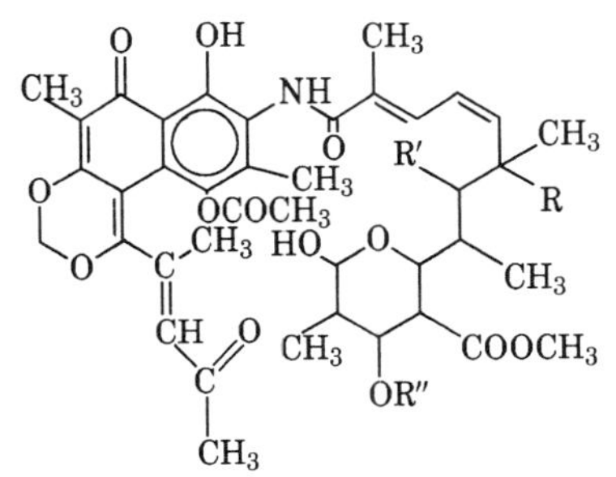						
streptoval A	[*60760-75-0*]	$C_{42}H_{51}NO_{16}$	(**14**)	OH	OH	$COCH_3$
streptoval B	[*60735-99-1*]	$C_{42}H_{51}NO_{15}$	(**15**)	H	OH	$COCH_3$
streptoval C	[*54955-14-5*]	$C_{40}H_{49}NO_{14}$	(**16**)	H	OH	H
streptoval E	[*60736-00-7*]	$C_{40}H_{47}NO_{14}$	(**17**)	H	O	H
streptoval G	[*60736-02-9*]	$C_{40}H_{49}NO_{15}$	(**18**)	OH	OH	H
streptoval J	[*60736-03-0*]	$C_{42}H_{51}NO_{15}$	(**19**)	H	$COCH_3$	H
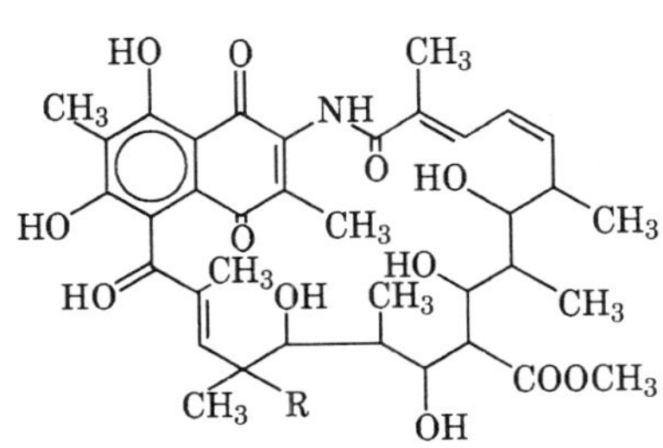						
Damavaricins						
damavaricin C	[*58849-86-8*]	$C_{37}H_{47}NO_{13}$	(**20**)	OH		
damavaricin D	[*59556-95-5*]	$C_{37}H_{47}NO_{12}$	(**21**)	H		

3). The damavaricins are related to the naturally occurring rifamycin W, and damavaricin D has been isolated from the streptovaricin complex. Besides damavaricins C and D, the basic methanolysis yields the corresponding atropisodamavaricins and lactonized damavaricins. The streptovaricins form acetonides and are hydrogenated to mixtures of di- and tetrahydro derivatives (4). Although few derivatives of the streptovaricins have been prepared, a large number of 19-*O*-substituted damavaricins (75) and lactonized damavaricins (76–78) have been reported. The synthesis of the naphthalene core of streptovaricin D has been accomplished (79) whereas several routes to the ansa-chain of the streptovaricins have been reported (80–83).

Biological Activity. The streptovaricins are mainly active against gram-positive organisms, and have very good activity against *Mycobacterium tuberculosis, M. leprae,* and *Staphylococcus aureus* infections in animals (60,84). The totally acetylated streptovaricins lose their antibacterial activity. Streptovaricin U has no antibacterial activity. Lactonization greatly reduces any biological activity a streptovaricin possesses. The antibacterial activities of the atropisostreptovaricins are about the same as the natural isomers. Although the damavaricins retain biological activity, they are not as active as the corresponding streptovaricins. However, many of the 19-*O*-substituted damavaricins exhibit greater antibacterial activity than the parent streptovaricins (76). The streptovals are also inactive as antibacterial agents (4).

The streptovaricins inhibit the reverse transcriptase of some RNA oncogenic viruses that may be involved in the process of viral transformation (see ANTIVIRAL AGENTS). The atropisostreptovaricins again have similar activities to the corresponding natural isomers. The streptovals and streptovarone exhibit greatly improved activity against reverse transcriptase relative to the streptovaricins (85), but their *in vitro* activities were low (86). The damavaricins also inhibit reverse transcriptase (4) as well as tumor cell growth (87).

Clinical studies with streptovaricins have been limited. The most significant toxicity reported in the clinical studies has been gastrointestinal disturbances. However, indications are that the toxicity does not result from the individual streptovaricins but from some impurity in some fermentation lots (85).

Assay Methods. The primary assay for the streptovaricins is the microbiological assay using the agar diffusion method or a turbidimetric procedure (60). The streptovaricins can also be identified by paper (60,88) or thin-layer chromatography (3).

Rifamycins. The rifamycins were first isolated from a broth of *Nocardia mediterranei* (the producing organism was originally identified as *Streptomyces mediterranei*). The rifamycins, the structures of which are shown in Figure 3, were originally designated as rifomycins. Only rifamycin B (**22**), which accounts for 10–15% of the crude complex, can be isolated easily as a stable, crystalline compound (6,89,90).

Difficulties encountered in the separation and stability of the individual rifamycins led to studies to increase the yield of individual components of the complex. The addition of sodium diethylbarbiturate to the fermentation medium of *N. mediterranei* resulted in the formation of rifamycin B as practically the only substance formed (91,92).

Rifamycin B (**22**) is oxidized by air to the more active rifamycin O (**23**) upon

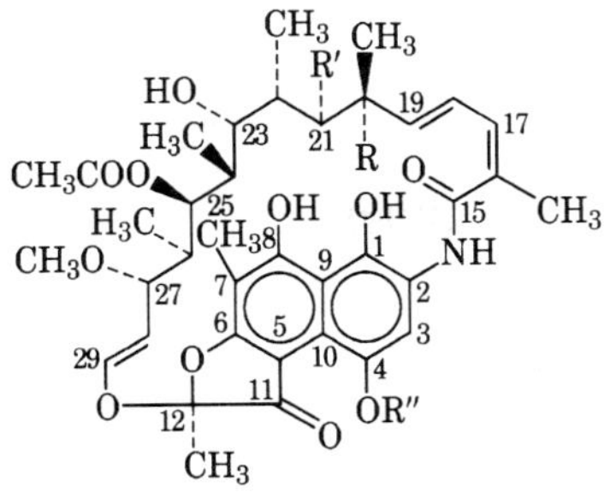

rifamycin	R	R′	R″
B (**22**)	H	OH	CH_2COOH
SV (**25**)	H	OH	H
L (**28**)	H	OH	$COCH_2OH$
Y (**29**)	OH	=O	CH_2COOH

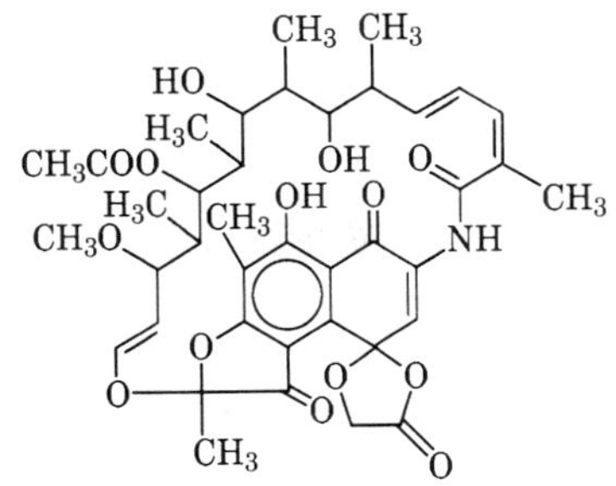

rifamycin O (**23**)

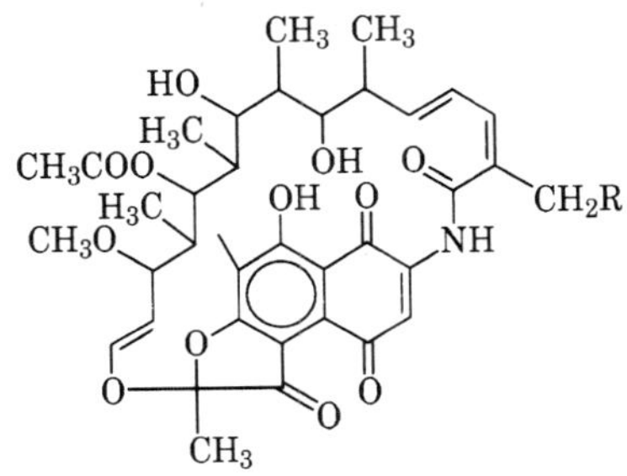

rifamycin S (**24**) R = H
rifamycin R (**27**) R = OH

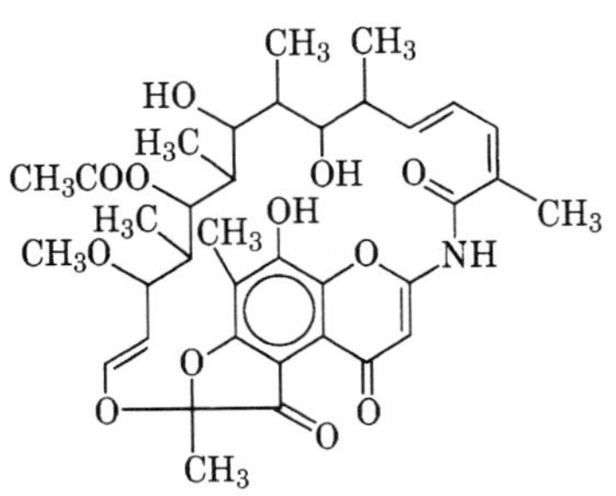

rifamycin G (**26**)

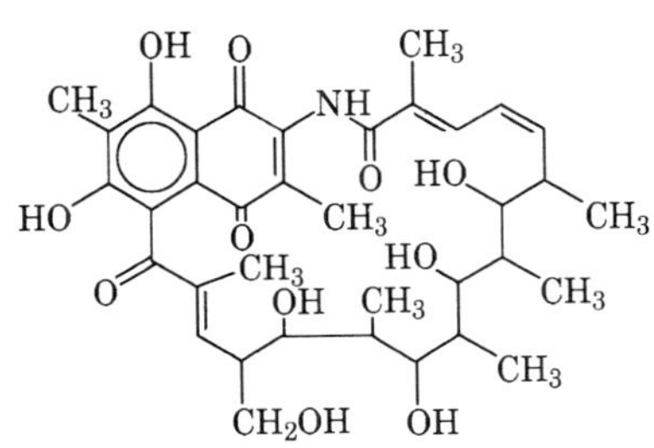

rifamycin W (**30**)

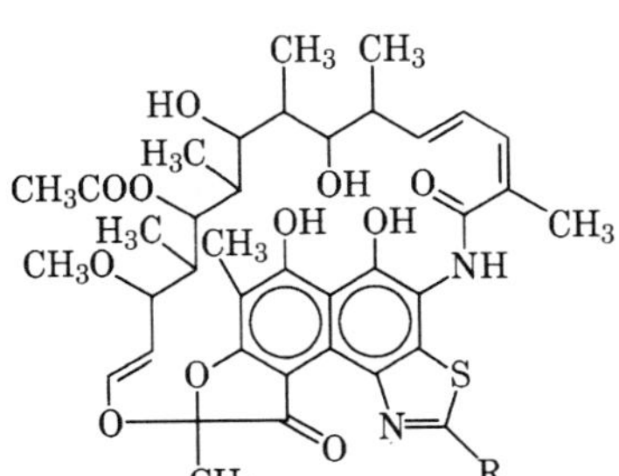

rifamycin P (**31**) R = H
rifamycin Q (**32**) R = CH_2OH

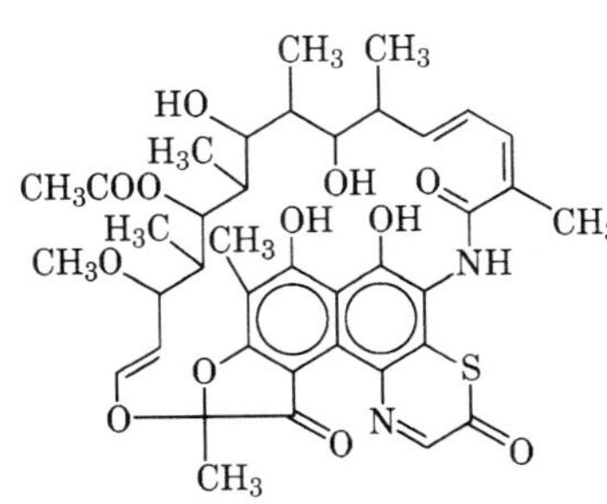

rifamycin verde (**33**)

Fig. 3. Structures of the rifamycins.

standing in solution. This reaction also occurs when rifamycin B is treated with a variety of oxidizing agents such as hydrogen peroxide. Rifamycin O can be reconverted to rifamycin B by treatment with ascorbic acid. Rifamycin O in turn can be hydrolyzed to the even more active rifamycin S (**24**) by the expulsion of glycolic acid. Rifamycin S can be reduced, using ascorbic acid, to rifamycin SV (**25**) and rifamycin SV can be oxidized to rifamycin S (5,93,94). Although rifamycins O and S were originally chemical derivatives of rifamycin B, microorganisms have been found that produce these rifamycins naturally.

Rifamycin G (**26**) is also isolated from the *Nocardia mediterranei* fermentation and appears to be a metabolic product of rifamycin S (17). Rifamycin R (**27**) is produced by a mutant of *N. mediterranei* (ATCC 31066) and appears to be biosynthesized from rifamycin S by the oxidation of the C-2 methyl group (15). Rifamycin L (**28**) is an oxidation product formed by incubating rifamycin S with washed mycelia of *N. mediterranei* (12). Rifamycin Y (**29**) is produced along with rifamycin B when this organism is cultured in the presence of diethylbarbituric acid (11). The amount of rifamycin Y produced was originally high, but through strain improvements, it is now produced in only small amounts (95). Rifamycin W (**30**) is produced by mutant 126 of *N. mediterranei* (13,14). The sulfur containing rifamycins P (**31**), Q (**32**), and verde (**33**) are produced by the mutant *Nocardia mediterranei* ATCC 31685 (16).

The structures of the rifamycins were arrived at by chemical degradation studies (1,4,96,97) and confirmed by x-ray crystallography (98–100). The absolute configuration of the ansa-bridge is 6(*S*), 7(*S*), 8(*R*), 9(*R*), 10(*R*), 11(*S*), 12(*R*), and 13(*S*) (101). Studies of ^{13}C nmr (4,13,14,102–104) and ir (104) have been reported and mass spectra of the rifamycins have been obtained (105–107).

Chemical Properties and Derivatives. There have been thousands of rifamycin derivatives prepared in an attempt to obtain a broader spectrum antibiotic having good oral absorption. Some of these are shown in Figure 4. Rifamycins B, O, and S have served as starting materials for the preparation of numerous classes of derivatives. Several of the semisynthetic derivatives are more active, have a broader spectrum of biological activity, and are therapeutically more effective than the parent antibiotics.

Treatment of rifamycin B (**22**) with amines, hydrazines, and alcohols yields the corresponding amides, hyrazides, and esters (**34**), respectively (4,6,108–110). The most active of this class is the *N,N*-diethyl derivative known as rifamide [*2750-76-7*], $C_{43}H_{58}N_2O_{13}$ (111). Rifamide is active against gram-positive bacteria and has very low toxicity (112–114). Good activity is exhibited against *M. tuberculosis*.

Rifamycin O (**23**) reacts with a variety of aromatic amines, hydrazines, amidrazones, and aminoguanidines to yield quinonimine derivatives (**35**). All of the derivatives show good activity against gram-positive bacteria. The condensation product with aminoguanidine, known as rifamycin AG, is a broad-spectrum antibiotic possessing good activity against gram-positive and gram-negative bacteria and *M. tuberculosis* (6,109,115).

Reaction of rifamycin S (**24**) with a variety of *o*-phenylenediamines and *o*-aminophenols produces a series of phenazines (**36**) and phenoxazines (**37**), respectively. Rifazine [*10238-70-7*] (**37**, R = H), $C_{43}H_{49}N_3O_{11}$, is the simplest of the phenazines (116–119).

Rifamycin S also undergoes conjugate addition reactions to the quinone ring by a variety of nucleophiles including ammonia, primary and secondary amines, mercaptans, carbanions, and enamines giving the C-3 substituted derivatives (**38**) of rifamycin SV (117,120,121). Many of the derivatives show excellent antibacterial properties (109,118,122,123). The 3-cyclic amino derivatives of rifamycin SV also inhibit the polymerase of RNA tumor viruses (123,124).

Treatment of rifamycin S with formaldehyde and secondary amines yields the 3-aminomethyl derivatives (**38**, X = CH_2NR_2) which are not of therapeutic interest (125). However, upon oxidation in acidic medium, the aminomethyl derivatives yield 3-formylrifamycin SV, also known as rifaldehyde [*13292-22-3*] (**38**, X = CHO), $C_{38}H_{47}NO_{13}$. Treatment of rifaldehyde with amines, hydrazines, hydroxylamines, and hydrazides yields a number of derivatives of the 3-formyl group having excellent biological activity (109,126–128). The most therapeutically useful derivative is the *N*-amino-*N*′-methylpiperazine hydrazone of rifaldehyde known as rifampin in the United States and rifampicin elsewhere.

Rifampicin [*13292-46-1*] (**39**), $C_{43}H_{58}N_4O_{12}$, is active against a variety of gram-positive and gram-negative bacteria; however, resistance develops rapidly if it is used alone (129–131). Rifampicin finds its greatest use in the treatment of tuberculosis (130,132–134). Because resistance develops rapidly, rifampicin is normally given in combination with other antituberculosis drugs, eg, isoniazid [*54-85-3*], $C_6H_7N_3O$ (130,132,135). Rifampicin has been used to treat nontuberculosis infections (136–138), with especially good activity against *Staphylococcus aureus* and *Staphylococcus epidermidis* (139).

Rifampicin has also shown antiviral activity but at levels 500–1000 times greater than required for antibacterial activity (130,140–142). Rifampicin shows promise in the treatment of leprosy (130,143). A large number of rifampicinlike derivatives are potent inhibitors of reverse transcriptase (123,144–148).

Rifapentine (DL473) [*61379-65-5*] (**40**), $C_{47}H_{64}N_4O_{12}$, a piperazinyl hydrazone of 3-formyl rifamycin SV, is similar to rifampicin. Rifapentine has activity similar to rifampicin, but its half-life is much longer (149,150).

A number of oxime derivatives of rifaldehyde have been prepared. Many of these derivatives exhibit good activity against rifampicin-resistant organisms (151,152).

Another class of active derivatives prepared from rifaldehyde, are the 3-aminomethylazinomethyl rifamycins (**41**). This class of compounds is prepared by treating rifaldehyde with hydrazine to yield hydrazones of the 3-aldehyde of rifamycin SV (**38**, X = $CH{=}NNH_2$), which is further treated with a suitable chloroformiminium chloride or the dimethyl acetals of tertiary amides to yield (**41**) (153–155). The piperidine analogue known as FCE 22250 (**42**) was chosen for further studies because of low toxicity coupled with high activity and a long half-life.

Three new classes of rifamycin derivatives were prepared from the common intermediate 3-amino-rifamycin S [*51756-80-0*] (**43**), $C_{37}H_{46}N_2O_{12}$. Reduction of (**43**) in the presence of an aldehyde yields derivatives of the general formula (**44**). Reaction of (**43**) and ammonia in THF solution yields (**45**) [*62041-01-4*] which in turn can be converted to imidazo-rifamycins (**46**) or to spiro-piperidyl-rifamycins (**47**) by treatment with aldehydes under reducing conditions or with ketones, respectively. All three classes of derivatives have *in vitro* antibacterial activities

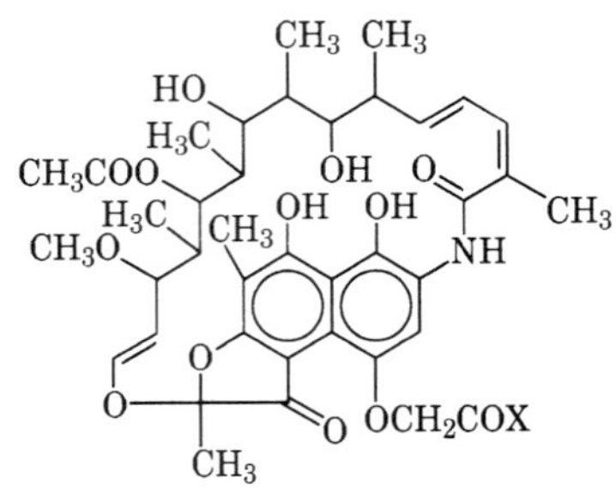

(34)

X = NR_2, $NHNR_2$, OR
X = $N(CH_2CH_3)_2$ = rifamide

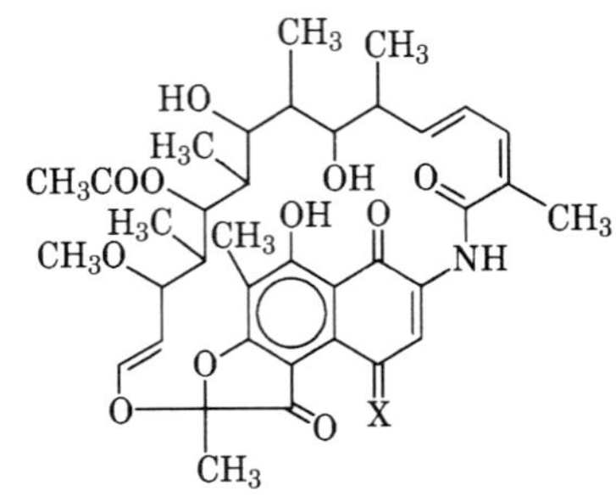

(35)

X = N-aryl
X = NNHCO-aryl
X = NNHC(NH)-aryl
X = NNHC(NH)-NHR
(rifamycin AG; R = H)

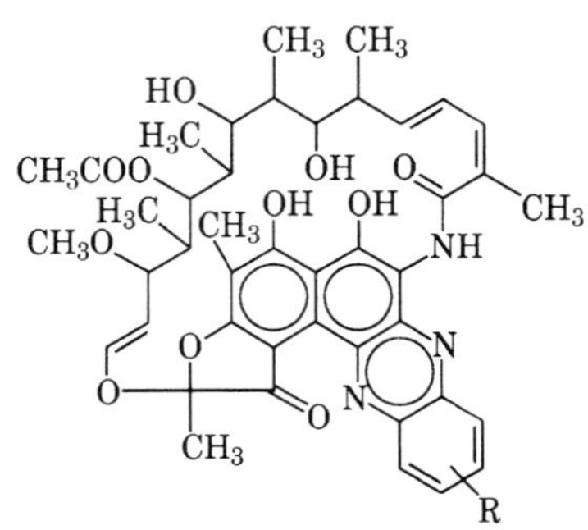

(36)

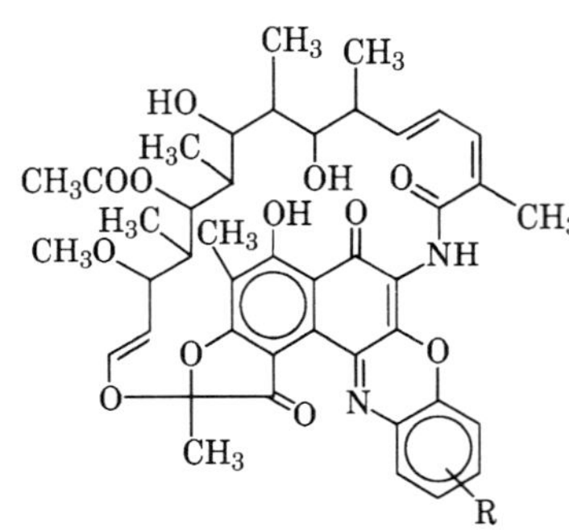

(37)

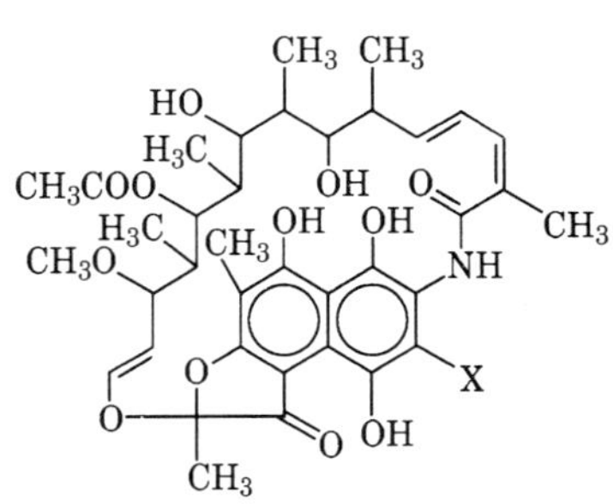

(38)

X = NRR′
X = SR
X = CHRR′
X = Cl, Br, CN

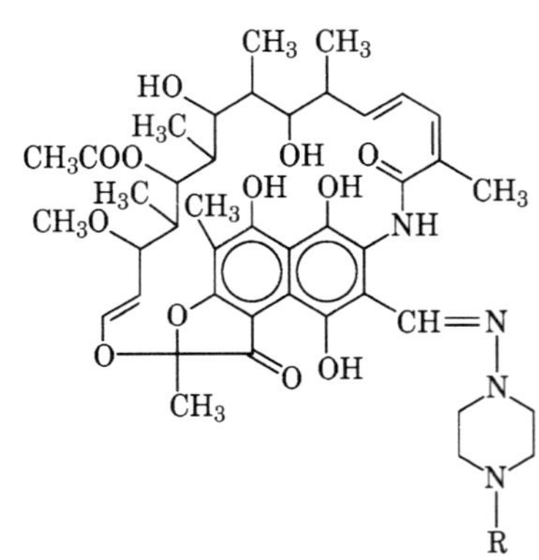

(39) R = CH_3
(40) R = cyclopentyl

Fig. 4. Structures of rifamycin derivatives.

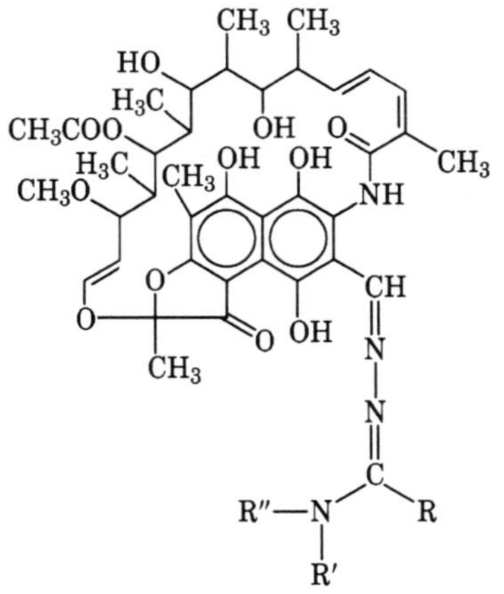

(**41**)

(**42**) R = H;
R′—R″ = $CH_2CH_2CH_2CH_2CH_2$

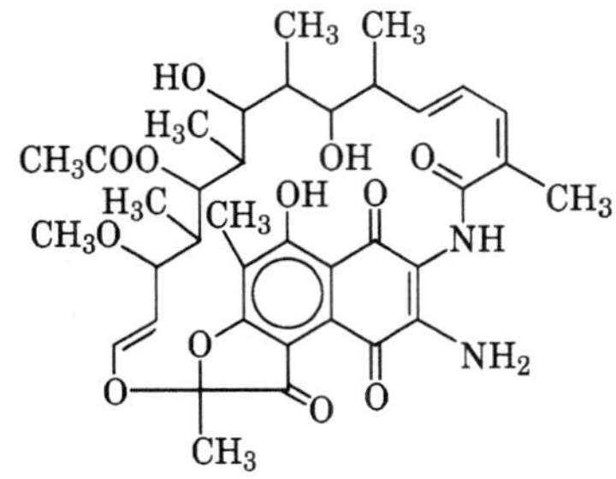

(**43**)

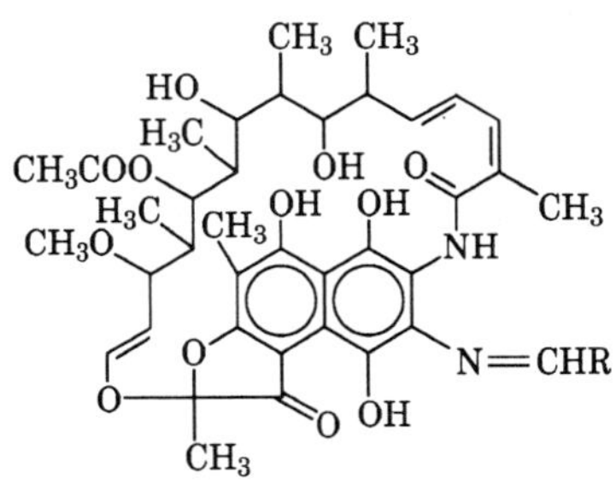

(**44**)

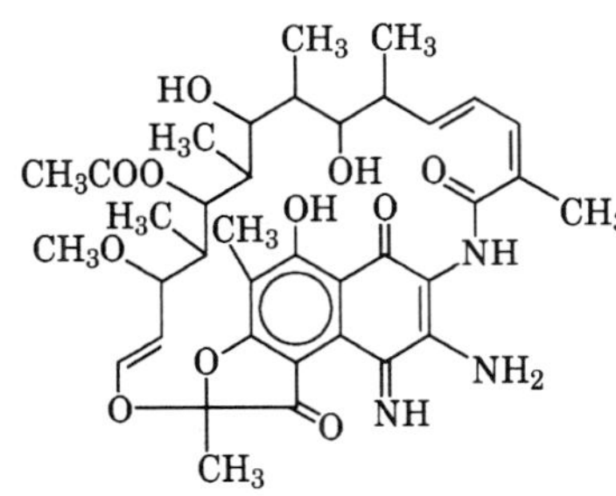

(**45**)

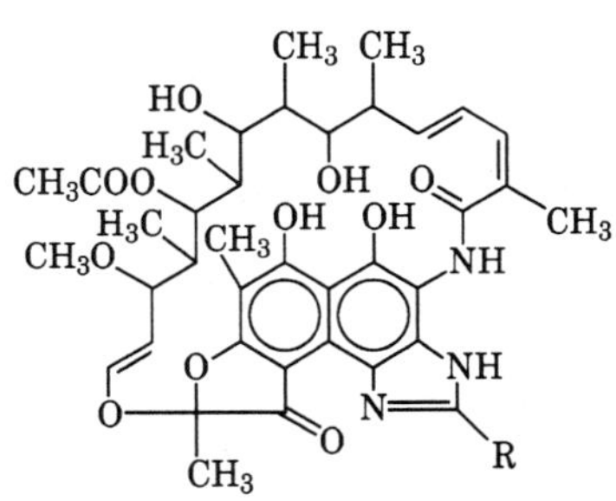

(**46**)

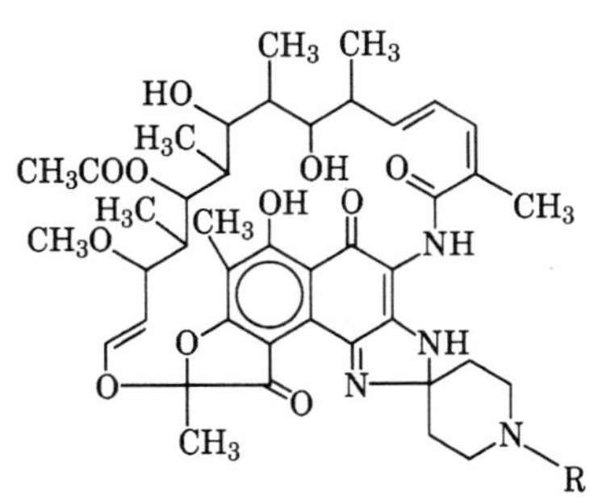

(**47**)

Fig. 4. *(Continued)*

comparable to those of rifampicin (156). The spiro-piperidyl-rifamycin rifabutin (LM 427) [*72559-06-9*] (**47**, R = $CH_2CH(CH_3)CH_3$), $C_{46}H_{62}N_4O_{11}$, showed good *in vitro* and *in vivo* activity against *Mycobacterium tuberculosis* (157) as well as displaying a broad spectrum of biological activity (158). It was also shown to be active against *M. aviumintracellulare* (159), which is frequently encountered in patients having acquired immune deficiency syndrome (AIDS). Rifabutin, used to treat AIDS patients having *Mycobacterium* infections, was shown to inhibit replication of human immunodeficiency virus type 1 (HIV-1) (160).

Modifications to the ansa-bridge of the rifamycins decreases the activity against gram-positive bacteria. Hydrogenation of the ansa-bridge double bonds increases the activity towards gram-negative bacteria as does removal of the C-11 acetate group (109).

Synthesis of the aromatic portion of rifamycin S has been accomplished (161) as has the ansa-bridge portion (162–164). Joining of these fragments resulted in rifamycin S total synthesis (165,166).

Biological Activity. The rifamycins are biologically active against gram-positive microorganisms and mycobacteria, particularly *M. tuberculosis* (6,167–170). At higher concentrations, the rifamycins are also active against gram-negative bacteria. The rifamycins show antiviral activity (140,141) and inhibit reverse transcriptase (171–173).

Rifamycin B is not biologically active but is spontaneously converted in aqueous solution to the active rifamycins O, S, and SV. Rifamycin SV was chosen for further studies because of its good *in vivo* activity, low toxicity, and solubility properties. Rifamycin SV is effective against a variety of infections as well as being active against tuberculosis and leprosy (168). Rifamycin P is the most active of the naturally occurring rifamycins (174).

Manufacture and Processing. Although fermentation procedures have not been reported, assumptions concerning fermentation media and optimal conditions have been made (95,174). The transformation of the biologically inactive rifamycin B to the biologically active rifamycin S is usually accomplished chemically. Several rifamycin B oxidases have been isolated that can enzymatically transform rifamycin B to rifamycin O, which is hydrolyzed in the fermentation medium to rifamycin S. The enzymes from *Monocillium* spp. ATC 20621 and *Humicola* spp. ATCC 20620 are intracellular (175,176) whereas the enzyme from *Curvularia lunata* var. *aeri* is extracellular (177). The use of a fluidized bed reactor containing immobilized whole cells of *Humicola* for the transformation of rifamycin B to rifamycin S has been described (178). Rifamycin SV producing strains have been isolated, but it is not known if these strains are used commercially.

Assay Methods. A large number of assays exist for the determination of the various rifamycins. Rifamycin SV and rifampicin can be determined by a microbiological assay using *Sarcina lutea* ATTC 9341 as the test organism (130,168), and rifampicin can be determined using *S. aureus* 560 (179). Rifamycins B, S, and SV can be separated by electrophoresis on agar gel and determined microbiologically using *B. subtilis* or *S. lutea* (180). Spectrophotometric assays exist for the rifamycins (130,168,181,182) and for rifampicin (183–185). Rifamycins B, O, S, and SV can be determined via polarography (186) or by amperometric titration (187). Rifamycins B, O, S, and SV can be separated by thin-layer chromatography on silica gel or by paper chromatography (188). Fluorimetric assays exist for rifamycin B (189) and rifampicin (179). High performance liquid chromatographic (hplc) procedures exist for rifamycins B, O, S, and SV (190), for rifampicin in formulations (191,192), in body fluids (193–197), in mixtures of antibiotics (198,199), and for rifapentine in plasma (200).

The *British Pharmacopoeia* specifies a biological assay for the sodium salt of rifamycin SV [*14897-39-3*]. It also specifies a spectrophotometric assay for rifampicin (201). The *United States Pharmacopeia* requires an hplc assay for rifampin (202).

Tolypomycins. The addition of small amounts of iron salts to the fermentation medium increases the production of tolypomycin Y (**48**) (7,203,204), the structure of which was arrived at by chemical degradation (205,206) and confirmed by x-ray crystallographic analysis (207) (Fig. 5). Mild acid hydrolysis of tolypomycin Y yields tolypomycinone [*22356-23-6*] (**49**, R = H), $C_{37}H_{43}NO_{13}$, and tolyposamine [*34174-76-0*], $C_{16}H_{13}NO_2$, (**50**). Further hydrolysis of tolypomycinone using acid yields tolyponone [*24317-12-2*] (**51**), $C_{14}H_{10}O_7$, which is also formed upon mild acid hydrolysis of rifamycin S. Reduction of tolypomycin Y produces tolypomycin R [*33889-22-4*] (**52**), $C_{43}H_{56}N_2O_{14}$ (208).

Tolypomycin Y (**48**) shows strong antibacterial activity against gram-positive bacteria and *Neisseria gonorrheae*. When administered by subcutaneous, intraperitoneal, and intravenous routes, tolypomycin Y is effective in mice infected with *Staphylococcus aureus, Streptococcus pyrogenes,* and *Diplococcus pneumoniae*. Cross-resistance is observed with rifampicin but not with other antibiotics. Resistance to tolypomycin Y develops rapidly. The bioactivity of tolypomycin R (**52**) is similar to that of tolypomycin Y (**48**) (208).

Tolypomycinone reacts with primary amines to form 3-*N*-substituted derivatives (209). The most interesting representative of this class of derivatives is 3-[β-(*N*-morpholyl)ethylamino]tolypomycinone (M/045) [*63793-73-7*] (**49**, R = $NHCH_2CH_2N(CH_2CH_2)_2O$), $C_{43}H_{55}N_3O_{14}$, which has greater activity against gram-positive bacteria than tolypomycin Y, whereas its activity towards gram-negative bacteria is similar to that of rifampicin. Because of greater stability in acidic solution, M/045 has a greater therapeutic effect than tolypomycin Y in mice (210).

A differential bioassay was developed to distinguish tolypomycin Y from rifamycin B (211).

Halomycins. The halomycins are a group of four antibiotics produced by *Micromonospora halophytica* and separated by partition chromatography on Chromosorb W coated with formamide (19). Further purification was accomplished using preparative tlc (212).

Treatment of halomycin B (**55**) using nitrous acid yields rifamycin S (**24**) and the pyrrolidine (**57**) as shown in Figure 6. The halomycin B structure was confirmed by heating rifamycin O (**23**) and (**57**) in tetrahydrofuran to yield halomycin B (20) which can also be converted to rifamycin S by electrochemical oxidation (213). Upon treatment with nitrous acid, halomycin A (**54**) yields rifamycin S along with the pyrrolidine (**58**). The structure for halomycin C (**56**) was determined to be 20-hydroxy halomycin B based on mass spectral data (212).

The halomycins are active against gram-positive bacteria. The halomycin complex exhibited high activity against bacterial strains resistant to penicillin G. Halomycin C appears to be the most active component of the complex (19). X-ray crystallographic studies indicated that the pyrrolidine ring alters the conformation of the ansa-chain relative to rifamycin SV by changing the spatial arrangements of the C-21 and C-23 hydroxyl groups, which are important for biological activity against DNA-dependent RNA polymerase (214).

Naphthomycins, Naphthoquinomycins, Actamycin, and Diastovaricins. The naphthomycins, shown in Figure 7, are a group of closely related antibiotics differing in the substituent at C-2 and C-30, and in the geometry about

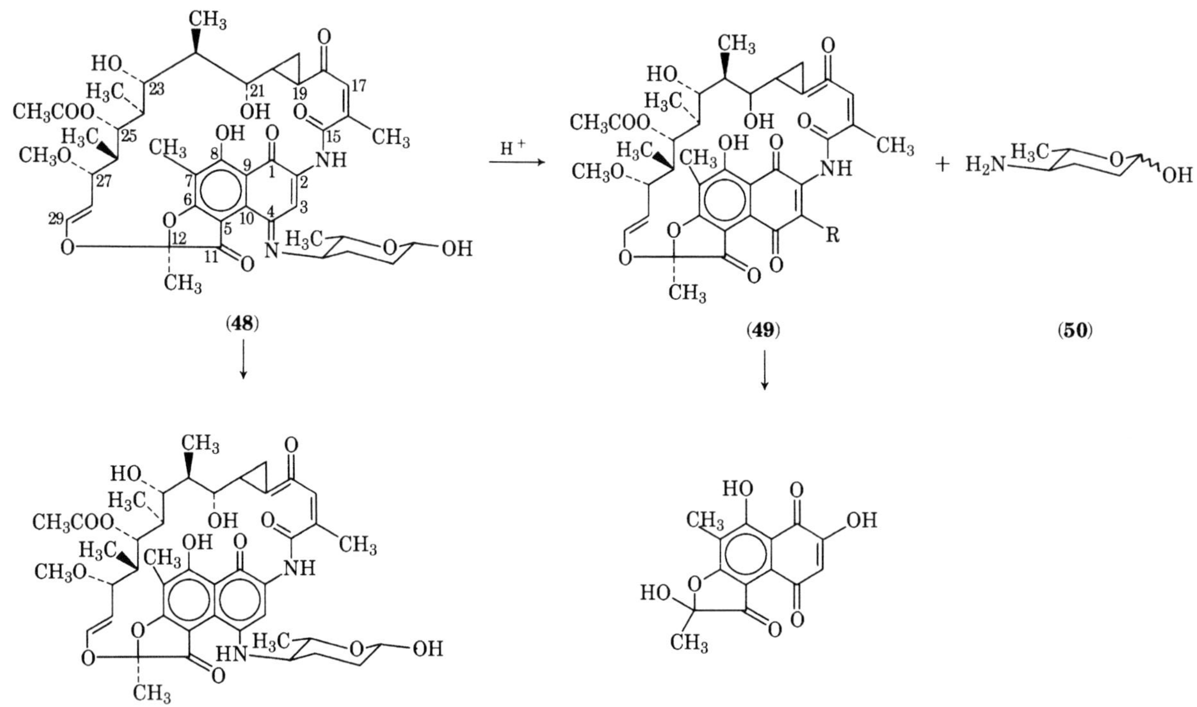

(48) (49) (50) (51) (52)

Fig. 5. Structures of the tolypomycins and degradation products.

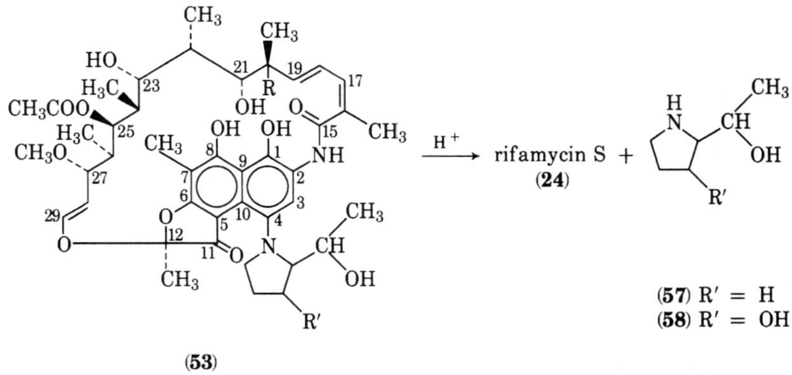

Fig. 6. Structures of the halomycins (**53**) and degradation products where for halomycin A (**54**) R = H, R′ = OH; for halomycin B (**55**) R = R′ = H; and for halomycin C (**56**) R = OH, R′ = H.

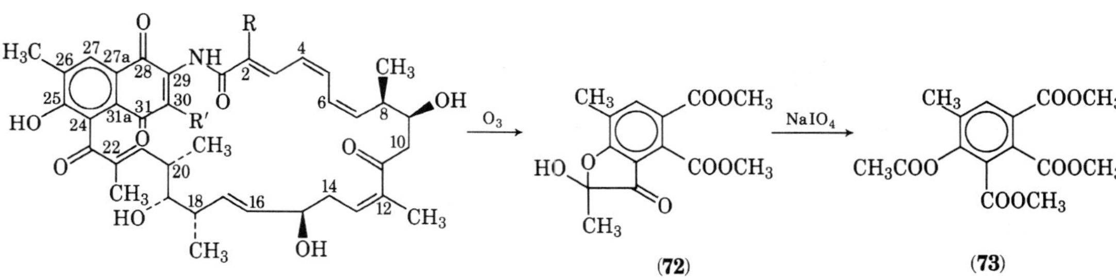

Fig. 7. Structures of the naphthomycins, naphthoquinomycins, actamycins, and diastovaricins.

Ansamacrolide	Substituent		Double-bond geometry					
	R	R′	C-2	C-4	C-6	C-12	C-16	C-21
naphthomycin A (**59**)	CH_3	Cl	Z	Z	E	E	E	E
naphthomycin B (**60**)	H	Cl	Z	E	Z	E	E	E
naphthomycin C (**61**)	H	H	Z	E	Z	E	E	E
naphthomycin D (**62**)	CH_3	OH	Z	Z	E	E	E	E
naphthomycin E (**63**)	CH_3	H	Z	Z	E	E	E	E
naphthomycin F (**64**)	CH_3	$SCH_2CH(NHCOCH_3)COOCH_3$	Z	Z	E	E	E	E
naphthomycin G (**65**)	CH_3	$SCH_2CH(NHCOCH_3)COOH$	Z	Z	E	E	E	E
naphthomycin H (**66**)	H	Cl	Z	Z	E	E	E	E
naphthoquinomycin A (**67**)	H	OCH_3	Z	Z	E	E	E	E
naphthoquinomycin B (**68**)	H	SCH_3	Z	Z	E	E	E	E
actamycin (**69**)	H	OH	Z	Z	E	E	E	E
diastovaricin I (**70**)	H	OH	Z	E	Z	E	E	E
diastovaricin II (**71**)	H	$SCH_2CH(NHCOCH_3)COOH$	Z	E	Z	E	E	E

the C-4 and C-6 double bonds. The naphthoquinomycins, diastovaricins, and actamycin are all closely related to the naphthomycins. Naphthomycin A (**59**) is isolated from a fermentation beer of *Streptomyces collinus* (Tü 105) (21). The structure originally proposed, based solely on nmr data, placed the phenolic group at C-27 rather than at C-25 as shown (215). However, spectral data for streptovaricin precursors and the reported value for the chemical shift of the phenolic proton, led to the proposal that the phenolic group was at C-25 (216). This was confirmed by degradation studies (217). Ozonolysis of (**59**) followed by treatment with diazomethane resulted in the isolation of (**72**) [*72033-45-5*], which, by reaction with sodium periodate, yielded (**73**) [*72033-46-6*] after treatment with diazomethane. The isolation of (**72**) and (**73**) confirmed the placement of the phenolic group at C-25. X-ray crystallographic studies confirmed the geometries assigned to the double bonds based on nmr studies and established the configuration as 8(*S*), 9(*S*), 15(*S*), 18(*S*), 19(*S*), and 20(*S*) (218). Naphthomycin A is active against gram-positive bacteria, but this activity is reversed by the addition of cysteine. It is also a vitamin K antagonist. Several sulfur-containing derivatives of naphthomycin A have been prepared (219) where the chlorine atom is replaced by an SR group. The biological activity of the C-3 *S*-substituted derivatives is similar to that of naphthomycin A.

Naphthomycin B (**60**) is produced by *Streptomyces galbus* (Tü 353) whereas naphthomycin C (**61**) is produced by *Streptomyces diastatochromogenes* (Tü 1892) (220). The structures for (**60**) and (**61**) were arrived at on the basis of nmr data in comparison with those of (**59**). The antibacterial activity of naphthomycin B is similar to that of naphthomycin A. Naphthomycin C is nearly inactive against bacteria and fungi. Naphthomycins D (**62**), E (**63**), F (**64**), and G (**65**) are isolated from a fermentation beer of *Streptomyces aurantiogriseus* (Tü 2357) (221). Structures were determined by nmr and degradation studies using ozone. Naphthomycin F exhibits some activity against gram-positive bacteria and fungi while naphthomycins D, E, and G are biologically inactive. Naphthomycin H (**66**) is isolated from *Streptomyces* Y-83,40369 and was shown to be a geometrical isomer of naphthomycin B (222). The biological activity of naphthomycin H is similar to that of naphthomycins A and B.

Naphthoquinomycins A (**67**) and B (**68**) are isolated from *Streptomyces S*-1998 (223) and the structures for (**67**) and (**68**) assigned on the basis of spectral data. Naphthoquinomycins A and B inhibit fatty acid synthesis in *E. coli*. Actamycin (**69**) is obtained from *Streptomyces* sp. E/784, and its structure arrived at on the basis of spectral data and degradation studies (224,225).

Diastovaricins I (**70**) and II (**71**) are produced by *Streptomyces diastochromogenes*. Diastovaricins I and II are active against Friend mouse leukemia cells. Spectral data were used to determine the structures (226).

Kanglemycin. Kanglemycin (**74**) (Fig. 8) is isolated from the fermentation broth filtrate of *Nocardia mediterranei* var *kanglensis* (1747-64) and its structure determined by x-ray crystallographic studies. The antibiotic is active against gram-positive bacteria (28).

Awamycin and Ansathiazin. Awaymycin (**75**) and ansathiazin (**76**) (Fig. 8) are produced by *Streptomyces albolongus* C-46366 (29); awamycin is also produced by *Streptomyces* sp. No. 80-217 (227). A mutant of *S. albolongus* (CM-11) formed by uv-mutagenesis produced 10 times more antibiotic than the parent strain. The structures for awamycin and ansathiazin were assigned on the basis of spectral data. Both antibiotics are active against gram-positive bacteria and awamycin is reported to have antitumor activity.

Benzoquinoids

Geldanamycin. Geldanamycin (**77**) (Fig. 9) is isolated from the filtered beer of *Streptomyces hygroscopicus* var. *geldanus* var. *nova* (30). This organism

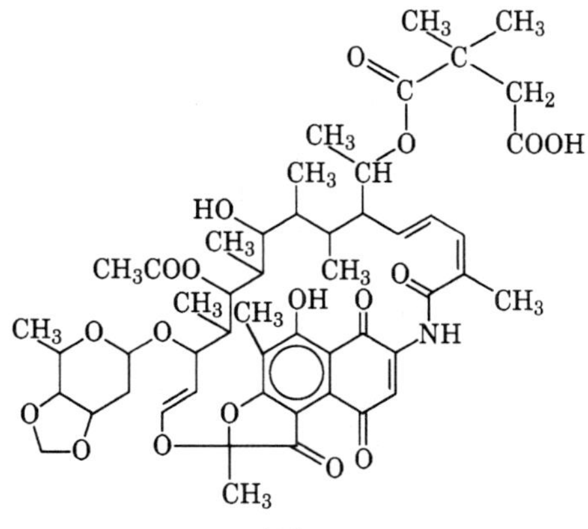

(74)

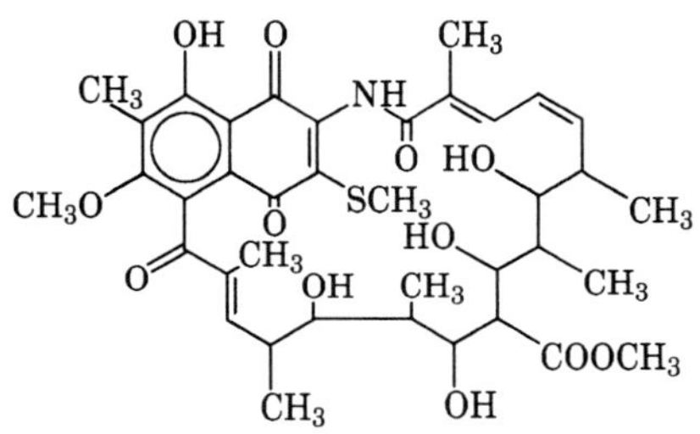

(75)

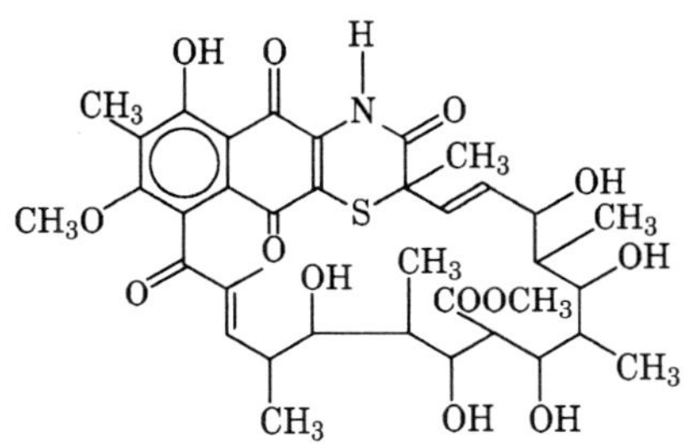

(76)

Fig. 8. Structures of kanglemycin (**74**), awamycin (**75**), and ansathiazin (**76**).

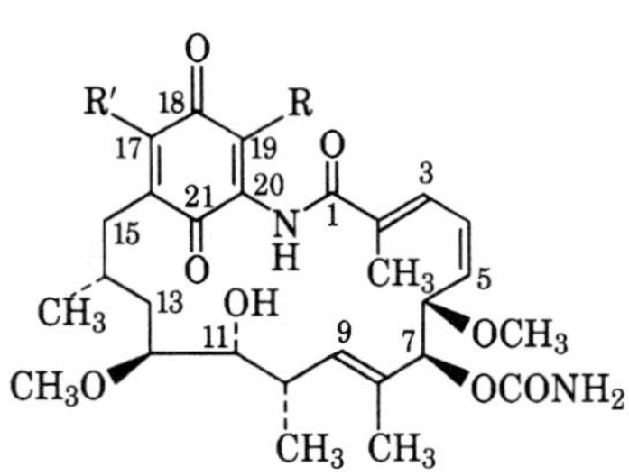

(**77**) R = H; R′ = OCH_3
(**80**) R = CHNR″; R′ = OCH_3
(**81**) R = H; R′ = NHR‴

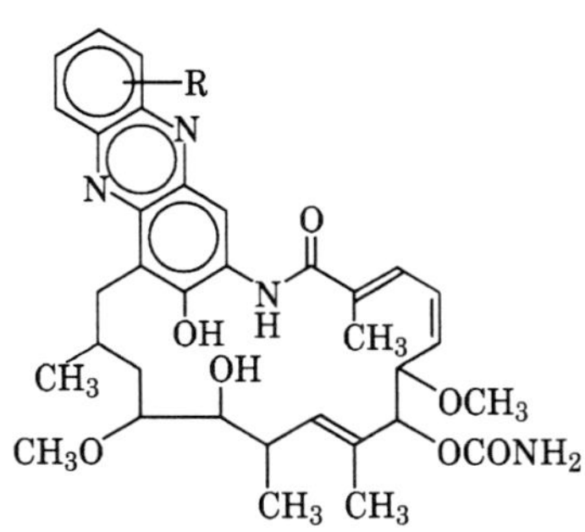

(**78**)

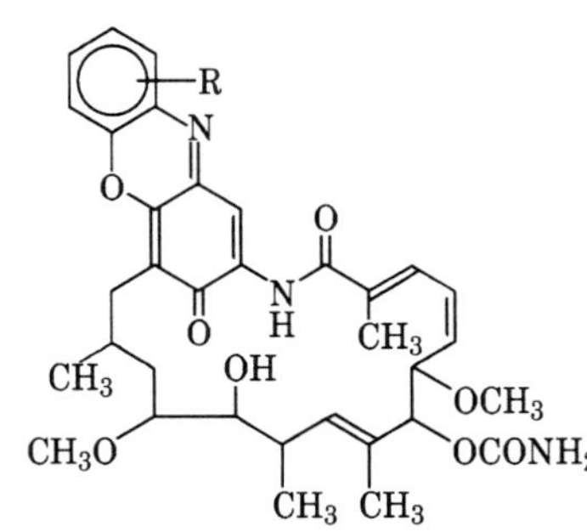

(**79**)

Fig. 9. Structures of geldanamycin (**77**) and derivatives.

also produces nigericin, nocardamine, and a libanamycinlike activity depending on the composition of the fermentation medium (228). The structure of geldanamycin was assigned in great part on the basis of nmr studies (4,229). Unlike the naphthoquinoid ansamacrolides, geldanamycin has little antibacterial activity, being primarily active against protozoa and fungi, especially *Tetrahymena pyriformis* and *Crithidia fasciculata* (see ANTIPARASITIC AGENTS) (30). Geldanamycin also has herbicidal activity (see HERBICIDES) (230,231).

Geldanamycin undergoes reaction with *o*-phenylenediamines and *o*-aminophenols to give compounds similar to rifazine (**36**, R = H). The corresponding geldanazines (**78**) and geldanoxazines (**79**) are active *in vitro* against reverse transcriptase but have no significant *in vivo* activity (232,233). Oxime and hydrazone derivatives of geldanaldehyde (**80**) similar to rifampicin (**39**) have been prepared and found to inhibit reverse transcriptase (234). Geldanamycin derivatives in which the 17-methoxy group has been replaced by a 17-alkylamino group (**81**) were found to have antitumor activity (235).

Herbimycins. Herbimycins A (**82**), B (**83**), and C (**84**), shown in Figure 10 along with some derivatives, are isolated from the fermentation broth of *Streptomyces hygroscopicus* AM-3672 (31). The structure of herbimycin A was assigned on the basis of spectral data (32) and confirmed by x-ray crystallographic studies (236). The structures for herbimycins B and C were derived by comparing spectral data to those for herbimycin A (34,35). The herbimycins possess strong herbicidal activity and exhibit some antitumor and antiviral activity.

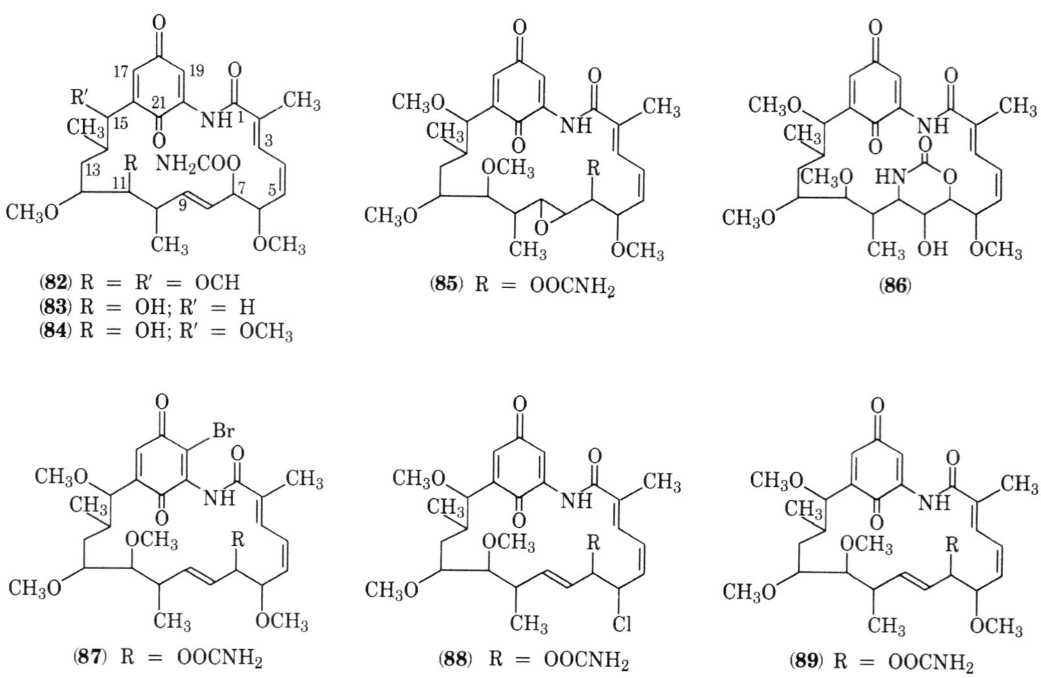

Fig. 10. Structures of the herbimycins and derivatives.

Several derivatives of herbimycin A have been prepared that possess greater antitumor activity than the parent. Treatment of herbimycin A with *m*-chloroperbenzoic acid yields 8,9-epoxyherbimycin A [*94513-90-3*] (**85**), $C_{30}H_{42}N_2O_{10}$, which upon treatment with boron trifluoride etherate is converted to herbimycin A-7,9-cyclic carbamate [*94513-91-4*], $C_3H_{42}N_2O_{10}$ (**86**). Treatment of (**82**), (**85**), or (**86**) with substituted amines yields the corresponding 17- or 19-amino derivatives. Most of the 19-amino derivatives examined, along with (**85**) and (**86**), have increased antitumor activity relative to herbimycin A (237). Other derivatives possessing antitumor activity include: 19-bromoherbimycin A [*103224-24-4*] (**87**),

$C_{30}H_{41}BrN_2O_9$, which is formed by treating herbimycin A with pyridinium hydrobromide perbromide; 6-chloro-6-demethoxyherbimycin A [*103224-10-8*] (**88**), $C_{29}H_{39}ClN_2O_8$, which is formed by treating herbimycin A with boron trichloride; and, 2,3,4,5-tetrahydroherbimycin A [*103224-30-2*] (**89**), $C_{30}H_{46}N_2O_9$, formed by catalytic hydrogenation of herbimycin A (238).

Macbecins. Macbecin I (**90**) and II (**91**) (Fig. 11) are isolated from the fermentation broth of *Nocardia* sp. C-14919 (35,239). The structures were assigned on the basis of ^1H nmr studies on the intact antibiotics as well as on several degradation products. The assigned structures were confirmed by x-ray crystallographic studies (36). The macbecins are active against gram-positive bacteria, fungi, and protozoa and exhibit *in vitro* antitumor activity against murine leukemia P 388 and melanoma B 16 (35) (see CHEMOTHERAUPETICS, ANTICANCER).

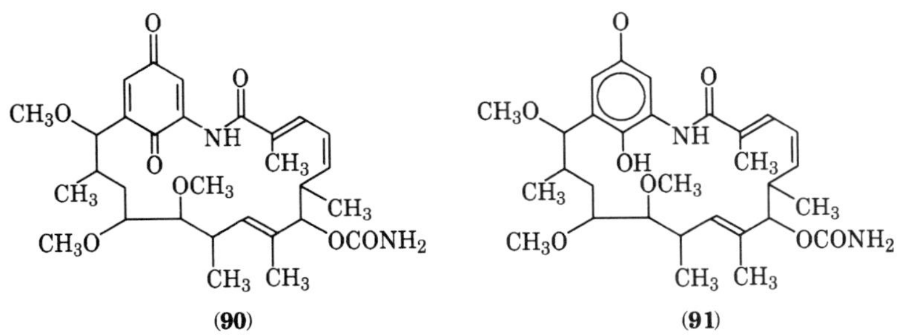

Fig. 11. Structures of macbecin I (**90**) and macbecin II (**91**).

Mycotrienins, Mycotrienols, Trienomycins, and Ansatrienins. The mycotrienins are produced by *Streptomyces rishiriensis* T-23 (38). The structures for mycotrienins I (**92**) and II (**93**) are shown in Figure 12. They were assigned primarily on the basis of nmr spectral analysis (37,240). Mycotrienins I and II are interconvertible via a redox reaction and acid hydrolysis of either compound yields D-alanine. Minor amounts of 20-*O*-methylmycotrienin II (**94**) and 23-deoxymycotrienin II [*97955-33-4*], $C_{36}H_{50}N_2O_7$, which is identical to trienomycin A (**97**), are also isolated from the *Streptomyces rishiriensis* T-23 fermentation broth (41). The structures are also shown in Figure 12.

Streptomyces rishiriensis T-23 also produces mycotrienols I (**95**) and II (**96**), the structures of which were based on spectral analysis. Treatment of mycotrienin II (**93**) with lithium aluminum hydride yields a mixture of mycotrienols I and II (40). The mycotrienols are also interconvertible via a redox reaction. The mycotrienins possess no antibacterial activity but are active against fungi and yeasts (qv), and exhibit weak antitumor activity. The mycotrienols are an order of magnitude less active than the mycotrienins, suggesting that the cyclohexanecarbonylalanine group is important for biological activity.

The trienomycins are isolated from *Streptomyces* sp. 83-16 (43,44). The assigned structures (Fig. 12) were based on spectral data. Acid hydrolysis of trienomycin A yielded D-alanine (42,44). The trienomycins have no antimicrobial activity but have good antitumor activity. Trienomycin A is the most active, exhibiting good *in vivo* antitumor activity against sarcoma 180 and P 388 leukemia in mice (241).

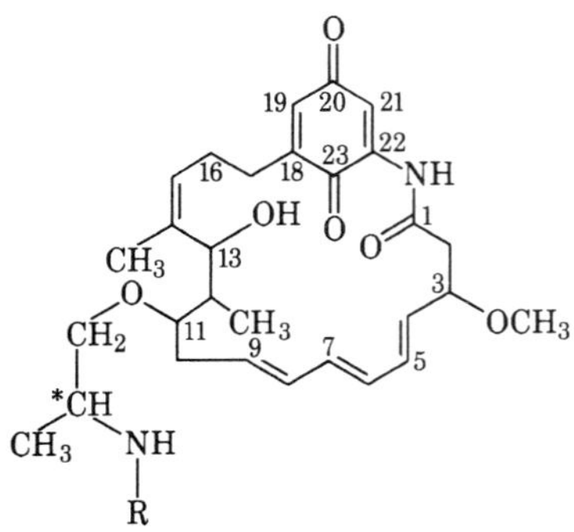

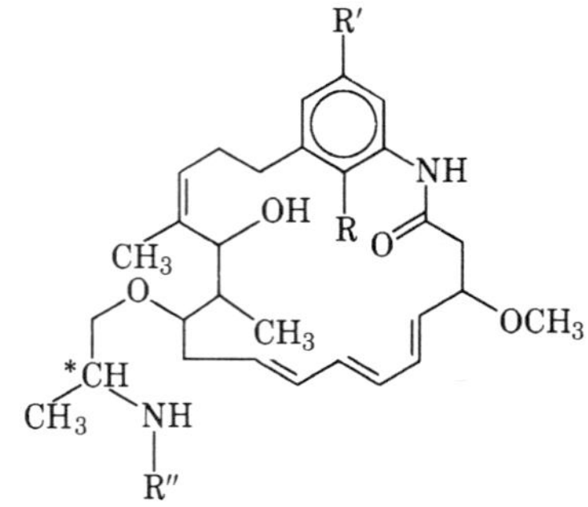

	R	*
mycotrienin I (**92**)	$CO(CH_2)_6$	D
ansatrienin A (**100**)	$CO(CH_2)_6$	L
ansatrienin A_2 (**101**)	$COCH(CH_3)CH_2CH_3$	L
ansatrienin A_3 (**102**)	$COCH_2CH(CH_3)_2$	L

	R	R′	R″	*
mycotrienin II (**93**)	OH	OH	$CO(CH_2)_6$	D
20-*O*-methylmycotrienin II (**94**)	OH	OCH_3	$CO(CH_2)_6$	D
trienomycin A (**97**)	H	OH	$CO(CH_2)_6$	D
trienomycin B (**98**)	H	OH	$COCH_2CH(CH_3)_2$	D
trienomycin C (**99**)	H	OH	$COCH(CH_3)CH_2CH_3$	D
ansatrienin B (**103**)	OH	OH	$CO(CH_2)_6$	L

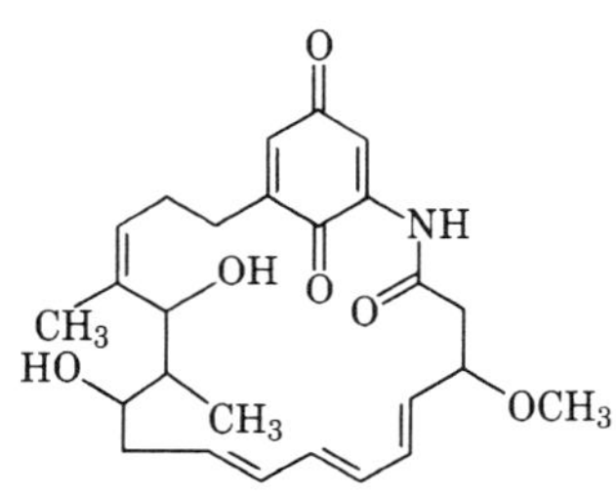

mycotroenol I (**95**)

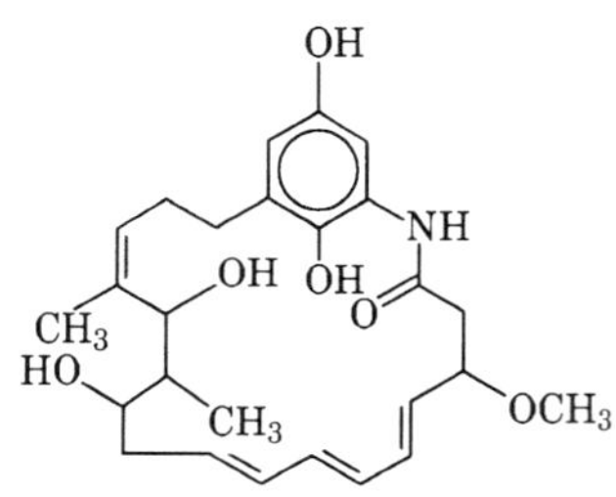

mycotrienol II (**96**)

Fig. 12. Structures of the mycotrienins, mycotrienols, trienomycins, and ansatrienins where * represents configuration of the asymmetric center.

The ansatrienins are produced by *Streptomyces collinus* Tü 1982. The structures (Fig. 12) were assigned on the basis of spectral data of the intact antibiotics as well as several derivatives. Acid hydrolysis of the ansatrienins yields L-alanine. The structures of ansatrienins A and B are the same as those for mycotrienins I and II, respectively, except for the configuration of the alanine moiety. The ansatrienins are active against fungi (45,46).

Maytansinoids and Maytansides (Ansamitocins). *Isolation and Structure Proof.* The maytansinoids were the first ansamacrolides to be found in plants. The term "maytansinoids" refers to those ansamacrolides related to maytansine (**104**), shown in Figure 13, whereas the term "maytansides" refers to maytansinoids lacking the ester side chain at C-3 as well as the corresponding elimination products (50). Maytansine (**104**) was first isolated from the alcoholic extract of *Maytenus ovatus* Loes. (47). Several other maytansinoids and maytansides have been isolated from this species (Table 1). The structure of maytansine was established by x-ray crystallographic analysis (47,242), and the structures of the other maytansinoids and maytansides were arrived at by com-

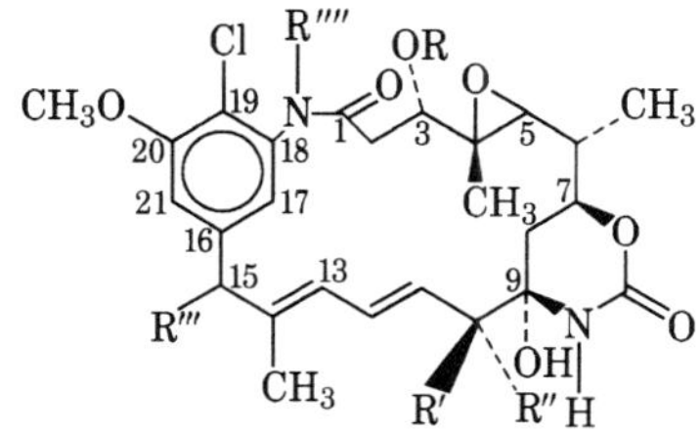

	R	R′	R″	R‴	R⁗
maytansine (**104**)	$COCHCH_3NCH_3COCH_3$	H	OCH_3	H	CH_3
maytanvaline (**105**)	$COCHCH_3NCH_3COCH_2CH(CH_3)_2$	H	OCH_3	H	CH_3
maytanprine (**106**)	$COCHCH_3NCH_3COCH_2CH_3$	H	OCH_3	H	CH_3
maytanbutine (**107**)	$COCHCH_3NCH_3COCH(CH_3)_2$	H	OCH_3	H	CH_3
maytanbutacine (**108**)	$COCH(CH_3)_2$	H	OCH_3	$OCOCH_3$	CH_3
maytanacine (**109**)	$COCH_3$	H	OCH_3	H	CH_3
maytansinol (**113**)	H	H	OCH_3	H	CH_3
colubrinol (**114**)	$COCHCH_3NCH_3COCH(CH_3)_2$	H	OCH_3	OH	CH_3
colubrinol acetate (**115**)	$COCHCH_3NCH_3COCH(CH_3)_2$	H	OCH_3	$OCOCH_3$	CH_3
normaytansine (**116**)	$COCHCH_3NCH_3COCH_3$	H	OCH_3	H	H
trewiasine (**117**)	$COCHCH_3NCH_3COCH(CH_3)_2$	H	OCH_3	OCH_3	CH_3
dehydrotrewiasine (**118**)	$COCHCH_3NCH_3COCCH_3{=}CH_2$	H	OCH_3	OCH_3	CH_3
demethyltrewiasine (**119**)	$COCHCH_3NHCOCH(CH_3)_2$	H	OCH_3	OCH_3	CH_3
10-epitrewiasine (**123**)	$COCHCH_3NCH_3COCH(CH_3)_2$	OCH_3	H	OCH_3	CH_3
nortrewiasine (**124**)	$COCHCN_3NCH_3COCH(CH_3)_2$	H	OCH_3	OCH_3	CH_3
normaytancyprine (**125**)	$COCHCH_3NCH_3C$)-dimethyl-cyclopropyl	H	OCH_3	H	H

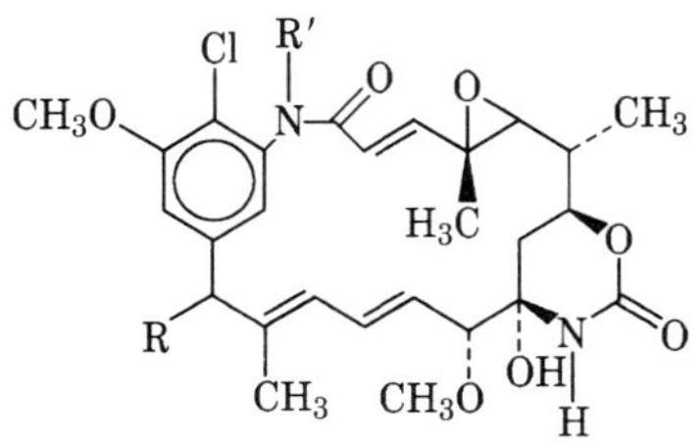

	R	R′
maysine (**110**)	H	CH_3
normaysine (**111**)	H	H

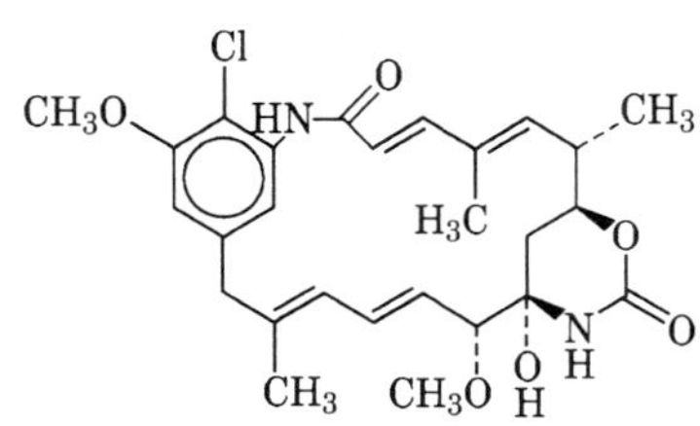

maysenine (**112**)

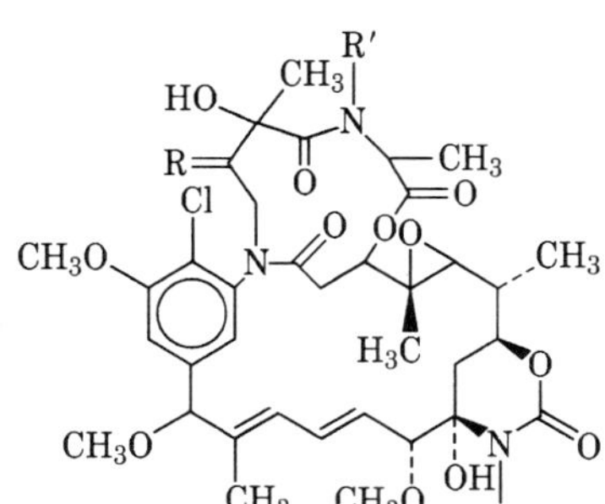

	R	R′
treflorine (**120**)	H, H	H
trenudine (**121**)	H, OH	H
N-methyltrenudone (**122**)	O	CH_3

Fig. 13. Structures of the maytansinoids from plant sources.

parative nmr studies using maytansine (48–52). The absolute configuration of maytansine is 3(*S*), 4(*S*), 5(*S*), 6(*R*), 7(*S*), 10(*R*), and 2′(*S*). Maytanvaline (**105**) is converted to maysine (**110**) and *N*-isovaleryl-*N*-methyl-L-alanine by basic methanolysis. Basic methanolysis of maytansine (**104**) also yields maysine alone with *N*-acetyl-*N*-methyl-L-alanine. Normaysine (**111**) is converted to maysenine (**112**) by treatment with chromous chloride in acetic acid. Maytanbutine (**107**) and maytanacine (**109**) are converted to maytansinol (**113**) upon reduction with lithium aluminum hydride. Maytansinol can be converted to maytanacine by treatment with acetic anhydride–pyridine.

Colubrinol (**114**) and colubrinol acetate (**115**) are isolated from *Colubrina texensis* Gray (Rhamnaceae) along with maytanbutine (51). Colubrinol is also isolated from *Trewia nudiflora* (Euphorbiaceae) (52). The structures for colubrinol and colubrinol acetate (Fig. 13) were established by high resolution ms and the comparison of their nmr spectra with that of the known maytanbutine.

Normaytansine (**116**) is isolated from *Maytenus buchananii* (54) and the maytansinoids trewiasine (**117**), dehydrotrewiasine (**118**), and demethyltrewiasine (**119**) are isolated from the ethanolic extract of the seed from *Trewia nudiflora* L. (Euphorbiaceae) (55). Also isolated from *Trewia nudiflora* are the maytansinoids treflorine (**120**), trenudine (**121**), and *N*-methyltrenudone (**122**), all of which contain an additional macrocyclic ring linking C-3 to the aromatic amide nitrogen (56). Several minor constituents were also isolated from the extract of *Trewia nudiflora* and include 10-epitrewiasine (**123**), nortrewiasine (**124**), and colubrinol (52). As with the other maytansinoids, these structures were elucidated by nmr and mass spectral studies. Normaytancyprine (**125**) is isolated along with maytansine from *Putterlickia verrucosa* Szyszyl. (Celastracea) (57).

Another large group of maytansinoids are produced by the microorganism *Nocardia* sp. C-25003 (N-1) and are designated ansamitocins (243). The structures of the ansamitocins, shown in Figure 14, were determined by spectral analysis. By comparison of reported physical data, it was concluded that ansamitocins P-0 (**113**), P-1 (**109**), and P-2 (**126**) were identical to maytansinol, maytanacine, and maytansinol propionate, respectively. Reductive cleavage of each ansamitocin yielded maytansinol. Alkaline hydrolysis of ansamitocins P-3 (**127**), P-3′ (**128**), and P-4 (**129**) yielded isobutyric, butyric, and isovaleric acids, respectively.

A mutant of *Nocardia* sp. C-14482 (N-1001) was prepared by treatment with ethidium bromide to produce the strain N-1231 which subsequently produced ansamitocins P-3 and P-4 in high yield. This mutant also produces sixteen minor ansamitocins isolated following the scheme used for the parent strain. The structures of the newer ansamitocins were determined on the basis of spectroscopic evidence and were designated PHO-3 (**130**), PND-0 (**131**), PND-1 (**132**), PND-2 (**133**), PND-3 (**134**), PND-4 (**135**), PHM-1 (**136**), PHM-2 (**137**), PHM-3 (**138**), PHM-4 (**139**), P4-βHY (**140**), P4-γHY (**141**), PND-4βHY (**142**), deClQND-0 (**143**), QND-0 (**144**), and deClQ-0 (**145**) (59,244) (Fig. 14). There is speculation that microorganisms such as *Nocardia* may be involved in the production of maytansinoids in plants (243). Ansamitocin P-3 has been isolated from two thudiaceous mosses, *Claopodium crispifolium* (Hook.) Ren. and Card. and *Anomodon attenuatus* (Hedw.) Hueb. (245).

Factors affecting the accumulation of ansamitocins P-2, P-3, and P-4 in *Nocardia* sp. C-15003 have been studied (246): the addition of isoleucine, propionate, propionaldehyde, or *n*-propyl alcohol to the fermentation medium resulted

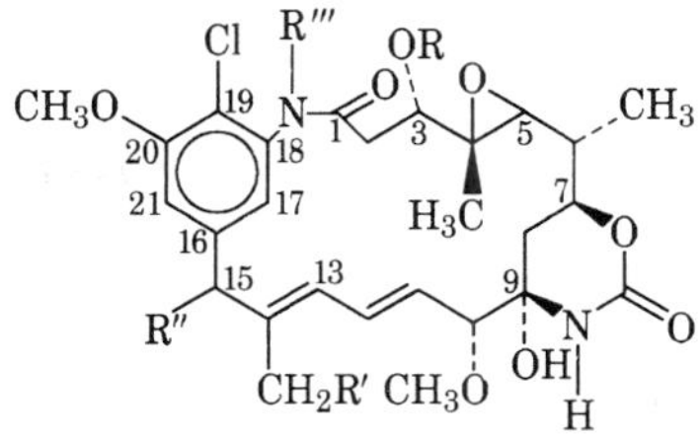

	R	R′	R″	R‴
ansamitocin P-0 (maytansinol) (**113**)	H	H	H	CH_3
ansamitocin P-1 (maytanacine) (**109**)	$COCH_3$	H	H	CH_3
ansamitocin P-2 (**126**)	$COCH_2CH_3$	H	H	CH_3
ansamitocin P-3 (**127**)	$COCH(CH_3)_2$	H	H	CH_3
ansamitocin P-3′ (**128**)	$COCH_2CH_2CH_3$	H	H	CH_3
ansamitocin P-4 (**129**)	$COCH_2CH(CH_3)_2$	H	H	CH_3
ansamitocin PHO-3 (**130**)	$COCH(CH_3)_2$	H	OH	CH_3
ansamitocin PND-0 (**131**)	H	H	H	H
ansamitocin PND-1 (**132**)	$COCH_3$	H	H	H
ansamitocin PND-2 (**133**)	$COCH_2CH_3$	H	H	H
ansamitocin PND-3 (**134**)	$COCH(CH_3)_2$	H	H	H
ansamitocin PND-4 (**135**)	$COCH_2CH(CH_3)_2$	H	H	H
ansamitocin PHM-1 (**136**)	$COCH_3$	OH	H	CH_3
ansamitocin PHM-2 (**137**)	$COCH_2CH_3$	OH	H	CH_3
ansamitocin PHM-3 (**138**)	$COCH(CH_3)_2$	OH	H	CH_3
ansamitocin PHM-4 (**139**)	$COCH_2CH(CH_3)_2$	OH	H	CH_3
ansamitocin P4-βHY (**140**)	$COCH_2C(OH)(CH_3)_2$	H	H	CH_3
ansamitocin P4-γHY (**141**)	$COCH_2CH(CH_2OH)CH_3$	H	H	CH_3
ansamitocin PND-4βHY (**142**)	$COCH_2C(OH)(CH_3)_2$	H	H	H

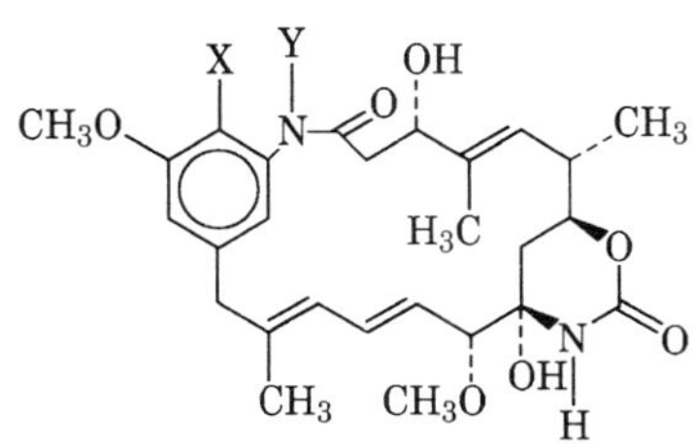

	X	Y
ansamitocin deClQND-0 (**143**)	H	H
ansamitocin QND-0 (**144**)	Cl	H
ansamitocin deClQ-0 (**145**)	H	CH_3

Fig. 14. Structures of the ansamitocins (maytansinoids from microorganisms).

in the increased production of P-2; the addition of valine, isobutyrate, isobutyraldehyde, or isobutyl alcohol increased the production of P-3, reaching more than 90% of the total ansamitocins produced; and the addition of leucine, isovalerate, isovaleraldehyde, or isoamyl alcohol increased the production of P-4.

Large-scale isolation of maytansine (**104**), maytanprine (**106**), and maytanbutine (**107**) from the seeds of *Maytenus rothian* (Walp) Lobreau-Callen from India is possible using preparative lc. As much as 75 g of crude extract can be chromatographed to yield maytansinoids in gram quantities. *Maytenus rothian* is the best known source of maytansinoids (247). A preparative separation using hplc has also been developed for the ansamitocins, capable of handling 5 g of material per run. A quantitative hplc assay has also been developed to determine the amount of ansamitocins in the fermentation of beer (248).

Compilations of the physical properties of the maytansinoids are found in two reviews (249,250).

Chemical Properties and Derivatives. Procedures for the total synthesis of several of the maytansinoids have been thoroughly reviewed (249,250). A variety of bacteria, actinomycetes, yeasts, and fungi were screened for their ability to modify the ansamitocins. Four general types of transformations were performed by many of the strains examined, most of which were actinomycetes. *Bacillus megaterium* IFO 12108, *Streptomyces flavotricini* IFO 12770, and *Streptomyces platensis* IFO 12901 converted ansamitocins P-4, P-3, P-2, P-1, and P-0 to the corresponding 20-*O*-demethyl derivatives (designated PMD-4, PMD-3, PMD-2, PMD-1, and PMD-0, respectively). PMD-3 was found to have better antitumor activity against P 388 and L 1210 than P-3. Maytansine was also 20-*O*-demethylated by the two *Streptomyces* strains. *Streptomyces coelicolor* IFO 3807 and *Streptomyces galbus* IFO 13399 deacylated P-4, P-3, P-2, and P-1 to yield maytansinol (P-0). Maytansine was not deacylated by either organism. *Streptomyces castaneus* IFO 13670 converts P-4, P-3, P-2, and P-1 to the corresponding 15-hydroxylated derivatives (designated PHO-4, PHO-3, PHO-2, and PHO-1, respectively). Upon further study, it was discovered that ansamitocin P-3 was converted into two 15-hydroxyl derivatives designated PHO-3 and epi-PHO-3. The configuration at C-15 was established as (*R*) for PHO-3 and (*S*) for epi-PHO-3 by x-ray analysis. *Streptomyces minutiscleroticus* IFO 13361 converted P-4, P-3, P-2, P-1, and P-0 to the corresponding *N*-demethylated derivatives designated PND-4 (**135**), PND-3 (**134**), PND-2 (**133**), PND-1 (**132**), and PND-0 (**131**), respectively (251–253).

Several semisynthetic maytansinoids have been prepared by acylating the C-3 hydroxyl group of maytansinol. Some of these derivatives have antiprotozoal and antitumor activity similar to maytansine (**104**) and ansamitocin P-3 (**127**) (52,254). 3-Epimaytansinoids have been synthesized and were not biologically active (255).

Biological Activity. The maytansinoids possess antitumor activity, particularly against P 388 lymphocytic leukemia, B 16 melanocarcinoma, and Lewis lung carcinoma. A number of semisynthetic esters of maytansinol have been prepared and exhibit good antileukemic activity (52,255). The maytansides lack antitumor activity, indicating that the ester at C-3 is a requirement for activity (50,52). The carbinolamide also appears to be necessary for antitumor activity. The maytansinoids do not inhibit bacterial RNA polymerase as do the other ansamacrolides. Besides having antitumor activity, the ansamitocins have antiprotozoal and antifungal activity (243). Maytansine has undergone Phase I and II clinical studies and does not appear to be effective (249,250,256).

Mode of Action and Biosynthesis

The mode of action of the naphthoquinoid ansamacrolides was established through studies using the rifamycins and streptovaricins (84,141,257,258). The ansamacrolides inhibit bacterial growth by inhibiting RNA synthesis. This is accomplished by forming a tight complex with DNA-dependent RNA polymerase.

This complex is between the ansamacrolide and the β-unit of RNA polymerase. The formation of the complex inhibits the initation step of RNA synthesis (259,260). The ansamacrolides form no such complex with mammalian RNA polymerase and thus have low mammalian toxicity.

The antiviral activity of the ansamacrolides does not result from inhibition of RNA polymerase but rather from the inhibition of the assembly of the virus particles (141,258).

The antitumor activity of geldanamycin and its derivatives appears to result from inhibition of DNA synthesis whereas RNA synthesis is not affected (261). The antitumor activity of the maytansinoids also appears to result from the inhibition of DNA synthesis. The mechanism of action of maytansine (**104**) has been hypothesized to be the acid catalyzed loss of water from the C-9 hydroxyl group of the carbinolamide to form a reactive acyl imine intermediate, which reacts rapidly with nucleophiles on the bases of DNA (262).

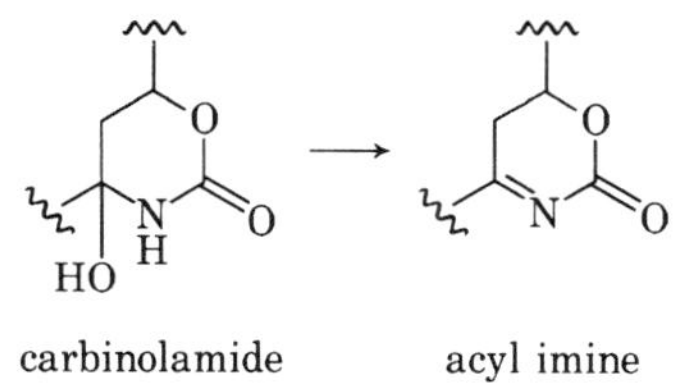

The ansa-chain of the ansamycins streptovaricins (4), rifamycins (263), geldanamycin (4), and herbimycin (32) has been shown to be polyketide in origin, being made up of propionate and acetate units with the *O*-methyl groups coming from methionine. The remaining aromatic C_7N portion of the ansamacrolides is derived from 3-amino-5-hydroxybenzoic acid (264–266) which is formed via shikimate precursors. Based on the precursors of the rifamycins and streptovaricins isolated from mutant bacteria strains, a detailed scheme for the biosynthesis of most of the ansamacrolides has been proposed (95,263).

Commercially Available Ansamacrolides

Rifampicin, the only commercially available ansamacrolide, is manufactured by Merrell Dow under the tradename Rifadin, and by CIBA under the trade name Rimactane. Rifampicin is also supplied in combination with isoniazid or pyrazinamide [*98-96-4*]. The rifampicin–isoniazid combination is known as Rifamate (Merrell Dow), Rifinah (Merrell Dow), and Rimactazid (CIBA); the rifampicin–pyrazinamide as Rifater (Merrell Dow). Several other rifamycin derivatives including rifabutin and rifapentine are undergoing clinical studies.

BIBLIOGRAPHY

"Antibiotics, Ansamacrolides" in *ECT* 3rd ed., Vol. 2, pp. 852–870, by F. J. Antosz, The Upjohn Company.

1. V. Prelog, *Pure Appl. Chem.* **7,** 551 (1963).
2. A. Lüttringhaus and H. Gralheer, *Justus Liebigs Ann. Chem.* **550,** 67 (1942).

3. K. L. Rinehart and co-workers, *Biochemistry* **13,** 186 (1974).
4. K. L. Rinehart, Jr. and L. S. Shield, *Fortschr. Chem. Org. Naturst.* **33,** 231 (1976).
5. W. M. J. Knöll, K. L. Rinehart, Jr., P. F. Wiley, and L. H. Li, *J. Antibiot.* **33,** 249 (1980).
6. P. Sensi, *Prog. Org. Biol. Med. Chem.* **1,** 337 (1964).
7. T. Kishi and co-workers, *J. Antibiot.* **25,** 11 (1972).
8. S. Sugawara, K. Karasawa, M. Watanabe, and T. Hidaka, *J. Antibiot. Ser. A* **17,** 29 (1964).
9. W. Kump and H. Bickel, *Helv. Chim. Acta* **56,** 2323 (1973).
10. G. Lancini and C. Hengeller, *J. Antibiot.* **22,** 637 (1969).
11. J. Leitich, V. Prelog, and P. Sensi, *Experientia* **23,** 505 (1967).
12. G. C. Lancini, G. G. Gallo, G. Sartori, and P. Sensi, in Ref. 10, p. 369.
13. E. Martinelli, G. G. Gallo, P. Antonini, and R. J. White, *Tetrahedron* **30,** 3087 (1974).
14. R. J. White, E. Martinelli, and G. Lancini, *Proc. Nat. Acad. Sci. U.S.A.* **71,** 3260 (1974).
15. E. Martinelli, P. Antonini, R. Cricchio, G. Lancini, and R. J. White, *J. Antibiot.* **31,** 949 (1978).
16. R. Cricchio and co-workers, *Tetrahedron* **36,** 1415 (1980).
17. G. Lancini and G. Sartori, *J. Antibiot.* **29,** 466 (1976).
18. A. K. Ganguly, Y. T. Liu, O. Z. Sarre, and S. Szmulewicz, *J. Antibiot.* **30,** 625 (1977).
19. M. J. Weinstein, G. M. Luedman, E. M. Oden, and G. H. Wagman, *Antimicrob. Agents Chemother.,* 435 (1968).
20. A. K. Ganguly and co-workers, *J. Chem. Soc. Chem. Commun.* 395 (1964).
21. M. Balerna, W. Keller-Schierlein, C. Martius, H. Wolf, and H. Zähner, *Arch. Mikrobiol.* **65,** 303 (1969).
22. M. Mayer, W. Keller-Schierlein, S. Megahed, H. Zähner, and A. Segre, *Helv. Chim. Acta* **69,** 1356 (1986).
23. W. Keller-Schierlein and co-workers, *J. Antibiot.* **36,** 384 (1983).
24. T. Mukhopadhyay, C. M. M. Franco, G. C. S. Reddy, B. N. Ganguli, and H. W. Fehlhaber, *J. Antibiot.* **38,** 948 (1985).
25. J. Mochizuki and co-workers, *J. Antibiot.* **39,** 157 (1986).
26. M. S. Allen, I. A. McDonald, and R. W. Rickards, *Tetrahedron Lett.* **22,** 1145 (1982).
27. M. Hotta and co-workers, in Ref. 25, p. 311.
28. C.-F. Xu and C.-H. He, *J. Antibiot.* **41,** 264 (1988).
29. S. Tanida and co-workers, *Experientia* **42,** 1167 (1986).
30. C. DeBoer, P. A. Meulman, R. J. Wnuk, and D. H. Peterson, *J. Antibiot.* **23,** 442 (1970).
31. S. Omura and co-workers, *J. Antibiot.* **32,** 255 (1979).
32. S. Omura, A. Nakagawa, and N. Sadakane, *Tetrahedron Lett.* **20,** 4323 (1979).
33. Y. Iwai and co-workers, *J. Antibiot.* **33,** 1114 (1980).
34. K. Shibata, S. Satsumabayashi, A. Nakagawa, and S. Omura, in Ref. 25, p. 1630.
35. M. Muroi, M. Izawa, Y. Kosai, and M. Asai, in Ref. 33, p. 205.
36. M. Muroi, K. Haibara, M. Asai, and T. Kishi, *Tetrahedron Lett.* **21,** 309 (1980).
37. M. Sugita, K. Furihata, H. Seto, N. Otake, and T. Sasaki, *Agric. Biol. Chem.* **46,** 1111 (1982).
38. M. Sugita and co-workers, *J. Antibiot.* **35,** 1460 (1982).
39. C. Coronelli, R. C. Pasqualucci, J. E. Thiemann, and G. Tamoni, *J. Antibiot.* **20,** 329 (1967).
40. M. Sugita and co-workers, in Ref. 38, p. 1474.
41. S. Hiramoto and co-workers, in Ref. 24, p. 1103.
42. S. Funayama, K. Okada, K. Komiyama, and I. Umezawa, in Ref. 24, p. 1107.
43. I. Umezawa and co-workers, in Ref. 24, p. 699.
44. S. Funayama, K. Okada, K. Iwasaki, K. Komiyama, and I. Umezawa, in Ref. 24, p. 1677.
45. M. Damberg, P. Russ, and A. Zeeck, *Tetrahedron Lett.* **23,** 59 (1982).

46. G. Lazar, H. Zähner, M. Damberg, and A. Zeeck, in Ref. 23, p. 187.
47. S. M. Kupchan and co-workers, *J. Am. Chem. Soc.* **94,** 1354 (1972).
48. S. M. Kupchan and co-workers, *J. Org. Chem.* **42,** 2349 (1977).
49. S. M. Kupchan, Y. Komoda, G. J. Thomas, and H. P. J. Hintz, *J. Chem. Soc. Chem. Commun.* 390 (1973).
50. S. M. Kupchan, Y. Komoda, A. R. Branfman, R. G. Dailey, Jr., and V. A. Zimmerly, *J. Am. Chem. Soc.* **96,** 3706 (1974).
51. M. C. Wani, H. L. Taylor, and M. E. Wall, in Ref. 49.
52. R. G. Powell, C. R. Smith, Jr., R. D. Plattner, and B. E. Jones, *J. Nat. Prod.* **46,** 660 (1983).
53. S. M. Kupchan and co-workers, *J. Am. Chem. Soc.* **97,** 5294 (1975).
54. A. T. Sneden and G. L. Beemsterboer, *J. Nat. Prod.* **43,** 637 (1980).
55. R. G. Powell, D. Weisleder, and C. R. Smith, Jr., *J. Org. Chem.* **46,** 4398 (1981).
56. R. G. Powell, D. Weisleder, C. R. Smith, Jr., J. Kozlowski, and W. K. Rohwedder, *J. Am. Chem. Soc.* **104,** 4929 (1982).
57. A. T. Sneden, W. C. Sumner, Jr., and S. M. Kupchan, *J. Nat. Prod.* **45,** 624 (1982).
58. M. Asai, E. Mizuta, M. Izawa, K. Haibara, and T. Kishi, *Tetrahedron* **35,** 1079 (1979).
59. M. Izawa, S. Tanida, and M. Asai, *J. Antibiot.* **34,** 496 (1981).
60. P. Siminoff, R. M. Smith, W. T. Sokolski, and G. M. Savage, *Am. Rev. Tuber. Pulm. Dis.* **75,** 576 (1957).
61. U.S. Pat. 3,116,202 (Dec. 31, 1963), A. Dietz and co-workers (to The Upjohn Co.).
62. G. B. Whitfield and co-workers, *Am. Rev. Tuber. Pulm. Dis.* **75,** 584 (1957).
63. R. C. Pandey, R. Misra, and K. L. Rinehart, Jr., *J. Chromatogr.* **169,** 129 (1979).
64. H. Yamazaki, *J. Antibiot.* **21,** 204 (1968).
65. K. L. Rinehart, Jr., *Acc. Chem. Res.* **5,** 57 (1972).
66. K. Sasaki, K. L. Rinehart, Jr., and F. J. Antosz, in Ref. 7, p. 68.
67. A. H.-J. Wang, I. C. Paul, K. L. Rinehart, Jr., and F. J. Antosz, *J. Am. Chem. Soc.* **93,** 6275 (1971).
68. A. H.-J. Wang and I. C. Paul, *J. Am. Chem. Soc.* **98,** 4612 (1976).
69. R. S. Cohn, C. Ingold, and V. Prelog, *Angew. Chem. Int. Ed. Eng.* **5,** 385 (1966).
70. K. L. Rinehart, Jr. and J. C. Cook, Jr., *J. Antibiot.* **27,** 1 (1974).
71. K. L. Rinehart, Jr., in A. L. Burlingame and N. Castagnoli, eds., *Mass Spectrometry in the Health and Life Sciences,* Elsevier, Amsterdam, the Netherlands, 1985, pp. 119–148.
72. K. Kakinuma, B. I. Milavetz, and K. L. Rinehart, Jr., *J. Org. Chem.* **41,** 1358 (1976).
73. K. L. Rinehart, Jr. and co-workers, *J. Am. Chem. Soc.* **97,** 196 (1975).
74. K. L. Rinehart, Jr. and co-workers, in Ref. 17, p. 201.
75. K. Sasaki, T. Naito, and K. Onodera, in Ref. 17, p. 147.
76. Jpn. Pat. 64 003 188 (1989), F. Yagihashi, A. Yamamoto, and T. Motomiya.
77. Jpn. Pat. 64 003 188 (1989), F. Yagihashi, A. Yamamoto, and T. Motomiya.
78. Jpn. Pat. 64 003 189 (1989), F. Yagihashi, A. Yamamoto, and T. Motomiya.
79. B. M. Trost and W. H. Pearson, *Tetrahedron Lett.* **24,** 269 (1983).
80. D. R. Mootoo and B. Fraser-Reid, *J. Org. Chem.* **52,** 4511 (1987).
81. S. L. Schreiber, Z. Wang, and G. Schulte, *Tetrahedron Lett.* **29,** 4085 (1988).
82. W. R. Roush and A. D. Palkowitz, *J. Org. Chem.* **54,** 3009 (1989).
83. D. R. Mootoo and B. Fraser-Reid, in Ref. 82, p. 5548.
84. D. M. Byrd and W. A. Carter in W. A. Carter, ed., *Selective Inhibitors of Viral Function,* Chemical Rubber Co. Press, Cleveland, Ohio, 1973, p. 329.
85. L. H. Li and co-workers, *J. Natl. Cancer Inst.* **58,** 239 (1977).
86. L. H. Li and co-workers, in Ref. 85, p. 245.
87. K. Onodera, Y. Aoi, and K. Sasaki, *Agric. Biol. Chem.* **40,** 1381 (1976).
88. K. L. Rinehart, Jr., P. K. Martin, and C. E. Coverdale, *J. Am. Chem. Soc.* **88,** 3149 (1966).

89. P. Sensi, A. M. Greco, and R. Ballotta, *Antibiot. Annu.* 1959–1960, 262 (1960).
90. Ger. Pat. 1,089,513 (Sept. 22, 1960), P. Sensi and P. Margalith (to Lepetit SpA).
91. P. Margalith and H. Pagani, *Appl. Microbiol.* **9,** 320 (1961).
92. D. Kluepfel, G. C. Lancini, and G. Sartori, *Appl. Microbiol.* **13,** 600 (1965).
93. P. Sensi, R. Ballota, and A. M. Greco, *Farmaco Ed. Sci.* **15,** 228 (1960).
94. P. Sensi, R. Ballota, A. M. Greco, and G. G. Gallo, *Farmaco Ed. Sci.* **16,** 165 (1961).
95. O. Ghisalba, J. A. L. Auden, T. Schupp, and J. Nüesch, in E. J. VanDamme, ed., *Biotechnology of Industrial Antibiotics,* New York, 1984, Chapt. 9.
96. W. Oppolzer and V. Prelog, *Helv. Chim. Acta* **56,** 2287 (1973).
97. W. Oppolzer, V. Prelog, and P. Sensi, *Experientia* **20,** 336 (1964).
98. M. Brufani, W. Fedeli, G. Giacomello, and A. Vaciago, in Ref. 97, p. 339.
99. M. Brufani, W. Fedeli, G. Giacomello, and A. Vaciago, in Ref. 11, p. 508.
100. M. Brufani, S. Cerrini, W. Fedeli, and A. Vaciago, *Mol. Biol.* **87,** 409 (1974).
101. J. Leitich, W. Oppolzer, and V. Prelog, in Ref. 97, p. 343.
102. H. Fuhrer, *Helv. Chim. Acta* **56,** 2377 (1973).
103. E. Martinelli, R. J. White, G. G. Gallo, and P. J. Beynon, *Tetrahedron Lett.* 1367 (1974).
104. E. Martinelli, R. J. White, and G. G. Gallo, *Tetrahedron* **29,** 3441 (1973).
105. L. F. Zerilli, M. Landi, G. G. Gallo, K. H. Maurer, and U. Rapp, *Biomed. Mass Spectrom.* **2,** 307 (1975).
106. P. Bravo and co-workers, *Biomed. Environ. Mass Spectrom.* **18,** 301 (1989).
107. K. Vekey, D. M. F. Edwards, and L. F. Zerilli, *J. Chromatogr.* **474,** 317 (1989).
108. P. Sensi and co-workers, *J. Med. Chem.* **7,** 586 (1964).
109. G. Lancini and W. Zanichelli, in D. Perlman, ed., *Structure-Activity Relationships Among the Semisynthetic Antibiotics,* Academic Press, New York, 1977, pp. 531–600.
110. F. R. Quinn, J. S. Driscoll, and C. Hansch, *J. Med. Chem.* **18,** 332 (1975).
111. N. Maggi, G. G. Gallo, and P. Sensi, *Farmaco Ed. Prat.* **20,** 147 (1965).
112. R. Pallanza, S. Füresz, M. T. Timbal, and G. Carniti, *Arzneim. Forsch.* **15,** 800 (1965).
113. S. Füresz, V. Arioli, and R. Scotti, in Ref. 112, p. 802.
114. V. Dezulian, M. G. Serralunga, and G. Maffii, *Toxicol. Appl. Pharmacol.* **8,** 126 (1966).
115. P. Sensi, M. T. Timbal, and A. M. Greco, *Antibiot. Chemother.* **12,** 488 (1962).
116. P. Sensi, N. Maggi, S. Füresz, and G. Maffii, *Antimicrob. Agents Chemother.* 699 (1966).
117. H. Bickel, F. Knüsel, W. Kump, and L. Neipp, *Antimicrob. Agents Chemother.* 352 (1966).
118. Neth. Pat. 6,411,694 (Apr. 12, 1965), (to CIBA Ltd.).
119. G. G. Gallo, C. R. Pasqualucci, N. Maggi, R. Ballotta, and P. Sensi, *Farmaco Ed. Sci.* **21,** 68 (1966).
120. W. Kump and H. Bickel, *Helv. Chim. Acta* **56,** 2348 (1973).
121. U.S. Pat. 3,923,791 (Dec. 2, 1975), W. D. Celmer (to Pfizer Inc.).
122. F. Kradolfer, L. Neipp, and W. Sackmann, *Antimicrob. Agents Chemother.* 359 (1966).
123. M. Green, J. Bragdon, and A. Rankin, *Proc. Natl. Acad. Sci. U.S.A.* **69,** 1294 (1972).
124. M. F. Dampier and H. W. Whitlock, Jr., in Ref. 53, p. 6254.
125. N. Maggi, V. Arioli, and P. Sensi, *J. Med. Chem.* **8,** 790 (1965).
126. N. Maggi, R. Pallanza, and P. Sensi, *Antimicrob. Agents Chemother.* 765 (1965).
127. S. Füresz, V. Arioli, and R. Pallanza, *Antimicrob. Agents Chemother.* 770 (1965).
128. Neth. Pat. 6,509,961 (Feb. 1, 1966), (to Lepetit SpA).
129. H. D. Cohn, *J. Clin. Pharmacol. J. New Drugs* **9,** 118 (1969).
130. G. Binda and co-workers, *Arzneim. Forsch.* **21,** 1907 (1971).
131. W. Wehrli, *Rev. Infectious Dis.* **5** (Suppl. 3), S407 (1983).
132. A. Vall-Spinosa, W. Lester, T. Moulding, P. T. Davidson, and J. K. McClatchy, *N. Eng. J. Med.* **283,** 616 (1970).

133. G. S. Avery, ed., *Drugs* **1,** 354 (1971).
134. W. Lester, *Ann. Rev. Microbiol.* **26,** 85 (1972).
135. D. B. Radner, *Chest* **64,** 213 (1973).
136. C. Thornberry, B. C. Hill, J. M. Swenson, and L. K. McDougal, *Rev. Infectious Dis.* **5** (Suppl. 3), S412 (1983).
137. B. Farr and G. L. Mandell, in B. C. Cunha, ed., *The Medical Clinics of North America,* Vol. 66/No. 1, W. B. Saunders, Philadelphia, Pa., 1982, pp. 157–168.
138. J. A. Kapusnik and M. A. Sande, in P. K. Peterson and J. Verhoef, eds., *Antimicrobial Agents Annual 3,* Elsevier, New York, 1988, Chapt. 16.
139. J. E. Kapusnik, F. Parenti, and M. A. Sandee, *J. Antimicrob. Chemother.* **13** (Suppl. C), 61 (1984).
140. E. A. C. Follett and T. H. Pennington, in M. A. Lauffer, F. B. Bang, K. Maramorosch, and K. M. Smith, eds., *Advances in Virus Research,* Vol. 18, Academic Press, New York, 1973, p. 105.
141. B. Moss, in W. A. Carter, ed., *Selective Inhibitors of Viral Functions,* Chemical Rubber Co. Press, Cleveland, Ohio, 1973, p. 313.
142. R. J. Clark, *N. Eng. J. Med.* **284,** 675 (1971).
143. W. E. Bullock, *Rev. Infectious Dis.* **5** (Suppl. 3), S606 (1983).
144. P. Sensi, *Pure Appl. Chem.* **35,** 383 (1973).
145. A. N. Tischler, U. R. Joss, F. M. Thompson, and M. Calvin, *J. Med. Chem.* **16,** 1071 (1973).
146. F. M. Thompson, A. N. Tischler, J. Adams, and M. Calvin, *Proc. Natl. Acad. Sci. U.S.A.* **71,** 107 (1974).
147. A. N. Tischler, F. M. Thompson, L. J. Libertini, and M. Calvin, *J. Med. Chem.* **17,** 948 (1974).
148. A. M. Fietta and L. G. Silvestri, *Eur. J. Biochem.* **52,** 391 (1975).
149. V. Arioli and co-workers, in Ref. 59, p. 1026.
150. H. C. Neu, *Antimicrob. Agents Chemother.* **24,** 457 (1983).
151. R. Cricchio, G. Lancini, G. Tamborini, and P. Sensi, *J. Med. Chem.* **17,** 396 (1974).
152. G. Pelizza, G. C. Allievi, and G. G. Gallo, *Farmaco Ed. Sci.* **31,** 31 (1974).
153. L. Marsili and co-workers, *J. Antibiot.* **37,** 1209 (1984).
154. A. Vigevani and co-workers, *Il Farmaco Ed. Pr.* **40,** 131 (1985).
155. T. Bruzzese, A. Cedro, E. Dell'Acqua, A. DiNardo, and A. Goi, *Il Farmaco Ed. Sc.* **41,** 196 (1986).
156. L. Marsili and co-workers, in Ref. 59, p. 1033.
157. A. Sanfilippo and co-workers, in Ref. 33, p. 1193.
158. C. D. Bruna and co-workers, in Ref. 23, p. 1502.
159. C. L. Woodley and J. O. Kilburn, *Am. Rev. Respir. Dis.* **126,** 586 (1982).
160. R. Anand, J. L. Moore, J. W. Curran, and A. Srinivasan, *Antimicrob. Agents Chemother.* **32,** 684 (1988).
161. H. Nagada, G. Schmid, H. Iio, and Y. Kishi, *Tetrahedron Lett.* **22,** 899 (1981).
162. H. Nagaoka and co-workers, *J. Am. Chem. Soc.* **102,** 7962 (1980).
163. H. Nagaoka and Y. Kishi, *Tetrahedron* **37,** 3873 (1981).
164. S. Masamune, B. Imperiali, and D. S. Garvey, *J. Am. Chem. Soc.* **104,** 5528 (1982).
165. H. Iio, H. Nagaoka, and Y. Kishi, in Ref. 162, p. 7965.
166. Y. Kishi, *Pure Appl. Chem.* **53,** 1163 (1981).
167. M. T. Timbal, *Antibiot. Annu.* **1959–1960,** 271 (1960).
168. N. Bergamini and G. Fowst, *Arzneim. Forsch.* **15,** 951 (1965).
169. W. Wehrli and M. Staehelin, *Bacteriol. Rev.* **35,** 290 (1971).
170. S. Riva and L. G. Silvestri, *Ann. Rev. Microbiol.* **26,** 199 (1972).
171. R. C. Ting, S. S. Yang, and R. C. Gallo, *Nature (London) New Biol.* **236,** 163 (1972).

172. R. G. Smith, J. Whang-Peng, R. C. Gallo, P. Levine, and R. C. Ting, *Nature (London) New Biol.* **236,** 166 (1972).
173. R. G. Smith and R. C. Gallo, *Life Sci.* **15,** 1711 (1974).
174. G. Lancini, in H. Pape and H.-J. Rehm, eds., *Biotechnology,* Vol. 4, *Microbial Products II,* VCH, Deerfield Beach, Fla., 1986, Chapt. 14.
175. B. L. Seong, H. J. Son, T.-I. Mheen, and M. H. Han, in Ref. 23, p. 1402.
176. B. L. Seong, H. J. Son, T.-I. Mheen, Y. H. Park, and M. H. Han, *J. Ferment. Technol.* **63,** 515 (1985).
177. R. M. Vohra, U. C. Banerjee, S. Das, and S. Dube, *Biotechnol. Lett.* **11,** 85 (1989).
178. Y. B. Hwana, B. H. Chung, H. N. Chang, and M. H. Han, *Bioprocess Eng.* **3,** 159 (1988).
179. J. M. Finkel, R. F. Pittillo, and L. B. Melleit, *Chemotherapy (Basel)* **16,** 380 (1971).
180. J. W. Lightbrown and P. deRossi, *Analyst* **90,** 89 (1965).
181. G. G. Gallo, P. Sensi, and P. Radaelli, *Farmaco Ed. Prat.* **15,** 283 (1960).
182. C. R. Pasqualucci, A. Vigevani, P. Radaelli, and G. G. Gallo, *J. Pharm. Sci.* **59,** 685 (1970).
183. T. P. Gandhi, A. A. Patel, and V. C. Patel, *Indian Drugs* **16,** 10 (1978).
184. R. B. Patel, T. P. Gandhi, A. A. Patel, V. C. Patel, and R. N. Gilbert, *Indian J. Pharm. Sci.* **41,** 40 (1979).
185. K. K. Mahalanabis, D. Basu, and B. Ray, *Analyst* **114,** 1311 (1989).
186. G. G. Gallo, L. Chiesa, and P. Sensi, *Anal. Chem.* **34,** 423 (1962).
187. G. G. Gallo, L. Chiesa, and P. Sensi, *Farmaco Ed. Sci.* **17,** 668 (1962).
188. P. Sensi, C. Coronelli, and B. J. R. Nicolaus, *J. Chromatogr.* **5,** 519 (1961).
189. P. Sensi, C. Coronelli, and A. Binaghi, *Farmaco Ed. Prat.* **15,** 283 (1960).
190. R. M. Vohra and S. Dube, *J. Chromatogr.* **477,** 463 (1989).
191. V. Vlasakova, J. Benes, and K. Zivny, *J. Chromatogr.* **151,** 199 (1978).
192. K. C. Graham, M. J. LeBelle, and W. L. Wilson, *J. Liq. Chromatogr.* **2,** 365 (1979).
193. J. B. Lecaillon, N. Febvre, J. P. Metayer, and C. Souppart, *J. Chromatogr.* **145,** 319 (1978).
194. B. Ratti, R. R. Parenti, A. Toselli, and L. F. Zerilli, *J. Chromatogr.* **225,** 526 (1981).
195. A. Weber, K. E. Opheim, A. L. Smith, and K. Wong, *Rev. Infectious Dis.* **5** (Suppl. 3), S433 (1983).
196. S. Oldfield, J. D. Berg, H. J. Stiles, and B. M. Buckley, *J. Chromatogr.* **377,** 423 (1986).
197. M. Ishii and H. Ogata, *J. Chromatogr.* **426,** 412 (1988).
198. P. S. Mandal, S. P. Tyagi, and S. K. Talwar, *Indian J. Pharm. Sci.* **48,** 183 (1986).
199. J. Woo, C. L. Wong, R. Teoh, and K. Chan, *J. Chromatogr.* **420,** 73 (1987).
200. B. Ratti, A. Toselli, and R. R. Parenti, *Il Farmaco Ed. Prat.* **36,** 319 (1981).
201. *British Pharmacopoeia 1988,* Her Majesty's Stationery Office, London, 1988, pp. 493–494.
202. *The United States Pharmacopeia, Twenty-Second Revision,* United States Pharmacopeial Convention, Rockville, Md., 1990, pp. 1226–1228.
203. M. Shibata, T. Hasegawa, and E. Higashide, *J. Antibiot.* **24,** 810 (1971).
204. T. Hasegawa, E. Higashide, and M. Shibata, in Ref. 203, p. 817.
205. T. Kishi and co-workers, *Tetrahedron Lett.* **1969,** 91 (1969).
206. T. Kishi, S. Harada, M. Asai, M. Muroi, and K. Mizuno, *Tetrahedron Lett.* **1969,** 97 (1969).
207. K. Kamiya, T. Sugino, Y. Wada, M. Nishikawa, and T. Kishi, *Experientia* **25,** 901 (1969).
208. M. Kondo, T. Oishi, and K. Tsuchiya, in Ref. 7, p. 16.
209. P. Belloma and co-workers, *J. Med. Chem.* **20,** 1287 (1977).
210. L. Stanzani, A. P. Venturini, and M. Mantovani, in Ref. 15, p. 1195.
211. T. Hasegawa, in Ref. 7, p. 25.

212. A. K. Ganguly, Y. T. Liu, O. Z. Sarre, and S. Szmulweicz, in Ref. 18, p. 625.
213. A. K. Ganguly and co-workers, in Ref. 72, p. 1258.
214. S. K. Arora and A. M. Kook, in Ref. 80, p. 1530.
215. T. H. Williams, *J. Antibiot.* **28,** 85 (1975).
216. P. V. Deshmukh and co-workers, in Ref. 68, p. 870.
217. M. Brufani, L. Cellai, and W. Keller-Schierlein, in Ref. 31, p. 167.
218. W. Keller-Schierlein and co-workers, in Ref. 153, p. 1357.
219. T. Okabe and co-workers, in Ref. 25, p. 316.
220. W. Keller-Schierlein and co-workers, in Ref. 23, p. 484.
221. M. Mayer, W. Keller-Schierlein, S. Megahed, H. Zähner, and A. Segre, *Helv. Chim. Acta* **69,** 1356 (1986).
222. T. Mukhopadhyay, C. M. M. Franco, G. C. S. Reddy, B. N. Ganguli, and H. W. Fehlhaber, in Ref. 24, p. 948.
223. J. Mochizuki and co-workers, in Ref. 25, p. 127.
224. M. S. Allen, I. A. McDonald, and R. W. Richards, in Ref. 161, p. 1145.
225. I. A. McDonald and R. W. Rickards, in Ref. 161, p. 1149.
226. M. Hotta and co-workers, in Ref. 25, p. 311.
227. I. Umezawa and co-workers, in Ref. 23, p. 1144.
228. C. DeBoer and A. Dietz, in Ref. 17, p. 1182.
229. K. Sasaki, K. L. Rinehart, Jr., G. Slomp, M. F. Grostic, and E. C. Olson, *J. Am. Chem. Soc.* **92,** 7591 (1970).
230. C. S. Rothrock and D. Gottlieb, *Can. J. Microbiol.* **30,** 1440 (1984).
231. R. M. Heisey and A. R. Putnam, *J. Natural Prod.* **49,** 859 (1986).
232. K. L. Rinehart, Jr. and co-workers, *Bioorg. Chem.* **6,** 353 (1977).
233. L. H. Li, T. D. Clark, C. H. Cowie, and K. L. Rinehart, Jr., *Cancer Treatment Rep.* **61,** 815 (1977).
234. K. L. Rinehart, Jr. and co-workers, *Bioorg. Chem.* **6,** 341 (1977).
235. K. Sasaki, H. Yasuda, and K. Onodera, in Ref. 31, p. 849.
236. A. Furusaki, T. Matsumoto, A. Nakagawa, and S. Omura, in Ref. 33, p. 781.
237. S. Omura and co-workers, in Ref. 153, p. 1264.
238. K. Shibata and co-workers, in Ref. 25, p. 415.
239. S. Tanida, T. Hasegawa, and E. Higashide, in Ref. 33, p. 199.
240. M. Ugita, T. Sasaki, K. Furihata, H. Seto, and N. Otake, in Ref. 38, p. 1467.
241. K. Komiyama and co-workers, *J. Antibiot.* **40,** 1768 (1987).
242. R. F. Bryan, C. J. Gilmore, and R. C. Haltiwanger, *J. Chem. Soc. Perkin Trans.* **2,** 897 (1973).
243. S. Tanida, T. Hasegawa, K. Hatano, E. Higashide, and M. Yoneda, in Ref. 33, p. 192.
244. S. Tanida, M. Izawa, and T. Hasegawa, in Ref. 59, p. 489.
245. K. Suwanborirux, C.-J. Chang, R. W. Spjut, and J. M. Cassady, *Experientia* **46,** 117 (1990).
246. K. Hatano, E. Higashide, S. Akiyama, and M. Yoneda, *Agric. Biol. Chem.* **48,** 1721 (1984).
247. D. E. Nettleton, Jr., D. M. Balitz, M. Brown, J. E. Moseley, and R. W. Myllymaki, *J. Natural Prod.* **44,** 340 (1981).
248. M. Izawa, K. Haibara, and M. Asai, *Chem. Pharm. Bull.* **28,** 789 (1980).
249. P. J. Reider and D. M. Roland, in A. Brossi, ed., *The Alkaloids* Vol. XXIII, Academic Press, New York, 1984, Chapt. 2.
250. C. R. Smith, Jr. and R. G. Powell, in S. W. Pelletier, ed., *Alkaloids: Chemical and Biological Perspectives,* Vol. 2, John Wiley & Sons, Inc., New York, 1984, Chapt. 4.
251. K. Nakahama, M. Izawa, M. Asai, M. Kida, and T. Kishi, in Ref. 59, p. 1581.
252. M. Izawa, K. Nakahama, F. Kasahara, M. Asai, and T. Kishi, in Ref. 59, p. 1587.

253. M. Izawa, Y. Wada, F. Kasahara, M. Asai, and T. Kishi, in Ref. 59, p. 1591.
254. A. Kawai and co-workers, *Chem. Pharm. Bull.* **32,** 3441 (1984).
255. H. Akimoto, A. Kawai, N. Hashimoto, and H. Nomura, in Ref. 254, p. 2565.
256. B. F. Issell and S. T. Crooke, *Cancer Treatment Rev.* **5,** 199 (1978).
257. W. Wehrli and M. Staehelin, *Bacteriol. Rev.* **35,** 290 (1971).
258. W. Wehrli and M. Staehelin, in J. W. Corcoran and F. E. Hahn, eds., *Antibiotics,* Vol. 3, Springer-Verlag, New York, 1975, p. 252.
259. W. Stender and K. H. Scheit, *Eur. J. Biochem.* **65,** 333 (1976).
260. L. R. Yarbrough, F. Y.-H. Wu, and C.-W. Wu, *Biochemistry* **15,** 2669 (1976).
261. H. Yamaki, H. Suzuki, E. C. Choi, and N. Tanaka, in Ref. 38, p. 886.
262. J. W. Lown, K. C. Majumdar, A. I. Meyers, and A. Hecht, *Bioorg. Chem.* **6,** 453 (1977).
263. O. Ghisalba, *Chimia* **39,** 79 (1985).
264. O. Ghisalba and J. Nüesch, in Ref. 59, p. 64.
265. J. J. Kirby, I. A. McDonald, and R. W. Rickards, *J. Chem. Soc. Chem. Commun.* 768 (1980).
266. K. Hatano, S. Akiyama, M. Asai, and R. W. Rickards, in Ref. 38, p. 1415.

General References

Ansamacrolides (Ansamycins)

K. L. Rinehart, Jr. and L. S. Shield, *Fortschr. Chem. Org. Naturst.* **33,** 231 (1976).
W. Wehrli, in *Topics in Current Chemistry: Medicinal Chemistry,* Vol. 72, Springer-Verlag, New York, 1977, pp. 21–51.
M. Brufani, in P. G. Sammes, ed., *Topics in Antibiotic Chemistry,* Vol. 1, Halsted Press, a division of John Wiley & Sons, Inc., New York, 1977, Part B.
A. K. Ganguly, in M. J. W. Weinstein and G. H. Wagman, eds., *Antibiotics: Isolation, Separation, and Purification,* Vol. 15, Elsevier, New York, 1978, pp. 39–68.
V. S. Sethi, in A. Aszalos, ed., *Antitumor Compounds of Natural Origin: Chemistry and Biochemistry,* Vol. II, CRC Press, Boca Raton, Fla., 1981, Chapt. 3.

Streptovaricins

B. I. Milavetz and W. A. Carter, *Pharmac. Ther. A* **1,** 289 (1977).

Rifamycins

W. Wehrli and M. Staehelin, in J. W. Corcoran and F. E. Hahn, eds., *Antibiotics: Mechanism of Action of Antimicrobial and Antitumor Agents,* Vol. III, Springer-Verlag, New York, 1975, pp. 252–268.
G. Lancini and W. Zanichelli, in *Structure-Activity Relationships Among Semisynthetic Antibiotics,* Academic Press, New York, 1977, pp. 531–600.
P. Sensi, in J. S. Bindra and D. Lednicer, eds., *Chronicles of Drug Discovery,* Vol. 1, John Wiley & Sons, Inc., New York, 1982, Chapt. 9.
C. Gurgo, S. Bridges, and M. Green, in P. E. Came and L. A. Caliguiri, eds., *Handbook of Experimental Pharmacology,* Vol. 61, Springer-Verlag, New York, 1982, Chapt. 13.
O. Ghisalba, J. A. L. Auden, T. Schupp, and J. Nüsch, in E. J. VanDamme, ed., *Biotechnology of Industrial Antibiotics,* Marcel Dekker, New York, 1984, Chapt. 9.
Ref. 263.
G. Lancini, in H. Pape and H.-J. Rehm, eds., *Biotechnology: Microbial Products II,* Vol. 4, VCH, 1986, Chapt. 14.

Maytansinoids

Y. Komoda and T. Kishi, in J. M. Cassady and J. D. Douros, eds., *Anticancer Agents Based on Natural Products Models,* Academic Press, New York, 1980, Chapt. 10.

P. J. Reider and D. M. Roland, in A. Brossi, ed., *The Alkaloids: Chemistry and Pharmacology,* Vol. XXIII, Academic Press, New York, 1984, Chapt. 2.
C. R. Smith, Jr. and R. G. Powell, in S. W. Pelletier, ed., *Alkaloids: Chemical and Biological Perspectives,* John Wiley & Sons, Inc., New York, 1984, Chapt. 4.

FREDERICK J. ANTOSZ
The Upjohn Company

CHLORAMPHENICOL AND ANALOGUES

Chloramphenicol [*56-76-7*] (**1**, R = NO_2), $C_{11}H_{12}Cl_2N_2O_5$, is a commercially significant antibacterial agent and its status in clinical practice has been reviewed (1–6). Although widespread use of this antibiotic declined in the United States in the 1960s because of reports of serious toxic effects, this situation changed a decade later when ampicillin-resistant Hemophilus influenzae emerged on the clinical scene (3,6). The appearance of Bacteroides species and of Streptococcus pneumoniae (6) resistant to β-lactam antibiotics contributed further to the resurgence. In the 1970s, chloramphenicol also became important in the treatment of serious *Salmonella* invasive gastroenteritis in infants less than three months of age (5). Because chloramphenicol crosses the blood–brain barrier, it is indicated in infections of the central nervous system caused by susceptible organisms (5,7). The antibacterial activities of chloramphenicol and thiamphenicol [*15318-45-3*] (**1**, R = SO_2CH_3), $C_{12}H_{15}Cl_2NO_5S$, a close analogue, against a number of chloramphenicol-sensitive gram-positive and gram-negative organisms are given in Table 1.

Table 1. Minimum Inhibitory Concentration (MIC) of Chloramphenicol and Thiamphenicol Against Sensitive Organisms

	MIC[a], μg/mL	
Organism	Chloramphenicol	Thiamphenicol
Enterobacter ridant	2	32
E. coli ATCC 10536	1	32
Klebsiella 0217604	4	64
Proteus mirabilis 12453	2	4
Proteus valgaris Napolitano	2	32
Proteus rettgeri Hewitt 104	4	8
Serratia 0213605	16	>128
B. subtilis 66333	2	4
Staphylococcus aureus 209P	4	8
Streptococcus pyogenes Bolden	4	4
Streptococcus 2040	4	4
Salmonella Gr. B. typhimurium	4	64
Shigella 1313	0.5	0.5

[a]Agar dilution, 48 h, Mueller-Hinton agar.

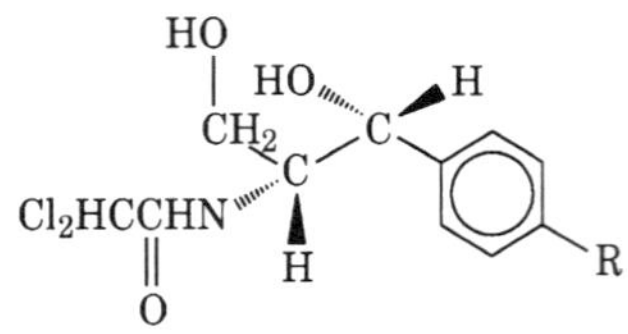

(**1**)

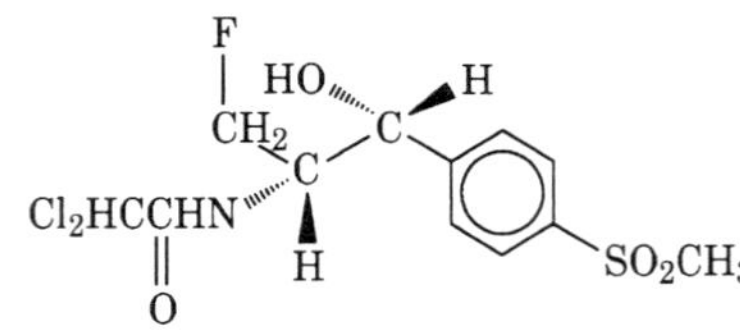

(**2**)

The emergence of quinolones (see ANTIBACTERIAL AGENTS, SYNTHETIC) and other antibiotics is expected to curtail the use of chloramphenicol in the future, but this drug is relatively inexpensive, orally active, and the toxicity, except for the rare idiosyncratic aplastic anemia (1–6), can be managed through monitoring of blood levels by sensitive modern analytical procedures (3). However, clinical use is being further curtailed by the emergence of chloramphenicol-resistant organisms. In Table 2, the median *in vitro* susceptibilities of chloramphenicol, thiamphenicol, and a fluoroanalogue, florfenicol [*76231-34-2*] (**2**), $C_{12}H_{14}Cl_2FNO_4S$, against a host of chloramphenicol-sensitive and -resistant organisms are given. Bacteria resistant to chloramphenicol are also resistant to thiamphenicol.

Table 2. ***In Vitro*** **Susceptibilities of Amphenicols**

Organism strain	No. of strains tested	Susceptibility[b]	MIC, μg/mL[a]		
			Chloramphenicol	Thiomphenicol	Florfenicol[c]
Enterobacter	4	S	4	64	4
	14	R	512	1024	8
Citrobacter	3	S	4	32	8
	3	R	512	1024	128
E. coli	9	S	4	64	8
	20	R	256	1024	8
Klebsiella	9	S	4	64	8
	20	R	512	1024	4
Providencia	4	S	16	128	8
	12	R	128	1024	8
Pseudomonas	13	R	128	128	256
Serratia	6	S	16	512	64
	18	R	512	1024	64
Salmonella	15	S	4	32	8
	7	R	256	1024	8
Shigella	9	S	1	2	2
Proteus	23	R	256	512	8
Acinetobacter	4	R	64	512	128
Staphylococcus	9	S	4	8	8
aureus	7	R	64	512	8
Streptococcus pneumonae	3	R	8	64	4

[a]Agar dilution, 24 h, Mueller-Hinton agar.
[b]S = susceptible; R = resistant.
[c]Florfenicol data are from Ref. 8.

Both chloramphenicol and thiamphenicol cause reversible bone marrow suppression (9). The irreversible, often fatal, aplastic anemia, however, is only seen for chloramphenicol (9). This rare (1 in 10,000–45,000) chloramphenicol toxicity has been linked to the nitroaromatic function (1,9). Thiamphenicol, which is less toxic than chloramphenicol in regard to aplastic anemia, lacks potency as can be seen in Table 1, and thiamphenicol has never found much usage in the United States. An analogue of thiamphenicol having antimicrobial potencies equivalent to chloramphenicol was sought. Florfenicol (**2**) was selected for further development from a number of closely related structures.

Bacterial Resistance of Amphenicol

Of the many mechanisms of bacterial resistance to chloramphenicol and thiamphenicol, the plasmid-mediated transmissible resistance conferred by the presence in resistant bacteria of chloramphenicol-acetyltransferases (CAT) is the most important. This enzyme catalyzes the acetyl-CoA dependent acetylation of chloramphenicol and thiamphenicol (1,10–12). CAT is a cytoplasmic enzyme of which there are three main types: I, II, and III, type III being the most catalytically active (13). The most commonly observed variant of CAT appears to be type I, and type I and type III proteins are known to associate with one another to form hybrids possessing properties of both (13,14). The type III variant of CAT has been studied in detail by steady-state kinetics (13). The data indicate the formation of a ternary complex in the rate determining step involving the enzyme, acetyl-CoA, and chloramphenicol, but the tightness of substrate binding is not a contributing factor. Four types of CAT, type A, B, C, and D, have been characterized in *Staphylococcus aureus* (15) and type C seems to be the most common variant in this species.

Plasmid-mediated bacterial inactivation of chloramphenicol and thiamphenicol can potentially lead to three products, the 3-*O*-acetyl (**3**), 1-*O*-acetyl (**4**), and 1,3-di-*O*-acetyl (**5**) derivatives as shown in Figure 1.

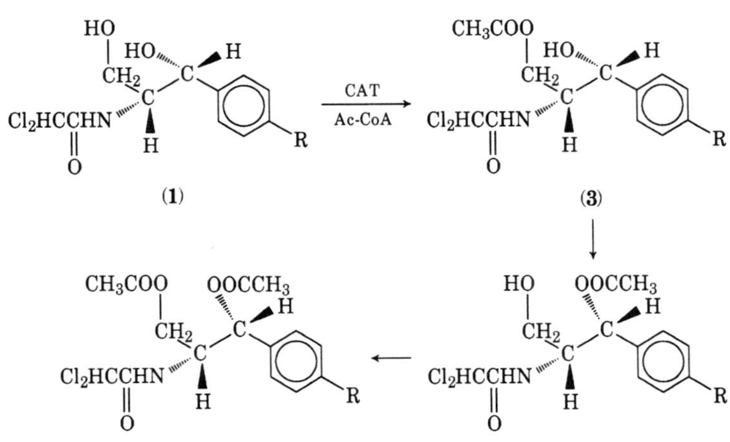

Fig. 1. CAT (catalyzed acetylation) of chloramphenicol (**1**, R = NO_2), and thiamphenicol (**1**, R = CH_3SO_2) where Ac-CoA is acetyl coenzyme A [*72-89-9*].

It was postulated (16–18) and later shown (10) that acetylation of the 3-hydroxyl group was the only enzymatic step in the inactivation process leading to products (**3**) and (**5**), and that formation of the 1-*O*-acetyl derivatives (**4**) resulted from a nonenzymatic intramolecular migration of the acetyl group from the 3- to the 1-position. Thus both the parent amphenicols (**1**) and the 1-*O*-acetyl derivatives (**4**) would seem to be substrates for the CAT enzyme (15,16,19) and prevention of enzymatic *O*-acetylation at the 3-position by replacement of that hydroxyl group using a suitable nonacylable function was proposed to block both modes of antibiotic inactivation (16). Extensive work on the modification of chloramphenicol at the 3-position had however, produced no therapeutically useful derivatives. Additionally, structure-activity relationship studies using chloramphenicol led to the conclusion that the 1,3-propanediol moiety was absolutely essential for amphenicol-type activity (20,21).

Structure-Activity Relationship of Chloramphenicol

Structure-activity and mechanism of action studies indicate that the requirements for chloramphenicol activity are: the D-threo-configuration, the 1,3-propanediol moiety, and a strong electron withdrawing group on the aromatic ring. The L-threo, the mirror image of (**1**), and the D-erythro and L-erythro isomers are not biologically active. Thus the speculation arose that certain specific intramolecular dipolar attractive interactions must exist in chloramphenicol leading to greater stabilization of one particular conformer over the others (16) where biological activity results from the most stable conformer.

The three basic conformational isomers of chloramphenicol are shown in Figure 2. In solution, the rotamers (**a**), (**b**), and (**c**) are expected to be in equilibrium and the concentration of any one of the three species at any given time is dependent on the nature of the solvent and temperature. Chloramphenicol has a specific rotation, $[\alpha]_D$, of $+18.5°$ in ethanol and $-30°$ in dimethyl formamide (22).

In solution, as in the crystalline state (23), the amide carbonyl bond is expected to be in a near syn-periplanar orientation with the C-2 hydrogen bond for all three rotameric forms (24–26). In the crystalline state, chloramphenicol exists in the rotameric form (**a**) (23) where the carbonyl group and the aromatic ring are in close proximity and the carbonyl atom is directed toward the π-electron system of the nitroaromatic ring suggesting strong dipolar attractive forces between the two moieties. In the rotameric form (**b**) such a dipolar interaction is not possible. Although, in conformer (**c**) a dipolar attractive interaction similar to the one seen in (**a**) is possible, it would require that the amide carbonyl bond depart from its thermodynamically preferred near syn-periplanar orientation to become near anti-periplanar. Therefore, rotamer (**a**) is the only conformation accommodating the preferred orientation of the amide carbonyl group with respect to both the C-2 methine hydrogen and the nitroaromatic group. Theoretical calculations (27) and molecular modeling (28) have also suggested that (**a**) is the most preferred conformation. Although the possibility of an intramolecular hydrogen bond between the 1,3 hydroxyl groups is implied in the x-ray crystal

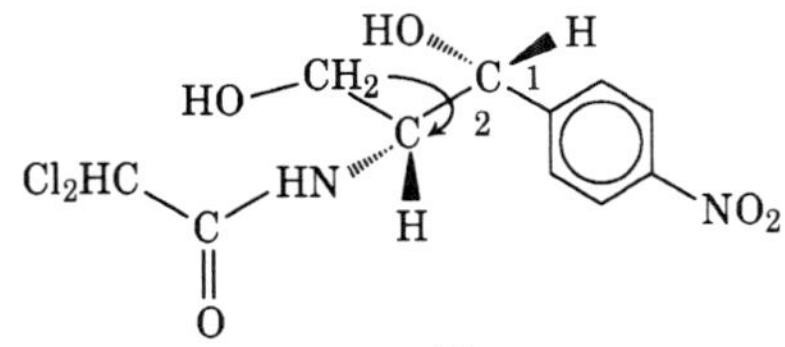

(1)

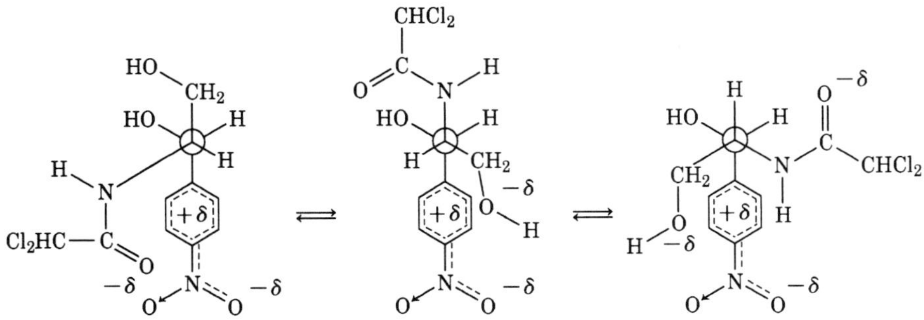

(a) (b) (c)

Fig. 2. Conformers (**a**), (**b**), and (**c**) of chloramphenicol (**1**, R = NO_2).

structure of chloramphenicol, the existence of such a bond in solution has been ruled out (27,29).

Fluoroanalogues

Because the lack of biological activity of 3-substituted chloramphenicols reported previously might result from the inability to exist in the "active" (**a**) type conformation, it was speculated that the size and nature of the C-3 substituent, maintenance of a low barrier to rotation about the C-2–C-3 bond, and the length of the carbon-substitutent atom bond at C-3 were highly critical for achieving a conformational preference of the (**a**) type. Thus, on the basis of the van der Waals radii of fluorine and oxygen being the same (0.14 nm) and the average C–O and C–F bond lengths being close (0.131 nm, 0.138 nm, respectively), the C-3-hydroxyl group of chloramphenicol was replaced by a fluorine atom. Optical rotation measurements of 3-fluoro-chloramphenicol [*73212-55-2*], (**6**, R = NO_2), $C_{11}H_{11}Cl_2FN_2O_4$, in ethanol, gave $[\alpha]_D = +24.4°$ and in dimethylformamide gave $-23.4°$. Thus, as in the case of chloramphenicol, the optical rotation changed from a positive to a negative value on going from a protic to a dipolar aprotic solvent. The solid-state conformation was determined by single crystal x-ray structure analysis (30) and the crystals contained two rotameric structures in the asymmetric unit. The first conformer (Fig. 3**a**), corresponds to rotamer (**a**) of chloramphenicol where all the chloramphenicol conformational features are maintained. The other conformer, shown in Figure 3**b**, corresponds to the extended form (**b**) seen in Figure 2. The fluorine atom in this case is located close to the plane of the aromatic ring system

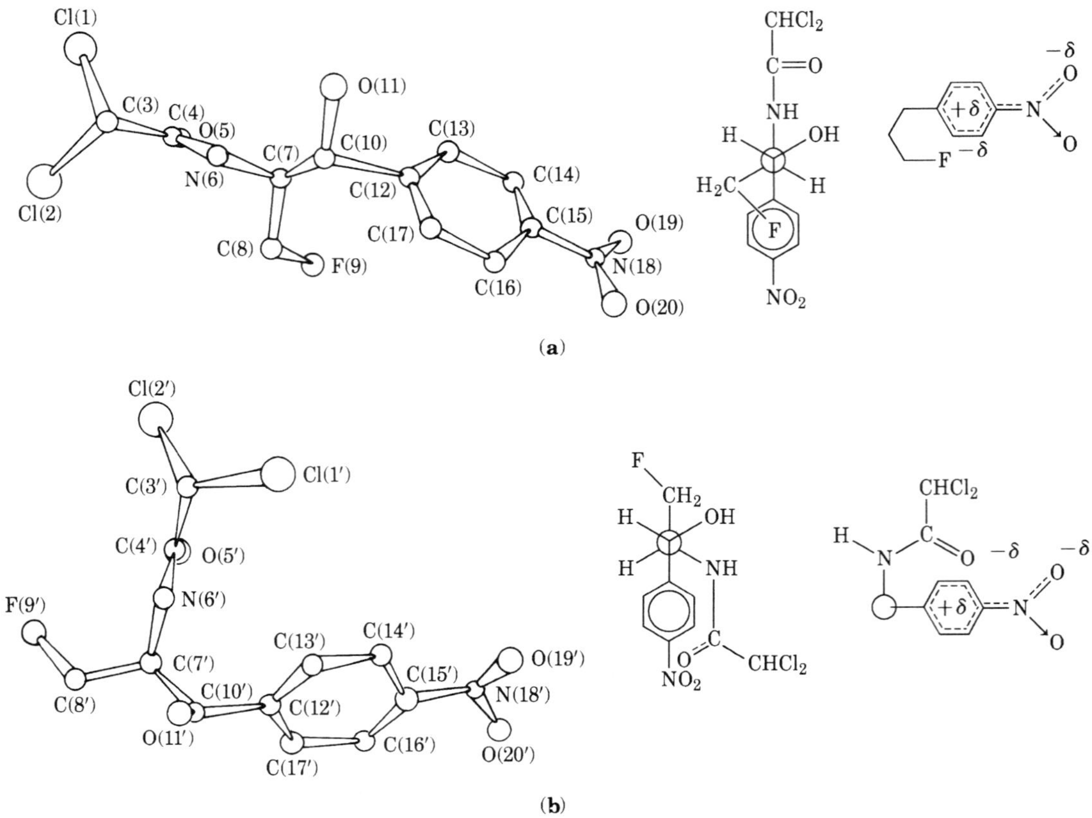

Fig. 3. The solid-state conformations of 3-fluoro-chloramphenicol (**6**, R = NO_2) showing the corresponding rotamers and schematics of the stabilizing attractive forces.

supporting the explanation given for the stability of conformer (**b**) of chloramphenicol.

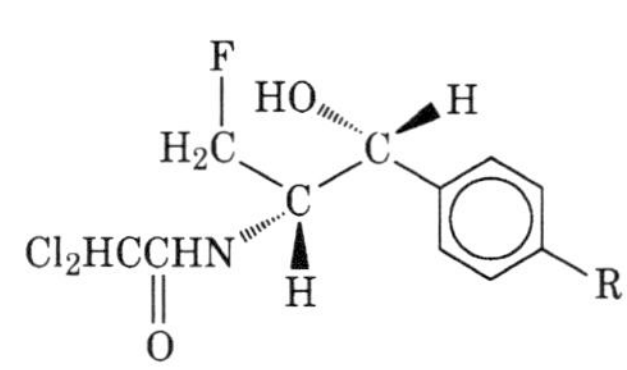

(**6**)

In solution, it is to be expected that rotamers (**a**) and (**b**) of structure (**6**) would exist in equilibrium with a third rotamer comparable to (**c**) of Figure 2, and that the thermodynamically preferred conformer (**6a**) would be responsible for biological activity. Removal of the nitro group from the aromatic ring was expected to have a destabilizing effect on the carbonyl-aromatic ring attraction in (**6a**) and in the solid state, the desnitro analogue (**6**, R = H) exists only in an extended form, comparable to rotamer (**b**), where the fluorine atom is no longer close to the plane of the aromatic ring. This desnitro compound is biologically inactive and it is therefore concluded that for activity a conformation of type (**a**) is required. The absence of the aromatic nitro group is thought to induce repulsive forces between the flourine atom and the electron system resulting in the deflection of the flourine away from the plane of the aromatic ring.

Biological Activity. The biological activity of 3-fluoro-chloramphenicol (**6**, R = NO_2) against chloramphenicol-sensitive and -resistant organisms was determined and is given in Table 3. Potencies against sensitive strains are similar to those of chloramphenicol. Additionally, this fluoroderivative is highly active against chloramphenicol-resistant organims having MICs ranging from 1 to 16 μg/L. This result prompted the synthesis and biological evaluation of a number of amphenicols containing a fluorine atom at the 3-position. The activity of the most promising, florfenicol (**2**), 3-fluorothiamphenicol, is also given in Table 3. Florfenicol is not only active against the chloramphenicol–thiamphenicol-resistant strains, but the potency of florfenicol against sensitive organisms is also superior to any of the other amphenicols.

The antimicrobial properties of florfenicol have been evaluated by several laboratories (8,31). As seen in Table 2, the median MICs of florfenicol against chloramphenicol-susceptible strains of *Enterobacter, Citrobacter, Providenia, Serratia, Salmonella,* and *Shigella* are similar to chloramphenicol whereas against *Klebsiella* the median MIC is half. However, against sensitive *Staphylococci* florfenicol is twice as potent as chloramphenicol. Thiamphenicol is considerably less potent than florfenicol against all of the above strains. Against 103 resistant strains, the MICs of chloramphenicol and thiamphenicol ranged from 64 to 1,026 μg/mL. Mean values were 256 and 512 μg/mL, respectively. For florfenicol, the MICs ranged from 1 to 64 μg/mL against the same resistant organism; the median value was 8 μg/mL.

Florfenicol inhibited 91% of the 399 bacterial isolates at a concentration of 12.5 μg/mL (31). At the same concentration, chloramphenicol and thiamphenicol inhibited only 70% and 24% of the isolates, respectively. Other work has also confirmed the superior activity of florfenicol against chloramphenicol-resistant strains (32–35). More recently it has been shown that florfenicol is active against *E. coli* strains that produce type I, II, or III CAT enzymes (36).

Veterinary Potential or Florfenicol. The absolute ban on the use of chloramphenicol in food producing animals in the United States and Canada has accentuated the need for an effective broad spectrum antibiotic in animal food medicine. Florfenicol and other antibiotics commonly used in veterinary medicine have been evaluated *in vitro* against a variety of important veterinary and aquaculture pathogens. Some of these data are shown in Tables 4 and 5, respectively. Florfenicol was broadly active having MICs lower than those of chloramphenicol in each of the genera tested (Table 4). Florfenicol was also superior to chloramphenicol, thiamphenicol, oxytetracyline [*79-57-2*], ampicillin [*69-53-4*], and oxolinic acid [*14698-29-4*] against the most commonly isolated bacterial pathogen of fish in Japan (Table 5) (37).

Florfenicol (**2**) has been approved in Japan for the treatment of pseudotuberculosis caused by *Pasteurella piscicida* and streptococcosis in yellowtail fish. The recommended dose is 10 mg/kg for up to one week and the drug withdrawal time is five days after cessation of treatment. Florfenicol is active in bovine

Table 3. Activity of Chloramphenicol and Analogues Against Chloramphenicol-Sensitive and -Resistant Organisms

Organism	MIC, μg/mL[a]			
	Chloramphenicol	Thiamphenicol	3-Fluoro-chloramphenicol	Florfenicol
Ent. aerogenes Ridant	2	8	4	0.5
Ent. aerogenes Jackson	>128	>128	8	4
E. coli ATCC 10536	1	32	0.5	0.5
E. coli 0128604	>128	>128	16	16
Kleb. pneumoniae 0217604	4	64	8	8
Kleb. pneumoniae 1117501	>128	>128	2	2
Prov. stuartll Rahal	16	64	2	1
Prov. stuartll 4GR	>128	>128	16	2
Serr. marcescens 0213605	8	>128	8	8
Serr. marcescens Brooke 4	>128	>128	4	2
Staph. aureus 209P	4	4	4	0.5
Staph. aureus Ziegler	8	4	4	0.5
Staph. aureus 59N	4	4	1	0.5
Staph. aureus 1613	2	4	2	0.25
Strep. pyogenes Bolden	0.5	4	2	0
Strep. pyogenes Cruz	2		4	4
Salm. Gr. B. typhimurium	2	32	4	4
Salm. Gr. C_2 Newport	2	8	4	2
Shig. dysenteriae ATCC 13213	0.5	0.25	0.25	0.25
Shig. dysenteriae SS1NO1	0.25	0.25	0.25	0.25
Prot. mirabilis ATCC 12453	4	1	4	1
Prot. mirabilis ATCC 12453	4	1	4	1
Prot. morgani Daly	8	32	8	1
Prot. mirabilis Charlot. Va.	32	32	8	0.5
Prot. morgani Garro	128	32	4	1
Prot. rettgeri 120	128	>128	16	2

[a]Mueller-Hinton broth, 24 h.

Table 4. Activity of Florfenicol (2) and Chloramphenicol (1, R = NO_2) Against Veterinary Bacterial Pathogens

Organism	No. of strains tested	MIC_{90}[a] (μg/mL) Florfenicol	Chloramphenicol
Streptococcus spp.	81	4.0	8.0
Staphylococcus spp.	23	5.0	64.0
Staphylococcus spp.	57	8.0	16.0
Corynebacterium spp.	14	2.0	16.0
Clostridium spp.	21	8.0	4.0
Pasteurella multocida	32	1.0	16.0
E. coli	272	16.0	>128.0
Salmonella spp.	68	16.0	>128.0
Klebsiella spp.	54	32.0	>128.0
Proteus spp.	26	16.0	>128.0
Pseudomonas spp.	35	>128.0	>128.0

[a]MIC_{90} is the antibiotic concentration at which 90% of the bacteria tested are inhibited.

Table 5. Minimum Inhibitory Concentrations (MIC_{90}[a]) of Florfenicol (2) and Other Antibiotics Against Bacterial Pathogens Isolated from Fish in Japan, μg/mL

Antibiotic	Organism (No. of strains tested) *Pasteurella piscicida* (50)	*Edwardsiella tarda* (50)	*Vibrio anguillarum* (35)
florfenicol	0.4	0.8	0.8
chloramphenicol	12.5	50.0	25.0
thiamphenicol	>100.0	>100.0	>100.0
oxytetracycline	6.3	50.3	
ampicillin	100.0		
oxolinic acid		1.6	

[a]MIC_{90} is the antibiotic concentration at which 90% of the bacteria tested are inhibited.

respiratory disease caused by *Pasteurella* species and mastitis caused by *Staphylococci* and *Streptococci*. It is also effective in neonatal colibacillosis caused by *E. coli*. The drug is being developed worldwide by Schering-Plough Animal Health for the treatment of aquatic and bovine diseases.

Structure–Activity Relationships of 3-Fluoro-amphenicols. A number of analogues of 3-fluorochloramphenicol (**6**, R = NO_2) and florfenicol (**2**) have been synthesized and the biological activities examined. Replacement of the dichloroacetyl group by a difluoroacetyl function in both series led basically to retention of potency and the spectrum of activity of the parent structures. However, changing the difluoroacetyl to a trifluoro- or a chlorodifluoroacetyl group abolished the antimicrobial activity almost completely. Reduced level of potency was also seen when the dichloroacetyl group was changed to a chlorofluoroacetyl group. Other amide functions such as methoxyacetyl or methylsulfonylacetyl did not give any appreciable activity. In the florfenicol structure, changing the methylsulfonyl group to methylsulfoxide greatly reduced the potency and the methylthio analogue was practically inactive.

Mechanism of Action of Florfenicol. The inhibitory activities of chloramphenicol (**1**, R = NO_2), thiamphenicol (**1**, R = SO_2CH_3), and florfenicol (**2**) against a sensitive *E. coli* strain have been studied (36). In two different liquid media, both chloramphenicol and florfenicol allowed only 20–30% residual growth at a drug concentration of 2 mg/L, whereas a thiamphenicol concentration of 25 mg/L was required to produce a similar effect. Florfenicol was also found to be a selective inhibitor of prokaryotic cells. At concentrations of 1 mg/L chloramphenicol and florfenicol, and at a concentration of 25 mg/L, thiamphenicol, inhibited protein synthesis. Florfenicol inhibited peptidyl transferase selectively on 70S ribosomes and florfenicol, like chloramphenicol and thiamphenicol, was inactive against *E. coli* strain A19-CM which is resistant at the ribosome level. The binding site for the three amphenicols on the prokaryotic 70S ribosome is different from that on the eukaryotic 80S ribosomes which accounts for the observed selective action. Both chloramphenicol and thiamphenicol have a higher affinity than florfenicol for a common ribosomal-receptor site that represents the peptidyl transferase domain. Although florfenicol is the most potent of these three drugs as an inhibitor in a cell-free transcription system, it is least effective in inhibiting the puromycin [*53-79-2*] reaction which is a measure of inhibition of puromycin-induced release of ribosome-associated nascent peptides.

***In Vivo* Effects of Florfenicol.** Comparative acute toxicities of florfenicol, chloramphenicol, and thiamphenicol in mice are given in Table 6. As can be seen, florfenicol is similar to thiamphenicol in acute toxicity by oral and subcutaneous (sc) administration, but is comparable to chloramphenicol by intraperitoneal (ip) and intravenous (iv) routes. Serum levels in mice following either a single or subcutaneous dose of 200 mg/kg of amphenicol have been determined and the results are given in Figure 4. The serum drug levels attained following oral administration were similar to those obtained by parenteral administration for all compounds. Serum levels of chloramphenicol and thiamphenicol were generally similar whereas florfenicol levels were much higher.

Table 6. Comparative Toxicities in Mice for Amphenicols

	Route of Administration[a]			
Compound	Oral	Ip	Sc	Iv[b]
chloramphenicol	2000	700	2450	90
thiamphenicol	>3000	>2500	>3000	370
florfenicol	>3000	1500	3000	100

[a]The vehicle employed, unless otherwise noted, was 50% propylene glycol plus 50% of a biological vehicle where complete solubility of the drugs was not attained at the concentrations employed.
[b]The vehicle employed was 25% propylene glycol plus 75% water and complete solubility was attained at the indicated concentrations.

In rats, the observed serum drug levels were generally much lower than those seen at the same dose in mice which would seem to indicate a greater metabolic degradation of these compounds in rats. Once again, levels after oral administration were very similar to those given parenterally for all of the compounds. Serum levels attained using florfenicol were similar to those seen using

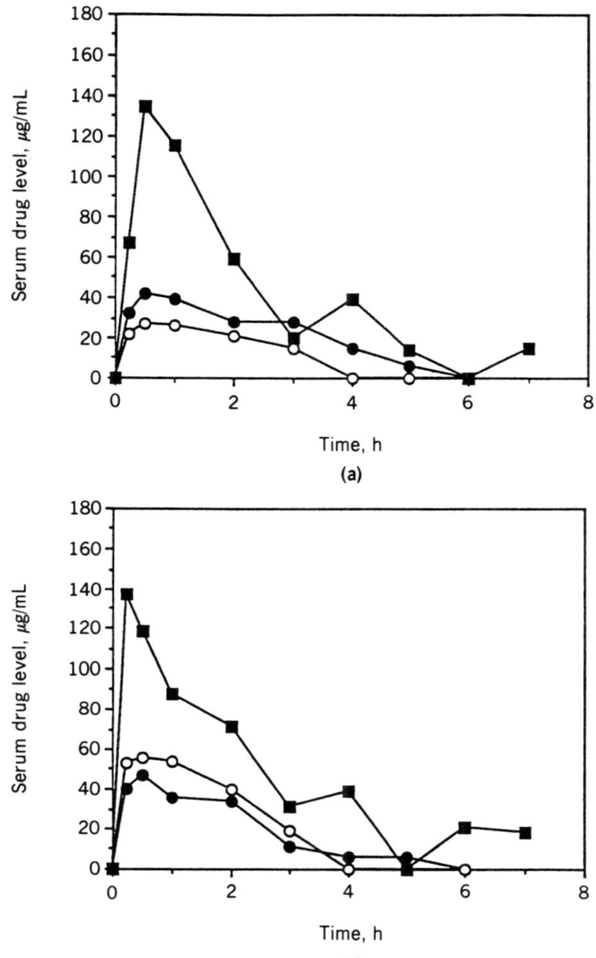

Fig. 4. Serum levels in mice of chloramphenicol (**1**, R = NO_2) ●, thiamphenicol (**1**, R = CH_3SO_2) ○, and florfenicol (**2**) ■, following a single dose of 200 mg/kg given (**a**) subcutaneously, and (**b**) orally.

thiamphenicol, however, these levels were lower after parenteral administration than after oral administration. Binding of florfenicol to serum protein is substantially higher than thiamphenicol binding at 60% and 16%, respectively, but is similar to chloramphenicol binding which is 47%.

The efficacy of florfenicol *in vivo* was determined by measuring the dose required to obtain values for protection from infection in 50% of the animals (PD_{50}) against 10 chloramphenicol-resistant strains and two chloramphenicol-sensitive isolates. Florfenicol, chloramphenicol, and thiamphenicol were evaluated concurrently against each strain. Against sensitive *Enterobacter*, PD_{50} by the subcutaneous and oral routes were similar for florfenicol and chloramphenicol (25 mg/kg sc, 5 mg/kg oral), but higher for thiamphenicol (30 mg/kg sc, 20 mg/kg oral). A dramatic effect was seen for florfenicol against *Shigella* (3 mg/kg sc, 2 mg/kg oral) as compared to chloramphenicol and thiamphenicol (100 mg/kg by both routes). Against resistant strains of *Enterobacter, Klebsiella, Providencia,*

Serratia, Salmonella, and *Staphylococcus,* the PD_{50} values for florfenicol ranged from 5 to 60 mg/kg whereas chloramphenicol and thiamphenicol were practically ineffective.

Pharmacokinetics in Nonrodents

The pharmacokinetic disposition of florfenicol (**2**) was studied in preruminant veal calves after administration of a single 22 mg/kg dose intravenously, orally after a 12-h fast, and orally 5 min postfeeding. The disposition of florfenicol in veal calves following a single iv dose was adequately described by a two-compartment open model where there was no significant effect of the animal's age on the pharmacokinetic parameters. Calves given the oral doses had a complex absorption pattern and delayed absorption. Administering florfenicol with milk delayed the onset of absorption and therefore the time to peak concentration. The disposition of the serum concentration of florfenicol in veal calves given by either oral method could be adequately described by a one-compartment pharmacokinetic model with first-order drug absorption and first-order drug elimination (38). The bioavailability of florfenicol was significantly less when given with milk replacer than when given on an empty stomach: after a 12-h fast median bioavailability was 88% of the dose; and given 5 min postfeeding, median bioavailability of the drug was 65%.

The elimination half-life of florfenicol after a single iv dose of 22 mg/kg (138–204 min) compares well with the elimination half-life of chloramphenicol (**1**, $R = NO_2$) reported in cattle, except in very young calves. Half-lives of 350 min (39) and 302 min (40) in 1-week-old calves, 207 min (40) in 4-week-old calves, 210 min in 6-week-old calves (40), and 264 min (39) and 210 min (41) in adult cows have been reported. The apparent volume of distribution (V_z) for florfenicol following iv administration ranged from 0.68 to 0.84 L/kg as compared to V_z values following iv administration of chloramphenicol in calves of from 0.905 to 1.23 L/kg (39,40). The total body clearance (CL) for florfenicol, 2.77–4.00 mL/(kg · min), is also similar to the CL of chloramphenicol, 1.9–4.03 mL/(kg · min), in calves (39,40).

Florfenicol concentrations in tissues and body fluids of male veal calves were studied after 11 mg/kg intramuscular doses administered at 12-h intervals (42). Concentrations of florfenicol in the lungs, heart, skeletal muscle, synovia, spleen, pancreas, large intestine, and small intestine were similar to the corresponding serum concentrations indicating excellent penetration of florfenicol into these tissues. Because the florfenicol concentration in these tissues decreased over time as did the corresponding serum concentrations, it was deemed that florfenicol equilibrated rapidly between these tissues and the blood. Thus serum concentrations of florfenicol can be used as an indicator of drug concentrations in these tissues.

High florfenicol concentrations were found in the kidney, urine, bile, and small intestine of three calves and in the large intestine of one calf. High florfenicol concentrations in the kidneys and urine indicate that florfenicol may be an excellent drug for treating urinary tract infection caused by susceptible organisms. On the basis of the high concentrations of florfenicol in the bile and the good absorption of the drug after oral administration, florfenicol may undergo some degree of enterohepatic recirculation.

Florfenicol concentrations in the brain, cerebrospinal fluid (CSF), and aqueous humor were one-fourth to one-half the corresponding serum concentrations. Concentrations in these tissues and fluids did not decrease as rapidly, maintaining a low, but fairly constant value. Because the brain, CSF, and aqueous humour are separated from the blood by specialized barriers, florfenicol can seemingly only cross these barriers to a limited extent.

Florfenicol has a wide tissue distribution, similar to that reported for chloramphenicol in calves and thiamphenicol in humans (43,44). Chloramphenicol attains concentrations higher than the corresponding plasma concentrations in bile and urine, as does florfenicol (43). Unlike florfenicol, chloramphenicol concentrations in the liver, kidney, spleen, and lungs are less than corresponding plasma concentrations. However, chloramphenicol penetrates the brain and CSF much better than does florfenicol, reaching values equal to plasma concentrations in the brain. The distribution of thiamphenicol into the kidney, urine, and muscles of humans compared with corresponding plasma concentration is similar to what was observed for florfenicol in calves (44). The penetration of thiamphenicol into the CSF is much smaller than that of florfenicol in calves.

Synthesis

The first syntheses of florfenicol (**2**), 3-fluorochloramphenicol, (**6**, R = NO_2) and other fluoroanalogues were accomplished beginning with thiamphenicol (**1**, R = SO_2CH_3) according to the reaction sequence

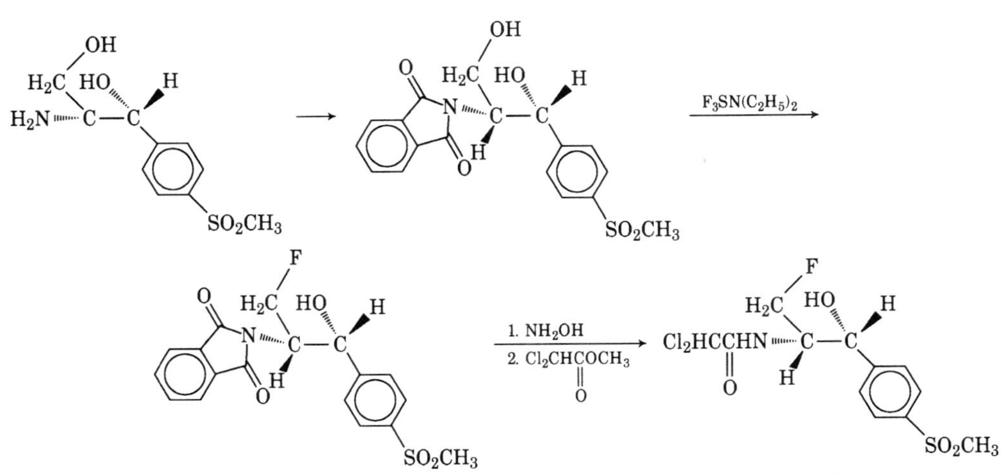

(**2**)

(16,22). Because the starting materials were optically active, the products were all pure enantiomers. Later, the synthetic scheme shown in Figure 5 was developed (22,45). Resolution of the racemic mixture was accomplished at the penultimate stage and the optically active D-threo-amine (**7**) was converted to florfenicol (**2**). This synthetic process also resulted in the synthesis of thiamphenicol shown in Figure 6 using 1,1,2,3,3,3-hexafluoropropyl diethylamine (FPA) (46). More recently an improved method of synthesis of florfenicol has been developed (47).

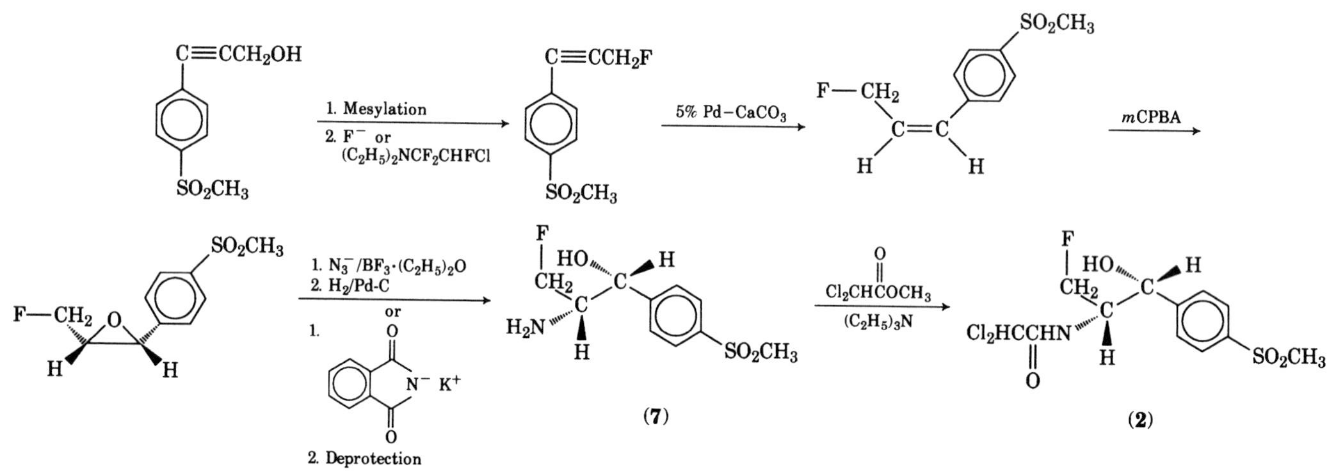

Fig. 5. Alternative synthesis of florfenicol (**2**) where *m*CPBA is *m*-chloroperbenzoic acid.

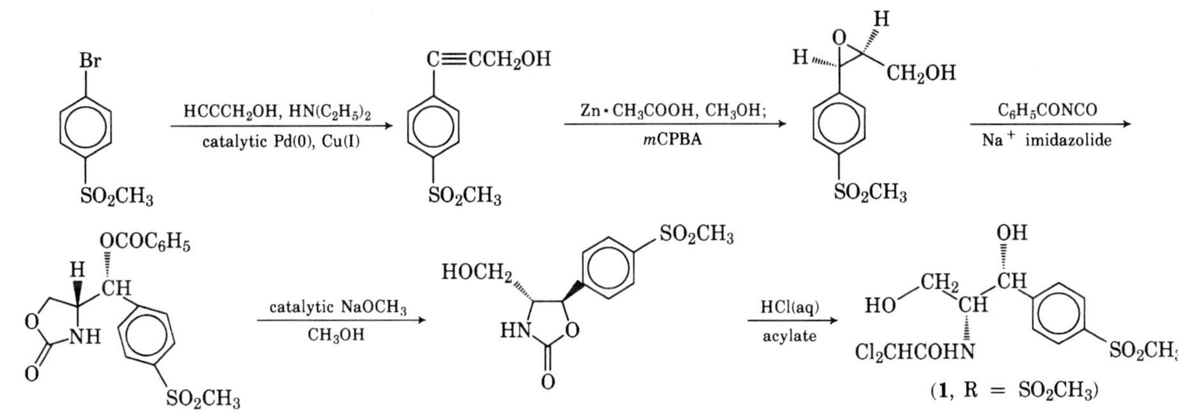

(**1**, R = SO_2CH_3)

Fig. 6. Synthesis of thiamphenicol (**1**, R = SO_2CH_3) (47) where *m*CPBA is *m*-chloroperbenzoic acid.

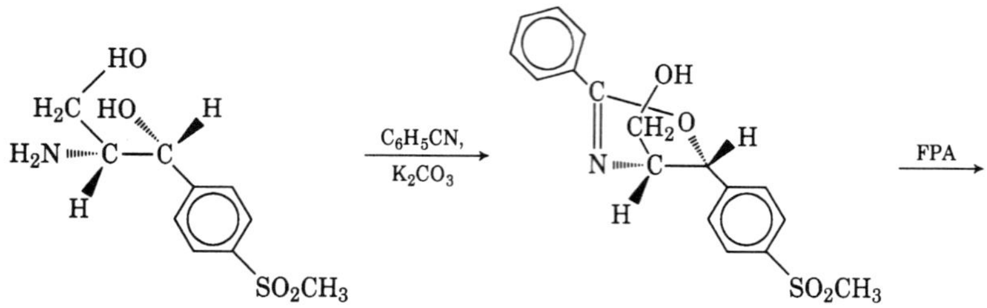

(**1**, R = SO_2CH_3)

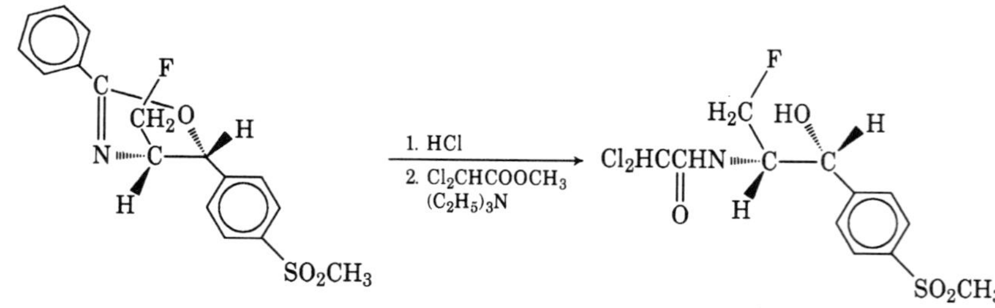

(**2**)

The C-H bond of the dichloracetyl group in chloramphenicol is readily oxidized by liver enzymes and the resulting oxalylchloride derivative is presumed to be implicated in chloramphenicol's toxic manifestation (48). Because such an oxidative mechanism might also be operative in bacterial systems, leading to inactivation of the antibacterial activity, the C–H bond was replaced by deuterium exchange under mildly basic conditions, in both the amphenicol and the 3-fluoroamphenicol series, to see if the rate of oxidation might be retarded. The deuterio analogues have been observed to be consistently twice as potent as the hydrogen analogues (49).

BIBLIOGRAPHY

"Chloromycetin" in *ECT* 1st ed., Vol. 3, p. 870, by H. B. Woodruff, Merck & Co., Inc.; "Streptomyces Antibiotics (Chloramphenicol)," Vol. 13, pp. 81–89, by John Ehrlich, Parke, Davis & Company; "Chloramphenicol" in *ECT* 2nd ed., Vol. 4, pp. 928–937, by John Ehrlich, Parke, Davis & Company; "Antibiotics (Chloramphenicol and its Derivatives)" in *ECT* 3rd ed., Vol. 2, pp. 920–930, by John Ehrlich, Detroit Institute of Technology.

1. A. A. Yunis, *Ann. Rev. Pharmacol. Toxicol.* **28,** 83–100 (1988).
2. H. S. Rosenkranz, *Mutat. Res.* **196,** 1–16 (1988).
3. E. G. Smyth and A. P. Pallett, *Br. J. Hosp. Med.* 424–428 (1988).
4. H. Knotte, *Internist* **30,** 32–37 (1989).
5. U. G. Bodhankar and M. S. Rawat, *Indian Pediatr.* **25,** 77–81 (1988).
6. E. L. Francke and H. C. Neu, *Med. Clin. North Am.* **71,** 1155–1168 (1987).
7. P. W. Kramer, R. S. Griffith, and R. C. Campbell, *J. Neurosurg.* **31,** 295–302 (1969).
8. T. W. Schafer, E. L. Moss, T. L. Nagabhushan, and G. H. Miller, *Curr. Chemother. Infect. Dis.* **1,** 444–446 (1980).

9. I. M. Skolimowski, R. C. Knight, and D. I. Edwards, *J. Antimicrob. Chemother.* **12,** 535–542 (1983).
10. Y. Nakagawa, Y. Nitahara, and S. Miyamura, *Antimicrob. Agents Chemother.* **16,** 719–723 (1979).
11. V. Ferrari and D. Della Bella, *Postgrad. Med. J. Suppl.* **50,** 17–22 (1974).
12. W. V. Shaw, *CRC Crit. Rev. Biochem.* **14,** 1–46 (1983).
13. C. Kleanthous and W. V. Shaw, *Biochem. J.* **223,** 211–220 (1984).
14. L. C. Packman and W. V. Shaw, *Biochem. J.* **193,** 541–552 (1981).
15. J. M. Liddell, W. V. Shaw, and I. D. A. Swan, *J. Mol. Biol.* **124,** 285–286 (1978).
16. T. L. Nagabhushan, D. Kandasamy, H. Tsai, W. N. Turner, and G. H. Miller, *Proc. 11th ICC 19th ICAAC Am. Soc. Microbiol.* **1,** 442–443 (1980); U.S. Pat. 4,235,892 (1980), T. L. Nagabhushan.
17. Y. Suzuki and S. Okamato, *J. Biol. Chem.* **242,** 4722–4730 (1967).
18. W. V. Shaw, *J. Biol. Chem.* **242,** 687–693 (1967).
19. Y. Nakagawa, *Jpn. J. Bacteriol.* **36,** 58 (1981).
20. F. E. Hahn, in D. Gottlieb and P. D. Shaw, eds., *Antibiotics,* Vol. 1, Springer-Verlag, New York, 1967, p. 321.
21. O. Pongs, in F. E. Hahn, ed., *Antibiotics,* Vol. 5, Part 1, Springer-Verlag, New York, 1979, pp. 26–42.
22. T. L. Nagabhushan, *Int. Chem. Cong. Pac. Basin Soc.* Honolulu, Hawaii, 1989; Abstract 420.
23. J. D. Dunitz, *J. Am. Chem. Soc.* **74,** 995 (1952).
24. A. M. Mathieson, *Tetrahedron Let.* **46,** 4137 (1965).
25. R. U. Lemieux, T. L. Nagabhushan, and R. Bruce, *Ann. N.Y. Acad. Sci.* **222,** 927–928 (1973).
26. R. U. Lemieux and M. A. Barton, *Ann. N.Y. Acad. Sci.* **222,** 929–932 (1973).
27. T. M. Bustard, R. S. Egan, and T. J. Perin, *Tetrahedron Lett.* **29,** 1961–1967 (1967).
28. T. L. Nagabhushan, B. Bauer, and M. Czarniecki, unpublished data, 1983.
29. O. Jardetzky, *J. Biol. Chem.* **238,** 2498–2508 (1963).
30. T. L. Nagabhushan and A. T. McPhail, unpublished data, 1979.
31. H. C. Neu, K. P. Fu, and K. Kong, *Curr. Chemother. Infect. Dis.* **1,** 446–447 (1980).
32. A. L. Smith, personal communication, 1979.
33. W. E. Sanders, Jr. and C. C. Sanders, personal communication, 1979.
34. J. Rahal, personal communication, 1978.
35. C. L. Weisseman, Jr., personal communication, 1978.
36. M. Cannon, S. Harford, and J. Davies, *J. Antimicrob. Chemother.* **28,** 307–317 (1990).
37. H. Fukui, Y. Fujihara, and K. Ferumasa, *Fish Pathol.* **22,** 201–207 (1987).
38. K. J. Varma, P. E. Adams, and T. E. Powers, *J. Vet. Pharmacol. Therap.* **9,** 412–425 (1986).
39. R. Reiche, M. Mulling, and H.-H. Frey, *J. Vet. Pharmacol. Therap.* **3,** 95–106 (1980).
40. G. E. Burrows, P. B. Barto, B. Martin, and M. L. Tripp, *Am. J. Vet. Res.* **44,** 1053–1057 (1983).
41. M. Pilloud, *Res. Vet. Sci.* **15,** 231–238 (1973).
42. P. Adams, K. J. Varma, T. E. Powers, and J. F. Lamendola, *Am. J. Vet. Res.* **12,** 1725–1732 (1987).
43. C. S. Sisodia and co-workers, *Am. J. Vet. Res.* **34,** 1147–1151 (1973).
44. T. A. Plomp, K. M. Schalkhauser, and R. A. A. Maes, *Arzneimittelforschung* **31,**1165–1168 (1981).
45. T. L. Nagabhushan, S. W. McCombie, W. A. Szarek, and D. Matsuura, unpublished data, 1983.
46. S. W. McCombie and T. L. Nagabhushan, *Tetrahedron Lett.* **28**(45), 5395–5398 (1987).

47. D. P. Schumacher, J. E. Clark, B. L. Murphy, and P. A. Fischer, *J. Org. Chem.* **55,** 5291–5294 (1990).
48. L. R. Pohl and G. Krishna, *Biochem. Pharmacol.* **27,** 335–341 (1978).
49. U.S. Pat. 4,226,808 (1980), T. L. Nagabhushan.

TATTANAHALLI NAGABHUSHAN
GEORGE H. MILLER
KANWAL J. VARMA
Schering-Plough Research

ELFAMYCINS

The elfamycins are so named because they exhibit antimicrobial activity through the inhibition of protein biosynthesis via binding to the *el*ongation *fa*ctor Tu (1). All of the known elfamycins are listed in Table 1 and the structures are given in Figure 1. These antibiotics are distinguished by low mammalian toxicity, narrow-range antimicrobial activity, and positive effects on feed utilization and growth promotion in farm animals. Elfamycins also improve milk production in lactating ruminants.

Elfamycins are natural products. Aurodox (**1**, R = CH_3) and efrotomycin (**2**, R = CH_3) have been synthesized chemically.

Properties

Elfamycins are slightly acidic because of the 4-hydroxy-2-pyridone or the carboxylic acid moiety. They are soluble in most polar organic solvents and the alkali and ammonium salts are water-soluble. The extractability of the free acids from aqueous solution into solvents such as dichloromethane and ethyl acetate is utilized in their isolation from fermentation broths.

Purifications of elfamycins have been described in the literature using Craig distribution (2,34), chromatography on Sephadex LH-20 (2,14,26) and Amberlite XAD-2 (10,17,19,26), supercritical fluid extraction (37), and chromatography on an Ito multilayer coil planet centrifuge (26,38). ^{1}H and ^{13}C nmr assignments of most elfamycins have been accomplished (3,24,26,32). The characteristic uv spectra permits some differentiation (12) and bathochromic shifts associated with Al^{3+} complexation have been used to quantify efrotomycin (**2**, R = CH_3) in feed premixes (39,40).

Structures. The first structure to be elucidated was that of aurodox (**1**, R = CH_3), the oldest member of the elfamycin family. The elfamycins are amorphous and crystalline derivatives are as yet unknown. A combination of degradation reactions and the spectroscopic and Roentgen crystallographic analyses of the products was used to determine structure (4). The discovery that the amide linkage in aurodox could be disconnected under extremely mild conditions permitted a sequence of controlled cleavages, the products of which provided the structural and stereochemical information required to permit complete stereochemical assignments of the parent molecule. As shown in Figure 2**a**, treatment of

Table 1. Elfamycins

Antibiotic	CAS Registry Number	Molecular formula	Producing organism	Structure no.	Refs.
aurodox	[*12704-90-4*]	$C_{44}H_{62}N_2O_{12}$	*Streptomyces goldiniensis* var. *goldiniensis*	(**1**, R = CH_3)	2–4
kirromycin[a]	[*50935-71-2*]	$C_{43}H_{60}N_2O_{12}$	*S. collinus* Tü 365	(**1**, R = H)	3,5–7
azdimycin	[*57035-45-7*]	unknown	*S. diastatochromogenes* ATCC 31013		8
efrotomycin	[*56592-32-6*]	$C_{59}H_{88}N_2O_{20}$	*S. lactamdurans* NRRL 3802	(**2**, R = CH_3)	9–13
dihydromocimycin	[*61994-74-9*]	$C_{43}H_{62}N_2O_{12}$	*S. ramocissimus* CBS190.69	(**3**)	12,14
heneicomycin	[*66170-37-4*]	$C_{44}H_{62}N_2O_{11}$	*S. filipinensis* NRRL 11044	(**4**, R = CH_3, R′ = C_2H_5, R″ = CH_3)	12,15,16
kirrothricin	[*79190-00-4*]	$C_{44}H_{64}N_2O_{10}$	*S. cinnamomeus* Tü 89	(**5**)	12,17,18
factumycin	[*84600-89-5*]	$C_{44}H_{62}N_2O_{10}$	*S. lavendulae* ATCC 31312	(**6**)	12,19
MSD A63A	[*81209-82-7*]	unknown	*Streptoverticillum hiroshimense*		20
L681,217	[*93522-10-2*]	$C_{36}H_{53}N_1O_{10}$	*S. cattleya* ATCC 39203	(**7**)	12,21,22
SB22484, factor 3	[*97328-80-8*]	$C_{41}H_{56}N_2O_{11}$	*S.* strain NRRL 15496	(**4**, R = H, R′ = CH_3, R″ = H)	12,23,24
SB22484, factor 4	[*97328-81-9*]	$C_{42}H_{58}N_2O_{11}$	*S.* strain NRRL 15496	(**4**, R = H, R′ = C_2H_5, R″ = H)	12,23,24
phenelfamycin A	[*118498-91-2*]	$C_{51}H_{71}N_1O_{15}$	*S. violaceoniger* str. AB999F-80, AB1047T-33	(**8**, R = H, R′ = $COCH_2C_6H_5$)	12,25,26
phenelfamycin B	[*118498-91-3*]	$C_{61}H_{71}N_1O_{15}$	*S. violaceoniger* str. AB999F-80, AB1047T-33	(**8**, R = $COCH_2C_6H_5$, R′ = H)	12,26,27
phenelfamycin C	[*118498-91-4*]	$C_{58}H_{83}N_1O_{18}$	*S. violaceoniger* str. AB999F-80, AB1047T-33	(**9**, R = H, R′ = $COCH_2C_6H_5$)	12,26,27
phenelfamycin D	[*118498-91-5*]	$C_{58}H_{83}N_1O_{18}$	*S. violaceoniger* str. AB999F-80, AB1047T-33	(**9**, R = $COCH_2C_6H_5$, R′ = H)	12,26,27
phenelfamycin E	[*114451-31-9*]	$C_{65}H_{95}N_1O_{21}$	*S. violaceoniger* str. AB999F-80, AB1047T-33	(**10**, R = H, R′ = $COCH_2C_6H_5$)	12,26,27
phenelfamycin F	[*114451-30-8*]	$C_{65}H_{95}N_1O_{21}$	*S. violaceoniger* str. AB999F-80, AB1047T-33	(**10**, R = $COCH_2C_6H_5$, R′ = H)	12,26,27
unphenelfamycin	[*118117-42-3*]	$C_{43}H_{65}N_1O_{14}$	*S. violaceoniger* str. AB999F-80, AB1047T-33	(**8**, R = R′ = H)	12,26,27
LL-E19020α	[*114451-31-9*]	$C_{65}H_{95}N_1O_{21}$	*S. lydicus* sp. *tanzanius* NRRL 18036	(**11**, R = H, R′ = $COCH_2C_6H_5$)	28–32
LL-E19020β	[*114451-30-8*]	$C_{65}H_{95}N_1O_{21}$	*S. lydicus* sp. *tanzanius* NRRL 18036	(**11**, R = $COCH_2C_6H_5$, R′ = H)	28–32
UK-69,753	[*118117-42-3*]	$C_{58}H_{86}N_2O_{18}$	*Amycolatopsis orientalis* ATCC 53550	(**12**)	12,31,33,34
N-demethylefrotomycin	[*124846-41-9*]	$C_{58}H_{86}N_2O_{20}$	*Nocardia* ATCC 53758	(**2**, R = H)	35

[a]Identical to mocimycin which was isolated from *S. ramocissimus* CB5190.69 (6,36).

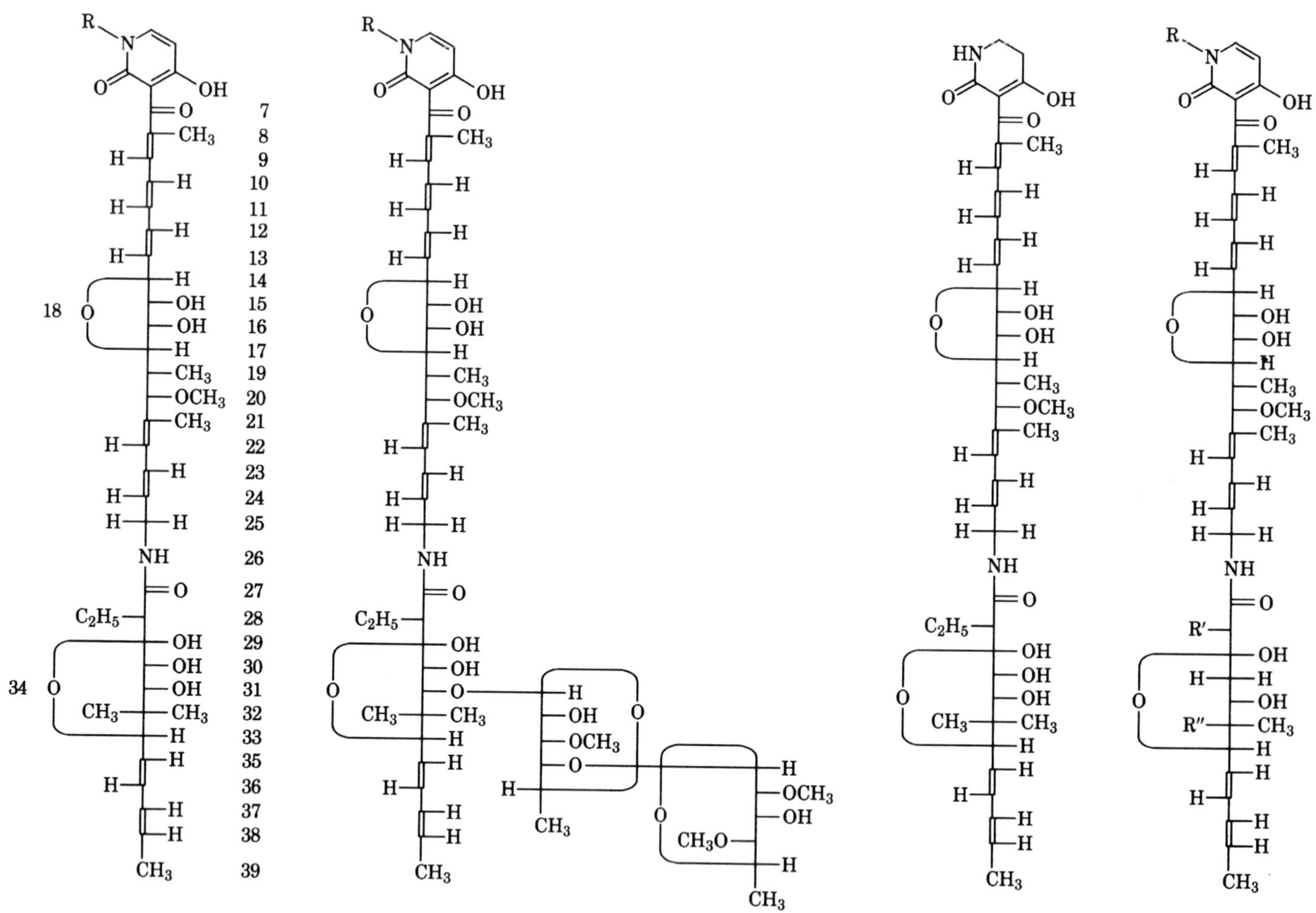

(1) (2) (3) (4)

Fig. 1. Structures of the elfamycins given in Table 1 where the carbon skeleton numbering scheme is shown for structure (**1**) only.

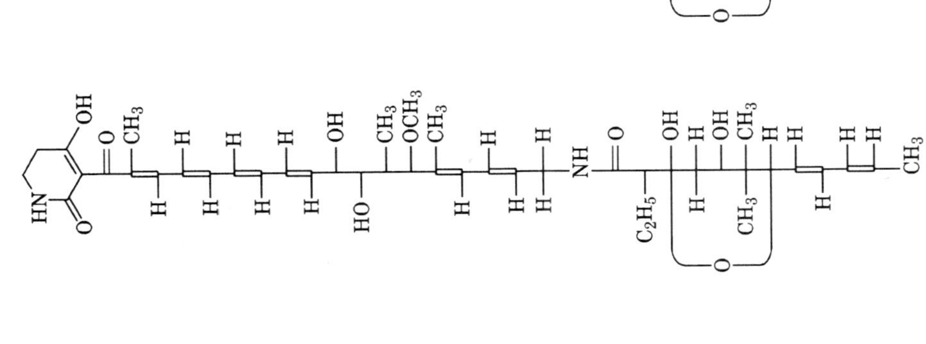

(5)

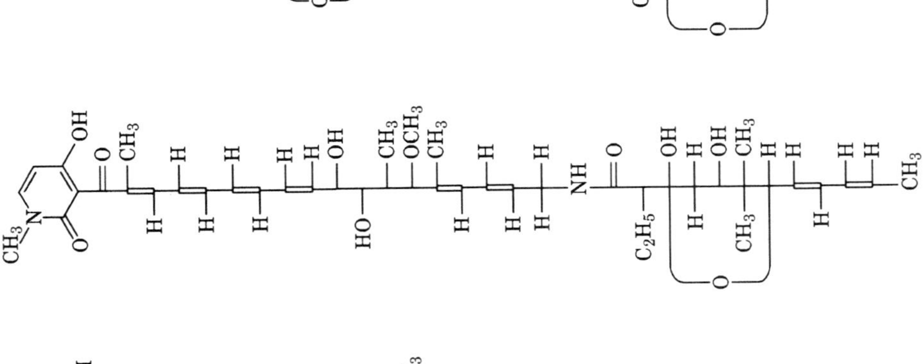

(6)

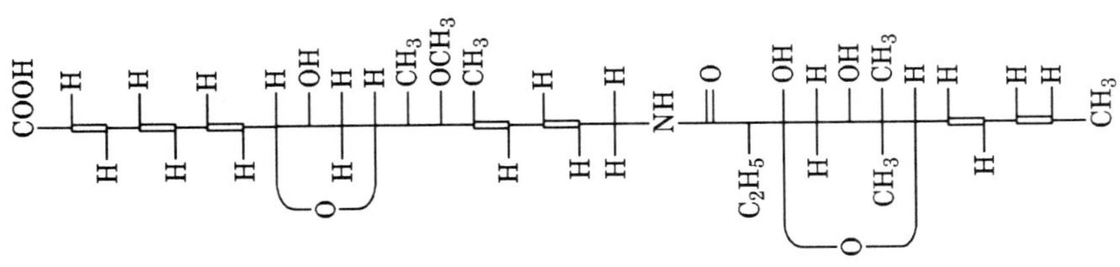

(7)

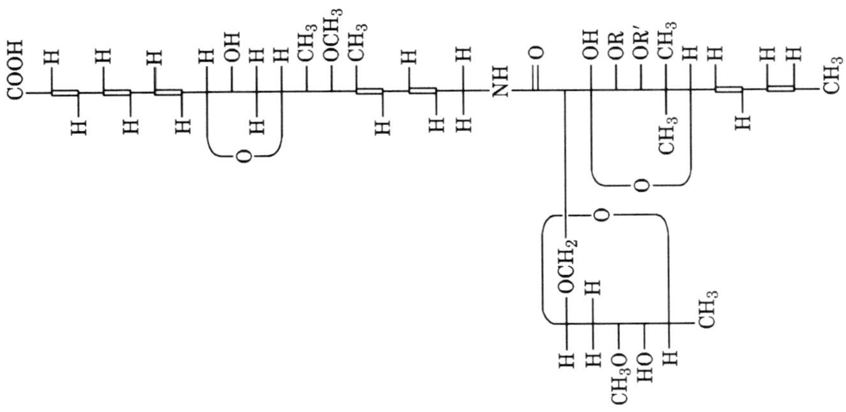

(8)

Fig. 1. *(Continued)*

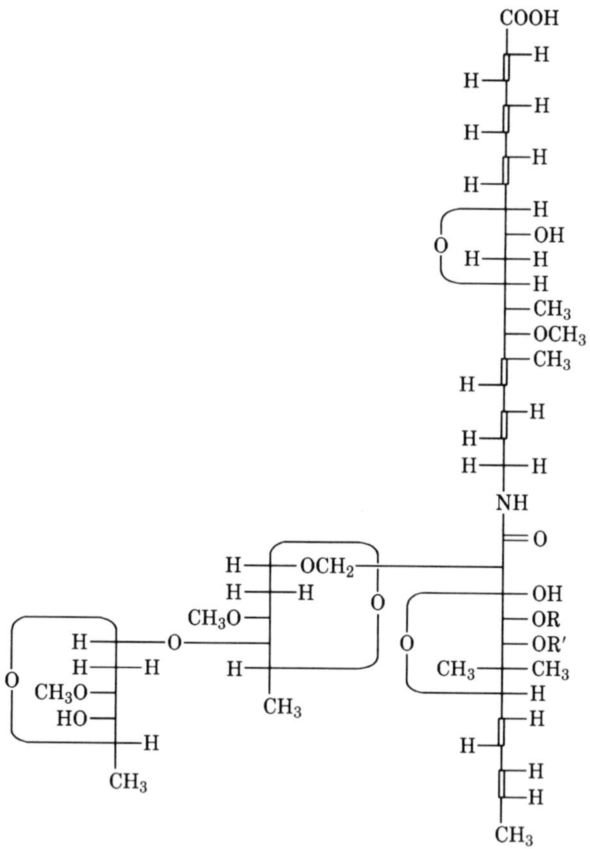

(9)

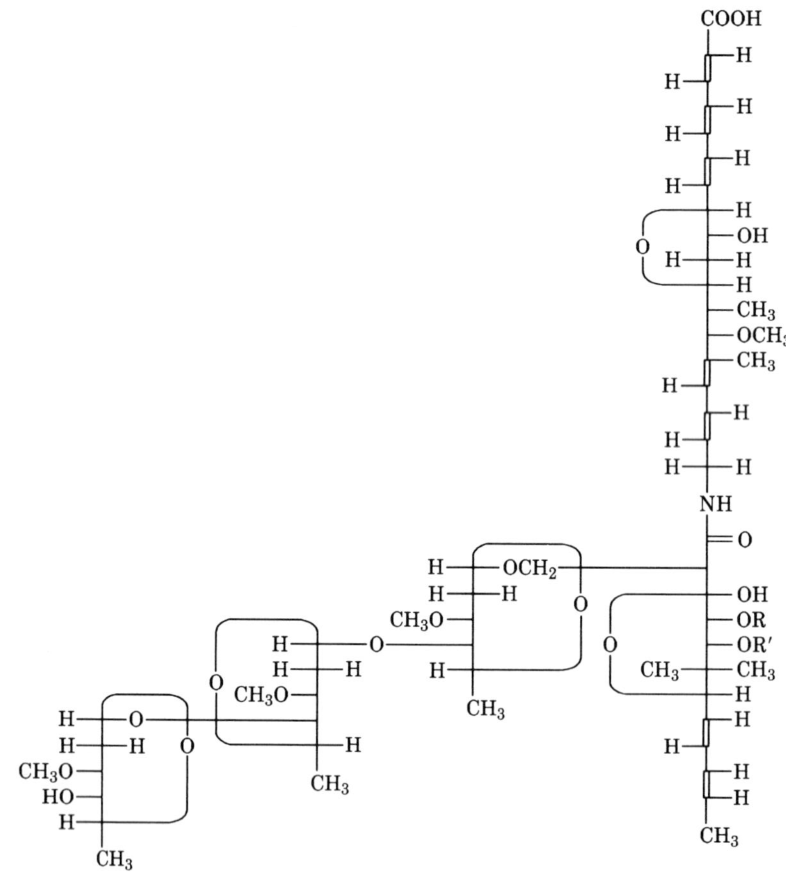

(10)

Fig. 1. *(Continued)*

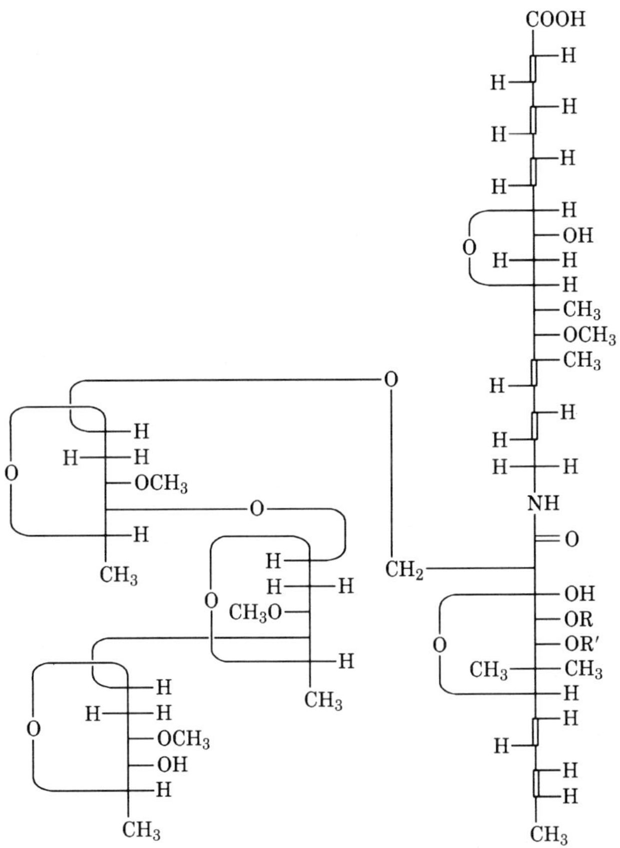

(11)

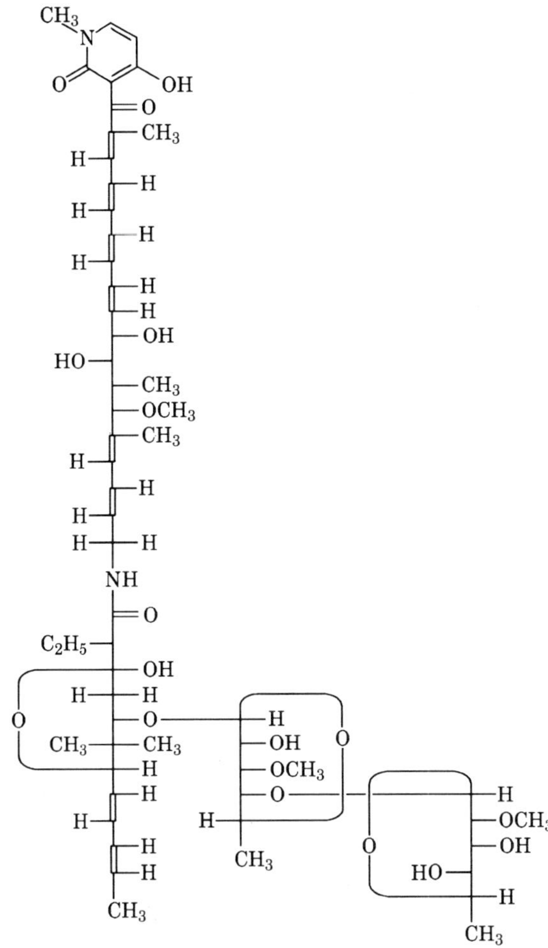

(12)

Fig. 1. *(Continued)*

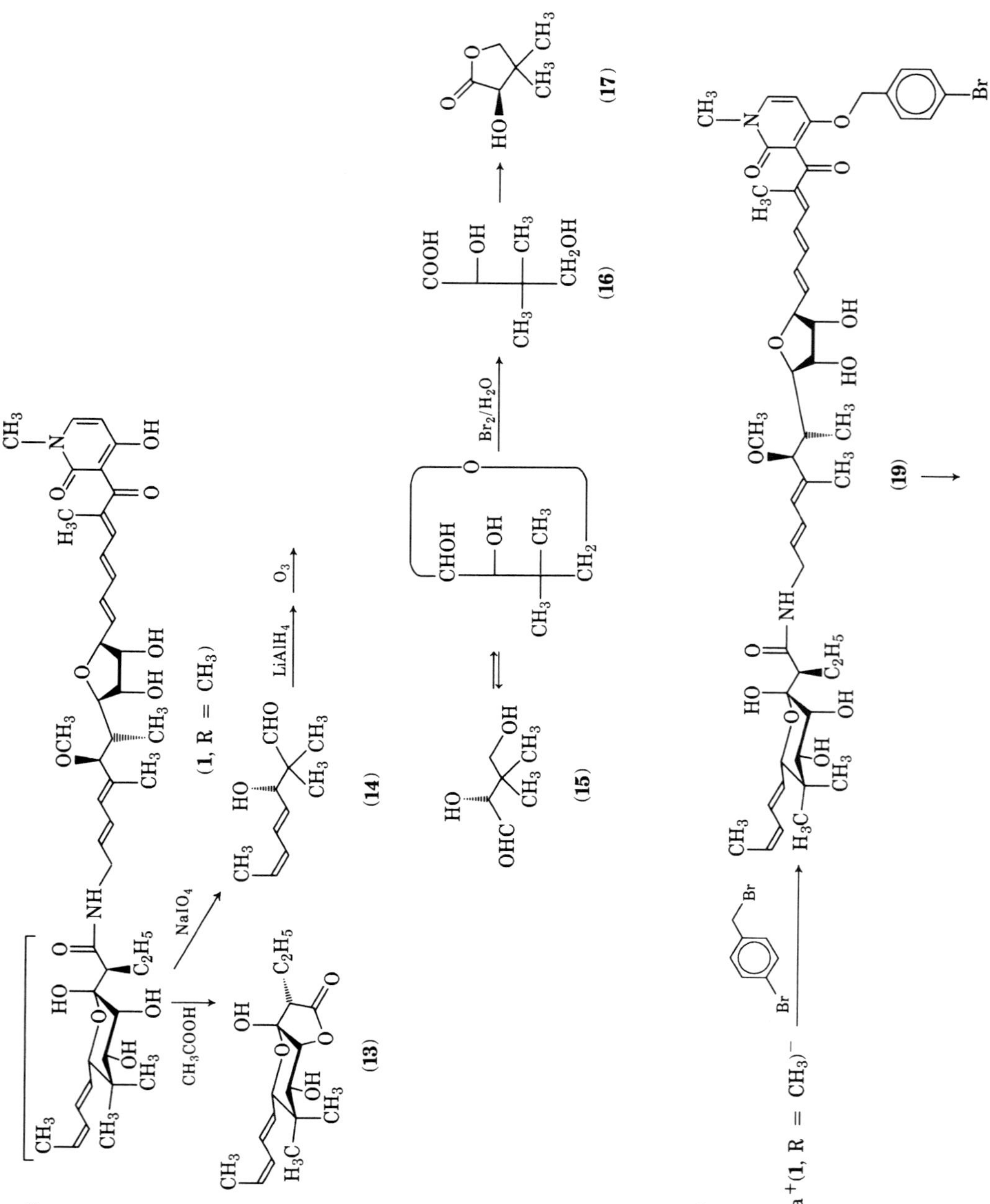
(a)
(b)
CH3COOH
NaIO4
LiAlH4
O3
Br2/H2O
(1, R = CH3)
Na+(1, R = CH3)−
(13)
(14)
(15)
(16)
(17)
(19)

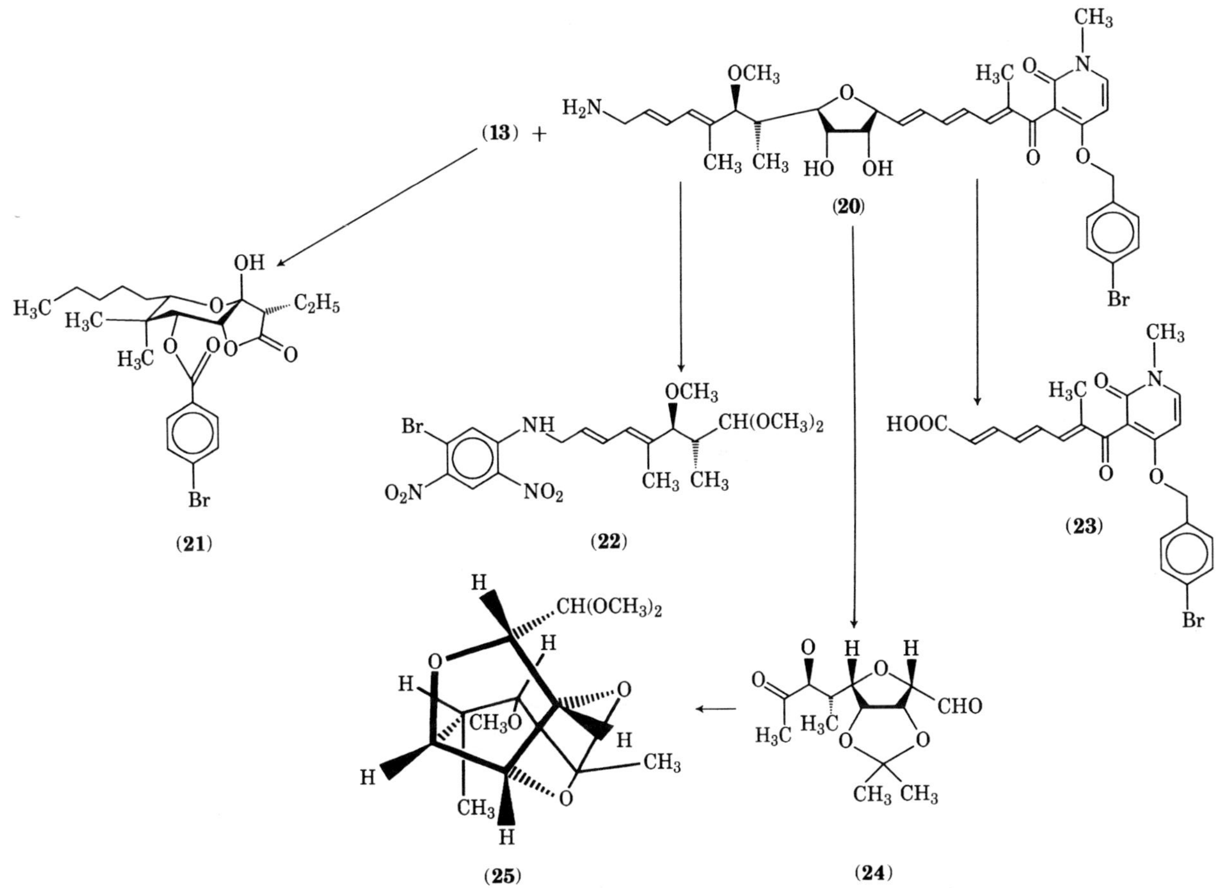

Fig. 2. Chemical degradation of aurodox: (**a**), goldinonic acid; (**b**), goldinamine.

aurodox using acetic acid at room temperature gave several products. One, termed goldinonic acid in view of the aurodox-producing microorganism's name, represented the carboxylic acid component of the amide and was isolated as the γ-lactone (**13**) (41). Periodate oxidation gave 3(*S*)-4(*E*)-6(*Z*)-3-hydroxy-2,2-dimethyl-4,6-octadienal (**14**), the absolute configuration of which was ascertained crystallographically (40) and by further degradation via (**15**) and (**16**) to the retrosynthetically significant (*R*)-(−)-pantolactone (**17**). Repetition of the acetic acid treatment using aurodox (4-bromobenzyl)ether (**19**) as substrate minimized side reactions and gave goldinonolactone (**13**) and the amine component, goldinamine (4-bromobenzyl)ether (**20**), thus accounting for all skeletal elements of the antibiotic.

The goldinamine bromobenzyl ether (**20**) gave rise to a series of useful degradation products. The most significant fragments are summarized in Figure 2**b**. The spectroscopic analysis of the trienone moiety was supplemented crystallographically via its derivative (**23**) (42), and the absolute configuration of (**13**) was confirmed by crystallographic analysis of the 4-bromobenzoate (**21**) (4,41).

The diene side of (**20**) was first obtained from its *N*-acetyl derivative by periodate oxidation in the form of 2(*R*)-3(*S*)-4(*E*)-6(*E*)-8-amino-3-methoxy-2,4-dimethyl-4,6-octadienal (4,43). The threo configuration of this product was ascertained by ir spectroscopy (4) and by crystallographic analysis of the *N*-dinitrophenyl (DNP) derivative of its dimethyl acetal (43). Anomalous scatter in the analysis of (**22**) made the determination of absolute configuration possible (44). The configuration of the central tetrahydrofuran ring in (**20**) was originally determined spectroscopically (43,45) and its topography was revealed by several degradation products (4). Most significant was the isolation of (**24**) which included not only the two stereocenters of (**22**), but also the four carbons in the tetrahydrofuran ring. Aldehyde (**24**) was converted to the crystalline tricyclic acetal (**25**). The topographic assignment of (**25**) also resulted from a crystallographic analysis (46), so that aurodox (**1**, R = CH_3) became known in complete stereochemical detail.

Similar degradation reactions were used to establish the absolute configuration of mocimycin (kirromycin) (**1**, R = H) (46), the constitution of which had been described previously (7,47). The chemical structures of most other subsequently discovered elfamycins have been determined spectroscopically and assignments of absolute configurations are not complete. The elfamycin structures shown in Figure 1 have complete stereochemical details that have been in part ascertained experimentally and in part are assumed to correspond to the aurodox topography.

Production, Biosynthesis, and Chemistry

The production of elfamycins is described in the references cited in Table 1. Fermentation yield improvements with aurodox (**1**, R = CH_3) proved difficult because of feedback inhibition (48). Aurodox-resistant strains (49), however, responded positively to conventional mutagenic methods leading to yield increases from 0.4 to 2.5 g/L (50). Scale-up of efrotomycin (**7**, R = CH_3) fermentations were found to be particularly sensitive to small changes in sterilization conditions of the oil-containing medium used (51).

Biosynthetic studies using acetate (Ac), propionate (Pr), and butyrate (Bu) revealed the polyketide nature of aurodox which has the composition $Pr(Ac)_8$ for the goldinamine skeleton C-7 to C-25 and the composition $Bu(Ac)_5$ for the C-27 to C-39 carbon chain of goldinonic acid. In contrast to the methyl branch at C-8, those at C-19 and C-21 are methionine-derived as are all remaining methyl groups (52,53). The biogenetic origin of the pyridone moiety is not clear.

Desmethylefrotomycin (**2**, R = H) is a product of microbial transformation of mocimycin (**1**, R = H) by *Nocardia* ATCC 53758 (35). Efrotomycin (**2**, R = CH_3) biosynthesis was studied in a resting cell system of *Nocardia lactamdurans* by glycosylation of aurodox (**1**, R = CH_3) via 6′-deoxyallosyl aurodox (54).

Mocimycin has been chemically converted to aurodox by protection of the 4-hydroxy group at the pyridone moiety as the benzoylformate, followed by *N*-methylation and hydrolytic removal of the protective group (1,55). Whereas aurodox esters are active growth promotors in animals, goldinamines that are *N*-acylated by acids other than goldinonic acid, such as acetic, benzoic, or arylsulfonic acids, lack useful antimicrobial or growth-promoting activity (1).

Mocimycin was prepared from dihydromocimycin (**3**) by oxidation with selenium dioxide (14,56). 4-Amino-4-dehydroxyefrotomycins were obtained by aminolysis of efrotomycin-4-*O*-phenylchlorophosphate (57).

Elfamycins having 4-hydroxy-2-pyridone moieties (**1–6**,**12**) readily undergo reversible internal cyclizations by conjugate addition of either oxygen functionality on the pyridone ring at C-9. These products can be isolated as exemplified by isoefrotomycin (58).

Aurodox and efrotomycin have been chemically synthesized (59). In analogy to the degradation studies shown in Figure 2, goldinonolactone (**13**) was synthesized from (**17**) by enantioselective reduction of ketolactone (**26**) followed by a sequence of acyclic stereoselective reactions shown in Figure 3. The stereocenters C-30 and C-31 in goldinonic acid (**1**) were generated by a Sharpless epoxidation of (**29**) followed by inversion at C-2 in epoxyalcohol (**31**). The (*E*),(*Z*)-geometry of the pentadienyl substituent was constructed from (**35**) by oxidation and addition of the anion derived from (*Z*)-crotyl diphenylphosphine oxide. Addition of butyrate to lactone (**36**) afforded a mixture of lactols which, after acetal hydrolysis, recyclization, and base-catalyzed epimerization, gave a mixture of C-3-epimeric dihydroxylactols from which pure goldinonolactone (**13**) could be isolated chromatographically (60).

The all-cis tetrahydrofuran ring of goldinamine was synthesized by the consecutive epoxide opening of the ester enolate derived from (**42**) according to an old concept as shown in Figure 4 (61). The five adjacent stereocenters in (**42**) were assembled by a kinetic resolution of alcohol (**38**) via its epoxide (**39**). One of the newly gained stereocenters was sacrificed to gain the allylic alcohol (**40**) which was subjected to a Sharpless epoxidation to introduce the third stereocenter. An epoxidation of the homoallylic alcohol (**41**) using *m*-chloroperbenzoic acid generated the two additional stereocenters with a diastereoselectivity of 94%. Hydrogenolysis of the benzyl ether, hydrolysis of the silyl ether, protection of the resulting diol, and oxidation of the primary alcohol gave a carboxylic acid which, after

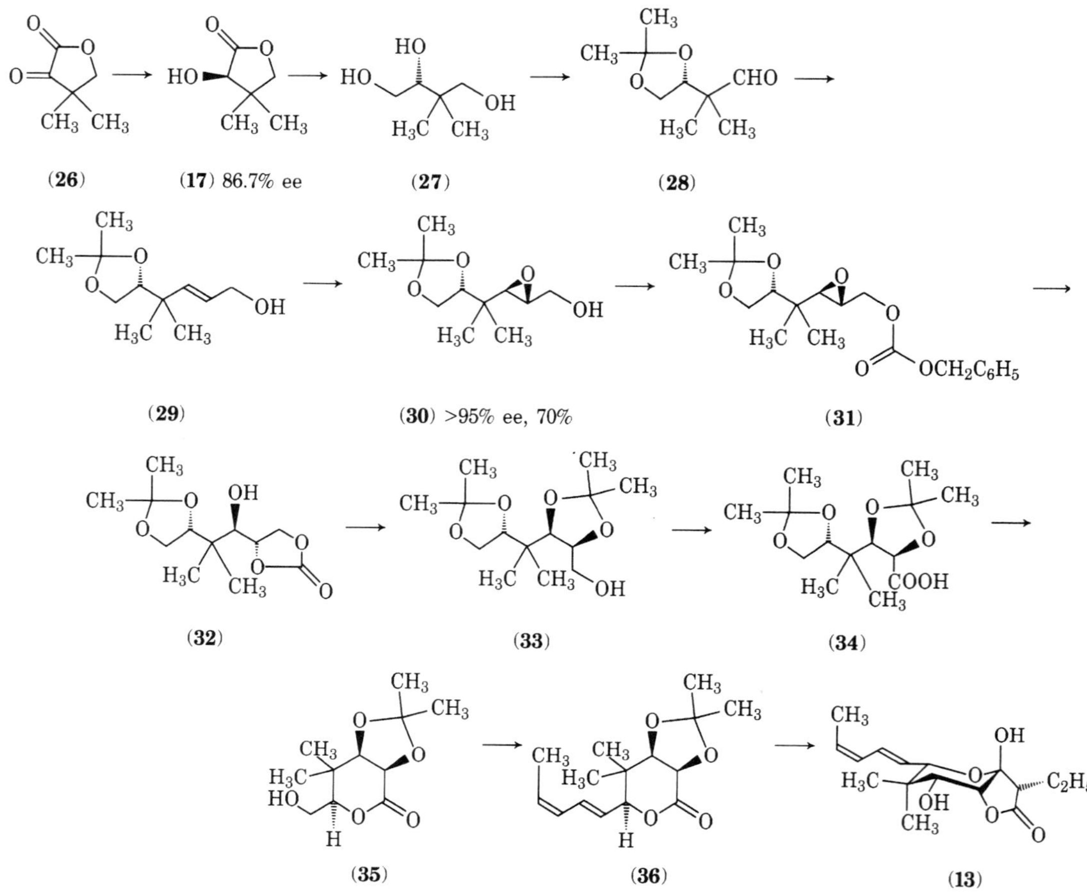

Fig. 3. Synthesis of goldinonolactone (**13**).

deprotonation, generated an alkoxide ion that opened the terminal epoxide to furnish the desired tetrahydrofuran (**43**). Further elaboration via (**44**) and (**45**) gave the terminal alkene (**46**) which was subjected to a stereoselective hydroboration to introduce the fifth consecutive stereocenter. A Swern oxidation of the resulting alcohol, followed by reaction with lithium di(α-methoxyvinyl)cuprate, led to the vinyl ether (**48**), the hydrolysis of which gave (**49**) with the sixth stereocenter in 73% de. A Wittig-Horner reaction using trimethyl phosphonoacetate established the (E)-geometry of the alkene with a stereoselectivity of 78%. An additional reduction step led to aldehyde (**50**).

Condensation of (**50**) with diisopropyl phosphonoacetonitrile established the second (E)-alkene with 82% stereoselectivity as shown in Figure 5. The third (E)-alkene, as shown in (**52**), was prepared from (**51**) after desilylation and oxidation of the resulting primary alcohol, followed by an additional Wittig-Horner reaction with 75% stereoselectivity. The missing (E),(E)-diene portion was generated by a Wittig-Horner condensation where one of the double bonds was provided by the reactant (E)-3-methyl-4-[4-(benzyloxy)-1,2-dihydro-1-methyl-2-oxo-3-pyridyl]-4-oxo-2-butenyl-1-triphenyl phosphonium bromide. Goldinamine deriva-

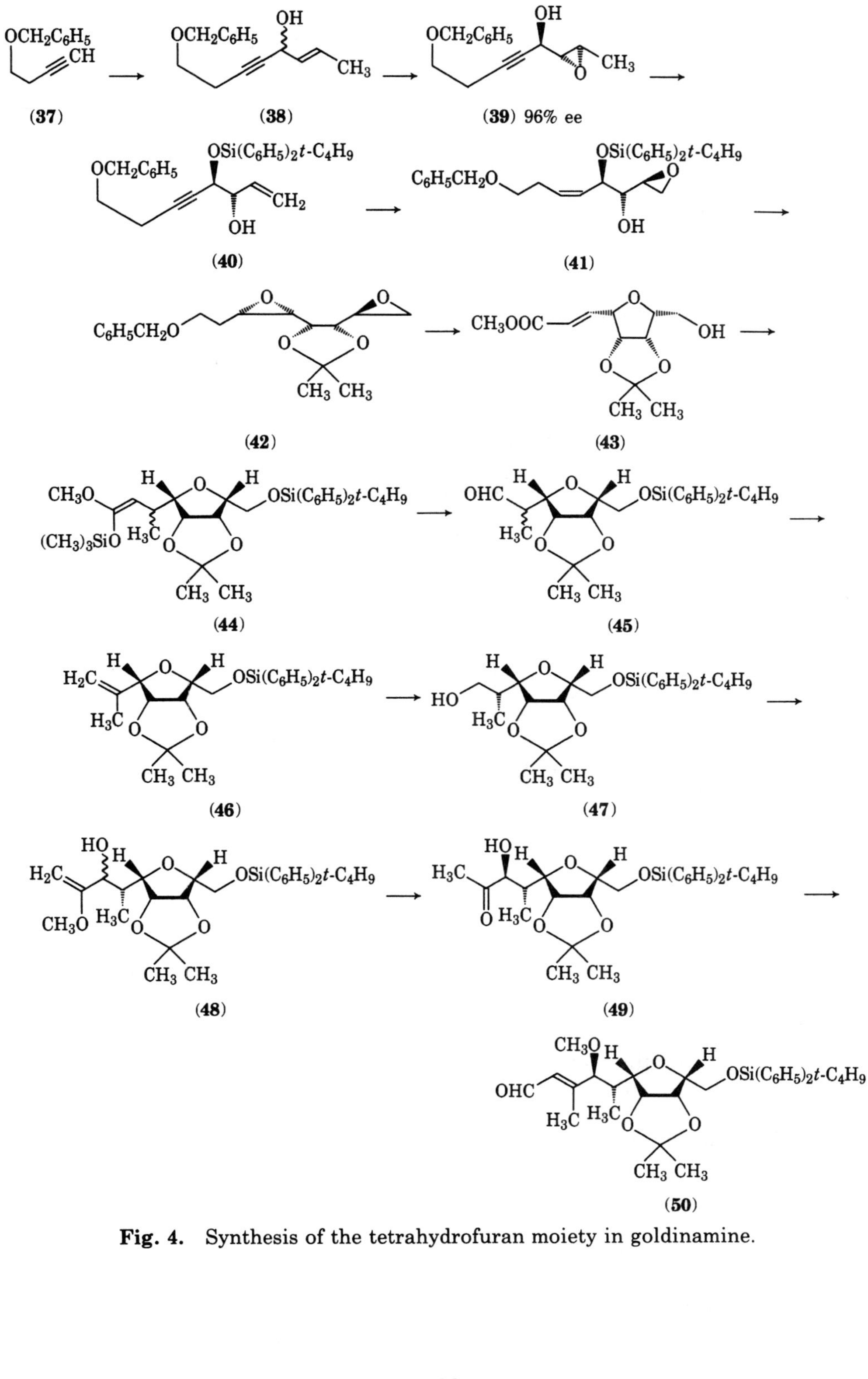

Fig. 4. Synthesis of the tetrahydrofuran moiety in goldinamine.

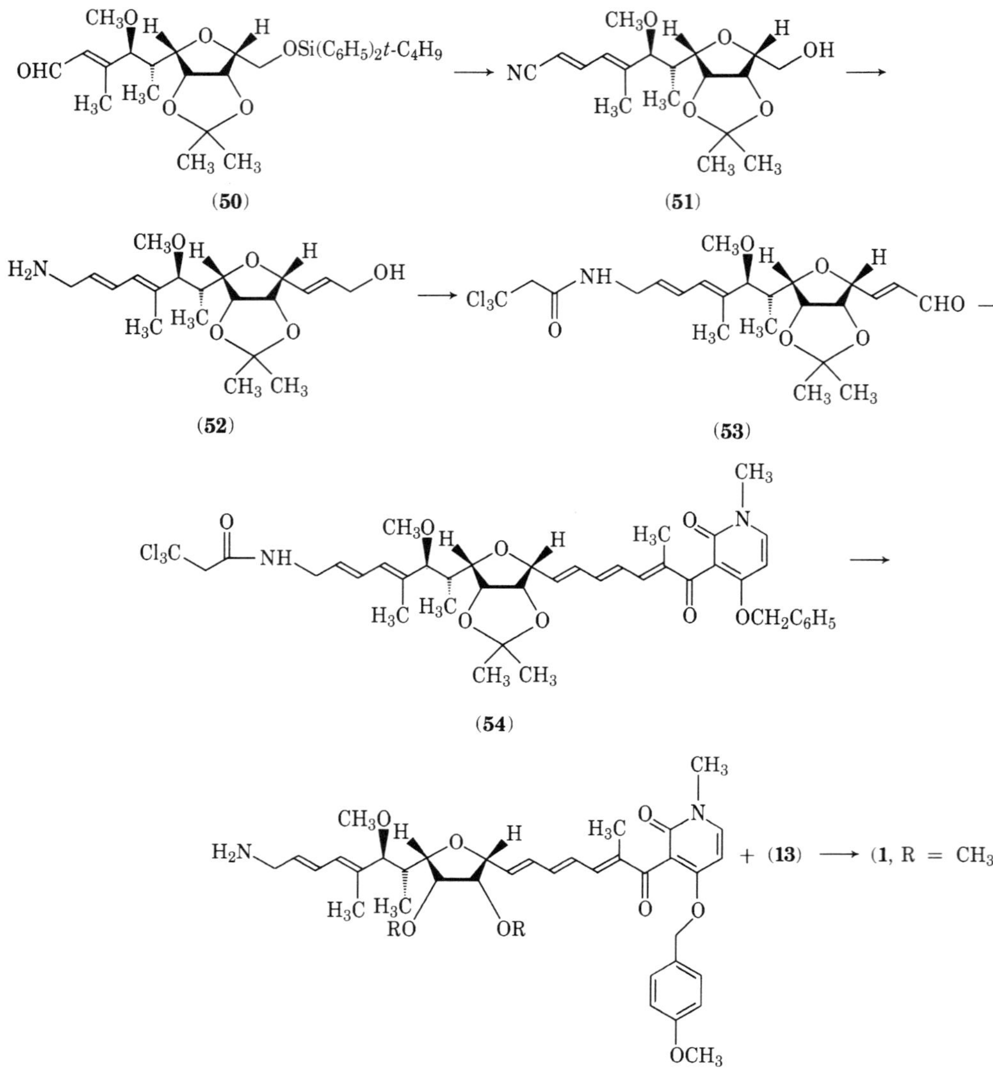

Fig. 5. Synthesis of goldinamine derivatives and aurodox where R′ is trimethylsilane.

tive (**55**), derived from (**54**) by changing the protective groups, served as a suitable substrate for the final condensation with (**13**) to furnish aurodox (**1**, R = CH_3), after removal of the silyl and benzyl ether groups (62).

Goldinonolactone and the tetrahydrofuran moiety of goldinamine were also synthesized from L- and D-mannose, respectively (63). The synthesis of efrotomycin has also been effected (62).

Biological Properties

Mode of Action. Elfamycins block bacterial protein biosynthesis at the level of elongation factor Tu (EF-Tu) (64–66). The biosynthesis of protein proceeds in four major stages, commencing with activation, which involves the synthesis of aminoacyl-*t*RNA (aa-*t*RNA) in the cytoplasm, followed by initiation, elongation, and termination. Binding of the ternary complex EF-Tu·GTP·aa-*t*RNA, where GTP is guanosine triphosphate, in the ribosomal *A*-site initiates the elongation phase. After hydrolysis of GTP to guanosine diphosphate (GDP) and the subsequent release of EF-Tu, peptide bond formation takes place. Elfamycins prevent the release of the EF-Tu·GDP complex from the ribosome and hence debilitate peptidyl transfer and peptide synthesis. This mode of action stands in unique contrast to that of other antibiotics which also interfere with the elongation phase of protein biosynthesis (60).

Antimicrobial Activity. The elfamycins' antimicrobial specificity and lack of toxicity in animals can be explained in view of species-dependent specificity of elfamycin binding to EF-Tu. Inefficient cellular uptake or the presence of a nonresponding EF-Tu were cited as responsible factors for the natural resistance in *Halobacterium cutirubrum* (67), *Lactobacillus brevis* (68), and in actinomycetes (5,69). The low activity of elfamycins against *S. aureus* was also attributed to an elfamycin-resistant EF-Tu system (70). However, cross-resistance with other antibacterial agents has not been observed (71).

Elfamycins have similar *in vitro* antimicrobial spectra and the activity against *Moraxella, Pasteurella, Yersinia, Haemophilus, Streptococcus, Corynebacterium,* and *Neisseria* appears to be common (2,23,72). Aurodox (**1**, R = CH_3) (2), azdimycin (8), LL-E19020α (**11**, R = H, R′ = $COCH_2C_6H_5$), LL-E19020β (**11**, R = $COCH_2C_6H_5$, R′ = H) (29), efrotomycin (**2**, R = CH_3), and heneicomycin (**4**, R = R″ = CH_3, R′ = C_2H_5) (72) have shown activity *in vivo* against *Streptococcus pyogenes* and the latter two antibiotics exhibited similar activity against *Moraxella bovis* (73,74).

Kirrothricin (**5**), aurodox, and efrotomycin were found to be active as protozoacids against the parasite *Eucoccidium dinophili* (75), and LL-E19020α and LL-E19020β were active against *Babesia* (see ANTIPARASITIC AGENTS). The latter two antibiotics acted as nematocides toward *Caenorhabditis elegans* and as anthelmintics (see ANTIPARASITIC AGENTS, ANTHELMINTICS) toward *Trichostrongylus colubiformis* (28).

Mocimycin (kirromycin) (**1**, R = CH_3) and tylosin [*1401-69-0*] (see ANTIBIOTICS, MACROLIDES) exhibited similar activity against *Treponema hyodysenteria,* the causative microorganism of dysentery in swine. Dihydromocimycin (**3**), however, was more active against this microorganism than either mocimycin or tylosin (14). Moreover, dihydromocimycin was active against tylosin-resistant *treponema* strains and was used as a feed additive at levels of 20–40 ppm (see FEEDS AND FEED ADDITIVES) (14). The growth-promoting activity of mocimycin was not observed for dihydromocimycin (14).

Efrotomycin (**2**, R = CH_3) is active against mycoplasma (PPLO) of chicks, pigs, and cattle (9). A combination of efrotomycin and bottromycin [*1393-68-6*], a complex of macrolides, has been used as a therapeutic and prophylactic agent to control mycoplasma infections and dysentery in swine (76). *Mycoplasma gallisepticum S*-6 was the most susceptible organism tested against heneicomycin (**4**, R = R″ = CH_3, R′ = C_2H_5) (15,16). Additionally, *M. gallisepticum* infected chicks receiving feed containing 50 ppm of LL-E19020α (**11**, R = H, R′ = $COCH_2C_6H_5$) had better weight gain, improved feed conversion, and less mortality than infected, nonmedicated chicks (30).

Efrotomycin is effective against *Clostridium perfringens* (77) and is much more active against 29 clinical isolates of *Clostridium difficile* than ciprofloxacin [*85721-33-1*] (78,79) (see ANTIBACTERIAL AGENTS, SYNTHETIC, QUINOLONES). Activity against gram-positive anaerobes, especially *Clostridium difficile,* was also noted for all phenelfamycins and unphenelfamycin (**8**, R = R′ = H). Phenelfamycin A (**8**, R = H, R′ = $COCH_2C_6H_5$) prolonged the survival of *C. difficile enterocolitis* in infected hamsters and was active against *Neisseria gonorrhoeae* and *streptococci* (80). Aurodox (**1**, R = CH_3) exhibited similar activity in the hamster model (81).

Antibiotic LL-E19020α and LL-E19020β are described as useful agents for the treatment of chronic respiratory disease, fowl cholera, and necrotic enteritis in birds (76) and as anthelmintics in monogastric and ruminant animals (28).

Growth Promotion. Elfamycins, in general, enhance the growth of farm animals (see GROWTH REGULATORS, ANIMAL). Growth improvement and feed conversion was studied using aurodox (**1**, R = CH_3) in chicks and turkeys. Effective drug levels were established to be 1–10 mg/kg of feed for chicks. The effective dose for turkeys was 5–10 times higher (82). The growth promoting ability of aurodox (**1**, R = CH_3) and mocimycin (**1**, R = H) was compared with other antibiotics and hormonal anabolic agents in chicks (83). Maximum weight improvement of 23% and enhanced feed efficiency of 13% in chicks was obtained with efrotomycin (**2**, R = CH_3) levels of 11 ppm (77). Antibiotics LL-E19020α and LL-E19020β proved to be more effective growth promotors in chicks than bacitracin [*1405-87-4*] and virginiamycin (see ANTIBIOTICS, PEPTIDES) (32) and also improved feed efficiency for meat-producing animals and fish. When given at 100 ppm in feed they caused weight gain by 6.1% in chicks. Feed efficiency was also increased, and during the first week of feeding an 8% increase was observed (84). Antibiotics LL-E19020α and LL-E19020β are more effective growth promotors than aurodox (32). The growth promoting effect of zinc bacitracin was potentiated by mocimycin (85). The improvement of weight gain and feed conversion by subtherapeutic antibiotic levels were shown to occur concomitantly with decreased cholyltaurine hydrolase activity suggesting that the inhibition of this enzyme is responsible for weight gain and improvement in feed utilization (see ENZYME INHIBITION) (86).

Efrotomycin (**2**, R = CH_3) is being developed as a growth promoting agent for swine. Pigs on efrotomycin diets gained weight 5.9–8.9% faster and utilized feed 1.7–4.0% more efficiently than the control animals. Growth improvement was linear from 2 to 16 ppm of drug, and the plateau of feed efficiency was reached at 4 ppm of efrotomycin (87). The effect on the shedding of *Salmonella,* however, is a matter of primary concern in swine and fowl, not only with efrotomycin (88), but with other elfamycins as well.

Stable growth promoting feed additives were prepared by granulation of the drug with alginic acid [*9005-32-7*] and magnesium hydroxide and adding it to oiled rice hulls (89) or by adsorption of elfamycins onto corn cob grits and coating with 10% tristearin [*555-43-1*] (90). A molecular efrotomycin dispersion with (2-vinylpyridine)-styrene copolymer served as a postrumen effective dosage form for oral administration (91).

Improvement of Lactation. Qualitative and quantitative improvements in milk production (see MILK AND MILK PRODUCTS) are dependent upon changes of ruminal volatile fatty acid (VFA) production and VFA composition. Increased propionate production at the expense of acetate and butyrate is responsible for growth enhancement and is achieved by additives such as polyether antibiotics (see ANTIBIOTICS, POLYETHERS). Relatively high levels of acetate and butyrate, however, are required to maintain adequate fat content in milk. Oral administration of elfamycins at concentrations of 1–10 mg/kg of animal body weight per day has been shown to increase milk volume from 2–15% relative to untreated animals without compromising fat content. The methods of administration to ruminants such as dairy cows and goats were similar to those employed for improving feed utilization and growth promotion (92).

Economic Aspects

The potential usefulness of elfamycins as growth promotors and feed-conversion enhancers is now generally recognized. Low original fermentation yields and difficulties in yield improvements discouraged early attempts to develop aurodox (**1**, R = CH_3) and mocimycin (kirromycin) (**1**, R = H) commercially. A development program for efrotomycin (**2**, R = CH_3), however, is ongoing as of this writing. Some of the newer elfamycins, such as the LL-E19020 pair, are considerably more active growth promotors than either aurodox or mocimycin, pointing toward the emergence of a second generation of elfamycins.

BIBLIOGRAPHY

1. H. Maehr, M. Leach, L. Yarmchuk, and M. Mitrovic, *J. Antibiot.* **32,** 361 (1979).
2. Ger. Pat. 2,140,322 (Feb. 17, 1972), J. Berger (to Hoffmann-LaRoche); Pat. 3,657,421 (Apr. 18, 1972), reissue Jan. 27, 1976, J. Berger (to Hoffmann-LaRoche); J. Berger, H. H. Lehr, S. Teitel, H. Maehr, and E. Grunberg, *J. Antibiot.* **26,** 15 (1972).
3. J. Barber, A. E. Derome, T. D. Howard, L. Lian, and G. Tebb, *Magn. Reson. Chem.* **27,** 748 (1989).
4. H. Maehr, M. Leach, T. H. Williams, and J. F. Blount, *Can. J. Chem.* **58,** 501 (1980).
5. H. Wolf and H. Zähner, *Arch. Microbiol.* **83,** 147 (1972).
6. U.S. Pat. 3,923,981 (Dec. 2, 1975), C. Vos and J. den Admirant; U.S. Pat. 3,927,211 (Dec. 16, 1975), C. Vos and J. den Admirant.
7. C. Vos and P. E. J. Verwiel, *Tetrahedron Lett.* 5173 (1973).
8. U.S. Pat. 3,898,327 (Aug. 5, 1975), M. W. Nimeck, E. Meyers, and W.-C. Liu.
9. Ger. Pat. 2,450,813 (Apr. 30, 1985), R. G. Wax and W. M. Maiese.
10. R. Wax, W. Maise, R. Weston, and J. Birnbaum, *J. Antibiot.* **29,** 670 (1976).
11. J. D. Strong, *Analyst* **111,** 853 (1986).
12. E. Selva and co-workers, *J. Antibiot.* **43,** 1349 (1990).
13. R. S. Dewey, B. H. Arison, J. Hannah, D. H. Shih, and G. Albers-Schoenberg, *J. Antibiot.* **38,** 1691 (1985).
14. Ger. Pat. 2,621,615 (Nov. 25, 1976), H. M. Jongsma, H. J. Kooreman, J. L. van Os, and C. Vos.
15. U.S. Pat. 4,071,631 (Jan. 31, 1978), S. B. Zimmerman and J. H. Chalmers, Jr.
16. S. B. Zimmerman, J. H. Chalmers, Jr., R. S. Dewey, E. O. Stapley, and S. Hernandez, *J. Antibiot.* **32,** 665 (1979).
17. I. Thein-Schranner, H. Zähner, H.-U. Hoppe, I. Hummel, and A. Zeeck, *J. Antibiot.* **35,** 948 (1982).

18. A. Zeeck, H. U. Hoppe, and I. Hummel, *Tetrahedron Lett.* **22,** 2357 (1981).
19. V. P. Gullo, S. B. Zimmerman, R. S. Dewey, O. Hensens, and P. J. Cassidy, *J. Antibiot.* **35,** 1705 (1982).
20. U.S. Pat. 4,311,693 (Jan. 19, 1982), S. Hernandez, S. B. Zimmerman, V. P. Gullo, and R. S. Dewey.
21. U.S. Pat. 4,476,139 (Oct. 9, 1984), A. J. Kempf and co-workers.
22. A. J. Kempf and co-workers, *J. Antibiot.* **39,** 1361 (1986).
23. Eur. Pat. 139,158 (May 2, 1985), E. Selva and co-workers.
24. P. Ferrari, D. Edwards, G. G. Gallo, and E. Selva, *J. Antibiot.* **43,** 1359 (1990).
25. D. R. Williams, G. T. Carter, F. Pinho, and D. B. Borders, *J. Chromatogr.* **484,** 381 (1989).
26. J. E. Hochlowski and co-workers, *J. Antibiot.* **41,** 1300 (1988).
27. M. Jackson, J. P. Karwowski, R. J. Theriault, M. L. Maus, and W. L. Kohl, *J. Antibiot.* **41,** 1293 (1988).
28. Eur. Pat. 252,380 (Jan. 13, 1988), G. T. Carter and co-workers.
29. W. M. Maiese and co-workers, *J. Antibiot.* **24,** 1489 (1989).
30. Eur. Pat. 313,846 (May 3, 1989), G. T. Carter and co-workers.
31. M. R. Jefson, J. Bordner, C. P. Reese, and E. B. Whipple, *J. Antibiot.* **42,** 1610 (1989).
32. G. T. Carter, D. W. Phillipson, J. J. Goodman, T. S. Dunne, and D. B. Borders, *J. Antibiot.* **41,** 1511 (1988).
33. U.S. Pat. 4,847,245 (July 11, 1989), M. R. Jefson, K. Kaneda, S. Nishiyama, and J. Tone.
34. M. S. Pacey and co-workers, *J. Antibiot.* **42,** 1453 (1989).
35. U.S. Pat. 4,847,245 (July 11, 1989), M. R. Jefson, J. Kaneda, S. Nishiyama, and J. Tone.
36. Ger. Pat. 2,140,674 (May 18, 1972), C. Vos and J. den Admirant.
37. K. A. Larson and M. L. King, *Biotechnol. Prog.* **2,** 73 (1986).
38. G. M. Brill, J. B. McAlpine, and J. E. Hochlowski, *J. Liq. Chromatogr.* **8,** 2259 (1985).
39. D. W. Fink and J. D. Stong, *Anal. Chem.* **56,** 2339 (1984).
40. L. Kaplan, D. W. Fink, and C. Halsey, *Anal. Chem.* **56,** 360 (1984).
41. H. Maehr and co-workers, *Helv. Chim. Acta* **55,** 3051 (1972).
42. H. Maehr, J. F. Blount, M. Leach, A. Stempel, and G. Büchi, *Helv. Chim. Acta* **55,** 3054 (1972).
43. H. Maehr and co-workers, *J. Am. Chem. Soc.* **95,** 8448 (1973).
44. H. Maehr, J. F. Blount, M. Leach, and A. Stempel, *Helv. Chim. Acta* **57,** 936 (1974).
45. H. Maehr, T. H. Williams, M. Leach, and A. Stempel, *Helv. Chim. Acta* **57,** 212 (1974).
46. H. Maehr, M. Leach, J. F. Blount, and A. Stempel, *J. Am. Chem. Soc.* **96,** 4034 (1974).
47. H. Maehr, M. Leach, L. Yarmchuk, and A. Stempel, *J. Am. Chem. Soc.* **95,** 8449 (1973).
48. C.-M. Liu, T. Hermann, and P. A. Miller, *J. Antibiot.* **30,** 244 (1977).
49. C.-M. Liu, T. Hermann, and P. A. Miller, *J. Antibiot.* **32,** 1025 (1979).
50. J. Unowsky and D. C. Hoppe, *J. Antibiot.* **31,** 662 (1978).
51. D. Jain and B. C. Buckland, *Bioprocess Eng.* **3,** 31 (1988).
52. C.-M. Liu, H. Maehr, M. Leach, M. Liu, and P. A. Miller, *J. Antibiot.* **30,** 616 (1977).
53. C.-M. Liu, T. H. Williams, and R. Pitcher, *J. Antibiot.* **32,** 414 (1979).
54. J. B. K. Nielsen and L. Kaplan, *J. Antibiot.* **42,** 944 (1989).
55. U.S. Pat. 4,112,216 (Sept. 5, 1978), H. Maehr; Neth. Pat. 7,607,573 (Jan. 10, 1976).
56. Ger. Pat. 2,658,977 (May 5, 1977), H. Jongsma, H. J. Kooreman, and J. L. van Os.
57. U.S. Pat. 4,515,783 (May 7, 1985), B. O. Linn and A. Lusi.
58. U.S. Pat. 4,482,545 (Nov. 13, 1984), R. J. Bochis and R. S. Dewey.
59. K. C. Nicolaou, *Chem. Br.* **21,** 813 (1985).
60. R. E. Dolle and K. C. Nicolaou, *J. Am. Chem. Soc.* **107,** 1691 (1985).
61. G. Ohloff, K.-H. Schulte-Elte, and B. Willhalm, *Helv. Chim. Acta.* **47,** 602 (1964).
62. R. E. Dolle and K. C. Nicolaou, *J. Am. Chem. Soc.* **107,** 1695 (1985).
63. R. E. Dolle and K. C. Nicolaou, *J. Chem. Soc., Chem. Commun.,* 1016 (1985).

64. D. L. Miller and H. Weissbach, *Arch. Biochem. Biophys.* **141,** 26 (1970).
65. A. Parmeggiani and G. W. M. Swart, *Ann. Rev. Microbiol.* **39,** 557 (1985).
66. A. Parmeggiani and G. Sander, *Top. Antibiot. Chem.* **5,** 159 (1980).
67. M. Kessel and F. Klink, *Eur. J. Biochem.* **114,** 481 (1981).
68. W. Wörner and H. Wolf, *Regard. Biochim.* **3,** 57 (1981).
69. C. Gloeckner and H. Wolf, *FEMS Microbiol. Lett.* **25,** 121 (1984).
70. C. C. Hall, J. D. Watkins, and N. H. Georgopapadakou, *Antimicrob. Agents Chemother.* **33,** 322 (1989).
71. T. M. Jacks, E. G. Frazier, G. K. Abruzzo, A. C. Graham, and R. A. Fromtling, *Methods Find. Exp. Clin. Pharmacol.* **11,** 697 (1989).
72. B. M. Frost, M. E. Valiant, B. Weissberger, and E. L. Dulaney, *J. Antibiot.* **29,** 1083 (1976).
73. U.S. Pat. 4,024,251 (May 17, 1977), W. M. Maiese and R. G. Wax.
74. B. M. Frost, M. E. Valiant, and E. L. Dulaney, *J. Antibiot.* **32,** 626 (1979).
75. L. Tejmar-Kolar and H. Zähner, *FEMS Microbiol. Lett.* **24,** 21 (1984).
76. Jpn. Pat. 61,165,332 (July 26, 1986), T. Konishi and K. Matsumoto.
77. M. W. Stutz, S. L. Johnson, F. R. Judith, and B. M. Miller, *Poult. Sci.* **62,** 1612 (1983).
78. S. W. B. Newsom, J. Matthews, and A. M. Rampling, *J. Antimicrob. Chemother.* **15,** 648 (1985).
79. C. R. Clabots, C. J. Shanholtzer, L. R. Peterson, and N. D. Gerding, *Diagn. Microbiol. Infect. Dis.* **6,** 49 (1987).
80. R. N. Swanson and co-workers, *J. Antibiot.* **42,** 94 (1989).
81. J. Unowsky, unpublished data, 1987.
82. M. L. Marusich, E. F. Ogrinz, and M. Mitrovic, *Poult. Sci.* **53,** 636 (1974).
83. W. L. Marusich, E. F. Ogrinz, N. Camerlengo, and M. Mitrovic, *Poult. Sci.* **57,** 1297 (1978).
84. U.S. Pat. 4,704,276 (Nov. 3, 1987), S. Kantor.
85. Ger. Pat. 2,406,164 (Aug. 14, 1974), T. van Eek and R. Beukers; Neth. Pat. 7,401,731 (Aug. 13, 1974).
86. S. D. Feighner and M. P. Dashkevicz, *Appl. Environ. Microbiol.* **53,** 331 (1987).
87. A. G. Foster, D. H. Baker, T. R. Cline, G. L. Cromwell, and T. Veum, *J. Anim. Sci.* **65,** 877 (1987).
88. T. M. Jacks, E. Frazier, F. R. Judith, and G. Oslon, *Am. J. Vet. Res.* **49,** 1832 (1988).
89. Eur. Pat. 91,767 (Oct. 19, 1983), R. J. Yarwood, H. C. Maxfield, and A. J. Phillips.
90. S. Afr. Pat. 8,706,440 (Apr. 27, 1988), M. H. Fisher and H. H. Mrozik.
91. U.S. Pat. 4,808,412 (Feb. 28, 1989), E. P. Smith and S. H. Wu.
92. U.S. Pat. 4,336,250 (June 22, 1982), C. C. Scheifinger.

H. Maehr
Hoffmann-LaRoche Inc.

GLYCOPEPTIDES (DALBAHEPTIDES)

The vancomycin–ristocetin family of glycopeptides (1,2) is a subclass of linear sugar-containing peptides composed of seven amino acids cross-linked to generate a specific stereochemical configuration. This configuration forms the basis of a particular mechanism of action, eg, complexation with the D-alanyl-D-alanine terminus of bacterial cell wall components (3,4). Because the mechanism of action is the distinguishing feature of these peptides, the term dalbaheptide, from

D-al(anyl-D-alanine)b(inding)a(ntibiotics) having hept(apept)ide structure, has been proposed to distinguish them within the larger and diverse groups of glycopeptide antibiotics (5).

About 40 different naturally produced dalbaheptides have been reported (Table 1), which correspond to a larger number of chemical entities (Tables 2–6).

Table 1. Naturally Occurring Dalbaheptides

Antibiotic	Producing organism	Year[a]	Company or Institute	Refs.
ristocetin A, B	*Nocardia lurida* NRRL 2430	1953	Abbott	1,2
vancomycin	*Streptomyces orientalis* NRRL 2450	1955	Lilly	1,6,10–13
actinoidin A, B	*Proactinomyces actinoides*	1956	Research Institute for the Discovery of New Antibiotics, Moscow	1,18,19
K-288	*Streptomyces haranomachiensis* n.sp.	1961	Tohoku University, Sendai	20
ristomycin A, B	*Proactinomyces fructiferi* var. *ristomycini*	1962	Research Institute for the Discovery of New Antibiotics, Moscow	1,2
avoparcin (LL-AV290 complex)	*Streptomyces candidus* NRRL 3218	1966	American Cyanamid	21–24
AM374	*Streptomyces eburosporeus* NRRL 3582	1969	American Cyanamid	25
A477	*Actinoplanes* sp. NRRL 3884	1971	Lilly	26
actaplanin (A4696 complex)	*Actinoplanes missouriensis* ATCC 23342	1971	Lilly	27–29
AB-65	*Saccharomonospora viride* T-80	1973	Dainippon Pharmaceuticals	30
teicoplanin (teichomycin complex)	*Actinoplanes teichomyceticus* ATCC 31121	1975	Lepetit	7,31–33
A35512 complex	*Streptomyces candidus* NRRL 8156	1976	Lilly	34–36
OA-7653A, B	*Streptomyces hygroscopicus* subsp. *hiwasaensis* ATCC 31613	1978	Otsuka Pharmaceutical	37,38
A51568A (*N*-demethyl-vancomycin), B	*Nocardia orientalis* NRRL 15232	1982	Lilly	39,40
A41030 complex	*Streptomyces virginiae* NRRL 15156	1982	Lilly	41–43
A47934	*Streptomyces toyocaensis* NRRL 15009	1982	Lilly	43–45
ardacin (AAD-216 complex, aridicin)	*Kibdelosporangium aridum* ATCC 39323	1983	Smith, Kline and French	15, 46–48

Table 1. ***(Continued)***

Antibiotic	Producing organism	Year[a]	Company or Institute	Refs.
chloropolysporin A, B, C	*Faenia interjecta* sp. nov. FERM BP-538	1983	Sankyo	49–51
A40926 complex	*Actinomadura* sp. ATCC 39727	1984	Lepetit	52–54
M43 complex	*Nocardia orientalis* NRRL 2450	1984	Lilly	55
kibdelin (AAD-609 complex)	*Kibdelosporangium aridum* subsp. *largum* ATCC 39922	1985	Smith, Kline and French	56,57
izupeptin A, B	*Nocardia* sp. FERM P-8656	1986	Kitasato Institute, Tokyo	58
parvodicin (AAJ-271)	*Actinomadura parvosata* ATCC 53463	1986	Smith, Kline and French	59
synmonicin A, B, C (CWI-785)	*Synnemomyces mamnoorii* ATCC 53296	1986	Eskayef Ltd.; Smith, Kline and French	60
actinoidin A_2	*Nocardia* sp. SKF-AAJ-193	1987	Smith, Kline and French	61,62
orienticin (PA42867A,B,C,D)	*Nocardia orientalis* PA-42867	1987	Shionogi	63,64
eremomycin	*Actinomyces* sp. INA-238	1987	Institute of New Antibiotics, Academy of Medical Sciences, Moscow	8,65
A42867	*Nocardia* sp. nov. ATCC 53492	1987	Lepetit	66
A82846 A, B, C	*Amycolatopsis orientalis* NRRL 18098	1987	Lilly	67–70
UK-68,597	*Actinoplanes* sp. ATCC 53533	1987	Pfizer	71,72
helvecardin A, B	*Pseudonocardia compacta* subsp. *helvetica* SANK 65185	1988	Sankyo	73
chloroorienticins A, B, C, D, E (PA-45052)	*Amycolatopsis orientalis* PA-45052	1988	Shionogi	69,74,75
A80407 A, B	*Kibdelosporangium philippinensis* NRRL 18198	1989	Lilly	76
MM 45289, MM 47756	*Amycolatopsis orientalis* NCIB 12531	1989	Beecham	77
MM 47761, MM 49721	*Amycolatopsis orientalis* NCIB 12608	1989	Beecham	78
MM 47766, MM 47767, MM 55256, MM 55260	*Amycolatopsis orientalis* NCIB 40011	1989	Beecham	79
MM 49728, MM 55266, MM 55267, MM 55268	*Amycolatopsis* sp. NCIB 40089	1989	Beecham	80
decaplanin	*Actinomyces* sp. DSM 4763	1990	Hoechst	81
UK-72,051	*Amycolatopsis orientalis* n.sp.	1990	Pfizer	82

[a]Year of patent or first publication.

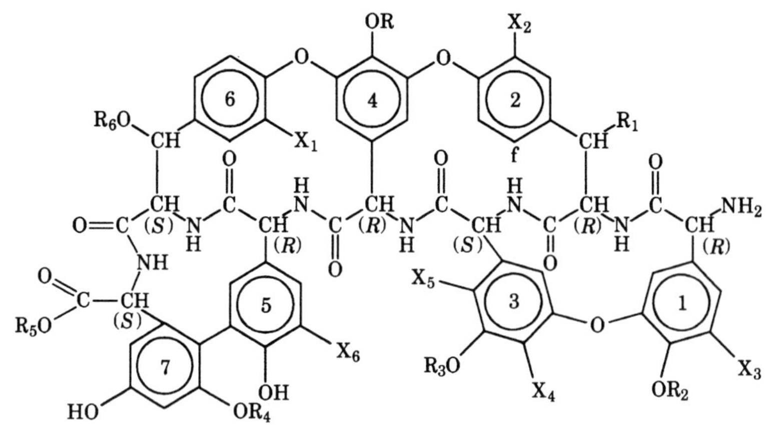

Table 2. Naturally Occurring Ristocetin-type Dalbaheptides

Antibiotic	CAS Registry Number	Structure[a] No.	R	R_1	R_2	R_3	R_4	R_5	R_6	X_1	X_2	X_3	X_4	X_5	X_6
ristocetin															
A[b]	[*11021-66-2*]	(1)	L-rha→D-glu←D-man←D-ara	OH	H	H	D-man	CH_3	L-ria	H	H	H	CH_3	H	H
B[c]	[*1405-59-0*]	(2)	L-rha→D-glu	OH	H	H	D-man	CH_3	L-ria	H	H	H	CH_3	H	H

actaplanin															
A	[*88357-81-7*]	**(3)**	D-glu←man	H	H	D-man	D-man	CH_3	L-ria	Cl	H	H	CH_3	H	H
B_1	[*88357-82-8*]	**(4)**	D-glu←rha	H	H	D-man	D-man	CH_3	L-ria	Cl	H	H	CH_3	H	H
B_2	[*88357-83-9*]	**(5)**	D-glu	H	H	D-man	D-man	CH_3	L-ria	Cl	H	H	CH_3	H	H
B_3	[*88357-84-0*]	**(6)**	D-glu←man	H	H	H	D-man	CH_3	L-ria	Cl	H	H	CH_3	H	H
C_1	[*88357-85-1*]	**(7)**	D-glu←L-rha	H	H	H	D-man	CH_3	L-ria	Cl	H	H	CH_3	H	H
C_3	[*88357-88-4*]	**(8)**	D-glu	H	H	D-man	H	CH_3	L-ria	Cl	H	H	CH_3	H	H
G	[*88381-73-1*]	**(9)**	D-glu	H	H	H	man	CH_3	L-ria	Cl	H	H	CH_3	H	H
A35512 B	[*63849-30-9*]	**(10)**	glu←rha	OH	H	fuc	man	CH_3	3-*epi*-L-van	H	H	H	H	Cl	H
A41030															
A	[*89139-41-3*]	**(11)**	H	H	H	H	H	H	H	Cl	Cl	H	H	H	Cl
B	[*89139-42-4*]	**(12)**	H	H	H	H	H	H	H	Cl	Cl	H	H	H	H
C	[*89140-21-6*]	**(13)**	H	H	gal	H	H	H	H	Cl	Cl	H	H	H	Cl
E	[*89139-43-5*]	**(14)**	H	H	H	H	H	H	H	Cl	H	H	H	H	H
F	[*89140-20-5*]	**(15)**	H	H	gal→gal	H	H	H	H	Cl	Cl	H	H	H	Cl
A47934	[*90039-80-8*]	**(16)**	H	H	SO_3H	H	H	H	H	Cl	Cl	H	H	H	Cl
UK-68,597[d]	[*121377-12-6*]	**(17)**	glu←van	H	H	SO_3H	H	H	H	Cl	Cl	Cl	H	H	Cl

[a]Abbreviations are ria = ristosamine, van = vancosamine, man = mannose, rha = rhamnose, glu = glucose, gal = galactose, ara = arabinose, and fuc = fucose.
[b]Identical to ristomycin A.
[c]Identical to ristomycin B.
[d]A C=O group is present instead of the terminal CH—NH_2.

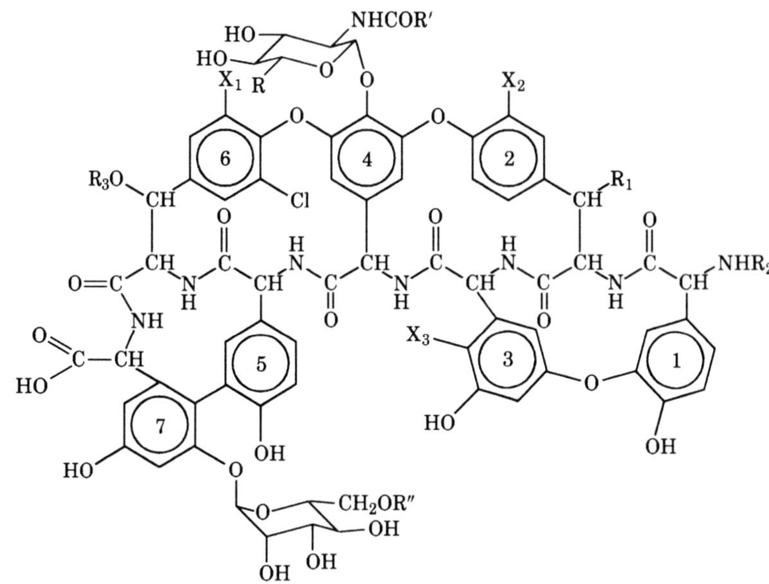

Table 3. Naturally Occurring Ristocetin-type Lipodalbaheptides

Antibiotic	CAS Registry Number	Structure[a]									
		No.	R	R′	R″	R_1	R_2	R_3	X_1	X_2	X_3
teicoplanin											
T-A2-1	[*91032-34-7*]	**(18)**	CH_2OH	$(CH_2)_2CH{=}CH(CH_2)_4CH_3$	H	H	H	Ac-*N*-D-glu	H	Cl	H
T-A2-2	[*91032-26-7*]	**(19)**	CH_2OH	$(CH_2)_6CH(CH_3)_2$	H	H	H	Ac-*N*-D-glu	H	Cl	H

T-A2-3	*[91032-36-9]*	**(20)**	CH_2OH	$(CH_2)_8CH_3$	H	H	H	Ac-*N*-D-glu	H	Cl	H
T-A2-4	*[91032-37-0]*	**(21)**	CH_2OH	$(CH_2)_6CH(CH_3)CH_2CH_3$	H	H	H	Ac-*N*-D-glu	H	Cl	H
T-A2-5	*[91032-38-1]*	**(22)**	CH_2OH	$(CH_2)_7CH(CH_3)_2$	H	H	H	Ac-*N*-D-glu	H	Cl	H
ardacin											
A	*[95935-21-0]*	**(23)**	COOH	$(CH_2)_8CH_3$	H	OH	CH_3	H	Cl	Cl	Cl
B	*[95935-22-1]*	**(24)**	COOH	$(CH_2)_7CH(CH_3)_2$	H	OH	CH_3	H	Cl	Cl	Cl
C	*[95935-23-2]*	**(25)**	COOH	$(CH_2)_8CH(CH_3)_2$	H	OH	CH_3	H	Cl	Cl	Cl
A40926											
A[b]	*[102961-73-9]*	**(26)**	COOH	$(CH_2)_9CH_3$	H	H	CH_3	H	H	H	Cl
B[c]	*[102961-74-0]*	**(27)**	COOH	$(CH_2)_8CH(CH_3)_2$	H	H	CH_3	H	H	H	Cl
PA	*[102961-76-2]*	**(28)**	COOH	$(CH_2)_9CH_3$	$COCH_3$	H	CH_3	H	H	H	Cl
PB[d]	*[102961-77-3]*	**(29)**	COOH	$(CH_2)_8CH(CH_3)_2$	$COCH_3$	H	CH_3	H	H	H	Cl
kibdelin											
A	*[103528-50-3]*	**(30)**	CH_2OH	$(CH_2)_8CH_3$	H	OH	CH_3	H	Cl	Cl	Cl
B	*[103528-49-0]*	**(31)**	CH_2OH	$(CH_2)_7CH(CH_3)_2$	H	OH	CH_3	H	Cl	Cl	Cl
C_1	*[103549-47-9]*	**(32)**	CH_2OH	$(CH_2)_8CH(CH_3)_2$	H	OH	CH_3	H	Cl	Cl	Cl
C_2	*[105997-85-1]*	**(33)**	CH_2OH	$(CH_2)_{10}CH_3$	H	OH	CH_3	H	Cl	Cl	Cl
D	*[105997-86-2]*	**(34)**	CH_2OH	$(CH_2)_2CH{=}CH(CH_2)_4CH_3$	H	OH	CH_3	H	Cl	Cl	Cl
parvodicin											
A	*[114797-09-9]*	**(35)**	COOH	$(CH_2)_8CH_3$	H	H	CH_3	H	H	H	Cl
B_1	*[110882-82-1]*	**(36)**	COOH	$(CH_2)_7CH(CH_3)_2$	H	H	CH_3	H	H	H	Cl
C_2	*[110882-85-4]*	**(37)**	COOH	$(CH_2)_{10}CH_3$	H	H	CH_3	H	H	H	Cl
C_4	*[110882-87-6]*	**(38)**	COOH	$(CH_2)_{10}CH_3$	$COCH_3$	H	CH_3	H	H	H	Cl

[a] Ac-*N*-D-glu = *N*-acetyl-D-glucosamine.
[b] Identical to parvodicin B_2.
[c] Identical to parvodicin C_1.
[d] Identical to parvodicin C_3.

Table 4. Naturally Occurring Vancomycin-type Dalbaheptides

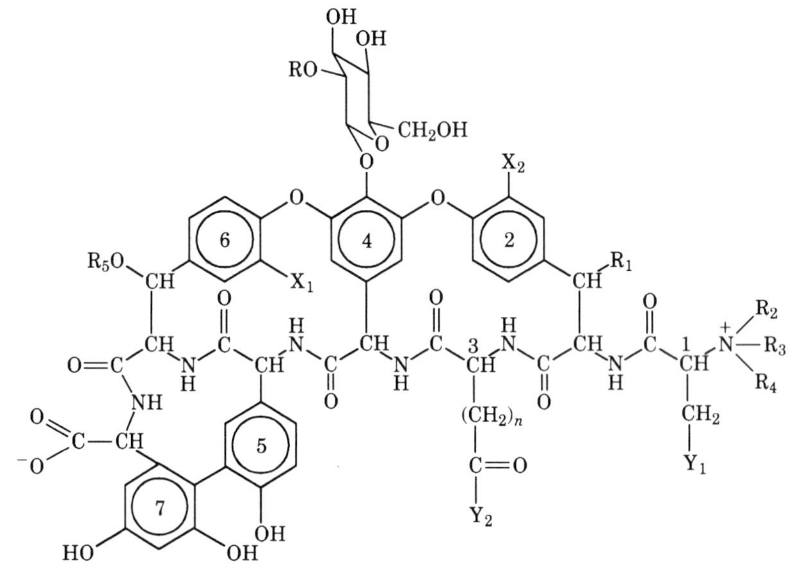

Antibiotic	CAS Registry Number	Structure[a] No.	n	R	R_1	R_2	R_3	R_4	R_5	X_1	X_2	Y_1	Y_2
vancomycin[b]	*[1404-90-6]*	**(39)**	1	L-van	OH	H	H	CH_3	H	Cl	Cl	iC_3H_7	NH_2
OA-7653													
A	*[73699-70-4]*	**(40)**	2	[c]	H	H	CH_3	CH_3	D-glu	Cl	Cl	H	NH_2
B	*[112219-76-8]*	**(41)**	2	[c]	H	H	CH_3	CH_3	D-glu	Cl	Cl	H	OH
A51568													
A	*[92182-33-7]*	**(42)**	1	L-van	OH	H	H	H	H	Cl	Cl	iC_3H_7	NH_2
B	*[92182-32-6]*	**(43)**	2	L-van	OH	H	H	H	H	Cl	Cl	iC_3H_7	NH_2

M43													
A	[*99759-15-6*]	**(44)**	1	L-van	OH	CH_3	CH_3	CH_3	H	Cl	Cl	iC_3H_7	NH_2
B	[*99776-55-3*]	**(45)**	1	L-van	OH	CH_3	CH_3	CH_3	H	Cl	Cl	iC_3H_7	OH
C	[*99759-14-5*]	**(46)**	1	H	OH	CH_3	CH_3	CH_3	H	Cl	Cl	iC_3H_7	NH_2
D	[*99759-13-4*]	**(47)**	1	L-van	OH	H	CH_3	CH_3	H	Cl	Cl	iC_3H_7	NH_2
orienticin													
A[d]	[*111073-20-2*]	**(48)**	1	4-*epi*-van	OH	H	H	CH_3	4-*epi*-van	Cl	H	iC_3H_7	NH_2
B	[*111073-19-9*]	**(49)**	1	L-oli	OH	H	H	CH_3	4-*epi*-van	Cl	H	iC_3H_7	NH_2
C[e]	[*112848-47-2*]	**(50)**	1	4-*epi*-van	OH	H	H	CH_3	4-*epi*-van	H	H	iC_3H_7	NH_2
D	[*112848-46-1*]	**(51)**	1	4-*epi*-van	OH	H	CH_3	CH_3	4-*epi*-van	Cl	H	iC_3H_7	NH_2
eremomycin[f]	[*110865-90-2*]	**(52)**	1	4-*epi*-van	OH	H	H	CH_3	4-*epi*-van	H	Cl	iC_3H_7	NH_2
A42867	[*114540-35-1*]	**(53)**	1	D-rha	OH	H	H	CH_3	van	H	Cl	iC_3H_7	NH_2
chloroorienticin													
A[g]	[*118395-73-6*]	**(54)**	1	4-*epi*-van	OH	H	H	CH_3	4-*epi*-van	Cl	Cl	iC_3H_7	NH_2
B	[*118373-81-2*]	**(55)**	1	H	OH	H	H	CH_3	4-*epi*-van	Cl	Cl	iC_3H_7	NH_2
C	[*118373-82-3*]	**(56)**	1	[c]	OH	H	CH_3	CH_3	4-*epi*-van	Cl	Cl	iC_3H_7	NH_2
D	[*118373-83-4*]	**(57)**	1	4-*epi*-van	OH	H	CH_3	CH_3	4-*epi*-van	Cl	Cl	iC_3H_7	NH_2
E	[*118373-84-5*]	**(58)**	1	H	OH	H	CH_3	CH_3	4-*epi*-van	Cl	Cl	iC_3H_7	NH_2
MM 47761[h]	[*126985-51-1*]	**(59)**	1	rha	OH	H	H	CH_3	4-*epi*-van	H	Cl	iC_3H_7	NH_2
MM 49721	[*126985-52-2*]	**(60)**	1	rha	OH	H	H	CH_3	4-*epi*-van	H	H	iC_3H_7	NH_2

[a]Abbreviations are van = vancosamine, glu = glucose, oli = olivose, 4-*epi*-van = 4-*epi*-vancosamine = eremosamine.
[b]Identical to K-288.
[c]The hydroxyl on ring 4 does not carry any sugar.
[d]Identical to UK-72,051.
[e]Identical to A82846C, MM 47756.
[f]Identical to A82846A, MM 45289.
[g]Identical to A82846B.
[h]Identical to decaplanin.

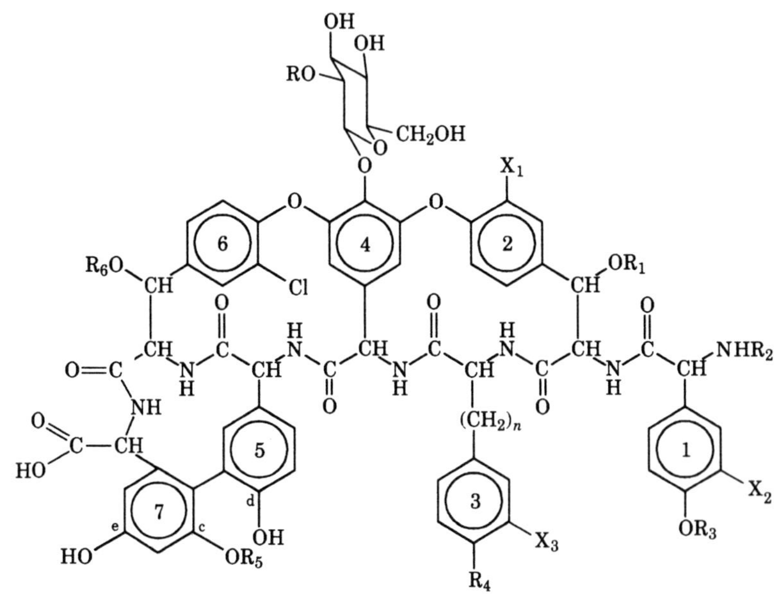

Table 5. Naturally Occurring Actinoidin-type Dalbaheptides

Antibiotic	CAS Registry Number	Structure[a]											
		No.	n	R	R_1	R_2	R_3	R_4	R_5	R_6	X_1	X_2	X_3
actinoidin													
A	[*60382-78-7*]	**(61)**	1	L-aco	H	H	H	H	D-man	L-aca	H	H	H
B	[*60382-79-8*]	**(62)**	1	L-aco	H	H	H	H	D-man	L-aca	H	Cl	H

avoparcin													
α	[*73957-86-5*]	(**63**)	0	L-ria	D-man	CH_3	L-rha	OH	H	L-ria	H	H	H
β	[*73957-87-6*]	(**64**)	0	L-ria	D-man	CH_3	L-rha	OH	H	L-ria	H	H	Cl
ϵ	[*88899-52-9*]	(**65**)	0	L-ria	H	CH_3	L-rha	OH	H	L-ria	H	H	Cl
chloropolysporin													
B	[*105650-11-1*]	(**66**)	0	H	man	CH_3	rha	OH	H	ria	Cl	H	Cl
C	[*105650-12-2*]	(**67**)	0	H	man	CH_3	H	OH	H	ria	Cl	H	Cl
actinoidin A_2	[*108905-13-1*]	(**68**)	1	L-rha	H	H	H	H	D-man	L-aca	H	H	H
helvecardin													
A	[*119979-33-8*]	(**69**)	0	ria	man	CH_3	*O*-Me-rha	OH	H	ria	H	H	Cl
B	[*119979-34-9*]	(**70**)	0	ria	H	CH_3	*O*-Me-rha	OH	H	ria	H	H	Cl
MM 47767[b]	[*129772-77-6*]	(**71**)	1	aco	H	CH_3	H	H	H	aca	H	Cl	H
MM 55256[b]	[*129652-26-2*]	(**72**)	1	aco	H	CH_3	H	H	H	aca	H	Cl	H

[a]Abbreviations are aco = acosamine, aca = actinosamine, ria = ristosamine, man = mannose, rha = rhamnose, and *O*-Me-rha = 2-*O*-methylrhamnose.
[b]These two dalbaheptides differ with regard to the steric position of the terminal NCH_3 group. One *O*-mannosyl group is present in position 5d, 7c, or 7e.

Table 6. Naturally Occurring Dalbaheptides: Synmonicin

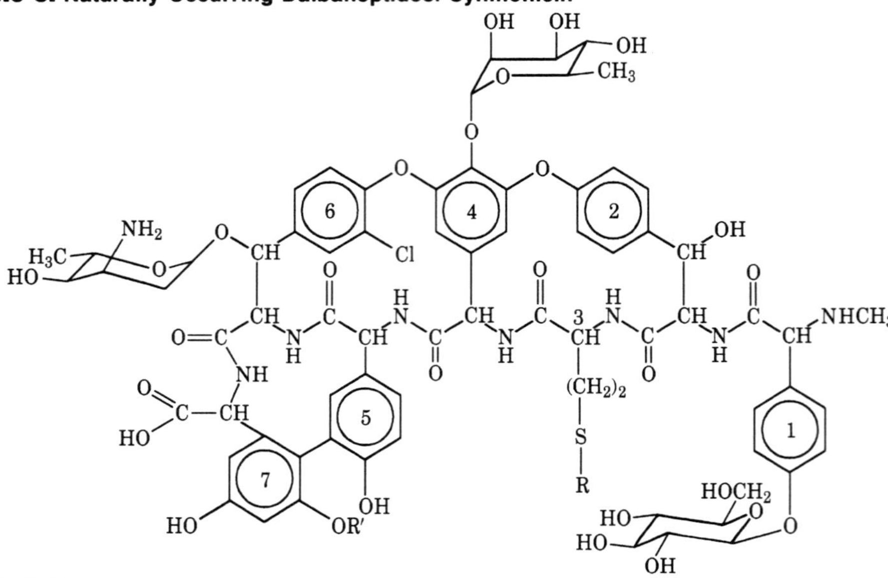

Synmonicin	CAS Registry Number	Structure[a] No.	R	R′	Empirical formula
A	[*114836-24-7*]	(**73**)	(O)CH_3	man	$C_{80}H_{86}ClN_8O_{34}S$
B[b]	[*114836-22-5*]	(**74**)	CH_3	man	$C_{80}H_{86}ClN_8O_{33}S$
C	[*114836-68-3*]	(**75**)	CH_3	H	$C_{74}H_{76}ClN_8O_{28}S$

[a]Abbreviation is man = mannose.
[b]Mixture of two epimers at terminal HC–NH, B′ is (*R*); B″ is (*S*).

Many of these antibiotics are groups of strictly related factors called complexes. Among them, vancomycin (**39**) (Table 4) (6) and the teicoplanins (**18–22**) (Table 3) (7) are used clinically as a result of high activity against gram-positive pathogens such as many coagulase-negative *Staphylococci* (CNS), corynebacteria, *Clostridium difficile,* multiresistant *Staphylococcus aureus,* and highly gentamicin-resistant Enterococci which are refractory to established drugs. Eremomycin (**52**) (Table 4) is under clinical evaluation (8). Many patents claim feed-utilization efficiency increase and growth promotion in domestic animals for several dalbaheptides but only avoparcin (**63**) (Avotan) (Table 5) is commercially used. The platelet aggregation ability of ristocetin A (**1**) (Table 2) renders it a suitable diagnostic agent for von Willebrand's disease, a hematologic disorder of genetic origin (9).

Knowledge of the mechanism of action and investigations on the physicochemical characteristics of the therapeutically used dalbaheptides has stimulated the transformation of natural antibiotics into new derivatives using both chemical and biosynthetic modification.

Naturally Occurring Dalbaheptides

Vancomycin (**39**) was introduced into clinical practice much earlier than its structure was elucidated. Structure investigations lasted 25 years (10–12), the structure being definitely assigned in 1983 (13). Structure elucidation of new dalbaheptides is done much faster by the instrumental techniques utilized in the 1990s. Especially helpful are: hplc on new stationary phases (14) (see CHROMATOGRAPHY); high field nmr including two-dimensional (2-D) experiments for structure elucidation (15,59) (see MAGNETIC SPIN RESONANCE); and soft ionization by fast atom bombardment mass spectrometry (qv) (fab/ms), which allows determinations of mol wt up to 2500 d with high accuracy (16,17).

Many labeling systems have been used for dalbaheptide structures. The one used herein, where each of the seven amino acids is identified by a number (see Table 2) and each atom by a letter, is widely applied because it permits easy comparison of ^{1}H and ^{13}C nmr data (31). The IUPAC system, utilized in *Chemical Abstracts* and generally in the description of semisynthetic derivatives, requires decodification for comparison of different dalbaheptides (83).

Descriptive Chemistry. Five of the seven amino acids that form the peptidic skeleton are common to all dalbaheptides: amino acid **2** is a tyrosine facultatively β-hydroxylated; amino acids **4** and **5** are *p*-hydroxyphenylglycines; **6** is a β-hydroxytyrosine and **7**, the carboxy terminal, is a *m,m′*-dihydroxyphenylglycine. Oxygen bonds between the phenyl moieties of amino acids **2** and **4** and **4** and **6** form two adjacent rings. Another ring is formed by a C–C bond between the phenyl moieties of amino acids **5** and **7**. The remaining two amino acids **1** and **3**, help to classify all known dalbaheptides into four types. The ristocetin-type (Tables 2 and 3) where amino acid **1**, the terminal NH_2, is a *p*-hydroxyphenylglycine and **3** is a *m,m′*-dihydroxyphenylglycine; an ether bond between these phenyl moieties generates an extra ring. The vancomycin-type (Table 4) where amino acids **1** and **3** are aliphatic, usually leucine and asparagine; in OA-7653A,B amino acid **1** is alanine; amino acid **3** is glutamine in OA-7653A (**40**), A51568B (**43**), glutamic acid in OA-7653B (**41**) and aspartic acid in M43B (**45**). The actinoidin-type (Table 5) where amino acid **1** is always a *p*-hydroxyphenylglycine and amino acid **3** is a phenylalanine or *p*-hydroxyphenlylglycine. Only a few dalbaheptides belong to this group. And synmonicin (Table 6) which is the only example of a dalbaheptide where amino acid **1** is aromatic (*p*-hydroxyglycine) and amino acid **3** is a thioamino acid such as methionine.

Dalbaheptides of natural origin have small variations in the nature and position of substituents. Additional chemical features are (*1*) chlorine atoms (up to four), methyl, and additional hydroxyl groups that can be present in different positions in the phenyl residues. No chlorine is present in ristocetins (**1**), (**2**), orienticin C (**50**), or MM 49721 (**60**). (*2*) Simple or complex carbohydrates that are linked to the core peptide in one or more positions through hemiacetalic bonds. Sugars found in dalbaheptides can be neutral, basic, or acidic. Vancosamine, *epi*-vancosamine, ristosamine, actinosamine, and acosamine were first isolated from dalbaheptides. A41030A (**11**), B (**12**), E (**14**), and A47934 (**16**) do not carry any sugar; minor amounts of partially (pseudoaglycones) or totally (aglycones) deglycosylated dalbaheptides are often present in many normally recovered batches. These compounds can be natural metabolites or artifacts of the recovery and purification processes. (*3*) A phenolic hydroxyl is esterified as a sulfate monoester in A47934 (**16**) and UK-68,597 (**17**). (*4*) The terminal carboxyl in ristocetin and actaplanin is a methyl ester and the terminal amino group can be methylated. In UK-68,597 (**17**) the *N*-terminal residue is replaced by a ketoamide functionality. (*5*) The amino group of glucosamine or 2-amino-2-deoxyglucuronic acid linked to the *p*-hydroxyphenyl residue of amino acid **4** of some ristocetin-type dalbaheptides is acylated by 10–12 carbon fatty acid residues. The acyl chains may be linear or branched (iso or anteiso), only exceptionally unsaturated. Acyl derivatives are present in teicoplanins, ardacins, A40926, kibdelins, and parvodicins and are therefore subgrouped as lipodalbaheptides (Table 3). The side chains differentiate the components of each complex.

Producing Organisms. All the dalbaheptides so far described are produced by strains belonging to the order of *Actinomycetales* (Table 1). Additionally, co-fermentation of *Actinoplanes missiouriensis* NRRL 15647 and strain NRRL 15646, which was obtained by chemical mutagenesis of an actaplanin-producing strain, yielded antibiotic CUC/CSV, an actaplaninlike dalbaheptide having a C=O function instead of the terminal HC–NH (84). An *Actinoplanes teichomyceticus* mutant yielded high levels of two teicoplaninlike compounds, designated RS-3 and RS-4, carrying 6-methyloctanoic and *n*-nonanoic residues as fatty acid chains (85,86).

Screening. The dalbaheptides were at first isolated by conventional screening procedures. More recently mechanistically based screens have been devised. For example, bioselective adsorption on affinity resins made by immobilizing *N*-acetyl-D-Ala-D-Ala [*19993-26-1*] on an activated support (87,88) was used in a screening campaign in which 72 strains were identified as producers of dalbaheptides, 60% of which produced ristocetin (89). The other characterized dalbaheptides were teicoplanins (Table 3), avoparcins (Table 5), actaplanins (Table 2), and the novel molecules A40926 (Table 3) and A42867 (**53**).

Other specific discovery assays have been used such as differential inhibition of a vancomycin resistant *S. aureus* strain and its susceptible parent, and an assay based on antagonism of the antibacterial activity by *N,N*-diacetyl-L-Lys-D-Ala-D-Ala [*24570-39-6*], a tripeptide analogue of the dalbaheptides receptor. Application of this latter test to 1936 cultures (90) led to the isolation of 42 dalbaheptides, six of which, including kibdelin (Table 3), parvodicin (Table 3), and actinoidin A_2 (**68**) were novel. A colorimetric assay based on competition between horseradish peroxidase bound teicoplanin and the putative dalbaheptide for a D-Ala-D-Ala peptide attached to the test tube wall has also been utilized (89,91) as has a polyclonal antibody against vancomycin, an enzyme-linked immunosorbant analysis (ELISA), which led to the discovery of A82846 (68).

Fermentation. Dalbaheptides are produced by submerged fermentation (qv) and this topic has been reviewed (92). In a few cases addition of a biosynthetic precursor of the core aglycones such as tyrosine and *p*-hydroxyphenylglycine, increased the yields (39). It has been reported (93) that the fatty acid moieties of teicoplanin (Table 3) biosynthetically derive from the degradation of long-chain fatty acids present in the membranes of the producing microorganisms. Addition to fermentation media of precursors that alter the membrane fatty acid composition led to an altered ratio of the components of teicoplanin (94,95) and A40926 (96).

Recovery and Purification. The dalbaheptides are present in both the fermentation broth and the mycelial mass, from which they can be extracted with acetone or methanol, or by raising the pH of the harvested material, eg, to a pH of 10.5–11 for A47934 (**16**) (44) and A41030 (41) and actaplanin (Table 2) (28). A detailed review on the isolation of dalbaheptides has been written (14). Recovery from aqueous solution is made by ion pair (avoparcin) or butanol (teicoplanin) extraction. The described isolation schemes use ion-exchange matrices such as Dowex and Amberlite IR, acidic alumina, cross-linked polymeric adsorbents such as Diaion HP and Amberlite XAD, cation-exchange dextran gel (Sephadex), and polyamides in various sequences. Reverse-phase hplc, ion-exchange, or affinity resins may be used for further purification (14,89).

Physical Properties. The molecular weight of dalbaheptides ranges from about 1150 to 2200. Pure dalbaheptides are obtained as colorless or whitish amorphous powders that usually retain water and solvent. Dalbaheptides are generally water-soluble. Teicoplanin can be obtained as an internal salt or as a partial monoalkaline (sodium) salt depending on the pH of the aqueous solution in the final purification step. Other dalbaheptides are obtained as acidic salts, such as hydrochlorides (vancomycin, actaplanin) or sulfates (ristocetin A, avoparcin, eremomycin). The presence of amino, carboxyl, benzylic, and phenolic hydroxyl functions, sugars, and aliphatic chains influence both water solubility and total charge. Aqueous solutions of vancomycin (**39**) at pH 3–5 and of teicoplanins (Table 3) at pH 7.4 are stable. The isoelectric points (pI) of dalbaheptides, determined by electrofocusing, vary from 3.2 for A47934 (**16**) to 8.1 for ristocetin (Table 2), A477, and A42867 (**53**). The lipid–water partition coefficient in *n*-octanol buffer is -1.1 at pH 5.1 for teicoplanin (Table 3) and -2.5 for vancomycin (**39**) indicating that teicoplanin is 5–10 times less hydrophilic than vancomycin.

Dalbaheptides are levorotatory. The absolute configuration of vancomycin was determined by x-ray analysis of degradation product CDP-I [*79517-31-0*] (11). The configurations of the seven amino acids are **1**(*R*), **2**(*R*), **3**(*S*), **4**(*R*), **5**(*R*), **6**(*S*), and **7**(*S*) (Table 2). Nmr experiments provided evidence that the stereochemistry of the chiral centers and the overall three-dimensional conformation is the same for most dalbaheptides. Natural and chemically produced epimers are exceptions. The anomeric conformation and position of the sugars have also been determined by nmr. In teicoplanin five of the conformations of the amide linkages are trans, the **5–6** linkage is cis. The stereostructure of teicoplanin aglycone is shown in Figure 1.

Biosynthesis. Biochemical studies on dalbaheptides have been reviewed (92,97). Experiments with ^{13}C and ^{2}H have shown that in vancomycin (**39**), D-tyrosine is the precursor of D-*p*-hydroxyphenylglycine and β-hydroxy-*m*-chlorotyrosine, and acetate the precursor of the two *m,m'*-dihydroxyphenylglycines (98). Similar results using either ^{13}C or radioactively labeled material have been reported for avoparcin (Table 5) (23), ristocetin (Table 2) (99,100), ardacin (Table 3) (101), and A47934 (102).

Kibdelins (Table 3) are converted into the corresponding ardacins by cul-

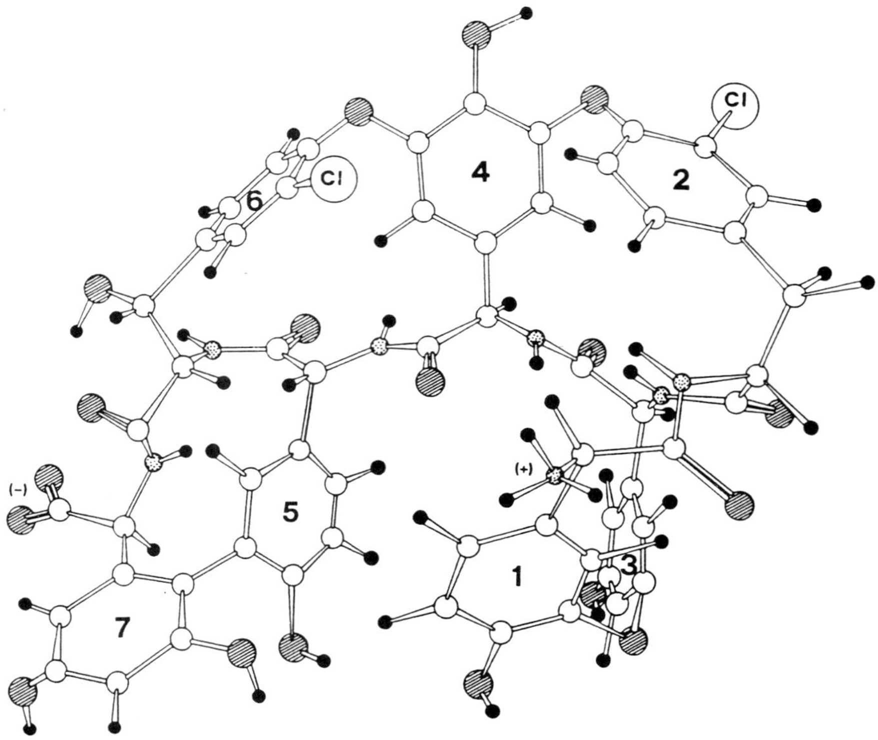

Fig. 1. Stereostructure of teicoplanin aglycone [*89139-42-4*] obtained from a Dreiding model (83) where ● is hydrogen, ◍ is nitrogen, ○ is carbon, ◍ is oxygen. The numbers represent the seven amino acids.

tures of *Kibdelosporangium aridum* indicating that the oxidation of carbon-6 of glucosamine to a carboxy group is the last biosynthetic step (100). Addition of ^{14}C-labeled ardacin aglycone to cultures of *K. aridum* gave origin to labeled ardacins but the yields were so low it is doubtful that the aglycone is the precursor of the glycosylated product. Other studies showed that the sulfate ester on phenolic hydroxyl in the production of A47934 (**16**) occurs prior to the formation of microbiologically active intermediates (102).

Biological Properties

Mechanism of Action. The basis of the antibacterial action of dalbaheptides is the ability to form a complex with the terminal D-Ala-D-Ala residues of growing peptidoglycan chains, thus preventing transglycosylation, eg, chain elongation, and transpeptidation, eg, cross-linking. The consequent defective cell wall stops bacterial growth and eventually leads to cell death. The mechanism of action has been reviewed both at the molecular (3) and at the biochemical level (4). Addition of a dalbaheptide to a growing culture results in rapid inhibition of incorporation of acetylglucosamine, a specific precursor of cell wall synthesis, and in accumulation of UDP-*N*-acetylmuramyl pentapeptide. Direct interaction studies of dalbaheptides and D-Ala-D-Ala-containing synthetic peptides have been carried out. Binding of dalbaheptides to D-Ala-D-Ala models has been measured by uv differential spectroscopy (3), affinity chromatography (87), microcalorimetry (103), nmr spectrometry (104), displacement of polyacrylamide-bound vancomycin (**39**) (105), and displacement of radiolabeled ristocetin bound to bacterial cells (106). These studies indicate that a free carboxyl and a D-configuration of the terminal amino acid of the model peptide are absolute requirements for formation of the complex. Stereostructure of the dalbaheptide skeleton forms a sort of receptor pocket which accomodates the D-Ala-D-Ala terminus of the target (3,107).

Antibacterial Activity. *In Vitro Properties.* The antibacterial spectrum of most dalbaheptides is known (Table 1). Vancomycin (**39**) and/or teicoplanin (Table 3) are generally introduced as reference compounds. Direct comparative data for some dalbaheptides tested under the same experimental conditions are given in Table 7 (54).

Dalbaheptides are active almost exclusively against gram-positive microorganisms including all major pathogens such as staphylococci, coagulase-positive and negative, streptococci of all groups, enterococci including *E. faecalis* and *faecium, corynebacteria* including JK and *acnes, Clostridia* including *C. perfringens* and *C. difficile,* anaerobic cocci eg, peptococci and streptostreptococci, and *Listeria monocytogenes*. Gram-positive organisms naturally resistant to dalbaheptides include Lactobacilli, *Leuconostoc, Pediococcus,* and *Nocardia.* Lipodalbaheptides are in general less active than other dalbaheptides against CNS. For example, the MIC against *S. haemolyticus* is from 4 to 50 μg/mL.

Gram-negative organisms are unsusceptible except *Gardnerella vaginalis* and *Branhamella catarrhalis* which are susceptible to teicoplanin, and *Neisseria gonorrhoeae* (Table 7). However, A47934 and A40926 (54) are exceptionally active against *N. gonorrhoeae,* having MICs of 8 and 2 μg/mL, respectively.

In the parvodicin and teicoplanin series (Table 3) the nature and length of the fatty acid portion of the glycolipid moiety only slightly influence the activity *in vitro*. Dalbaheptides are bactericidal against actively growing, but not against resting, bacteria (33).

In Vivo Properties. The efficacy of dalbaheptides has been assessed in various models of experimental septicemia in mice. In general there was good correlation between the ED_{50}s (effective doses which prevent death in 50% of test animals) and the MICs on test strains. Teicoplanin was very effective, ED_{50} values ranged from 0.11 to 0.72 mg/kg sc administration for septicemias caused by *S. pyogenes, S. pneumoniae,* and *S. aureus,* whereas for vancomycin ED_{50}s were from 0.58 to 7.2 mg/kg (33). Eremomycin (**52**) had therapeutic activity 2–3 times greater than vancomycin. Therapeutic indices for staphylococcal–streptococcal sepsis exceeded ten times those of vancomycin (8). Ardacin and kibdelin are less active *in vivo* than vancomycin, although the MIC on the tested strains are comparable. This happens also for A40926 in respect to teicoplanin (54). In fact, the long lasting ardacin was more active than vancomycin when administered as early as 18 h before infection (48). Teicoplanin and vancomycin were also found to be active against experimentally induced endocarditis in rabbits (108) and rats (109), and against localized infection in mice (109).

Pharmacokinetics. Pharmacokinetic studies in mice via iv administration have been done mainly on vancomycin (**39**), eremomycin (**52**), and lipodalbaheptides (Table 3). A systematic investigation of the relationship between pI, lipophilicity, and pharmacokinetic parameters of ardacin and its hydrolysis

Table 7. Antibacterial Activity of Dalbaheptides, MICs in μg/mL[a]

	Organism and strain					
Antibiotic	*Staphylococcus aureus* TOUR	*Staphyloccus epidermidis* ATCC 12228	*Streptococcus pyogenes* C 203	*Enterococcus faecalis* ATCC 7080	*Clostridium perfringens* ISS 30543	*Neisseria gonorroheae* ISM 68/126
ristocetin[b]	4	2	0.25	1	2	64
vancomycin[c]	0.25	0.5	0.13	0.5	0.13	32
avoparcin[d]	2	2	0.25	0.25	0.5	128
actaplanin[b]	1	2	0.13	0.5	1	>128
teicoplanin[e]	0.13	0.13	0.06	0.13	0.03	32
A35512B[b]	1	0.5	0.13	0.5	2	64
A41030[b]	0.03	0.008	0.13	0.13		64
A47934[b]	0.06	0.03	0.13	0.13	0.25	8
ardacin[e]	1	4	0.13	2	0.06	64
A40926[e]	0.06	0.06	0.06	0.06	0.004	2

[a]From Ref. 54 and the Author's Laboratory.
[b]Structures shown in Table 2.
[c]Structure (**39**).
[d]Structures shown in Table 5.
[e]Structures shown in Table 3.

products in comparison to those of ristocetin (Table 2), teicoplanin, and vancomycin has been carried out (110). Ardacin and its pseudoaglycone having pIs ≈ 3.8 yielded high and prolonged serum concentrations and the half-life ranged from 226 to 492 min. In contrast, vancomycin and ristocetin, which have pIs ≈ 8, had $t_{1/2} = 20$ and 62 min, respectively; teicoplanin and ardacin aglycone, pIs ≈ 5, had intermediate elimination rates, $t_{1/2}$ ranged from 118 to 155 min. For dalbaheptides having similar pIs, clearance decreased and half-life increased as lipophilicity increased.

Mechanism of Resistance. Recently, resistance to high levels of dalbaheptides has been described in both *E. faecium* and *E. faecalis* (111). This resistance, inducible by low concentrations of dalbaheptides, is plasmid mediated and is transferable. Concomitant with the induction of resistance is the appearance or increased expression of a protein having a molecular weight of either 39,500 or 39,000. The enzymatic activity of this material has been postulated (112). Although the mechanism of resistance induction by dalbaheptides is unknown, different dalbaheptides have different induction capacity. Vancomycin (**39**) is the most powerful inducer; teicoplanin is a very weak inducer.

Chemical Modifications

Although it has been known for some time that hydrolysis of ristocetin leads to derivatives having different or increased antimicrobial activity (113), little chemical modification work was reported in the literature until the mid-1980s. Selective removal of sugar moieties and glycosylation, deacylation, deamination, dechlorination, introduction of bromine or chlorine atoms, esterification or amidation of the terminal carboxyl, acylation or alkylation of the terminal (or sugar) amino groups have all since been described, mainly in patent applications. Some of the aglycones and pseudoaglycones showed a weak activity against selected gram-negative bacteria (83,114,115). Among the most interesting compounds produced was SKF 104662 [*121089-14-3*], obtained from the B′ epimer of synmonicin (Table 6) after selective removal of the neutral sugars and glycosylation of the hydroxyl on ring **4** (116). The *in vitro* activity (117) and therapeutic efficacy in experimental infection in mice for SKF 104662 were similar to vancomycins or teicoplanins. SKF 104662 was less toxic than vancomycin in mice and the pharmacokinetic profile in mice, rats, and dogs indicated that the half-life was also longer than that of vancomycin (118).

The -$NH(CH_2)_3N(CH_3)_2$ amide of teicoplanin factor A2-2, coded MDL 62,873 [*122173-74-4*], was also prepared. The combined effect of a moderate basicity and a slightly increased lipophilicity at neutral pH probably led to a better penetration through the cell wall. MDL 62,873 was consistently more active than teicoplanin against CNS clinical isolates (119,120). No semisynthetic dalbaheptide is under clinical evaluation at this writing.

Economic Aspects

Only three dalbaheptides are commercialized: vancomycin (**39**) and teicoplanin (**18–22**) for human health, and avoparcin (**63–65**) for animal usage. Vancomycin, the main trademark of which is Eli Lilly's Vancocin, had 1990 sales around $160 million. Total annual production is in the vicinity of 8 t. Teicoplanin, trademarked Targocid, had 1990 sales of $35 million corresponding to 200 kg. Teicoplanin is commercialized in Europe, Hong Kong, Korea, and the Middle East. It is at the late developmental clinical phase in North America and Japan. Avoparcin is used as a growth promoting feed additive (see FEEDS AND FEED ADDITIVES; GROWTH REGULATORS).

Clinical Use. Vancomycin and teicoplanin as formulated drugs are lyophilized powders to be reconstituted with sterile water for injection. Vancomycin hydrochloride [*1404-93-9*], is presented in vials of 500 mg that give 100–200 mL solution of pH 2.5–4.2. It is administered by slow (60 min) infusion at a dose of 500 mg every 6 h or 1 g every 12 h/d. The teicoplanin contains the five factors (87%) plus 13% of the pseudoaglycone T-A3-1. It is presented in vials containing 200 mg of lyophilized power that after dissolution with 3 mL of solvent gives a solution at pH 7.5. The dose regimen is 200–800 mg/d by iv bolus administration.

BIBLIOGRAPHY

1. F. Sztaricskai and R. Bognár in R. Bognár and C. Szántay, eds., *Recent Developments in the Chemistry of Natural Carbon Compounds,* Vol. 10, Akadémiai Kiadó, Budapest, 1984, pp. 91–201.
2. G. S. Katrukha and A. B. Silaev in W. Voelter, E. Bayer, Y. A. Ovchinnikov, and V. T. Ivanov, eds., *Chemistry of Peptides and Proteins,* Vol. 3, W. de Gruyter and Co., Berlin, 1986, pp. 289–306.
3. J. C. J. Barna and D. H. Williams, *Ann. Rev. Microbiol.* **38,** 339 (1984).
4. P. E. Reynolds, *Eur. J. Clin. Microbiol. Infect. Dis.* **8,** 943 (1989).
5. F. Parenti and B. Cavalleri, *J. Antibiot.* **42,** 1882 (1989).
6. R. Wise and D. Reeves, eds., *J. Antimicrob. Chemother.* **14** (Suppl. D), 1 (1984); R. P. F. Cheung and J. T. DiPiro, *Pharmacotherapy* **6,** 153 (1986); M. J. Rybak and S. C. Boike, *Drug Intell. Clin. Pharm.* **20,** 757 (1986); P. E. Hermans and M. P. Wilhelm, *Mayo Clin. Proc.* **62,** 901 (1987).
7. A. Del Favero and F. Menichetti, *Drugs of Today* **24,** 641 (1988); D. Speller, D. Greenwood, and P. J. Daly, eds., *J. Antimicrob. Chemother.* **21** (Suppl. A), 1 (1988); F. Fraschini, *J. Chemother.* **1** (Suppl. 4), 590 (1989); D. M. Campoli-Richards, R. N. Brogden, and D. Faulds, *Drugs* **40,** 449 (1990).
8. M. G. Brazhnikova, N. N. Lomakina, and T. F. Berdnikova in M. E. Bushell and U. Gräfe, eds., *Bioactive Metabolites from Microorganisms, Progress in Industrial Microbiology,* Vol. 27, Elsevier, Amsterdam, 1989, pp. 163–175.
9. H. Gralnick, *J. Am. Med. Assoc.* **238,** 1625 (1977).
10. F. J. Marshall, *J. Med. Chem.* **8,** 18 (1965).
11. G. M. Sheldrick, P. G. Jones, O. Kennard, D. H. Williams, and G. A. Smith, *Nature (London)* **271,** 223 (1978).
12. M. P. Williamson and D. H. Williams, *J. Am. Chem. Soc.* **103,** 6580 (1981).
13. C. M. Harris, H. Kopecka, and T. M. Harris, *J. Am. Chem. Soc.* **105,** 6915 (1983).
14. R. D. Sitrin and G. Folena-Wasserman in G. H. Wagman and R. Cooper, eds., *Natural Products Isolation,* Elsevier, Amsterdam, 1989, pp. 111–152.
15. P. W. Jeffs, L. Mueller, C. DeBrosse, S. L. Heald, and R. Fisher, *J. Am. Chem. Soc.* **108,** 3063 (1986).
16. G. D. Roberts, S. A. Carr, S. Rottschaefer, and P. W. Jeffs, *J. Antibiot.* **38,** 713 (1985).

17. S. Naylor, N. J. Skelton, and D. H. Williams, *J. Chem. Soc. Chem. Commun.* 1619 (1986).
18. G. Batta, F. Sztaricskai, J. Csanádi, I. Komáromi, and R. Bognár, *J. Antibiot.* **39,** 910 (1986).
19. T. F. Berdnikova, N. N. Lomakina, and N. P. Potapova, *Antibiotiki* **27,** 252 (1982); *Chem. Abstr.* **97,** 72754z (1982).
20. K. Matsumoto, *J. Antibiot. Ser. A.* **14,** 141 (1961).
21. M. P. Kunstmann, L. A. Mitscher, J. N. Porter, A. J. Shay, and M. A. Darken, *Antimicrob. Agents Chemother.,* 242 (1968).
22. G. S. Redin and A. C. Dornbush, *Antimicrob. Agents. Chemother.,* 246 (1968).
23. W. J. McGahren and co-workers, *J. Am. Chem. Soc.* **102,** 1671 (1980).
24. W. J. McGahren and co-workers, *J. Antibiot.* **36,** 1671 (1983).
25. U.S. Pat. 3,803,306 (Apr. 9, 1974), M. P. Kunstmann and J. N. Porter (to American Cyanamid Co.).
26. U.S. Pat. 3,780,174 (Dec. 18, 1973), R. L. Hamill, M. E. Haney, Jr., and W. M. Stark (to Lilly).
27. U.S. Pat. 4,064,233 (Dec. 20, 1977), R. L. Hamill, W. M. Stark, and D. C. DeLong (to Lilly).
28. M. Debono and co-workers, *J. Antibiot.* **37,** 85 (1984).
29. A. H. Hunt, T. K. Elzey, K. E. Merkel, and M. Debono, *J. Org. Chem.* **49,** 641 (1984).
30. A. Tamura and I. Takeda, *J. Antibiot.* **28,** 395 (1975).
31. A. H. Hunt, R. M. Molloy, J. L. Occolowitz, G. G. Marconi, and M. Debono, *J. Am. Chem. Soc.* **106,** 4891 (1984); J. C. J. Barna, D. H. Williams, D. J. M. Stone, T.-W. C. Leung, and D. M. Doddrell, *J. Am. Chem. Soc.* **106,** 4895 (1984); S. L. Heald, L. Mueller, and P. W. Jeffs, *J. Magn. Reson.* **72,** 120 (1987).
32. C. Coronelli, G. G. Gallo, and B. Cavalleri, *Farm. Ed. Sci.* **42,** 767 (1987).
33. R. Pallanza and co-workers, *J. Antimicrob. Chemother.* **11,** 419 (1983).
34. U.S. Pat. 4,122,168 (Oct. 24, 1978), K. H. Michel and C. E. Higgens (to Lilly).
35. K. H. Michel, R. M. Shah, and R. L. Hamill, *J. Antibiot.* **33,** 1397 (1980); M. Debono, R. M. Molloy, M. Barnhart, and D. E. Dorman, *J. Antibiot.* **33,** 1407 (1980).
36. D. A. Preston and P. W. Ensminger, *17th Interscience Conference on Antimicrobial Agents and Chemotherapy,* New York, 1977, Abstr. No. 43.
37. T. Kamogashira, T. Nishida, and M. Sugawara, *Agric. Biol. Chem.* **47,** 499 (1983).
38. P. W. Jeffs, B. Yellin, L. Mueller, and S. L. Heald, *J. Org. Chem.* **53,** 471 (1988); S-G. Ang, M. P. Williamson, and D. H. Williams, *J. Chem. Soc. Perkin Trans. 1* **1988,** 1949 (1988).
39. L. D. Boeck, F. P. Mertz, R. K. Wolter, and C. E. Higgens, *J. Antibiot.* **37,** 446 (1984).
40. A. H. Hunt, G. G. Marconi, T. K. Elzey, and M. M. Hoehn, *J. Antibiot.* **37,** 917 (1984).
41. L. D. Boeck, F. P. Mertz, and G. M. Clem, *J. Antibiot.* **38,** 1 (1985).
42. A. H. Hunt, D. E. Dorman, M. Debono, and R. M. Molloy, *J. Org. Chem.* **50,** 2031 (1985).
43. J. H. Eggert, K. H. Michel, L. D. Boeck, W. M. Nakatsukasa, and R. E. Kastner, *23rd Interscience Conference on Antimicrobial Agents and Chemotherapy,* Las Vegas, Nev., 1983, Abstr. No. 440, and F. T. Counter and co-workers, Abstr. No. 444.
44. L. D. Boeck and F. P. Mertz, *J. Antibiot.* **39,** 1533 (1986).
45. A. H. Hunt, J. L. Occolowitz, M. Debono, R. M. Molloy, and G. M. Maciak, *23rd Interscience Conference on Antimicrobial Agents and Chemotherapy,* Las Vegas, Nev., 1983, Abstr. No. 441.
46. M. C. Shearer and co-workers, *J. Antibiot.* **38,** 555 (1985).
47. R. D. Sitrin and co-workers, *J. Antibiot.* **38,** 561 (1985).
48. S. F. Grappel, A. J. Giovenella, L. Phillips, D. H. Pitkin, and L. J. Nisbet, *Antimicrob. Agents Chemother.* **28,** 660 (1985).

49. T. Okazaki, R. Enokita, H. Miyaoka, T. Takatsu, and A. Torikata, *J. Antibiot.* **40,** 917 (1987).
50. T. Takatsu and co-workers, *J. Antibiot.* **40,** 924 (1987).
51. T. Takatsu and co-workers, *J. Antibiot.* **40,** 933, 946 (1987).
52. Eur. Pat. Appl. 177,822 (Apr. 16, 1986), E. Selva and co-workers (to Lepetit).
53. J. P. Waltho, D. H. Williams, E. Selva, and P. Ferrari, *J. Chem. Soc. Perkin Trans. 1* 2103 (1987).
54. B. P. Goldstein and co-workers, *Antimicrob. Agents Chemother.* **31,** 1961 (1987).
55. Eur. Pat. Appl. 159,180 (Apr. 11, 1985), H. M. Higgins, Jr., M. H. McCormick, K. E. Merkel, and K. H. Michel (to Lilly).
56. M. C. Shearer and co-workers, *J. Antibiot.* **39,** 1386 (1986).
57. G. Folena-Wasserman and co-workers, *J. Antibiot.* **39,** 1395 (1986).
58. P. Spiri-Nakagawa and co-workers, *J. Antibiot.* **39,** 1719 (1986).
59. S. B. Christensen and co-workers, *J. Antibiot.* **40,** 970 (1987).
60. V. Arjuna Rao and co-workers, *26th Interscience Conference on Antimicrobial Agents and Chemotherapy,* New Orleans, La., 1986, Abstr. No. 939, and A. K. Verma, R. Prakash, S. A. Carr, G. D. Roberts, and R. D. Sitrin, Abstr. No. 940.
61. J. J. Dingerdissen and co-workers, *J. Antibiot.* **40,** 165 (1987).
62. S. L. Heald, L. Mueller, and P. W. Jeffs, *J. Antibiot.* **40,** 630 (1987).
63. Eur. Pat. Appl. 231,111 (Aug. 5, 1987), N. Tsuji and co-workers (to Shionogi).
64. N. Tsuji, M. Kobayashi, T. Kamigauchi, Y. Yoshimura, and Y. Terui, *J. Antibiot.* **41,** 819 (1988).
65. G. F. Gause and co-workers, *J. Antibiot.* **42,** 1790 (1989).
66. E. Riva and co-workers, *J. Antibiot.* **42,** 497 (1989).
67. Eur. Pat. Appl. 265,071 (Apr. 27, 1988), R. L. Hamill, J. A. Mabe, D. F. Mahoney, W. M. Nakatsukasa, and R. C.-F. Yao (to Lilly).
68. R. C. Yao and co-workers, *28th Interscience Conference on Antimicrobial Agents and Chemotherapy,* Los Angeles, 1988, Abstr. No. 974, R. L. Hamill and co-workers, Abstr. No. 975, A. H. Hunt, J. L. Occolowitz, M. Debono, and R. M. Molloy, Abstr. No. 976, and F. T. Counter, P. W. Ensminger, and C. Y. E. Wu, Abstr. No. 977.
69. R. Nagarajan, D. M. Berry, and A. A. Schabel, *J. Antibiot.* **42,** 1438 (1989).
70. R. Nagarajan, D. M. Berry, A. H. Hunt, J. L. Occolowitz, and A. A. Schabel, *J. Org. Chem.* **54,** 983 (1989).
71. Eur. Pat. Appl. 265,143 (Apr. 27, 1988), K. S. Holdom, H. Maeda, J. C. Ruddock, and J. Tone (to Pfizer Ltd.).
72. J. Horsman, J. C. Ruddock, and D. A. Wakefield, *29th Interscience Conference on Antimicrobial Agents and Chemotherapy,* Houston, Tex., 1989, Abstr. No. 415; N. J. Skelton, D. H. Williams, R. A. Monday, and J. C. Ruddock, *J. Org. Chem.* **55,** 3718 (1990).
73. Jpn. Pat. Kokai 63 152,397A (June 24, 1988), M. Inukai and co-workers (to Sankyo).
74. Eur. Pat. Appl. 287,110 (Oct. 19, 1988), E. Kondo and co-workers (to Shionogi).
75. N. Tsuji, T. Kamigauchi, M. Kobayashi, and Y. Terui, *J. Antibiot.* **41,** 1506 (1988).
76. Eur. Pat. Appl. 299,707 (Jan. 18, 1989), L. E. Doolin and co-workers (to Lilly).
77. Int. Pat. Appl. WO 02441 (Mar. 23, 1989), M. Athalye, A. Elson, M. L. Gilpin, and L. R. Jeffries (to Beecham); V. M. Good, M. N. Gwynn, and D. J. C. Knowles, *J. Antibiot.* **43,** 550 (1990).
78. Int. Pat. Appl. WO 07612 (Aug. 24, 1989), M. Athalye, A. L. Elson, and M. L. Gilpin (to Beecham); S. J. Box, A. L. Elson, M. L. Gilpin, and D. J. Winstanley, *J. Antibiot.* **43,** 931 (1990).
79. Eur. Pat. Appl. 339,982 (Nov. 2, 1989), M. Athalye, N. J. Coates, and P. H. Milner (to Beecham).
80. Eur. Pat. Appl. 375,448 (June 27, 1990), N. J. Coates, C. J. Davis, L. M. Curtis, and R. Sykes (to Beecham); S. J. Box and co-workers, *J Antibiot.* **44,** 807 (1991).

81. Eur. Pat. Appl. 356,894 (Mar. 7, 1990), C. M. M. Franco and co-workers (to Hoechst).
82. N. J. Skelton, D. H. Williams, M. J. Rance, and J. C. Ruddock, *J. Chem. Soc. Perkin Trans. 1* 77 (1990).
83. A. Malabarba, P. Ferrari, G. G. Gallo, J. Kettenring, and B. Cavalleri, *J. Antibiot.* **39,** 1430 (1986).
84. Eur. Pat. Appl. 159,436 (Oct. 30, 1985), C. L. Hershberger (to Lilly).
85. M. Zanol, A. Cometti, A. Borghi, and G. Lancini, *Chromatographia* **26,** 234 (1988).
86. A. Borghi and co-workers, *J. Antibiot.* **42,** 361 (1989).
87. A. Corti and G. Cassani, *Appl. Biochem. Biotech.* **11,** 101 (1985).
88. G. Folena-Wasserman, R. D. Sitrin, F. Chapin, and K. M. Snader, *J. Chromatography* **392,** 225 (1987).
89. G. Cassani in Ref. 8, p. 221.
90. J. B. Rake and co-workers, *J. Antibiot.* **39,** 58 (1986).
91. A. Corti, C. Rurali, A. Borghi, and G. Cassani, *Clin. Chem.* **31,** 1606 (1985).
92. G. C. Lancini in Ref. 8, p. 283.
93. A. Borghi, D. Edwards, L. F. Zerilli, and G. C. Lancini, *J. Gen. Microbiol.* **137,** 587 (1991).
94. Eur. Pat. Appl. 204,179 (Dec. 10, 1986), F. Assi, G. C. Lancini, and A. Gianantonio (to Lepetit).
95. G. Cassani, G. C. Lancini, B. Cavalleri, and F. Parenti, *Chimica Oggi,* **1987**(5), 11 (1987).
96. Eur. Pat. Appl. 259,781 (Mar. 16, 1988), E. Selva, L. Gastaldo, M. Denaro, G. Cassani, and F. Parenti (to Lepetit).
97. G. C. Lancini and B. Cavalleri in H. Kleinkauf and H. von Doehren, eds., *Biochemistry of Peptide Antibiotics, Recent Advances in the Biotechnology of β-Lactams and Microbial Bioactive Peptides,* W. de Gruyter and Co., Berlin, 1990, pp. 159–178.
98. S. J. Hammond, M. P. Williamson, D. H. Williams, L. D. Boeck, and G. G. Marconi, *J. Chem. Soc. Chem. Commun.* 344 (1982).
99. S. J. Hammond, D. H. Williams, and R. V. Nielsen, *J. Chem. Soc. Chem. Commun.* 116 (1983).
100. C. A. Townsend and A. M. Brown, *J. Am. Chem. Soc.* **105,** 913 (1983).
101. S. K. Chung, P. Taylor, Y. K. Oh, C. DeBrosse, and P. W. Jeffs, *J. Antibiot.* **39,** 642 (1986).
102. M. J. Zmijewski, Jr., B. Briggs, R. Logan, and L. D. Boeck, *Antimicrob. Agents Chemother.* **31,** 1497 (1987).
103. A. Rodríguez-Tebar, D. Vázquez, J. L. Pérez Velázquez, J. Laynez, and I. Wadsö, *J. Antibiot.* **39,** 1578 (1986).
104. S. W. Fesik, I. M. Armitage, G. A. Ellestad, and W. J. McGahren, *Molec. Pharmacol.* **25,** 275, 281 (1984); M. P. Williamson, D. H. Williams, and S. J. Hammond, *Tetrahedron* **40,** 569 (1984); M. P. Williamson and D. H. Williams, *J. Chem. Soc. Perkin Trans. 1,* 949 (1985); G. E. Hawkes, H. Molinari, S. Singh, and L.-Y. Lian, *J. Magn. Reson.* **74,** 188 (1987).
105. P. W. Smith, G. Chang, and W. C. Still, *J. Org. Chem.* **53,** 1587 (1988).
106. K-H. Kim, Y. Martin, E. Otis, and J. Mao, *J. Med. Chem.* **32,** 84 (1989).
107. S. H. Heald, L. Mueller, and P. W. Jeffs, *J. Magn. Reson.* **72,** 120 (1987); H. Molinari, G. Hawkes, and K. Sales, *XVI Congresso Nazionale di Chimica,* Bologna, 1988, Abstr. No. 101; J. P. Waltho, J. Cavanagh, and D. H. Williams, *J. Chem. Soc. Chem. Commun.* 707 (1988); L. Mueller, S. L. Heald, J. C. Hempel, and P. W. Jeffs, *J. Am. Chem. Soc.* **111,** 496 (1989).
108. H. F. Chambers and M. A. Sande, *Antimicrob. Agents Chemother.* **26,** 61 (1984); P. M. Sullam, M. G. Täuber, C. J. Hackbarth, and M. A. Sande, *Antimicrob. Agents Chemother.* **27,** 135 (1985).
109. M. Berti, G. Romanò, R. Pallanza, and V. Arioli, *Chemioterapia* **4** (Suppl. 2), 688 (1985);

V. Arioli, M. Berti, G. C. Candiani, and R. Pallanza, *Chemioterapia* **4** (Suppl. 2), 694 (1985).
110. D. H. Pitkin, B. A. Mico, R. D. Sitrin, and L. J. Nisbet, *Antimicrob. Agents Chemother.* **29,** 440 (1986).
111. R. Leclercq, E. Derlot, J. Duval, and P. Courvalin, *N. Engl. J. Med.* **319,** 157 (1988); R. Leclercq, E. Derlot, M. Weber, J. Duval, and P. Courvalin, *Antimicrob. Agents Chemother.* **33,** 10 (1989); T. I. Nicas, C. Y. E. Wu, J. N. Hobbs, Jr., D. A. Preston, and N. E. Allen, *Antimicrob. Agents Chemother.* **33,** 1121 (1989).
112. S. Al-Obeid, E. Collatz, and L. Gutmann, *Antimicrob. Agents Chemother.* **34,** 252 (1990).
113. J. E. Philip, J. R. Schenck, M. P. Hargie, J. C. Holper, and W. E. Grundy, *Antimicrob. Agent Ann.* 10 (1960).
114. A. Malabarba and co-workers, *J. Antibiot.* **37,** 988 (1984).
115. R. D. Sitrin and co-workers, *J. Antibiot.* **39,** 68 (1986).
116. Eur. Pat. Appl. 273,727 (July 6, 1988), S. B. Christensen, S. K. Chung, and P. W. Jeffs (to Smith, Kline and Beckman).
117. J. H. Jorgensen, J. S. Redding, and L. A. Maher, *Antimicrob. Agents Chemother.* **33,** 560 (1989); J. D. C. Yao, G. M. Eliopoulos, and R. C. Moellering, Jr., *Antimicrob. Agents Chemother.* **33,** 965 (1989).
118. H. S. Allaudeen and co-workers, *28th Interscience Conference on Antimicrobial Agents and Chemotherapy,* Los Angeles, Calif., 1988, Abstr. No. 982, and S. B. Christensen, W. R. Hewitt, and B. A. Mico, Abstr. No. 984.
119. A. Malabarba and co-workers, *J. Med. Chem.* **32,** 2450 (1989).
120. P. R. Hunter, R. C. George, and J. W. Griffiths, *Antimicrob. Agents Chemother.* **34,** 1769 (1990).

B. Cavalleri
F. Parenti
Merrell-Dow Research Institute

β-LACTAMS

CARBAPENEMS AND PENEMS

In the period up to 1970 most β-lactam research was concerned with the penicillin and cephalosporin group of antibiotics (1). Since that time, however, a wide variety of new mono- and bicyclic β-lactam structures have been described. The carbapenems, characterized by the presence of the bicyclic ring system (**1**, X = CH_2) originated from natural sources; the penem ring (**1**, X = S) and its derivatives are the products of the chemical synthetic approach to new antibiotics. The chemical names are: 7-oxo-(*R*)-1-azabicyclo[3.2.0]hept-2-ene-2-carboxylic acid [*78854-41-8*], $C_7H_7NO_3$, and 7-oxo-(*R*)-4-thia-1-azabicyclo[3.2.0]hept-2-ene-2-carboxylic acid [*69126-94-9*], $C_6H_5NO_3S$, respectively.

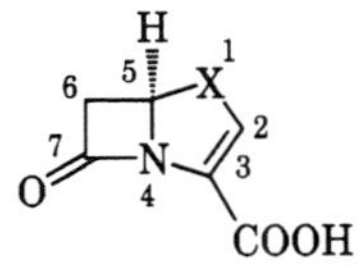

(**1**)

Carbapenems

During the screening of soil samples for inhibitors of peptidoglycan biosynthesis, thienamycin (**2**), shown in Table 1, and other related derivatives were isolated from the microorganism *Streptomyces cattleya* (2). At the same time, while screening for natural inhibitors of *β*-lactamases, a group of interrelated metabolites known as the olivanic acids, obtained from *Streptomyces olivaceus,* was found (3). The first three compounds isolated were the sulfated derivatives having structures (**3**) and (**4**); subsequently, several nonsulfated analogues were also isolated (4). Thienamycin has the (*R*)-configuration at the C-8 hydroxy group with a trans-arrangement of protons on the *β*-lactam ring. In the olivanic acids the stereochemistry is 8(*S*)- with a cis-substituted *β*-lactam in the sulfated series, and both *cis-* and *trans-β*-lactams in the case of the C-8 hydroxy compounds. These compounds represented a completely new family of *β*-lactam structures based on the 7-oxo-1-azabicyclo[3.2.0]hept-2-ene-2-carboxylic acid ring system, and which are now generally referred to as the 1-carbadethiapen-2-em or simply 1-carbapenem family of antibiotics.

Subsequently, other structural variations were reported encompassing compounds such as PS-5 (**5**) (5), carpetimycin A (**6**) (6), asparenomycin A (**7**) (7), and pluracidomycin A (**8**) (8), from a wide variety of streptomycete strains. Following these structures the simplest member of the series, having the completely unsubstituted nucleus (**1**, X = CH_2), was isolated from bacterial strains of *Serratia* and *Erwinia* (9). All other natural products reported have substituents at both the C-6 and C-2 positions of the bicyclic ring system. Differences in the nature and stereochemistry of these substituents has provided a wide variety of structures, and over forty variations have been reported and comprehensively listed (10).

Occurrence, Fermentation, and Biosynthesis. Although a large number of *Streptomyces* species have been shown to produce carbapenems, only *S. cattleya* (2) and *S. penemfaciens* (11) have been reported to give thienamycin (**2**). Generally the antibiotics occur as a mixture of analogues or isomers and are often co-produced with penicillin N and cephamycin C. Yields are low compared to other *β*-lactams produced by streptomycetes, and titres are of the order of 1–20 μg/mL despite, in many cases, a great deal of effort on the optimization of the media and fermentation conditions. The rather poor stability of the compounds also contrib-

Table 1. Carbapenems

Name	CAS Registry Number	Molecular formula	Structure number	Structure
thienamycin	*[59995-64-1]*	$C_{11}H_{16}N_2O_4S$	**(2)**	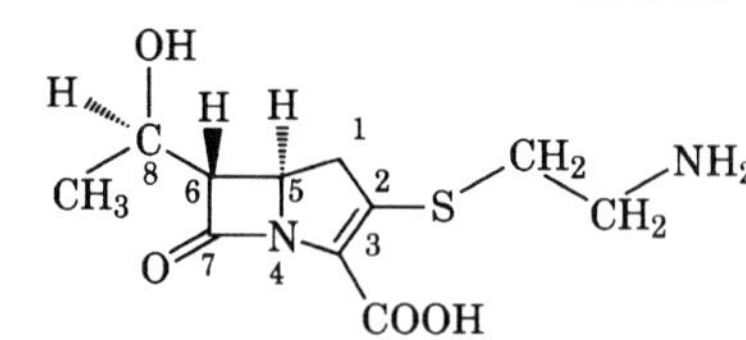
MM 4550 MM 13902	*[12795-21-0]* *[57459-82-2]*	$C_{13}H_{16}N_2O_9S_2$ $C_{13}H_{16}N_2O_8S_2$	(**3**, n = 1) (**3**, n = 0)	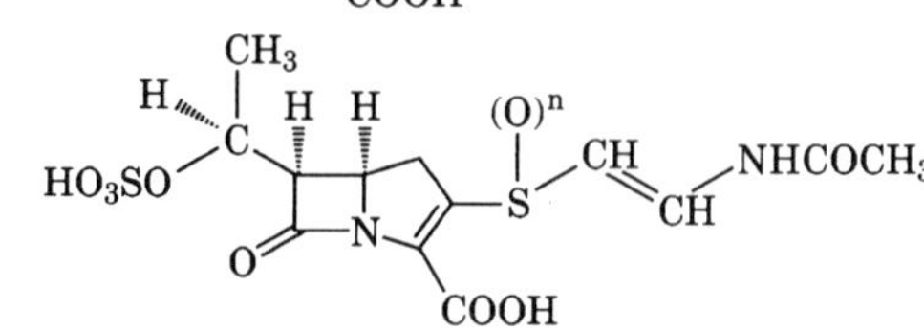
MM 17880	*[61036-81-5]*	$C_{13}H_{18}N_2O_8S_2$	**(4)**	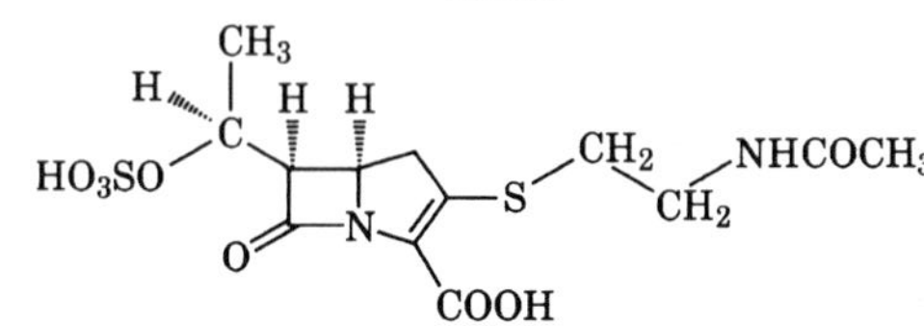

PS–5	[*67007-29-8*]	$C_{13}H_{18}N_2O_4S$	(**5**)	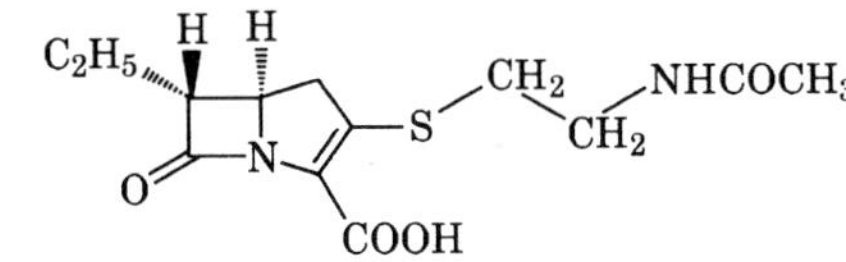
carpetimycin A	[*77209-15-5*]	$C_{14}H_{18}N_2O_6S$	(**6**)	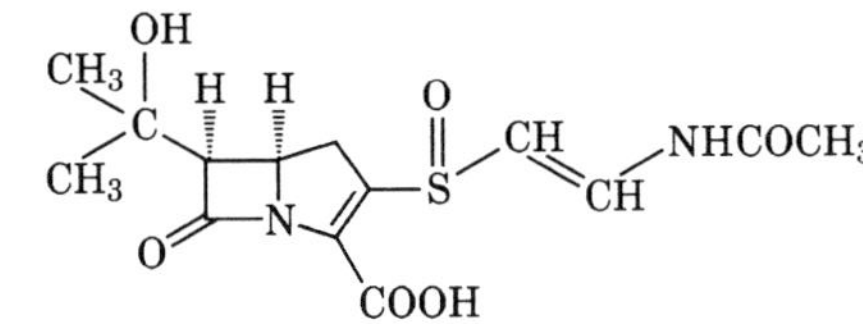
asparenomycin A	[*76466-24-5*]	$C_{14}H_{16}N_2O_6S$	(**7**)	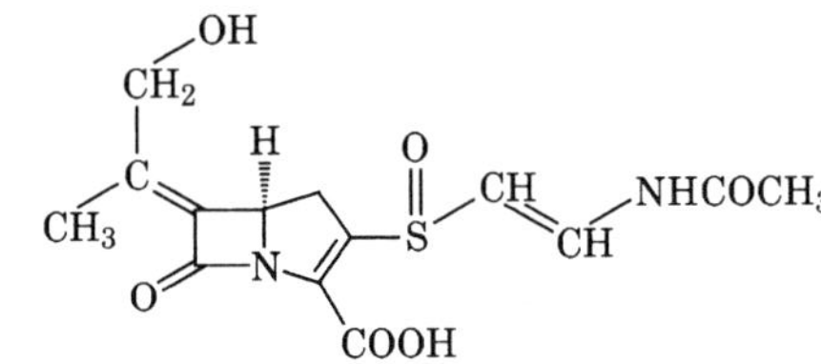
pluracidomycin A	[*82138-64-5*]	$C_9H_{11}NO_{10}S_2$	(**8**)	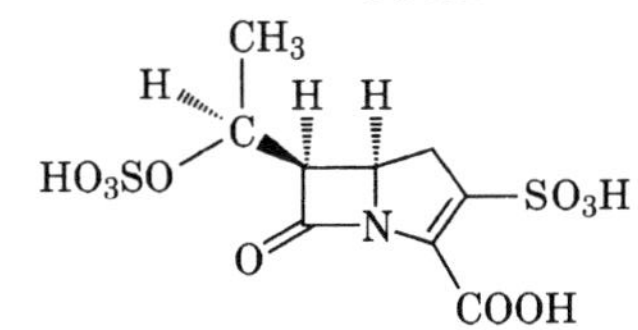

utes to a low recovery in the isolation procedures. The fermentation and isolation processes for thienamycin and the olivanic acids has been reviewed in some detail (12).

Early studies on the biosynthesis of thienamycin (**2**) indicated that the pyrroline ring was derived from glutamic acid (13,14). Further reports demonstrated that the C-6 and C-7 atoms of the β-lactam ring are acetate derived, that the C-2 cysteaminyl side chain is from cysteine, and the two carbon atoms of the hydroxyethyl side chain originate from methionine (15). The incorporation of glutamate into the parent ring system (**1**, X = CH_2) has also been reported (16); in this study bacterial production of various isomers of the saturated carbapenam ring structure was observed including one isomer having stereochemistry opposite to that of thienamycin at the C-5 position. Some streptomycete carbapenem producers have also been found to produce structures in which the pyrroline ring is reduced as in (**9**) (17). The microbiological interconversion of a number of the olivanic acids has been demonstrated (4), as well as the role of carbapenems, such as (**10**, R = H, OH, or OSO_3H) possessing a pantothenyl side chain, as early precursors of other carbapenem antibiotics (18). The possible biosynthetic pathways leading to both carbapenem and carbapenam ring structures have been discussed in some detail (14,16,18).

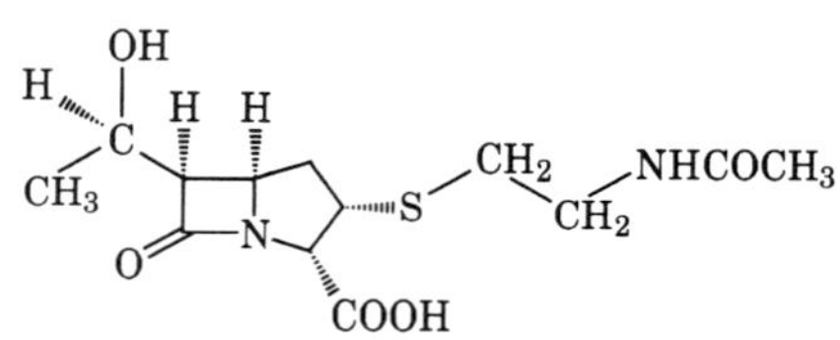

(**9**)

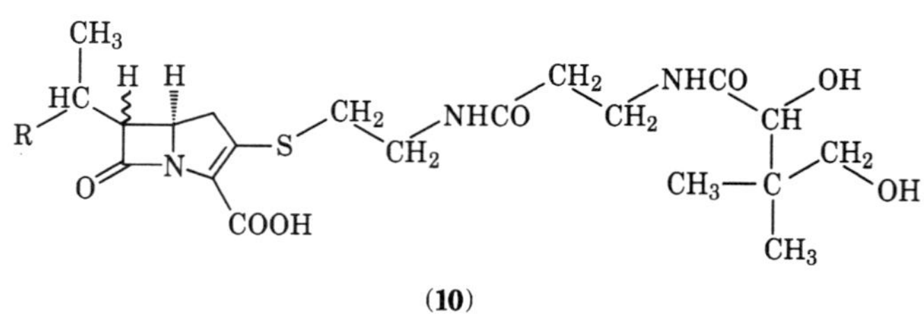

(**10**)

Properties. Thienamycin is isolated as a colorless, hygroscopic, zwitterionic solid, although the majority of carbapenems have been obtained as sodium salts and, in the case of the sulfated olivanic acids, as disodium salts (12). Concentrated aqueous solutions of the carbapenems are generally unstable, particularly at low pH. All the substituted natural products have characteristic uv absorption properties that are often used in assay procedures. The ir frequency of the β-lactam carbonyl is in the range 1760–1790 cm^{-1}.

Structure Determinations. The structural elucidation of the early carbapenems, thienamycin, and the olivanic acids, followed a fairly similar sequence making use of both spectroscopic and degradation studies (19–22). Infrared absorption spectra suggested the presence of a β-lactam ring (ν_{max} 1765 cm^{-1}) and in the case of thienamycin (**2**) a trans-arrangement of β-lactam protons was indicated by the small coupling constant ($J_{5,6}$ <3 Hz) for the β-lactam hydrogens in the nmr. For the sulfated olivanic acids, (**3**) and (**4**), the coupling constant ($J_{5,6} \approx 6$ Hz) indicated the more familiar *cis*-β-lactam stereochemistry found in the penicillins and cephalosporins. One key reaction in confirming the structure of MM 13902 (**3**, $n = 0$) is shown in Figure 1**a**. The rearrangement of the monomethyl ester (**11**) produced the pyrrole (**12**); subsequent esterification followed by elimination of the sulfate residue gave a mixture of isomers (**13**) (20). Elimination of the sulfate moiety of the diester (**14**) gave solely the (*E*)-isomer of the ethylidene derivative (**15**) indicating a stereospecific (E)-$_2$ elimination from which it was possible to assign the configuration at C-8 as (*S*). The (*E*)-geometry of the C-2 acetamidoethenyl substituent was clearly apparent from the coupling constant ($J = 14$ Hz) of the double bond protons.

In the case of thienamycin (Fig. 1**b**) the absolute stereochemistry at C-5 was unambiguously determined from the ene-lactam (**16**). The resultant (*R*)-aspartic acid (**17**) demonstrated that the absolute stereochemistry at C-5 of thienamycin is (*R*), corresponding to that found in the C-5 position of both penicillins and cephalosporins. Confirmation of the stereochemical assignments in both thienamycin (**2**) and the olivanic acid MM 13902 (**3**, $n = 0$) has been confirmed by x-ray crystallography (19,21,22). The structural determination of the nonsulfated derivatives from *S. olivaceus* (23), PS-5 (**5**) (5), the carpetimycins (6), and the asparenomycins (7) followed a similar pattern.

Reactions. Although carbapenems are extremely sensitive to many reaction conditions, a wide variety of chemical modifications have been carried out. Many derivatives of the amino, hydroxy, and carboxy group of thienamycin (**2**) have been prepared primarily to study structure–activity relationships (24). The most interesting class of *N*-derivatives are the amidines which are usually obtained in good yield by reaction of thienamycin with an imidate ester at pH 8.3. Introduction of this basic but less nucleophilic moiety maintains or improves the potency of the natural material while greatly increasing the chemical stability. Thus *N*-formimidoyl thienamycin [*64221-86-9*] (MK 0787) (**18**), $C_{12}H_{17}N_3O_4S$, (25) was chosen for clinical evaluation and development. Another reaction of thienamycin involves isomerization to the chemically stable, but biologically inactive, Δ-1 isomer (26). Reductive removal of the cysteaminyl side chain has also been reported (27).

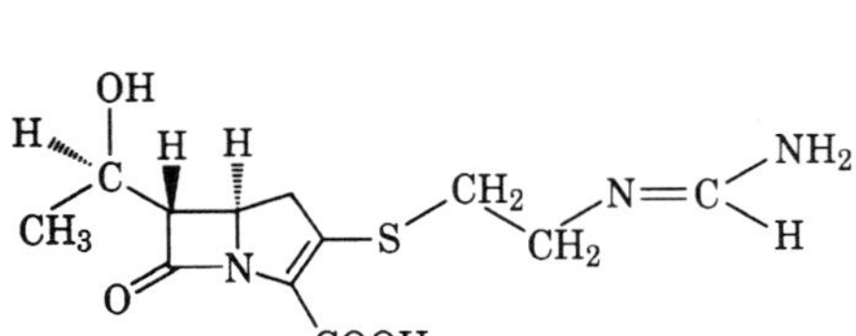

(**18**)

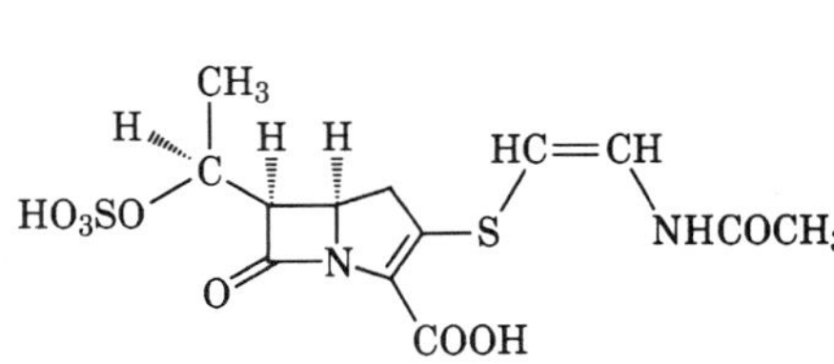

(**19**)

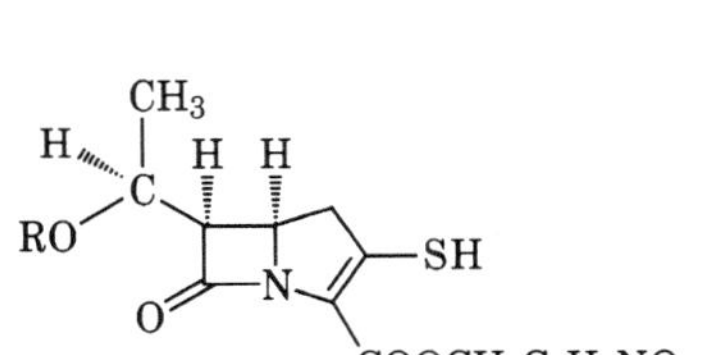

(**20**)

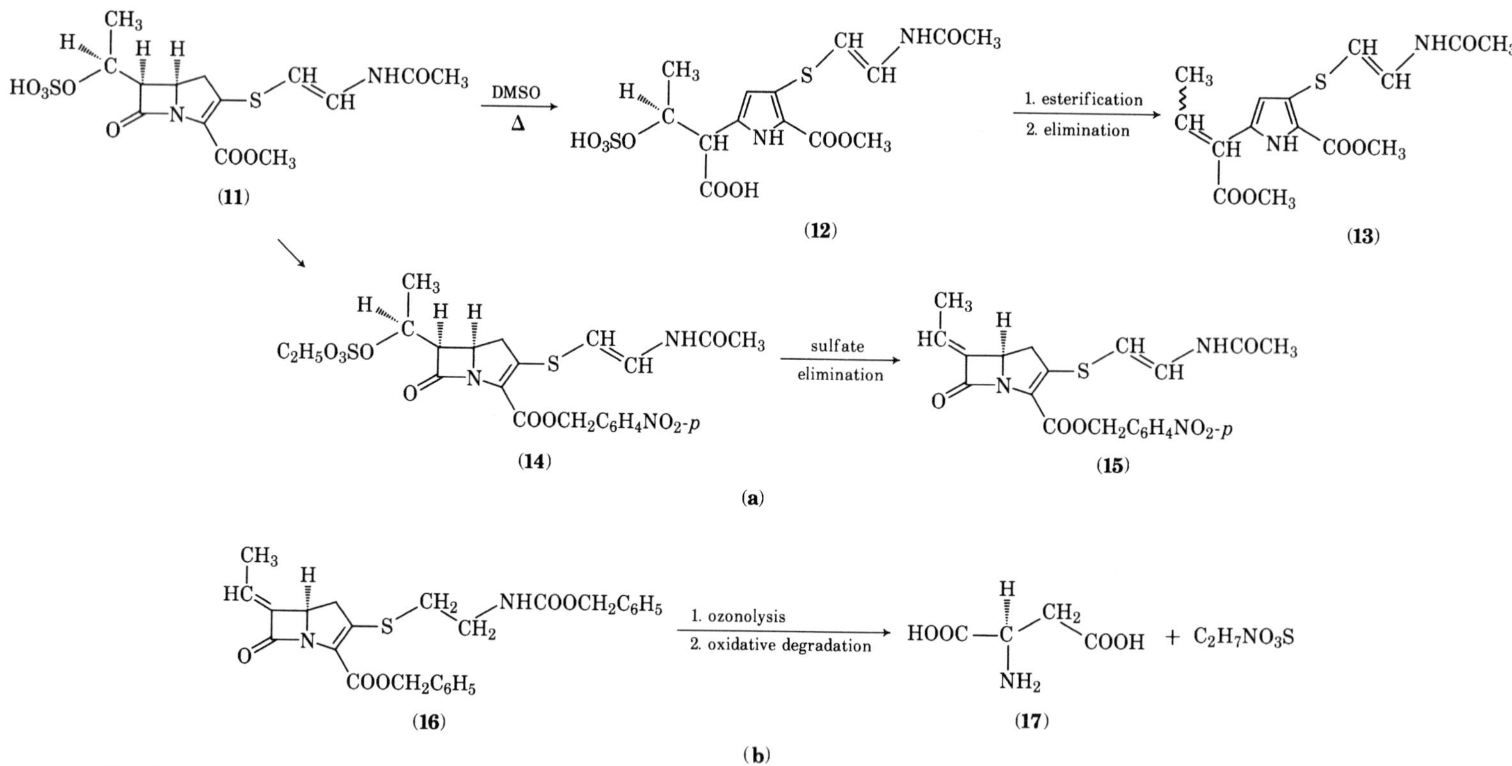

Fig. 1. Reaction sequences for steps in the structure determination of **(a)** MM 13902 (**3**, $n = 0$) and **(b)** thienamycin (**2**).

In the olivanic acid series of carbapenems the (*E*)-acetamidoethenyl grouping can be isomerized to the (*Z*)-isomer (**19**) (22) and reaction with hypobromous acid provides a bromohydrin that fragments to give a thiol of type (**20**) when R = H, SO_3H, or $COCH_3$. The thiol is not isolated but can react to provide new alkyl or alkenyl C-2 substituents (28). In the case of the nonsulfated olivanic acids, inversion of the stereochemistry at the 8(*S*)-hydroxyl group by way of a Mitsunobu reaction affords an entry to the 8(*R*)-thienamycin series (29). An alternative method for introducing new sulfur substituents makes use of a displacement reaction of a carbapenem (*S*)-oxide with a thiol (30). Microbial deacylation of the acylamino group in PS-5 (**5**) has been used to obtain the aminoethyl analogue (31).

Synthesis. One consequence of the discovery of the carbapenem natural products has been the development of new synthetic methods, the impetus for which was provided by the exceptional antibacterial potential of the compounds coupled with the extremely poor fermentation yields. Only chemical synthesis could provide the quantities of MK 0787 (**18**) necessary for clinical trials and commercial production.

The racemic form of the unsubstituted nucleus (**1**, X = CH_2) was synthesized by several groups (32–34) prior to the disclosure of the natural material. One reaction path involved an azetidinone (**21**) where R = CH_2OH or $CH{=}CH_2$ converted to the corresponding phosphorane (**22**) where R′ = *o*-NO_2-C_6H_4 when R = CH_2OH and R′ = $COCH_3$ when R=CH=CH_2.

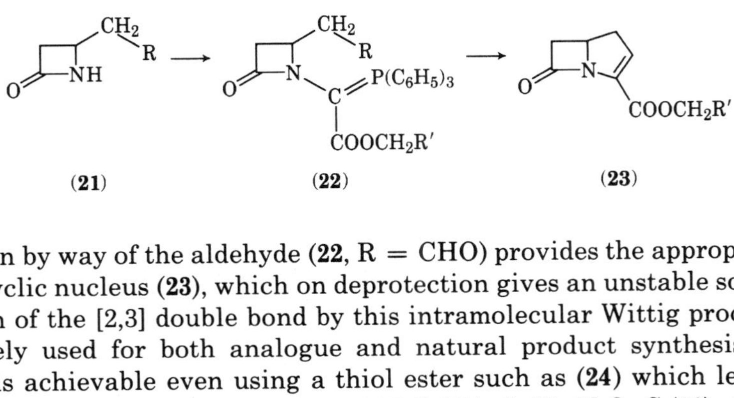

(**21**) (**22**) (**23**)

Cyclization by way of the aldehyde (**22**, R = CHO) provides the appropriate ester of the bicyclic nucleus (**23**), which on deprotection gives an unstable sodium salt. Formation of the [2,3] double bond by this intramolecular Wittig procedure has been widely used for both analogue and natural product synthesis (35). Cyclization is achievable even using a thiol ester such as (**24**) which leads to the protected olivanic acid MM 22383 [*74819-56-0*] (**25**), $C_{28}H_{26}N_4O_{11}S$ (36). Addition of thiols to the double bond of phosphorane-derived derivatives lacking a C-2 substituent (**26**) has been used for the synthesis of PS-5 (**5**) (37).

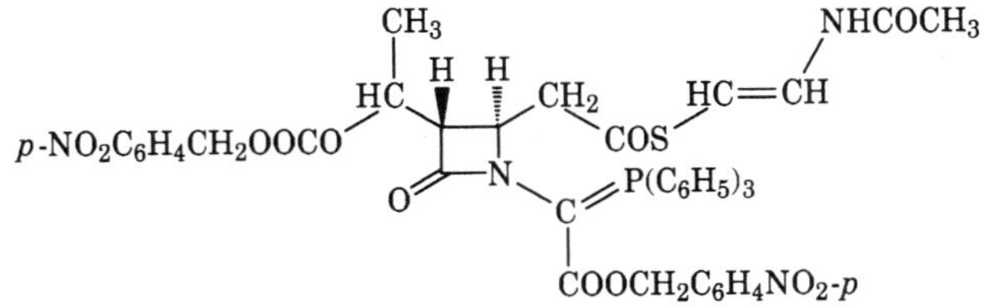

(**24**)

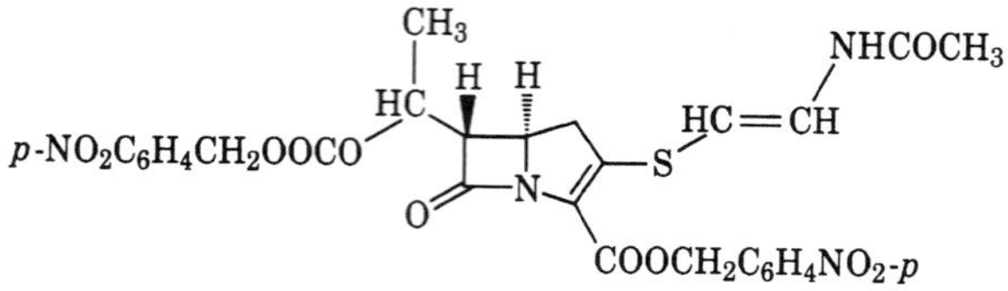

(25)

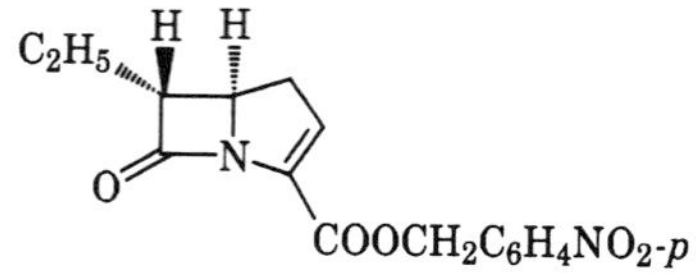

(26)

A synthetic approach that involves the [3,4] bond formation using a carbene insertion reaction has been highly successful and is illustrated by the enantioselective synthesis of (+)-thienamycin starting from L-aspartic acid [*56-84-8*], $C_4H_7NO_4$ (38). Over several steps the acid is converted to a protected form of the azetidinone (**27**) having all three contiguous chiral centers with the correct stereochemistry. Progression to the diazoketone (**28**) and cyclization using a catalytic amount of rhodium acetate provides the bicyclic ester (**29**, R = H) in high yield. Introduction of the cysteaminyl side via an activated enol derivative of (**29**) and deprotection gives (+)-thienamycin;

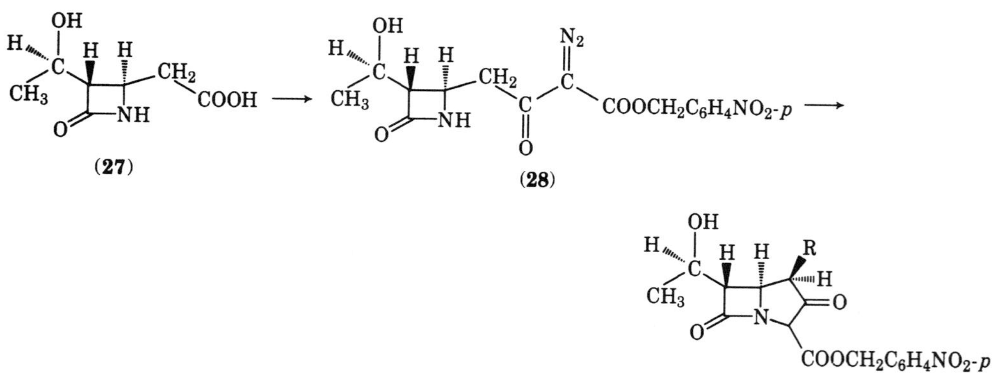

(27) (28) (29)

using the amidine side chain (**30**) MK 0787 (**18**) is directly accessible (39).

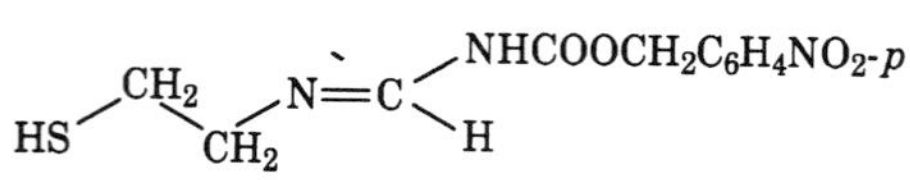

(30)

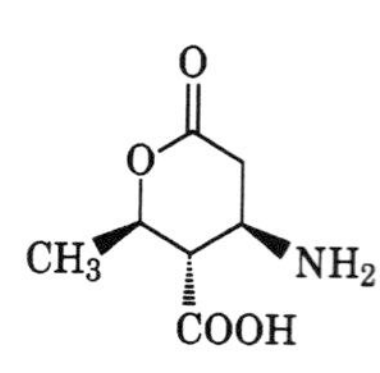

(31)

A second method makes use of the lactone (**31**) from acetone dicarboxylate (40) and for which a synthesis from (−)-carvone has been reported (41). Displacement of chlorine from the 6-aminopenicillanic acid (6-APA) derived *β*-lactam (**32**) by (**33**) illustrates yet another approach to the diazoketone (**28**) (42).

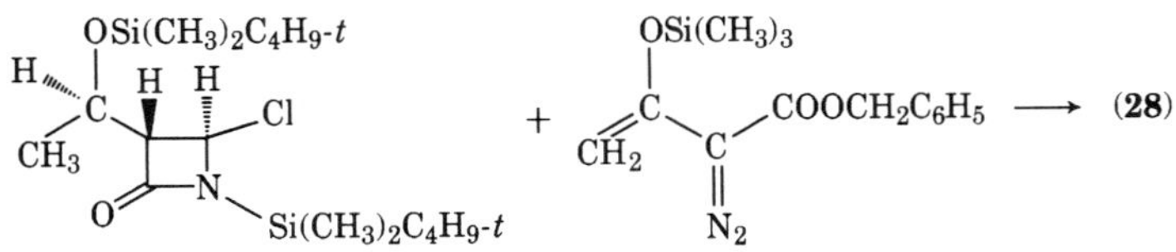

(**32**) (**33**)

Formal syntheses of thienamycin (**2**) from precursors such as carbohydrates (43–45), amino acids (46,47), isoxazolidines (48), and tricarbonyliron lactam complexes (49) have also been reported. Many other methods for carbapenem synthesis have been widely reviewed (10,50–52).

In addition to variable chemical stability the carbapenems are susceptible to *β*-lactam cleavage by a dehydropeptidase enzyme (DHP-I) located on the brush borders of the kidney (53). Clinically, MK 0787 (**18**) is used with an inhibitor of this enzyme, cilastatin [*78852-98-9*] (MK 0791) (**34**), $C_{16}H_{26}N_2O_5S$, which has a dramatic effect not only on the urinary recovery of the drug, but also reduces any nephrotoxic potential (52) (see ENZYME INHIBITION).

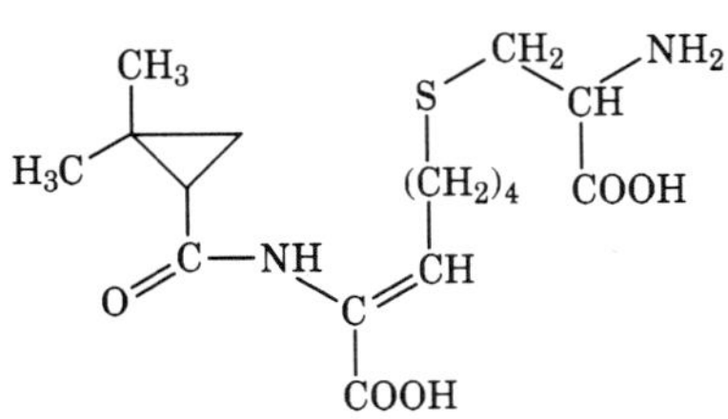

(**34**)

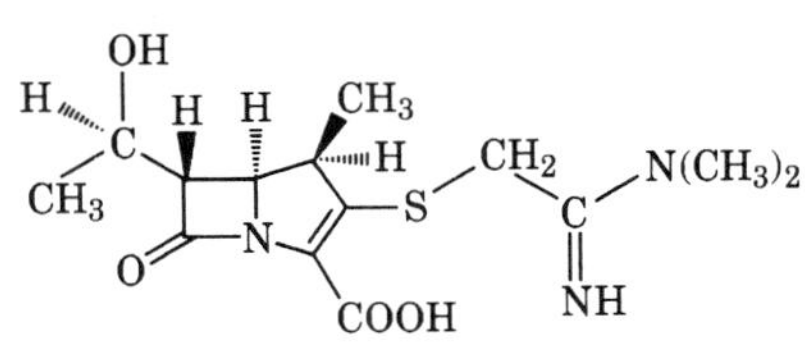

(**35**)

The disclosure that the 1*β*-methylcarbapenem [*90822-26-7*] (**35**), $C_{14}H_{21}N_3O_4S$, exhibits greatly increased chemical and enzymatic stability over MK 0787 (**18**) (54), while retaining the antibacterial potency of thienamycin (**2**), has caused attention to be focused on methods leading to structural analogues. The synthesis of (**35**) was by alkylation of (**27**) and elaboration via the ketoester (**29**, R = CH_3), but little stereocontrol was attainable by this procedure. Many newer approaches use the acid (**36**) derived by diastereoselective displacement at C-4 using the chiral 4-acyloxy *β*-lactams (**37**, R = CH_3 or C_6H_5) (55,56). Examples make use of the tin or the boron enolates producing (**36**) in yields of 72–79% with a ratio of *β*:*α* isomers ranging from 24:1 to 60:1 (57–59).

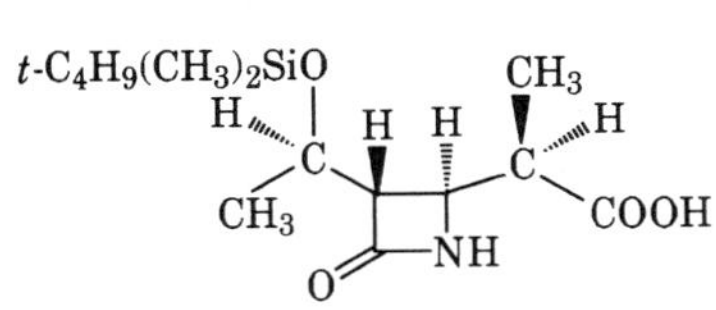

(**36**)

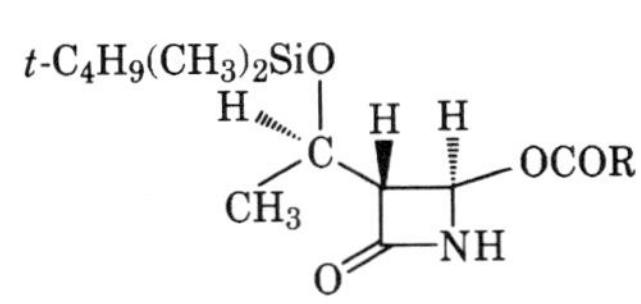

(**37**)

Other approaches to (**36**) make use of (**37**, R = CH_3) and reaction with a tributylstannyl allene (60) or 3-siloxypentadiene (61). A chemicoenzymatic synthesis for both thienamycin (**2**) and 1β-methyl analogues starts from the chiral monoester (**38**), derived by enzymatic hydrolysis of the dimethyl ester, and proceeding by way of the β-lactam (**39**, R = H or CH_3) (62,63). (*S*)-Methyl-3-hydroxy-2-methylpropanoate [*80657-57-4*] (**40**), $C_5H_{10}O_3$, has also been used as starting material for (**36**) (64), whereas 1,3-dipolar cycloaddition of a chiral nitrone with a crotonate ester affords the oxazolidine (**41**) which again can be converted to a suitable β-lactam precursor (65).

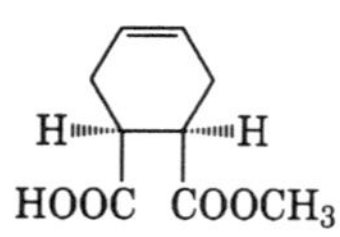

(**38**)

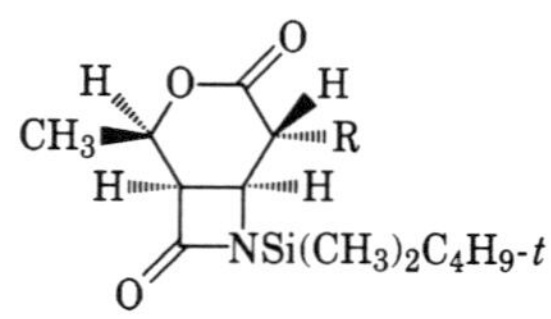

(**39**)

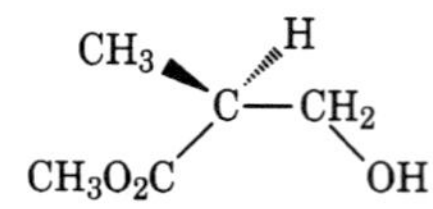

(**40**)

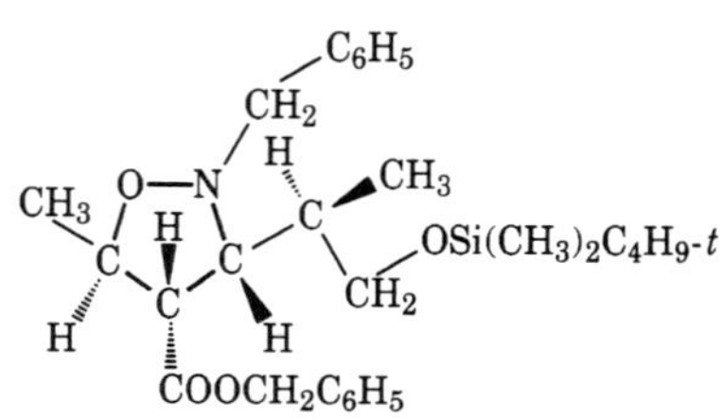

(**41**)

A number of highly potent DHP-I stable 1β-methylcarbapenems having a variety of C-2 substituents have now been described (60,66–69) including SM 7338 [*96036-03-2*] (**42**), $C_{17}H_{25}N_3O_5S$. An acylamino compound (66) and a 1β-methoxy analogue (70) provide other variations. The pyrrolidine substituted 1β-methylcarbapenem SM 7338 (**42**) is being developed as a broad-spectrum parenteral antibiotic under the name meropenem; the synthesis of (**42**) is by way of the lactone (**43**) derived by a novel Diels-Alder approach to dihydropyran precursors of (**43**) (71).

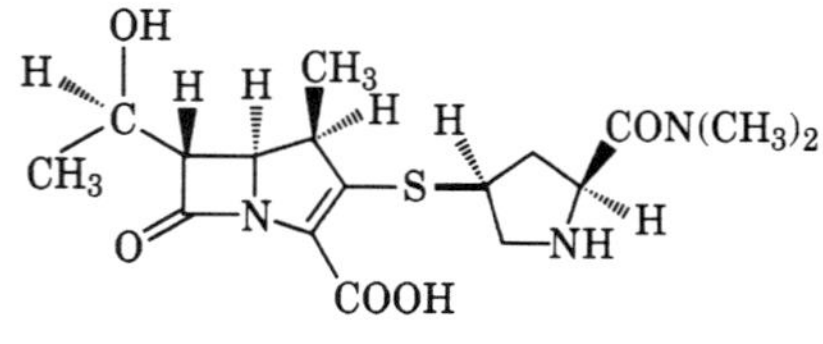

(**42**)

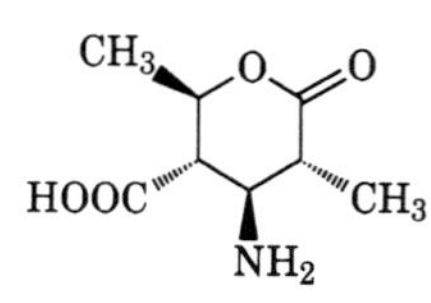

(**43**)

Biological Properties. Thienamycin, the olivanic acids, and the majority of carbapenems are highly active broad-spectrum antibiotics having good stability to β-lactamases. Of the natural products thienamycin (**2**) is the most potent having a spectrum of activity encompassing both aerobic and anaerobic gram-positive and gram-negative bacteria including *Pseudomonas* species (72). The latter activity is attributed to the presence of the basic cysteaminyl substituent at C-2. All of the *N*-acylated derivatives have much reduced antipseudomonal activity (24).

The sulfated compounds MM 13902 (**3**, $n = 0$) and MM 17880 (**4**) are also broad-spectrum agents, but not as potent as thienamycin and all lack any significant activity against *Pseudomonas* (73). Many carbapenems are excellent inhibitors of isolated β-lactamases, particularly the olivanic acid sulfoxide MM 4550 (**3**, $n = 1$) (3). The possible mechanism of action of the carbapenems as inhibitors of β-lactamases has been discussed in some detail (74). Other carbapenems such as PS-5 (**5**) (75), the carpetimycins (76), asparenomycins (77), and pluracidomycins (8) are all highly active as antibiotics or β-lactamase inhibitors. The parent nucleus itself (**1**, $X = CH_2$) is intrinsically active, but chemically unstable (9).

Unlike the classical penicillins and cephalosporins, a rigid adherence to the cis-stereochemistry around the β-lactam ring is not necessary for biological activity as both trans-isomers such as (**2**), and cis-isomers (**3**, $n = 1$) are active. The hydroxy olivanic acids are not as potent as the sulfates, although in this case the cis-β-lactams are more active than the trans-isomers (73). Thienamycin (**2**) is the most stable to β-lactamases, a property attributed to the presence of the 8(*R*)-hydroxy group. Several synthetic examples lacking the C-6 hydroxyethyl side chain are much more susceptible to hydrolysis by β-lactamases (78). The order of antibiotic activity in the nonsulfated hydroxyethyl series appears to be trans-substituted β-lactam with 8(*R*) stereochemistry ie, thienamycin > cis 8(*S*) > trans 8(*R*) (25). As in the case of other β-lactam antibiotics, the carbapenems inhibit bacterial growth by interfering with peptidoglycan synthesis in the cell wall. In *E. coli* thienamycin has the greatest affinity for the penicillin binding protein PBP-2 (79); most of the newer cephalosporins bind principally to PBP-3.

The amidine derivative MK 0787 (**18**), named as imipenem, shows greatly improved solution stability as compared to thienamycin (**2**). Against some 800 isolates of gram-positive and gram-negative bacteria, imipenem (**18**) inhibited the majority of organisms at concentrations below 1 μg/mL and was not hydrolyzed by plasmid mediated or chromosomal β-lactamases (80). The plasma half-life of imipenem in humans was considered satisfactory (1 h), but urinary recovery of antibiotic was extremely variable ranging from 6–40% (81) resulting from degradation by the DHP-I enzyme. This led to the development of the imipenem-cilastatin combination as the clinical product (82). The urinary recovery of the combination was established consistently at some 70% (83). The imipenem-cilastatin combination is used as a parenteral broad-spectrum antibiotic against a wide variety of bacterial infections. No other carbapenem has yet been developed as a commercial product although effort is being directed towards the DHP-I stable 1β-methyl analogues of which meropenem (**42**) would appear to be at the most advanced stage of development. This is a broad-spectrum agent similar to imipenem but with greatly enhanced stability to dehydropeptidase and should therefore not require concomitant administration of an enzyme inhibitor (69).

Penems

Historically, the development of penems is contemporary with that of the naturally occurring carbapenems and the direction of penem research has clearly been influenced by the structures of the closely related natural products. The origins of the two groups of compounds is, however, quite different. Unlike carbapenems, no penems have been found in nature. When first described (84,85) they were viewed as hybrid molecules combining structural features of penicillins and cephalosporins.

Synthesis. *Woodward's Phosphorane Route.* The first penem synthesis, shown in Figure 2, utilized an intramolecular Wittig reaction to form the [2,3] double bond of the thiazoline ring (84). Reductive acylation of the penicillin derived disulfide (**44**) gave the thioester (**45**). Ozonolysis of the latter provided the oxalimide (**46**) which on mild methanolysis gave the azetidinone (**47**). Well established methods were applied to convert (**47**) to the phosphorane (**48**) which underwent thermal cyclization to the penem ester (**49**). Catalytic hydrogenation gave the penem acid [*64370-39-4*] (**50**) which was shown to possess antibacterial activity in spite of its rather limited stability.

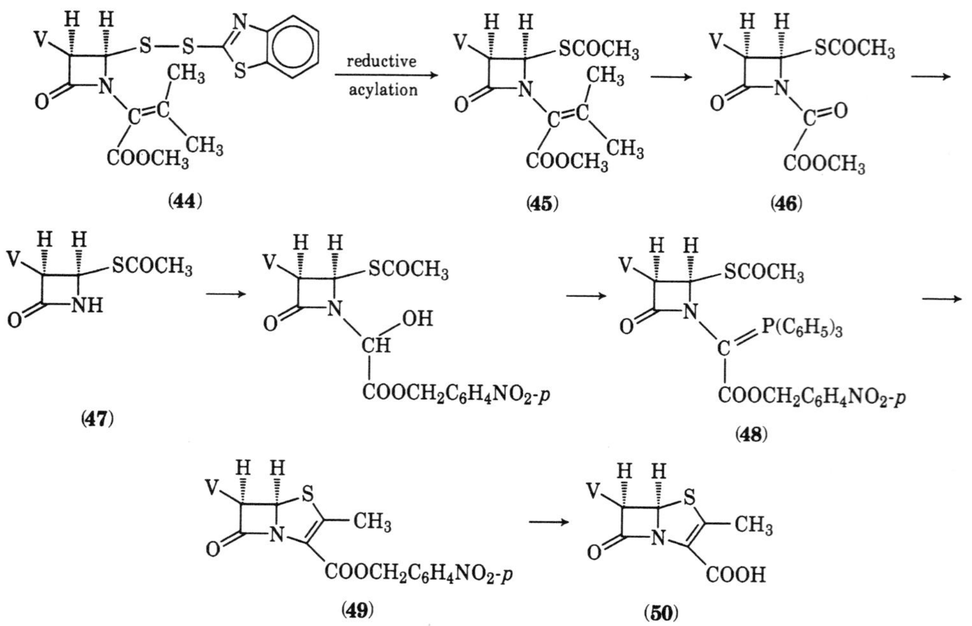

(**44**) (**45**) (**46**) (**47**) (**48**) (**49**) (**50**)

Fig. 2. Reaction scheme for the first penem synthesis where V = $C_6H_5OCH_2CONH$.

Extension of the Phosphorane Route. Ample evidence of the versatility of the phosphorane synthesis strategy is provided by the proliferation of penems that followed. Nucleophilic displacement of the acetate function of the acetoxyazetidinone (**51**, R = $OCOCH_3$) [*28562-53-0*] (86) provided azetidinones where R = $SCOCH_3$, $SCSSC_2H_5$, and $SCSOC_2H_5$, which on elaboration gave the penems (**52**, R = CH_3) (87), (**52**, R = SC_2H_5) (88), and (**52**, R = OC_2H_5) (89). Similar treatment of 3-substituted (or disubstituted) acetoxyazetidinones allowed the synthesis of a number of 2-substituted-6-alkyl- and 6,6-dialkylpenems (90).

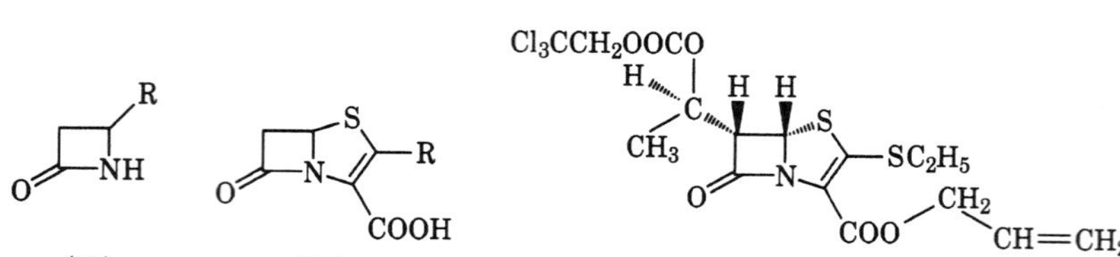

(**51**) (**52**) (**55**)

The naturally occurring carbapenems provided the impetus for the synthesis of the four racemates of 6-(1-hydroxyethyl)-2-ethylthiopenem-3-carboxylate (91). As seen in Figure 3, conversion of (**51**, R = $OCOCH_3$) to the *N*-protected azetidinone followed by aldol type addition gave a mixture of alcohols. Further manipulation gave the four isomers of the trithiocarbonate (**53**) each of which was progressed separately to the corresponding penems (**54**) by means of the phosphorane route. During the thermal cyclization each product gives rise to a second isomer by virtue of a novel epimerization at C-5; eg, thermolysis of the rel-5(*R*)-, 6(*S*)-, 8(*R*)-isomer [*76431-42-0*] gives (**55**) which could be shown to be the rel-5(*R*)-, 6(*R*)-, 8(*S*)-isomer [*80575-20-8*]. Double deprotection of (**54**) gave the corresponding penem salts (**56**).

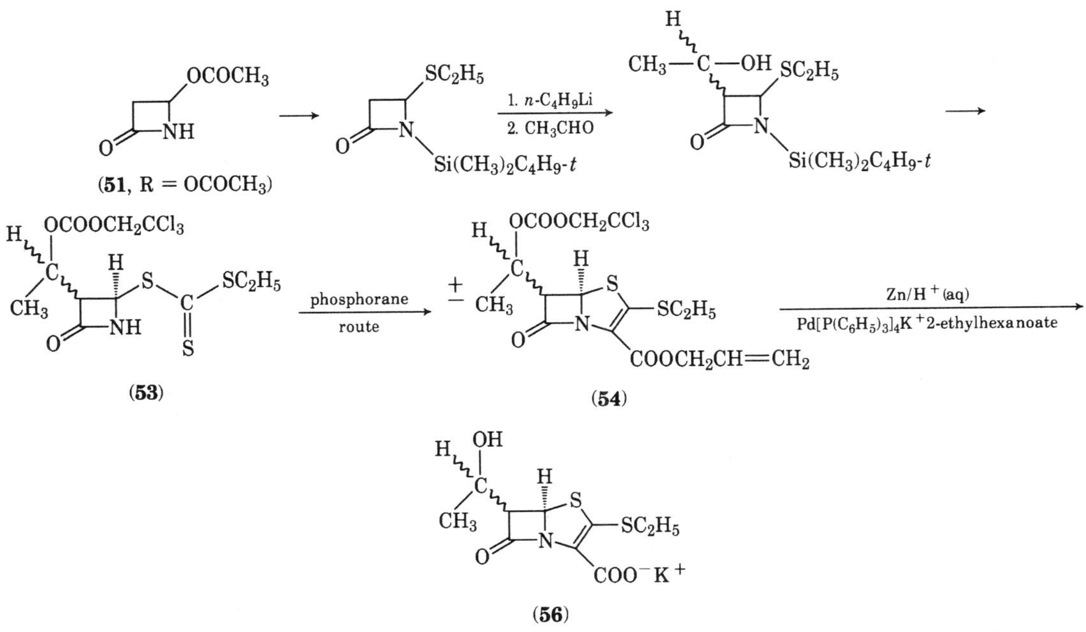

Fig. 3. Reaction scheme for the synthesis of the 6-(1-hydroxyethyl)-2-ethylthiopenem-3-carboxylates (92).

A useful modification of the phosphorane route (92) allowed the synthesis of a wide range of penems from common advanced intermediates. Cleavage of tritylthioazetidinones of the type (**57**) when R = H, $CH_2OCOCH_2C_6H_4NO_2$-*p*, etc, using silver nitrate–methanol afforded the silver mercaptides (**58**) which on acylation–cyclization gave a wide range of penems in which the 2-substituent was linked through carbon or sulfur.

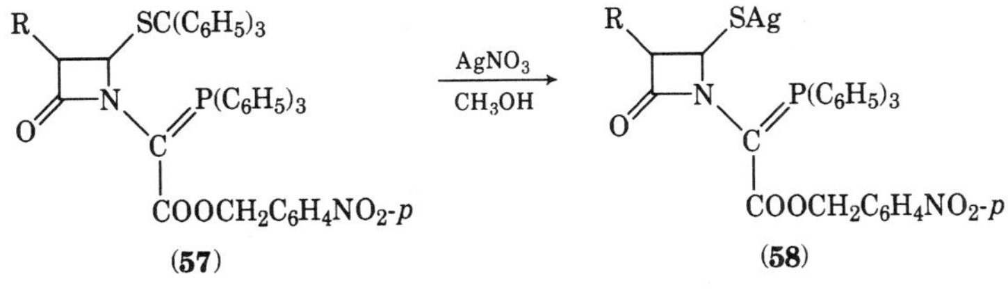

In common with the naturally occurring carbapenem thienamycin (**2**), the introduction of the *trans*-6-[1-(*R*)-hydroxyethyl] group had a profound effect on the biological properties of the penems. This, together with an indication from an early study (93) that, as with other *β*-lactams, the 5(*R*)-enantiomer was solely responsible for antibacterial activity, provided impetus for the development of methods for the synthesis of chiral penems.

The stereochemical outcome of aldol reactions of 6,6-dibromopenicillanates was described in 1977 (94) making 6-aminopenicillanic acid [*551-16-6*] (6-APA), $C_8H_{12}N_2O_3S$, an attractive starting material for the synthesis of chiral 6-(1-hydroxyethyl)penems (Fig. 4). Metal-halogen exchange on (**59**) followed by treat-

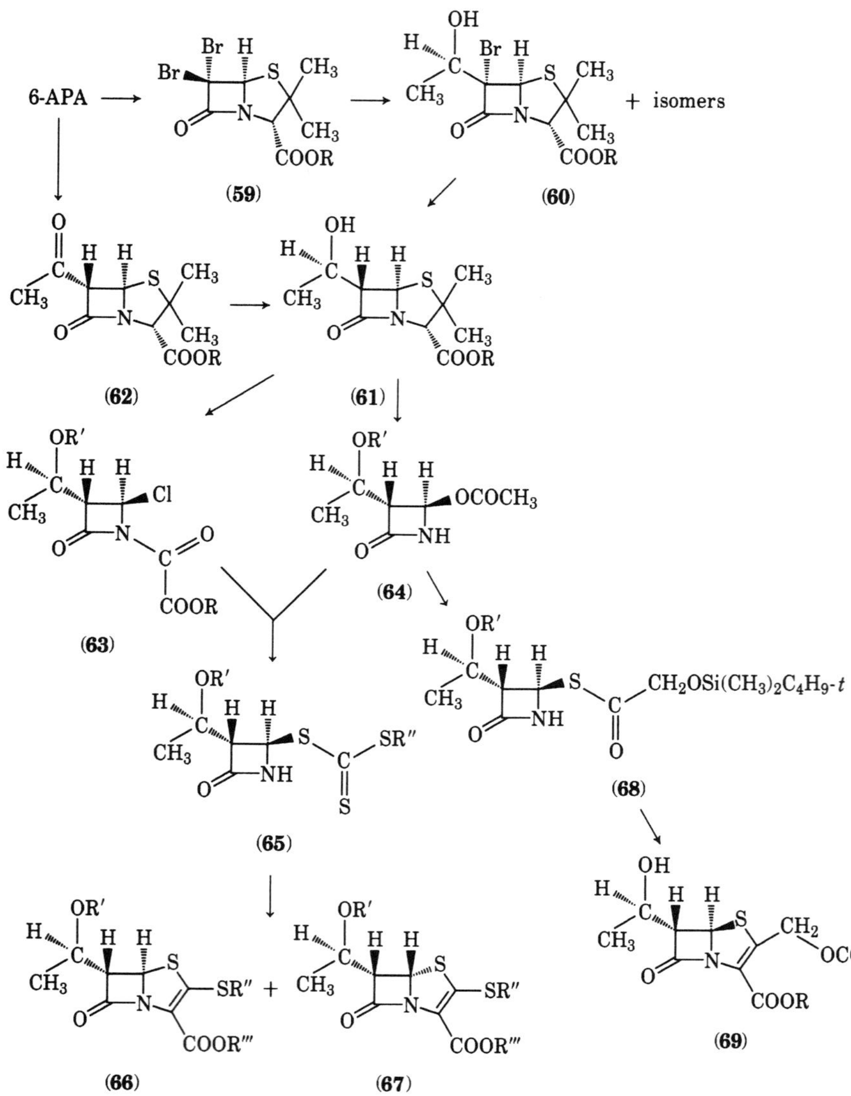

Fig. 4. Reaction scheme for the synthesis of chiral 6-(1-hydroxyethyl) penems where R′ = H, $OCOCH_2CCl_3$, $Si(CH_3)_2C_4H_9$-*t*; R″ = C_2H_5; and R‴ = $CH_2CH{=}CH_2$, $CH_2C_6H_5$.

ment with acetaldehyde gave a mixture of isomers from which the bromohydrin (**60**) having a desired 8(*R*) stereochemistry could be isolated as major product. Reductive removal of the halogen atom using zinc then gave the (5(*R*)-, 6(*S*)-, 8(*R*))-penem (**61**). Alternatively, diazotization followed by Lewis acid catalyzed reaction using acetaldehyde gave the unstable *trans*-6-acetylpenicillanate (**62**) which could be reduced stereoselectively to the alcohol (**61**) (95).

Several groups have built on this work and prepared intermediates capable of conversion to penems using the phosphorane route. In one method (96), chlorinolysis was used to prepare the chloroazetidinone (**63**), in another (97) the thiazolidine ring was cleaved using mercuric acetate, obtaining the acetoxyazetidinone (**64**) after oxidative removal of the nitrogen substituent. Reaction of either (**63**) (96) or (**64**) (98) with a sodium trithiocarbonate proceeded with retention of configuration to give the azetidinones (**65**) from which the (5(*R*)-, 6(*S*)-, 8(*R*))-penems (**66**) were obtained. Some of the unwanted *cis*-5(*S*)-penems (**67**) were also formed during cyclization. Moreover, when either pure (**66**) or (**67**) was heated the same equilibrium mixture was obtained (96,98), possibly involving the intermediacy of a betaine (98). The acetoxyazetidinone (**64**) has also been used to prepare intermediates of the type (**68**) for the synthesis of the Farmitalia clinical candidate FCE 22101 [*84845-58-9*] (**69**, R = Na), $C_{10}H_{12}N_2O_6SNa$ (99).

Other methods for cleavage of the thiazolidine ring of penicillanates with retention of the sulfur atom have been described. Silver assisted cleavage afforded mercaptides (100) whereas the sulfenic acids generated by the thermolysis of penicillanate sulfoxides have been intercepted by acetylenes (101–103) and mercaptans (104,105). All these methods have provided useful intermediates for the synthesis of penems. Total synthesis starting from L-threonine [*72-19-5*], $C_4H_9NO_3$, (106,107) and ethyl-3-(*S*)-hydroxybutanoate [*56816-01-4*], $C_6H_{12}O_3$, (108) has also been used to prepare intermediates for the preparation of penems having defined stereochemistry.

Alternative Methods for Formation of the Penem Ring. The reactivity of the carbonyl group in readily available oxalimides of the type (**70**) (Fig. 5) provides a useful alternative to the original phosphorane route. Treatment of oxalimides (**70**, X = O, R = CH_2OCONH_2) and (**70**, X = S, R = SC_2H_5) with a trialkyl phosphite at elevated temperatures gives the corresponding penems (**72**) (100,109) and (**74**) (110,111). In a detailed study (112) it was shown that for thiolesters (**70**, X = 0) carbene generation/trapping gives intermediate trialkyoxyphosphoranes (**71**) which then cyclize thermally. In contrast, cyclization of trithiocarbonates (**70**, X = S, R = SC_2H_5) does not proceed via phosphorane intermediates. Instead it has been proposed (110,111) that the carbene generated added intramolecularly to the more reactive thiocarbonyl group to form an episulfide (**73**) which was then desulfurized to give the penem. The lower temperature and shorter reaction times required for cyclization of trithiocarbonates avoided the troublesome C-5 epimerization experienced with the conventional phosphorane route. The high reactivity of the oxalimide carbonyl group has also been demonstrated by the more recently reported (113) cyclization of phosphoranes of the type (**70**, X = $P(C_6H_5)_3$).

Although the phosphorus mediated ring closures are by far the most widely used, other methods for [2,3] bond formation have been described. Treatment of the dithiocarbamates (**75**, R = H or $Si(CH_3)_3$; R′ = 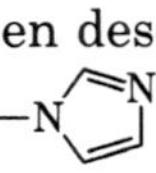) with a strong base

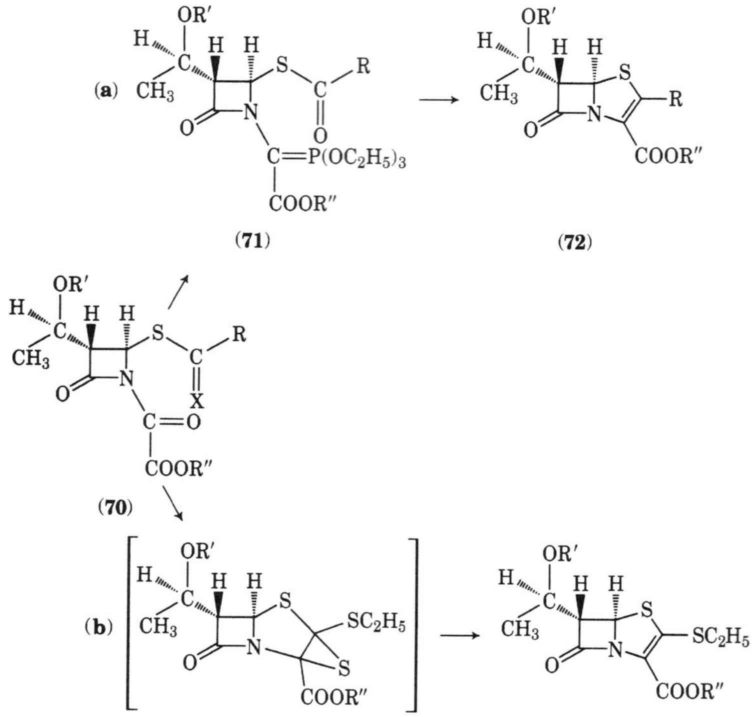

(71) (72) (70) (73) (74)

Fig. 5. An alternative method for the formation of the penem ring where R′ = $OCOCH_2CCl_3$, $Si(CH_3)_2$, or C_4H_9-*t* and R″ = CH_2CCl_3, or $CH_2C_6H_4NO_2$-*p*, CH_2CHCH_2; (**a**) X = O, R = CH_2OCONH_2, (**b**) X = S, R = SC_2H_5.

at low temperature gives thioxopenams (**76**) which are readily alkylated to give 2-alkylthiopenems (114).

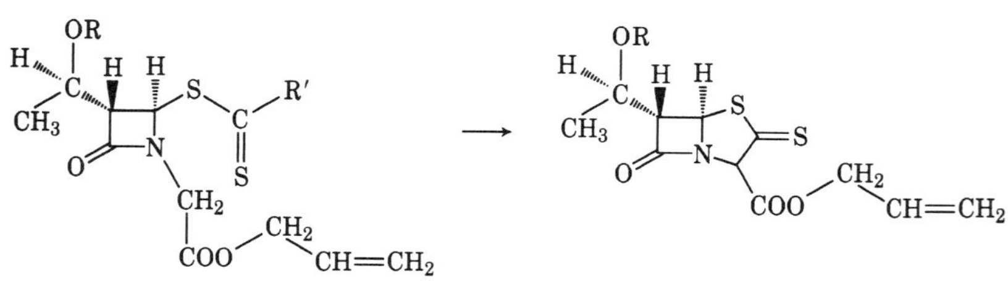

(**75**) (**76**)

An analogous preparation of thioxopenams from dithiocarbonates (**75**, R = t-$C_4H_9(CH_3)_2Si$, R′ = OC_6H_5) has also been described (115). Additionally, an intramolecular Michael addition reaction to form the [2,3] bond has been exploited in penem synthesis to prepare FCE 22101 (**69**) (116).

Synthesis of 2-thioxopenams (**76**) has also been realized using a [sulfur, C-5] ring closure (117). Cyclization of 4-(*S*)-chloroazetidinone [*88816-44-8*] (**77**, R = R′ = $SCOCH_3$), $C_{19}H_{19}ClN_2O_8S_2$, using imidazole in aqueous dioxane proceeded stereospecifically to give the 5-(*R*)-thioxopenam [*83362-57-6*] (**78**). Using the same approach, the azetidinone (**77**, R = 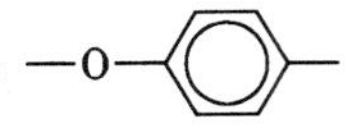Cl; R′ = $SCOC_4H_9$-*t*), was cyclized to the corresponding 2-aryloxypenem [*85768-34-9*] (118).

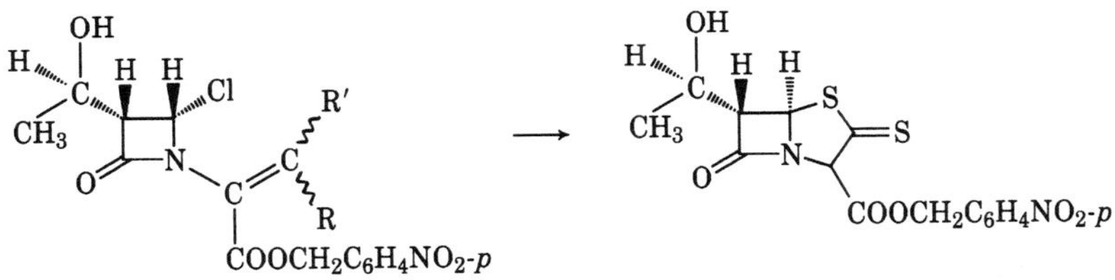

(**77**) (**78**)

Only one successful penem synthesis involving a [nitrogen, C-3] ring closure has been described (119). Cyclization of the ketomalonate (**79**) using hydrofluoric acid–pyridine afforded the penams (**80**) which were converted to the penem (**81**). Attempts to adapt the versatile diazoketoester–carbapenem cyclization for penem synthesis failed (120).

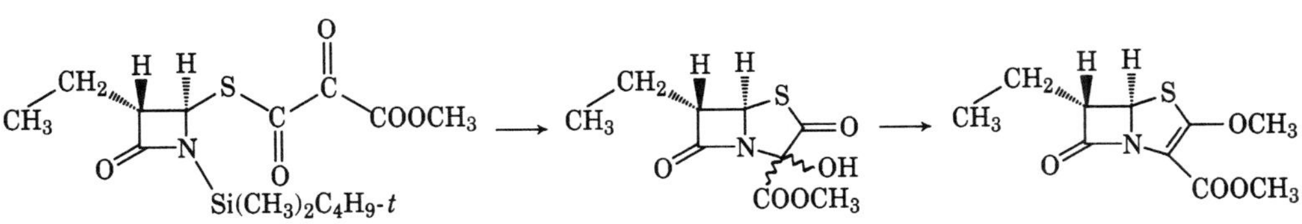

(**79**) (**80**) (**81**)

Ring contraction of 2-thiocephems has also been examined as a route to penems. Desulfurization of (**82**, n = 0) using triphenylphosphine gave mixtures of 5(*R*)- and 5(*S*)-penems (121). The stereochemical problem was neatly overcome by regioselective oxidation to the thiosulfonate (**82**, n = 2) which underwent stereospecific thermal extrusion of sulfur dioxide (122) to give the 5(*R*)-penem (**83**).

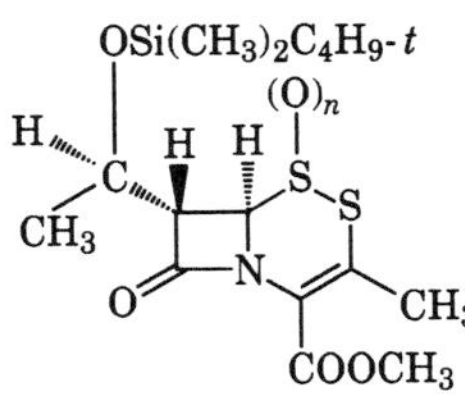

(**82**)

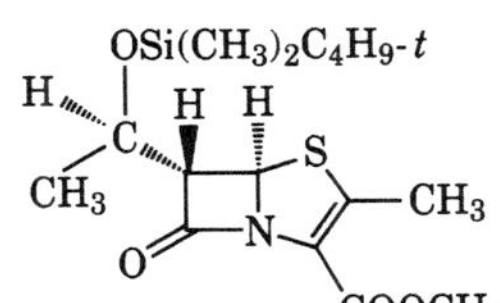

(**83**)

Modification of Intact Penems. Functional group modification has been used by a number of researchers (123–125) to synthesize a wide range of 2-substituted penems. For example, activation of the hydroxyl group of (**84**, R = OH) followed by displacement reactions provided 2-heterocyclylthiomethylpenems (**84**, R = (*S*)-heterocyclyl) (125) and 2-(quaternary ammonio)methylpenems (**84**, R = NR'_3X) (125).

Displacement of the sulfinyl group of penems (**85**), obtained by regioselective oxidation of (**74**, R′ = $Si(CH_3)_2C_4H_7$-*t*, R″ = CH_2CHCH_2) (Fig. 5) using thiolates,

Table 2. Penems

Name	CAS Registry Number	Molecular formula	Structure number	Structure	Reference
SCH 29482 SCH 34343	*[77646-83-4]* *[94392-37-7]*	$C_{10}H_{13}NO_4S_2$ $C_{11}H_{14}N_2O_6S_2$	(**94**, R = C_2H_5) (**94**, R = $(CH_2)_2OCONH_2$)	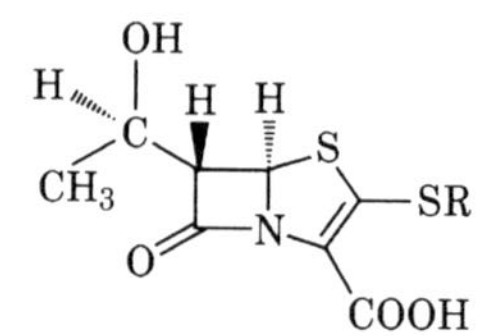	145 146
FCE 221201 FCE 22891	*[84845-58-9]* *[87238-52-6]*	$C_{10}H_{12}N_2O_6S$ $C_{13}H_{16}N_2O_8S$	(**69**, R = Na) (**69**, R = CH_2OCOCH_3)	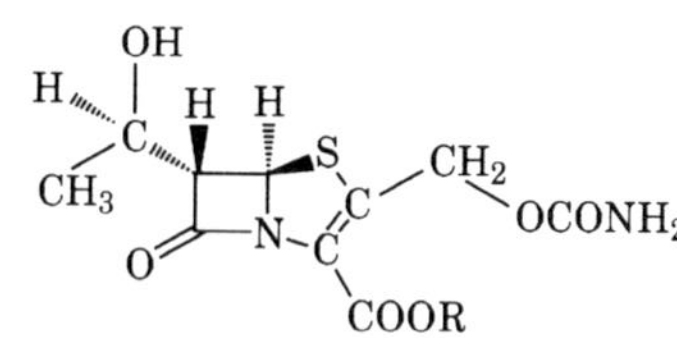	146 147
HRE 664	*[109888-39-3]*	$C_{15}H_{14}N_2O_6S$	(**95**)	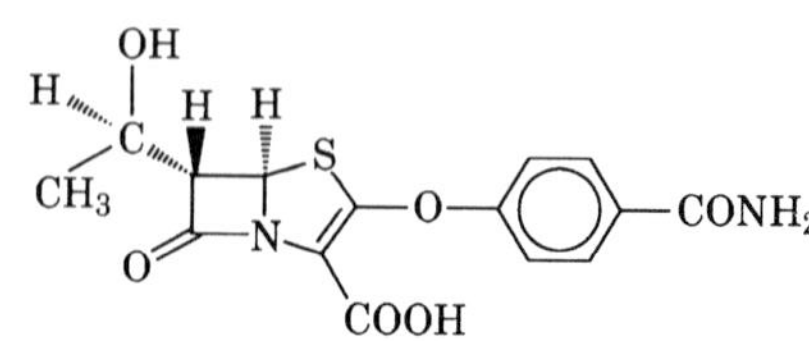	148

SUN 5555	[*106560-14-9*]	$C_{12}H_{15}NO_5S$	(**96**, R = 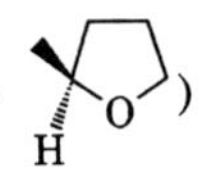)	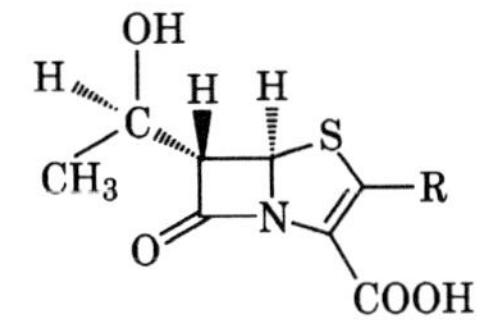	149
CGP 31608	[*97644-04-7*]	$C_9H_{12}N_2O_4S$	(**96**, R = CH_2NH_2)		150
CP 65207	[*120788-07-0*]	$C_{12}H_{15}NO_5S_3$	(**96**, R = 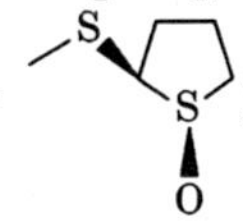)		151
FCE 25199	[*120796-34-1*]	$C_{15}H_{17}NO_7S$	(**97**)	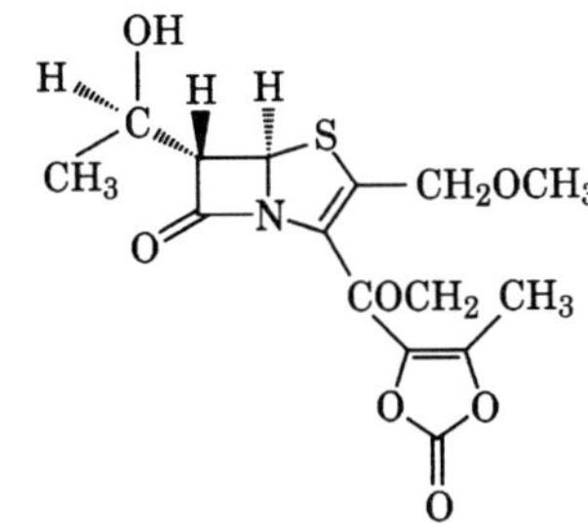	152

allowed the synthesis of a range of thio-substituted penems from a common intermediate (126). A conceptually similar approach (127) enables the displacement of the phenolic leaving group in penems (**86**, R = CN) [*85768-38-3*] or (**86**, R = NO_2) [*114707-30-1*] using amines and thus provides a general route to 2-aminopenems (**87**) where R can be CH_3 or n-C_3H_7; R′ can be H, CH_3, $CH_2COOC_2H_5$, or 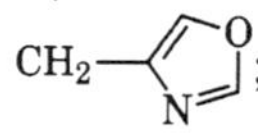; or both R and R′ can be 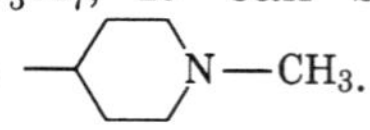.

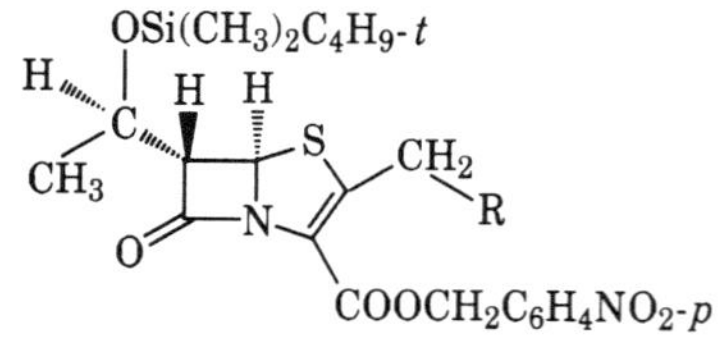

(**84**)

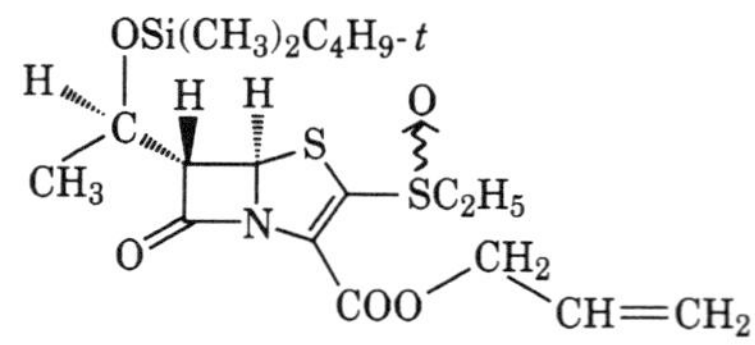

(**85**)

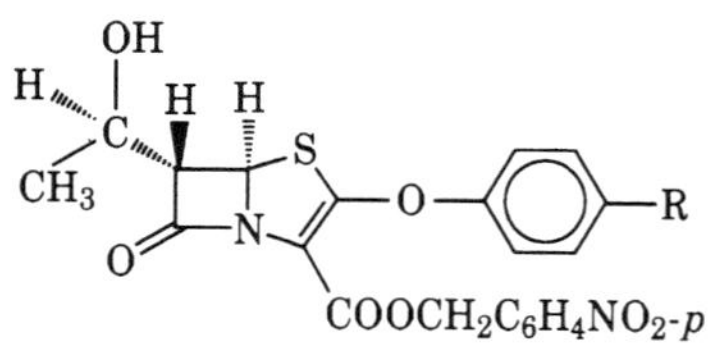

(**86**)

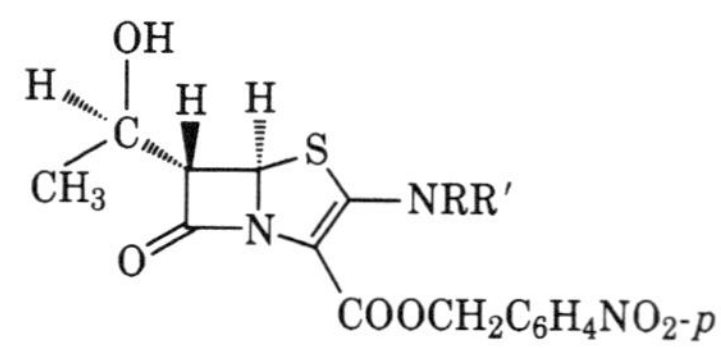

(**87**)

Although less researched than the 2-position, modifications at the 6-position of intact penems have been reported. Generation of the dianion of the penem (**52**, R = CH_3) using a strong base such as n-butyllithium or lithium diisopropylamide, followed by reaction with electrophiles yields 6-substituted 2-methylpenems in moderate yield (128). The enhanced acidity of the 6-proton in the bromopenem (**88**) [*114409-16-4*] has been exploited to prepare the Beecham β-lactamase inhibitor BRL 42715 [*102209-75-6*] (**89**, R = Na), $C_{10}H_8N_4O_3SNa$ (105). Lithium diphenylamide, a weaker base, was used to generate the anion of (**88**) which on sequential treatment with 1-methyl-1,2,3-triazole-4-carbaldehyde and acetic anhydride gives a mixture of diastereomers of the bromoacetate (**90**). Reductive elimination then provided the (Z)-penem (**89**, R = $CH_2C_6H_4OCH_3$-p) as major product which on Lewis acid mediated deprotection gave BRL 42715 (**89**, R = Na).

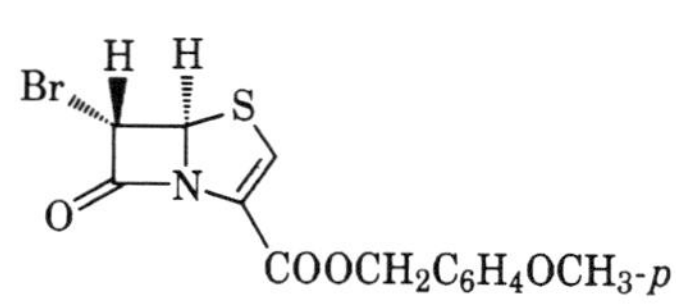

(**88**)

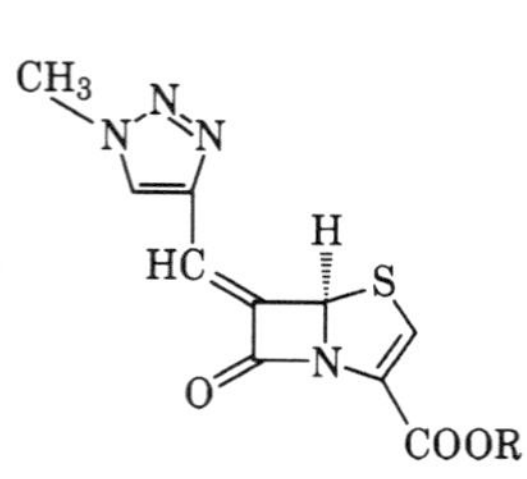

(**89**)

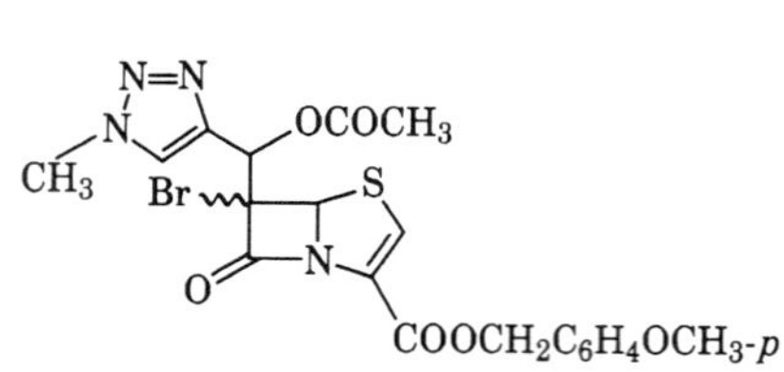

(**90**)

Biological Properties. In marked contrast to the antibacterially inactive penicillanic and cephalosporanic acids, 6-unsubstituted penems such as (**52**, R = CH_3) [*69077-00-5*] exhibit good activity against both gram-positive and gram-negative bacteria (87). The introduction of a 6β-acylamino substituent had an adverse effect on the stability of the penem nucleus, and compounds such as (**50**) were shown to possess only weak antibacterial properties (84,85). Attempts to stabilize 6β-acylaminopenems by the introduction of a 6α-methoxy or 6α-methyl group was only marginally successful in the latter case and the compounds still showed only weak activity (129). A similar reduction in stability and activity was also noted for the 6α-methoxypenem (**91**, R = OCH_3) [*78839-71-1*], whereas the 6α-methylpenem (**91**, R = CH_3) [*73660-04-5*], although considerably more stable, still only exhibited modest antibacterial activity (130). 6,6-Dialkylpenems have been shown to be extremely stable but are devoid of antibacterial reactivity (130).

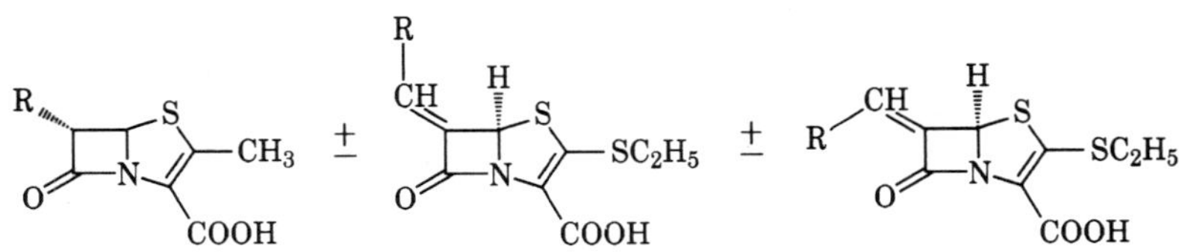

(**91**) (**92**) (**93**)

The introduction of the 1-hydroxyethyl group found in the naturally occurring carbapenems proved to be crucial for providing penems with broad-spectrum antibacterial activity and stability to β-lactamases. As with their carbapenem counterparts, biological activity showed a marked dependence on the relative stereochemistry at the three chiral centers. Evaluation of the four racemic penems (**56**) revealed that the *trans*-8(*R*)-isomer [*76431-47-5*], corresponding to thienamycin (**2**), was 20- to 30-fold more potent than the *trans*-8(*S*)-isomer [*76431-46-4*] whereas the two cis-penems [*80629-72-7*] and [*80629-98-7*] were of intermediate potency (131). Further structural modification at C-8 resulted in adverse effects on antibacterial activity. Removal of the C-8 methyl group resulted in loss of activity especially against β-lactamase producing bacteria (131) whereas an additional methyl group as in the 6-hydroxyisopropylpenem [*79583-53-2*], (**91**, R = $COH(CH_3)_2$), $C_{10}H_{13}NO_4S$, resulted in almost complete loss of activity (132). The increased bulk produced by substitution of the C-8 methyl group was also detrimental, 1-hydroxy-1-propylpenem [*75940-79-3*] (**91**, R = $CHOHC_2H_5$) and 1-hydroxy-2-phenylethylpenem [*75940-91-9*] (**91**, R = $CHOHCH_2C_6H_5$) being devoid of useful activity (132).

The introduction of unsaturation at [6,8] resulted in loss of all antibacterial activity except that against *staphylococcal* species. Both the (*Z*)- (**92**, R = CH_3) [*81463-33-4*] and the (*E*)- (**93**, R = CH_3) [*81463-37-8*] isomers of the 6-ethylidenepenem sodium salt showed potent broad spectrum β-lactamase inhibitory properties and were capable of reducing the minimum inhibitory concentration (MIC) values of sensitive β-lactams, eg, Amoxycillin, when used in combination (133). Unlike the saturated counterparts, the 6-methylenepenems were tolerant to profound changes at C-8. Improved inhibitory properties and synergism resulted from the introduction of a furan ring. However, in the case of this larger C-8 substituent, a preferred geometry for the double bond was identified and the (*Z*)-isomer (**92**, R = 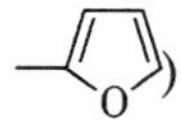) [*93853-68-0*] was more active than the (*E*)-isomer (**93**, R = 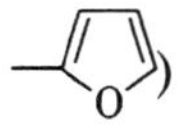) (134). Structure–activity relationships for a wide range of (*Z*)-6-heterocyclylmethylenepenems have been reported (135).

With a few exceptions the 2-substituent can be seen as a means of fine-tuning the activity, pharmacokinetics, and stability properties of the penems. The great number of structural variations (98,124,125,136–144) that can be accommodated without loss of the potent activity conveyed by the 6-[1-(*R*)-hydroxyethyl] group bears witness to the undemanding nature of this region of the molecule. These penems typically exhibit excellent activity against gram-positive bacteria including β-lactamase producing strains. The activity against gram-negative organisms, including those producing β-lactamase, is generally good although of a lower order. Except for compounds possessing a basic amino function in the 2-substituent (98,144) penems are devoid of useful activity against *pseudomonas* species.

Research carried out in a number of laboratories has revealed several members of the penem family having biological properties worthy of further investigation. Some of these compounds are given in Table 2. Extensive studies (135) have also identified BRL 42715 [*102209-75-6*] (**89**, R = Na), $C_{10}H_7N_4O_3SNa$, as a potentially useful β-lactamase inhibitor capable of potentiating the activity of a number of clinically important β-lactam antibiotics against resistant strains (153).

In terms of activity there seems little to prevent some of these compounds finding a place in therapy, especially those such as SCH 29482, SUN 5555, and FCE 25199 which have oral absorption properties. However, as is the case for the carbapenems, some penems are extensively metabolized by human renal dehydropeptidase-I enzyme (144). Although no penem has received approval for clinical use as of this writing, expectations are high that future research and development will change that.

Economic Aspects

Extensive carbapenem and penem antibiotic research has been ongoing since thienamycin was discovered in 1978. However, only the imipenem-cilastatin combination has become a commercial product. Launched in 1985 in the United States as a broad-spectrum hospital product under the name Primaxin, this product had worldwide sales of some $300 million in 1988. Sales were predicted to rise to $345 million for the year ending 1989 (154).

BIBLIOGRAPHY

"Penicillin" in *ECT* 1st ed., Vol. 9, pp. 922–943, by W. E. Brown, Squibb Institute for Medical Research; "Penicillins" in *ECT* 2nd ed., Vol. 14, pp. 652–707, by Kenneth Butler, Chas. Pfizer & Co., Inc.; "Antibiotics, β-Lactams" in *ECT* 3rd ed., Vol. 2, pp. 871–919, by J. R. E. Hoover, and C. H. Nash, Smith Kline & French Laboratories.

1. "Cephalosporins and Penicillins" in E. H. Flynn, ed., *Chemistry and Biology,* Academic Press, New York and London, 1972.
2. J. S. Kahan and co-workers, *J. Antibiot.* **32,** 1 (1979).
3. D. Butterworth, M. Cole, G. Hanscomb, and G. N. Rolinson, *J. Antibiot.* **32,** 287 (1979); J. D. Hood, S. J. Box, and M. S. Verall, *J. Antibiot.* **32,** 295 (1979).
4. S. J. Box, J. D. Hood, and S. R. Spear, *J. Antibiot.* **32,** 1239 (1979).
5. K. Yamamoto and co-workers, *J. Antibiot.* **33,** 796 (1980).
6. M. Nakayama and co-workers, *J. Antibiot.* **34,** 818 (1981).
7. N. Tsuji and co-workers, *J. Antibiot.* **35,** 24 (1982).
8. N. Tsuji and co-workers, *J. Antibiot.* **35,** 536 (1982).
9. W. L. Parker and co-workers, *J. Antibiot.* **35,** 653 (1982).
10. R. Southgate and S. Elson, *Fortschr. Chem. Org. Naturst.* (Progress in the *Chemistry of Organic Natural Products*), Springer-Verlag, New York, **47,** 1 (1985).
11. Eur. Pat. Appl. 38,534 (Apr. 15, 1981), K. Tanaka, N. Tsuji, E. Kondo, and Y. Kawamura (to Shionogi and Co., Ltd.).
12. D. Butterworth, J. D. Hood, and M. S. Verall in A. Mizrahi, ed., *Advances in Biotechnological Processes,* Vol. 1, A. R. Liss, New York, 1982, pp. 251–282.
13. G. Albers-Schönberg and co-workers, "Abstract 228" in *Abstracts of the 16th Interscience Conference on Antimicrobial Agents and Chemotherapy, 1976,* American Society for Microbiology, Washington, D.C., 1976.
14. J. M. Williamson, *CRC Crit. Rev. Biotechnol.* **4,** 111 (1986).
15. J. M. Williamson and co-workers, *J. Biol. Chem.* **260,** 4637 (1985); D. R. Houk, K. Kobayashi, J. M. Williamson, and H. G. Loss, *J. Am. Chem. Soc.* **108,** 5365 (1986).
16. B. W. Bycroft in P. H. Bentley and R. Southgate, eds., *Recent Advances in the Chemistry of β-Lactam Antibiotics,* RSC Special Publication No. 70, London, 1989, pp. 23–32.
17. T. Haneishi and co-workers, *J. Antibiot.* **36,** 1581 (1983).
18. Y. Fukagawa, M. Okabe, T. Yoshioka, and T. Ishikura in A. G. Brown and S. M. Roberts, ed., *Recent Advances in the Chemistry of β-Lactam Antibiotics,* Special Publication No. 52, London, 1985, pp. 163–182.
19. G. Albers-Schönberg and co-workers, *J. Am. Chem. Soc.* **100,** 6491 (1978).
20. A. G. Brown, D. F. Corbett, A. J. Eglington, and T. T. Howarth, *J. Chem. Soc., Chem. Commun.,* 523 (1977); 953 (1977).
21. A. G. Brown, D. F. Corbett, A. J. Eglington, and T. T. Howarth in G. I. Gregory, ed., *Recent Advances in Chemistry of β-Lactam Antibiotics,* RSC Special Publication No. 38, London, 1981, pp. 255–268.
22. A. G. Brown, D. F. Corbett, A. J. Eglington, and T. T. Howarth, *J. Antibiot.* **39,** 2551 (1983).
23. A. G. Brown, D. F. Corbett, A. J. Eglington, and T. T. Howarth, *J. Antibiot.* **32,** 961 (1979).
24. W. J. Leanza and co-workers in ref. 21, pp. 240–254.
25. W. J. Leanza, K. J. Wildonger, T. W. Miller, and B. G. Christensen, *J. Med. Chem.* **22,** 1435 (1979).
26. D. H. Shih and R. W. Ratcliffe, *J. Med. Chem.* **24,** 639 (1981).
27. D. H. Shih, J. Hannah, and B. G. Christensen, *J. Am. Chem. Soc.* **100,** 8004 (1978).
28. D. F. Corbett, *J. Chem. Soc., Chem. Commun.,* 803 (1981).
29. D. F. Corbett, S. Coulton, and R. Southgate, *J. Chem. Soc. Perkin I,* 3011 (1982).
30. K. Yamamoto and co-workers, *Tetrahedron Lett.* **23,** 897 (1982).
31. Y. Fukagawa, K. Kubo, T. Ishikura, and K. Kounu, *J. Antibiot.* **33,** 543 (1980).
32. L. D. Cama and B. G. Christensen, *J. Am. Chem. Soc.* **100,** 8006 (1978).
33. J. H. Bateson and co-workers, *J. Chem. Soc. Perkin I,* 3242 (1981).
34. H. R. Pfaendler, J. R. Gosteli, R. B. Woodward, and G. Rihs, *J. Am. Chem. Soc.* **103,** 4526 (1981).
35. J. H. Bateson and co-workers in ref. 21, pp. 291–313.

36. R. J. Ponsford and R. Southgate, *J. Chem. Soc., Chem. Commun.,* 1085 (1980).
37. J. H. Bateson and co-workers, *J. Chem. Soc., Chem. Commun.,* 1084 (1980).
38. T. N. Salzman, R. W. Ratcliffe, B. G. Christensen, and F. A. Bouffford, *J. Am. Chem. Soc.* **102,** 6161 (1980).
39. I. Shinkai and co-workers, *Tetrahedron Lett.* **23,** 4903 (1982).
40. D. G. Melillo and co-workers, *Tetrahedron Lett.* **22,** 913 (1981).
41. T. Kametani and co-workers, *J. Chem. Soc., Chem. Commun.,* 646 (1989).
42. S. Karady, J. S. Amato, R. A. Reamer, and L. M. Weinstock, *J. Am. Chem. Soc.* **103,** 6765 (1981).
43. M. Miyashita, M. N. Chida, and A. Yoshikoshi, *J. Chem. Soc., Chem. Commun.,* 1354 (1982).
44. N. Ikota, O. Yoshino, and K. Koga, *Chem. Pharm. Bull.* **30,** 1929 (1982).
45. A. Knierzinger and A. Vasella, *J. Chem. Soc., Chem. Commun.,* 9 (1985).
46. H. Maruyama, M. Shiozaki, and T. Hiraoka, *Bull. Chem. Soc. Jap.* **58,** 3264 (1985).
47. M. Shiozaki, N. Ishida, H. Maruyama, and T. Hiraoka, *Tetrahedron* **39,** 2399 (1983).
48. T. Kametani, T. Nagahara, and T. Honda, *J. Org. Chem.* **50,** 2327 (1985).
49. S. T. Hodgson, D. M. Hollinshead, and S. V. Ley, *J. Chem. Soc., Chem. Commun.,* 494 (1984).
50. T. Kametani, *Heterocycles* **17,** 463 (1982).
51. T. Nagahara and T. Kametani, *Heterocycles* **25,** 729 (1987).
52. F. M. Kahan, H. Kropp, J. G. Sundelof, and J. Birnbaum, *J. Antimicrob. Chem.,* Supplement D, **12,** 1 (1983).
53. H. Kropp, J. G. Sundelof, R. H. Hajdu, and F. M. Kahan, *Antimicrob. Agents Chemother.* **22,** 62 (1982).
54. D. H. Shih, F. Baker, L. D. Cama, and B. G. Christensen, *Heterocycles* **21,** 29 (1984).
55. A. Afonso and co-workers in ref. 16, pp. 295–302.
56. D. M. Tschaen and co-workers, *Tetrahedron Lett.* **29,** 2779 (1988).
57. R. Deziel and D. Faureau, *Tetrahedron Lett.* **47,** 5687 (1986).
58. F. Shirai and T. Nakai, *J. Org. Chem.* **52,** 5491 (1987).
59. T. Shibata and Y. Sugimura, *J. Antibiot.* **42,** 374 (1989).
60. J. Haruta and co-workers, *Chem. Pharm. Bull.* **37,** 2338 (1989).
61. T. J. Sowin and A. I. Meyers, *J. Org. Chem.* **53,** 4154 (1988).
62. S. Kobayashi, K. Kamijama, T. Iimori, and M. Ohno, *Tetrahedron Lett.* **25,** 2557 (1984); **29,** 1057 (1988).
63. H. Kaga, S. Kobayashi, and M. Ohno, *Tetrahedron Lett.* **30,** 113 (1989).
64. F. Shirai and T. Nakai, *Chem. Lett.,* 445 (1989).
65. M. Ihara, M. Takahashi, M. Fukumoto, and T. Kametani, *Heterocycles* **27,** 327 (1988).
66. T. N. Salzmann and co-workers in ref. 16, pp. 171–189.
67. S. Schmitt, T. N. Salzmann, D. M. Shih, and B. G. Christensen, *J. Antibiot.* **41,** 780 (1988).
68. C. Kim, B. Y. Luh, P. F. Misco, and M. J. Hitchcock, *J. Med. Chem.* **32,** 601 (1989).
69. P. Davey and co-workers, eds., *J. Antimicrob. Chemother.* **24,** Supplement A (1989).
70. Y. Nagao and co-workers, *J. Chem. Soc., Chem. Commun.* 821 (1989).
71. R. Bayles and co-workers in ref. 16, pp. 190–195.
72. H. C. Neu, *Am. J. Med.* **78** (Supplement 6A), 33 (1985).
73. M. J. Basker, R. J. Boon, and P. A. Hunter, *J. Antibiot.* **33,** 878 (1980).
74. J. R. Knowles, *Acc. Chem. Res.* **18,** 97 (1985).
75. M. Sakamoto and co-workers, *J. Antibiot.* **32,** 272 (1979).
76. F. Kobayashi and co-workers, *Antimicrob. Agents Chemother.* **21,** 536 (1982).
77. Y. Kimura and co-workers, *J. Antibiot.* **35,** 32 (1982).
78. M. J. Basker and co-workers, *J. Antibiot.* **34,** 1224 (1981).
79. B. G. Spratt, V. Jobanputra, and W. Zimmerman, *Antimicrob. Agents Chemother.* **12,** 406 (1977).

80. H. C. Neu and P. Labthavikul, *Antimicrob. Agents Chemother.* **21,** 180 (1982).
81. S. R. Norrby, B. Björnegard, F. Ferber, and K. H. Jones, *J. Antimicrob. Chemother.* (Suppl. D), **12,** 109 (1983).
82. J. Birnbaum, F. M. Kahan, H. Kropp, and J. S. MacDonald in H. C. Neu, *Am. J. Med.* **78** (Suppl. 6A), 33 (1985).
83. S. R. Norrby and co-workers, *Antimicrob. Agents Chemother.* **23,** 300 (1983).
84. R. B. Woodward in J. Elks, ed., *Recent Advances in the Chemistry of β-Lactam Antibiotics,* The Chemical Society, 1977, pp. 167–180.
85. I. Ernest and co-workers, *J. Am. Chem. Soc.* **100,** 8214 (1978).
86. K. Claues, D. Grimm, and G. Prossel, *Justus Liebigs Ann. Chem.,* 539 (1974).
87. M. Lang and co-workers, *Am. Chem. Soc.* **101,** 6296 (1979).
88. M. Lang, K. Prasad, J. Gosteli, and R. B. Woodward, *Helv. Chim. Acta* **63,** 1093 (1980).
89. Brit. Pat. 2,042,508A (1980), E. G. Brain (to Beecham Group, Ltd.).
90. Eur. Pat. 0,000,258 (1978), J. Gosteli, I. Ernest, M. Lang, and R. B. Woodward.
91. S. W. McCombie, *Tetrahedron Lett.* **22,** 3489 (1981).
92. U.S. Pat. 4,272,437 (1981), M. Ménard and A. Martel (to Bristol-Myers).
93. H. R. Pfaendler, J. Gosteli, and R. B. Woodward, *J. Am. Chem. Soc.* **101,** 6306 (1979).
94. F. DiNinno, T. R. Beattie, and B. G. Christensen, *J. Org. Chem.* **42,** 2960 (1977).
95. S. Karady, J. S. Amato, R. A. Reamer, and L. M. Weinstock, *J. Am. Chem. Soc.* **103,** 6765 (1981).
96. V. M. Girijavallabhan and co-workers, *Tetrahedron Lett.* **22,** 3485 (1981).
97. A. Yoshida and co-workers, *Chem. Pharm. Bull.* **29,** 2899 (1981).
98. T. Hayashi and co-workers, *Chem. Pharm. Bull.* **29,** 3158 (1981).
99. G. Franceschi and co-workers, *J. Antibiot.* **36,** 938 (1983).
100. M. Alpegiani and co-workers, *J. Am. Chem. Soc.* **107,** 6398 (1985).
101. M. Foglio and co-workers, *Heterocycles* **20,** 1491 (1983).
102. M. Foglio, G. Franceschi, C. Scarafile, and P. Zini, *Heterocycles* **16,** 1919 (1981).
103. M. Foglio, C. Battistini, F. Zarini, and G. Franceschi, *Heterocycles* **19,** 485 (1982).
104. M. Alpegiani and co-workers, *Tetrahedron Lett.* **24,** 1627 (1983).
105. N. F. Osborne and co-workers, *J. Chem. Soc., Chem. Commun.,* 371 (1989).
106. H. Yanagisawa, A. Ando, M. Shiozaki, and T. Hiraoki, *Tetrahedron Lett.* **24,** 1037 (1983).
107. S. Chackalamannil and co-workers, *J. Org. Chem.* **53,** 450 (1988).
108. G. I. Georg, J. Kant, and H. S. Gill, *J. Am. Chem. Soc.* **109,** 1129 (1987).
109. C. Battistini, C. Scarafile, M. Foglio, and G. Franceschi, *Tetrahedron Lett.* **25,** 2395 (1984).
110. A. Afonso and co-workers, *J. Am. Chem. Soc.* **104,** 6138 (1982).
111. A. Yoshida and co-workers, *Chem. Pharm. Bull.* **31,** 768 (1983).
112. E. Perrone and co-workers, *Tetrahedron Lett.* **25,** 2399 (1984).
113. A. J. Baker and M. J. Jenkins in ref. 16, pp. 259–272.
114. V. M. Girijavallabhan, A. K. Ganguly, P. Pinto, and R. Versace, *J. Chem. Soc., Chem. Commun.,* 908 (1983).
115. W. J. Leanza and co-workers, *Tetrahedron* **39,** 2505 (1983).
116. S. Hanessian, A. Bedeschi, C. Battistini, and N. Mongilli, *J. Am. Chem. Soc.* **107,** 1438 (1985).
117. N. Daniels, G. Johnson, and B. C. Ross, *J. Chem. Soc., Chem. Commun.,* 1006 (1983).
118. M. D. Cooke, K. W. Moore, B. C. Ross, and S. E. Turner, *J. Chem. Soc., Chem. Commun.,* 1005 (1983).
119. H. H. Wasserman and W. T. Han, *J. Am. Chem. Soc.* **107,** 1444 (1985).
120. J. Marchand-Brynaert, L. Ghosez, and E. Cossement, *Tetrahedron Lett.* **21,** 3085 (1980).
121. E. Perrone and co-workers, *Tetrahedron Lett.* **24,** 1631 (1983).
122. E. Perrone and co-workers, *J. Org. Chem.* **51,** 3413 (1986).
123. S. Oida and co-workers, *Chem. Pharm. Bull.* **28,** 3258 (1980).

124. G. Franceschi and co-workers, *J. Antibiot.* **37,** 685 (1984).
125. E. Perrone and co-workers, *J. Antibiot.* **39,** 1351 (1986).
126. F. DiNinno, D. A. Muthard, R. W. Ratcliffe, and B. G. Christensen, *Tetrahedron Lett.* **23,** 3535 (1982).
127. A. J. Baker, M. R. Teall, and G. Johnson, *Tetrahedron Lett.* **28,** 2283 (1987).
128. Y. Ueda and co-workers, *Can. J. Chem.* **60,** 904 (1982).
129. J. Banville, P. La Pointe, B. Belleau, and M. Ménard, *Can. J. Chem.* **66,** 1390 (1988).
130. J. Gosteli, W. Holick, M. Land, and R. B. Woodward in ref. 21, pp. 359–367.
131. S. W. McCombie and co-workers, *21st Interscience Conference on Antimicrobial Agents and Chemotherapy,* Abstract 830, Chicago, 1981, American Society for Microbiology, Washington, D.C., 1981.
132. Brit. Pat. Appl. 2,042,515 A (Sept. 24, 1980), M. Ménard and A. Martel (to Bristol-Myers).
133. M. J. Basker and N. F. Osborne, *J. Antibiot.* **43,** 70 (1990).
134. N. J. P. Broom, K. Coleman, P. A. Hunter, and N. F. Osborne, *J. Antibiot.* **43,** 76 (1990).
135. N. J. P. Broom and co-workers in ref. 16, pp. 247–258.
136. V. M. Girijavallabhan and co-workers, *J. Antibiot.* **39,** 1182 (1986).
137. V. M. Girijavallabhan and co-workers, *J. Antibiot.* **39,** 1187 (1985).
138. G. Emmer and co-workers, *J. Antibiot.* **38,** 1371 (1985).
139. M. Lang and co-workers, *J. Antibiot.* **40,** 217 (1987).
140. H. G. Capraro and co-workers, *J. Antibiot.* **41,** 759 (1988).
141. M. Ishiguro and co-workers, *J. Antibiot.* **41,** 1685 (1988).
142. M. Alpegiani, E. Perrone, and G. Franceschi, *Heterocycles* **27,** 49 (1988).
143. M. Alpegiani and co-workers, *Heterocycles* **27,** 1329 (1988).
144. O. Zak and co-workers, *J. Clin. Pharmacol.* **28,** 128 (1988).
145. I. Phillips, R. Wise, and H. C. Neu, eds., *J. Antimicrob. Chemother.* **9** (Suppl. C) (1982).
146. R. Wise and I. Phillips, eds., *J. Antimicrob. Chemother.* **15** (Suppl. C) (1985).
147. D. Reeves, D. Speller, R. Spencer, and P. J. Daly, eds., *J. Antimicrob. Chemother.* **23** (Suppl. C) (1989).
148. G. Seibert and co-workers, *J. Antibiot.* **40,** 660 (1987); N. Kesel and co-workers, *J. Antibiot.* **40,** 1184 (1987).
149. T. Nishino and co-workers, *J. Antibiot.* **42,** 977 (1989).
150. *26th Interscience Conference on Antimicrobial Agents and Chemotherapy,* Abstracts 630–645, New Orleans, 1986, American Society for Microbiology, Washington, D.C., 1986.
151. T. Gootz and co-workers, *Antimicrob. Agents Chemother.* **33,** 1160 (1989).
152. *29th Interscience Conference on Antimicrobial Agents and Chemotherapy,* Abstracts 85–87, Houston, 1989, American Society for Microbiology, Washington, D.C., 1989.
153. K. Coleman, D. R. J. Griffin, J. W. J. Page, and P. A. Upshon, *Antimicrob. Agents Chemother.* **33,** 1580 (1989).
154. *Scrip Review Issue,* 1989, p. 12.

General Reference

R. W. Ratcliffe and G. Albers-Schönberg, "The Chemistry of Thienamycin and Other Carbapenem Antibitotics" and I. Ernest, "The Penems" in R. B. Morin and M. Gorman, eds., *Chemistry and Biology of β-Lactam Antibiotics,* Academic Press, New York, 1982, Chapts. 4 and 5.

ROBERT SOUTHGATE
NEAL F. OSBORNE
SmithKline Beecham Pharmaceuticals (UK)

CEPHALOSPORINS

The cephalosporins, a subgroup of *β*-lactam antibiotics, consist of a 4-membered lactam ring fused through the nitrogen and the adjacent tetrahedral carbon atom to a second heterocycle forming a 6-membered dihydrothiazine ring. Other structural features common to all the cephalosporins are a carboxyl group on the dihydrothiazine ring on the carbon next to the ring nitrogen and a functionalized amino group on C-7, the carbon of the *β*-lactam ring opposite the nitrogen. These features are evidenced in 7-aminocephalosporanic acid [*957-68-6*] (7-ACA), $C_{10}H_{12}N_2O_5S$ (**1**). Cephalosporins, like all *β*-lactam antibiotics, exert their antibacterial effect by interfering with the synthesis of the bacterial cell wall. These antibiotics tend to be "irreversible" inhibitors of cell wall biosynthesis and they are usually bactericidal at concentrations close to their bacteriostatic levels. Cephalosporins are widely used for treating bacterial infections. They are highly effective antibiotics and have low toxicity.

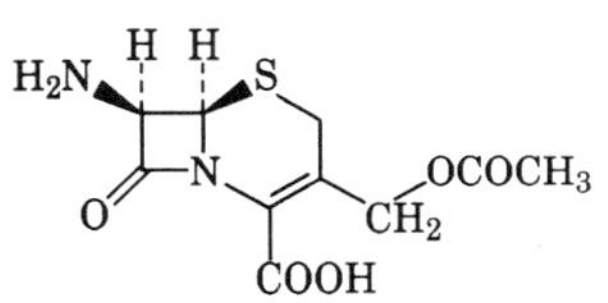

(**1**)

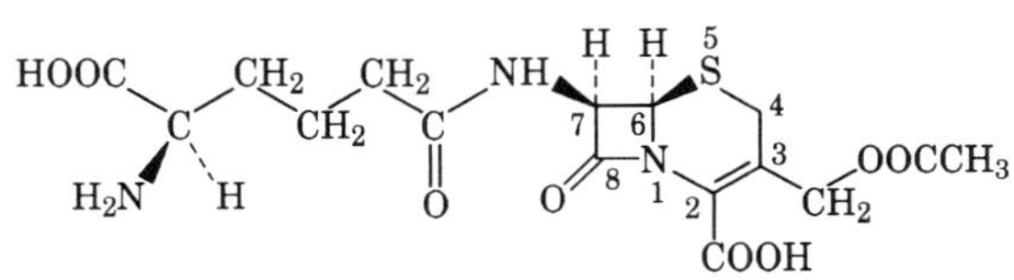

(**2**)

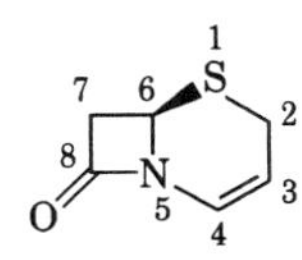

(**3**)

Nomenclature and Stereochemistry

Chemical Abstracts indexes cephalosporins as 5-thia-1-azabicyclo[4.2.0]oct-2-enes. Using this system then, cephalosporin C [*61-24-5*], $C_{16}H_{21}N_3O_8S$, (**2**) is 3-[(acetyloxy)methyl]-7-[(5-amino-5-carboxy-1-oxopentyl)amino]-8-oxo-5-thia-1-azabicyclo[4.2.0]oct-2-ene-2-carboxylic acid. Such names are too cumbersome for general use. One simplification defines the ring system of the cephalosporins as cepham (**3**) (1). Hence cephalosporins become 3-acetoxymethyl-7-acylamino-3-cephem-4-carboxylic acids. In this widely used nomenclature system, the numbering, as shown in structure (**3**), differs from that used by *Chemical Abstracts* shown in structure (**2**). In both systems, however, the important C-7, C-6, and C-3 positions remain unchanged. The cephem numbering system (**3**) is used herein.

The largest group of cephalosporins are derived from the cephem nucleus containing a 3-(substituted)methyl group. It is sometimes useful to designate the methyl group as C-3′. Thus, for example, a 3-acetoxymethyl, present in structures (**1**) and (**2**), becomes the 3′-acetoxy functionality. Another simplification in nomenclature is the use of the term cephalosporanic acid (2) as the parent name. Although convenient, this system is somewhat restrictive for cephalosporins in which the 3-acetoxymethyl grouping has been replaced. The clinically important cephalosporins all have a generic name which, by convention, starts with ceph- or cef-.

Structures of the naturally occurring cephalosporins, cephamycins, and the 7-formamido cephalosporins are as indicated in Tables 1 and 2. The amide bearing carbon, C-7, has the *R*-configuration; the nonplanar fused rings are folded along the C-6 to ring nitrogen bond; and the configuration of the fused rings is such that

Table 1. Naturally Occurring Cephalosporins

Name	CAS Registry Number	Molecular formula	R	R′	Ref.
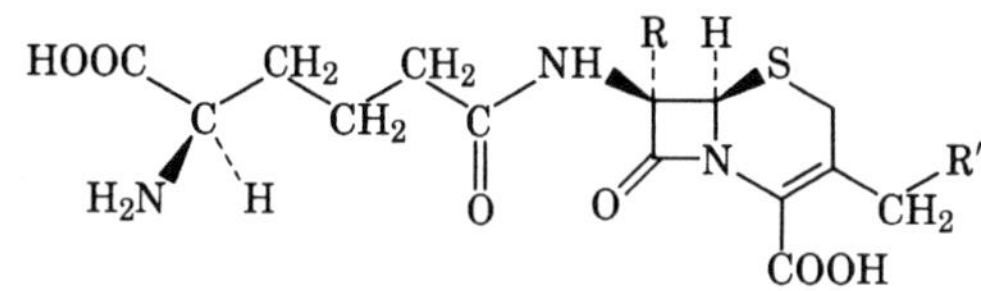					
deacetoxycephalosporin C	[*26924-74-3*]	$C_{14}H_{19}N_3O_8S$	—H	—H	3
deacetylcephalosporin C	[*1476-46-6*]	$C_{14}H_{19}N_3O_7S$	—H	—OH	3
cephalosporin C	[*61-24-5*]	$C_{16}H_{12}N_2O_5S$	—H	$—OCOCH_3$	4–6
O-carbamoyldeacetylcephalosporin C	[*32178-83-9*]	$C_{15}H_{19}N_4O_8S$	—H	$—OCONH_2$	7
3′-methylthiodeacetoxycephalosporin C (F_1)	[*60890-86-0*]	$C_{15}H_{21}N_3O_6S_2$	—H	$—SCH_3$	8
3′-sulfothiodeacetoxycephalosporin C (F_2)	[*65580-83-8*]	$C_{14}H_{19}N_3O_9S_3$	—H	$—SSO_2OH$	9
C43-219	[*60857-71-8*]	$C_{19}H_{28}N_4O_8S_2$	—H	$—SC(CH_3)_2CH(NH_2)CO_2H$	10
7α-methoxycephalosporin C	[*32178-82-8*]	$C_{17}H_{23}N_3O_9S$	$—OCH_3$	$—OCOCH_3$	7
cephamycin C	[*34279-55-1*]	$C_{16}H_{22}N_4O_9S$	$—OCH_3$	$—OCONH_2$	7,11
cephamycin A	[*34279-78-2*]	$C_{25}H_{29}N_3O_{14}S_2$	$—OCH_3$	$—OCOC(OCH_3){=}CHC_6H_4OSO_2OH\text{-}p$	11
cephamycin B	[*34729-77-1*]	$C_{25}H_{29}N_3O_{11}S$	$—OCH_3$	$—OCOC(OCH_3){=}CHC_6H_4OH\text{-}p$	11
Takeda C2801X	[*62851-50-7*]	$C_{25}H_{29}N_3O_{12}S$	$—OCH_3$	$—OCOC(OCH_3){=}CHC_6H_3(OH)_2\text{-}m,p$	12
SF-1623	[*56487-86-6*]	$C_{15}H_{21}N_3O_{10}S_3$	$—OCH_3$	$—SSO_2OH$	13
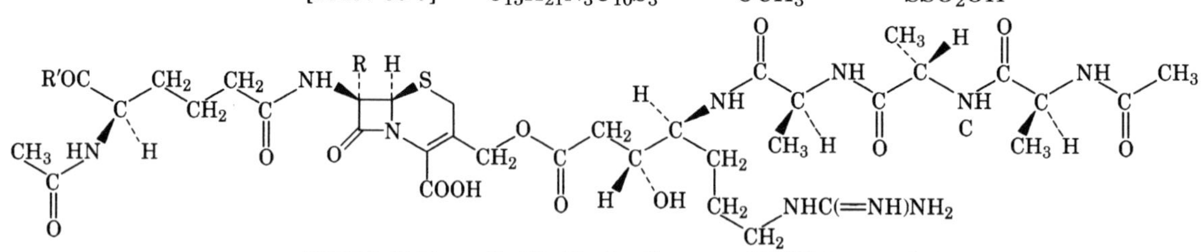					
SQ 28,516	[*92121-46-5*]	$C_{36}H_{55}N_{11}O_{15}S_1$	—NHCHO	—OH	14
SQ 28,517	[*92131-67-4*]	$C_{36}H_{56}N_{12}O_{14}S_1$	—NHCHO	$—NH_2$	14

Table 2. Naturally Occurring Cephabacin Cephalosporins[a]

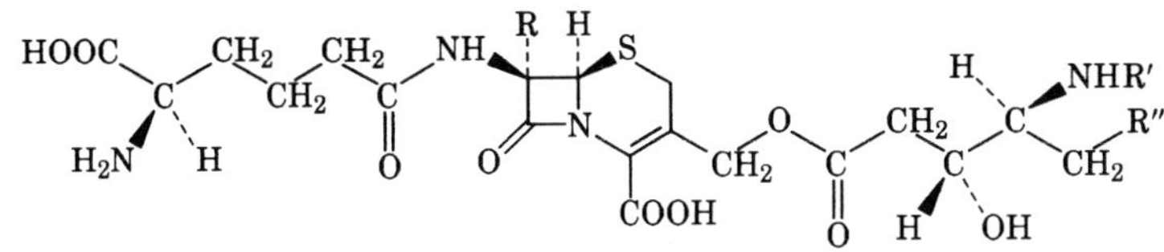

Name	CAS Registry Number	Molecular formula	R	R′	R″	Refs.
cephabacin F_1	*[95041-98-8]*	$C_{26}H_{41}N_9O_{11}S$	—NHCHO	-L-Ala	—$CH_2CH_2NHC(=NH)NH_2$	15
or chitinovorin A						16
cephabacin F_2	*[95722-76-2]*	$C_{29}H_{46}N_{10}O_{12}S$	—NHCHO	-L-Ala-L-Ala	—$CH_2CH_2NHC(=NH)NH_2$	15
or chitinovorin B						16
cephabacin F_3	*[95042-00-5]*	$C_{32}H_{51}N_{11}O_{13}S$	—NHCHO	-L-Ala-L-Ala-L-Ala	—$CH_2CH_2NHC(=NH)NH_2$	15
cephabacin F_4	*[95042-01-6]*	$C_{26}H_{41}N_9O_{12}S$	—NHCHO	-L-Ser	—$CH_2CH_2NHC(=NH)NH_2$	15
cephabacin F_5	*[95042-02-7]*	$C_{29}H_{46}N_{10}O_{14}S$	—NHCHO	-L-Ser-L-Ser	—$CH_2CH_2NHC(=NH)NH_2$	15
cephabacin F_6	*[95042-03-8]*	$C_{32}H_{51}N_{11}O_{15}S$	—NHCHO	-L-Ser-L-Ser-L-Ala	—$CH_2CH_2NHC(=NH)NH_2$	15
cephabacin F_7	*[95230-87-8]*	$C_{26}H_{41}N_7O_{12}S$	—NHCHO	-L-Ser	—$CH_2CH_2CH_2NH_2$	15
cephabacin F_8	*[95230-88-9]*	$C_{29}H_{46}N_8O_{14}S$	—NHCHO	-L-Ser-L-Ser	—$CH_2CH_2CH_2NH_2$	15
cephabacin F_9	*[95230-89-0]*	$C_{32}H_{51}N_9O_{15}S$	—NHCHO	-L-Ser-L-Ser-L-Ala	—$CH_2CH_2CH_2NH_2$	15
cephabacin H_1	*[95042-07-2]*	$C_{25}H_{40}N_8O_{10}S$	—H	-L-Ala	—$CH_2CH_2NHC(=NH)NH_2$	15
cephabacin H_2	*[95042-08-3]*	$C_{28}H_{45}N_9O_{11}S$	—H	-L-Ala-L-Ala	—$CH_2CH_2NHC(=NH)NH_2$	15
cephabacin H_3	*[95042-09-4]*	$C_{31}H_{50}N_{10}O_{12}S$	—H	-L-Ala-L-Ala-L-Ala	—$CH_2CH_2NHC(=NH)NH_2$	15
cephabacin H_4	*[95042-10-7]*	$C_{25}H_{40}N_8O_{11}S$	—H	-L-Ser	—$CH_2CH_2NHC(=NH)NH_2$	15
cephabacin H_5	*[95042-11-8]*	$C_{28}H_{45}N_9O_{13}S$	—H	-L-Ser-L-Ser	—$CH_2CH_2NHC(=NH)NH_2$	15
cephabacin H_6	*[95042-12-9]*	$C_{31}H_{50}N_{10}O_{14}S$	—H	-L-Ser-L-Ser-L-Ala	—$CH_2CH_2NHC(=NH)NH_2$	15
cephabacin M_1	*[99332-96-4]*	$C_{31}H_{50}N_8O_{13}S$	—OCH_3	-L-Val-L-Orn	—$CONH_2$	17
cephabacin M_2	*[99313-76-5]*	$C_{34}H_{55}N_9O_{15}S$	—OCH_3	-L-Val-L-Orn-L-Ser	—$CONH_2$	17
cephabacin M_3	*[99313-75-4]*	$C_{37}H_{60}N_{10}O_{16}S$	—OCH_3	-L-Val-L-Orn-L-Ser-L-Ala	—$CONH_2$	17
cephabacin M_4	*[99313-74-3]*	$C_{41}H_{69}N_{11}O_{15}S$	—OCH_3	-L-Val-L-Orn-L-Val-L-Orn	—$CONH_2$	17
cephabacin M_5	*[99332-95-3]*	$C_{44}H_{74}N_{12}O_{17}S$	—OCH_3	-L-Val-L-Orn-L-Val-L-Orn-L-Ser	—$CONH_2$	17
cephabacin M_6	*[99313-73-2]*	$C_{47}H_{79}N_{13}O_{18}S$	—OCH_3	-L-Val-L-Orn-L-Val-L-Orn-L-Ser-L-Ala	—$CONH_2$	17

[a]Structure as shown where R, R′, and R″ are as indicated.

the hydrogen on C-6 is cis to the hydrogen (or the methoxyl or formamido moiety) on C-7. Thus the absolute stereochemistry is 6*R*:7*R*. The stereochemistry of substituents attached to the rings is designated using α and β. For cephalosporins, the hydrogens at C-7 and C-6 are α, the acylamino group at C-7 is β; in the cephamycins, the C-7 methoxyl is α. These stereochemical relationships and the absolute stereochemistry are also essential for antibacterial activity in the synthetic and semisynthetic cephalosporins. Antibacterial activity is lost whenever either or both of the asymmetric centers are inverted.

Discovery and Historical Development

In 1945 it was noted that a strain of *Cephalosporium acremonium* produced antibiotic material that was active against gram-negative as well as gram-positive organisms (18). The main hydrophilic component, originally called cephalosporin N, and active against both gram-negative and gram-positive organisms was found, in fact, to be a penicillin and was renamed penicillin N. A second hydrophilic antibiotic, present only in minute amounts in the original fermentation mixture and stable to acidic hydrolysis, was named cephalosporin C (**2**) (19). Chemical studies demonstrated that, like penicillin N, cephalosporin C contained a fused β-lactam ring system and a D-α-aminoadipamido substituent. Additional experiments led to the correct structure which was confirmed by x-ray crystallographic studies (4–6,20).

All cephalosporins found in nature (Tables 1 and 2) have the D-α-aminoadipic acid 7-acyl side chain (21). All of these compounds can be classified as having rather low specific activity. A substantial amount of the early work in the cephalosporin area was unsuccessfully directed toward replacing the aminoadipic acid side chain or modifying it appropriately by fermentation or enzymatic processes (6,22). A milestone in the development of cephalosporins occurred in 1960 with the discovery of a practical chemical process to remove the side chain to afford 7-ACA (**1**) (1). Several related processes were subsequently developed (22,23). The ready availability of 7-ACA opened the way to thousands of new semisynthetic cephalosporins. The cephalosporin structure offers more opportunities for chemical modification than does that of penicillins. There are two side chains that especially lend themselves to chemical manipulation: the 7-acylamino and 3-acetoxymethyl substituents.

The cephamycins, described almost simultaneously by two different groups (7,11) in 1971, are produced by the *Streptomyces* species and have the same basic cephalosporin ring system (Table 1) (24). The natural products have the 7β-D-α-aminoadipamido substituent and a methoxy substituent instead of a hydrogen in the 7α-position. These cephamycins have only weak antibacterial activity, but the methoxy group imparts good β-lactamase stability to these compounds. Semisynthetic analogues (25) have been obtained either by chemically introducing a methoxy group at the 7α-position of the cephalosporin ring system or by manipulating the 3- and 7β-substituents of the natural cephamycins (26,27). Some 7α-methoxylated cephalosporins have also been reported which have amino acid or oligopeptides at the C-3 position (Table 2) (17). The compounds are designated herein as cephabacin M (from *ceph*em *a*ntibiotics of *bac*terial orig*in* with the M designating a 7-methoxyl substituent).

In 1984 a new class of naturally occurring cephalosporins containing a 7α-formamido group, was isolated (14–16). These compounds (Table 2) called chitinovorins (16) or cephabacin F (15) where F designates a formamido group in the 7-position, also have amino acids or oligopeptides at the C-3 position. An analogous 7α-hydrogen series, named cephabacin H, containing the same oligopeptides at C-3 has also been isolated (15). The derivatives all have weak antibacterial activity. The 7α-formamido group imparts good β-lactamase stability (15) as is the case for the cephamycins. Chemical manipulation afforded a series of semisynthetic analogues having excellent broad-spectrum activity (28–30).

Biogenesis

The biosynthesis of cephalosporins and penicillins (Fig. 1) both start from the amino acids and proceed via δ-(L-α-aminoadipyl)-L-cysteinyl-D-valine [*32467-88-2*] (LLD-ACV) (**9**), $C_{14}H_{25}N_3O_6S$, (31–34), often referred to as the Arnstein tripeptide (35). Because LLD-ACV is not transported into intact cells, a cell-free system was required to determine that this intermediate is the precursor of the penicillins (36). The cell-free system, obtained fron *C. acremonium,* converts LLD-ACV into isopenicillin N [*525-94-0*] $C_{14}H_{21}N_3O_6S$ (**10**). Isopenicillin N (IPN) synthetase, the enzyme which catalyses this conversion, requires oxygen, Fe^{2+}, a reducing agent such as ascorbate, and a thiol group such as that of dithiothreitol for high activity (37,38). Isopenicillin N synthetase is present in *P. chrysogenum* and in species of *Streptomyces* as well as in *C. acremonium* (39,40). In the *Cephalosporium* species and the *Streptomyces* species, an epimerase is present that converts (**10**) into penicillin N [*58678-43-6*] (**11**), $C_{14}H_{21}N_3O_6S$, (41,42) which then undergoes a ring expansion to deacetoxycephalosporin C (**4**) (43–46). The ring expansion enzyme (REX) from *C. acremonium* appears to be bifunctional (47); it also catalyzes (48,49) the subsequent hydroxylation of (**4**) to deacetylcephalosporin C (**5**), itself a precursor of cephalosporin C (**2**) (50). The introduction of the methoxyl group is also a two-step process involving molecular oxygen (51,52). Using cephalosporin C (**2**) or the corresponding carbamoyl derivative (**6**) as a substrate, another dioxygenase catalyzes the incorporation of a 7α-hydroxy function (53). The resulting 7α-hydroxycephalosporin is then methylated using *S*-adenosylmethionine to form the corresponding 7α-methoxycephalosporin (53). The details of these steps are discussed in depth in the literature (21,39,42,48,54,55). Rapid advances in this area have been made possible by the successful cloning and expression of isopenicillin N synthetase (IPNS) and ring expansion-hydroxylase (REX) (39,48,56,57).

Physical Properties

Most cephalosporin antibiotics are white, off-white, tan, or pale yellow solids that are usually amorphous, but can sometimes be obtained crystalline. The cephalosporins do not usually have sharp melting points, but rather decompose upon

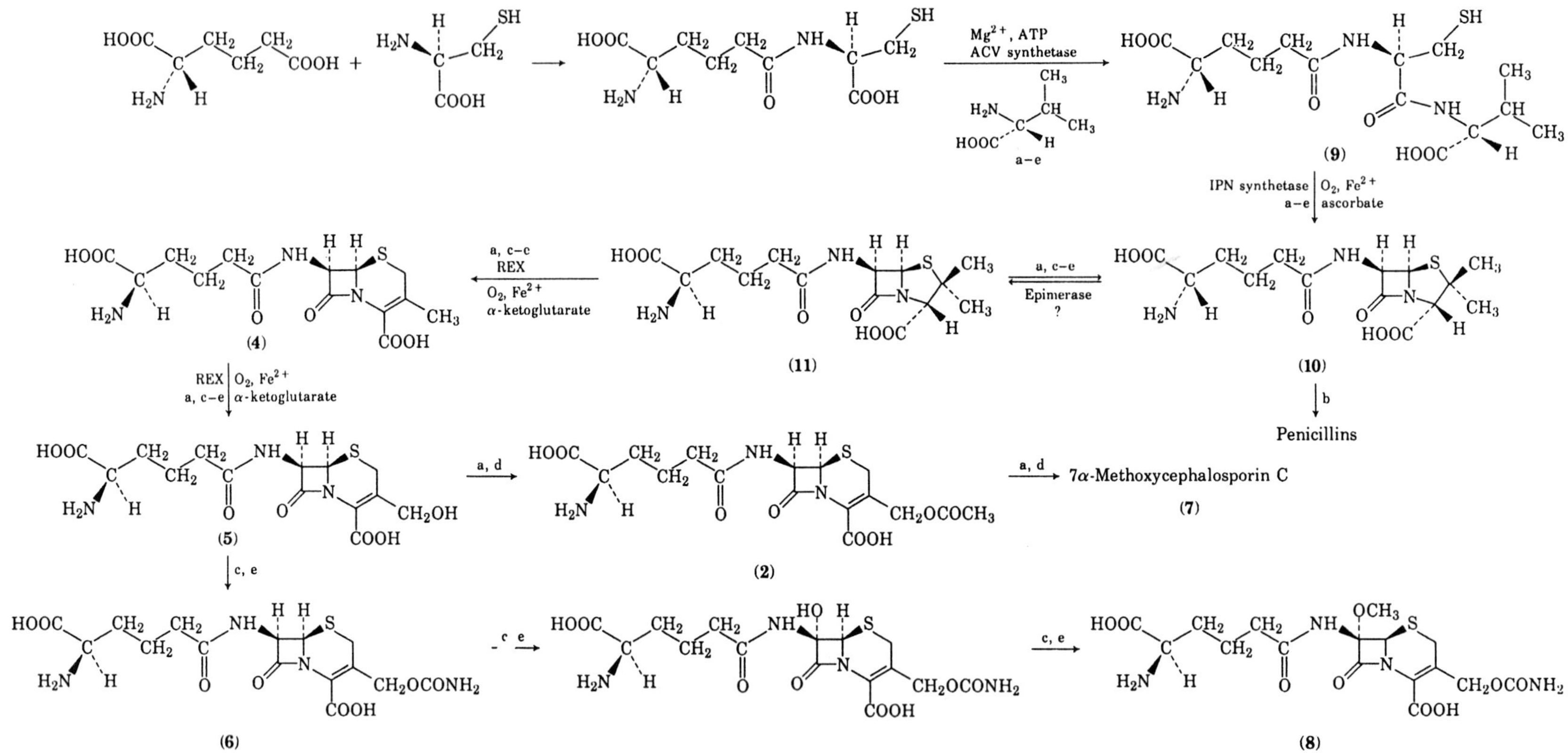

Fig. 1. Biosynthesis of cephalosporins and cephamycins. a, *Cephalosporium acremonium;* b, *Penicillium chrysogenum;* c, *Streptomyces clavuligerus;* d, *Streptomyces lipmanii;* e, *Streptomyces wadayamensis;* REX is a ring expansion enzyme (deacetoxycephalosporin C synthethase).

heating at elevated temperatures. The acid strength, pK_a, of the carboxyl group on the dihydrothiazine ring depends on environment (58). Representative pK_a values are given in Table 3, as are other physical properties.

Table 3. Properties of Cephalosporins[a]

Name	pK_a[b]	β-Lactam ir stretching frequency, cm^{-1}[c]	Uv absorption, λ max, nm (ε, $cm^{-1}M^{-1}$)[d]	Nmr absorption, ppm[e]		
				H-7[f]	H-6[g]	$J_{6,7}$, Hz[h]
7-ACA	1.75, 4.63	1806	261(8500)	5.53d 4.83d	5.13d	4.5
cephalosporin C	<2.6, 3.1, 9.8	1780	260(8900)	5.66d	5.15d	4.7
cephalothin	5.0	1760	265(9000)	5.70d	5.14d	4.5
cephalexin	5.2, 7.3	1775	260(7750)	6.10d	5.45d	4.2
cephamycin C	4.2, 5.6, 10.4[i]	1770	264(6900) 242(5700)		5.19s	

[a]Refs. 7, 11, 58–62, 64, 65.
[b]Values for aqueous solutions unless otherwise noted, either by direct determination or by extrapolation from mixed solvents.
[c]Nujol mull.
[d]In aqueous solution.
[e]In D_2O relative to external standard tetramethyl silane (TMS); s = singlet, d = doublet.
[f]Range 5.23–6.21.
[g]Range 4.24–5.46.
[h]$J_{6,7}(cis)$ = 4–5 Hz, $J_{6,7}(trans)$ = 1.5–2 Hz, $J_{7\text{-}NH}$ = 8–11 Hz.
[i]In 66% DMF.

One of the distinguishing physical characteristics of the cephalosporins is the infrared stretching frequency of the β-lactam carbonyl. This absorption occurs at higher frequencies (1770–1815 cm^{-1}) than those of either normal secondary amides (1504–1695 cm^{-1}) or ester carbonyl groups (1720–1780 cm^{-1}). The nuclear magnetic resonance signals of the protons attached to the β-lactam ring provide information about ring integrity, the attachments to the ring, and the relative stereochemistry of substituents in these positions. These protons form an ABX system with the vicinal proton on the side-chain amide nitrogen (Table 3) (58–62). The cephalosporins display uv absorption maxima at about 260 nm, which is at a longer wavelength than expected for a normal α,β-unsaturated amide, usually about 220–230 nm. This shift may result from the presence of low lying sulfur atom d-orbitals (63).

Chemical Properties

Much of the chemical reactivity of the β-lactam antibiotics is associated with the β-lactam moiety. The geometry and the accompanying increased ring strain results in very little, if any, amide-resonance stabilization leading to a marked increase in chemical reactivity when compared to a normal amide. In fact, in many instances the reactivity of the lactam carbonyl is analogous to that of a carboxylic acid anhydride. Fused β-lactam antibiotics are readily attacked by nucleophiles with resultant ring opening and loss of biological activity. The cephalosporins are more resistant to ring opening than the penicillins. For example, although alcohols readily attack the penicillin β-lactam, cephalosporins are sufficiently stable to permit the use of methanol as a recrystallization solvent.

Cephalosporins are attacked by nucleophiles such as alkoxide or hydroxylamine, but because of the nature of the dihydrothiazine ring, products equivalent to the penicilloic or penilloic acids derived from the penicillins under similar conditions, are not obtained. Rather, studies indicate involvement of the 3′-substituent as a leaving group (66). Thus 3-deacetoxycephalosporins are found to be more stable than cephalosporins having 3′-substituents that can be readily lost. The 3′-substituents exert an electronic influence on the chemical reactivity of the β-lactam carbonyl group (67,68) which can be correlated with antibiotic activity within a series (69,70) and it has been claimed that cephalosporin antibacterial activity is related to the ease of C-3′ group expulsion (71). The deacetoxy analogues generally show lower levels of activity *in vitro* than their 3′-functionalized analogues.

It has also been proposed that reactivity is enhanced because nucleophilic attack on the β-lactam carbonyl is concerted with expulsion of the leaving group (69). However, evidence is now available which strongly suggests that in alkaline hydrolysis and alcoholysis fission of the cephalosporin β-lactam C–N bond occurs before that of the C-3′ leaving group (71–73). This would seem to be the case for interaction of cephalosporins with β-lactamase enzymes (72,74). Thus cephalosporin reactivity is probably best represented by the sequence outlined in Figure 2. In the case where the attacking nucleophile is part of the active site of an enzyme (serine hydroxyl), the acyl–enzyme complex may break down before expulsion of the leaving group or may expel the leaving group to afford a new acyl–enzyme complex (72). The chemistry of cephalosporins, as utilized in synthetic processes and manufacturing, such as deacylation and acylation of the C-7 amino, esterification of the carboxyl, modification at C-3, and introduction of functionality into

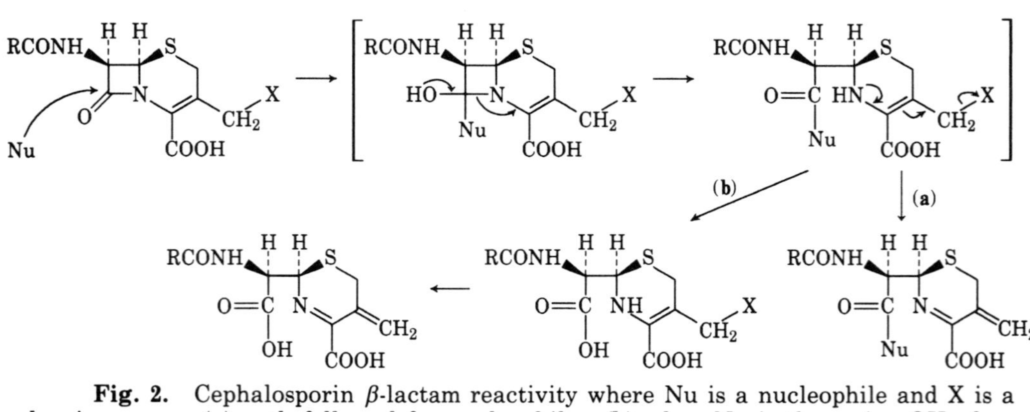

Fig. 2. Cephalosporin β-lactam reactivity where Nu is a nucleophile and X is a leaving group; (**a**) path followed for nucleophiles; (**b**) when Nu is the serine OH of an enzyme (ie, Nu = Enz–Ser–OH) deacylation may precede expulsion of the leaving group and both pathways (**a**) and (**b**) may be observed.

the C-7α position are discussed later. A discussion of other aspects of cephalosporin chemistry, such as deamination or epimerization at C-7, reactions at sulfur, isomerization of and addition to the double bond, and modification on the allylic methylene group at C-2, can be found in leading reviews (75,76).

Biological Properties

The clinical effectiveness of the cephalosporins depends on a number of properties. The antibiotic must inhibit, or preferably kill, bacteria at acceptable concentrations of the drug (*in vitro* activity); it must be capable of achieving host serum and tissue levels greater than those required to inhibit the pathogenic organism; and the selective toxicity profile must allow for safe administration to the host. Much of the chemical work that has been done on the cephalosporins has been directed toward modifying and improving the antibacterial and pharmacokinetic properties of these substances. Pharmacokinetic properties include the efficiency and rate of adsorption, the rate of metabolism and excretion, tissue distribution, serum binding, and serum and tissue inactivation. Chemical studies are thus closely coupled to extensive biological evaluation which examines *in vitro* and *in vivo* antibacterial activity expressed as minimum inhibitory concentration (MIC), minimum bactericidal concentration (MBC), and protective effectiveness in laboratory animals (PD_{50} or ED_{50}) as well as pharmacokinetic characteristics. New compounds are also tested for the ability to resist inactivation by β-lactamases (penicillinases, cephalosporinases) produced by bacteria.

Classification. As of 1991, there were approximately fifty different cephalosporins in clinical use or at an advanced stage of evaluation and development (77–79). Cephalosporins may be classified for convenience by their clinical pharmacology, β-lactamase resistance, chemical structure, or their antibacterial spectrum. The most common classification, which is somewhat arbitrary, divides the cephalosporins into three groups or "generations" (Tables 4–8), based primarily on their antibacterial spectrum (80,81). First-generation cephalosporins are characterized by good gram-positive activity and modest to weak gram-negative activity. They are usually active against aerobic gram-positive cocci, with the exception of enterococci, methicillin-resistant *Staphylococcus aureus*, and *Staphlylococcus epidermidis,* and are also active against some gram-negative bacilli such as *Escherichia coli, Klebsiella pneumoniae,* and *Proteus mirabilis,* and most anaerobic cocci and bacilli other than *Bacteroides fragilis* (80–83).

The second-generation cephalosporins have a broader spectrum and generally are more active against aerobic gram-negative organisms than the first-generation cephalosporins. Most second-generation cephalosporins have some activity against indole-positive *Proteus, Enterobacter* (except *E. cloacae*), and *Haemophilus influenzae.* Two agents included in this group, cefotetan (**26**) and cefoxitin (**23**), which are both cephamycin analogues, have significant activity against *Bacteroides fragilis* (80–83).

Third-generation cephalosporins have an expanded gram-negative spectrum and are the most active against enteric gram-negative bacilli, including penicillinase-producing strains, as well as *Serratia* and *Citrobacter.* They are also highly active against *Haemophilus influenzae* and *Neisseria gonorrhoeae, Streptococcus pneumoniae* and *pyogenes,* moderately active against *Pseudomonas aeruginosa* and anaerobes, but somewhat less active against staphylococci than the first-generation cephalosporins (80–84). A subset of the third-generation cephalosporins has somewhat improved, albeit still not ideal, activity against *Pseudomonas aeruginosa.* This group includes cefoperazone (**34**), cefpiramide (**42**), ceftazidime (**40**), and cefsulodin (**41**) (80), although cefsulodin is relatively inactive against most other organisms. Cefixime (**16**), a new oral cephalosporin touted as the first oral third-generation cephalosporin, has a spectrum between those of the second- and third-generation compounds. It has enhanced activity against *Enterobacteriaceae,* and against most common respiratory and urinary tract pathogens, but lacks activity against *Staphylococcus aureus,* enterococci, and *Pseudomonas aeruginosa.* Cefixime (**16**) would be better compared to its direct competition in the oral market, cefaclor (**13**), and the oral prodrug cefuroxime axetil (**20**).

Some other compounds tentatively labeled as examples of the as yet undefined "fourth-generation" have appeared in the literature, but these are probably best thought of as third-generation cephalosporins, having slight advantages over earlier examples of this group. This group includes cefpirome (**43**), cefepime (**45**), and others undergoing clinical trials.

It should be noted that the classification into generations is not chronological, and so some second-generation compounds came to the market relatively recently, after the third-generation was firmly established. Within each class there is also a further classification as to whether the compounds are administered orally or parenterally. A contemporary variation of the generation classification has been proposed, but has not found widespread use (85). Another proposal groups cephalosporins according to clinical indication or application (86).

The structures of selected cephalosporins on the U.S. market, or in the final stages of development, are shown in Tables 4–8 (see also 78, 87). For every cephalosporin which has made it to the marketplace, literally thousands of analogues were synthesized in order to establish the structure-activity profile and allow selection of a clinical candidate. In addition to these compounds, there is a tremendous number of cephalosporin compounds currently at various stages of development. A more extensive listing of the newer cephalosporins under preclinical or clinical evaluation may be found in a number of reviews (79,88).

***In Vitro* Antibacterial Activity and Structure-Activity Relationships.** The *in vitro* antibacterial activity of any particular cephalosporin is a combination of the degree and type of activity at the target site, the ease with which it can penetrate to the target, and the ability to resist the attack of destructive enzymes. The nature and complexity of the biochemical target(s) for β-lactam antibiotics is fairly well-established and the mechanisms of penetration are also understood to some extent. However, many factors are involved, including pharmacokinetic and pharmacodynamic properties, and the antibiotic may not perform as predicted. The *in vitro* antibacterial activity of a representative selection of the cephalosporins given in Tables 4–8 against a variety of bacterial organisms is depicted in Table 9. More extensive comparative listings of the antibacterial activity of cephalosporins may be found in the literature (77,80,81,130–132).

Structure-activity relationships can be inferred by comparison of the antibacterial properties of the clinical agents and related compounds. Different acyl side chains can result in significant changes in the antibacterial activity, both with respect to potency and to breadth of spectrum. The highest activities are observed when the acylamino side chain at C-7 is a substituted acetic acid. Homologation of the acetic acid moiety lowers activity dramatically as exemplified by the naturally occurring cephalosporins which all have weak activity. Compared to the penicillins, there is somewhat more latitude for the type of substituent(s) attached to the α-carbon of the side chain. The phenylacetyl and phenoxyacetyl side chains are less effective in conferring activity on the cephalosporins than the pencillins. More effective are acetic acid side chains attached to heterocyclic rings, such as thiophene, pyridine, tetrazole, furan, and aminothiazole. Much smaller and nonaromatic substituents can also confer high levels of *in vitro* activity. Such substituents include the cyano, methylthio, methylsulfonyl, and trifluoromethylthio moieties. The aromatic ring seems to only be associated with effective activity when there is a second substituent on the α-carbon of the acyl group. The effect of this second substituent on the cephalosporin activity is different to what is observed for the penicillins. The carboxyl and sulfonic acid groups do not particularly improve activity overall for the cephalosporins, whereas the amino and hydroxy group seem effective in this regard. A second substituent on the α-carbon introduces a new chiral center. The D-configuration (*R*) is preferred for the phenylglycine and mandelic acid derived side chains as in both cases the L-configuration (*S*)-analogue is considerably less active.

That the aminothiazole group imparts high potency to the cephalosporins against gram-negative bacteria was first demonstrated in 1977 with the synthesis of cefotiam [*61625-34-2*] (SCE-963), $C_{18}H_{22}N_9O_4Na$ (**50**) (133). The oxime group in cefuroxime (**35**) and most of the aminothiazoyl containing cephalosporins is especially effective in further increasing potency and conferring resistance to certain β-lactamases. A marked stereochemical preference is observed: the syn-isomer where the alkoxy and acyl groups are cis to each other, is considerably more active than the corresponding antiisomer.

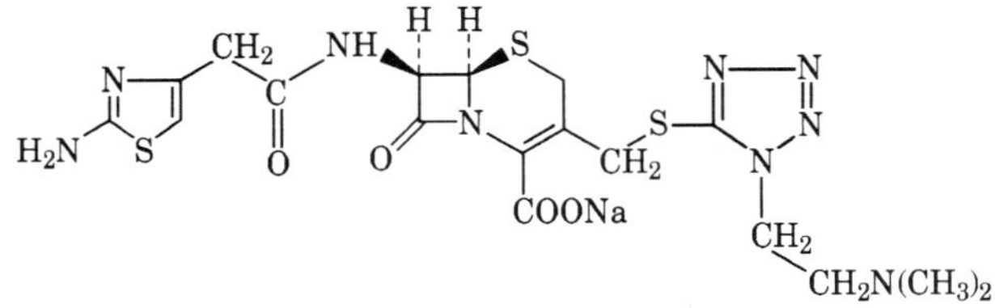

(**50**)

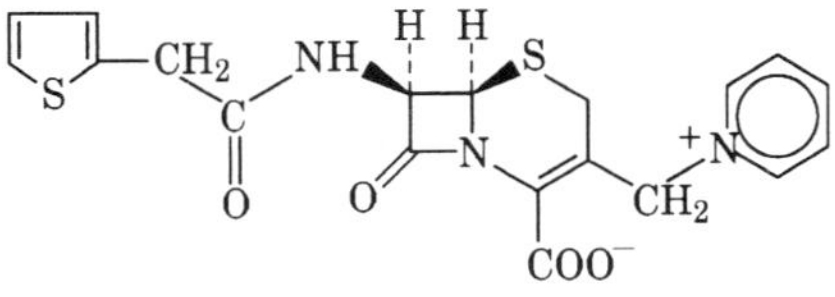

(**51**)

One of the principal deficiencies of the older cephalosporins was the lack of resistance to β-lactamases. Compounds with improved β-lactamase resistance have one or more of the following characteristics: a second, monovalent substituent on the α-carbon of the C-7 acyl group such as found in cefamandole (**30**) and cefoperazone (**34**); a syn-oxime substituent, eg, cefuroxime (**35**) and ceftriaxone (**39**); a methoxy, or formamido, substituent on the β-lactam ring at the 7α-position such as in cefoxitin (**23**). However, these various substituents have different effects, and increased resistance to one enzyme does not indicate resistance to all β-lactamases. The 7α-methoxy, or formamido, group is the most effective in im-

Table 4. Oral Cephalosporins

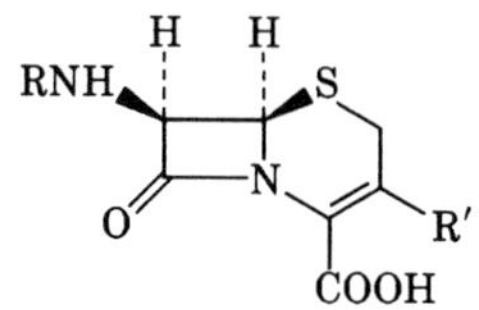

Generic name	CAS Registry Number	Molecular formula	Generation	Trade name	Company	Structure number	R	R′	Year[a]	Refs.
cephalexin	[*15686-71-2*]	$C_{16}H_{17}N_3O_4S$	[1]	Keflex	Lilly	**(12)**	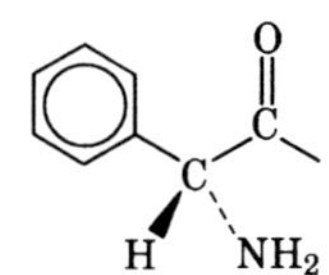	—CH_3	1970	77,89
cefaclor	[*53994-73-3*]	$C_{15}H_{14}N_3O_4S_1Cl$	[2]	Ceclor	Lilly	**(13)**	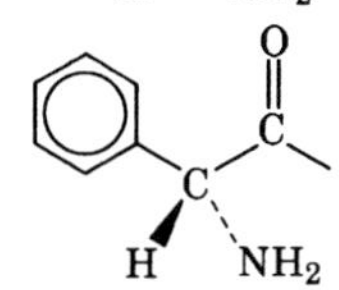	—Cl	1979	77,90
cephradine	[*3882-53-3*]	$C_{16}H_{19}N_3O_4S$	[1]	Velosef Anspor	Squibb SKF-Beecham	**(14)**	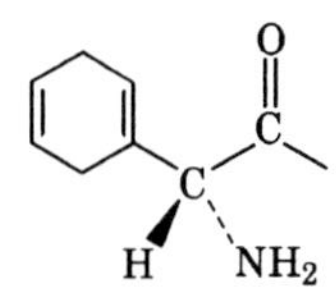	—CH_3	1972	77,91–93
cefadroxil	[*50370-12-2*]	$C_{16}H_{17}N_3O_5S$	[1]	Duricef Ultracef	Mead-Johnson Bristol-Myers Squibb	**(15)**	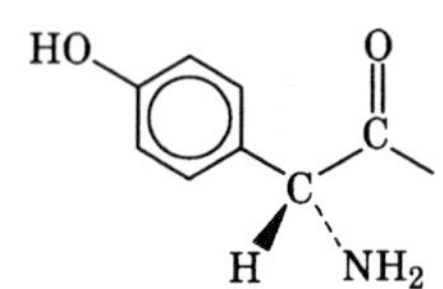	—CH_3	1980	77,94,95

cefixime	[*79350-37-1*]	$C_{16}H_{15}N_5O_7S_2$	[2–3]	Suprax Cefspan[b]	Lederle Fujisawa	**(16)**	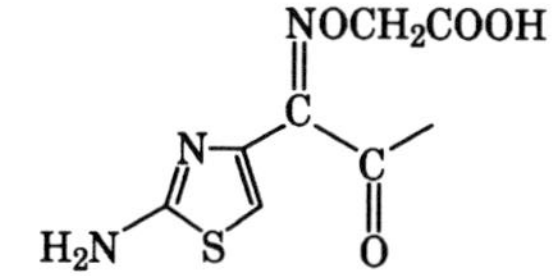	—CH=CH$_2$	1989 1987[c]	96
ceftibuten	[*97519-39-6*]	$C_{15}H_{14}N_4O_6S_2$	[2–3]	Sch 39720 7432-S	Schering Shionogi	**(17)**	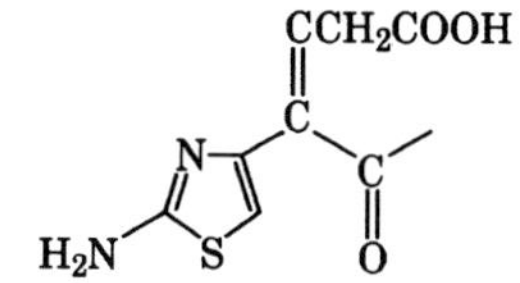	—H	Ph-III	97
cefprozil	[*92665-29-7*]	$C_{18}H_{19}N_3O_5S$	[2]	BMY28100	Bristol-Myers Squibb	**(18)**	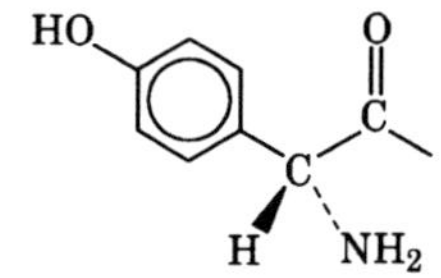	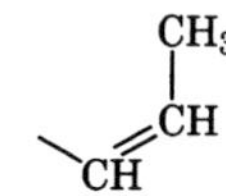	Ph-III	98
—	[*107888-46-0*]	$C_{15}H_{15}N_5O_6S_2$	[2–3]	BMY28232	Bristol-Myers Squibb	**(19)**	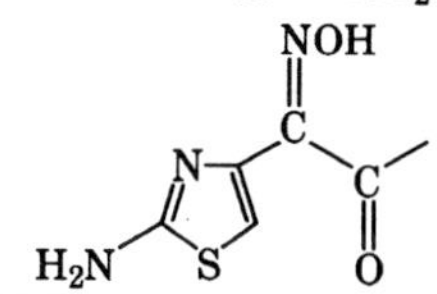	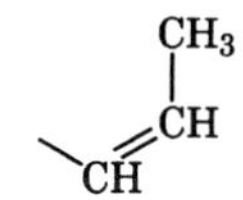	Pre-Cl	99

[a]Year is year of launch in the U.S. market unless indicated. Pre-Cl = preclinical, Ph III = phase III clinical trials (status as of Feb. 1990).
[b]Japanese trade name.
[c]Launch date in Japan.

Table 5. Oral Cephalosporins-Prodrugs

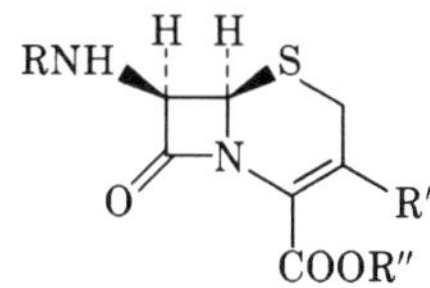

Generic name[a]	CAS Registry Number	Molecular formula	Trade name	Company	Structure number	R	R′	R″	Year[b]	Refs.
cefuroxime-axetil [2]	*[64544-07-6]*	$C_{20}H_{22}N_4O_{10}S_1$	Ceftin	Glaxo Roche	**(20)**	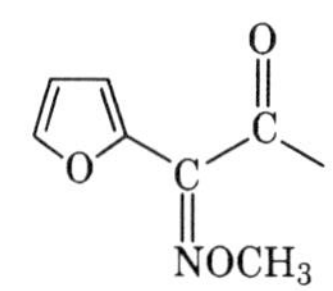	$—CH_2OCONH_2$	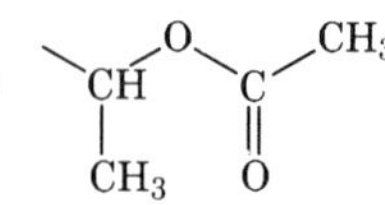	1988	100
cefpodoxime-proxetil [2–3]	*[87239-81-4]*	$C_{21}H_{27}N_5O_9S_2$	Banan	Upjohn Sankyo	**(21)**	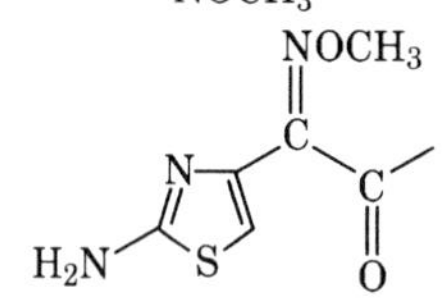	$—CH_2OCH_3$	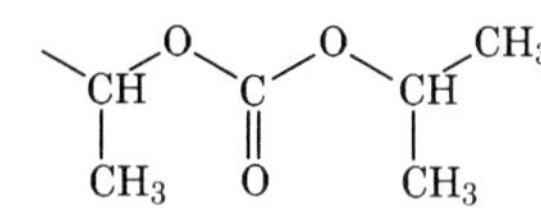	Ph-III	101
[2–3]	*[107888-50-6]*	$C_{18}H_{19}N_5O_7S_2$	BMY28271	Bristol-Myers Squibb	**(22)**	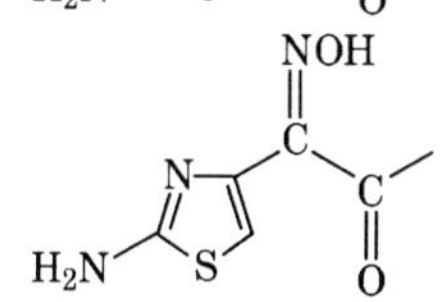	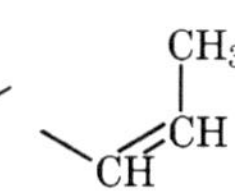	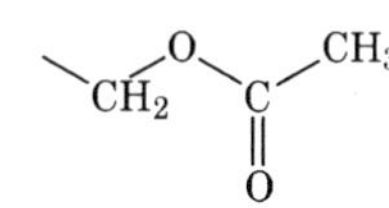	Pre-Cl	99,102

[a]Drug generation is given in brackets.
[b]Year is year of launch in the U.S. market unless indicated. Pre-Cl = preclinical and Ph III = phase III clinical trials (status as of Feb. 1990).

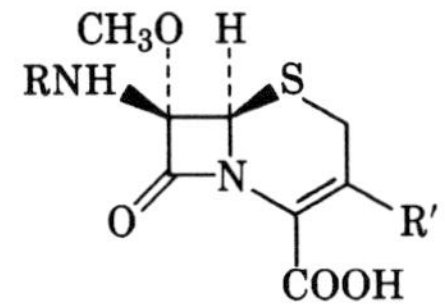

Table 6. Cephamycins

Generic name[a]	CAS Registry Number	Molecular formula	Trade name	Company	Structure number	R	R′	Year[b]	Refs.
cefoxitin [2]	*[35607-66-0]*	$C_{15}H_{15}N_3O_6$	Mefoxin	Merck	**(23)**	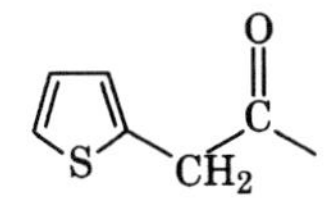	$-CH_2OCONH_2$	1978	103
cefmetazole [2]	*[56796-20-4]*	$C_{14}H_{15}N_7O_4S_3$	Zefazone Cefmetazon[c]	Upjohn Sankyo	**(24)**	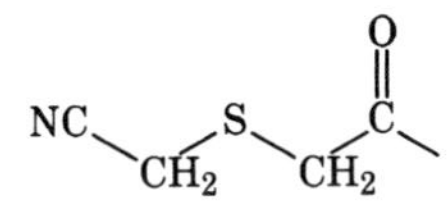	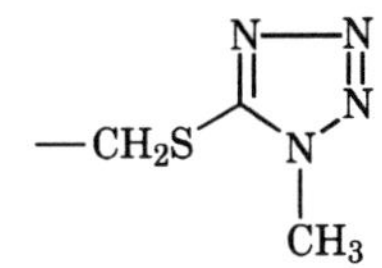	1989 1980[d]	104
cefminox [2]	*[75221-13-0]*	$C_{15}H_{19}N_7O_6S_3$	Meicelin[c]	Meiji Seika Bristol-Myers Squibb	**(25)**	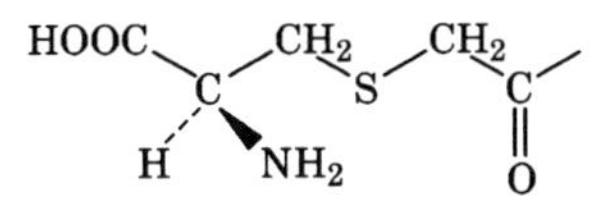	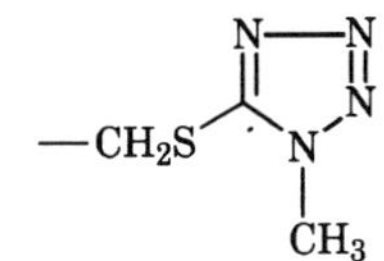	1987[d]	105
cefotetan [2]	*[69712-56-7]*	$C_{16}H_{15}N_7O_7S_4$	Ceftan Apacef[c] Yamatetan	Stuart Yamanouchi	**(26)**	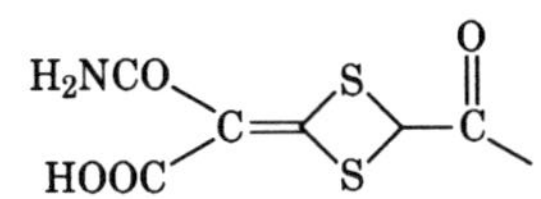	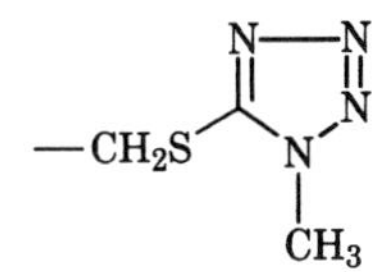	1986 1984[d]	106

[a]Drug generation is given in brackets.
[b]Year is year of launch in the U.S. market unless indicated.
[c]Japanese trade name.
[d]Launch date in Japan.

Table 7. Parenteral Cephalosporins

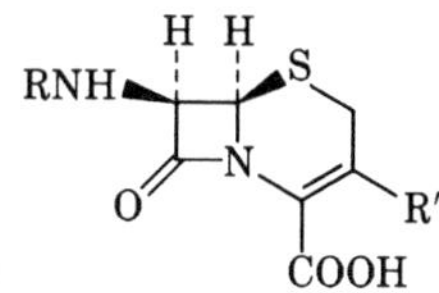

Generic name[a]	CAS Registry Number	Molecular formula	Trade name	Company	Structure number	R	R′	Year[b]	Refs.
cephalothin [1]	*[153-61-7]*	$C_{16}H_{16}N_2O_6S_2$	Keflin	Lilly	**(27)**	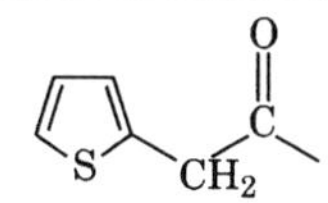	$—CH_2OCOCH_3$	1965	87,107
cephacetrile [1]	*[23239-41-0]*	$C_{13}H_{13}N_3O_6S$	Celospor	Ciba	**(28)**	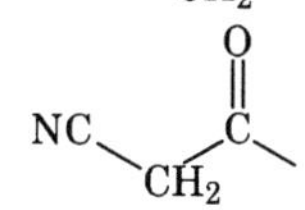	$—CH_2OCOCH_3$	1979[c]	87,108
cephapirin [1]	*[21593-23-7]*	$C_{17}H_{17}N_3O_6S_2$	Cefadyl	Bristol-Myers Squibb	**(29)**	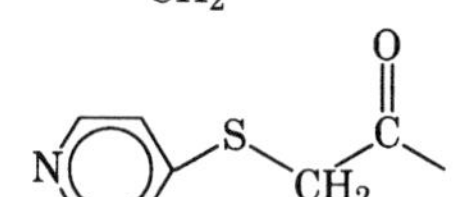	$—CH_2OCOCH_3$	1974	87,91,109
cefamandole [2]	*[34444-01-4]*	$C_{11}H_{18}N_6O_5S_2$	Mandol	Lilly	**(30)**	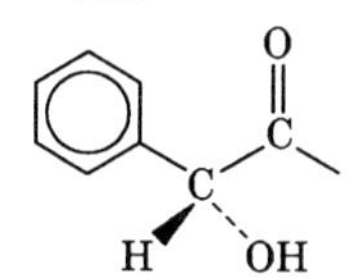	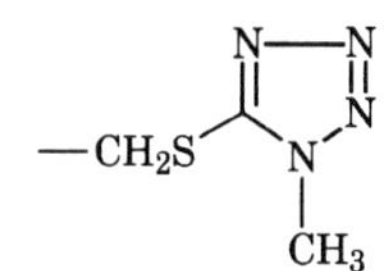	1978	87,110
cefonicid [2]	*[61270-58-4]*	$C_{18}H_{17}N_6O_8S_3Na$	Monicid	SKF-Beecham	**(31)**	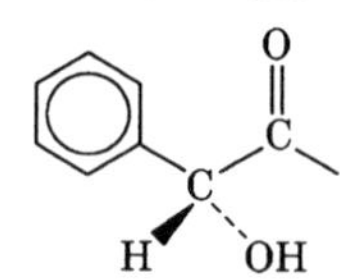	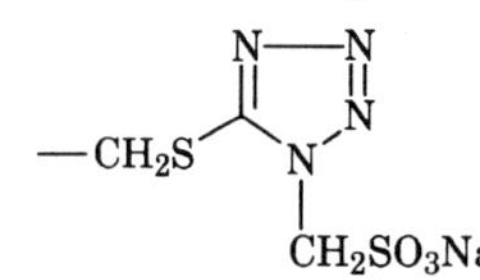	1984	87,111
cefazolin [1]	*[25953-19-9]*	$C_{14}H_{14}N_8O_4S_3$	Ancef Kefzol	SKF-Beecham Lilly	**(32)**	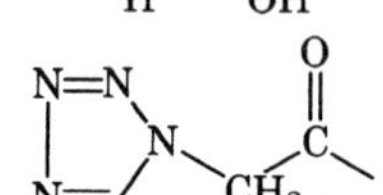	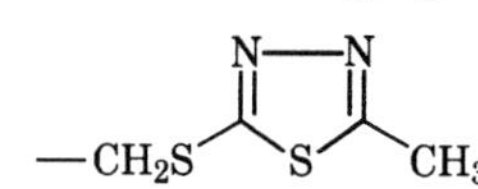	1973	87,112

ceforanide [2]	[*40851-79-4*]	$C_{20}H_{19}N_7O_6S_2$	Precef	Bristol-Myers Squibb	**(33)**	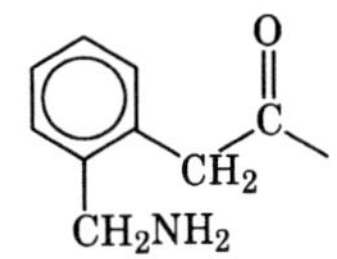	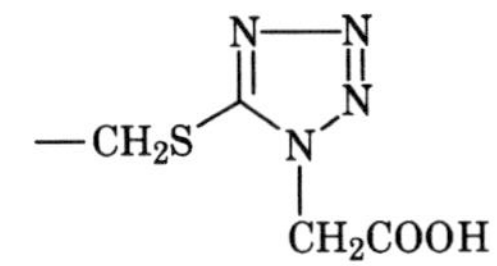	1984	87,113
cefoperazone [3]	[*62893-19-0*]	$C_{25}H_{27}N_9O_8S_2$	Cefobid	Pfizer Roerig	**(34)**	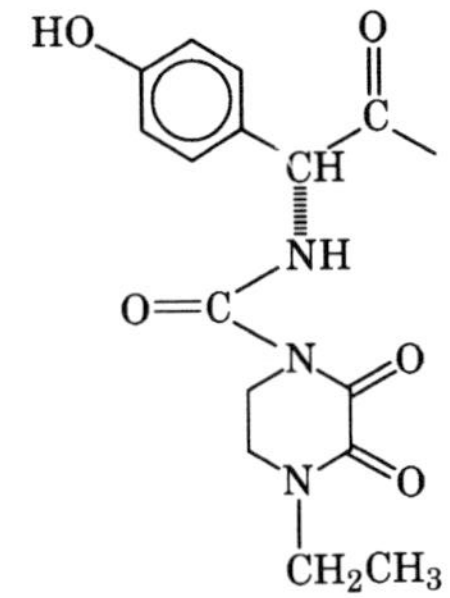	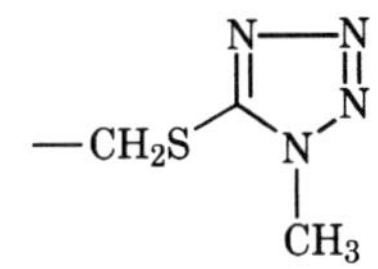	1983	87,114
cefuroxime [2]	[*55268-75-2*]	$C_{16}H_{16}N_4O_8S$	Zinacef Kefurox	Glaxo Lilly	**(35)**	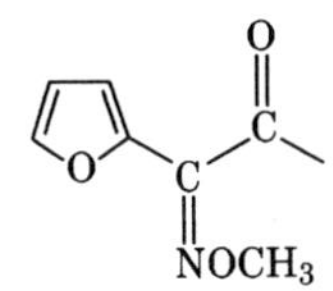	$—CH_2OCONH_2$	1983	87,115
cefotaxime [3]	[*63527-52-6*]	$C_{16}H_{17}N_5O_7S_2$	Claforan	Hoechst-Roussel	**(36)**	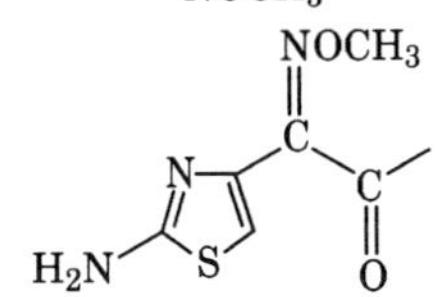	$—CH_2OCOCH_3$	1981	87,116
ceftizoxime [3]	[*68401-81-0*]	$C_{13}H_{13}N_5O_5S_2$	Cefizox	SKF-Beecham	**(37)**	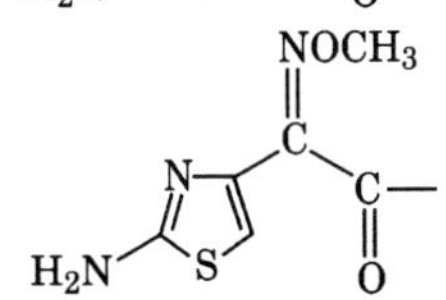	—H	1983	87,117

Table 7. ***(Continued)***

Generic name[a]	CAS Registry Number	Molecular formula	Trade name	Company	Structure number	R	R′	Year[b]	Refs.
cefmenoxime [3]	[*65085-01-0*]	$C_{16}H_{17}N_9O_5S_3$	Cefmax Bestcall[e]	Abbott Takeda-Roche	**(38)**	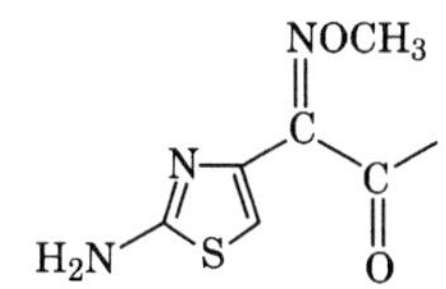	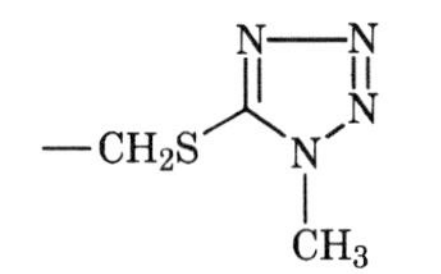	1987[d] 1983[f]	87,118
ceftriaxone [3]	[*73384-59-5*]	$C_{18}H_{17}N_8O_7S_3Na$	Rocephin	Roche	**(39)**	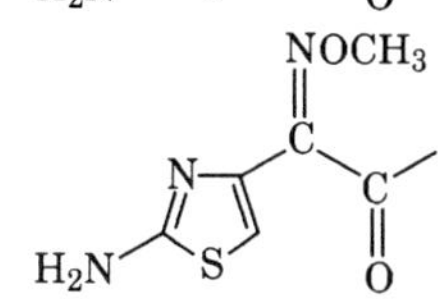	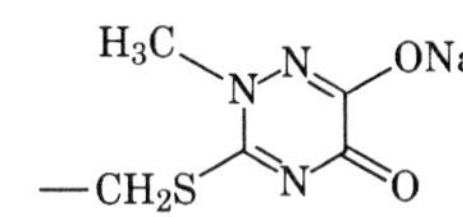	1985	87,119
ceftazidime [3]	[*72558-82-8*]	$C_{22}H_{22}N_6O_7S_2$	Fortaz Tazicef Tazidime	Glaxo SKF-Beecham Lilly	**(40)**	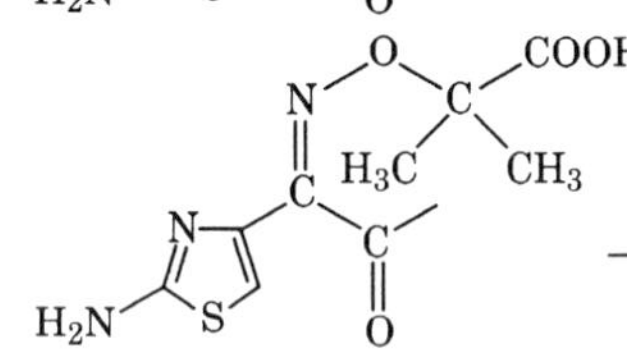	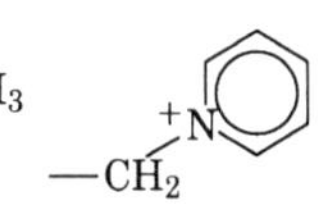	1985	87,120
cefsulodin	[*52152-93-9*]	$C_{22}H_{20}N_4O_8S_2$	Cefomonil	Takeda-Abbott	**(41)**	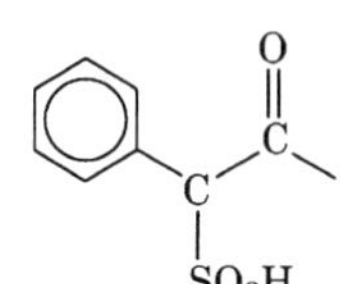	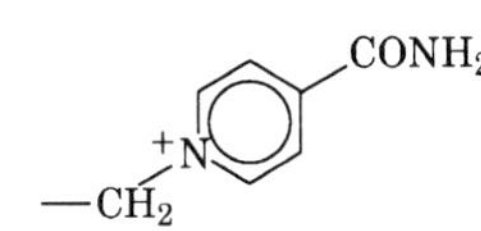	1986[d]	121

cefpiramide [3]	[*70797-11-4*]	$C_{25}H_{24}N_8O_7S_2$	Cefpiran Sepatren[e] Suncefal[e]	Wyeth-Ayerst Sumitomo Yamanouchi Banyu	**(42)**	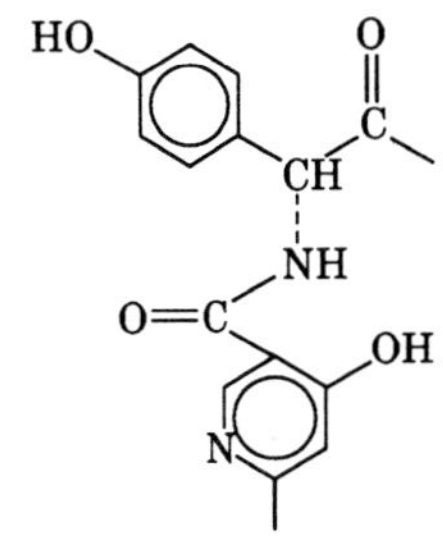	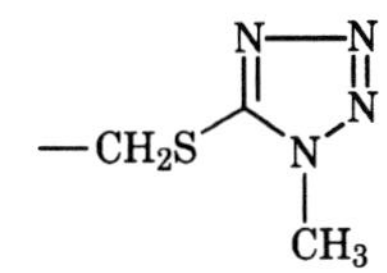	1989[d] 1985[f] 1985[f]	122
cefpirome [3–"4"]	[*84957-29-9*]	$C_{22}H_{22}N_6O_5S_2$	HR810	Hoechst-Roussel	**(43)**	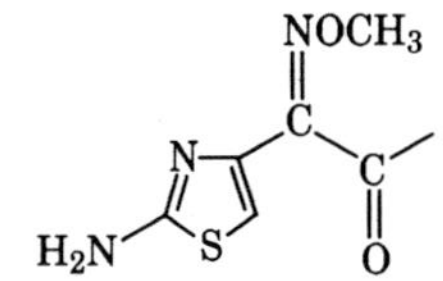	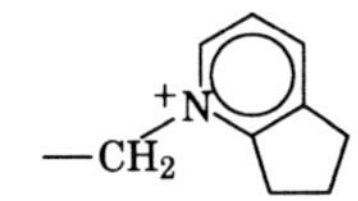	Ph-III	123
cefpimizole [3]	[*84572-33-8*]	$C_{28}H_{26}N_6O_{10}S_2$	U-63,196 Ajicef[e] Renilan[e]	UpJohn Ajinomoto Mochida	**(44)**	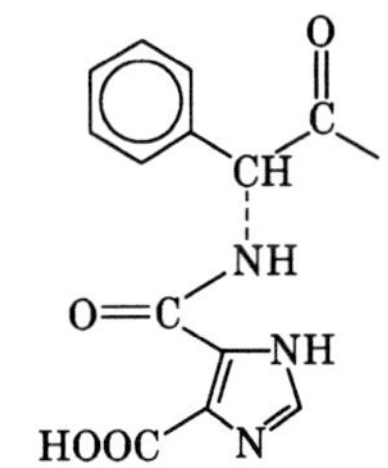	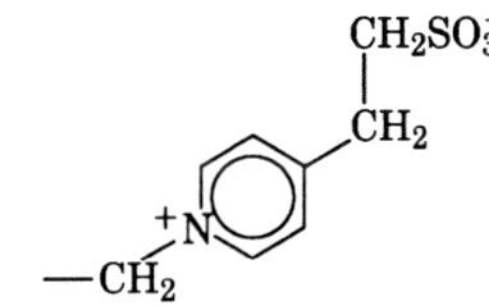	Ph-III 1987[f] 1987[f]	124
cefepime [3–"4"]	[*88040-23-7*]	$C_{19}H_{24}N_6O_5S_2$	BMY28142	Bristol-Myers Squibb	**(45)**	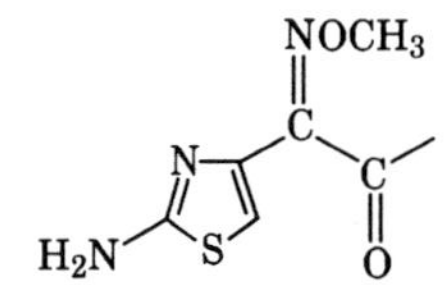	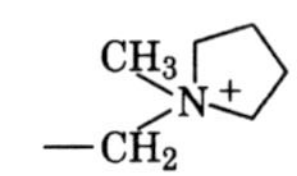	Ph-II	125

Table 7. (*Continued*)

Generic name[a]	CAS Registry Number	Molecular formula	Trade name	Company	Structure number	R	R′	Year[b]	Refs.
[3–"4"]	[*106772-82-1*]	$C_{28}H_{21}N_9O_{11}S_3Na_2$	M14659	Mochida	**(46)**	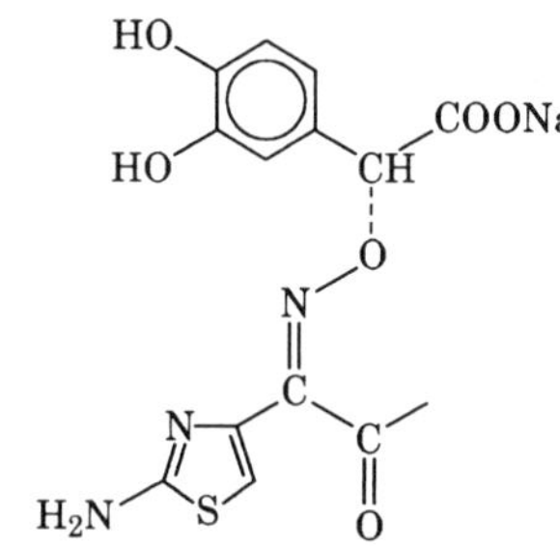	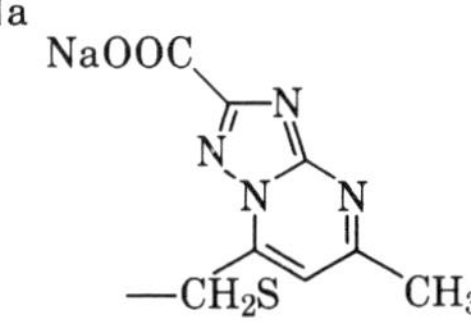	Pre-Cl	126
["4"]	[*115622-58-7*]	$C_{31}H_{31}N_8O_8S_2F_3$	Ro 23-9424	Roche	**(47)**	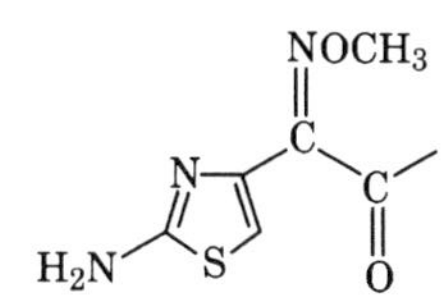	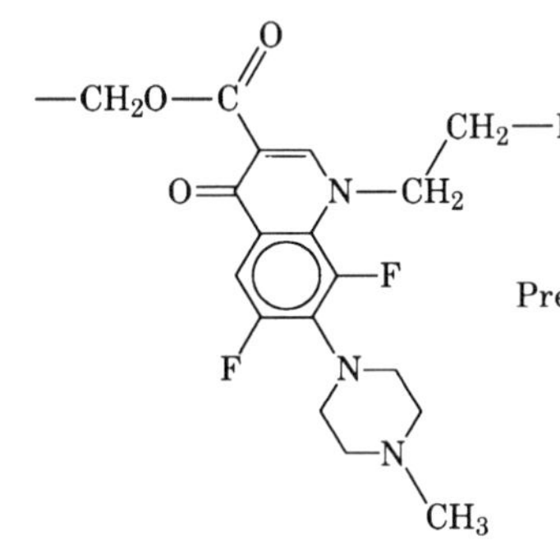	Pre-Cl	127

[a]Drug generation is given in brackets.
[b]Year is year of launch in the U.S. market unless indicated. Pre-Cl = preclinical, Ph III = phase III clinical trials, and Ph II = phase II clinical trials (status as of Feb. 1990).
[c]Launch date in Italy.
[d]New drug application (NDA) filing or approval.
[e]Japanese trade name.
[f]Launch date in Japan.

Table 8. Oxadethiacephalosporins

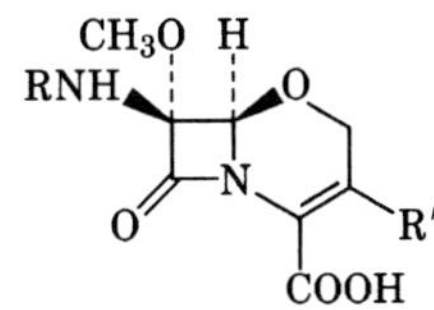

Generic name	CAS Registry Number	Molecular formula	Generation	Trade name	Company	Structure number	R	R′	Year[b]	Refs.
moxalactam latamoxef	*[64952-97-2]*	$C_{20}H_{20}N_6O_8S$	[3]	Moxam	Lilly	**(48)**	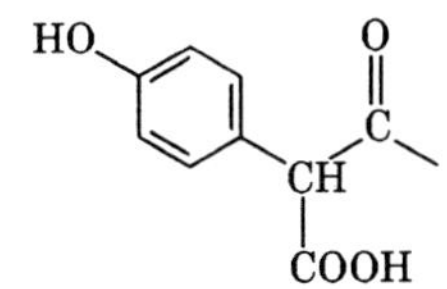	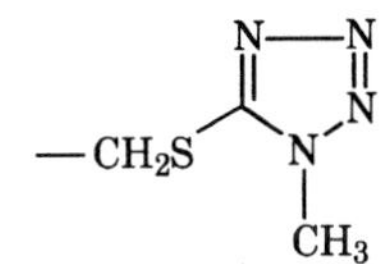	1981	128
flomoxef	*[99665-00-6]*	$C_{15}H_{18}N_6O_6S_2F_2$	[3]	Flumarin[b]	Shionogi	**(49)**	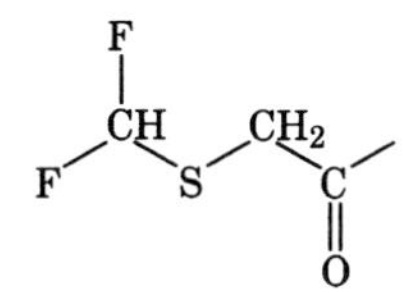	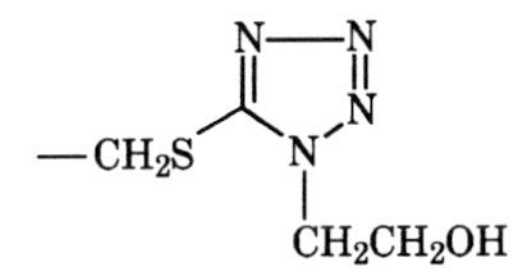	1988[c]	129

[a]Year is year of launch in the U.S. market unless indicated.
[b]Japanese trade name.
[c]Launch date in Japan.

Table 9. Cephalosporin *In Vitro* Antibacterial Activity, μg/mL, and Pharmacokinetic Data[a,b]

	(12)	(27)	(13)	(31)	(35)	(16)	(36)	(39)	(40)	(23)	(48)	(47)
Antibacterial activity												
Staph. aureus[c]	2	0.5	1	4	1	32	2	4	4	2	8	1
Strep. pyogenes	0.5	0.12	0.25	0.5	0.03	0.25	0.03	0.03	0.12	1	1	0.06
Strep. pneumoniae	2	0.12	1	1	0.12	0.5	0.12	0.25	0.25	2	1	<0.06
E. coli[d]	4	4	1	1	1	0.25	0.03	0.12	0.12	4	0.12	0.06
Kleb. pneumoniae[e]	4	4	1	4	2	0.12	0.03	0.06	0.12	2	0.12	0.06
Pr. mirabilis[d]	8	2	1	0.5	1	0.03	0.03	0.03	0.06	2	0.12	0.25
Proteus sp (ind+)	>128	>128	>128	16	8		0.12	0.12	0.12	4	0.25	0.12
Enterobacter sp[f]	>128	>128	>128	32	16	64	0.12	0.25	0.25	>128	0.12	0.06
Citrobacter sp	>128	64	>128		8		0.25	0.5	0.5		0.25	0.12
Serratia marcescens	>128	>128	>128	64	64		0.12	0.25	0.12	16	0.25	<0.12
Ps. aeruginosa	>128	>128	>128	>128	>128	>128	32	32	2	>128	16	4
H. influenzae	8	4	1	1	0.5	0.03	0.03	0.03	0.12	2	0.12	0.03
B. fragilis[g]	64	64	>128	128	32		32	64	64	4	4	128
N. gonorrhoeae[c]	2	0.5	0.5	1	0.06	0.06	<0.01	<0.01	0.06	0.5	0.06	<0.01
Pharmacokinetic data												
plasma half life, h	0.8	0.6	0.6	4.0	1.3	0.7	1.2	8.0	1.8	0.8	2.5	
protein bound, %	10	65	25	98	35	25	35	95	17	70	50	
urinary excretion, %	90	70[h]	60[i]	96	95	65	70[h]	65	85	80	80	
biliary excretion, %	0.3	0.03	0.05	1	0.5		1	35	<3	<2	20	

[a]Refs. 77, 80, 81, 130–132. Cefixime Ref. 96; Ro 24-9424 Ref. 127.
[b]Note that values are approximated from references because specific strains are not specified.
[c]Penicillin resistant.
[d]Ampicillin sensitive.
[e]Cefazolin sensitive.
[f]Cefotaxime sensitive.
[g]Penicillin sensitive.
[h]Plus 30% deacetylcephalosporin.
[i]±30% inactive.

parting a broad spectrum of resistance to β-lactamases, followed by acyl side-chain oxime ether substituents. Thus cefuroxime (**35**), the first oximino–cephalosporin, has a β-lactamase resistance spectrum similar to that of cefotaxime (**36**) whereas the less resistant cefamandole (**30**) is similar to cefoperazone (**34**).

The nature of the C-3 substituent influences the pharmacokinetic properties, as well as the intrinsic antibacterial activity. Analogues having a 3-methyl substituent generally have a low level of antibacterial activity except when the 7-acyl group is D-mandelic acid, D-phenylglycine, or a related moiety. Thus an α-hydroxyl or amino group on the side chain compensates, at least in part, for the lack of a nucleophilic leaving group on the C-3 methyl. The 7-phenylglycine analogues were the only known cephalosporins that combined *in vitro* activity and efficient oral adsorption. However, more recent work has shown that a C-3 vinyl substituent permits reasonable oral adsorption even for the oximino–aminothiazole side chain, eg, cefixime (**16**), and BMY28232 (**19**), and it appears likely that other C-3 substituents will be found to facilitate oral adsorption. Changes in the 3′-position are thought to primarily effect pharmacologic activity. However some improvement in antibacterial activity can also result from such substitution. Of particular note is the improvement seen using pyridinium and other quaternary-ammonium moieties, especially against the gram-positive organisms, and certain heterocyclic thio groups. A number of the newer cephalosporins, such as M14659 (**46**), have been rationally designed to take advantage of the so called *ton B* mediated iron transport system (*vide infra*) and show considerably improved activity against *Enterobacteriaceae* and *Pseudomonas aeruginosa* in particular (126).

Another interesting approach to obtaining potent, broad-spectrum activity has been reported (127). The "dual-action" antibacterial concept involves incorporation of two moieties having complimentary antibacterial modes of action into the same molecule, and uses the mode of action of one part to release the second antibacterial at the site of action. This approach is exemplified in Ro 23–9424 (**47**) (127), which uses the mode of action and reactivity of the cephalosporin moiety (Fig. 2) to release the quinolone portion.

Intrinsic Activity and Biochemical Targets. β-Lactam antibiotics affect sensitive bacteria by inhibiting late stages in the biosynthesis of their cell wall peptidoglycan (134–136). The cell wall contains a matrix of complex macromolecules that provide rigidity and mechanical stability by way of cross-linked latticeworklike structures. The peptidoglycan is one of these polymers, and is made up of glycan strands (consisting of alternating *N*-acetylglucosamine and *N*-acetylmuramic acid units) which are interconnected by peptide cross-links in a three-dimensional array. The last step in the construction of the peptidoglycan cross-link involves a transpeptidase-catalyzed reaction in which an amide bond is formed. In the case of the peptidoglycan from *Escherichia coli* (type A1γ-peptidoglycan) the cross-linking-peptide chain is a pentapeptide attached to the *N*-acetylmuramic acid residue of the glycan repeating subunit via the lactic acid side chain. The third amino acid in the pentapeptide is *meso*-diaminopimelic acid (DAP) which is anchored into the peptide via its L-center. The cross-linking step, referred to as transpeptidation, which the β-lactam antibiotics inhibit, is the process whereby the amino group of the D-center of *meso*-DAP forms an amide link to the penultimate D-alanine of another peptide with concomitant expulsion of the final D-alanine residue. Not all peptide chains are involved in the formation of cross-links, in *E. coli* the extent of cross-linking is about 33%, and the noncross-linked chains may be modified by the removal of the terminal D-alanine unit without transpeptidation. The enzyme responsible for this "pruning" is a D,D-carboxypeptidase (135,136).

β-Lactam antibiotics inhibit the cross-linking process by forming a covalent acyl-enzyme complex with transpeptidase or carboxypeptidase (Fig. 2). A serine hydroxyl function in the active site of these enzymes is implicated in the formation of the acyl-enzyme bond. The stability of the acyl-enzyme bond in the case of the transpeptidases is such that for all intents and purposes the enzyme is irreversibly inhibited. The carboxypeptidases, although intrinsically more sensitive to the *β*-lactam antibiotics in general, are reversibly inhibited because the acyl-enzyme bond is readily hydrolyzed to regenerate free enzyme (135,136).

Cell wall biosynthesis is a dynamic process and during cell division new cell wall must be synthesized (135–137). Thus there must also be some cleavage of certain peptide cross-links and/or glycan chains to allow for new cell growth and division. Hence the picture is rather more complicated than the basic mechanism just described and there is a group of enzymes which has an affinity for *β*-lactam antibiotics. These enzymes, the so-called penicillin-binding proteins (PBPs), are embedded in the inner, or plasma, membrane. These PBPs are designated according to decreasing molecular weight (Table 10) and are species specific. Hence PBP1 of *Escherichia coli* may not have anything in common with PBP 1 of any other organism. There are invariably multiple PBPs varying in number (from 3 in gonococci to 10 in *Escherichia coli*), molecular weight, and number of copies per cell. In the case of *Escherichia coli* there are seven important PBPs, of which four, namely 1A, 1B, 2, and 3 are considered essential for cell viability (134–136,138).

In a general sense all cephalosporins are thought to form a covalent acyl-enzyme complex involving the *β*-lactam carbonyl and a serine hydroxyl in the enzyme active site. However, as shown by competitive binding experiments, cephalosporins differ in their affinities for the various PBPs and it is well-established that preferential binding to the various PBPs produces different morphological effects (135–138). These effects are summarized in Table 10. Preferential binding to PBP 1B of *Escherichia coli,* as exhibited by cephaloridine [*50-59-9*] (**51**), $C_{19}H_{17}N_3O_4S_2$, produces rapid lysis with little malformation of the cells. Septum cross-wall formation is controlled by PBP 3, and cephalosporins, such as cefuroxime (**35**), that primarily inhibit this protein leave growing cells incapable of cell division. Each cephalosporin inhibits the various proteins to varying extents, and the intrinsic activity depends largely on the concentration of the compound required to saturate the most susceptible PBP. The 7*α*-methoxycephalosporins have considerable affinity for the nonessential PBPs 4, 5, and 6, whereas cephalosporins with no 7*α*-substituent bind very little if at all to these proteins. The binding of selected cephalosporins to the PBPs of *Escherichia coli* is detailed in Table 11.

Resistance. Resistance to the cephalosporins may result from the alteration of target pencillin-binding sites (PBPs), decreased permeability of the bacterial cell wall and outer membrane, or by inactivation via enzyme mediated hydrolysis of the lactam ring (80,81,138–140). This resistance can be either natural or acquired. Although resistance is often attributed specifically to one of these

Table 10. Penicillin Binding Proteins (PBPs) of *E. coli*[a]

PBP	Mol wt	Number per cell[b] (%)	Proposed *in vitro* functions	Effects of inactivation
1A	92,000	200 (6.5)	transglycosylase, minor transpeptidase, can partially compensate for loss of 1B activity; essential for cylindrical cell wall synthesis	rapid lysis if both 1A and 1B are inactivated
1B	90,000	250 (8)	transglycosylase, major transpeptidase of cell elongation; essential for cylindrical cell wall synthesis	rapid lysis
2	66,000	20 (0.7)	maintenance of cylindrical rod shape; requires transpeptidase?	spherical cell formation
3	60,000	50 (1.6)	transglycosylase–transpeptidase required septum cross-wall synthesis	nonseptate cells; filament formation
4	49,000	110 (3.6)	secondary transpeptidase, D,D-carboxypeptidase, or D,D-endopeptidase acting on maturing peptidoglycan	controls transpeptidation delayed transpeptidation is absent
5	42,000	1800 (59)	D,D-carboxypeptidase	none apparent
6	40,000	600 (19.6)	D,D-carboxypeptidase	none apparent

[a]Refs. 134–136, and 138.
[b]Number in parenthesis is the % of the total PBPs per cell.

Table 11. Selective Binding of Cephalosporins to the PBPs of *E. coli*[a,b]

Antibiotic	PBP-1A	PBP-1B	PBP-2	PPB-3	PBP-4	PBP-5	PBP-6
cephalexin (**12**)	4	240	250	8	30	250	250
cephalothin (**27**)	<0.25	16	37	1	60	125	130
cephaloridine (**51**)	0.25	2.5	50	8	17	250	250
cefotaxime (**36**)	0.05	0.7	5	0.05	30	50	50
cefoxitin (**23**)	0.1	3.9	>250	5.8	7.0	0.6	0.9

[a]Refs. 134, 136, and 138.
[b]Concentration required for 50% inhibition of binding of [^{14}C]pencillin G, μg/mL.

factors, in reality it reflects the interplay of several factors. In most instances, however, resistance results from the production of a β-lactamase enzyme, which opens the β-lactam ring as depicted in Figure 2.

Bacteria produce chromosomally and R-plasmid (resistance factor) mediated β-lactamases. The plasmid-mediated enzymes can cross interspecific and intergeneric boundaries. This transfer of resistance via plasmid transfer between strains and even species has enhanced the problems of β-lactam antibiotic resistance. Many species previously controlled by β-lactam antibiotics are now resistant. The chromosomal β-lactamases are species specific, but can be broadly classified by substrate profile, sensitivity to inhibitors, analytical isoelectric focusing, immunological studies, and molecular weight determination. Individual enzymes may inactivate primarily penicillins, cephalosporins, or both, and the substrate specificity predetermines the antibiotic resistance of the producing strain. Some β-lactamases are produced only in the presence of the β-lactam antibiotic (inducible) and others are produced continuously (constitutive).

Because of the highly permeable nature of the cell wall of gram-positive organisms, they produce β-lactamases which are not only found throughout the cell wall, but also in the extracellular environment. Hence the extracellular β-lactamases can act on the antibiotic before the cell is entered. Gram-negative organisms produce cell-bound β-lactamases which reside in the periplasmic space. Thus, for gram-negative bacteria, the antibiotic must penetrate the outer cell membrane/wall before coming in contact with a β-lactamase (80,139,140).

In general, the first-generation cephalosporins are readily hydrolyzed by the inducible, chromosomally mediated β-lactamases, and to a lesser extent, by the plasmid mediated enzymes such as the common TEM enzymes. Second- and third-generation cephalosporins have greater stability to hydrolysis by these better enzymes, although the emergence of resistant organisms during therapy using second- and third-generation cephalosporins is an increasingly serious medical problem (141). Resistance is most frequently reported in cases involving species which characteristically possess inducible β-lactamase enzymes (such as the so-called "overproducers" *Enterobacter* and *Serratia*) (142). Although it has also been proposed that some β-lactamases may also protect the enzyme-producing organisms from the effects of certain, mostly third-generation, cephalosporins by "trapping" or binding the drugs without actually hydrolyzing them (141,143), experiments have cast doubts as to the validity of this proposal (144).

Resistance can also arise when target enzymes, ie, the PBPs, and in particular the transpeptidases, are modified. Target-mediated cephalosporin resistance can involve either a reduced affinity for an existing PBP, or the acquisition of a supplementary, β-lactam insensitive PBP (139).

The third general mechanism for resistance is permeability barriers, or inaccessibility of the target because of transport problems. In gram-positive organisms, the PBPs are located outside of the cytoplasmic membrane and problems from permeability barriers is minimal. However the situation with gram-negative bacteria is considerably more complex, and penetration to the target can be a problem.

Transport and Cell Penetration. One of the causes of bacterial resistance to the cephalosporins is poor transport of the antibiotic through the outer membrane of gram-negative bacteria. This lipid-bilayer membrane carries receptor proteins for the recognition and transport of essential nutrients, but provides an effective barrier to large molecules. In the case of the cephalosporins there can be a considerable difference between the concentration required to inhibit intact cells and the concentrations required to saturate the target enzymes in broken cell preparations. For example, the concentration of cefoxitin (**23**) required to inhibit the growth of whole organisms is ca 100-fold that needed to saturate the essential PBPs, whereas using cefuroxime (**35**) the minimum inhibitory concentration (MIC) is only four times the concentration required to saturate the PBPs (145).

Many small molecules can penetrate the outer cell membrane by diffusion through channels created by outer-membrane proteins called porins. Porin channels are implicated in the transport of cephalosporins because cells deficient in porins are much more impermeable than are cells that are rich in porins. The porins appear to function as a molecular sieve, allowing molecules of relatively low molecular weight to gain access to the periplasmic space by passive diffusion. In enterobacteria, a clear correlation exists between porin quantity and cephalosporin resistance, suggesting that the outer membrane is the sole barrier to permeability. However, such a relationship is not clearly defined for *Pseudomonas aeruginosa,* where additional barriers may be involved (139,144,146).

Bacteria have evolved efficient acquisition systems to meet specific needs. In the case of the transport of iron, an essential trace element, an iron-chelating ligand (siderophore) is used to sequester the iron, and this siderophore-iron complex is then recognized by specific receptors on the outer membrane. This is known as the *ton B*-dependent iron transport pathway because it appears to be under the general control of the *ton B* gene in *Escherichia coli.* The most common siderophores are catechols, α-hydroxy carboxylic acids or *N*-alkylated hydroxamic acids. In an attempt to exploit the iron-transport pathway a number of cephalosporins have been made containing catechol or the isosteric 3-hydroxypyrid-4-one group (147–150). These compounds, such as E-0702 [*77768-57-1*] (**52**), $C_{29}H_{21}N_7O_{12}S_2Na_2$, (148) and M14659 (**46**) (150), have good activity against *Pseudomonas aeruginosa* and *Enterobacteriaceae* and experiments do indeed implicate the iron-transport mechanism in the ability of these catechol cephalosporins to penetrate to their targets.

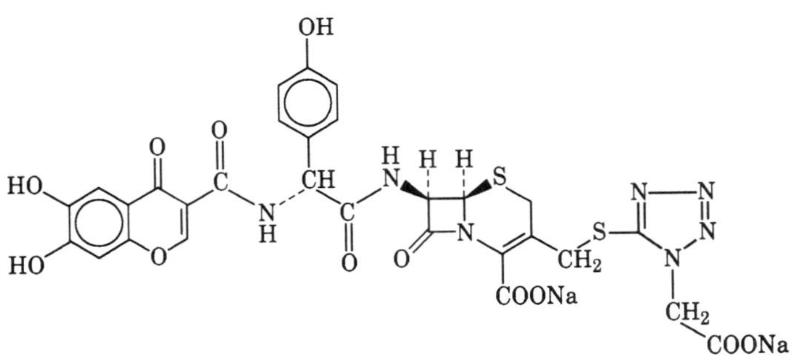

(**52**)

Pharmacokinetics. The pharmacokinetic properties of the cephalosporins depend to a large extent on the substituent at C-3. The 3′-acetoxy group is metabolized in the body to the less active 3′-alcohol. Most other substituents, including the 3′-carbamate, are metabolically stable. Most cephalosporins are eliminated rapidly, having serum half-lives of 1–2 hours. A number of the clinically useful cephalosporins contain the 1-*N*-substituted tetrazol-5-ylmercaptomethyl moiety in the C-3 position, eg, cefoperazone (**34**), moxalactam (**48**), cefmenoxime (**38**), cefamandole (**30**), and cefotetan (**26**). However, these compounds do not all have identical pharmacokinetic properties because there is always some interaction with, or contribution from, the 7-acyl group in determining the overall pharmacokinetic profile. When the tetrazoylmercaptomethyl 3′-substituent bears an acidic group such as found in cefonicid (**31**) or ceforanide (**33**), the rate of urinary excretion is reduced and higher, longer-lasting serum levels are obtained. Ceftriaxone (**39**), having an acidic group in its C-3 substituent, has extremely long-lasting serum levels. It appears that acidic C-3 substituents are associated with a very high level of serum binding which may contribute to the long-lasting serum levels. Most cephalosporins, both parenteral and oral, are renally excreted by glomerular filtration (free unbound drug) or tubular secretion (total drug). Some cephalosporins are readily secreted and have renal clearances via tubular secretion essentially independent of protein binding; others are eliminated predominantly by filtration, eg, ceftazidime (**40**), and exhibit renal clearance that is highly dependent on free drug concentration. A few of the cephalosporins, such as cefoperazone (**34**) and ceftriaxone (**39**) are also eliminated by biliary excretion. In such cases of dual excretion modes, the biliary excretion increases when renal function is impaired and, conversely, renal excretion increases when biliary excretion is impaired (80,151,152).

Manufacture and Chemical Synthesis

At present all of the cephalosporins are manufactured from one of four β-lactams, cephalosporin C (**2**), penicillin V [*87-08-1*], penicillin G [*113-98-4*], and cephamycin C (**8**), which are all produced in commercial quantities by fermentation (87). The manufacturing process consists of three steps: fermentation, isolation, and chemical modification.

Fermentation. The commercial β-lactam antibiotics which act as starting material for all of the cephalosporins are produced by submerged fermentation. The organisms used for the commercial production of the penicillins and cephalosporins are mutants of *Penicillin chrysogenum* and *Cephalosporium acremonium,* respectively (3,153,154). Both are true fungi (eucaryotes). In contrast, the cephamycins are produced by certain species of procaryotic *Streptomyces,* including *Streptomyces clavuligerus* and *Streptomyces lipmanii* (21,103).

Superior penicillin producing cultures are capable of producing in excess of 30 mg/mL of penicillin G (154). Cephalosporin producing strains, however, generally grow poorly and cephalosporin C production is not as efficient as is that of penicillin. Factors such as strain maintenance, strain improvement, fermentation development, inoculum preparation, and fermentation equipment requirements are discussed in the literature (3,154).

The β-lactam antibiotics are produced by secondary metabolic reactions that differ from those responsible for the growth and reproduction of the microorganism. In order to enhance antibiotic synthesis, nutrients must be diverted from the primary pathways to the antibiotic biosynthetic sequences. Although most media for the production of penicillins and cephalosporins are similar, they are individually designed for the specific requirements of the high yielding strains and the fermentation equipment used.

A typical fermentation medium for penicillin production contains lactose, corn steep liquor, and calcium carbonate (3,153,154). In most industrial processes the carbohydrate source, glucose, beet molasses, or lactose, is continuously added to the fermentation. The rate of glucose addition must be carefully monitored, by pH or rate of oxygen depletion, because the synthesis of penicillin is markedly reduced in the presence of excess glucose. Alternative sources of nitrogen, such as cottonseed, peanut, linseed, or soybean meal, have been substituted for corn steep liquor. Other ingredients that help to optimize penicillin production are monopotassium phosphate, for additional phosphorus and pH control, and sodium, ammonium or magnesium sulfate, for additional sulfur. Media for the large scale fermentative production of cephalosporin C generally contains fish meal, peanut meal or soybean meal, corn steep liquor, beet molasses, lard oil or methyl oleate, glucose, and methionine (155). In the cephalosporin medium, lipids, rather than glucose, provide the main source of carbon and energy. The sulfur for cephalosporin C is derived more efficiently from methionine than from sulfate. The yields of cephalosporin C are potentiated by both methionine and methyl oleate. The media and fermentation conditions for cephamycin C production are briefly discussed (3,103).

Isolation. Isolation procedures rely primarily on solubility, adsorption, and ionic characteristics of the β-lactam antibiotic to separate it from the large number of other components present in the fermentation mixture. The penicillins are monobasic carboxylic acids which lend themselves to solvent extraction techniques (154). Pencillin V, because of its improved acid stability over other penicillins, can be precipitated directly from broth filtrates by addition of dilute sulfuric acid (154,156). The separation process for cephalosporin C is more complex because the amphoteric nature of cephalosporin C precludes direct extraction into organic solvents. This antibiotic is isolated through the use of a combination of ion-exchange and precipitation procedures (157). The use of neutral, macroporous resins such as XAD-2 or XAD-4, allows for a more rapid elimination of impurities in the initial steps of the isolation (158). The isolation procedure for cephamycin C also involves a series of ion exchange treatments (103).

Chemical Modification. The chemistry and synthetic strategies used in the commercial synthesis of cephalosporins have been reviewed (87) and can be broadly divided into: (*1*) Selection of starting material; penicillin precursors must be rearranged to the cephalosporin nucleus; (*2*) cleavage of the acyl side chain of the precursor; (*3*) synthesis of the C-7 and C-3 side-chain precursors; (*4*) acylation of the C-7 amino function to introduce the desired acylamino side chain; (*5*) introduction of the C-3 substituent; and (*6*) protection and/or activation of functional groups that may be required.

For a viable commercial process, the selection of materials and the choice of synthetic route is governed primarily by cost, not by overall yield. The selection of starting material is dictated usually by the desired C-3 substituent. For cephalosporins containing 3-acetoxymethyl or 3-(substituted)methyl such as 3-thiomethyl and 3-aminomethyl derived moieties, the most direct synthetic route is from cephalosporin C, whereas pencillin V or G is the preferred starting material for the synthesis of the C-3 methyl cephalosporins. The three chemical transformations (*3*), (*5*), and (*6*) can potentially be carried out in a variety of ways, the precise sequence being determined by a balance of competing factors such as cost and optimization of yield (87).

Substituents at C-3. *3-Methyl Cephalosporins.* Direct reduction of the 3′-acetoxy group of compounds derived from cephalosporin C, either by catalytic hydrogenation (159) or by acid catalyzed reduction using trialkyl-silanes (160), leads to 3-methyl cephalosporins (87). However, the most economical route to this class of compounds involves the acid-catalyzed rearrangement of penicillin sulfoxides (161,162). Since the disclosure of this rearrangement in 1963, a tremendous number of papers dealing with the nature of the catalyst, solvent, optimal reaction parameters, and the reaction mechanism have appeared and been thoroughly reviewed (87,163–165). In addition, over twenty U.S. patents dealing with this process have been issued (87). The process requires that the carboxylic acid be protected, and high yields can be obtained by the correct choice of protecting group, solvent, and acid catalyst. The relatively stable *p*-nitrobenzyl, trichloroethyl, and benzhydryl esters of pencillin V or G sulfoxide undergo ring expansion best in either 1,4 dioxane or 1,1,2-trichloroethane with pyridine salts of phosphoric acid as the catalyst (166). This affords the cephalosporin with a carboxylic acid protecting group suitable for subsequent C-7 side-chain manipulation. Alternatively the carboxylic acid protecting group can be chosen such that deprotection occurs during isolation of the ring-expanded product. The best example is the use of trimethylsilyl protection. Thus reaction of penicillin G sulfoxide [*34104-15-9*] (**53**), $C_{16}H_{18}N_2O_5S$, with pyridinium hydrobromide (PyHBr) and bistrimethylsilylurea [*18297-63-7*] (BSU) in toluene followed by hydrolysis of the silyl ester affords the cephalosporin [*27255-72-7*] (**54**) in 97% yield (87,167).

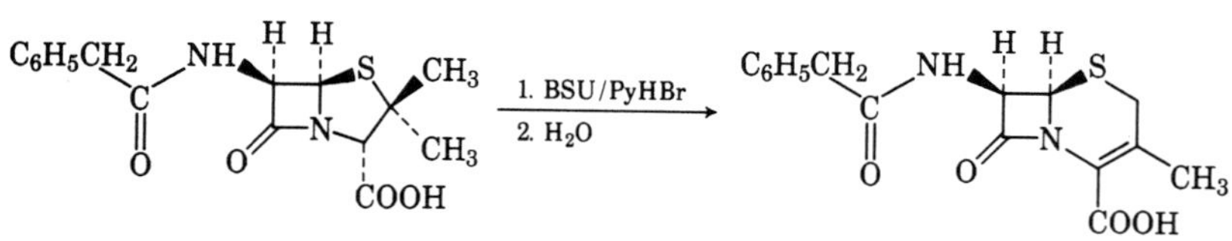

(**53**) (**54**)

3′-Heteroatom Cephalosporins. The most direct synthetic route to this class of compounds utilizes cephalosporin C (**2**) as the starting material, and thus requires the introduction of a new 3′-substituent. The 3′-acetoxy group can be readily displaced using a variety of nitrogen, sulfur, and carbon nucleophiles under both aqueous and nonaqueous conditions (75,87,168). However, oxygen nucleophiles do not readily displace the acetoxy group. These displacement reactions have been extensively studied, and tend to follow S_N kinetics (75,168). In fact under aqueous conditions, mechanistic studies show that displacement occurs via an S_N process with carboxylate stabilization of the intermediate carbonium ion (75,87,169,170). These displacement reactions generally cannot be carried out on cephalosporin esters, lactones, or sulfoxides.

Nitrogen nucleophiles used to diplace the 3′-acetoxy group include substituted pyridines, quinolines, pyrimidines, triazoles, pyrazoles, azide, and even aniline and methylaniline if the pH is controlled at 7.5. Sulfur nucleophiles include alkylthiols, thiosulfate, thio and dithio acids, carbamates and carbonates, thioureas, thioamides, and most importantly, from a biological viewpoint, heterocyclic thiols. The yields of the displacement reactions vary widely. Two general approaches for improving 3′-acetoxy displacement have been reported. One approach involves initial, or *in situ,* conversion of the acetoxy moiety to a more facile leaving group. The other approach utilizes Lewis or Brønsted acid activation (87).

Conversion to a more facile, sulfur-derived, leaving group can be achieved by treatment with sodium thiosulfate or salts of thio and dithio acids (75,87).

Under anhydrous conditions, boron tribromide converts the 3′-acetoxy group to a bromide whereas trimethylsilyl iodide gives good yields of the 3′-iodide (87,171,172). These 3′-halides are much more reactive, even when the carboxyl group is esterified, and can be displaced readily by cyano and by oxygen nucleophiles (127).

Methane sulfonic acid, trifluoroacetic acid, hydrogen iodide, and other Brønsted acids can facilitate 3′-acetoxy displacement (87,173). Displacement yields can also be enhanced by the addition of inorganic salts such as potassium thiocyanate and potassium iodide (174). Because initial displacement of the acetoxy by the added salt does not appear to occur, the role of these added salts is not clear. Under nonaqueous conditions, boron trifluoride complexes of ethers, alcohols, and acids also facilitate displacement (87,175).

An alternative to the displacement of a 3′-acetoxy group for entry into the 3′-heteroatom cephalosporins is functionalization of a 3-methyl cephalosporin. The low cost of the 3-methyl cephalosporins, derived from penicillin via the Morin rearrangement, makes this approach potentially competitive and attractive. Allylic bromination, using *N*-bromosuccinimide [*128-08-5*] (NBS) and photochemical initiation, of the esters of 3-methyl-Δ-3-cephem sulfoxides affords the desired 3′-bromo compound (87,176,177), although this reaction does not appear to work on the corresponding sulfide (75). Bromination can be carried out with a variety of substituents on the C-7 amino, but can suffer from significant bromination at C-2. However, the competing bromination at C-2 appears to be related to the nature of the substituent on the C-7 amino. Of particular note, the trichloroacetylamino group at C-7 completely suppresses the undesired C-2 bromination (176). Figure 3 outlines a particularly interesting application of this chemistry. Overbrominating, to afford a mixture of the 3′-monobromo compound [*71543-00-5*] (**55**) and corresponding C-2,3′-dibromo compound [*71542-99-9*] (**56**) is deliberate. This mixture is then treated with trimethylphosphite to effect selective reduction at C-2, and the resulting 3′-monobromo cephalosporin (**55**) can then be subjected to nucleophilic displacement (87,178).

3-Hydroxymethyl Cephalosporins. The 3-hydroxymethyls are a special case of the 3′-heteroatom cephalosporins. These deacetylcephalosporins serve as precursors for the 3′-carbamoyl cephalosporins, such as cefuroxime (**35**) and cefoxitin (**23**). Treatment of a cephalosporin acid or ester to hydrolytic deacylation under basic conditions often suffers from the problems of double bond migration and/or *β*-lactam cleavage. However, deacylation of 7-aminocephalosporanic acid (7-ACA) (**1**), derived from cephalosporin C (**2**) by treatment with sodium hydroxide at 0°C affords the corresponding 3′-hydroxy compound in good yield (64%) (87,179,180). This deacylation can also be achieved by enzymatic hydrolysis, using a variety of esterases and substrates (75,179–182). The mild selective conditions of enzyme catalyzed cleavage make this the process of choice for industrial production and such a process using a yeast, *Rhodotorula rubra* as the source of esterase activity is employed in the commercial synthesis of cefuroxime (**35**) (87,182).

3-Methylene Cephalosporins and 3-Heteroatom Cephalosporins. A number of clinically important cephalosporins have a hydrogen or a heteroatom at C-3 instead of a methyl or substituted methyl, requiring removal of the C-3 methyl from either of the commercially available starting materials. Removal is conven-

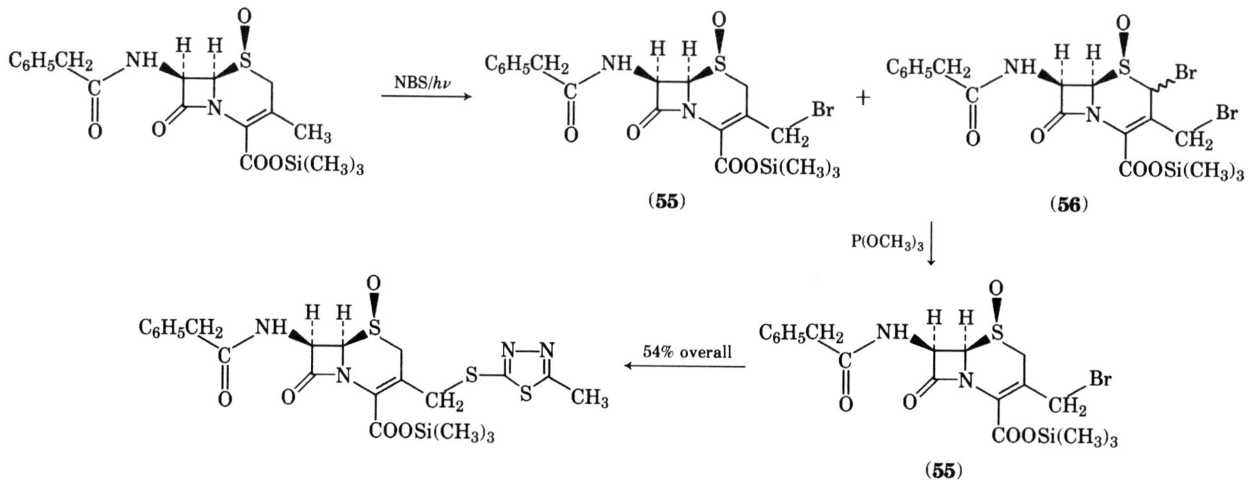

Fig. 3. Allylic bromination at C-3′; NBS = *N*-bromosuccinimide.

iently achieved via ozonolysis of an exocyclic methylene group, and a number of routes have been developed to produce such 3-methylene cephalosporins from both penicillins and cephalosporins derived from cephalosporin C. Industrially, the most viable routes to 3-methylenecephams are via penicillins, and thus the penicillins serve as the precursors for the C-3 heteroatom cephalosporins (87). This chemistry has been reviewed rather extensively, along with the subsequent conversion into heteroatom substituted cephalosporins (164,183–185). One of these routes takes advantage of advances in penicillin sulfoxide chemistry and involves the rearrangement of the sulfoxide to the 3-exomethylenecepham. This is exemplified in the synthesis of cefaclor (**13**) shown in Figure 4 (185,186).

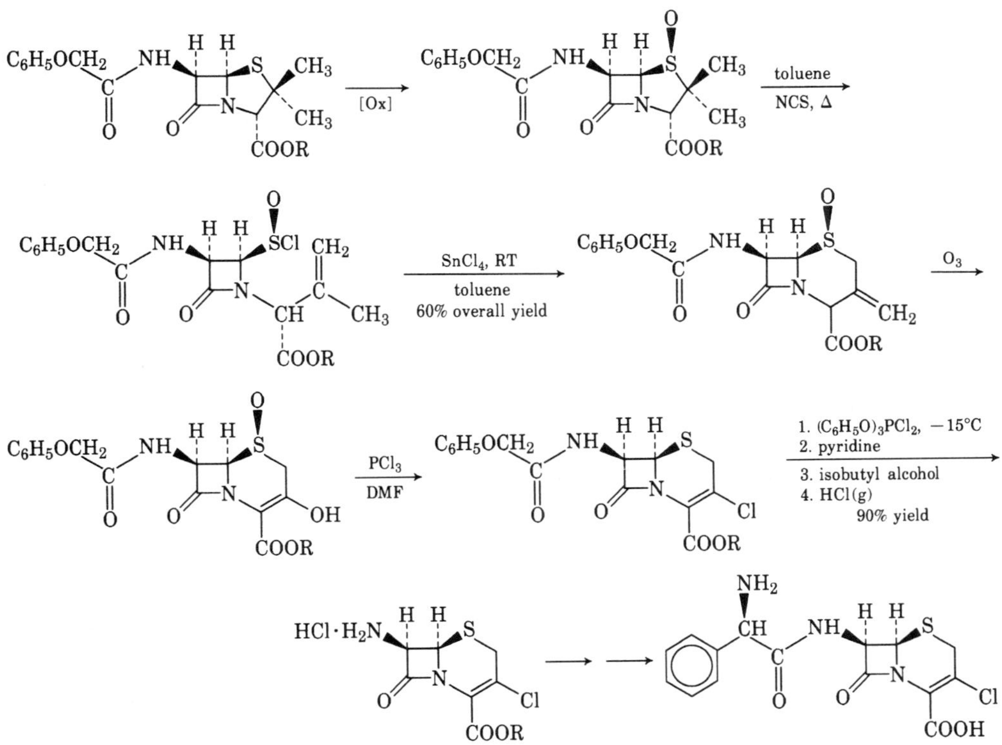

(**13**)

Fig. 4. Synthesis of cefaclor (**13**) where DMF is dimethylformamide.

7β-Substituents. C-7 Side-Chain Cleavage. 7-ACA (**1**), the key intermediate for the synthesis of a large number of the clinical cephalosporins, cannot be made by fermentation, nor can it be readily obtained by enzymatic removal (187) of the aminoadipic acid side chain from cephalosporin C. Because substantial quantities of 7-ACA are used, an efficient method for the chemical cleavage of the acyl side chain of cephalosporin C was sought. The first efficient method involved diazotization using nitrosyl chloride, followed by hydrolysis or alcoholysis, and gave 7-ACA in 50% yield (22,23,87,188). The reaction proceeds via the formation of the cyclic imino ether (**57**) which is easily hydrolyzed under very mild conditions.

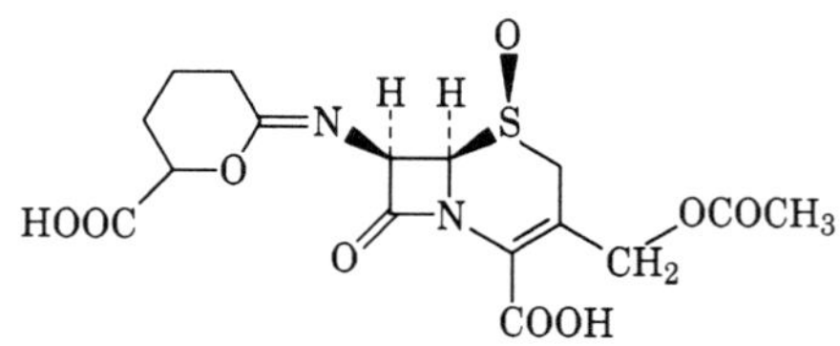

(**57**)

Subsequently, a general procedure for amide side-chain cleavage was developed that involves initial conversion of the amide linkage into an imino chloride, achieved under anhydrous conditions using phosphorus trichloride, or more preferably phosphorus pentachloride. Alcoholysis to the imino ether, followed by hydrolysis affords 7-ACA (22,23,87,189). The process requires protection of both the free carboxyl and the amino group. This can be achieved either by reaction with acetyl chloride before cleavage or by silylation of the anhydrous sodium salt as shown in Figure 5. In the latter case, the overall yield of 7-ACA is 92.5%. A procedure for the removal of the acyl-amino side chain of cephalosporin C, using an immobilized enzyme has been reported (187). However, this process requires either the prior removal, or protection, of the adipoyl side-chain amino group, before enzymatic cleavage of the amide bond can be effected.

C-7 Amino Acylation. Essentially all the known methods for amine acylation, used in the peptide field, have been applied to cephalosporins. These procedures generally involve acyl group activation prior to the acylation (76). Commercially, the most common activation probably involves either mixed anhydride or acid chloride formation (87). Acylations via acid chloride activation are usually carried out under Schotten-Baumman conditions eg, aqueous acetone, 0°C, sodium bicarbonate, pH 7–8. A wide variety of active esters have also been used in cephalosporin chemistry, especially heterocyclic-thio esters and *o*-acyl substituted hydroxylamines, either preformed or generated *in situ,* as well as coupling methods employing dicyclohexylcarbodiimide (DCC) and similar condensing agents. The use of active esters in the acylation step is exemplified by the synthesis of ceftriaxone (**39**). In order to circumvent the poor solubility of 7-ACA in organic solvents, prior formation of a soluble derivative, usually an ester, is required for most activation methods using nonaqueous conditions. Silylation of 7-ACA meets this requirement and the silylated moiety has the advantage of easy removed during the workup (87). Other commonly used esters include benzhydryl, *t*-butyl, and less frequently *p*-nitrobenzyl, *p*-methoxybenzyl, trichloroethyl, and allyl. However, the use of esters in the acylation step, under conditions requiring an organic base, often leads to the problematic isomerization of the ring double bond (75).

7α-Substituents. A large number of cephalosporins with a variety of substituents in the 7α-position have been prepared and their biological properties and antibacterial activities evaluated. The substituents introduced into the 7α-position include groups such as CH_3–, CH_3CH_2–, $-CH_2OH$, $-CH_2F$, $-CH_2NH_2$, $-COOH$, $-CHO$, $-COCH_3$, $-OCH_3$, $-OCH_2CH_3$, $-OCH_2CN$, $-OCH_2CH_2OCH_3$, $-SCH_3$, $-CN$, $-Br$, $-SCH_3$, $-CH(COOCH_2CH_3)_2$, $-N_3$, $-NHCHO$, $-PO(OCH_3)_2$, and $-SCH_3$ (25–30). Of these 7α-substituted cephalosporins, only those with the methoxyl and formamido group have interesting antibacterial activity. A number of

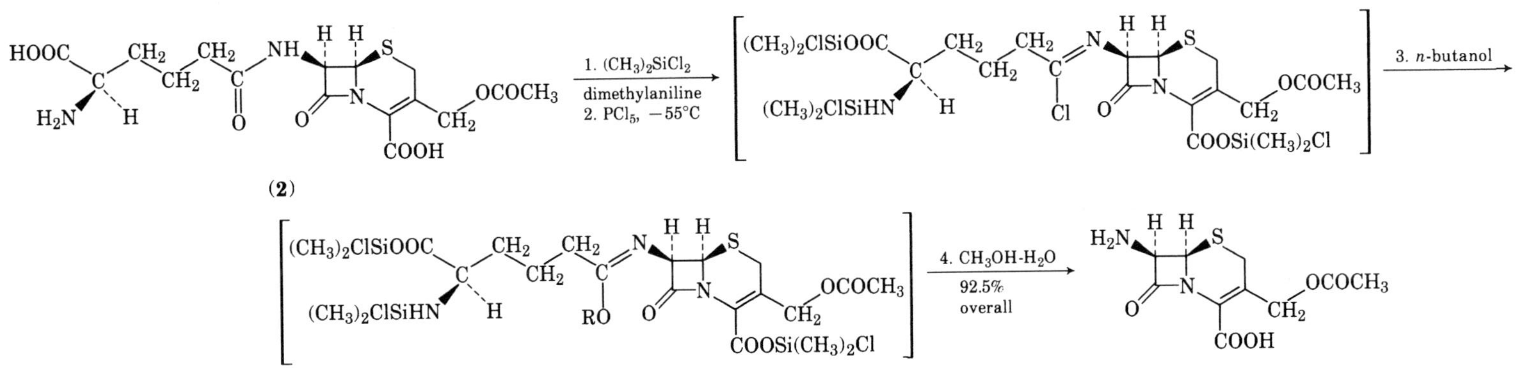

(**2**) (**1**)

Fig. 5. C-7 side-chain cleavage.

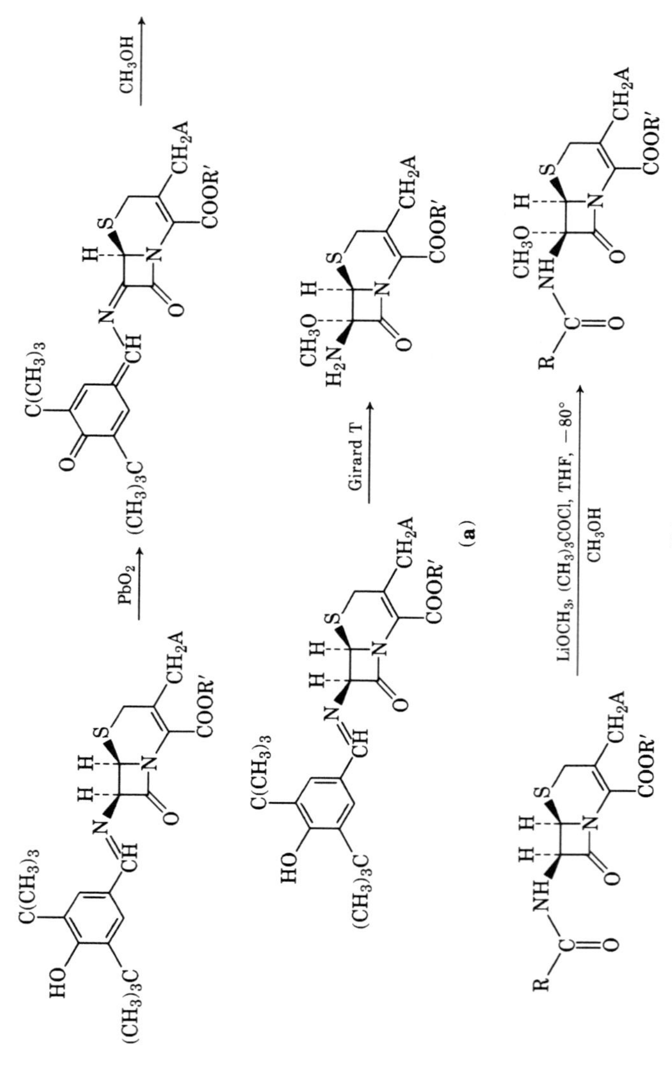
PbO$_2$
CH$_3$OH
Girard T
(a)
LiOCH$_3$, (CH$_3$)$_3$COCl, THF, −80°
CH$_3$OH
(b)

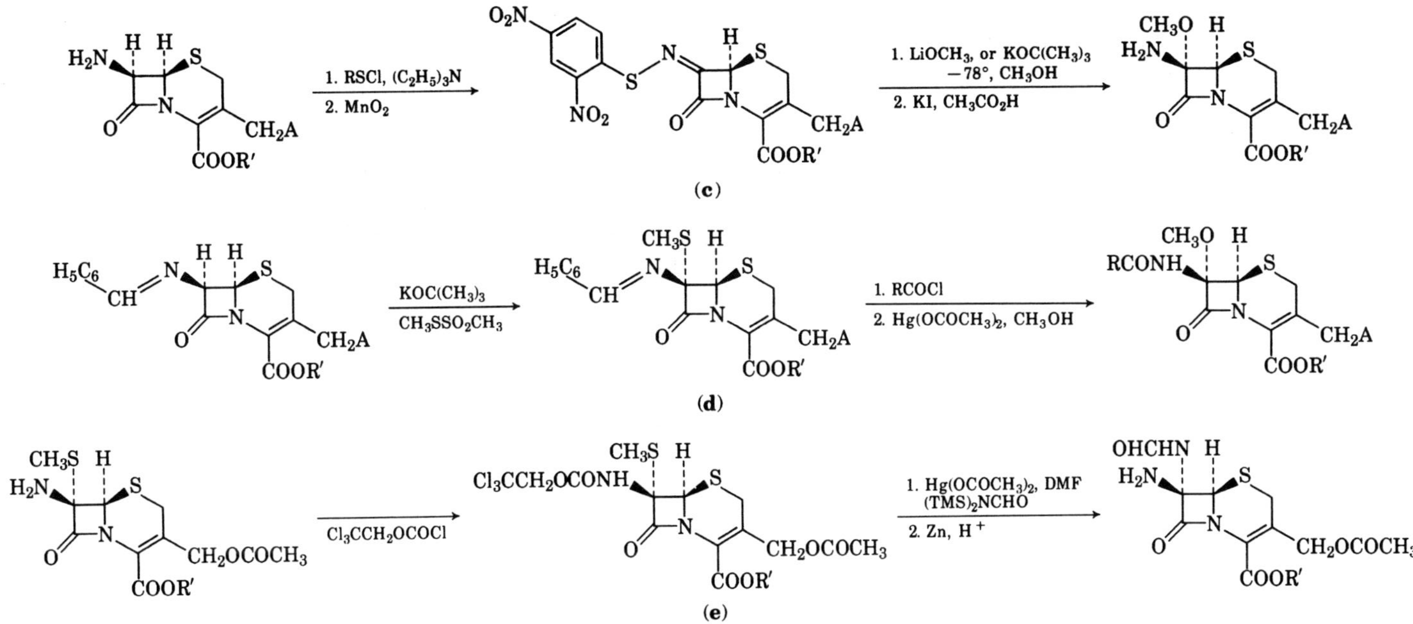

Fig. 6. Synthetic routes to **(a)**–**(d)** 7α-methoxycephalosporins and **(e)** 7α-formamidocephalosporins where THF = tetrahydrofuran, and DMF = dimethylformamide, and TMS = trimethylsilane.

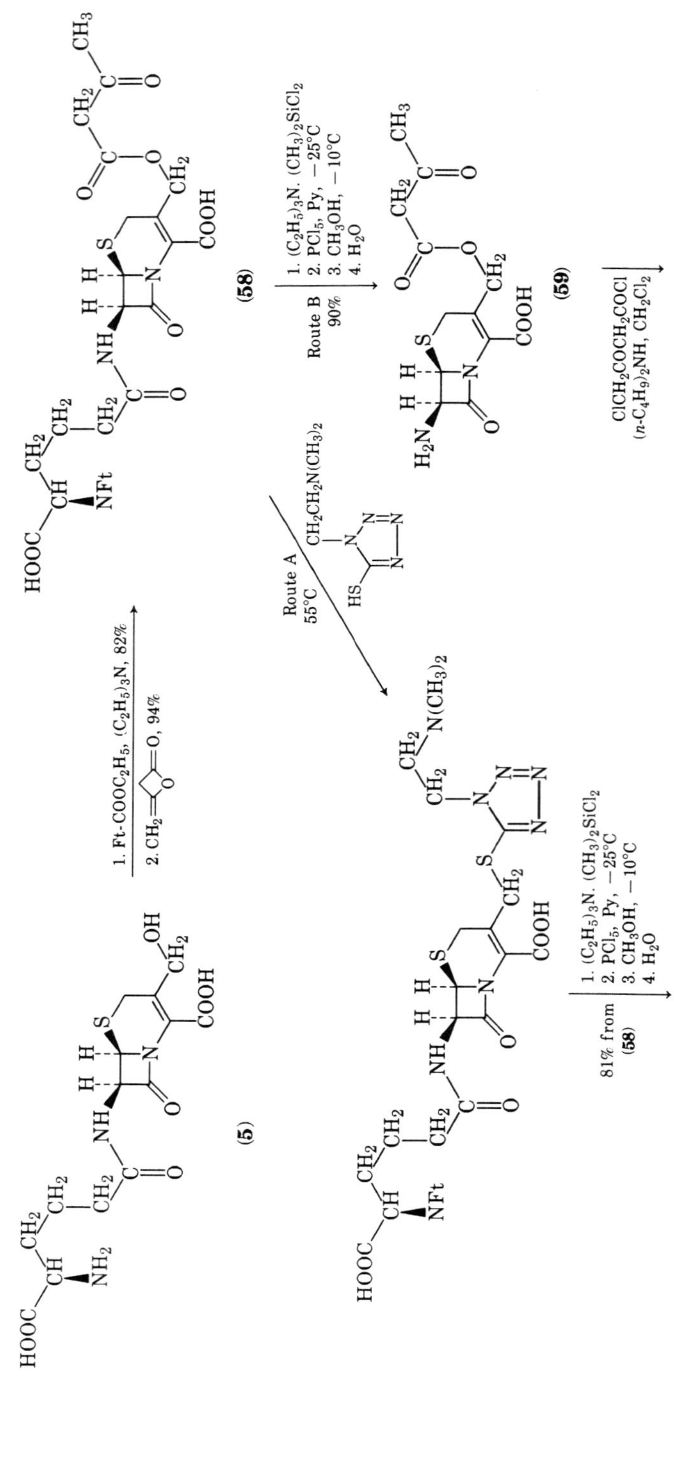

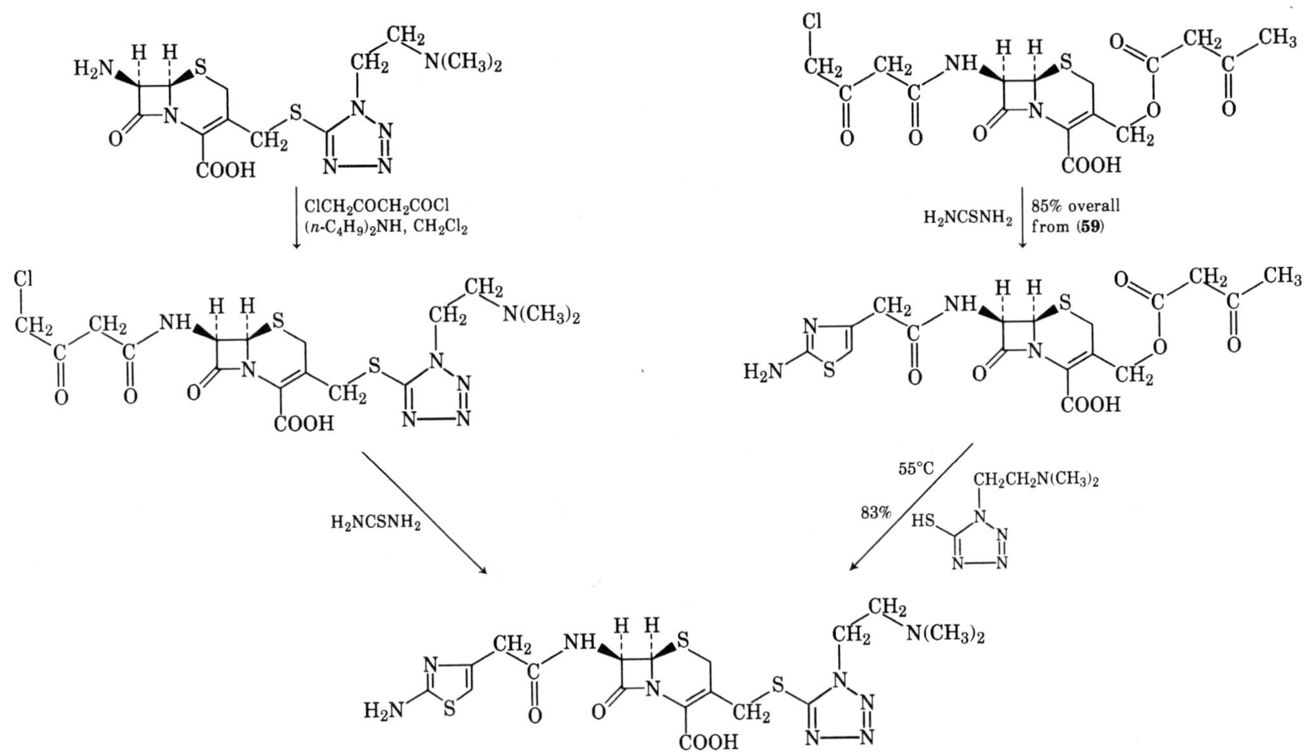

Cefotiam (**50**)

Fig. 7. Synthesis of cefotiam (**50**) where Ft is phthalimido and Py is pyridine, C_5H_5N.

synthetic routes to such compounds have been described and reviewed (27). The key steps for the introduction of the 7α-substituent in three of these approaches to the 7α-methoxy substituted cephalosporins are depicted in Figure 6 (27). Figure 6e outlines the introduction of the 7α-formamido substituent (28–30).

Synthetic Strategy. For a commercial process the deciding factor in a synthetic strategy is overall cost, although other factors, such as patent rights and material and facility availability also play a role. Generally the goal is to introduce the more expensive materials as late in the total synthesis as possible (87), and to use a convergent rather than a long linear approach. These factors generally favor complete construction of side chains before attachment to the cephalosporin nucleus and most routes follow this approach. The variety of routes to some of the more common, newer C-3 and C-7 side chains have been compared in a review (87). The side chains may also be built up on the cephalosporin nucleus, generally 7-ACA, and furthermore, the sequence by which side chains are either built up or introduced intact, can also be varied. The syntheses of cefotiam (**50**), ceftazidime (**40**), and ceftriaxone (**39**) are described by the reaction sequences given in Figures 7–9, respectively. These are not necessarily the commercial, or the most efficient, routes.

The schemes depicted in Figure 7 contrast two complimentary approaches to cefotiam (**50**) in which the timing of the introduction of the C-3 substituent differs. In Route A the heterocyclic thiol C-3 substituent is introduced even before the removal of the aminoadipoyl acyl side chain. The acetonylacetyl C-3 substituent was introduced because it gave considerably higher yields and cleaner product in the nucleophilic displacement step than the corresponding acetoxy, and the starting material, deacetylcephalosporin C (**5**) was readily available from a fermentation process (190,191).

In the synthesis of ceftazidime (**40**) (Fig. 8), the protected, preassembled aminothiazole side chain [*68672-66-2*] (**60**) is coupled to a protected 7-ACA first and the C-3 displacement step carried out last. By way of contrast, in the synthesis of ceftriaxone (**39**) (Fig. 9), the preformed C-3 substituent is introduced onto the cephalosporin nucleus in the first step and then the acyl-amino side chain is introduced. This last step is noteworthy for two reasons in that it demonstrates the use of an activated thio ester in the coupling step and that no protecting group chemistry is required (192,193).

The substitution of oxygen for sulfur, as found in the 1-oxadethiacephems such as moxalactam (**48**), introduces a new synthetic problem in that the oxadethiacephem nucleus is not naturally occurring. Synthetic routes to the oxadethiacephems have been investigated fairly extensively (194,195) and in all the routes published the key step is the stereocontrolled introduction of the ring oxygen function. The synthesis of moxalactam (**48**), depicted in Figure 10, efficiently retains all the carbons of its starting material, 6-aminopenicillanic acid [*551-16-6*] (6-APA) (**61**), $C_8H_{12}N_2O_2S$ (194–196).

Nuclear Analogues of Cephalosporins

In the search for improved antibacterials not only has the effect produced by the variation of the C-7 amido side chain and the 3′ substituent been studied, but so also has the more synthetically challenging question of the effect of changes in

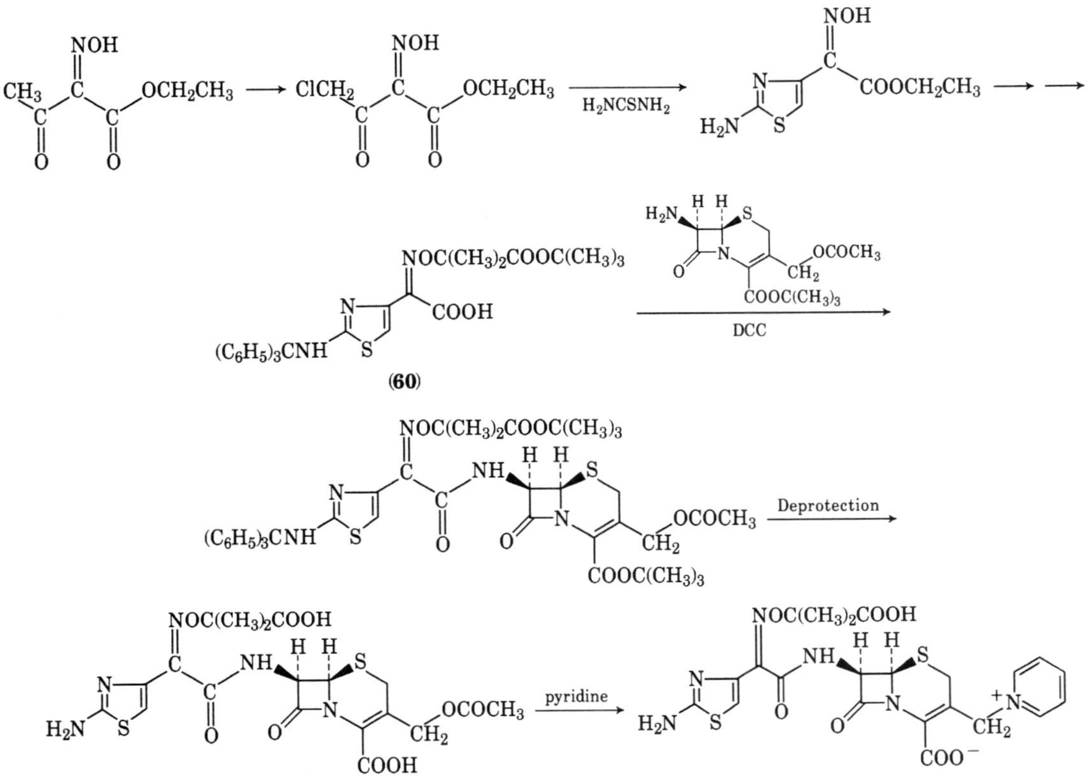

Ceftazidime (**40**)

Fig. 8. Synthesis of ceftazidime (**40**); DCC = *N,N*-dicyclohexyl-carbodiimide.

the cephem nucleus (194,197,198). Nuclear analogues have been studied since the early 1970s but only the oxacephem class has reached the marketplace.

Oxadethiacephalosporins. The first compound in which the sulfur of a cephalosporin was replaced by oxygen was 1-oxa-dethiacephalothin [*54214-83-4*] (**64**), $C_{16}H_{16}N_2O_7S$, (199) which is approximately twice as active as cephalothin (**27**) taking into account that the synthetic material was racemic. The oxygen analogue [*62504-53-4*], (**65**, X = OH, R = 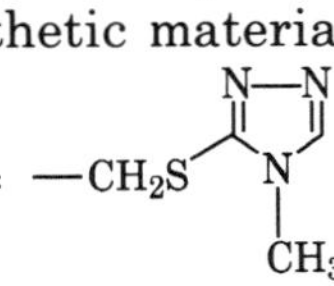) of cefamandole (**30**) was as good as, or better than, cefamandole itself, except for activity against *Proteus mirabilis* (200). Other studies have demonstrated that the 1-oxa-dethiacephalosporins are for the most part comparable to, or slightly better than, the cephalosporin counterpart in antibacterial activity, although there are some exceptions, such as 1-oxa-dethiacephalexin [*66204-07-7*] (**65**, X = NH_2, R = H), $C_{15}H_{15}N_3O_5$, which was inactive (201). Both the 1-oxa-dethiacephalosporins having a mandelamido C-7 side chain and either a methyl (**65**, X = OH, R = CH_3) or a hydrogen (**65**, X = OH, R = H) at C-3 had properties comparable to cephalexin (**12**) (201). The effect of various substituents at C-3, C-7β, acylamino side chains, and C-7α, methoxy and hydrogen, in the 1-oxa-dethiacephalosporin system have been exten-

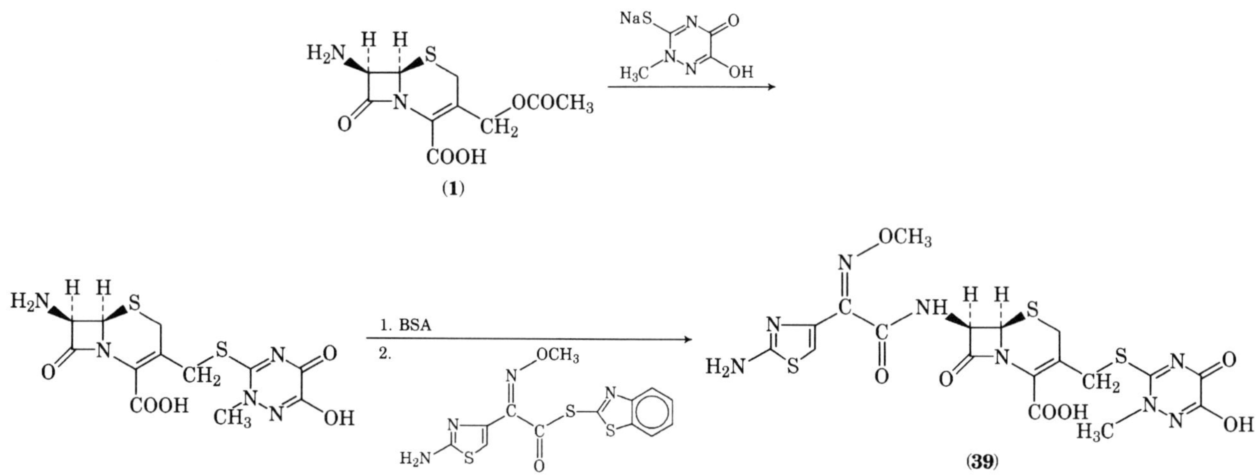

Fig. 9. Synthesis of ceftriaxone (**39**) where BSA is *N,O*-bis (trimethylsilyl) acetamide.

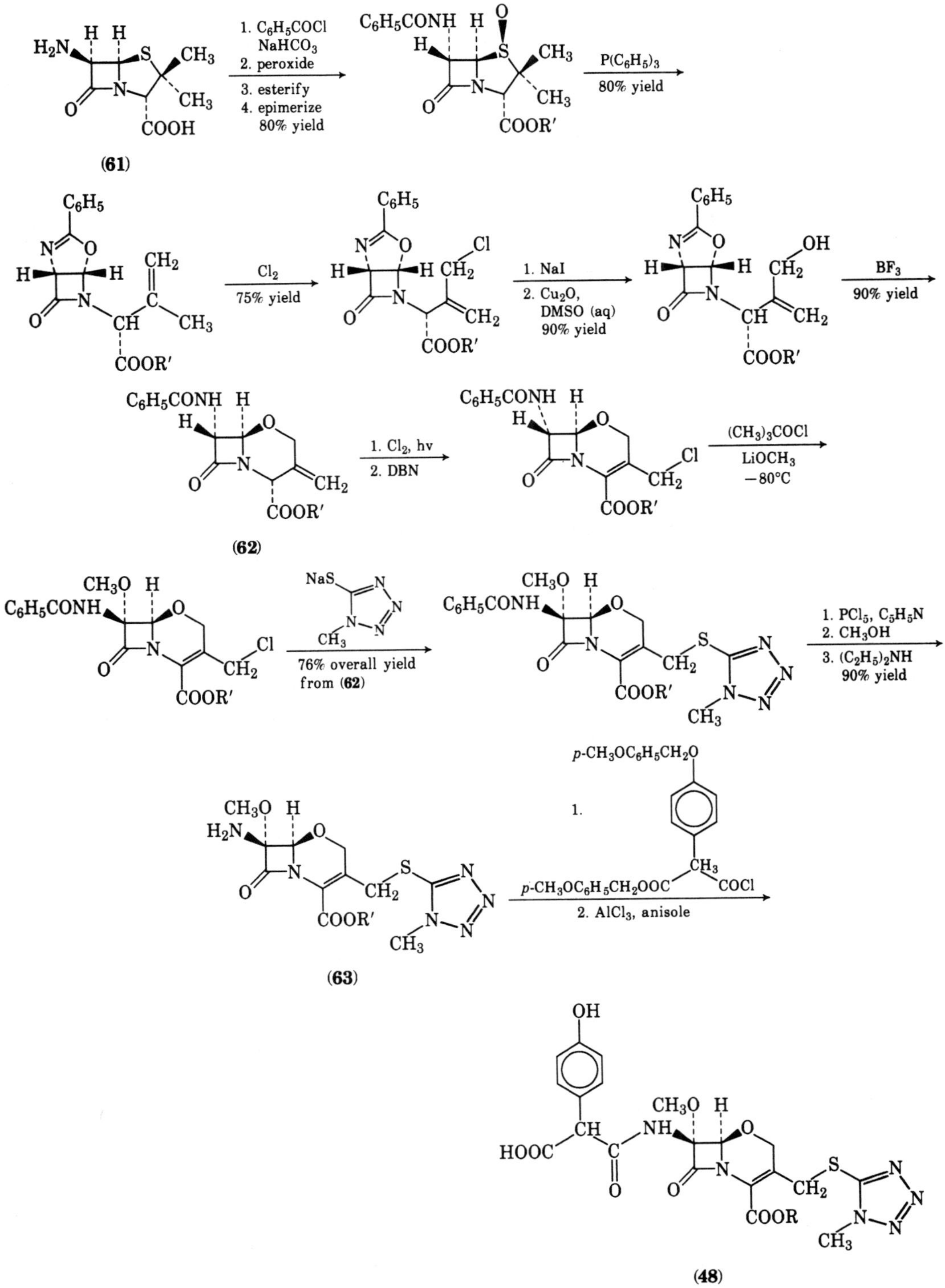

Fig. 10. Synthesis of moxalactam **(48)** from 6-aminopenicillanic acid (6-APA) where R′ is $CH(C_6H_5)_2$, DMSO is dimethyl sulfoxide and DBN is 1,5-diazabicyclo[4.3.0]non-5-ene. The overall yield of **(63)** is 26% from 6-APA in the 13 steps shown.

sively studied (194). These structure-activity investigations led to the development of moxalactam (**48**) (194,195). The substitution of oxygen for sulfur in moxalactam (**48**) affords a several-fold increase in intrinsic activity, but introduces the synthetic challenge of the oxacephem nucleus.

The antibacterial spectrum of moxalactam (Table 9) is similar in breadth and potency to that of cefotaxime (**36**). Hence, moxalactam (**48**) is classified with the third-generation cephalosporins. In general 1-oxacephalosporins are considerably more susceptible to β-lactamases than their sulfur counterparts (194). However, moxalactam is highly resistant to a broad spectrum of β-lactamases owing to the effect of the 7α-methoxy group (194,195).

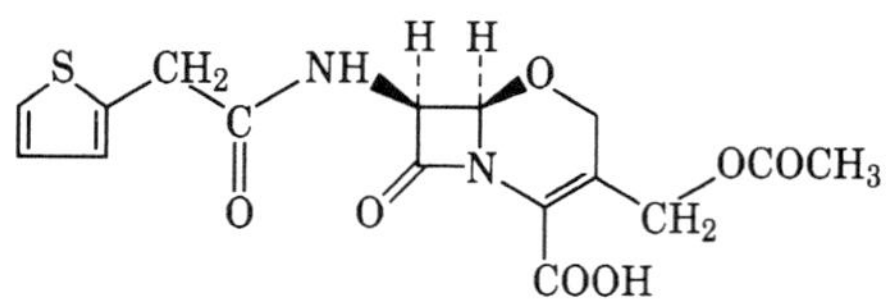

(**64**)

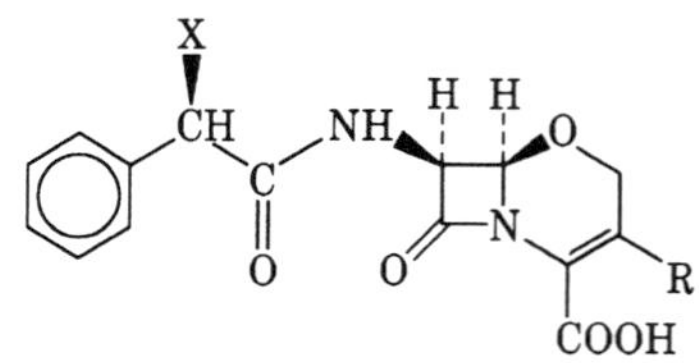

(**65**)

Isocephems. 1-Dethia-2-thia-cephalosporins, more conveniently known as isocephems, have attracted interest mostly from a synthetic perspective (197), but analogues have been found, eg, (**66**, X = S, Y = 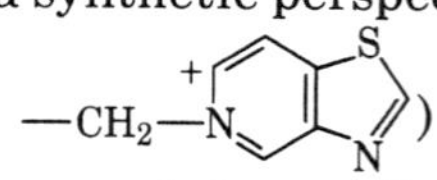) which exhibit good antibacterial activity *in vitro* (198). Structure-activity studies in the series clearly demonstrate the need for a polar, preferably positively charged potential

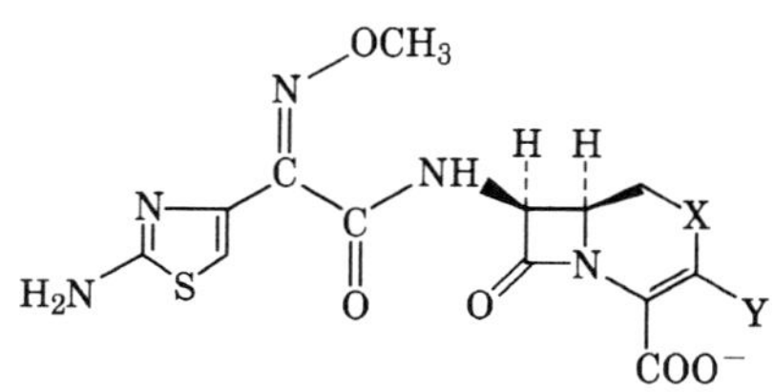

(**66**)

leaving group in the 3′-position to obtain good antibiotic activity. The insertion of a *z*-double bond between the 3- and 3′-position of the isocephems, eg, [*121037-46-5*] (**66**, X = S, Y = 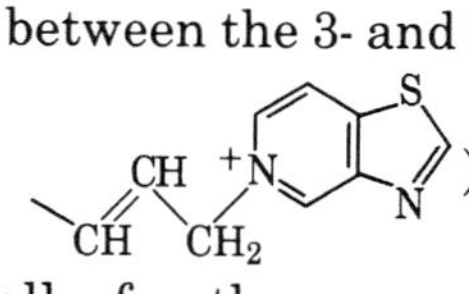), profoundly effects the antibacterial potency (198,202) especially for the gram-positive bacteria where the improvement in potency can be 10-fold. The corresponding 2-oxa-isocephems have also been investigated, eg, [*123581-13-5*] (**66**, X = O, Y = 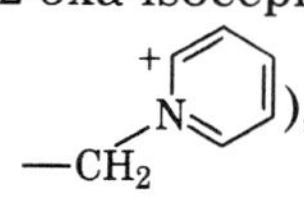), but were found to not be quite as potent as the corresponding isocephem and more susceptible to β-lactamase inactivation (198).

Carbacephems. The carbacephems in which the sulfur group is replaced by a methylene, were introduced in the 1970s (197,200,203,204). Generally these 1-carba analogues possess antibacterial activity that is similar to the cephalosporin counterpart (197,204), and have the advantage of possessing greater chemical stability (204). Thus, for example, locarbacef [*76470-66-1*] (LY163892 or KT3777) (**67**), $C_{11}H_{16}N_2O_4Cl$, has antibacterial activity similar to that of cefaclor (**13**) (205), but is 140 times more stable than cefaclor in pH 7.4 buffer (206). Furthermore, human pharmacokinetic studies demonstrate that locarbacef has a longer half life and better oral bioavailability than cefaclor (207). A new series of 3-acyl-1-carbacephems with promising activity and pharmacokinetic properties has been reported (208). Carbacephems having an iminoaminothiazole acyl side chain and a quaternary nitrogen group directly attached to the ring in the C-3 position, eg, LY258360 [*125615-06-7*] (**68**), $C_{18}H_{19}N_7O_5S$, are reported to exhibit improved potency compared to the third-generation cephalosporins, with the exception of their poor activity against *Pseudomonas* (209).

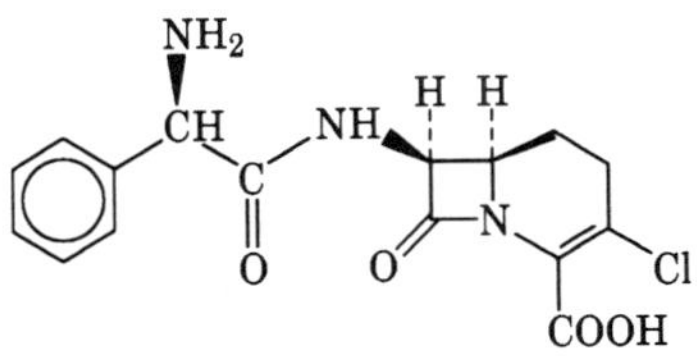

(**67**)

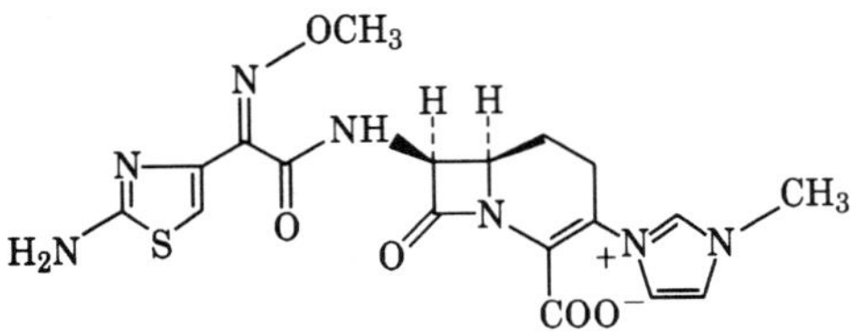

(**68**)

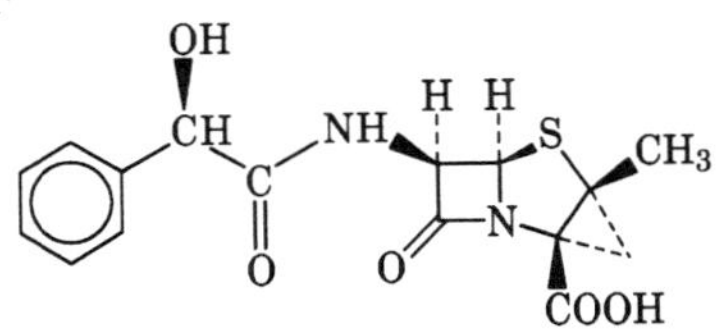

(**69**)

α-Cyclopropylpenams. Although at first glance (2,3)-α-methylenepenams such as [*109279-36-9*] (**69**), look like penicillins, these compounds can be considered as nuclear analogues of the cephalosporins. The profile of their antibacterial activity and their mode of action support this view. The penicillin nucleus in these compounds is locked in the open conformation, which is the proposed biologically active form for pencillins. However, the absolute configuration at the carboxylic acid bearing carbon (C-2) is opposite to the naturally occurring penicillins, and places the carboxyl group in a position similar to that found in cephalosporins. Thus the compounds exhibit an antibacterial profile closer to those of their cephalosporin analogues than to that of the penicillin counterpart. Consistent with this, the compounds in this series show a marked selectivity for PBP-3 of *Escherichia coli,* and are more susceptible to cephalosporinases than penicillinases (210).

Cephalosporins with Special Properties

Chromogenic Cephalosporins. A 3-substituted pyridinium cephalosporin, known as PADAC [*77449-91-3*] (**70**), $C_{27}H_{26}N_6O_4S_2$, is purple in color but on hydrolysis or treatment with β-lactamase releases the 3′-pyridinium group with concomitant loss of the purple color (211). PADAC is an example of the chromogenic cephalosporins which are useful in studying interactions with β-lactamases. Another example is nitrocefin [*41906-86-9*] (**71**), $C_{21}H_{16}N_4O_8S_2$, which is light yellow in solution, $\lambda_{max} = 386$ nm. On hydrolysis, or treatment with β-lactamase, the color changes to a deep red, $\lambda_{max} = 482$ nm, but unlike the situation with PADAC the nitrocefin 3′ group is not expelled (212) (see CHROMOGENIC MATERIALS).

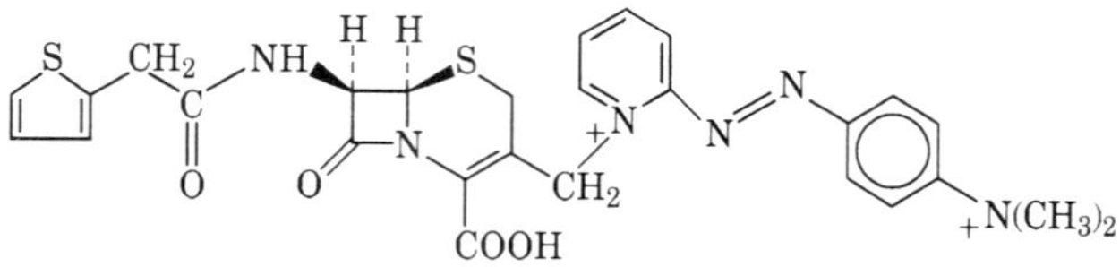

(**70**)

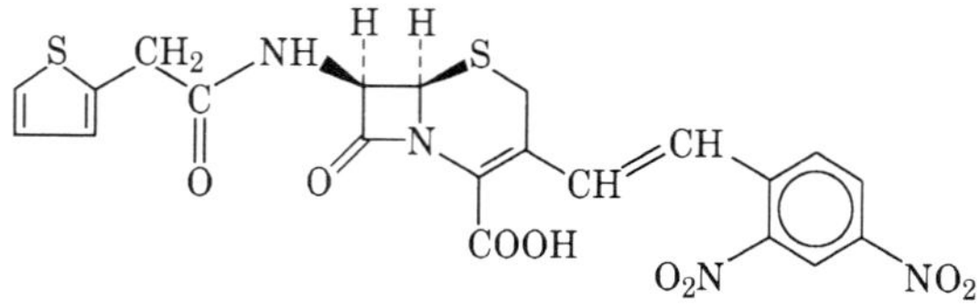

(**71**)

Uses

The cephalosporins are used for treating infectious diseases of bacterial origin in both humans and animals (80,132,213–218). First-generation cephalosporins such as cephalothin (**27**) and cephalexin (**12**) are the most active against staphylococci and nonenterococcal streptococci and are effective alternatives to the penicillins in patients with endocarditis, osteomyelitis, septic arthritis, and cellulitis (213). They are especially useful for treating patients who are allergic to the pencillins or who have mixed infections from gram-positive and gram-negative bacteria. Although these drugs have proved useful in treating infections such as bacteremias, urinary tract infections, and pneumonias, caused by gram-negative bacilli, their use as single agents in this regard is not recommended, because activity against gram-negative organisms is somewhat weak and unpredictable. The first-generation cephalosporins have been widely used for prophylaxis in cardiovascular, orthopedic, biliary, pelvic, and intra-abdominal surgery (132,214). Cefazolin (**32**), which has a longer half-life than the other first-generation compounds, is the first-generation agent of choice for surgical prophylaxis (132,213,214).

Cefuroxime (**35**) is effective against community-acquired pneumonia in which ampicillin-resistant *Haemophilus influenzae* is the probable etiologic agent. Cefoxitin (**23**) is used to treat mixed aerobic–anaerobic infections including pelvic infections, intra-abdominal infections, and nosocomial aspiration pneumonia. Cefonicid (**31**), because of its long half-life has been used in a once-a-day regimen to treat a variety of mild to moderate infections including community-acquired pneumonias, urinary tract infections, and infections of the skin and soft tissue (132,215).

Whereas third-generation cephalosporins do have some coverage against gram-positive infections, they are not the agents of choice. Similarly, most community-acquired infections are better treated with drugs other than the third-generation cephalosporins. The treatment of meningitis is an important exception (132,216). Cefotaxime (**36**), ceftriaxone (**39**), and ceftazidime (**40**) have proven effective in treating meningitis, especially in children where *Haemophilus influenzae, Streptococcus pneumoniae,* and *Neisseria meningitidis* are the leading etiologic agents. In meningitis caused by enteric gram-negative bacilli (community-acquired or nosocomial) the third-generation cephalosporins are clearly the drugs of choice. Ceftazidime (**40**) has even been used to treat patients with *Pseudomonas aeruginosa* meningitis with an 80% response rate.

The third-generation cephalosporins are effective in the treatment of bacteremias, pneumonias, urinary tract infections, intra-abdominal infections, and skin and soft tissue infections. Because of its long half-life, cetriaxone (**39**) has been used for outpatient treatment of skin and soft tissue infections. Single dose ceftriaxone (**39**) has also been effectively used in the treatment of urinary tract infections (132,215). Ceftriaxone is now the agent of choice for treating uncomplicated urethral, anorectal, or pharyngeal gonorrhea, and gonococcal opthalmia, including infections caused by penicillin-resistant strains (132,217). There is no clearly defined role that cephalosporins play in the treatment of syphilis, mycoplasma or clamydial infections, or bacterial vaginosis, but ceftriaxone may be effective in the treatment of chancroid.

Toxicity

The cephalosporins generally cause few side effects (80,132,219–221). Thrombophlebitis occurs as a result of intravenous administration of all cephalosporins. Hypersensitivity reactions related to the cephalosporins are the most common side effects observed, but these are less common than found with the penicillins. Clinically only about 5–10% of patients with allergic reactions to the penicillins manifest the same reactions to the cephalosporins, and data would contradict any true cross-reactivity to cephalosporins in patients who have previously reacted to penicillin (80,132,219).

Although immediate reactions of anaphylaxis, bronchospasm, and urticaria have been reported, most commonly patients exhibiting an adverse reaction develop a maculopapular rash, usually after several days of therapy. They may also develop fever and eosinophilia (80,219). Cefoperazone (**34**) and ceftriaxone (**39**), having greater biliary excretion than other cephalosporins, are associated with an increased risk of diarrhea, which may be caused by selection of cytotoxin producing stains of *Clostridium difficile* (219).

Generally, nephrotoxicity is not a problem. Some cephalosporins, especially those with the 3-methylthiotetrazole side chain, such as moxalactam (**48**), show a tendency to promote bleeding. This appears to be due to a reduction in the synthesis of prothrombin and can be a problem especially in elderly patients, patients with renal insufficiency, or patients suffering from malnutrition (219). The same side chain seems to promote a disulfiramlike reaction in patients consuming alcohol following a cephalosporin dose (80,219).

Economic Aspects

Cephalosporins first entered the marketplace in 1964, when cephalothin (**27**) and cephaloridine (**51**), which are both injectable, were launched. By the late 1970s, the injectable cephalosporins had become important therapeutic agents in the hospitals. Also in 1964 the first oral cephalosporin, cephaloglycin [*3577-01-3*], $C_{18}H_{19}N_3O_6S$, was launched only to be displaced by the end of the year by cephalexin (**12**). For years cephalexin was the leading oral cephalosporin on the market. It has since been displaced by cefaclor (**13**). With the advent of the more β-lactamase stable cephalosporins such as cefoxitin (**23**) and cefuroxime (**35**), and the more potent agents such as cefotaxime (**36**) and other third-generation compounds, cephalosporins now dominate the antibiotic market worldwide.

On a worldwide basis, total pharmaceutical sales for 1983 were $71 billion, antibiotic sales totaled $9.2 billion, and cephalosporins accounted for $1.2 billion, or 13% of the antibacterial market (87). By 1986 the worldwide antibiotic market was valued at around $11 billion and cephalosporins represented approximately 41% of the total. Marketing projections for antibiotics from 1986 to 1991 show an average annual growth rate of 3.7% to $13,144 million. Cephalosporin sales are projected to reach 40% or $5,234 million (222). Worldwide, in 1988, the top selling cephalosporin was the oral cephalosporin Ceclor (**13**) which had 1988 sales of $605 million and projected 1989 sales of $696 million. The second best selling, and top parenteral, cephalosporin was Rocephin (**39**) which had 1988 sales of $475.8 million and projected sales of $597 million for 1989. The third cephalosporin on the list was Claforan (**36**) which had 1988 sales of $376.2 million and a projected decline in sales for 1989 to $353 million as it is displaced by the newer cephalosporins (223).

BIBLIOGRAPHY

"Antibiotics (β-Lactams)" in *ECT* 3rd ed., Vol. 2, pp. 871–919, by J. R. E. Hoover and C. H. Nash, Smith Kline and French Laboratories; "Antibiotics, β-Lactams (Cephalosporins)" in *ECT* 3rd ed., Supplement, pp. 64–83, by C. O'Callaghan and J. Elks, Glaxo Pharmaceuticals Ltd.

1. R. B. Morin, B. G. Jackson, E. H. Flynn, and R. W. Roeske, *J. Am. Chem. Soc.* **84,** 3400 (1962).
2. B. Loder, G. G. F. Newton, and E. P. Abraham, *Biochem. J.* **79,** 408 (1961).
3. R. P. Elander and H. Aoki in R. B. Morin and M. Gorman, eds., *Chemistry and Biology of β-Lactam Antibiotics,* Vol. 3, Academic Press, New York, 1982, p. 84.
4. E. P. Abraham and G. G. F. Newton, *Biochem. J.* **79,** 377 (1961).
5. E. P. Abraham, *Pharm. Rev.* **14,** 473 (1962); E. P. Abraham, *Quart. Rev. Chem. Soc.* **21,** 231 (1967); R. B. Morin and B. E. Jackson, *Fortschr. Chem. Org. Naturst.* **28,** 343 (1970).
6. E. P. Abraham and P. B. Loder in E. H. Flynn, ed., *Cephalosporins and Penicillins: Chemistry and Biology,* Academic Press, New York, 1972, p. 1.
7. R. Nagarajan and co-workers, *J. Am. Chem. Soc.* **93,** 2308 (1971).
8. T. Kanzaki and co-workers, *J. Antibiot.* **27,** 361 (1974).
9. T. Kanzaki and Y. Fujisawa in D. Perlman, ed., *Advances in Applied Microbiology,* Vol. 20, Academic Press, New York, 1976, p. 159; Ger. Offen. 2,723,463 (1977), T. Niwa and co-workers (to Meiji Seika Kaisha Ltd.).

10. K. Kitano and co-workers, *Ann. Meet. Soc. Ferment. Tech. Jpn.*, 1975, Abstract p. 244; T. Kanzaki and co-workers, *Hakko Kogaku Zasshi* **54,** 720 (1976); Jpn. Kokai 76,488 (1976), T. Kanzaki and co-workers (to Takeda Chemical Industries).
11. E. O. Stapley and co-workers, *Eleventh Interscience Conference on Antimicrobial Agents and Chemotherapy,* 1971, Atlantic City, N.J., American Society of Microbiology, Washington, 1971, Abstract 15; E. O. Stapley and co-workers, *Antimicrob. Agents Chemother.* **2,** 122 (1972); G. Albers-Schonberg, B. H. Arison, and J. L. Smith, *Tetrahedron Lett.* **29,** 2911 (1972).
12. H. Fukase and co-workers, *J. Antibiot.* **29,** 113 (1976).
13. Jpn. Kokai 76 76,487 (1976), S. Inoue and co-workers (to Meiji Confectionary Co., Ltd.); Ger. Offen. 2,455,992 (1975), S. Inoue and co-workers (to Meiji Confectionary Readings); see also U.S. Pat. 4,010,155 (1977).
14. P. D. Singh and co-workers, *J. Antibiot.* **37,** 773 (1984).
15. H. Ono, Y. Nozaki, N. Katayama, and H. Okazaki, *J. Antibiot.* **37,** 1528 (1984); S. Harada, S. Tsubotani, H. Ono, and H. Okazaki, *J. Antibiot.* **37,** 1536 (1984); S. Tsubotani and co-workers, *J. Antibiot.* **37,** 1546 (1984); Y. Nozaki and co-workers, *J. Antibiot.* **37,** 1555 (1984).
16. J. Shoji and co-workers, *J. Antibiot.* **37,** 1486 (1984).
17. Y. Nozaki and co-workers, *J. Antibiot.* **38,** 1141 (1985); S. Tsubotani, T. Hida, H. Ono, and S. Harada, *J. Antibiot.* **38,** 1152 (1985).
18. C. G. Brotzu, *Richerche su di Nuovo Antibiotico,* Lav. Ist. Igiene Cagliari, 1948.
19. G. G. F. Newton and E. P. Abraham, *Nature (London)* **175,** 548 (1955); G. G. F. Newton and E. P. Abraham, *Biochem. J.* **62,** 651 (1956).
20. D. C. Hodgkin and E. N. Maslen, *Biochem. J.* **79,** 393 (1961).
21. S. W. Queener and N. Neuss in Ref. 3, p. 2. The only naturally occurring cephalosporins isolated not containing an aminoadipoyl side chain would appear to be metabolites arising from enzymatic oxidative deamination followed by oxidative decarboxylation. See p. 62 of this reference.
22. F. M. Huber, R. R. Chauvette, and W. G. Jackson in Ref. 6, p. 27.
23. L. D. Hatfield and co-workers in G. I. Gregory, ed., *Recent Advances in the Chemistry of β-Lactam Antibiotics: Proceedings of the Second International Symposium, 1980,* The Royal Society of Chemistry, London, 1981, p. 109.
24. A. K. Miller and co-workers, *Antimicrob. Agents Chemother.* **2,** 287 (1972) and prior articles in this series; H. Fukase and co-workers, *J. Antibiot.* **29,** 113 (1976) and related papers.
25. H. R. Onishi and co-workers, *Antimicrob. Agents Chemother.* **5,** 38 (1974) and prior articles in this series; H. Nakao and co-workers, *J. Antibiot.* **29,** 554 (1976) and references cited therein.
26. T. Hiraoka and co-workers, in J. Elks, ed., *Recent Advances in the Chemistry of β-Lactam Antibiotics: Proceedings of the First International Symposium, 1976,* The Royal Society of Chemistry, London, 1977, p. 123.
27. E. M. Gordon and R. B. Sykes in R. B. Morin and M. Gorman, eds., *Chemistry and Biology of β-Lactam Antibiotics,* Vol. 1, Academic Press, New York, 1982, p. 199.
28. A. W. Guest and co-workers, *J. Chem. Soc., Perkin Trans. 1,* 45 (1987); P. H. Milner and co-workers, *J. Chem. Soc., Chem. Commun.,* 1335 (1984).
29. C. L. Branch and co-workers, *J. Antibiot.* **40,** 646 (1987).
30. M. J. Basker and co-workers, *Twenty Seventh Interscience Conference on Antimicrobial Agents and Chemotherapy 1987,* New York, American Society for Microbiology, Washington, 1987, Abstract 810.
31. P. B. Loder and E. P. Abraham, *Biochem. J.* **123,** 471 (1971).
32. P. Adriaens and co-workers, *Antimicrob. Agents Chemother.* **8,** 638 (1975).
33. P. A. Fawcett and co-workers, *Biochem. J.* **157,** 651 (1976).

34. P. B. Loder and E. P. Abraham, *Biochem. J.* **123,** 477 (1971).
35. H. R. V. Arnstein and D. Morris, *Biochem. J.* **76,** 357 (1960).
36. P. A. Fawcett and co-workers, *J. Gen. Microbiol.* **79,** 293 (1973).
37. C.-P. Pang and co-workers, *Biochem. J.* **222,** 789 (1984).
38. J. E. Baldwin, R. L. White, E.-M. M. John, and E. P. Abraham, *Biochem. J.* **203,** 791 (1982).
39. J. E. Baldwin and E. P. Abraham, *Nat. Prod. Rep.,* 129 (1988).
40. S. A. Goulden and F. W. Chattaway, *Biochem. J.* **110,** 55 (1968).
41. T. Konomi and co-workers, *Biochem. J.* **184,** 427 (1979).
42. E. P. Abraham and co-workers in G. I. Gregory, ed., *Recent Advances in the Chemistry of β-Lactam Antibiotics: Proceedings of the Second International Symposium,* The Royal Society of Chemistry, London, 1981, p. 125.
43. M. Yoshida and co-workers, *Pro. Natl. Acad. Sci. USA* **75,** 6253 (1978).
44. J. E. Baldwin, J. W. Keeping, P. D. Singh, and C. A. Vallejo, *Biochem. J.* **194,** 649 (1981).
45. D. J. Hook, L. T. Chang, R. P. Elander, and R. B. Morin, *Biochem. Biophys. Res. Commun.* **87,** 258 (1979).
46. J. E. Baldwin and co-workers, *Biochem. J.* **245,** 831 (1987).
47. A. Scheidegger, M. T. Kuenzi, and J. Nuesch, *J. Antibiot.* **37,** 522 (1984).
48. J. E. Baldwin in P. H. Bentley and R. Southgate, eds., *Recent Advances in the Chemistry of β-Lactam Antibiotics: Proceedings of the Fourth International Symposium,* The Royal Society of Chemistry, London, 1989, p. 1.
49. M. K. Turner, J. E. Farthing, and S. J. Brewer, *Biochem. J.* **173,** 839 (1978).
50. M. Liersch, J. Nuesch, and H. J. Treichler in *Proceedings of the Second International Symposium on Genetics of Industrial Micro-organisms,* Academic Press, London, 1976, p. 179.
51. J. D. Hood and co-workers in A. G. Brown and S. M. Roberts, eds., *Recent Advances in the Chemistry of β-Lactam Antibiotics: Proceedings of the Third International Symposium,* The Royal Society of Chemistry, London, 1985, p. 52.
52. J. O'Sullivan, R. T. Aplin, C. M. Stevens, and E. P. Abraham, *Biochem. J.* **179,** 47 (1979).
53. J. O'Sullivan and E. P. Abraham, *Biochem. J.* **186,** 613 (1980).
54. J. E. Baldwin in A. G. Brown and S. M. Roberts, eds., *Recent Advances in the Chemistry of β-Lactam Antibiotics: Proceedings of the Third International Symposium,* The Royal Society of Chemistry, London, 1985, p. 62.
55. E. P. Abraham in S. F. Queener, J. A. Webber, and S. W. Queener, eds., *Beta-Lactam Antibiotics for Clinical Use,* Marcel Dekker, New York, 1986, p. 103.
56. R. D. G. Cooper in P. H. Bentley and R. Southgate, eds., *Recent Advances in the Chemistry of β-Lactam Antibiotics: Proceedings of the Fourth International Symposium,* The Royal Society of Chemistry, London, 1989, p. 196.
57. S. M. Samson and co-workers, *Nature (London)* **318,** 191 (1985).
58. P. V. Demarco and R. Nagarajan in Ref. 6, p. 311.
59. K. D. Barrow and T. M. Spotswood, *Tetrahedron Lett.,* 3325 (1965).
60. J. L. Luche and co-workers, *Tetrahedron* **24,** 1275 (1968).
61. D. A. Johnson, D. Mania, C. A. Panetta, and H. H. Silvestri, *Tetrahedron Lett.,* 1903 (1968).
62. G. E. Gutowski, *Tetrahedron Lett.,* 1779 (1970).
63. D. M. Green, A. G. Long, P. J. May, and A. F. Turner, *J. Chem. Soc.,* 766 (1964).
64. M. Windholz, ed., *The Merck Index,* 9th ed., Merck Corporation, Rahway, N.J., 1976.
65. J. P. Hou and J. W. Poole, *J. Pharm. Sci.* **60,** 503 (1971); G. F. H. Green, J. E. Page, and S. E. Staniforth, *J. Chem. Soc.,* 1595 (1965); B. Fechtig, H. Peter, H. Bickel, and E. Vischer, *Helv. Chim. Acta* **51,** 1108 (1968).

66. J. M. T. Hamilton-Miller, G. G. F. Newton, and E. P. Abraham, *Biochem. J.* **116,** 371 (1970); J. M. T. Hamilton-Miller, E. Richards, and E. P. Abraham, *Biochem. J.* **116,** 385 (1970); H. Bundgaard, *Arch. Pharm. Chem. Sci. Ed.* **3,** 94 (1975); C. H. O'Callaghan and co-workers, *J. Bacteriol.* **110,** 988 (1970).
67. J. Nishikawa and K. Tori, *J. Med. Chem.* **27,** 1657 (1984).
68. B. Coene, A. Schanck, J.-M. Dereppe, and M. Van Meerssche, *J. Med. Chem.* **27,** 694 (1984).
69. D. B. Boyd, D. K. Herron, W. H. W. Lunn, and W. A. Spitzer, *J. Am. Chem. Soc.* **102,** 1812 (1980).
70. D. B. Boyd in Ref. 27, p. 437.
71. D. B. Boyd, R. B. Hermann, E. E. Presti, and M. M. Marsh, *J. Med. Chem.* **18,** 408 (1975); D. B. Boyd and W. H. W. Lunn, *J. Med. Chem.* **22,** 778 (1979).
72. W. S. Faraci and R. F. Pratt, *J. Am. Chem. Soc.* **106,** 1489 (1984); W. S. Faraci and R. F. Pratt, *Biochemistry* **24,** 903 (1985); W. S. Faraci and R. F. Pratt, *Biochemistry* **25,** 2934 (1986); W. S. Faraci and R. F. Pratt, *J. Am. Chem. Soc.* **108,** 5328 (1986).
73. E. J. J. Grabowski, A. W. Douglas, and G. B. Smith, *J. Am. Chem. Soc.* **107,** 267 (1985).
74. D. Agathocleous, S. Buckwell, P. Proctor, and M. I. Page, in A. G. Brown and S. M. Roberts, eds., *Recent Advances in the Chemistry of β-Lactam Antibiotics: Proceedings of the Third International Symposium,* The Royal Society of Chemistry, London, 1985, p. 18; M. I. Page and P. J. Proctor, *J. Am. Chem. Soc.* **106,** 3829 (1984).
75. C. F. Murphy and J. A. Webber in Ref. 6, p. 134.
76. G. V. Kaiser and S. Kukolja in Ref. 6, p. 74.
77. G. N. Rolinson, *J. Antimicrob. Chemother.* **17,** 5 (1986).
78. J. Elks, *Drugs* **34** (Suppl. 2), 240 (1987).
79. *Ibid.,* 247 (1987).
80. G. L. Mandell in G. L. Mandell, R. G. Douglas, Jr., and J. E. Bennett, *Anti-infective Therapy,* John Wiley & Sons, Inc., New York, 1985, p. 76.
81. E. Squires and R. Cleeland, *Microbiology and Pharmacokinetics of Parenteral Cephalosporins,* Roche Laboratories, Nutley, N.J., 1985.
82. *Med. Lett.* **25,** 57 (June 10, 1983).
83. A. J. Weinstein, *Drugs* **19,** 137 (1980).
84. P. Garzone, J. Lyon, and V. L. Yu, *Drug Intel. Clin. Pharm.* **17,** 615 (1983).
85. J. D. Williams, *Drugs* **34**(Suppl. 2), 15 (1987).
86. R. Quintiliani, *Scand. J. Infect. Dis.* Suppl. **42,** 99 (1984).
87. C. A. Bunnell, W. D. Luke, and F. M. Perry, Jr., in Ref. 55, p. 255.
88. J. V. Heck, *Ann. Rep. Med. Chem.* **24,** 101 (1989); T. F. Walsh, *Ann. Rep. Med. Chem.* **23,** 121 (1988); G. L. Dunn, *Ann. Rep. Med. Chem.* **21,** 131 (1986); G. L. Dunn, *Ann. Rep. Med. Chem.* **20,** 127 (1985).
89. C. W. Ryan, R. L. Simon, and E. M. VanHeyningen, *J. Med. Chem.* **12,** 310 (1969); U.S. Pat. 3,946,002 (1976), M. Davison, D. B. Frankham, and T. W. M. Spence (to Lilly Industries Ltd., London); U.S. Pat. 3,957,773 (1976), B. Burton and W. Graham (to Glaxo Laboratories, Ltd.).
90. U.S. Pat. 3,925,372 (1975), R. Chauvette (to Eli Lilly and Company).
91. U.S. Pat. 3,503,967 (1970), H. H. Sivestri and D. Aaron (to Bristol Myers Co.).
92. J. E. Dolfini and co-workers, *J. Med. Chem.* **14,** 117 (1971).
93. U.S. Pat. 3,912,726 (1975), P. Diassi and M. S. Atwal (to E. R. Squibb & Sons).
94. U.S. Pat. 3,489,752 (1970), L. B. Crast, Jr. (to Bristol Myers Co.).
95. U.S. Pat. 3,985,741 (1976), L. B. Crast, Jr., and W. Gottstein (to Bristol Myers Co.).
96. *Drugs Future* **8,** 682 (1983); **13,** 790 (1988); *Drugs* **38,** 524 (1989).
97. *Drugs Future* **13,** 1096 (1988).
98. *Drugs Future* **14,** 507 (1989).
99. K. Tomatsu and co-workers, *Twenty-Seventh Interscience Conference on Antimicrobial*

Agents and Chemotherapy, New York, American Society for Microbiology, Washington, 1987, Abstract 806.
100. *Drugs Future* **10,** 112 (1985); **13,** 171 (1988).
101. *Drugs Future* **13,** 16 (1988); **14,** 73 (1989).
102. S. Masuyoshi and co-workers, *Twenty-Eighth Interscience Conference on Antimicrobial Agents and Chemotherapy 1988,* Los Angeles, American Society for Microbiology, Washington, 1988, Abstract 527.
103. E. O. Stapley, E. S. Inamine, and C. M. Martin in Ref. 55, p. 411.
104. *Drugs Future* **2,** 517 (1977); **6,** 505 (1981); *Drugs Today* **17,** 11 (1981).
105. *Drugs Future* **7,** 395 (1982); **11,** 502 (1986).
106. *Drugs Future* **6,** 609 (1981); **9,** 781 (1984); *Drugs Today* **19,** 335 (1983).
107. R. R. Chauvette and co-workers, *Antimicrob. Agents Chemother.* **7,** 687 (1962).
108. Review Issue, *Arneimittel Forsch.* **24,** 1445–1534 (1974); *Drugs Today* **10,** 242 (1974); **9,** 51 (1973).
109. U.S. Pat. 3,422,100 (1969), L. B. Crast, Jr. (to Bristol Myers Co.); U.S. Pat. 3,578,661 (1971), R. E. Havranek and L. B. Crast, Jr. (to Bristol Myers Co.); *Drugs Today* **10,** 299 (1974); **9,** 56 (1973).
110. U.S. Pat. 4,110,338 (1978), T. Kamiya, K. Tanaka, and Y. Shiokawa (to Fujisawa Pharmaceutical Co.); U.S. Pat. 3,641,021 (1972), C. W. Ryan (to Eli Lilly and Co.); *Drugs Today* **10,** 50 (1974); *Drugs Future* **2,** 646 (1977).
111. U.S. Pat. 4,159,373 (1979), D. A. Berges (to SmithKline Corp.); *Drugs Today* **21,** 12 (1985); *Drugs Future* **4,** 634 (1979); **10,** 777 (1985).
112. U.S. Pat. 4,338,461 (1982), K. Unuma, H. Saito, J. Inomata, and S. Takizawa (to Nippon Kasei Chemical Co., Ltd.); U.S. Pat. 3,767,667 (1973), T. Kamiya and Y. Saito (to Fujisawa Pharmaceutical Co.); U.S. Pat. 3,962,272 (1976), A. S. Katner (to Eli Lilly and Company); U.S. Pat. 3,798,229 (1974), K. Kariyone and co-workers (to Fujisawa Pharmaceutical Co., Ltd.); U.S. Pat. 3,824,246 (1974), G. W. Huffman (to Eli Lilly and Company).
113. *Drugs Future* **2,** 643 (1977); **10,** 864 (1985); *Drugs Today* **20,** 615 (1984).
114. *Drugs Future* **4,** 675 (1979); **6,** 569 (1981); *Drugs Today* **17,** 535 (1981); U.S. Pat. 4,304,909 (1981), I. Saikawa and co-workers (to Toyama Chemical Co., Ltd.).
115. *Drugs Future* **3,** 266 (1978); **4,** 302 (1979); Eur. Pat. Appl. 88,183 (1983) (to Glaxo Laboratories Ltd.); U.S. Pat. 4,093,803 (1978), M. C. Cook, G. I. Gregory, and J. Bradshaw (to Glaxo Laboratories, Ltd.); U.S. Pat. 4,013,680 (1977), M. G. Johnson, J. P. Turnbull, and H. A. Crisp (to Glaxo Laboratories, Ltd.).
116. *Drugs Future* **3,** 905 (1978); **5,** 519 (1980); *Drugs Today* **17,** 16 (1981); U.S. Pat. 4,098,888 (1978), M. Ochiai and co-workers, (to Takeda Chemical Industries, Ltd.).
117. *Drugs Future* **2,** 291 (1977); **8,** 445 (1983); *Drugs Today* **19,** 133 (1983); U.S. Pat. 4,166,115 (1979), T. Takaya and co-workers, (to Fujisawa Pharmaceutical Co., Ltd.); H. Tankasugi and co-workers, *J. Antibiot.* **37,** 846 (1983).
118. *Drugs Today* **19,** 335 (1983); *Drugs Future* **5,** 146 (1980).
119. *Drugs Future* **6,** 132 (1981); **8,** 254 (1983); *Drugs Today* **19,** 653 (1983); U.S. Pat. 4,349,672 (1982), M. Montavon and R. Reiner (to Hoffmann-La Roche, Inc.).
120. *Drugs Future* **6,** 612 (1981); **9,** 782 (1984); *Drugs Today* **20,** 555 (1984); U.S. Pat. 4,331,661 (1982), W. E. Marki, M. R. Brown, and J. E. F. Rivier (to The Salk Institute for Biological Studies).
121. *Drugs Future* **5,** 67 (1980); **6,** 110 (1981); *Drugs Today* **17,** 542 (1981).
122. *Drugs Future* **5,** 254 (1980); **11,** 417 (1986).
123. *Drugs Future* **9,** 252 (1984); **14,** 371 (1989).
124. *Drugs Future* **7,** 373 (1982); **12,** 581 (1987).
125. *Drugs Future* **10,** 805 (1985); **14,** 994 (1989).
126. *Drugs Future* **13,** 1047 (1988).

127. H. A. Albrecht and co-workers, *J. Med. Chem.* **33,** 77 (1990); R. N. Jones, A. L. Barry, and C. Thornsberry, *Antimicrob. Agents Chemother.* **33,** 944 (1989); G. Beskid and co-workers, *Antimicrob. Agents Chemother.* **33,** 1072 (1989).
128. *Drugs Future* **4,** 667 (1979); **6,** 578 (1981); *Drugs Today* **18,** 134 (1982).
129. *Drugs Future* **11,** 452 (1986); **13,** 572 (1988).
130. J. D. Williams and F. Moosdeen, *Drugs* **34**(Suppl. 2), 44 (1987).
131. R. R. Kohler, L. J. Wheat, and A. C. White in Ref. 55, p. 351.
132. G. R. Donowitz and G. L. Mandell, *N. Engl. J. Med.* **318,** 490 (1988).
133. H. Numata and co-workers, *J. Antibiotics* **31,** 1262 (1978).
134. A. Tomaz in R. K. Root and M. A. Sande, eds., *New Dimensions in Antimicrobial Therapy,* Churchill Livingstone, New York, 1984, p. 1.
135. D. J. Tipper in Ref. 55, p. 17.
136. D. J. Waxman and J. L. Strominger in Ref. 3, p. 210.
137. G. D. Shockman, L. Daneo-Moore, T. D. McDowell, and W. Wong in Ref. 3, p. 304.
138. N. H. Georgopapdakou in P. K. Peterson and J. Verhoef, eds., *Antimicrobial Agents Annual 3,* Elsevier Science Publishers, New York, 1988, p. 409.
139. D. M. Livermore, *Drugs* **34**(Suppl. 2), 64 (1987); T. D. Gootz in D. M. Bailey, ed., *Annual Reports in Medicinal Chemistry,* Vol. 20, Academic Press, New York, 1985, p. 137.
140. A. A. Medeiros and G. A. Jacoby in Ref. 55, p. 49.
141. C. C. Sanders, *J. Infect. Dis.* **147,** 585 (1983).
142. C. C. Sanders and W. E. Sanders, Jr., *Rev. Infect. Dis.* **5,** 639 (1983).
143. R. L. Then and P. Angehrn, *Antimicrob. Agents Chemother.* **21,** 711 (1982).
144. H. Nikaido, *J. Antimicrob. Chemother.* **22**(Suppl. A), 17 (1988).
145. M. H. Richmond and W. Wotten, *Antimicrob. Agents Chemother.* **10,** 219 (1976).
146. H. Nikaido in M. Stalton and G. A. Shockman, eds., *β-Lactam Antibiotics: Mode of Action, Recent Developments and Future Prospects,* Academic Press, New York, 1981, p. 249; H. Nikaido, E. Y. Rosenberg, and J. Foulds, *J. Bacteriology* **153,** 232 (1983).
147. I. Chopra in R. C. Allen, ed., *Annual Reports in Medicinal Chemistry,* Vol. 24, Academic Press, New York, 1989, p. 139.
148. N. Watanabe, T. Nagasu, K. Katsu, and K. Kitoh, *Antimicrob. Agents Chemother.* **31,** 497 (1987).
149. N. A. C. Curtis and co-workers, *Antimicrob. Agents Chemother.* **32,** 1879 (1988).
150. H. Mochizuki and co-workers, *Twenty-Eighth Interscience Conference on Antimicrobial Agents and Chemotherapy 1988,* Los Angeles, American Society for Microbiology, Washington, 1988, Abstract 428; H. Mochizuki and co-workers, *J. Antibiot.* **41,** 377 (1988).
151. K. I. Plaisance, C. H. Nightingale, and R. Quintiliani in Ref. 55, p. 285.
152. T. Bergan, *Drugs* **34**(Suppl. 2), 89 (1987).
153. P. A. Lemke and D. R. Brannon in E. H. Flynn, eds., *Cephalosporins and Penicillins: Chemistry and Biology,* Academic Press, New York, 1972, p. 370; A. L. Demain, *Lloydia* **37,** 147 (1974).
154. D. A. Lowe in Ref. 55, p. 117.
155. J. L. Ott and co-workers, *Appl. Microbiol.* **10,** 515 (1962).
156. U.S. Pat. 4,354,971 (1982), J. M. Edmundowicz and E. E. Allanson (to Eli Lilly and Co.).
157. U.S. Pat. 3,184,454 (1965), E. P. Abraham, G. G. F. Newton, and C. W. Hale (to National Research Development Corp.).
158. U.S. Pat. 3,725,400 (1973), W. Vosser (to CIBA, Ltd.).
159. R. J. Stedman, K. Swered, and J. R. E. Hoover, *J. Med. Chem.* **7,** 117 (1964).
160. U.S. Pat. 4,346,219 (1982), W. C. Vladuchick (to Eli Lilly and Co.).
161. R. B. Morin and co-workers, *J. Am. Chem. Soc.* **85,** 1896 (1963).

162. R. B. Morin and co-workers, *J. Am. Chem. Soc.* **91,** 1401 (1969).
163. R. D. G. Cooper and D. O. Spry in E. H. Flynn, ed., *Cephalosporins and Penicillins: Chemistry and Biology,* Academic Press, New York, 1972, p. 181.
164. R. D. G. Cooper and G. A. Koppel in R. B. Morin and M. Gorman, eds., *Chemistry and Biology of β-Lactam Antibiotics,* Vol. 1, Academic Press, New York, 1982, p. 1.
165. R. D. G. Cooper, L. D. Hatfield, and D. O. Spry, *Accounts Chem. Res.* **6,** 32 (1973); P. G. Sammes, *Chem. Rev.* **76,** 113 (1976); R. J. Stoodley, *Tetrahedron* **31,** 2321 (1975).
166. U.S. Pat. 3,725,399 (1973), N. V. Ellerton, W. F. Paradise, and P. E. Sanford (to Glaxo Laboratories, Ltd.); U.S. Pat. 3,953,440 (1976), L. D. Hatfield (to Eli Lilly and Co.); U.S. Pat. 3,879,379 (1975), U. Stache and co-workers (to Farbwerke Hoechst Aktiengesellschaft Vormals Meister Lucius and Bruning).
167. U.S. Pat. 4,003,894 (1977), J. Verweij, H. S. Tan, and H. J. Kooreman (to Gist-Brocades).
168. L. D. Hatfield and co-workers, *Philos. Trans. R. Soc. London* **B289,** 173 (1980).
169. J. D. Cocker and co-workers, *J. Chem. Soc.* 5015 (1965).
170. A. B. Taylor, *J. Chem. Soc.* 7020 (1965).
171. H. Yazawa, H. Nakamura, K. Tanaka, and K. Kariyone, *Tetrahedron Lett.* 3991 (1974).
172. R. Bonjouklian and M. L. Phillips, *Tetrahedron Lett.* **22,** 3915 (1981); U.S. Pat. 4,266,049 (1981), R. Bonjouklian (to Eli Lilly and Co.).
173. U.S. Pat. 4,014,874 (1977), H. Peter, B. Muller, W. Sibral, and H. Bickel (to CIBA-GEIGY Corp.).
174. U.S. Pat. 3,270,012 (1966), H. M. Higgins, Jr. (to Eli Lilly and Co.); J. L. Spencer and co-workers, *J. Org. Chem.* **32,** 500 (1967).
175. U.S. Pat. 4,317,907 (1982), I. Saikawa and co-workers, (to Toyama Chemical Co., Ltd.).
176. I. Miskolczi, F. Sztariczai, and R. Bognar, *Org. Prep. Proc. Int.* **13,** 315 (1981).
177. U.S. Pat. 4,182,870 (1979), C. A. Bruynes and J. J. DeKoning (to Gist-Brocades); U.S. Pat. 4,107,431 (1978), J. C. Clark, J. Kennedy, A. G. Long, and N. G. Weir (to Glaxo Labs. Ltd.); U.S. Pat. 4,113,591 (1978), B. Laudon. B. R. Cowley, and D. C. Humber (to Glaxo Lab. Ltd.).
178. Eur. Pat. Appl. 34,394 (1981), C. A. Bruynes and T. K. Jurriens (to Gist-Brocades).
179. N. Nomura and co-workers, *J. Med. Chem.* **17,** 1312 (1974).
180. T. Takaya, H. Takasugi, T. Murakawa, and H. Nakano, *J. Antibiot.* **34,** 1300 (1981).
181. J. D. Jeffrey, E. P. Abraham, and G. G. F. Newton, *Biochem. J.* **81,** 591 (1961).
182. E. M. Wilson, *Chem. Ind. London* 217 (1984).
183. F. A. Jung, W. R. Pilgrim, J. P. Poysen, and P. J. Siret, in P. G. Sammes, ed., *Topics in Antibiotic Chemistry,* Vol. 4, John Wiley & Sons, Inc., New York, 1980.
184. M. Gorman, R. R. Chauvette, and S. Kukolja, in J. S. Bindra and D. Lednicer, eds., *Chronicles of Drug Discovery,* Vol. 2, John Wiley & Sons, Inc., New York, 1983, p. 49.
185. S. Kukolja and R. R. Chauvette in R. B. Morin and M. Gorman, eds., *Chemistry and Biology of β-Lactam Antibiotics,* Vol. 1, Academic Press, New York, 1982, p. 93.
186. S. Kukolja and co-workers, *J. Org. Chem.* **41,** 2276 (1976).
187. Y. Shibuya, K. Matsumoto, and T. Fujii, *Agric. Biol. Chem.* **45,** 1561 (1981).
188. R. B. Morin and co-workers, *J. Am. Chem. Soc.* **91,** 1396 (1969).
189. B. Fechtig, H. Peter, and E. Vischer, *Helv. Chim. Acta.* **51,** 1108 (1968).
190. S. Tsushima and co-workers, *Chem. Pharm. Bull. Jpn.* **27,** 696 (1979).
191. K. Morita and co-workers, *Phil. Trans. R. Soc. London* **B289,** 181 (1980).
192. R. Reiner and co-workers, *J. Antibiot.* **33,** 783 (1980).
193. Eur. Pat. 037,380 (1981), G. Ascher (to Biochemie G.m.b.H.).
194. W. Ngata, M. Narisada, and Y. Yoshida in R. B. Morin and M. Gorman, eds., *Chemistry and Biology of β-Lactam Antibiotics,* Vol. 2, Academic Press, New York, 1982, p. 1.
195. B. G. Jackson, R. B. Kammer, M. Yoshioka, and H. Yamada in Ref. 55, p. 371.

196. M. Yoshioka and co-workers, *Tetrahedron Lett.* **21,** 351 (1980).
197. K. G. Holden in R. B. Morin and M. Gorman, eds., *Chemistry and Biology of β-Lactam Antibiotics,* Vol. 2, Academic Press, New York, 1982, p. 101.
198. J. Aszodi and co-workers in P. H. Bentley and R. Southgate, eds., *Recent Advances in the Chemistry of β-Lactam Antibiotics: Proceedings of the Fourth International Symposium,* The Royal Society of Chemistry, London, 1989, p. 350.
199. L. D. Cama and B. G. Christensen, *J. Am. Chem. Soc.* **96,** 7582 (1974).
200. R. A. Firestone and co-workers, *J. Med. Chem.* **20,** 551 (1977).
201. C. L. Branch and M. J. Pearson, *J. Chem. Soc. Perkins Trans. 1,* 2268 (1979).
202. J. Aszodi and co-workers, Abstract 123, and J. F. Chantot, Abstract 436, *Twenty-Eighth Interscience Conference on Antimicrobial Agents and Chemotherapy, 1988,* Los Angeles, American Society for Microbiology, Washington, 1988.
203. R. N. Guthikonda, L. D. Cama, and B. G. Christensen, *J. Am. Chem. Soc.* **96,** 7584 (1974).
204. J. A. Eudaly and co-workers in P. H. Bentley and R. Southgate, eds., *Recent Advances in the Chemistry of β-Lactam Antibiotics: Proceedings of the Fourth International Symposium,* The Royal Society of Chemistry, London, 1989, p. 333.
205. F. T. Counter and co-workers, *Twenty-Seventh Interscience Conference on Antimicrobial Agents and Chemotherapy,* New York, American Society for Microbiology, Washington, 1987, Abstract 1188.
206. J. F. Quay and co-workers in Ref. 205, Abstract 1205.
207. H. R. Black and co-workers in Ref. 202, Abstract 123.
208. J. A. Eudaly and co-workers in P. H. Bentley and R. Southgate, eds., *Recent Advances in the Chemistry of β-Lactam Antibiotics: Proceedings of the Fourth International Symposium,* The Royal Society of Chemistry, London, 1989, p. 333.
209. G. K. Cook and co-workers in Ref. 202, Abstract 532.
210. J. Christenson and co-workers in P. H. Bentley and R. Southgate, eds., *Recent Advances in the Chemistry of β-Lactam Antibiotics: Proceedings of the Fourth International Symposium,* The Royal Society of Chemistry, London, 1989, p. 33; J. G. Christenson, D. L. Pruess, M. K. Talbot, and D. D. Keith, *Antimicrob. Agents Chemother.* **32,** 1005 (1988); N. H. Georgopapadakou and co-workers, *Antimicrob. Agents Chemother.* **31,** 1069 (1987).
211. P. Schindler and P. Huber in U. Brodbeck, ed., *Enzyme Inhibitors,* Verlag Chemie, Weinheim, FRG, 1980, p. 169.
212. R. B. Sykes and M. Matthew in J. M. T. Hamilton-Miller and J. T. Smith, eds., *Beta-Lactamases,* Academic Press, New York, 1980, p. 17.
213. J. Symonds and A. M. Geddes, *Drugs* **34**(Suppl. 2), 121 (1987).
214. D. W. McEniry and S. L. Gorbach, *Drugs* **34**(Suppl. 2), 216 (1987).
215. L. O. Gentry, *Drugs* **34**(Suppl. 2), 154 (1987).
216. H. C. Neu, *Drugs* **34**(Suppl. 2), 135 (1987).
217. I. Phillips, *Drugs* **34**(Suppl. 2), 164 (1987).
218. R. Finch, *Drugs* **34**(Suppl. 2), 180 (1987).
219. S. R. Norrby, *Drugs* **34**(Suppl. 2), 105 (1987).
220. B. R. Meyers, *Am. J. Med.* **79**(Suppl. 2A), 96 (1985).
221. R. Platt, *J. Antimicrob. Chemother.* **10**(Suppl. C), 135 (1982).
222. *Chemistry and Industry* **22,** 769 (Nov. 16, 1987). Note that the figure given in this article of $1,100 million is in error. The correct figure is $11,000 million.
223. *Scrip,* Review issue, 12 (1989).

JOHN ROBERTS
Hoffmann-La Roche, Inc.

β-LACTAMASE INHIBITORS

The antibacterial effectiveness of penicillins, cephalosporins, and other β-lactam antibiotics depends upon selective acylation and consequently, inactivation, of transpeptidases involved in bacterial cell wall synthesis. This acylating ability is a result of the reactivity of the β-lactam ring (**1**). Bacteria that are resistant to β-lactam antibiotics often produce enzymes called β-lactamases that inactivate the antibiotics by catalyzing the hydrolytic opening of the β-lactam ring to give products (**2**) devoid of antibacterial activity.

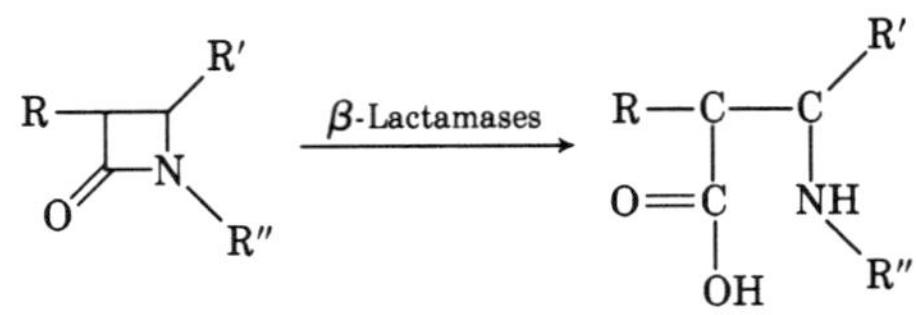

(**1**) active (**2**) inactive

A book (1) and several general reviews (2–4) on β-lactamases have been published. Based on sequence data, it has been suggested that β-lactamases evolved from the enzymes involved in bacterial cell wall synthesis (5–7).

One approach to combating antibiotic resistance caused by β-lactamase is to inhibit the enzyme (see ENZYME INHIBITION). Effective combinations of enzyme inhibitors with β-lactam antibiotics such as penicillins or cephalosporins, result in a synergistic response, lowering the minimal inhibitory concentration (MIC) by a factor of four or more for each component. However, inhibition of β-lactamases alone is not sufficient. Pharmacokinetics, stability, ability to penetrate bacteria, cost, and other factors are also important in determining whether an inhibitor is suitable for therapeutic use. Almost any class of β-lactam is capable of producing β-lactamase inhibitors. Several reviews have been published on β-lactamase inhibitors, detection, and properties (8–15).

Classification and Occurrence

Several classification schemes for β-lactamases have been used to describe the activity of β-lactamase inhibitors (3,16,17). The β-lactamases from gram-positive *Staphylococcal* organisms have been divided into four types A, B, C, D on the basis of serological data and are referred to in general as penicillinases or penases. Gram-negative bacteria produce a much greater diversity of β-lactamases and therefore require a more complex classification scheme. The Richmond-Sykes system (18) is most frequently used to describe the activity of β-lactamase inhibitors. This system defines five classes (I–V) of β-lactamases; a sixth (VI) has been added for the β-lactamases produced by *Bacteroides* species (19). The many plasmid-derived β-lactamases of gram-negative bacteria have been separated by isoelectric focusing. They have been given names depending on plasmid origin. The clinically most important plasmid-derived β-lactamases and their prevalence, indicated in parenthesis, are (19): TEM-1 (75%), TEM-2 (3%), OXA-1 (5%), OXA-2

(3%), OXA-3 (2%), SHV-1 (2%), and others (10%). The TEM-1 enzyme stands out as an important target for a β-lactamase inhibitor.

Table 1 shows the clinically most important bacteria that produce β-lactamases, separated into gram-positive and gram-negative organisms. The prevalence of β-lactamase is indicated as is the origin and Richmond-Sykes classification. Based on these data the most important β-lactamases to inhibit clinically are the gram-positive penases, the gram-negative TEM, which are Richmond-Sykes type III, and the gram-negative chromosomal cephalosporinases–cephases which are Richmond-Sykes type I. These enzymes are subsequently referred to as penase, TEM(III), and cephase(I).

Table 1. β-Lactamase Producing Bacteria[a]

Organism	β-Lactamase producers, %[b]	Enzyme origin, %		Richmond-Sykes classification
		Chromosomal	Plasmid	
Gram-positive				
Staphylococcus aureus	80		penase	
Staphylococcus epidermidis	80		penase	
Gram-negative				
Escherichia coli	25 (16–76)	15	TEM, 80; OXA-1, 7.5	I,III,V
Haemophilus influenzae	25–60		TEM, 92; ROB-1, 8	III
Neisseria gonorrhea	1–10		TEM, 100	III
Salmonella	6		TEM, 100	III
Shigella			TEM	III
Klebsiella pneumoniae	60–90		TEM, 24; SHV-1, 76	III,IV
Enterobacter	25–30	73	TEM, 27	I,III
Citrobacter	20–50	77	TEM, 23	I,III
Pseudomones aeruginosa	23	44	TEM, 9; PSE, 10; others	I,III
Bacillus catarrhalis	87	90^+	BRO-1	
Bacillus fragilis	87	~100		VI

[a]Refs. 16, 19–21.
[b]Clinically resistant bacteria, virtually 100% of the *Enterobacteriaceae,* produce a low level of chromosomal enzyme that can clinically be selected for higher levels.

Mechanistic Aspects of β-Lactamase Inhibition

The clinically important β-lactamases, eg, the penases, TEM(III), and cephases(I), are serine proteases that form an acyl enzyme intermediate with β-lactam substrates and β-lactam-derived β-lactamase inhibitors (22–25). Mechanistic studies using several β-lactamase inhibitors have been extensively reviewed (6,26,27) and a general inhibition scheme is illustrated in Figure 1. Following Michaelis complex formation (**3**), the enzyme's serine hydroxyl reacts with the β-lactam carbonyl, forming an acyl enzyme intermediate (**4**), the structure of which depends on the inhibitor. This acyl enzyme can then follow Route B and hydrolyze to products and active enzyme, react further to give an irreversibly inactivated enzyme (Route A) or a transiently inactivated enzyme species (Route C). The transient inhibition can result from a conformational change in the enzyme or formation of a species resistant to hydrolysis. In the latter case, amino acrylatelike esters are commonly encountered. A suitably long-lived transient species can occupy the β-lactamase and protect a β-lactam from hydrolysis resulting in a synergistic effect. The transiently inhibited species can then hydrolyze to products, revert to acyl enzyme, or react further to give irreversibly inactivated enzyme. β-Lactamase inhibitors that branch away from acyl enzyme (**4**) formation have been termed branched pathway inhibitors. The exact mechanistic pathway is dependent on both the enzyme and the inhibitor.

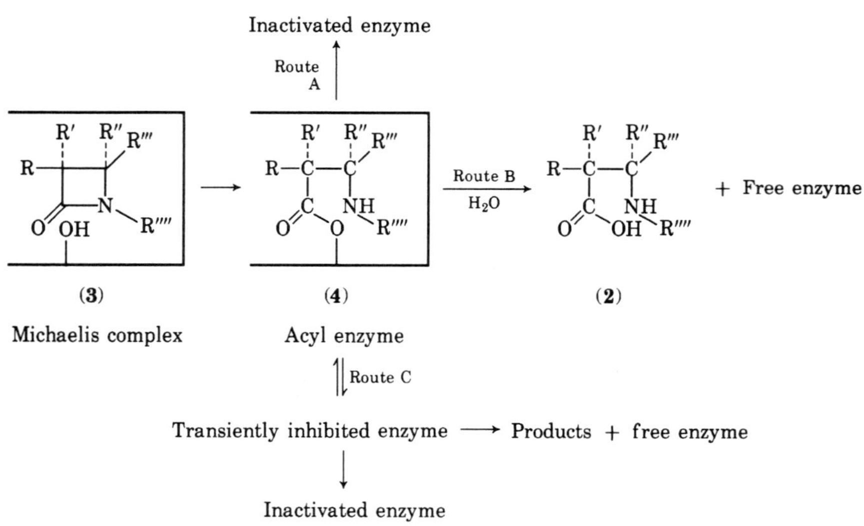

Fig. 1. Scheme for the interaction of β-lactamase inhibitors and β-lactamases where the enzyme is represented by 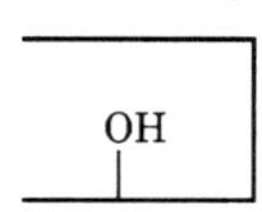

Active site directed β-lactam-derived inhibitors have a competitive component of inhibition, but once in the active site they form an acyl enzyme species which follows one or more of the pathways outlined in Figure 1. Compounds that follow Route C and form a transiently inhibited enzyme species and are subsequently hydrolyzed to products have been termed inhibitory substrates or competitive substrates. Inhibitors that give irreversibly inactivated β-lactamase (Route A) are called suicide inactivators or irreversible inhibitors. The term progressive inhibitor has also been used. An excellent review has appeared on inhibitor interactions with β-lactamases (28).

The activity of β-lactamase inhibitors is often expressed as an IC_{50} value, which is defined as the concentration of inhibitor that causes 50% inhibition of enzyme activity for a given set of conditions. IC_{50} values, which vary widely according to substrate, time of incubation, and other factors, are presented herein solely to give an indication of potency and enzyme inhibitor specificity. Values that decrease with preincubation are indicative of irreversible inhibitors.

β-Lactam-Based Inhibitors

Penicillins, Cephalosporins, and Monobactams. Early attempts at inhibiting β-lactamases using inorganics or penicillin fragments were not successful (29,30). The use of reactive penicillin species, such as diazo species from 6-aminopenicillanic acid (6-APA) or ampicillin (31) and penicillin isocyanates (32), was successful, but not practical because of the reactivity of the molecules. The discovery that cephalosporin C (33) and methicillin (34) inhibited β-lactamases resulted in the screening of numerous antibiotics and a number of β-lactamase-resistant penicillins and cephalosporins were found to be β-lactamase inhibitors. These compounds act by forming a transiently inhibited acyl enzyme species as a result of conformational change (35–37) and are inhibitory substrates (Fig. 1, Route C). No clinically useful inhibitors have been identified from this class. These efforts have been extensively reviewed (8,10,13,14,37).

Modern β-lactamase-resistant cephalosporins, primarily of the aminothiazole oxime structural type shown in Figure 2, have been reported to be inhibitors of type I cephases (38–42). β-Lactamase inhibition occurs through the transiently inhibited enzyme species (**7**), which requires a good C-3 leaving group. Cephalosporins without a leaving group at C-3 are poor inhibitors (43,44). None of the aminothiazole oxime structural types is an effective synergist.

A series of activated cephalosporins having electron withdrawing groups at C-3, such as [*128474-86-2*] (**8**, R = CHO), $C_{10}H_{11}NO_5S$, and [*123036-47-5*] (**8**, R = CN), $C_{10}H_{10}N_2O_4S$, were prepared in an attempt to increase β-lactamase inhibition (45,46). These compounds inhibited type I cephases and synergy was demon-

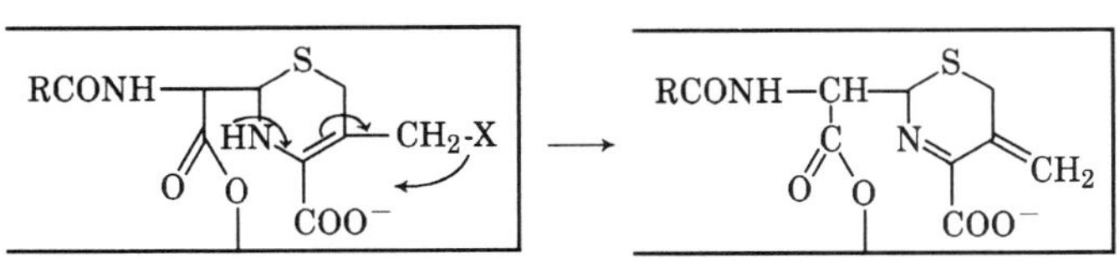

(**6**) Acyl enzyme

(**7**) Transient inhibition

Fig. 2. Transiently inhibited species for cephalosporins bound to β-lactamase. R, an aminothiazole oxime, =

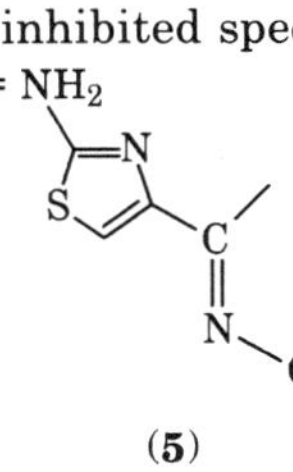

(**5**)

strated for (**8**, R = CN) with ceftizoxime. A series of 7-*β*-[2-(1,3-dithiolan-2-ylidene)acetamido] cephalosporins represented by [*71908-34-4*] (**9**), $C_{18}H_{20}N_5O_6S_4$, are potent inhibitors of type I cephase and give synergy with cephaloridine (47). Several monobactams have been reported to be inhibitory substrates for type I cephases (48–51). Interesting exceptions are the monophosphams such as represented by [*120020-34-0*] (**10**), $C_{12}H_{15}N_5O_8PS$, and some oligopeptide monobactams which are reported to be irreversible inhibitors (Fig. 1, Route A) (52,53). The monocyclic *β*-lactam [*123039-89-4*] (**11**), $C_9H_8N_2O_2$, inhibits a broad range of *β*-lactamases, but is too unstable for practical use (54).

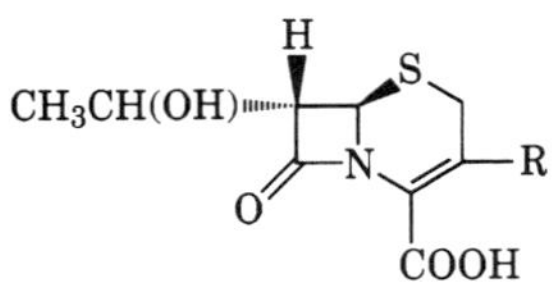

(**8**)

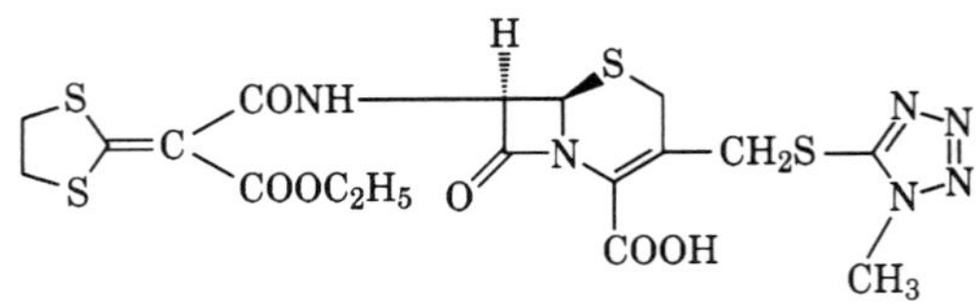

(**9**)

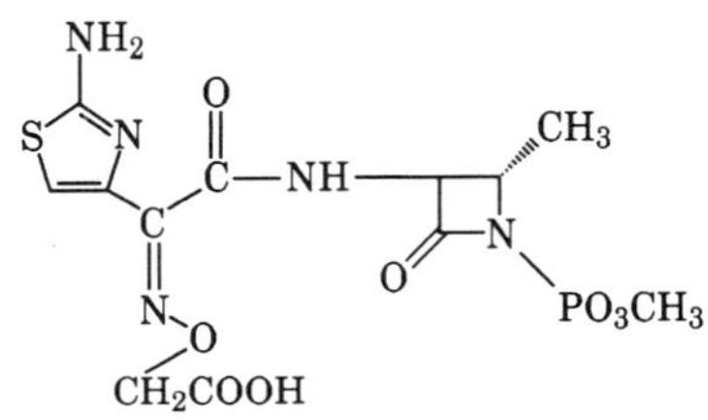

(**10**)

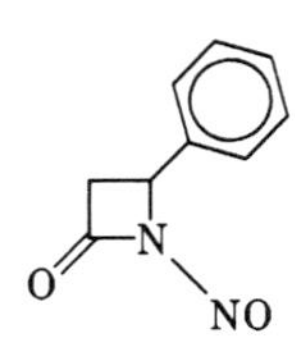

(**11**)

Clavulanic Acid Class of *β*-Lactamase Inhibitors. The discovery of clavulanic acid [*58001-44-8*] (**12**, R = CH_2OH, R′ = H), $C_8H_9NO_5$, in 1976 marked a turning point in *β*-lactamase inhibitor research (55). This natural product isolated from *Streptomyces clavuligerus,* ATCC 27064, using a cleverly designed screen, had an unusual *β*-lactam structure compared to classical penicillins and cephalosporins. The nucleus has the trivial name clavam: an oxygen replaces the ring sulfur and there is no C-6 substituent. The x-ray structure of clavulanic acid, the chemical name of which is [2(*R*)-(2*α*,3(*Z*),5*α*)]-3-(2-hydroxy-ethylidine)-7-oxo-4-oxa-1-azabicyclo[3.2.0.]heptane-2-carboxylic acid, has been published (56).

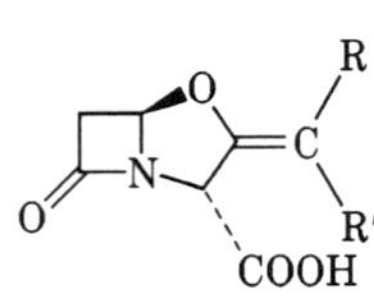

(**12**)

Clavulanic acid has only weak antibacterial activity, but is a potent irreversible inhibitor for many clinically important *β*-lactamases (10–14,57,58) including penases, and Richmond-Sykes types II, III, IV, V, VI (*Bacteroides*). Type I cephases are poorly inhibited. Clavulanic acid synergizes the activity of many penicillins and cephalosporins against resistant strains. The chemistry (59–63), microbiology (64,65), structure activity relationships (10,13,60–62,66), biosynthesis (67–69), and mechanism of action (6,26,27,67) have been reviewed.

Mechanistic studies (6,26,27,67) have shown that the acyl enzyme species is the ring opened compound (**13**), which can tautomerize to the transiently inhibited amino acrylate (**14**), and both of these species can react further to give irreversibly inactivated enzyme. Three inactivated forms of the enzyme have been detected. Two, according to labeling studies, retain the complete clavulanate skeleton and the other retains only the carbon chain of the *β*-lactam ring. Structure (**15**) has been suggested as one possible inactivated form.

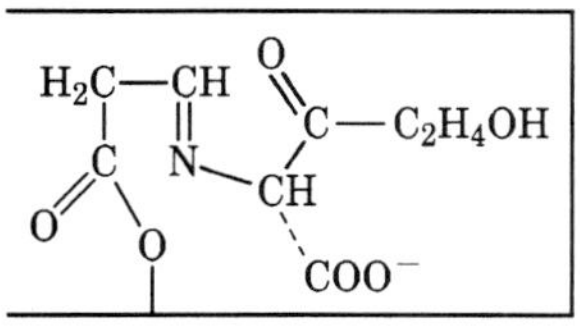

(**13**)

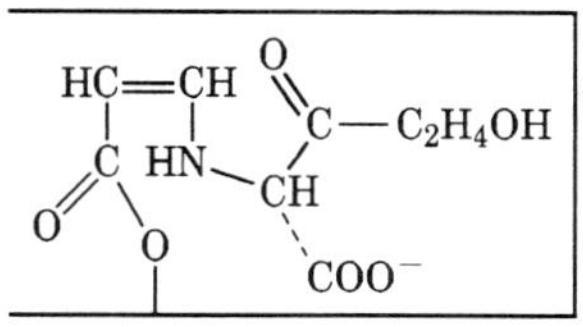

(**14**)

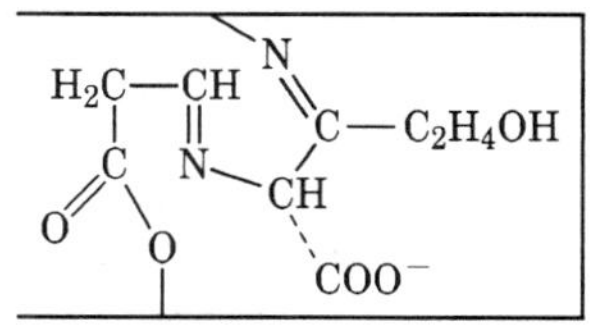

(**15**)

Consistent with the proposed mechanism is the finding that deoxyclavulanate [*69779-62-0*] (**12**, R = CH_3, R′ = H), $C_8H_9NO_4$, and the oxapenem [*78808-19-2*] (**16**), $C_8H_9NO_4$, are potent inhibitors whereas the dihydroclavulanic acid [*59897-36-2*] (**17**), $C_8H_{11}NO_5$, is devoid of useful activity. The oxapenem (**16**) is a more potent inhibitor than clavulanic acid (**12**, R = CH_2OH, R′ = H) but (**16**) is unstable.

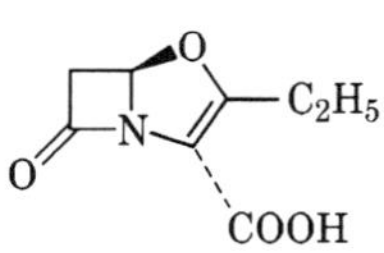

(**16**)

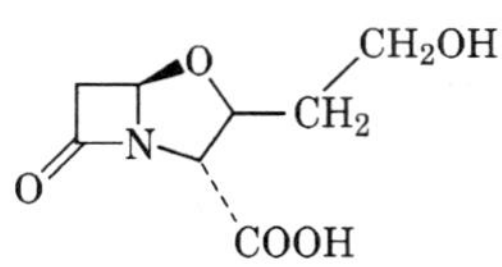

(**17**)

Analogues of clavulanic acid have been reviewed (10,11,13,60–62,66). Inhibitory activity for selected compounds is given in Table 2. A number of clavulanic acid analogues are more potent than the parent compound. Additionally, the natural (*Z*)-isomer (**12**, R = CH_2OH, R′ = H) is more potent than the (*E*)-isomer (**12**, R = H, R′ = CH_2OH) which is a general trend. One surprising feature is that the carboxyl group of clavulanic acid is not required for potent inhibition of *β*-lactamase (10,13,62,70,71). For antibacterial activity *β*-lactam antibiotics normally require an analogously positioned carboxylic acid or other acidic group. Descarboxy compounds [*86408-58-4*] (**18**, R = H, R′ = SC_6H_5), $C_{12}H_{11}NO_2S$, and [*79908-54-6*] (**18**, R = $COOCH_3$, R′ = H), $C_8H_8NO_4$, demonstrate *β*-lactamase inhibition. Few derivatives have been reported that have substitution at C-6. The hydroxyalkyl compounds [*66825-93-2*] (**19**) and (**20**), $C_{10}H_{13}NO_6$, have been disclosed as *β*-lactamase inhibitors (13) where the cis epimer (**19**) is more active than the trans (**20**).

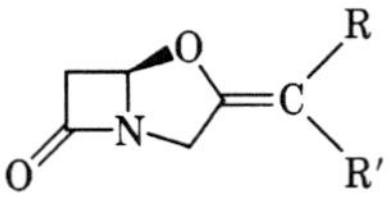

(**18**)

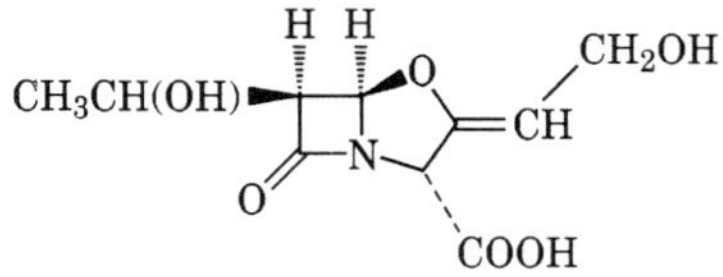

(**19**)

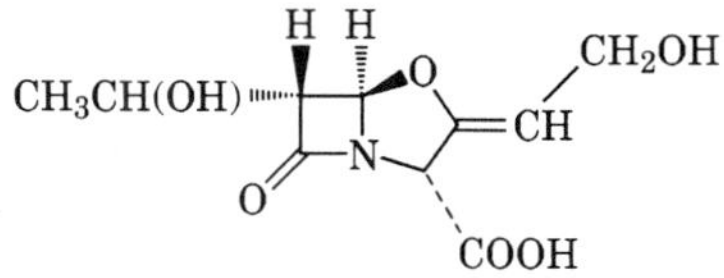

(**20**)

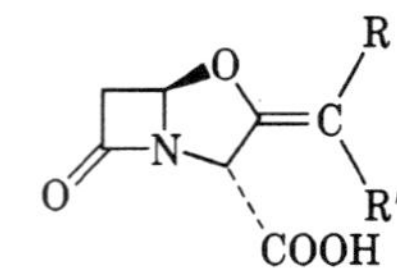

(12)

Table 2. *β*-**Lactamase Inhibitory Activity for Clavulanic Acid and Analogues**[a]

CAS Registry Number	Molecular formula	R	IC$_{50}$, μg/mL[b] Penase	TEM(III)	Cephase	Refs.
[*58001-44-8*]	$C_8H_6NO_5$	CH_2OH	8.2[c]	1.35[c]	10[c]	60,66
			0.03	0.06		13
[*69779-62-0*]	$C_8H_9NO_4$	H	0.12	0.09	5	66
[*78739-05-6*]	$C_{10}H_{11}NO_6$	CH_2OOCCH_3	0.04		0.4	10,66
[*63617-41-4*]	$C_{10}H_{12}N_2O_6$	$CH_2OOCNHCH_3$	1.5	2.5	0.45	10,13,66
[*67529-21-9*]	$C_{10}H_{13}NO_5$	$CH_2OC_2H_5$	0.03	0.05	60	13
[*130097-27-7*]	$C_8H_9NO_8S$	CH_2OSO_3H		3.4		10
[*78739-06-7*]	$C_9H_{10}NO_4S$	CH_2SCH_3	1.6[c]	1.1[c]		10,60,62,66
[*64832-72-0*]	$C_{10}H_{14}N_2O_4S$	$CH_2SCH_2CH_2NH_2$	0.23[c]	0.05[c]		10,60,62
[*78739-07-8*]	$C_9H_9N_2O_4S_2$	$CH_2S_2CNH_2$	3.8[c]	0.08[c]		10,60,62
[*65788-55-8*]	$C_8H_{10}N_2O_4$	CH_2NH_2	0.8[c]	0.08[c]		10,60,62,72
[*75167-01-0*]	$C_9H_{12}N_2O_4$	CH_2NHCH_3	0.03	0.007	>50	13
[*67065-13-8*]	$C_9H_{10}N_2O_5$	CH_2NHCHO	0.04[c]	0.01[c]		10,60,62
[*78739-12-5*]	$C_{15}H_{15}N_3O_5$	$CH_2NHCONHC_6H_5$	0.12[c]	0.06[c]		10,60
[*78739-15-8*]	$C_9H_{12}N_2O_6S$	$CH_2NHSO_2CH_3$	1.15[c]	0.33[c]		10,60,62
[*87131-73-5*]	$C_{10}H_{13}NO_5$	$CH_2CH_2CH_2OH$	0.02	0.1		13
[*64475-40-7*]	$C_{13}H_{11}NO_4$	C_6H_5	<0.08[c]	0.02[c]		10,73
[*62319-53-3*]	$C_8H_9NO_5$	H[e]	0.6[d]	1[d]		10,13,62

[a]Structure (**12**) where R is defined and R′ is H unless otherwise indicated.
[b]The substrate is nitrocefin unless otherwise indicated.
[c]The substrate is ampicillin.
[d]The substrate is penicillin G.
[e]R′ = CH_2OH.

Carbapenem *β*-Lactamase Inhibitors. Carbapenems are another class of natural product *β*-lactamase inhibitors discovered about the same time as clavulanic acid. Over forty naturally occurring carbapenems have been identified; many are potent *β*-lactamase inhibitors. Carbapenem is the trivial name for the 1-azabicyclo[3.2.0]hept-2-ene ring system (**21**) shown in Table 3. The synthesis (74), biosynthesis (75), and *β*-lactamase inhibitory properties (13,14,66) of carbapenems have been reviewed. Carbapenems are often more potent than clavulanic acid and include type I cephases in the spectrum of inhibition. Table 3 lists the available *β*-lactamase inhibition data. Synergy is frequently difficult to demonstrate because the compounds are often potent antibacterials.

The mechanism of *β*-lactamase inhibition has been studied (6,26,27,83). Reversible inhibition is believed to result from the Δ^1-pyrroline (**22**). Certain sulfate ester containing compounds, such as SF-2103A (**21**, R = R″ = SO_3H, R′ = H), give long-lived transiently inhibited species suggesting possible conformational changes or additional ionic bonds to the enzyme (6,83,86,89). MM-22381 (**21**, R = H, R′ = CH_3, R″ = $SCH_2CH_2NHCOCH_3$) and MM-22383 (**21**, R = R′ = H, R″ =

SCHCHNHCOCH$_3$), are the only compounds having trans β-lactam stereochemistry for which IC$_{50}$ data are available. Other trans compounds, such as thienamycin, imipenem, and the PS-5 series, have been reported to be β-lactamase inhibitors (90–97), but only limited data have been published. Several totally synthetic carbapenems have also been reported to be potent β-lactamase inhibitors (98,99). Asparenomycin [*76466-24-5*] (**23**), $C_{14}H_{16}N_2O_6S$, and its analogues are reported to be irreversible inhibitors (14,100). These compounds have functionality at C-6 that could react further in a Michael fashion. However, no commercial β-lactamase inhibitor product has been developed from this class of compounds, in part because of low fermentation titers (high synthetic cost), poor stability (82,84,87), short half-life in humans (101), and degradation by renal dipeptidase (102–106).

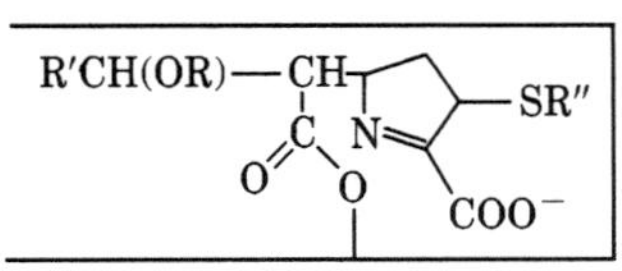

(**22**)

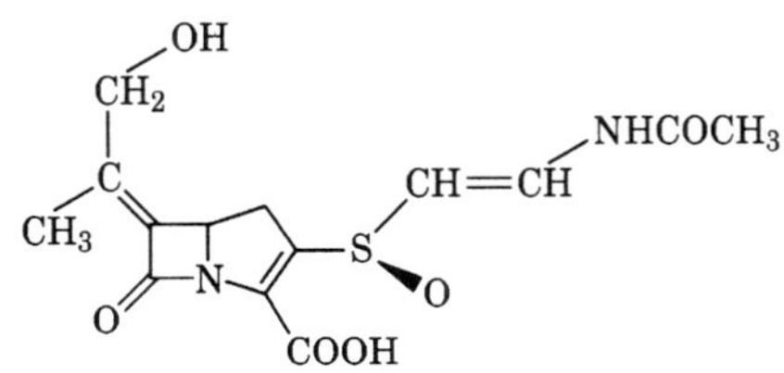

(**23**)

Penem β-Lactamase Inhibitors. The synthesis and antibacterial properties of penems, the trivial name for the 4-thia-1-azabicyclo[3.2.0]hept-2-ene ring system (**24**), have been reviewed (107,108). Like the closely related carbapenems, many of the penems are potent antibacterials. Additionally, penems are also susceptible to degradation by renal dipeptidase, but to a lesser extent. The limited β-lactamase inhibitory data available for penems are presented in Table 4. SCH-29,482 [*77646-83-4*] (**24**, R = H, R′ = CH(OH)CH$_3$, R″ = SC$_2$H$_5$), $C_{10}H_{13}NO_4S_2$, is reported to be an inhibitor of type I cephases and the OXA-2 enzyme (109). Compounds [*101803-54-7*] and [*101914-68-5*] (**24**, R = H, R′ = CH$_3$CH(OH), R″ = 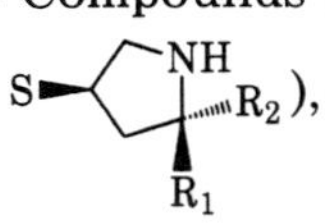), and [*130097-28-8*] (**24**, R = R′ = R″ = CH$_3$) are also reported to inhibit cephase(I) and some TEM inhibition was reported for the first two. 6-Methylene-substituted penems (R = R$_1$CH=) are potent, better β-lactamase inhibitors than clavulanic acid, and many of the compounds are potent synergists. No data have been published on the mechanism of β-lactamase inhibition. Comparative data for [*102209-75-6*] the (*Z*)-isomer of (**24**, R = 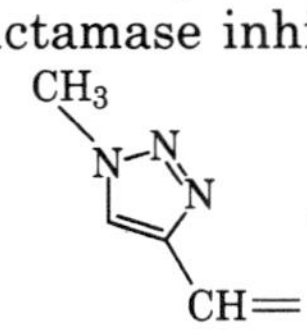, R″ = H) with other other β-lactamase inhibitors (113), and structure activity relationships have been reported (115).

Penam Sulfone β-Lactamase Inhibitors. Natural product discoveries stimulated the rational design of β-lactamase inhibitors based on the readily accessible penicillin nucleus. An early success was penicillanic acid sulfone, (2(*S*)-cis)-3,3-dimethyl-7-oxo-4,4-dioxide-4-thia-1-azabicyclo [3.2.0]heptane-2-car-

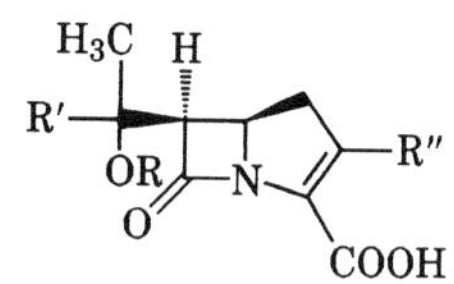

(21)

Table 3. Activity of Carbapenem β-Lactamase Inhibitors[a]

Name	CAS Registry Number	Molecular formula	R	R′	R″	IC$_{50}$, μg/mL[b] Penase	TEM(III)	Cephase(I)	Refs.
trans stereochemistry in side chain R″									
C-19393S$_2$	[*76025-74-6*]	$C_{14}H_{18}N_2O_9S_2$	SO_3H	CH_3	$OSCHCHNHCOCH_3$	0.21	0.00027	0.043	76–80
C-19393S$_2$M$_1$	[*82795-81-1*]	$C_{14}H_{18}N_2O_8S_2$	SO_3H	CH_3	$SCHCHNHCOCH_3$	2.4	0.025	0.0024	76–78
MM-4550	[*76985-32-5*]	$C_{13}H_{16}N_2O_9S_2$	SO_3H	H	$OSCHCHNHCOCH_3$	0.003	0.0004	0.001	76,81,82
MM-13902	[*79057-46-8*]	$C_{13}H_{16}N_2O_8S_2$	SO_3H	H	$SCHCHNHCOCH_3$	0.008	0.003	0.0001	14,76,81–83
C-19393H$_2$	[*83916-44-3*]	$C_{14}H_{18}N_2O_6S$	H	CH_3	$OSCHCHNHCOCH_3$	0.004	0.003	0.034	76,78–80
C-19393H$_2$M$_1$	[*82502-20-3*]	$C_{14}H_{18}N_2O_6S$	H	CH_3	$SCHCHNHCOCH_3$	0.6	0.01	0.005	76
C-19393E$_5$	[*83310-72-9*]	$C_{13}H_{16}N_2O_6S$	H	H	$OSCHCHNHCOCH_3$	0.0003	0.027	0.015	76,84
epithienamycin B (MM-22382)	[*65376-20-7*]	$C_{13}H_{16}N_2O_5S$	H	H	$SCHCHNHCOCH_3$	0.012	0.32	0.001	14,76,83
cis stereochemistry in side chain R″									
	[*83916-46-5*]	$C_{14}H_{18}N_2O_6S$	H	CH_3	$OSCHCHNHCOCH_3$	0.59		0.014	79
	[*83916-36-3*]	$C_{14}H_{18}N_2O_9S_2$	SO_3H	CH_3	$OSCHCHNHCOCH_3$	0.24		0.0005	79

	[83916-40-9]	$C_{13}H_{16}N_2O_9S_2$	SO_3H	H	$OSCHCHNHCOCH_3$	0.003		0.0002	79
	[83916-39-6]	$C_{14}H_{18}N_2O_8S_2$	SO_3H	CH_3	$SCHCHNHCOCH_3$	1		0.004	79
	[83916-42-1]	$C_{13}H_{16}N_2O_8S_2$	SO_3H	H	$SCHCHNHCOCH_3$	0.014		0.0004	79,85
	[83916-38-5]	$C_{14}H_{18}N_2O_6S$	H	CH_3	$OSCHCHNHCOCH_3$	0.055		0.001	79
	[83916-41-0]	$C_{13}H_{16}N_2O_6S$	H	H	$OSCHCHNHCOCH_3$	0.0006		0.019	79
	[83916-37-4]	$C_{14}H_{18}N_2O_5S$	H	CH_3	$SCHCHNHCOCH_3$	0.36		0.017	79
	[75443-31-1]	$C_{13}H_{16}N_2O_5S$	H	H	$SCHCHNHCOCH_3$	0.065		1.5	79
no stereochemistry in side chain R″									
C-19393S_2M_2	*[82795-82-2]*	$C_{14}H_{20}N_2O_8S_2$	SO_3H	CH_3	$SCH_2CH_2NHCOCH_3$	2.3	0.042	0.006	76
MM-17880	*[79057-45-7]*	$C_{13}H_{18}N_2O_8S_2$	SO_3H	H	$SCH_2CH_2NHCOCH_3$	0.021	0.005	0.0003	76,81,82
C-19393H_2M_3	*[82889-91-6]*	$C_{14}H_{20}N_2O_5S$	H	CH_3	$SCH_2CH_2NHCOCH_3$	0.56	0.0045	0.0054	76
epithienamycin A (MM-22380)	*[63582-78-5]*	$C_{13}H_{18}N_2O_5S$	H	H	$SCH_2CH_2NHCOCH_3$	0.037	0.69	0.003	76
pluracidomycin A (SF-2103A)	*[82138-64-5]*	$C_9H_{11}N_{10}S_2$	SO_3H	H	SO_3H	4	0.05	0.0015	14,86–88
trans β-Lactam stereochemistry									
MM-22381	*[96193-21-4]*	$C_{14}H_{20}N_2O_5S$	H	CH_3	$SCH_2CH_2NHCOCH_3$	5^c	0.03^c	0.2	14
MM-22383	*[65322-98-7]*	$C_{13}H_{16}N_2O_5S$	H	H	$SCHCHNHCOCH_3$	0.04^c	0.04^c	10	14

[a]Structure (**21**) where R, R′, and R″ are as defined.
[b]The substrate is amicillin for penicillinases and cephalothin for cephases unless otherwise indicated.
[c]The substrate is nitrocefin.

Table 4. Activity of Penem β-Lactamase Inhibitors[a]

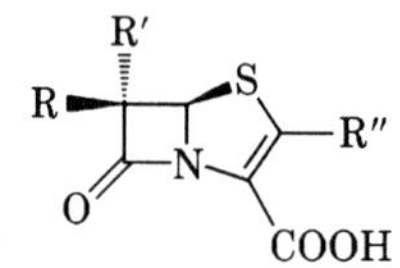

CAS Registry Number	Molecular formula	R	R′	R″	IC$_{50}$, μg/mL[b] Penase	TEM(III)	Cephase(I)	Refs.
[77646-83-4][c]	$C_{10}H_{13}NO_4S_2$	H	$CH_3CH(OH)$	SC_2H_5			inhib.	109
[101803-54-7]	$C_{13}H_{16}N_2O_6S_2$	H	$CH_3CH(OH)$	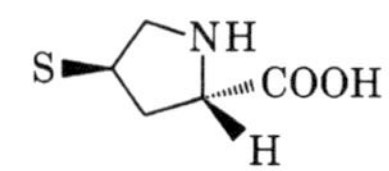	no	0.2–1.8[d]	0.07–0.25[d]	110
[101914-68-5]	$C_{13}H_{18}N_2O_5S_2$	H	$CH_3CH(OH)$	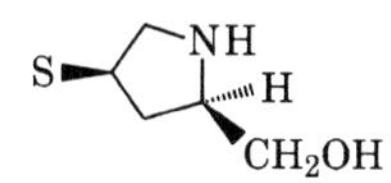				
[130097-28-8]	$C_9H_{11}NO_3S$	CH_3	CH_3	CH_3	no	no	inhib.	111
[81519-84-8]	$C_{10}H_9NO_4S$	$CH_3COCH=$	Z isomer	CH_3	claim broad spectrum			112
[102209-75-6]	$C_{10}H_8N_4O_3S$	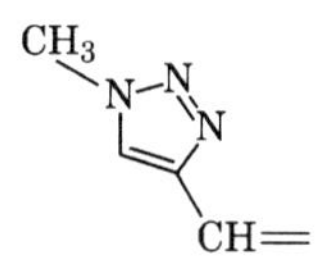	Z isomer	H	0.016	0.002	0.002	113–115
[81519-81-5]	$C_8H_7NO_3S$	$CH_3CH=$[e]	Z isomer	H	0.6	0.04	1	115–117
[128658-07-1]	$C_8H_7NO_3S$	$CH_3CH=$	E isomer	H	3.2	0.2	0.4	115,117
[93853-93-0]	$C_{11}H_7NO_3S_2$	2-thienyl-CH=[e]	Z isomer	H	0.4	0.008	0.02	116
[93853-93-1]	$C_{11}H_7NO_3S_2$	2-thienyl-CH=	E isomer	H	>10	3.5	3	116
[93854-02-5]	$C_{11}H_7NO_3S_2$	3-thienyl-CH=	Z isomer	H	0.012	0.13	0.003	115,116
[93854-42-3]	$C_{11}H_7NO_4S$	2-furyl-CH=	Z isomer	H	0.013	0.003	0.005	115,116
[93853-72-6]	$C_{11}H_7NO_4S$	3-furyl-CH=	Z isomer	H	0.4	0.1	0.025	116

[a]Structure (**24**) where R, R′, and R″ are as defined.
[b]The substrate is nitrocefin.
[c]Compound SCH-29482.
[d]Data is from a series of compounds where substituents on R″ vary. Best compounds are the one shown and where R″ = 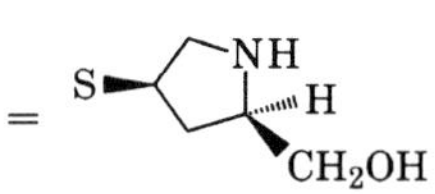
[e]Represents a series of compounds.

boxylic acid [*68373-14-8*] (sulbactam) (**25**, R = R′ = H, R″ = R‴ = CH_3), $C_8H_{11}NO_5S$. The synthesis (118), microbiology (119–121), and clinical use (122–124) of sulbactam have been reviewed. Sulbactam, with minor exceptions, is a weak antibacterial, but is a potent irreversible inactivator of many β-lactamases including penases, and Richmond-Sykes type II,III,IV,V; VI(*Bacteroides*) β-lactamases. Sulbactam is better than clavulanic acid against type I cephases and synergy is observed for combinations of many penicillins and cephalosporins. Because sulbactam is not well absorbed orally, prodrug forms have been developed (125–129). Mechanism of action studies (26,27), including labeling experiments, are consistent with formation of a thiazolidine opened acyl enzyme species (**26**) that tautomerizes to the transiently inhibited form (**27**). These intermediates, shown in Figure 3, can react further to give inactivated enzyme species such as (**28**).

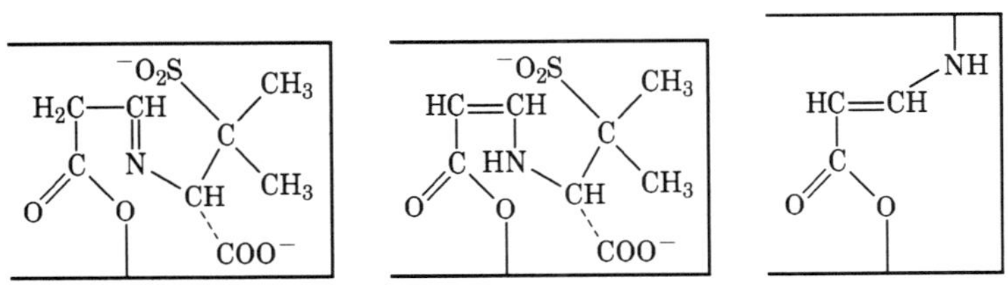

(**26**) (**27**) (**28**)

Fig. 3. Species in the inhibition pathway for sulbactam with β-lactamase.

Numerous other penicillin sulfones have been reported to be β-lactamase inhibitors, as illustrated in Table 5. The effect of C-6 substituents has been extensively explored starting with 6-APA sulfone (**25**, R = NH_2, R′ = H, R″ = R‴ = CH_3), which has modest activity. Mechanistic considerations led to preparation of sulfones of poor substrates, compounds such as methicillin, cloxacillin, nafacillin, and quinacillin sulfone (**25**, R = an aromatic moiety, R′ = H, and R″ = R‴ = CH_3), which were expected to have a long-lived acyl enzyme species (**26**) that would react further to give inactivated enzyme. The compounds all act as β-lactamase inhibitors; but they are poor synergists. Aged solutions of cloxacillin sulfone, where R = 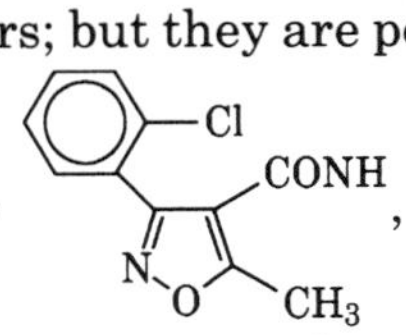, in phosphate buffer gave a product that inhibited *C. freundi* β-lactamase (153,154). Both the C-6 phthalamido [*82087-00-1*] and sulfonamido [*83053-95-6*] compounds, for example, are reported to inhibit *B. cereus* β-lactamase through a conformational change in the enzyme (138).

C-6 halo-substituted sulfones (**25**, R = H, Cl or Br, R′ = Cl or H, R″ = R‴ = CH_3) have been shown to be weak β-lactamase inhibitors and a series of C-6-heteroarylmethylene derivatives, where R = 2-pyridyl-CH or 2-thiazoyl-CH, are potent suicide β-lactamase inhibitors and synergists that are superior to clavulanic acid. Structure activity relationships are discussed in the literature (141) and based on reactions with methanol, used to mimic the serine hydroxyl of β-lactamases, the acyl enzyme species (**29**) has been suggested as being responsi-

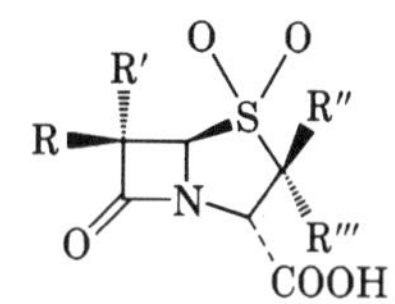

Table 5. Activity of Penam Sulfone β-Lactamase Inhibitors[a] (**25**)

Compound	CAS Registry Number	Molecular formula	R	R''	IC$_{50}$, μg/mL[b,c]			Refs.
					Penase	TEM(III)	Cephase(I)	
sulbactam	[*68373-14-8*]	$C_8H_{11}NO_5S$	H	CH_3	1.4	1.7	5	118–120, 130–132
6-APA sulfone	[*36091-15-3*]	$C_8H_{12}N_2O_5S$	NH_2	CH_3	0.4	40	50	13
methicillin sulfone	[*76350-41-9*]	$C_{17}H_{20}N_2O_8S$	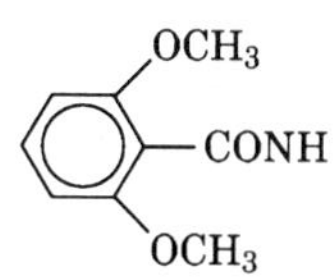	CH_3	>50	16	20	13,33,40
cloxacillin sulfone	[*76788-83-5*]	$C_{19}H_{18}ClN_3O_7S$	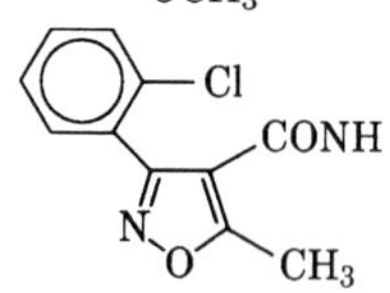	CH_3	+	+	+	133–135
nafacillin sulfone	[*77724-76-6*]	$C_{25}H_{21}N_2O_7S$	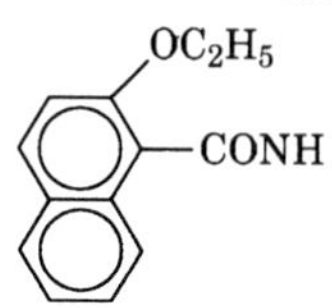	CH_3		+		133

quinacillin sulfone	[*76788-82-4*]	$C_{19}H_{17}N_3O_8S$	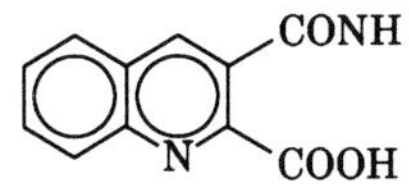	CH_3	+	+		136,137
	[*82087-00-1*]	$C_{16}H_{14}N_2O_7S$	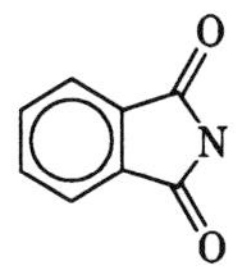	CH_3	+			138
d	[*83053-95-6*]	$C_9H_{11}F_3N_2O_7S_2$	CF_3SO_2NH	CH_3	+	+		137,138
	[*76350-36-2*]	$C_8H_{10}ClNO_5S$	H	CH_3	50	30	30	13,139
e	[*76613-60-8*]	$C_8H_{10}ClNO_5S$	Cl	CH_3	50	13	>50	13
	[*75527-87-6*]	$C_8H_{10}BrNO_5S$	Br	CH_3	4	1.5	>50	13
d,f	[*108636-76-6*]	$C_{14}H_{14}N_2O_5S$	2-pyridyl-CH=	CH_3	+	+	+	140,141
f	[*110088-75-7*]	$C_{12}H_{12}N_2O_5S_2$	2-thiazolyl-CH=	CH_3	+	0.0009[g]	+[g]	141
f	[*123361-44-4*]	$C_{11}H_{15}NO_5S$	CH_2CHCH_2	CH_3	claim broad spectrum[c]			142
	[*78494-75-4*]	$C_{15}H_{17}NO_6S$	$C_6H_5CH(OH)$	CH_3	no[g]		+[g]	142
	[*76517-56-1*]	$C_9H_{13}NO_6S$	$HOCH_2$	CH_3	+[g]	+[g]	+[g]	143
	[*87579-77-9*]	$C_8H_9NO_5S$	H	CH_2[h]	1.5	0.25	15.5	144
BLP-2013	[*79886-07-0*]	$C_8H_{10}ClNO_5S$	H	CH_2Cl	9.6	+		145–147
i	[*115095-95-9*]	$C_8H_{10}ClNO_5S$	H	CH_3	15.3[j]			146
	[*95123-11-8*]	$C_8H_{10}BrNO_5S$	H	CH_2Br	+[j]	+[j]		147
	[*95123-10-7*]	$C_8H_{11}NO_6S$	H	CH_2OH	no[j]	+[j]		147
d	[*95123-06-1*]	$C_{15}H_{15}NO_7S$	H	$CH_2OOCC_6H_5$	no[j]	+[j]		147
	[*112110-80-2*]	$C_9H_{10}N_2O_5S_2$	H	CH_2SCN	2.8[j]	+[j]	+[j]	148
	[*89051-51-4*]	$C_{10}H_{16}N_5O_5S_2$	H	CH_2N_3	1.1[j]			131
	[*112110-81-3*]	$C_{10}H_{16}N_5O_5S_2$	H	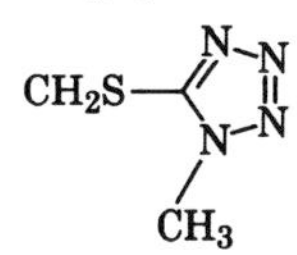	0.335[j]			148

Table 5. ***(Continued)***

Compound	CAS Registry Number	Molecular formula	R	R″	IC$_{50}$, μg/mL[b,c] Penase	TEM(III)	Cephase(I)	Refs.
	[112110-83-5]	$C_{15}H_{15}N_5O_5S_2$	H	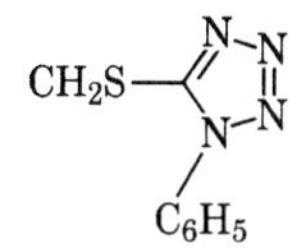	0.354[j]			148
d	*[98382-76-4]*	$C_{12}H_{16}N_4O_5S$	H	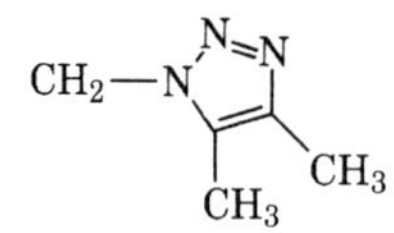	0.026[j]			149
tazobactam	*[89786-04-9]*	$C_{10}H_{12}N_4O_5S$	H	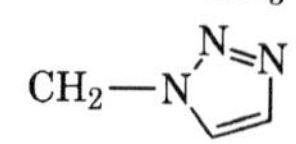	0.27	0.04[j]	0.93	131,148,150, 151,152
d	*[89785-79-5]*	$C_{11}H_{11}N_4O_7SK$	H	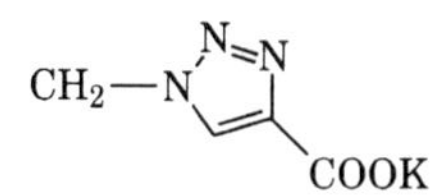	0.23[j]			131
	[89786-01-6]	$C_{12}H_{14}N_4O_7S$	H	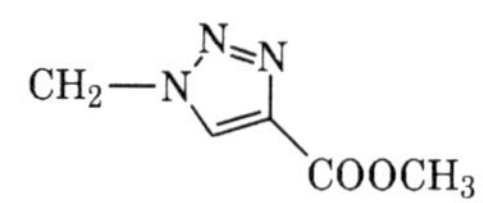	0.018[f]			131
	[98382-75-3]	$C_{10}H_{13}N_5O_5S$	H	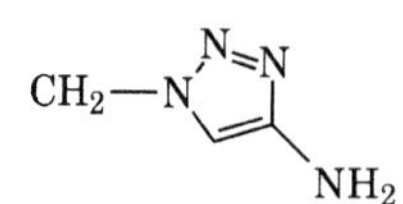	0.22[f]			131

[a]Structure (**25**) where R, R″ are as defined, and R′ = H and R‴ = CH_3 unless otherwise indicated.
[b]Substrate is nitrocefin unless otherwise indicated.
[c]+ Indicates enzyme inhibition, but no available IC$_{50}$ data.
[d]Represents series of compounds.
[e]R′ = Cl.
[f]R′ = (*Z*)-isomer.
[g]Substrate is ampicillin.
[h]Cyclopropane ring bonded also to C-2.
[i]R‴ = CH_2Cl.
[j]Substrate is penicillin G.

ble for inactivation of the enzyme, at least for the R = 2-pyridyl-CH series. Compounds having a leaving group such as acetoxy or fluoro, which can undergo elimination to form the 6-thiazolyl methylene penam sulfone have been reported to be β-lactamase inhibitors and synergists (155). The allyl sulfone (**25**, R = CH_2CHCH_2, R′ = H, R″ = R‴ = CH_3) was reported to be a potent β-lactamase inhibitor, although no IC_{50} was provided (142). The C-6 hydroxyalkyl sulfones (**25**, R = C_6H_5CHOH or $HOCH_2$, R′ = H, R″ = R‴ = CH_3) have been reported as β-lactamase inhibitors. The compound where R = C_6H_5CHOH was primarily active against cephase(I), but the simpler species, R = $HOCH_2$, is reported to be a broad spectrum inhibitor and synergist.

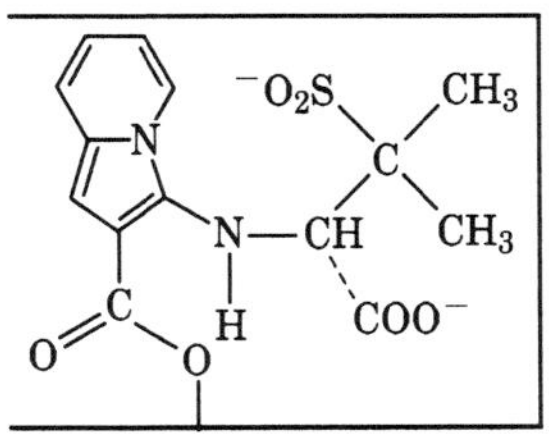

(**29**)

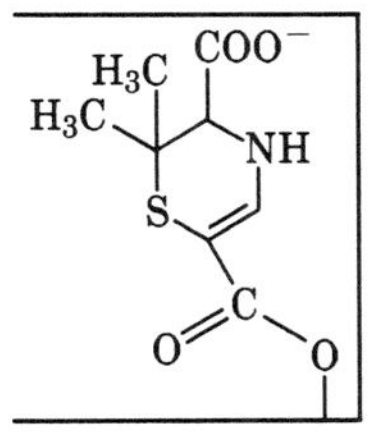

(**30**)

Substitution of the β-methyl group of penicillanic acid sulfone has also been extensively investigated. The cyclopropyl sulfone (**25**, R = R′ = H, R″ = CH_2, R‴ = CH_3) resembles sulbactam, but is overall a weaker inhibitor. BLP-2013 (**25**, R = R′ = H, R″ = CH_2Cl, R‴ = CH_3) has activity similar to that of sulbactam (145,147). The α-chloromethyl analogue (**25**, R = R′ = H, R″ = CH_3, R‴ = CH_2Cl) is less active. Potent activity has been demonstrated for a series of heteroaryl-substituted methyl sulfones. The compound most studied is tazobactam (YTR-830) (**25**, R = R′ = H, R″ = CH_2—N 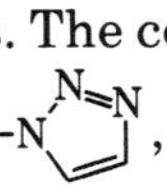 , R‴ = CH_3) which has activity similar to that of clavulanic acid and inhibits many of the type I cephases. The synthesis (156), microbiological activity (152,156–162), and stability (152) of tazobactam have been reported.

Penam β-Lactamase Inhibitors. Penam is the trivial name given derivatives of the penicillin nucleus (**31**) the chemical name of which is 4-thia-1-azabicyclo[3.2.0]heptane. Table 6 gives activity data for a diverse group of penams. The report that 6-β-bromopenicillanic acid [*26631-90-3*], [2(*S*)-(2α,5α,6β)]-6-bromo-3,3-dimethyl-7-oxo-4-thia-1-azabicyclo[3.2.0]heptane-2-carboxylic acid, (**31**, R = Br, R′ = H, R″ = R‴ = CH_3) is a potent inhibitor led to intense study both of this compound and analogues. The microbiology profile of 6-β-bromopenicillanic acid has been reported (165) and the compound has progressed to clinical trials (6). Mechanistic studies have demonstrated that the dihydrothiazine derivative (**30**) is responsible for inactivation of β-lactamases (6,26,27). The iodo analogue of the bromopenicillinic acid (**31**, R = I, R′ = H, R″ = R‴ = CH_3) has comparable activity, but the chloro compound (**31**, R = Cl, R′ = H, R″ = R‴ = CH_3) is substantially less active. Mutual prodrugs of the 6-β-bromo and 6-β-iodo compounds with ampicillin have been reported (189). Analogues of 6-β-bromopenicillanic acid having a substituent in the β-methyl position have

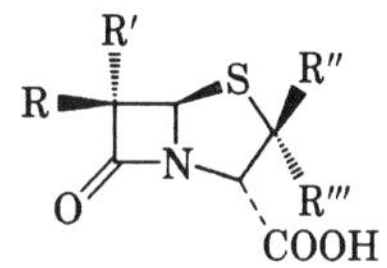

Table 6. Activity of Penam-Based β-Lactamase Inhibitors[a] (**31**)

CAS Registry Number	Molecular formula	R	R′	R″	R‴	IC$_{50}$, μg/mL[b] Penase	TEM(III)	Cephase(I)	Refs.
[*26631-90-3*]	$C_8H_{10}BrNO_3S$	Br	H	CH_3	CH_3	1.5/0.005[c]	0.15/0.005[c]	1.5/0.32[c]	13,163–167,168
[*74772-32-0*]	$C_8H_{10}INO_3S$	I	H	CH_3	CH_3	0.7	0.06	5.5	13,164,165
[*74326-78-6*]	$C_8H_{10}ClNO_3S$	Cl	H	CH_3	CH_3	0.3	50	50	13,163,164,167, 169,170
[*108934-65-2*]	$C_8H_9BrFNO_3S$	Br	H	CH_2F	CH_3		0.13	0.2	168
[*108934-66-3*]	$C_8H_9BrN_4O_3S$	Br	H	CH_2N_3	CH_3	0.2	0.04	8.1	168
[*105433-74-7*]	$C_{10}H_{12}BrNO_3S$	Br	H	$CH_2O_2CCH_3$	CH_3	3.2	0.4	50	168
[*105433-78-1*]	$C_9H_{12}BrNO_4S$	Br	H	CH_2OCH_3	CH_3	1.3	0.2	6.3	168
[*108934-67-4*]	$C_{10}H_{14}BrNO_4S$	Br	H	$CH_2OC_2H_5$	CH_3	3.2	0.4	50	168
[*105458-19-3*]	$C_9H_9BrN_2O_3S_2$	Br	H	CH_2SCN	CH_3	0.4	0.03	50	168
[*83116-77-2*][g]	$C_{11}H_{13}NO_4S$	$CH_3COCH=$	Z isomer	CH_3	CH_3	0.00015	0.0002	1.2	171–177
[*83116-56-7*]	$C_{11}H_{13}NO_4S$	$CH_3COCH=$	E isomer	CH_3	CH_3		0.002	0.012	171
[*104127-98-2*]	$C_{10}H_{13}NO_5S_2$	$CH_3SO_2CH=$	Z isomer	CH_3	CH_3	+[d]		+[d]	178
[*104127-97-1*]	$C_{15}H_{15}NO_5S_2$	$C_6H_5SO_2CH=$	Z isomer	CH_3	CH_3	+[d]		+[d]	178
[*86492-69-5*]	$C_{11}H_{14}N_2O_4S$	$CH_3CN(OH)CH=$	Z isomer	CH_3	CH_3		0.2[e]		176
[*93040-42-7*]	$C_{10}H_{13}NO_4S$	$CH_3OCH=$	Z isomer	CH_3	CH_3		+[d]		179,180
[*73707-58-1*]	$C_{15}H_{16}N_2O_3S_2$	p-CH_3-C_6H_5S-$N=$	Z isomer	CH_3	CH_3			0.5	181

[*82954-44-7*]	$C_{10}H_{10}F_6N_2O_7S_3$	$(CF_3SO_2)_2N$	H	CH_3	CH_3	+[d]			182
[*83670-99-9*]	$C_8H_{12}N_2O_6S_2$	HO_3SNH	H	CH_3	CH_3			+[d]	183
[*89012-63-5*]	$C_{12}H_{18}N_2O_5S$	$C_2H_5O_2CCH_2NH$	H	CH_3	CH_3	no[f]	no[f]	0.25[f]	184
[*89012-69-1*]	$C_{12}H_{20}N_2O_4S$	$CH_3CH(OH)(CH_2)_2NH$	H	CH_3	CH_3	no	no	0.32	184
[*89012-67-9*]	$C_{12}H_{18}N_2O_4S$	$CH_3CO(CH_2)_2NH$	H	CH_3	CH_3	no	no	0.40	184
[*118499-99-3*]	$C_{16}H_{17}N_3O_4S$	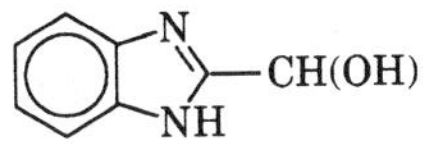	H	CH_3	CH_3				185
[*118498-71-8*]	$C_{18}H_{19}N_3O_4S$	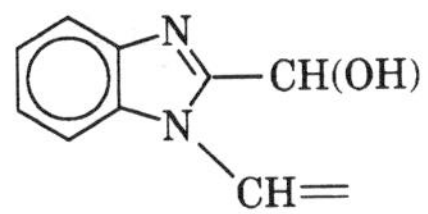	H	CH_3	CH_3				185
[*113656-11-4*]	$C_{17}H_{19}NO_5S$	$C_6H_5CH_2COCH(OH)$	H	CH_3	CH_3	+[f]	+[f]		186
[*113656-13-6*]	$C_{16}H_{17}NO_5S$	$C_6H_5COCH(OH)$	H	CH_3	CH_3	+	+		186
[*109990-44-5*]	$C_{20}H_{19}NO_5S$	1-naphthylCOCH(OH)	H	CH_3	CH_3	+	+	+	187
[*94723-14-5*]	$C_{18}H_{17}NO_6S_2$	2-naphthylSO_2O[f]	H	CH_3	CH_3				188
[*94723-16-7*]	$C_{24}H_{22}NO_6PS$	2-naphthyl$OP(C_6H_5)O_2$	H	CH_3	CH_3				188
[*94723-39-4*]	$C_9H_{11}NO_4S$	epoxide		CH_3	CH_3				188
[*49628-20-8*]	$C_{17}H_{19}NO_5S$	$C_6H_5OCH_2COCH_2$	H	CH_3	CH_3				188

[a]Structure (**30**) where R, R′, R″, and R‴ are defined as indicated.
[b]The substrate is nitrocefin unless otherwise indicated.
[c]These data correspond to set of β-methyl analogues.
[d]Compound is active but no IC_{50} data are available.
[e]Substrate is penicillin G.
[f]Represents a series of compounds.
[g]Compound Ro 15-1903.

been explored, (**31**, R = Br, R′ = H, R″ = CH_2F, R‴ = CH_3) is an example, and in general offer no advantages.

The (*Z*)-isomer of 6-acetylmethylene penicillanic acid, RO 15-1903 [*83116-77-2*] (**31**, R = CH_3COCH, (*Z*)-isomer, R″ = R‴ = CH_3), [2(*S*)-(2α,5α,6(*Z*))]-3,3-dimethyl-7-oxo-6-(2-oxo-propylidene)-4-thia-1-azabicyclo[3.2.0]-heptane-2-carboxylic acid, has been reported to be a more potent β-lactamase inhibitor than clavulanic acid. Despite excellent β-lactamase inhibition, the compound gave poorer than expected *in vitro* and *in vivo* synergy results (172,173). This was subsequently attributed to poor uptake of drug by bacteria (172) and instability in serum (177). Mechanistic studies with the TEM enzyme indicate that one molecule of compound inactivates one molecule of enzyme. There is essentially no turnover as substrate. Four inactivated forms of the β-lactamase enzyme have been detected in inhibition studies using RO 15-1903. Structures (**32**) and (**33**) have been implicated in this inactivation. The corresponding (*E*)-isomer (**31**, R = CH_3COCH = (*E*)-isomer, R″ = R‴ = CH_3) and the related structures, (*Z*)-isomers where R = CH_3SO_2CH or $C_6H_5SO_2CH$, are less potent.

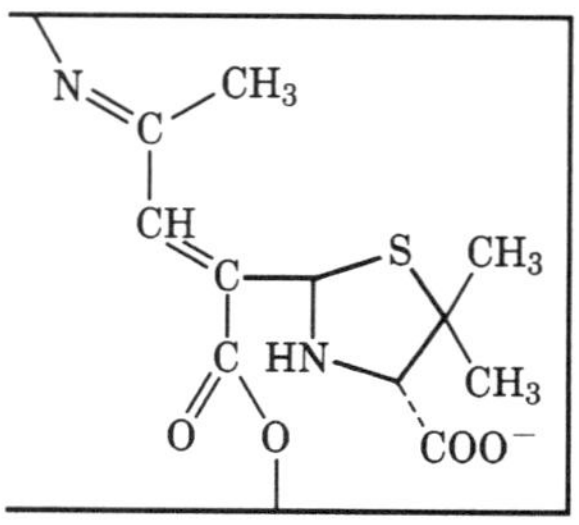

(**32**)

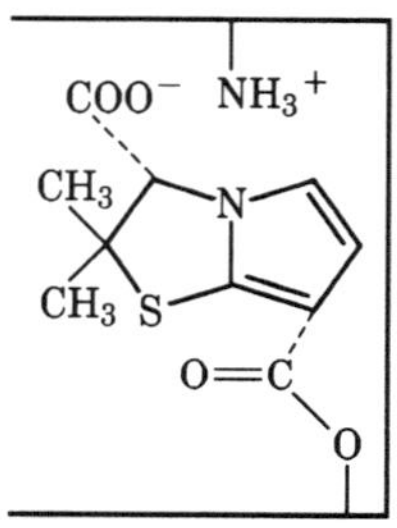

(**33**)

The 6-methoxymethylene penicillanic acid [*93040-42-7*] (**31**, R = CH_3OCH_1 (*Z*)-isomer, R″ = R‴ = CH_3) was designed to mimic the aminoacrylate species found using clavulanic acid and sulbactam. Upon the reaction of this compound with the enzyme, the potential exists for further Michael addition to inactivate the enzyme. The compound is indeed a β-lactamase inhibitor but no synergy data have been reported. The related imine structure [*73707-58-1*], R = *p*-$CH_3C_6H_5N$, (*Z*)-isomer, has also been reported to inhibit type I cephases. The 6-β-[bistrifluoromethane sulfonyl]amidopenicillanic acid [*82954-44-7*] (**31**, R = $(CF_3SO_2)_2N$), R′ = H, R″ = R‴ = CH_3) was speculated to inhibit the penase from *B. cereus* by triflation or generation of an imine that reacts further with the enzyme. Several C-6 substituted amino penams have been reported to inhibit type I cephases; the complex hydroxyalkyl penicillanic acid derivatives are reported to be potent β-lactamase inhibitors and synergists. In combination with amoxicillin, compound (**31**, R = 1-naphthylCOCH(OH), R′ = H, R″ = R‴ = CH_3) protected animals from infection. Synergy has been reported for the usual penam structures where R involves a 2-naphthylsulfoxide or a 2-naphthylphosphate against a β-lactamase producing strain of *Klebsiella*. An unusual series of ferrocenyl penicillins (190) were reported to inhibit β-lactamases.

Other Unusual β-Lactam Based Inhibitors. There are a number of other unusual β-lactams reported to have β-lactamase inhibition activity (191–194). In general these compounds are not very potent and are not irreversible inhibitors. Data are also very limited.

Economic Aspects

Although a broad range of β-lactamase inhibitors has been discovered, only clavulanic acid and sulbactam have been commercialized. Clavulanic acid (**12**, R = CH_2OH, R′ = H), manufactured by SmithKline Beecham, is sold as an oral and parenteral product in combination with amoxicillin under the trade name Augmentin. A parenteral product in combination with ticarcillin [*34787-01-4*], $C_{15}H_{16}N_2O_6S$, has the trade name, Timentin. In 1990 worldwide sales of clavulanic acid containing products were about $725 million.

Sulbactam (**25**, R = R′ = H, R″ = R‴ = CH_3) is produced by Pfizer. The oral version of sulbactam in combination with ampicillin is called Unasyn Oral which is the mutual prodrug sultamicillin. Two sulbactam parenteral products are sold, a combination product with ampicillin called Unasyn and a combination with cefoperazone [*62893-19-0*] called Sulperazon. In addition, sulbactam is sold alone for parenteral use with any β-lactam antibiotic as Betamaze. In 1990 worldwide sales of sulbactam containing products were over $280 million.

BIBLIOGRAPHY

"Antibiotics, β-Lactams, Clavulanic Acid, Thienamycin, and Others," in *ECT* 3rd ed., Supplement, pp. 83–131, by Allan Brown, Beecham Research Laboratories.

1. J. M. T. Hamilton Miller and J. T. Smith, eds., *Beta-Lactamases,* Academic Press, New York, 1979.
2. M. Richmond in M. Salton and G. D. Shockman, eds., *β-Lactam Antibiotics (Mode of Action, New Developments, and Future Prospects),* Academic Press, New York, 1981, pp. 261–274.
3. A. A. Medeiros and G. A. Jacoby in S. F. Queener, J. A. Webber, and S. W. Queener, eds., *Clinical Pharmacology 4, β-Lactam Antibiotics for Clinical Use,* Marcel Dekker Inc., New York, 1986, pp. 49–84.
4. N. H. Georgopapadakou and R. B. Sykes in A. L. Demain and N. A. Solomon, eds., *Antibiotics Containing the β-Lactam Structure,* Part II, Springer-Verlag, New York, 1983, pp. 1–78.
5. J.-M. Ghuysen, *Rev. Inf. Dis.* **10,** 726 (1988).
6. R. F. Pratt in M. Sandler and H. J. Smith, eds., *Design of Enzyme Inhibitors as Drugs,* Oxford University Press, New York, 1989, pp. 178–205.
7. R. A. Nicholas and J. L. Strominger, *Rev. Inf. Dis.* **10,** 733 (1988).
8. M. Cole in Ref. 1, pp. 205–289.
9. R. B. Sykes and K. Bush in R. B. Morin and M. Gorin, eds., *Chemistry and Biology of β-Lactam Antibiotics,* Vol. 3, Academic Press, New York, 1982, pp. 155–207.
10. J. R. E. Hoover in Ref. 4, pp. 212–226.
11. M. Cole, *Phil. Trans. R. Soc. London* **B289,** 207 (1980).
12. M. Cole, *Drugs of the Future* **6,** 697 (1980).
13. C. Reading and M. Cole, *J. Enzyme Inhibition* **1,** 83 (1986).
14. S. J. Cartwright and S. G. Waley, *Med. Res. Rev.* **3,** 341 (1983).
15. C. E. Newall in Ref. 2, pp. 287–300.

16. A. Bauernfeind, *Rev. Inf. Dis.* **8,** 5470 (1986).
17. K. Bush, *Rev. Inf. Dis.* **10,** 681 (1988).
18. R. H. Richmond and R. B. Sykes, *Adv. Microb. Physiol.* **9,** 31 (1973).
19. H. C. Neu, *Am. J. Med.* **79,** (Suppl. B.), 2 (1985).
20. B. Wiedemann, C. Kliebe, and M. Kresken, *J. Antimicrob. Chemother.* **24,** (Suppl. B), 1 (1989).
21. J. D. Williams, *Drugs* **35,** (Suppl. 7), 3 (1988).
22. A. L. Fink, *Pharm. Res.* **2,** 55 (1985).
23. S. J. Cartwright, A. K. Tan, and A. L. Fink, *Biochem. J.* **263,** 905 (1989).
24. R. Virden, A. K. Tan, and A. L. Fink, *Biochemistry* **29,** 145 (1990).
25. V. Knott-Hunziker, S. Petursson, S. G. Waley, B. Jaurin, and T. Grundstrom, *Biochem. J.* **207,** 315 (1982).
26. J. R. Knowles, *Acc. Chem. Res.* **18,** 97 (1985).
27. J. R. Knowles in F. E. Hahn, ed., *Antibiotics VI (Modes and Mechanisms of Microbial Growth Inhibitors),* Springer-Verlag, New York, 1983, pp. 90–107.
28. K. Bush and R. B. Sykes, *J. Antimicrob. Chemother.* **11,** 97 (1983).
29. R. Reid, L. C. Felton, and M. A. Pittroff, *Proc. Soc. Exp. Biol. Med.* **63,** 438 (1946).
30. O. K. Behrens and L. Garrison, *Arch. Biochem. Biophys.* **27,** 94 (1950).
31. G. V. Patil and R. A. Day, *Biochem. Biophys. Acta* **293,** 490 (1973).
32. H. Ogawara and H. Umezawa, *Biochimica Biophys. Acta.* **327,** 481 (1973).
33. E. P. Abraham and G. G. F. Newton, *Biochem. J.* **63,** 628 (1956).
34. G. N. Rolinson, S. Stevens, F. R. Batchelor, J. Cameron-Wood, and E. B. Chain, *Lancet* (2), 564 (1960).
35. N. Citri, A. Samuni, and N. Zyk, *Proc. Nat. Acad. Sci. USA* **73,** 1048 (1976).
36. A. L. Fink, K. M. Behner, and A. K. Tan, *Biochem.* **26,** 4248 (1987).
37. M. Cole in J. D. Williams, ed., *Antibotic Interactions,* Academic Press, New York, 1979, pp. 99–135.
38. M. Arisawa, J. Ohshima, and H. B. Maruyama, *Chem. Pharm. Bull.* **30,** 3333 (1982).
39. K. P. Fu and H. C. Neu, *J. Antibiot.* **32,** 909 (1979).
40. M. H. Richmond, *J. Antimicrob. Chemother.* **6,** 445 (1980).
41. K. P. Fu and H. C. Neu, *Antimicrob. Agents Chemother.* **14,** 322 (1978).
42. K. Murakami and T. Yoshida, *Antimicrob. Agents Chemother.* **27,** 727 (1985).
43. R. L. Charnas and R. L. Then, *Rev. Inf. Dis.* **10,** 752 (1988).
44. W. S. Faraci and R. F. Pratt, *Biochemistry* **24,** 903 (1985).
45. S. Nishimura and co-workers, *J. Antibiot.* **42,** 159 (1989).
46. S. Nishimura, N. Yasuda, H. Sasaki, Y. Matsmoto, and T. Kamimura, *J. Antibiot.* **43,** 114 (1990).
47. S. Ohya, T. Miyadera, and M. Yamazaki, *Antimicrob. Agents. Chemother.* **21,** 613 (1982).
48. H. C. Neu and N.-X. Chin, *Antimicrob. Agents Chemother.* **31,** 505 (1987).
49. R. L. Then, *Chemotherapy* **30,** 398 (1984).
50. K. Bush, J. S. Freudenberger, and R. B. Sykes, *Antimicrob. Agents Chemother.* **22,** 414 (1982).
51. R. B. Sykes and co-workers, *Nature* **291,** 489 (1981).
52. K. Bush, S. A. Smith, S. K. Tanaka, and D. B. Bonner, *J. Antimicrob. Chemother.* **22,** 801 (1988).
53. R. Cooper, K. Bush, P. A. Principe, W. H. Trejo, J. S. Wells, and R. B. Sykes, *J. Antibiot.* **36,** 1252 (1983).
54. K. Tanizawa, K. Santoh, and Y. Kanaoka, *FEBS Lett.* **250,** 218 (1989).
55. A. G. Brown, D. Butterworth, M. Cole, G. Hanscomb, J. D. Hood, and C. Reading, *J. Antibiot.* **29,** 668 (1976).
56. T. T. Howarth, A. G. Brown, and T. J. King, *J. Chem. Soc., Chem. Commun.,* 266 (1976).

57. C. Reading and M. Cole, *Antimicrob. Agents Chemother.* **11,** 852 (1977).
58. H. C. Neu and K. P. Fu, *Antimicrob. Agents Chemother.* **14,** 650 (1978).
59. P. C. Cherry and C. E. Newall in R. B. Morin and M. Gorman, eds., *Chemistry and Biology of β-Lactam Antibiotics,* Academic Press, New York, 1982, pp. 362–402.
60. C. E. Newall in G. I. Gregory, ed., *Recent Advances in the Chemistry of β-Lactam Antibiotics,* Royal Society of Chemistry, London, 1980, pp. 151–167.
61. P. H. Bently, P. D. Berry, G. Brooks, M. L. Gilpin, E. Hunt, and I. I. Zomaya in Ref. 60, pp. 175–183.
62. R. D. G. Cooper in P. G. Sammes, ed., *Topics in Antibiotic Chemistry,* Vol. 3, John Wiley & Sons, Inc., New York, 1980, pp. 43–73.
63. G. Brooks and co-workers in A. G. Brown and S. M. Roberts, eds., *Recent Advances in the Chemistry of β-Lactam Antibiotics,* Royal Society of Chemistry, London, 1984, pp. 222–241.
64. C. Reading and B. Slocombe in Ref. 3, pp. 527–592.
65. *J. Antimicrob. Chemother.* **24,** (Suppl. B), (1989).
66. A. G. Brown, *J. Antimicrob. Chemother.* **7,** 15 (1981).
67. C. Reading and P. Hepburn, *Biochem. J.* **179,** 67 (1979).
68. A. Basak, S. P. Salowe, and C. A. Townsend, *J. Am. Chem. Soc.* **112,** 1654 (1990).
69. S. W. Elson in P. H. Bently and R. Southgate, eds., *Recent Advances in the Chemistry of β-Lactam Antibiotics,* Royal Society of Chemistry, London, 1989, pp. 303–320.
70. C.-P. Mak, K. Prasad, and F. Turnowsky, *J. Antibiot.* **36,** 398 (1983).
71. E. Hunt, P. H. Bently, G. Brooks, and M. L. Gilpin, *J. Chem. Soc., Chem. Commun.,* 906 (1977).
72. S. W. Elson, J. Gillett, N. H. Nicholoson, and J. W. Tyler, *J. Chem. Soc., Chem. Commun.,* 979 (1988).
73. P. H. Bently, G. Brooks, M. L. Gilpin, and E. Hunt, *J. Chem. Soc., Chem. Commun.,* 905 (1977).
74. R. W. Ratcliffe and G. Alber-Schonberg in R. B. Morin and M. Gorman, eds., *Chemistry and Biology of β-Lactam Antibiotics,* Academic Press, New York, 1982, pp. 227–313.
75. Y. Fukagawa, M. Okabe, T. Yoshioka, and T. Ishikura in Ref. 63, pp. 163–182.
76. K. Okonogi, S. Harada, S. Shinagawa, A. Imada, and M. Kuno, *J. Antibiot.* **35,** 963 (1982).
77. F. Kobayashi and co-workers, *Antimicrob. Agents Chemother.* **21,** 536 (1982).
78. K. Okonogi, Y. Nozaki, A. Imada, and M. Kuno, *J. Antibiot.* **34,** 212 (1981).
79. S. Harada, S. Tsubotani, M. Asai, K. Okonogi, and M. Kondo, *J. Med. Chem.* **26,** 753 (1983).
80. M. Nakayama and co-workers, *J. Antibiot.* **33,** 1388 (1980).
81. D. Butterworth, M. Cole, G. Hanscomb, and G. N. Rolinson, *J. Antibiot.* **32,** 287 (1979).
82. J. D. Hood, S. J. Box, and M. S. Verrall, *J. Antibiot.* **32,** 295 (1979).
83. R. L. Charnas and J. R. Knowles, *Biochemistry* **20,** 2732 (1981).
84. S. Harada, Y. Nozaki, and S. Shinagawa, *J. Antibiot.* **35,** 957 (1982).
85. Y. Nakamura and co-workers, *J. Antibiot.* **39,** 956 (1986).
86. A. Yamaguchi, T. Hirata, and T. Sawai, *Antimicrob. Agents Chemother.* **25,** 348 (1984).
87. T. Yoshida, Y. Kazuno, T. Shomura, S. Murata, S. Inouye, and T. Ito, *J. Antibiot.* **39,** 956 (1986).
88. T. Ito and co-workers, *J. Antibiot.* **35,** 533 (1982).
89. T. Niwa, T. Yoshida, A. Tamura, Y. Kazuno, S. Inouye, and T. Ito, *J. Antibiot.* **39,** 943 (1986).
90. R. H. Richmond, *J. Antimicrob. Chemother.* **7,** 279 (1981).
91. M. Toda, K. Sato, H. Nakazawa, M. Inoue, and S. Mitsuhashi, *Antimicrob. Agents Chemother.* **18,** 837 (1980).

92. K. Okamura, M. Sakamoto, Y. Fukagawa, T. Isahikura, and J. Lein, *J. Antibiot.* **32,** 280 (1979).
93. K. Okamura, M. Sakamoto, and T. Isahikura, *J. Antibiot.* **33,** 293 (1980).
94. K. Okamura, M. Sakamoto, and T. Isahikura, *Biochem. J.* **185,** 177 (1980).
95. N. Shibamoto, M. Nishino, K. Okamura, Y. Fukagawa, and T. Ishikura, *J. Antibiot.* **35,** 763 (1982).
96. M. Sakamoto, T. Ishikura, and Y. Fukagawa, *J. Antibiot.* **37,** 1414 (1984).
97. H. Hirai, K. Sawada, M. Aratani, and M. Hashimoto, *Tetrahedron. Lett.* **25,** 5075 (1984).
98. H. Natsugari and co-workers, *J. Antibiot.* **36,** 855 (1983).
99. N. Tamura and co-workers, *Chem. Pharm. Bull.* **35,** 996 (1987).
100. K. Murakami, M. Doi, and T. Sawai, *Antimicrob. Agents Chemother.* **25,** 348 (1984).
101. M. J. Basker, R. J. Boon, S. J. Box, E. A. Prestige, G. M. Smith, and S. R. Specer, *J. Antibiot.* **36,** 416 (1983).
102. J. Birnbaum, F. M. Kahan, H. Kropp, and J. S. MacDonald, *Am. J. Med.* **78,** (Suppl. 6A), 3 (1985).
103. H. Mikami, M. Ogashiwa, Y. Saini, M. Inoue, and S. Mitsuhashi, *Antimicrob. Agents Chemother.* **22,** 693 (1982).
104. S. R. Norrby and co-workers, *Antimicrob. Agents Chemother.* **23,** 300 (1983).
105. N. Shibamoto, M. Sakamoto, Y. Fukagawa, and T. Ishikura, *J. Antibiot.* **35,** 729 (1982).
106. P. Schindler, G. Huber, and W. Konig, *J. Antibiot.* **36,** 876 (1983).
107. I. Ernest in R. B. Morin and M. Gorman, eds., *Chemistry and Biology of β-Lactam Antibiotics,* Academic Press, New York, 1982, pp. 315–360.
108. S. W. McCombie and A. K. Ganguly, *Med. Res. Rev.* **8,** 393 (1988).
109. J. C. Pechere, R. Letarte, R. Guay, C. Asselin, and C. Morin, *J. Antimicrob. Chemother.* **9,** Suppl. C, 123–132 (1982).
110. G. Emmer and co-workers, *J. Antibiot.* **38,** 1371 (1985).
111. R. B. Woodward, *Phil. Trans. R. Soc. London* **B289,** 239 (1980).
112. S. Adams, *Heterocycles* **22,** 1509 (1984).
113. K. Coleman, D. R. J. Griffin, J. W. J. Page, and P. A. Upshon, *Antimicrob. Agents Chemother.* **33,** 1580 (1989).
114. N. F. Osborne, N. J. P. Broom, S. Coulton, J. B. Harbridge, M. A. Harris, I. Stirling-Francois, and G. Walker, *J. Chem. Soc., Chem. Commun.,* 371 (1989).
115. N. J. P. Broom, S. Coulton, I. Francois, J. B. Harbridge, J. H. C. Nayler, and N. F. Osborne in Ref. 69, pp. 247–258.
116. N. J. P. Broom, K. Coleman, P. A. Hunter, and N. F. Osborne, *J. Antibiot.* **43,** 76 (1990).
117. M. J. Basker and N. F. Osborne, *J. Antibiot.* **43,** 70 (1990).
118. A. J. Sweetman, *Drugs of the Future* **5,** 23 (1980).
119. A. R. English, J. A. Retsema, A. E. Girard, J. G. Lynch, and W. E. Barth, *Antimicrob. Agents Chemother.* **14,** 414 (1978).
120. J. A. Retsema, A. R. English, and A. E. Girard, *Antimicrob. Agents Chemother.* **17,** 615 (1980).
121. J. Klastersky and P. Van der Auwera, *Diagn. Microbial. Inf. Dis.* **12,** 165S (1989).
122. D. M. Campoli-Richards and R. N. Brogden, *Drugs* **33,** 577 (1987).
123. D. L. Hemsell and H. H. Werner, eds., *Drugs* **31,** (Suppl. 2), 1–28 (1986).
124. *Diagn. Microbiol. Inf. Dis.* **12,** (Suppl.) 107S–215S (1989).
125. G. Foulds and co-workers, *Proceedings of the 11th International Congress and the 19th Interscience Conference on Antimicrobial Agents and Chemotherapy,* Boston, Mass., 1980, p. 353.
126. J. G. Stam and co-workers, *22nd Interscience Conference on Antimicrobial Agents and Chemotherapy,* Miami Beach, Fla., 1982, p. 510.

127. *Chemotherapy* **33,** (Suppl. 3), (1984).
128. B. Baltzen and co-workers, *J. Antibiot.* **33,** 1183 (1980).
129. A. R. English, D. Girard, V. J. Jasys, R. J. Martingano, and M. S. Kellogg, *J. Med. Chem.* **33,** 344 (1990).
130. K. P. Fu and H. C. Neu, *Antimicrob. Agents Chemother.* **15,** 171 (1979).
131. R. G. Micetich and co-workers, *J. Med. Chem.* **30,** 1469 (1987).
132. J. A. Retsema, W. U. Schelky, A. E. Girard, and A. R. English, *Drugd. Expt. Clin. Res.* **7,** 255 (1981).
133. A. L. Fink, L. M. Ellerby, and P. M. Bassett, *J. Am. Chem. Soc.* **111,** 6871 (1989).
134. T. Swai, A. Yamaguchi, and K. Tsukamato, *Rev. Inf. Dis.* **10,** 721 (1988).
135. A. Yamaguchi, A. Adachi, T. Hirata, H. Adachi, and T. Sawai, *J. Antibiot.* **38,** 83 (1985).
136. J. Fisher, R. L. Charnas, S. M. Bradley, and J. R. Knowles, *Biochemistry* **20,** 2726 (1981).
137. P. S. F. Mezes, A. J. Clark, G. I. Dmitrienko, and T. Viswanatha, *FEBS Lett.* **143,** 265 (1982).
138. G. I. Dmitrienko, C. R. Copeland, L. Arnold, M. E. Savard, A. J. Clarke, and T. Viswanatha, *Bioorg. Chem.* **13,** 34 (1985).
139. S. J. Cartwright and A. F. W. Coulson, *Nature* **278,** 360 (1979).
140. Y. L. Chen, C.-W. Chang, and K. Hedberg, *Tetrahedron Lett.* **27,** 3449 (1986).
141. Y. L. Chen and co-workers, *J. Antibiot.* **40,** 803 (1987).
142. S. Hanessian and M. Alepegiani, *Tetrahedron* **45,** 941 (1989).
143. M. G. Waters, J. A. Retsema, and M. S. Kellogg, *ASM Meeting,* Lake George, New York, 1983.
144. D. D. Kieth, J. Tengi, P. Rossman, L. Todaro, and M. Weigele, *Tetrahedron* **39,** 2445 (1983).
145. W. J. Gottstein, L. B. Crest, R. G. Graham, U. J. Haynes, and D. N. McGregor, *J. Med. Chem.* **24,** 1531 (1981).
146. S. N. Maiti, P. Spevak, K. Ogawa, and R. G. Micetich, *J. Org. Chem.* **53,** 3803 (1988).
147. W. J. Gottstein, U. J. Hayes, and D. N. McGreggor, *J. Med. Chem.* **28,** 518 (1985).
148. H. Tanaka, M. Tanaka, A. Nakai, S. Yamada, N. Ishida, T. Otani, and S. Torii, *J. Antibiot.* **41,** 579 (1988).
149. R. G. Micetich and co-workers, *Proceedings of the 14th International Congress on Chemotherapy,* Kyoto, Japan, 1985, pp. 249–250.
150. L. Gutmann, M.-D. Kitzis, S. Yamabe, and J. F. Acar, *Antimicrob. Agents Chemother.* **29,** 955 (1986).
151. F. Moosdeen, J. D. Williams, and S. Yamabe in Ref. 149, pp. 1272–1273.
152. N. Ishida, A. Hyodo, C. Hanehara, Y. Miyake, Y. Kawaguchi, and S. Yamabe in Ref. 149, pp. 1274–1275.
153. A. Yamaguchi, A. Adachi, T. Hirata, H. Adachi, and T. Sawai, *J. Antibiot.* **38,** 83 (1985).
154. A. Yamaguchi, T. Inaba, and T. Sawai, *J. Antibiot.* **41,** 690 (1988).
155. Y. L. Chen, K. Hedberg, J. F. Barrett, and J. A. Retsema, *J. Antibiot.* **41,** 134 (1988).
156. M. N. Serradell and J. Castaner, *Drugs of the Future* **11,** 588 (1986).
157. S. C. Aronoff, M. R. Jacobs, S. Johenning, and S. Yamabe, *Antimicrob. Agents Chemother.* **26,** 580 (1984).
158. M. R. Jacobs, S. C. Arnoff, S. Johenning, and S. Yamabe, *J. Antimicrob. Chemother.* **18,** 177 (1986).
159. M. R. Jacobs, S. C. Arnoff, S. Johenning, D. M. Shales, and S. Yamabe, *Antimicrob. Agents Chemother.* **29,** 980 (1986).
160. P. C. Appelbaum, M. R. Jacobs, S. K. Spangler, and S. Yamabe, *Antimicrob. Agents Chemother.* **30,** 789 (1986).

161. S. C. Arnoff, M. R. Jacobs, P. H. Labrozzi, and S. Yamabe, *J. Antimicrob. Chemother.* **18,** 271 (1986).
162. *Diagn. Microb. Inf. Dis.,* 477 (1990).
163. P. J. Claes, H. Vanderhaeghie, E. Roets, and J. Thomis, *J. Antibiot.* **37,** 75 (1985).
164. W. von Daehne, *J. Antibiot.* **33,** 451 (1980).
165. H. C. Neu, *Antimicrob. Agents Chemother.* **23,** 63 (1983).
166. R. F. Pratt and M. J. Loosemore, *Proc. Natl. Acad. Sci. USA* **75,** 4145 (1978).
167. R. Wise, J. M. Andrews, and N. Patel, *J. Antimicrob. Chemother.* **7,** 531 (1981).
168. W. von Daehne, E. T. Hansen, and N. Rastrup-Andersen, in Ref. 63, pp. 375–380.
169. J. E. G. Kemp, M. D. Closier, S. Narayanaswami, and M. H. Stefaniak, *Tetrahedron Lett.* **21,** 2991 (1980).
170. B. A. Moore and K. W. Brammer, *Antimicrob. Agents Chemother.* **20,** 327 (1981).
171. S. Adam, R. L. Then, and P. Angehrn, *J. Antibiot.* **40,** 108 (1987).
172. N.-X. Chin, M. J. McElrath, and H. C. Neu, *Chemotherapy* **34,** 318 (1988).
173. P. Angehrn and M. Arisawa, *J. Antibiot.* **35,** 1584 (1982).
174. M. Arisawa and R. L. Then, *J. Antibiot.* **35,** 1578 (1982).
175. M. Arisawa and R. L. Then, *Biochem. J.* **209,** 609 (1983).
176. M. Arisawa and S. Adam, *Biochem. J.* **211,** 447 (1983).
177. M. Arisawa and R. L. Then, *J. Antibiot.* **36,** 1372 (1983).
178. D. Habich and K. Metzger, *Heterocycles* **24,** 289 (1986).
179. D. G. Brenner, *J. Org. Chem.* **50,** 18 (1985).
180. D. G. Brenner and J. R. Knowles, *Biochemistry* **23,** 5839 (1984).
181. E. M. Gordon, H. W. Chang, C. M. Cimarusti, B. Toeplitz, and J. Z. Gougoutas, *J. Am. Chem. Soc.* **102,** 1690 (1980).
182. P. S. F. Mezes, R. W. Friesen, T. Viswanatha, and G. I. Dmitrienko, *Heterocycles* **19,** 1207 (1982).
183. M. Yamashita, S. Hashimoto, and M. Ezaki, *J. Antibiot.* **36,** 1774 (1983).
184. M. J. Calverley and M. Begtrup, *J. Antibiot.* **36,** 1507 (1983).
185. Y. L. Chen, K. G. Hedberg, and K. J. Guarino, *Tetrahedron Lett.* **30,** 1067 (1989).
186. L. A. Reed, D. A. Charleson, and R. A. Volkmann, *Tetrahedron Lett.* **28,** 3431 (1987).
187. J. A. Retsema and co-workers, *28th Interscience Conference on Antimicrobial Agents and Chemotherapy,* 122 (1988).
188. J. C. Sheehan, E. Chacko, T. J. Commons, Y. S. Lo, D. R. Ponzi, and A. Schwabacher, *J. Antibiot.* **37,** 1441 (1984).
189. W. von Daehne, B. H. Nielsen, and I. Roholt, in Ref. 149, pp. 1280–1281.
190. E. I. Edwards, R. Epton, and G. Marr, *J. Organomet. Chem.* **85,** C23 (1975).
191. R. Joyeau, H. Molines, R. Labia, and M. Wakselman, *J. Med. Chem.* **31,** 370 (1988).
192. D. Davies and M. J. Pearson, *J. Chem. Soc., Perkin Trans. 1,* 2539 (1981).
193. D. Habich, P. Naab, and K. Metzger, *Tetrahedron Lett.* **24,** 2559 (1983).
194. R. Joyeau, L. D. S. Yadav, and M. Wakselman, *J. Chem. Soc., Perkin Trans. 1,* 1899 (1987).

J. G. Stam
Pfizer Central Research

MONOBACTAMS

β-Lactam antibiotics are one of the best established classes of antimicrobial agents for the treatment of infectious diseases. Since the discovery of penicillin in 1929 and cephalosporin C in 1955, a large number of semisynthetic analogues has become available. In addition, more sophisticated screening and isolation techniques have afforded a variety of new naturally-occurring β-lactams such as the cephamycins, clavulanic acid, nocardicin, and the carbapenems. In late 1979 another entirely new class of β-lactams, characterized by the 2-oxoazetidine-1-sulfonic acid moiety (**1**, R = H or OCH_3, R′ = acyl), was discovered (1,2). To distinguish this class of monocyclic, bacterially-produced β-lactam antibiotics from the bicyclic cephalosporins and cephamycins (**2**, R = H or OCH_3, R′ = γ-linked α-aminoadipoyl), the term monobactam was introduced (1).

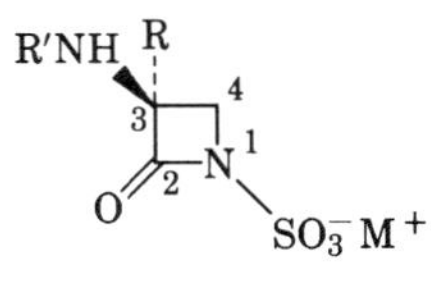

(**1**)

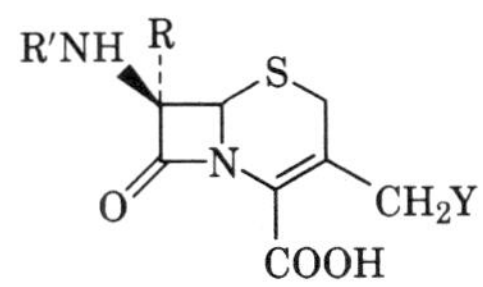

(**2**)

By screening strains of bacteria that were specifically responsive to β-lactam antibiotics, monobactams, varying in substitution at the C-3 position, were identified. The structures of the compounds along with their producing organisms are shown in Table 1. More recently, the 4β-methyl analogue (**4**, R″ = CH_3) of SQ 26,445/sulfazecin (**4**, R″ = H) was isolated using a differential antibacterial assay (3).

Structure Determination and Biological Properties

The structure and absolute stereochemistry of SQ 26,445 (sulfazecin) (**4**, R″ = H) was unequivocally established by x-ray crystallography (4) and that of SQ 26,180 (**3**, R = OCH_3) was determined by synthesis from a homochiral precursor of defined structure. Starting with 7-aminocephalosporanic acid (7-ACA), stereospecific introduction of the methoxyl group and degradation of the thiazine ring ultimately afforded the corresponding N-1 unsubstituted azetidinone. Subsequent sulfonation confirmed the structure of SQ 26,180 (**8**, R = CH_3) (Fig. 1) (5). The structure and absolute configuration of the nonmethoxylated naturally-occurring monobactam (**6**, X = OH, Y = R = H, M = K) was also proven by total synthesis (6). The monobactams (**1**) correspond in configuration to the naturally-occurring cephalosporins and cephamycins (**2**). And, as is the case with regard to bicyclic β-lactams, monobactams of opposite absolute configuration to those encountered in nature are inactive. The enantiomeric pair of monobactams shown in Table 2 provide a striking illustration of this specificity.

In penicillins, incorporation of a 6α-methoxyl substituent increases chemical stability three- to fivefold as measured by basic hydrolysis of the β-lactam moiety (7). Similar substitution of a 7α-methoxyl substituent in cephalosporins also results in a modest increase in chemical stability. In contrast, methoxylated monobactams having simple side chains are much less chemically stable than the corresponding nonmethoxylated compounds. However, the methoxylated monobactams show an increased stability to β-lactamases when compared to

Table 1. Naturally Occurring Monobactams

Organism	Monobactam	CAS Registry Number	Molecular formula	Structure number	Structure
Chromobacterium violaceum	SQ 26,180	[*79720-08-4*]	$C_6H_{10}N_2O_6S{\cdot}K$	(**3**, R = OCH_3)	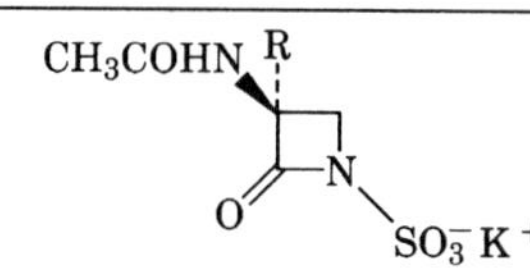
Gluconobacter oxydans	SQ 26,445	[*77912-79-9*]	$C_{12}H_{20}N_4O_9S$	(**4**, R″ = H)	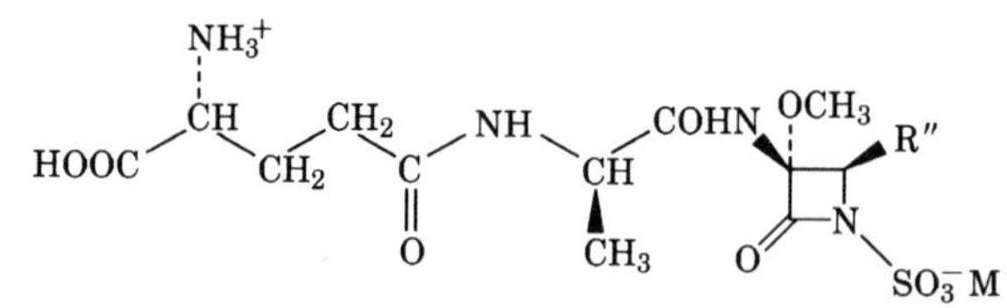
Pseudomonas acidophila	sulfazecin	[*77912-79-9*]	$C_{12}H_{20}N_4O_9S$	(**4**, R″ = H)	
Pseudomonas mesoacidophila	isosulfazecin	[*77900-75-5*]	$C_{12}H_{20}N_4O_9S$	(**5**)	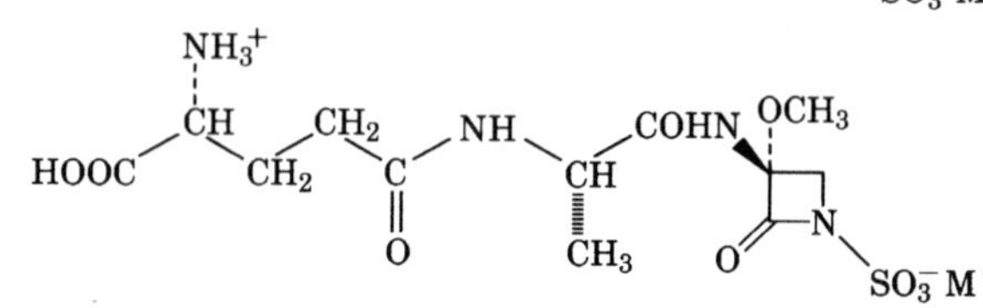

Agrobacterium radiobacter			(**6**) 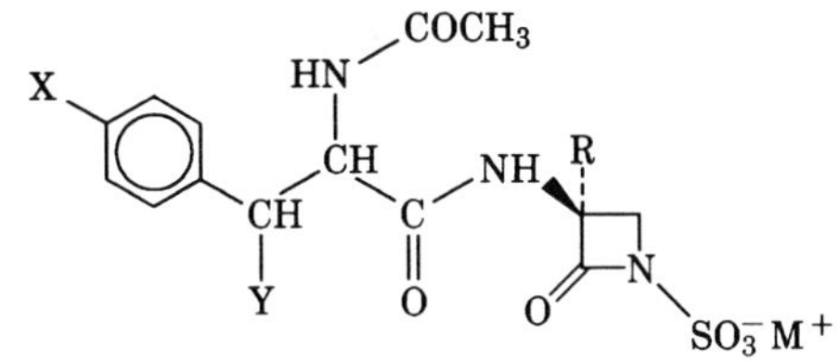
	[*79720-13-1*]	$C_{15}H_{19}N_3O_7S{\cdot}Na$	(**6**, X = H, Y = H, R = OCH_3, M = Na)
	[*79720-14-2*]	$C_{15}H_{19}N_3O_8S{\cdot}K$	(**6**, X = OH, Y = H, R = OCH_3, M = K)
	[*81919-28-0*]	$C_{14}H_{17}N_3O_7S{\cdot}K$	(**6**, X = OH, Y = H, R = H, M = K)
	[*79720-16-4*]	$C_{15}H_{19}N_3O_{12}S{\cdot}2Na$	(**6**, X = $OSO_3^-Na^+$, Y = OH, R = OCH_3, M = Na)
	[*79720-17-5*]	$C_{15}H_{19}N_3O_{15}S_3{\cdot}3Na$	(**6**, X = $OSO_3^-Na^+$, Y = $OSO_3^-Na^+$, R = OCH_3, M = Na)

Table 2. Biological Activity of (*S*)- and (*R*)-3-{[(2-amino-4-thiazolyl)(methoxyimino)acetyl]amino}-2-oxo-1-azetidinesulfonic acid

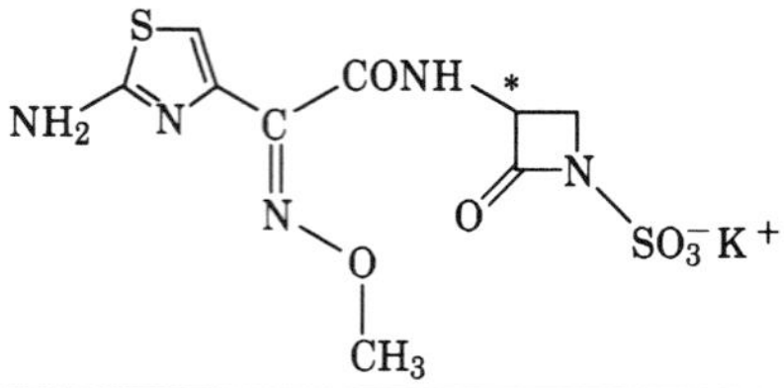

Absolute configuration[a]	CAS Registry Number	Molecular formula	MIC, μg/mL[b]
(*S*)-	[*78626-19-4*]	$C_9H_{11}N_5O_6S_2$	<0.05
(*R*)-	[*78612-61-0*]	$C_9H_{11}N_5O_6S_2$	>100

[a]At C-3.
[b]For *P. rettgeri*.

their unsubstituted analogues. This relationship is demonstrated by comparing the V_{max}, the rate at saturating substrate concentration for SQ 26,180 (**3**, R = OCH_3) with that of the nonmethoxylated counterpart SQ 26,396 [*79720-10-8*] (**3**, R = H), $C_5H_8N_2O_5SK$, against the clinically important RTEM lactamase (8). The relative V_{max} (RTEM) are <0.01 and 12, respectively compared to cephaloridine at 100. Similarly, increased β-lactamase stability was also observed for the 7-methoxy-cephalosporins (9).

The monobactams, like pencillins and cephalosporins, interfere with the synthesis of bacterial cell walls. β-Lactam antibiotics bind to a series of penicillin-binding proteins (PBPs) on the cytoplasmic membrane and their antibacterial effect is believed to result from inhibition of a subset of these PBPs known as peptidoglycan transpeptidases. These enzymes are responsible for cross-linking adjacent peptide strands within the peptidoglycan structure of the cell wall, thus improving the integrity of the bacterial protective coat. Aztreonam [*78110-38-0*], $C_{13}H_{17}N_5O_8S_2$, a clinically efficacious monobactam, specifically binds to PBP-3 thereby preventing septation and subsequently leading to cell death (10).

The biosynthetic origin of monobactams has been elucidated by fermentation experiments using radioactively labeled amino acids (qv). The monocyclic ring of the naturally-occurring C-4 unsubstituted monobactams is derived from serine. Similar techniques have shown that the methyl moiety of the methoxyl group in 3α-methoxylated monobactams is derived from methionine (11).

All of the naturally-occurring monobactams discovered as of this writing have exhibited poor antibacterial activity. However, as in the case of the penicillins and cephalosporins, alteration of the C-3 amide side chain led to many potent new compounds (12). Furthermore, the monobactam nucleus provides a unique opportunity to study the effect of structural modifications at the N-1 and C-4 positions of the azetidinone ring on biological activity. In contrast to the bicyclic β-lactams, these positions on the monocyclic ring system are readily accessible by synthesis.

Synthesis

Initial syntheses employed the sulfonation of an N-1 unsubstituted azetidinone as the key step. The natural product SQ 26,180 (**3**, R = OCH_3) (**8**, R = CH_3) as well as other methoxylated monobactams were synthesized, starting from either 7-aminocephalosporanic acid [*957-68-6*] (7-ACA) or 6-aminopenicillanic acid [*551-16-6*] (6-APA), via sulfonation of the N-1 unsubstituted intermediates (**7** or **10**) as shown in Figure 1 (5,13). Subsequently, many more C-3 side-chain analogues were prepared by this method. Examples include the sulfonation of the N-1 unsubstituted azetidinones (**10**) that carry protecting groups on the nitrogen atom attached to C-3. 3-Aminomonobactamic acid [*79720-18-6*] (3-AMA) (**12**), $C_3H_6N_2O_4S$, was readily formed by removal of the protecting group and could subsequently be acylated to afford a wide variety of side-chain monobactams (**13**). 6-APA proved to be a readily available source for 3-AMA (14,15) which alternatively could be prepared from the *O*-methylhydroxamate (**9**) of L-serine (16).

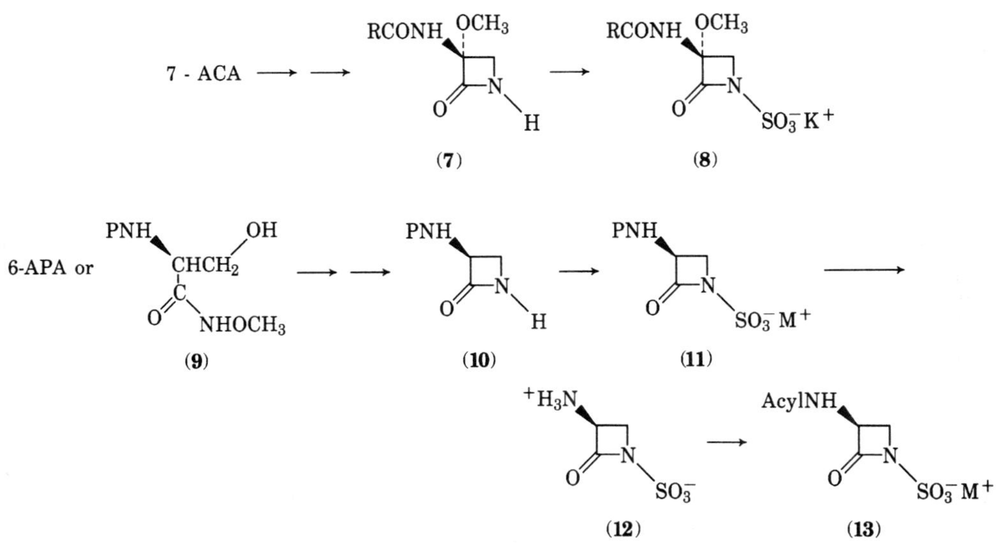

Fig. 1. Synthesis of C-3 side chain analogues of monobactams, where P is an amine protecting group and other terms are defined in text.

A second, conceptually distinct chiral synthesis of monobactams was developed from *β*-hydroxy amino acids. As shown in Figure 2, cyclization of the acylsulfamate of an amino-protected *O*-mesylserine derivative (**14**, R = H) leads directly to the monobactam (**15**). This methodology was also applied to the synthesis of 4*α*- (**15**, R = CH_3) and 4*β*-methyl monobactams from L-threonine and allothreonine, respectively (17). The 3(*S*)-(*trans*)-3-amino-4(*S*)-methylmonobactamic acid [*80082-65-1*] (**16**, R = CH_3), $C_4H_8N_2O_4S$, is the key intermediate to aztreonam [*78110-38-0*] (**17**), $C_{13}H_{17}N_5O_8S_2$, the first monobactam to achieve the status of a clinically useful antibiotic. A similar synthesis was utilized for the preparation of carumonam [*89638-04-8*] (**19**), $C_{12}H_{14}N_6O_{10}S_2$, the second mono-

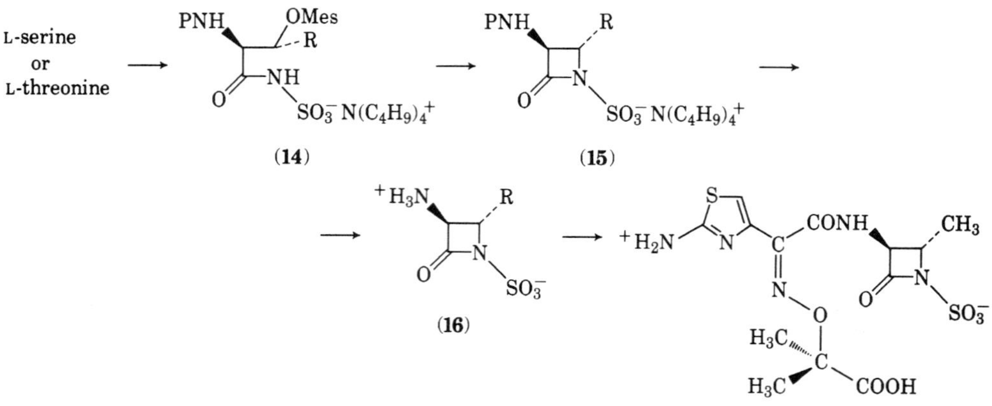

(14) (15) (16) (17)

Fig. 2. Synthesis of clinically useful monobactams where R = H, CH_3; P is an amino protecting group, and Mes = mesyl is methanesulfonyl.

bactam to undergo clinical trials. The absolute stereochemistry at the C-3 and C-4 positions, as well as the functionality on the C-4 position, were controlled by starting with either L-threonate or L-(+)-tartaric acid. Thus the appropriately substitued acyl sulfamate (**18**), analogous to (**14**), was obtained for conversion to carumonam (**19**) (18).

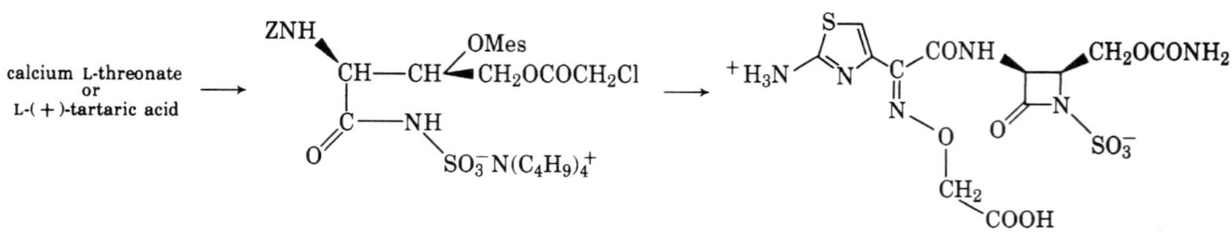

(**18**) (**19**)

Although the activity of methoxylated monobactams could be improved by appropriate side-chain modifications, difficulty of synthesis and poor chemical stability focused attention on the nonmethoxylated analogues. Both high intrinsic activity and excellent β-lactamase stability are exhibited by monobactams that combine C-3 aminothiazole oxime side chains and 4-alkyl, 4-alkenyl, and 4-alkynyl groups (19).

Aztreonam. Aztreonam (**17**) is a totally synthetic compound having an antibacterial spectrum that is unique among β-lactam antibiotics. It exhibits potent and specific activity against a wide range of both β-lactamase-producing and nonproducing aerobic gram-negative bacteria, including *Pseudomonas aeruginosa,* but displays minimal inhibition against anaerobic and gram-positive aerobic, bacteria, eg, staphylococci and streptococci. When tested against strains of *Enterobacteriaceae,* aztreonam inhibited 50% of these isolates at concentrations <0.3 μg/mL (20). Particularly sensitive were members of *Escherichia coli, Klebsiella pneumoniae, Proteus mirabilis, Morganella morganii, Serratia marcescens,* and many *Providencia sp.*. The activity of aztreonam is comparable to that of the aminoglycosides and third generation cephalosporins against these *Enterobacteriaceae*. Similarly, 50% of strains of ampicillin-sensitive and resistant *Haemophilis influenzae* and *Neisseria gonorrhoeae* were inhibited at <0.1 μg/mL.

Most dramatic were results with strains of *Pseudomonas aeruginosa* where aztreonam inhibited 50% of the strains at 4 μg/mL, and 90% at 16 μg/mL. Cessation of growth is not the sole consequence of exposure of these organisms to aztreonam as the inhibitory concentration is also generally the lethal concentration. Aztreonam does not inhibit the majority of strains of *Acinetobacter sp.* and many strains of *Pseudomonas sp.*, eg, *P. maltophilia* and *P. cepacia*. Also little or no activity was seen against *Bacteroides fragilis*. Minimal inhibitory concentrations for the anaerobic and gram-positive aerobic bacteria were usually >100 μg/mL.

Many of the gram-negative isolates tested were multiresistant β-lactamase-producing strains that are becoming more common and are gaining increasing clinical importance, particularly in nosocomial infections. Generally resistant to first- and second-generation cephalosporins and most of the penicillins, these organisms elaborate a diverse array of β-lactamases targeted at destroying β-lactam antibiotics. Aztreonam exhibits a high degree of resistance to enzymatic hydrolysis by most of the common β-lactamases (21). Particularly insidious are the plasmid-mediated TEM β-lactamases carried by a wide range of gram-negative bacteria. The plasmid-mediated genes responsible for enzyme production pass freely among the *Enterobacteriaceae,* pseudomonads, *Neisseria gonorrhoeae,* and *Haemophilus influenzae*. Other β-lactamases capable of this same promiscuous behavior are the OXA, SHV, and PSE enzymes. Aztreonam, similar to many of the third-generation aminothiazoleoxime cephalosporins, such as cefotaxime and ceftizoxime, is stable to these β-lactamases. In contrast, compounds like cephaloridine and cefoperazone are readily hydrolyzed. Aztreonam, cefotaxime, and ceftizoxime are also generally stable to the chromosomally mediated β-lactamases, which are widely distributed among specific gram-negative bacteria, but are not transferable. These enzymes, which may coexist in the same organisms with a plasmid-mediated β-lactamase, are found in organisms such as *Klebsiella, Providencia, Enterobacter, Serratia, Proteus,* and *Bacteriodes*. The only enzyme in this group showing any appreciable destruction of aztreonam is the relatively uncommon K-1 β-lactamase produced by some strains of *Klebsiella oxytoca*.

In vivo, aztreonam shows excellent pharmacokinetic properties. One hour after parenteral administration of a 1 g dose, serum levels are 45–50 μg/mL. After a 2 g dose, levels are 90 μg/mL at 1 h and there is no apparent accumulation of the drug after multiple dosing. Aztreonam is excreted primarily by the kidneys with two-thirds of an administered dose eliminated unchanged in the urine. Renal secretion, glomerular filtration, and nonrenal mechanisms are involved in the elimination of aztreonam. Serum half-life for elimination is 1.7 h. Aztreonam penetrates well into peritoneal fluid, pleural fluid, and blister fluid. Furthermore, aztreonam undergoes biliary secretion with peak biliary concentrations greater than 40 μg/mL at 2.5 h.

In the clinic, aztreonam has demonstrated efficacy in urinary tract infections (UTI), lower respiratory tract infections, bacteremias, skin and soft tissue infections, intraabdominal and pelvic infections, leukopenic patients, and bone and joint infections resulting from susceptible gram-negative pathogens, allowing for selective therapy or combination therapy if mixed pathogens are involved. Aztreonam markedly decreases the fecal aerobic gram-negative bacteria, selectively eradicating pathogenic bacteria from the intestinal tract without significantly altering the gut anaerobes or impairing their colonization-resistance function.

Overall, aztreonam appears to be a safe agent having toxicity side effects similar to those of other β-lactams. The safety profile suggests that aztreonam may be useful as a replacement for aminoglycoside therapy (22). The biological properties of aztreonam have been extensively reviewed (23).

Alternative N-1 Activating Groups

β-Lactam antibiotics exert their antibacterial effects via acylation of a serine residue at the active site of the bacterial transpeptidases. Critical to this mechanism of action is a reactive β-lactam ring having a proximate anionic charge that is necessary for positioning the ring within the substrate binding cleft (24).

All of the naturally-occurring monobactams and aztreonam are characterized by the presence of the N-1 sulfonate group, which serves a dual function. The electron withdrawing sulfonate moiety renders the β-lactam ring more reactive toward nucleophilic attack and at the same time provides the anionic charge necessary for binding. A variety of monobactam subclasses bearing N-1 activating groups having the necessary physical properties for antibacterial activity are listed in Table 3 (25).

Monophosphams. The similar tetrahedral geometry and bond lengths of tetracoordinate phosphorus(V) compared to those of sulfur(VI) suggested that phosphonic and phosphinic acid groups might act as bioisosteres for the sulfonic acid moiety in the parent monobactams. The 2-oxoazetidine-1-phosphonates and phosphinates (monophosphams) were prepared, as shown in Figure 3**a**, by lithiation of the 3-acylamino-2-azetidinones **(20)** followed by treatment with a phosphoryl or phosphonyl halide to yield the phosphonylated, or phosphinylated, azetidinones **(21)** or **(23)**. Cleavage of these esters using trimethylsilyl bromide [*2857-97-8*] in the presence of bistrimethylsilylacetamide [*10416-59-8*] (BSA), which serves as an acid scavenger, affords the diacidic and monoacidic monophosphams **(22)** and **(24)**, respectively (26,27). Alternatively, Figure 3**b** reaction of lithiated azetidinones with phosphoryl dichlorides yields azetidinone-1-phosphonyl chlorides **(26)**, which are solvolyzed to give the monoacidic ester **(27)** or aminolized to ultimately afford the phosphonamide **(29)**. Cleavage of the urethane protecting groups by catalytic hydrogenation, when R = $C_6H_5CH_2O$, or by trifluororacetic acid, when R = *t*-C_4H_9O, gives 3-aminomonophosphamic acids. Coupling using *N,N'*-dicyclohexylcarbodiimide (DCC)-*N*-hydroxybenzotriazole or mixed anhydride with carboxylic acids formed 3-acylamino derivatives for screening (27).

The monophosphams are less intrinsically active, but more stable to β-lactamases, than their monobactam counterparts. SQ 27,327 [*84486-64-6*] **(30)**, $C_{14}H_{20}N_5O_8PS$, the most active monophospham, is generally two- to eightfold less active against aerobic gram-negative organisms than is its sulfonic acid analogue, aztreonam **(17)**. SQ 27,327, however, is substantially more stable to the K-1

Table 3. Biologically Active Monobactam Subclasses

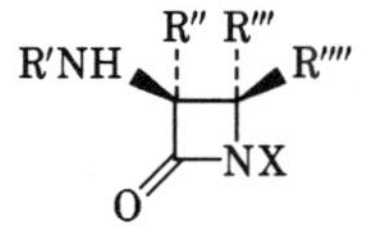

Subclass	Substituent, X
monobactams	SO_3H
monophosphams	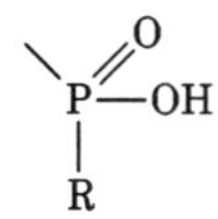
monosulfactams	OSO_3H
monophosphatams	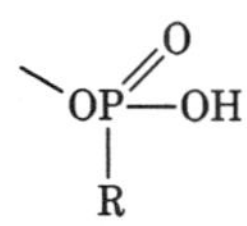
monoxacetams (oxamazins)	OCH_2COOH
monocarbams	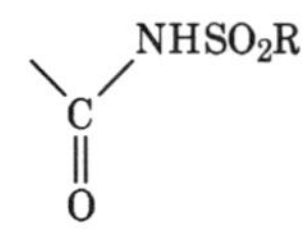
tetrazole-activated monobactams	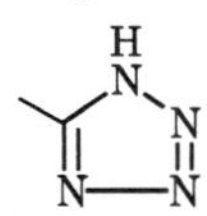

β-lactamase than aztreonam and this is reflected in the >250-fold difference in antimicrobial activity against the enzyme-producing *Klebsiella* strain (Table 4) (27).

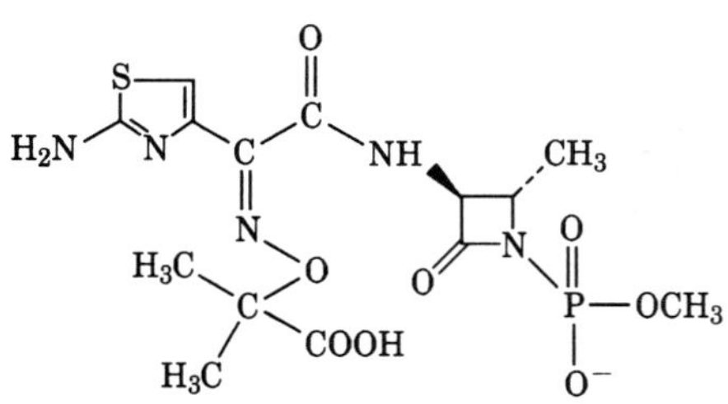

(30)

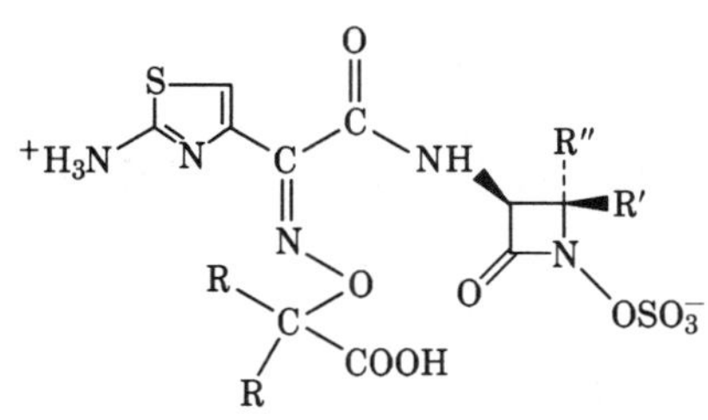

(31)

Monosulfactams. The monosulfactams are a class of monobactams activated by an *N*-1-OSO_3^- group (28) resulting in good intrinsic activity against gram-negative bacteria. However, a C-4 unsubstituted (**31**, R = CH_3, R′ = R″ = H) [*109895-38-7*] and the monomethylated [*87577-16-0*] (**31**, R = CH_3, R′ = H, R″ = CH_3), $C_{13}H_{17}N_5O_9S_2$ and [*102572-29-2*] (**31**, R = CH_3, R′ = CH_3, R″ = H),

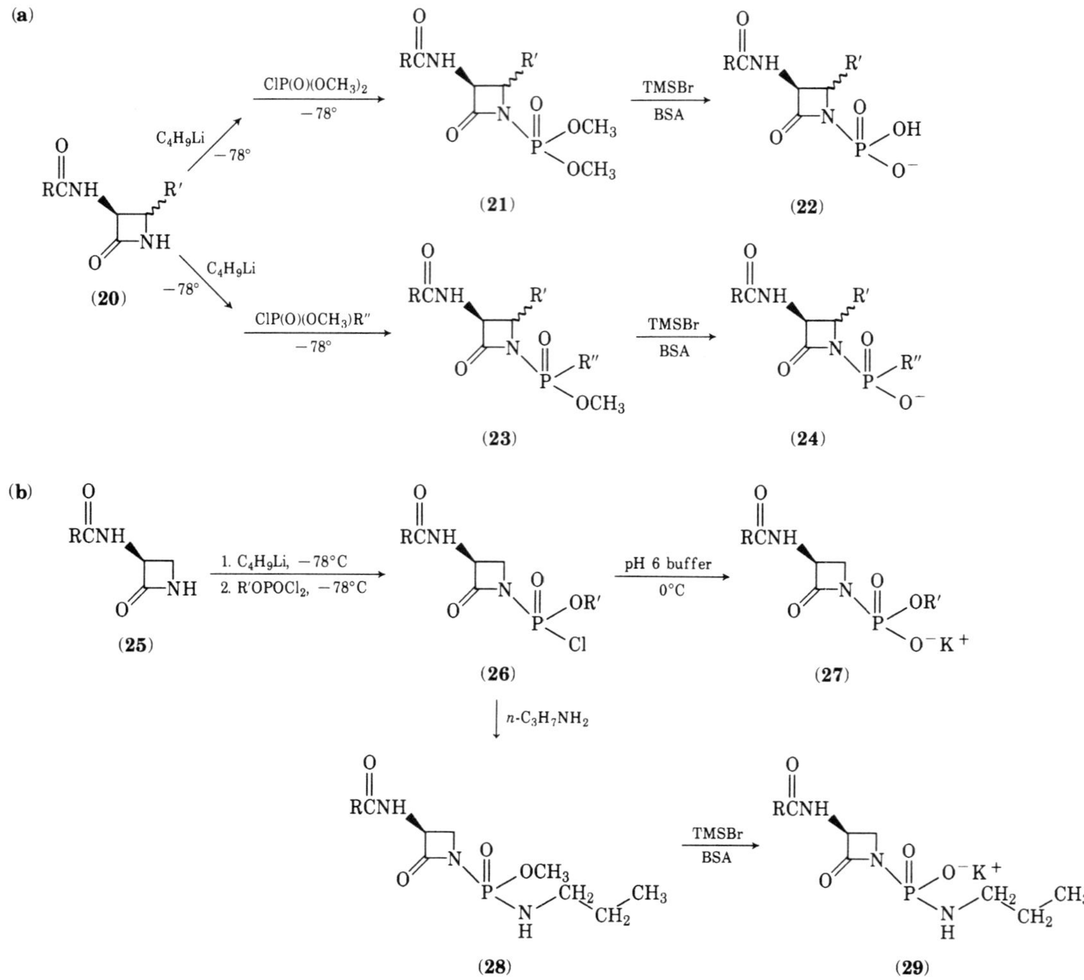

Fig. 3. Synthesis of diacidic and monoacidic N-1 azetidinone phosphonates and phosphinates where TMSBr is trimethylsilyl bromide; BSA is bis-trimethylsilylacetamide; (**a**) R = $C_6H_5CH_2O$, t-C_4H_9O, or 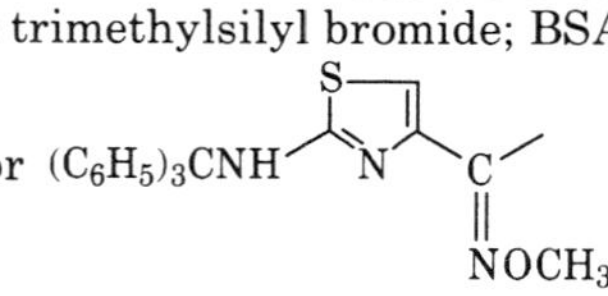 ; R′ = H or CH_3; and R″ = CH_3 or C_6H_5; and (**b**) R = t-C_4H_9O or $C_6H_5CH_2O$; and R′ = CH_3, CH_2CH_2F, CH_2CF_3, n-C_4H_9, or C_6H_5.

$C_{13}H_{47}N_5O_9S_2$, monosulfactams have poor chemical and β-lactamase stability. Dimethylation at the C-4 position (**31**, R = R′ = R″ = CH_3) [*109885-32-7*] improves the chemical and β-lactamase stability remarkably, while maintaining a high degree of antibacterial activity. Furthermore, removal of the *gem*-dimethyl groups from the C-3 side chain of this C-4 dimethyl compound minimally affects β-lactamase stability (29), but affords an orally-absorbed monosulfactam SQ 30,213 [*102507-71-1*] (tigemonam) (**31**, R = H, R′ = R″ = CH_3), $C_{12}H_{15}N_5O_9S_2$ (30). The high degree of oral absorption (80% in mice) is atypical for monobactams. The

Table 4. Biological Activities of Aztreonam and N-1 Substituted Analogues

Gram-negative organisms	SC#[b]	MIC, μg/mL[a] Aztreonam Low[c]	Aztreonam High[c]	SQ 27,327 Low[c]	SQ 27,327 High[c]	Tigemonam Low[c]	Tigemonam High[c]	SQ 28,112 Low[c]	SQ 28,112 High[c]
E. coli TEM+	10,404	0.1	0.2	0.4	0.8	0.4	0.8	0.4	0.8
E. coli TEM−	10,439	0.1	0.2	0.8	0.8	0.8	0.8	0.8	0.8
Ent. cloacae P99+	10,435	12.5	50	6.3	50	6.3	50	6.3	50
Ent. cloacae P99−	10,441	0.1	0.2	0.8	1.6	0.8	1.6	0.8	1.6
K. aerogenes K1+	10,436	100	>100	0.4	0.4	0.4	0.4	>100	0.4
K. aerogenes K1−	10,440	<0.05	0.1	0.4	0.8	0.4	0.8	0.8	0.8
K. pneumo.	11,066	0.4	0.8	1.6	1.6	0.8	1.6	3.1	25
C. freundii	10,204	0.1	25	0.4	3.1	0.2	0.4	0.8	50
Prot. rettgeri	8,217	<0.05	<0.05	0.2	0.2	<0.05	<0.05	<0.5	0.4
Prot. vulgaris	10,951B	<0.05	0.2	0.4	0.4	<0.05	<0.05	0.4	1.6
Ser. marcescens	9,782	0.2	0.2	1.6	3.1	0.8	1.6	0.8	3.1
Ps. aeruginosa	8,329	3.1	12.5	>100	>100	50	>100	3.1	12.5
Ps. aeruginosa	9,545	0.4	0.8	6.3	6.3	1.6	1.6	0.8	1.6

[a]Minimum Inhibitory Concentration measured by dilution of test compound in agar inoculated with microorganism.
[b]Squibb culture number.
[c]Low and high refer to the number of colony forming units (CFU) used to inoculate the media. Comparing MICs at low (10^4) and high (10^6) CFU, a large increase in MIC at the high inoculum level vs a β-lactamase producer is an indication of instability via enzymatic hydrolysis.

oral absorption and efficacy of tigemonam in animals have been confirmed in humans in Phase I and Phase II clinical studies (31).

Syntheses of the C-4 unsubstituted and monomethylated monosulfactams utilized hydroxamates derived from *β*-hydroxy amino acids such as serine or threonine, as precursors to the *β*-lactam ring (28). By analogy, synthesis of C-4 dimethylated monobactams (Fig. 4) required a protected form of (*S*)-*β*-hydroxyvaline [*102509-19-7*] (**33**). This unusual amino acid was prepared by reaction of the dianion of Boc-glycine benzyl ester [*54244-89-8*] (**32**) with acetone. Cleavage of the benzyl ester afforded the Boc-protected racemic valine [*102507-13-1*], $C_{10}H_{19}NO_5$, which was resolved as its phenethylamine salt to give Boc-(*S*)-*β*-hydroxyvaline. Coupling *O*-benzylhydroxylamine with amino acid (**33**) afforded the hydroxamate (**34**). Selective sulfonation of the hydroxyl group of (**34**) and subsequent heating of the resulting sulfate (**35**) [*107982-89-8*] in the presence of base effected cyclization at the sterically hindered carbon atom and gave the *β*-lactam intermediate (**36**) [*10250-25-5*]. The final steps, benzyl cleavage, sulfonation, and coupling of the side chain, yielded tigemonam (**38**) (**31**, R = H, R′ = R″ = CH_3), and related analogues (32).

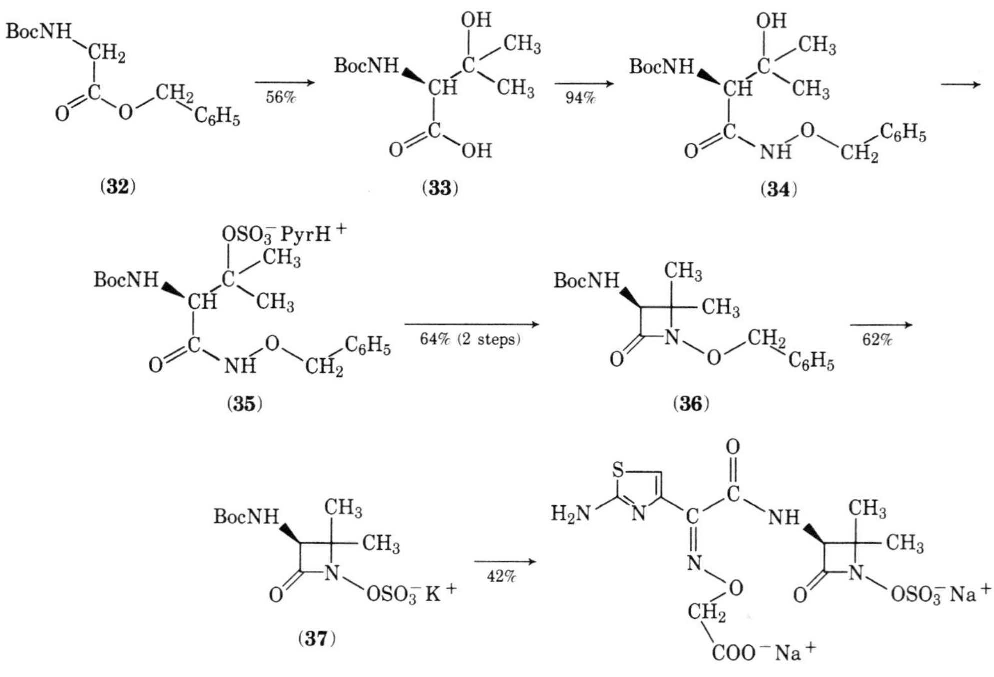

Fig. 4. Synthesis of tigemonam where Boc is the *t*-butoxycarbonyl group; $PyrH^+$ is the pyridinium ion; and the numbers given are percent yield of product.

As shown in Table 4, tigemonam is also relatively inactive vs gram-positive bacteria and anaerobes, yet exhibits a high degree of activity against gram-negative organisms as well as resistance to hydrolysis by *β*-lactamases. The activity of tigemonam is cidal against the *Enterobacteriaceae, Hemophilus influ-*

enzae, and *N. gonorrhoeae* but unlike aztreonam, the activity is weak against the pseudomonads (33,34).

Monophosphatams. The preparation of monophosphatams, the phosphorus analogues of the monosulfactams, is exemplified by the reaction shown in Figure 5 (35). Phosphorylation of the *N*-hydroxyazetidinone [*80542-48-9*] (**39**) using methylphosphonic or methylphosphoric dichlorides in the presence of 2,6-lutidine or triethylamine at low temperatures gave the chloro intermediates (**40**), which were hydrolyzed *in situ* at pH 3–4 to give the stable potassium salts (**41**) in from 12 to 40% yield. Removal of the *t*-butoxycarbonyl group under acidic conditions at −10°C for 1 h gave the amine salts (**42**) in good yield. The desired acyl side chains were then coupled to the amino group using an activated ester of the side chain to afford various acyl derivatives such as SQ 28,112 [*91603-73-5*] (**43**), $C_{14}H_{20}N_5O_8PS$.

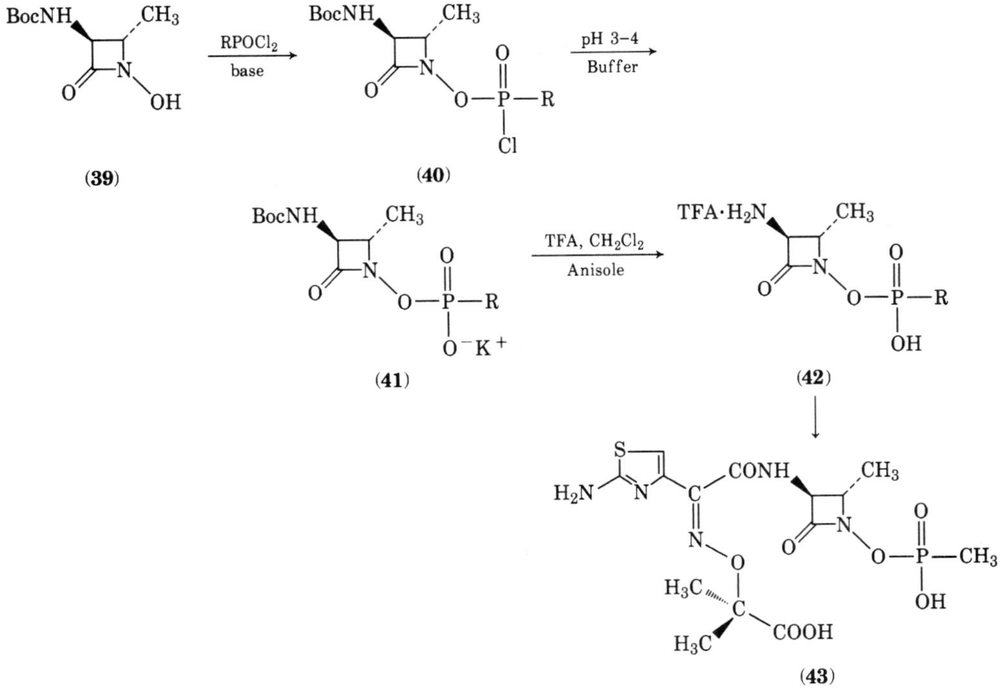

Fig. 5. Synthesis of monophosphatams where Boc is *t*-butoxycarbonyl; R is either CH_3 or OCH_3; and TFA is trifluoroacetic acid.

The biological activity of SQ 28,112 (**43**), shown in Table 4, is typical of the monophosphatams which possess the same spectrum of activity as aztreonam, but have comparatively lower intrinsic activity and *β*-lactamase stability.

Monoxacetams. Once the structural criteria for ring activation and binding of monobactams to the transpeptidases were established, researchers set out to determine if an oxygen atom could provide the requisite *β*-lactam ring reactivity. It was first necessary to insulate the anionic binding moiety from the oxygen atom by a "spacer" group, and using glycolate structures appended to the *N*-1 position of the *β*-lactam ring, structures (**44**) were synthesized (36). These 3-acylamino-1-azetidinyloxyacetic acids are known variously as monoxacetams and as oxamazins.

The oxyacetic acid residue of the monoxacetams is bioisosteric with the sulfate moiety of the monosulfactams. In addition, the presence of the carboxylate moiety provided opportunity for the preparation of esters that could act as orally absorbed prodrugs. Based on structure-activity relationships, SQ 82,291 (**45**) [*90898-90-1*], $C_{12}H_{15}N_5O_6S$, the nonacidic methoxime side chain of which was necessary to maintain oral absorption of the prodrugs, was prepared. SQ 82,291 has a high, specific affinity for the PBP-3 transpeptidase of gram-negative bacteria. However, it lacks the isobutyric acid moiety of aztreonam (**17**) on the oxime residue and whereas the activity of SQ 82,291 vs the *Enterobacteriaceae* was maintained, antipseudomonal activity was significantly diminished (37).

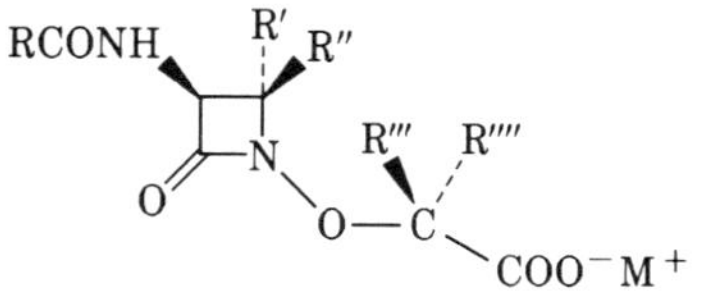

(**44**)

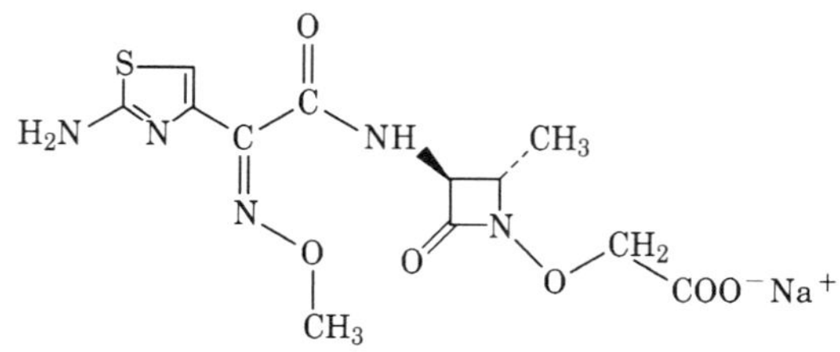

(**45**)

To synthesize the monoxacetam structures (Fig. 6), alkylation of *N*-protected 1-hydroxyazetidinones (**46**) with the appropriate haloacetic acid derivatives provided (**47**). Alternatively, (**47**) could be prepared from the acyclic hydroxamate ester (**48**). Deprotection of (**47**) furnished the zwitterionic intermediate (**49**) [*90849-16-4*], $C_6H_{10}N_2O_4$, which subsequently underwent acylation using the C-3 aminothiazole oxime side chain to afford SQ 82,291 (**45**) also known as oximonam (37).

The first oral monobactam to reach clinical trials was SQ 82,531 [*90850-05-8*] (gloximonam), (**50**), $C_{18}H_{25}N_5O_8S$, the *t*-butyl glycolate ester of oximonam. This prodrug ester could be formed in high yield by direct alkylation of SQ 82,291 or by carrying the ester from the earlier synthetic stages (**47** or **48**, R′ = $—CH_2COO$*t*-C_4H_9). On a multikilogram scale, the overall yield via the route from *N*-hydroxyazetidinone (**46**) was ca 25% (38).

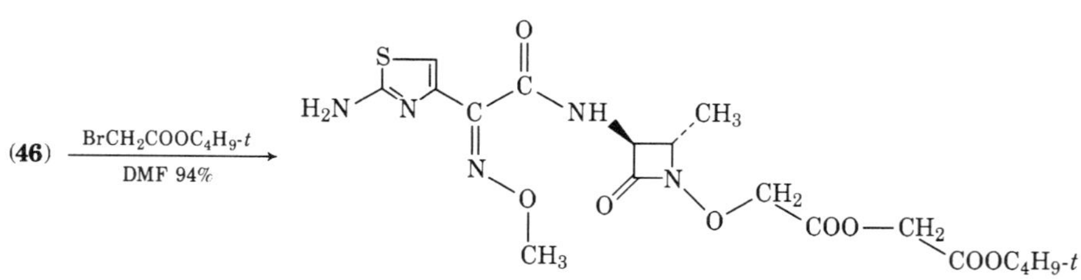

(**50**)

Gloximonam [*90850-05-8*] (**50**) was selected for clinical studies because of its desirable physical properties such as crystallinity and solubility and because it was orally absorbed in mice, rats, and monkeys, thus resulting in therapeutic serum levels of the active oximonam. Gloximonam was orally efficacious in a variety of murine gram-negative systemic infections (39) and in Phase I clinical studies gloximonam was well-absorbed. However, further clinical development

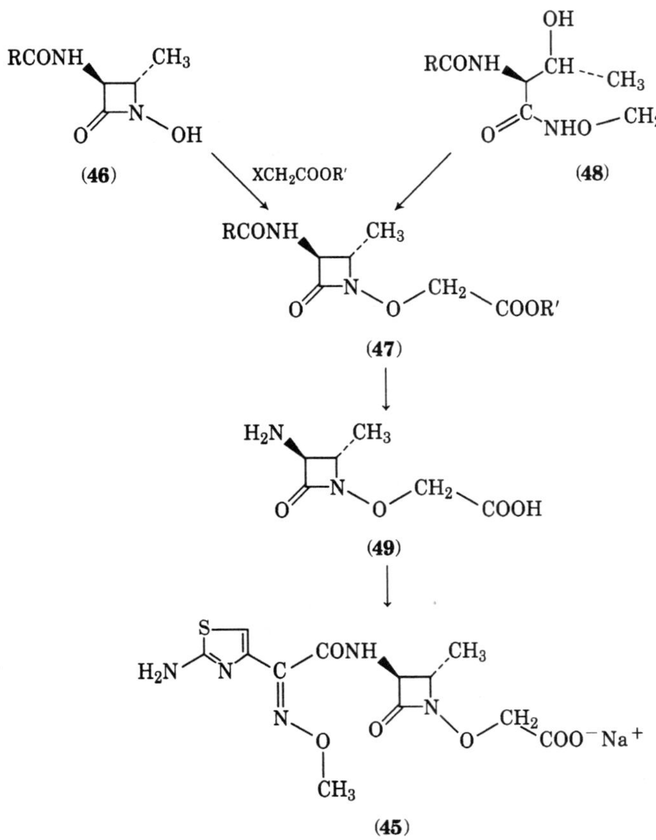

Fig. 6. Synthesis of monoxacetams.

was discontinued once it was determined that in humans the β-lactam ring underwent significant hydrolysis.

The *N*-aza (40) and *N*-thia (41) analogues of the monoxacetams have also been prepared. Within the nitrogen series, the imino acids [*121142-86-7*] (**51**, X = N=CHCOOH, R = H, CH_3) the imine stereochemistry of which is unknown, have moderate, notably less than oximonam anti-gram-negative activity. Both the *N*-glycyl [*121142-76-5*] (**51**, X = $NHCH_2COOH$, R = H) and *N*-thioacetic acid [*102652-83-5*] (**52**) derivatives were devoid of any significant biological activity.

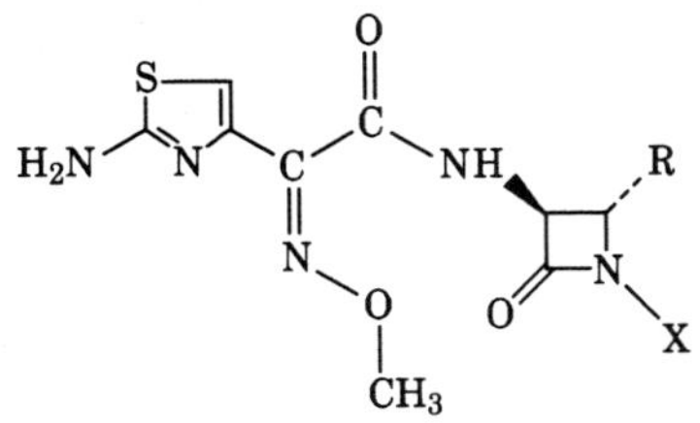

(**51**)

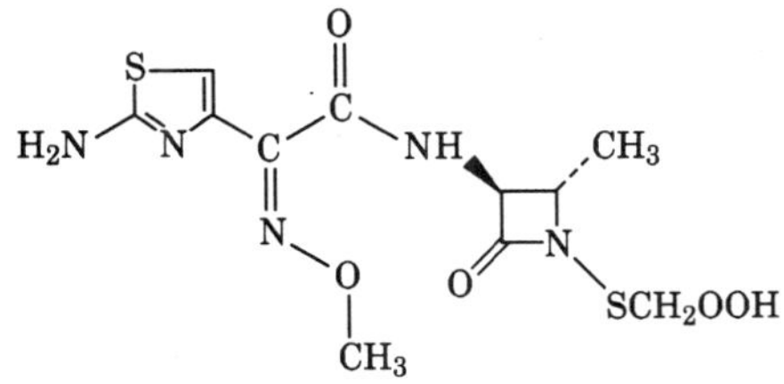

(**52**)

Monocarbams. The reaction sequence for monocarbam preparation is shown in Figure 7. Simple carboxyl *N*-1 substitution, representing one of the simplest carbonyl activating groups, led, as expected, to a carbamic acid that was hydrolytically unstable. Several monocarbams (**55**) and (**56**), having acidic sulfonylaminocarbonyl activating groups, were produced (42). The monocarbams (**55**) were prepared by sequential acylation of 3-aminoprotected 2-azetidinones (**53**) using alkyl or aryl isocyanates, removal of the protecting group, and acylation with activated side-chain acids. Reaction of (**53**) and chlorosulfonyl isocyanate (CSI) afforded the key intermediate (**54**) from which the aminosulfonyl derivatives (**56**) were prepared.

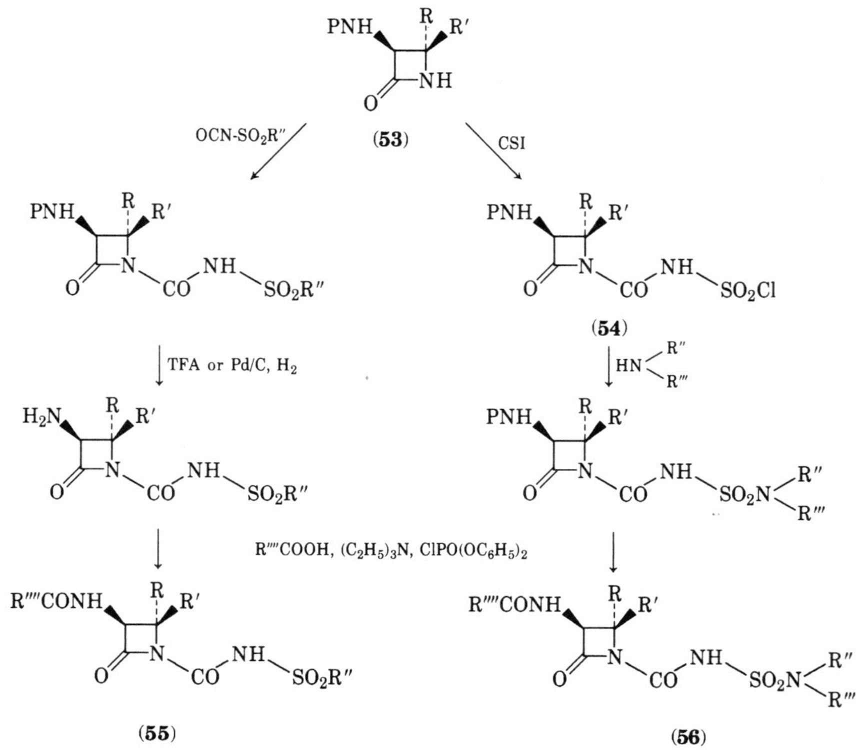

Fig. 7. Synthesis of monocarbams where P is an amino protecting group and CSI is chlorosulfonyl isocyanate. TFA is trifluoroacetic acid.

Monocarbams of the general formula (**56**), in which the terminal nitrogen is incorporated in a heterocyclic group, represent a class of potent β-lactamase stable antibacterials which are specifically active against gram-negative microorganisms. SQ 82,228 [*109895-39-8*] (**57**), featuring an *N*-ethylpiperazinedione terminus, is one of the most active (42), displaying favorable pharmacokinetic properties in experimental animals and good efficacy when tested parenterally in model infections.

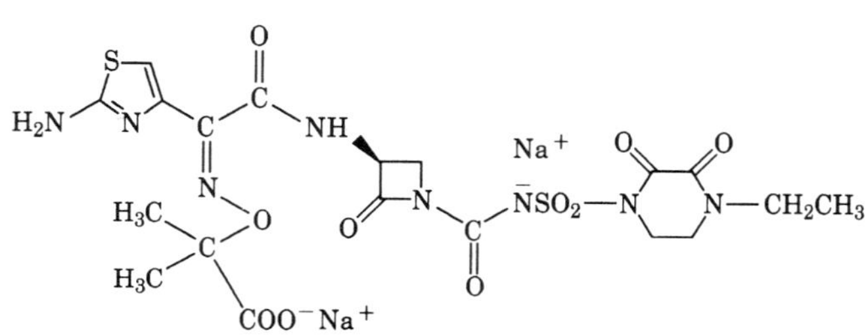

(**57**)

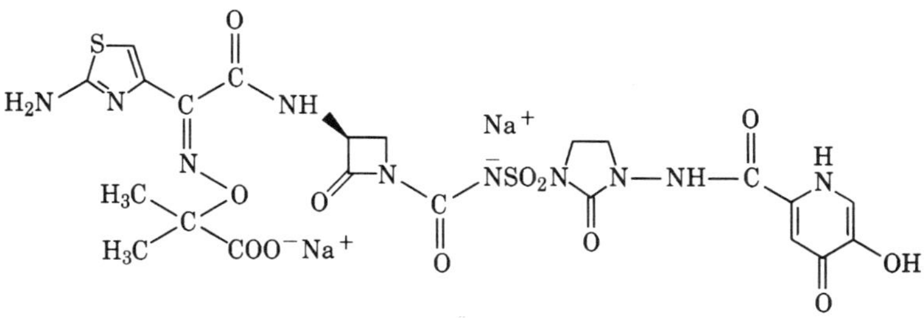

(**58**)

Continued efforts to improve the activity of the monobactams against nonfermenting gram-negative rods such as *Pseudomonas aeruginosa,* led to the discovery of SQ 83,360 [*104393-00-2*] (**58**), $C_{22}H_{24}N_{10}O_{12}S_2{\cdot}2Na$, a 3-hydroxy-4-pyridone containing monocarbam. The enhanced activity of SQ 83,360 is a direct consequence of the hydroxypyridone moiety, a catechol surrogate recognized by the bacterial *ton* B-dependent ion transport pathway. Uptake of SQ 83,360 by this mechanism increases the concentration of drug in the periplasmic space where the transpeptidases are located (43).

The synthesis of SQ 83,360 is shown in Figure 8. The heterocyclic terminus (**60**) [*112333-97-8*] of the *N*-1 substituent was constructed in five steps from kojic acid [*501-30-4*] (**59**). After persilylation, the reaction of intermediate (**60**) and the chlorosulfonyl isocyanate adduct of Z-protected 3-amino-2-azetidinone to afford monocarbam [*112333-98-9*] (**61**) occurred. Hydrogenolysis of the Z-group formed (**62**) [*12335-07-6*], followed by coupling and deprotection of the side chain in (**63**) [*112334-01-7*] to yield SQ 83,360 [*108319-07-9*] (**58**) in the acid form. A ready supply of Z-azetidinone [*80082-81-1*], $C_{11}H_{12}N_2O_3$, used in the synthesis of SQ 83,360, was obtained starting with the mesylate of Z-protected serine amide (17). The azetidinone was formed in 53–57% overall yield (44). SQ 83,360 (**58**) demonstrated exceptional activity *in vitro* as can be seen in Table 5. Against numerous clinical *Enterobacteriaceae* strains, SQ 83,360 was similar to aztreonam, yet against nonfermenting gram-negative bacilli, particularly *Ps. aeruginosa,* SQ 83,360 was significantly superior.

Table 5. Comparative *In Vitro* Activity of SQ 83,360 and Aztreonam

		MIC$_{90}$[b], μg/mL	
Organism[a]	Number of strains	SQ 83,360	Azetreonam
Enterobacteriaceae	236	<0.03–1.0	<0.03–12.5
Ps. aeruginosa	35	0.05	15.3
Pseudomonas sp.	58	6.3	90.4
Acinetobacter sp.	45	14.7	46.0

[a]All are nonfermentors except the *Enterobacteriaceae.*
[b]Minimum inhibitory concentration for 90% of the strains used.

Since the discovery of SQ 83,360, compounds such as U-78,608 [*123444-35-9*] (**64**) having different linker groups between the hydroxypyridone group and the sulfonyl residue have been reported. U-78,608 and SQ 83,360 have similar *in vitro* and *in vivo* activity (45).

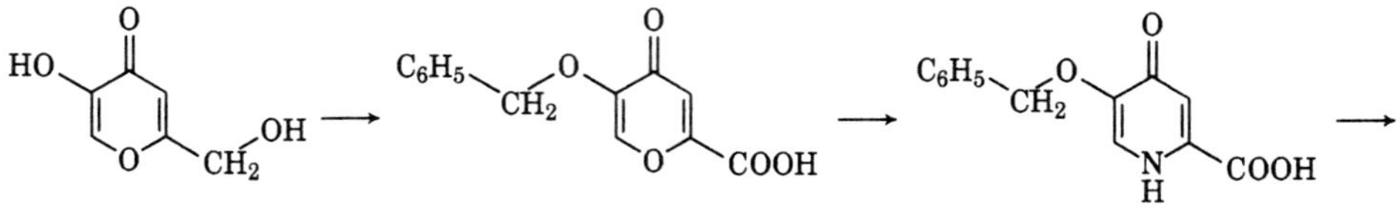

(59)

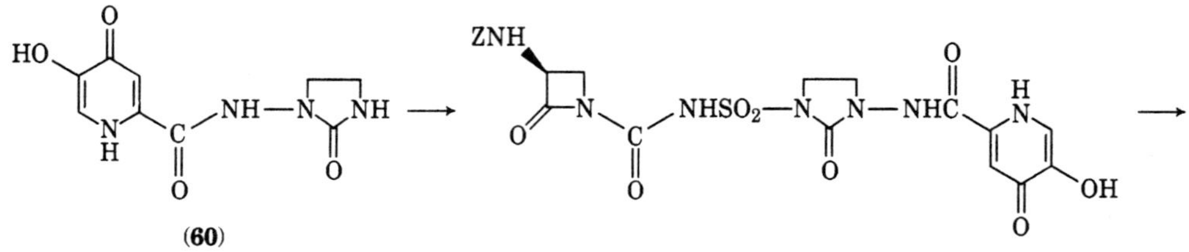

(60)

(61)

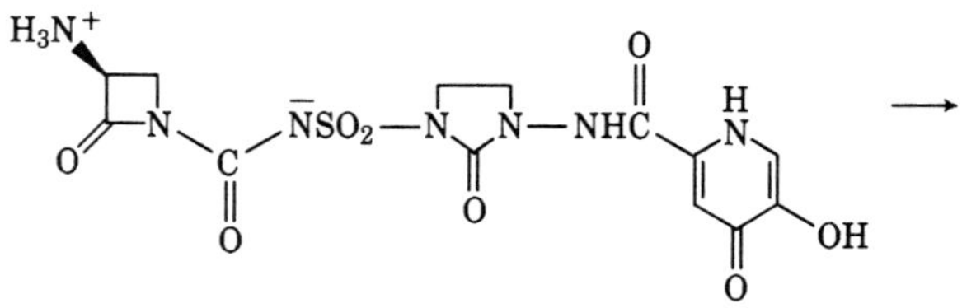

(62)

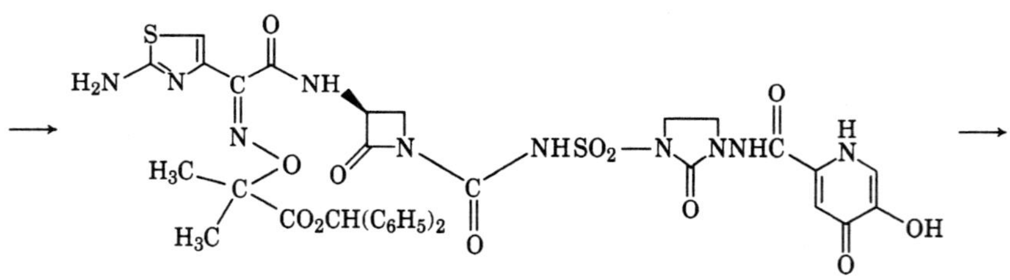

(63)

Fig. 8. Synthesis of SQ 83,360.

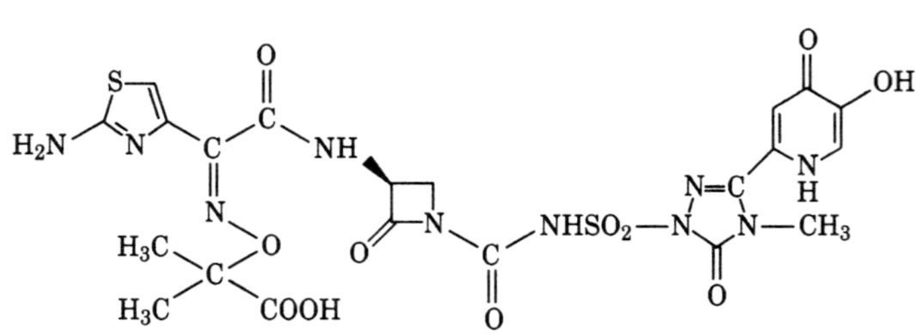

(64)

Tetrazole-Activated Monobactams. The tetrazole moiety is one of the closest isosteres to a carboxylic acid, having similar electron-withdrawing capability and acidity. The syntheses of homochiral tetrazole derivatives (**65**) from serine and threonine have been reported (46). As shown in Figure 9, cyclization of amides (**66**, R = H) [*90208-25-6*] and (**66**, R = CH_3) [*90181-42-3*] was effected using an intramolecular Mitsunobu reaction. Deprotection afforded the amine salts (**67**, R = H) [*90181-47-8*] and (**67**, R = CH_3) [*90181-49-0*], which were then coupled to afford the side-chain analogues (**65**, R′ = CH_3 or $C(CH_3)_2COOH$). Using similar methodology, other C-4 substituted monobactams were prepared whereby the configuration of R in structure (**65**) was either cis or trans and R = CH_2F, $COOCH_3$, CH_2OH, or CH_2OCONH_2 (47).

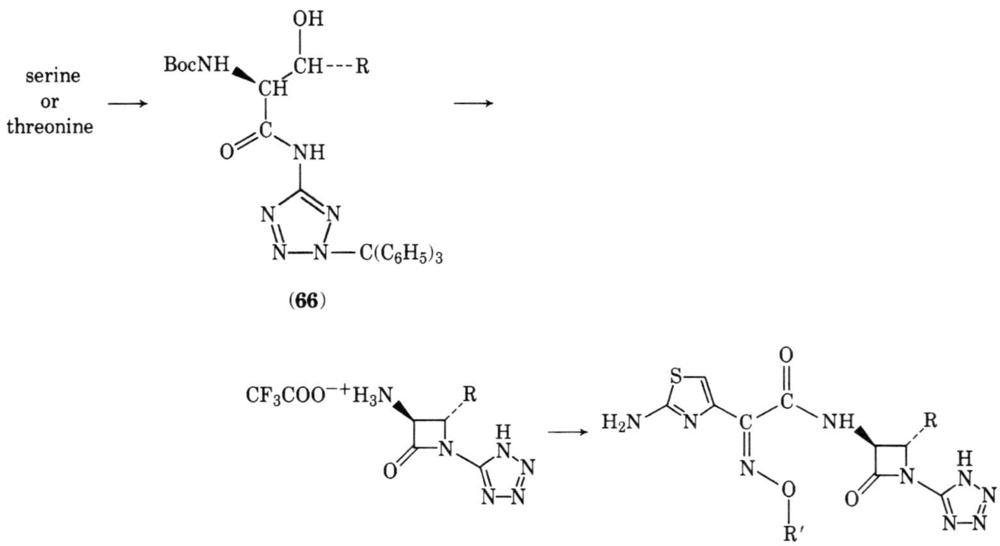

(**66**)

(**67**) (**65**)

Fig. 9. Synthesis of *N*-(tetrazolyl-5-yl) azetidinones where Boc is *t*-butoxycarbonyl; R = H or CH_3; and R′ = CH_3 or $C(CH_3)_2COOH$.

Alternatively, various 4-substituted derivatives have been prepared via synthesis of amino acid (**68**) by reaction of the anion formed from protected glycine and an appropriately substituted Schiff base.

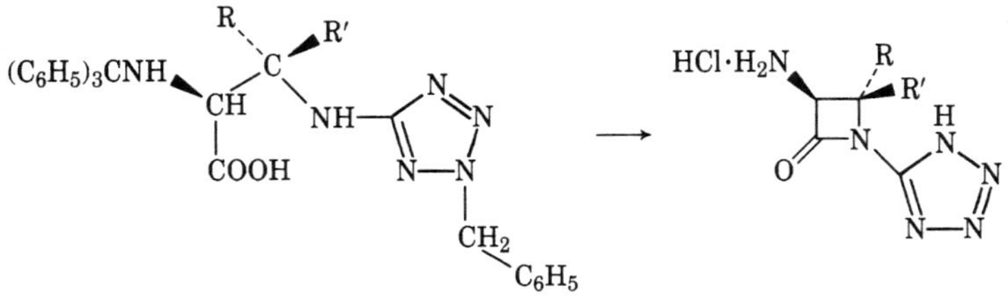

(**68**) (**69**)

Cyclization of (**68**) and deprotection then yield the monobactam nucleus (**69**) which may be coupled with various C-3 side chains (48). Direct alkylation of the *N*-unsubstituted azetidinone (**70**) using fluorotetrazole [*93607-94-4*], $C_8H_7FN_4$, also produced (**69**) after deprotection of (**71**) (49).

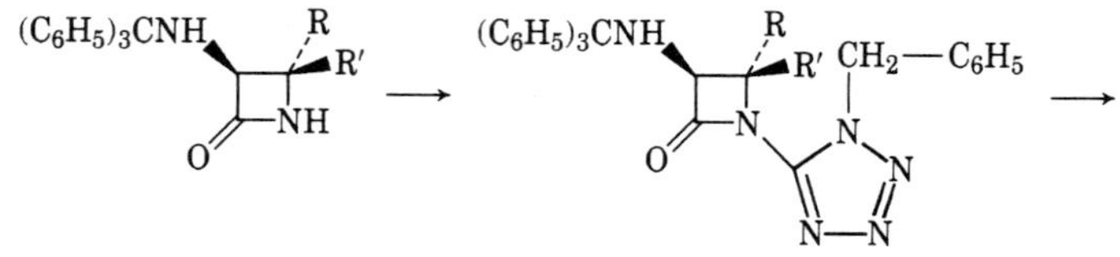

(70) (71)

(69) →

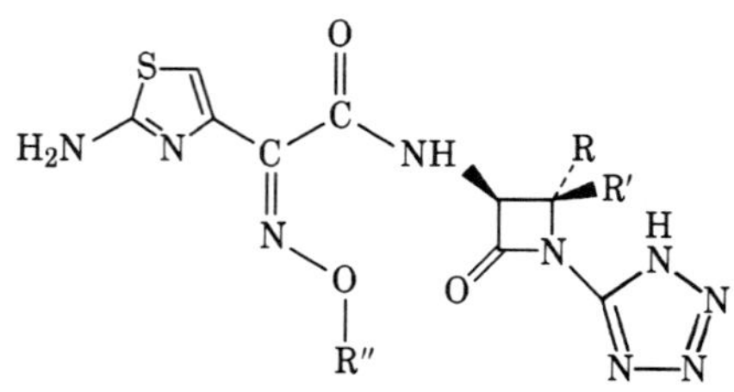

(72)

Structure (**72**) exemplifies the C-4 derivatives prepared by these routes. From this class of monobactams, RU-44790 [*110012-78-7*] (**72**, R = H, R′ = CH_2F, R″ = 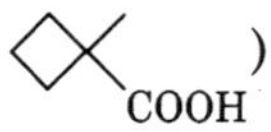) was identified as having improved *β*-lactamase stability and activity vs gram-negative bacilli similar to that of aztreonam (50).

Economic Aspects

Two monobactams were in clinical use as of 1990. Aztreonam (**17**) manufactured by Bristol-Myers Squibb, has the worldwide trademark of Azactam. Aztreonam has received regulatory approval for use in humans in 67 countries and has pending filings in seven additional countries. The global experience with aztreonam and its clinical acceptance signifies that the monobactams are recognized as an important new class of antibacterial agents. Carumonam, manufactured by Takeda, received approval for human usage in 1988 in Japan. Carumonam (**19**) has the trademark, Amasulin.

BIBLIOGRAPHY

"Antibiotics, *β*-Lactams-Monobactams" in *ECT* 3rd ed., Supplement Vol., pp. 131–144, by C. M. Cimarusti and R. B. Sykes, Squibb Institute for Medical Research.

1. R. B. Sykes and co-workers, *Nature (London)* **291,** 489 (1981).
2. A. Imada and co-workers, *Nature (London)* **289,** 590 (1981).
3. W. L. Parker, J. O'Sullivan, and R. B. Sykes, *Advances in Applied Microbiology,* Vol. 31, Academic Press, New York, 1986, pp. 181–205; A. G. Brown, *Pure Appl. Chem.* **59,** 475 (1987).
4. K. Kamiya, M. Takamoto, Y. Wada, and M. Asai, *Acta Cryst.* **B37,** 1626 (1981).
5. W. L. Parker and co-workers, *J. Antibiot.* **35,** 189 (1982).
6. W. L. Parker and M. L. Rathnum, *J. Antibiot.* **35,** 300 (1982).

7. J. M. Indelicato and W. L. Wilham, *J. Med. Chem.* **17,** 528 (1974).
8. R. B. Sykes, D. P. Bonner, N. H. Georgopapadakou, and J. S. Wells, *J. Antimicrob. Chemother.* (Suppl. E) **8,** 1 (1981).
9. C. M. Cimarusti, *J. Med. Chem.* **27,** 247 (1984).
10. N. H. Georgopapadakou, S. A. Smith, and R. B. Sykes, *Antimicrob. Agents Chemother.* **21,** 950 (1982).
11. J. O'Sullivan and co-workers, *J. Antimicrob. Agents Chemother.* **21,** 558 (1982).
12. D. P. Bonner and R. B. Sykes in C. S. F. Easmon and J. Jeljaszewicz, eds., *Medical Microbiology,* Vol. 4, Academic Press, New York, 1984, pp. 171–197; D. P. Bonner and R. B. Sykes, *J. Antimicrob. Chemother.* **14,** 313 (1984).
13. T. Matsuo and co-workers, *Chem. Pharm. Bull.* **31,** 2200 (1983).
14. W. H. Koster, C. M. Cimarusti, and R. B. Sykes in R. B. Morin and M. Gorman, eds., *Chemistry and Biology of β-Lactam Antibiotics,* Vol. 3, Academic Press, New York, 1982, pp. 339–375.
15. T. Matsuo and co-workers, *Chem. Pharm. Bull.* **31,** 1874 (1983).
16. D. M. Floyd and co-workers, *J. Org. Chem.* **47,** 5160 (1982).
17. D. M. Floyd, A. W. Fritz, and C. M. Cimarusti, *J. Org. Chem.* **47,** 176 (1982).
18. P. S. Manchand and co-workers, *J. Org. Chem.* **53,** 5507 (1988).
19. C. M. Cimarusti and co-workers, *Tetrahedron* **39,** 2577 (1983); H. Breuer, *J. Antimicrob. Chemother* (Suppl. E) **8,** 21 (1981).
20. R. B. Sykes, D. P. Bonner, K. Bush, and N. H. Georgopapadakou, *Antimicrob. Agents Chemother* **21,** 85 (1982).
21. K. Bush, J. Freudenberger, and R. S. Sykes, *Antimicrob. Agents Chemother.* **22,** 414 (1982).
22. R. E. Reese, D. E. Sentochnik, R. G. Douglas, and R. F. Betts, *Handbook of Antibiotics,* Little, Brown and Company, Boston, 1988, pp. 134–141.
23. *Reviews of Infectious Disease,* **1** (Suppl. 4) (1985); D. P. Bonner and R. B. Sykes, *J. Antimicrob. Chemother.* **14,** 313 (1984); R. B. Sykes and D. P. Bonner, *Amer. J. Med.* **78**(2A), 2 (1985); R. B. Sykes, D. P. Bonner, and E. A. Swabb, *Pharmac. Ther.* **29,** 321 (1985); *Chemotherapy* **35,** (Suppl. 1) (1989).
24. N. C. Cohen, *J. Med. Chem.* **26,** 259 (1983); J. B. Bartolone, G. J. Hite, J. A. Kelly, and J. R. Knox, in A. G. Brown and S. M. Roberts, eds., *Recent Advances in the Chemistry of β-Lactam Antibiotics,* The Royal Society of Chemistry, 1985, pp. 318–327; J. A. Kelley, J. C. Boyington, P. C. Moews, and J. R. Knox, in H. Umezawa, ed., *Frontiers of Antibiotic Research,* Academic Press, New York, 1987, pp. 327–337.
25. C. M. Cimarusti and R. B. Sykes, *Med. Res. Rev.* **4,** 1 (1984).
26. W. H. Koster and co-workers, *J. Am. Chem. Soc.* **105,** 3743 (1983).
27. W. H. Koster and co-workers, *Abstracts, 22nd Interscience Conference on Antimicrobial Agents and Chemotherapy* No. 674, American Society for Microbiology, Miami, Fla., 1982; Ref. 25, pp. 15–17.
28. E. M. Gordon and co-workers, *J. Am. Chem. Soc.* **104,** 6053 (1982).
29. S. K. Tanaka and co-workers, *Antimicrob. Agents. Chemother.* **31,** 219 (1987).
30. J. M. Clark and co-workers, *Antimicrob. Agents Chemother.* **31,** 226 (1987).
31. D. A. Conrad, A. L. Lentnek, and the Tigemonan Study Group, *29th Interscience Conference on Antimicrobial Agents and Chemotherapy,* Abstract No. 921, Houston, Tex., 1989.
32. W. A. Slusarchyk and co-workers, *Tetrahedron Lett.* **27,** 2789 (1986); J. D. Godfrey, Jr., R. H. Mueller, and D. J. Von Langen, 2793 (1986).
33. W. H. Koster and D. P. Bonner in H. Umezawa, ed., *Frontiers of Antibiotic Research,* Academic Press, New York, 1987, pp. 211–226.
34. R. B. Sykes, W. H. Koster, and D. P. Bonner, *J. Clin. Pharmacol.* **28,** 113 (1988).
35. W. A. Slusarchyk and co-workers, *Heterocycles* **21,** 191 (1984).

36. F. R. Atherton and R. W. Lambert, *Tetrahedron* **40,** 1039 (1984); H. Breuer and coworkers, *Abstract No. 135, 24th Interscience Conference on Antimicrobial Agents and Chemotherapy,* Washington, D.C., 1984; S. R. Woulfe and M. J. Miller, *Tetrahedron Lett.* **25,** 3293 (1984); S. R. Woulfe and M. J. Miller, *J. Med. Chem.* **28,** 1447 (1985).
37. H. Breuer and co-workers, *J. Antibiot.* **38,** 813 (1985).
38. C. M. Cimarusti, *Gazz. Chim. Ital.* **116,** 169 (1986).
39. W. H. Koster and co-workers, Abstract No. 136, and J. M. Clark and co-workers, Abstract No. 139, *24th Interscience Conference on Antimicrobial Agents and Chemotherapy,* Washington, D.C., 1984.
40. W. V. Curran and R. H. Lenhard, *J. Med. Chem.* **32,** 1749 (1989).
41. S. R. Woulfe and M. J. Miller, *J. Org. Chem.* **51,** 3133 (1986).
42. Ref. 25, pp. 17–20.
43. Ref. 33, pp. 211–218; H. Nikaido and E. Y. Rosenberg, *J. Bacteriol.* **172,** 1361 (1990).
44. U.S. Pat. 4,587,051 (May 6, 1986), W. H. Koster (to E. R. Squibb and Sons, Inc.); ref. 33, pp. 214–216.
45. M. R. Barbachyn and T. C. Tuominen, Abstract No. 231, and G. E. Zurenko and B. Hannon, Abstract No. 233, *Interscience Conference on Antimicrobial Agents and Chemotherapy,* Los Angeles, Cal., 1988.
46. A. Andrus, B. Partridge, J. V. Heck, and B. G. Christensen, *Tetrahedron Lett.* **25,** 911 (1984).
47. C. Yoshida and co-workers, *J. Antibiot.* **39,** 215 (1986).
48. M. Klich and G. Teutsch, *Tetrahedron Lett.* **25,** 3849 (1984).
49. M. Klich and G. Teutsch, *Tetrahedron* **42,** 2677 (1986).
50. G. Teutsch, M. Klich, and J. F. Chantot, *Abstracts, Interscience Conference on Antimicrobial Agents and Chemotherapy,* No. 843, New Orleans, La., 1986; Ref. 47.

KLAUS R. LINDNER
DANIEL P. BONNER
WILLIAM H. KOSTER
Bristol-Myers Squibb Pharmaceutical Research Institute

PENICILLINS AND OTHERS

The basic structural features of the penicillin nucleus (**1**) include a β-lactam ring fused through nitrogen and the adjacent tetrahedral carbon to a second heterocycle which, in natural penicillin is a 5-membered thiazolidine ring. Biologically active penicillins are generally characterized by a functionalized amino group in the 6β-position of the β-lactam ring and a carboxyl group in the 3-position in the thiazolidine ring.

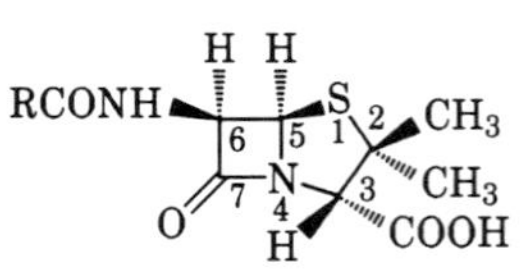

(**1**)

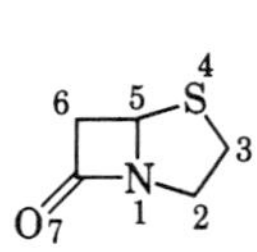

(**2**)

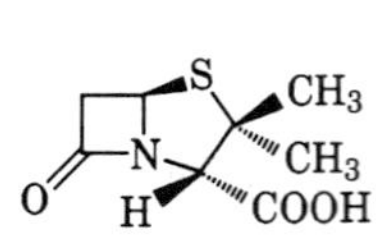

(**3**)

In general, penicillins exert their biological effect, as do the other β-lactams, by inhibiting the synthesis of essential structural components of the bacterial cell wall. These components are absent in mammalian cells so that inhibition of the synthesis of the bacterial cell wall structure occurs with little or no effect on mammalian cell metabolism. Additionally, penicillins tend to be irreversible inhibitors of bacterial cell-wall synthesis and are generally bactericidal at concentrations close to their bacteriostatic levels. Consequently penicillins have become widely used for the treatment of bacterial infections and are regarded as one of the safest and most efficacious classes of antibiotics.

Nomenclature

Chemical Abstracts indexes most penicillins as 4-thia-1-azabicyclo[3.2.0]heptane-7-ones (**2**). Using this system, penicillin G [*161-33-6*] (**1**, R = $C_6H_5CH_2$), $C_{16}H_{18}N_2O_4S$, is 6-(2-phenyl-acetamido)-3,3-dimethyl-7-oxo-4-thia-1-azabicyclo[3.2.0]heptane-2-carboxylic acid. The unsubstituted bicyclic ring system of the penicillins is designated as penam [*53908-04-6*], C_5H_7NOS, (**2**) (1) and the penicillins (**1**) are generally 6-acylamino-2,2-dimethylpenam-3-carboxylic acids. A further simplification is the use of the term penicillanic acid [*87-53-6*], $C_8H_{11}NO_3S$, to designate the penicillin ring system having the substituents indicated in (**3**). Thus the penicillins are named as the appropriate acylaminopenicillanic acid. Because the great majority of the variations of the penicillin structure are in the acyl side chain, the carbonyl of the acyl group is included in the basic moiety name penicillin.

Landmarks in the development of penicillin are its discovery in 1929 (2) and the subsequent recognition in 1940 of the potential utility of penicillin for controlling antibacterial infections in animals (3) and shortly afterward in humans (4,5). The realization that penicillin was a useful drug occurred during World War II (6).

Penicillin was first obtained as a product of the fungus *Penicillium notatum*, but the early manufacture utilizing this organism was not amenable to large-scale handling. Modifications were soon introduced which improved yields and quality of the product (7). Although there was an intense effort from 1940 on, it took until 1945 before the combined results of chemical degradation and x-ray crystallography made it possible to be certain that the penicillin structure contained the β-lactam thiazolidine ring system (8).

In 1959 isolation of the penicillin nucleus, 6-aminopenicillanic acid [*551-16-6*] (6-APA) (**1**, RCO = H), $C_8H_{12}N_2O_3S$, from fermentations deficient in side-chain precursors was reported (9). It was soon found that 6-APA could be produced more efficiently by enzymatically removing the acyl group from various penicillins using penicillin acylases. Subsequently this intermediate has also been made by chemically removing the 6-acyl group after first protecting the C-3 carboxyl group. However, for manufacturing purposes, the majority of 6-APA is currently produced by enzymatic deacylation of penicillin G or penicillin V [*87-08-1*]. The isolation of 6-aminopenicillanic acid has resulted in the synthesis and biological evaluation of many thousands of penicillin analogues derived primarily by acylation of the 6-amino group. From this vast chemical investment an important range of semisynthetic penicillins have found wide clinical use (Table 1).

Table 1. Penicillins[a] in Clinical Use

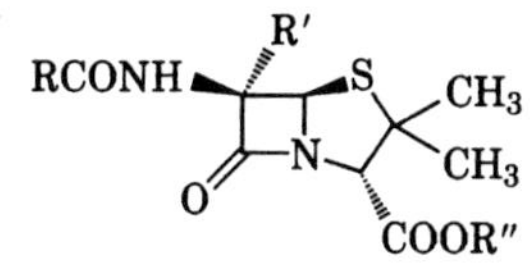

(**4**)

Name	CAS Registry Number	Molecular formula	R	R′	R″[b]	Administration: Route[c]	Administration: Av dose, g/d
Limited spectrum							
benzyl penicillin (penicillin G)[b]	[*61-33-6*]	$C_{16}H_{18}N_2O_4S$	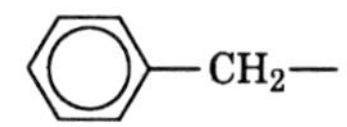	H	H	iv im po	3–12[d] 1–7[e] 0.6–6[f]
phenoxymethyl penicillin (penicillin V)	[*87-08-1*]	$C_{16}H_{18}N_2O_5S$	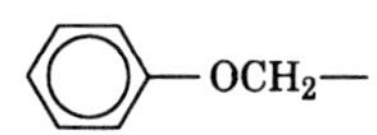	H	H	po	0.3–6
phenethicillin	[*132-93-4*]	$C_{17}H_{20}N_2O_5S{\cdot}K$	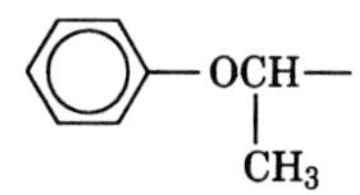	H	K	po	0.5–1
β-Lactamase stable							
methicillin	[*132-92-3*]	$C_{17}H_{20}N_2O_6S{\cdot}Na$	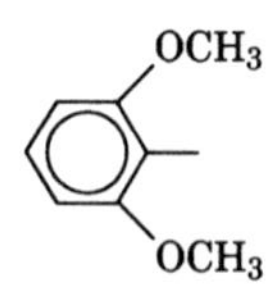	H	Na	im iv	6–12 4–6
			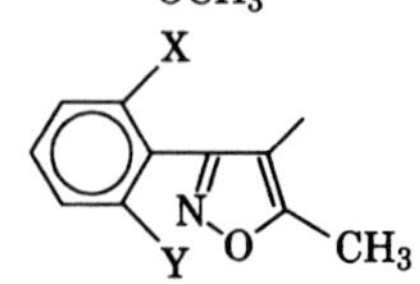				

oxacillin	*[66-79-5]*	$C_{19}H_{19}N_3O_5S$	X = H, Y = H	H	H	po, im, iv	1–6
cloxacillin	*[61-72-3]*	$C_{19}H_{18}ClN_3O_5S$	X = H, Y = Cl	H	H	po	1–6
dicloxacillin	*[3116-76-5]*	$C_{19}H_{17}Cl_2N_3O_5S$	X = Cl, Y = Cl	H	H	po	1–6
flucloxacillin	*[5250-39-5]*	$C_{19}H_{17}ClFN_3O_5S$	X = Cl, Y = F	H	H	po	1–6
nafcillin	*[985-16-0]*	$C_{21}H_{22}N_2O_5SNa$	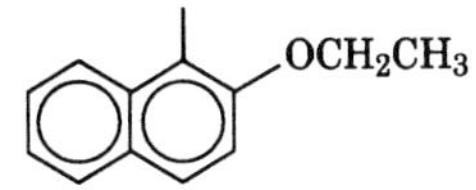	H	Na	po	1–2
Broad spectrum							
ampicillin	*[69-53-4]*	$C_{16}H_{19}N_3O_4S$	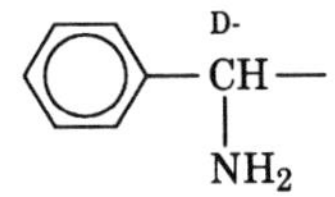	H	H	po, iv, im	1–4
hetacillin[g]	*[3511-16-8]*	$C_{19}H_{23}N_3O_4S$	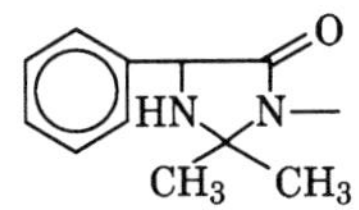	H	H		1–4
pivampicillin	*[33817-20-8]*	$C_{22}H_{29}N_3O_6S$	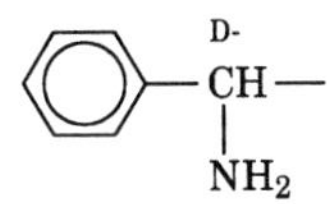	H	$CH_2OCOC(CH_3)_3$		1–4
talampicillin	*[47747-56-8]*	$C_{24}H_{23}N_3O_6S$	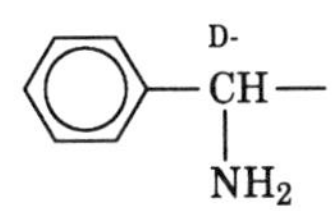	H	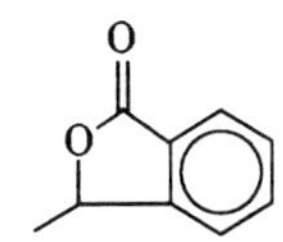		1–4
bacampicillin	*[50972-17-3]*	$C_{21}H_{27}N_3O_7S$	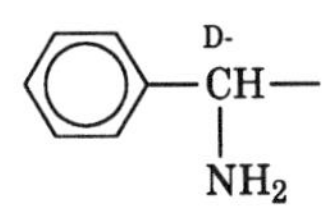	H	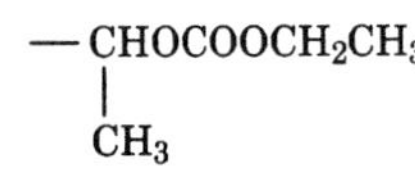	po, iv, im	1–4
ciclacillin	*[3485-14-1]*	$C_{15}H_{23}N_3O_4S$		H	H	po, iv	1–4

Table 1. ***(Continued)***

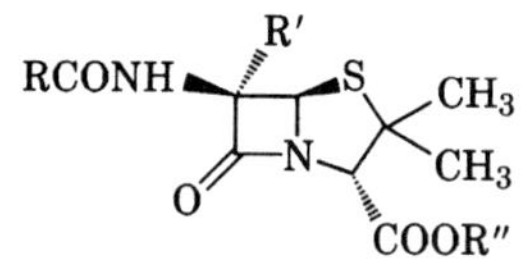

(**4**)

Name	CAS Registry Number	Molecular formula	R	R′	R″[b]	Administration: Route[c]	Administration: Av dose, g/d
amoxicillin	[*26787-78-0*]	$C_{16}H_{19}N_3O_5S$	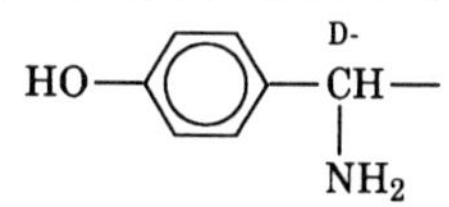	H	H	po	0.75–3
carbenicillin[h]	[*4697-36-3*]	$C_{17}H_{18}N_2O_6S$	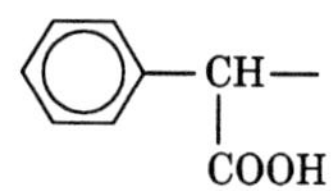	H	H	iv	to 30
ticarcillin	[*34787-01-4*]	$C_{15}H_{16}N_2O_6S_2$	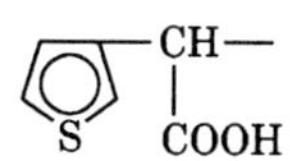	H	H	iv	to 20
sulbenicillin	[*41744-40-5*]	$C_{16}H_{18}N_2O_7S_2$	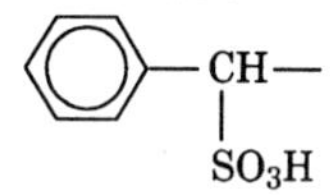	H	H	iv	to 20
azlocillin	[*37091-66-0*]	$C_{20}H_{23}N_5O_6S$	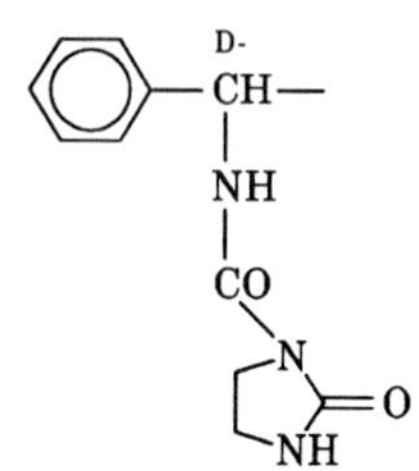	H	H	iv	to 20

mezlocillin	*[51481-65-3]*	$C_{21}H_{25}N_5O_8S_2$	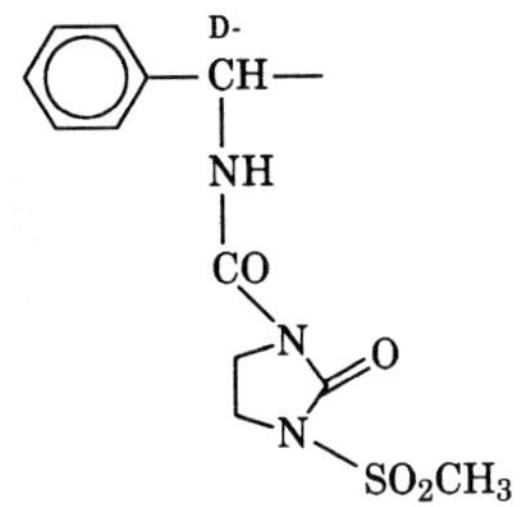	H	H	iv	to 20
piperacillin	*[61477-96-1]*	$C_{23}H_{27}N_5O_7S$	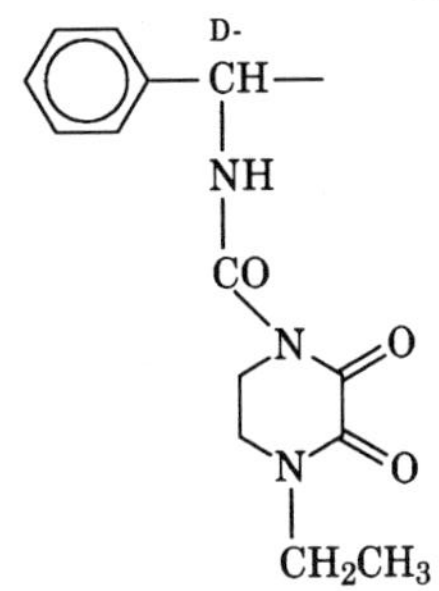	H	H	iv	to 20
Directed spectrum β-lactamase stable							
temocillin	*[66148-78-5]*	$C_{16}H_{18}N_2O_7S$	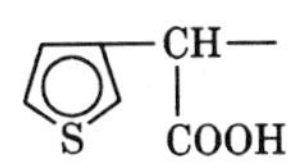	OCH_3	H	iv	to 20

[a]Structure (**4**) where R, R′, and R″ are as indicated.
[b]Whereas the carboxylic acid is shown for many of these compounds, the majority of the commercial penicillins are sodium salts. Penicillin G is available as the calcium salt *[973-53-5]*, $C_{16}H_{17}N_2O_4S\cdot\frac{1}{2}Ca$, and as the potassium salt *[113-98-4]*, $C_{16}H_{17}N_2O_4S\cdot K$.
[c]Where iv = intravenously, im = intramuscularly, and po = orally.
[d]Corresponds to 5–20 million units.
[e]Corresponds to 2–12 million units.
[f]Corresponds to 1–10 million units.
[g]The R group given is equivalent to the RCONH of (**4**) and attaches directly to the C-6 of the penam ring (**2**).
[h]Also available as the indanyl ester *[26605-69-6]*, $C_{26}H_{26}N_2O_6SNa$.

Physical Properties

Penicillins have several properties that are characteristic of β-lactam antibiotics (10). They are obtained in relatively pure form as off-white, tan, or yellow freeze-dried or spray-dried solids that are usually amorphous. Alternatively they are sometimes obtained as crystalline solids, often as hydrates. Penicillins do not usually have sharp melting points, but decompose upon heating to elevated temperatures. Most natural members have a free carboxyl group and commercial preparations are generally either supplied as salts, most frequently as sodium salts, or in zwitterionic form as hydrates, eg, amoxicillin trihydrate [*61336-70-7*]. The acid strength of the carboxyl group in aqueous solution varies from pK_{a1} = 2.73 for oxacillin to pK_{a1} = 3.06 for carbenicillin. For zwitterionic penicillins amoxicillin, having pK_{a1} = 2.67 and pK_{a2} = 7.11, is typical (11).

Spectral Characteristics. The infrared stretching frequency of the penicillin β-lactam carbonyl group normally occurs at relatively high frequencies (1770–1815 cm^{-1}) as compared to the absorptions for the secondary amide (1504–1695 cm^{-1}) and ester (1720–1780 cm^{-1}) carbonyl groups. There is little difference between solution and KBr spectra. The nuclear magnetic resonance spectrum of penicillins invariably provides information about the integrity of the ring, attachments to the ring, and the stereochemistry of those attachments. The disposition of the C-5 and C-6 resonances are also characteristic. Thus the proton on the amide-bearing carbon appears as a quartet that collapses to a doublet on D_2O exchange, where δ = 5.42–6.15 ppm and J_{AB}(cis) = 4–5 Hz; J_{AB}(trans) = 1.5–2 Hz. The proton on the bridgehead carbon appears as a doublet, δ = 4.66–5.60; the C-3 proton absorbs in the range δ = 3.88–4.58 (10,12). More recently nmr spectroscopy has also proved to be a valuable tool for detecting and characterizing penicillin metabolites in biofluids (13).

The carbon-13 nmr features of penicillins have been studied and documented (10,14). The four sp^3 ring carbons give rise to resonances in order of decreasing chemical shift, C-3, C-5, C-2, and C-6. For C-3 the chemical shift ranges are 73.5–74.4 ppm for alkali metal salts in D_2O and 70.9–71.6 ppm for HCl salts having nonionizable carboxylate functions. The C-5 resonance ranges from 66.0 to 68.7 ppm, and the C-6 resonance from 57.4 to 59.1 ppm, or 60.0 to 64.6 ppm for examples with protonated side chains. The C-2 ranges from 63.9 to 66.0 ppm. The range for the 2α,2β-methyl signals are both narrow: for the high field 2-methyl the range is 26.5–28.0 ppm; and for the low field 2-methyl the range is 29.6–32.5 ppm. The carbonyl carbons give rise to resolved signals in most spectra in the range 167–176 ppm.

The analysis of penicillins by mass spectrometry (qv) has developed with the advent of novel techniques such as fast atom bombardment. The use of soft ionization techniques has enabled the analysis of thermally labile nonvolatile compounds. These techniques have proven extremely valuable in providing abundant molecular weight information from underivatized penicillins, both as free acids and as metal salts (15).

Stereochemistry. The absolute stereochemistry of the penicillins is 3(S):5(R):6(R) and the stereochemistry of the substituents attached to the ring is designated by the α and β-notations. Thus the β-lactam hydrogens are α, the acylamino group is β, and the penicillin carboxyl is α as in (**1**). A study of derivatives for which crystal structures have been determined has shown that the derivatives belong to two classes, either open or closed, depending on the conformation of the thiazolidine ring as shown in Figure 1 (16). In both cases puckering of the ring system is relatively pronounced and it is suggested that changes in the nature of the conformation influence biological activity.

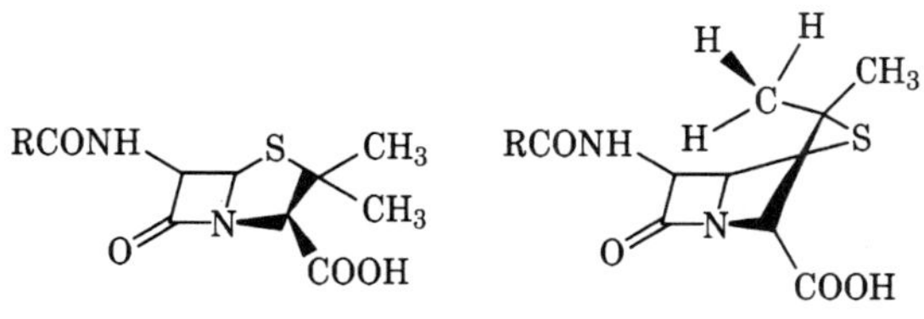

(**a**) (**b**)

Fig. 1. The penicillin thiazolidine ring: (**a**) open conformation; (**b**) closed conformation.

Chemical Properties

Penicillin to Other β-Lactam Conversions. Penicillin G (**1**, R = $C_6H_5CH_2$), penicillin V (**1**, R = $C_6H_5OCH_2$), and 6-APA (**1**, RCO = H) have proven to be cheap and versatile starting materials for a number of conversions to novel β-lactam containing systems. The commercial process for the preparation of 7-aminodeacetoxycephalosporanic acid (7-ADCA) and the deacetoxycephalosporins, cephalexin and cephradine, utilizes penicillin V (17) and the processes for conversion of penicillins into cephalosporins are well documented (18). Moxalactam [*64952-97-2*] (**5**) and related oxacephems are similarly derived from 6-APA (19).

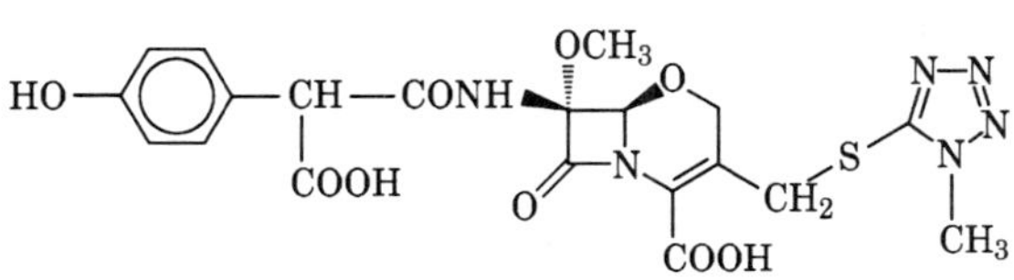

(**5**)

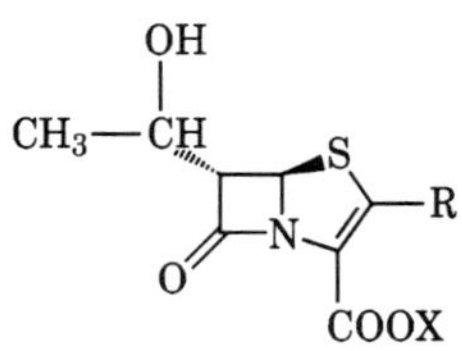

(**6**)

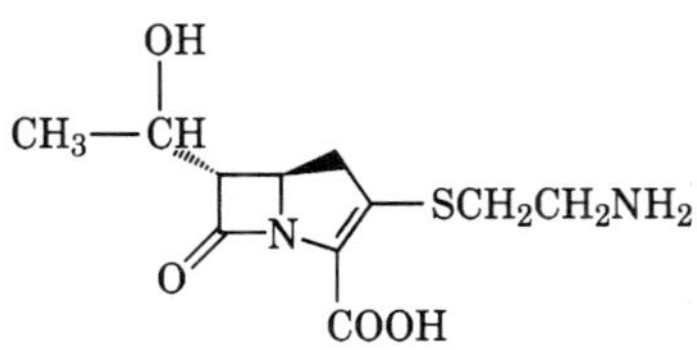

(**7**)

Synthesis of the penem ring system (**6**) from 6-APA has been described (20) and subsequently syntheses of novel penems having potential biological utility such as FCE 22101 (**6**, R = CH_2OCONH_2, X = Na) [*84845-58-9*] (21), SCH 29482 (**6**, R = SCH_2CH_3, X = Na) [*77646-84-5*], and SCH 34343 (**6**, R = $SCH_2CH_2OCONH_2$, X = Na) [*94392-35-5*] (22) have been documented. Similarly, synthesis of the

Table 2. α-Amino, Acyl, or Ureido Penicillins[a] Under Investigation or Exhibiting Limited Clinical Utility

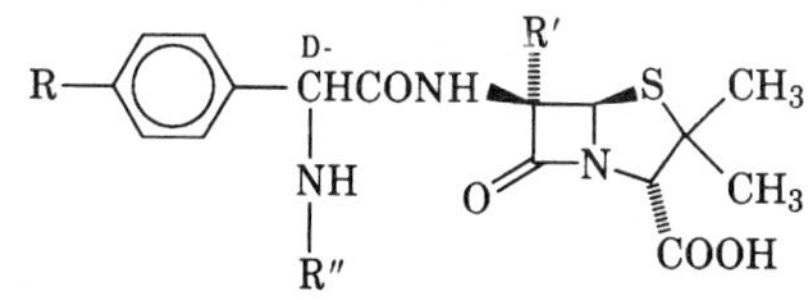

(**8**)

Name	CAS Registry Number	Molecular formula	R	R′	R″	Originating company	Status[b]
apalcillin	[*63469-19-2*]	$C_{25}H_{23}N_5O_6S$	H	H	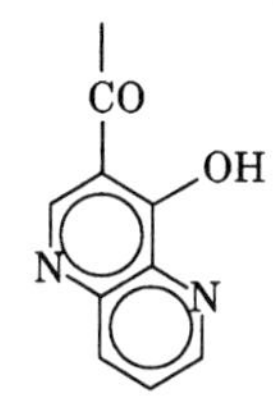	Sumitomo	C
piridacillin (Cl-867)	[*69414-41-1*]	$C_{32}H_{35}N_5O_{11}S_2$	OH	H	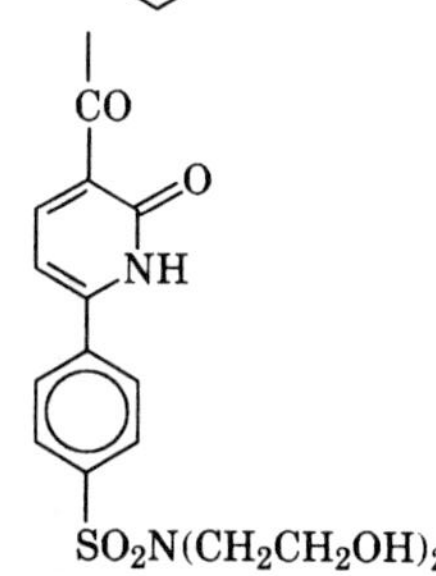	Warner-Lambert	N

PL-385	[*85550-66-9*]	$C_{29}H_{30}N_8O_8S$	OH	H	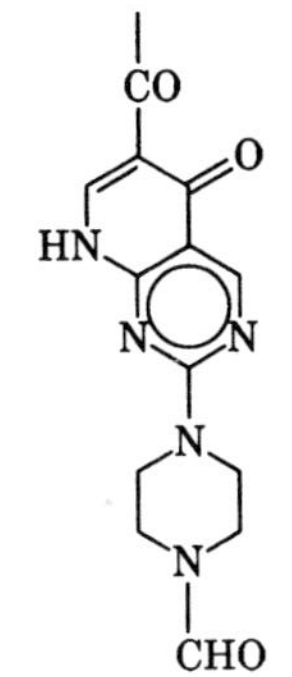	Dainippon	N
BLP-1908	[*65742-38-3*]	$C_{20}H_{20}N_6O_8S$	OH	H	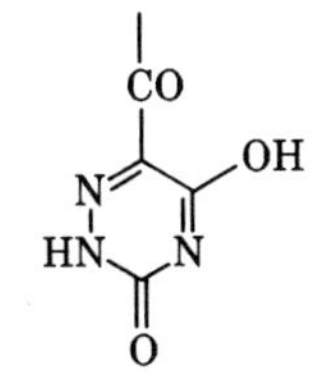	Bristol	N
					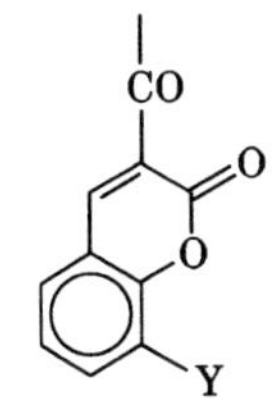		
TEI 1194	[*67609-13-6*]	$C_{26}H_{23}N_3O_7S{\cdot}Na$	H	H	Y = H	Teijin	N
TEI 2012	[*71344-34-8*]	$C_{26}H_{23}N_3O_8S{\cdot}Na$	H	H	Y = OH	Teijin	N
aspoxicillin (TA058)	[*63358-49-6*]	$C_{21}H_{27}N_5O_7S$	OH	H	—CO—$CHNH_2$—$CH_2CONHCH_3$	Tanabe	I

Table 2. ***(Continued)***

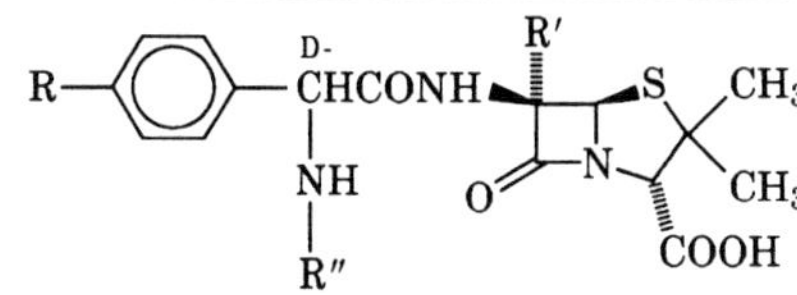

(**8**)

Name	CAS Registry Number	Molecular formula	R	R′	R″	Originating company	Status[b]
furazlocillin	[*66327-51-3*]	$C_{25}H_{26}N_6O_8S$	H	H	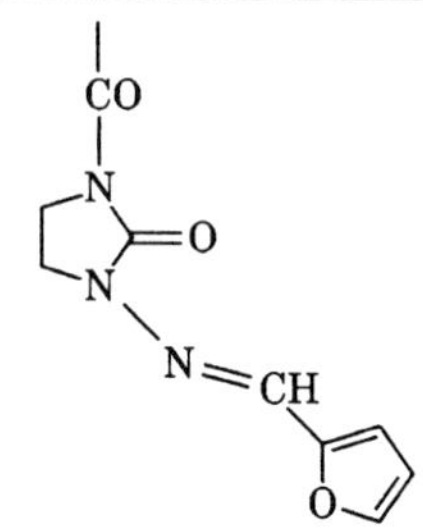	Bayer	C
EMD 39734		$C_{29}H_{29}N_7O_9S$	OH	H	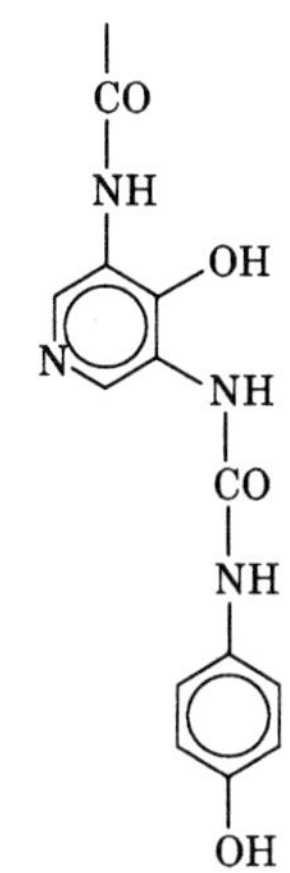	E. Merck	N

VX-VC-43	[*82509-56-6*]	$C_{27}H_{28}N_8O_9S_2$	H	H	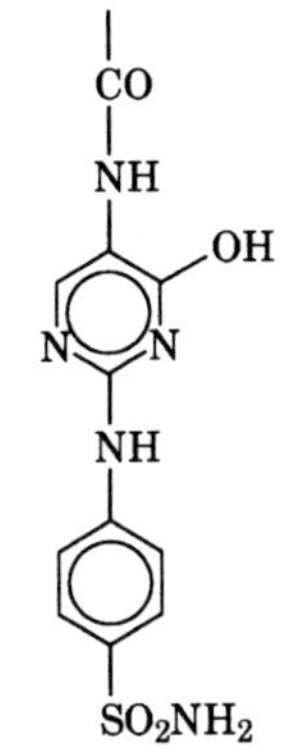	Thomae	N
lenampicillin[c]	[*86273-18-9*][c]	$C_{21}H_{23}N_3O_7S^c$	H	H	H	Kanebo	C
formidacillin (BRL 36650)	[*94425-13-5*]	$C_{24}H_{28}N_6D_{10}S{\cdot}Na$	3,4 di-OH	NHCHO	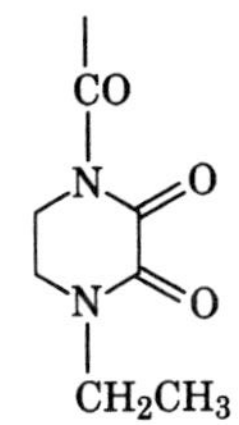	Beecham	N

[a]Structure (8) where R, R′, and R″ are as indicated.
[b]Drug status where C = in clinical use, I = investigational, and N = not progressed.
[c]Represents the 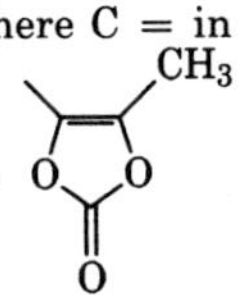 ester.

carbapenem antibiotic thienamycin [*59995-64-1*] (**7**), $C_{11}H_{16}N_2O_4S$, has been achieved from 6-APA (23).

Synthesis

The only penicillins used in their natural form are benzylpenicillin (penicillin G) and phenoxymethylpenicillin (penicillin V). The remainder of penicillins in clinical use are derived from 6-APA and most penicillins having useful biological properties have resulted from acylation of 6-APA using standard procedures.

A variety of coupling methods have been employed including: acid chorides, mixed anhydrides, mixed sulfonic acid anhydrides, *N,N*-dicyclohexylcarbodiimide and similar condensing agents, activated esters with *N*-hydroxysuccinimide, and *N*-hydroxybenzotriazole together with other acylating agents commonly used in peptide synthesis. The choice of coupling reaction is invariably dependent on the side chain. The efficiency of the reaction varies considerably between side chains so that optimization of the coupling procedure is often a matter of trial and error.

The use of protecting groups is common in penicillin chemistry: the amino function is normally protected by a trityl, benzyloxycarbonyl, *p*-nitrobenzyloxycarbonyl, trichloroethyloxycarbonyl, or trimethylsilyl group; and the carboxylic acid is usually protected as a benzyl, *p*-nitrobenzyl, *p*-methoxybenzyl, or trichloroethyl ester. Acylations may thus be carried out in aqueous or nonaqueous media with subsequent removal of the protecting group as required. Use of a tertiary base such as triethylamine is common in anhydrous reactions. Aqueous systems frequently utilize acetone–water mixtures in the presence of sodium bicarbonate. The resulting penicillins are usually isolated by the partitioning of the product as its free acid into a water-immiscible organic solvent and the salts or other impurities into water. The penicillins can be precipitated from organic solution as amine salts, or as alkali metal salts. Potassium or sodium ethylhexanoate dissolved in lower aliphatic alcohols may be used to effect purification. Alkali metal salts may be further purified by use of ion-exchange resins, eg, Diaion HP-20SS, and the final product isolated by freeze-drying (lyophilization), spray-drying, precipitation, or occasionally by crystallization.

After the influx of several α-amino penicillins such as ampicillin (Table 1) and related analogues, and the β-lactamase stable isoxazolyl penicillins in the 1950s and early 1960s, attention turned to identifying derivatives having a broader spectrum of activity. Carbenicillin and later, ticarcillin, reached the marketplace by the early 1970s. Effort then focused on the acyl and ureido ampicillin derivatives and culminated in the commercialization first of azlocillin and mezlocillin, then of piperacillin. Many thousands of derivatives were synthesized and some of the more active examples that have made little or no commercial impact are shown in Tables 2 and 3.

Examples of the acyl ampicillin series found in Table 2 are apalcillin (24), piridacillin (25), PL-385 (26), BLP-1908 (27), the Teijin examples TEI 1194 and TEI 2012 (28), and aspoxicillin (29). Other members of the ureido series of penicillins, of which azlocillin, mezlocillin, and piperacillin (Table 1) are commercial examples, are furazlocillin, EMD 39734 (30), and VX-VC-43 (31). During the 1970s the

Table 3. Other Penicillins[a] Under Investigation or Exhibiting Limited Clinical Utility

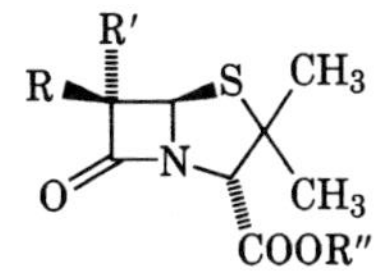

(**9**)

Name	CAS Registry Number	Molecular formula	R	R′	R″	Originating company	Status[b]
mecillinam	[*32887-01-7*]	$C_{15}H_{23}N_3O_2S$	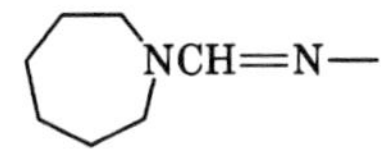	H	H	Leo	C
bacmecillinam	[*50846-45-2*]	$C_{20}H_{31}N_3O_6S$	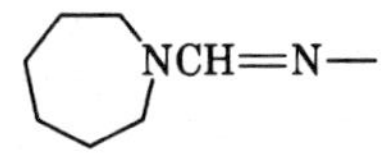	H	$-CH(CH_3)OCOC_2H_5$ (with C=O)	Kyowa	I
BRL 28917	[*85621-68-7*]	$C_{17}H_{20}N_2O_{10}S_2$	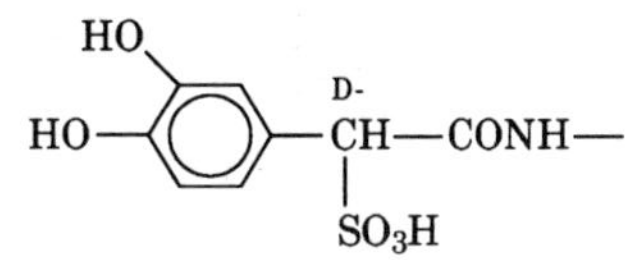	OCH_3	H	Beecham	N
BRL 20330	[*78968-21-5*]	$C_{23}H_{24}N_2O_7S_2$	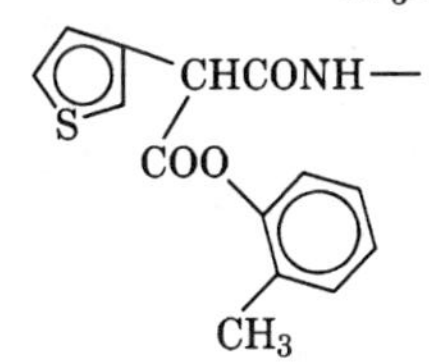	OCH_3	H	Beecham	N
BRL 44154	[*110717-44-7*]	$C_{19}H_{24}N_5O_6S_2$	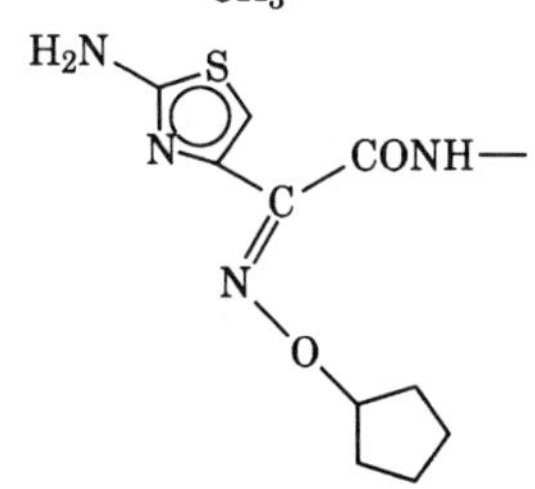	H	H	Beecham	I

[a]Structures (**9**), where R, R′, and R″ are as indicated.
[b]Drug status where C = in clinical use, I = investigational, and N = not progressed.

amidino penicillin mecillinam (Table 3), which possesses an unusual nonacyl side-chain substituent and a spectrum of activity limited to gram-negative organisms, was prepared. More recently a further example of a different type of acylated β-lactamase stable 6-APA derivative, BRL 44154, has been documented (32). BRL 44154 exhibits high activity against gram-positive organisms and methicillin-resistant staphylococci (MRSA).

Chemical Modification. Chemical modification of most positions in the penicillin nucleus have been carried out and these are summarized in Table 4. Apart from acylation of 6-APA, few of these modifications have proven profitable in terms of improving the biological properties of the derived penicillins. However, one of the modifications that has led to beneficial properties is substitution at the 6α-position.

Table 4. **Chemical Modification of Penicillins**

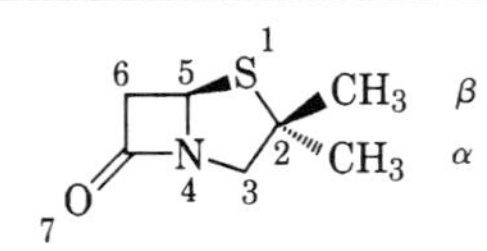

Position	Modification	
	From	To
1	S	SO, SO_2, $\overset{+}{S}CH_3$
2	$(CH_3)_2$	$(H)_2$, cycloalkyl, 4-piperidinyl, $=CH_2$ α-CH_2 halogen, β-CH_2 halogen, α-CHF_2, β-CHF_2 β-$COCH_3$, β-CH_2OR, β-CH_2OH, β-CHO, β-vinyl β-$CON(CH_3)_2$, β- $\overbrace{\text{—CH—CH}}^{\text{O}}$, β-CH=CHCOOR β-CH=CHCN, β-CH=CH-tetrazole
3	COOH	H, COOR, $COOCH_2OCOR$, COR, COOOCOR, CO_2SiR_3, CON_3, CH_2OH, CH_2OR, CH_2COOH, $COCHN_2$, CN, $PO(OCH_3)OH$, $PO(OCH_3)_2$, CHO, OH, NCO, CH_2N_3, CH_2NR_2, COSH, CH_2COOH, CH_2CH_2COOH, tetrazolyl, diketopiperazinyl
2,3		cyclopropyl
6β	RCONH	H, NH_2, R′CONH, R″N=C=N, ArC=N, R_2N, RNH, $CH(OH)CH_3$, CH_2OH, CH_2NH_2, OH, halogen, 6-oxopenam
6α	H	CH_3, OCH_3, NH_2, RCONH, OH, CH_2OH, CH_2Cl, CH_2F, D, CH(OH)F, CH_2NH_2, SCH_3, $S(O)CH_3$, $CH_2CO_2CH_3$, CH_2Ar, COOH, COOR, NC, $NHCH_3$, $CH=CH_2$, CN, OCH_2CH_3, CH=CH-CN, CH=CHCHO, NHC_6H_5, $NHNHCOOCH_3$, $NHOCH_3$, N(CHO)NHCHO, $NHCOOCH_3$, $NHCOCF_3$, $NHSO_2CH_3$, $N(OCH_3)CHO$, $NHCONHCH_3$, $N(CH_3)_2$, $N(CH_3)CHO$, NHNHCHO, $N(OH)CH_3$, $COCH_3$, CH_2OCH_3, $CH_2OSO_2CH_3$, CHO, ONHCHO, ONH_2, C_6H_5, tetrazolyl, succinimido-oxy, $CH(OH)OCH_3$, $CONH_2$, CH(OH)Ar, CH_2CH_2CN
7	O	S

6α-Substituents. In the early 1970s the discovery of the cephamycins, stabilized to attack by β-lactamase through the presence of 7α-methoxy substituents, stimulated a parallel investigation of penicillins substituted by methoxy at the 6α-position (33). This study led to the identification of the directed spectrum β-lactamase stable temocillin, (Table 1) (34), the stability of which results from the combination of the 6α-methoxy group and the dicarboxylic acid functionality. Temocillin possessed high and prolonged serum levels in human volunteers but was directed in its spectrum of activity to gram-negative organisms. Whereas temocillin was active only by the parenteral route, the *o*-methylphenyl ester prodrug BRL 20330 (Table 3) was found to be absorbed orally (35). Replacement of the side-chain carboxylic acid by sulfonic acid led to BRL 28917 (Table 3), a β-lactamase stable penicillin with high activity against *Pseudomonas sp.* (36).

Substitution of penicillins by 6α-methoxy was found to be compatible with an α-acidic side chain in terms of antibacterial activity, but less beneficial when the side chain contained an α-acyl or α-ureido substituent. However, analogues of the ureido penicillin VX-VC-43 (Table 2) containing a 6α-methoxy substituent (**10**) were found to combine good stability to β-lactamase and relatively high antibacterial activity (37). Following an extensive program to identify other 6α-substituents that would stabilize the acyl and ureido series of penicillins, the 6α-formamido series (**11**) represented by formidacillin (BRL 36650) (Table 2) was developed (38).

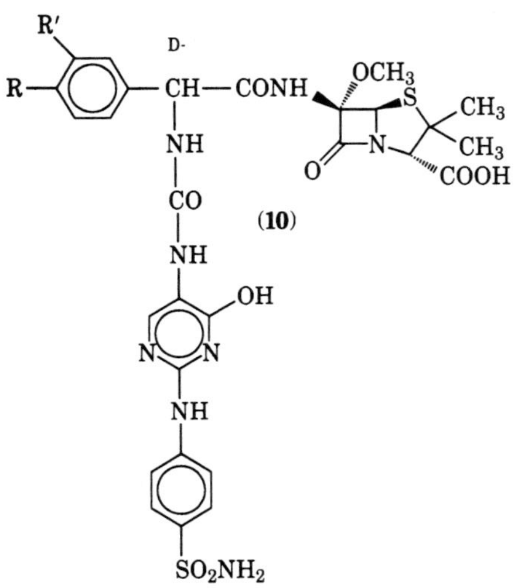

(**10**)

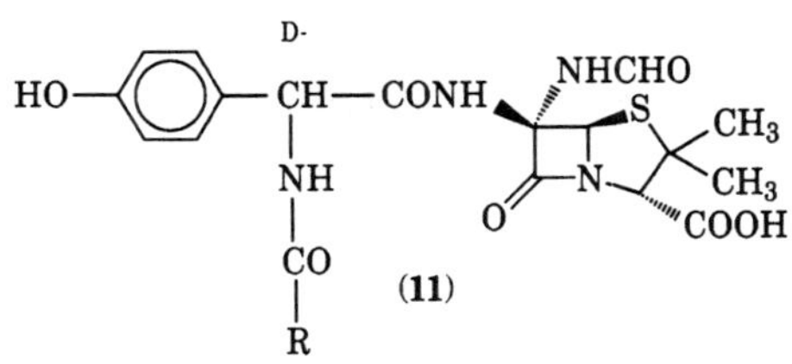

(**11**)

The early chemistry leading to these derivatives was originally carried out via the 6α-(methylthio) derivative (**17**) which was prepared by way of a Schiff's base (39). The 6α-thiomethyl group could then be displaced by various nucleophiles giving rise to 6α-methoxy or other 6α-substituted penicillins. A stereospecific one-step introduction of a methoxy group at C-6 in penicillins provided a simple entry to 6α-methoxy penicillins (40) in yields ranging from 50–62%.

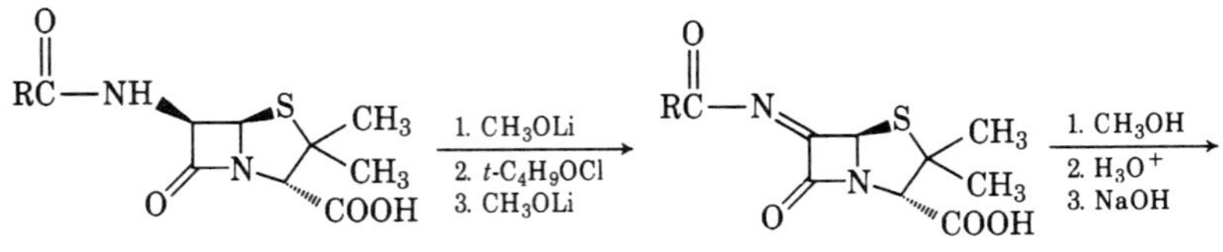

(1) (12)

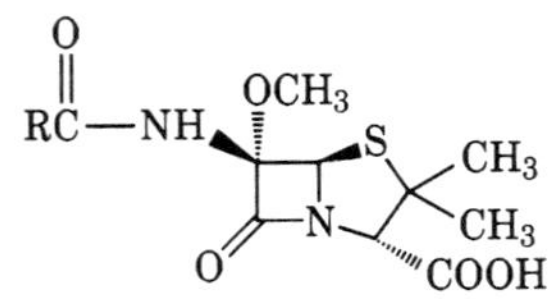

(13)

After formation of the acylimine (**12**), methanol adds to the less sterically hindered α-face of the molecule with high selectivity to provide (**13**). A further direct incorporation of a 6α-methoxy group (41) and subsequently a 6α-formamido group into penicillin has been achieved using trifluoromethanesulfonamides of type (**14**) (42).

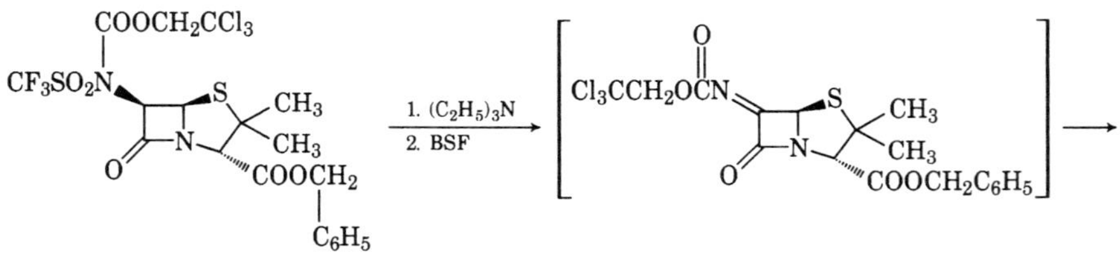

(14) (15)

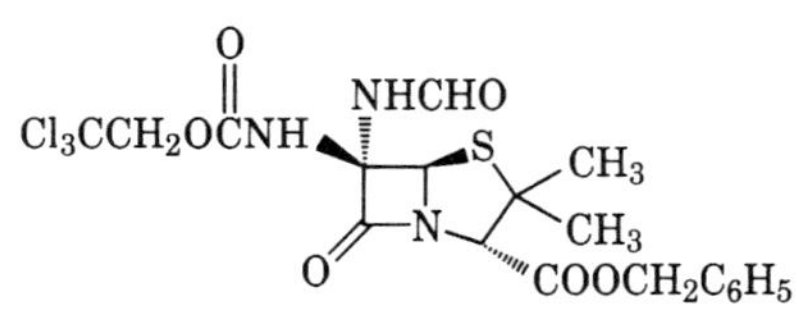

(16)

Preparation of (**14**) and treatment with *N,N*-bis(trimethylsilyl)formamide (BSF) and triethylamine provided (**16**) in 84% yield via the putative imine intermediate (**15**). The trifluoromethyl group could be replaced by other moieties such as 2,4,5 trichlorophenyl, pentafluorophenyl, or nonafluorobutyl with increasing effectiveness.

Further versatility was added to the range of substituents available for introduction into the 6α-position by use of the 6α(succinimido-oxy) derivative (**18**) prepared by treatment of the 6α-(methylthio) derivative (**17**) with *N*-hydroxysuccinimide and silver(I) acetate in dimethylformamide in virtually quantitative yield. In this way the 6α-cyanopenicillin (**19**, X = CN), 6α-vinylpenicillin (**19**, X = $CH{=}CH_2$) and 6α-phenylpenicillin (**19**, X = C_6H_5) could be prepared in high yield (43).

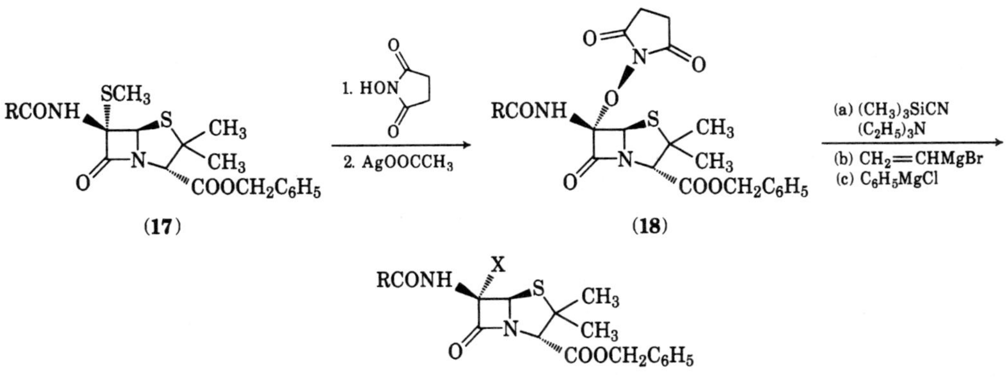

(17) (18)

(19)

2α-2β-Substituents. Modifications at the 2-position of the penicillin nucleus has led to benefits in the biological properties of the modified penicillins. The series that has generated the most interest is the bisnorpenicillins depicted by (**22**). Synthesis of these derivatives employs as a key intermediate the 4-mercaptoazetidin-2-one (**20**) which in turn is generated from penicillin V (**1**, R = $C_6H_5OCH_2$) (44).

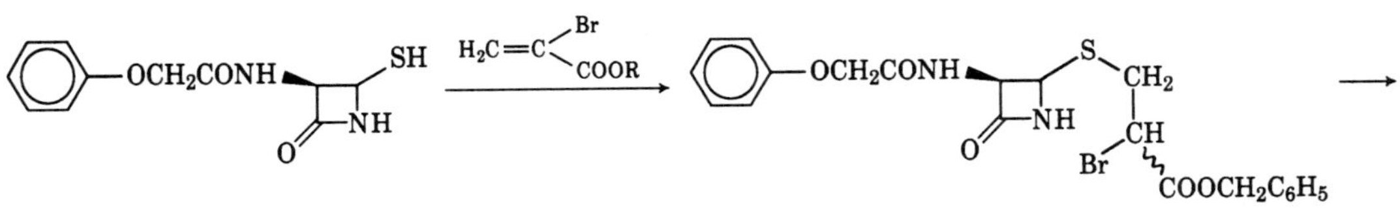

(20) (21)

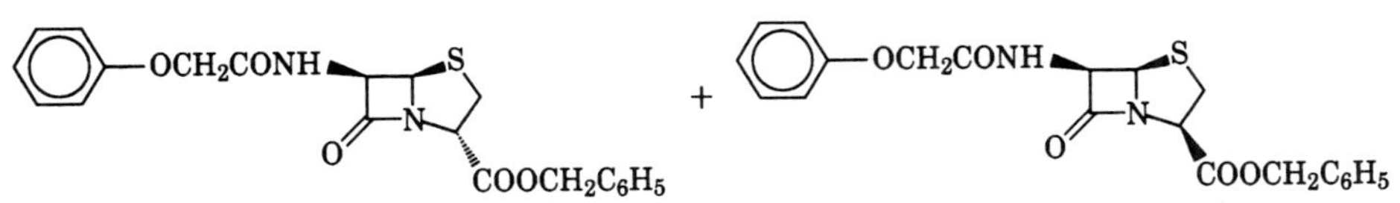

(22) (23)

Reaction of (**20**) with ethyl 2-bromoacrylate in hexamethylphosphorous triamide (HMPT) in the presence of 1,5-diazabicyclo[4.3.0]non-5-ene (DBN) yielded (**22**) in moderate yield. Use of potassium carbonate in DMF afforded a mixture of C-3 epimeric bisnorpenicillanates (**22**) and (**23**) (45). Removal of the phenoxyacetyl side chain was effected by Delft cleavage namely, treatment with phosphorus pentachloride in methylene chloride containing *N*-methylmorpholine followed by methanol and water. Acylation was then carried out using standard procedures. Bisnorampicillin was found to be less active than ampicillin, but bisnorpiperacillin proved to be similar in activity to piperacillin. Bisnorpiperacillin possesses a slightly broader spectrum of activity than the unmodified compound because of increased stability towards bacterial β-lactamases (46). Studies have shown that enhanced β-lactamase stability is a general property of acyl and ureido bisnorpenicillin derivatives (47). Synthesis of cycloalkanespiro-2-bisnorpenicillins led to a further series of nuclear modified penicillins:

spirocyclohexane, spirocyclopentane, spirocyclobutane, and spirocyclopropane analogues have been prepared. Only those containing the spirocyclopropane ring system provided analogues that had antibacterial properties comparable to those of the corresponding penicillins (48). Other modifications of the C-2 position are shown in Table 4.

2,3-Methylenepenams. The penam nucleus is also able to assume both of the ring conformations depicted in Figure 1 and the (2,3)-α-and (2,3)-β-methylenepenams shown in Figure 2 may be regarded as penicillin analogues (49). Studies involving the (2,3)-methylenepenams, when R = $CH_2C_6H_5$, and analogues revealed that the (2,3)α-methylenepenam penetrated the outer membrane of *E. coli* more readily than did the penicillin G analogue (50). Intrinsically, the alpha compound was found to have similar activity to that of penicillin G. By comparison the β-methylene analogue had poor activity.

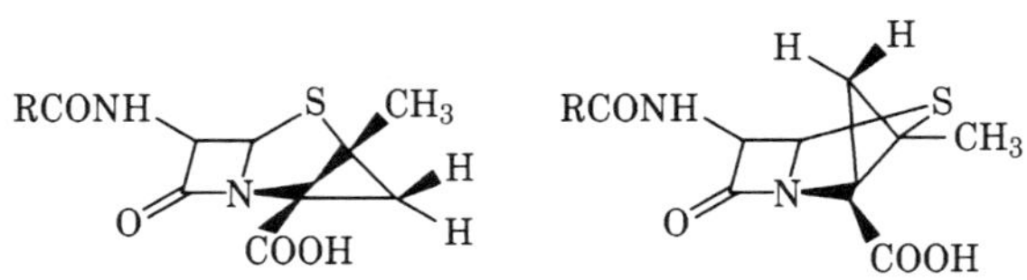

Fig. 2. Structures of (**a**) (2,3)-α-methylenepenams, and (**b**) (2,3)-β-methylenepenams.

Later investigations led to improved synthetic routes and analogues such as those shown in Figure 3 (51). Conversion of the mixed imide (**24**) to the bromide (**25**) provided a mixture of α- and β-bromomethylpenicillins in which the α-isomer predominated. 1,8-Diazabicyclo[5,4,0]undec-7-ene(DBU)-mediated cyclization led to the (2,3)-methylenepenicillanic acid (**26**) which was converted by catalytic hydrogenation followed by reverse phase chromatography to the potassium salt (**27**). The preparation of analogues (**29**) was carried out by reaction of (**26**) with 1.1 equivalents of PCl_5 and 1.9 equivalents of dry pyridine in chloroform followed by treatment with *n*-propanol and saturated aqueous sodium chloride solution. The resulting amine (**28**) was used without further purification and then transformed into derivatives (**29**) using suitably activated acids followed by hydrogenolysis using 10% palladium on carbon.

In general the (2,3)-α-methylenepenams were found to be chemically reactive and biologically active in contrast to the β-methylene counterparts supporting the hypothesis that the open conformation of penicillins is the biologically active form. In (2,3)-α-methylenepenams the absolute configuration at C-2 is opposite that of the naturally occurring penicillins placing the carboxyl group in a position more akin to that found in the cephalosporins. The compounds were found to be more susceptible to cephalosporinase than penicillinase. The (2,3)-α-methylene penicillin G analogue was found to be more selective in the inhibition of PBP 3 than was the β-analogue. It also bound more effectively to PBP 2 of *S. aureus* than the β-isomer (52).

Biochemical studies showed that the α-aminoadipoyl analogue derived from (2,3)-β-methylenepenam was not a substrate for expandase activity but rather, it was a potent reversible inhibitor of the ring expansion of α-aminoadipoyl-penicillin into deacetoxycephalosporanic acid by the expandase enzyme (53).

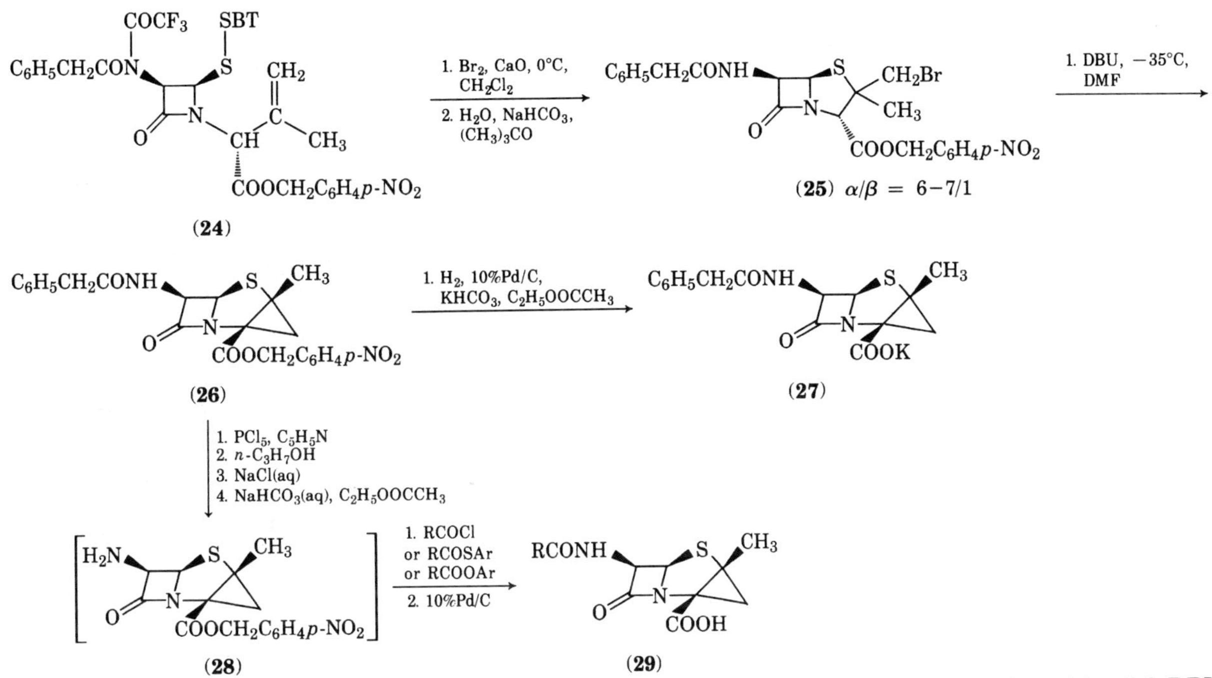

Fig. 3. Reaction scheme for the synthesis of (2,3)-α-methylenepenams where BT is 2-benzthiazolyl, DBU is 1,8-diazabicyclo[5.4.0]undec-7-ene.

Degradation. Penicillins are rapidly hydrolyzed by aqueous alkali to the corresponding penicilloic acids (**30**) which are stable as salts, but which decarboxylate on acidification to yield penilloic acids (**31**).

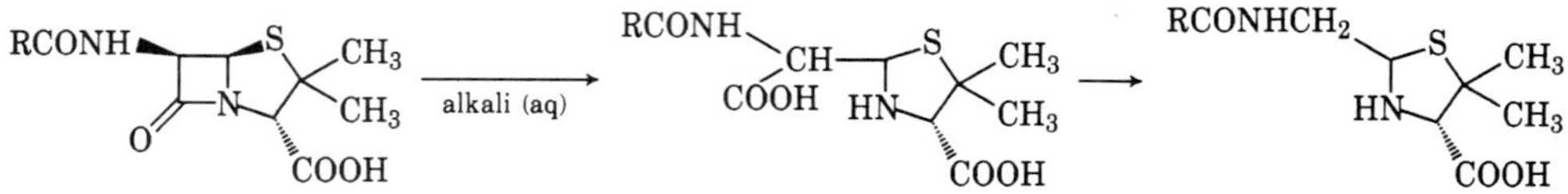

(**1**) (**30**) (**31**)

Penicillins are also degraded by aqueous acids via initial reaction of the side-chain carbonyl group with the *β*-lactam. Penicillenic acids (**33**) are obtained when hydrolysis is carried out at pH 4, penillic acids (**34**) at pH 2.

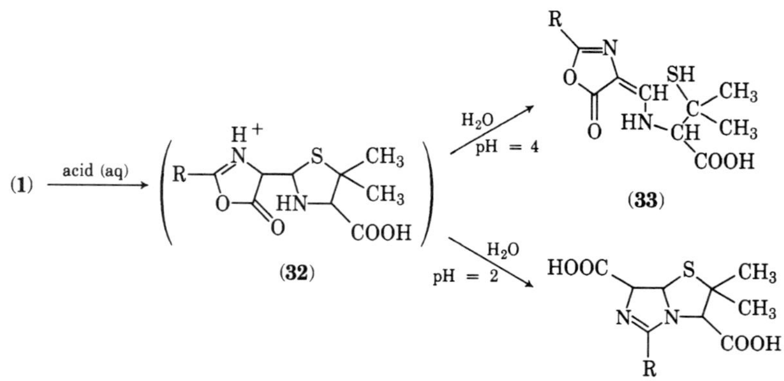

(**32**) (**33**) (**34**)

Electron-withdrawing groups at the *α*-position of the 6-acyl substituent stabilize the compounds to attack by acids. Thus phenoxymethylpenicillin and ampicillin are more stable to acid and more efficiently absorbed by the oral route than is benzyl penicillin. Degradation of temocillin (Table 1), a 6*α*-methoxypenicillin, occurs in a way similar to that shown for (**31**), (**33**), or (**34**). Although temocillin exhibits good stability in mild aqueous acid or base, in stronger acid the corresponding methoxypenillic acid is formed and under alkaline conditions the methoxypenicilloic acid is generated together with the C-5 epimer (54). Degradation of BRL 36650 (**35**) shown in Figure 4, containing a 6*α*-formamido substituent, occurs quite readily under mildly acidic aqueous conditions. Cleavage of the C-5–C-6 bond leads to two fragments, the formamido acid (**39**) and *N*-formyl penicillamine (**40**) presumably generated via the putative penicillenic acid intermediate (**36**) (55). The presence of the electron-withdrawing 6*α*-formamido group serves to accelerate degradation and leads to products that have been detected for penicillins in aqueous solution left over a longer period of time.

Biological Properties

Structure-Activity Relationships. Biological evaluation of penicillins yields information such as *in vitro* and *in vivo* antibacterial activities, minimum inhibitory concentration (MIC), minimum bactericidal concentration (MBC),

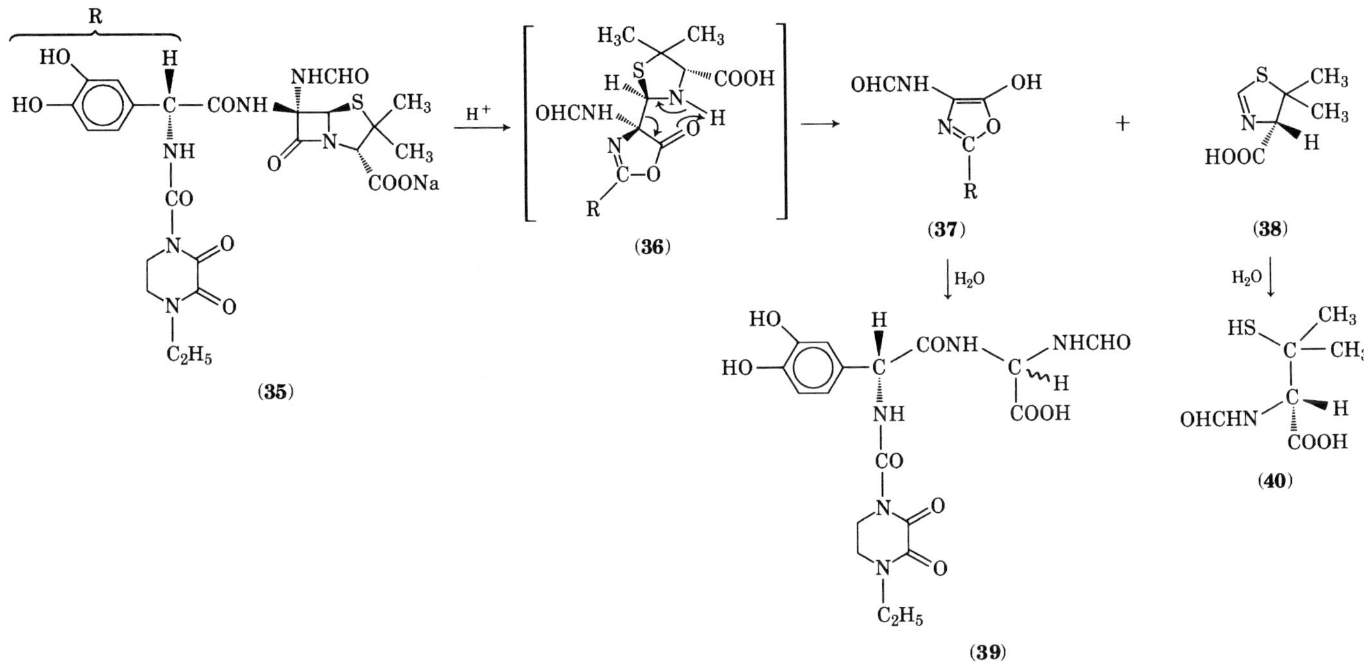

Fig. 4. Acid degradation of BRL 36650.

protective effectiveness in laboratory animals (PD_{50}), and pharmacokinetic characteristics including efficiency of absorption, serum levels, tissue distribution, urinary excretion, metabolism, serum binding, serum and tissue activation, biliary excretion, recycling, etc. Penicillins are also tested for ability to resist inactivation by β-lactamase produced by both gram-positive and gram-negative bacteria.

Table 5 gives a range of activities for some modified penicillins. Penicillin G remains probably the most active penicillin against gram-positive organisms. However, the majority of *Staphylococcus aureus* strains are resistant to penicillin by virtue of β-lactamase production. The β-lactamase resistant penicillins such as oxacillin, cloxacillin, flucloxacillin, and nafcillin are active against most penicillin-resistant *Staphylococci* but lack activity against methicillin-resistant *Staphylococci* (MRSA) and the majority of gram-negative organisms.

Ampicillin and its congeners amoxicillin, bacampicillin, and ciclacillin, have largely similar antibacterial spectra, exhibiting activity against both penicillin sensitive gram-positive and gram-negative microorganisms. The susceptibility of ampicillin or amoxicillin to β-lactamase may be overcome by combination with a β-lactamase inhibitor. Clavulanic acid and sulbactam are two such β-lactamase inhibitors in clinical use. The products Augmentin, a combination of clavulanic acid and amoxycillin for oral use primarily, and Unasyn, a combination of sulbactam and ampicillin for use by injection, are highly active against both gram-positive aerobic and anaerobic organisms in addition to many important gram-negative pathogens. Carbenicillin and ticarcillin possess moderate broad-spectrum activity and were the first penicillins to show activity against *Pseudomonas* species. Azlocillin, mezlocillin, and piperacillin possess largely the same characteristics but have slightly greater potency against many gram-negative species. Temocillin is the first penicillin to possess a high level of activity against gram-negative organisms, notably the *Enterobacteriaceae,* as well as excellent β-lactamase stability. The introduction of the 6α-methoxy group and the concomitant β-lactamase stability compromises activity against both gram-positive organisms and *Pseudomonas* species. Similarly BRL 36650, having the 6α-formamido substituent, exhibits high bacterial β-lactamase stability. Some activity is retained against most species of *Streptococci* but the compound is inactive against *Staphylococci*. BRL 36650 is highly potent against most gram-negative organisms including refractory gram-negative species such as *Acinetobacter* and *Pseudomonas*. The level of potency is much greater than either that of ticarcillin or piperacillin (56).

The pharmacology of penicillins differs markedly from compound to compound but has been well reviewed (57). The majority of derivatives, including penicillin G and the antipseudomonal penicillins, are unstable in gastric acid and are not available orally. The isoxazolyl penicillins are relatively acid stable but not consistently well absorbed by the oral route. Nafcillin and oxacillin are poorly absorbed orally; cloxacillin, dicloxacillin, and flucloxacillin are more reliable. Penicillin V, ampicillin, and particularly amoxicillin are relatively well absorbed orally. Esters of ampicillin such as bacampicillin, pivampicillin, and talampicillin improve the level of oral absorption of ampicillin to that achieved by amoxicillin. Absorption can be diminished by food after oral administration, however, and peak blood levels, usually achieved after 1 to 2 h, are somewhat delayed after ingestion of food.

Table 5. Antibacterial Activity[a] of Modified Penicillins

Organism	Penicillin G	Ampicillin	Amox+clav[b]	Ox/Naf[c]	BRL 44154	Carb/Ticar[d]	Azlo/Mezlo/Piper[e]	Temocillin	BRL 36650
Gram-positive									
Staphylococcus (−)[f]	++	++	++	++	++	+	++	−	−
Staphylococcus (+)[g]	−	−	++	++	++	−	−	−	−
Staphylococcus (MRSA)[h]	−	−	−	−	+	−	−	−	−
Streptococcus pyogenes	++	++	++	+	++	+	++	−	+
Streptococcus pneumoniae	++	++	++	+	++	+	++	−	+
Enterococcus	++	++	++	−	−	+	++	−	−
Bacterioides fragilis	−	+	++	−	−	+	+	−	−
other anaerobes	−	+	++	−	−	+	+	−	−
Gram-negative									
E. coli	−	+	++	−	+	+	++	++	++
Klebsiella	−	−	++	−	−	−	+	++	++
Proteus mirabilis	−	++	++	−	++	++	++	++	++
other *Proteus*	−	−	−	−	−	+	++	++	++
Enterobacter	−	−	−	−	−	+	++	++	++
Pseudomonas aeruginosa	−	−	−	−	±	+	+±	−	++
other *Pseudomonas* species	−	−	−	−	−	−	+	−	++

[a] ++ = very active at low concentrations; + = moderately active against most strains or active only at high concentrations; and − = inactive against the majority of strains.
[b] A mixture of amoxicillin and clavulanic acid.
[c] Activity represents that of both oxacillin and nafcillin.
[d] Activity represents that of both carbenicillin and ticarcillin.
[e] Activity represents that of azlocillin, mezlocillin, and piperacillin.
[f] *Staphylococcus* (−) = β-lactamase negative.
[g] *Staphylococcus* (+) = β-lactamase positive.
[h] *Staphylococcus* (MRSA) = methicillin-resistant *Staphylococcus aureus*.

Parenteral penicillins generally are virtually 100% available and half-lives vary from one-half hour for nafcillin and oxacillin to over an hour for amoxicillin and ticarcillin. The exception to this rule is temocillin which has an unusually long half-life of 4.5–5 h. Long-acting preparations of penicillin G are available as the procaine and benzathine derivatives. When these derivatives are administered intramuscularly, they provide effective serum levels for 12–24 and 3–4 weeks, respectively. Penicillins are eliminated by active secretion by renal tubular epithelial cells. This secretion can be blocked by probenicid [*57-66-9*], $C_{13}H_{19}NO_4S$, resulting in higher blood levels and prolonged serum half-lives. Most penicillins distribute well to body cavities and kinetics are usually linear. Urinary concentrations of penicillins are normally high, even in the presence of renal failure. Intracellular penetration of penicillins because of their lipid insolubility is invariably low. They also penetrate poorly into the central nervous system and the eye. Biliary concentrations of penicillins generally exceed serum concentrations in the presence of adequate bile flow.

The penicillins in general, are renowned for their lack of toxicity. The most common adverse effect of the use of penicillins is an allergic reaction which can change from a mild rash to fatal anaphylactic shock in rare cases. All penicillins cross the placenta and are excreted in maternal milk. However, the relative freedom from toxicity renders these compounds valuable agents during pregnancy and lactation.

Biosynthesis. The microbial synthesis of penicillins from eukaryotes such as *C. acremonium* and *P. chrysogenum* has been comprehensively reviewed (58). In essence the biosynthesis of penicillins is described in Figure 5 although certain stages have yet to be fully characterized. Products are isopenicillin N, penicillin G, and penicillin N [*525-94-0*], $C_{14}H_{21}N_3O_6S$.

The key enzyme in this sequence, isopenicillin N synthase (IPNS), has been purified from *E. coli* (59) and the recombinant enzyme shown to be a single polypeptide of 336 amino acids containing two cysteines, numbers 106 and 255 from the *N*-terminus, and probably a ferrous ion in a nonheme environment. The enzyme has been crystallized and studies undertaken to obtain suitably sized crystals for diffraction studies.

The IPNS enzyme has also been shown to recognize modified tripeptides. The synthesis of a range of tripeptides, other than aminoadipoyl cysteinyl valine (ACV) (Table 6), has given rise to a selection of modified penicillins using IPNS as a means of cyclizing the tripeptide (58).

Mode of Action. Penicillins exert their antibacterial effect by inhibiting the high molecular weight penicillin binding proteins (PBPs) that are implicated in the final stages of peptidoglycan synthesis (60). These PBPs have been extensively studied in *E. coli* in terms of the relative affinity of the drug for the three physiologically important proteins. Inhibition of PBP-1 causes rapid cell lysis whereas inhibition of PBP-2 and PBP-3 results in the formation of spherical and filamentous cells, respectively. PBP-1 may be further subdivided into PBPs 1*a* and 1*b*. PBP-1*a* is usually highly sensitive to penicillins and it is the effective inhibition of PBP-1*b* that is regarded as the more crucial stage leading to cell lysis. The low molecular weight PBPs 4, 5, and 6 of *E. coli* are believed to be nonessential for cell survival and therefore not considered of major importance in the killing mechanism of penicillins (61).

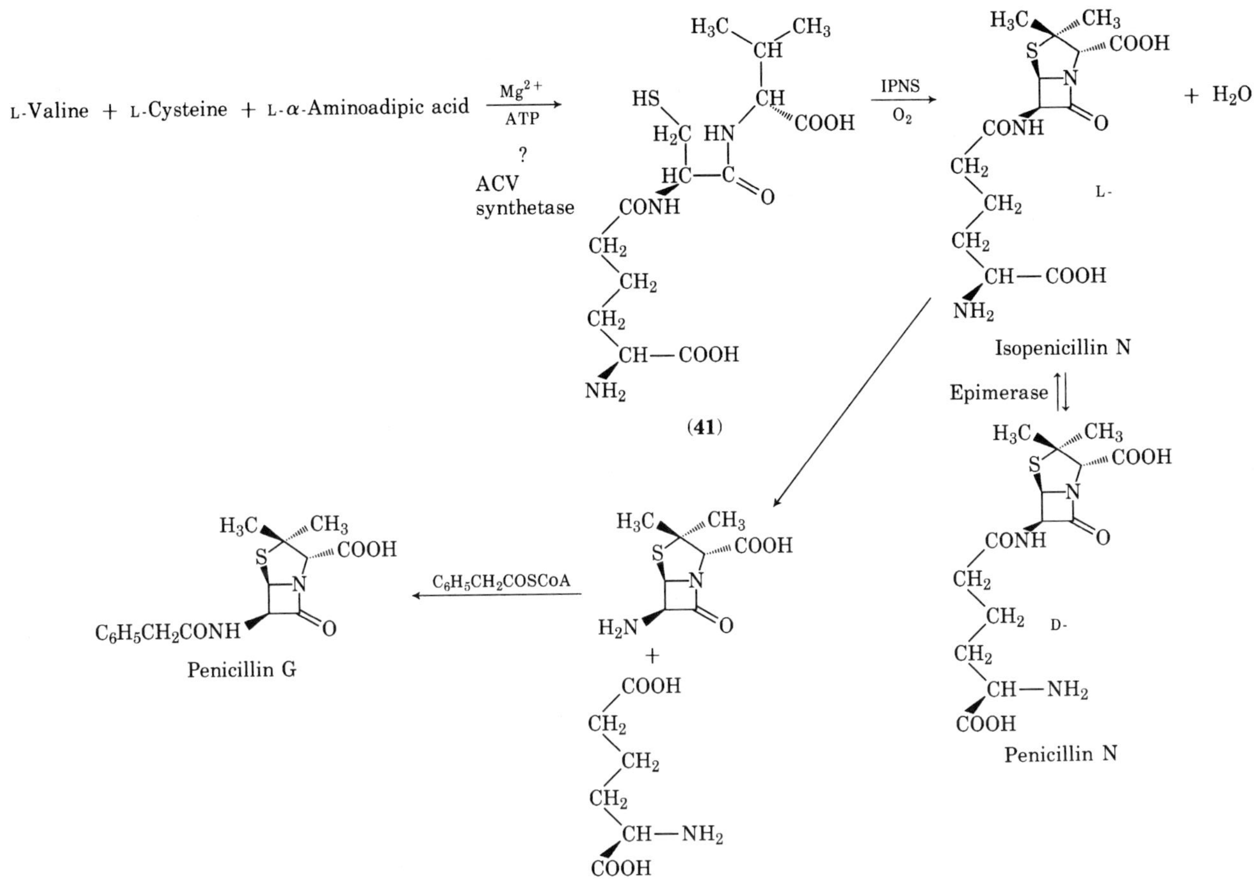

Fig. 5. Biosynthesis of penicillins when ACV is aminoadipoly cysteinyl valine and IPNS is isopenicillin N synthase and $C_6H_5CH_2COSCoA$ represents benzyl coenzyme A. ACV synthetase is thought to catalyze the first step of this reaction sequence. Courtesy of J. E. Baldwin, Royal Society of Chemistry.

Table 6. Modified Penicillins[a] Prepared Using IPNS[b]

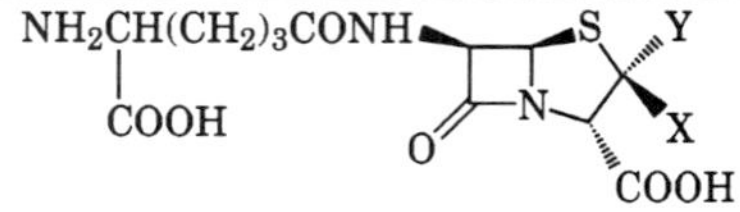

(**42**)

X	Y	X	Y
CH_3	H	CH_3	CH_2OH
H	CH_3	H	CH_2OH
CH_3	C_2H_5	CH_3	OCH_3
C_2H_5	CH_3		
$-CH{=}CH_2$	H	H	C_2H_5
H	$-CH{=}CH_2$	H	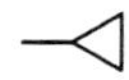
CH_3	$-CH{=}CH_2$	H	$-C{=}C{=}CH_2$

[a]Structure (**42**) where X and Y are as indicated.
[b]Ref. 58.

The changes that occur when an organism is exposed to different concentrations of penicillin provides a characteristic morphological response that can be particularly well demonstrated in the scanning electron microscope. Penicillin G binds preferentially to PBP 3 causing filamentation. Most penicillins bind principally to PBPs 1 and 3 leading to filamentous forms. Exceptions are amoxicillin and mecillinam. Amoxicillin binds to PBPs 1 and 3 but at low concentrations forms spheroplasts and at higher concentrations, filaments. Mecillinam binds only to PBP 2 producing ovoid and later spherical cells. The novel penicillin BRL 44154 has been shown to bind to PBPs 1 and 3 and also to PBP 2′, the protein responsible for the resistance in methicillin-resistant *Staphylococcus aureus* (MRSA) (62).

Rapid progress is being made in the protein crystallography of low molecular weight PBPs and demonstration of the feasibility of obtaining water-soluble forms of high molecular weight PBPs suggests that crystallization of the latter proteins can be achieved. These enzymes are inserted into the cytoplasmic membrane only at their amino termini and water-soluble forms are expected to be suitable for crystallization and x-ray analysis (57). Preliminary investigations involving a *β*-lactam-sensitive, bifunctional D-alanyl-carboxypeptidase–transpeptidase (C Pase–T Pase) from *Streptomyces* R61 have identified the three-dimensional structure and catalytic site of interaction with penicillins (63).

Production, Manufacture, and Processing

Most methods used for the production of the commercially important *α*-amino penicillins, such as ampicillin and amoxicillin, are based on modifications of an enamine process employing the appropriate phenylglycine and methylacetoacetate followed by coupling with 6-APA (64). Other aspects of the fermentation, strain maintenance, equipment, inoculum development, media, and procedures used in the production of penicillin are well covered in previous editions of the *Encyclopedia*. Developments in these areas have been reviewed (65).

Economic Aspects

Worldwide retail antibiotic sales in 1986 were approximately $11 billion of which penicillins comprised approximately $2.5 billion (66). Preparations containing ampicillin are estimated at $800 million whereas those containing amoxicillin are estimated at $1,020 million. Total sales include sales of ampicillin esters and the ureidopenicillins derived from ampicillin. Total sales of penicillins, including semisynthetic penicillins in the United States in 1987, were $25 million (67). Sales of amoxicillin, the largest single selling penicillin, were approximately $350 million.

The volume of bulk ampicillin produced was 3,700 t, valued at $260 million, and of bulk amoxicillin, 2,200 t valued at $190 million, both as the trihydrate. It was predicted that the market for bulk ampicillin was growing at 3% annually and would reach 4,160 t by 1990. Sales of amoxicillin, growing at 10% annually were predicted to reach a consumption of 2,800 t in 1990 with an average growth rate of 8%. The demand for both products is expected to reach 4,500 t by 2000 (68).

The cost of 6-APA, ampicillin, and amoxicillin is invariably linked to that of penicillin G. 6-APA remains one of the cheapest raw starting materials available for the preparation of semisynthetic penicillins and other related β-lactam antibiotics.

BIBLIOGRAPHY

"Penicillin," in *ECT* 1st ed., Vol. 9, pp. 922–943, by W. E. Brown, Squibb Institute for Medical Research; in *ECT* 2nd ed., Vol. 14, pp. 652–707, K. Butler, Chas. Pfizer and Co.; "Antibiotics (β-Lactams)," in *ECT* 3rd ed., Vol. 2, pp. 871–919, by J. R. E. Hoover and C. H. Nash, SmithKline and French Laboratories.

1. J. C. Sheehan, K. R. Henery-Logan, and D. A. Johnson, *J. Am. Chem. Soc.* **75,** 3292 (1953).
2. A. Fleming, *Br. J. Exp. Pathol.* **10,** 226 (1929).
3. E. Chain and co-workers, *Lancet* (II), 177 (1941).
4. E. P. Abraham, and co-workers, *Lancet* **1977**(II) (1941).
5. H. W. Florey and co-workers, *Antibiotics,* Vol. 11, Oxford University Press, London, 1949.
6. H. T. Clarke, J. R. Johnson, and R. Robinson, eds., *The Chemistry of Penicillin,* Princeton University Press, N.J., 1949.
7. A. J. Moyer and R. D. Coghill, *J. Bacteriol.* **53,** 329 (1947).
8. D. Crowfoot, C. W. Bunn, B. W. Rogers-Low, and A. Turner-Jones in Ref. 6, p. 310; H. T. Clarke, J. R. Johnson, and R. Robinson in Ref. 6, p. 5; Ref. 5, p. 667.
9. F. R. Batchelor, F. P. Doyle, J. H. C. Nayler, and G. N. Rolinson *Nature (London)* **183,** 257 (1959), F. R. Batchelor, E. B. Chain, T. L. Hardy, K. R. L. Mansford, and G. N. Rolinson, *Proc. R. Soc. London* Ser.B. **154,** 498 (1961).

10. E. H. Flynn, ed., *Cephalosporins and Penicillins, Chemistry and Biology,* Academic Press, Inc., New York, 1972.
11. M. Tsuji and Y. Yamada in S. Mitsuhashi, ed., *β-Lactam Antibiotics,* Japan Scientific Society's Press, Tokyo, 1981, p. 235.
12. S. K. Branch, A. F. Casy, and E. M. A. Ominde, *J. Pharm. Biomed. Anal.* **5,** 73 (1987).
13. J. R. Everett, J. W. Tyler, and G. Woodnutt, *J. Pharm. Biomed. Anal.* **7,** 397 (1989).
14. S. K. Branch, A. F. Casy, A. Lipczynski, and E. M. A. Ominde, *Magn. Reson. Chem.* **24,** 465 (1986).
15. M. Barber and co-workers, *Biomed. Mass Spectrom.* **9,** 11 (1982).
16. A. Balsamo, P. Domiano, B. Macchia, F. Macchia, and N. Nardelli, *Eur. J. Med. Chem.* **15,** 559 (1980); N. C. Cohen, *J. Med. Chem.* **26,** 259 (1983).
17. R. D. G. Cooper and D. O. Spry in Ref. 10, p. 181.
18. F. A. Jung, W. R. Pilgrim, J. P. Poyser, and P. Sirel in P. G. Sammes, ed., *Topics in Antibiotic Chemistry,* Vol. 4, John Wiley & Sons, Inc., New York, 1980, Chapt. 4.
19. Y. Hamashima and co-workers, in G. I. Gregory, ed., *Recent Advances in the Chemistry of β-Lactam Antibiotics,* Special Publication No. 38, The Royal Society of Chemistry, 1980, p. 57.
20. I. Ernest and co-workers, *J. Am. Chem. Soc.* **100,** 8214 (1978).
21. M. Foglio, C. Battistini, F. Zarini, C. Scarafile, and G. Francheschi, *Heterocycles* **20,** 1491 (1983).
22. A. Alfonso, A. K. Ganguly, V. Girijavallabhan, and S. McCombie in A. G. Brown and S. M. Roberts, eds., *Recent Advances in the Chemistry of β-Lactam Antibiotics,* The Royal Society of Chemistry, Special Publication No. 52, 1984, p. 266.
23. M. Aratani, K. Sawada, and M. Hashimoto, *Tetrahedron Lett.,* 3921 (1982); H. Maruyama and T. Hiraoka, *J. Org. Chem.* **51,** 399 (1986).
24. H. Tobiki and co-workers, *Yakugaku Zasshi* **100,** 49 (1980). *Chem. Abstr.* **93,** 95179x (1980).
25. J. S. Kaltenbronn and co-workers, *J. Antibiot.* **32,** 621 (1979).
26. S. Nakamura and co-workers, *18th Interscience Conference on Antimicrobial Agents and Chemotherapy,* Atlanta, Ga., 1978, Abstract No. 30.
27. P. C. Fuchs and co-workers, *J. Antibiot.* **30,** 1098 (1977).
28. Y. Suzuki and co-workers, *J. Antibiot.* **32,** 711 (1979).
29. T. Nishino, N. Ishii, T. Tanino, S. Ohshima, and T. Yamaguchi, *19th Interscience Conference on Antimicrobial Agents and Chemotherapy,* Boston, Mass., 1979, Abstr. No. 241.
30. R. Gottschlich, V. Rudolph, H. Wahling, and E. Poetsh, *20th Interscience Conference on Antimicrobial Agents and Chemotherapy,* New Orleans, La., 1980, Abstr. No. 146.
31. B. Wetzel and co-workers in Ref. 19, p. 26.
32. P. C. A. Chapman and co-workers, *29th Interscience Conference on Antimicrobial Agents and Chemotheraphy,* Abstracts 89–91, Houston, Tex., 1989.
33. J. E. Baldwin, F. J. Urban, R. D. G. Cooper, and F. L. Jose. *J. Amer. Chem. Soc.* **95,** 2401 (1973); G. A. Koppel and R. E. Koehler, *J. Amer. Chem. Soc.,* **95,** 2403 (1973); R. A. Firestone and B. G. Christensen, *J. Org. Chem.* **38,** 1436 (1973); W. A. Slusarchyk and co-workers, *J. Org. Chem.* **38,** 943 (1973).
34. B. Slocombe and co-workers, *Antimicrob. Agents Chemother.* **20,** 38 (1981).
35. R. J. Ponsford, D. J. Merrikin, B. Slocombe, M. J. Basker, and T. C. G. Tasker, *23rd Interscience Conference on Antimicrobial Agents and Chemotherapy,* Las Vegas, Nev., 1983, Abstr. No. 415; M. J. Basker, D. J. Merrikin, R. J. Ponsford, B. Slocombe, and T. C. G. Tasker, *J. Antimicrob. Chemother.* **18,** 399 (1986).
36. G. Burton, D. J. Best, R. A. Dixon, R. F. Kenyon, and A. G. Lashford, *J. Antibiot.* **39,** 1419 (1986).
37. D. J. Best, G. Burton, F. P. Harrington, and J. Lowther, *J. Antibiot.* **39,** 483 (1986).

38. R. J. Ponsford and co-workers, in Ref. 22, p. 32; M. J. Basker and co-workers, *Antimicrob. Agents Chemother.* **26,** 734 (1984).
39. T. Jen, J. Frazee, and J. R. E. Hoover, *J. Org. Chem.* **38,** 2857 (1973).
40. P. Feyen and W. Schröck, *Angew. Chem. Int. Ed.* **20,** 808 (1981).
41. P. S. F. Mézes, R. W. Friesen, T. Viswanatha, and G. I. Dmitrienko, *Heterocycles* **19,** 1207 (1982).
42. C. L. Branch, M. J. Pearson, and T. C. Smale, *J. Chem. Soc. Perkin 1,* 401 (1988).
43. A. V. Stachulski, *J. Chem. Soc. Chem. Commun.* 401 (1986).
44. N. F. Osborne, *J. Chem. Soc., Perkin 1,* 146 (1980).
45. Ref. 44, p. 150.
46. M. J. Driver and co-workers, *J. Antibiot.* **37,** 297 (1984).
47. M. J. Driver, J. Lowther, and A. W. Taylor, *J. Antibiot.* **38,** 550 (1985).
48. N. F. Osborne, *J. Chem. Soc., Perkin 1,* 1435 (1982).
49. D. D. Keith, J. Tengi, P. Rossman, L. Todaro, and M. Weigele, *Tetrahedron* **39,** 2445 (1983).
50. J. S. Chapman and N. H. Georgopapadakou, *Antimicrob. Agents Chemother.* **31,** 1994 (1987).
51. J. Christenson and co-workers in P. H. Bentley and R. Southgate, eds., *Recent Advances in the Chemistry of β-Lactam Antibiotics,* The Royal Society of Chemistry, Special Publication No. 70, 1988 p. 33; J. G. Christenson, D. L. Pruess, M. K. Talbot, and D. D. Keith, *Antimicrob. Agents Chemother.* **32,** 1005 (1988).
52. N. F. Georgopapadakou ánd co-workers, *Antimicrob. Agents Chemother.* **31,** 1069 (1987).
53. J. E. Baldwin and co-workers, *Biochem. J.* **245,** 831 (1987).
54. G. Burton, C. R. Dobson, J. R. Everett, *J. Pharm. Pharmacol.* **38,** 758 (1986).
55. E. A. Cutmore and co-workers, *J. Chem. Soc. Chem. Commun.* 21 (1987).
56. M. J. Basker and co-workers, *Antimicrob. Agents Chemother.* **26,** 734 (1984).
57. M. F. Parry, *Med. Clin. North Am., Update on Antibiotics 1, The Penicillins* **71,** 1093 (1987).
58. J. E. Baldwin in Ref. 51, p. 1; R. D. G. Cooper in Ref. 51, p. 196; J. E. Baldwin and E. P. Abraham, *Natural Product Rpts.* **5,** 129 (1988).
59. S. M. Samson and co-workers, *Nature* **318,** 191 (1985).
60. W. B. Pratt and R. Fekety, eds., *The Antimicrobial Drugs,* Oxford University Press, London, 1986.
61. B. W. Bycroft and R. E. Shute, *Pharm. Res.* **1,** 3 (1985).
62. B. G. Spratt and K. D. Cromie, *Revs. Inf. Dis.* **10,** 699 (1988).
63. J. B. Bartolone, G. J. Hite, J. A. Kelly, and J. R. Knox in Ref. 22, p. 318; J. R. Knox and J. A. Kelly, in S. M. Roberts, ed., *Chemical and Biochemical Problems in Molecular Recognition,* The Royal Society of Chemistry, London, 1989; J. A. Kelly, J. R. Knox, H. Zhao, J.-M. Frère, and J-M. Ghuysen, *J. Mol. Biol.* **209,** 281 (1989).
64. A. A. W. Long, J. H. C. Nayler, H. Smith, T. Taylor, and N. Ward, *J. Chem. Soc.* **(C)**, 1920 (1971).
65. J. O. Neway, ed., *Fermentation Process Development of Industrial Organisms,* Marcel Dekker Inc., New York, 1989.
66. Scrip No. **1238,** 24 (1987).
67. *Synthetic Organic Chemicals, United States Production and Sales 1987,* U.S. International Trade Commission Publication 2118, U.S. Government Printing Office, Washington, D.C., 1987.
68. Scrip No. **1206,** 24 (1987).

R. J. PONSFORD
SmithKline Beecham Pharmaceuticals

LINCOSAMINIDES

Lincomycin [*154-21-2*] (**1**, R = OH, R′ = H), $C_{18}H_{34}N_2O_6S$, the first lincosaminide antibiotic to which a structure was assigned, is defined chemically as methyl 6,8-dideoxy-6-(1-methyl-*trans*-4-propyl-L-pyrrolidin-2-ylcarbonylamino)-1-thio-D-erythro-D-galacto-octopyranoside. Both lincomycin and the semisynthetic clindamycin [*18323-44-9*] (**1**, R = H, R′ = Cl), $C_{18}H_{33}ClN_2O_5S$, are widely used in clinical practice. The trivial name of the sugar fragment of this antibiotic, methyl α-thiolincosaminide, has lent itself to the other members of this family, whether produced as secondary metabolites of soil microorganisms or derived semisynthetically by chemical modification. Thus celesticetin [*2520-21-0*] (**2**, R = 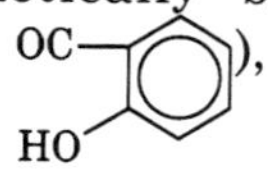), $C_{24}H_{36}N_2O_9S$, and desalicetin [*19246-70-9*] (**2**, R = H), $C_{17}H_{32}N_2O_7S$, are also lincosaminides.

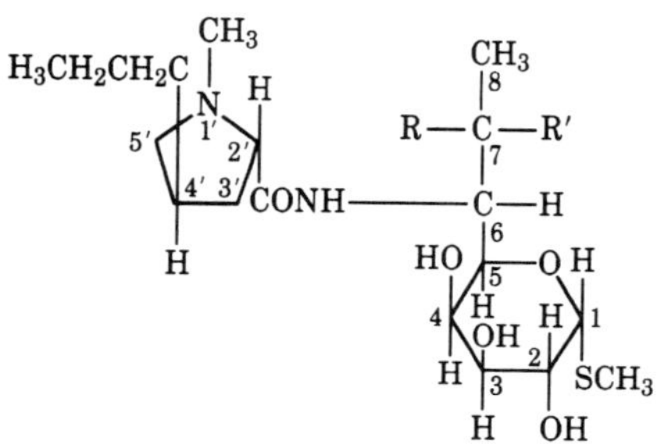

(**1**)

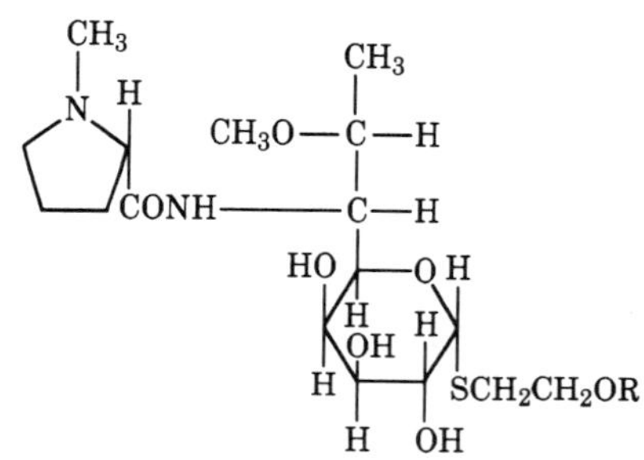

(**2**)

Lincomycin

The discovery and biological properties of lincomycin (**1**, R = OH, R′ = H) were described in 1962 (1). This antibiotic is active *in vitro* and *in vivo* against most of the common gram-positive pathogens. Resistance by Staphylococci is developed slowly in a stepwise manner, based on *in vitro* serial subculture experiments, and its activity is not influenced by body fluids up to concentrations of 50% in the assay medium (2).

Lincomycin was produced originally by the actinomycete *Streptomyces lincolnensis* var. *lincolnensis,* NRRL 2936. Subsequently, it has been produced by a variety of *Streptomyces* strains (3–8) and by strain 1146 of *Actinomyces roseolus* (9). Extraction by standard procedures using *Sarcina lutea* as the assay organism on agar trays (10) leads to a crystalline hydrochloride having molecular formula $C_{18}H_{34}N_2O_6S{\cdot}HCl{\cdot}\frac{1}{2}\,H_2O$ (11). A second, more dense form exists as the monohydrate (12), and this latter form of lincomycin hydrochloride is the Upjohn trademarked Lincocin available commercially. This salt is highly soluble in water, and moderately soluble in methanol and ethanol. It has one basic function, $pK_a = 7.5$, and the specific rotation $[\alpha]_D^{25} = +137°$ (1.0 mg/mL in H_2O). The structure of lincomycin was determined by classical chemical degradation studies, together with the interpretation of proton magnetic resonance and mass spectra (13–18).

Biosynthesis. The terminal *C*-methyl of the propyl side chain, the *S*-methyl, and the *N*-methyl groups are derived from methionine (19). *trans*-4-Propyl-L-proline [*31101-27-6*] was shown to accumulate when *Streptomyces lincolnensis* is grown in media deficient in sulfur, and the addition of L-tyrosine or L-dihydroxyphenylalanine (DOPA) was shown to stimulate this production. From a comparison of the incorporation of label from L-[1-^{14}C]tyrosine and L[U-^{14}C]tyrosine, it was demonstrated that all of the carbon atoms of the propylproline other than the *C*-terminal methyl group were derived from tyrosine and, by the use of L-[^{15}N]tyrosine, it was determined that the proline nitrogen atom was derived from the amino–nitrogen atom of tyrosine (20).

Later work (21), using ^{2}H- and ^{13}C-labeled precursors, in combination with ^{13}C nmr and mass spectral analyses, showed that glucose is converted into tyrosine and thence into DOPA which, via an initial 2,3-extradiol cleavage, cyclization, loss of two carbon atoms, and introduction of a *C*-methyl group from *S*-adenosylmethionine, gives propylproline. The biosynthesis of methyl α-thiolincosaminide [*14810-93-6*], $C_9H_{19}NO_5S$, has also been demonstrated to proceed from glucose via a C-5 and a C-3 unit, using both specifically ^{13}C-labeled substrates and uniformily labeled D-(^{13}C) glucose followed by the analysis of the ^{13}C spin coupling patterns (22).

Mechanism of Action. The earliest studies on the mechanism of action of lincomycin showed that lincomycin had the immediate effect on *Staphylococcus aureus* of complete inhibition of protein synthesis (23). This inhibition results from the blocking of the peptidyltransferase site of the 50S subunit of the bacterial ribosome (24). Little effect on DNA and RNA synthesis was observed.

Resistance to Lincomycin. Resistance to lincomycin is developed slowly, and is usually caused by modification of 23S ribosomal RNA, which leads to co-resistance to macrolide, lincosaminide, and streptogramin B antibiotics (25). Inactivation of lincomycin by clinical isolates of strains of *Staphylococcus aureus* and *Staphylococcus haemolyticus,* though retention of sensitivity to macrolides (see ANTIBIOTICS, MACROLIDES) and streptogramins (see ANTIBIOTICS, PEPTIDES), has been found to be the consequence of the conversion of the antibiotic into its 3-(5′-adenylate) (26).

Pharmacology and Uses. Lincomycin hydrochloride (Lincocin) is available in oral dosage forms and as a sterile solution for injection. After oral administration to adults, mean peak serum levels were achieved between 2 and 6 h. Single 500 and 1000 mg doses gave peak levels of 1.8–5.3 and 2.5–6.7 μg/mL, respectively (27–31). The half-life is 4.2–5.4 h (32). Intramuscular administration gave peak levels at 0.5–2 h. After 300 and 600 mg doses, mean peak levels were 11.7–15 and 9.3–18.5 μg/mL, respectively (27,28,31,33). Lincomycin activity is widely distributed in human body tissues and fluids. Significant concentrations were found in bile, peritoneal fluid, pleural fluid, the eye, bone, and brain (31,34–37). It is excreted both in the urine and in the feces (27–31).

Lincomycin has found use in the treatment of diseases of the ear, throat, nose, respiratory tissue, skin and soft tissue, bone, joint, dental, and septicemic infections caused by staphylococci, pneumonococci, and streptococci (other than enterococci). It has also been used in the treatment of diphtheria and a variety of anaerobic infections, including actinomycosis (38).

Celesticetin and Other Lincosaminide Metabolites

The production (39) and isolation (40) of the antibiotic celesticetin (**2**, R = 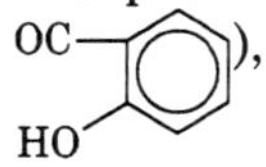), a salicylate ester of the β-hydroxyethylthio-substituent was reported as early as the 1950s, although its structure was not determined until 1968 (41). Reexamination of the fermentation showed the presence of desalicetin (**2**, R = H) and a variety of other β-hydroxyethylthio esters, together with similar esters of 7-de-*O*-methyldesalicetin [*39032-07-0*], (**3**) $C_{16}H_{30}N_2O_7S$ (42,43). One auxotrophic mutant of *S. caelestis* produced 7-de-*O*-methylcelesticetin [*39032-05-8*] (**4**, R = CH_3, R′ = H), $C_{23}H_{34}N_2O_9S$ (44), whereas a second produced both de-*N*-methylcelesticetin [*40736-31-0*] (**4**, R = H, R′ = CH_3), $C_{23}H_{34}N_2O_9S$, and de-*N*-methyl-7-de-*O*-methylcelesticetin, (**4**, R = R′ = H), $C_{22}H_{32}N_2O_9S$ (45).

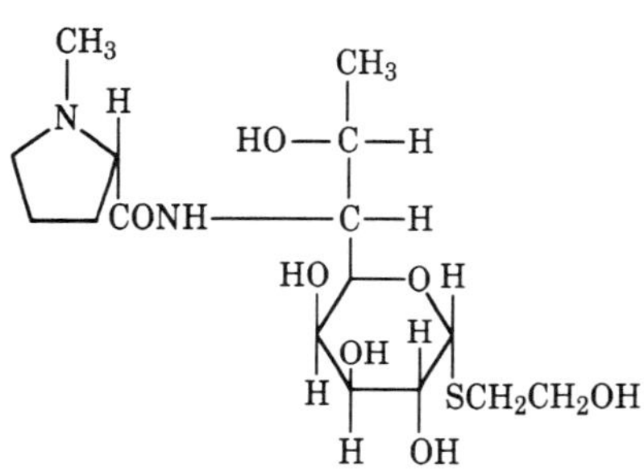

(**3**)

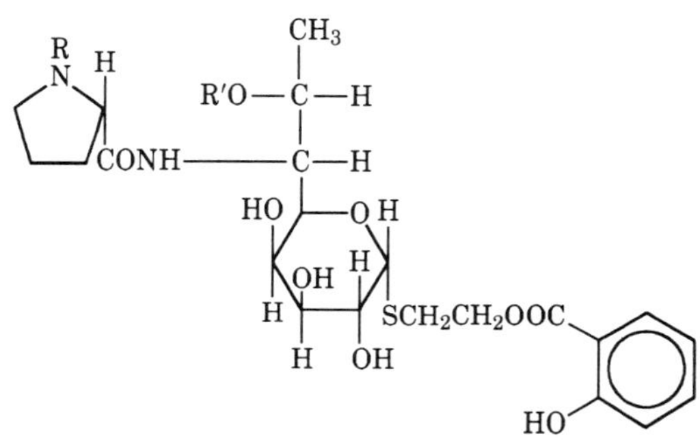

(**4**)

Antibiotic Bu-2545, 7-*O*-methyl-4′-depropyllincomycin [*75007-09-9*] (**1**, R = OCH_3, R′ = H but lacking the 4′-propyl group), $C_{16}H_{30}N_2O_6S$, produced by *Streptomyces* strain No. H 230-5, possesses structural features in common with both celesticetin and lincomycin (46,47).

Accompanying lincomycin in *S. lincolnensis* fermentations is a small amount of the analogue 4′-depropyl-4′-ethyllincomycin [*2520-24-3*], $C_{17}H_{32}N_2O_6S$ (48), which has considerably lower antibacterial activity than the parent compound. Extension of the normal six-day fermentation to twelve days resulted in the formation of lincomycin sulfoxide [*23444-78-2*], $C_{18}H_{34}N_2O_7S$, and 1-demethylthio-1-hydroxylincomycin [*22099-01-0*], (lincomycose) $C_{17}H_{32}N_2O_7$, (**5**, R = CH_3, R′ = H, R″ = OH) (49), both of which have greatly reduced antibacterial activity. The addition of D,L-ethionine and of methyl α-thiolincosaminide, the sugar fragment of lincomycin, produces the *S*-ethyl analogue of lincomycin, [*14042-43-4*], $C_{19}H_{36}N_2O_6S$ (50,51) and de-*N*-methyllincomycin [*16843-70-2*] (**5**, R = R′ = H, R″ = SCH_3), $C_{17}H_{32}N_2O_6S$, (52) respectively, whereas the addition of ethyl α-thiolincosaminide [*6734-39-0*], $C_{10}H_{21}NO_5S$, the sugar fragment of the *S*-ethyl analogue, produces the de-*N*-methyl-*S*-ethyl analogue (**5**, R = R′ = H, R″ = SC_2H_5), $C_{18}H_{34}N_2O_6S$ (50).

Rapid inactivation of added lincomycin was found to result from the growth of *Streptomyces rochei* in a synthetic medium; the antibiotic was converted into lincomycin 3-phosphate [*23670-99-7*] (**5**, R = CH_3, R′ = PO_3H_2, R″ = SCH_3), $C_{18}H_{35}N_2O_9PS$, readily cleaved back to the antibiotic upon treating with alkaline phosphatase (53).

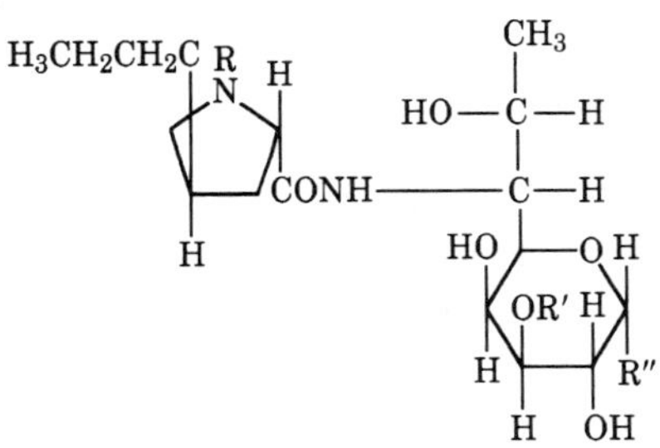

(**5**)

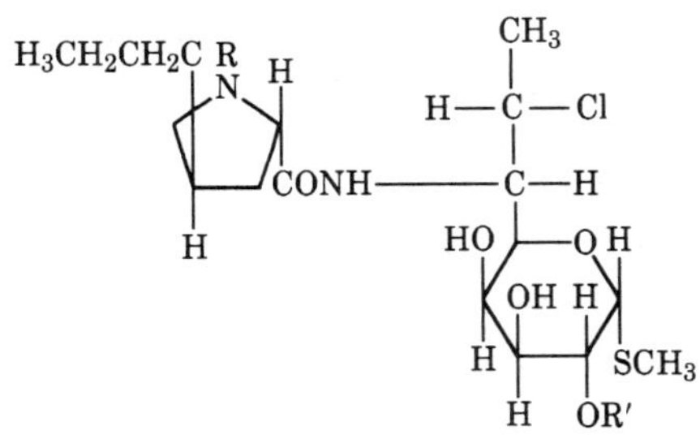

(**6**)

Clindamycin

Clindamycin, 7(*S*)-7-chloro-7-deoxylincomycin [*18323-44-9*], (**1**, R = H, R′ = Cl), also known as Cleocin, first resulted from the reaction of lincomycin and thionyl chloride (54); improved synthetic methods involve the reaction of lincomycin and triphenylphosphine dichloride or triphenylphosphine in carbon tetrachloride (55). Clindamycin is significantly more active than lincomycin against gram-positive bacteria *in vitro,* and is absorbed rapidly following oral administration. Clindamycin 2-palmitate [*36688-78-5*], (**6**, R = CH_3, R′ = $OC(CH_2)_{14}CH_3$), $C_{34}H_{63}ClN_2O_8S$, the 2-palmitate ester of clindamycin, is tasteless and antibacterially inactive. However, following oral administration, esterase cleavage occurs to give good blood levels of clindamycin (56), and this ester has been developed as a pediatric formulation of the antibiotic (Cleocin Pediatric). Given intramuscularly, clindamycin hydrochloride causes pain at the site of injection; the 2-phosphate ester (**6**, R = CH_3, R′ = PO_3H_2) (57) (Cleocin Phosphate), antibacterially inactive, is much better tolerated on injection, and is hydrolyzed by phosphatase, giving good levels of clindamycin.

Pharmacology and Uses. Oral administration of clindamycin hydrochloride gives peak serum levels of 1.9–2.7, 3.6, and 5.6 μg/mL one hour after single doses of 150, 300, and 450 mg, respectively. The half-life is 2.0–3.8 h (58–61). Clindamycin palmitate was also absorbed rapidly, following oral administration, but serum concentrations were lower than after clindamycin hydrochloride (58). Intramuscular injection of single 300, 450, and 600 mg doses of clindamycin phosphate gave mean peak serum levels of 3.8–4.9, 5.3, and 6.2–6.3 μg/mL respectively, 2–4 h after administration (59,62,63). Applied topically, clindamycin 2-phosphate [*24729-96-2*] (Cleocin T), $C_{18}H_{34}ClN_2O_8PS$, is effective in the treatment of acne vulgaris (64). Clindamycin, like lincomycin, is distributed widely in human body tissues, except that no significant levels are found in the brain or eye (38).

Antibacterial activity of clindamycin is found both in urine and feces after administration of clindamycin. This activity is a consequence of the presence of both clindamycin and its metabolite, de-*N*-methylclindamycin [*22431-45-4*] (**6**, R = R′ = H). Unlike de-*N*-methyllincomycin, the de-*N*-methyl analogue is as active *in vitro* as clindamycin. The analogue has been isolated from the urine of humans who had received clindamycin, and its presence in serum has been detected (65).

Clindamycin has found use in the treatment of common infections caused by gram-positive cocci. It is also efficacious in the treatment of anaerobic infections, including actinomycosis (38). Clindamycin has been shown to be active against strains of *Plasmodium* in animals (66–68).

Resistance to Clindamycin. Cross-resistance between lincomycin and clindamycin is complete (64), and co-resistances of lincomycin also apply to clindamycin. However, the inactivation of clindamycin by clinical isolates of *Staphylococcus haemolyticus* and *Staphylococcus aureus* is caused by adenylylation at the 4-position to form clindamycin 4-(5′-adenylate) [*29752-38-3*] (**7**) in contrast to the lincomycin 3-(5′-adenylate) [*117785-83-8*] (**8**) that forms (26).

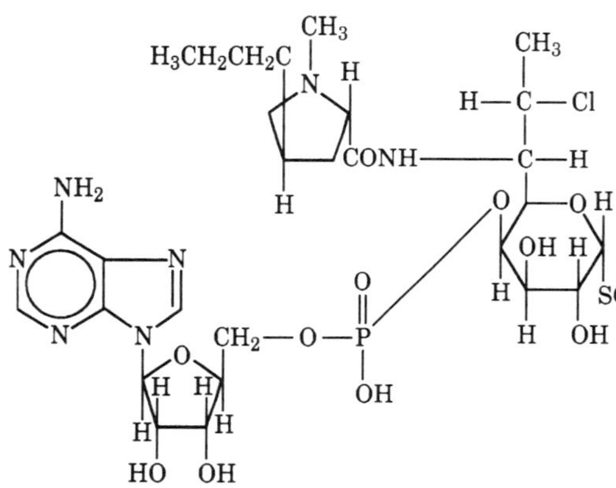

(**7**)

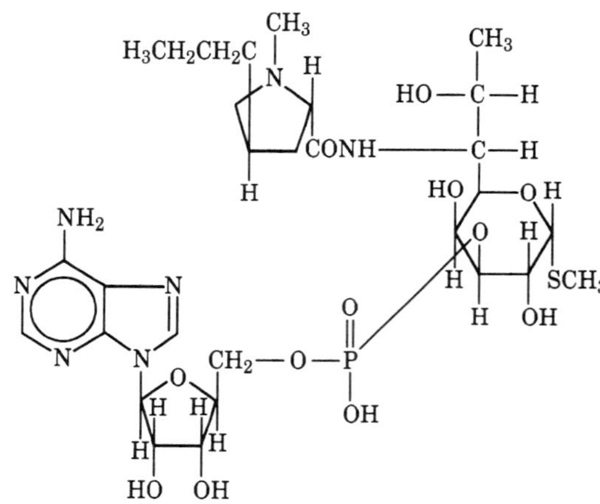

(**8**)

Biomodification of Clindamycin. When added to fermentations of *Streptomyces punipalus,* clindamycin is converted into de-*N*-methylclindamycin (**6**, R = R′ = H). However, when clindamycin is incubated with *Streptomyces armentosus,* clindamycin sulfoxide [*68366-52-9*], $C_{18}H_{33}ClN_2O_6S$, which has low antibacterial activity, is formed (69). Clindamycin 3-phosphate [*28708-34-1*], antibacterially inactive *in vitro,* and the ribonucleotides clindamycin 3-(5′-cytidylate) [*31186-90-0*], clindamycin 3-(5′-adenylate) [*31186-91-1*], clindamycin 3-(5′-uridylate) [*36010-69-2*], and clindamycin 3-(5′-guanylate) [*36010-70-5*], all inactive *in vitro,* can be generated (70). All of these derivatives protect mice infected with *Staphylococcus aureus,* however, presumably because of biotransformation into clindamycin (71–73).

Other Chemically Modified Lincosaminides

Modification of the Thioglycoside. Reaction of lincomycin (**1**, R = OH, R′ = H) and bromine in aqueous solution results in the replacement of the methylthio moiety with a hydroxy group producing lincomycose (**5**, R = CH_3, R′ = H, R″ = OH). Treatment of lincomycose with alkanethiols in the presence of strong acids affords a mixture of dialkyl dithioacetals (**9**) which are inactive and de-*S*-methyl-*S*-alkyllincomycins (**10**), which are equivalent to lincomycin in activity if the alkyl groups are small, but are less active when larger alkyl substituents are present (54). Inversion of the configuration of the methylthio-substituent from α- to β- results in the total loss of activity (74).

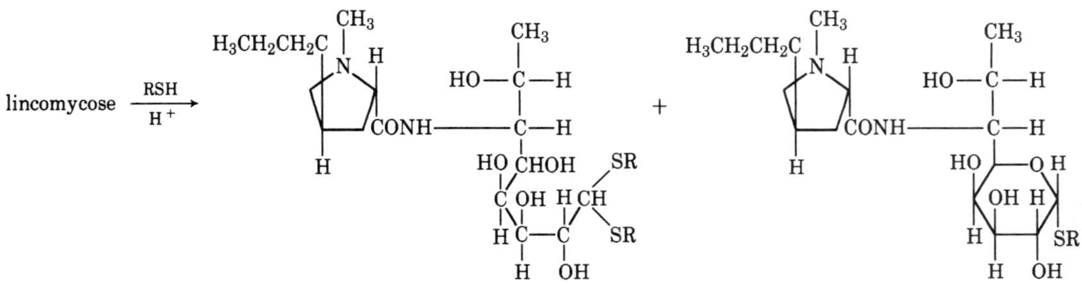

(**9**) (**10**)

Modification of the Carbohydrate Ring Substituents. Replacement of lincomycin's 2-hydroxy group by methoxy or by hydrogen (74), or inversion of the configuration of the hydroxy group at C-4 or at C-2 destroys activity (75).

Modification of the Carbohydrate Side Chain. The importance to the antibacterial activity of the C-7 substituents in lincosaminides has been shown not only by the increased activity of the 7(*S*)-chloro-analogue, clindamycin, (**1**, R = H, R′ = Cl) and of the corresponding 7(*S*)-bromo-7-deoxylincomycin [*18464-22-7*] (**1**, R = H, R′ = Br), $C_{18}H_{33}BrN_2O_5S$, and iodo, $C_{18}H_{33}IN_2O_5S$ [*26390-05-6*] (**1**, R = H, R′ = I) analogues, but also by the minimal activity of 7-deoxylincomycin [*72496-56-1*] (**1**, R = R′ = H), $C_{18}H_{34}N_2O_5S$, and 7-deoxy-7-ketolincomycin [*17057-59-9*] (**1**, R = R′ = O), $C_{18}H_{32}N_2O_6S$, and the diminished activity of 7-epilincomycin [*17017-22-0*] (**1**, R = H, R′ = OH), $C_{18}H_{34}N_2O_6S$, and of the 7(*R*)-7-chloro-deoxylincomycin [*78788-75-7*] (**1**, R = Cl, R′ = H), $C_{18}H_{33}ClN_2O_5S$, derived from it (76). 7-*O*-Methyllincomycin [*39679-82-8*] (**1**, R = OCH_3, R′ = H), $C_{19}H_{36}N_2O_6S$, possesses somewhat enhanced activity (77) and the 7(*S*)-methoxy analogue (**1**,R = H, R′ = OCH_3) is significantly more active than lincomycin, albeit with the same spectrum of activity, but activity decreases rapidly as the size of the alkoxy-substitutent increases (78).

Whereas introducing a thiol moiety at C-7 markedly reduced the antibacterial activity relative to lincomycin (79), the 7(*S*)-7-deoxy-7-alkylthiolincomycins exhibited considerably enhanced antibacterial activity without apparent regard for the size of the alkyl group (80–82). A marked increase in gram-negative activity was shown when the 7(*S*)-substitutent contained a 2- or 3-hydroxy or amino group, but this activity was insufficient to be effective in infected mice (83–85).

Modification of the Proline Fragment. Besides the 4′-ethyl analogue of lincomycin, co-produced in fermentations of *S. lincolnensis* (48), other 4′-alkyl analogues were chemically synthesized. A peak of activity is reached with the *trans*-4′-pentyl and hexyl analogues (**11**), the cis-epimers (**12**) possessing about one-half of the activity of the trans-. Equivalent activities are shown by *N*-ethyl lincosaminides, but larger *N*-alkyl substituents show decreased activity (86).

Similar structure activity relationships were found in the 4′-alkyl analogues of clindamycin (87). The de-*N*-methylclindamycin intermediates including de-*N*-methylclindamycin itself, but unlike de-*N*-methyllincomycin, are also highly active antibacterially. In addition, they are active *in vivo* as antimalarial agents (66–68).

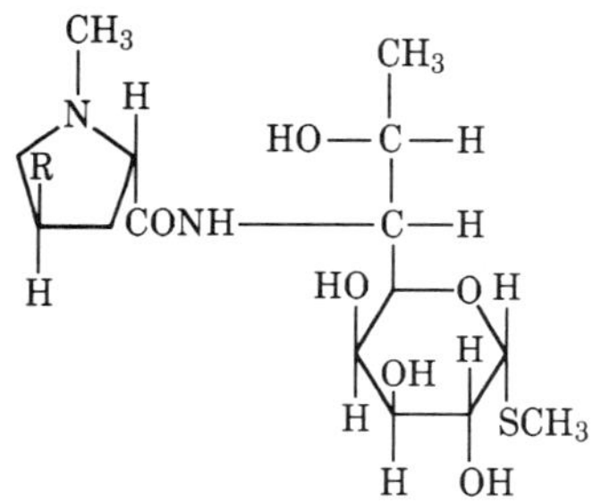

(**11**)

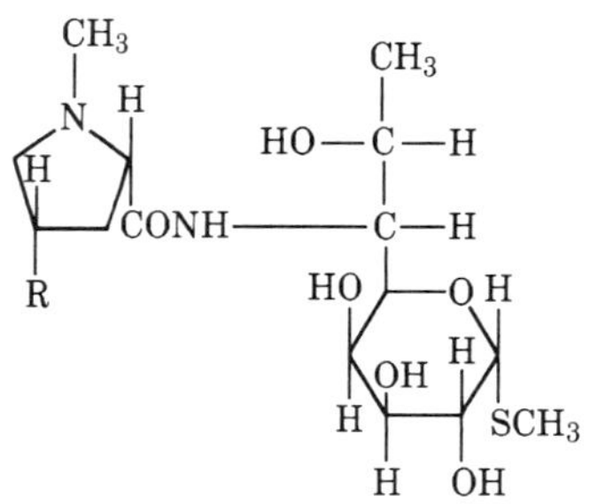

(**12**)

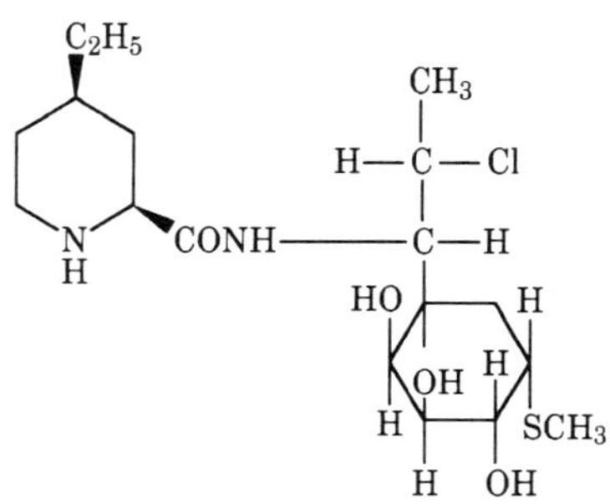

(**13**)

Other Changes in the Amino Acid Fragment

The influence of the size of the cyclic amino acid in analogues of clindamycin has been examined. The (2-(*S*)-*cis*)-4-ethylpipecolic acid analogue (6-ring), $C_{17}H_{31}ClN_2O_5S$, [*78822-40-9*] (**13**) was highly active both *in vitro* and *in vivo,* but the *cis*-(*R*)-isomer [*80081-63-6*] had minimal activity. Significant activity was shown for the azepine (7-ring) analogue, $C_{16}H_{29}ClN_2O_5S$, [*88015-22-9*], but the azetidine (4-ring) analogue, $C_{13}H_{23}ClN_2O_5S$ [*88000-11-7*], showed little activity (88).

Synthesis of the Carbohydrate Fragment of Lincomycin. The first formal synthesis of methyl α-thiolincosaminide started with D-galactose and was reported in 1970 (89). Other syntheses beginning with carbohydrates may also be found in the literature (90–96). Additionally, the total synthesis of methyl β-lincosaminide, albeit racemic, has been reported, utilizing the cyclocondensation of activated dienes and aldehydes, catalyzed by Lewis acids, to yield functionalized pyran rings (97). The synthesis of methyl 1-thio-α-lincosaminide from methyl α-D-galactopyranoside involving an elegant stereocontrolled introduction of the amino–alcohol substituents at C-6 and C-7 has also been reported (98).

Economic Aspects

It is estimated (99) that U.S. sales of clindamycin in 1989 were $45 million. Clindamycin was the seventh most widely used of all prescription products in U.S. hospitals (100). Figures are not available for worldwide sales of clindamycin, or for sales of lincomycin, which are virtually all outside the United States.

The composition of matter patents in the United States issued to The Upjohn Company on clindamycin phosphate and hydrochloride expired at the end of 1986 and in early 1987, respectively. Since then, these compounds have been available generically from more than two dozen companies in the United States alone (101,102).

BIBLIOGRAPHY

"Antibiotics (Lincosaminides)" in *ECT* 3rd ed., Vol. 2, pp. 930–936, by Thomas E. Eble, The Upjohn Co.

1. D. J. Mason, A. Dietz, and C. DeBoer, *Antimicrob. Agents Chemother.* **1962,** 544 (1963).
2. C. Lewis, H. W. Clapp, and J. E. Grady, *Antimicrob. Agents Chemother.* **1962,** 570 (1963).
3. E. L. Patterson, J. H. Hash, M. Lincks, P. A. Miller, and N. Bohonos, *Science* **146,** 1691 (1964).
4. U.S. Pat. 3,697,380 (Oct. 10, 1972), A. D. Argoudelis, J. H. Coats, and T. R. Pyke (to The Upjohn Co.).
5. U.S. Pat. 3,833,475 (Sept. 3, 1974), F. Reusser and A. D. Argoudelis (to The Upjohn Co.).
6. U.S. Pat. 3,726,766 (Apr. 10, 1973), A. D. Argoudelis and J. H. Coats (to The Upjohn Co.).
7. U.S. Pat. 3,812,014 (May 21, 1974), A. D. Argoudelis and J. H. Coats (to The Upjohn Co.).
8. Belg. Pat. 830,068 (Dec. 10, 1975), M. E. Bergy, J. H. Coats, and V. S. Malik (to The Upjohn Co.).
9. USSR Pat. 287,743 (Feb. 18, 1975), G. F. Gause and co-workers (to The University of Moscow).
10. L. J. Hanka, D. J. Mason, M. R. Burch, and R. W. Treick, *Antimicrob. Agents Chemother.* **1962,** 565 (1963).
11. R. R. Herr and M. E. Bergy, *Antimicrob. Agents Chemother.* **1962,** 560 (1963).
12. U.S. Pat. 3,313,695 (Apr. 11, 1967), D. VanOverloop (to The Upjohn Co.).
13. H. Hoeksema, *Abstracts, Papers, 149th Meeting of the American Chemical Society,* Detroit, Mich., 1965, p. 9-C.
14. H. Hoeksema and co-workers, *J. Am. Chem. Soc.* **86,** 4223 (1964).
15. R. R. Herr and G. Slomp, *J. Am. Chem. Soc.* **89,** 2444 (1967).
16. W. Schroeder, B. Bannister, and H. Hoeksema, *J. Am. Chem. Soc.* **89,** 2448 (1967).
17. G. Slomp and F. A. MacKellar, *J. Am. Chem. Soc.* **89,** 2454 (1967).
18. B. J. Magerlein, R. D. Birkenmeyer, R. R. Herr, and F. Kagan, *J. Am. Chem. Soc.* **89,** 2459 (1967).
19. A. D. Argoudelis, T. E. Eble, J. A. Fox, and D. J. Mason, *Biochemistry* **8,** 3408 (1969).
20. D. F. Witz, E. J. Hessler, and T. L. Miller, *Biochemistry* **10,** 1128 (1971).
21. N. M. Brahme and co-workers, *J. Am. Chem. Soc.* **106,** 7873 (1984).
22. N. M. Brahme and co-workers, *J. Am. Chem. Soc.* **106,** 7878 (1984).

23. J. J. Josten and P. M. Allen, *Biochem. Biophys. Res. Commun.* **14,** 241 (1964).
24. E. F. Gale, ed., *The Molecular Basis of Antibiotic Action,* John Wiley & Sons, Inc., 2nd ed., New York, 1981, pp. 478–480.
25. B. Weisblum, *J. Antimicrob. Chemother. Suppl. A,* **16,** 63 (1985).
26. A. Brisson-Noël, P. Delrieu, D. Samain, and P. Courvalin, *J. Biol. Chem.* **263,** 15,880 (1988).
27. J. J. Vavra, W. T. Sokolski, and J. B. Lawson, *Antimicrob. Agents Chemother.* **1963,** 176 (1964).
28. K. Kaplan, W. H. Chew, and L. Weinstein, *Am. J. Med. Sci.* **250,** 137 (1965).
29. C. E. McCall, N. H. Steigbigel, and M. Finland, *Am. J. Med. Sci.* **254,** 144 (1967).
30. R. F. McGehee, Jr., C. B. Smith, C. Wilcox, and M. Finland, *Am. J. Med. Sci.* **256,** 279 (1968).
31. A. Medina, N. Fiske, I. Hjelt-Harvey, C. D. Brown, and A. Prigot, *Antimicrob. Agents Chemother.* **1963,** 189 (1964).
32. J. G. Wagner, J. I. Northam, and W. T. Sokolski, *Nature (London)* **207,** 201 (1965).
33. P. Ma, M. Lim, and J. H. Nodine, *Antimicrob. Agents Chemother.* **1963,** 183 (1964).
34. P. A. Thomas and P. C. Jolly, *Am. Rev. Respir. Dis.* **96,** 1044 (1967).
35. E. F. Becker, *Am. J. Ophthalmol.* **67,** 963 (1969).
36. W. McL. Davis, Jr., *Oral Surg.* **27,** 688 (1969).
37. V. J. Nemecek, M. Hejzlar, and V. Vacek, *Arzneim. Forsch.* **19,** 1760 (1969).
38. R. J. Fass in E. M. Kagan, ed., *Antimicrobial Therapy,* 2nd ed., W. B. Saunders Co., Philadelphia, Pa., 1974, p. 83.
39. C. DeBoer, A. Dietz, J. R. Wilkins, C. Lewis, and G. M. Savage, *Antibiot. Ann.* 831 (1954–1955).
40. H. Hoeksema, G. F. Crum, and W. H. DeVries, *Antibiot. Ann.* 837 (1954–1955).
41. H. Hoeksema, *J. Am. Chem. Soc.* **90,** 755 (1968).
42. A. D. Argoudelis and T. F. Brodasky, *J. Antibiot.* **25,** 194 (1972).
43. A. D. Argoudelis and T. F. Brodasky, *J. Antibiot.* **27,** 642 (1974).
44. A. D. Argoudelis, J. H. Coats, P. G. Lemaux, and O. K. Sebek, *J. Antibiot.* **25,** 445 (1972).
45. A. D. Argoudelis, J. H. Coats, P. G. Lemaux, and O. K. Sebek, *J. Antibiot.* **26,** 7 (1973).
46. M. Hanada, M. Tsunakawa, K. Tomita, H. Tsukiwa, and H. Kawaguchi, *J. Antibiot.* **33,** 751 (1980).
47. S. Toda, S. Nakagawa, T. Naito, and H. Kawaguchi, *J. Antibiot.* **34,** 596 (1981).
48. A. D. Argoudelis, J. A. Fox, and T. E. Eble, *Biochemistry* **4,** 698 (1965).
49. A. D. Argoudelis and D. J. Mason, *J. Antibiot.* **22,** 289 (1969).
50. A. D. Argoudelis, J. A. Fox, D. J. Mason, and T. E. Eble, *J. Am. Chem. Soc.* **86,** 5044 (1964).
51. A. D. Argoudelis and D. J. Mason, *Biochemistry* **4,** 704 (1965).
52. A. D. Argoudelis, J. A. Fox, and D. J. Mason, *Biochemistry* **4,** 710 (1965).
53. A. D. Argoudelis and J. H. Coats, *J. Antibiot.* **22,** 341 (1969).
54. R. D. Birkenmeyer, *Abstract, 5th Interscience Conference on Antimicrobial Agents and Chemotherapy,* 1965, p. 18.
55. R. D. Birkenmeyer and F. Kagan, *J. Med. Chem.* **13,** 616 (1970).
56. A. A. Sinkula, W. Morozowich, and E. L. Rowe, *J. Pharm. Sci.* **62,** 1106 (1973).
57. W. Morozowich, D. J. Lamb, R. M. DeHaan, and J. E. Gray, *Abstracts, Papers, American Pharmaceutical Association Meeting,* Washington, D.C., 1970, p. 63.
58. R. M. DeHaan, C. M. Metzler, D. Schellenberg, W. D. VandenBosch, and E. L. Masson, *Int. J. Clin. Pharmacol.* **62,** 105 (1972).
59. C. M. Metzler, R. M. DeHaan, D. Schellenberg, and W. D. VandenBosch, *J. Pharm. Sci.* **62,** 591 (1973).
60. B. R. Meyers, K. Kaplan, and L. Weinstein, *Appl. Microbiol.* **17,** 653 (1969).

61. J. G. Wagner, E. Novak, N. C. Patel, C. G. Chidester, and W. L. Lummis, *Am. J. Med. Sci.* **256,** 25 (1968).
62. R. M. DeHaan, C. M. Metzler, D. Schellenberg, and W. D. VandenBosch, *J. Clin. Pharmacol.* **13,** 190 (1973).
63. R. J. Fass and S. Saslaw, *Am. J. Med. Sci.* **263,** 369 (1972).
64. J. L. LeFrock, A. Molavi, and R. A. Prince, *Med. Clin. N. Am.* **66,** 103 (1982).
65. T. F. Brodasky, A. D. Argoudelis, and T. E. Eble, *J. Antibiot.* **21,** 327 (1968).
66. C. Lewis, *J. Parasitol.* **54,** 169 (1968).
67. K. G. Powers, *Amer. J. Trop. Med. Hyg.* **18,** 485 (1969).
68. L. H. Schmidt, J. Harrison, R. Ellison, and P. Worcester, *Amer. J. Trop. Med. Hyg.* **19,** 1 (1970).
69. A. D. Argoudelis, J. H. Coats, D. J. Mason, and O. K. Sebek, *J. Antibiot.* **22,** 309 (1969).
70. J. H. Coats and A. D. Argoudelis, *J. Bacteriol.* **108,** 459 (1971).
71. A. D. Argoudelis and J. H. Coats, *J. Am. Chem. Soc.* **93,** 534 (1971).
72. A. D. Argoudelis, J. H. Coats, and S. A. Mizsak, *J. Antibiot.* **30,** 474 (1977).
73. V. P. Marshall, T. E. Patt, and A. D. Argoudelis, *J. Indust. Microbiol.* **1,** 17 (1986).
74. B. Bannister, *J. Chem. Soc. Perkin Trans.* **1,** 3025 (1972).
75. B. Bannister, *J. Chem. Soc. Perkin Trans.* **1,** 3031 (1972).
76. B. J. Magerlein, *Adv. Appl. Microbiol.* **14,** 203 (1971).
77. B. Bannister, *J. Chem. Soc. Perkin Trans.* **1,** 1676 (1973).
78. B. Bannister, *J. Chem. Soc. Perkin Trans.* **1,** 360 (1974).
79. B. J. Magerlein and F. Kagan, *J. Med. Chem.* **12,** 974 (1969).
80. B. Bannister, *J. Chem. Soc. Perkin Trans.* **1,** 1057 (1977).
81. B. Bannister, *J. Chem. Soc. Perkin Trans.* **1,** 274 (1978).
82. B. Bannister, *J. Chem. Soc. Perkin Trans.* **1,** 967 (1978).
83. B. Bannister, *J. Chem. Soc. Perkin Trans.* **1,** 540 (1980).
84. B. Bannister, *Tetrahedron* **40,** 1633 (1984).
85. B. Bannister, *J. Chem. Res.* (*S*) 90 (1989); *J. Chem. Res.* (*M*) 701 (1989).
86. B. J. Magerlein, R. D. Birkenmeyer, and F. Kagan, *J. Med. Chem.* **10,** 355 (1967).
87. B. J. Magerlein and F. Kagan, *J. Med. Chem.* **12,** 780 (1969).
88. R. D. Birkenmeyer, S. J. Kroll, C. Lewis, K. F. Stern, and G. E. Zurenko, *J. Med. Chem.* **27,** 216 (1984).
89. B. J. Magerlein, *Tetrahedron Lett.* **33** (1970).
90. G. B. Howarth, W. A. Szarek, and J. K. N. Jones, *J. Chem. Soc.* (*C*) 2218 (1970).
91. H. Saeki and E. Ohki, *Chem. and Pharm. Bull. (Japan)* **18,** 789 (1970).
92. T. Alsumi, T. Fukumaru, T. Ogawa, and M. Matsui, *Agr. Biol. Chem.* **37,** 2621 (1973).
93. G. R. Woolard, E. B. Rathbone, W. A. Szarek, and J. K. N. Jones, *J. Chem. Soc. Perkin Trans.* **1,** 950 (1976).
94. A. Gateau-Olesker, A. M. Sepulchre, G. Vass, and S. D. Gero, *Tetrahedron* **33,** 393 (1977).
95. I. Hoffe and V. Schöllkopf, *Liebigs Ann. Chem.* 1474 (1980).
96. S. M. David and J.-C. Fischer, *Carbohydrate Res.* **38,** 147 (1974).
97. S. J. Danishefsky, E. Larson, and J. P. Springer, *J. Am. Chem. Soc.* **107,** 1274 (1985).
98. S. Knapp and P. J. Kukkola, *J. Org. Chem.* **55,** 1632 (1990).
99. M. I. Smith, Pharmaceutical Data Services of Scottsdale, Ariz., personal communication, 1990.
100. *Scrip,* No. 1500, March 28th, 1990, p. 18.
101. *American Druggist Blue Book, 1989–1990,* Hearst Corp., New York, 1990, pp. 128, 129.
102. *Drug Topics Red Book, 1989,* Medical Economics Co., Oradell, N.J., 1989, p. 238.

BRIAN BANNISTER
The Upjohn Company

MACROLIDES

Macrolide antibiotics are well established antimicrobial agents in both clinical and veterinary medicine. These agents can be administered orally and are generally used to treat infections in the respiratory tract, skin and soft tissues, and genital tract caused by gram-positive organisms, *Mycoplasma* species, and certain susceptible gram-negative and anaerobic bacteria.

The macrolide class is large and structurally diverse. Macrolides are produced by fermentation of soil microorganisms. Additionally, structural modifications using both chemical and microbiological means have yielded biologically active semisynthetic derivatives. A book devoted to macrolides includes work to 1984 (1). Monographs are available for older macrolides such as erythromycin (2–5) and spiramycin (6) and newer ones such as roxithromycin (7–9), azithromycin (10), and erythromycin acistrate (11). Reviews have covered the field generally (12–15) and specifically with regard to chemical modification (15,16), antimicrobial properties (17,18), and pharmacokinetics and pharmacology (19–23). These complex molecules are also targets for total synthesis (24–26).

The term macrolide was introduced to denote the class of substances produced by *Streptomyces* species containing a macrocyclic lactone ring (27). The generalized structure is a highly substituted monocyclic lactone (aglycone) to which is attached one or more saccharides glycosidically linked to hydroxyl groups on either the aglycone or another saccharide. The aglycones are derived via similar polyketide biosynthetic pathways and thus share many structural features in terms of pattern and stereochemistry of substituents (28). Traditional macrolide antibiotics are divided into three families according to the size of the aglycone which can be 12-, 14-, or 16-membered.

Because one or more aminosugars are usually present, these compounds are basic and can form acid addition salts. In addition, one or more neutral sugars are often present. A few macrolides possess no aminosugar. The saccharides share some common features: they tend to be highly deoxygenated and *N*- and/or *O*-methylated; and the amino groups are located at either positions 3 or 4. The most common sugars found in macrolides are given in Table 1.

12-Membered Ring Macrolides

The few macrolides having 12-membered rings are listed in Table 2. Methymycin (**12**, R = OH, R′ = H), isolated from culture broths of a *Streptomyces* species (29), was the first macrolide structure elucidated (30). It is comprised of the aglycone methynolide (**13**, R = OH, R′ = H) and the aminosugar desosamine (**1**, R = OH, R′ = H) (31,32). Methymycin was also the first conventional macrolide made by total synthesis (33).

Neomethymycin (**12**, R = H, R′ = OH), an isomer co-produced with methymycin, is the product of hydroxylation at C-12 rather than C-10 of the lactone (34,35). The corresponding aglycone, neomethynolide (**13**, R = H, R′ = OH), was isolated with methynolide from broths of *S. venezuelae* (36). The stereochemistry of 12(*R*)- for neomethynolide was established by total synthesis (37). YC-17 (**12**, R = R′ = H), also found in broths of *S. venezuelae,* is a possible

Table 1. Saccharides Found in Macrolide Antibiotics

Saccharide	CAS Registry Number	Molecular formula	Structure number	Structure
		Aminosaccharides		
D-angolosamine	[*14702-57-9*]	$C_8H_{17}NO_3$	(**1**, R = H, R′ = OH)	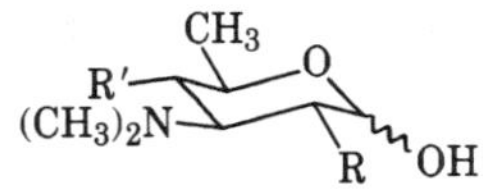
D-desosamine	[*5779-39-5*]	$C_8H_{17}NO_3$	(**1**, R = OH, R′ = H)	
D-mycaminose	[*519-21-1*]	$C_8H_{17}NO_4$	(**1**, R = R′ = OH)	
D-forosamine	[*18423-27-3*]	$C_8H_{17}NO_2$	(**2**)	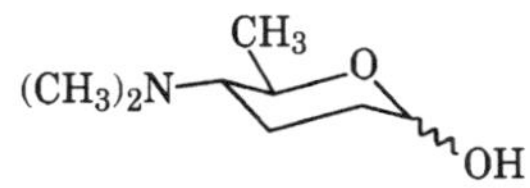
L-megosamine	[*65832-97-5*]	$C_8H_{17}NO_3$	(**3**)	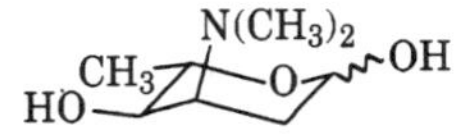
		Neutral saccharides		
D-aldgarose	[*26428-87-5*]	$C_9H_{14}O_6$	(**4**)	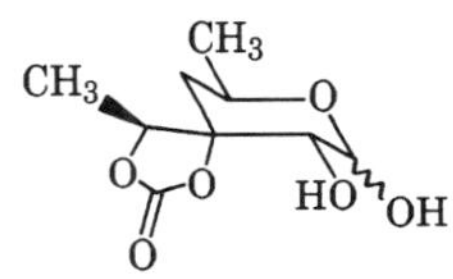
L-amicetose	[*67528-19-2*]	$C_6H_{12}O_3$	(**5**, R = H, R′ = OH)	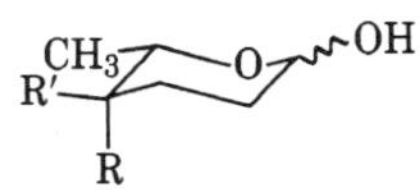
L-cinerulose	[*33985-39-6*]	$C_6H_{10}O_3$	(**5**, R + R′ = O)	
L-rhodinose	[*35903-48-1*]	$C_6H_{12}O_3$	(**5**, R = OH, R′ = H)	
L-arcanose	[*26548-40-3*]	$C_8H_{16}O_4$	(**6**)	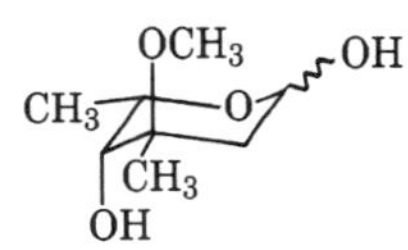
D-chalcose	[*3150-28-5*]	$C_7H_{14}O_4$	(**7**)	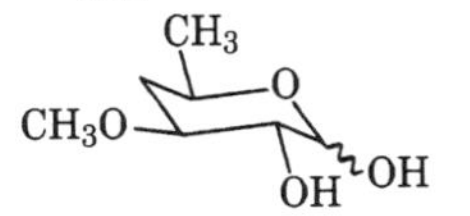
L-cladinose	[*470-12-2*]	$C_8H_{16}O_4$	(**8**, R = CH_3)	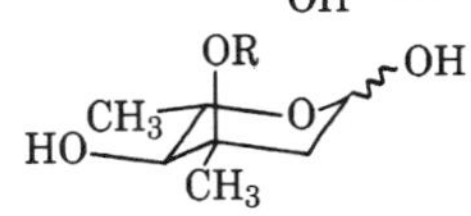
L-mycarose	[*6032-92-4*]	$C_7H_{14}O_4$	(**8**, R = H)	
D-6-deoxyallose	[*4348-84-9*]	$C_6H_{12}O_5$	(**9**, R = R′ = H)	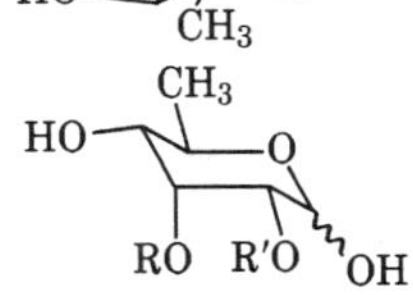
D-javose	[*921-90-4*]	$C_7H_{14}O_5$	(**9**, R = H, R′ = CH_3)	
D-mycinose	[*21967-31-7*]	$C_8H_{16}O_5$	(**9**, R = R′ = CH_3)	
D-4,6-dideoxy-3-dehydroallose		$C_6H_{10}O_4$	(**10**)	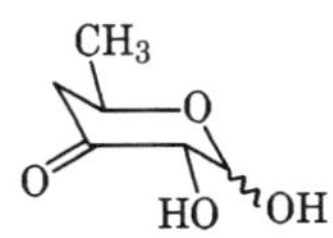
L-oleandrose	[*6786-76-1*]	$C_7H_{14}O_4$	(**11**, R = CH_3)	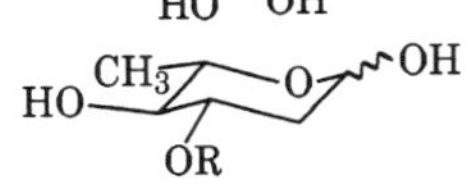
L-olivose	[*25029-50-9*]	$C_6H_{12}O_4$	(**11**, R = H)	

Table 2. Macrolides Having a 12-Membered Ring

Macrolide	CAS Registry Number	Molecular formula	Structure number	Structure
methymycin	[*497-72-3*]	$C_{25}H_{43}NO_7$	(**12**, R = OH, R′ = H)	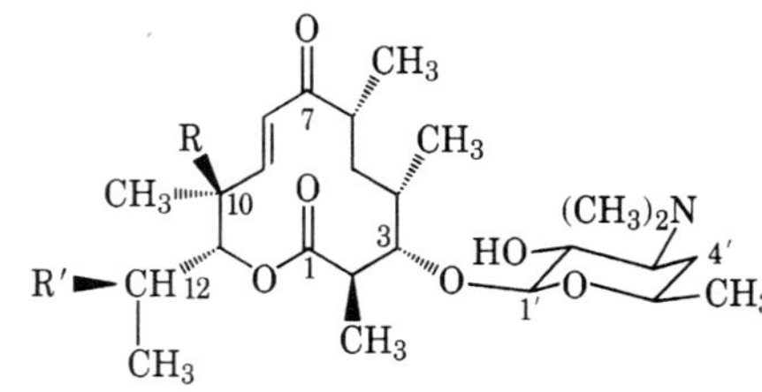
neomethymycin	[*497-73-4*]	$C_{25}H_{43}NO_7$	(**12**, R = H, R′ = OH)	
YC-17	[*11091-33-1*]	$C_{25}H_{43}NO_6$	(**12**, R = R′ = H)	
methynolide	[*534-32-7*]	$C_{17}H_{28}O_5$	(**13**, R = OH, R′ = H)	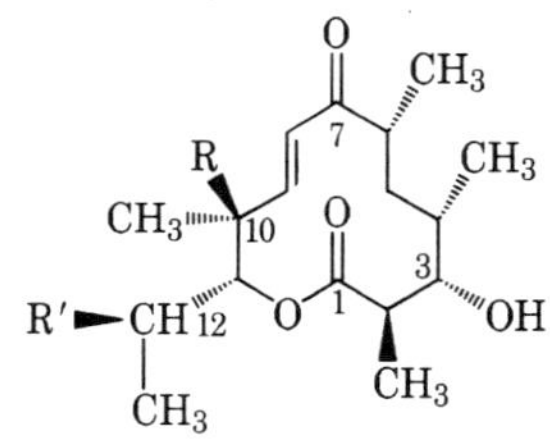
neomethynolide	[*497-51-8*]	$C_{17}H_{28}O_5$	(**13**, R = H, R′ = OH)	

precursor of methymycin and neomethymycin. The hydroxyl groups at C-12 and C-10 are probably added as late steps in the biosynthesis (38).

14-Membered Ring Macrolides

Natural Products. Naturally occurring 14-membered macrolides are given in Table 3.

Erythromycins. Erythromycin A (**14**, R = OH, R′ = CH_3, R″ = H), the most widely used macrolide antibiotic, was the principal product found in culture broths of *Streptomyces erythreus* (39), now reclassified as *Saccharopolyspora erythraea* (40). It contains a highly substituted aglycone, erythronolide A, (**16**, R = R′ = OH) to which desosamine (**1**, R = OH, R′ = H) and cladinose (**8**, R = CH_3) are attached (41). The complete stereochemistry of erythromycin A was established by x-ray analysis of its hydroiodide dihydrate (42); total synthesis of erythromycin A was a landmark achievement (43), a task previously considered hopeless (44).

Several minor factors are co-produced with erythromycin A. Erythromycin B (**14**, R = R″ = H, R′ = CH_3) lacks a hydroxyl group at C-12 (45). Erythromycin C (**14**, R = OH, R′ = R″ = H) contains mycarose (**8**, R = H) as the neutral sugar (46). Erythromycin D (**14**, R = R′ = R″ = H), which simultaneously lacks the C-12 hydroxyl group and contains mycarose (47), plus three of its 3″- and 4″-esters were isolated from culture broths of a *Nocardia* species (48). Erythromycin F (**14**, R = R″ = OH, R′ = CH_3), which contains a 2-hydroxymethyl group (49,50), may be the biosynthetic precursor of erythromycin E (**15**), which possesses an ortho ester linkage (51,52). Several possible biosynthetic intermediates or shunt metabolites of erythromycin have been obtained after single-step fermentations or two-step, precursor-feeding experiments utilizing biosynthetically blocked mutants of the erythromycin-producing organism (53–59). Additional compounds related to erythromycin have been discovered (50,52,60–62).

Chemical degradation of erythromycin A yielded its aglycone, erythronolide A (**16**, R = R′ = OH), whereas erythronolide B (**16**, R = H, R′ = OH) was obtained from fermentation (63,64). Biosynthesis of erythromycin proceeds via 6-deoxyerythronolide B (**16**, R = R′ = H) and then erythronolide B (64,65). The first total synthesis of erythromycin-related compounds was erythronolide B (66); syntheses of erythronolide A and 6-deoxyerythronolide B soon followed (67,68).

Oleandomycin. Oleandomycin (**17**, R = CH_3) was the primary factor in culture broths of *S. antibioticus* (69,70). Its aglycone, oleandolide (**18**, X = $\overset{O}{\blacktriangle}CH_2$), differs from the erythronolides at several centers, most notably by the epoxide at C-8. In addition, the neutral sugar in oleandomycin is oleandrose (**11**, R = CH_3) (71–73). A by-product, oleandomycin Y (**17**, R = H), contains the neutral sugar L-olivose (**11**, R = H) (74). Oleandolide has been obtained by total synthesis and chemical degradation (75,76). The aglycone lacking the epoxide function, 8,8a-deoxyoleandolide (**18**, X = $\overset{H}{\blacktriangle}CH_3$), was produced by a mutant of the erythromycin-producing organism (59).

Pikromycin. Pikromycin (**19**, R = OH, R′ = H), the first macrolide discovered (77,78), is produced by *S. felleus*. Pikromycin is identical to amaromycin and albomycetin (79–81) and may be identical to proactinomycin (82–84). The structure of pikromycin was determined from chemical degradation, mass spectrometry, nmr, and x-ray crystallography (85–90). Its aglycone, pikronolide (**20**, R = OH), was produced by *S. venezuelae* (36). A derivative, kromycin (**22**, R = OH), was formed from pikromycin under acidic conditions (87,88,91) and more drastic conditions produced an intramolecular spiroketal of pikronolide named kromin (89,91). 10,11-Dihydropikromycin (**21**, R = OH, R′ = H), was also produced by *S. venezuelae* (92).

Narbomycin (**19**, R = R′ = H), produced by *S. narbonensis* (93), is 12-deoxypikromycin (94). Its aglycone, narbonolide, (**20**, R = H) is produced by *S. venezuelae* and bioconverted into narbomycin and then pikromycin (95–97). The stereochemistry of narbonolide was established by total synthesis (98). 12-Deoxykromycin (**22**, R = H) was formed from narbonolide in a manner analogous to that yielding kromycin (99). Two derivatives of narbonolide containing mycaminose (**1**, R = R′ = OH) rather than desosamine (**1**, R = OH, R′ = H) were produced by species of *Nocardiopsis:* kayamycin (**21**, R = H, R′ = OH) (100) and 5-*O*-mycaminosylnarbonolide (**19**, R = H, R′ = OH). This latter macrolide and its 9-dihydro derivative had previously been obtained by feeding narbonolide to strains of *S. platensis* (101,102).

Megalomicins. The megalomicins, the first macrolides obtained from *Micromonospora* species (*M. megalomicea*) (103), were later obtained from culture broths of *M. inositola* (104,105). Structures were determined by nmr and x-ray crystallographic studies (106,107). Megalomicins are the only 14-membered macrolides containing two aminosugars: desosamine and a novel aminosugar, megosamine (**3**) glycosidically linked to the 6-hydroxyl group. The megalomicins also differ from erythromycin in that the neutral sugar is either mycarose (**8**, R =

Table 3. Naturally Occurring 14-Membered Macrolides

Macrolide	CAS Registry Number	Molecular formula	Structure number	Structure
Erythromycins and erythronolides				
erythromycin A	*[114-07-8]*	$C_{37}H_{67}NO_{13}$	(**14**, R = OH, R′ = CH_3, R″ = H)	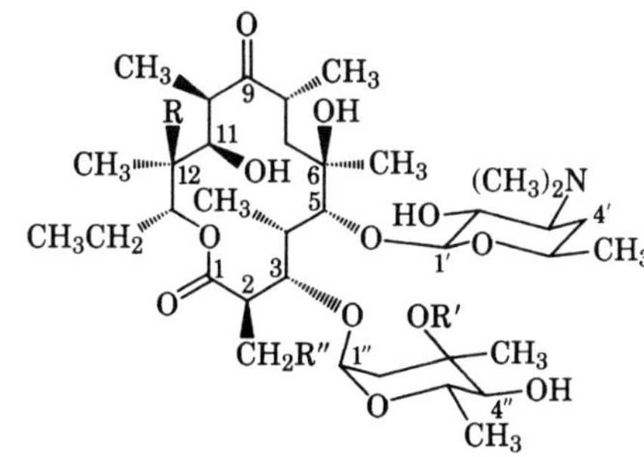
erythromycin B	*[527-75-3]*	$C_{37}H_{67}NO_{12}$	(**14**, R = R″ = H, R′ = CH_3)	
erythromycin C	*[1675-02-1]*	$C_{36}H_{65}NO_{13}$	(**14**, R = OH, R′ = R″ = H)	
erythromycin D	*[33442-56-7]*	$C_{36}H_{65}NO_{12}$	(**14**, R = R′ = R″ = H)	
erythromycin F	*[82230-93-1]*	$C_{37}H_{67}NO_{14}$	(**14**, R = R″ = OH, R′ = CH_3)	
erythromycin E	*[41451-91-6]*	$C_{37}H_{65}NO_{14}$	(**15**)	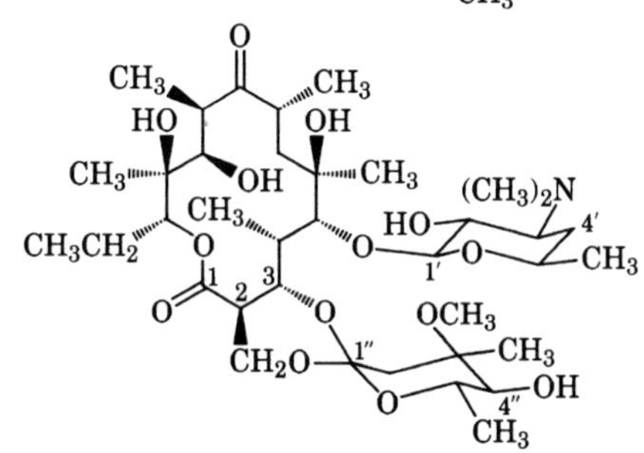
erythronolide A	*[26754-37-0]*	$C_{21}H_{38}O_8$	(**16**, R = R′ = OH)	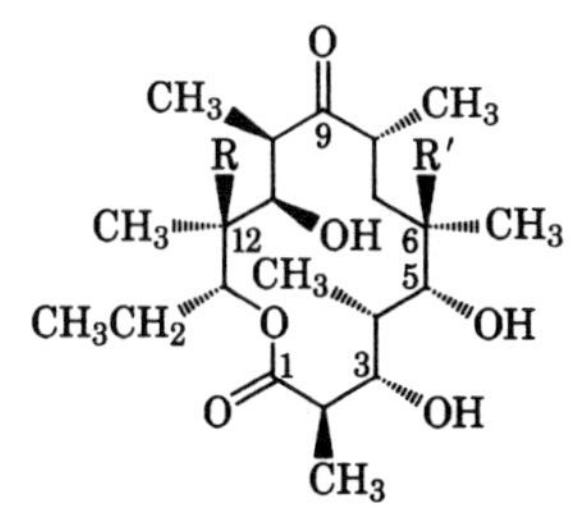
erythronolide B	*[3225-82-9]*	$C_{21}H_{38}O_7$	(**16**, R = H, R′ = OH)	
6-deoxyerythronolide B	*[15797-36-1]*	$C_{21}H_{38}O_6$	(**16**, R = R′ = H)	

Oleandomycin and related compounds

oleandomycin	[*3922-90-5*]	$C_{35}H_{61}NO_{12}$	(**17**, R = CH_3)
oleandomycin Y	[*64743-16-4*]	$C_{34}H_{59}NO_{12}$	(**17**, R = H)

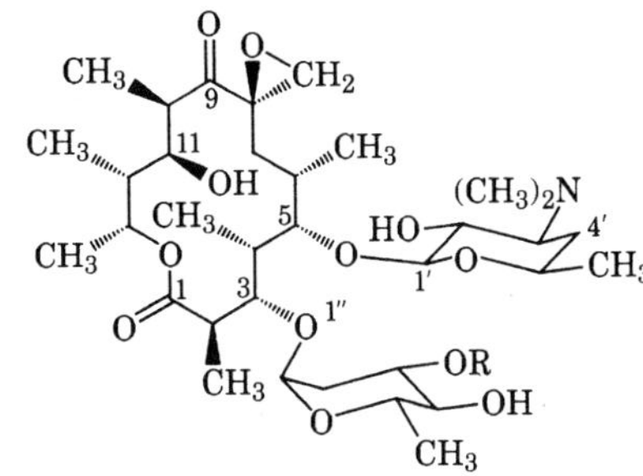

oleandolide	[*68540-16-9*]	$C_{20}H_{34}O_7$	(**18**, X = )
8,8a-deoxyoleandolide	[*53428-54-9*]	$C_{20}H_{36}O_6$	(**18**, X = 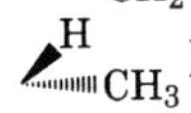)

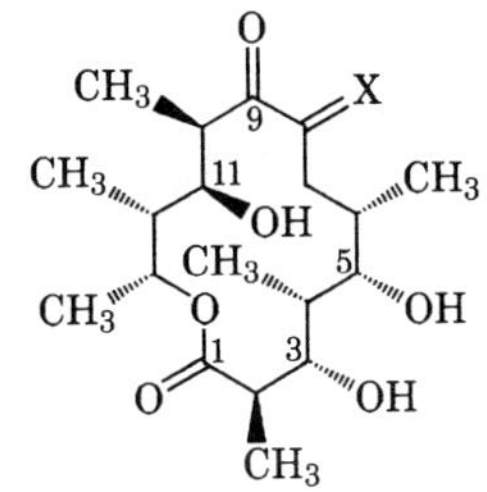

Pikromycin and related compounds

pikromycin	[*19721-56-3*]	$C_{28}H_{47}NO_8$	(**19**, R = OH, R′ = H)
narbomycin	[*6036-25-5*]	$C_{28}H_{47}NO_7$	(**19**, R = R′ = H)
5-*O*-mycaminosylnarbonolide	[*61435-20-9*]	$C_{28}H_{47}NO_8$	(**19**, R = H, R′ = OH)

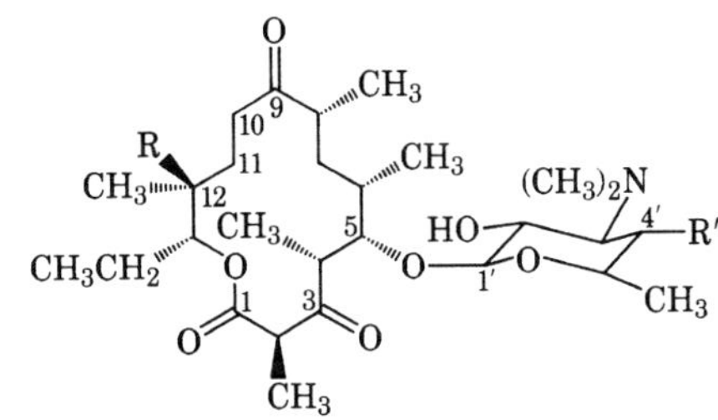

Table 3. (*Continued*)

Macrolide	CAS Registry Number	Molecular formula	Structure number	Structure
pikronolide narbonolide	[*51724-55-1*] [*32885-75-9*]	$C_{20}H_{32}O_6$ $C_{20}H_{32}O_5$	(**20**, R = OH) (**20**, R = H)	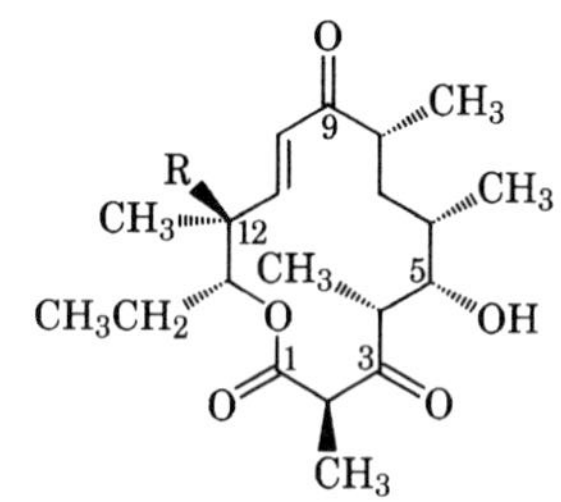
10,11-dihydropikromycin kayamycin[a]	[*27656-56-0*] [*102907-96-0*]	$C_{28}H_{49}NO_8$ $C_{28}H_{49}NO_8$	(**21**, R = OH, R′ = H) (**21**, R = H, R′ = OH)	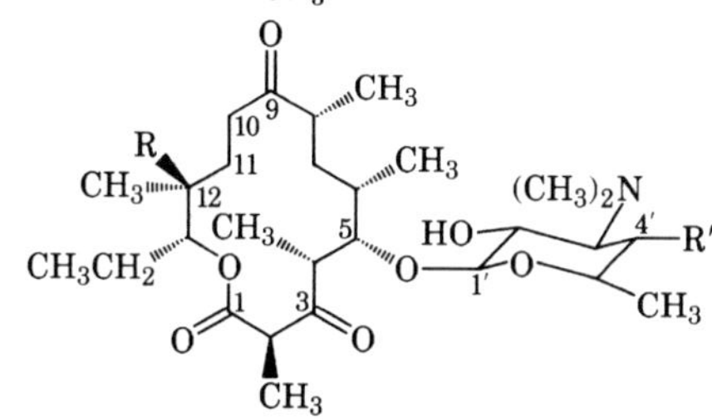
12-deoxykromycin kromycin	[*55527-99-6*] [*20509-23-3*]	$C_{20}H_{30}O_4$ $C_{20}H_{30}O_5$	(**22**, R = H) (**22**, R = OH)	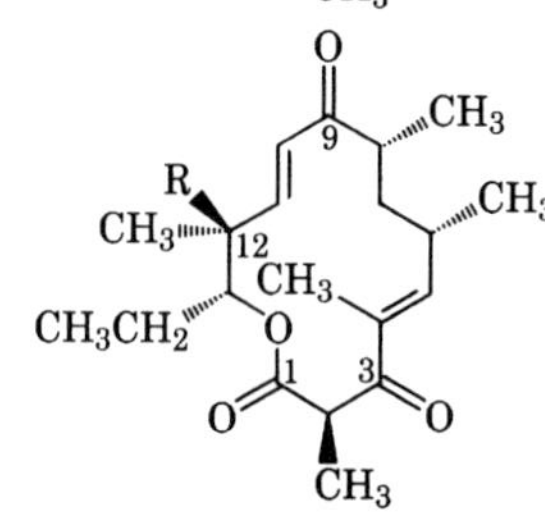

Megalomicins

megalomicin A[b]	*[28022-11-9]*	$C_{44}H_{80}N_2O_{15}$	(**23**, R = R′ = H)
megalomicin B[c]	*[49669-75-2]*	$C_{46}H_{82}N_2O_{16}$	(**23**, R = H, R′ = OCCH$_3$)
megalomicin C$_1$[d]	*[49669-76-3]*	$C_{48}H_{84}N_2O_{17}$	(**23**, R = R′ = OCCH$_3$)
megalomicin C$_2$[e]	*[49669-77-4]*	$C_{49}H_{86}N_2O_{17}$	(**23**, R = OCCH$_3$, R′ = OCCH$_2$CH$_3$)
XK-41-B$_2$	*[50692-34-7]*	$C_{47}H_{84}N_2O_{16}$	(**23**, R = H, R′ = OCCH$_2$CH$_3$)

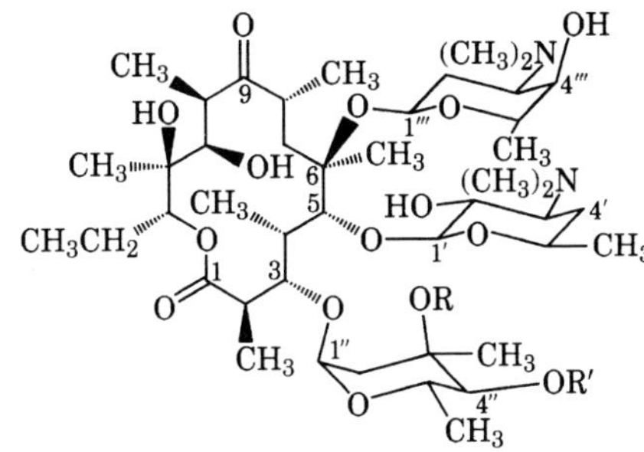

Lankamycin and related compounds

lankamycin (kujimycin B)	*[30042-37-6]*	$C_{42}H_{72}O_{16}$	(**24**, R = OCCH$_3$, R′ = OH, R″ = H)
kujimycin A	*[33955-27-0]*	$C_{40}H_{70}O_{15}$	(**24**, R = H, R′ = OH, R″ = H)
15-dehydrolankamycin	*[65423-08-7]*	$C_{42}H_{70}O_{16}$	(**24**, R = OCCH$_3$, R′ + R″ = O)
15-*O*-(4-*O*-acetylarcanosyl)-lankamycin	*[65423-07-6]*	$C_{52}H_{88}O_{20}$	(**24**, R = OCCH$_3$, R′ = *O*-4-*O*-acetylarcanosyl, R″ = H)

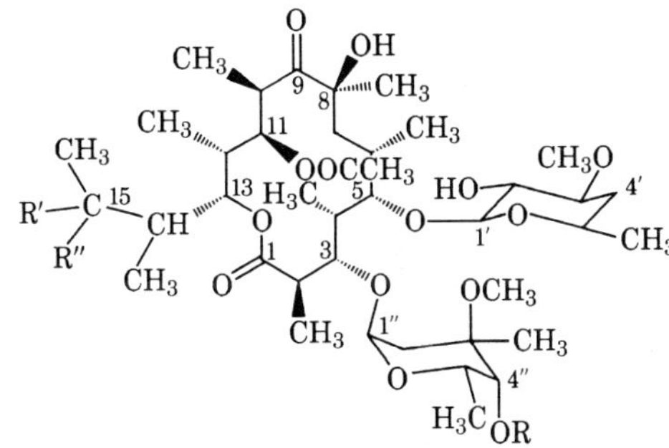

3″-*O*-demethyl-2″,3″-anhydrolankamycin	*[60462-94-4]*	$C_{41}H_{68}O_{15}$	(**25**)

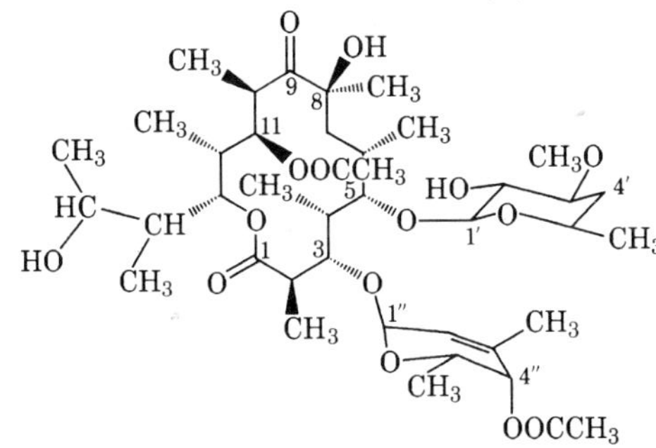

Table 3. (*Continued*)

Macrolide	CAS Registry Number	Molecular formula	Structure number	Structure
23672 RP	[*37359-09-4*]	$C_{48}H_{82}O_{20}$	(**26**, R = α-hydroxyisovaleryl)	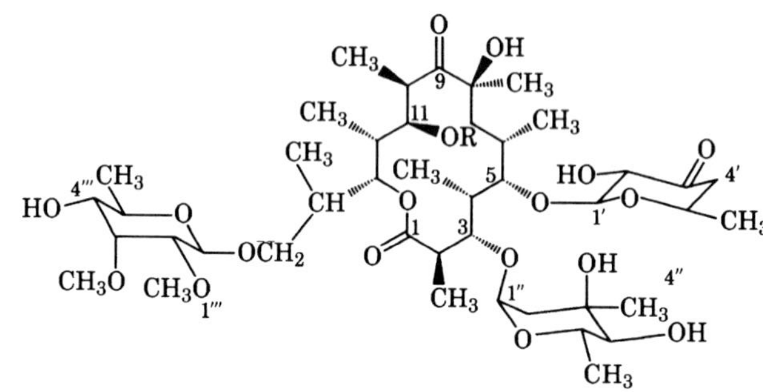

[a]Also known as A1-R6-4.
[b]Also known as XK-41-C.
[c]Also known as XK-41-B$_1$.
[d]Also known as XK-41-A$_2$.
[e]Also known as XK-41-A$_1$.

H) or its acyl derivative instead of cladinose (**8**, R = CH_3). The aglycones of megalomicin and erythromycin A are identical (104).

Lankamycins/Kujimycins. The lankamycin/kujimycin group lack amino-sugars but contain the neutral sugars D-chalcose (**7**) and L-arcanose (**6**) or their derivatives. Lankamycin (**24**, R = $OCCH_3$, R′ = OH, R″ = H) was found in culture broths of *S. violaceoniger* (108) and its structure revised after detailed nmr studies (109,110). Kujimycin A (**24**, R = H, R′ = OH, R″ = H) and kujimycin B, identical to lankamycin, were isolated from broths of *S. spinichromogenes* (111,112). Related factors have been found which possess an unsaturated sugar (**25**), or a C-15 ketone or a 4-*O*-acetylarcanosyl moiety on the C-15 hydroxyl group (113,114). In 23672 RP (**26**) the C-11 hydroxyl group is acylated by an α-hydroxyvaleryl group (115).

Semisynthetic Derivatives. Erythromycin has been the principal subject of modification of 14-membered macrolides; some of the derivatives are being commercially launched (116).

Decomposition of erythromycin under acidic conditions generates erythromycin 8,9-anhydro-6,9-hemiketal [*33396-29-1*] (**27**) and erythromycin 6,9;9,12-spiroketal [*23893-13-2*] (**28**) (117–119) as shown in Figure 1. The ring-contracted product [*105882-69-7*] (**29**), derived from trans-acylation in (**27**), was a minor component in mother liquors from which erythromycin was isolated (60), and was also found in pharmaceutical formulations of erythromycin (50). Syntheses of (**29**) have been published (120,121). A related pair of ring contractions was observed in oleandomycin, yielding a 10-membered bicyclic product (122). In all of these examples, the bicyclic product apparently stabilizes the 10- and 12-membered lactones. In contrast, translactonization of 14-hydroxy-6-*O*-meth-

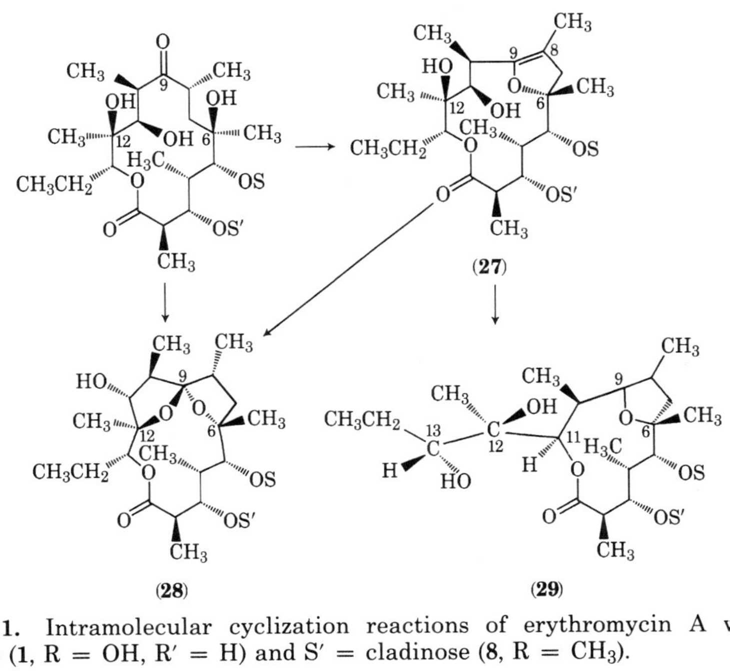

(**27**) (**28**) (**29**)

Fig. 1. Intramolecular cyclization reactions of erythromycin A where S = desosamine (**1**, R = OH, R′ = H) and S′ = cladinose (**8**, R = CH_3).

ylerythromycin A yielded a ring-expanded 15-membered macrolide which existed in two tautomeric forms possessing either a 9-ketone or a 9,12-hemiketal (123).

The acid-instability of erythromycin makes it susceptible to degradation in the stomach to intramolecular cyclization products lacking antimicrobial activity. Relatively water-insoluble, acid-stable salts, esters, and/or formulations have therefore been employed to protect erythromycin during passage through the stomach, to increase oral bioavailability, and to decrease the variability of oral absorption. These various derivatives and formulations also mask the very bitter taste of macrolides.

Ester derivatives of erythromycin, shown in Table 4, were synthesized soon after its discovery (124–126). They are readily prepared by acylation of the 2′-hydroxyl group; the neighboring 3′-dimethylamino group directs acylation to this site. Commonly used esters of erythromycin are propionate, acetate, ethyl succinate, and ethyl carbonate (**30**). 2′-Esters do not bind bacterial ribosomes, however, and consequently must be hydrolyzed back to the parent to exert antibacterial activity (127). Ester derivatives are still being prepared (128,129). 2′-*O*-Acetylerythromycin stearate, also known as erythromycin acistrate, is an example undergoing clinical trial (11,130,131). Two salts of 2′-*O*-propionylerythromycin, *N*-acetylcysteinate (erythromycin stinoprate) and mercaptosuccinate, were prepared to combine antibiotic and mucolytic properties in a single agent (132–134).

Ester derivatives of oleandomycin (**17**, R = CH_3) were also pursued leading to commercial development of (**31**) also known as troleandomycin (TOA), which improved oral bioavailability and taste as compared to the parent (135).

Another successful strategy for derivatization of erythromycin employed modification of functional groups involved in intramolecular cyclizations. The C-9 ketone, C-6 hydroxyl group, C-8 proton, and/or C-11,12-diol of erythromycin were converted into functional groups which participate poorly, if at all, in intramolecular cyclizations. Some derivatives which have been extensively evaluated in preclinical and clinical trials exhibit such desirable properties as better stability under acidic conditions, greater oral bioavailability, and higher and more prolonged concentrations of antibiotic in serum and tissues.

The 11,12-carbonate of erythromycin (**32**) is an older cyclic ester which had greater stability and antibiotic activity by diminishing formation of intramolecular enol ether (**27**) (136,137). A later analogue, the 11-*N*-12-*O*-cyclic carbamate of 11-deoxy-11-(2-dimethylamino)ethylamino-6-*O*-methylerythromycin (A-62514), was selected for further study (138,139).

Several structural modifications of the C-9 ketone of erythromycin have been explored; oximes and hydrazones are less prone to intramolecular cyclization, but they often have less antibiotic activity than erythromycin (140). Synthesis of more complex oxime derivatives resulted in the development of roxithromycin, the 9-[*O*-(2-methoxyethoxy)methyl]oxime (**33**) (141). Reduction of the oximes and hydrazones produced 9(*S*)-erythromycylamine (**34**) as the principal product, with minor amounts of the 9(*R*)-isomer (140); however, clinical studies showed that 9(*S*)-erythromycylamine and its *N*-benzylidene derivative were poorly absorbed in humans (142). Evaluation of more complex oxazine derivatives of erythromycylamine led to dirithromycin, the 2-(2-methoxyethoxy)ethylidene oxazine derivative (**35**) (143). A third route to modification of the ketone utilized a

Beckmann rearrangement of the 9-oxime to expand the 14-membered ring to a 15-membered intermediate, which was subsequently reduced and *N*-methylated to yield azithromycin (**36**) (144,145). The term azalide was proposed to denote these 15-membered azalactones (10,145).

Other approaches to inhibiting intramolecular cyclizations of erythromycin have also proven successful. From a series of *O*-alkyl derivatives of erythromycin, clarithromycin (6-*O*-methylerythromycin) (**37**) was selected for clinical development (146,147). Another approach replaced the C-8 proton of erythromycin with fluorine, which was accomplished by both chemical and bioconversion methods to yield flurithromycin (**38**) (148).

A few bioconversions of erythromycin have been reported (149). Certain modifications of erythromycin at both its 9-ketone and 11-hydroxyl groups inhibited hydrolysis of the lactone ring by an esterase isolated from *Escherichia coli* which might be involved in bacterial resistance to macrolides (150). The 9-methoxime-11-[(2-dimethylaminoethyl)oxymethyl] derivative of erythromycin, ER 42859, showed good antibacterial activity and pharmacokinetics in animals, but it gave lower blood levels than erythromycin when administered to humans (151). Other 9,11-cyclic derivatives of erythromycin have been synthesized (152,153). Series of 9-*N*-alkyl derivatives of both 9(*S*)- and 9(*R*)-erythromycylamine have been prepared which improved activity *in vivo* over erythromycin (154,155).

A novel series of 9,12-epoxy derivatives synthesized from (**27**) showed unexpectedly good antibiotic activity; A-69334, which possesses an 11-amino-9,12-epoxy moiety, exhibited pharmacokinetic advantages in animals (156). Some 4″-*O*-acyl derivatives of erythromycin did not induce resistance to other macrolide, lincosamide (see ANTIBIOTICS, LINCOSAMINIDES), and streptogramin B (see ANTIBIOTICS, PEPTIDES) antibiotics (MLS-resistance) in bacteria which were inducibly-resistant to erythromycin (157,158); furthermore, certain derivatives of erythromycin modified at the 4″-hydroxyl and/or 11,12-diol groups demonstrated activity against constitutively MLS-resistant bacteria (159,160). Some 4″-*O*-acyl derivatives promoted less gastrointestinal motility in dogs compared to erythromycin, and 4″-*O*-carbamoyl-6-*O*-methylerythromycin (A-63075) was selected for further study (161).

16-Membered Ring Macrolides

Natural Products. 16-Membered macrolides are divided into leucomycin- and tylosin-related groups, which differ in the substitution pattern of their aglycones. Multi-factor complexes are usually produced and some compounds have been isolated from culture broths of different organisms and then been given different names.

The leucomycin complex of ten factors, Type I in Table 5, occasionally called kitasamycin, was produced by *S. kitasatoensis* (162), later reclassified as *Streptoverticillium* (163). After extensive studies, a 16-membered lactone substituted by an aminosugar and a neutral sugar was proposed, with individual factors differing in acylation of the 3- and 4″-hydroxyl groups (164–172). X-ray crystallography of demycarosylisoleucomycin A_3 hydrobromide confirmed these structures

Table 4. Derivatives of Erythromycin and Oleandonycin

Macrolide	CAS Registry Number	Molecular formula	Structure number	Structure
			Erythromycin esters	
2′-*O*-acetylerythromycin	*[992-69-8]*	$C_{39}H_{69}NO_{14}$	(**30**, R = $COCH_3$)	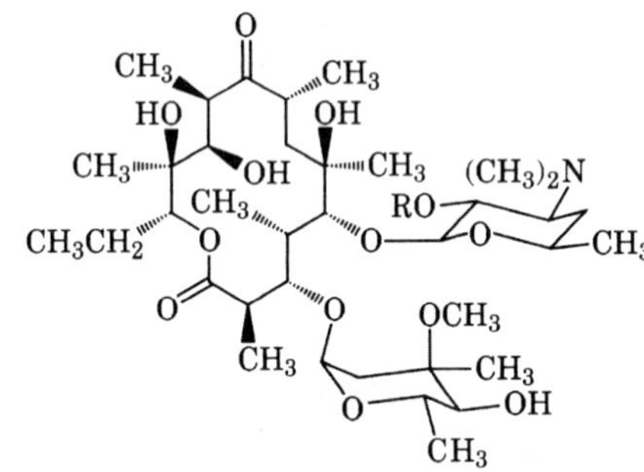
2′-*O*-propionylerythromycin	*[134-36-1]*	$C_{40}H_{71}NO_{14}$	(**30**, R = $COCH_2CH_3$)	
erythromycin ethyl carbonate	*[7218-80-6]*	$C_{40}H_{71}NO_{15}$	(**30**, R = $COOCH_2CH_3$)	
erythromycin ethyl succinate	*[1264-62-6]*	$C_{43}H_{75}NO_{16}$	(**30**, R = $CO(CH_2)_2COOCH_2CH_3$)	
			Other derivatives[a]	
tri-*O*-acetyloleandomycin	*[2751-09-9]*	$C_{41}H_{67}NO_{15}$	(**31**)	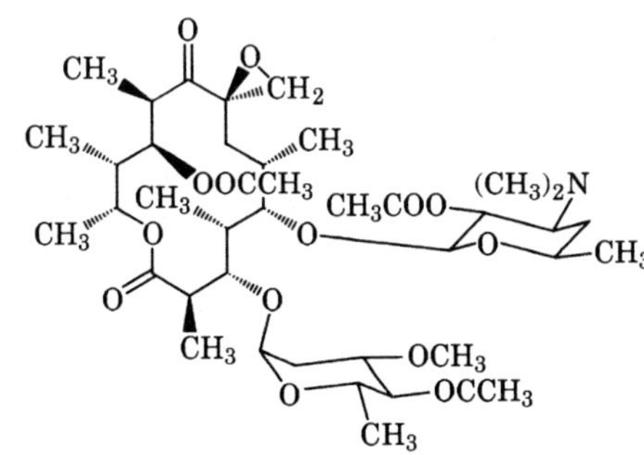

Compound	CAS Registry Number	Molecular formula		Structure
erythromycin-11,12-carbonate	[*55224-05-0*]	$C_{38}H_{65}NO_{14}$	**(32)**	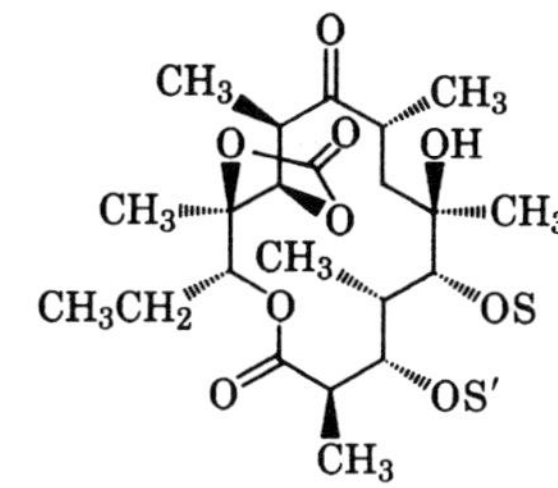
roxithromycin	[*80214-83-1*]	$C_{41}H_{76}N_2O_{15}$	**(33)**	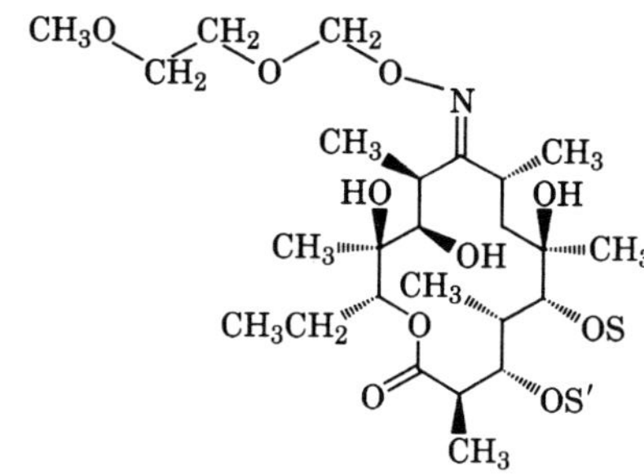
9(*S*)-erythromycylamine	[*26116-56-3*]	$C_{37}H_{70}N_2O_{12}$	**(34)**	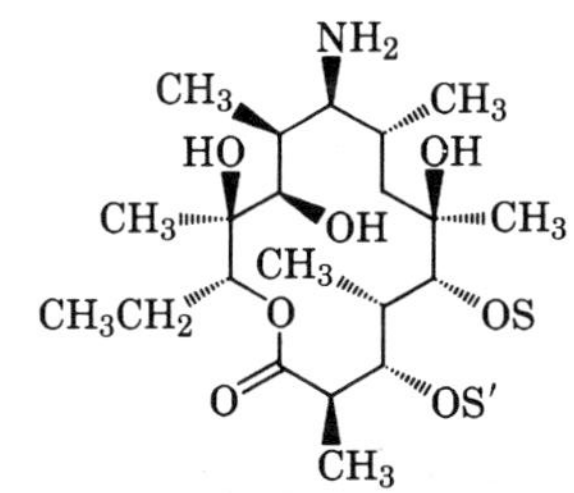
dirithromycin	[*62013-04-1*]	$C_{42}H_{78}N_2O_{14}$	**(35)**	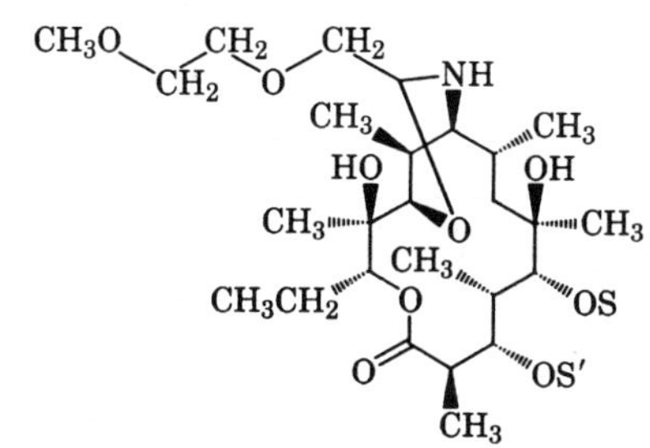

Table 4. (*Continued*)

Macrolide	CAS Registry Number	Molecular formula	Structure number	Structure
azithromycin	[*83905-01-5*]	$C_{38}H_{72}N_2O_{12}$	(**36**)	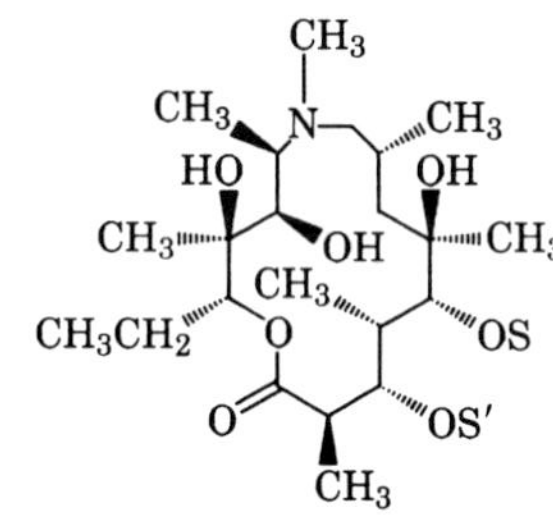
clarithromycin	[*81103-11-9*]	$C_{38}H_{69}NO_{13}$	(**37**)	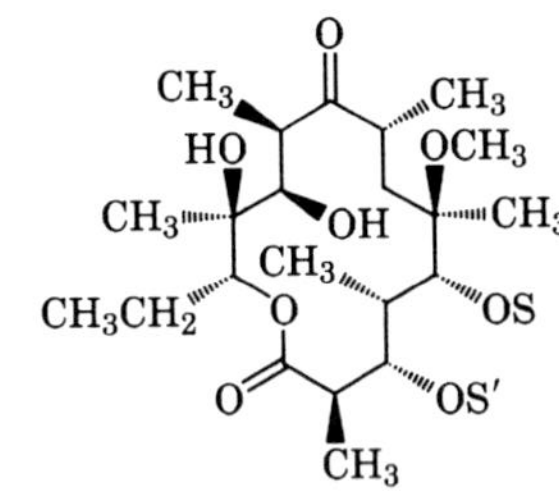 (**37**)
flurithromycin	[*82664-20-8*]	$C_{37}H_{66}FNO_{13}$	(**38**)	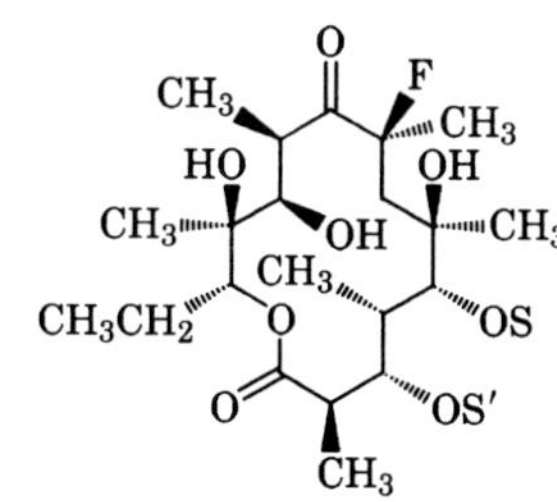 (**38**)

[a]S = desosamine (**1**, R = OH, R′ = H); S′ = cladinose (**8**, R = CH_3).

Table 5. Leucomycins and Related Compounds

Macrolide	CAS Registry Number	Molecular formula	R	R′
	Type I: C-9 to C-13 is dienol			
	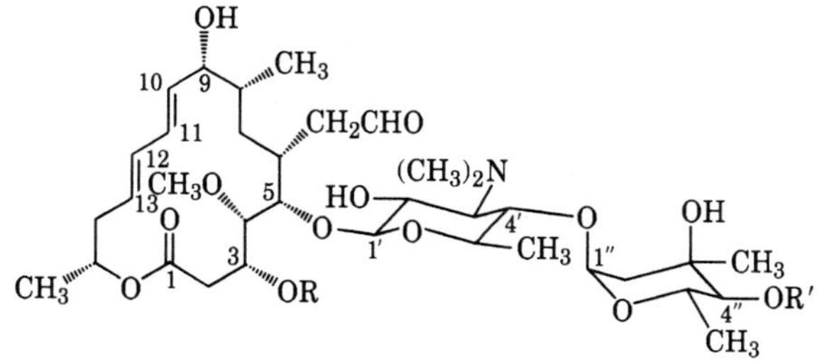			
	(39)			
leucomycin A_1	[*16846-34-7*]	$C_{40}H_{67}NO_{14}$	H	isovaleryl
leucomycin A_5	[*18361-45-0*]	$C_{39}H_{65}NO_{14}$	H	butyryl
leucomycin A_7	[*18361-47-2*]	$C_{38}H_{63}NO_{14}$	H	propionyl
leucomycin A_9	[*18361-49-4*]	$C_{37}H_{61}NO_{14}$	H	acetyl
leucomycin V	[*22875-15-6*]	$C_{35}H_{59}NO_{13}$	H	H
leucomycin A_3 (josamycin)	[*16846-24-5*]	$C_{42}H_{69}NO_{15}$	acetyl	isovaleryl
leucomycin A_4	[*18361-46-1*]	$C_{41}H_{67}NO_{15}$	acetyl	butyryl
leucomycin A_6	[*18361-48-3*]	$C_{40}H_{65}NO_{15}$	acetyl	propionyl
leucomycin A_8	[*18361-50-7*]	$C_{39}H_{63}NO_{15}$	acetyl	acetyl
leucomycin U	[*31642-61-2*]	$C_{37}H_{61}NO_{14}$	acetyl	H
platenomycin A_1	[*40615-47-2*]	$C_{43}H_{71}NO_{15}$	propionyl	isovaleryl
midecamycin A_2	[*35457-81-9*]	$C_{42}H_{69}NO_{15}$	propionyl	butyryl
midecamycin A_1 (espinomycin A_1), (platenomycin B_1)	[*35457-80-8*]	$C_{41}H_{67}NO_{15}$	propionyl	propionyl
platenomycin C_2 (espinomycin A_3)	[*35867-32-4*]	$C_{40}H_{65}NO_{15}$	propionyl	acetyl
DHP	[*35906-56-0*]	$C_{38}H_{63}NO_{14}$	propionyl	H
espinomycin A_2	[*40867-12-7*]	$C_{42}H_{69}NO_{15}$	propionyl	isobutyryl
platenomycin A_0	[*52310-62-0*]	$C_{44}H_{73}NO_{15}$	butyryl	isovaleryl
	Type II: C-9 to C-13 is epoxyenol			
	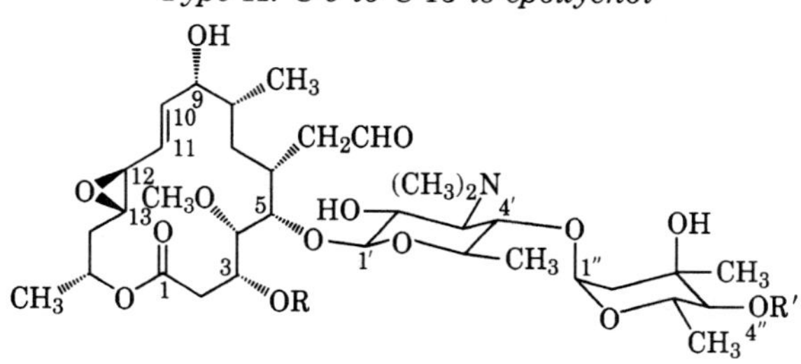			
	(40)			
maridomycin I (platenomycin C_3)	[*35908-44-2*]	$C_{43}H_{71}NO_{16}$	propionyl	isovaleryl
maridomycin VII	[*56078-81-0*]	$C_{42}H_{69}NO_{16}$	propionyl	butyryl

Table 5. ***(Continued)***

Macrolide	CAS Registry Number	Molecular formula	R	R′
maridomycin III (platenomycin C_1)	[*35775-82-7*]	$C_{41}H_{67}NO_{16}$	propionyl	propionyl
maridomycin V	[*35942-57-5*]	$C_{40}H_{65}NO_{16}$	propionyl	acetyl
maridomycin II (platenomycin C_4)	[*35908-45-3*]	$C_{42}H_{69}NO_{16}$	acetyl	isovaleryl
maridomycin IV	[*35942-56-4*]	$C_{40}H_{65}NO_{16}$	acetyl	propionyl
maridomycin VI	[*35775-66-7*]	$C_{39}H_{63}NO_{16}$	acetyl	acetyl

Type III: C-9 to C-13 is epoxyenone

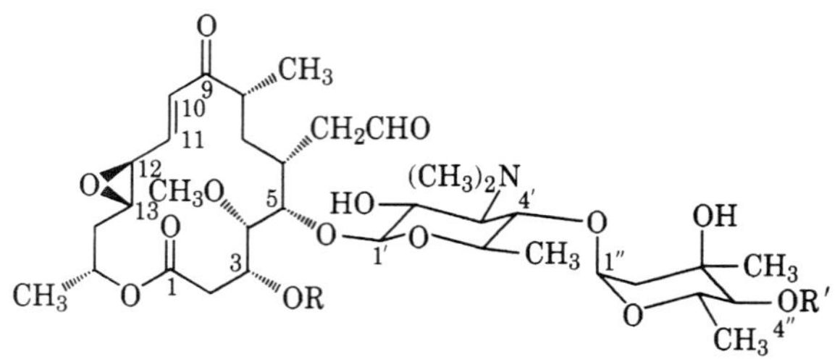

(**41**)

Macrolide	CAS Registry Number	Molecular formula	R	R′
carbomycin A (deltamycin A_4)	[*4564-87-8*]	$C_{42}H_{67}NO_{16}$	acetyl	isovaleryl
deltamycin A_3	[*58880-24-3*]	$C_{41}H_{65}NO_{16}$	acetyl	butyryl
deltamycin A_2	[*58880-23-2*]	$C_{40}H_{63}NO_{16}$	acetyl	propionyl
deltamycin A_1	[*58880-22-1*]	$C_{39}H_{61}NO_{16}$	acetyl	acetyl
deltamycin X, (EOA)	[*40625-70-5*]	$C_{37}H_{59}NO_{15}$	acetyl	H
EOP	[*41688-43-1*]	$C_{38}H_{61}NO_{15}$	propionyl	H

Type IV: C-9 to C-13 is dienone

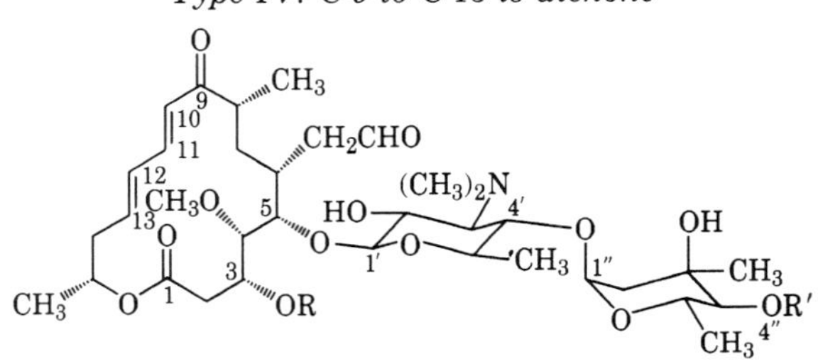

(**42**)

Macrolide	CAS Registry Number	Molecular formula	R	R′
carbomycin B	[*21238-30-2*]	$C_{42}H_{67}NO_{15}$	acetyl	isovaleryl
platenomycin W_1	[*35867-31-3*]	$C_{43}H_{69}NO_{15}$	propionyl	isovaleryl
platenomycin W_2	[*52310-61-9*]	$C_{44}H_{71}NO_{15}$	butyryl	isovaleryl
niddamycin	[*20283-69-6*]	$C_{39}H_{63}NO_{14}$	H	butyryl
midecamycin A_4	[*36083-82-6*]	$C_{42}H_{67}NO_{15}$	propionyl	butyryl
midecamycin A_3	[*36025-69-1*]	$C_{41}H_{65}NO_{15}$	propionyl	propionyl
DOA	[*52278-66-7*]	$C_{37}H_{59}NO_{14}$	acetyl	H
DOP	[*40580-80-1*]	$C_{38}H_{61}NO_{14}$	propionyl	H

except in stereochemistry at C-9 (173,174). Interconversion between leucomycin and isoleucomycin had suggested a β configuration for the C-9 hydroxyl group (175,176), but this was later revised to α (177) and confirmed by additional x-ray studies (178). Conformational studies of the leucomycins have been published (179). Josamycin (**39**, R = acetyl, R′ = isovaleryl) was obtained from culture broths of *S. narbonensis* var. *josamyceticus* (180). Although initially reported as a new macrolide, it was later shown to be identical to leucomycin A_3 (181). Josamycin was commercially launched in Japan in 1970 and is now one of the most widely used 16-membered macrolides (116).

The maridomycin complex of seven factors (**40**), Type II in Table 5, was obtained from culture broths of *S. hygroscopicus* (182–188). The principal difference from the leucomycins is a 12,13-epoxide. Confirmation of structure was obtained from x-ray crystallography and spectroscopy of 9-*O*-acyl derivatives of maridomycin III (188,189).

Carbomycin A (**41**, R = acetyl, R′ = isovaleryl), Table 5, Type III, the first 16-membered macrolide discovered, was isolated from culture broths of *S. halstedii* (also called magnamycin or M-4209) (190–192). The 17-membered lactone originally proposed was later revised to a 16-membered lactone (27,193,194). Its absolute stereochemistry was established by correlations with the maridomycins (187,188). Carbomycin is also produced by *S. thermotolerans* (195,196). The deltamycins, produced by *S. halstedii* subsp. *deltae*, are five related factors which differ in their 4″-*O*-acyl group (197,198). Deltamycin A_4 is the same as carbomycin A. Deltamycins differ from the leucomycins and maridomycins by having a 12,13-epoxy-9-ketone. De-epoxidation and isomerization products have been prepared by chemical and microbiological methods (199). Carbomycin B (**42**, R = acetyl, R′ = isovaleryl), Type IV, Table 5, a minor component in the fermentation broths of *S. halstedii,* contains a C-9 to C-13 dienone (200).

Many related complexes, such as the platenomycin complex (initially YL-704) from *S. platensis* subsp. *malvinus,* have been obtained (201–206). A series designated by abbreviations (DHP, DOA) was obtained from another strain of *S. platensis* (207); the midecamycin complex (initially SF-837) was produced by *S. mycarofaciens* (208–212); the espinomycin factors were produced by *S. fungicidicus* var. *espinomyceticus* (213,214); the turimycin series of fifteen components, isolated from culture broths of *S. hygroscopicus,* are identical to other members of the leucomycin group (215). Niddamycin (**42**, R = H, R′ = butyryl) was isolated from cultures of *S. djakartensis* (216).

Macrolides containing only neutral sugars were obtained from a platenomycin-producing organism (217,218). Four demycarosyl derivatives of platenomycins were isolated from biosynthetically blocked mutants of *S. platensis;* two of these possessed a methyl group rather than an aldehyde (219). A pair of novel compounds related to carbomycin were isolated in which one contained an unusual 10,11-dihydro-12,13-diol moiety, the other a 14-hydroxy-epoxyenone moiety (220).

The spiramycin complex, also discovered as foromacidine, was isolated from culture broths of *S. ambofaciens* (221–224). The spiramycins, shown in Table 6, are distinguished by a second aminosugar, forosamine (**2**), attached to the 9-hydroxyl group by a β-glycosidic linkage (225). The three primary factors, spiramycin I, II, and III, differ in their 3-*O*-acyl substituent. Two minor factors containing a

Table 6. Spiramycins and 16-Membered Aglycones

Macrolide	CAS Registry Number	Molecular formula	Structure number	Structure
spiramycin I (foromacidine A)	*[24916-50-5]*	$C_{43}H_{74}N_2O_{14}$	(**43**, R = H, R′ = mycarosyl)	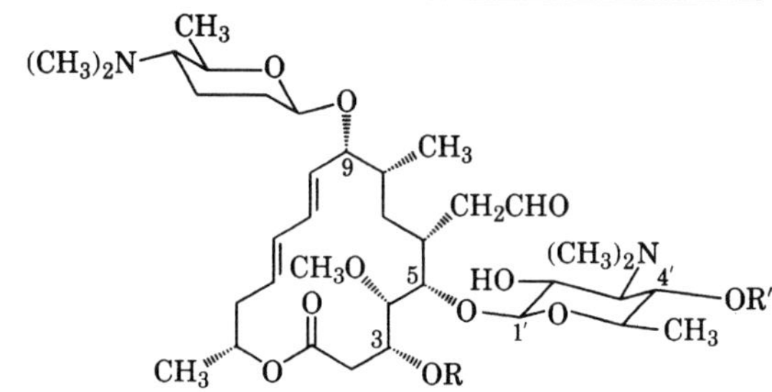
spiramycin II (foromacidine B)	*[24916-51-6]*	$C_{45}H_{76}N_2O_{15}$	(**43**, R = acetyl, R′ = mycarosyl)	
spiramycin III (foromacidine C)	*[24916-52-7]*	$C_{46}H_{78}N_2O_{15}$	(**43**, R = propionyl, R′ = mycarosyl)	
neospiramycin I	*[70253-62-2]*	$C_{36}H_{62}N_2O_{11}$	(**43**, R = R′ = H)	
leuconolide A_1	*[56201-85-5]*	$C_{20}H_{32}O_7$	(**44**)	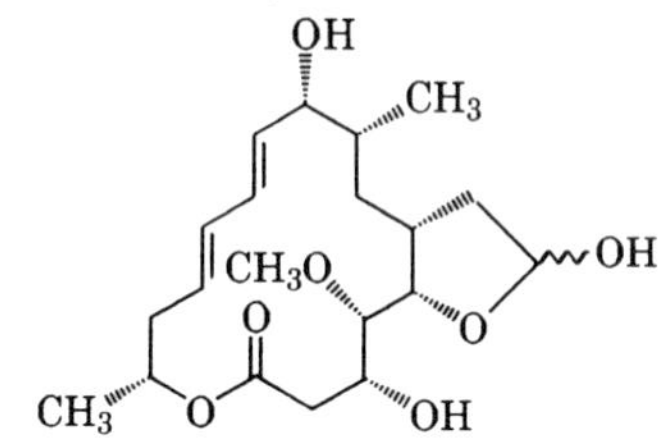
spiramycin U	*[87708-39-2]*	$C_{42}H_{71}NO_{16}$	(**45**, R = CHO, R′ = mycarosyl)	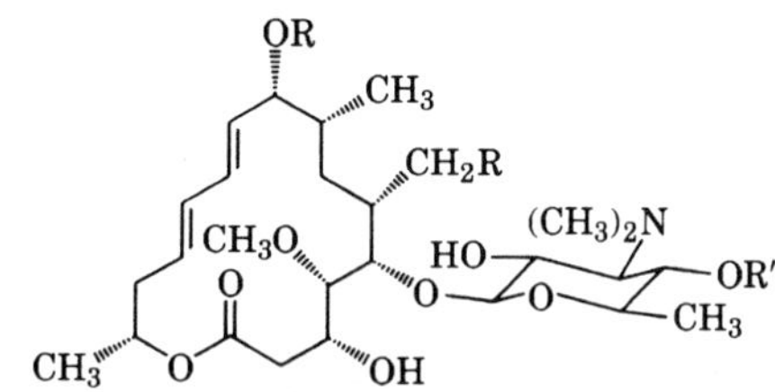
spiramycin S	*[87718-67-0]*	$C_{42}H_{73}NO_{16}$	(**45**, R = CH_2OH, R′ = mycarosyl)	
forocidin I	*[70282-40-5]*	$C_{28}H_{47}NO_{10}$	(**45**, R = CHO, R′ = H)	
platenolide I	*[52212-94-9]*	$C_{20}H_{32}O_6$	(**46**, R, R′ = O)	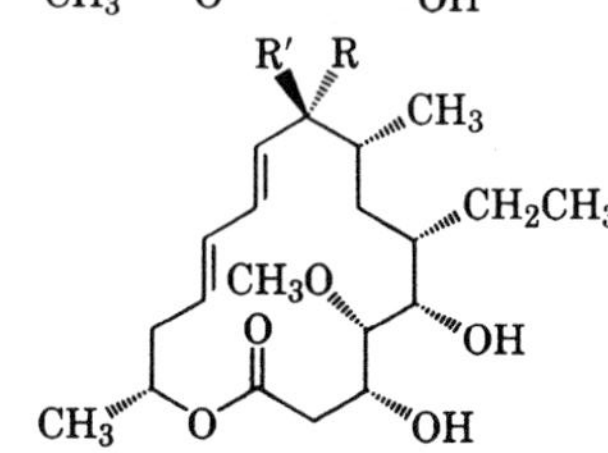
platenolide II	*[52212-95-0]*	$C_{20}H_{34}O_6$	(**46**, R = OH, R′ = H)	

neutral sugar instead of forosamine, spiramycin U and S, have been found (226). Selective hydrolysis of the terminal sugar yields the respective neospiramycin factors (227); subsequent hydrolysis of forosamine produces the forocidin factors (228).

Aglycones of 16-membered macrolides exist as intramolecularly cyclized hemiketals as a result of the close proximity of the aldehyde and 5-hydroxyl group. Aglycones can differ in their oxidation level at C-9 and C-12,13, types I–IV, Table 5, and their 3-*O*-substituent. The names parallel those of the parent macrolides; for example, leuconolide A_1 (**44**) is the aglycone from any of five leucomycin factors. The Polonovski reaction is generally used to cleave the aminosugar; acid hydrolysis destroys the aglycone before the sugar is hydrolyzed (229–231). Two reduced aglycones (**46**) were isolated from either a mutant strain or producing strain of *S. platensis* (232,233); these compounds are biosynthetic precursors of platenomycin; oxidation to the aldehyde occurs as a late step in the fermentation (234).

A second large group of 16-membered macrolides differs from the leucomycins in the substitution pattern of the aglycone. One difference is a methyl or hydroxymethyl group at C-14. If hydroxymethyl is present, it may be glycosidically substituted by a neutral sugar such as mycinose (**9**, R = R′ = CH_3), or an analogue. The most prominent member of this group is tylosin shown in Table 7, an important veterinary antibiotic produced by *S. fradiae* (235,236). The structure was deduced by chemical methods (237,238) and nmr (225). Absolute configuration was established by spectroscopic comparison with leucomycins (239) and later confirmed by correlations based upon x-ray crystallography of biosynthetic precursors (240,241).

Relomycin (Table 7) was found in culture broths of *S. hygroscopicus* (242) and shown to be 20-dihydrotylosin by chemical and microbial reductions (242,243). Macrocin (3‴-*O*-demethyltylosin), was found as a minor factor in tylosin (244). It is the immediate biosynthetic precursor of tylosin and can be produced by a biosynthetically blocked mutant of *S. fradiae* (245–248). 6-Deoxy-2-*O*-methyl-D-allose [*921-90-4*], a neutral sugar in macrocin, is a constituent of cardiac glycosides from *Antiaris toxicaria* and named javose (**9**, R = H, R′ = CH_3) (249). 2‴-*O*-Demethylmacrocin (DOMM) obtained from another mutant of *S. fradiae* (246,250–252), is the penultimate precursor of tylosin (245–248). Other mutant strains of *S. fradiae* yield shunt metabolites of tylosin in which mycinose is altered (251,252). CP-56,064 [*88378-53-4*] is an analogue of tylosin in which mycarose is replaced by amicetose (**5**, R = H, R′ = OH) (253).

The 2-deoxy sugar, mycarose, is selectively hydrolyzed from tylosin under acidic conditions to produce demycarosyltylosin known as desmycosin (236). Although analogous hydrolyses have been performed in the leucomycin–spiramycin group, desmycosin possesses greater antibiotic activity than demycarosyl compounds from the other group. Hydrolyses of mycarose from macrocin and 3‴-*O*-demethylmacrocin produce lactenocin and 2″-*O*-demethyllactenocin (DOML), respectively (246,250). Desmycosin was also produced by mutant strains of *S. fradiae* possessing a biosynthetic block in synthesis and/or attachment of mycarose (246). Other mutants of *S. fradiae* which are blocked in oxidation of the C-23 methyl group, attachment of a neutral sugar onto the C-23 hydroxyl group, or oxidation of the C-20 methyl group provide a convenient source of tylosin-related

Table 7. Tylosin and Related Compounds

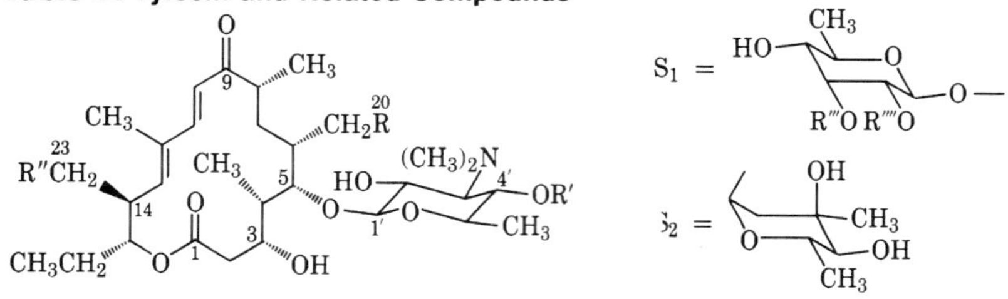

(**47**)

Macrolide	CAS Registry Number	Molecular formula	R	R′	R″	R‴	R⁗
tylosin	[*1401-69-0*]	$C_{46}H_{77}NO_{17}$	CHO	S_2	S_1	CH_3	CH_3
relomycin (20-dihydrotylosin)	[*62902-06-1*]	$C_{46}H_{79}NO_{17}$	CH_2OH	S_2	S_1	CH_3	CH_3
macrocin	[*11049-15-3*]	$C_{45}H_{75}NO_{17}$	CHO	S_2	S_1	H	CH_3
O-demethylmacrocin	[*79404-98-1*]	$C_{44}H_{73}NO_{17}$	CHO	S_2	S_1	H	H
desmycosin	[*11032-98-7*]	$C_{39}H_{65}NO_{14}$	CHO	H	S_1	CH_3	CH_3
lactenocin	[*110469-05-1*]	$C_{38}H_{63}NO_{14}$	CHO	H	S_1	H	CH_3
O-demethyllactenocin (DOML)	[*81557-35-9*]	$C_{37}H_{61}NO_{14}$	CHO	H	S_1	H	H
23-*O*-demycinosyltylosin	[*79592-92-0*]	$C_{38}H_{63}NO_{13}$	CHO	S_2	OH		
23-(demycinosyloxy)tylosin	[*79404-97-0*]	$C_{38}H_{63}NO_{12}$	CHO	S_2	H		
GS-77-1	[*86293-46-1*]	$C_{38}H_{65}NO_{11}$	CH_3	S_2	H		
GS-77-3	[*81661-90-7*]	$C_{38}H_{65}NO_{12}$	CH_3	S_2	OH		

macrolides in which mycarose is present and mycinose is absent (246,250,251). Some of these are shown in Table 7.

Other macrolides in this group have a 12,13-epoxide. Angolamycin (**48**, S = α-L-mycarosyl), produced by *S. eurythermus* (254) and identical to shincomycin A from *S. flavochromogenes* (255), contains the 2-deoxy aminosugar, angolosamine (**1**, R = H, R′ = OH) (256,257). Its structure is given in Table 8. Demycarosylangolamycin was isolated from cultures of *Streptomyces* sp. AS-NG-16 and named staphcoccomycin (**48**, S = H) (258). CP-56,063 [*88378-52-3*] and its 20-dihydro derivative CP-56,678, contain amicetose (**5**, R = H, R′ = OH) as the terminal neutral sugar (253). Acumycin (**49**, S = α-L-cinerulosyl) was isolated from culture broths of *S. griseoflavus* (259) and shown by x-ray crystallography to contain cinerulose (**5**, R = O) as the terminal sugar (260). This same compound was isolated from culture broths of *S. fradiae* var. *acinicolor* and designated cirramycin B or B-58941 (261–263). A6888C, produced by *S. flocculus,* contains dihydrocinerulose; its 20-dihydro derivative, A6888X, was also isolated (264). Cirramycin F-1, identical to A6888C, established that the neutral sugar was L-rhodinose (**5**, R = OH, R′ = H) (265). Cirramycin F-2 possesses the 4-epimeric sugar, L-amicetose (**5**, R = H, R′ = OH) (265). M119-a (**49**, S = L-cladinosyl) produced by an alkalophilic actinomycete, differs from the others in the terminal neutral sugar (266).

Table 8. Angolamycin and Related Compounds

Macrolide	CAS Registry Number	Molecular formula	Structure number	Structure
angolamycin (shincomycin A)	*[1402-83-1]*	$C_{46}H_{77}NO_{17}$	(**48**, S = α-L-mycarosyl)	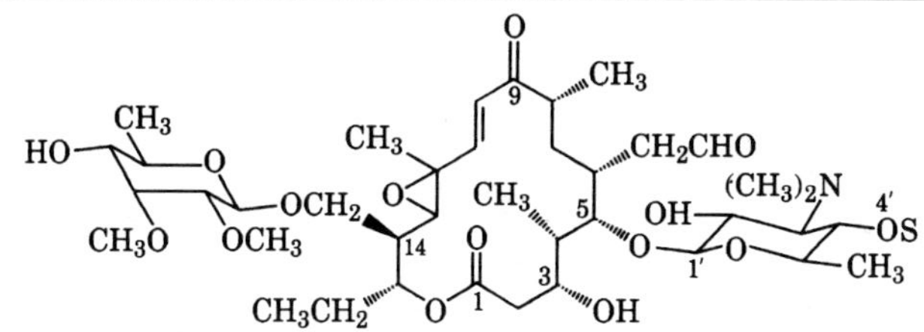
staphcoccomycin	*[73047-31-1]*	$C_{39}H_{65}NO_{14}$	(**48**, S = H)	
acumycin (cirramycin B)	*[25999-30-8]*	$C_{37}H_{59}NO_{12}$	(**49**, S = α-L-cinerulosyl)	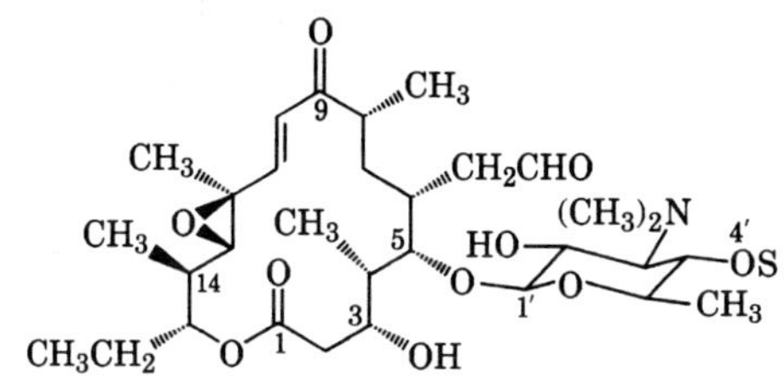
cirramycin F-1 (A6888C)	*[74918-31-3]*	$C_{37}H_{61}NO_{12}$	(**49**, S = α-L-rhodinosyl)	
cirramycin F-2	*[120851-46-9]*	$C_{37}H_{61}NO_{12}$	(**49**, S = α-L-amicetosyl)	
M119-a	*[111205-12-0]*	$C_{39}H_{65}NO_{13}$	(**49**, S = α-L-cladinosyl)	

Many 16-membered macrolides possess a tylosin-type aglycone with only one saccharide, an aminosugar that is either mycaminose or desosamine. These compounds, given in Table 9, also differ in their degree of oxidation at C-20, C-23, and C-9 to C-13 (dienone or epoxyenone). The first macrolide found in this group was cirramycin A_1 (**50**, R = CHO, R′ = OH, R″ = H), obtained from a complex of

Table 9. Cirramycin A_1, Rosaramicin, and Related Compounds

C-9 to C-13, epoxyenone

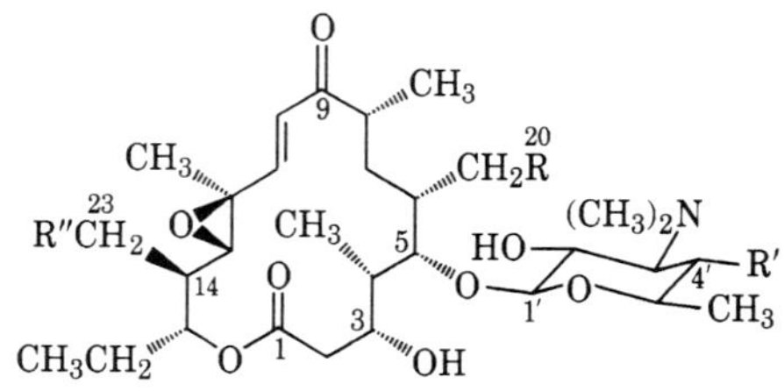

(**50**)

Macrolide	CAS Registry Number	Molecular formula	R	R′	R″
cirramycin A_1	*[25339-90-6]*	$C_{31}H_{51}NO_{10}$	CHO	OH	H
rosaramicin	*[35834-26-5]*	$C_{31}H_{51}NO_9$	CHO	H	H
izenamicin A_3	*[86132-03-8]*	$C_{31}H_{51}NO_{10}$	CHO	H	OH
juvenimicin A_4	*[61475-37-4]*	$C_{31}H_{53}NO_9$	CH_2OH	H	H
M-4365 A_1	*[59227-84-8]*	$C_{31}H_{53}NO_8$	CH_3	H	H
juvenimicin A_2	*[61417-47-8]*	$C_{30}H_{51}NO_8$	H	H	H

C-9 to C-13, dienone

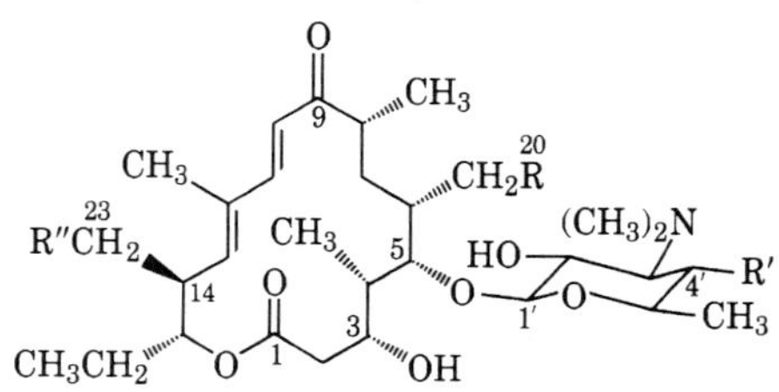

(**51**)

Macrolide	CAS Registry Number	Molecular formula	R	R′	R″
5-*O*-mycaminosyl-tylonolide (OMT)	*[61257-02-1]*	$C_{31}H_{51}NO_{10}$	CHO	OH	OH
23-deoxy-5-*O*-mycaminosyltylonolide (DOMT)	*[115033-64-2]*	$C_{31}H_{51}NO_9$	CHO	OH	H
izenamicin B_3	*[80240-61-5]*	$C_{31}H_{51}NO_9$	CHO	H	OH
M-4365 G_2	*[56689-42-0]*	$C_{31}H_{51}NO_8$	CHO	H	H
juvenimicin B_3	*[61417-48-9]*	$C_{31}H_{53}NO_9$	CH_2OH	H	OH
juvenimicin B_1	*[58947-83-4]*	$C_{31}H_{53}NO_8$	CH_2OH	H	H
GS-77-4	*[80830-18-8]*	$C_{31}H_{53}NO_9$	CH_3	OH	OH
GS-77-2	*[80830-17-7]*	$C_{31}H_{53}NO_8$	CH_3	OH	H
M-4365 G_1	*[59227-83-7]*	$C_{31}H_{53}NO_7$	CH_3	H	H
izenamicin B_2	*[86131-94-4]*	$C_{30}H_{51}NO_8$	H	H	OH

several factors produced by *S. cirratus* (267–269). It was later prepared by hydrolysis of cinerulose from cirramycin B (**49**, S = α-L-cinerulosyl) (261). Rosaramicin, formerly rosamicin, was produced by *M. rosaria* and shown to be 4′-deoxycirramycin A_1 (270,271). Its x-ray crystal structure has been published (272). Structures of four related compounds from the juvenimicin complex of eight factors, A_1–A_4 and B_1–B_4, isolated from culture broths of *M. chalcea* var. *izumensis* (273,274), have been determined. Shortly thereafter, the M-4365 complex of six factors (A_1–A_3 and G_1–G_3), produced by *M. capillata* (275,276), was reported. From the izenamicin complex of seven factors produced by a *Micromonospora* species, three were new fermentation products (277). Many compounds isolated are identical: juvenimicin A_3, M-4365 A_2, and izenamicin A_1 are the same as rosaramicin. Structures have been proven by chemical interconversions (261,277,278) and microbial transformations (279).

Hydrolysis of both neutral sugars from tylosin yields 5-*O*-mycaminosyltylonolide (OMT) (**51**, R = CHO, R′ = R″ = OH) (238). However, this compound is more conveniently obtained by hydrolysis of mycarose from the fermentation-derived demycinosyltylosin. Analogous hydrolyses of mycarose from DMOT, GS-77-3, and GS-77-1 yield, respectively, 23-deoxy-5-*O*-mycaminosyltylonolide (DOMT), 20-deoxo-5-*O*-mycaminosyltylonolide (GS-77-4), and 5-*O*-mycaminosyltylactone (GS-77-2), all given in Table 9 (246,250,280).

The mycinamicin complex, also found as AR-5 complex, was isolated from culture broths of *M. griseorubida* (281–284). These compounds are listed in Table 10. They contain a 2,3-double bond and a methyl group rather than a two-carbon substituent at C-6 of the aglycone. A hydroxyl group is present at C-14 in mycinamicins II and V. X-ray crystallography established the stereochemistry and absolute configuration of the aglycone as identical to tylosin (**47**) (284–287). Minor factors were isolated where the C-14 hydroxymethyl group is either unsubstituted or glycosidated (288), and where the aminosugar was lacking (289). These products were also found after either acid hydrolysis or Polonovski reaction of the parent mycinamicins (290,291). Mycinamicin II (**52**, R = R′ = CH_3, R″ = OH), now named miporamicin, is being investigated in pre-clinical studies (292).

A few neutral 16-membered macrolides have been isolated. The structure of chalcomycin (**54**, R = CH_3, R′ = OH), produced by *S. bikiniensis,* was established as a 16-membered lactone having two neutral sugars, D-chalcose (**7**) and D-mycinose (**9**, R = R′ = CH_3) (293–295). Its aglycone resembles that of the mycinamicins, but differs by a hydroxyl group, the stereochemistry of which is uncertain, at C-8 and a methyl group at C-15 (296–298). X-ray crystallographic data has been reported (299). The closely-related neutramycin (**54**, R = H, R′ = OH), produced by *S. rimosus* (300–302), lacks a substituent at C-6 of the aglycone. These compounds are given in Table 11. CP-70,662, produced by *S. hirsutus,* may be 8-deoxychalcomycin (303,304). CP-70,663, its *de*-desosaminyl derivative, was also obtained from this culture (303).

Also shown in Table 11 are the aldgamycins, produced by *S. lavendulae,* containing a novel bicyclic sugar, aldgarose (**4**) (305). Structures of aldgamycin F (**55**, R = OH) and E (**56**) were established by chemical degradation and mass spectrometry and differ by the presence or absence, respectively, of a hydroxyl group at C-8 and a double bond at C-10,11 of the aglycone (306–308). 8-Deoxyaldgamycin F (**55**, R = H), was named aldgamycin G, from *S. avidinii,* or

Table 10. The Mycinamicins

Macrolide	CAS Registry Number	Molecular formula	R	R′	R″
		C-9 to C-13, epoxyenone			

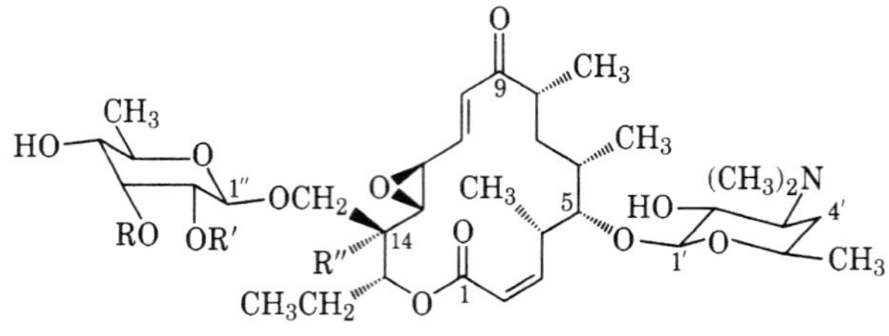

(**52**)

Macrolide	CAS Registry Number	Molecular formula	R	R′	R″
mycinamicin I	[*73665-15-3*]	$C_{37}H_{61}NO_{12}$	CH_3	CH_3	H
mycinamicin II (miporamicin)	[*73684-69-2*]	$C_{37}H_{61}NO_{13}$	CH_3	CH_3	OH

C-9 to C-13, dienone

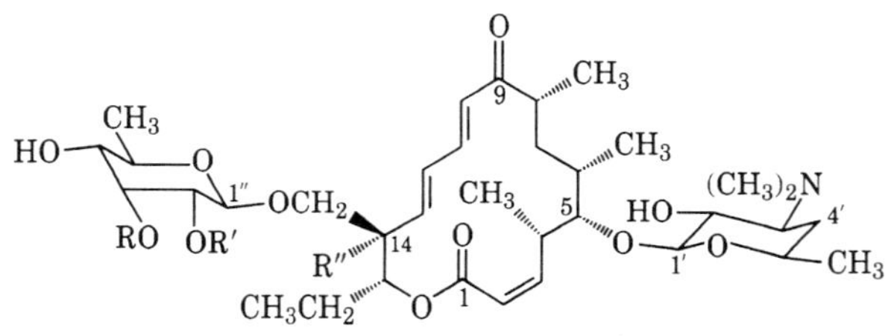

(**53**)

Macrolide	CAS Registry Number	Molecular formula	R	R′	R″
mycinamicin III	[*73684-70-5*]	$C_{36}H_{59}NO_{11}$	H	CH_3	H
mycinamicin IV	[*73684-71-6*]	$C_{37}H_{61}NO_{11}$	CH_3	CH_3	H
mycinamicin V	[*73684-72-7*]	$C_{37}H_{61}NO_{12}$	CH_3	CH_3	OH
mycinamicin VI	[*85687-36-1*]	$C_{35}H_{57}NO_{11}$	H	H	H
mycinamicin VII[a]	[*77704-58-6*]	$C_{29}H_{47}NO_7$			H

[a]Mycinamicin VII has a hydrogen where the sugar is at the C-14 position.

CP-70,661, from *S. hirsutus* (303,309). Its *des*-epoxy analogue was isolated from culture broths of *S. anandii* subsp. *swalpus* and named swalpamycin (**57**) (310,311).

As is the case for the leucomycin group, several aglycones exist in the tylosin group which differ in the pattern of oxidation and substitution of the lactone. Some of these are listed in Table 12. If the aglycone contains an aldehyde, cleavage of the aminosugar yields the hemiketal, such as tylonolide (**58**) and rosaranolide (**60**) (231). The aldehyde-reduced aglycone (**59**), is a biosynthetic intermediate to these macrolides (240,246); it was synthesized or produced from biosynthetically blocked mutants of *S. fradiae* (246,312). Four possible mycinolides and corresponding protomycinolides (deoxymycinolides) are exemplified by mycinolide IV (**61**, R = OH) and protomycinolide IV (**61**, R = H) (313).

Table 11. Chalcomycin, Aldgamycin, and Related Compounds

Macrolide	CAS Registry Number	Molecular formula	Structure number	Structure
chalcomycin	*[20283-48-1]*	$C_{35}H_{56}O_{14}$	(**54**, R = CH_3, R′ = OH)	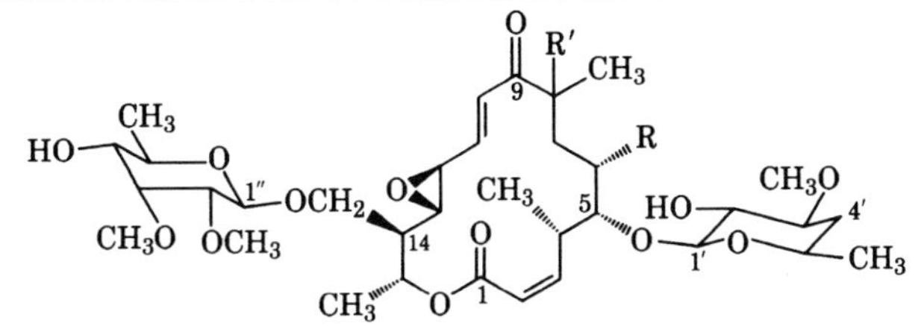
neutramycin	*[1404-08-6]*	$C_{34}H_{54}O_{14}$	(**54**, R = H, R′ = OH)	
CP-70,662	*[125225-83-4]*	$C_{35}H_{56}O_{13}$	(**54**, R = CH_3, R′ = H)	
aldgamycin F	*[55141-41-8]*	$C_{37}H_{56}O_{16}$	(**55**, R = OH)	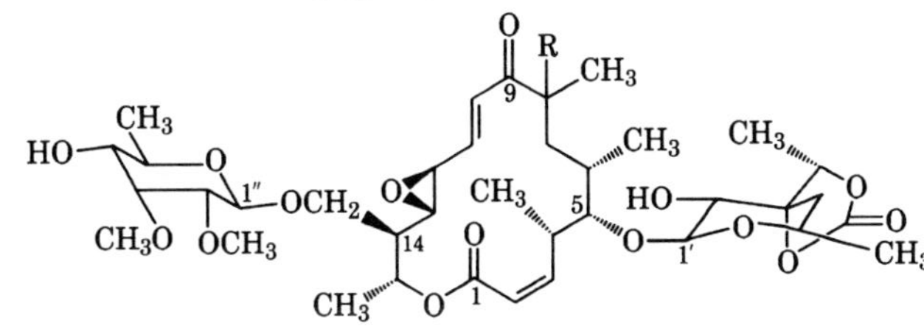
aldgamycin G (8-deoxyaldgamycin F)	*[107745-56-2]*	$C_{37}H_{56}O_{15}$	(**55**, R = H)	
aldgamycin E	*[11011-06-6]*	$C_{37}H_{58}O_{15}$	(**56**)	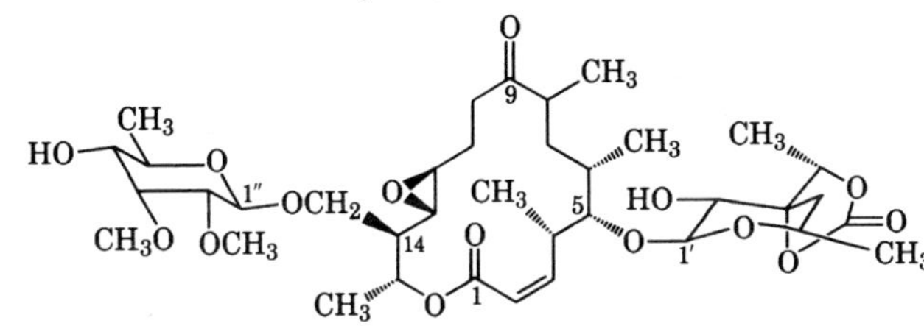
swalpamycin	*[112008-27-2]*	$C_{37}H_{56}O_{14}$	(**57**)	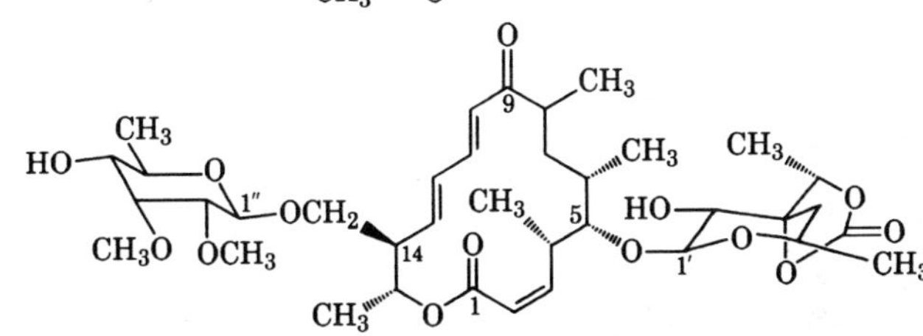

Table 12. **Tylosin-Type Aglycones**

Macrolide	CAS Registry Number	Molecular formula	Structure number	Structure
tylonolide	[*61219-81-6*]	$C_{23}H_{36}O_7$	(**58**)	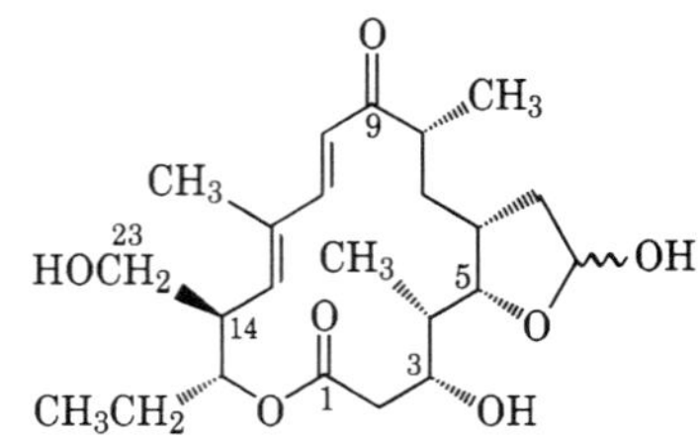
tylactone (protylonolide)	[*74758-60-4*]	$C_{23}H_{38}O_5$	(**59**)	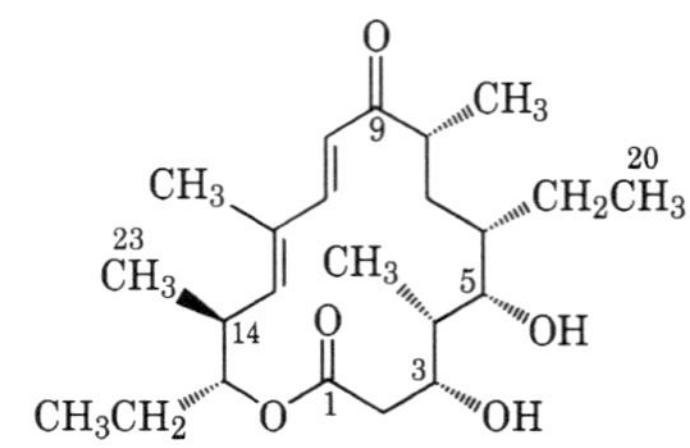
rosaranolide	[*81560-69-2*]	$C_{23}H_{36}O_7$	(**60**)	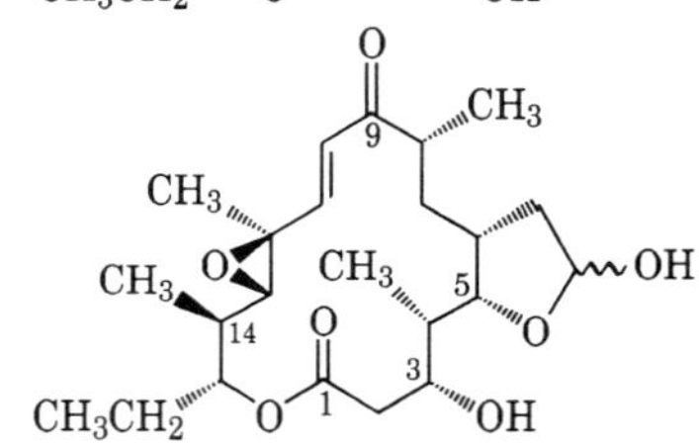
mycinolide IV protomycinolide IV	[*77704-61-1*] [*79495-87-7*]	$C_{21}H_{32}O_5$ $C_{21}H_{32}O_4$	(**61**, R = OH) (**61**, R = H)	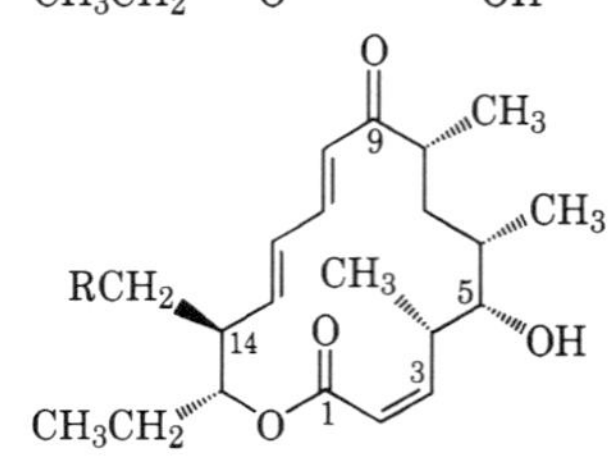

Semisynthetic Derivatives. No significant improvements in activity have resulted from modifications of the 3-, 9-, 2′- and/or 4″-hydroxyl groups (314). 3″-*O*-Acyl derivatives have not been found via fermentation, but chemical acylation of the 3″-hydroxyl group yields products having good antibiotic activity and better pharmacokinetics than the parent macrolides. Two such compounds have been developed (315,316): 3″-*O*-propionyl-leucomycin A_5 (rokitamycin) [*74014-51-0*], $C_{42}H_{69}NO_{15}$, formerly TMS-19-Q, (317,318); and 9,3″-di-*O*-acetylmidecamycin (miokamycin) [*55881-07-7*], $C_{45}H_{71}NO_{17}$, (319–321). At least part of the *in vivo* improvement was attributed to slower elimination of active metabolites from serum (322–324). These compounds are shown in Figure 2. More recently, 3″- and 4″-*O*-substituted derivatives of spiramycin were synthesized which showed similar *in vivo* improvements in activity (325–327). Rokitamycin and miokamycin have been commercially launched in Japan (116).

The enhanced activity from acylation of the 3- and 4″-hydroxyl groups of

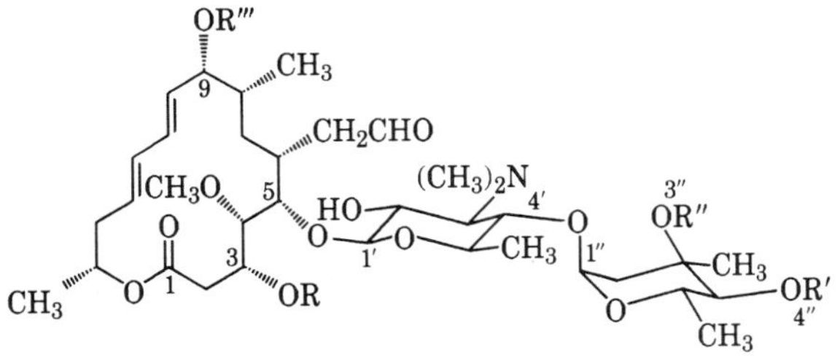

(62)

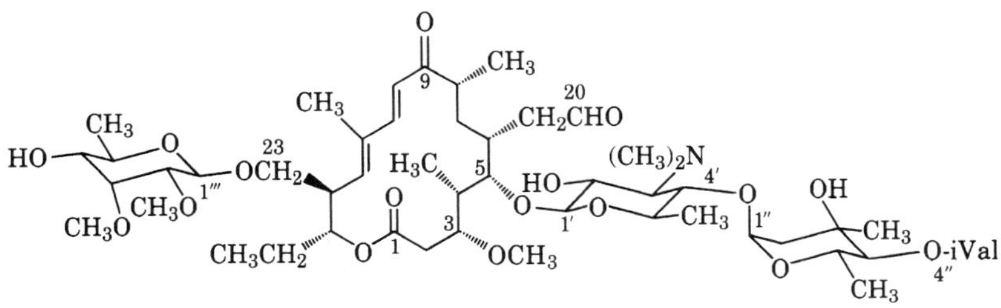

(63)

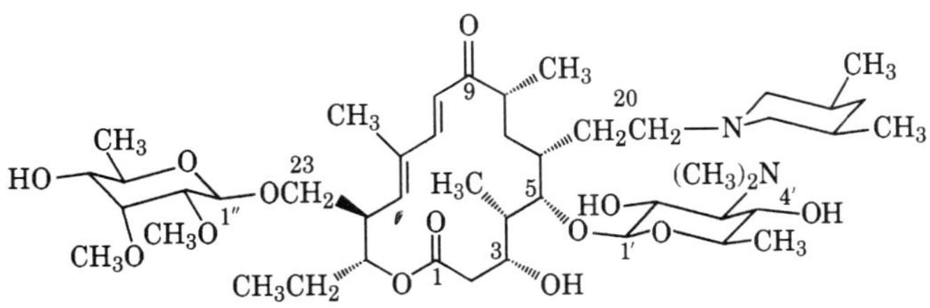

(64)

Fig. 2. Semisynthetic derivatives of 16-membered macrolides rokitamycin (**62**, R = R‴ = H, R′ = butyryl, R″ = propionyl); miokamycin (**62**, R = R′ = propionyl, R″ = R‴ = acetyl); AIV-tylosin (**63**); and tilmicosin (**64**).

leucomycin prompted analogous studies of tylosin. Bioconversion of tylosin by *S. thermotolerans* yielded 3- and/or 4″-*O*-acyl derivatives possessing increased activity against certain resistant microorganisms and higher concentrations of antibiotic in serum after oral administration (328–331). From this study, 3-*O*-acetyl-4″-*O*-isovaleryltylosin (AIV-tylosin) [*63409-12-1*], $C_{53}H_{87}NO_{19}$, (**63**) was developed as a new veterinary antibiotic (332,333).

Analogous bioconversions of tylosin-related macrolides have been reported (334). This work was extended by cloning the 4″-*O*-acylase gene and expressing it in *S. lividans* by bioconversion of spiramycin to 4″-*O*-isovalerylspiramycin [*129204-11-1*], $C_{48}H_{82}N_2O_{15}$, (335,336). Chemical methods for selective 4″-*O*-substitution of tylosin were also developed (337–339). Derivatives more stable to liver esterases have been prepared, such as 4″-*O*-(4-methoxyphenylacetyl)tylosin, [*103360-00-5*], $C_{55}H_{85}NO_{19}$, which is undergoing preclinical studies (338–340). Other studies have explored synthesis and structure–activity relations of 4″-*O*-acyl derivatives of tylosin-related macrolides (334,341,342). Modification of the

4″-*O*-acyl group in the deltamycins also produces derivatives having good antibiotic activity (343,344).

Another successful approach utilized the modification of the tylosin aldehyde moiety. Although reduction of tylosin reduced activity, other modifications yielded compounds having good antibiotic potency, better oral bioavailability, and higher and more prolonged concentrations of antibiotic activity in serum (345). These modifications were especially promising when applied to desmycosin (Table 7) (345–348). Similar pharmacokinetic improvements were observed for the structurally related mycinamicins (282). As a result of these advantages, mycinamicin II (miporamicin) (**52**, R = R′ = CH_3, R″ = H) and 19-deformyl-4′-deoxydesmycosin (TMC-016) [*85382-79-2*], $C_{38}H_{65}NO_{12}$, are being studied for clinical development (292,347,349). Reductive amination of the aldehyde in desmycosin yielded another useful series (350–354), from which 20-deoxo-20-(3,5-dimethylpiperidinyl)desmycosin (tilmicosin) [*108050-54-0*] (**64**), $C_{46}H_{80}N_2O_{13}$, was selected as a therapeutic agent to treat pneumonia in cattle and pigs because of its activity against *Pasteurella* species, oral bioavailability, and prolonged concentrations *in vivo* (355–357).

Derivatives of 5-*O*-mycaminosyltylonolide (**51**, R = CHO, R′ = R″ = OH) have been extensively investigated. Modifications of its 23-hydroxyl group, especially replacements by secondary amino groups, yielded derivatives having excellent activity *in vitro* but lacking good oral efficacy or bioavailability in animals (346,358–361). Clinical trials of the structurally related macrolide, rosaramicin (**50**, R = CHO, R′ = R″ = H), were unsuccessful (362). 4′-Deoxy-5-*O*-mycaminosyltylonolide and 4′-deoxydesmycosin were twice as potent as the corresponding compounds containing a 4′-hydroxyl group (363,364). Tylosin was modified by substituting or replacing mycinose (342,349,365,366) and some unusual 14,15-cyclopropyl derivatives were prepared (361,365). Modifications of the dienone of tylosin and desmycosin also yielded active derivatives (348,367–369).

Hybrid Macrolides

Other macrolides have been prepared which represent hybrids of structures within the 14-membered family, within the 16-membered family, or between the two families. These hybrids have been made by chemical, bioconversion, or genetic manipulations. The 9-*O*-[(2-methoxyethoxy)methyl]oxime of tylosin (Table 7) was synthesized, using the oxime found in roxithromycin (**33**) (369). 3-*O*-Cladinosyl derivatives of 16-membered macrolides were synthesized in which the neutral sugar of erythromycin (Table 3) was attached to tylosin derivatives at their 3-hydroxyl group, analogous to its position in erythromycin (370).

An aglycone obtained from one macrolide has been bioconverted by a microorganism producing either the same or a different macrolide; such studies have been important for production of new compounds and elucidation of biosynthetic pathways (102,279,289,291,371–378). These bioconversions were pioneered by use of cerulenin [*17397-89-6*], $C_{12}H_{17}NO_3$, an inhibitor of fatty acid synthesis, to block biosynthesis of the aglycone normally produced by the bioconverting organism (379). For example, feeding tylactone (protylonolide) (**59**) to the spiramycin-producer, *S. ambofaciens,* yielded four 9-dihydro-9-*O*-forosaminyl derivatives named chimeramycins (371). A summary of microbial conversions of macrolides has been published (149).

The advent of molecular biology has opened new possibilities for producing hybrid macrolides. Insertion of DNA from the oleandomycin-producer into a mutant strain of the erythromycin-producer produced derivatives of erythromycin lacking the 2-methyl group (380). Genetic manipulations of biosynthetic pathways in macrolide-producing microorganisms complement traditional chemical and microbiological approaches (381–386). For example, inactivation of the *Ery F* gene coding for a 6-hydroxylase in *Sa. erythraea* produced 6-deoxyerythromycin (385).

Biological Properties

Antimicrobial Properties. Macrolides inhibit growth of gram-positive bacteria, *Mycoplasma* species, and certain gram-negative and anaerobic bacteria. Susceptible gram-positive bacteria include many species of *Staphylococcus* and *Streptococcus;* susceptible gram-negative bacteria include *Bordetella pertussis, Legionella pneumophila, Moraxella catarrhalis* (formerly *Branhamella*), and *Haemophilus ducreyi.* Although erythromycin has some *in vitro* activity against *Haemophilus influenzae,* it does not exhibit a high level of *in vivo* efficacy. An important susceptible anaerobe is *Propionibacterium acnes.* Comparative evaluations have been published (387–392), and monographs devoted to particular macrolides and reviews are available (2–14).

Azithromycin (**36**) expanded the traditional *in vitro* spectrum because of increased activity against gram-negative bacteria, including *H. influenzae* (393). *In vivo* efficacy against this organism was demonstrated in a middle ear infection in gerbils (394); however, azithromycin failed to eradicate *H. influenzae* in an initial study of chronic bronchitis (395). Miokamycin (**62**), rokitamycin (**62**), clarithromycin (**37**), and roxithromycin (**33**) showed increased activity *in vitro* against *L. pneumophila* (387,388); the latter two were more efficacious than erythromycin against *Legionella* in gerbils (396–398). *Helicobacter pylori* (formerly *Campylobacter*) is very susceptible to macrolides *in vitro,* but treatment of gastric disorders with macrolides has not been established (399,400). Spiramycin (Table 6) is used to treat *Toxoplasma gondii,* but more effective agents are being sought (401). Some macrolides were active at high doses against acute toxoplasmosis or toxoplasmic encephalitis in mice (402–404), and new methods for *in vitro* testing have been reported (405). Azithromycin (**36**) was active *in vivo* against *Borrelia burgdorferi,* causative agent of Lyme disease (406,407). *In vitro* inhibition of *Chlamydia pneumoniae* (formerly *Chlamydia* TWAR) (408) and several species of mycobacteria, including *M. avium,* have been reported (409–414). These examples represent efforts to expand the spectrum of macrolides and establish clinical efficacy against organisms susceptible *in vitro.*

Macrolides inhibit growth of bacteria by inhibiting protein synthesis on ribosomes (17,415,416). Bacterial resistance to macrolides is often accompanied by cross-resistance to lincosamide and streptogramin B antibiotics (MLS-resistance), which can be either inducible or constitutive (417). 14-Membered macrolides generally induce resistance to themselves, whereas 16-membered macrolides do not; consequently, one advantage of the latter is their activity against bacteria which are inducibly resistant to erythromycin. Both 14- and 16-membered macrolides lack activity against constitutively resistant strains (387,388).

Bacterial resistance to antibiotics usually results from modification of a target site, enzymatic inactivation, or reduced uptake into bacterial cells. Modification of the macrolide's ribosomal binding site involves methylation of the 23S fraction of ribosomal RNA (418). Modifications of 23S rRNA are under investigation (419–421). Derivatives of erythromycin have been found which do not induce resistance and also inhibit strains of both inducibly and constitutively resistant bacteria (157–160). Enzymatic inactivations of macrolides include esterases (422), phosphorylases (423,424), and glycosidases (425), and reduced uptake into cells has been reported (426,427). The possible origin of bacterial inactivating enzymes from macrolide-producing organisms has been proposed (422).

Pharmacokinetics and Pharmacology. Older macrolides such as erythromycin (Table 7) exhibit relatively low serum concentrations, short *in vivo* half-lives, highly variable oral absorption, and low oral bioavailability. Improvements in these pharmacokinetic parameters have been accomplished for newer derivatives. Pharmacokinetic parameters for individual compounds are available (2–14,19,20,428–430). A beneficial feature of macrolides is the high concentrations in certain tissues such as pulmonary and prostate that far exceed serum concentrations (430). Another capability of these agents is intracellular penetration, a necessary condition for activity against intracellular pathogens. Cellular uptake, intracellular antibacterial activity, and interactions of macrolides with components of the host defense system are being investigated (431–443).

The principal route of macrolide excretion is by way of the liver. Effects of macrolides on hepatic metabolic enzymes, particularly cytochrome P-450, have been studied in order to identify and reduce potential interference with metabolism of other drugs (21–23,444–447). Several macrolides are initially metabolized to products which retain antibiotic activity, thereby contributing to more persistent *in vivo* concentrations of antibiotic. Clarithromycin (**37**) is oxidized to its 14-hydroxy derivative; the 14(*R*) isomer has activity comparable to clarithromycin (448) and is readily prepared by biotransformation using *Mucor circinelloides* (449). Dirithromycin (**35**) is a pro-drug converted to erythromycylamine (**34**) *in vivo* (450). Miokamycin and rokitamycin (**62**) undergo deacylations to other compounds in the leucomycin group (322–324,451,452).

The principal side effects of macrolides are gastrointestinal problems, such as pain, indigestion, diarrhea, nausea, and vomiting (453). Erythromycin (**14**) and oleandomycin (**17**, R = CH_3) stimulate motility of the gastrointestinal tract in dogs, whereas 16-membered macrolides do not (454,455). This model has been used to select derivatives of erythromycin having reduced potential for causing gastrointestinal upset. More recently, erythromycin was shown to exert a variety of effects on the human gastrointestinal tract, including induction of migrating motor complex and increased lower esophageal sphincter pressure and gastric emptying (456,457). At least part of its mechanism of action was proposed as an agonist of motilin, an endogenous peptide in the gastrointestinal tract responsible for interdigestive contractility (458–460). The dog contractility model was also used to select and evaluate macrolide derivatives for the treatment of gastrointestinal motility disorders (461–463).

Biosynthetic Patterns and Conformational Analysis

Macrolides are obtained by controlled submerged aerobic fermentations of soil microorganisms. Although species of *Streptomyces* have dominated, species of *Saccharopolyspora, Micromonospora,* and *Streptoverticillium* are also well represented. New techniques such as enzyme-linked immunosorbent assay (ELISA) based assays may prove beneficial for discovering new structures (464).

The biosynthetic steps which convert the initially produced aglycone to final product have been elucidated by bioconversions, blocked mutants, and radiolabeled precursors (465,466). The regular pattern of methyl substituents on alternating carbon atoms (propionate or Gerzon's rule) results from methyl groups of propionate (27,467). Many studies have established that aglycones are assembled from acetate and propionate units, with an occasional butyrate, and sugars are derived from glucose (28,465,466). Enzymes involved in these processes have been reviewed (383,468), and factors regulating production have been summarized (469). Biosynthesis of aglycones and fatty acids have some common features (383,470); this analogy was exploited by blocking aglycone biosynthesis with cerulenin, an inhibitor of fatty acid biosynthesis (379). A common pattern found in the stereochemistry of substituents on aglycones (Celmer's model or rules) has been used to predict stereochemistry and identify incorrect assignments (471,472). The sequence of biosynthetic intermediates leading to aglycones (383,473–477) and deoxysugars (478,479) is under investigation.

Conformational analysis has been used to find and predict conformations which maximize antibiotic activity, using x-ray crystal structures coupled with nmr and cd spectra. An early approach utilized the Dale diamond lattice conformational model (480), which was extended to other diamond lattice models (472,481–483). Other studies have relied on nmr data (225,484–491). However, extensive correlations between conformation and biological activity have not been successful (486,492).

Economic Aspects

Macrolide antibiotics are used clinically to treat infections resulting from susceptible organisms in the upper and lower respiratory tract, skin and soft tissues, and genital tract. They are generally used orally, although they can be given intravenously. For the latter purpose, a water-soluble salt is used, because macrolides are poorly soluble in water as free bases. Commonly employed acid-addition salts include the lactobionate, gluceptate, tartrate, and phosphate. For oral administration, acid-stable esters, salts, formulations, and coatings are necessary for erythromycin, which is unstable as its unprotected free base under acidic conditions such as those in the stomach. They are not administered by intramuscular injection because of severe pain.

Reported worldwide sales of macrolides are estimated as $800,000,000 annually; approximately 25% of this figure are U.S. sales. Table 13 provides a partial list of products commercially available for clinical use (493–495).

Macrolides are regarded as among the safest of antibiotics (496). The principal side effects, which in some cases are sufficiently severe to require cessation of the drug, are gastrointestinal.

Table 13. Commercial Macrolide Products[a]

Macrolide	Trade name[b]	Clinical route	Manufacturer
erythromycin[c]	Ery-tab	oral	Abbott
erythromycin[d]	Erythromycin base filmtab	oral	Abbott
erythromycin[e]	Erythromycin delayed-release capsules	oral	Abbott
erythromycin[f]	E-Mycin	oral	Boots
erythromycin[c]	Ilotycin	oral	Dista
erythromycin[e]	Eryc	oral	Parke-Davis
erythromycin topical solution	Eryderm 2%	topical	Abbott
erythromycin topical solution	ETS-2%	topical	Paddock
erythromycin ophthalmic ointment	Ilotycin ointment	topical	Dista
erythromycin stearate	Erythrocin stearate	oral	Abbott
erythromycin lactobionate	Erythrocin lactobionate-iv	iv	Abbott
	Erythrocin piggyback	iv	Abbott
erythromycin gluceptate	Ilotycin gluceptate	iv	Dista
erythromycin ethyl succinate	E.E.S.	oral	Abbott
	Eryped	oral	Abbott
erythromycin estolate	Ilosone	oral	Dista
erythromycin acistrate	Erasis (Finland)	oral	Orion
erythromycin stinoprate	Erythrocist (Italy)	oral	Refarmed
propionylerythromycin mercaptosuccinate	Zalig (Italy)	oral	Pierrel
erythromycin-11,12-carbonate	Davercin (Poland)	oral	Tarchomin
roxithromycin	Rulid (France)	oral	Roussel-Uclaf
clarithromycin	Klaricid (Ireland)	oral	Abbott
azithromycin	Sunamed (Yugoslavia)	oral	Pliva
triacetyloleandomycin	TAO	oral	Roerig
josamycin	Josamycin (Japan)	oral	Yamanouchi
miokamycin	Miocamycin (Japan)	oral	Meiji Seika
rokitamycin	Ricamycin (Japan)	oral	Toyo Jozo

[a]Abstracted from refs. 116, 495; updates of 116 and 495, and 494.
[b]Country in parenthesis represents launch outside the United States.
[c]Enteric-coated tablets.
[d]Film-coated tablets.
[e]Enteric-coated pellets.
[f]Coated tablets.

Relatively few macrolides are used in veterinary medicine (497). The most important is tylosin (Tylan, Elanco Products), which is used to control chronic respiratory disease caused by *Mycoplasma gallisepticum* in poultry. It is also used to treat and control infections in pigs resulting from gram-positive bacteria and *Mycoplasma,* and as a growth promotant for pigs and poultry. Other macrolides in veterinary use include erythromycin, oleandomycin, and spiramycin (497). Newer macrolides being developed for veterinary applications are 3-*O*-acetyl-4″-*O*-isovaleryltylosin and tilmicosin. The pharmacokinetics of macrolides in animals have been reviewed (428,429), and a book devoted to pharmacokinetics of veterinary antibiotics is scheduled (498).

BIBLIOGRAPHY

"Antibiotics" in *ECT* 1st ed., Vol. 2, pp. 7–37, by J. S. Kiser and H. B. Woodruff; in *ECT* 2nd ed., Vol. 2, pp. 533–540, by W. E. Brown, The Squibb Institute for Medical Research; "Macrolide Antibiotics" in *ECT* 2nd ed., Vol. 12, pp. 632–660, by R. Morin and M. Gorman, Eli Lilly and Company; "Antibiotics, Macrolides" in *ECT* 3rd ed., Vol. 2, pp. 937–957, by A. K. Mallams, Schering-Plough Corporation.

1. S. Omura, ed., *Macrolide Antibiotics: Chemistry, Biology, and Practice,* Academic Press, Orlando, Fla., 1984.
2. D. C. Brittain, *Med. Clin. North Am.* **71,** 1147 (1987).
3. J. D. Nelson, ed., *Pediatr. Infect. Dis.* **5,** 118 (1986).
4. J. A. Washington and W. R. Wilson, *Mayo Clin. Proc.* **60,** 189 (1985).
5. *Ibid.,* p. 271.
6. P. Davey, J.-C. Pechere, and D. Speller, eds., *J. Antimicrob. Chemother.* **22**(Suppl. B) (1988).
7. H. C. Neu and J. F. Acar, eds., *Brit. J. Clin. Pract.* **42**(Suppl. 55) (1988).
8. I. Phillips, J.-C. Pechere, A. Davies, and D. Speller, eds., *J. Antimicrob. Chemother.* **20**(Suppl. B) (1987).
9. J. P. Butzler, ed., *Macrolides: A Review with an Outlook on Future Developments,* Excerpta Medica, Amsterdam, The Netherlands, 1986.
10. D. A. Leigh and G. L. Ridgway, eds., *J. Antimicrob. Chemother.* **25**(Suppl. A) (1990).
11. P. Davey and R. Williams, eds., *J. Antimicrob. Chemother.* **21**(Suppl. D) (1988).
12. H. A. Kirst, *Ann. Rep. Med. Chem.* **25,** Chapt. 13 (1990).
13. H. A. Kirst and G. D. Sides, *Antimicrob. Agents Chemother.* **33,** 1413 (1989).
14. *Ibid.,* p. 1419.
15. Y. Fukagawa, K. Kiyoshima, and T. Yoshioka, *Life Chemistry Reports* **6,** 267 (1988).
16. H. A. Kirst, in G. Lukacs and M. Ohno, eds., *Recent Progress in the Chemical Synthesis of Antibiotics,* Springer-Verlag, Heidelberg, Germany, 1990, p. 39.
17. I. Phillips and J. D. Williams, eds., *J. Antimicrob. Chemother.* **16**(Suppl. A) (1985).
18. R. W. Auckenthaler, A. Zwahlen, and F. A. Waldvogel, in P. K. Peterson and J. Verhoef, eds., *Antimicrobial Agents Annual 3,* Elsevier Science Publishers, Amsterdam, The Netherlands, 1988, p. 122.
19. P. Periti, T. Mazzei, E. Mini, and A. Novelli, *Clin. Pharmacokinetics* **16,** 193 (1989).
20. *Ibid.,* p. 261.
21. T. M. Ludden, *Clin. Pharmacokinetics* **10,** 63 (1985).
22. J. Descotes, P. Andre, and J. C. Evreux, *J. Antimicrob. Chemother.* **15,** 659 (1985).
23. H. C. Neu, in J. C. Petrie, ed., *Gastrointestinal, Haematological and Infectious Disease Therapy,* Elsevier Science Publishers, Amsterdam, The Netherlands, 1985, p. 215.
24. K. Tatsuta in Ref. 16, p. 1.
25. R. K. Boeckman and S. W. Goldstein, in J. ApSimon, ed., *The Total Synthesis of Natural Products,* Vol. 7, John Wiley & Sons, Inc., New York, 1988, p. 1.
26. I. Paterson and M. M. Mansuri, *Tetrahedron* **41,** 3569 (1985).
27. R. B. Woodward, *Angew. Chem.* **69,** 50 (1957).
28. D. O'Hagan, *Natl. Prod. Reports* **6,** 205 (1989).
29. M. N. Donin, J. Pagano, J. D. Dutcher, and C. M. McKee, *Antibiot. Annu. 1953–1954,* 179 (1953).
30. C. Djerassi and J. A. Zderic, *J. Am. Chem. Soc.* **78,** 2907 (1956).
31. *Ibid.,* p. 6390.
32. D. G. Manwaring, R. W. Rickards, and R. M. Smith, *Tetrahedron Lett.,* 1029 (1970).
33. S. Masamune, H. Yamamoto, S. Kamata, and A. Fukuzawa, *J. Am. Chem. Soc.* **97,** 3513 (1975).
34. C. Djerassi and O. Halpern, *J. Am. Chem. Soc.* **79,** 2022 (1957).

35. C. Djerassi and O. Halpern, *Tetrahedron* **3,** 255 (1958).
36. I. Maezawa, A. Kinumaki, and M. Suzuki, *J. Antibiot.* **27,** 84 (1974).
37. J. Inanaga, Y. Kawanami, and M. Yamaguchi, *Chem. Lett.,* 1415 (1981).
38. A. Kinumaki and M. Suzuki, *J. Chem. Soc., Chem. Commun.,* 744 (1972).
39. J. M. McGuire and co-workers, *Antibiot. Chemother.* **2,** 281 (1952).
40. D. P. Labeda, *Int. J. Syst. Bacteriol.* **37,** 19 (1987).
41. P. F. Wiley and co-workers, *J. Am. Chem. Soc.* **79,** 6062 (1957).
42. D. R. Harris, S. G. McGeachin, and H. H. Mills, *Tetrahedron Lett.,* 679 (1965).
43. R. B. Woodward and co-workers, *J. Am. Chem. Soc.* **103,** 3215 (1981).
44. R. B. Woodward, in A. Todd, ed., *Perspectives in Organic Chemistry,* Interscience Publishers, New York, 1956, p. 160.
45. P. F. Wiley and co-workers, *J. Am. Chem. Soc.* **79,** 6070 (1957).
46. P. F. Wiley, R. Gale, C. W. Pettinga, and K. Gerzon, *J. Am. Chem. Soc.* **79,** 6074 (1957).
47. J. Majer, J. R. Martin, R. S. Egan, and J. W. Corcoran, *J. Am. Chem. Soc.* **99,** 1620 (1977).
48. J. Tone and co-workers, *Proc. 24th Intersci. Conf. Antimicrob. Agents Chemother.,* Oct. 8–10, 1984, Washington, D.C., Abstr. No. 796.
49. J. R. Martin and co-workers, *J. Antibiot.* **35,** 426 (1982).
50. Th. Cachet, E. Roets, J. Hoogmartens, and H. Vanderhaeghe, *J. Chromatogr.* **403,** 343 (1987).
51. J. R. Martin, R. S. Egan, A. W. Goldstein, and P. Collum, *Tetrahedron* **31,** 1985 (1975).
52. Th. Cachet and co-workers, *J. Chromatogr.* **445,** 290 (1988).
53. J. R. Martin, T. J. Perun, and R. L. Girolami, *Biochemistry* **5,** 2852 (1966).
54. J. R. Martin and W. Rosenbrook, *Biochemistry* **6,** 435 (1967).
55. J. R. Martin and T. J. Perun, *Biochemistry* **7,** 1728 (1968).
56. J. R. Martin and R. S. Egan, *Biochemistry* **9,** 3439 (1970).
57. J. R. Martin, T. J. Perun, and R. S. Egan, *Tetrahedron* **28,** 2937 (1972).
58. J. R. Martin, R. S. Egan, T. J. Perun, and A. W. Goldstein, *Tetrahedron* **29,** 935 (1973).
59. J. R. Martin and co-workers, *J. Antibiot.* **27,** 570 (1974).
60. I. O. Kibwage and co-workers, *J. Antibiot.* **40,** 1 (1987).
61. I. O. Kibwage and co-workers, *J. Chromatogr.* **346,** 309 (1985).
62. Th. Cachet and co-workers, *J. Chromatogr.* **409,** 91 (1987).
63. R. A. LeMahieu and co-workers, *J. Med. Chem.* **17,** 953 (1974).
64. P. P. Hung, C. L. Marks, and P. L. Tardrew, *J. Biol. Chem.* **240,** 1322 (1965).
65. S. Masamune, G. S. Bates, and J. W. Corcoran, *Angew. Chem. Int. Eng. Ed.* **16,** 585 (1977).
66. E. J. Corey and co-workers, *J. Am. Chem. Soc.* **100,** 4620 (1978).
67. E. J. Corey and co-workers, *J. Am. Chem. Soc.* **101,** 7131 (1979).
68. S. Masamune and co-workers, *J. Am. Chem. Soc.* **103,** 1568 (1981).
69. B. A. Sobin, A. R. English, and W. D. Celmer, *Antibiot. Annu. 1954–1955,* 827 (1955).
70. W. D. Celmer, H. Els, and K. Murai, *Antibiot. Annu. 1957–1958,* 476 (1958).
71. F. A. Hochstein and co-workers, *J. Am. Chem. Soc.* **82,** 3225 (1960).
72. W. D. Celmer, *J. Am. Chem. Soc.* **87,** 1797 (1965).
73. H. Ogura, K. Furuhata, Y. Harada, and Y. Iitaka, *J. Am. Chem. Soc.* **100,** 6733 (1978).
74. W. D. Celmer, *Pure Appl. Chem.* **28,** 413 (1971).
75. K. Tatsuta, Y. Kobayashi, and H. Gunji, *J. Antibiot.* **41,** 1520 (1988).
76. K. Tatsuta, Y. Kobayashi, H. Gunji, and H. Masuda, *Tetrahedron Lett.* **29,** 3975 (1988).
77. H. Brockmann and W. Henkel, *Naturwissenschaften* **37,** 138 (1950).
78. H. Brockmann and W. Henkel, *Chem. Ber.* **84,** 284 (1951).
79. T. Hata and co-workers, *J. Antibiot. Ser. A* **8,** 9 (1955).
80. B. Takahashi, *J. Antibiot. Ser. A* **7,** 149 (1954).
81. H. Ogura, A. Otagoshi, Y. Sano, and T. Hata, *Chem. Pharm. Bull.* **15,** 682 (1967).

82. A. D. Gardner and E. Chain, *Brit. J. Exp. Pathol.* **23,** 123 (1942).
83. H. Brockmann, H. Genth, and R. Strufe, *Chem. Ber.* **85,** 426 (1952).
84. S. A. Waksman, *The Actinomycetes,* Vol. 2, Williams & Wilkins, Baltimore, Md., 1961, p. 215.
85. H. Muxfeldt, S. Shrader, P. Hansen, and H. Brockmann, *J. Am. Chem. Soc.* **90,** 4748 (1968).
86. R. W. Rickards, R. M. Smith, and J. Majer, *J. Chem. Soc., Chem. Commun.,* 1049 (1968).
87. R. E. Hughes, H. Muxfeldt, C. Tsai, and J. J. Stezowski, *J. Am. Chem. Soc.* **92,** 5267 (1970).
88. C. Tsai, J. J. Stezowski, and R. E. Hughes, *J. Am. Chem. Soc.* **93,** 7286 (1971).
89. H. Ogura, K. Furuhata, and H. Kuwano, *Tetrahedron Lett.,* 4715 (1971).
90. A. Furusaki, T. Matsumoto, K. Furuhata, and H. Ogura, *Bull. Chem. Soc. Japan* **55,** 59 (1982).
91. H. Brockmann and R. Oster, *Chem. Ber.* **90,** 605 (1957).
92. J. Majer, J. B. McAlpine, R. S. Egan, and J. W. Corcoran, *J. Antibiot.* **29,** 769 (1976).
93. R. Corbaz and co-workers, *Helv. Chim. Acta* **38,** 935 (1955).
94. V. Prelog, A. M. Gold, G. Talbot, and A. Zamojski, *Helv. Chim. Acta* **45,** 4 (1962).
95. T. Hori, I. Maezawa, N. Nagahama, and M. Suzuki, *J. Chem. Soc., Chem. Commun.,* 304 (1971).
96. I. Maezawa, T. Hori, and M. Suzuki, *Agric. Biol. Chem.* **38,** 91 (1974).
97. *Ibid.,* p. 539.
98. T. Kaiho, S. Masamune, and T. Toyoda, *J. Org. Chem.* **47,** 1612 (1982).
99. H. Ogura, K. Furuhata, H. Kuwano, and N. Harada, *J. Am. Chem. Soc.* **97,** 1930 (1975).
100. Jpn. Pat. Appl. Kokai 86-234,791 (Oct. 20, 1986).
101. *Meiji Seika Kenkyu Nenpo* (24), 52 (1985); *Chem. Abstr.* **106,** 152654 (1987).
102. I. Maezawa, A. Kinumaki, and M. Suzuki, *J. Antibiot.* **29,** 1203 (1976).
103. J. Marquez and co-workers, *J. Antibiot.* **22,** 259 (1969).
104. I. Kawamoto and co-workers, *J. Antibiot.* **27,** 493 (1974).
105. R. S. Egan and co-workers, *J. Antibiot.* **27,** 544 (1974).
106. T. T. Thang and co-workers, *J. Am. Chem. Soc.* **100,** 663 (1978).
107. P. Bartner and co-workers, *J. Chem. Soc., Perkin Trans. 1,* 1600 (1979).
108. E. Gaumann and co-workers, *Helv. Chim. Acta* **43,** 601 (1960).
109. G. Roncari and W. Keller-Schierlein, *Helv. Chim. Acta* **49,** 705 (1966).
110. R. S. Egan and J. R. Martin, *J. Am. Chem. Soc.* **92,** 4129 (1970).
111. S. Omura and co-workers, *J. Antibiot.* **22,** 500 (1969).
112. S. Omura and co-workers, *J. Antibiot.* **22,** 629 (1969).
113. J. R. Martin and co-workers, *Helv. Chim. Acta* **59,** 1886 (1976).
114. J. R. Martin and co-workers, *Helv. Chim. Acta* **60,** 2559 (1977).
115. B. Arnoux, C. Pascard, L. Raynaud, and J. Lunel, *J. Am. Chem. Soc.* **102,** 3605 (1980).
116. J. Hardcastle, ed., *Pharmaprojects (Therapeutic Categories, Main Volume),* Vol. 10, PJB Publications, Richmond, UK, 1989, p. a616, and monthly updates.
117. P. Kurath, P. H. Jones, R. S. Egan, and T. J. Perun, *Experientia* **27,** 362 (1971).
118. Th. Cachet and co-workers, *Int. J. Pharm.* **55,** 59 (1989).
119. C. Vinckier and co-workers, *Int. J. Pharm.* **55,** 67 (1989).
120. I. O. Kibwage and co-workers, *J. Org. Chem.* **52,** 990 (1987).
121. H. A. Kirst, J. A. Wind, and J. W. Paschal, *J. Org. Chem.* **52,** 4359 (1987).
122. A. A. Nagel and co-workers, *J. Org. Chem.* **51,** 5397 (1986).
123. T. Adachi, *J. Org. Chem.* **54,** 3507 (1989).
124. H. W. Murphy, *Antibiot. Annu. 1953–1954,* 500 (1953).
125. V. C. Stephens, *Antibiot. Annu. 1953–1954,* 514 (1953).
126. R. K. Clark and E. L. Varner, *Antibiot. Chemother.* **7,** 487 (1957).

127. P. L. Tardrew, J. C. H. Mao, and D. Kenney, *Appl. Microbiol.* **18,** 159 (1969).
128. L. Dall'Asta and co-workers, *J. Antibiot.* **41,** 139 (1988).
129. B. Steffansen and H. Bundgaard, *Int. J. Pharm.* **56,** 159 (1989).
130. A. Gordin and co-workers, *Antimicrob. Agents Chemother.* **32,** 1019 (1988).
131. P. Saloranta, L. Roos, E. Elonen, and H. Allonen, *J. Antimicrob. Chemother.* **24,** 455 (1989).
132. M. DeBernardi, F. Feletti, G. Gazzani, and G. B. Fregnan, *Int. J. Clin. Pharm. Ther. Tox.* **26,** 444 (1988).
133. G. Ricevuti and co-workers, *Chemotherapy (Basel)* **34,** 374 (1988).
134. E. Concia and co-workers, *J. Int. Med. Res.* **14,** 137 (1986).
135. W. D. Celmer, *Antibiot. Annu. 1958–1959,* 277 (1959).
136. A. Neszmelyi and H. Bojarska-Dahlig, *J. Antibiot.* **31,** 487 (1978).
137. P. T. Mannisto and co-workers, *Proc. 13th Int. Cong. Chemother.*, Vienna, Austria, Aug. 28–Sept. 2, 1983, Vol. 5, Part 107, p. 5.
138. W. R. Baker and co-workers, *Proc. 27th Intersci. Conf. Antimicrob. Agents Chemother.*, New York, Oct. 4–7, 1987, Abstr. No. 221.
139. W. R. Baker, J. D. Clark, R. L. Stephens, and K. H. Kim, *J. Org. Chem.* **53,** 2340 (1988).
140. H. Sakakibara and S. Omura, in Ref. 1, p. 89.
141. J.-F. Chantot, A. Bryskier, and J.-C. Gasc, *J. Antibiot.* **39,** 660 (1986).
142. E. H. Massey, B. S. Kitchell, L. D. Martin, and K. Gerzon, *J. Med. Chem.* **17,** 105 (1974).
143. P. Luger and R. Maier, *J. Cryst. Mol. Struct.* **9,** 329 (1979).
144. S. Djokic and co-workers, *J. Chem. Res. (S),* 152 (1988).
145. G. M. Bright and co-workers, *J. Antibiot.* **41,** 1029 (1988).
146. S. Morimoto and co-workers, *J. Antibiot.* **43,** 286 (1990).
147. S. Morimoto and co-workers, *J. Antibiot.* **43,** 295 (1990).
148. L. Toscano, G. Fioriello, S. Silingardi, and M. Inglesi, *Tetrahedron* **40,** 2177 (1984).
149. P. F. Wiley, *Dev. Indust. Microbiol.* **26,** 97 (1985).
150. J. Wilson, J. Durodie, and M. Foulstone, *J. Antimicrob. Chemother.* **22,** 84 (1988).
151. J. M. Wilson, P. C. T. Hannan, C. Shillingford, and D. J. C. Knowles, *J. Antibiot.* **42,** 454 (1989).
152. J. S. Davies, E. Hunt, and I. I. Zomaya, *J. Chem. Soc., Perkin Trans. 1,* 1409 (1990).
153. E. Hunt and co-workers, *J. Antibiot.* **42,** 293 (1989).
154. H. A. Kirst and co-workers, *Proc. 29th Intersci. Conf. Antimicrob. Agents Chemother.*, Sept. 17–20, 1989, Houston, Tex., Abstr. No. 1021.
155. R. N. Swanson and co-workers, in Ref. 154, Abstr. No. 1025.
156. L. A. Freiberg and co-workers, *Proc. 9th Internat. Symp. Future Trends Chemother.*, March 26–28, 1990, Palexpo (Geneva), Switzerland, p. 30.
157. N. E. Allen, *Antimicrob. Agents Chemother.* **11,** 669 (1977).
158. S. Kadam, C. C. Doran, and R. C. Goldman, *Proc. 88th Mtg. Amer. Soc. Microbiol.*, May 8–13, 1988, Miami, Fla., Abstr. No. A-11.
159. P. B. Fernandes, W. R. Baker, L. A. Freiberg, D. J. Hardy, and E. J. McDonald, *Antimicrob. Agents Chemother.* **33,** 78 (1989).
160. R. C. Goldman and S. K. Kadam, *Antimicrob. Agents Chemother.* **33,** 1058 (1989).
161. L. A. Freiberg and co-workers, in Ref. 138, Abstr. No. 224.
162. T. Hata and co-workers, *J. Antibiot.* **6,** 87 (1953).
163. T. Arai, S. Kuroda, and Y. Mikami, in T. Arai, ed., *Actinomycetes: The Boundary Microorganisms,* Toppan Co. Ltd., Tokyo, Japan, 1976, p. 632.
164. S. Omura, H. Ogura, and T. Hata, *Tetrahedron Lett.,* 609 (1967).
165. *Ibid.,* p. 1267.
166. T. Hata and co-workers, *Chem. Pharm. Bull.* **15,** 358 (1967).
167. S. Omura, M. Katagiri, and T. Hata, *J. Antibiot.* **20,** 234 (1967).

168. S. Omura, M. Katagiri, H. Ogura, and T. Hata, *Chem. Pharm. Bull.* **15,** 1529 (1967).
169. S. Omura, M. Katagiri, and T. Hata, *J. Antibiot.* **21,** 199 (1968).
170. *Ibid.,* p. 272.
171. S. Omura, M. Katagiri, H. Ogura, and T. Hata, *Chem. Pharm. Bull.* **16,** 1167 (1968).
172. *Ibid.,* p. 1181.
173. M. Hiramatsu and co-workers, *Bull. Chem. Soc. Japan* **40,** 2982 (1967).
174. M. Hiramatsu and co-workers, *Bull. Chem. Soc. Japan* **43,** 1966 (1970).
175. S. Omura and co-workers, *Chem. Pharm. Bull.* **16,** 1402 (1968).
176. S. Omura and co-workers, *Chem. Pharm. Bull.* **18,** 1501 (1970).
177. L. A. Freiberg, R. S. Egan, and W. H. Washburn, *J. Org. Chem.* **39,** 2474 (1974).
178. A. Ducruix, C. Pascard, A. Nakagawa, and S. Omura, *J. Chem. Soc., Chem. Commun.,* 947 (1976).
179. S. Omura and co-workers, *Tetrahedron* **28,** 2839 (1972).
180. T. Osono and co-workers, *J. Antibiot. Ser. A* **20,** 174 (1967).
181. S. Omura, Y. Hironaka, and T. Hata, *J. Antibiot.* **23,** 511 (1970).
182. M. Muroi, M. Izawa, and T. Kishi, *Experientia* **28,** 129 (1972).
183. M. Muroi and co-workers, *Experientia* **28,** 501 (1972).
184. M. Muroi and co-workers, *Experientia* **28,** 878 (1972).
185. H. Ono, T. Hasegawa, E. Higashide, and M. Shibata, *J. Antibiot.* **26,** 191 (1973).
186. M. Muroi and co-workers, *J. Antibiot.* **26,** 199 (1973).
187. M. Muroi, M. Izawa, and T. Kishi, *Chem. Pharm. Bull.* **24,** 450 (1976).
188. *Ibid.,* p. 463.
189. K. Kamiya, Y. Wada, and K. Kondo, *Chem. Pharm. Bull.* **26,** 1753 (1978).
190. F. W. Tanner, A. R. English, T. M. Lees, and J. B. Routien, *Antibiot. Chemother.* **2,** 441 (1952).
191. J. F. Pagano, M. J. Weinstein, and C. M. McKee, *Antibiot. Chemother.* **3,** 899 (1953).
192. R. L. Wagner and co-workers, *J. Am. Chem. Soc.* **75,** 4684 (1953).
193. M. E. Kuehne and B. W. Benson, *J. Am. Chem. Soc.* **87,** 4660 (1965).
194. R. B. Woodward, L. S. Weiler, and P. C. Dutta, *J. Am. Chem. Soc.* **87,** 4662 (1965).
195. R. Okamoto and co-workers, *J. Ferment. Technol.* **57,** 519 (1979).
196. J. K. Epp, S. G. Burgett, and B. E. Schoner, *Gene* **53,** 73 (1987).
197. Y. Shimauchi and co-workers, *J. Antibiot.* **31,** 261 (1978).
198. Y. Shimauchi and co-workers, *J. Antibiot.* **32,** 878 (1979).
199. M. Sakamoto and co-workers, *J. Antibiot.* **37,** 130 (1984).
200. F. A. Hochstein and K. Murai, *J. Am. Chem. Soc.* **76,** 5080 (1954).
201. M. Suzuki and co-workers, *Tetrahedron Lett.,* 435 (1971).
202. M. Suzuki and co-workers, *J. Antibiot.* **24,** 904 (1971).
203. T. Furumai and co-workers, *J. Antibiot.* **27,** 95 (1974).
204. A. Kinumaki and co-workers, *J. Antibiot.* **27,** 102 (1974).
205. A. Kinumaki and co-workers, *J. Antibiot.* **27,** 107 (1974).
206. A. Kinumaki and co-workers, *J. Antibiot.* **27,** 117 (1974).
207. Jpn. Pat. Appl. Kokai 73-10,288 (Feb. 2, 1973).
208. T. Niida and co-workers, *J. Antibiot.* **24,** 319 (1971).
209. T. Tsuruoka and co-workers, *J. Antibiot.* **24,** 452 (1971).
210. S. Inouye and co-workers, *J. Antibiot.* **24,** 460 (1971).
211. T. Tsuruoka and co-workers, *J. Antibiot.* **24,** 476 (1971).
212. T. Tsuruoka and co-workers, *J. Antibiot.* **24,** 526 (1971).
213. U.S. Pat. 3,819,834 (June 25, 1974).
214. Jpn. Pat. Appl. Kokai 72-25,384 (Oct. 20, 1972).
215. I. Haupt, H. Fricke, J. Cerna, and I. Rychlik, *J. Antibiot.* **29,** 1314 (1976).
216. G. Huber and co-workers, *Arzneim.-Forsch.* **12,** 1191 (1962).
217. U. Grafe and co-workers, *J. Antibiot.* **33,** 566 (1980).

218. U. Grafe and co-workers, *J. Antibiot.* **33,** 574 (1980).
219. T. Furumai and M. Suzuki, *J. Antibiot.* **28,** 775 (1975).
220. Jpn. Pat. Appl. Kokai 86-257,198 (Nov. 14, 1986).
221. S. Pinnert-Sindico, L. Ninet, J. Preud'Homme, and C. Cosar, *Antibiot. Annu. 1954–1955,* 724 (1955).
222. R. Corbaz and co-workers, *Helv. Chim. Acta* **39,** 304 (1956).
223. R. Paul and S. Tchelitcheff, *Bull. Soc. Chim. France,* 650 (1965).
224. S. Omura and co-workers, *J. Am. Chem. Soc.* **91,** 3401 (1969).
225. S. Omura and co-workers, *J. Am. Chem. Soc.* **97,** 4001 (1975).
226. Jpn. Pat. Appl. Kokai 83-74,695 (May 6, 1983).
227. R. Paul and S. Tchelitcheff, *Bull. Soc. Chim. France,* 443 (1957).
228. R. J. Adamski and co-workers, *J. Med. Chem.* **9,** 932 (1966).
229. S. Omura and co-workers, *J. Antibiot.* **27,** 147 (1974).
230. N. N. Girotra and N. L. Wendler, *Tetrahedron Lett.,* 227 (1975).
231. S. Masamune, Y. Hayase, W. K. Chan, and R. L. Sobczak, *J. Am. Chem. Soc.* **98,** 7874 (1976).
232. T. Furumai and M. Suzuki, *J. Antibiot.* **28,** 783 (1975).
233. U. Grafe and co-workers, *J. Antibiot.* **33,** 663 (1980).
234. T. Furumai, K. Takeda, and M. Suzuki, *J. Antibiot.* **28,** 789 (1975).
235. J. M. McGuire and co-workers, *Antibiot. Chemother.* **11,** 320 (1961).
236. R. L. Hamill, M. E. Haney, M. Stamper, and P. F. Wiley, *Antibiot. Chemother.* **11,** 328 (1961).
237. R. B. Morin and M. Gorman, *Tetrahedron Lett.,* 2339 (1964).
238. R. B. Morin, M. Gorman, R. L. Hamill, and P. V. Demarco, *Tetrahedron Lett.,* 4737 (1970).
239. S. Omura, A. Nakagawa, M. Machida, and H. Imai, *Tetrahedron Lett.,* 1045 (1977).
240. S. Omura and co-workers, *J. Antibiot.* **33,** 915 (1980).
241. N. D. Jones and co-workers, *J. Antibiot.* **35,** 420 (1982).
242. H. A. Whaley and co-workers, *Antimicrob. Agents Chemother.,* 45 (1963).
243. L. I. Feldman and co-workers, *Antimicrob. Agents Chemother.,* 54 (1963).
244. R. L. Hamill and W. M. Stark, *J. Antibiot., Ser. A* **17,** 133 (1964).
245. A. J. Kreuzman, J. R. Turner, and W.-K. Yeh, *J. Biol. Chem.* **263,** 15,626 (1988).
246. R. H. Baltz and E. T. Seno, *Antimicrob. Agents Chemother.* **20,** 214 (1981).
247. E. T. Seno and R. H. Baltz, *Antimicrob. Agents Chemother.* **20,** 370 (1981).
248. R. H. Baltz, E. T. Seno, J. Stonesifer, and G. M. Wild, *J. Antibiot.* **36,** 131 (1983).
249. P. Muhlradt, E. Weiss, and T. Reichstein, *Helv. Chim. Acta* **47,** 2164 (1964).
250. H. A. Kirst and co-workers, *J. Antibiot.* **36,** 376 (1983).
251. R. Okamoto and co-workers, *J. Antibiot.* **35,** 921 (1982).
252. K. Kiyoshima and co-workers, *J. Antibiot.* **40,** 1123 (1987).
253. Y. Yamauchi and co-workers, *Proc. 7th Int. Symp. Biol. Actinomycetes,* May 22–26, 1988, Tokyo, Japan, Abstr. No. P4-14.
254. R. Corbaz and co-workers, *Helv. Chim. Acta* **38,** 1202 (1955).
255. N. Nishimura and co-workers, *J. Antibiot.* **18,** 251 (1965).
256. M. Brufani and W. Keller-Schierlein, *Helv. Chim. Acta* **49,** 1962 (1966).
257. A. Kinumaki and M. Suzuki, *J. Antibiot.* **25,** 480 (1972).
258. I. R. Shimi, S. Shoukry, and F. T. Ali, *J. Antibiot.* **32,** 1248 (1979).
259. H. Bickel and co-workers, *Helv. Chim. Acta* **45,** 1396 (1962).
260. J. Clardy, J. Finer-Moore, L. Weiler, and D. C. Wiley, *Tetrahedron* **37**(Suppl. 1), 91 (1981).
261. T. Suzuki, *Bull. Chem. Soc. Jpn.* **43,** 292 (1970).
262. T. Suzuki, N. Sugita, and M. Asai, *Chem. Lett.,* 789 (1973).
263. T. Suzuki, E. Mizuta, and N. Sugita, *Chem. Lett.,* 793 (1973).

264. S. M. Nash and co-workers, in J. D. Nelson and C. Grassi, eds., *Current Chemotherapy and Infectious Disease,* Vol. 1, American Society for Microbiology, Washington, D.C., 1980, p. 462.
265. Y. Sawada and co-workers, *J. Antibiot.* **42,** 242 (1989).
266. T. Kawasaki and co-workers, in Ref. 138, Abstr. No. 227.
267. H. Koshiyama and co-workers, *J. Antibiot. Ser. A* **16,** 59 (1963).
268. H. Koshiyama and co-workers, *J. Antibiot.* **22,** 61 (1969).
269. H. Tsukiura, M. Konishi, M. Saka, T. Naito, and H. Kawaguchi, *J. Antibiot.* **22,** 89 (1969).
270. G. H. Wagman and co-workers, *J. Antibiot.* **25,** 641 (1972).
271. H. Reimann and R. S. Jaret, *J. Chem. Soc., Chem. Commun.,* 1270 (1972).
272. A. K. Ganguly and co-workers, *Tetrahedron Lett.,* 4699 (1980).
273. K. Hatano, E. Higashide, and M. Shibata, *J. Antibiot.* **29,** 1163 (1976).
274. T. Kishi, S. Harada, H. Yamana, and A. Miyake, *J. Antibiot.* **29,** 1171 (1976).
275. T. Furumai and co-workers, *J. Antibiot.* **30,** 443 (1977).
276. A. Kinumaki, K. Harada, T. Suzuki, M. Suzuki, and T. Okuda, *J. Antibiot.* **30,** 450 (1977).
277. H. Imai and co-workers, *J. Antibiot.* **42,** 1000 (1989).
278. H. Tsukiura and co-workers, *J. Antibiot.* **22,** 100 (1969).
279. S. Omura, H. Ikeda, H. Matsubara, and N. Sadakane, *J. Antibiot.* **33,** 1570 (1980).
280. H. A. Kirst and co-workers, *J. Antibiot.* **35,** 1675 (1982).
281. M. Hayashi, M. Ohno, and S. Satoi, *J. Chem. Soc., Chem. Commun.,* 119 (1980).
282. S. Satoi, N. Muto, M. Hayashi, T. Fujii, and M. Otani, *J. Antibiot.* **33,** 364 (1980).
283. M. Hayashi and co-workers, *J. Antibiot.* **34,** 276 (1981).
284. A. K. Ganguly, Y-T. Liu, A. T. McPhail, and O. Sarre, *J. Chem. Soc., Chem. Commun.,* 853 (1982).
285. M. Hayashi, K. Kinoshita, S. Satoi, and K. Nakatsu, *J. Antibiot.* **35,** 1243 (1982).
286. K. Kinoshita and co-workers, *J. Antibiot.* **38,** 522 (1985).
287. K. Kinoshita, S. Satoi, M. Hayashi, and K. Nakatsu, *J. Antibiot.* **42,** 1003 (1989).
288. M. Hayashi and co-workers, *J. Antibiot.* **36,** 175 (1983).
289. J. Lotvin and co-workers, *J. Antibiot.* **35,** 1407 (1982).
290. M. Hayashi and co-workers, *J. Antibiot.* **34,** 346 (1981).
291. A. K. Ganguly and co-workers, *J. Chem. Soc., Chem. Commun.,* 855 (1982).
292. M. Sasaki and co-workers, *Jpn. J. Antibiot.* **42,** 2412 (1989).
293. P. W. K. Woo, H. W. Dion, and Q. R. Bartz, *J. Am. Chem. Soc.* **83,** 3352 (1961).
294. H. W. Dion, P. W. K. Woo, and Q. R. Bartz, *J. Am. Chem. Soc.* **84,** 880 (1962).
295. P. W. K. Woo, H. W. Dion, and L. F. Johnson, *J. Am. Chem. Soc.* **84,** 1066 (1962).
296. P. W. K. Woo, H. W. Dion, and Q. R. Bartz, *J. Am. Chem. Soc.* **84,** 1512 (1962).
297. P. W. K. Woo, H. W. Dion, and Q. R. Bartz, *J. Am. Chem. Soc.* **86,** 2724 (1964).
298. *Ibid.,* p. 2726.
299. J. Krc and R. B. Scott, *Microscope* **23,** 15 (1975).
300. D. V. Lefemine and co-workers, *Antimicrob. Agents Chemother.,* 41 (1963).
301. M. P. Kunstmann and L. A. Mitscher, *Experientia* **21,** 372 (1965).
302. L. A. Mitscher and M. P. Kunstmann, *Experientia* **25,** 12 (1969).
303. W. Cullen and co-workers, in Ref. 253, Abstr. No. P4-13.
304. J. R. Hauske, J. Dibrino, M. Guadliana, and G. Kostek, *J. Org. Chem.* **51,** 2808 (1986).
305. M. P. Kunstmann, L. A. Mitscher, and E. L. Patterson, *Antimicrob. Agents Chemother.,* 87 (1964).
306. H. Achenbach and W. Karl, *Chem. Ber.* **108,** 759 (1975).
307. *Ibid.,* p. 772.
308. *Ibid.,* p. 780.

309. S. Mizobuchi, J. Mochizuki, H. Soga, H. Tanba, and H. Inoue, *J. Antibiot.* **39,** 1776 (1986).
310. C. M. M. Franco, J. N. Gandhi, S. Chatterjee, and B. N. Ganguli, *J. Antibiot.* **40,** 1361 (1987).
311. S. Chatterjee and co-workers, *J. Antibiot.* **40,** 1368 (1987).
312. S. Omura, C. Kitao, and H. Matsubara, *Chem. Pharm. Bull.* **28,** 1963 (1980).
313. M. Hayashi and co-workers, *J. Antibiot.* **34,** 1075 (1981).
314. S. Omura and co-workers, *J. Antibiot.* **21,** 532 (1968).
315. J. R. Prous, ed., *Drugs of the Future* **10,** 486 (1985).
316. J. R. Prous, ed., *Drugs of the Future* **8,** 790 (1983).
317. H. Sakakibara, O. Okekawa, T. Fujiwara, M. Otani, and S. Omura, *J. Antibiot.* **34,** 1001 (1981).
318. H. Sakakibara, O. Okekawa, T. Fujiwara, M. Aizawa, and S. Omura, *J. Antibiot.* **34,** 1011 (1981).
319. S. Omoto, K. Iwamatsu, S. Inouye, and T. Niida, *J. Antibiot.* **29,** 536 (1976).
320. S. Inouye, S. Omoto, K. Iwamatsu, and T. Niida, *J. Antibiot.* **33,** 61 (1980).
321. K. Kawaharajo, Y. Sekizawa, and M. Inoue, *J. Antibiot.* **34,** 436 (1981).
322. K. Fukaya and co-workers, *Chemotherapy (Japan)* **29,** 448 (1981).
323. T. Shomura, S. Someya, S. Murata, K. Umemura, and M. Nishio, *Chem. Pharm. Bull.* **29,** 2413 (1981).
324. T. Yoshida and co-workers, *Jpn. J. Antibiot.* **35,** 1462 (1982).
325. H. Sano and co-workers, *J. Antibiot.* **37,** 750 (1984).
326. *Ibid.,* p. 760.
327. H. Sano and co-workers, *J. Antibiot.* **38,** 1350 (1985).
328. R. Okamoto and co-workers, *J. Antibiot.* **33,** 1300 (1980).
329. R. Okamoto and co-workers, *J. Antibiot.* **33,** 1309 (1980).
330. M. Tsuchiya and co-workers, *J. Antibiot.* **34,** 305 (1981).
331. M. Tsuchiya, T. Sawa, T. Takeuchi, H. Umezawa, and R. Okamoto, *J. Antibiot.* **35,** 673 (1982).
332. B. J. Skelly, D. Andersen, M. Pruss, and R. Pellegrino, *Avian Diseases* **30,** 505 (1986).
333. T. M. Jacks, F. R. Judith, S. D. Feighner, and R. O. Likoff, *Am. J. Vet. Res.* **47,** 2325 (1986).
334. H. A. Kirst and co-workers, *J. Antibiot.* **39,** 1724 (1986).
335. J. K. Epp, M. L. Huber, J. R. Turner, and B. E. Schoner, in C. L. Hershberger, S. W. Queener, and G. Hegeman, eds., *Genetics and Molecular Biology of Industrial Microorganisms,* American Society for Microbiology, Washington, D.C., 1989, p. 35.
336. J. K. Epp, M. L. B. Huber, J. R. Turner, T. Goodson, and B. E. Schoner, *Gene* **85,** 293 (1989).
337. M. Tsuchiya and co-workers, *J. Antibiot.* **35,** 661 (1982).
338. T. Yoshioka and co-workers, *J. Antibiot.* **41,** 1617 (1988).
339. K. Kiyoshima and co-workers, *Chem. Pharm. Bull.* **37,** 861 (1989).
340. T. Takeuchi and co-workers, *J. Antibiot.* **40,** 1358 (1987).
341. K. Kiyoshima and co-workers, *J. Antibiot.* **42,** 1661 (1989).
342. A. G. Fishman, A. K. Mallams, and R. R. Rossman, *J. Chem. Soc., Perkin Trans. 1,* 787 (1989).
343. Y. Shimauchi and co-workers, *J. Antibiot.* **33,** 284 (1980).
344. Y. Shimauchi, M. Sakamoto, K. Hori, T. Ishikura, and J. Lein, *J. Antibiot.* **34,** 245 (1981).
345. H. A. Kirst and co-workers, *J. Med. Chem.* **31,** 1631 (1988).
346. H. Matsubara, K. Miyano, A. Nakagawa, and S. Omura, *Chem. Pharm. Bull.* **30,** 97 (1982).

347. T. Fujiwara, H. Watanabe, Y. Kogami, Y. Shiritani, and H. Sakakibara, *J. Antibiot.* **42,** 903 (1989).
348. A. G. Fishman, A. K. Mallams, M. S. Puar, R. R. Rossman, and R. L. Stephens, *J. Chem. Soc., Perkin Trans. 1,* 1189 (1987).
349. T. Fujiwara, A. Sakai, and H. Sakakibara, *J. Antibiot.* **43,** 327 (1990).
350. H. A. Kirst and co-workers, *J. Antibiot.* **42,** 1673 (1989).
351. M. Debono and co-workers, *J. Antibiot.* **42,** 1253 (1989).
352. H. A. Kirst and co-workers, *J. Antibiot.* **41,** 938 (1988).
353. H. Matsubara, J. Inokoshi, A. Nakagawa, H. Tanaka, and S. Omura, *J. Antibiot.* **36,** 1713 (1983).
354. S. Omura, K. Miyano, H. Matsubara, and A. Nakagawa, *J. Med. Chem.* **25,** 271 (1982).
355. E. E. Ose, *J. Antibiot.* **40,** 190 (1987).
356. E. E. Ose and L. V. Tonkinson, *Vet. Rec.* **123,** 367 (1988).
357. R. N. Gourlay, L. H. Thomas, S. G. Wyld, and C. J. Smith, *Res. Vet. Sci.* **47,** 84 (1989).
358. A. Tanaka, T. Tsuchiya, S. Umezawa, M. Hamada, and H. Umezawa, *J. Antibiot.* **34,** 1377 (1981).
359. A. Tanaka and co-workers, *J. Antibiot.* **35,** 113 (1982).
360. H. A. Kirst and co-workers, *J. Antibiot.* **39,** 1108 (1986).
361. H. A. Kirst and co-workers, *J. Antibiot.* **40,** 823 (1987).
362. R. C. Brunham, C-C. Kuo, C. E. Stevens, and K. K. Holmes, *Ann. Int. Med.* **97,** 216 (1982).
363. A. Tanaka, T. Tsuchiya, S. Umezawa, and H. Umezawa, *J. Antibiot.* **34,** 1374 (1981).
364. A. Tanaka, A. Watanabe, T. Tsuchiya, S. Umezawa, and H. Umezawa, *J. Antibiot.* **34,** 1381 (1981).
365. A. K. Mallams and R. R. Rossman, *J. Chem. Soc., Perkin Trans. 1,* 775 (1989).
366. J. P. Leeds and H. A. Kirst, *Synth. Commun.* **18,** 777 (1988).
367. S. Omura, H. Matsubara, K. Tsuzuki, and A. Nakagawa, *J. Antibiot.* **37,** 1007 (1984).
368. K. Tsuzuki, H. Matsubara, A. Nakagawa, and S. Omura, *J. Antibiot.* **39,** 1784 (1986).
369. C. Ruggeri and co-workers, *J. Antibiot.* **42,** 1443 (1989).
370. A. K. Mallams and R. R. Rossman, *J. Chem. Soc., Perkin Trans. 1,* 799 (1989).
371. S. Omura, N. Sadakane, Y. Tanaka, and H. Matsubara, *J. Antibiot.* **36,** 927 (1983).
372. S. Omura, N. Sadakane, and H. Matsubara, *Chem. Pharm. Bull.* **30,** 223 (1982).
373. N. Sadakane, Y. Tanaka, and S. Omura, *J. Antibiot.* **35,** 680 (1982).
374. N. Sadakane, Y. Tanaka, and S. Omura, *J. Antibiot.* **36,** 921 (1983).
375. S. Omura, C. Kitao, and N. Sadakane, *J. Antibiot.* **33,** 911 (1980).
376. S. Omura, N. Sadakane, C. Kitao, H. Matsubara, and A. Nakagawa, *J. Antibiot.* **33,** 913 (1980).
377. B. K. Lee and co-workers, *J. Antibiot.* **36,** 742 (1983).
378. L. Toscano, G. Fioriello, R. Spagnoli, and L. Cappelletti, *J. Antibiot.* **36,** 1585 (1983).
379. S. Omura, *Bacteriol. Rev.* **40,** 681 (1976).
380. J. B. McAlpine and co-workers, *J. Antibiot.* **40,** 1115 (1987).
381. D. A. Hopwood, *Phil. Trans. Royal Soc. London, B* **324,** 549 (1989).
382. C. R. Hutchinson, C. W. Borell, S. L. Otten, K. J. Stutzman-Engwall, and Y. Wang, *J. Med. Chem.* **32,** 929 (1989).
383. J. A. Robinson, *Chem. Soc. Rev.* **17,** 383 (1988).
384. E. T. Seno and R. H. Baltz, in S. Shapiro, ed., *Regulation of Secondary Metabolism in Actinomycetes,* CRC Press, Boca Raton, Fla., 1989, p. 1.
385. S. Donadio and co-workers, in Ref. 335, p. 53.
386. J. M. Weber, B. Schoner, and R. Losick, *Gene* **75,** 235 (1989).
387. D. J. Hardy and co-workers, *Antimicrob. Agents Chemother.* **32,** 1710 (1988).
388. P. B. Fernandes and D. J. Hardy, *Drugs Exptl. Clin. Res.* **14,** 445 (1988).

389. A. L. Barry, R. N. Jones, and C. Thornsberry, *Antimicrob. Agents Chemother.* **32,** 752 (1988).
390. J. E. Hoppe and A. Eichhorn, *Eur. J. Clin. Microbiol. Infect. Dis.* **8,** 653 (1989).
391. P. Le Noc, J. Croize, A. Bryskier, D. Le Noc, and J. Robert, *Pathol. Biol.* **37,** 553 (1989).
392. N. Bornstein, C. Roudier, and J. Fleurette, *J. Antimicrob. Chemother.* **15,** 17 (1985).
393. J. Retsema and co-workers, *Antimicrob. Agents Chemother.* **31,** 1939 (1987).
394. A. E. Girard and co-workers, *Antimicrob. Agents Chemother.* **31,** 1948 (1987).
395. B. I. Davies, F. P. V. Maesen, and R. Gubbelmans, *J. Antimicrob. Chemother.* **23,** 743 (1989).
396. P. B. Fernandes, R. N. Swanson, D. J. Hardy, E. J. McDonald, and N. Ramer, *Drugs. Exptl. Clin. Res.* **14,** 441 (1988).
397. S. Kohno and co-workers, *J. Antimicrob. Chemother.* **24,** 397 (1989).
398. R. B. Fitzgeorge and A. S. R. Featherstone, *J. Antimicrob. Chemother.* **23,** 462 (1989).
399. D. J. Hardy, C. W. Hanson, D. M. Hensey, J. M. Beyer, and P. B. Fernandes, *J. Antimicrob. Chemother.* **22,** 631 (1988).
400. J. A. Garcia-Rodriguez, J. E. Garcia Sanchez, M. I. Garcia Garcia, E. Garcia Sanchez, and J. L. Munoz Bellido, *Antimicrob. Agents Chemother.* **33,** 1650 (1989).
401. R. McCabe and J. S. Remington, *N. Eng. J. Med.* **318,** 313 (1988).
402. B. J. Luft, *Eur. J. Clin. Microbiol. Infect. Dis.* **6,** 479 (1987).
403. H. R. Chang, F. C. Rudareanu, and J.-C. Pechere, *J. Antimicrob. Chemother.* **22,** 359 (1988).
404. F. G. Araujo, D. R. Guptill, and J. S. Remington, *Antimicrob. Agents Chemother.* **32,** 755 (1988).
405. S. Chamberland, H. A. Kirst, and W. L. Current, in Ref. 154, Abstr. No. 1036.
406. V. Preac-Mursic, B. Wilske, G. Schierz, E. Suss, and B. Gross, *Eur. J. Clin. Microbiol. Infect. Dis.* **8,** 651 (1989).
407. R. C. Johnson, C. Kodner, and M. Russell, in Ref. 138, Abstr. No. 235.
408. K. Chirgwin, P. M. Roblin, and M. R. Hammerschlag, *Antimicrob. Agents Chemother.* **33,** 1634 (1989).
409. N. Khardori, K. Rolston, B. Rosenbaum, S. Hayat, and G. P. Bodey, *J. Antimicrob. Chemother.* **24,** 667 (1989).
410. S. Naik and R. Ruck, *Antimicrob. Agents Chemother.* **33,** 1614 (1989).
411. C. B. Inderlied, P. T. Kolonoski, M. Wu, and L. S. Young, *J. Infect. Dis.* **159,** 994 and 1095 (1989).
412. P. B. Fernandes, D. J. Hardy, D. McDaniel, C. W. Hanson, and R. N. Swanson, *Antimicrob. Agents Chemother.* **33,** 1531 (1989).
413. N. Ramasesh, J. L. Krahenbuhl, and R. C. Hastings, *Antimicrob. Agents Chemother.* **33,** 657 (1989).
414. E. A. Gorzynski, S. I. Gutman, and W. Allen, *Antimicrob. Agents Chemother.* **33,** 591 (1989).
415. M. DiGiambattista, G. Chinali, and C. Cocito, *J. Antimicrob. Chemother.* **24,** 485 (1989).
416. M. A. Arevalo, F. Tejedor, F. Polo, and J. P. G. Ballesta, *J. Med. Chem.* **32,** 2200 (1989).
417. S. Mitsuhashi and M. Inoue, in L. E. Bryan, ed., *Antimicrobial Drug Resistance,* Academic Press, Orlando, Fla., 1984, p. 279.
418. B. Weisblum, *Brit. Med. Bull.* **40,** 47 (1984).
419. S. Douthwaite, T. Powers, J. Y. Lee, and H. F. Noller, *J. Mol. Biol.* **209,** 655 (1989).
420. M. Zalacain and E. Cundliffe, *J. Bacteriol.* **171,** 4254 (1989).
421. S. K. Kadam, *J. Bacteriol.* **171,** 4518 (1989).
422. M. Arthur, A. Brisson-Noel, and P. Courvalin, *J. Antimicrob. Chemother.* **20,** 783 (1987).

423. V. P. Marshall, W. F. Liggett, and J. I. Cialdella, *J. Antibiot.* **42,** 826 (1989).
424. K. O'Hara, T. Kanda, K. Ohmiya, T. Ebisu, and M. Kono, *Antimicrob. Agents Chemother.* **33,** 1354 (1989).
425. M.-S. Kuo and co-workers, *Antimicrob. Agents Chemother.* **33,** 2089 (1989).
426. B. C. Lampson, W. von David, and J. T. Parisi, *Antimicrob. Agents Chemother.* **30,** 653 (1986).
427. G. Palu, S. Valisena, M. F. Barile, and G. A. Meloni, *Eur. J. Epidemiology* **5,** 146 (1989).
428. G. E. Burrows, *J. Am. Vet. Med. Assoc.* **176,** 1072 (1980).
429. G. S. Duthu, *J. Pharm. Sci.* **74,** 943 (1985).
430. F. Fraschini, M. Falchi, and V. Copponi, *New Perspectives Clin. Microbiol.* **4,** 79 (1981).
431. F. O'Grady, *Symp. Soc. Gen. Microbiol.* **38,** 341 (1985).
432. L. Madgwick, S. Mayer, and P. Keen, *J. Antimicrob. Chemother.* **24,** 709 (1989).
433. M. Raghoebar, E. Lindeyer, W. B. Van den Berg, and C. A. M. Van Ginneken, *Biochem. Pharm.* **37,** 3221 (1988).
434. M. L. Villa, F. Valenti, F. Scaglione, M. Falchi, and F. Fraschini, *J. Antimicrob. Chemother.* **24,** 765 (1989).
435. A. M. Cuffini, N. A. Carlone, V. Tullio, and M. Borsotto, *Microbios* **58,** 27 (1989).
436. R. P. Gladue, G. M. Bright, R. E. Isaacson, and M. F. Newborg, *Antimicrob. Agents Chemother.* **33,** 277 (1989).
437. R. Anderson, G. Joone, and C. E. J. van Rensburg, *J. Antimicrob. Chemother.* **22,** 923 (1988).
438. M. Ishiguro and co-workers, *J. Antimicrob. Chemother.* **24,** 719 (1989).
439. M. T. Labro and C. Babin-Chevaye, *J. Antimicrob. Chemother.* **24,** 731 (1989).
440. B. Scorneaux and P. M. Tulkens, in Ref. 154, Abstr. No. 157.
441. R. Anderson, *J. Infect. Dis.* **159,** 966 (1989).
442. A. Naess and C. O. Solberg, *Acta Pathol. Microbiol. Immunol. Scand.* **96,** 503 (1988).
443. K. Takeshita and co-workers, *Drugs Exptl. Clin. Res.* **15,** 527 (1989).
444. D. Pessayre, *Int. J. Clin. Pharm. Res.* **3,** 449 (1983).
445. E. Sartori, M. Delaforge, and D. Mansuy, *Biochem. Pharm.* **38,** 2061 (1989).
446. M. Tinel and co-workers, *J. Pharmacol. Expt. Ther.* **250,** 746 (1989).
447. T. Miura and co-workers, *J. Antimicrob. Chemother.* **24,** 551 (1989).
448. T. Adachi and co-workers, *J. Antibiot.* **41,** 966 (1988).
449. J. Sasaki, K. Mizoue, S. Morimoto, T. Adachi, and S. Omura, *J. Antibiot.* **41,** 908 (1988).
450. G. Bozler, G. Heinzel, U. Lechner, K. Schumacher, and U. Busch, *Proc. 28th Intersci. Conf. Antimicrob. Agents Chemother.,* Oct. 23–26, 1988, Los Angeles, Calif., Abstr. No. 924.
451. T. Watanabe and co-workers, *Drugs Exptl. Clin. Res.* **10,** 371 (1984).
452. A. Sakai and co-workers, in Ref. 450, Abstr. No. 913.
453. B. L. Carter, J. C. Woodhead, K. J. Cole, and G. Milavetz, *Drug Intelligence and Clin. Pharm.* **21,** 734 (1987).
454. Z. Itoh, T. Suzuki, M. Nakaya, M. Inoue, and S. Mitsuhashi, *Antimicrob. Agents Chemother.* **26,** 863 (1984).
455. X. Y. Qin, M-A. Pilot, H. H. Thompson, and J. P. Maskell, *J. Antimicrob. Chemother.* **18,** 747 (1986).
456. T. Tomomasa, T. Kuroume, H. Arai, K. Wakabayashi, and Z. Itoh, *Dig. Dis. Sci.* **31,** 157 (1986).
457. AGA Abstracts, *Gastroenterology* **98,** A64, A342, A345, A358, and A375 (1990).
458. Y. Kondo, K. Torii, S. Omura, and Z. Itoh, *Biochem. Biophys. Res. Commun.* **150,** 877 (1988).
459. T. Peeters and co-workers, *Am. J. Physiol.* **257,** G470 (1989).
460. C. H. S. McIntosh and J. C. Brown, *Adv. Metabolic Disorders* **11,** 439 (1988).

461. K. Tsuzuki and co-workers, *Chem. Pharm. Bull.* **37,** 2687 (1989).
462. T. Sunazuka and co-workers, *Chem. Pharm. Bull.* **37,** 2701 (1989).
463. N. Inatomi and co-workers, *J. Pharm. Expt. Ther.* **251,** 707 (1989).
464. R. C. Yao and D. F. Mahoney, *Appl. Environ. Microbiol.* **55,** 1507 (1989).
465. S. Omura and Y. Tanaka, in Ref. 1, p. 199.
466. S. Omura and A. Nakagawa, in J. W. Corcoran, ed., *Antibiotics,* Vol. 4, Springer-Verlag, Berlin, Germany, 1981, p. 175.
467. K. Gerzon and co-workers, *J. Am. Chem. Soc.* **78,** 6396 (1956).
468. J. Neuzil and Z. Hostalek, *Folia Microbiol.* **31,** 402 (1986).
469. S. Omura and Y. Tanaka, in H. Pape and H.-J. Rehm, volume eds., *Biotechnology,* Vol. 4, VCH Publishers, Weinheim, Germany, 1986, p. 359.
470. D. A. Hopwood and D. H. Sherman, *Ann. Rev. Genetics* **24,** 1990.
471. W. D. Celmer, *J. Am. Chem. Soc.* **87,** 1801 (1965).
472. W. D. Celmer, *Pure Appl. Chem.* **28,** 413 (1971).
473. S. Yue, J. S. Duncan, Y. Yamamoto, and C. R. Hutchinson, *J. Am. Chem. Soc.* **109,** 1253 (1987).
474. D. E. Cane and C-C. Yang, *J. Am. Chem. Soc.* **109,** 1255 (1987).
475. K. Kinoshita, S. Takenaka, and M. Hayashi, *J. Chem. Soc., Chem. Commun.,* 943 (1988).
476. S. Takano, Y. Sekiguchi, Y. Shimazaki, and K. Ogasawara, *Tetrahedron Lett.* **30,** 4001 (1989).
477. M. L. B. Huber and co-workers, *Antimicrob. Agents Chemother.* **34** (1990).
478. J. Vara, M. Lewandowska-Skarbek, Y-G. Wang, S. Donadio, and C. R. Hutchinson, *J. Bacteriol.* **171,** 5872 (1989).
479. O. Han and H-W. Liu, *J. Am. Chem. Soc.* **110,** 7893 (1988).
480. J. Dale, *Angew. Chem. Int. Eng. Ed.* **5,** 1000 (1966).
481. R. S. Egan, J. R. Martin, T. J. Perun, and L. A. Mitscher, *J. Am. Chem. Soc.* **97,** 4578 (1975).
482. H. Ogura, K. Furuhata, H. Kuwano, and M. Suzuki, *Tetrahedron* **37**(Suppl. 1), 165 (1981).
483. T. H. Keller, E. G. Neeland, S. Rettig, J. Trotter, and L. Weiler, *J. Am. Chem. Soc.* **110,** 7858 (1988).
484. J. R. Everett, I. K. Hatton, and J. W. Tyler, *Magn. Res. Chem.* **28,** 114 (1990).
485. J. R. Everett, I. K. Hatton, E. Hunt, J. W. Tyler, and D. J. Williams, *J. Chem. Soc., Perkin Trans. 2,* 1719 (1989).
486. J. R. Everett and J. W. Tyler, *J. Chem. Soc., Perkin Trans. 2,* 325 (1988).
487. *Ibid.,* 1987, p. 1659.
488. M. S. Puar, R. Brambilla, and R. Jaret, *J. Antibiot.* **35,** 381 (1982).
489. M. S. Puar, *J. Antibiot.* **34,** 602 (1981).
490. J. G. Nourse and J. D. Roberts, *J. Am. Chem. Soc.* **97,** 4584 (1975).
491. D. J. Ager and C. K. Sood, *Mag. Res. Chem.* **25,** 948 (1987).
492. K. H. Kim and Y. C. Martin, *Prog. Clin. Biol. Res.* **291,** 325 (1989).
493. I. Nakayama, in Ref. 1, p. 261.
494. *Physicians Desk Reference,* 44th ed., Medical Economics Co., Oradell, N.J., 1990.
495. *Scrip's Antibacterial Report,* PJB Publications, Richmond, UK, 1989.
496. R. E. Brummett and K. E. Fox, *Antimicrob. Agents Chemother.* **33,** 791 (1989).
497. R. C. Wilson, in Ref. 1, p. 302.
498. J. E. Riviere, A. L. Craigmill, and S. F. Sundlof, eds., *Comparative Pharmacokinetics and Residues of Veterinary Antimicrobials,* CRC Press, Boca Raton, Fla., 1990, Chapt. 5.

H. A. Kirst
Eli Lilly and Company

NUCLEOSIDES AND NUCLEOTIDES

The naturally occurring nucleoside and nucleotide antibiotics listed alphabetically in Table 1 exist as either the *C*- or *N*-glycosides (see NUCLEIC ACIDS). These antibiotics contain a variety of purine and pyrimidine rings, including the diazepin, maleimide, indole, imidazole, pyrrolopyrimidine, pyrazolpyrimidine, pyrazole, oxazine, triazene, hydantoin, and the purine ring having a 6-*N*-phosphoramidate substituent. The structures of the carbohydrate moieties also vary. Some nucleosides have additional carbon atoms attached to the ribosyl moiety and in some cases ribose has been replaced by other sugars such as allulose, olefinic sugars, 2-, 3-, 4-, or 5-amino sugars, 4-fluororibose, 4-aminohexoses, aminouronic acids, disaccharides, or a tricyclic dodecose.

The naturally occurring nucleoside/nucleotide antibiotics, which have been isolated from bacteria, fungi, blue-green algae, and marine sponges, have proven to be useful biochemical probes in eucaryotic, procaryotic, viral, fungal, and plant systems. For example, the tunicamycins [*11089-65-9*] and mureidomycins are valuable because of their importance in peptidoglycan synthesis, mucopolysaccharide synthesis, and inhibition of the synthesis of the oligosaccharide moiety of the glycoproteins. Antibiotic A201A, which is structurally similar to puromycin, is an inhibitor of protein synthesis. 2′-Deoxycoformycin has found dramatic application in the treatment of hairy cell leukemia (see CHEMOTHERAPEUTIC AGENTS, ANTICANCER) and as a means to increase the therapeutic efficacy of cordycepin, formycin, and arabinofuranosyladenine (ara-A) through inhibition of adenosine deaminase. The 4-aminohexose nucleoside antibiotics and 2′-amino-2′-deoxyguanosine have been useful in elucidating the steps in protein synthesis, whereas psicofuranine and decoyinine have found use as inducers of sporulation in bacteria. Showdomycin is an excellent probe of membrane transport. The arabinosylpyrimidine nucleosides, ara-T and ara-U, have antiviral activities (see ANTIVIRAL AGENTS). Several nucleoside analogues have herbicidal activity (see HERBICIDES). Some excellent reviews on the nucleoside/nucleotide antibiotics are available (1–4) (see also ANTIPARASITIC AGENTS).

C-Nucleosides

The naturally occurring *C*-nucleosides containing *C*-glycosyl linkages are shown in Table 1.

Ezomycins. The ezomycins [*58572-97-7*] (**1**–**6**) are naturally occurring pyrimidine nucleosides produced by the *Streptomyces* (1–4). The ezomycins, which are comprised of ezomycin A [*60182-24-3*], ezomycin B [*60182-25-4*], and ezomycin C, inhibit the phytopathogenic fungi, *Sclerotinia* and *Botrytis* sp., but not bacteria and yeast. Ezomycins B_1, B_2, C_1, and C_2 are pseudouridine-type *C*-nucleosides, whereas ezomycins A_1 and A_2 are *N*-nucleosides. Ezomycins A_1, B_1, and C_1, but not A_2, B_2, and C_2, contain the L-cystathionine moiety. Ezomycins A_1, A_2, B_1, and B_2 are β-nucleosides; C_1 and C_2 are α-nucleosides.

Showdomycin. Showdomycin (2-β-D-ribofuranosylmaleimide) (**7**) is a maleimide *C*-nucleoside antibiotic synthesized by *S. showdoensis*; isoshowdomycin (**8**) and maleimycin (**9**) have also been isolated (1–6). Showdomycin is not phospho-

Table 1. *C*-Nucleoside Antibiotics

Name	CAS Registry Number	Molecular formula	Structure number	R	Structure
ezomycin A_1	*[39422-19-0]*	$C_{26}H_{38}N_8O_{15}S$	**(1)**	*β*-1-cytosine	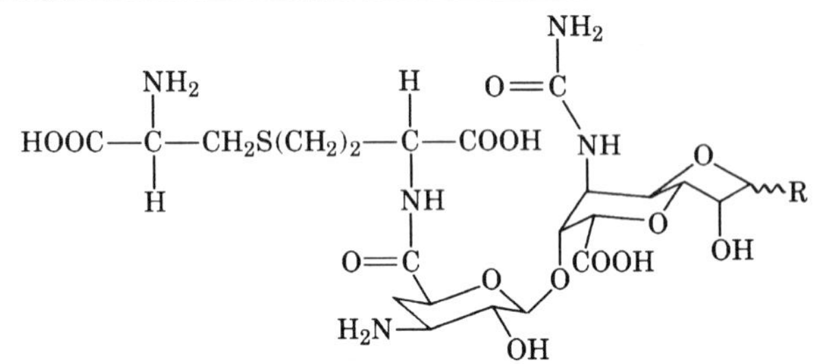
ezomycin B_1	*[39422-20-3]*	$C_{26}H_{39}N_7O_{17}S$	**(2)**	*β*-5-uracil	
ezomycin C_1	*[58002-06-5]*	$C_{26}H_{37}N_7O_{16}S$	**(3)**	*α*-5-uracil	
ezomycin A_2	*[54328-22-2]*	$C_{19}H_{26}N_6O_{12}$	**(4)**	*β*-1-cytosine	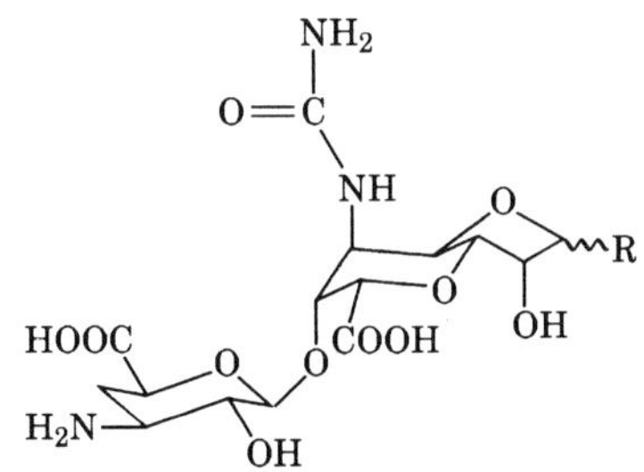
ezomycin B_2	*[57973-16-7]*	$C_{19}H_{25}N_5O_{13}$	**(5)**	*β*-5-uracil	
ezomycin C_2	*[57973-14-5]*	$C_{19}H_{25}N_5O_{13}$	**(6)**	*α*-5-uracil	
showdomycin	*[16755-07-0]*	$C_9H_{11}NO_6$	**(7)**	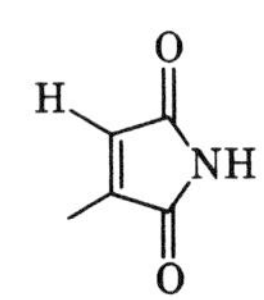	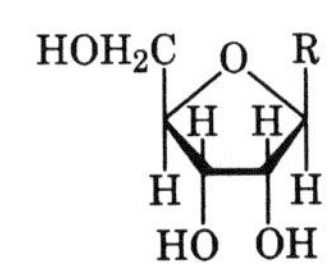

isoshowdomycin	[*37760-84-2*]	$C_9H_{11}NO_6$	**(8)**	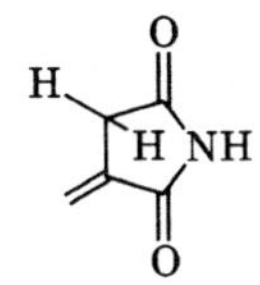	
maleimycin	[*50988-16-4*]	$C_7H_7NO_3$	**(9)**		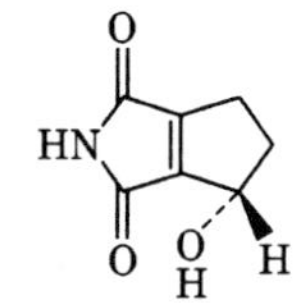
oxazinomycin	[*32388-21-9*]	$C_9H_{11}NO_7$	**(10)**	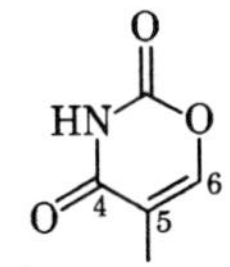	
pyrazomycin (pyrazofurin)	[*30868-30-5*]	$C_9H_{13}N_3O_6$	**(11)**	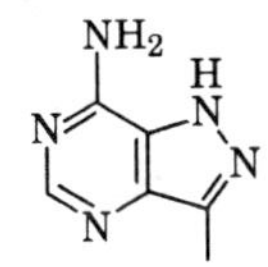	
formycin formycin B	[*6742-12-7*] [*13877-76-4*]	$C_{10}H_{13}N_5O_4$ $C_{10}H_{12}N_4O_5$	**(12)** **(13)**	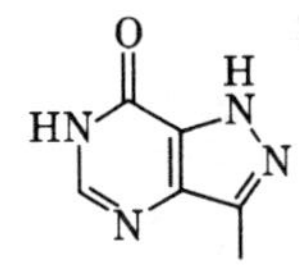	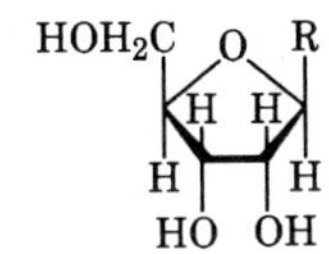
oxoformycin B	[*19246-88-9*]	$C_{10}H_{12}N_4O_6$	**(14)**	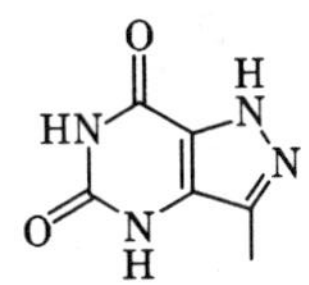	

rylated by nucleoside kinase and is not a substrate for nucleoside phosphorylase. Once (**7**) enters the cell, it blocks the uptake of glucose and other nutrients.

Oxazinomycin. Oxazinomycin (5-β-D-ribofuranosyl-1,3-oxazine-2,4-dione) (**10**), also known as minimycin, has been isolated from *S. hygroscopicus* and *S. mediocidicus* (1–4). It inhibits gram-positive and gram-negative bacteria and tumor growth, but not yeast or fungi. The biosynthesis of (**6**) proceeds from [5-^{14}C,3-^{3}H]glutamic acid with carbons 4, 5, and 6 of the oxazine ring of (**10**) arising from carbons 5, 4, and 3 of glutamic acid (4). The naturally occurring *C*-nucleoside antibiotics showdomycin (**7**), formycin (**12**), oxazinomycin (**10**), and pyrazomycin (**11**) utilize acetate and glutamate as the carbon source for their aglycons (4,7–10).

Pyrazomycin. Pyrazomycin (**11**), 3-(1-β-D-ribofuranosyl)-4-hydroxypyrazole 5-carboxamide, is isolated from *S. candidus* (1–4,9,10). The incorporation of [2-^{13}C]acetate and [1- and U-^{14}C]glutamate into the four contiguous carbons of pyrazomycin has been reported (11,12). Pyrazomycin 5′-phosphate inhibits orotidylic acid decarboxylase. Pyrazomycin inhibits adenosine phosphorylation and decreases the incorporation of deoxyuridine into DNA of Novikoff hepatoma cells in culture. It also inhibits the growth of tumor cells and the cytopathic effects of vaccinia, herpes simplex, vesicular stomatitis, Newcastle disease, measles, Sindbis, polio, hepatitis A, and coxsackie viruses (13,14). The inhibitory action of (**11**) on viral multiplication is reversed by uridine.

Pyrazolopyrimidine Nucleosides. Formycin (**12**), formycin B (**13**), and oxoformycin B (**14**) are naturally occurring pyrazolopyrimidine nucleosides isolated from *Nocardia interforma, S. lavendulae,* and *S. gunmaences* (1–4,15). Coformycin, which inhibits adenosine deaminase, is also isolated from *N. interforma*. Glutamate is the precursor for the biosynthesis of formycin (10). Compound (**12**), but not (**13**), is converted to its 5′-mono-, di-, and triphosphates (3). Compound (**12**) inhibits DNA and protein synthesis, but not RNA synthesis (1–4). Formycin 5′-triphosphate (FTP) is incorporated into RNA and codes as an adenosine 5′-triphosphate (ATP) with bacterial, viral, plant, RNA, and DNA polymerases (16). It is also an important probe of nucleoside transport (17–21). The tight binding of formycin 5′-phosphate to adenosine monophosphate (AMP) deaminase has been reported (22). Compound (**12**) has been used to elucidate ribosome-transfer RNA (tRNA) interactions and it has been reported that the stacking of $tRNA^{PheCCF}$, where the tail is phenylalanine–cytosine–cytosine–formycin, is weaker than that of $tRNA^{PheCCA}$ (23,24). Donor activity for peptide bond formation of formycin-containing tRNA is much lower than that of normal tRNA. Compound (**12**) is rapidly deaminated by erythrocytic adenosine deaminase; the absence of this deaminase in erythrocytes of adenosine deaminase (ADA)-deficient patients results in the conversion of (**12**) to FTP in quantities three times greater than ATP (25). An analogue of inosine [*58-63-9*] (**13**), inhibits tumors and viruses, but does not appear to act via formation of the 5′-phosphates. It has been reported that (**12**) inhibits L5178Y cell growth as a competitive inhibitor of the nicotinamide adenine dinucleotide cation (NAD^+) and showed that (**7**) and (**13**) inhibit poly(ADP-ribose) polymerase (26,27). The formycin analogue of NAD^+ has been shown to inhibit DNA synthesis (28).

Most recently the structure of pyrrolosine has been shown to be an isomeric *C*-nucleoside analogue of 9-deazainosine, which is 7-(β-D-ribofuranosyl)-4-oxo-

3*H*,5*H*-pyrrolo[3,2-*d*]pyrimidine (29). Pyrrolosine inhibits development of starfish embryos.

N-Nucleosides

The naturally occurring nucleoside analogues discussed in this section contain the *N*-glycosyl linkage and either purine, pyrimidine, imidazole, diazepin, or indole rings. The purine nucleosides inhibit protein synthesis, RNA and DNA synthesis, and methyltransferases; they have antimycoplasmal, antiviral, hypotensive, antifungal, antimycobacterial, and antitumor activities and induce sporulation (1–4). The pyrimidine nucleosides inhibit protein synthesis, virus replication, RNA and DNA synthesis, and cAMP phosphodiesterase. The imidazole nucleosides inhibit nucleic acid synthesis. The diazepin nucleosides inhibit adenosine deaminase (ADA). The indole nucleosides inhibit bacteria, yeast, fungi, and viruses.

Purine *N*-Nucleosides. The purine *N*-nucleoside antibiotics are given in Tables 2 and 3.

2′-Amino-2′-deoxypurines. 2′-Amino-2′-deoxyadenosine (**15**) is a naturally occurring *N*-nucleoside isolated from *Actinomadura* that shows antimycoplasmal activity (1,4). Adenosine is the direct precursor for its biosynthesis (30). 2′-Amino-2′-deoxyguanosine (**16**), isolated from a strain of *Enterobacter cloacae* (1,4), shows the growth of HeLa S3 cells and Sarcoma 180 *in vivo* and has been tested for antibacterial activity.

3′-Amino-3′-deoxyadenosine. 3′-Amino-3′-deoxyadenosine (**17**) is elaborated by *Cordyceps militaris, Aspergillus nidulans,* and *Helminthosporium* (3,4). The biosynthesis proceeds directly from adenosine. Compound (**17**) inhibits RNA polymerase, but not DNA polymerase, and replaces the adenosyl residue at the 3′-terminus of tRNA. Phenylalanyl-(3′-amino-3′-deoxyadenosyl)-tRNA has acceptor but not donor activity (31,32). Compound (**17**) also inhibits retroviral RNA-dependent DNA polymerase (33).

Cordycepin (3′-Deoxyadenosine). Cordycepin (**18**), also known as 3′-deoxyadenosine, is isolated from *C. militaris* and *A. nidulans* (1–4). The biosynthesis utilizes adenosine. It was originally proposed to be an analogue of adenine having the branched chain sugar, cordycepose. A functional analogue of adenosine, (**18**) is a most versatile biological probe for studying many reactions in procaryotes, eucaryotes, plants, and viruses. Initial studies indicated that (**18**) was converted to its 5′-phosphates and inhibits RNA, but not DNA, synthesis. A chain terminator in RNA synthesis, (**18**) inhibits poly(A) and tRNA methylation of the 5′-terminus of mRNA and blocks the production of RNA viruses. It inhibits 2′-*O*-methylation, forms the 3′-deoxy analogues of *S*-adenosylmethionine (SAM) and *S*-adenosylhomocysteine (SAH), and inhibits initiation of HSV DNA synthesis (34). Both SAM and SAH analogues inhibit nucleic acid methylation. When (**18**) replaces adenosine at the 3′-end of the $tRNA^{Phe}$, the addition of phenylalanine at the 2′-hydroxyl position results in a charged tRNA that is not a substrate for poly(U)-directed polyphenylalanine synthesis. 3′*d*ATP inhibits RNA and DNA synthesis in permeabilized *E. coli*. 3′*d*ATP has been reported to be a substrate for 2′,5′-oligoadenylate (2-5A) synthetase. The resulting 2′,5′-cordycepin trimer

Table 2. Purine *N*-Nucleoside Antibiotics[a]

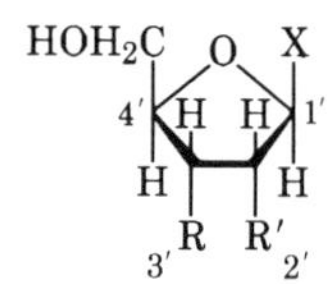

Name	CAS Registry Number	Molecular formula	Structure number	R	R′	X^b
2′-amino-2′-deoxy-adenosine	[*10414-81-0*]	$C_{10}H_{14}N_6O_3$	(**15**)	OH	NH_2	A
2′-amino-2′-deoxy-guanosine	[*60966-26-9*]	$C_{10}H_{14}N_6O_4$	(**16**)	OH	NH_2	G
3′-amino-3′-deoxy-adenosine	[*2504-55-4*]	$C_{10}H_{14}N_6O_3$	(**17**)	NH_2	OH	A
cordycepin (3′-deoxyadenosine)	[*73-03-0*]	$C_{10}H_{13}N_5O_3$	(**18**)	H	OH	A
puromycin	[*53-79-2*]	$C_{22}H_{29}N_7O_5$	(**19**)	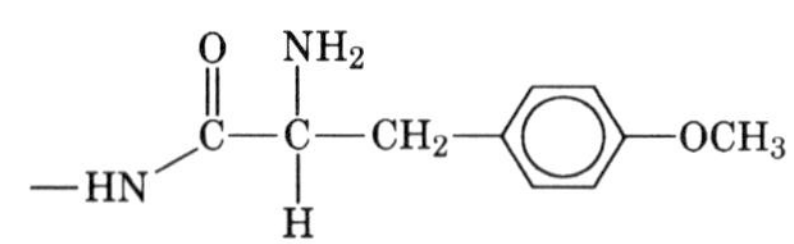	OH	*N*,*N*-dimethyl-A
homocitrullylamino-adenosine	[*59204-62-5*]	$C_{17}H_{27}N_9O_5$	(**20**)	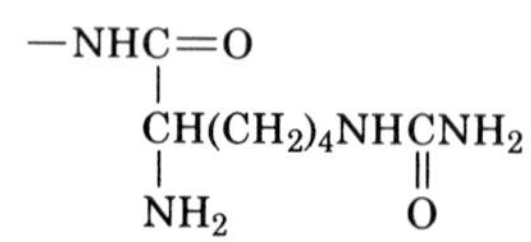	OH	A

lysylaminoadenosine	[*31518-64-6*]	$C_{16}H_{26}N_8O_4$	**(21)**	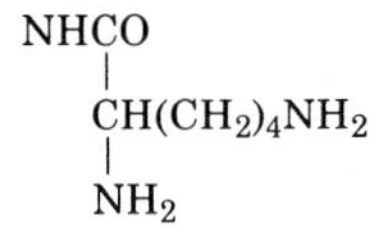	OH	A
chryscandin	[*86936-90-5*]	$C_{20}H_{23}N_7O_6$	**(22)**	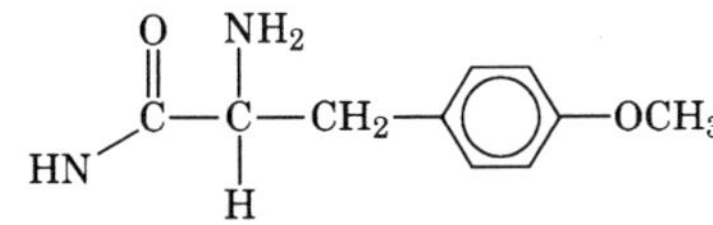	OH	A
A201A		$C_{37}H_{50}N_6O_{14}$	**(23)**	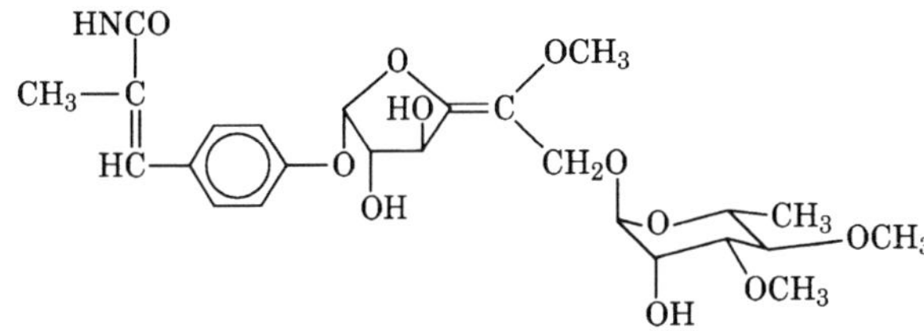	OH	*N*,*N*-dimethyl-A
psicofuranine[c]	[*1874-54-0*]	$C_{11}H_{15}N_5O_5$	**(24)**	OH	OH	A
decoyinine[c,d] (angustmycin A)	[*2004-04-8*]	$C_{11}H_{13}N_5O_4$	**(25)**	OH	OH	A

[a]Structure as shown unless otherwise indicated where R, R′, and X are as defined.

[b]A = adenine 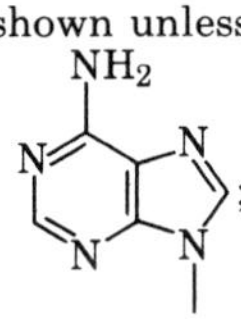; G = guanine 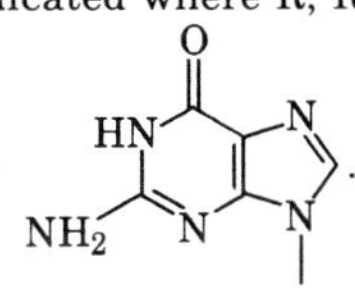.

[c]A CH_2OH group replaces the H at position 1′.

[d]The CH_2OH and H at position 4′ are replaced by a CH_2 group.

Table 3. Other Purine *N*-Nucleoside Antibiotics, Sugar-1′-X

Name	CAS Registry Number	Molecular formula	Structure number	Sugar[a]	X[b]
arabinofuranosyladenine (ara-A, vidarabine)	[*5536-17-4*]	$C_{10}H_{13}N_5O_4$	**(26)**	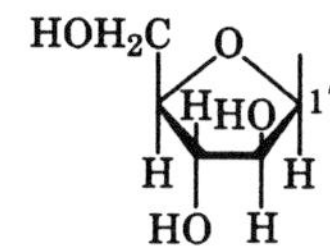	A
sinefungin	[*58944-73-3*]	$C_{15}H_{23}N_7O_5$	**(27)**	S, R = $CH_2CHNH_2(CH_2)_2CHNH_2COOH$	A
A9145A			**(28)**	S, R = $CH_2CHNH_2(CH_2)_2CHNH_2CONH_2$	A
A9145C	[*66753-47-7*]	$C_{15}H_{21}N_7O_5$	**(29)**	S, R = $CHCHNH_2(CH_2)_2CHNH_2COOH$	A
herbicidins	[*77699-46-8*]			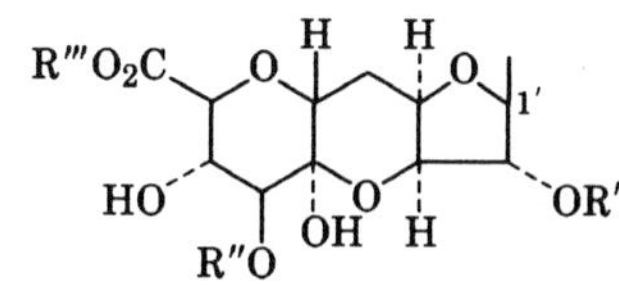	
				R′ R″ R‴	
herbicidin A	[*55353-31-6*]		**(30)**	CH_3 $CO(CH_2OH))C{=}CHCH_3$ CH_3	A
herbicidin B	[*55353-32-7*]		**(31)**	CH_3 H CH_3	A
herbidicin E	[*72067-15-3*]	$C_{23}H_{30}N_5O_{10}$	**(32)**	CH_3 $COCH(CH_3)_2$ CH_3	A
herbicidin F	[*72283-61-5*]		**(33)**	CH_3 $CO(CH_3)C{=}CHCH_3$ CH_3	A
herbicidin G	[*72283-62-6*]		**(34)**	H $CO(CH_3)C{=}CHCH_3$ H	A
doridosine	[*73027-05-1*]	$C_{11}H_{15}N_5O_5$	**(35)**	S, R = CH_2OH	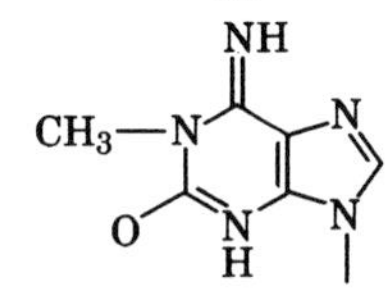
aristeromycin	[*19186-33-5*]	$C_{11}H_{15}N_5O_3$	**(36)**	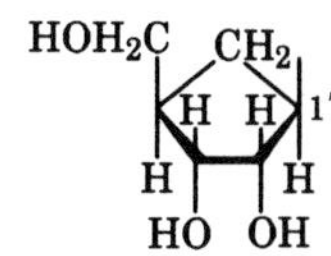	A
neplanocin A	[*72877-50-0*]	$C_{11}H_{13}N_5O_3$	**(37)**	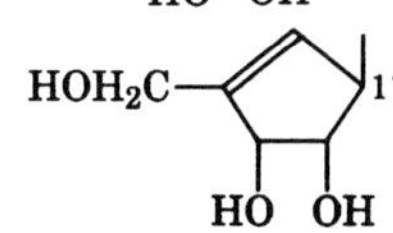	A

neplanocin B	[*72877-49-7*]	$C_{11}H_{13}N_5O_4$	**(38)**	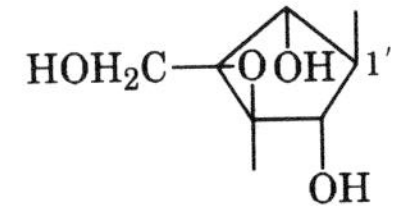	A
neplanocin C	[*72877-48-6*]	$C_{11}H_{13}N_5O_4$	**(39)**	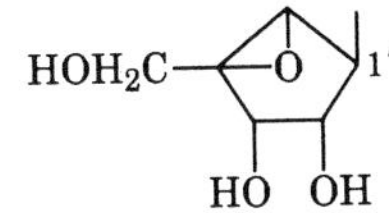	A
neplanocin D	[*72877-47-5*]	$C_{11}H_{12}N_4O_4$	**(40)**	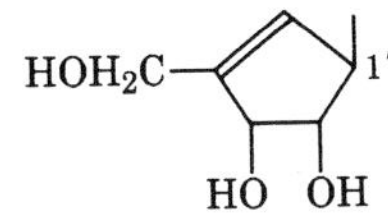	G
neplanocin F	[*72877-51-1*]	$C_{11}H_{13}N_5O_3$	**(41)**	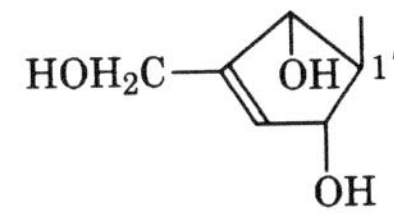	A
nucleocidin	[*24751-69-7*]	$C_{10}H_{13}FN_6O_6S$	**(42)**	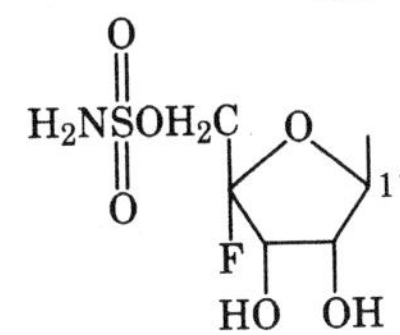	A
ascamycin	[*91432-48-3*]	$C_{13}H_{18}ClN_7O_7S$	**(43)**	S, R = 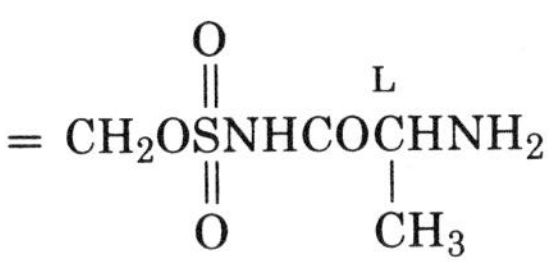	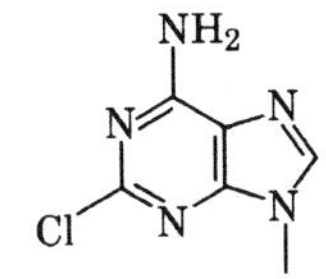
nebularine	[*550-33-4*]	$C_{10}H_{12}N_4O_4$	**(44)**	S, R = CH_2OH	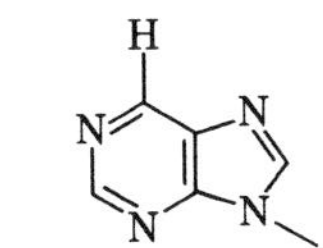
crotonoside	[*3373-53-3*]	$C_5H_5N_5O$	**(45)**	S, R = CH_2OH	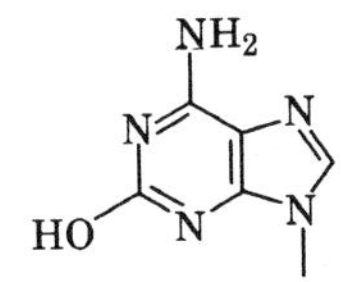

Table 3. (*Continued*)

Name	CAS Registry Number	Molecular formula	Structure number	Sugar[a]	X[b]
griseolic acid	[*79030-08-3*]	$C_{14}H_{13}N_5O_8$	**(46)**	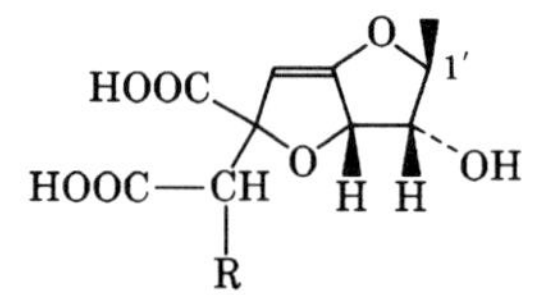 , R = H	A
griseolic acid B	[*98808-01-8*]	$C_{14}H_{13}N_5O_9$	**(47)**	, R = OH	A
griseolic acid C	[*100242-49-7*]	$C_{14}H_{15}N_5O_7$	**(48)**	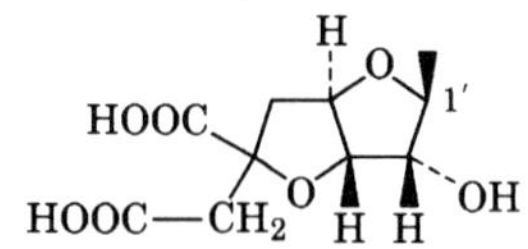	A
oxetanocin A	[*103913-16-2*]	$C_{10}H_{13}N_5O_3$	**(49)**	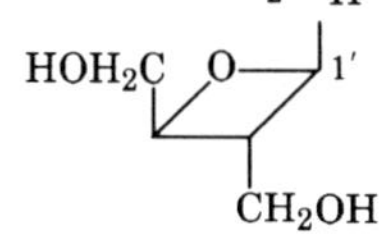	A
oxanosine	[*80394-72-5*]	$C_{10}H_{12}N_4O_6$	**(50)**	S, R = CH_2OH	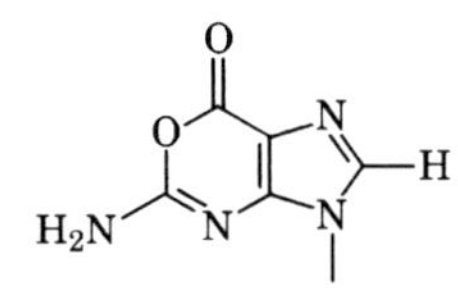

[a]S = 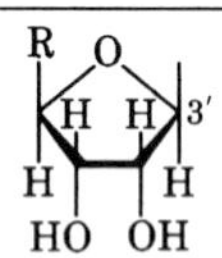 and R is as indicated.

[b]A = adenine 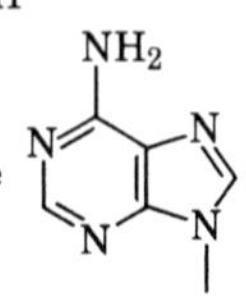; G = guanine 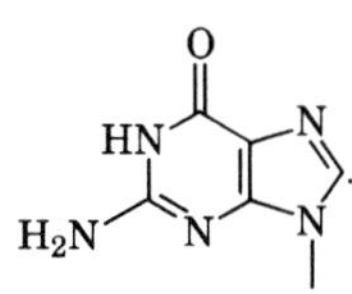.

5′-triphosphate inhibits translation in lysed rabbit reticulocytes better than authentic 2-5A (35). The inhibition of HIV-1 replication in infected MT-2 cells, and of HIV-1 reverse transcriptase by the cordycepin and phosphorothioate analogues of 2-5A has also been reported (36,37). Compound (**18**) damages messenger RNA (mRNA) processing in brain tissue, affects mRNA processing in yeast and *Xenopus* oocytes, increases mRNA stability in HL60 cells, affects mobility of nucleoside transport in cells, disrupts microtubule networks, and has been used to detect mRNAs in nonheat shocked thymic lymphocytes that code for noncordycepin inhibited heat shock proteins (38–47).

Puromycin. Puromycin (**19**), elaborated by *S. alboniger* (1–4), inhibits protein synthesis by replacing aminoacyl-tRNA at the A-site of peptidyltransferase (48,49). Photosensitive analogues of (**19**) have been used to label the A-site proteins of peptidyltransferase and $tRNA^{Phe}$ (50). Compound (**19**) and its carbocyclic analogue have been used to study the accumulation of glycoprotein-derived free sialooligosaccharides, accumulation of mRNA, methylase activity, enzyme transport, rat embryo development, the acceptor site of human placental 80S ribosomes, and gene expression in mammalian cells (51–60).

Three other 3′-aminoacyl-3′-deoxyadenosine nucleosides, homocitrullylaminoadenosine (**20**), lysylaminoadenosine (**21**), and chryscandin (**22**), have been reported (61). Unlike (**19**), chryscandin has strong inhibitory activity against *C. albicans* (61). Antibiotic FR900403, which is structurally similar to (**22**), has been isolated from the fungus, *Kernia* sp. F-19849; it inhibits *C. albicans* (62).

Antibiotic A201A. Antibiotic A201A (**23**), produced by *S. capreolus,* is an N^6-dimethyladenine nucleoside structurally similar to puromycin (**19**). Compound (**23**) which contains an aromatic acid and monosaccharide residues (1,4), inhibits the incorporation of amino acids into proteins but has no effect on RNA or DNA synthesis. Compound (**23**) does not accept polypeptides as does (**19**), and does appear to block formation of the initiation complex of the 50S subunit. It may block formation of a puromycin-reactive ribosome.

Ketohexose Nucleosides. The ketohexose nucleosides, psicofuranine (**24**) and decoyinine (**25**), are isolated from *S. hygroscopicus* (1–4). Psicofuranine, 6-amino-9-(β-D-psicofuranosyl)purine, and decoyinine, 9-β-D-(5,6-psicofuranoseenyl)-6-aminopurine, angustmycin A, inhibit XMP aminase noncompetitively. Unlike many other nucleoside antibiotics, phosphorylation is not necessary for inhibitory action. Compound (**25**) effectively induces sporulation; (**24**) inhibits GMP synthesis and subsequently AMP synthesis. Therefore (**24**) is sufficient to cause sporulation of *B. subtilis*. Guanine and adenine can reverse this inhibition. Compounds (**24**) and (**25**) regulate sporulation by induction of the alarmones in bacteria, ie, lowering levels of nucleotides and/or folate, enzyme promoters, and early sporulation genes (63–70).

Arabinosyladenine. Three naturally occurring arabinosyl purine and pyrimidine nucleoside antibiotics (ara-A, ara-T, and ara-U) have been isolated from the sponge, *Cryptotethia crypta* (1–4). 9-β-D-Arabinofuranosyladenine (**26**), also known as ara-A and Vidarabine, has also been isolated from *S. antibioticus* and *S. herbaceus*. (The adenosine deaminase inhibitor, 2′-deoxycoformycin (**59**), is also isolated from *S. antibioticus*). The structural elucidation and chemical syntheses of (**26**), its phosphate triester, and 2-halo and 5′-*O*-acyl derivatives have been reported (1–4,71–73). An enzymatic synthesis of (**26**) via transarabinosylation of ara-U has been reported (74) as has the biosynthesis of (**26**) from [^{14}C, ^{3}H, 2′-^{18}O and 3′-^{18}O]adenosine has been reported (75). Ara-A is a potent antiviral drug that shows great selectivity in its influence on host cell metabolism and virus growth (76–78). Ara-A and its 5′-monophosphate are effective antiviral analogues with known activity against herpes simplex keratitis, herpes simplex encephalitis, neonatal herpes, herpes zoster, and mucocutaneous herpes in humans. Ara-ATP is a competitive inhibitor of 2′*d*ATP. Compound (**26**) is incorporated into HSV DNA and it is an S-phase specific inhibitor of tumor cell growth (79). Many studies have reported the inhibition of DNA and RNA synthesis in procaryotes, eucaryotes, and viruses by ara-A and ara-ATP (1–4,80). Eucaryotic and procaryonuclear poly(A) polymerase, poly(ADP-ribose) polymerase, and nuclear poly(A) polymerase are not inhibited by ara-ATP (26,81,82). Although ara-ATP does not inhibit the DNA-dependent DNA polymerase from *E. coli,* it does inhibit the DNA-dependent DNA polymerase from mammalian cells. HSV-1 induced DNA polymerase is more sensitive to ara-ATP than is DNA polymerase-α in uninfected rabbit kidney cells (83). In L5178Y cells, (**26**) is deaminated to ara-hypoxanthine which is converted to ara-inosinetic phosphate ara-ITP; ara-ITP does not affect RNA or protein synthesis, but does inhibit DNA polymerase-α in L5178Y cells. In permeabilized mouse leukemic cells, the DNA replicating complex is competitively inhibited by ara-ATP (4). Ara-A inhibits DNA synthesis by blocking formation of new replicons and inhibits DNA repair, nucleoside uptake, and DNA strand elongation (84–87). 2′-Deoxycoformycin (**59**) potentiates the growth inhibition of (**26**) in cultured human cell lines (88). Inhibition of virus replication occurs when (**26**) is converted to ara-ATP (89).

Sinefungins. *S. griseolus* produces the nucleosides, sinefungin, A9145, A9145A, and A9145C (**27**–**29**) (1,4) which are structural analogues of SAM and SAH and inhibit methyltransferase reactions (90). The biosynthesis of (**27**) was reported to utilize L-arginine and ATP (91). It was demonstrated that adenosine and ornithine are converted to (**27**) (92) and it has been shown that adenine, ribose, and ornithine are the direct precursors of (**27**) (93). Regulation of the biosynthesis of (**27**) has been studied using auxotrophic mutants of *S. incarratus;* mutants with blocks in arginine synthesis were not able to produce (**27**) (94). Compounds (**27**–**29**) have antifungal, antiparasitic, and antiviral activity, and inhibit transformed cells in culture (95–97). They are potent inhibitors of the SAM:*O*-methyltransferases I and III (98,99). Compound (**27**) is the most potent inhibitor of protein methylase I activity and is comparable to SAH with respect to its inhibitory effect (100). It inhibits photolabeling of restriction enzyme EcoR-II methyltransferase, protein methylation, EcoR-I adenine DNA methylase, and myelin basic protein (arginine) methyltransferase (101–104). Methyltransferase inhibitors may provide a therapeutic strategy in myasthenia gravis (105).

Herbicidins. Five adenine nucleosides, herbicidins A, B, E, F, and G (**30**–**34**), have been isolated from the culture filtrates of *S. sagamonensis* No. 4075 (1,4) which also produces ara-A and 2′-deoxycoformycin. The herbicidins contain a tricyclic dodecose. They inhibit dicotyledonous plants, germination of plant seeds, growth of blue-green algae, and fungi.

Doridosine. Doridosine, N^1-methylisoguanosine, (**35**) was isolated from the dorid nudibranchs of *Anisodoris nobilis* and the sponge, *Tedania* (106,107). The injection of (**35**) into the saphenous vein of anesthetized rats produces hypotension and bradycardia almost immediately. The observed changes in the electrocardiograms are minor and indicate little interference with conduction of the impulse within the heart (see CARDIOVASCULAR AGENTS).

Thraustomycin. Thraustomycin and β-thraustomycin are isolated from *S. exfoliatus* (4). Although their structures have not been totally elucidated, hydrolysis of thraustomycin shows that it contains equimolar quantities of adenine, L-leucine, and a tetrahydroxymonocarboxylic acid. Thraustomycin is a potent inhibitor of the fungus, *M. hiemallis* (+), but does not inhibit bacteria.

Aristeromycin. Aristeromycin (**36**), the first carbocylic analogue of adenosine, was isolated from the culture filtrates of *S. citricolor* as part of a search for inhibitors of bacterial leaf blight (1–4). A herbicidally active hypoxanthine analogue of (**36**), coaristeromycin, has also been isolated (108). Several chemical syntheses of (**36**) have appeared (1–4,109). It inhibits *Xanthomonas oryzae* and *Pyricularia oryzae,* bacterial leaf blight, blast disease of rice plants, and leaf mold of tomato plants, by inhibiting both cell division and elongation. An inhibitor of *S*-adenosylmethionine (SAM) hydrolase, (**36**) also affects *de novo* synthesis of purine nucleotides and RNA. Adenosine and, to a lesser degree, adenine, inosine and 2′-deoxyadenosine reverse inhibition by (**36**). Aristeromycin diphosphate binds to polynucleotide phosphorylase. The copolymer poly(A, aristeromycin) is formed when aristeromycin 5′-diphosphate is incubated with an excess of ADP. Aristeromycin derivatives of *S*-adenosyl homocysteine (SAH) have been synthesized in an effort to characterize the SAH binding site on mRNA (guanine-7)-methyltranferases that are required for efficient translation and mRNA-ribosome binding. Whereas the aristeromycin derivatives of SAH showed only marginal inhibitory activity, *S*-tubercidinyl-L-homocysteine was a potent inhibitor. The metabolism of analogues of (**36**) and neplanocin A (**37**) have been compared (110). Both nucleosides are converted to the 5′-triphosphates; however, only (**36**) was converted to a guanosine nucleotide derivative. A carbocyclic inosine analogue of (**36**) has potent antileishmanial activity (111). The 3-deazaaristeromycin derivative is an inhibitor of SAH hydrolase and has been widely used in studies related to RNA synthesis in transformed cells and viruses and cell differentiation (112–118). 3-Deazaaristeromycin-resistant clones of human β-lymphoblasts have been reported to display elevated SAM levels (119). The biosynthesis of the cyclopentane ring of (**36**) proceeds by carbon–carbon bond formation between C-2 and C-6 of D-glucose (120).

Neplanocins. Neplanocins A–D and F (**37**–**41**) are carbocyclic nucleoside antibiotic products of *Ampullariella regularis* (1,4) that are structurally related to (**36**) in that they contain either a cyclopentene or epoxy cyclopentane ring (121,122). The chemical syntheses of (**37**–**41**) and the 3-deazaneplanocins have been reported (123–126). Compound (**37**), which is converted to its 5′-triphosphate, has potent antitumor and antiviral activities (127–129). It strongly inhibits SAM in cells and viruses (128–131) and is converted to the 3′-keto derivative by *S*-adenosylhomocysteine hydrolase (132,133).

Nucleocidin and Ascamycin. Nucleocidin, 4′-fluoro-5′-*O*-sulfamoyladenosine, (**42**) is isolated from the culture filtrates of *S. clavus* (1–4). Among the nucleoside antibiotics (**42**) is unique in that it contains a fluoride group at carbon-4′. Compound (**42**) and its chemically synthesized analogues are inhibitors of bacteria and are extremely toxic to mammals. Compound (**42**) inhibits protein

Table 4. Pyrrolopyrimidine, Diazepin, and Other *N*-Nucleoside Antibiotics, Sugar-1′-X

Name	CAS Registry Number	Molecular formula	Structure number	Sugar[a]	X
Pyrrolopyrimidine N-nucleosides					
tubercidin	*[69-33-0]*	$C_{11}H_{14}N_4O_4$	**(51)**	S	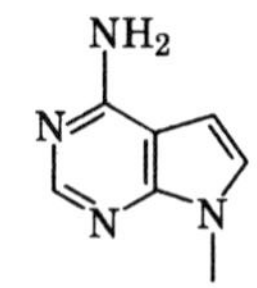
toyocamycin	*[606-58-6]*	$C_{12}H_{13}N_5O_4$	**(52)**	S	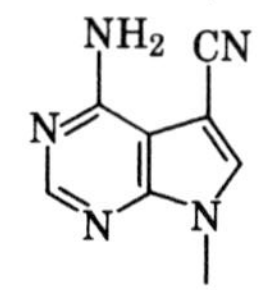
sangivamycin	*[18417-89-5]*	$C_{12}H_{15}N_5O_5$	**(53)**	S	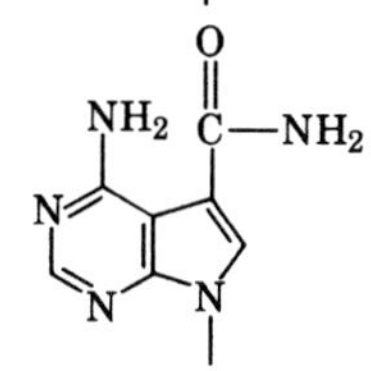
cadeguomycin	*[81645-08-1]*	$C_{12}H_{14}N_4O_7$	**(54)**	S	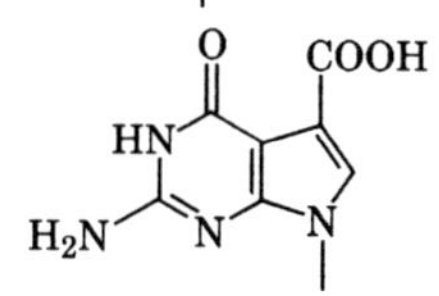
kanagawamicin	*[84873-16-5]*	$C_{13}H_{17}N_5O_6$	**(55)**	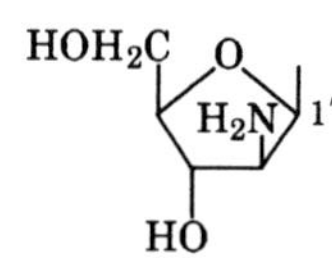	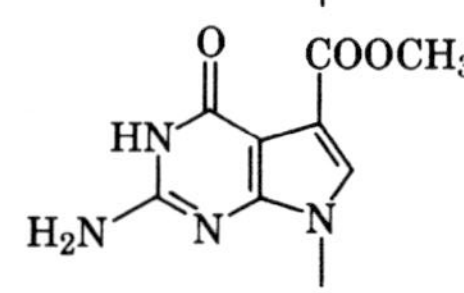
mycalisine A mycalisine B	*[98890-73-4]* *[98890-72-3]*	$C_{13}H_{13}N_5O_3$ $C_{13}H_{12}N_4O_4$	**(56)** **(57)**	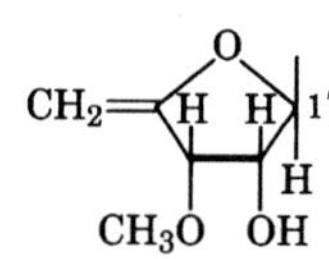	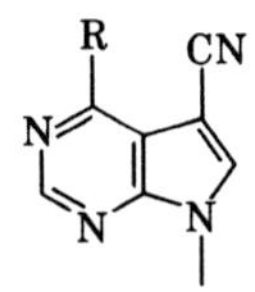 R = NH_2 R = OH

Diazepin N-nucleosides					
coformycin	*[11033-22-0]*	$C_{11}H_{16}N_4O_5$	**(58)**	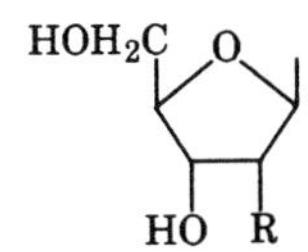 R = OH	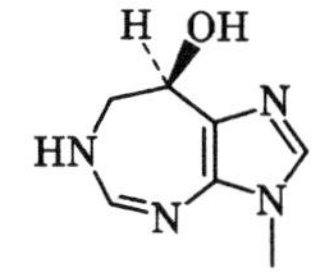
2′-deoxycoformycin (co-vidarabine, pentostatin)	*[53910-25-1]*	$C_{11}H_{16}N_4O_4$	**(59)**	R = H	
adechlorin (2′-chloro-2′-deoxycoformycin)	*[96328-17-5]*	$C_{11}H_{15}ClN_4O_4$	**(60)**	R = Cl	
adecypenol	*[104493-13-2]*	$C_{12}H_{16}N_4O_4$	**(61)**	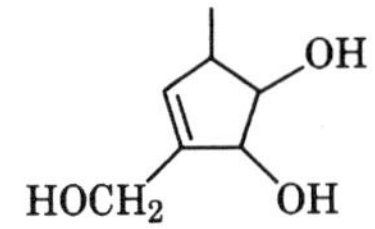	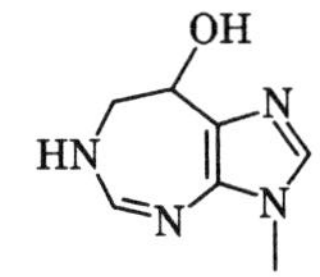
		Others			
bredinin	*[50924-49-7]*	$C_{19}H_{13}N_3O_6$	**(62)**	S	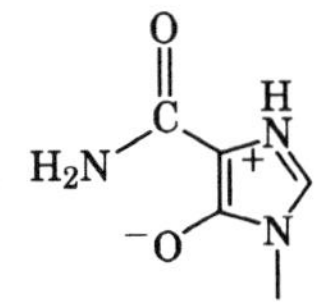
neosidomycin	*[72033-44-4]*	$C_{17}H_{20}N_2O_6$	**(63)**	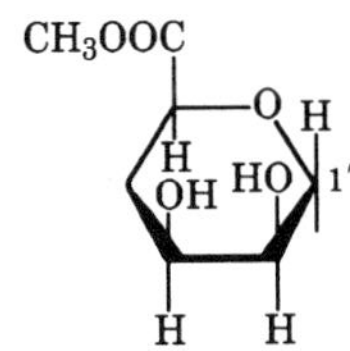	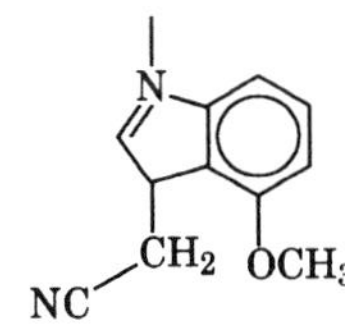
SF-2140	*[91284-30-9]*	$C_{18}H_{20}N_2O_6$	**(64)**	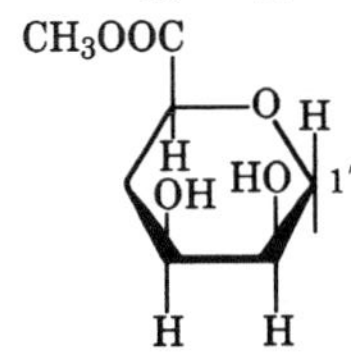	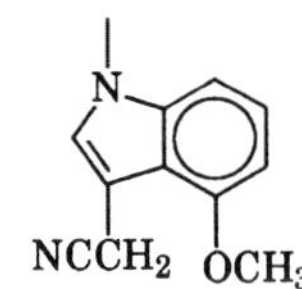

[a]S = 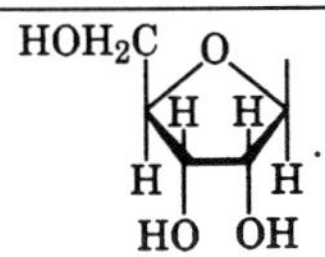.

synthesis by blocking peptide elongation. Ascamycin, 5′-O-L-alanylsulfamoyl-2-chloroadenosine, (**43**) is similar to (**42**) in that it contains a sulfamoyl group (134,135). A specific amino-peptidase that hydrolyzes the alanyl group of (**43**) has been purified (136). Two additional 5′-O-sulfamoylnucleosides that are isolated from the *Streptomyces* are 5′-O-sulfamoyladenosine and 5′-O-sulfamoyltubercidin (137).

Nebularine. Nebularine (**44**) is a naturally occurring purine riboside isolated from *S. yokosukanensis* (1,3,4). It is phosphorylated, and inhibits purine biosynthesis and RNA synthesis, but is not incorporated into RNA by *E. coli* RNA polymerase. It has also found application as a transition state analogue for treatment of schistosomiasis and as a substrate for the restriction endonuclease, HindII (138–141).

Crotonoside. Crotonoside, also called isoguanosine, (**45**) has been isolated from *Croton tiglium* (4). Ip administration of [2-^{14}C]crotonoside to rats results in incorporation into nucleic acids (qv). It inhibits the inducible binding sites, inhibits glutamic acid dehydrogenase, and accumulates cAMP.

Griseolic Acids. The three griseolic acids, A, B, and C (**46**–**48**), have been isolated from *S. griseoaurantiacus* (1,142). They are potent inhibitors of cAMP phosphodiesterase (143,144). Compound (**46**) stimulates glycogen degradation in mice.

Oxetanocins. Oxetanocin A (**49**), formerly oxetanocin, is the first naturally occurring oxetanose derivative and is isolated from *Bacillus megaterium* (1,145). It inhibits gram-positive bacteria, herpes viruses, and human immunodeficiency virus (HIV) (146). The chemical synthesis of (**49**) and several derivatives has been reported (147).

Oxanosine. Oxanosine (**50**), which is isolated from *S. capreolus* (1,148), is a potent inhibitor of GMP synthetase and nucleic acid synthesis (149,150).

Pyrrolopyrimidine Nucleosides. The pyrrolopyrimidine *N*-nucleoside antibiotics are listed in Table 4. Tubercidin (**51**), toyocamycin (**52**), sangivamycin (**53**), cadeguomycin (**54**), and kanagawamicin (**55**) are pyrrolopyrimidine nucleoside antibiotics isolated from strains of *Streptomyces* (1–4,151). In addition, the 5′-α-D-glucopyranosyl derivatives of tubercidin and toyocamycin have been isolated from blue-green algae (152). The structural elucidation, physical and chemical properties, chemical syntheses, and biological activities of these nucleosides and their derivatives have been reported (1–4,153–172). GTP, but not ATP, is a precursor for the biosynthesis of the pyrrolopyrimidine ring of (**51**–**53**) (161). The nitrile group of (**52**) is enzymatically converted to the carboxamide group in the formation of (**53**) (162). Tubercidin 5′-triphosphate is not a substrate for either adenosine deaminase or nucleoside phosphorylase. Compound (**51**) is incorporated into RNA, inhibits the processing of RNA, but does not inhibit the transcription of 45S precursor RNA to 28S and 18S RNA; DNA synthesis is inhibited. The importance of (**51**) as a methylase inhibitor has been studied in detail (163,164). *S*-Tubercidinyl homocysteine inhibits methylation (163,164). It has been used to elucidate the requirements for the catalytic site of RNA polymerase with respect to initiation, elongation, and termination. Rous sarcoma virus and SV40 replication are inhibited by 5-bromotubercidin. Another promising application of (**51**) is as an anthelmintic agent. It inhibits purine nucleotide synthesis in *Schistosoma mansoni* and xylotubercidin inhibits HSV-2 in mice (165,166).

Toyocamycin (**52**), a cytotoxic analogue of adenosine, is selectively incorporated into the 45S ribosomal RNA precursor and prevents cleavage into 28S and 18S RNA (1,4). When Ehrlich ascites cells are incubated with (**52**) in a medium rich in amino acids, the high transcription rate of rRNA is lowered. Toyocamycin is incorporated into HeLa cell 45S RNA, blocks polyadenylation of heterogenous nuclear RNA (hnRNA), and inhibits Friend, Rous Sarcoma, and vesicular stomatitis virus replication. Sangivamycin (**53**), which is phosphorylated but not deaminated, is incorporated into RNA and DNA of normal tissue, inhibits *de novo* purine synthesis in L1210 cells in culture, competitively inhibits amino acid activation, charging of tRNA, and can replace ATP in the synthesis of RNA catalyzed by *E. coli* RNA polymerase. The molecular effects of (**53**) and its thio analogue on RNA and DNA synthesis in L1210 cells and as an inhibitor of rhodopsin kinase have been studied (167,168). Compound (**54**) potentiates the cytotoxicity of ara-C and has been used in studies on the T-cell immunostimulatory properties of certain peptides (169–171). Compound (**55**) has antitumor and antibacterial activity (172).

Two other pyrrolopyrimidine nucleoside antibiotics, mycalisines A and B (**56**, **57**), have been isolated from the sponge *Mycale* sp. (173). These bioactive marine metabolites inhibit cell division of starfish eggs.

Diazepin Nucleosides. Four naturally occurring diazepin nucleosides, coformycin (**58**), 2′-deoxycoformycin (**59**), adechlorin or 2′-chloro-2′-deoxycoformycin (**60**), and adecypenol (**61**), have been isolated (1–4,174,175). The biosynthesis of (**59**) and (**60**) have been reported to proceed from adenosine and C-1 of D-ribose (30,176,177). They are strong inhibitors of adenosine deaminase and AMP deaminase (178). Compound (**58**) protects adenosine and formycin (**12**) from deamination by adenosine deaminase. Advanced hairy cell leukemia has shown rapid response to (**59**) with or without α- or β-interferon treatment (179–187). In addition, (**59**) affects interleukin-2 production, receptor expression on human T-cells, DNA repair synthesis, immunosuppression, natural killer cell activity, and cytokine production (188–194).

Bredinin, Neosidomycin, and SF-2140. Bredinin (**62**), isolated from the culture filtrates of *Eupenicillium brefeldianum* (1,4), inhibits the multiplication of L5178Y, HeLa S3, RK-13, mouse L-cells, and Chinese hamster cells. GMP can reverse the inhibition by (**62**), but (**62**) is not incorporated into the nucleic acids. The inhibition of nucleic acid synthesis and chromosomal damage in the S and G_2 phases that is caused by (**62**), is reversed by GMP. It blocks the conversion of IMP to XMP and XMP to GMP. In combination with GMP, (**62**) interferes with intracellular cAMP levels and thereby inhibits cell division.

Neosidomycin (**63**) and SF-2140 (**64**) are indole *N*-glycosides produced by *S. hygroscopicus* and *Actinomadura*, respectively (195,196). A revised structure for (**63**) has appeared (197). Compound (**64**) contains an α-*N*-glycoside linkage. Both (**63**) and (**64**) show activity against gram-positive bacteria, yeast, fungi, and viruses.

Pyrimidine–Nucleoside Antibiotics. *Fatty Acyl Nucleosides.* The nucleoside antibiotics with fatty acyl groups containing adenine, uracil, and dihydrouracil aglycons, the tunicamycins [*11089-65-9*] (**65**–**74**), streptovirudins (**75**–**84**), corynetoxins (**85**–**97**), are given in Table 5. The structure of liposidomycin (**98**), septacidins [*62362-59-8*] (**99**), spicamycin [*62362-59-8*] (**100**), and FR900848 are given in Figure 1. The chemistry of the tunicamycins has been extensively reviewed (198). The structural elucidation of (**75**–**78**) has been described (199). The isolation and structural elucidation of liposidomycins A, B, and C, septacidins, corynetoxins, and spicamycins have been described (200–205). Compounds (**85**–**97**), produced by *Corynebacterium rathayi,* infect rye grass which is toxic to grazing animals (205). A recently discovered fatty acyl nucleoside antibiotic with antifungal activity, antibiotic FR900848 [*120500-69-8*] (**101**), contains one isolated and four serial cyclopropane rings (206).

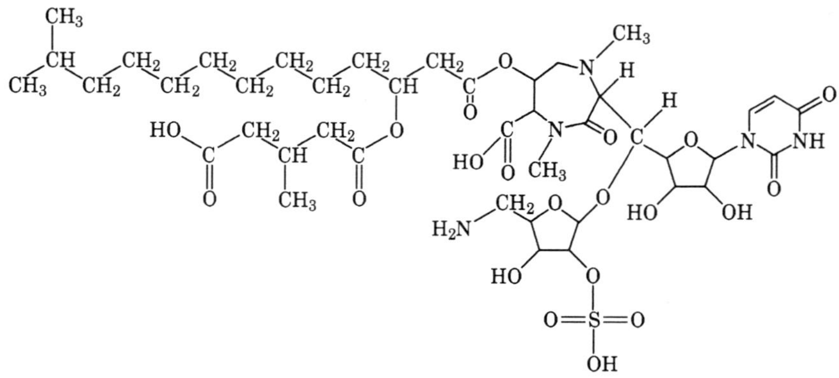

Liposidomycin B (**98**)

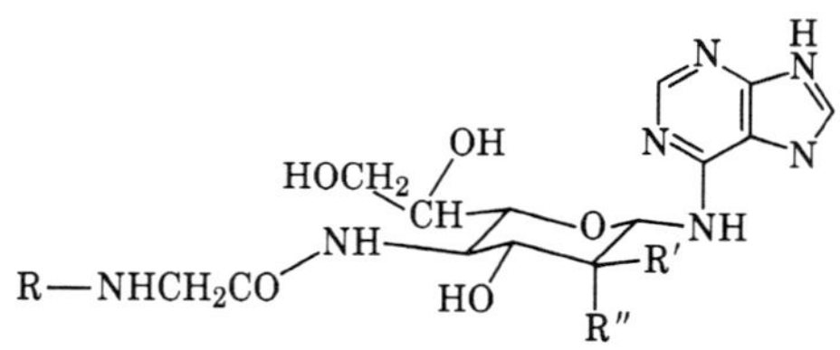

Septacidin (**99**, R′ = OH, R″ = H) R = $(CH_3)_2CH(CH_2)_nCO-$,
Spicamycin (**100**, R′ = H, R″ = OH) $CH_3CH_2CH_2(CH_2)_nCO-$
n = 8–14

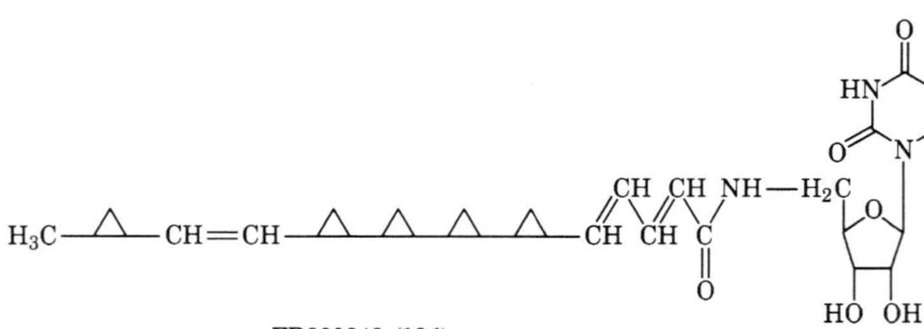

FR900848 (**101**)

Fig. 1. Structures of fatty acyl *N*-nucleosides.

The tunicamycins, streptovirudins, corynetoxins, and liposidomycins inhibit the formation of glycoproteins and peptidoglycans. Oligosaccharide synthesis for glycoproteins requires a polyisoprenoid carrier followed by transfer to the nascent polypeptide (207–209). The tunicamycins and streptovirudins are

Table 5. Pyrimidine Fatty Acyl *N*-Nucleoside

Name	CAS Registry Number	Molecular formula	Structure number	R
		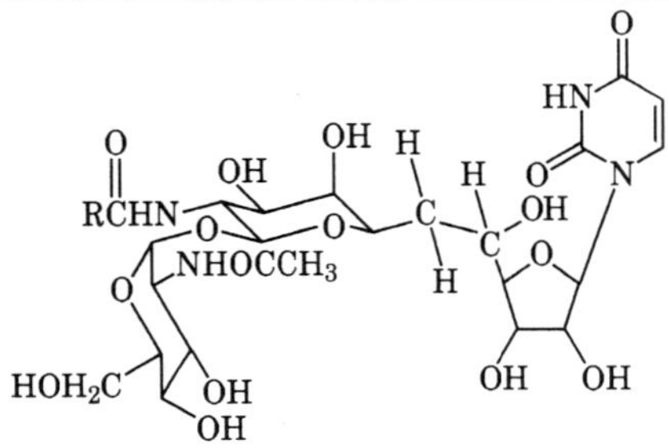		
Tunicamycins [11089-65-9]				
I(A_0)	[*73942-10-6*]	$C_{36}H_{58}N_4O_{16}$	**(65)**	$(CH_3)_2CH(CH_2)_7\overset{H}{C}=\underset{H}{C}-$
II(C)	[*66081-37-6*]	$C_{37}H_{60}N_4O_{16}$	**(66)**	$(CH_3)_2CH(CH_2)_8\overset{H}{C}=\underset{H}{C}-$
III(A_2)	[*76544-45-1*]	$C_{37}H_{60}N_4O_{16}$	**(67)**	$CH_3(CH_2)_{10}\overset{H}{C}=\underset{H}{C}-$
IV(B_2)	[*73942-09-3*]	$C_{38}H_{62}N_4O_{16}$	**(68)**	$CH_3(CH_2)_{11}\overset{H}{C}=\underset{H}{C}-$
V(A)	[*66054-36-2*]	$C_{38}H_{62}N_4O_{16}$	**(69)**	$(CH_3)_2CH(CH_2)_9\overset{H}{C}=\underset{H}{C}-$
VI	[*88263-43-8*]	$C_{38}H_{62}N_4O_{16}$	**(70)**	$(CH_3)_2CH(CH_2)_{11}-$
VII(B)	[*66081-36-5*]	$C_{39}H_{64}N_4O_{16}$	**(71)**	$(CH_3)_2CH(CH_2)_{10}\overset{H}{C}=\underset{H}{C}-$
VIII(C_2)	[*73942-07-1*]	$C_{39}H_{64}N_4O_{16}$	**(72)**	$CH_3(CH_2)_{12}\overset{H}{C}=\underset{H}{C}-$
IX(D_1)	[*73942-08-2*]	$C_{40}H_{66}N_4O_{16}$	**(73)**	$CH_3(CH_2)_{13}\overset{H}{C}=\underset{H}{C}-$
X(D)	[*66081-38-7*]	$C_{40}H_{66}N_4O_{16}$	**(74)**	$(CH_3)_2CH(CH_2)_{11}\overset{H}{C}=\underset{H}{C}-$
Streptovirudins (A_1–D_1)[a]				
A_1	[*51330-28-0*]	$C_{35}H_{58}N_4O_{16}$	**(75)**	$(CH_3)_2CH(CH_2)_6CH=CH-$

Table 5. (*Continued*)

Name	CAS Registry Number	Molecular formula	Structure number	R
B_1	[*51330-30-4*]	$C_{36}H_{60}N_4O_{16}$	**(76)**	$CH_3CH_2\overset{CH_3}{\overset{\vert}{C}}H(CH_2)_6CH{=}CH-$
B_{1a}	[*81093-26-7*]	$C_{36}H_{60}N_4O_{16}$	**(77)**	$(CH_3)_2CH(CH_2)_7CH{=}CH-$
C_1	[*51330-32-6*]	$C_{37}H_{62}N_4O_{16}$	**(78)**	$(CH_3)_2CH(CH_2)_8CH{=}CH-$
D_1	[*51330-34-8*]	$C_{38}H_{64}N_4O_{16}$	**(79)**	$CH_3CH_2\overset{CH_3}{\overset{\vert}{C}}H(CH_2)_8CH{=}CH-$
Streptovirudins $(A_2–D_2)$[b]				
A_2	[*51330-29-1*]	$C_{35}H_{56}N_4O_{16}$	**(80)**	$(CH_3)_2CH(CH_2)_6CH{=}CH-$
B_2	[*51330-31-5*]	$C_{36}H_{58}N_4O_{16}$	**(81)**	$CH_3CH_2\overset{CH_3}{\overset{\vert}{C}}H(CH_2)_6CH{=}CH-$
B_{2a}	[*73942-10-6*]	$C_{36}H_{58}N_4O_{16}$	**(82)**	$(CH_3)_2CH(CH_2)_7CH{=}CH-$
C_2	[*51330-33-7*]	$C_{37}H_{60}N_4O_{16}$	**(83)**	$(CH_3)_2CH(CH_2)_9CH{=}CH-$
D_2	[*51330-35-9*]	$C_{38}H_{62}N_4O_{16}$	**(84)**	$CH_3CH_2\overset{CH_3}{\overset{\vert}{C}}H(CH_2)_5CH{=}CH-$
Corynetoxins[c]				
H16*i*	[*82138-73-6*]	$C_{39}H_{66}N_4O_{17}$	**(85)**	$(CH_3)_2CH(CH_2)_{10}\overset{OH}{\overset{\vert}{C}}HCH_2-$
H18*i*	[*82138-74-7*]	$C_{41}H_{70}N_4O_{17}$	**(86)**	$(CH_3)_2CH(CH_2)_{12}\overset{OH}{\overset{\vert}{C}}HCH_2-$
H17*a*	[*82151-55-1*]	$C_{40}H_{68}N_4O_{17}$	**(87)**	$CH_3CH_2\overset{CH_3}{\overset{\vert}{C}}H(CH_2)_{10}\overset{OH}{\overset{\vert}{C}}HCH_2-$
H19*a*	[*82151-56-2*]	$C_{42}H_{72}N_4O_{17}$	**(88)**	$CH_3CH_2\overset{CH_3}{\overset{\vert}{C}}H(CH_2)_{12}\overset{OH}{\overset{\vert}{C}}HCH_2-$
U16*i*	[*66081-36-5*]	$C_{39}H_{64}N_4O_{16}$	**(89)**	$(CH_3)_2CH(CH_2)_{10}CHCH-$
U18*i*	[*82138-75-8*]	$C_{41}H_{68}N_4O_{16}$	**(90)**	$(CH_3)_2CH(CH_2)_{12}CHCH-$
U17*a*	[*82151-57-3*]	$C_{40}H_{66}N_4O_{16}$	**(91)**	$CH_3CH_2\overset{CH_3}{\overset{\vert}{C}}H(CH_2)_{10}CHCH-$
U19*a*	[*82138-76-9*]	$C_{42}H_{70}N_4O_{16}$	**(92)**	$CH_3CH_2\overset{CH_3}{\overset{\vert}{C}}H(CH_2)_{12}CHCH-$
S16*i*	[*82138-77-0*]	$C_{39}H_{66}N_4O_{16}$	**(93)**	$(CH_3)_2CH(CH_2)_{12}-$
S18*i*	[*82138-78-1*]	$C_{41}H_{70}N_4O_{16}$	**(94)**	$(CH_3)_2CH(CH_2)_{14}-$
S15*a*	[*82138-79-2*]	$C_{38}H_{64}N_4O_{16}$	**(95)**	$CH_3CH_2\overset{CH_3}{\overset{\vert}{C}}H(CH_2)_{10}-$
S17*a*	[*82138-80-5*]	$C_{40}H_{68}N_4O_{16}$	**(96)**	$CH_3CH_2\overset{CH_3}{\overset{\vert}{C}}H(CH_2)_{12}-$
S19*a*	[*82138-87-6*]	$C_{42}H_{72}N_4O_{16}$	**(97)**	$CH_3CH_2\overset{CH_3}{\overset{\vert}{C}}H(CH_2)_{14}-$

[a]Dihydrouracil incorporated, ie, uracil C=C bond is saturated.
[b]Uracil as shown in structure.
[c]The H stands for β-hydroxy, S for saturated, and U for α–β unsaturated; i = iso; a = anteiso.

lipophilic nucleoside analogues of UDP-*N*-acetylglucosamine and inhibit the first transferase reaction in the lipid-linked saccharide pathway, ie, the formation of dolichyl pyrophosphoryl-*N*-acetylglucosamine, by inhibiting dolichyl-P:*N*-acetylglucosamine-1-*P* transferase. Compounds (**65**–**74**) are also biological probes for cell wall polymer synthesis in bacteria, fungi, immunoglobulin synthesis, and glycoprotein biosynthesis in insects and plants. By blocking synthesis of the phosphomannosyl residues on lysosomal enzymes with tunicamycin, the secretion of the lysosomal enzymes is inhibited as is the synthesis of rhodopsin and the conversion of procollagen to collagen (210–212). The effect of the tunicamycins and streptovirudins on adherence of group B streptococci to influenza virus-infected kidney cells has been investigated (213). In hen oviduct microsomes, tunicamycin abolished *N*-glycosylation, but had no effect on *O*-glycosylation (214). Tunicamycin inhibits peptidoglycan and 1,3-poly(glycerol phosphate) teichoic acid in *B. licheniformis,* the formation of undecaprenylacetylglucosaminyl-lipid and the formation of the peptidoglycan and teichuronic acid, *N*-acetylglucosamine-lipid formation in plants, glycosylation of human interferon, glycosaminoglycan synthesis in cultured embryonic chick fibroblasts, and polysaccharide chain formation of corneal keratan sulfate (215–218). Tunicamycin has also been used to investigate virus assembly. The reduction of glycosamine incorporation caused by tunicamycin is accompanied by a decrease in infectious virus particle production (219). In a model system for *N*-glycosylation of the intracellular transport of rabies virus, tunicamycin blocks surface expression and accumulation of G protein (220), the first step in the lipid cycle of *E. coli* and *S. cerevisiae* (221) and IgE binding protein (222). Tunicamycin inhibits the incorporation of *O*-linked sugars on the CD45 phosphotyrosine phosphatase (223), glycosylation of tumor necrosis factor (224), causes accumulation of dolicholphosphoryl-mannose, an essential anchor to CHO cell mutants (225), and inhibits lipoprotein lipase complex mannose *N*-linked oligosaccharides (226). Monoclonal antibodies to HSV glycoproteins did not react with cell extracts of tunicamycin treated cells (227). The inhibition of *N*-glycosylation processing and transport surface expression also has been reported (227–231).

Peptidyl *N*-Nucleoside Antibiotics. *Polyoxins and Neopolyoxins.* The polyoxins [*11113-80-7*] (**102**–**113**), and neopolyoxins (**114**–**116**), shown in Figure 2, are peptidylpyrimidine nucleoside antibiotics that have achieved use as agricultural fungicides (1–4,232) (see FUNGICIDES, AGRICULTURAL). Polyoxins A–M, which have the pyrimidine chromophore, have been isolated from *S. cacaoi* var *asoensis* and *S. piomogenus*. Polyoxin N [*37362-29-1*] (**113**), $C_{16}H_{23}N_5O_{12}$, neopolyoxin A [*75044-69-8*] [*72864-26-7*] (**114**), $C_{20}H_{25}N_5O_{10}$, and neopolyoxin B [*75005-71-9*] (**115**), $C_{20}H_{25}N_5O_{11}$, have the imidazole chromophore and neopolyoxin C [*75044-70-1*] (**116**), $C_{20}H_{25}N_5O_{10}$, contains uracil and a 3-hydroxypyridine rings (1–4). The thymine chromophore of the polyoxins is formed by a pathway independent of thymidylate synthetase (233). The aminouronic acid moiety is not directly synthesized from glucose, but rather via phosphoenolpyruvate and the 5′-aldehyde of uridine to give the intermediate octofuranose uronic acid nucleoside (1). Oxidative elimination of the two terminal carbons (carbon 7′ and 8′) followed by the addition of the amino group on carbon-5′ results in the formation of the polyoxin nucleoside. Azetidine carboxylic acid (polyoximic acid) is formed from the carbon–nitrogen skeleton of isoleucine and carbamoylpolyoxamic acid, is formed

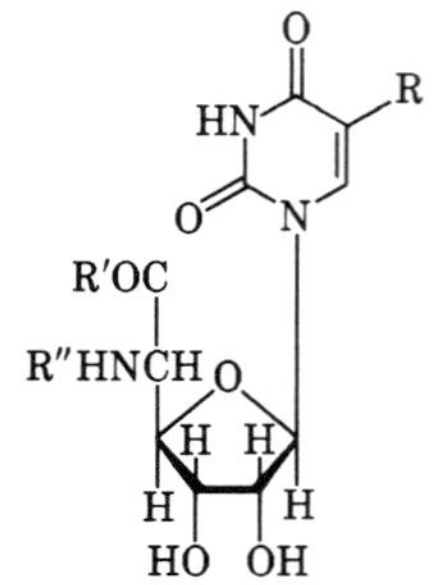

Polyoxins A–M (**102–112**)

R = -CH_2OH, -COOH, -CH_3, -H

R′ = -OH, 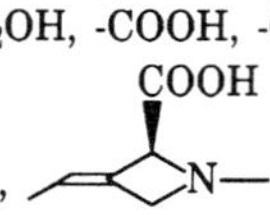

R″ = -H, 5-*O*-carbamoyl-2-amino-2-deoxy-L-xylonic acid or the 3-deoxy derivative

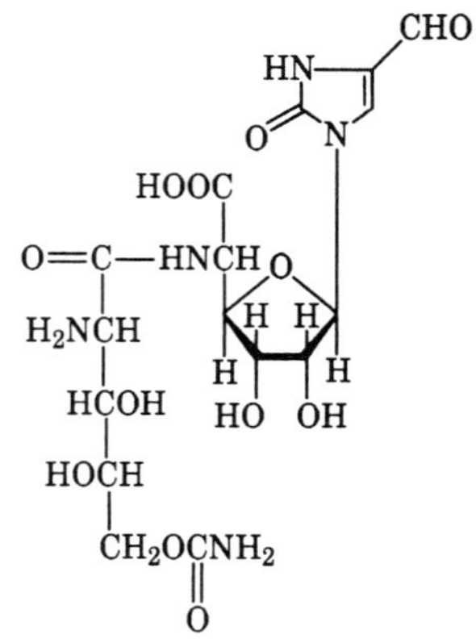

Polyoxin N (**113**)

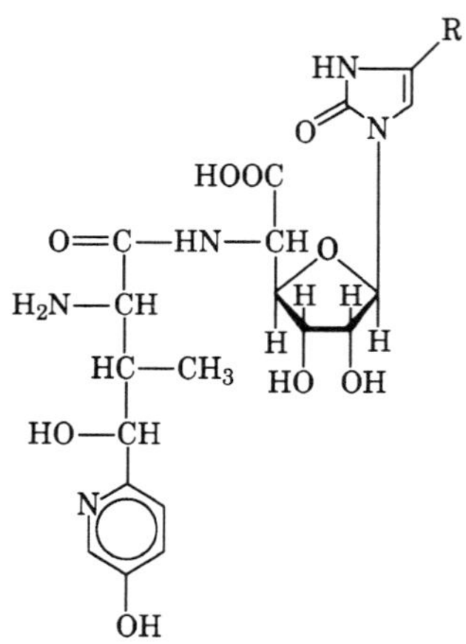

Neopolyoxin A (**114**, R = CHO)
Neopolyoxin B (**115**, R = COOH)

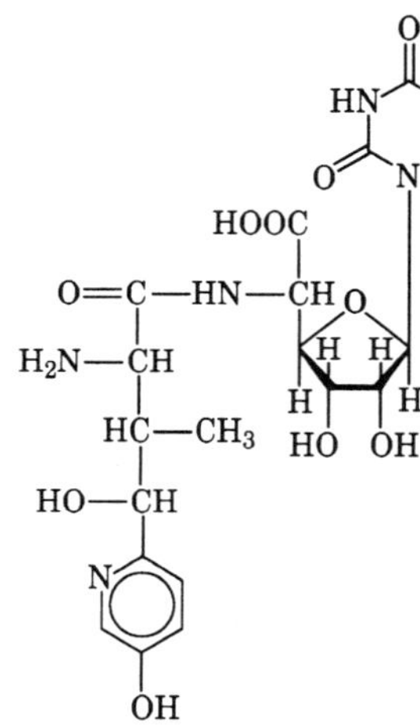

Neopolyoxin C (**116**)

Fig. 2. Structures of the polyoxins. Polyoxin A [*19396-03-3*] [*11016-31-2*], $C_{23}H_{32}N_6O_{14}$; polyoxin B [*19396-06-6*] [*11016-32-3*] [*24695-50-9*], $C_{17}H_{25}N_5O_{13}$; polyoxin C [*21027-33-8*] [*11043-74-6*], $C_{11}H_{15}N_3O_8$; polyoxin D [*22976-86-9*] [*33401-46-6*] [*11043-75-7*] [*29762-31-0*], $C_{17}H_{23}N_5O_{14}$; polyoxin E [*22976-87-0*] [*11043-76-8*], $C_{17}H_{23}N_5O_{13}$; polyoxin F [*23116-76-9*] [*11043-77-9*], $C_{23}H_{30}N_6O_{15}$; polyoxin G [*22976-88-1*] [*11043-78-0*], $C_{17}H_{25}N_5O_{12}$; polyoxin H [*24695-54-3*] [*11043-79-1*], $C_{23}H_{32}N_6O_{13}$; polyoxin I [*22886-33-5*] [*11043-80-4*], $C_{17}H_{22}N_4O_9$; polyoxin J [*22976-89-2*] [*25546-71-8*], $C_{17}H_{25}N_5O_{12}$; polyoxin K [*22886-46-0*], $C_{22}H_{30}N_6O_{13}$; polyoxin L [*22976-90-5*] [*100566-82-3*] [*25546-73-0*] [*59519-82-3*], $C_{16}H_{23}N_5O_{12}$; and polyoxin M [*34718-88-2*], $C_{16}H_{23}N_5O_{11}$.

from glutamate. Compounds (**102–116**) inhibit sheath-blight disease of rice crops, ie, the pathogenic fungus *Pellicularia philamentosa f. sasakii.* The polyoxins, which are structurally similar to UDP-*N*-acetylglucosamine, inhibit chitin synthetase (1). Polyoxin D (**104**) is a competitive inhibitor of UDP-*N*-acetylglucosamine. Polyoxin-resistant mutants of *A. kikuchiana* have a decreased uptake of the polyoxins. To overcome the resistance of these mutants, three approaches have been used: transnucleosidation, biosynthesis of a polyoxin with the 5-fluorouracil moiety, and decarboxylation of the 5-carboxyuracil polyoxins.

Mureidomycins and Pacidamycins. The mureidomycins and pacidamycins are listed in Table 6. The four peptidylnucleosides, mureidomycins A–D (**117–120**), are produced by *S. flavidovirens* (234). They contain two moles of *m*-tyrosine, one mole of 2-amino-3-*N*-methylaminobutyric acid, and one mole of methionine (235). The mureidomycins inhibit *Pseudomonas aeruginosa* (236). Treatment of *P. aeruginosa* with mureidomycin C results in cell lysis. The mureidomycins show low toxicity to mice and protect them from infection with *P. aeruginosa*. Mureidomycin A inhibits peptidoglycan synthesis and lipid-intermediate formation from UDP-*N*-acetylmuramyl (MurNAc)-pentapeptide and UDP-*N*-

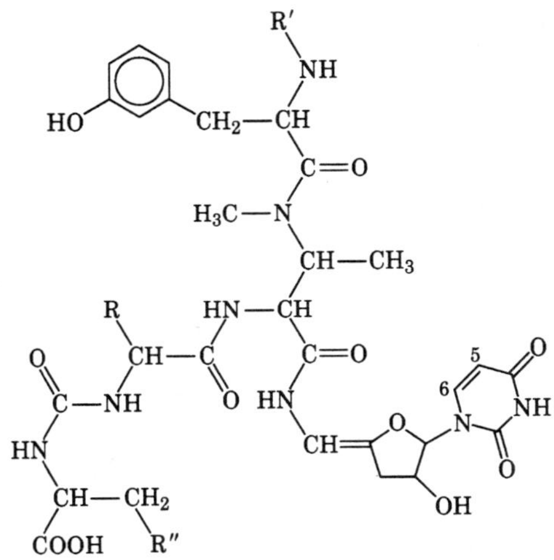

Table 6. The Mureidomycins[a] and the Pacidamycins[b]

Name	CAS Registry Number	Molecular formula	Structure number	R′	R″
Mureidomycins					
mureidomycin A	[*114797-04-5*]	$C_{38}H_{48}N_8O_{12}S$	(**117**)	H	*m*-hydroxyphenyl
mureidomycin B[c]	[*114797-05-6*]	$C_{38}H_{50}N_8O_{12}S$	(**118**)	H	*m*-hydroxyphenyl
mureidomycin C	[*114797-06-7*]	$C_{40}H_{51}N_9O_{13}S$	(**119**)	glycyl	*m*-hydroxyphenyl
mureidomycin D[c]	[*114797-07-8*]	$C_{40}H_{53}N_9O_{13}S$	(**120**)	glycyl	*m*-hydroxyphenyl
Pacidamycins					
pacidamycin 1	[*121264-05-9*]		(**121**)	alanyl	indol-3-yl
pacidamycin 2	[*121264-06-0*]	$C_{39}H_{49}N_9O_{12}$	(**122**)	alanyl	phenyl
pacidamycin 3	[*121280-49-7*]	$C_{39}H_{49}N_9O_{13}$	(**123**)	alanyl	*m*-hydroxyphenyl
pacidamycin 4			(**124**)	H	indol-3-yl
pacidamycin 5	[*121255-43-0*]	$C_{36}H_{44}N_8O_{11}$	(**125**)	H	phenyl
pacidamycin 6			(**126**)	glycyl	indol-3-yl
pacidamycin 7		$C_{38}H_{47}N_9O_{12}$	(**127**)	glycyl	phenyl

[a] R is $(CH_2)_2SCH_3$.
[b] R is CH_3.
[c] Positions 5, 6 of the uracil moiety are saturated to give dihydrouracil.

acetylglucosamine (237). Mureidomycin A inhibits translocase, thereby blocking formation of undecaprenyl-pyrophosphoryl-MurNAc-pentapeptide.

The seven pacidamycins (**121–127**), which are isolated from the culture filtrates of *S. coeruleorubidus,* are structurally similar to the mureidomycins. These seven peptidyl nucleoside antibiotics differ in the terminal amino acid residues (238). The biosynthesis of these nucleoside antibiotics is markedly affected by the amino acids added to the culture medium (239).

4-Aminohexose Nucleosides. The 4-aminohexose nucleosides (**128–140**) are listed in Table 7 (1–4,240–242). A biosynthetic relationship between the 4-aminohexose peptidyl nucleoside antibiotics and the pentopyranines has been proposed (1). The 4-aminohexose pyrimidine nucleoside antibiotics block peptidyl transferase activity and inhibit transfer of amino acids from aminoacyl-tRNA to polypeptides. Hikizimycin, gougerotin, amicetin, and blasticidin S bind to the peptidyl transferase center at overlapping sites (243).

Nikkomycins. The nikkomycins (**141–159**), isolated from *S. tendae,* are nucleoside-peptide antibiotics (1,4,244,245) as shown in Table 8. Nikkomycins X and Z are structurally identical to neopolyoxins A and C, respectively. Compound (**141**) is a competitive inhibitor of chitin synthetase. Two new nikkomycins, nikkomycin pseudo-Z and pseudo-J (**158, 159**), contain a *C*-glycosidic bond between C-5 of uracil and C-1′ of 5-amino-5-deoxy-D-*allo*furanuronic acid (245). Other peptidyl nucleosides, the albomycins [*1414-39-7*] (**160–162**), amipurimycin (**163**), and miharamycins A and B (**164, 165**), are shown in Table 9 (246–248). These nucleosides are either iron chelators or inhibitors of plant pathogens (244,246–248).

Glycosyl Nucleosides. There are five glycosyl antibiotics with either the adenine, uracil, or 7-deazaguanine aglycon given in Table 10. The glycosylated moieties are either *N*-linked disaccharides, unsaturated uronic acid, or a sulfated sugar. The antibacterial and antifungal activities of these compounds have been described (249–253).

Octosyl Acids. Three octosyl uronic acid nucleosides, produced by *S. cacaoi* sub sp. *asoensis* are shown in Figure 3. The biosynthesis of (**172**) and (**173**) has been reported (1). The replacement of the pyrimidine chromophore of (**171**) with adenine results in a nucleoside analogue that is a competitive inhibitor of cAMP.

Arabinosylpyrimidine Nucleosides. 1-*β*-D-Arabinofuranosylthymine [*605-23-2*] (ara-T) (**174**), $C_{10}H_{14}N_2O_6$, and 1-*β*-D-arabinofuranosyluracil [*3083-77-0*] (ara-U) (**175**), $C_9H_{12}N_2O_6$, also known as spongouridine, were first isolated from

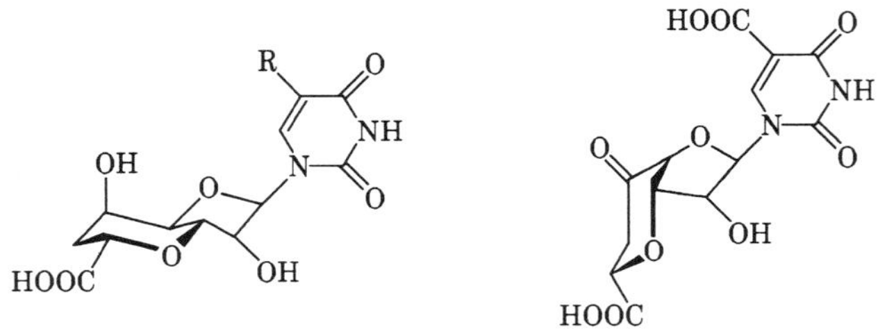

Octosyl acid A (**171**, R = COOH)
Octosyl acid B (**172**, R = CH_2OH)

Octosyl acid C (**173**)

Fig. 3. Octosyl acids.

the sponge *C. crypta* (1–4). The structures of these compounds are shown in Figure 4. Compound **(174)** moderately inhibits *E. coli* and is not toxic to mammalian cells. It inhibits the replication of HSV-1 and varicella zoster virus and is converted to the 5′-triphosphate which inhibits HSV-induced deoxypyrimidine kinase. The chemical synthesis of **(174)** from thymidine has been reported (254). Ara-TTP competitively inhibits DNA polymerase-α and -β isolated from L5178Y cells with respect to dTTP; thymidine and uridine reverse the inhibition by **(174)**. RNA polymerases are not inhibited by ara-TTP. It is phosphorylated to its 5′-triphosphate in noninfected and HSV-infected cells. Ara-U **(175)** can satisfy the uracil requirement of an auxotroph of *E. coli* and can be phosphorylated to its 5′-phosphate. Although **(175)** has little or no antiviral activity, the 5-alkyl derivatives of **(175)** show antiherpesviral activity and inhibition of cell growth in human

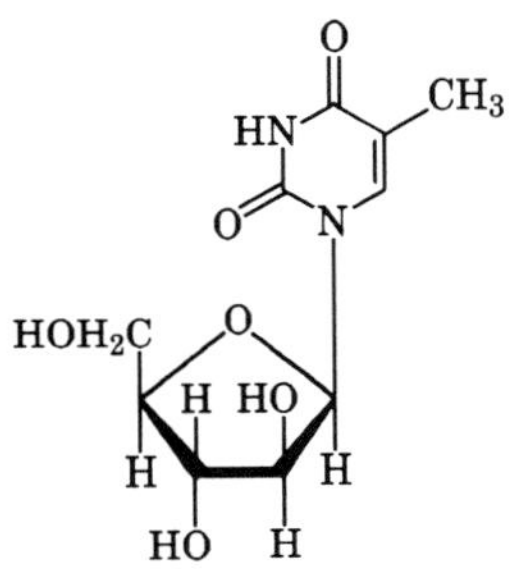

Ara-T **(174)**

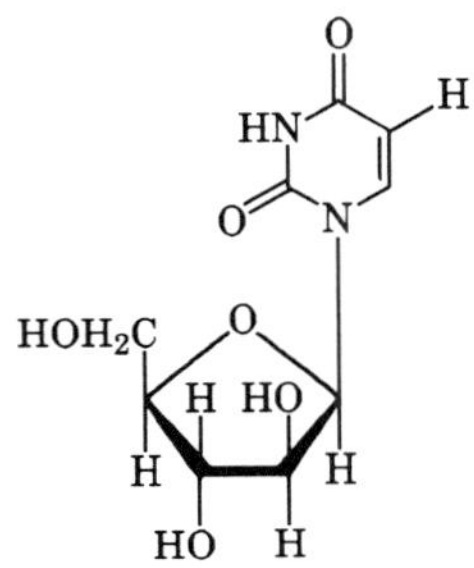

Ara-U **(175)**

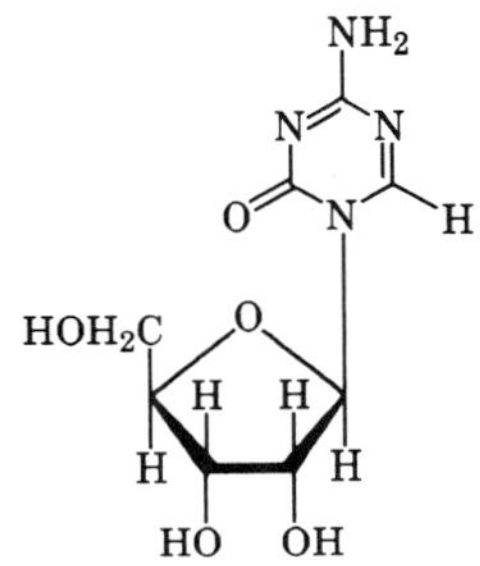

5-Azacytidine **(176)**

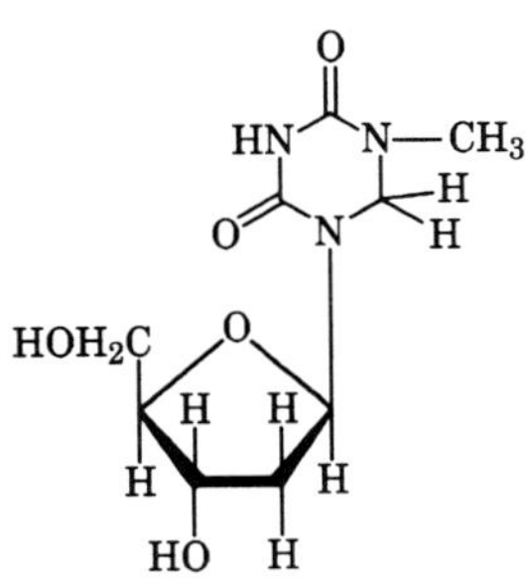

5,6-Dihydro-5-azathymidine **(177)**

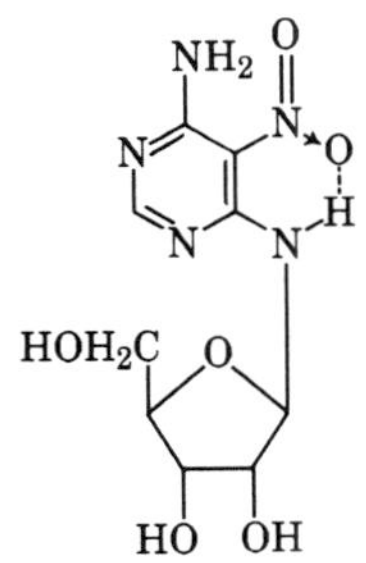

Clitocine **(178)**

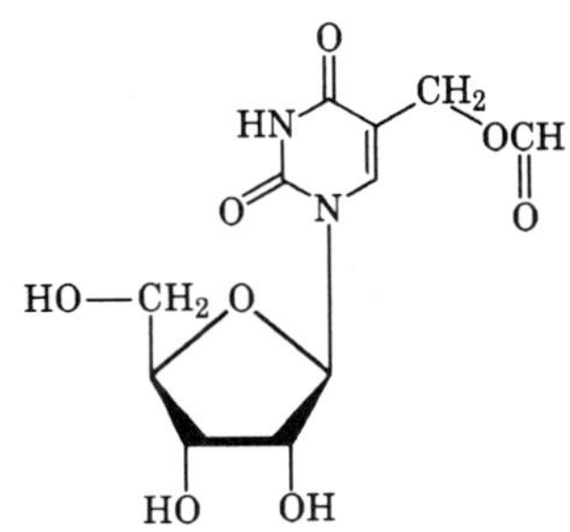

5-Formyloxymethyluridine **(179)**

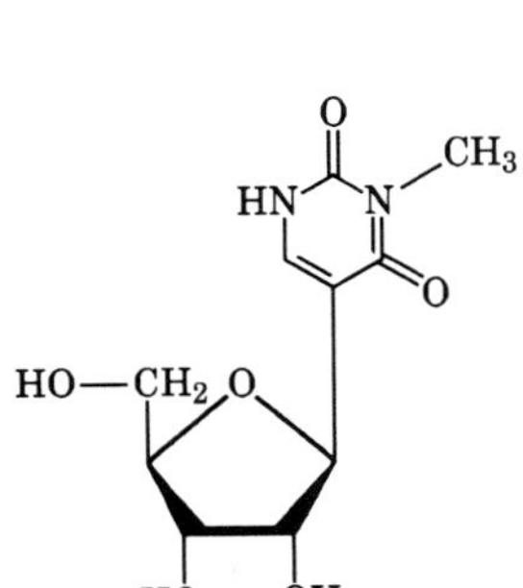

3-Methylpseudouridine **(180)**

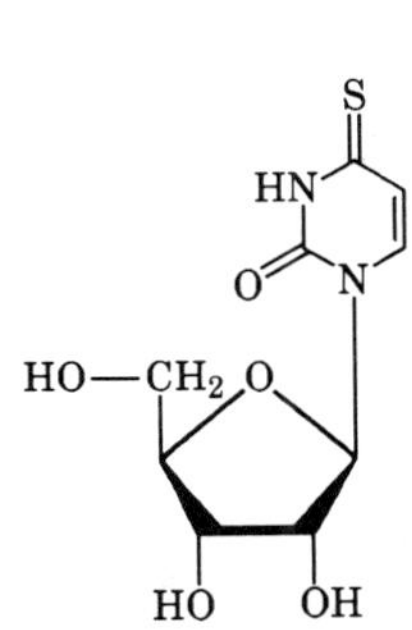

4-Thiouridine **(181)**

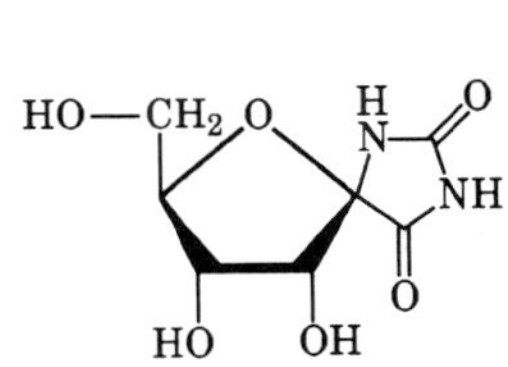

Hydantocidin **(182)**

Fig. 4. Structures of pyrimidine nucleosides.

Table 7. 4-Aminohexose Nucleosides

Name	CAS Registry Number	Molecular formula	Structure number	Structure
gougerotin	[*2096-42-6*]	$C_{16}H_{25}N_7O_8$	**(128)**	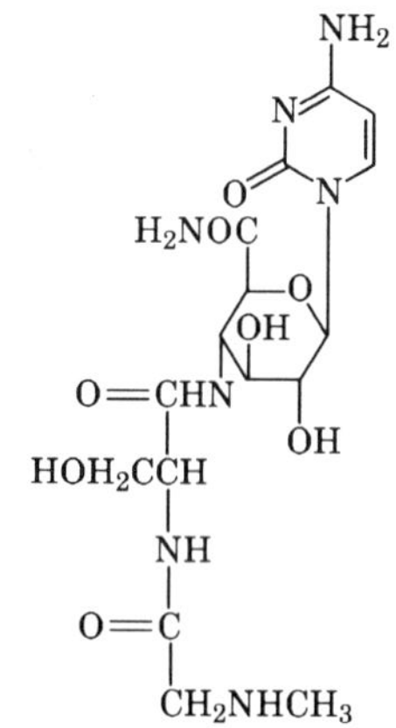
blasticidin S	[*2079-00-7*]	$C_{17}H_{26}N_8O_5$	**(129)**	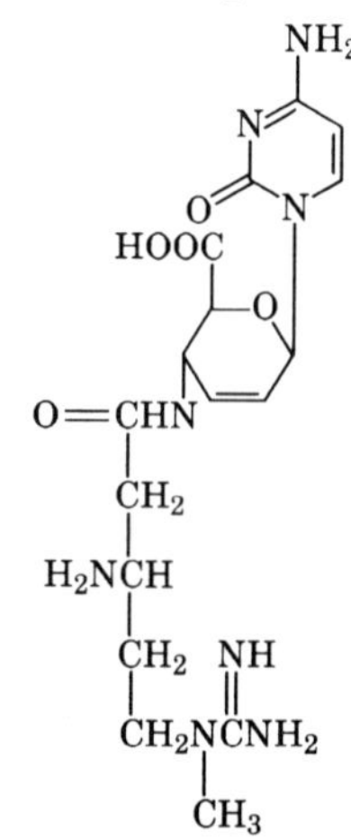

				R	R′	R″
amicetin	[*17650-86-1*]	$C_{29}H_{42}N_6O_9$	**(130)**	α-methylseryl	H	$N(CH_3)_2$
bamicetin	[*43043-14-7*]	$C_{28}H_{40}N_6O_9$	**(131)**	α-methylseryl	H	$NHCH_3$
oxamicetin	[*52665-75-5*]	$C_{29}H_{42}N_6O_{10}$	**(132)**	α-methylseryl	OH	$N(CH_3)_2$
plicacetin	[*43043-15-8*]	$C_{25}H_{35}N_5O_7$	**(133)**	H	H	$N(CH_3)_2$
norplicacetin	[*64040-47-7*]	$C_{24}H_{33}N_5O_7$	**(134)**	H	H	$NHCH_3$

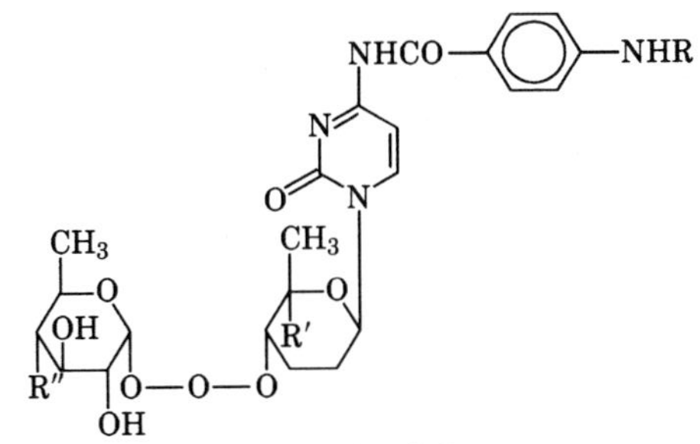

hikizimycin (anthelmycin)	[*12706-94-4*]	$C_{21}H_{37}N_5O_{14}$	**(135)**

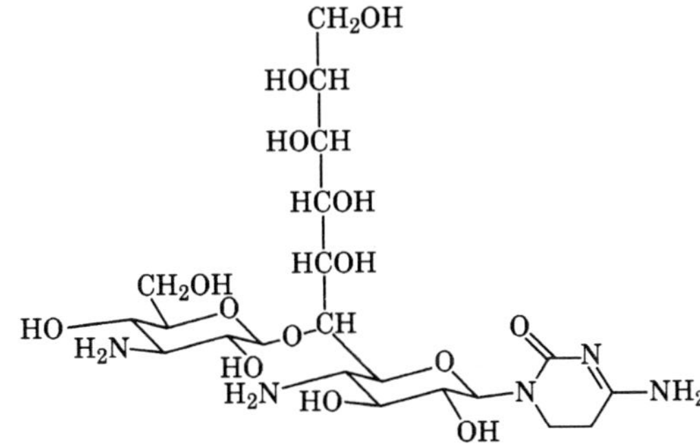

Table 7. ***(Continued)***

Name	CAS Registry Number	Molecular formula	Structure number	Structure
bagougeramine A	[*104840-35-9*]	$C_{17}H_{28}N_{10}O_7$	**(136)**	R = H 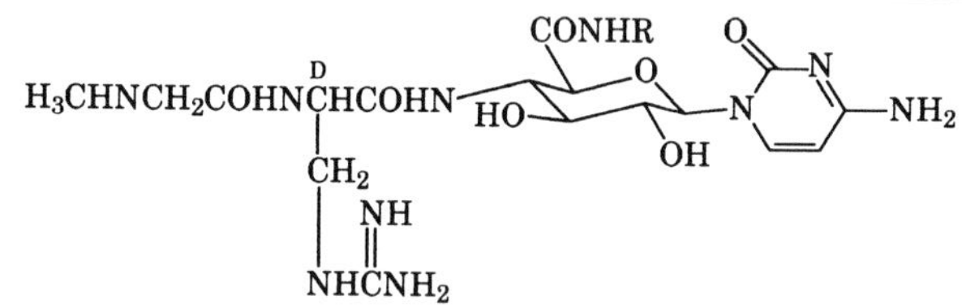
bagougeramine B	[*104840-34-8*]	$C_{24}H_{44}N_{12}O_7$	**(137)**	R = $(CH_2)_3NH(CH_2)_4NH_2$
arginomycin	[*106133-33-9*]	$C_{18}H_{28}N_8O_5$	**(138)**	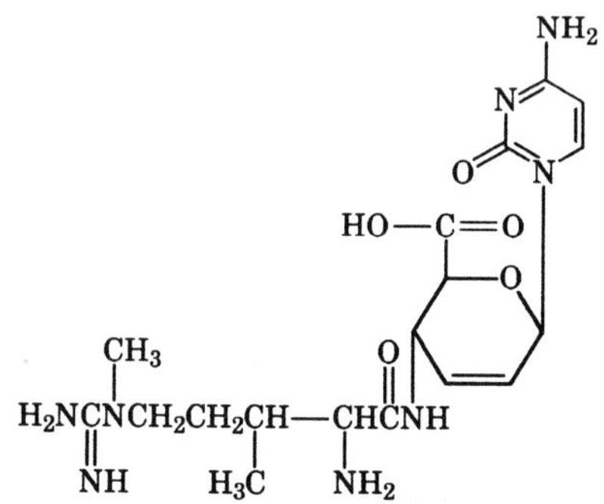

mildiomycin	*[67527-71-3]*	$C_{19}H_{30}N_8O_9$	**(139)**	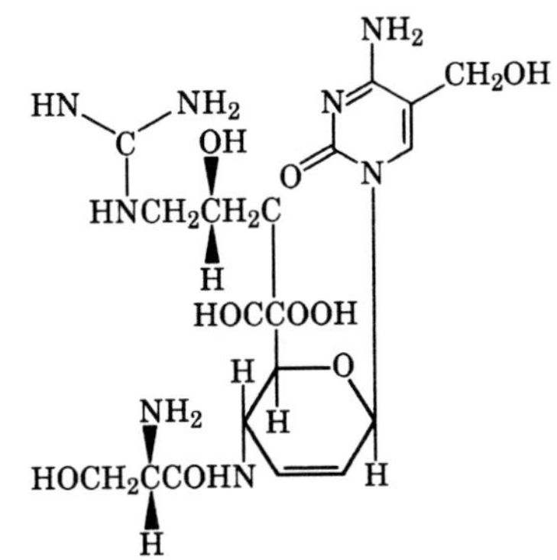
SCH 36605 (rodaplutin)	*[113667-06-4]*	$C_{24}H_{39}N_9O_7$	**(140)**	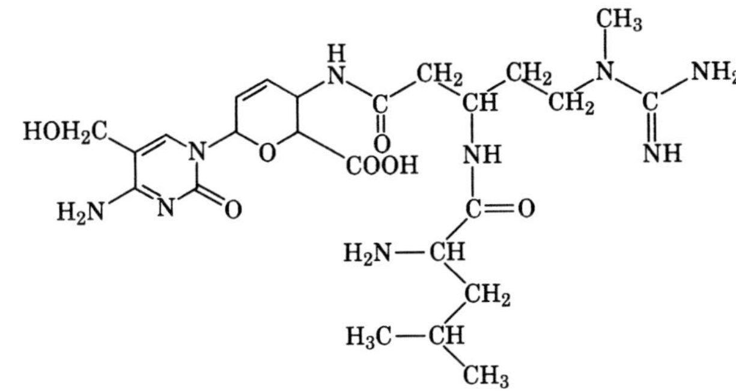

Table 8. The Nikkomycins

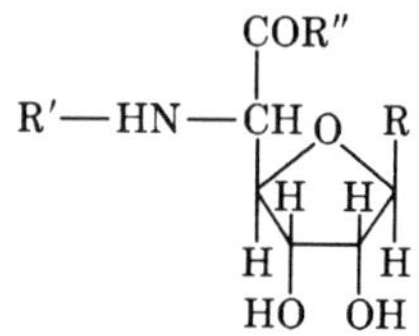

Nikkomycin	CAS Registry Number	Molecular formula	Structure number	R	R′	R″
Z	*[59456-70-1]* *[75044-70-1]*	$C_{20}H_{25}N_5O_{10}$ $C_{20}H_{25}N_5O_{10}$	**(141)**	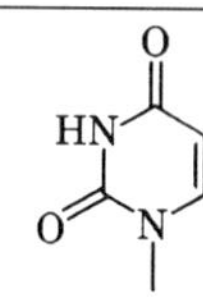	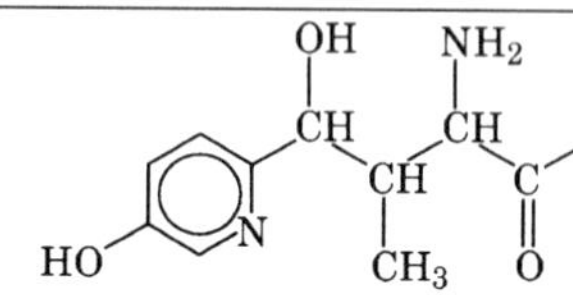	OH
X	*[72864-26-7]*	$C_{20}H_{25}N_5O_{10}$	**(142)**	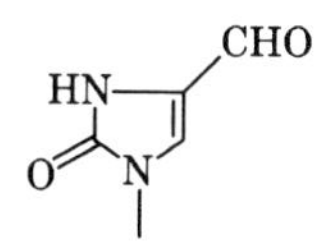	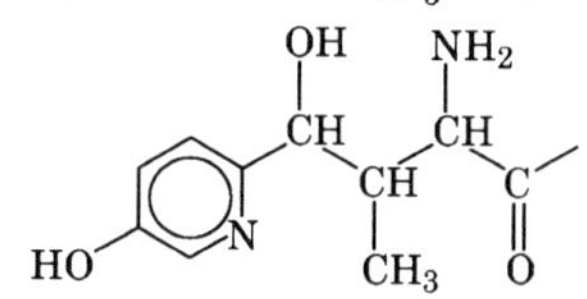	OH
J	*[77368-59-3]*	$C_{25}H_{32}N_6O_{13}$	**(143)**	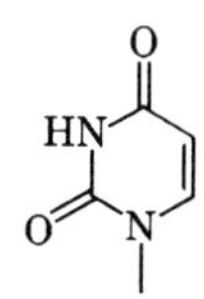	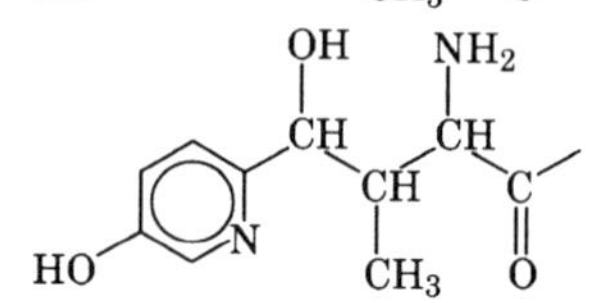	Glu
I	*[77368-60-6]*	$C_{25}H_{32}N_6O_{13}$	**(144)**	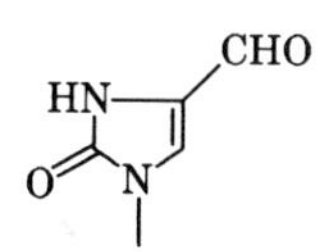	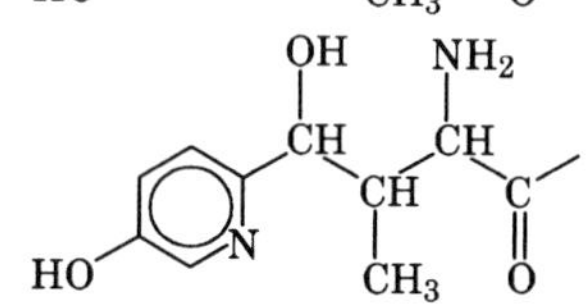	Glu

B_z			**(145)**	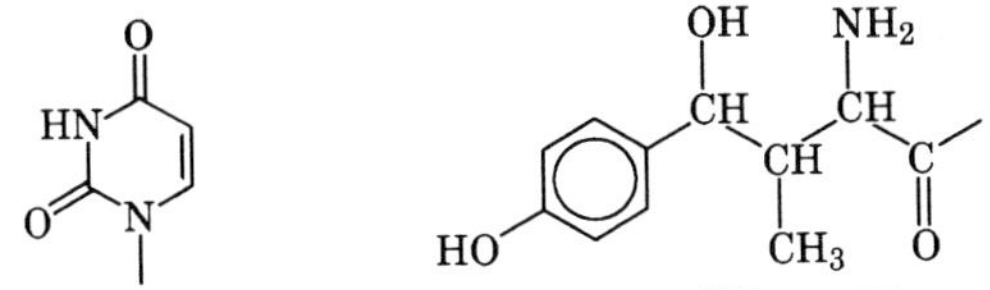	OH
B_x	*[75410-71-8]*	$C_{21}H_{26}N_4O_{10}$	**(146)**	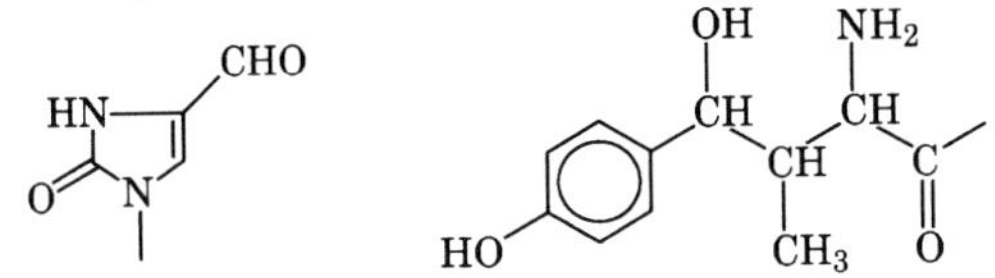	OH
K_z	*[95259-45-3]*	$C_{19}H_{23}N_5O_9$	**(147)**	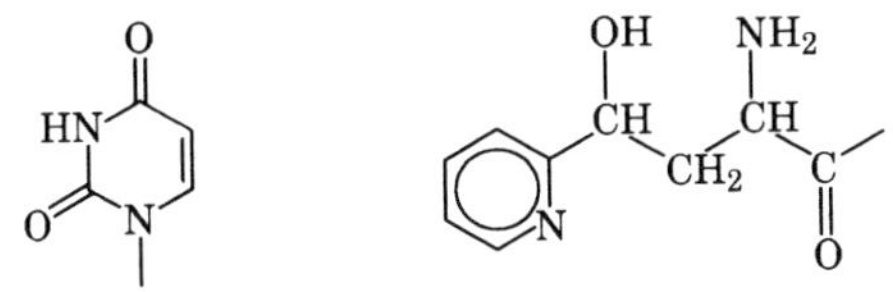	OH
K_x	*[95259-46-4]*	$C_{19}H_{23}N_5O_9$	**(148)**	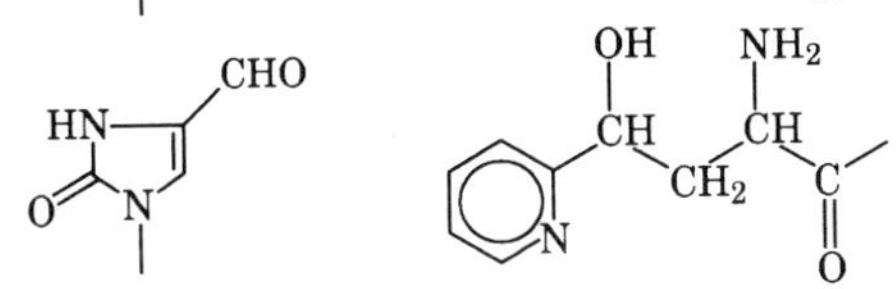	OH
O_z	*[95259-47-5]*	$C_{19}H_{23}N_5O_{10}$	**(149)**	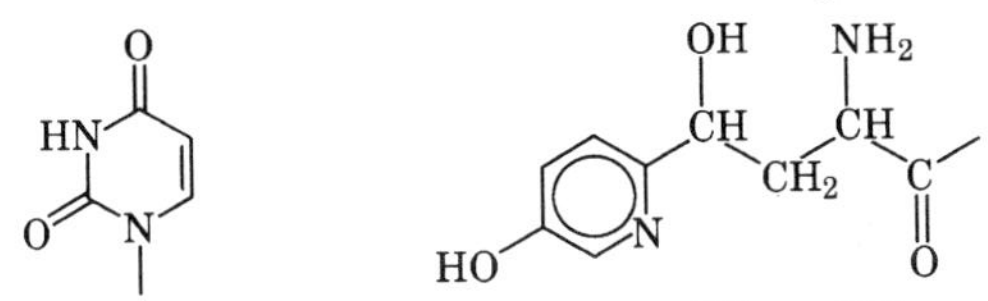	OH
O_x	*[95259-49-7]*	$C_{19}H_{23}N_5O_{10}$	**(150)**	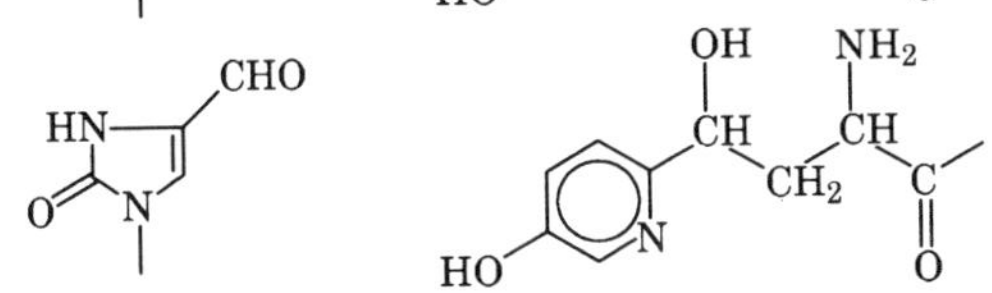	OH

Table 8. *(Continued)*

Nikkomycin	CAS Registry Number	Molecular formula	Structure number	R	R′	R″
Q_z	*[95259-50-0]*	$C_{24}H_{32}N_6O_{12}$	**(151)**	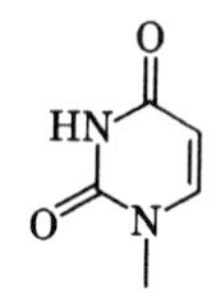	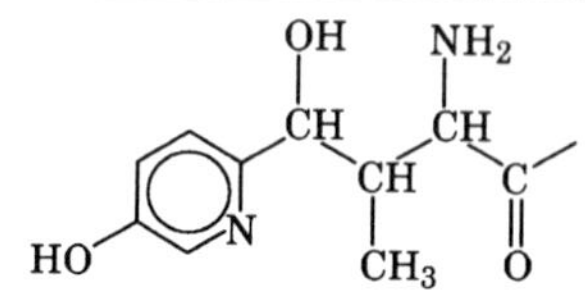	Homo-Ser
Q_x	*[95259-48-6]*	$C_{24}H_{32}N_6O_{12}$	**(152)**	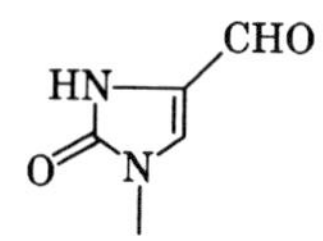	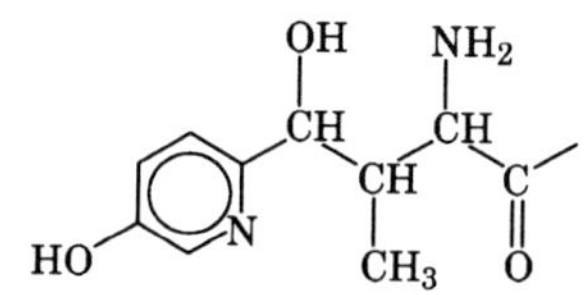	Homo-Ser
P_x		$C_{20}H_{25}N_5O_9$	**(153)**	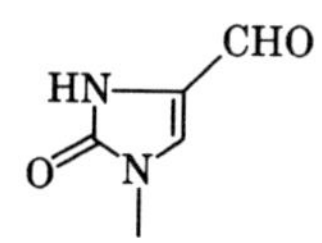	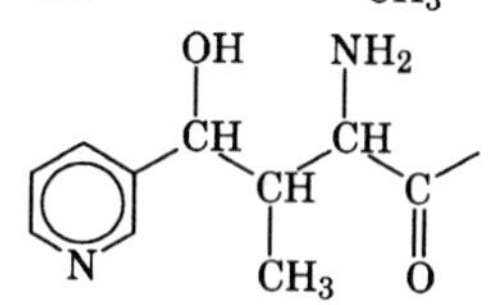	OH
R_z		$C_{25}H_{32}N_6O_{12}$	**(154)**	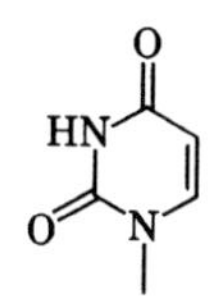	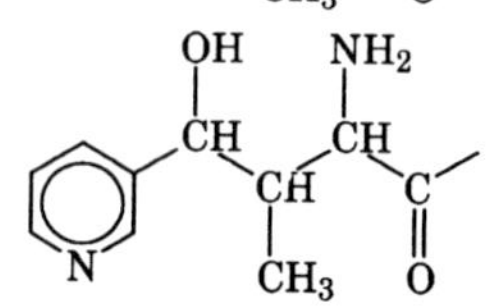	Glu

R_x		$C_{25}H_{32}N_6O_{12}$	**(155)**	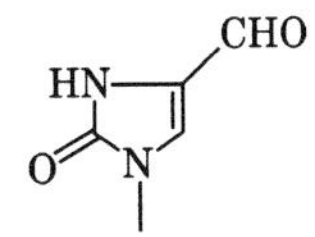	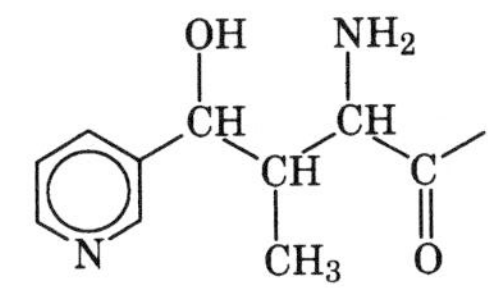	Glu
W_z		$C_{19}H_{21}N_4O_9$	**(156)**	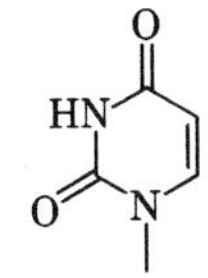	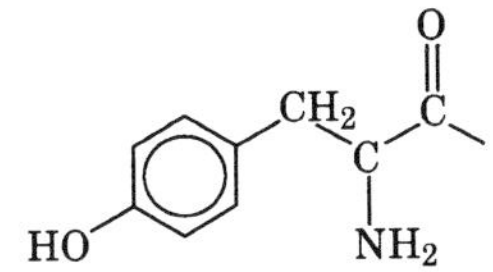	OH
W_x		$C_{19}H_{21}N_4O_9$	**(157)**	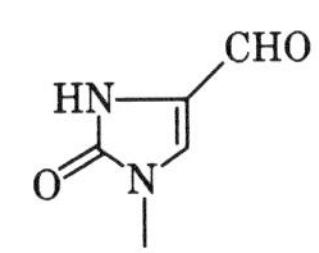	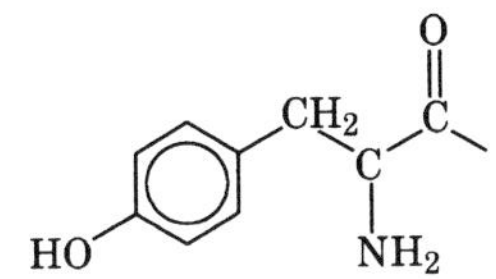	OH
pseudo-Z	[*120796-21-6*]	$C_{20}H_{25}N_5O_{10}$	**(158)**	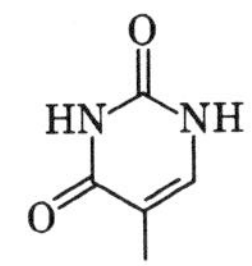	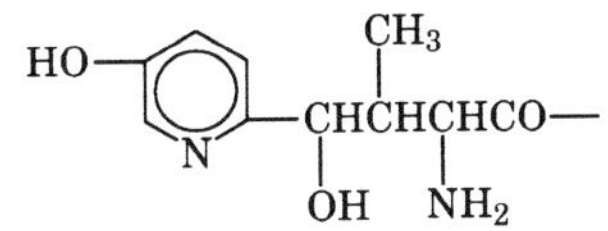	OH
pseudo-J	[*120796-22-7*]	$C_{25}H_{32}N_6O_{13}$	**(159)**	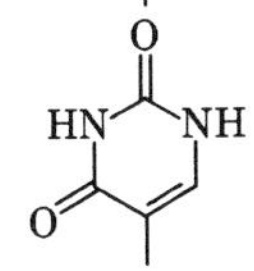	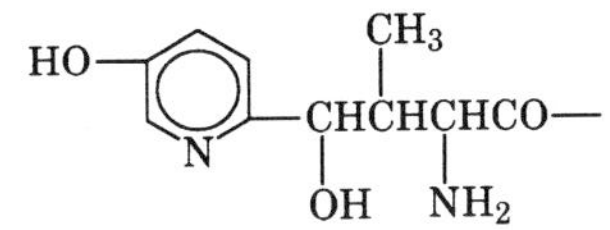	NHCHCOOH (with $(CH_2)_2COOH$ attached to CH)

Table 9. Other Peptidyl Nucleosides

Name	CAS Registry Number	Molecular formula	Structure number	X	Structure
albomycin δ_2	*[82113-67-5]*	$C_{37}H_{60}N_{12}O_{18}S$	**(160)**	$NCONH_2$	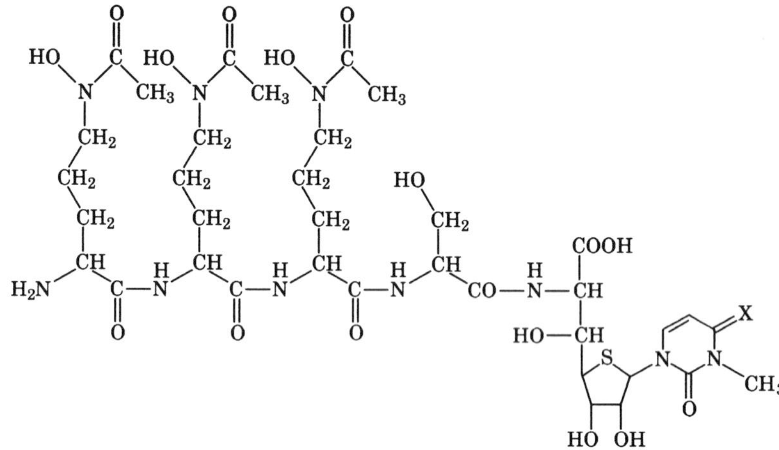
albomycin ϵ	*[82113-68-6]*	$C_{36}H_{59}N_{11}O_{17}S$	**(161)**	NH	
albomycin δ_1	*[82113-69-7]*	$C_{36}H_{58}N_{10}O_{18}S$	**(162)**	O	
amipurimycin	*[61991-08-0]*	$C_{20}H_{29}N_7O_8$	**(163)**		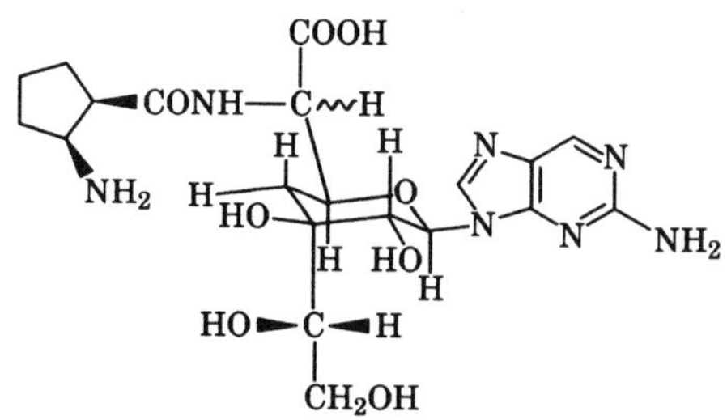
miharamycin A	*[12626-16-3]*	$C_{20}H_{30}N_{10}O_9$	**(164)**	OH	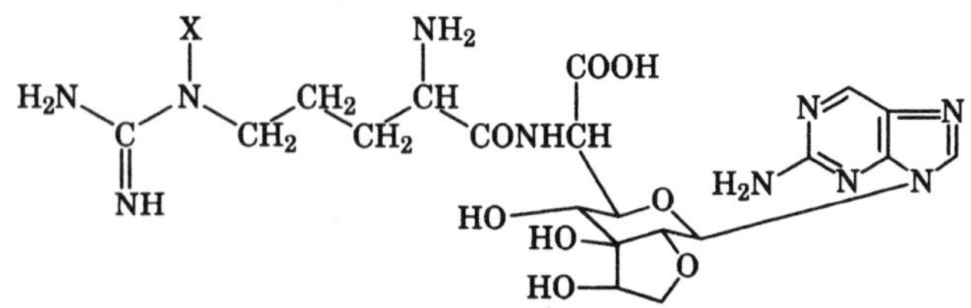
miharamycin B	*[12626-15-2]*	$C_{20}H_{30}N_{10}O_8$	**(165)**	H	

Table 10. Glycosyl Nucleosides

Name	CAS Registry Number	Molecular formula	Structure number	R	Structure
dapiramicin A	[*67298-15-1*]	$C_{21}H_{29}N_5O_{10}$	**(166)**		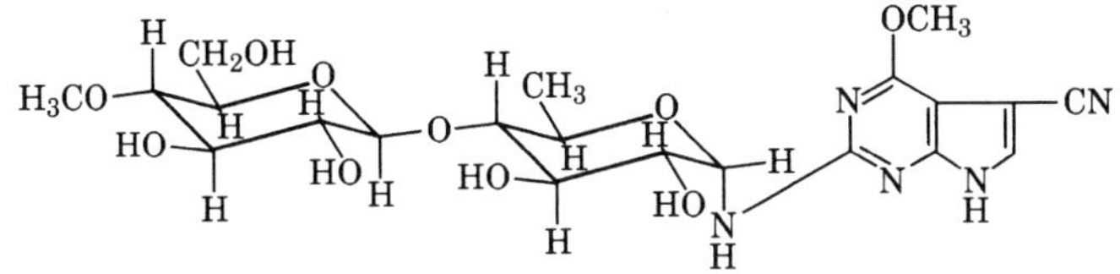
epidapiramicin A dapiramicin B	[*90044-19-2*] [*90044-18-1*]	$C_{21}H_{29}N_5O_{10}$ $C_{21}H_{29}N_5O_{11}$	**(167)** **(168)**	H OH	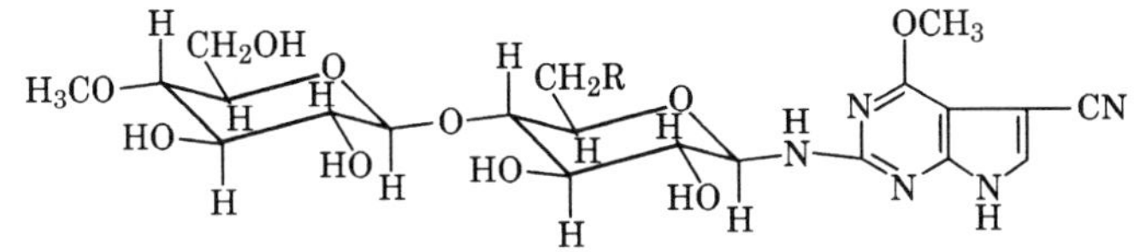
capuramycin	[*102770-00-3*]	$C_{23}H_{31}N_5O_{12}$	**(169)**		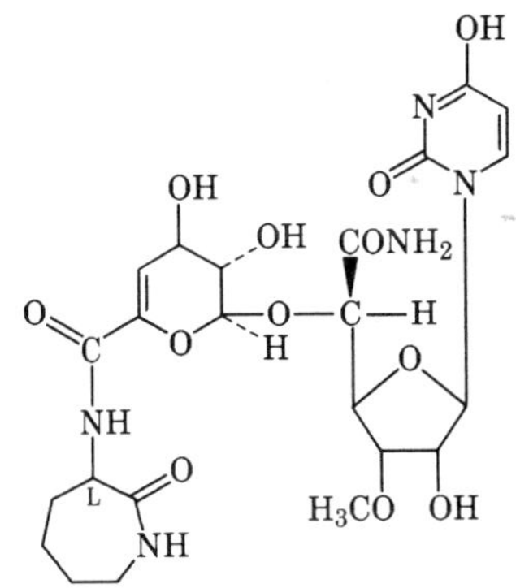
adenomycin	[*76174-56-6*]	$C_{25}H_{39}N_7O_{18}S$	**(170)**	$-\overset{O}{\overset{\Vert}{C}}\underset{\underset{NH_2}{\vert}}{C}HCH_2OH$	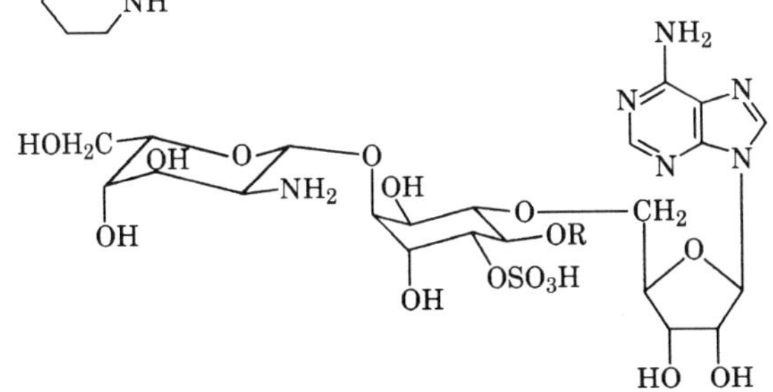

Table 11. ***N*-Nucleoside Antibiotics**

Name	CAS Registry Number	Molecular formula	Structure number	Structure
agrocin 84	[*59111-78-3*]	$C_{21}H_{34}N_6O_{16}P_2$	**(183)**	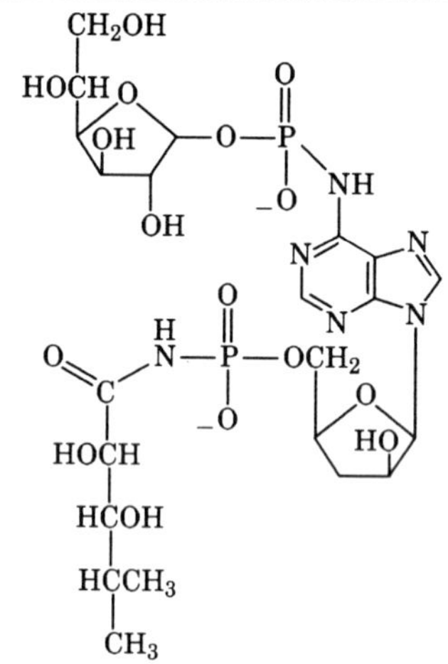
thuringiensin	[*23526-02-5*]	$C_{22}H_{30}N_5O_{19}P$	**(184)**	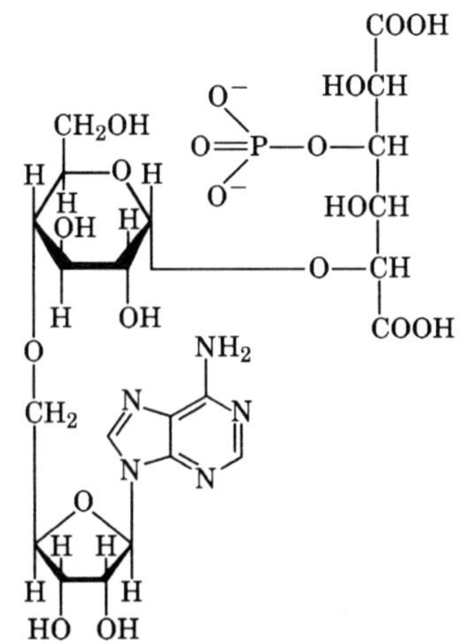

phosmidosine		$C_{16}H_{24}N_7O_8P$	(**185**)	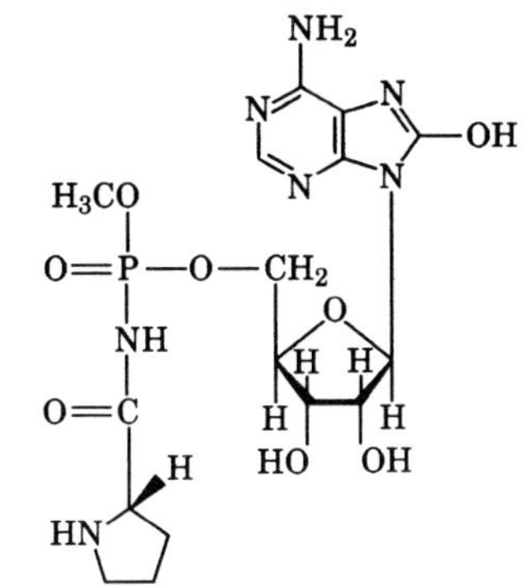
fosfadecin	[*120909-50-4*]	$C_{13}H_{19}N_5O_{10}P_2$	(**186**)	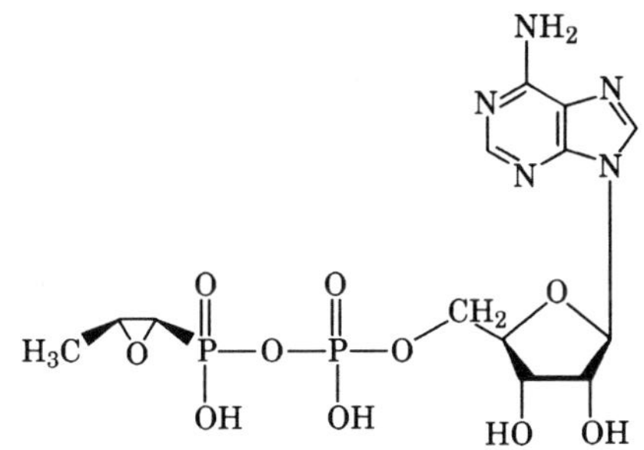
fosfocytocin	[*126986-25-2*]	$C_{12}H_{20}N_4O_{13}P_2$	(**187**)	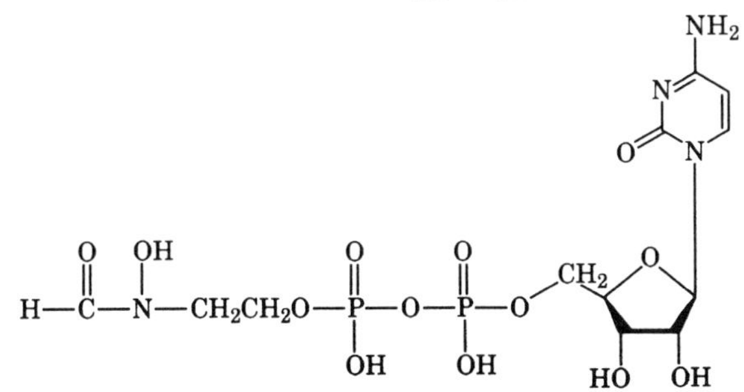

lung embryonic fibroblasts (255). The 5-halogenated derivatives of ara-UTP inhibit DNA polymerase-α, -β, and -γ and reverse transcriptase.

5-Azacytidine. 5-Azacytidine [*320-67-2*] (**176**), $C_8H_{12}N_4O_5$, 4-amino-1-β-D-ribofuranosyl-*S*-triazine-2-(1*H*-one), was chemically synthesized in 1964 and subsequently isolated from culture filtrates of *S. ladakanus* (1–4). Compound (**176**) and 2′-deoxy-5-azacytidine, which are cytotoxic analogues of cytidine, are potent anticancer agents that inhibit DNA methylation (256). Compound (**176**) is incorporated into RNA and tRNA and subsequently inhibits protein synthesis. It inhibits the maturation of 28S and 18S RNA, but not 13S RNA. Polysome degradation increases in the presence of (**176**) with an accumulation of monosomes and an inhibition of protein synthesis. The incorporation of 2′-deoxy-5-azacytidine 5′-monophosphate into DNA causes chromosome breakage. 5-Azacytidine induces 2-5A synthetase in human lymphoid cells (257). This hypomethylating nucleoside is a useful probe to analyze specific gene products in the earliest stages of chemical carcinogenesis (258,259). It can retransform cells by a mechanism other than DNA methylation (260). In addition, (**176**) induces cytidine deaminase in HL-60 cells, is incorporated into RNA and DNA, inhibits HIV-1 replication, upregulates Epstein-Barr virus nuclear antigen-2, suppresses ADP-ribosylation, and causes chromosomal aberrations (261–268). Compound (**176**) has been used in the treatment of thalassemia (269). The biochemistry of (**176**) has been reviewed (270).

5,6-Dihydro-5-azathymidine. 5,6-Dihydro-5-azathymidine [*57350-36-4*] (**177**), $C_9H_{15}N_3O_5$, contains the *s*-triazine ring and is isolated from the culture filtrates of *S. platensis* var. *clarensis* (4). The inhibition of viral replication by (**177**) can be reversed by either thymidine, deoxyuridine, or deoxycytidine. Compound (**177**) protects mice following intracerebral inoculation with HSV-1 and appears to inhibit the phosphorylation of thymidine rather than its incorporation into DNA. Resistance to (**177**) by *E. coli* appears to result from a change in thymidine phosphorylase.

Clitocine. Clitocine [*105798-74-1*] (**178**), $C_9H_{13}N_5O_6$, isolated from the mushroom, *Clitocybe inversa,* is 5-nitro-4-(β-D-ribofuranosylamino)pyrimidine-6-amine (271). The crystal structure indicates that the nitro group is hydrogen bonded to the 4-amino hydrogen atom as shown in Figure 4 (272). The base is in the *anti* conformation and the sugar moiety is disordered. Chemical syntheses of (**178**) have been described (273,274). Clitocine shows strong insecticidal activity against *Pectinophora gossypiella* (271) and is a substrate and inhibitor of adenosine kinase (274).

Uridine Analogues. 5-Formyloxymethyluridine (**179**), $C_{11}H_{14}N_2O_8$, produced by *Serratia plymuthica* (274), inhibits bacterial growth. 3-Methylpseudouridine (**180**), $C_{10}H_{14}N_2O_6$, has been isolated and identified from the culture filtrates of *Nocardia lactamdurans* (275). 1-Methylpseudouridine was previously isolated from *S. platensis* (276). 4-Thiouridine (**181**), $C_9H_{12}N_2O_5S$, has been isolated from an actinomycete resembling *S. hygroscopicus* (277).

Hydantocidin. Hydantocidin (**182**), $C_7H_{10}N_2O_6$, is elaborated by *S. hygroscopicus* (278). It is unique in that the anomeric carbon of the ribosyl moiety forms the spiro bond of hydantoin (279). The ribofuranose moiety which has been reported to be in a C_2-*endo* conformation (279) has been synthesized (280,281). Hydantocidin is a herbicidal nucleoside with activity against monocotyledenous and dicotyledenous plants.

N-Nucleotides

The *N*-nucleotide antibiotics are given in Table 11. Agrocin 84 (**183**) is an adenine 6-*N*-phosphoramidate nucleotide analogue that contains adenine, phosphate, and D-glucose in a 1:2:1 ratio (1,4). It is produced by *Agrobacterium radiobacter* strain K-84. Crown gall tumors are induced in a number of dicotyledonous plants following inoculation at wound sites with virulent *A. tumefaciens*. Crown gall tumors are caused by the incorporation of part of a virulent plasmid carried by *A. tumefaciens*. Proof that a large plasmid was essential is that the virulent strain C-58 can be converted to a stable avirulent strain. Crown gall formation by *A. tumefaciens* represents the first case where there is a transfer of genetic information from *A. tumefaciens* into the genome of the plant. This type of interaction has been termed "genetic colonization."

Thuringiensin (**184**), produced by *B. thuringiensis* (1,4), is a β-exotoxin that exerts its toxic action on insects and mammals through the inhibition of RNA polymerases.

The structure of the nucleotide antibiotic, phosmidosine (**185**), isolated from the culture filtrates of *Streptomyces* sp. RK-16 has been elucidated (282). Phosmidosine inhibits pore formation of *Botrytis cinerea* and *Aspergillus niger*.

Fosfadecin (**186**) and fosfocytocin (**187**) are adenine and cytosine nucleotide antibiotics isolated from the culture filtrates of *Pseudomonas viridiflava* PK-5 and *P. fluorescens* PK-52, respectively (283). Hydrolysis produces fosfoxacin which is also isolated from the culture filtrates. Compounds (**186**) and (**187**) inhibit gram-positive and gram-negative bacteria.

BIBLIOGRAPHY

"Antibiotics (Nucleosides)," in *ECT* 3, Vol. 2, pp. 962–986, by A. Bloch, Roswell-Park Memorial Institute.

1. K. Isono, *J. Antibiot.* **41,** 1711 (1988).
2. K. Isono, *Pharmacol. Therapeut.,* in press, 1992.
3. R. J. Suhadolnik, *Nucleoside Antibiotics,* John Wiley & Sons, Inc., New York, 1970.
4. R. J. Suhadolnik, *Nucleosides as Biological Probes,* John Wiley & Sons, Inc., New York, 1979.
5. A. P. Reuter, J. A. Figueiredo, and I. M. Ismael, *Carbohydrate Res.* **188,** 19 (1989).
6. E. F. Elstner and co-workers, *Biochemistry* **12,** 4992 (1973).
7. K. Isono and R. J. Suhadolnik, *J. Antibiot.* **30,** 272 (1977).
8. E. F. Elstner, R. J. Suhadolnik, and A. Allerhand, *J. Biol. Chem.* **248,** 5385 (1973).
9. J. G. Buchanan, A. Stobie, and R. H. Wightman, *J.C.S. Chem. Commun.,* 916 (1980).
10. C. R. Petrie, III and co-workers, *J. Med. Chem.* **29,** 268 (1986).
11. R. J. Suhadolnik and N. L. Reichenbach, *Biochemistry* **20,** 7042 (1981).
12. J. G. Buchanan, M. R. Hamblin, G. R. Sood, and R. H. Wightman, *J.C.S. Chem. Commun.,* 917 (1980).
13. J. M. Crance, E. Biziagos, J. Passagot, H. van Cuyck-Gandrie, and R. Deloince, *J. Med. Virol.* **31,** 155 (1990).
14. M. E. Sant and co-workers, *Cancer Res.* **49,** 2645 (1989).
15. Y. Kang, S. B. Larson, R. K. Robins, and G. R. Revenkar, *J. Med. Chem.* **32,** 1547 (1989).

16. P. Laquel, M. Castrouviejo, and S. Litvak, *Nucleic Acids Res.* **18,** 4867 (1990).
17. C. R. Crawford, C. Y. Ng, L. D. Noel, and J. A. Belt, *J. Biol. Chem.* **265,** 9732 (1990).
18. P. G. Plagemann and C. Woffendin, *Biochim. Biophys. Acta* **1010,** 7 (1989).
19. D. Vijayalakshmi and J. A. Belt, *J. Biol. Chem.* **263,** 19,419 (1988).
20. P. G. Plagemann, J. M. Aran, and C. Woffendin, *Biochim. Biophys. Acta* **1022,** 93 (1990).
21. E. S. Jakobs, D. J. Van Os-Corby, and A. R. P. Paterson, *J. Biol. Chem.* **265,** 22,210 (1990).
22. V. L. Giranda, H. M. Berman, and V. L. Schramm, *Biochemistry* **27,** 5813 (1988).
23. A. Maelicke, M. Sprinzl, F. Von der Haar, T. A. Khwaja, and F. Cramer, *Eur. J. Biochem.* **43,** 617 (1974).
24. E. Baksht, N. de Groot, M. Sprinzl, and F. Cramer, *FEBS Letters* **55,** 105 (1975).
25. R. P. Agarwal and co-workers, *J. Clin. Invest.* **57,** 1025 (1976).
26. W. E. G. Müller and co-workers, *Cancer Res.* **35,** 3673 (1975).
27. W. E. G. Müller and R. K. Zahn, *Experientia* **31,** 1014 (1975).
28. R. J. Suhadolnik and co-workers, *J. Biol. Chem.* **252,** 4134 (1977).
29. B. A. Otter, S. A. Patil, R. S. Klein, and S. E. Ealick, *J. Am. Chem. Soc.* **114,** 668 (1992).
30. R. J. Suhadolnik, S. Pornbanlualap, D. C. Baker, K. N. Tiwari, and A. K. Hebbler, *Arch. Biochem. Biophys.* **270,** 374 (1989).
31. M. Sprinzl and F. Cramer, *Nature New Biol.* **245,** 3 (1973).
32. T. H. Frazer and A. Rich, *Proc. Natl. Acad. Sci. USA* **70,** 2671 (1973).
33. N. Van der Heyden, C. Rodi, and L. Ratner, *AIDS Res. Human Retroviruses* **5,** 647 (1989).
34. N. M. Kredich, *J. Biol. Chem.* **255,** 7380 (1980).
35. P. Doetsch, J. M. Wu, Y. Sawada, and R. J. Suhadolnik, *Nature* **291,** 355 (1981).
36. D. C. Montefiori and co-workers, *Proc. Natl. Acad. Sci. USA* **86,** 7191 (1989).
37. W. E. G. Müller and co-workers, *Biochemistry* **30,** 2027 (1991).
38. D. A. Kuznetsov and N. I. Musajev, *Int. J. Neurosci.* **51,** 53 (1990).
39. J. S. Butler, P. P. Sadhale, and T. Platt, *Mol. Cell. Biol.* **10,** 2599 (1990).
40. P. A. Mathew, L. K. Ellis, and G. P. Studzinski, *J. Cell. Physiol.* **140,** 212 (1989).
41. P. G. Plagemann, J. M. Aran, R. M. Wohlhueter, and C. Woffendin, *Biochim. Biophys. Acta* **1022,** 103 (1990).
42. L. L. McGrew, E. Dworkin-Rastl, and J. D. Richter, *Genes Dev.* **3,** 803 (1989).
43. J. McLauchlan, C. L. Moore, S. Simpson, and J. B. Clements, *Nuc. Acids Res.* **16,** 5323 (1988).
44. S. K. Pramanik, W. J. Meadus, and J. Bag, *Eur. J. Biochem.* **172,** 355 (1988).
45. G. W. Zieve, R. J. Feeney, and E. J. Roemer, *Cell Motility Cytoskeleton* **7,** 337 (1987).
46. R. A. Colbert and D. A. Young, *J. Biol. Chem.* **262,** 9939 (1987).
47. D. Zarkower and M. Wickens, *EMBO J.* **6,** 177 (1987).
48. J. Remme, T. Margus, R. Villems, and K. H. Nierhaus, *Eur. J. Biochem.* **183,** 281 (1989).
49. D. Synetos and C. Coutsogeorgopoulos, *Eur. J. Biochem.* **184,** 47 (1989).
50. C. C. Hall, D. Johnson, and B. S. Cooperman, *J. Biochem.* **27,** 3983 (1988).
51. S. Inoue, M. Iwasaki, K. Ishii, K. Kitajima, and Y. Inoue, *J. Biol. Chem.* **264,** 18,520 (1989).
52. A. Santos, A. Perez-Castillo, N. C. Wong, and J. H. Oppenheimer, *J. Biol. Chem.* **262,** 16,880 (1987).
53. K. S. Park and co-workers, *J. Biol. Chem.* **262,** 14,702 (1987).
54. T. J. Flynn, L. Friedman, and T. N. Black, *Experientia* **43,** 433 (1987).
55. U. Geigenmuller, T. P. Hausner, and K. H. Nierhaus, *Eur. J. Biochem.* **16,** 715 (1986).
56. L. Kaczmarek, E. Surmacz, and R. Baserga, *Cell Biol. Int. Rep.* **10,** 455 (1986).
57. R. Vince, S. Daluge, and J. Brownell, *J. Med. Chem.* **29,** 2400 (1986).
58. J. A. Vara, A. Portela, J. Ortin, and A. Jimenez, *J. Nuc. Acids Res.* **14,** 4617 (1986).

59. C. J. Weitzmann and B. S. Cooperman, *Biochemistry* **29,** 3450 (1990).
60. D. M. Graifer, O. S. Fedorova, and G. G. Karpova, *FEBS Lett.* **277,** 4 (1990).
61. M. Yamashita and co-workers, *J. Antibiot.* **37,** 1279 (1984).
62. T. Iwamoto and co-workers, *J. Antibiot.* **43,** 1183 (1990).
63. A. Cumaraswamy and H. Henney, Jr., *Biochem. Cell Biol.* **68,** 769 (1990).
64. A. Fouet and A. L. Sonenshein, *J. Bacteriol.* **172,** 835 (1990).
65. U. Bai, M. Lewandoski, E. Dubnau, and I. Smith, *J. Bacteriol.* **172,** 5432 (1990).
66. M. A. Glazebrook, J. L. Doull, C. Stuttard, and L. C. Vining, *J. Gen. Microbiol.* **136,** 581 (1990).
67. W. L. Nicholson and G. H. Chambliss, *J. Bacteriol.* **169,** 5867 (1987).
68. K. Ochi, *J. Bacteriol.* **169,** 3608 (1987).
69. E. J. Dubnau, K. Cabane, and I. Smith, *J. Bacteriol.* **169,** 1182 (1987).
70. C. E. Rohlman and R. G. Matthews, *J. Bacteriol.* **172,** 7200 (1990).
71. C. McGuigan, S. M. Tollerfield, and P. A. Riley, *Nucleic Acids Res.* **17,** 6065 (1989).
72. J. A. Secrist, III, A. T. Shortnacy, and J. A. Montgomery, *J. Med. Chem.* **31,** 405 (1988).
73. D. C. Baker, T. H. Haskell, S. R. Putt, and B. J. Sloan, *J. Med. Chem.* **22,** 273 (1979).
74. T. Utagawa and co-workers, *FEBS Letters* **109,** 261 (1980).
75. R. J. Suhadolnik, S. Pornbanlualap, J. M. Wu, D. C. Baker, and A. K. Hebbler, *Arch. Biochem. Biophys.* **270,** 363 (1989).
76. W. E. G. Müller, R. K. Zahn, K. Bittlingmaier, and D. Falke, *Ann. N.Y. Acad. Sci.* **284,** 34 (1977).
77. J. P. Verheyden, *Rev. Infec. Dis.* **10,** Suppl. 477 (1988).
78. D. Pavan-Langston, R. A. Buchanan, and C. A. Alford, Jr., eds., *Adenine Arabinoside: An Antiviral Agent,* Raven Press, New York, 1975.
79. T. Takimoto and co-workers, *Leukemia* **2,** 443 (1988).
80. T. Yoshida, T. Suzutani, and M. Azuma, *Tohoku J. Exp. Med.* **156,** 279 (1988).
81. T. Spector, *J. Biol. Chem.* **250,** 7372 (1975).
82. W. E. G. Müller, *Experientia* **32,** 1572 (1976).
83. W. E. G. Müller and R. K. Zahn, in P. Chandra, ed., *Antiviral Mechanisms in the Control of Neoplasia,* Plenum Publishing Corp., New York, 1979, p. 553.
84. D. E. Bell and A. Fridland, *Biochim. Biophys. Acta* **606,** 57 (1980).
85. G. Iliakis and R. Seaner, *Rad. Environ. Biophys.* **27,** 295 (1988).
86. M. Kubota and co-workers, *Anticancer Res.* **8,** 339 (1988).
87. Y. Ohno, D. Spriggs, A. Matsukage, T. Ohno, and D. Kufe, *J. Cancer Res.* **49,** 2077 (1989).
88. Y. Kuroki, M. Shimoyama, S. Inaba, and M. Hirose, *Japanese J. Cancer Res.* **80,** 482 (1989).
89. J. Balzarini and E. DeClercq, *Biochem. Biophys. Res. Commun.* **173,** 781 (1990).
90. N. M. Mozier, K. P. McConnell, and J. L. Hoffman, *J. Biol. Chem.* **263,** 4527 (1988).
91. H. Malina, M. H. Tempete, and M. Robert-Gero, *J. Antibiot.* **40,** 505 (1987).
92. D. R. Berry and B. J. Abbott, *J. Antibiot.* **31,** 185 (1978).
93. A. Parry, *J. Am. Chem. Soc.* **113,** 4704 (1991).
94. H. Malina and M. Robert-Gero, *Applied Environ. Micro.* **54,** 1109 (1988).
95. W. K. Long, G. E. Fronko, R. G. Lindemeyer, B. Wu, and E. E. Henderson, *J. Virol.* **61,** 221 (1987); L. L. Nolan, *Antimicrob. Agents Chemother.* **31,** 1542 (1987).
96. R. A. Neal, M. V. Iwabi, and M. Robert-Gero, *Comptes Rendus Acad. Sci.* **308,** 485 (1989).
97. B. L. Mirkin, R. F. O'Dea, and H. P. C. Hogenkamp, *Cancer Res.* **47,** 3650 (1987).
98. R. T. Borchardt, L. E. Eiden, B-S. Wu, and C. O. Rutledge, *Biochem. Biophys. Res. Commun.* **89,** 919 (1979).
99. J. Najbauer and D. W. Oswald, *J. Biol. Chem.* **265,** 12,717 (1990).
100. W.-K. Paik and S. Kim, in A. Meister, ed., *Protein Methylation (Biochemistry: A Series of Monographs),* Vol. I, John Wiley & Sons, Inc., New York, 1980, p. 197.

101. S. Som and S. Friedman, *J. Biol. Chem.* **265,** 4278 (1990).
102. N. O. Reich and N. Mashhoon, *J. Biol. Chem.* **265,** 8966 (1990).
103. C. Barbis, J. Sanchez, M. Robert-Gero, and C. Hardisson, *FEMS Microbiol. Lett.* **57,** 239 (1990).
104. S. G. Amur, G. Shanker, J. M. Cochran, H. S. Ved, and R. A. Pieringer, *J. Neuroscience Res.* **16,** 367 (1986).
105. D. B. Drachman and R. Adams, *Proc. Natl. Acad. Sci. USA* **85,** 4032 (1988).
106. F. A. Fuhrman, G. J. Fuhrman, Y. H. Kim, L. A. Pavelka, and H. S. Mosher, *Science* **207,** 193 (1980).
107. R. P. Gregson, R. J. Quinn, and A. F. Cook, *Chem. Abst.* **91,** 39,792 (1979).
108. B. G. Issac, S. W. Ayer, L. J. Letendre, and R. J. Stonard, *J. Antibiot.* **44,** 729 (1991).
109. A. K. Saksena, *Tetrahedron Lett.* **21,** 133 (1980).
110. L. L. Bennett, Jr., P. W. Allan, L. M. Rose, R. M. Comber, and J. A. Secrist, III, *Molec. Pharmacol.* **29,** 383 (1986).
111. O. Hiraoko and co-workers, *Biochem. Biophys. Res. Commun.* **134,** 1114 (1986).
112. A. A. Fischer, K. Muller, and C. Scholtissek, *J. Virol.* **177,** 523 (1990).
113. R. Djurhuus, A. M. Svardal, and P. M. Ueland, *J. Cancer Res.* **49,** 324 (1989).
114. K. Dalaker and H. Prydz, *J. Biochem. Pharm.* **35,** 3433 (1986).
115. J. Aarbakke, G. A. Miura, and P. S. Prytz, *J. Cancer Res.* **46,** 5469 (1986).
116. P. S. Backlund, Jr., D. Carotti, and G. L. Cantoni, *Eur. J. Biochem.* **160,** 245 (1986).
117. A. M. Svardal, R. Djurhuus, H. Refsum, and P. M. Ueland, *J. Cancer Res.* **46,** 5095 (1986).
118. R. G. Liteplo, *Cancer Lett.* **39,** 319 (1988).
119. M. L. Greenberg, S. Chaffee, and M. S. Hershfield, *J. Biol. Chem.* **264,** 795 (1989).
120. R. J. Parry, V. Bornemann, and R. Subramanian, *J. Am. Chem. Sic.* **111,** 5819 (1989).
121. S. Yaginuma, N. Muto, T. Tsujino, and M. Hayashi, *J. Antibiot.* **34,** 359 (1981).
122. M. Hayashi, S. Yaginuma, H. Yoshioka, and K. Nakatsu, *J. Antibiot.* **34,** 675 (1981).
123. M.-L. Lim and V. E. Marquez, *Tetrahedron Lett.* **24,** 5559 (1983).
124. J. Michael, G. Oflenbaecher, and J. Retey, *Helv. Chim. Acta* **66,** 1915 (1983).
125. M. Arita, K. Adachi, Y. Ito, H. Sawai, and M. Ohno, *J. Am. Chem. Soc.* **105,** 4049 (1983).
126. C. K. Tseng and co-workers, *J. Med. Chem.* **32,** 1442 (1989).
127. E. DeClerq and co-workers, *Antimicrob. Agents Chemother.* **33,** 1291 (1989).
128. E. DeClerq, *Biochem. Pharmacol.* **36,** 2567 (1987).
129. R. M. Ransohoff, P. Narayan, D. F. Ayers, F. M. Rottman, and T. W. Nilsen, *J. Antiviral Res.* **7,** 317 (1987).
130. M. Cools and E. DeClerq, *Biochem. Pharmacol.* **40,** 2259 (1990).
131. D. L. Kramer, C. W. Porter, R. T. Borchardt, and J. R. Sufrin, *J. Cancer Res.* **50,** 3838 (1990).
132. S. D. Paisley, M. S. Wolfe, and R. T. Borchardt, *J. Med. Chem.* **32,** 1415 (1989).
133. M. Inaba, K. Nagashima, S. Tsukagoshi, and Y. Sakurai, *Cancer Res.* **46,** 1063 (1986).
134. K. Isono and co-workers, *J. Antibiot.* **37,** 670 (1985).
135. E. Takahashi and T. Beppu, *J. Antibiot.* **35,** 939 (1982).
136. M. Osada and K. Isono, *Antimicrob. Agents Chemother.* **27,** 230 (1985).
137. H. Osada and K. Isono, *Biochem. J.* **233,** 459 (1986).
138. W. M. Kati and R. Wolfenden, *Science* **243,** 1591 (1989).
139. M. H. el Kouni, N. J. Messier, and S. Cha, *J. Biochem. Pharm.* **36,** 3815 (1987).
140. M. H. el Kouni and S. Cha, *J. Biochem. Pharm.* **36,** 1099 (1987).
141. J. Jiricny, S. G. Wood, D. Martin, and A. Ubasawa, *Nuc. Acids Res.* **14,** 6579 (1986).
142. F. Nakagawa, T. Okazaki, A. Naito, Y. Iijima, and M. Yamazaki, *J. Antibiot.* **38,** 823 (1985).
143. S. Takahashi and co-workers, *J. Antibiot.* **38,** 830 (1985).
144. S. Takahashi, F. Nakagawa, and S. Sato, *J. Antibiot.* **41,** 705 (1988).
145. N. Shimada and co-workers, *J. Antibiot.* **39,** 1623 (1986).

146. H. Hoshino, N. Shimizu, N. Shimada, T. Takita, and T. Takeuchi, *J. Antibiot.* **40,** 1077 (1987).
147. F. X. Wilson and co-workers, *Tetrahedron Lett.* **31,** 6931 (1990).
148. N. Shimada and co-workers, *J. Antibiot.* **34,** 1216 (1981).
149. N. Yagisawa and co-workers, *J. Antibiot.* **35,** 755 (1982).
150. Y. Uehara, M. Hasegawa, M. Hori, and H. Umezawa, *Cancer Res.* **45,** 5230 (1985).
151. S. Naruto, H. Uno, A. Tanaka, H. Kotani, and Y. Takase, *Heterocycles* **20,** 27 (1983); B. D. Yuan and co-workers, *J. Antibiot.* **38,** 1588 (1985).
152. J. B. Stewart and co-workers, *J. Antibiot.* **41,** 1048 (1988).
153. C. R. Loomis and R. M. Bell, *J. Biol. Chem.* **263,** 1682 (1988).
154. T. T. Tsunoda and K. Isono, *J. Antibiot.* **42,** 102 (1989).
155. P. K. Gupta and co-workers, *J. Med. Chem.* **32,** 402 (1989).
156. E. DeClercq, J. Balzarini, D. Madej, F. Hansske, and M. J. Robins, *J. Med. Chem.* **30,** 481 (1987).
157. S. H. Krawczyk and co-workers, *J. Med. Chem.* **33,** 3160 (1990).
158. H. Osada, T. Sonoda, K. Tsunoda, and K. Isono, *J. Antibiot.* **42,** 102 (1989).
159. N. K. Saxena, L. A. Coleman, J. C. Drach, and L. B. Townsend, *J. Med. Chem.* **33,** 1980 (1990).
160. P. K. Gupta and co-workers, *J. Med. Chem.* **32,** 402 (1989).
161. E. F. Elstner and R. J. Suhadolnik, *J. Biol. Chem.* **246,** 6973 (1971).
162. T. Uematsu and R. J. Suhadolnik, *Arch. Biochem. Biophys.* **162,** 614 (1974).
163. M. Kaehler, J. Coward, and F. Rottman, *Biochemistry* **16,** 5770 (1977).
164. G. Seibert, A. Maidhof, R. K. Zahn, and W. E. G. Müller, *Gann* **69,** 739 (1978).
165. E. DeClercq and M. J. Robins, *Antimicrob. Agents Chemother.* **30,** 719 (1986).
166. E. DeClerq and M. J. Robins, *Nuc. Acids Res.* **15,** 3059 (1987).
167. R. I. Glazer and A. L. Peale, *Biochem. Pharmacol.* **29,** 305 (1980).
168. K. Palczewski, N. Kahn, and P. A. Hargrave, *Biochemistry* **29,** 6276 (1990).
169. K. Ramasamy, B. S. Sharma, W. B. Jolley, R. K. Robins, and G. R. Revankar, *J. Med. Chem.* **32,** 1905 (1989).
170. S. H. Kim, T. Okabe, N. Tanaka, and H. Suzuki, *J. Antibiot.* **40,** 1776 (1987).
171. H. Suzuki and co-workers, *Cancer Res.* **47,** 713 (1987).
172. S. Naruto, H. Uno, A. Tanaka, H. Kotani, and Y. Takase, *Heterocycles* **20,** 27 (1983).
173. Y. Kato, N. Fusetani, S. Matsunaga, and K. Hashimoto, *Tetrahedron Lett.* **26,** 3483 (1985).
174. S. Omura and co-workers, *J. Antibiot.* **38,** 1008 (1985).
175. S. Omura, H. Tanaka, H. Kuga, and N. Imamura, *J. Antibiot.* **39,** 309 (1986).
176. J. C. Hanvey, J. K. Hardman, R. J. Suhadolnik, and D. C. Baker, *Biochemistry* **23,** 904 (1984).
177. J. C. Hanvey and co-workers, *Biochemistry* **26,** 5636 (1987).
178. D. J. Merkler, M. Brenowitz, and V. L. Schramm, *Biochemistry* **29,** 8358 (1990).
179. M. R. Grever and co-workers, *J. Clin. Oncol.* **3,** 1196 (1985).
180. H. M. Golomb and M. J. Ratain, *N. Engl. J. Med.* **316,** 870 (1987).
181. A. S. Spiers and co-workers, *N. Engl. J. Med.* **316,** 825 (1987).
182. J. B. Johnston and co-workers, *Cancer Res.* **46,** 2179 (1986).
183. B. C. Lembersky, M. J. Ratain, C. Westbrook, and H. M. Golomb, *Am. J. Hematol.* **27,** 60 (1988).
184. A. D. Ho and co-workers, *Blood* **72,** 1884 (1988).
185. S. Durrleman, J. L. Grem, and B. D. Cheson, *Eur. J. Haematol.* **43,** 297 (1989).
186. J. P. Dutcher, K. M. Salva, and P. H. Wiernik, *Am. J. Clin. Oncol.* **13,** 290 (1990).
187. R. M. Lowenthal, J. S. Wiley, K. F. Rooney, D. R. Challis, and G. M. Woods, *Brit. J. Haematol.* **76,** 555 (1990).
188. T. J. Ruers, W. A. Buurman, and C. J. van der Linden, *J. Immunol.* **138,** 116 (1987).
189. S. S. Matsumoto, J. Yu, and A. L. Yu, *J. Biol. Chem.* **263,** 7153 (1988).

190. A. Begleiter, L. Pugh, L. G. Israels, and J. B. Johnston, *J. Cancer Res.* **48,** 3981 (1988).
191. F. Lamballe, P. V. LePrise, E. LeGall, and J. C. David, *J. Leukemia* **3,** 97 (1989).
192. E. H. Kraut, J. C. Neff, B. A. Bouroncle, D. Gochnour, and M. R. Grever, *J. Clin. Oncol.* **8,** 848 (1990).
193. N. Lahat and co-workers, *Scand. J. Immunol.* **32,** 205 (1990).
194. A. D. Ho and co-workers, *Leukemia* **4,** 584 (1990).
195. R. Furuta, R. S. Naruto, A. Tamura, and K. Yokogawa, *Tetrahedron Lett.,* 1701 (1979).
196. T. Ito and co-workers, *J. Antibiot.* **37,** 931 (1984).
197. J. G. Buchanan, J. Stoddart, and R. H. Wightman, *J. Chem. Soc., Chem. Commun.,* 823 (1989).
198. G. Tamura, ed., *Tunicamycin,* Japan Scientific Society Press, Tokyo, 1982.
199. E. Eckardt, W. Ihn, D. Tresselt, and D. Krebs, *J. Antibiot.* **34,** 1631, 1632 (1981).
200. K. Isono and co-workers, *J. Antibiot.* **38,** 1617 (1985).
201. M. Ubukata, K. Isono, K. Kimura, C. C. Nelson, and J. A. McCloskey, *J. Am. Chem. Soc.* **110,** 4416 (1988).
202. Y. Hayakawa and co-workers, *Agric. Biol. Chem.* **49,** 2685 (1985).
203. J. D. Dutcher, M. H. von Saltza, and F. E. Pansy, *Antimicrob. Agents Chemother.* **1963,** 83 (1964).
204. H. L. Agahigrian and co-workers, *J. Org. Chem.* **30,** 1085 (1965).
205. J. F. Edgar and co-workers, *J. Chem. Soc., Chem. Commun.,* 222 (1982).
206. M. Yoshida and co-workers, *J. Antibiot.* **43,** 748 (1990).
207. I. Tabas and S. Kornfeld, *J. Biol. Chem.* **254,** 11,655 (1979).
208. C. J. Waechter and W. J. Lennarz, *Annu. Rev. Biochem.* **45,** 95 (1976).
209. R. T. Schwarz and R. Datema, *Trends Biochem. Sci.* **5,** 65 (1980).
210. K. VonFigura, M. Rey, R. Prinz, B. Voss, and K. Ullrich, *Eur. J. Biochem.* **101,** 103 (1979).
211. E. L. Kean, J. J. Plantner, and L. Poncz, *Fed. Proc.* **39,** 1899 (1980).
212. D. Duksin and P. Bornstein, *J. Biol. Chem.* **252,** 955 (1977).
213. Y. T. Pan, J. W. Schmitt, B. A. Sanford, and A. D. Elbein, *J. Bacteriol.* **139,** 507 (1979).
214. J. A. Hanover, W. J. Lennarz, and J. D. Young, *J. Biol. Chem.* **255,** 6713 (1980).
215. G. E. Bettinger, A. N. Chaterjee, and F. E. Young, *J. Biol. Chem.* **252,** 4118 (1977).
216. H. R. Perkins, S. J. Thorpe, and C. A. Brown, *Biochem. Soc. Trans.* **8,** 163 (1980).
217. W. T. Forsee, G. Valkovich, and A. D. Elbein, *Arch. Biochem. Biophys.* **174,** 469 (1976).
218. J. Fujisawa, Y. Iwakura, and Y. Kawade, *J. Biol. Chem.* **253,** 8677 (1978).
219. G. W. Hart and W. J. Lennarz, *J. Biol. Chem.* **253,** 5795 (1978).
220. S. R. Burger and co-workers, *J. Gen. Virol.* **72,** 359 (1991).
221. M. Ikeda, M. Wachi, H. K. Jung, F. Ishino, and M. Matsuhashi, *J. Bacteriol.* **173,** 1021 (1991).
222. P. J. Walker, K. A. Byrne, D. H. Cybinski, D. L. Doolan, and Y. H. Wang, *J. Gen. Virol.* **72,** 67 (1991).
223. D. L. Hogue, K. C. Hodgson, and C. E. Cass, *Biochem. Cell Biol.* **68,** 199 (1990).
224. V. Gross and co-workers, *J. Biol. Chem.* **371,** 231 (1990).
225. A. Giraud and J. L. Franc, *Eur. J. Cell Biol.* **48,** 128 (1989).
226. R. Konig, G. Ashwell, and J. A. Hanover, *J. Biol. Chem.* **263,** 9502 (1988).
227. U. Blank, C. S. Ra, and J. P. Kinet, *J. Biol. Chem.* **266,** 2639 (1991).
228. B. Sherry, D. M. Jue, A. Zentella, and A. Cerami, *Biochem. Biophys. Res. Commun.* **173,** 1072 (1990).
229. N. Singh and A. M. Tartakoff, *Molec. Cell. Biol.* **11,** 391 (1991).
230. A. J. Hoogewerf and A. Bensadoun, *J. Biol. Chem.* **266,** 1048 (1991).
231. J. J. York, S. Sonza, M. R. Brandon, and K. J. Fahey, *Arch. Virol.* **115,** 147 (1990).
232. M. Uramoto, C.-J. Kim, K. Shin-ya, H. Kusakabe, and K. Isono, *J. Antibiot.* **44,** 375 (1991).
233. K. Isono and R. J. Suhadolnik, *Arch. Biochem. Biophys.* **173,** 141 (1976).

234. M. Inukai and co-workers, *J. Antibiot.* **42,** 662 (1989).
235. F. Isono, M. Inukai, S. Takahashi, and T. Haneishi, *J. Antibiot.* **42,** 667 (1989).
236. F. Isono, T. Katayama, M. Inukai, and T. Haneishi, *J. Antibiot.* **42,** 674 (1989).
237. F. Isono and M. Inukai, *Antimicrob. Agents Chemother.* **35,** 234 (1991).
238. R. H. Chen, A. M. Buko, D. N. Whittern, and J. B. McAlpine, *J. Antibiot.* **42,** 512 (1989).
239. J. P. Karwowski and co-workers, *J. Antibiot.* **42,** 506 (1989).
240. V. Gullo and co-workers, *J. Antibiot.* **41,** 20 (1988).
241. R. Cooper, M. Conover, and M. Patel, *J. Antibiot.* **41,** 123 (1988).
242. H. Dellweg and co-workers, *J. Antibiot.* **41,** 1145 (1988).
243. A. Gonzalez, D. Vazquez, and J. Jimenez, *Biochim. Biophys. Acta* **561,** 403 (1979).
244. H.-P. Fiedler, ed., *Chemische Biologie der Mikroorganismen, Ergebnisse des Sonderforschungsbreichs 76 der Deutschen Forschungsmeinschaft 1972–1985.*
245. H. Heitsch and co-workers, *J. Antibiot.* **42,** 711 (1989).
246. H. Maehr and R. G. Pitcher, *J. Antibiot.* **24,** 830 (1971).
247. G. Benz and co-workers, *Angew Chem. Int. Ed. Engl.* **21,** 527 (1982).
248. G. Benz, *Liebigs Ann. Chem.,* 1399 (1984).
249. T. Ogita and co-workers, *Tetrahedron Lett.* **21,** 3203 (1980).
250. H. Yamaguchi and co-workers, *J. Antibiot.* **39,** 1047 (1986).
251. H. Seto and co-workers, *Tetrahedron Lett.* **29,** 2343 (1988).
252. T. Shimura and co-workers, *J. Antibiot.* **36,** 1300 (1983).
253. N. Nishizawa and co-workers, *J. Antibiot.* **37,** 1 (1984).
254. J. J. McGowan, C. P. Allen, J. M. Barnett, and G. A. Gentry, *Virology* **101,** 516 (1980).
255. H. Machida and co-workers, *Antimicrob. Agents Chemother.* **16,** 158 (1979).
256. J. J. Harrison, A. Anisowicz, I. K. Gade, M. Raffeld, and R. Sager, *Proc. Natl. Acad. Sci. USA* **80,** 6606 (1983).
257. N. Fujii and K. Oguma, *J. Gen. Virol.* **67,** 2521 (1986).
258. B. Richardson, L. Kahn, E. J. Lovett, and J. Hudson, *J. Immunol.* **137,** 35 (1986).
259. S. Rainier and A. P. Feinberg, *Proc. Natl. Acad. Sci. USA* **85,** 6384 (1988).
260. N. Jahroudi, R. Foster, J. Price-Haughey, G. Beitel, and L. Gedamu, *J. Biol. Chem.* **265,** 6506 (1990).
261. R. L. Momparler and J. Lalibert, *Leukemia Res.* **14,** 751 (1990).
262. J. Bouchard and co-workers, *Antimicrob. Agents Chemother.* **34,** 206 (1990).
263. R. A. Winegar, J. W. Phillips, L. H. Lutze, and W. F. Morgan, *Somatic Cell Molec. Genet.* **16,** 251 (1990).
264. C. Bonnerot, S. Amigorena, W. H. Fridman, J. Even, and M. Dameron, *J. Immunol.* **144,** 323 (1990).
265. M. C. John and R. M. Amasino, *Molec. Cell. Biol.* **9,** 4298 (1989).
266. J. Bouchard, *Leukemia Res.* **13,** 715 (1989).
267. V. I. Avramis, W. C. Powell, and R. A. Mecum, *Cancer Chemother. Pharmacol.* **24,** 155 (1989).
268. J. Hoshino, J. Frahm, and H. Kroger, *Biochem. Biophys. Res. Commun.* **142,** 468 (1987).
269. C. Dunbar, W. Travis, Y. W. Kan, and A. Nienhuis, *Brit. J. Haematol.* **72,** 467 (1989).
270. A. B. Glover and B. Leyland-Jones, *Cancer Treatment Reports* **71,** 959 (1987).
271. I. Kubo and M. Kim, *Tetrahedron Lett.* **27,** 4277 (1986).
272. S. B. Larson, R. J. Moss, G. D. Kini, and R. K. Robins, *Acta Cryst.* **C44,** 1076 (1988).
273. T. Kamikawa, S. Fujie, Y. Yamagiwa, M. Kim, and H. Kawaguchi, *J. Chem. Soc. Chem. Commun.,* 195 (1988).
274. K. Isshiki and co-workers, *J. Antibiot.* **43,** 584 (1990).
275. J. B. K. Nielsen and B. H. Arison, *J. Antibiot.* **42,** 1248 (1989).
276. A. D. Argoudelis and S. A. Mizsak, *J. Antibiot.* **29,** 818 (1976).
277. T. Kurokawa and co-workers, *Japan Kokai Tokyo Koho,* JP 90312596 (1990).
278. M. Nakajima and co-workers, *J. Antibiot.* **44,** 293 (1991).

279. H. Haruyama and co-workers, *J. Chem. Soc. Perkin Trans. 1,* 1637 (1991).
280. S. Mio, R. Ichinose, K. Goto, S. Sugai, and S. Sato, *Tetrahedron* **47,** 2111 (1991).
281. S. Mio, Y. Kumagawa, and S. Sugai, *Tetrahedron* **47,** 2133 (1991).
282. N. Katayama, S. Tsubotani, Y. Nozaki, S. Harada, and H. Ono, *J. Antibiot.* **43,** 238 (1990).
283. M. Uramoto and co-workers, *J. Antibiot.* **44,** 375 (1991).

ROBERT J. SUHADOLNIK
NANCY L. REICHENBACH
Temple University School of Medicine

OLIGOSACCHARIDES

Oligosaccharide antibiotics represented by the everninomicins (**1**) (Table 1) (1–13), flambamycins (**2**) (Table 2) (22,23), avilamycins (**2**) (Table 2) (14,15), and curamycins (**2**) (Table 2) (18,19) have complex and unique structural features. They are sometimes referred to as orthosomycins because characteristically these oligosaccharides possess two acid sensitive ortho-ester linkages, cleavage of which results in the complete loss of antibiotic activity (3,22). Another structural feature common to all the oligosaccharide antibiotics is the presence of a substituted phenol ester derived from dichloroisoeverninic acid attached to the sugar ring B at C-13. This acidic phenolic group, which could form salts with organic or inorganic bases, eg, the sodium or *N*-methylglucamine salt, is also essential for the antibiotic activity (3).

An additional structural feature found in everninomicin D is the presence of a nitro sugar residue, evernitrose (**3**) linked to ring B at C-12. Everninomicins B (1), C (2), and 13-384 Component 1 (13) also have this unique nitro sugar. Chemical modification of the nitro group has led to the preparation of a number of highly active derivatives (6,7).

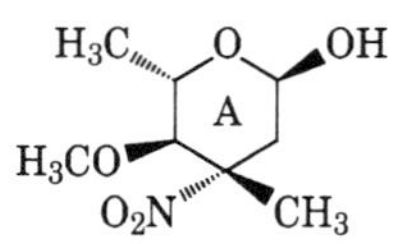

(**3**)

The first member of the oligosaccharide antibiotics to have its complete chemical structure elucidated was everninomicin D (5). The structure of everninomicin D then represented a template for the assignment of the structures of flambamycin (23), avilamycins (14), curamycins (19), and, more recently, everninomicin 13-384 Components 1 and 5 (13).

Properties

Physical Properties. Oligosaccharide antibiotics are colorless solids, which are often crystalline and have defined melting points and optical rotations, $[\alpha]_D$ (see Tables 1 and 2). They have a characteristic uv spectrum resulting from

Table 1. The Everninomicins[a]

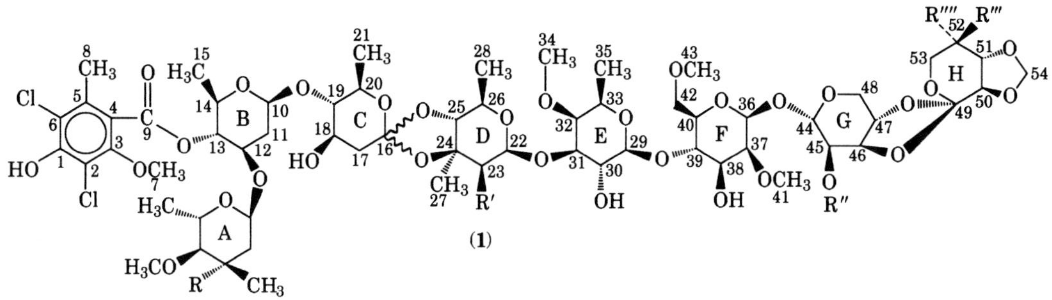

(**1**)

Compound	CAS Registry Number	Molecular formula	Mp, °C	$[\alpha]_D$	R	R′	R″	R‴	R⁗	Reference
everninomicin B	[*53296-30-3*]	$C_{66}H_{99}Cl_2NO_{36}$	184–185	−33.1	NO_2	OH	CH_3	(*S*)-$CH(CH_3)OCH_3$	OH	1
everninomicin C	[*53296-29-0*]	$C_{63}H_{93}Cl_2NO_{34}$	181–184	−33.7	NO_2	H	CH_3	OH	H	2
everninomicin D	[*39340-46-0*]	$C_{66}H_{99}Cl_2NO_{35}$	169–171	−34.2	NO_2	H	CH_3	(*S*)-$CH(CH_3)OCH_3$	OH	3–5,6–11
everninomicin 2	[*116217-29-9*]	$C_{58}H_{86}Cl_2O_{31}$	212–216	0.5	[b]	H	CH_3	(*S*)-$CH(CH_3)OCH_3$	OH	12

Table 1. The Everninomicins[a] (*Continued*)

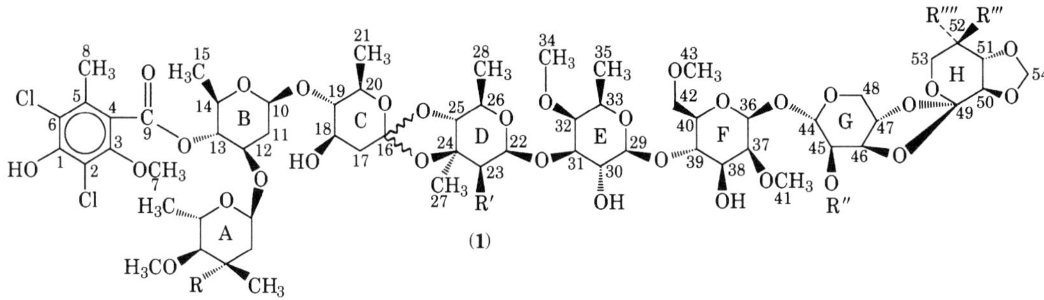

(1)

Compound	CAS Registry Number	Molecular formula	Mp, °C	$[\alpha]_D$	R	R′	R″	R‴	R⁗	Reference
everninomicin 3	[*64743-83-5*]	$C_{66}H_{98}Cl_2O_{33}$	157–158	−29.8	[c]	H	CH_3	(*S*)-$CHOCH_3$ with CH_3 on the CH	OH	11
everninomicin 7	[*64597-16-6*]	$C_{66}H_{100}Cl_2O_{34}$	173–175	−35.6	OH	H	CH_3	(*S*)-$CHOCH_3$ with CH_3 on the CH	OH	11
everninomicin-13-384-Component 1		$C_{70}H_{97}Cl_2NO_{38}$		−47.2	NO_2	OH	H	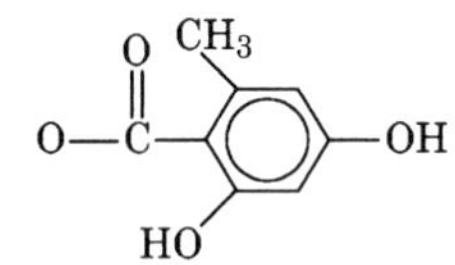	H	13
everninomicin-13-384-Component 5		$C_{70}H_{99}Cl_2NO_{36}$		−42.3	NH_2	OH	H	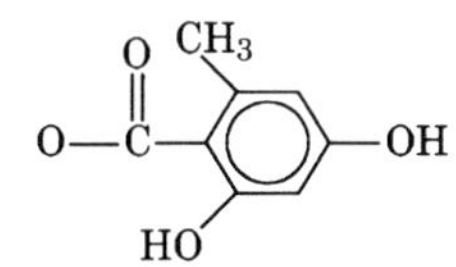	H	13

[a]The numbering is as in everninomicin (13).
[b]Ring A is absent in everninomicin 2, ie, Ring A = H.
[c]There is a double bond in everninomicin 3 at R.

Table 2. The Avilamycins (R = H), Curamycin (R = H), and Flambamycin (R = OH).

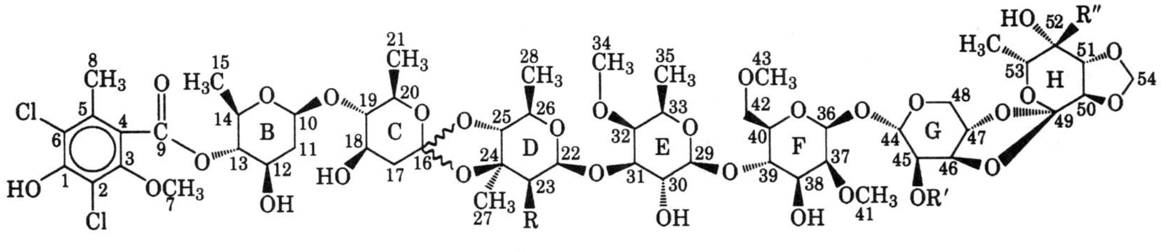

(2)

Compound	CAS Registry Number	Molecular formula	R′	R″	References
avilamycin A	*[69787-79-7]*	$C_{61}H_{88}Cl_2O_{32}$	$COCH(CH_3)_2$	$COCH_3$	14–17
avilamycin A_1			$COCH_2CH_3$	H	
avilamycin B[a]	*[73240-30-9]*	$C_{59}H_{84}Cl_2O_{32}$	$COCH_3$	$COCH_3$	
avilamycin C[b]	*[69787-80-0]*	$C_{61}H_{90}Cl_2O_{32}$	$COCH(CH_3)_2$	$CH(OH)CH_3$	
avilamycin D_1	*[82278-49-7]*	$C_{57}H_{82}Cl_2O_{31}$	H	$COCH_3$	
avilamycin D_2			$COCH_3$	$CH(OH)CH_3$	
avilamycin E			H	$CH(OH)CH_3$	
avilamycin F[c]	*[104748-55-2]*	$C_{60}H_{87}ClO_{32}$	$COCH(CH_3)_2$	$COCH_3$	
avilamycin G	*[104748-53-0]*	$C_{62}H_{90}Cl_2O_{32}$	COC_4H_9	$COCH_3$	
avilamycin H[d]	*[104748-54-1]*	$C_{61}H_{89}ClO_{32}$	$COCH(CH_3)_2$	$COCH_3$	
avilamycin I	*[104765-14-2]*	$C_{60}H_{86}Cl_2O_{32}$	$COCH_2CH_3$	$COCH_3$	
avilamycin J[e]	*[104748-48-3]*	$C_{60}H_{86}Cl_2O_{32}$	$COCH(CH_3)_2$	$COCH_3$	
avilamycin K[f]	*[104748-51-8]*	$C_{61}H_{88}Cl_2O_{33}$	$COCH(CH_3)_2$	$COCH_3$	
avilamycin L	*[104748-52-9]*	$C_{60}H_{86}Cl_2O_{32}$	$COCH(CH_3)_2$	CHO	
avilamycin M[g]	*[104748-50-7]*	$C_{60}H_{86}Cl_2O_{32}$	$COCH(CH_3)_2$	$COCH_3$	
avilamycin N[h]	*[104748-49-4]*	$C_{60}H_{86}Cl_2O_{32}$	$COCH(CH_3)_2$	$COCH_3$	
curamycin A[i,a]	*[73240-30-9]*	$C_{59}H_{84}Cl_2O_{32}$	$COCH_3$	$COCH_3$	18–21
flambamycin	*[42617-24-3]*	$C_{61}H_{88}Cl_2O_{33}\cdot H_2O$	$COCH(CH_3)_2$	$COCH_3$	22–26

[a]Avilamycin B and curamycin A appear to be identical.
[b]$[\alpha]_D = -4.8$.
[c]Avilamycin F has an H at C-2.
[d]Avilamycin H has an H at C-6.
[e]Avilamycin J has an H in place of the C_{43} CH_3 group.
[f]Avilamycin K has a CH_2OH in place of the C_{35} CH_3 group.
[g]Avilamycin M has an H in place of the C_{35} CH_3 group.
[h]Avilamycin N has an H in place of the C_{41} CH_3 group.
[i]$[\alpha]_D = 5.3$.

the phenolic ester residue, $\lambda_{max} \approx 285$ nm (ϵ ca 22), which shifts under basic conditions to $\lambda_{max} \approx 295$ nm (ϵ ca 80). The infrared spectrum shows absorptions for the hydroxyl and carbonyl moieties, and for other special features such as the nitro group which occurs at 1538 cm^{-1}. Molecular ions can be readily generated using modern mass spectral conditions such as fast atom bombardment (fab) (see MASS SPECTROMETRY) and the fingerprint fragmentation pattern has been employed for structure elucidation (8,13,14). X-ray analyses have been performed for the confirmation of partial structures of everninomicin (9,16) and avilamycin (27).

Modern nmr spectroscopic methods have also contributed heavily to the structure assignments (see MAGNETIC SPIN RESONANCE).

Chemical Properties and Structure. Oligosaccharide antibiotics are sensitive to acid pH because of the ortho-ester linkages. Yet all of them have an acidic phenolic group which makes the molecule relatively unstable to handling conditions. The ortho-ester connecting the C and D rings (Table 1) is comparatively more sensitive to acid pH than the one linking the G and H residues. Other chemically reactive groups in these compounds are the glycosidic linkages, the hydroxyl and carbonyl groups, and any nitro groups present.

Everninomicins

Everninomicin D is the principal component from cultures of *Micromonospora carbonacae* (10). Its structure (5) was elucidated using extensive chemical degradation coupled with spectroscopic analysis and it was the first reported instance of a natural product containing a tertiary nitrosugar. X-ray analyses of both the olgose residue (9) and the nitrosugar (16) have been reported as has a complete mass spectral analysis of everninomicin D (8).

Chemical modification (6) of the nitro group in everninomicin D has resulted in the formation of amino, mono- and dialkylamino, *N*-acylamino, and *N*-hydroxylamino (3) (and its nitrone derivatives) everninomicin D, all of which possess great antibiotic activity (6) against gram-positive bacteria. Aminoeverninomicin D can be converted to the known (11) everninomicins 3 and 7 under buffered conditions (6). Electrochemical reduction (11) of nitro and nitroso everninomicin D also produced everninomicins 3 and 7. Phosphite reduction of nitroso everninomicin causes fragmentation of sugar residue A leading to everninomicin 2 (12).

Everninomicin B (1), though a secondary component, has been extensively investigated because of its improved pharmacokinetic properties over those of everninomicin D (3,22). Everninomicin 13-384 Components 1 and 5, produced by *Micromonospora carbonacae* (24,28) are newer members of the everninomicin family. The structures (13) were assigned through extensive use of nmr and fab ms techniques. As in everninomicin B, C, and D, 13-384 Component 1 has a nitrosugar, whereas Component 5 has an amino sugar in its place, shown by conversion of Component 1 to 5 by hydrogenation (13). ^{13}C nmr confirmed the characteristic ortho-ester carbon atoms at 119.2 and 120.4 ppm and an extra aromatic residue in ring H at C-52. Nmr and high resolution fab ms confirmed this aromatic residue to be the 2,4-dihydroxy-6-methylbenzoyloxy group. The two aromatic residues on either end of this molecule appear much closer in conformational proximity, as observed by nmr (13) in $CDCl_3$ solution. Chemical modification of Components 1 and 5 has provided a number of active derivatives (7).

Flambamycin, Curamycin, and Avilamycins

Flambamycin. Flambamycin, produced by *Streptomyces hygroscopicus* DS 23230, also has undergone substantial chemical degradative experiments during structure elucidation (22,23,25). Nmr interpretation of the structure was reported later (20,26). The structure, shown in Table 2, chemical reactions, and biological properties have been thoroughly reviewed (22); it melts at 202–203°C.

Curamycin. Curamycin A, presented in Table 2, is the primary component of the culture *Streptomyces curacoi* (21) and preliminary chemical degradation studies (18) established it to be an oligosaccharide antibiotic. Later when the complete structure was assigned (17,19), the close relationship to flambamycin was evident. Curamycin A melts at 192–199°C.

Avilamycins. At least 16 avilamycins have been reported (14,15,29) and are listed in Table 2. Avilamycins A (mp 181–182) and C (mp 188–189) are the primary components (14) produced by the strain *Streptomyces viridochromogenes*. Structures have been established based on chemical degradation, nmr, and x-ray analysis. Structures for the rest of the compounds have been assigned based on nmr and fab ms (14) interpretations. Avilamycins F and H have only one chlorine atom in the phenolic ring. Relative hplc retention times for all the avilamycins have been recorded (14).

Other Oligosaccharides and Relevant Work

Sporacuracin A [*58205-91-7*], $C_{63}H_{94}Cl_2O_{35}$, mp 145–148°C, $[\alpha]_D$ −26.3, and sporacuracin B [*58205-92-8*], $C_{63}H_{94}Cl_2O_{35}$, mp 162–165°C, $[\alpha]_D$ −18.8, are other probable members of the oligosaccharide antibiotic family (22,30).

In all of the oligosaccharides discussed herein the stereochemistry of the ortho-ester in ring C of C-16 has not as of this writing been unequivocally established. Based on mechanistic studies using simple model reactions, a proposal has been put forward (31) to define the possible configuration at the ortho-ester linkage of oligosaccharide antibiotics. Experimental evidence suggests that the axially oriented oxygen atom (Fig. 1) with respect to ring C undergoes cleavage under acidic pH. The total synthesis of curacin [*20585-97-1*], $C_{15}H_{18}Cl_2O_7$, corresponding to the A B ring fragment, has been reported (32) and has been compared to its synthetic regioisomer, isocuracin.

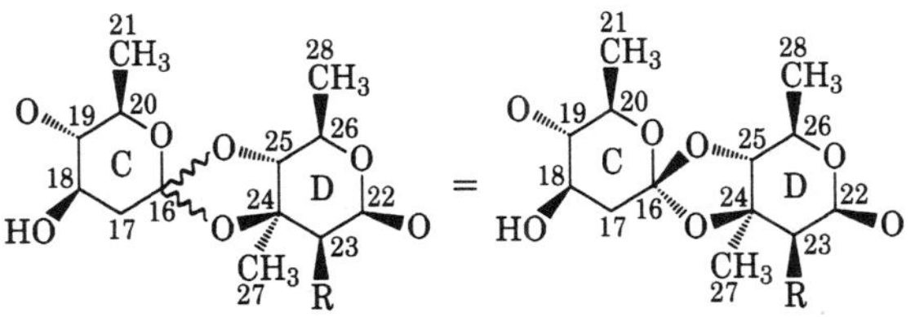

Fig. 1. Proposed ortho-ester configuration for ring C of the oligosaccharide antibiotics.

Biological Properties. The *in vitro* activity is such that oligosaccharides, in general, are highly potent, but are narrow-spectrum antibiotics (3,4,22). Everninomicins are active against a wide variety of gram-positive aerobes and anaerobes; *Neisseria, Mycoplasma,* and some *Mycobacteria*. These antibiotics are active against methicillin (see ANTIBIOTICS, PENICILLINS AND OTHERS): resistant *Staphylococci,* and compare to vancomycin (see ANTIBIOTICS, PEPTIDES). They lack cross-resistance with vancomycin because of different modes of action and their anaerobe activity is comparable to clindamycin (see ANTIBIOTICS, LINCOSAMINIDES). Everninomicin 13-384 Component 5 is said to have measurable *Pseudomonas* activity, but other members are devoid of any gram-negative activity. Comparatively, flambamycin is less potent than everninomicin D (22).

In Vivo Activity. Although highly active *in vitro,* everninomicin D posed pharmacokinetic problems when tested *in vivo* on mice and dogs. Everninomicin B, a close relative, on the other hand, gave better blood levels and *in vivo* efficacy. Similarly, Sch 23199, a semisynthetic *N*-acetyl analogue of everninomicin D also had useful activity and pharmacokinetic profile. In preliminary experiments in animals, everninomicin 13-384 Component 1 shows good *in vivo* activity. Finally, potential for use of oligosaccharide antibiotics in both human and animal health care has been claimed in various patents (7,24,28,30,33).

BIBLIOGRAPHY

"Antibiotics, Oligosaccharides" in *ECT* 3rd ed., Vol. 2, pp. 986–990, by A. K. Ganguly, Schering Corporation.

1. A. K. Ganguly and A. K. Saksena, *J. Antibiot.* **28,** 707 (1975).
2. A. K. Ganguly and S. Szmulewicz, *J. Antibiot.* **28,** 710 (1975).
3. A. K. Ganguly, in P. Sammes, ed., *Oligosaccharide Antibiotics, Topics in Antibiotic Chemistry,* Vol. 2, Ellis Horwood Publishers, Chichester, UK 1978, p. 49.
4. A. K. Ganguly, in M. Grayson, ed., *Oligosaccharides, Antibiotics, Chemotherapeutics, and Antibacterial Agents for Disease Control,* John Wiley & Sons, Inc., New York, 1982, p. 205.
5. A. K. Ganguly, O. Z. Sarre, D. Greeves, and J. Morton, *J. Am. Chem. Soc.* **97,** 1982 (1975).
6. A. K. Ganguly, V. M. Girijavallabhan, G. H. Miller, and O. Z. Sarre, *J. Antibiot.* **35,** 561 (1982).
7. U.S. Pat. 4,622,314 (Dec. 12, 1978), A. K. Ganguly, O. Z. Sarre, and V. M. Girijavallabhan (to Schering-Plough Corp.).
8. A. K. Ganguly and co-workers, *J. Antibiot.* **38,** 808 (1985).
9. A. K. Ganguly, O. Z. Sarre, A. T. McPhail, and K. D. Onan, *J. Chem. Soc. Chem. Commun.* **9,** 313 (1977).
10. M. J. Weinstein, G. M. Luedemann, E. M. Oden, and G. H. Wagman, *Antimicrob. Agents Chemother.* 24 (1964).
11. A. K. Ganguly and co-workers, *J. Chem. Soc. Chem. Commun.* **2,** 56 (1980).
12. A. K. Ganguly, S. Szmulewicz, O. Z. Sarre, and V. M. Girijavallabhan, *J. Chem. Soc. Chem. Commun.* **15,** 609 (1976).
13. A. K. Ganguly and co-workers, *Heterocycles* **28,** 83 (1989).
14. J. L. Mertz and co-workers, *J. Antibiot.* **39,** 877 (1986).
15. W. Keller-Schierlein, W. Heilman, W. D. Ollis, and C. Smith, *Helv. Chim. Acta* **62,** 7 (1979).
16. A. K. Ganguly, O. Z. Sarre, A. T. McPhail, and R. W. Miller, *J. Chem. Soc. Chem. Commun.* **1,** 22 (1979).
17. A. K. Ganguly, A. K. Bose, and N. F. Cappuccino, *J. Antibiot.* **32,** 1213 (1979).
18. E. G. Gros, V. Deulofeu, O. L. Galmarini, and B. Frydman, *Experientia* **24,** 323 (1968).
19. N. F. Capuccino, A. K. Bose, J. B. Morton, and A. K. Ganguly, *Heterocycles* **15,** 1621 (1981).

20. L. Valente and co-workers, *Tetrahedron Lett.* **13,** 1149 (1979).
21. O. L. Galmarini and V. Deulofeu, *Tetrahedron* **15,** 76 (1961).
22. D. E. Wright, *Tetrahedron,* Report No. 62, **35,** 1207 (1979).
23. W. D. Ollis, C. S. Smith, and D. E. Wright, *Tetrahedron* **35,** 105 (1979).
24. U.S. Pat. 4,735,903 (Apr. 5, 1988), J. A. Waitz and A. Horan (to Schering-Plough Corp.).
25. L. Ninet and co-workers, *Experientia* **30,** 1270 (1974).
26. W. D. Ollis and co-workers, *Tetrahedron* **35,** 993 (79).
27. E. Kupfer, K. Neupert-Laves, M. Dobler, and W. Keller-Schierlein, *Helv. Chim. Acta* **65,** 3 (1982).
28. U.S. Pat. 4,597,968 (July 1, 1986), J. A. Waitz and co-workers (to Schering-Plough Corp.).
29. F. Buzzetti and co-workers, *Experientia* **24,** 320 (1968).
30. Jpn. Pat. 75 125,094 (Oct. 1, 1975), A. Tamura and co-workers (to Dainippon); *J. Antibiot.,* Suppl. to Index 30, 77-111 and 77-112 (1977).
31. J-M Beau, G. Jaurand, J. Esnault, and P. Sinay, *Tetrahedron Lett.* **28,** 1105 (1987).
32. P. Jutten, J. Dornhagen, and H. Scharf, *Tetrahedron* **43,** 4133 (1987).
33. U.S. Pat. 4,436,734 (Mar. 13, 1984), E. E. Ose (to Eli Lilly Corp.).

V. M. Girijavallabhan
Ashit K. Ganguly
Schering-Plough Corporation

PEPTIDES

Peptide antibiotics are classified according to their overall shape, which can be linear or cyclic, and by the nature of the bonds joining the constituent and other amino acids (qv), which can be all amide bonds or amide plus ester bonds. Most peptide antibiotics are cyclic peptides that do not contain disulfide linkages. In homomeric cyclic peptides, all the atoms in the ring are contributed by amino acids; in heteromeric cyclic peptides such as enniatin [*11113-62-5*] and valinomycin [*2001-95-8*], $C_{54}H_{90}N_6O_{18}$, some atoms in the ring arise from nonamino acids. Homodetic homomeric cyclic peptides contain a ring of amino acids cyclized only through amide bonds. Amphomycin [*1402-82-0*], $C_{58}H_{91}N_{13}O_{20}$, and bacitracin [*1405-87-4*] are examples. The term heterodetic homomeric cyclic peptide applies if some amino acids in the ring are joined by ester bonds, eg, esterified serine or threonine hydroxyls as in virginiamycin S_1 [*23152-29-6*], $C_{43}H_{49}N_7O_{10}$. Heteromeric cyclic peptides, also called true depsipeptides, contain both amino acids and hydroxy acids in the ring. However, the term cyclic depsipeptide is often used in the literature to refer to heterodetic homomeric peptide lactones in which ester bonds to amino acid hydroxyls are present in the ring.

Peptide antibiotics differ in many respects from proteins (qv) and from peptides having hormonal or other functions in higher animals and plants (see Hormones). Among the significant differences are low molecular weight, where the vast majority of peptide antibiotics have molecular weights in the 500–1500 range. The average protein has mol wt of 40,000. Many peptide antibiotics have unusual fatty acids and amino acids, such as D-amino acids, *N*-methyl amino acids, or imino acids, and they usually lack methionine and histidine. Additionally, peptide antibiotics often have other nonamino acid moieties, such as the chromophore of dactinomycin [*50-76-0*]. Ring closure in cyclic peptide antibiotics is by amide or ester bonds, not by disulfide bonds as in normal proteins; or, if the peptide antibiotic has a tail, a diamino acid or a hydroxy amino acid provides the amide or ester bonds for ring closure. Peptide antibiotics are resistant to the usual proteases and peptidases. Also they are synthesized on multienzyme complexes much as are fatty acids, and not on ribosomes as are proteins (1). Nisin and related lantibiotics are exceptions. Most peptide antibiotics are synthesized, and are often marketed, as groups of closely related structures, usually reflecting a lack of specificity of the biosynthetic enzymes. Even the small fraction of peptide antibiotics that have therapeutic usefulness are quite toxic. For these useful peptide antibiotics the therapeutic index, the ratio of the minimum toxic dose to the maximum effective dose, is smaller than for most nonpeptide antibiotics.

Although historically most useful antibiotics have come from spore forming microorganisms, marine organisms have yielded the candidate antitumor peptide didemnin B [*77327-50-0*] (2) and cytostatic peptides such as the patellamides (3). Many of the marine peptides have little or no antimicrobial activity. Antibacterial peptides called magainins are found in frog skin (4) and antibacterial proteins called defensins are found in mammalian white blood cells (5). The commercially important insecticidal proteins from *Bacillus thuringensis* (6) are not discussed herein nor are the numerous peptide siderophores (7,8), which, except for the albomycins (9), are usually not antimicrobial.

Peptide antibiotics are not often the drugs of first choice for systemic therapy of important human disease. However, the World Health Organization, which chooses drugs especially for Third World use based on efficacy, safety, quality, price, and availability, includes as essential such peptide antibiotics as bleomycin, dactinomycin, and bacitracin (as an ointment containing neomycin), plus several β-lactams (see ANTIBIOTICS, β-LACTAMS) (10). Systemic use of peptide antibiotics is many times limited by nephrotoxicity and other toxicities. Semisynthesis or complete chemical synthesis of analogues of peptide antibiotics has most often not resulted in improved drugs. The complex structure of peptide antibiotics adds considerably to the problems of synthesis, but more recent efforts toward improved peptide antibiotics are encouraging (see FERMENTATION). Methods of bioassay and other laboratory use of economic antibiotics are available (11). Table 1 is a list of important peptide antibiotics. More extensive listings of minor peptide antibiotics have been published (32–36).

Peptide Antibacterials from Bacillus Species

Several economically useful peptide antibiotics are produced by members of the bacterial genus *Bacillus*.

Polymyxins. Polymyxin B [*1404-26-8*] and polymyxin E [*1066-17-7*] (colistin) are the only medically important antibacterials from a group of some thirty closely related cyclic decapeptide antibiotics in the polymyxin family, some of which are given in Table 2. Most family members contain a seven amino acid ring attached to a three amino acid tail, to which is attached a fatty acyl group. The octapeptins have a tail containing a single amino acid. Each antibiotic contains several L-α,γ-diaminobutyric acid [*1758-80-1*], $C_4H_{10}N_2O_2$, (Dab) residues, L-threonine [*72-19-5*], at least one D-amino acid, and one of a very limited list of fatty acids.

Polymyxin B sulfate [*1404-20-5*] (USP) and colistin sulfate [*1264-72-8*] (USP) are mixtures containing mostly polymyxins B_1 and B_2 and colistins A (polymyxin E_1) and B (polymyxin E_2), respectively. Polymyxin B sulfate, colistin sulfate (Coly-Mycin S), and colistin sodium methanesulfonate [*8068-28-8*], (colistimethate sodium, Coly-Mycin M), $C_{58}H_{105}N_{16}Na_5O_{28}S_5$, an injectable prodrug, at one time were important systemic agents against most gram-negative infections, especially those caused by *Pseudomonas aeruginosa*. However, in the 1990s most polymyxin is used, safely and effectively, as topical preparations. It is not absorbed through the skin. Polymyxin studies have been reviewed extensively; information regarding general basic science (37–41), chemistry (42), biosynthesis from multienzyme complexes and production (43,44), mechanism of action (43–45), and medical aspects (46,47) is available.

Polymyxin B sulfate, colistin sulfate, and colistimethate sodium are soluble in water and less soluble in organic solvents (48). Preliminary studies of polymyxin conformation in solution have been published (49). It took 15 years to elucidate the structures of the polymyxins following their discovery in 1947 (42). Before structures were elucidated, polymyxin A, later rediscovered as polymyxin M, polymyxins C and D, and the circulins were found to be more toxic than polymyxin B or E (the latter rediscovered as colistin). The structure determination of polymyxin B_1 was made possible by enzymic degradation, careful determination of D- and L-amino acids, and by final chemical synthesis of the molecule (42). There were limited but futile attempts to make medically useful improvements on polymyxins; exceptions were the sulfomethylation product of polymyxin B, no longer sold in the United States, and polymyxin E (colistimethate). Synthesis involved blocking the five amino groups of the DAB residues by treatment with formaldehyde and sodium bisulfite (50,51). The sodium salt of this *N*-blocked derivative, variously called the sodium methanesulfonate or the *N*-sulfomethyl derivative, has low antibiotic activity *in vitro* but is less nephrotoxic and less painful on injection. *In vivo,* hydrolytic removal of the methanesulfonate groups results in release of antibiotically active polymyxin B or E (colistin).

Most polymyxin B sold for human use in the United States is in dermatological, otic, and ophthalmic preparations that usually contain one or more other spectrum extending antibacterials such as bacitracin, neomycin sulfate [*1405-10-3*], $C_{23}H_{46}N_6O_{13}$, linear gramicidin, oxytetracycline [*79-57-2*], $C_{22}H_{24}N_2O$, or trimethoprim [*738-70-5*], $C_{14}H_{18}N_4O_3$. These preparations may also contain an antiinflammatory corticoid such as hydrocortisone [*50-23-7*], $C_{21}H_{30}O_5$, or hydrocortisone acetate [*50-03-3*], $C_{23}H_{32}O_6$, the analgesic diperidon [*537-12-2*], $C_{22}H_{28}ClN_3O_4$, or a detergent, eg, thonzonium bromide [*553-08-2*], $C_{32}H_{55}BrN_4O$.

Polymyxin B sulfate, whether as the bulk drug or as the final product, is sold by the unit and not by weight. One unit of polymyxin B sulfate is the weight of polymyxin B sulfate having the microbiological activity, usually against *Bordetella bronchoseptica,* of one unit of polymyxin B base. One milligram of standard polymyxin base has been arbitrarily set equal to 10,000 units. Polymyxin B sulfate USP must have the microbiological potency of at least 6000 units polymyxin B base per milligram. Most drug has a considerably higher purity.

Table 1. Peptide Antibiotics of Special Interest

Name	CAS Registry Number	Related to	Producing microbe	Antibiotic activity[a]	Mol wt (number of amino acids)	Number in family	Reference[b]
From bacteria of the genus Bacillus							
bacillomycin F	[*81689-97-6*]		*B. subtilis*	F	1080 (8)	2	12
bacilysin	[*29393-20-2*]		*B. subtilis, B. pumilis*	P, N	270 (2)		13
bacitracin (ayfivin)			*B. licheniformis*	P	1410 (11)	5	
circulins	[*9008-54-2*]	polymyxins	*B. circulans*	N	1150 (10)	2	
colistin (polymyxin E)	[*1066-17-7*]	polymyxins, circulins	*B. colistinus*	N	1150 (10)	10	
edeine A_1	[*27656-72-0*]		*B. brevis*	P, N, F	755 (5)	6	14
gramicidins, linear, Dubos			*B. brevis*	P	1900 (15)	6	
gramicidin S	[*113-73-5*]		*B. brevis*	P	1141 (10)	3	
iturins A	[*52229-90-0*]	bacillomycin B	*B. subtilis*	F	1050 (8)	9	15
micrococcin P_1	[*67401-56-3*]	thiocillins	*B. pumilis*	P	1144 (5)	4	16
mycobacillin	[*18524-67-9*]		*B. subtilis*	F	1528 (13)	1	17
polymyxins		circulins	*B. polymyxa*	N	1200 (10)	10	
subtilin	[*1393-38-0*]	nisin	*B. subtilis*	P	3321 (26)	1	
tyrocidines			*B. brevis*	P, N	1300 (10)	5	
tyrothricin			*B. brevis*	P, N	mixture		
From higher bacteria of the genus Streptomyces							
albomycin	[*1414-39-7*]		*S. subtropicus, S. griseus*	P	950 (5)	5	18
amphomycin (glumamycin)	[*1402-82-0*]		*S. canis, S. violaceus, S. albogriseolus*	P	1290 (10)	5	
bialaphos (SF 1293)	[*35597-43-4*]	phosphinothricin	*S. hydroscopicus*, other *S.* species	P, N,[c]	323 (3)	2	
bicozamycin (bicyclomycin)	[*38129-37-2*]		*S. sapporonensis, S. aizunensis*	N	302 (2)		
bleomycins		phleomycin, peplomycin, liblomycin	*S. verticillus*	P, N, M, F, T	1416 (12)		

bottromycin			*S. bottropensis*	P, N	810	(6)	5	19
capreomycins		viomycin, enviomycin, tuberoactinomycin	*S. capreolus*	P, N, M	670	(5)	4	
dactinomycin (actinomycin D)	[*50-76-0*]	cactinomycin	*S. parvullus*	P, T	1255	(10)	30	
daptomycin (deptomycin, LY 146032, semisynthetic)	[*103060-53-3*]		from A21978C of *S. roseosporus*	P	1619	(14)	6	
distamycin A (stallimycin)	[*636-47-5*]	netropsin	*S. distallicus*	F, T, V	482	(4)		20
enduracidins (enramycin)			*S. fungicidicus*	P, M	2355	(17)	4	
enviomycin (tuberactinomycin N)		capreomycin	*S. griseoverticilatus* var. *tuberacticus*	P, N, M	686	(6)	3	
ferrimycin A	[*15319-50-3*]	sideromycins	*S. griseoflavus*, other *S.* species	P, N	978	(6)	4	21
globomycin	[*67076-74-8*]		*S. halstedii*, *S. cinnamomeus*	N	657	(5)		22
ilamycin A (rufomycin A)	[*11006-41-2*]		*S. islandicus*, *S. atratus*	P, M	1026	(7)	5	23
neoviridogrisein II (etamycin B)	[*66002-40-2*]		*S. griseoviridis*	P	863	(7)	3	24
netropsin	[*1438-30-8*]	distamycin	*S. netropsis*, *S. chromogens*, others	P, N, M, T, V	430	(2)	8	20,25
quinomycin C	[*11001-74-4*]	triostin	*S. aureus*	P, N, T, V	1141	(8)	3	26
stendomycin	[*11006-78-3*]		*S. antimycoticus*	P, N, F	1850	(14)	3	27
streptogramin A (pristinamycin IIA, virginiamycin M1)	[*21411-53-0*]	virginiamycin M2	*S. virginiae*, others	P	527	(3)	2	
streptogramin B (pristinamycin IA)	[*3131-03-1*]	virginiamycin S_1	*S. virginiae*, others	P	867	(6)	12	
streptothricin A (polymycin A)	[*3484-67-1*]	noureothricins, racemomycins	*S. lavendulae*, others	P, N, M, V	1143	(6)	7	
telomycin	[*19246-24-3*]		*S. canus*, others	P	1272	(11)	1	28

Table 1. (*Continued*)

Name	CAS Registry Number	Related to	Producing microbe	Antibiotic activity[a]	Mol wt (number of amino acids)	Number in family	Reference[b]
thiopeptin			*S. tateyamensis*	P	1550 (6)	8	29
thiostrepton (bryamycin)	[*1393-48-2*]		*S. aureus*	P	1665 (9)	1	
valinomycin	[*2001-95-8*]		*S. fulvissimus*, *S. tsusimaensis*	P, N, F, T, V	1111 (6)		
viridogrisein (etamycin A)	[*299-20-7*]		*S. lavendulae*	P	879 (7)	2	30
zinostatin (neocarzinostatin)			*S. carcinostaticus*	P, T	11400 (113)	1	
From other bacteria							
nisin		subtilin	*Streptococcus lactis*	P	3354 (34)		
From fungi							
alamethicin	[*27061-78-5*]		*Trichoderma viridis*		1970 (19)	2	
cilofungin (semisynthetic)	[*79404-91-4*]	aculeacin B, sporiofungin	from echinocandin B of *Aspergillus* spp.	F	1029 (4)		
cyclosporine (cyclosporin A)		other cyclosporins	*Cylindrocarpon lucidium*, *Tolypocladium inflatum*	F, [c]	1203 (11)	25	
destruxin B			*Oospora destructor*, others	[d]	584 (5)		31
echinocandin B	[*54651-05-7*]		*Aspergillus rugulosus*, *A. nidulans* var. *roseus*	F	1060 (4)	8	
enniatin A	[*2503-13-1*]		*Fusarium orthoceras*, *F. sciroi*	P, M, F	682 (6)	3	
tentoxin	[*28540-82-1*]		*Alternaria tenuis*, *A. alternata*	[e]	415 (4)	2	

[a]Activity against gram-positive bacteria = P; gram-negative bacteria = N; mycobacteria = M; fungi = F; tumors = T; and viruses = V.
[b]Only peptides not discussed in the text of this article are given references.
[c]Active against the immune system.
[d]Active as an insecticide.
[e]Active as a herbicide.

Table 2. Polymyxins[a]

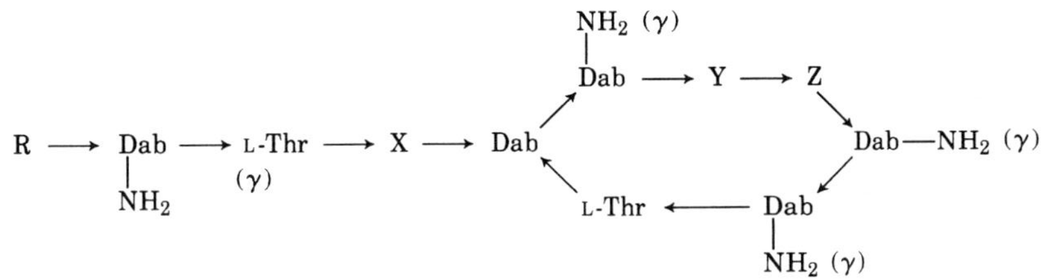

Name	CAS Registry Number	Molecular formula	R[b]	X	Y	Z
polymyxin A	[*1404-24-6*]	$C_{56}H_{98}N_{16}O_{13}$	MeOct	D-Dab	D-Leu	L-Thr
polymyxin B_1	[*4135-11-9*]	$C_{55}H_{96}N_{16}O_{13}$	MeOct	L-Dab	D-Phe	L-Leu
polymyxin B_2	[*34503-87-2*]	$C_{50}H_{93}N_{15}O_{15}$	MeHep	L-Dab	D-Phe	L-Leu
polymyxin D_1	[*10072-50-1*]	$C_{49}H_{91}N_{15}O_{15}$	MeOct	D-Ser	D-Leu	L-Thr
polymyxin D_2	[*34167-45-8*]	$C_{53}H_{100}N_{16}O_{13}$	MeHep	D-Ser	D-Leu	L-Thr
polymyxin E_1 (colistin A)	[*7722-44-3*]		MeOct	L-Dab	D-Leu	L-Leu
polymyxin E_2 (colistin B)	[*7239-48-7*]	$C_{52}H_{98}N_{16}O_{13}$	MeHep	L-Dab	D-Leu	L-Leu
circulin A	[*5854-98-8*]	$C_{53}H_{100}N_{16}O_{13}$	MeOct	L-Dab	D-Leu	L-*i*Leu
octapeptin B_1 [c,d]	[*50935-41-6*]		βOHMeDec		D-Leu	L-Phe

[a]Where the arrows represent bonds from the *C*-terminus to the *N*-terminus, and Dab = α,γ-diaminobutyric acid [*27252-32-0*], $C_4H_{10}N_2O_2$. Unless otherwise noted all Dab is L-Dab.

[b]MeHep = 6-methylheptanoic acid [*929-10-2*], $C_7H_{16}O_2$, MeOct = 6-methyloctanoic acid [*504-99-4*], $C_9H_{18}O_2$, MeDec = 8-methyldecanoic acid [*5601-60-5*], $C_{11}H_{22}O_2$, and βOHMeDec is the MeDec β-hydroxy derivative.

[c]For the octapeptins, omit the dipeptide, L-Thr-X, in the amino acid tail.

[d]The amino acid in the tail bound to R is D-Dab.

Some polymyxins are sold for second-line systemic therapy. Polymyxin B sulfate and colistimethate sodium can be used for intravenous, intramuscular, or intrathecal administration, especially for *Pseudomonas aeruginosa* infections, but also for most other gram-negative organisms, such as those resistant to first-line antibiotics. Nephrotoxicity and various neurotoxicities are common in parenteral, but not in topical, use. Resistance to polymyxins develops slowly, involves mutation and, at least in some bacteria, adaptation, a poorly understood type of resistance that is rapidly lost on transfer to a medium free of polymyxin. Resistance can involve changes in the proteins, the lipopolysaccharides, and lipids of the outer membrane of the cell (52). Polymyxin and colistin show complete cross-resistance.

The mode of action of the polymyxins is related to the effect these molecules have on membrane integrity. Polymyxins are detergentlike, highly positively charged molecules that damage membranes of gram-negative bacteria by binding to lipopolysaccharide (endotoxin) and phospholipids of the membranes. Bactericidal action by this mechanism, which allows leakage of low molecular weight metabolites, can occur on both growing and nongrowing cells (53). Polymyxin B nonapeptide [*86408-36-8*], prepared by enzymic removal of both the fatty acid and the terminal amino acid of the tail (54–56), is a less toxic polymyxin derivative retaining considerable membrane permeabilizing activity. It does not, however, appear to be medically useful (57). Other cleavage products of polymyxins have been prepared (58).

The ability of polymyxin and its cleavage products to bind to lipopolysaccharide (endotoxin) suggests usefulness in preventing endotoxic shock caused by endotoxemia in gram-negative bacteremias. Practical use in patients has been limited, however, because of the toxicity of these agents (57). Polymyxin immobilized on a solid support is useful for removing contaminating endotoxin from genetically engineered mammalian proteins made in *E. coli* (see GENETIC ENGINEERING) (59).

Bacitracin. Bacitracin, a cyclic peptide active against gram-positive bacteria, was discovered in 1943. Bacitracin received drug certification in 1949 (60–62). Whereas human usage of bacitracin is almost exclusively topical, the vast majority of bacitracin manufactured worldwide is used as an animal feed additive. Reviews of work on bacitracin include its chemistry (63–67), comprehensive aspects (62), medical aspects (62,68), biosynthesis on large enzyme complexes and genetics (69–71), and production (71,72).

Bacitracin is produced and marketed as a mixture of at least nine water-soluble peptides. The principal (60–80%) component is bacitracin A [*22601-59-8*], $C_{66}H_{103}N_{17}O_{16}S$. The most probable structure of bacitracin A is shown in Figure 1 (67,73). Bacitracin B [*1402-99-9*] is similar to bacitracin A except that one of the isoleucines of bacitracin A is replaced by a valine. Bacitracin F [*22601-63-4*], $C_{66}H_{98}N_{16}O_{17}S$, is a relatively inactive degradation product of bacitracin A. Many other bacitracin components have been identified (64–67,74), but structure determinations are incomplete. At least some of the bacitracin F and probably some of the other bacitracins are degradation products of bacitracins A and B produced during fermentation or in manufacture and storage. Pure bacitracins are prepared by hplc (74). Bacitracin A has not been crystallized, but hydrogen–tritium exchange and nmr studies show that there is at least one internal hydrogen bond

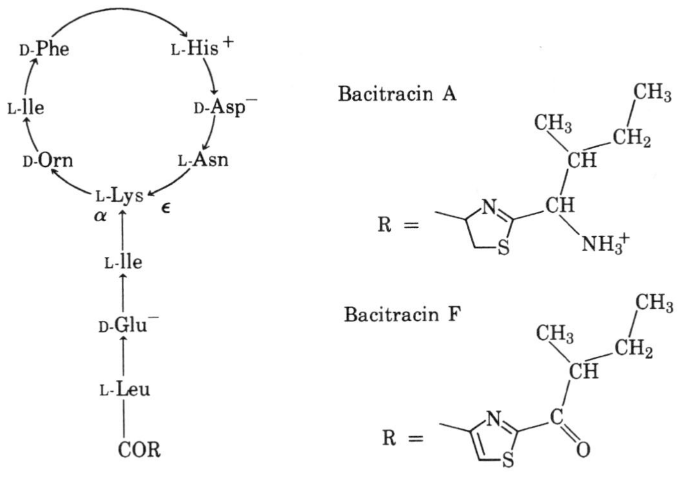

Fig. 1. Bacitracins A and F.

and that the peptide tail is folded over the ring. Only a partial chemical synthesis has been reported (75).

Bacitracin is a bitter, amorphous, white to pale buff, hygroscopic powder that is freely soluble in water and soluble in ethanol. When dry, bacitracin is stable from 25°C to 37°C. Several formulations of bacitracin have been developed. Anhydrous grease preparations of bacitracin are fairly stable, but water miscible creams are not. Zinc bacitracin, the form used in human medicine, is a white or pale yellowish gray or tan hygroscopic powder containing 2–10% zinc on a dry weight basis. It is sparingly soluble in both water and alcohol and more stable than bacitracin in troches, ointments, and tablets. Zinc may also enhance antibacterial activity. Bacitracin methylenedisalicylic acid (BMD) [*1405-88-5*], the form usually used in animal feed, is a white to grayish brown powder having a somewhat disagreeable odor. It is soluble in water, pyridine, and ethanol (67,76,77) and has increased stability over bacitracin. Both BMD and the zinc salt are used in animal feed (78,79,81,82).

One unit of bacitracin USP is defined as the activity given by 26 micrograms of the dried FDA master standard. The USP requirement for commercial bacitracin is a potency of at least 40 units per milligram in a bioassay with *Micrococcus flavus* or *Sarcina lutea*. Pure bacitracin A has a potency of about 100 units per milligram. Standards for animal feed grades of bacitracin are available (71,81,82) as are analytical methods (67).

Bacitracin given parenterally is sufficiently nephrotoxic that it is rarely used in human medicine for other than topical indications (80). Thus safe and effective use, especially as the zinc salt, is limited almost completely to ointments, sprays, and solutions for skin and ophthalmic use in concentrations of 250 to 1000 units per milliliter. Bacitracin is only rarely skin sensitizing. As in the case of polymyxin, bacitracin is usually combined with other antibiotics to enlarge its spectrum of activity, or with corticoids or analgesics to relieve pain or itching.

In the United States, 100 times more bacitracin by weight is used as a feed additive than for human medicine. Most of this is BMD but some is the zinc salt.

BMD can be used in subtherapeutic nutritional dosages for increase in feed efficiency and for growth promotion in poultry, swine, and feedlot cattle at concentrations of 2.6–33 g/t of feed (78,81,82). Prophylactic or therapeutic medicinal dosing at higher concentrations is used for necrotic enteritis in chickens, transmissable enteritis in turkeys, ulcerative enteritis in quail, dysentery in swine, and liver abcess in cattle (see FEEDS AND FEED ADDITIVES).

Bacitracin is bactericidal for many gram-positive cocci and rods, and a few gram-negative bacteria. Common gram-negative, enteric rod bacteria such as *E. coli* and *Pseudomonas,* and yeasts and other fungi are resistant. Resistant strains of bacteria do not normally arise during therapy, but an increasing number of *Staphylococcus aureus* strains are bacitracin-resistant. The mechanism of antibacterial action has been reviewed (69,70,83) and may be limited to a single step in bacterial cell wall biosynthesis. Bacitracin binds to the polyisoprenyl pyrophosphate used in bacterial cell wall biosynthesis and prevents its further cycling, thus blocking wall synthesis. There is a metal requirement for this binding and zinc is especially active. In mammalian metabolism, somewhat related polyisoprenyl pyrophosphates are found; the binding of bacitracin to these substances may be responsible for the parenteral toxicity of this antibiotic.

Tyrothricin and Gramicidin S. The linear pentadecapeptide gramicidins and the cyclic decapeptide tyrocidines, shown in Table 3, together comprise tyrothricin [*1404-88-2*], an antibiotic mixture active against many gram-positive and a few gram-negative bacteria. Tyrothricin was discovered in a strain of *Bacillus brevis* in 1939 (84). Gramicidin S [*113-73-7*], $C_{60}H_{92}N_{12}O_{10}$, which stands for gramicidin Soviet, is also called gramicidin J, for Japan. This material forms a group of cyclic decapeptides, also shown in Table 3, that is related to tyrocidines and obtained from another strain of *B. brevis* (85).

Although tyrothricin is too toxic for parenteral therapy, it was formerly sold in the United States as oral lozenges. Modern tyrothricin formulations are composed of 70–80% tyrocidines and 30–20% linear gramicidins. Tyrocidines are not as active as linear gramicidins and are too toxic for any therapeutic use by themselves. The bactericidal linear gramicidins are used in the United States solely as an ophthalmic solution in combination with polymyxin B sulfate and neomycin sulfate. The linear gramicidin is used in this aqueous product as a substitute for bacitracin, which lacks stability under such conditions.

The linear gramicidin of commerce is a mixture of at least eight closely related components, about 80% of which is valine–gramicidin A. Gramicidin USP is a white, odorless crystalline powder containing not less than 90% gramicidins on a dry weight basis. It is almost insoluble in water, but is readily soluble in methanol (86). Research on linear gramicidin's comprehensive aspects (86,87), conformations and mechanism of action (88–92), biosynthesis on multienzyme complexes (93), and chemistry (87–91) has been reviewed.

Linear gramicidins are neutral, uncharged, hydrophobic pentadecapeptides composed of alternating D- and L-amino acids. Glycine is classified here as a D-amino acid. This mixture is called gramicidin [*1405-97-6*] or gramicidin D (Dubos) [*1393-88-0*]. Gramicidin D is also used as a name for isoleucine–gramicidin A or as a name for an undefined polar fraction. Various naturally occurring and other gramicidins have been synthesized (94–96). Purified Val-gramicidin A has been shown to adopt different dimeric conformations depending on its physical

Table 3. Gramicidins and Tyrocidines

Name	CAS Registry Number	Molecular formula	R	R′
Linear gramicidins				
HCO—R—Gly—L-Ala—D-Leu—L-Ala—D-Val—L-Val—D-Val—L-Trp—D-Leu—R′—D-Leu—L-Trp—D-Leu—L-Trp—$NH_2CH_2CH_2OH$				
gramicidin A	*[11029-61-1]*			
Val-gramicidin A	*[4419-81-2]*	$C_{99}H_{140}N_{20}O_{17}$	L-Val	L-Trp
Ile-gramicidin A	*[5536-03-8]*	$C_{100}H_{142}N_{20}O_{17}$	L-Ile	L-Trp
gramicidin B	*[11041-38-6]*			
Val-gramicidin B	*[4422-52-0]*	$C_{97}H_{139}N_{19}O_{17}$	L-Val	L-Phe
Ile-gramicidin B			L-Ile	L-Phe
gramicidin C	*[9062-61-7]*			L-Tyr
Val-gramicidin C	*[58442-65-2]*	$C_{97}H_{139}N_{19}O_{18}$	L-Val	L-Tyr
Ile-gramicidin C			L-Ile	L-Tyr
Tyrocidines				
(cyclic) —L-Val—L-Orn—L-Leu—D-Phe—L-Pro—R—R′—L-Asn—L-Gln—L-Tyr—				
tyrocidine A	*[1481-70-5]*	$C_{66}H_{87}N_{13}O_{13}$	L-Phe	D-Phe
tyrocidine B	*[865-28-1]*	$C_{68}H_{88}N_{14}O_{13}$	L-Tyr	D-Phe
tyrocidine C	*[3252-29-7]*	$C_{70}H_{89}N_{15}O_{13}$	L-Tyr	D-Tyr
Gramicidin S				
(cyclic) —R—L-Orn—L-Leu—D-Phe—L-Pro—R′—L-Orn—L-Leu—D-Phe—L-Pro—				
gramicidin S_1	*[113-73-5]*	$C_{60}H_{92}N_{12}O_{14}$	L-Val	L-Val
gramicidin S_2	*[92278-12-1]*	$C_{59}H_{90}N_{12}O_{10}$	2-Aba[a]	L-Val
gramicidin S_3	*[92278-13-2]*	$C_{58}H_{88}N_{12}O_{10}$	2-Aba[a]	2-Aba[a]

[a]2-Aba is 2-aminobutanoic acid *[80-60-4]*, $C_4H_9NO_2$.

state. The pore forms seen in solvents appear to be parallel and antiparallel double helical forms. The pore forms viewed in the solid state by x-ray crystallography are antiparallel double helices (88–91,97). Dimeric gramicidin is also believed to produce channel forms that conduct monovalent cations in membrane models (88–91). The handedness of many of these helical forms is still under debate.

Tyrocidine *[8011-61-8]* is a mixture of three closely related components. Tyrocidine studies on mechanism of action (98), biosynthesis on multienzyme complexes (93,99,100), and chemistry (101) are available, and tyrothricin production is discussed (102). Although the mechanism of action of linear gramicidins has been well researched, such work on tyrocidine is more limited; it appears that tyrocidine damages membranes (103,104).

Gramicidin S is a fairly toxic mixture of at least three components, of which gramicidin S_1 is the primary component. Although it is not used in human medicine in the United States, there is some topical use of gramicidin S elsewhere. The biosynthesis of gramicidin S_1 on two multifunctional enzymes has been the subject of much work (99,100,105) and the mode of action of S_1 and analogues as membrane damaging agents also has elicited much attention (104). Gramicidin S studies are reviewed: mechanism of action (101), biosynthesis on multienzyme complexes (93,99,100), and chemistry (101).

Peptide Antibacterials and Antifungals from Other Microorganisms

Most peptide antibiotics have been isolated from *Actinomycetes* and especially from *Streptomyces*. Many β-lactams, however, and a few other useful peptide antibiotics have come from fungi. These streptomycete and fungal peptide antibiotics include many structural types having varied therapeutic indications and/or scientific applications.

Capreomycin, Viomycin, and Enviomycin. Capreomycin (Capastat, Lilly), a bacteriostatic, antimycobacterial peptide mixture isolated from *Streptomyces capreolus* was first reported in 1961 (106–108). This tuberactinomycin family member, shown in Table 4, was introduced into the U.S. market in 1971 where it has remained a useful but nephrotoxic and ototoxic second-line alternative to first-line tuberculosis therapies. Because capreomycin is somewhat less toxic than viomycin (tuberoactinomycin B [*32988-50-4*]), $C_{25}H_{43}N_{13}O_{10}$ (109,110), capreomycin has now displaced viomycin in the United States and most other markets. The structure of viomycin is shown in Figure 2. The related enviomycin (tuberactinomycin N [*33103-22-9*]), $C_{25}H_{43}N_{13}O_{10}$, is marketed in Japan (111–114). Available tuberactinomycin literature includes: capreomycin reviews (87,115, 116); capreomycin production (117); and viomycin synthesis on multienzyme complexes (118).

Capreomycin disulfate USP is a white solid, soluble in water and relatively

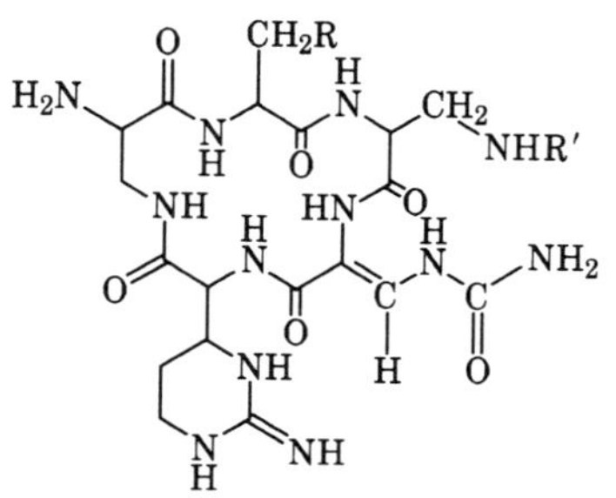

Table 4. Capreomycins

Name	CAS Registry Number	Molecular formula	R	R′
capreomycin IA	[*37280-35-6*]	$C_{25}H_{44}N_{14}O_8$	OH	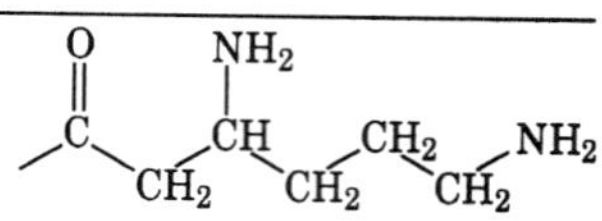
capreomycin IB	[*33490-33-4*]	$C_{25}H_{44}N_{14}O_7$	H	
capreomycin IIA	[*62639-89-8*]	$C_{19}H_{32}N_{12}O_7$	OH	H
capreomycin IIB	[*62639-90-1*]	$C_{19}H_{32}N_{12}O_6$	H	

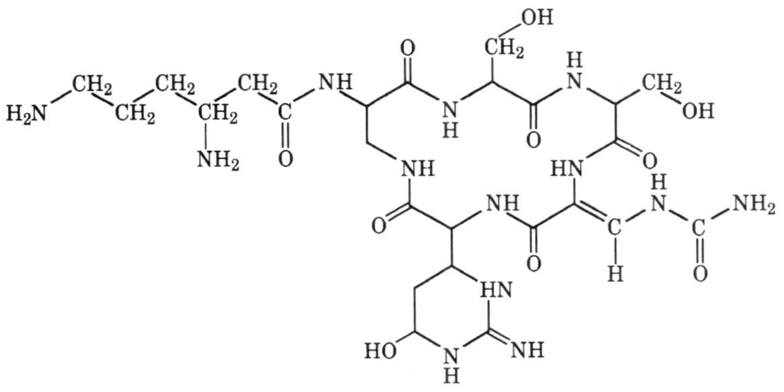

Fig. 2. Structure of viomycin (tuberoactinomycin B).

insoluble in most organic solvents. It must contain at least 90% capreomycins I. A representative lot contained 25% capreomycin IA, 67% IB, 3% IIA, and 6% IIB (119). The structures of capreomycin IA and IB have been determined and the compounds synthesized (120). The structures of capreomycin IIA and IIB are probably those shown in Table 4.

Capreomycin given intramuscularly is used to treat tuberculosis patients who do not respond to first-line drugs. Auditory, vestibular, and kidney function must be monitored during long-term therapy. Capreomycin shows cross-resistance to viomycin and some strains resistant to the more toxic second-line drug kanamycin [*8063-07-8*] (see ANTIBIOTICS, AMINOGLYCOSIDES) are also resistant to capreomycin. But capreomycin is not cross-resistant to streptomycin and other preferred therapies (121,122). As for many primary tuberculosis therapies, mutational resistance to capreomycin can develop during therapy. To slow the development of resistance, capreomycin is normally given with other drugs such as ethambutol [*74-55-5*], $C_{10}H_{24}N_2O_2$, or isoniazid [*54-85-3*], $C_6H_7N_3O$. Capreomycin is not effective with the majority of *Mycobacterium avium/intracellulare* strains that are often seen in autoimmune deficiency syndrome (AIDS) patients (123,124) (see ANTIVIRAL AGENTS; IMMUNOTHERAPEUTIC AGENTS). This latter resistance is apparently caused by poor uptake of drug into the bacterial cell. Capreomycin and other tuberactinomycins inhibit protein synthesis in sensitive bacteria at the translocation step and resistant strains have altered ribosomes (125).

Streptogramins. Virginiamycin (antibiotic 899, staphylomycin) was isolated in 1955 from a strain of *Streptomyces virginiae* and was shown to have several components inhibitory to gram-positive bacteria (126). Two years earlier a peptide antibiotic called streptogramin had been isolated (127). Streptogramin is a peptide antibiotic mixture similar to virginiamycin. Many virginiamycinlike antibiotics have since been discovered including the ostreogrycins, vernamycins, synergistins, PA 114, mikamycins, and pristinamycins. Streptogramin family antibiotics find some use in human medicines, eg, pristinamycin is used in France (128), but most streptogramin production is for use as animal feed additives. Virginiamycin is used in feeds in the United States and elsewhere and mikamycin is used in Japan (129). The literature on streptogramins, especially virginiamycins, has been reviewed (130–135).

As shown in Figure 3 and Table 5, these antibiotics are mixtures of two quite

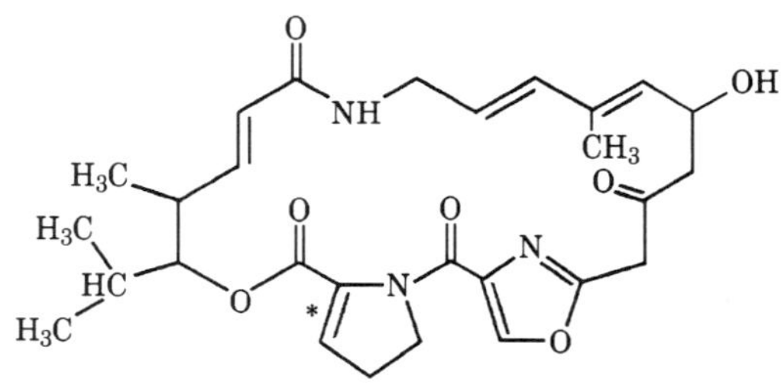

Fig. 3. Structure of streptogramin A, and when the bond denoted * is saturated, of ostreogrycin G. Synonyms for streptogramin A are mikamycin A, PA114A, pristinamycin IIA, vernamycin A, ostreogrycin A, virginiamycin M1, and staphylomycin M. Synonyms for ostreogrycin G are virginiamycin M2 and pristinamycin IIB (132).

different groups of chemical structures usually denoted A (or M) and B (or S) (136). The A components may be considered highly modified depsipeptides. Although they constitute a family of only two members, differing by one double bond, they have been given many different synonyms. The B component is a group of closely related cyclic hexadepsipeptides containing nonprotein aminoacids.

Members of the A group (Fig. 3) are less powerful antibiotics than members of the B group (Table 5), but they are powerfully synergistic with members of the B group. Singly, A- and B-group members act bacteriostatically; acting together, they are bactericidal. Group A members make up about 70% by weight, group B members about 30%, of the antibiotic yielded by a producer strain. This ratio also gives the maximum synergy and represents the approximate concentration ratio found in the marketed antibiotic. The biosynthesis of these A- and B-group substances is not well understood (137,138).

Virginiamycin is an amorphous white powder slightly soluble in water and very soluble in chloroform or methanol. Solubility and other physical data, including x-ray crystallographic, have been reviewed (180). Data for the separated components of virginiamycin and for pristinamycin are similar. The compounds that comprise virginiamycin were crystallized following separation by silica gel chromatography, and the structures were then determined using a variety of procedures (139). Virginiamycin is relatively stable at both acid and neutral pH, allowing oral dosage in animals and adequate shelf life in feed mixtures. Semisynthetic studies have produced more water-soluble derivatives of the A and B components of pristinamycin and these may allow development of an injectable formulation (140).

The production of virginiamycin by fermentation of *S. virginiae* has been detailed (133). Although crude, dry virginiamycin is relatively stable, there is a need to improve virginiamycin's stability when it is used as a feed additive, because of incompatibility with components in some feed mixtures. Virginiamycin is mixed with carboxymethylcellulose and chalk and the mixture is then granulated to a particle size of less than 1 mm. This preparation contains 50% virginiamycin and is known as Stafac 500 (SmithKline Beecham) (133).

Pristinamycin is used in France as an oral drug for human use, especially for staphylococcal infections (141); it has at least *in vitro* activity against a number of anaerobes and genital pathogens as well (142,143). U.S. rules for animal feed use permit virginiamycin in subtherapeutic nutritional dosages for increase in feed

Table 5. Streptogramin B Antibiotics

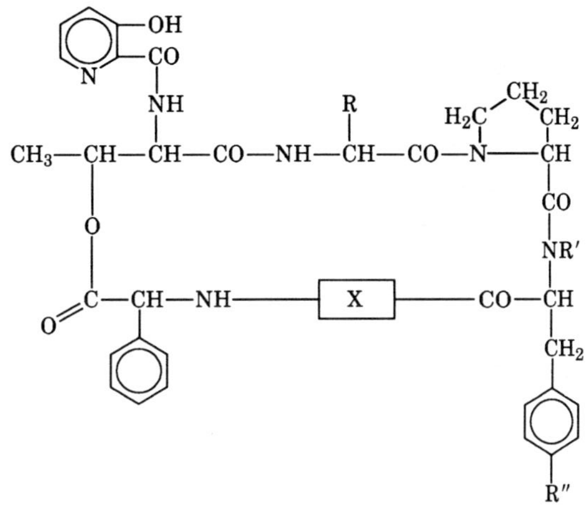

Name	CAS Registry Number	Molecular formula	R	R′	R″	X
streptogramin B[a]	[*3131-03-1*]	$C_{45}H_{54}N_8O_{10}$	CH_2CH_3	CH_3	$N(CH_3)_2$	4-oxopipecolic acid
vernamycin Bβ[b]	[*57206-54-9*]	$C_{44}H_{52}N_8O_{10}$	CH_2CH_3	CH_3	$NHCH_3$	4-oxopipecolic acid
vernamycin Bγ[c]	[*28979-74-0*]	$C_{44}H_{52}N_8O_{10}$	CH_3	CH_3	$N(CH_3)_2$	4-oxopipecolic acid
vernamycin Bδ	[*29139-17-1*]	$C_{43}H_{50}N_8O_{10}$	CH_3	CH_3	$NHCH_3$	4-oxopipecolic acid
vernamycin C[d]	[*14014-70-1*]	$C_{43}H_{52}N_8O_{11}$	CH_2CH_3	CH_3	$N(CH_3)_2$	aspartic acid
ostreogrycin B_3	[*31508-69-7*]	$C_{45}H_{54}N_8O_{11}$	CH_2CH_3	CH_3	$N(CH_3)_2$	3-hydroxy-4-oxopipecolic acid
patricin A	[*2105-09-1*]	$C_{42}H_{49}N_7O_9$	CH_2CH_3	CH_3	H	proline
patricin B	[*38188-91-9*]	$C_{43}H_{51}N_7O_9$	CH_2CH_3	CH_3	H	pipecolic acid
virginiamycin S_1[e]	[*23152-29-6*]	$C_{43}H_{49}N_7O_{10}$	CH_2CH_3	CH_3	H	4-oxopipecolic acid
virginiamycin S_2[e]	[*33477-38-2*]	$C_{42}H_{49}N_7O_{10}$	CH_2CH_3	H	H	4-hydroxypipecolic acid
virginiamycin S_3[e]	[*33477-39-3*]	$C_{43}H_{49}N_7O_{11}$	CH_2CH_3	CH_3	H	3-hydroxy-4-oxopipecolic acid
virginiamycin S_4[e]	[*33477-37-1*]	$C_{42}H_{47}N_7O_{10}$	CH_3	CH_3	H	4-oxopipecolic acid

[a]Also known as mikamycin B, PA114B, pristinamycin 1A, vernamycin Bα, ostreogrycin B, and synergistin B.
[b]Also known as pristinamycin 1B and ostreogrycin B2.
[c]Also known as pristinamycin 1C and ostreogrycin B1.
[d]Also known as doricin.
[e]The virginiamycins are also known as staphylomycins.

efficiency and for growth promotion in poultry (5.5–16.5 g/t of feed) and swine (5.5–11 g/t) (81,82,144). Higher concentrations are used for prophylactic or therapeutic medicinal dosing against necrotic enteritis in turkeys and dysentery in swine (81,82).

Virginiamycinlike antibiotics inhibit protein synthesis by binding to ribosomes. The molecular mechanism of the powerful synergistic action of these drugs is complicated and is the subject of continuing research (145). Resistance to these antibiotics can occur as the result of permeability decrease, ribosomal mutation, *O*-acetylation or hydrolysis of antibiotic, or by methylation of adenyl groups in ribosomal RNA (146–149). The mechanism of resistance varies. Macrolide–lincosamide–streptogramin B (MLS) resistance determines methylation of ribosomal RNA and resistance to the streptogramin B group member (see ANTIBIOTICS, LINCOSAMINIDES; ANTIBIOTICS, MACROLIDES). MLS resistance is often plasmid-borne and transferable, and sometimes induceable. Such MLS resistance in a strain may not cause loss of synergy between A- and B-group members or loss of bacteriostatic sensitivity to the mixture of the A and B members, but it does lead to a loss of cidal activity. One concern is that the subtherapeutic use of streptograminlike antibiotics in animal husbandry could generate and cause the spread of resistance to human pathogens, resulting in the development of resistance to erythromycin and clindamycin in human pathogens.

Thiostrepton and Related Antibiotics. Thiostrepton [*1393-48-2*] (thiactin, bryamycin), $C_{72}H_{85}N_{19}O_{18}S_5$, shown in Figure 4, is a large (mol wt 1660) antibiotic discovered in 1954 and produced by *Streptomyces azureus* (150,151). It is a member of a family of large, poorly water-soluble, bacteriostatic antibiotics that are active at very low concentrations against gram-positive bacteria, inactive against most

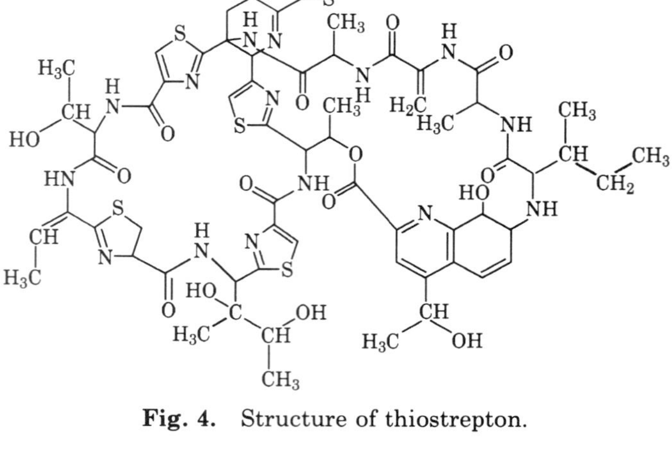

Fig. 4. Structure of thiostrepton.

gram-negative bacteria, and inactive against fungi. Thiostrepton inhibits sensitive bacteria by binding tightly to ribosomes and inhibiting protein synthesis (152,153). Other antibiotics in this family that contain multiple thiazole rings include nosiheptide [*56377-79-8*], $C_{51}H_{43}N_{13}O_{12}S_6$, (multhiomycin), actinotiocin [*39454-49-4*], micrococcin P, saramycetin [*11130-70-4*], siomycin [*11017-43-9*], $C_{64}H_{88}N_{16}O_{22}S_4$, (sporangiomycin) and the thiopeptins (154). The structure of thiostrepton was almost completely determined by a combination of chemical fragmentation and x-ray crystallography (155). The two dehydroalanines in the side chain were finally detected by ^{13}C nmr (156). The solution conformation of thiostrepton is similar to its crystalline conformation (157). Commercial thiostrepton is a white to light yellow crystalline powder, insoluble in water or diethyl ether, but very soluble in chloroform or dimethylformamide (158).

Thiostrepton family members are biosynthesized by extensive modification of simple peptides. Thus, from amino acid incorporation studies, the somewhat smaller (mol wt 1200) nosiheptide, which contains five thiazole rings, a trisubstituted indole, and a trisubstituted pyridine, is speculated to arise from a simple dodecapeptide. This work shows that the thiazole moieties arise from the condensation of serine with cysteine (159,160). Only a few reports on the biosynthesis of the thiostrepton family are available (159,160). Thiostrepton is presently used in the United States only as a polyantimicrobial veterinary ointment (Panalog, Squibb), but thiazole antibiotics have, in the past, been used as feed additives in various parts of the world. General (158) and mechanism of action (152) reviews on thiostrepton are available.

Enduracidin. Enduracidin (enramycin) is a mixture of some six cyclic lipopeptides isolated from *Streptomyces fungicidicus* in 1967 (161). As shown in Figure 5, enduracidins are composed of a ring of 16 D- and L-amino acids and an amino acid tail that is acylated with a branched-chain fatty acid. The enduracidin of commerce is a mixture of the hydrochlorides of enduracidin A [*34438-27-2*], $C_{107}H_{138}Cl_2N_{26}O_{31}$, enduracidin B [*34304-21-7*], $C_{108}H_{140}Cl_2N_{26}O_{31}$, enduracidin S_A [*40958-06-3*], $C_{107}H_{139}ClN_{26}O_{31}$, and enduracidin S_B [*40958-07-4*], $C_{108}H_{141}ClN_{26}O_{31}$ (161,162). This poorly soluble but bactericidal antibiotic is active *in vitro* on a variety of gram-positive bacteria such as *Staphylococcus aureus* and *Streptococcus pyogenes*. Mycobacteria are sensitive, but gram-negative bacteria and fungi are insensitive to enduracidin (161). Enduracidin does not show cross-resistance with other marketed antibacterials (161,163). It inhibits the transglycosylation step in bacterial cell wall synthesis (164). Enduracidin is not active when given orally, but does have a long half-life when given parenterally. It is painful at the injection site and is fairly toxic when given intravenously to animals (165). It is used only in veterinary medicine in Japan.

Bicozamycin. Bicozamycin (bicyclomycin) [*38129-37-2*], $C_{12}H_{18}N_2O_7$, shown in Figure 6, is a colorless, water-soluble diketopiperazine or cyclic dipeptide derived from leucine and isoleucine. It was simultaneously isolated in 1972 in Japan from cultures of two different *Streptomyces* strains (166). Bicozamycin is a relatively nontoxic but weak antibiotic having a minimum inhibitory concentration of 25–100 μg/mL for *Salmonella* and 5–50 μg/mL for *E. coli*. It inhibits bacterial cell wall synthesis and is bactericidial to gram-negative pathogens such as *E. coli, Salmonella,* and *Shigella. Proteus, Pseudomonas,* gram-positive bacteria, mycobacteria, and fungi are insensitive. This antibiotic is poorly absorbed when orally administered. Initial clinical trials for therapeutic

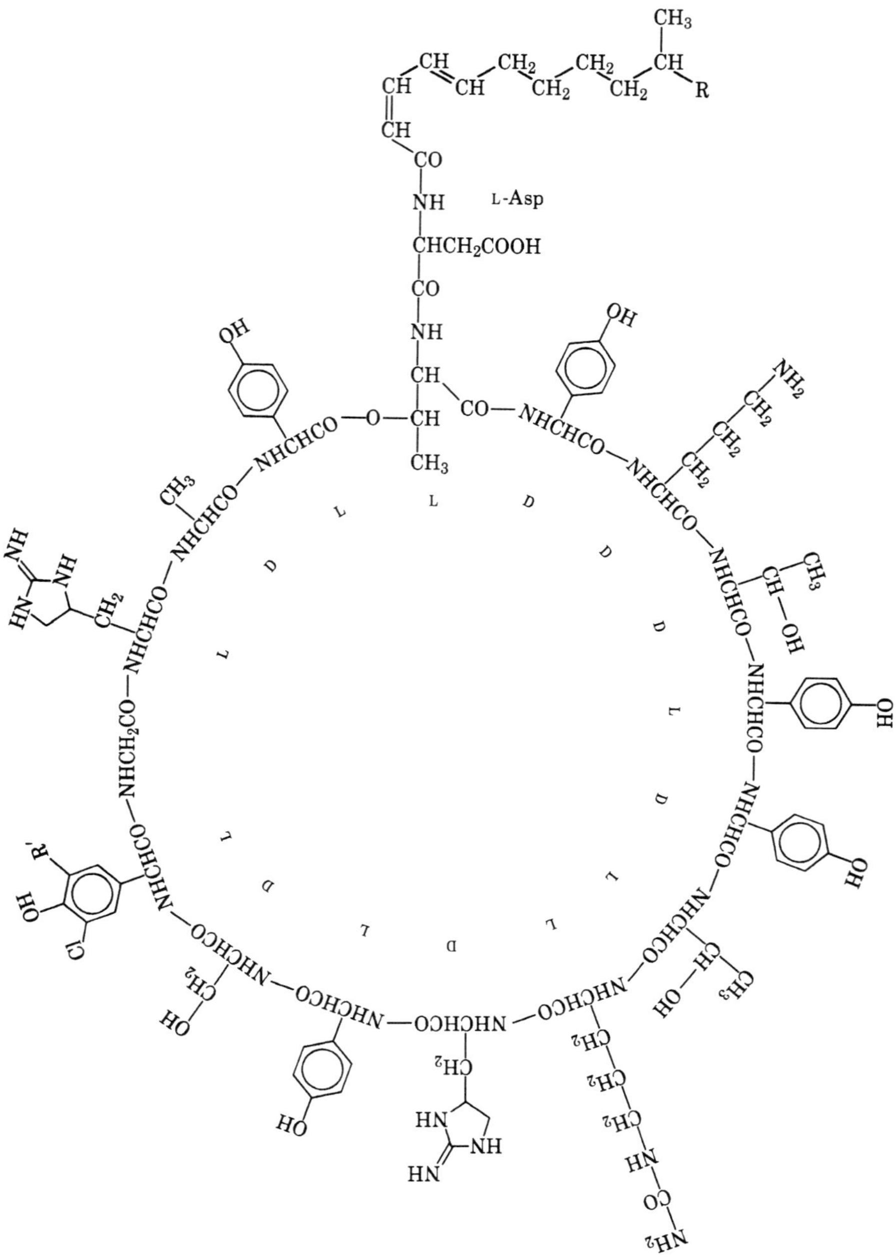

Fig. 5. Enduracidins where for enduracidin A, R = CH_3, R′ = Cl; enduracidin B, R = C_2H_5, R′ = Cl; enduracidin S_A, R = CH_3, R′ = H; and enduracidin S_B, R = C_2H_5, R′ = H.

and prophylactic use in travelers' diarrhea were promising (167), but human use was not pursued in the United States. This fermentation product is now sold in the Orient for bacterial diarrhea in calves and pigs. Chemical synthetic and analogue work, and other aspects of the bicozamycin literature, have been reviewed (168).

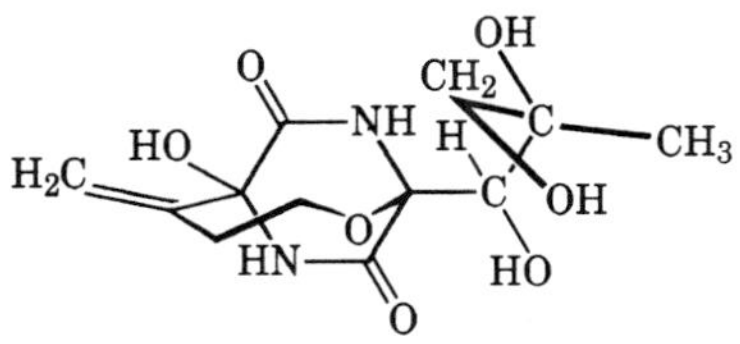

Fig. 6. Structure of bicozamycin (bicyclomycin).

Amphomycin (glumamycin). In 1953 a highly surface-active lipopeptide mixture of antibiotics, named amphomycin because of its amphoteric properties (169–172), was isolated from a strain of *Streptomyces canus.* Amphomycin [*1402-82-0*], $C_{58}H_{91}N_{13}O_{20}$, shown in Figure 7, is composed of two compounds, identical except for the different fatty acids. Aspartocin [*4117-65-1*], $C_{42}H_{64}N_{12}O_{12}S_2$, and tsushimycin [*11054-63-0*], $C_{59}H_{93}N_{13}O_{20}$, are similar to amphomycin (172). Amphomycin (glumamycin) has moderate activity toward gram-positive pathogens, but is inactive against gram-negative bacteria and fungi. Although amphomycin is toxic when given systemically, it was introduced in the United States in 1963 as the water-insoluble, crystalline calcium salt in ointments for topical use. However, it has been withdrawn, probably for economic reasons.

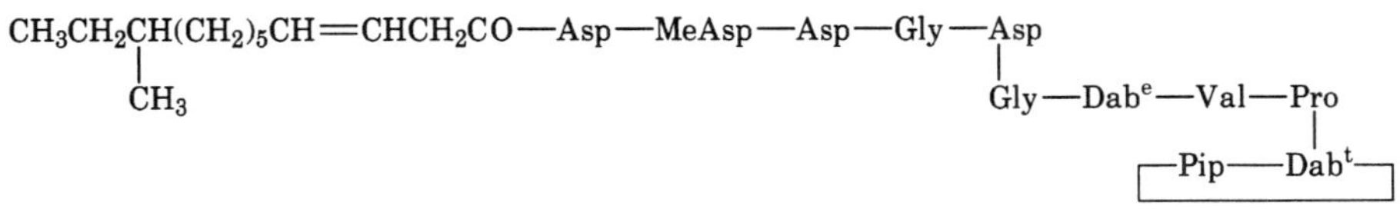

Fig. 7. Structure of one member of the amphomycin family where Dab^e = D-*erythro*-α,β-diaminobutyric acid; Dab^t = L-*threo*-α,β-diaminobutyric acid; Pip = D-pipecolic acid; MeAsp = *N*-methylaspartic acid. The amphomycin family member having (+)-3-anteisotridecenoic acid is shown; not shown is the second member containing (+)-3-isododecenoic acid.

Amphomycin inhibits bacterial cell wall synthesis at the translocase step by binding to the large isoprenoid lipid called undecaprenylphosphate (173,174). In eukaryotes, in a similar manner, amphomycin binds to the large isoprenoid lipid called dolicholphosphate thus blocking the transfer of mannose from its uridinediphosphate derivative to this lipid (173–177). Amphomycin has been patented for use as a feed additive (178).

Valinomycin and the Enniatins. Neutral ionophores such as the cyclic dodecadepsipeptide valinomycin [*2001-95-8*], $C_{54}H_{90}N_6O_{18}$, (Fig. 8) from *Streptomyces fulvissimus* (179), and the cyclic hexadepsipeptides enniatin [*11113-62-5*] and beauvericin [*26048-05-5*], $C_{45}H_{57}N_3O_9$, from fungi (180–182), are toxic to many forms of life. Valinomycin is active against gram-positive bacteria such as *Staphylococcus aureus,* whereas gram-negative bacteria are some 25 times less sensitive. Mammals are about as sensitive to valinomycin as gram-positive bacteria (179). In sharp contrast to the less toxic, charged polyether ionophores, neutral ionophores have little economic value, but they have been important in elucidating cell membrane function. Carrier ionophores, such as valinomycin and lasalocid, should be sharply distinguished from channel-forming antibiotics, such

(D-Hov—D-Val—L-Lac—L-Val)$_3$ (cyclic)

(a)

(D-Hov—L-*N*-methyl-amino acid)$_3$ (cyclic)

(b)

Fig. 8. Structure of (**a**) valinomycin and (**b**) the enniatins and beauvericin. Hov = α-hydroxy-isovaleric acid and Lac = lactic acid. The *N*-methylamino acid for enniatin A is isoleucine; enniatin B, valine; enniatin C, leucine; and beauvericin, phenylalanine.

as the linear gramicidins or alamethicin [*27061-78-5*], $C_{77}H_{128}N_{20}O_{22}$, that allow ion passage through membranes by forming channels through the whole width of the membrane (183). Neutral peptide ionophores have been reviewed (184–187).

Valinomycin is a cyclic dodecadepsipeptide (mol wt 1110) that forms a shell or bracelet of beta sheet that is held rigidly together by hydrogen bonds between carbonyl oxygens and amino hydrogens (188). Because these hydrogen bonds define a relatively inflexible cavity the size of a potassium ion, valinomycin is quite specific for potassium ion. Sodium binds with 3,000- to 10,000-fold lower avidity than potassium. Many valinomycin analogues have been chemically synthesized (189). The biosynthesis of valinomycin on a multienzyme complex and some genetic work have been described (190).

The enniatins enniatin A [*2503-13-1*], $C_{36}H_{63}N_3O_9$, enniatin B [*917-13-5*], $C_{33}H_{57}N_3O_9$, enniatin C [*19893-23-3*], $C_{36}H_{63}N_3O_9$, and beauvericin (Fig. 8) are 700–800 molecular weight cyclic hexadepsipeptides. They form valinomycinlike hydrophilic cavities surrounded by outer lipophilic regions, but they have more flexible structures than those seen with valinomycin and therefore have less specificity for potassium over sodium ion than valinomycin (186,187).

Nisin and Other Lantibiotics. Nisin [*1414-45-5*], $C_{143}H_{230}N_{42}O_{37}S_7$, shown in Figure 9, is the only economically useful member of a group of large (mol wt ca 3500) peptide antibiotics synthesized on ribosomes, as are normal proteins, rather than on large multienzyme complexes. These antibiotics are highly modified following translation; in particular some amino acid residues are converted

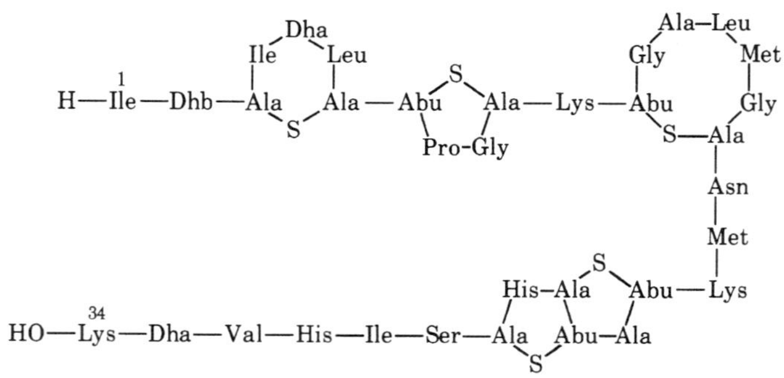

Fig. 9. Structure of nisin where Dha = dehydroalanine; Dhb = dehydrobutyrine; Abu = α-aminobutyric acid; Ala-S-Ala = *meso*-lanthionine; and Abu-S-Ala = *threo*-methyllanthionine. The first amino acid residue (Ala or Abu) of the lanthionine or methyllanthionine is a D residue; all other amino acids are L.

to lanthionine [*922-55-4*], $C_6H_{12}N_2O_4S$, and related amino acids, hence the group name, lantibiotics. The lantibiotics include nisin from *Streptococcus lactis* group N, subtilin from *Bacillus subtilis,* epidermin [*8022-54-4*], from *Staphylococcus epidermidis,* duramycin [*1391-36-2*], from *Streptoverticillium* sp., gallidermin [*117978-77-5*], ancovenin [*88201-41-6*], Ro 09-0198 [*110655-58-8*], $C_{89}H_{125}N_{25}O_{25}S_3$, pep 5 [*84931-86-2*], and cinnamycin [*1405-39-6*] (191,192).

Lantibiotics may also be classified as bacteriocins, a group of antibacterial proteins that are often bactericidal to bacteria closely related to the producing species (193). Nisin or nisin-producing bacteria have been used in Europe and elsewhere since 1960 as preservatives in canned and other foods, and as a protective agent during the manufacture of Swiss cheese. Nisin is surprisingly stable to the heat of canning processes, especially at acid pH. In 1988 the U.S. Food and Drug Administration (FDA) allowed nisin at 250 ppm to be a Generally Recognized As Safe (GRAS) additive for certain pasteurized cheese spreads to inhibit outgrowth of *Clostridium botulinum* spores, especially heat-treated ones, and toxin production (194). This may have been stimulated by the growing awareness of the undesirability of potentially carcinogenic nitrites that are added to foods for the same purpose (see FOOD ADDITIVES). Consumers normally eat small amounts of nisin because nisin-producing strains are ubiquitous in milk products. Nisin is about as toxic as sodium chloride with an oral LD_{50} in rats of 7 g/kg. Nisin literature has been reviewed (87,195).

Nisin was first isolated in 1947 (196,197). The structure was proposed in 1971 (198) and confirmed in 1988 by mass spectrometry studies (199). The first total chemical synthesis was reported in 1988 (200). Nisin is synthesized on ribosomes as a 57-amino acid prepeptide. It is then modified by enzymic elimination of water from the serines and the threonines of the prepeptide. Some of the resulting dehydroalanines and dehydrobutyrines combine with nearby cysteines in the molecule to form the thioether amino acids *meso*lanthionine and *threo*methyllanthionine. The *N*-terminal 23-amino acid leader peptide is probably cleaved off last, leaving nisin. The nisin produced by fermentation of *S. lactis* on skim milk is a semipurified mixture containing true nisin plus degradation products of nisin, some of which have antibacterial activity. The UK producer Aplin and Barrett sells a product, Nisaplin, that is 2.5% pure nisin (195,201). One gram of pure nisin contains 40,000 International or Reading units.

Nisin acts bactericidally primarily against gram-positive bacteria. It acts best at acid pH and is almost insoluble at physiological pH. Nisin and probably all lantibiotics appear to permeabilize the bacterial cell membrane to release small molecules, resulting in an immediate collapse of the membrane potential (202).

Semisynthetic Peptides in Development. Although the efforts to develop synthetic or semisynthetic improvements on naturally occurring peptide antibiotics have met with few successes, two lipopeptide antimicrobials are in development: the semisynthetic antibacterial daptomycin [*103060-53-3*], $C_{72}H_{101}N_{17}O_{26}$, and the semisynthetic antifungal cilofungin [*79404-91-4*], $C_{49}H_{71}N_7O_{17}$. As of 1991, cilofungin was less promising than originally expected.

Daptomycin (deptomycin, LY 146032), shown in Figure 10, is a cyclic peptide lactone containing a 10-amino acid ring and a 3-amino acid tail to which is attached a fatty acid. Daptomycin is derived from A21978C [*82800-76-8*], a complex of similar antibiotics produced by *Streptomyces roseosporus* (203,204). The A21978C complex is isolated and its ornithine amino groups are protected by

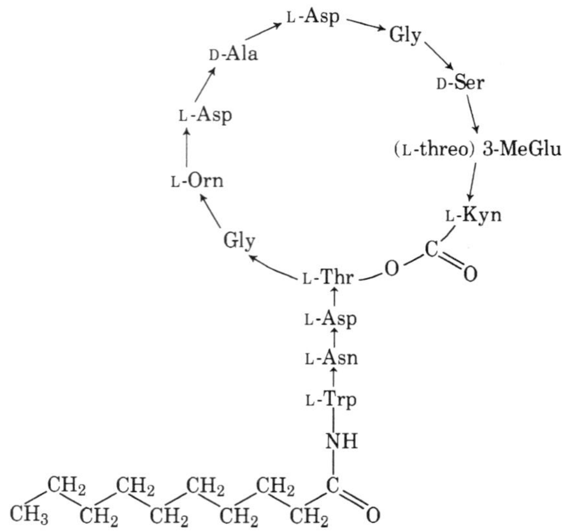

Fig. 10. Structure of daptomycin where L-Kyn is L-kynurenine and (L-*threo*) 3-MeGlu is L-*threo*-3-methylglutamic acid.

making the *tert*-butyloxycarbonyl (*t*-BOC) derivative. These blocked derivatives are then incubated with a culture of *Actinoplanes utahensis* to biologically remove the various fatty acids from the common polypeptide nucleus. The blocked polypeptide nucleus is then isolated and reacylated on the tryptophan amino group by an activated *n*-decanoyl group. The *t*-BOC group is then removed with trifluoroacetic acid, resulting in daptomycin (205,206).

Daptomycin is bactericidal to many gram-positive bacteria including *Streptococcus faecalis,* methicillin-resistant staphylococci, and certain vancomycin-resistant strains; most gram-negative bacteria are resistant (207). This antibiotic inhibits bacterial cell wall formation and the primary target, though not completely known, may be lipoteichoic acid formation (208). Daptomycin showed less toxicity in animals than the precursor A21978C (205,206) and was active in several animal models of infection (209). When it is administered once a day by injection, daptomycin may provide an alternative to vancomycin and other agents. Daptomycin was in clinical trials in the United States in 1991.

Cilofungin (LY121019), a semisynthetic cyclic lipohexapeptide shown in Figure 11, is being developed as an injectable anti-*Candida* drug. It is an analogue of the antifungal echinocandin B [*54651-05-7*], $C_{52}H_{81}N_7O_{18}$, (ECB, Lilly A30912A, Sandoz SF7810F) (see ANTIPARASITIC AGENTS, ANTIMYCOTICS). ECB, isolated from species of the fungus *Aspergillus* (210), is one of a closely related family of antifungals isolated from various *Aspergillus* species. This family includes echinocandin C [*71018-12-7*], $C_{52}H_{81}N_7O_{15}$, and echinocandin D [*71018-13-8*], $C_{52}H_{81}N_7O_{13}$, A30912H [*77368-57-1*], $C_{53}H_{83}N_7O_{16}$, aculeacin A [*58814-86-1*], mulundocandin [*108351-20-8*], $C_{48}H_{77}N_7O_{16}$, and sporiofungin [*82029-17-2*], $C_{50}H_{80}N_8O_6$ (210). ECB is produced by fermentation and is then deacylated biologically using *Actinoplanes utahensis*. The deacylated ECB is then isolated and reacylated chemically using an activated 4-*n*-octyl-oxo-benzoyl moiety (210). The resulting cilofungin is much less toxic than ECB.

Candida albicans is considerably more sensitive to cilofungin than to ECB

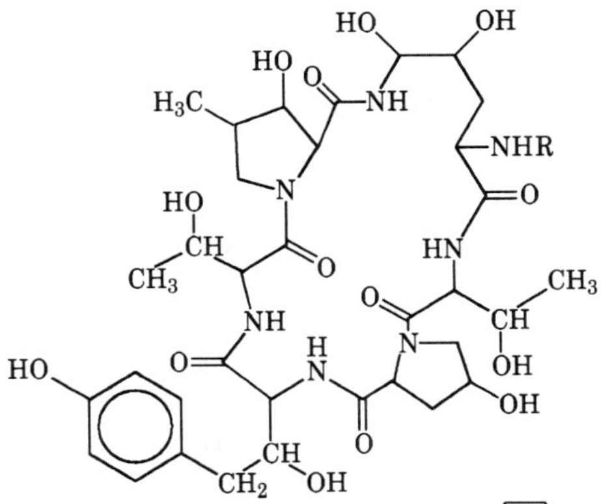

Fig. 11. Structure of cilofungin, R = —CO—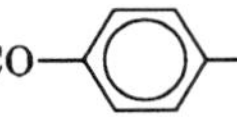—O(CH$_2$)$_7$CH$_3$, and echinocandin B, R = linoleoyl.

(211). Cilofungin has an *in vitro* spectrum of activity limited to *Candida albicans, C. tropicalis, Neurospora crassa,* and *Malassezia pachydermatis,* but most other fungi, including other *Candida* species are resistant (212). *Pneumocystis carinii,* an important cause of pneumonia in AIDS and other immunosuppressed patients, is sensitive to cilofungin (213). The mechanism of action of cilofungin in susceptible species appears to be by inhibition of the synthesis of 1,3,β-D-glucan, a wall component (214). Cilofungin is some twenty times less toxic than amphotericin B. In some animal infection models cilofungin is more effective than amphotericin B, but the fungicidal activity of cilofungin is not as reliable as desired (211,215). Cidal agents are especially valuable in therapy of immunosuppressed patients in whom systemic *Candida* infection is a significant problem. Cilofungin was in clinical trials in 1991. Unfortunately, initial trials showed excessive toxicity of the cilofungin preparation, much of it attributable to the vehicle required for this insoluble compound (216).

Antitumor Agents and Immunomodulators

Bleomycin and cyclosporine are the two economically most important streptomycete peptide antibiotics used as antitumor agents and immunomodulators, although dactinomycin is important medically for several tumors (see CHEMOTHERAPEUTICS, ANTICANCER; IMMUNOTHERAPEUTIC AGENTS).

Dactinomycin. Dactinomycin [*50-76-0*] (actinomycin D, actinomycin C1, Cosmegen), $C_{62}H_{86}N_{12}O_{16}$, is the only actinomycin in clinical use. Dactinomycin, an antineoplastic drug, was discovered in 1943 and is made in rather pure form by *Streptomyces parvullus.* Dactinomycin has some bacteriostatic antibacterial and antifungal activity, but high toxicity limits its use to antineoplastic therapy. It may be used alone or with other antineoplastics, with or without surgery and synergistic x-ray therapy. Dose limiting bone marrow toxicity may result in low white cell and platelet count. Intestinal mucosal damage also occurs. Reviews of more detailed chemotherapeutic information are available (217–222).

Dactinomycin forms bright red, rhomboid prisms from absolute ethanol.

Dactinomycin USP is a yellow lyophylized powder. It is soluble in water at 10°C, slightly soluble in water at 37°C, and soluble in ethanol. Dactinomycin USP must be protected from light and should be preserved in air-tight containers and stored below 40°C. Great care must be taken in handling this extremely toxic drug to avoid breathing the powder or getting dry powder or solutions on the skin. Dactinomycin is mutagenic, carcinogenic, and teratogenic. The LD_{50} in mice, when given intravenously, is about 0.7 mg/kg body weight. Other properties, fermentation, and purification methods (223–226), and studies on biosynthesis by multienzyme complexes can be found in the literature (227).

Almost all actinomycins have the same chromophore, a planar phenoxazinone dicarboxylic acid called actinocin. In dactinomycin, the structure of which is shown in Figure 12, the two pendent pentapeptide lactones are identical, but in other actinomycins these lactones may be different. In other actinomycins the first amino acid, amide linked with actinocin, is usually L-threonine, as in dactinomycin; the second position is sometimes D-allo-isoleucine instead of D-valine; the third position may be sarcosine or oxoproline; the fourth position is sarcosine; and the fifth position is sometimes *N*-methyl isoleucine instead of *N*-methylvaline. The lactone ring is always present.

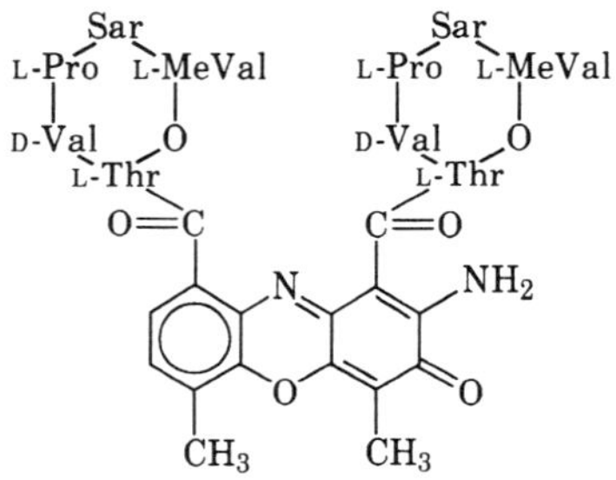

Fig. 12. Structure of dactinomycin (actinomycin D) where MeVal = *N*-methylvaline and Sar = sarcosine.

Total synthesis of dactinomycin has been accomplished, and at least thirty natural actinomycins and many synthetic and semisynthetic actinomycins have been tested (220,221,228). At one time cactinomycin [*8052-16-2*] (actinomycin C), a mixture of actinomycins D, C2, and C3, was sold as an antineoplastic.

Dactinomycin is useful against several tumors especially Wilms' tumor of the kidney and rhabdomyosarcoma. Wilms' tumor normally requires a combination of surgery, x-ray irradiation, and dactinomycin plus vincristine [*57-22-7*], $C_{46}H_{56}N_4O_{16}$, or other antitumor agents. Resistance of the multidrug type to dactinomycin can develop during therapy (229).

Dactinomycin intercalates into double stranded DNA inhibiting DNA-dependent RNA polymerase. Dactinomycin also can cause single strand breaks in DNA. Inhibition of DNA synthesis only takes place at higher concentrations of drug. A molecular model for dactinomycin-DNA interaction has been proposed based in part on x-ray crystallography of a complex containing two molecules of deoxyguanosine and one of dactinomycin (230). Conformations of dactinomycin–oligonucleotide complexes are being studied by nmr (231,232).

Bleomycin and Related Glycopeptides. Bleomycins (Fig. 13) comprise a group of related antineoplastic glycopeptide antibiotics isolated from a strain of *Streptomyces verticillus* in 1965 (233). Bleomycin sulfate [*9041-93-4*], (Blenoxane, Bristol-Myers Squibb) is mostly a mixture of bleomycins A_2 and B_2. Bleomycins differ from one another in having different R groups on the glycopeptide bleomycinic acid moiety as shown in Table 6. Hundreds of bleomycin analogues have been made either by adding different precursor amines during fermentation or by adding various amines to purified bleomycinic acid (234,235). Peplomycin (pepleomycin) (236–241) and liblomycin (238,239) are newer semisynthetic analogues having potentially useful properties. Peplomycin is marketed in Japan and was in clinical trials in the United States in 1990; liblomycin is at an early stage of development. Others include tallysomycin A [*65057-90-1*], reported in 1977, which was in clinical development at one time (240), and phleomycin [*11006-33-0*], reported in 1956 (241), which was too toxic for drug use.

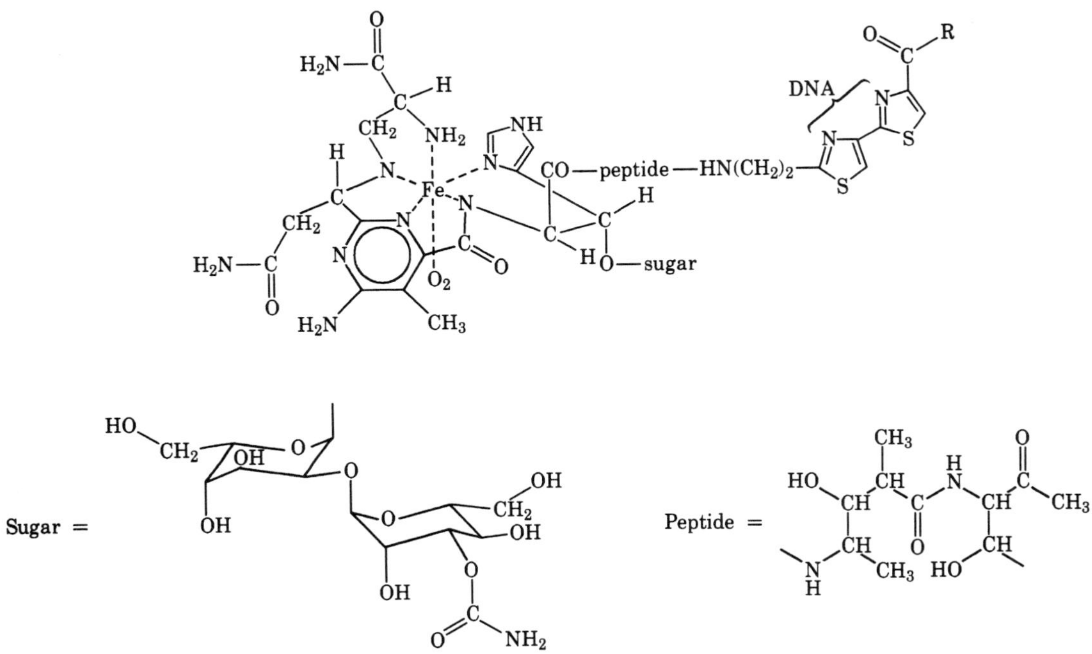

Fig. 13. Structure of the bleomycin–Fe(II)–O_2 complex showing the DNA binding region.

Bleomycin sulfate is used alone or in combination with other antineoplastic drugs. Using this DNA-cleaving drug, disease-free (possibly cured) survivors of some cancers have been seen, especially with the use of bleomycin in multidrug regimens. General reviews of bleomycin (234,242–245) and its mechanism of action (246–249) are available.

The structure of bleomycin shown in Figure 13 was reported in 1978 (250); total synthesis was reported in 1982 (251–253). The commercial fermentation and purification of bleomycin has been described (244) as has its nonribosomal synthesis using a large multienzyme complex (254).

Table 6. Bleomycins and Derivatives[a]

Name	CAS Registry Number	Molecular formula	R
bleomycinic acid	[*37364-66-2*]	$C_{50}H_{69}N_{15}O_{22}S_2$	$-OH$
bleomycin A_2	[*11116-31-7*]	$C_{55}H_{81}N_{16}O_{21}S_3$	$-NHCH_2CH_2CH_2-\overset{+}{S}(CH_3)_2$
bleomycin B_2	[*9060-10-0*]	$C_{55}H_{81}N_{19}O_{21}S_2$	$-NHCH_2CH_2CH_2CH_2NHC(=\overset{+}{N}H_2)NH_2$
peplomycin peplomycin sulfate	[*68247-85-8*] [*70384-29-1*]	$C_{61}H_{88}N_{18}O_{21}S_2$ $C_{61}H_{90}N_{18}O_{25}S_3$	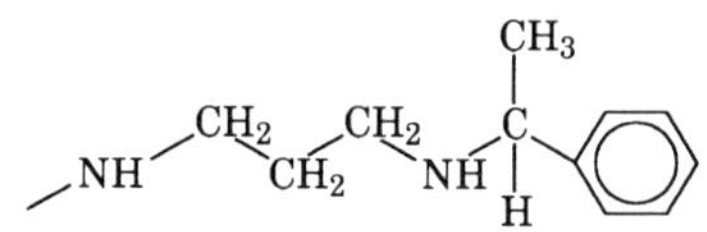
liblomycin	[*88266-67-5*]	$C_{99}H_{125}N_{19}O_{25}S_2$	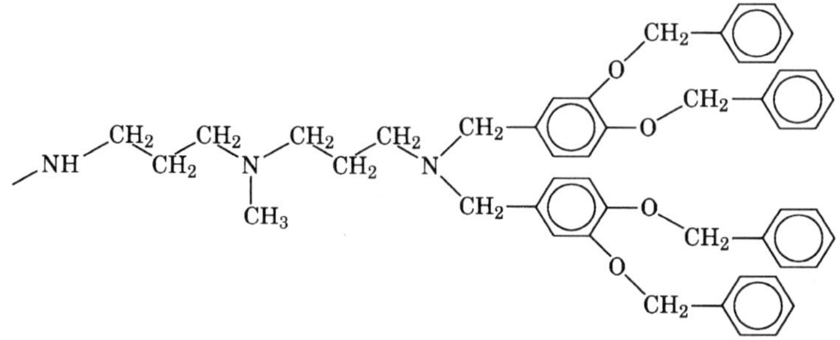

[a]See Figure 13.

Bleomycin sulfate USP, a white or yellowish powder, is very soluble in water and sparingly soluble in alcohol. It is 55–70% bleomycin A_2 and 25–32% bleomycin B_2 based on total bleomycin base content as determined by hplc. This combination was found to be more effective than use of either bleomycin alone (234).

Bleomycin has been used for the treatment of testicular cancer and malignant lymphomas and it has some activity against squamous cell carcinomas of various organs. Because of low bone marrow and immune system toxicities, bleomycin is often combined with bone marrow suppressive drugs such as vinblastine [*865-21-4*], $C_{46}H_{58}N_4O_9$, vincristine [*23214-92-8*], $C_{27}H_{29}NO_{11}$, cisplatin [*15663-27-1*], $N_2H_6Cl_2Pt$, doxorubicin [*23214-92-8*], $C_{27}H_{29}NO_{11}$, dacarbazine [*4342-03-4*], $C_6H_{10}N_{60}$, and mitomycin C [*50-07-7*], $C_{15}H_{18}N_4O_5$. Tumors often acquire resistance to bleomycin during therapy because of increased levels of bleomycin hydrolase or of DNA repair activities, or possibly by decreased transport (255). The hydrophilic bleomycin is not affected by the multidrug resistance system involving transport. The toxicity of bleomycin, which is high in lung and skin and low in bone marrow, correlates well with the level of a bleomycin hydrolase, a cysteine proteinase with cathepsin-*H*-like activity (256). This enzyme inactivates bleomycin by cleaving the alanine carboxamide moiety on the pyrimidoblamic unit.

Newer bleomycins such as peplomycin and especially liblomycin, are more resistant to bleomycin hydrolase. This results in less lung toxicity but more bone marrow toxicity, and allows for a different spectrum of antitumor action. Bleomycin is inactive orally; it is given intravenously, intramuscularly, subcutaneously, or directly into a cavity such as the pleural cavity. The majority of drug is excreted unchanged in the urine.

Much work has been done on the mechanism of action of bleomycin as an agent that causes single and double strand breaks in DNA. Bleomycin binds to cupric ions strongly, but the less strongly bound ferrous ions are probably the important ions *in vivo*. Ferrous bleomycin binds with considerable base specificity to DNA (257) probably in part by the intercalation of the bithiazole moiety between the DNA base pairs. Reductive activation of dioxygen by this antibiotic damages the strand of DNA in at least two separate ways (246–249). This mechanism of action of bleomycin is still speculative and has stimulated the design of other structures with similar action. An antitumor antibiotic called zinostatin [*9014-02-2*] (neocarzinostatin) acts somewhat similarly to bleomycin. It is composed of a protein (mol wt 11,000) having a tightly bound nonpeptide prosthetic group [*81604-85-5*] (258).

Cyclosporine. Cyclosporin A [*59865-13-3*], $C_{62}H_{111}N_{11}O_{12}$, called cyclosporine in the United States and called Sandimmun(e) by Sandoz, is a lipophilic cyclic undecapeptide, isolated from fungi, that has important immunosuppressant action. This action is directed mainly against T-lymphocytes resulting in inhibition of cell-mediated immunity and, to a lesser extent, antibody production. Cyclosporine has little effect on blood-forming tissues and phagocytic cells, thus preserving much antibacterial defense; it is not mutagenic. In 1970 culture extracts of the fungi *Cylindrocarpon lucidum* and *Tolypocladium inflatum* showed a narrow spectrum of antifungal activity *in vitro* but poor antifungal activity in whole animal models (259). Purified extracts, composed mainly of cyclosporin A, shown in Figure 14, and cyclosporin B, yielded high immunosuppressive and antiinflammatory activities, surprisingly not linked to general cytostatic effects (260). This initial work was followed by the separation and purification of some 25 naturally occurring cyclosporins, denoted A through Z, as well as the synthesis of analogues.

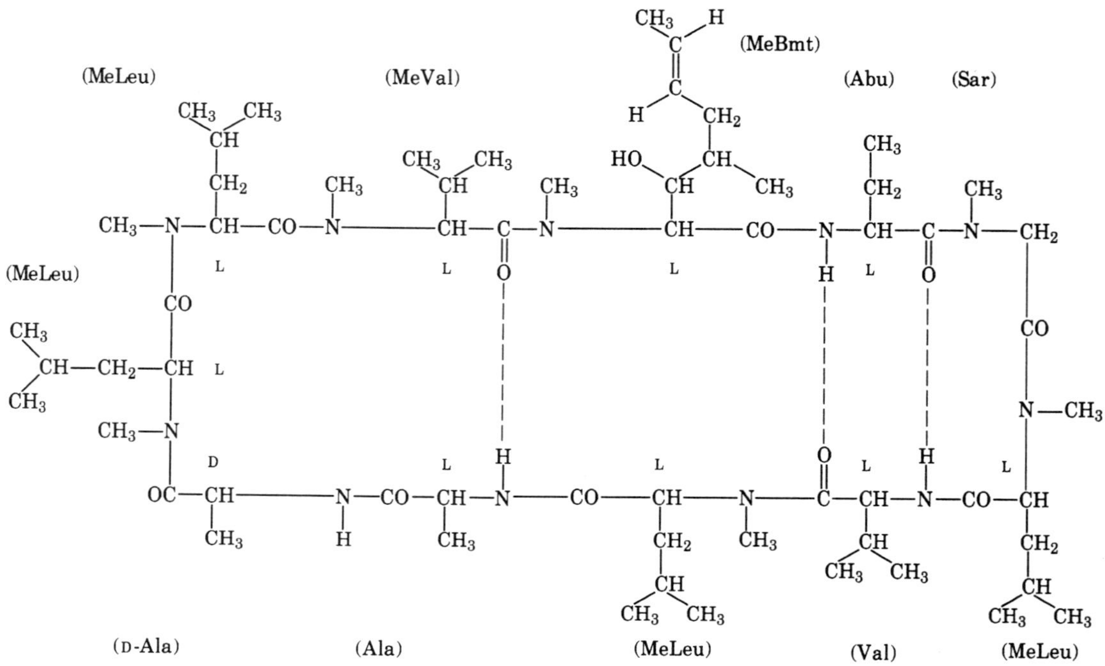

Fig. 14. Structure of cyclosporin A where MeBmt = 4-(2-butenyl)-4,*N*-dimethylthreonine; Sar = sarcosine; MeLeu = *N*-methylleucine; MeVal = *N*-methylvaline; and Abu = α-aminobutyric acid. The dotted lines indicate hydrogen bonds.

The first human kidney and bone marrow transplants using cyclosporine were reported in 1978. Oral or intravenous cyclosporine is an immunosuppressant for transplantation of these and other organs and investigations are underway for its possible use in a variety of autoimmune diséases including rheumatoid arthritis, severe psoriasis, and Crohn's disease. Dose-dependent nephrotoxicity (261–264) remains the primary limitation of the drug and necessitates close monitoring of patients, including measurement of drug levels in blood. Cyclosporine research has been reviewed (265–274).

Cyclosporin A forms white prismatic crystals from acetone and is only slightly soluble in water and saturated hydrocarbons, but is very soluble in methanol, ethanol, acetone, and diethyl ether. Optical and nmr data on cyclosporins and x-ray crystallographic data on cyclosporin A and an iododerivative have been reviewed (273,275).

Cyclosporin A contains no charged groups. Acid hydrolysis yields a mixture of amino acids. Sequence analysis and x-ray crystallography yielded the cyclic structure containing the several hydrogen bonds shown in Figure 14 (259,275).

The structures of some 25 naturally occurring cyclosporins have been determined (269). The probable conformation of cyclosporin A in aqueous media has been reviewed (273). The cyclosporins have been synthesized chemically and many analogues have been made, but no cyclosporin has been shown to be more medically useful than cyclosporin A (cyclosporine) (273,276). The production and purification of cyclosporine have been described (277). The nonribosomal biosynthesis of cyclosporine A in cell-free systems involves a single, very large multienzyme polypeptide that appears to catalyze at least 40 enzymic reactions (278).

The mechanism of action of cyclosporine is not fully known, but the main effects appear to act against T-helper cells and against the production and release of lymphokines including interleukin-2 and γ-interferon. Cyclosporine inhibits at least one form of the enzymic activity responsible for the cis-trans isomerization of the peptide bond involving the imino group of prolyl residues in proteins (274,279–282). This class of ubiquitous small enzymes, denoted peptidyl-prolyl-*cis-trans*-isomerases (rotamases or immunophilins), is composed of at least cyclophilin, to which cyclosporine binds (274,283), and FK-binding protein, to which two other immunosuppressive macrolides under development, FK506 and rapamycin, bind (see ANTIBIOTICS, MACROLIDES) (284–289). It has not yet been proven how inhibition of rotamase by cyclosporine explains the immunosuppressive action of the drug. It is hypothesized that rotamase acts as a foldase on one or more control proteins that modulate the rate of synthesis of messenger RNA for interleukins and other proteins, but this may not fully explain cyclosporine action.

Ubenimex. Ubenimex [*58970-76-6*] (bestatin), $C_{16}H_{24}N_2O_4$, is a dipeptide isolated from *Streptomyces olivoreticuli* that stimulates parts of the immune system, especially T-cell function including delayed-type hypersensitivity, and causes tumor regression in animals (290–292). It has a low toxicity but lacks antibiotic activity. Ubenimex, in the class of nonantiinfective fermentation products, has been marketed in Japan since 1987 as an oral anticancer agent to be used with other agents to maintain patients with adult acute nonlymphocytic leukemia who are in remission.

Ubenimex, [(2(*S*),3(*R*))-3-amino-2-hydroxy-4-phenylbutanoyl]-L-leucine, was isolated as an inhibitor of aminopeptidases, on which it acts as a strong, reversible transition-state analogue inhibitor (293). Analogues of ubenimex have been made and some other aminopeptidase inhibitors, not all of them peptides, have been isolated from streptomycetes (294–296).

Antimetabolite Peptides

An antimetabolite is defined as a special sort of enzyme inhibitor that acts as a structural analogue of a required metabolite such as a vitamin, amino acid, sugar, hormone, etc, and prevents production or use of that metabolite. Peptide antimetabolites have been useful as research tools in the past, especially those that are glutamine or glutamate analogues or that serve as prodrugs for such analogues. For instance, tabtoxin, 2-amino-4-(3-hydroxy-2-oxo-3-azetidinyl)-butanoyl-L-threonine [*40957-90-2*], $C_{11}H_{19}N_3O_6$, and bacilysin, L-alanyl-L-2,3-epoxy-

4-oxohexahydrophenylalanine [*29393-20-2*], $C_{12}H_{18}N_2O_5$, inhibit glutamine metabolism, as does bialaphos as a source of the amino acid L-phosphinothricin [*35597-44-5*], $C_5H_{12}NO_4P$.

Bialaphos, Phosphinothricin, and Glufosinate. In 1972 and 1973 two groups independently reported finding an antibacterial and antifungal tripeptide called bialaphos [*71048-99-2*], $C_{11}H_{22}N_3O_6P{\cdot}Na$, (formerly SF-1293) in *Streptomyces*

Table 7. Commercial Peptide Antibiotics[a]

Compound	CAS Registry Number	Company[b]	Use[c]
bacitracin (Zn salt) and methylenedisalicylate sodium salt	[*1405-89-6*] [*55852-84-1*]	A.L. Labs, Lundbeck, IMC, Krka, Kayaku	H (topical antibacterial), An (feed additive)
bialaphos		Meiji Seika	Ag (herbicide)
bicozamycin (bicyclomycin)		Fujisawa	An (feed additive)
bleomycin sulfate	[*9041-93-4*]	Nippon Kayaku	H (antitumor)
peplomycin sulfate	[*70384-29-1*]		
capreomycin sulfate	[*1264-72-8*]	Lilly	H (antimycobacterial)
colistin sulfate and colistimethate sodium derivative	[*1264-72-8*] [*8068-28-8*]	A.L. Labs, Kayaku, Banyu, Rhône Poulenc	H and An (topical and systemic antibacterial)
cyclosporine		Sandoz	H (antiimmune)
dactinomycin (actinomycin D)		Merck	H (antitumor)
enduracidin (enramycin)		Takeda	An (antibacterial)
enviomycin (tuberactinomycin N)	[*33103-22-9*]	Toyo Joso	H (antimyobacterial)
gramicidin (linear/ Dubos)		Biochemie, Lundbeck	H (topical antibacterial)
gramicidin S		Nikken	H (topical antibacterial)
mikamycin	[*11006-76-1*]	Kaneka	An (feed additive)
nisin		Applin and Barrett	H (food preservative)
phosphinothricin, (amino acid)		Hoechst	Ag (herbicide)
polymyxin B sulfate	[*1404-20-5*]	A.L. Labs, Pfizer	H and An (topical and systemic antibacterial)
pristinamycin	[*11006-76-1*]	Rhône Poulenc	H (systemic antibacterial)
thiopeptin	[*12609-84-6*]	Fujisawa	An (feed additive)
thiostrepton		CalBioChem	An (topical antibacterial)
tyrothricin	[*1404-88-2*]	Biochemie	H (topical antibacterial)
ubenimex (bestatin)	[*58970-76-6*]	Nippon Kayaku	H (immune stimulant)
virginiamycin		SmithKline Beecham	An (feed additive)
zinostatin (neocarzinostatin)	[*9014-02-2*]	Kayaku	H (antitumor)

[a]Japanese producers courtesy of Japan Antibiotics Research Association (306).
[b]Use in humans, H; in animals, An; plants, Ag.

viridochromogenes and in *S. hygroscopicus* (297,298). Bialaphos (L-phosphinothricyl-L-alanyl-L-alanine) yields the novel amino acid L-phosphinothricin, L-2-amino-4-methylphosphinylbutyric acid [*35597-44-5*], $C_5H_{12}NO_4P$. L-Phosphinothricin, an analogue of methionine sulfoximine, was found to be an irreversibly inactivating inhibitor, competitive with glutamate, for glutamine synthetase from a variety of sources (297–299). The inhibition of cells by L-phosphinothricin was reversed by glutamine [*6899-04-3*]. Susceptible cells must be able to hydrolyze bialaphos, because the tripeptide itself has no activity on glutamine synthetase. Papers on biosynthesis and genetics may be found in the literature (299,300).

Although the antibacterial and antifungal activities of bialaphos and phosphinothricin were not found to be useful, the two agents were later used as biodegradeable, relatively nonselective, postemergent herbicides. Glutamine synthetase inhibition is toxic to plants because the enzyme is key to ammonia assimilation. There is some selectivity for individual plant species as shown by the LD_{50} for bialaphos ranging from 0.125 to 8.5 kg/ha (301–303). Bialaphos, produced by fermentation of *S. hygroscopicus,* is sold in Japan. Racemic phosphinothricin [*53369-07-6*], $C_5H_{12}NO_4P$, produced chemically (304,305), is sold in the United States and elsewhere as the monoammonium salt called glufosinate-ammonium [*77182-82-2*], $C_5H_{12}NO_4P{\cdot}H_3N$, or Total or Basta (formerly Hoe 661, Hoe 39866).

Economic Aspects

Table 7 lists the peptide antibiotics of commercial interest and their producers. Annual U.S. sales of peptide antibiotics, as wholesale price to pharmacies, can be approximated from the following estimates: bleomycin, $25 million for 2 kg; bacitracin, polymyxin, and gramicidin D, $100 million total for 4 t, 10 kg, and 400 g respectively; and cyclosporine, $130 million (3 t). Minor peptides having less than $1 million sales each, such as capreomycin and dactinomycin, are sold at a loss. U.S. agricultural sales figures for bacitracin, $20 million (>200 t), and virginiamycin, $12–15 million (ca 70 t), are indicative of the 1991 market.

BIBLIOGRAPHY

"Polypeptide Antibiotics" in *ECT* 2nd ed., Vol. 16, pp. 306–345, by S. Wilkinson, The Wellcome Research Laboratories, Beckenham, Kent, UK; "Antibiotics (Peptides)" in *ECT* 3rd ed., Vol. 2, pp. 991–1036, by D. Perlman, University of Wisconsin.

1. H. Kleinkauf and H. von Döhren, eds., *Biochemistry of Peptide Antibiotics,* W. de Gruyter, Berlin, 1990.
2. P. Jouin and co-workers, *J. Med. Chem.* **34,** 486 (1991).
3. F. J. Schmitz and co-workers, *J. Org. Chem.* **54,** 3463 (1989).
4. C. L. Bevins and M. Zasloff, *Annu. Rev. Biochem.* **59,** 395 (1990).
5. P. Elsbach, *Trends Biotechnol.* **8,** 26 (1990); C. P. Hill, J. Yee, M. E. Selsted, and D. Eisenberg, *Science* **251,** 1481 (1991).
6. H. Hofte and H. R. Whiteley, *Microbiol. Rev.* **53,** 242 (1989).

7. J. B. Neilands, *Annu. Rev. Biochem.* **50,** 715 (1981).
8. P. Demange and co-workers, *Biochemistry* **29,** 11041 (1990).
9. E. K. Dolence and co-workers, *J. Med. Chem.* **34,** 956, 968 (1991).
10. *The Use of Essential Drugs,* 6th List, Technical Report Series No. 296, World Health Organization, Geneva, Switzerland, 1990.
11. V. Lorian, ed., *Antibiotics in Laboratory Medicine,* 3rd ed., Williams & Wilkins, Baltimore, Md., 1991.
12. F. Peypoux and co-workers, *Eur. J. Biochem.* **153,** 335 (1985).
13. J. E. Walker and E. P. Abraham, *Biochem. J.* **118,** 563 (1970).
14. H. Wojciechowska, A. Konitz, and E. Borowski, *Int. J. Pept. Protein Res.* **26,** 279 (1985).
15. F. Besson, F. Peypoux, M. J. Quentin, and G. Michel, *J. Antibiot.* **37,** 172 (1984).
16. B. W. Bycroft and M. S. Gowland, *J. Chem. Soc. Chem. Commun.* **1978,** 256 (1978).
17. S. K. Das, S. Majumdar, S. Basu, S. Mukherjee, and S. K. Bose, *J. Antibiot.* **39,** 581 (1986).
18. G. Benz and co-workers, *Justus Liebig's Ann. Chem.* **1984,** 1399, 1408, 1434 (1984).
19. D. Schipper, *J. Antibiot.* **36,** 1076 (1983).
20. J. W. Lown and K. Krowicki, *J. Org. Chem.* **50,** 3774 (1985).
21. H. Bickel and co-workers, *Tetrahedron* **1966**(Suppl. 8), 171 (1966).
22. M. Nakajima and co-workers, *J. Antibiot.* **31,** 427 (1978).
23. Y. Iitaka, H. Nakamura, K. Takada, and T. Takita, *Acta Crystallogr.* **30**(Sect. B), 2817 (1974).
24. Y. Okumura and co-workers, *J. Antibiot.* **32,** 1130 (1979).
25. F. E. Hahn, in J. W. Corcoran and F. E. Hahn, eds., *Antibiotics—Mechanism of Action of Antimicrobial and Antitumor Agents,* Vol. 3, Springer-Verlag, Berlin, 1975, p. 79.
26. D. G. Martin and co-workers, *J. Antibiot.* **28,** 332 (1975).
27. T. P. Pitner and D. W. Urry, *Biochemistry* **11,** 4132 (1972).
28. N. G. Kumar and D. W. Urry, *Biochemistry* **12,** 4392 (1973).
29. O. D. Hensens and G. Albers-Schönberg, *J. Antibiot.* **36,** 832 (1983).
30. I. Oberbaumer, E. Grell, F. Raschdorf, and W. J. Richter, *Helv. Chim. Acta* **65,** 2280 (1982).
31. J. P. Springer and co-workers, *J. Am. Chem. Soc.* **106,** 2388 (1984).
32. T. Korzybski, Z. Kowszyk-Gindifer, and W. Kurylowicz, *Antibiotics: Origin, Nature, and Properties,* American Society for Microbiology, Washington, D.C., 1978.
33. J. Bérdy, *Handbook of Antibiotic Compounds,* Vol. 4/1, Vol. 4/2, CRC Press, Boca Raton, Fla., 1980.
34. A. I. Laskin and H. A. Lechevalier, eds., *CRC Handbook of Microbiology,* 2nd ed., Vol. 9, Parts A and B, CRC Press, Boca Raton, Fla., 1988.
35. B. W. Bycroft, ed., *Dictionary of Antibiotics and Related Substances,* Chapman and Hall, London, 1988.
36. *Amino Acids and Peptides,* the Royal Society of Chemistry, Cambridge, UK, Vol. 22, 1991.
37. O. L. Sebek, in D. Gottlieb and P. D. Shaw, eds., *Antibiotics—Mechanism of Action,* Vol. 1, Springer-Verlag, Berlin, 1967, p. 142.
38. H. Paulus, in D. Gottlieb and P. D. Shaw, eds., *Antibiotics—Biosynthesis,* Vol. 2, Springer-Verlag, Berlin, 1967, p. 254.
39. E. Katz and A. L. Demain, *Bacteriol Rev.* **41,** 449 (1977).
40. D. R. Storm, K. S. Rosenthal, and P. E. Swanson, *Ann. Rev. Biochem.* **46,** 723 (1977).
41. E. F. Gale, E. Cundliffe, P. E. Reynolds, M. H. Richmond, and M. J. Waring, *The Molecular Basis of Antibiotic Action,* 2nd ed., John Wiley & Sons, Inc., London, 1981, p. 175.

42. K. Vogler and R. O. Studer, *Experientia* **22,** 345 (1966).
43. H. Kleinkauf and H. von Döhren, in L. C. Vining, ed., *Biochemistry and Genetic Regulation of Commercially Important Antibiotics,* Addison-Wesley, Reading, Mass., 1983, p. 95.
44. S. Nakamura, *Iyakuhin Kaihatsu Kiso Kosa* **14,** 678 (1972).
45. S. Komura and K. Kurahashi, *J. Biochem.* **97,** 1409 (1985).
46. J. Horton and G. A. Pankey, *Med. Clin. North Am.* **66,** 135 (1982).
47. R. Fekety, in G. L. Mandel and co-eds., *Principles and Practice of Infectious Diseases,* 3rd ed., Churchill Livingston, New York, 1990, p. 323.
48. P. J. Weiss, M. L. Andrew, and W. W. Wright, *Antibiot. Chemother.* **7,** 375 (1957).
49. S. Macura, N. G. Kumar, and L. R. Brown, *J. Magnet. Reson.* **60,** 99 (1984).
50. T. Suzuki, K. Hayashi, K. Fujikawa, and K. Tsukamoto, *Report of the Institute of Protein Research,* Osaka University, Japan, 1963.
51. M. Barnett, S. R. M. Bushby, and S. Wilkinson, *Brit. J. Pharmacol.* **23,** 552 (1964).
52. R. A. Moore, L. Chan, and R. E. W. Hancock, *Antimicrob. Agents Chemother.* **26,** 539 (1984); R. E. W. Hancock, *Ann. Rev. Microb.* **38,** 237 (1984); H. Nikaido, *Handbook of Experimental Pharmacology,* Vol. 91, Springer-Verlag, Berlin, 1989, p. 1.
53. D. S. Feingold, C-C. Hsu Chen, and I. J. Sud, *Ann. N.Y. Acad. Sci.* **235,** 480 (1974).
54. S. Chihara, T. Tobita, M. Yahata, A. Ito, and Y. Koyama, *Agric. Biol. Chem.* **37,** 2455 (1973).
55. M. Vaara and T. Vaara, *Nature* **303,** 526 (1983).
56. H. Nikaido and M. Vaara, *Microbiol. Rev.* **49,** 1 (1985).
57. R. L. Danner and co-workers, *Antimicrob. Agents Chemother.* **33,** 1428 (1989).
58. Y. Kimura and co-workers, *Proc. 9th Symp. Peptide Chem. Japan, Protein Res. Found.* **9,** 159 (1972); M. Ito-Kagawa and Y. Koyama, *J. Antibiol.* **33,** 1551 (1980); H. Kitamura-Matsunaga, Y. Kimura, and T. Araki, in N. Izumiya, ed., *Peptide Chemistry, 1984, Protein Res. Found., Osaka,* 1985, p. 333; N. Yasuda, Y. Kimura, M. Miyama, and H. Matsunaga, *Agric. Biol. Chem.* **53,** 3245 (1989).
59. K. Hanasawa and co-workers, *Prog. Clin. Biol. Res.* **264,** 337 (1988); M. A. F. A. Boons, M. L. P. M. Verhoeven, L. A. A. E. Sluyterman, and H. M. Buck, *Appl. Biochem. Biotechnol.* **22,** 95 (1989); K. W. Talmadge and C. J. Siebert, *J. Chromatog.* **476,** 175 (1989).
60. B. A. Johnson, H. S. Anker, and F. L. Meleney, *Science,* **102,** 376 (1945).
61. H. S. Anker, B. A. Johnson, J. Goldberg, and F. L. Meleney, *J. Bacteriol.* **55,** 249 (1948).
62. *Bacitracin,* S. B. Penick and Co., New York, 1952.
63. A. Arriagada and co-workers, *Brit. J. Exp. Pathol.* **30,** 425 (1949).
64. E. P. Abraham and G. G. F. Newton, *Ciba Foundation Symposium: Amino Acids and Peptides with Antimetabolic Activity,* 1958, p. 205.
65. L. C. Craig, W. Kronigsberg, and R. J. Hill, in Ref. 64, p. 226.
66. R. O. Studer, in G. P. Ellis and G. B. West, eds., *Progress in Medicinal Chemistry,* Vol. 5, Plenum Press, New York, 1967, p. 1.
67. G. A. Brewer, in K. Florey, ed., *Analytical Profiles of Drug Substances,* Vol. 9, Academic Press, New York, 1980, p. 1.
68. F. L. Meleney and B. A. Johnson, *Am. J. Med.* **7,** 794 (1949); E. Jawetz, *Polymyxin, Neomycin, and Bacitracin,* Antibiotic Monograph No. 5, Medical Encyclopedia, Inc., New York, 1965, p. 85; E. Jawetz, *Pediat. Clin. North Am.* **8,** 1057 (1961).
69. D. R. Storm and W. A. Toscano, Jr. in Ref. 37, p. 1.
70. W. A. Toscano, Jr. and D. R. Storm, *Pharmac. Ther.* **16,** 199 (1982).
71. O. Frøyshov, in E. J. Vandamme, ed., *Biotechnology of Industrial Antibiotics,* Marcel Dekker, New York, 1984, p. 665.

72. R. J. Hickey, *Prog. Ind. Microbiol.* **4,** 94 (1964).
73. C. Ressler and D. V. Kashelikar, *J. Am. Chem. Soc.* **88,** 2025 (1966); R. E. Galardy, M. P. Printz, and L. C. Craig, *Biochemistry* **10,** 2429 (1971).
74. K. Tsuji, J. H. Robertson, and J. A. Bach, *J. Chromatog.* **99,** 597 (1974); H. Oka, Y. Ikai, N. Kawamura, and M. Yamada, *J. Chromatog.* **462,** 315 (1989).
75. E. Munekata, Y. Masui, T. Shiba, and T. Kaneko, *Bull. Chem. Soc. Japan* **46,** 3187 (1973).
76. P. J. Weiss, M. L. Andrew, and W. W. Wright, *Antibiot. Chemother.* **7,** 374 (1957).
77. M. L. Andrew and P. J. Weiss, *Antibiot. Chemother.* **9,** 277 (1959).
78. B. L. Damron and H. R. Wilson, *Poultry Sci.* **64,** 1451 (1985); J. H. Choi and K. S. Ryu, *Brit. Poult. Sci.* **28,** 113 (1987).
79. P. W. Waldroup and co-workers, *Poult. Sci.* **67,** 103 (1988).
80. J. V. Scudi and W. Antopol, *Proc. Soc. Exptl. Biol. Med.* **64,** 503 (1947); H. A. Zintel, R. A. Ma, A. C. Nichols, and H. Ellis, *Am. J. Med. Sci.* **218,** 439 (1949).
81. S. F. Sundlof, J. E. Riviere, and A. L. Craigmill, *Food Animal Residue Avoidance Databank (FARAD) Trade Name File, A Comprehensive Compendium of Food Animal Drugs,* University of Florida, Gainsville, 1988.
82. S. Muirhead, ed., *1990 Feed Additive Compendium,* Miller Publishing Co., Minnetonka, Minn., 1989.
83. D. R. Storm, *Ann. N.Y. Acad. Sci.* **235,** 387 (1974).
84. R. J. Dubos, *J. Exptl. Med.* **70,** 1 (1939); R. D. Hotchkiss, *Adv. Enzymol.* **4,** 153 (1944); R. P. Crease, *Science* **246,** 883 (1989).
85. G. F. Gause and M. G. Brayhnikova, *Nature* **154,** 703 (1944).
86. G. A. Brewer, in K. Florey, ed., *Analytical Profiles of Drug Substances,* Vol. 8, Academic Press, New York, p. 179.
87. E. Gross, in M. J. Weinstein and G. H. Wagman, eds., *Antibiotics: Isolation, Separation, and Purification,* Elsevier, Amsterdam, The Netherlands, 1978, p. 415.
88. B. Cornell, *J. Bioenerg. Biomemb.* **19,** 655 (1987).
89. F. R. Salemme, *Science* **241,** 145 (1988); B. A. Wallace, *Curr. Topics Membr. Transp.* **33,** 35 (1988).
90. S.-S. Sung and P. C. Jordan, *J. Theor. Biol.* **40,** 369 (1989).
91. B. A. Wallace, *Annu. Rev. Biophys. Chem.* **19,** 127 (1990).
92. F. E. Hunter, Jr. and L. S. Schwartz, in Ref. 37, p. 643.
93. K. Kurahashi, in W. Corcoran, ed., *Antibiotics—Biosynthesis,* Vol. 4, Springer-Verlag, New York, 1981, p. 325.
94. R. Sarges and B. Witkop, *J. Am. Chem. Soc.* **87,** 2020 (1965).
95. P. Daumas, F. Heitz, L. Ranjalahy-Rasoloanijaq, and R. Lazaro, *Biochemie* **71,** 77 (1989).
96. D. Lelievre, Y. Trudelle, F. Heitz, and G. Spach, *Int. J. Peptide Protein Res.* **33,** 379 (1989).
97. B. A. Wallace and K. Ravikumar, *Science* **241,** 182 (1988); D. A. Langs, *Science* **241,** 188 (1988).
98. F. E. Hunter, Jr. and L. S. Schwartz, in Ref. 37, p. 636.
99. J. Vater, in *CRC Crit. Rev. Biotechnol.* **8,** 205 (1988).
100. G. Mittenhuber, R. Weckermann, and M. A. Marahiel, *J. Bacteriol.* **171,** 4881 (1989).
101. M. Waki and N. Izumiya, in Ref. 99, p. 206.
102. R. D. Hotchkiss and R. J. Dubos, *J. Biol. Chem.* **141,** 155 (1941).
103. A. Bohg and H. Ristow, *Eur. J. Biochem.* **170,** 253 (1987); F. J. Aranda and B. de Kruijff, *Biochim. Biophys. Acta* **937,** 195 (1988).
104. W. Danders and co-workers, *Antimicrob. Agents Chemother.* **22,** 785 (1982); K. Sato and U. Nagai, in Y. Kiso, ed., *Peptide Chemistry 1985,* Protein Research Foundation, 1986,

p. 217; H. Aoyagi, S. Ando, S. Lee, and N. Izumiya, *Tetrahedron* **44,** 877 (1988); T. Katsu and co-workers, *Biochim. Biophys. Acta* **983,** 135 (1989).
105. M. Kanda and co-workers, *J. Biochem.* **105,** 653 (1989); J. Vater and co-workers, *Biol. Chem. Hoppe-Seyler* **370,** 1013 (1989); H.-J. Skarpeid, T.-L. Zimmer, B. Shen, and H. von Döhren, *Eur. J. Biochem.* **187,** 627 (1990).
106. *Chem. Eng. News* **39,** 57 (1961).
107. W. B. Sutton, R. S. Gordee, W. E. Wick, and LaV. Stanfield, *Ann. N.Y. Acad. Sci.* **135,** 947 (1966).
108. M. Lucchesi, *Antibiot. Chemother.* **16,** 27 (1970).
109. P. G. Caltrider in Ref. 37, p. 677.
110. L. Trnka, in K. Bartmann, H. Iwainsky, and co-eds., *Antituberculosis Drugs,* Springer-Verlag, Berlin, 1988, p. 150.
111. A. Nagata and co-workers, *J. Antibiot. (Tokyo)* **24,** 681 (1968).
112. T. Ando and co-workers, *J. Antibiot.* **24,** 680 (1971).
113. A. Nagata and co-workers, in M. Hejzlar and co-workers, *Advances in Antimicrobial Antineoplastic Chemotherapy,* Vol. 1/2, University Park Press, Baltimore, Md., 1972, p. 1039.
114. U.S. Pat. 3,892,732 (July 1, 1975), J. Abe and co-workers (to Toyo Joso K.K.).
115. J. M. Bond, in R. J. Schnitzer and F. Hawking, eds., *Experimental Chemotherapy,* Vol. 4, Academic Press, New York, 1966, p. 560.
116. H. Otten, in Ref. 110, p. 191.
117. U.S. Pat. 3,143,468 (Aug. 4, 1964), E. B. Herr, Jr., R. L. Hamill, and J. M. McGuire (to Eli Lilly & Co.).
118. J. F. Carter II and co-workers, *Biochemistry (U.S.)* **13,** 1221, 1227 (1974).
119. E. B. Herr, Jr. and M. O. Redstone, *Ann. N.Y. Acad. Sci.* **135,** 940 (1966).
120. S. Nomoto, T. Teshima, T. Wakamiya, and T. Shiba, *Tetrahedron* **34,** 921 (1978).
121. J. K. McClatchy, W. Kanes, P. T. Davidson, and T. S. Moulding, *Tubercle* **58,** 29 (1977).
122. M. M. Tsukamura, *Microbiol. Immunol.* **24,** 11 (1980).
123. Y. Mizuguchi, T. Udou, and T. Yamada, *Microbiol. Immunol.* **27,** 425 (1983).
124 L. Heifets and P. Lindholm-Levy, *Antimicrob. Agents Chemother.* **33,** 1298 (1989).
125. P. Marrero, M. J. Cabañas, and J. Modelell, *Biochem. Biophys. Res. Commun.* **97,** 1047 (1980); T. Yamada, T. Teshima, T. Shiba, and K. H. Nierhaus, *Nucleic Acids Res.* **8,** 5767 (1980); I. Medow, S. Rüsch-Gerdes, and K.-H. Schröder, *Zentralbl. Bakteriol. Mikrobiol. Hyg.* **A266,** 359 (1987).
126. P. DeSomer and P. Van Dijck, *Antibiot. Chemother.* **5,** 632 (1955).
127. J. Charney, W. P. Fisher, C. Curran, R. A. Machlowitz, and A. A. Tytell, *Antibiot. Chemother.* **3,** 1283 (1953).
128. J. Preud'homme, P. Tarridec, and A. Belloc, *Bull. Soc. Chim. France* **2,** 585 (1968).
129. M. Arai and co-workers, *J. Antibiot.* **11**(Ser. A), 14 (1958).
130. D. Vazquez, in Ref. 37, p. 387.
131. C. Cocito, *Microbiol. Rev.* **43,** 145 (1979).
132. E. Cundliffe, in Ref. 41, p. 402.
133. A. M. Biot, in Ref. 71, p. 695.
134. C. Cocito, in F. E. Hahn, ed., *Antibiotics—Modes and Mechanisms of Microbial Growth Inhibitors,* Springer-Verlag, Berlin, 1983, p. 296.
135. E. F. Gale, in Ref. 41, p. 176.
136. P. Crooy and R. De Neyes, *J. Antibiot.* **25,** 371 (1972).
137. J. W. Reed and D. G. I. Kingston, *J. Natl. Prod.* **49,** 626 (1986).
138. H. S. Kim and co-workers, *J. Antibiot.* **42,** 769 (1989).
139. H. Vanderhaeghe, P. Van Dijck, G. Parmentier, and P. DeSomer, *Antibiot. Chemother.*

7, 606 (1957); H. Vanderhaeghe and G. Parmentier, *J. Am. Chem. Soc.* **82,** 4414 (1960); G. R. Delpierre and co-workers, *J. Chem. Soc. (C)* **1966,** 1653; D. G. I. Kingston, Lord Todd, and D. H. Williams, *J. Chem. Soc. (C)* **1966,** 1669; J. Vanderhaeghe, G. Janssen, and F. Compernolle, *Tetrahedron Lett.* **28,** 2687 (1971).
140. F. LeGoffic, *J. Antimicrob. Chemother.* **16**(Suppl. A), 13 (1985); J. M. Paris and co-workers, *Pathol. Biol.* **33,** 493 (1985); J. C. Barriere and co-workers, *30th Interscience Conference on Antimicrobial Agenys and Chemotherapy,* American Society for Microbiology, Atlanta, Ga., 1990, p. 214.
141. J. Modai, *J. Antimicrob. Chemother.* **16**(Suppl. A), 195 (1985).
142. J. C. Lefevre and R. Bauriaud, *Antimicrob. Agents Chemother.* **33,** 2152 (1989).
143. H. LaForest, M. Fourgeaud, H. Richet, and P. H. LaGrange, *Antimicrob. Agents Chemother.* **32,** 1094 (1988).
144. R. E. Buresh, R. D. Miles, and R. H. Harms, *Poultry Sci.* **64,** 1041 (1985); R. D. Miles, D. M. Janky, and R. H. Harms, *Poultry Sci.* **64,** 139 (1985); S. A. Woodward and co-workers, *Poultry Sci.* **67,** 1222 (1988).
145. M. Di Giambattista, G. Chinali, and C. Cocito, *J. Antimicrob. Chemother.* **24,** 485 (1989).
146. D. Dubnau, *CRC Crit. Rev. Biochem.* **16,** 103 (1984).
147. J. Duval, *J. Antimicrob. Chemother.* **16**(Suppl. A), 137 (1985).
148. B. Weisblum, *J. Antimicrob. Chemother.* **16,** 63 (1985).
149. E. Cundliffe, in L. E. Bryan, ed., *Handbook of Experimental Pharmacology,* Vol. 91, Springer-Verlag, Berlin, 1989, p. 227.
150. J. F. Pagano, M. J. Weinstein, H. A. Stout, and R. Donovick, *Antibiot. Ann.* **1955/1956,** 554 (1956).
151. J. Vandeputte and J. D. Dutcher, *Antibiot. Ann.* **1955/1956,** 560 (1956).
152. S. Pestka and J. W. Bodley, in Ref. 25, p. 551; E. Cundliffe, in Ref. 41, p. 329; 492; E. Cundliffe, in Ref. 149, p. 228.
153. T.-P. Hausner, V. Geigenmüller, and K. H. Nierhaus, *J. Biol. Chem.* **263,** 13,103 (1988).
154. F. Benazet and co-workers, *Experientia* **36,** 414 (1980).
155. M. Bodanszky and co-workers, *J. Am. Chem. Soc.* **86,** 2478 (1964); M. Bodansky, J. A. Scozzie, and I. Muramatsu, *J. Am. Chem. Soc.* **91,** 4934 (1969); B. Anderson, D. C. Hodgkin, and M. A. Viswamitra, *Nature* **225,** 233 (1970).
156. K. Tori and co-workers, *Tetrahedron Lett.* **3,** 185 (1976).
157. O. D. Hensens, G. Albers-Schönberg, and B. F. Anderson, *J. Antibiot.* **36,** 799 (1983).
158. K. Florey, in K. Florey, ed., *Analytical Profiles of Drug Substances,* Vol. 7, Academic Press, New York, 1978, p. 423.
159. D. R. Houck and co-workers, *J. Am. Chem. Soc.* **109,** 1250 (1987).
160. T. Frenzel, P. Zhou, and H. G. Floss, *Arch. Biochem. Biophys.* **278,** 35 (1990).
161. S. Goto, S. Kuwahara, N. Okubo, and H. Zenyoji, *J. Antibiot.* **21,** 119 (1968); E. Higashide, K. Hatano, M. Shibata, and K. Nakazawa, *J. Antibiot.* **21,** 126 (1968); K. Tsuchiya, M. Kondo, T. Oishi, and T. Yamazaki, *J. Antibiot.* **21,** 147 (1968).
162. M. Hori, N. Sugita, and M. Miyazaki, *Chem. Pharmaceut. Bull.* **21,** 1171 (1973); M. Hori and co-workers, *Chem. Pharmaceut. Bull.* **21,** 1175 (1973); H. Iwasaki and co-workers, *Chem. Pharmaceut. Bull.* **21,** 1184 (1973).
163. M. Kawakami, Y. Nagai, T. Fujii, and S. Mitsuhashi, *J. Antibiot.* **24,** 583 (1971).
164. T. Tamura, H. Suzuki, Y. Nishimura, J. Mizoguchi, and Y. Hirota, *Proc. Nat. Acad. Sci. USA* **77,** 4499 (1980); F. Ishino and co-workers, *J. Biol. Chem.* **261,** 7024 (1986).
165. H. Yokotani and co-workers, *Takeda Kenkyusho Nempo* **28,** 76 (1969).
166. T. Miyoshi and co-workers, *J. Antibiot.* **25,** 569 (1972); S. Miyamura and co-workers, *J. Antibiot.* **25,** 610 (1972).
167. C. D. Ericsson and co-workers, *Ann. Intern. Med.* **98,** 20 (1983); C. D. Ericsson and co-

workers, *Gastroenterology* **88,** 473 (1985); C. D. Ericsson and H. L. DuPont, in P. K. Peterson and J. Verhoef, eds., *Antimicrobial Agents Annual,* Vol. 3, Elsevier, New York, 1988, p. 541.

168. N. Tanaka, in F. E. Hahn, ed., *Antibiotics—Mechanism of Action of Antibacterial Agents,* Springer-Verlag, New York, 1979, p. 18; R. M. Williams and C. A. Durham, *Chem. Rev.* **88,** 511 (1988).
169. B. Heinemann, M. A. Kaplan, R. D. Muir, and I. R. Hooper, *Antibiot. Chemother.* **111,** 1239 (1953).
170. D. E. Tisch, J. B. Hoekstra, D. S. Fisher, and H. L. Dickison, *Antibiot. Ann.* **1954/1955,** 1011 (1955).
171. U.S. Pat. 3,126,317 (Mar. 24, 1964), B. Heinemann, I. R. Hooper, and M. A. Kaplan (to Bristol-Myers Co.).
172. M. Bodanszky, G. F. Sigler, and A. Bodanszky, *J. Am. Chem. Soc.* **95,** 2352 (1973).
173. H. Tanaka, R. Oiwa, S. Matsukara, and S. Omura, *Biochem. Biophys. Res. Commun.* **86,** 902 (1979).
174. H. Tanaka and co-workers, *J. Antibiot.* **35,** 1216 (1982).
175. M. S. Kang, J. P. Spencer, and A. D. Elbein, *J. Biol. Chem.* **253,** 8860 (1978).
176. D. K. Banerjee, *J. Biol. Chem.* **264,** 2024 (1989).
177. J. H. Lakey, R. Maget-Dana, and M. Ptak, *Biochem. Biophys. Res. Commun.* **150,** 384 (1988).
178. U.S. Pat. 4,414,206 (Nov. 8, 1983), M. Gordon and G. J. Christie (to Bristol-Myers Co.).
179. H. Brockmann and G. Schmidt-Kastner, *Chem. Ber.* **88,** 57 (1955); H. Brockmann and H. Geeren, *Anal. Chemie* **603,** 216 (1957).
180. E. Gäumann and co-workers, *Experientia* **3,** 202 (1947).
181. P. A. Plattner and U. Nager, *Experientia* **3,** 325 (1947).
182. R. L. Hamill, C. E. Higgens, H. E. Boaz, and M. Gorman, *Tetrahedron Lett.* **1969,** 4255 (1969).
183. T. M. Balasubramanian and co-workers, *J. Am. Chem. Soc.* **103,** 6127 (1981); B. F. Gisin and co-workers, *J. Am. Chem. Soc.* **103,** 6373 (1981); M. K. Mathew and P. Balaram, *Molec. Cell. Biochem.* **50,** 47 (1983); E. T. Kaiser and F. J. Kézdy, *Ann. Rev. Biophys. Biophys. Chem.* **16,** 561 (1987).
184. B. C. Pressman and M. Fahim, *Ann. Rev. Pharmacol. Toxicol.* **22,** 465 (1982).
185. E. P. Bakker in Ref. 37, p. 67.
186. R. Hilgenfeld and W. Saenger, *Top. Curr. Chem.* **101,** 1 (1982).
187. G. R. Painter and B. C. Pressman, *Top. Curr. Chem.* **101,** 83 (1982).
188. M. M. Shemyakin, N. A. Aldanova, E. I. Vinogradova, and M. Y. Feigina, *Tetrahedron Lett.* **28,** 1921 (1963); M. Pinkerton, L. K. Steinrauf, and P. Dawkins, *Biochem. Biophys. Res. Commun.* **35,** 512 (1969); V. T. Ivanov and co-workers, *Biochem. Biophys. Res. Commun.* **34,** 803 (1969).
189. Yu. A. Ovchinnikov, V. T. Ivanov, and A. M. Shkrob, *Membrane-Active Complexones, BBA Library,* Vol. 12, Elsevier, Amsterdam, The Netherlands, 1974.
190. T. Anke and F. Lipmann, *FEBS Lett.* **82,** 337 (1977); H. Ristow, J. Salnikow, and H. Kleinkauf, *FEBS Lett.* **42,** 127 (1974); J. B. Perkins and co-workers, *J. Bacteriol.* **172,** 3108 (1990).
191. N. Schnell and co-workers, *Nature* **333,** 276 (1988).
192. C. Kaletta and K.-D. Entian, *J. Bacteriol.* **171,** 1597 (1989).
193. T. R. Klaenhammer, *6th International Symposium on the Genetics of Industrial Microorganisms,* Vol. 1, Strasbourg, France, 1990, p. 433.
194. *Fed. Reg.* **53**(66), 11,247 (Apr. 1988).
195. E. Lipinska, in V. M. Woodbine, ed., *Antibiotics and Antibiosis in Agriculture with Special Reference to Synergism,* Butterworths, London, 1977, p. 103; A. Hurst, *Adv. Appl. Microbiol.* **27,** 85 (1981); K. Rayman and A. Hurst in Ref. 71, p. 807.
196. L. A. Rogers, *J. Bacteriol.* **16,** 321 (1928).

197. A. T. R. Mattick and A. Hirsh, *Nature* **554,** 551 (1944).
198. E. Gross and J. L. Morell, *J. Am. Chem. Soc.* **93,** 4634 (1971).
199. M. Barber, and co-workers, *Experientia* **44,** 266 (1988).
200. K. Fukase and co-workers, *Tetrahedron Lett.* **29,** 795 (1988).
201. U.S. Pat. 2,935,503 (May 3, 1960), H. B. Hawley and R. H. Hall (to Aplin and Barrett, Ltd.).
202. E. Ruhr and H.-G. Sahl, *Antimicrob. Agents Chemother.* **27,** 841 (1985).
203. M. Debono and co-workers, *J. Antibiot.* **40,** 761 (1987).
204. G. M. Eliopoulos, C. Thauvin, B. Gerson, and R. C. Moellering, Jr., *Antimicrob. Agents Chemother.* **27,** 357 (1985).
205. L. D. Boeck, D. S. Fukuda, B. J. Abbott, and M. Debono, *J. Antibiot.* **41,** 1085 (1988).
206. M. Debono and co-workers, *J. Antibiot.* **41,** 1093 (1988).
207. G. M. Eliopoulos and co-workers, *Antimicrob. Agents Chemother.* **30,** 532 (1986); R. N. Jones and A. L. Barry, *Antimicrob. Agents Chemother.* **31,** 625 (1987).
208. P. Canepari, M. Boaretti, M. del Mar Lleó, and G. Satta, *Antimicrob. Agents Chemother.* **34,** 1220 (1990).
209. J. T. Mader and K. Adams, *Antimicrob. Agents Chemother.* **33,** 689 (1989); S. Kennedy and H. F. Chambers, *Antimicrob. Agents Chemother.* **33,** 1522 (1989); C. S. Haworth, M. W. Sobieski, W. M. Scheld, and T. S. Park, *Antimicrob. Agents Chemother.* **34,** 245 (1990).
210. M. Debono and co-workers, *Ann. N.Y. Acad. Sci.* **544,** 152 (1988); L. D. Broeck, D. S. Fukuda, B. J. Abbott, and M. Debono, *J. Antibiot.* **42,** 382 (1989).
211. R. S. Gordee and co-workers, *Ann. N.Y. Acad. Sci.* **544,** 294 (1988).
212. M. Hobbs, J. Perfect, and D. Durack, *Eur. J. Clin. Microbiol. Infect. Dis.* **7,** 77 (1988); L. H. Hanson and D. A. Stevens, *Antimicrob. Agents Chemother.* **33,** 1391 (1989); R. P. Rennie and L. Hellman, *Mycoses* **32,** 145 (1989).
213. W. L. Current and C. J. Boylan, in Ref. 140, p. 229.
214. C. S. Taft and C. P. Selitrennikoff, *J. Antibiot.* **41,** 697 (1988).
215. A. Padula and H. F. Chambers, *Antimicrob. Agents Chemother.* **33,** 1822 (1989); J. R. Perfect, M. M. Hobbs, K. A. Wright, and D. T. Durack, *Antimicrob. Agents Chemother.* **33,** 1811 (1989); C. J. Morrison and D. A. Stevens, *Antimicrob. Agents Chemother.* **34,** 746 (1990).
216. B. N. Doebbeling and co-workers, in *30th Interscience Conference on Antimicrobial Agents and Chemotherapy,* American Society for Microbiology, 1990, p. 183.
217. S. A. Waksman, ed., *Ann. N.Y. Acad. Sci.* **89,** 283 (1960).
218. S. A. Waksman, ed., *Actinomycin: Nature, Formation, and Activities,* Wiley-Interscience, New York, 1968.
219. U. Hollstein, *Chem. Rev.* **74,** 625 (1974); "Selman A. Waksman Conference on Actinomycins: Their Potential for Cancer Chemotherapy," *Cancer Chemother. Rept.* **58** (1974).
220. J. Meinhofer and E. Atherton, in D. Perlman, ed., *Structure-Activity Relationships Among the Semisynthetic Antibiotics,* Academic Press, New York, 1977, p. 427.
221. A. B. Mauger, in P. G. Sammes, ed., *Topics in Antibiotic Chemistry,* Vol. 4, Ellis Horwood, Chicester, UK, 1980, p. 224.
222. M. J. Waring, in Ref. 41, p. 314.
223. *IARC Monographs* **12,** 29 (1976).
224. R. A. Manaker, F. Gregory, L. C. Vining, and S. A. Waksman, *Antibiot. Annu.* **1954–1955,** 853 (1955).
225. A. B. Mauger and E. Katz, *J. Chromatog. Lib.* **15,** 1 (1978).
226. U.S. Pat. 2,378,876 (June 19, 1945), S. A. Waksman and H. B. Woodruff (to Merck Co.).
227. E. Katz, in Ref. 38, p. 276; F. Fawaz and G. H. Jones, *J. Biol. Chem.* **263,** 4602 (1988); C. E. Barry, III, P. G. Nayar, and T. P. Begley, *Biochemistry* **28,** 6323 (1989).
228. S. K. Sengupta and co-workers, *J. Med. Chem.* **24,** 1052 (1981); M. Yamada, K.

Nakajima, and K. Okawa, in N. Izumiya, ed., *Peptide Chemistry 1984,* Protein Research Foundation, Osaka, Japan, 1985, p. 153.
229. G. Bradley, P. F. Juranka, and V. Ling, *Biochim. Biophys. Acta* **948,** 87 (1988); *Cancer Bull.* **41**(1), 1 (1989).
230. H. M. Sobell, *Prog. Nucleic Acid Res. Mol. Biol.* **13,** 153 (1973).
231. E. V. Scott, G. Zon, L. G. Marzilli, and W. D. Wilson, *Biochemistry* **27,** 7940 (1988).
232. M. Delepierre and co-workers, *J. Biomol. Struct. Dyn.* **7,** 557 (1989).
233. H. Umezawa, K. Maeda, T. Takeuchi, and Y. Okami, *J. Antibiot. Ser. A* **19,** 200 (1966); H. Umezawa, Y. Suhara, T. Takita, and K. Maeda, *J. Antibiot. Ser. A* **19,** 210 (1966).
234. H. Umezawa, in J. M. Cassady and J. D. Douros, eds., *Anticancer Agents Based on Natural Product Models,* Academic Press, New York, 1980, p. 147.
235. N. J. Oppenheimer and co-workers, *J. Biol. Chem.* **257,** 1606 (1982); B. J. Carter and co-workers, *J. Biol. Chem.* **265,** 4193 (1990).
236. S. Oka, *Rec. Results Cancer Res.* **74,** 163 (1980).
237. U.S. Pat. 4,195,018 (Mar. 25, 1980), T. Takita and co-workers (to Nippon Kayaku KK).
238. K. Takahashi and co-workers, *Cancer Treat. Rev.* **14,** 169 (1987).
239. R. A. Newman and co-workers, *Investig. New Drugs* **8,** 33 (1990).
240. S. T. Crooke and co-workers, *Rec. Results Cancer Res.* **76,** 83 (1981).
241. K. Maeda, H. Kosaka, K. Yagishita, and H. Umezawa, *J. Antibiot. Ser. A* **9,** 82 (1956).
242. S. K. Carter, T. Ichikawa, G. Mathé, and H. Umezawa, eds., *Gann Monogr. Cancer Res.* (19) (1976).
243. C. W. Haidle and R. S. Lloyd, in Ref. 38, p. 124; H. Umezawa, in J. M. Cassady and J. D. Douros, eds., *Anticancer Agents Based on Natural Product Models,* Academic Press, New York, 1980, p. 147.
244. T. Takita, in Ref. 71, p. 595.
245. B. I. Sikic, M. Rozencweig, and S. K. Carter, *Bleomycin Chemotherapy,* Academic Press, Orlando, Fla., 1985.
246. S. M. Hecht, *Fed. Proc.* **45,** 2784 (1986); S. M. Hecht, *Acc. Chem. Res.* **19,** 383 (1986).
247. I. H. Goldberg, in C. Chagas and B. Pullman, eds., *Molecular Mechanisms of Carcinogenic and Antitumor Activity,* The Vatican Press distributed by Adenine Press, Guilderland, N.Y., 1987, p. 425.
248. J. Stubbe and J. W. Kosarich, *Chem. Rev.* **87,** 1107 (1987).
249. D. H. Petering, R. W. Byrnes, and W. E. Antholine, *Chem.-Biol. Interac.* **73,** 133 (1990).
250. T. Takita and co-workers, *J. Antibiot.* **25,** 755 (1972).
251. T. Takita and co-workers, *Tetrahedron Lett.* **23,** 521 (1982).
252. Y. Aoyagi and co-workers, *J. Am. Chem. Soc.* **104,** 5237 (1982).
253. S. Saito and co-workers, *J. Antibiot.* **36,** 92 (1983).
254. T. Takita and Y. Muraoka, in Ref. 1, p. 289.
255. J. S. Lazo, C. J. Boland, and P. E. Schwartz, *Cancer Res.* **42,** 4026 (1982); P. R. Twentyman, *Pharmacol. Ther.* **23,** 417 (1984); B. I. Sikic, *Cancer Surv.* **5,** 81 (1986); J. E. Zuckerman and co-workers, *Cancer Res.* **46,** 1748 (1986).
256. S. M. Sebti and co-workers, *Biochemistry* **28,** 6544 (1989).
257. M. Takeshita, A. P. Grollman, E. Ohtsubo, and H. Ohtsubo, *Proc. Natl. Acad. Sci. USA* **75,** 5983 (1987).
258. N. Ishida, K. Miyazaki, K. Kumagai, and M. Rikimaru, *J. Antibiot. Ser. A* **18,** 68 (1965); K. Edo, K. Saito, Y. Akiyama-Murai, and M. Mizugaki, *J. Antibiot.* **41,** 554 (1988); P. C. Dedon and I. H. Goldberg, *J. Biol. Chem.* **265,** 14,713 (1990); B. Pullman, *Adv. Drug Res.* **18,** 1 (1989); D. H. Chin, C. H. Zeng, C. E. Costello, and I. H. Goldberg, *Biochemistry* **27,** 8106 (1988); L. F. Povirk and I. H. Goldberg, *Biochemie* **69,** 815 (1987); T. Sasaki, Y. Koide, and K. Yoshigaga, *Meth. Enzymol.* **178,** 422 (1989); F. Suzuki, R. B. Pollard, S. Uchimura, T. Munakata, and H. Maeda, *Cancer Res.* **50,** 3897 (1990); E. Otsuji, T. Takahashi, T. Yamaguchi, N. Yamaguchi, and J. Imanishi, *Gastroenterol. Japan* **25,** 244 (1990); L. C. Sieker and G. G. Gnanarajah, *Proteins* **3,** 252 (1988); P. Van Roey and T. A. Beerman, *Proc. Natl. Acad. Sci. USA* **86,** 6587 (1989).

259. A. Rüegger and co-workers, *Helv. Chim. Acta* **59,** 1075 (1976).
260. J. F. Borel, C. Feurer, H. V. Gubler, and H. Stähelin, *Agents Actions* **6,** 468 (1976).
261. B. Ryffel and co-workers, *Arch. Toxicol.* **53,** 107 (1983).
262. J. P. Scott and T. W. Higenbottam, *Med. Toxicol.* **3,** 107 (1988).
263. R. Maiorca and co-workers, *Adv. Exp. Med. Biol.* **252,** 273 (1989).
264. H. Dieperink, *Dan. Med. Bull.* **36,** 235 (1989).
265. B. D. Kahan, ed., *Cyclosporine,* Vol. 1, Grune and Stratton, Inc., Orlando, Fla., 1984.
266. R. Schindler, ed., *Cyclosporin in Autoimmune Diseases,* Springer-Verlag, Berlin, 1985.
267. J. F. Borel, *Prog. Allergy* **38,** 9 (1986).
268. M. M. A. Hassan and M. A. Al-Yahya, in K. Florey, ed., *Analy. Profiles Drug Subst.* **16,** 145 (1987).
269. A. von Wartburg and R. Traber, in G. P. Ellis and G. B. West, eds., *Prog. Med. Chem.* **25,** 1 (1988).
270. G. Feutren, *Curr. Op. Immunol.* **2,** 239 (1989).
271. M. J. Mihatsch, G. Thiel, and B. Ryffel, *Toxicol. Lett.* **46,** 125 (1989).
272. A. Aszalos, *J. Med.* **19,** 297 (1988).
273. H. G. Fliri and R. M. Wenger, in Ref. 1, p. 245.
274. S. L. Schreiber, *Science* **251,** 283 (1991).
275. T. J. Petcher, H.-P. Weber, and A. Rüegger, *Helv. Chim. Acta* **59,** 1480 (1976).
276. J. D. Aebi and co-workers, *J. Med. Chem.* **33,** 999 (1990).
277. U.S. Pat. 4,117,118 (Sept. 26, 1978), E. Härri and A. Rüegger (to Sandoz, Ltd.).
278. A. Lawen and R. Zocher, *J. Biol. Chem.* **265,** 11,355 (1990).
279. G. Fischer, H. Bang, and C. Mech, *Biomed. Biochim. Acta* **43,** 1101 (1984).
280. G. Fischer and H. Bang, *Biochim. Biophys. Acta* **828,** 39 (1985).
281. G. Fischer and F. X. Schmid, *Biochemistry* **29,** 2205 (1990).
282. J. Liu, M. W. Albers, C.-M. Chen, S. L. Schreiber, and C. T. Walsh, *Proc. Natl. Acad. Sci. USA* **87,** 2304 (1990).
283. N. Takahashi, T. Hayano, and M. Suzuki, *Nature* **337,** 473 (1989); G. Fischer and co-workers, *Nature,* **337,** 476 (1989).
284. A. W. Thomson, *Immunol. Today* **10,** 6 (1989).
285. T. E. Starzl and co-workers, *Lancet* **1989ii,** 1000 (1989).
286. J. J. Siekierka and co-workers, *Nature* **341,** 755 (1989).
287. M. W. Harding, A. Galat, D. E. Uehling, and S. L. Schreiber, *Nature* **341,** 758 (1989).
288. R. K. Harrison and R. L. Stein, *Biochemistry* **29,** 3813 (1990).
289. M. K. Rosen, R. F. Standaert, A. Galat, M. Nakatsuka, and S. L. Schreiber, *Science* **248,** 863 (1990).
290. H. Umezawa and co-workers, *J. Antibiot.* **29,** 97 (1976); H. Umezawa, *Drug Exptl. Clin. Res.* **10,** 519 (1984).
291. H.-U. Schorlemmer and co-workers, *Behring Inst. Mitt.* **74,** 157 (1984).
292. K. Ota and N. Ogawa, *Biomed. Pharmacother.* **44,** 93 (1990).
293. J. S. Ricci, Jr., A. Bousvaros, and A. Taylor, *J. Org. Chem.* **47,** 3063 (1982).
294. R. Nishizawa and co-workers, *J. Med. Chem.* **20,** 510 (1977).
295. Y. Mine and co-workers, *J. Antibiot.* **36,** 1045 (1983).
296. H. Umezawa, *Rev. Infect. Dis.* **6,** 412 (1984).
297. E. Bayer and co-workers, *Helv. Chim. Acta* **55,** 224 (1972).
298. Y. Kondo and co-workers, *Meiji Seika Kenkyusho Nempo* **13,** 34 (1973).
299. R. E. Hoagland, in H. G. Cutler, ed., *Biologically Active Natural Products-Potential Use in Agriculture,* American Chemical Society, Washington, D.C., 1988, p. 182.
300. T. Hidaka and co-workers, *J. Antibiot.* **42,** 491 (1989).
301. K. L. Carlson and O. C. Burnside, *Weed Sci.* **32,** 841 (1984).
302. R. R. Bellinder, K. K. Hatzios, and H. P. Wilson, *Weed Sci.* **33,** 779 (1985).
303. S. M. Ridley and S. F. McNally, *Plant Sci.* **39,** 31 (1985).
304. U.S. Pat. 4,168,963 (Sept. 25, 1979), W. Rupp and co-workers (to Hoechst).
305. L. Maier and P. J. Lea, *Phosphorus Sulfur* **17,** 1 (1983).

306. *KYM International Directory of Pharmaceutical Companies 1988/1989,* KYM Consultancy Ltd., Kent, UK; *Corporate Technology Directory, U.S. Edition,* Corporate Technology Information Services, Wellesley Hills, Mass., 1991.

EDMUND M. WISE, JR.
The Wellcome Research Laboratories

ANTIBIOTICS, POLYENES.

See ANTIPARASITIC AGENTS–ANTIMYCOTICS.

POLYETHERS

The polyether antibiotics were first recognized as a separate class with the publication of the structure of monensin in 1967 (1). Several members of the group have since found commercial application as anticoccidials in poultry farming and in improvement of feed efficiency for ruminants (see FEEDS AND FEED ADDITIVES).

These antibiotics are characterized by multiple tetrahydrofuran and tetrahydropyran rings connected by aliphatic bridges, direct C–C linkage, or spiro linkage. Other features include a free carboxyl function, many lower alkyl groups, and a variety of functional oxygen groups. These structural features enable the molecule to form a cyclic conformation with the oxygen functions at the center and the alkyl groups on the outer surface. This conformation results in lipid solubility, even for the salt forms, enabling transport of cations across lipid membranes. The potential for using these antibiotics as biochemical tools was recognized long before the structural features had been elucidated. Studies of antibiotics as inhibitors of mitochondrial enzymes (2) revealed the unique characteristics of nigericin and dianemycin, both later shown to be polyethers. The term ionophore was first used to describe compounds that can transport ions across artificial or natural membranes (3). The polyethers are sometimes referred to as carboxylic acid ionophores to distinguish these antibiotics from other compounds showing ionophoretic activity.

Individual polyethers exhibit varying specificities for cations. Some polyethers have found application as components in ion-selective electrodes for use in clinical medicine or in laboratory studies involving transport studies or measurement of transmembrane electrical potential (4). The methyl ester of monensin [*28636-21-7*] has been incorporated into a membrane slide assembly used for the assay of serum sodium (see BIOSENSORS) (5). Studies directed toward the design of a lithium selective electrode resulted in the synthesis of a derivative of monensin lactone that is highly specific for lithium (6).

Properties

A classification based first on ion specificity, then on structural features has been suggested for the polyethers (7). Another method uses the presence of unsaturation or of aromatic groups in the molecular skeleton (8). In this review the compounds are classified based on the number of carbons in the backbone according to the numbering system proposed in reference 9. The carbon backbone or skeleton refers to the longest chain of contiguous carbons between the carboxyl group and the terminal carbon.

Table 1 lists the polyether antibiotics arranged by the number of carbons in the skeleton. Many of these compounds were isolated independently in separate laboratories and thus have more than one designation. The groups are subdivided depending on the number of spiroketals. Two classes fall outside this scheme: the pyrrole ether type containing a heterocyclic ring, and the acyltetronic acid type, that has an acylidene tetronic acid instead of a carboxylic acid. These compounds are ionophores and because of their common features are included as polyethers.

Lysocellin (**1**, R = COOH, R′ = CH_2CH_3), X14889 A (**1**, R = R′ = CH_3), and X14889 C (**1**, R = COOH, R′ = CH_3) are examples of the shortest polyethers, those having 22 carbons in the backbone. X14889 A, a minor factor, is atypical in that it does not contain a carboxylic acid. The 24 carbon backbone group includes the commercially important lasalocid A (**2**), one of the first polyethers to be isolated. Monensin A (**3**, R = CH_3, R′ = CH_2CH_3, R″ = H), another widely used anticoccidial and feed efficiency enhancer, has one spiroketal and a 26 C backbone. Other closely related members of this subgroup are monensin B (**3**, R = R′ = CH_3, R″ = H) and laidlomycin (**3**, R = $COCH_2CH_3$, R′ = CH_3, R″ = H). The propionyl ester of laidlomycin [*84799-02-0*] (**3**, R = R″ = $COCH_2CH_3$, R′ = CH_3), $C_{40}H_{66}O_{13}K$, is currently at the preregistration stage of development for use as a feed additive.

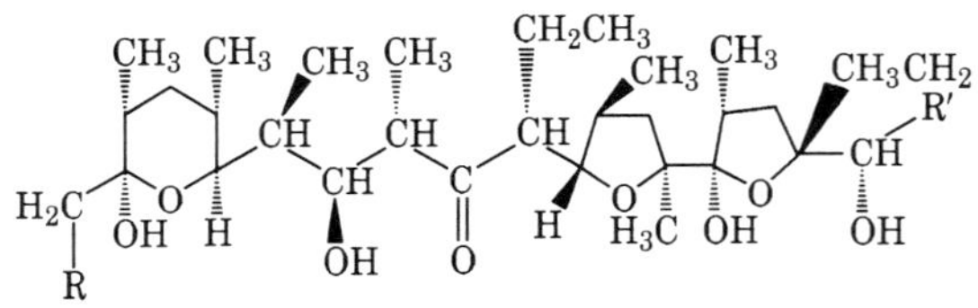

(**1**)

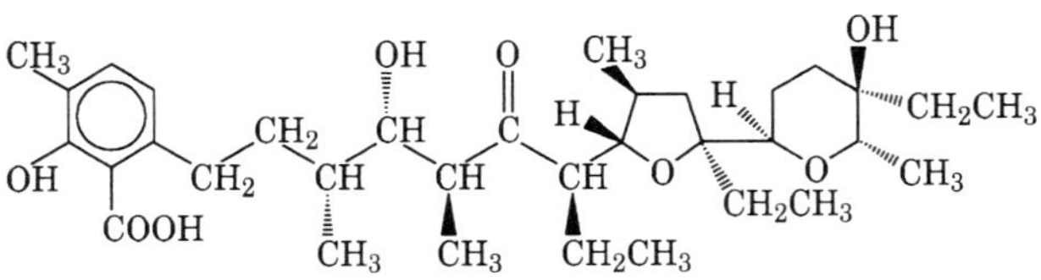

(**2**)

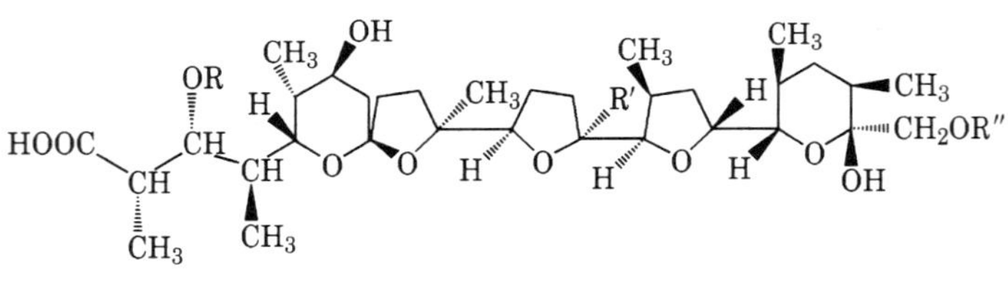

(**3**)

The C-30 skeleton group accounts for about 60% of the polyethers for which structures have been determined. Most of these contain a sugar moiety, usually 2,3,6-trideoxy-4-*O*-methyl-D-erythro pyranose [*65104-53-2*]. The single spiroketal subset is illustrated by two closely related compounds, semduramicin (**4**, R = H, R′ = sug 1) and maduramicin alpha (**4**, R = CH_3, R′ = sug 2). Maduramicin is marketed as an anticoccidial and semduramicin is under development for the same application.

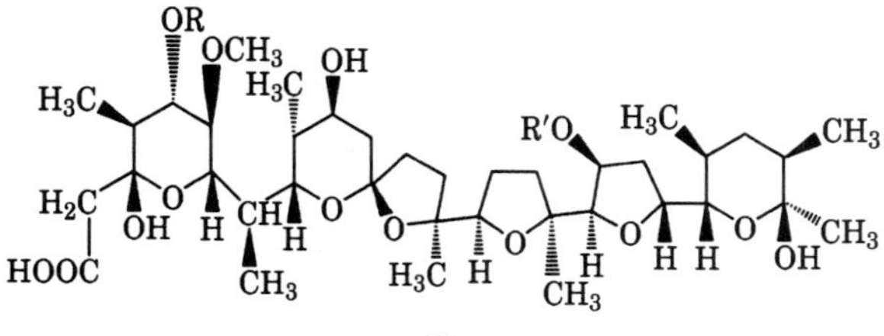

(**4**)

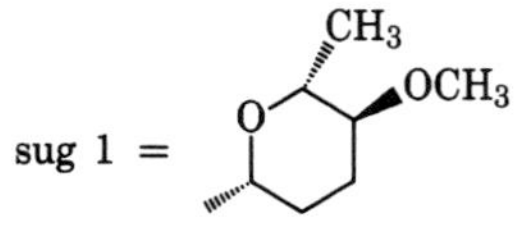

(**5**)

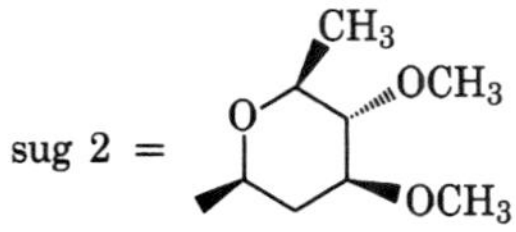

(**6**)

The group containing two spiroketals is also characterized by an alpha–beta unsaturated ketone. Representative structures are illustrated by CP53607 (**7**, R = CH_3, R′ = OH, R″ = R‴ = R⁗ = H), which has no sugar, dianemycin (**7**, R = CH_3, R′ = OH, R″ = R⁗ = H, R‴ = sug), which has one sugar, and A130 B (**7**, R = R‴ = H, R′ = R⁗ = sug, R″ = CH_3), which has two sugars.

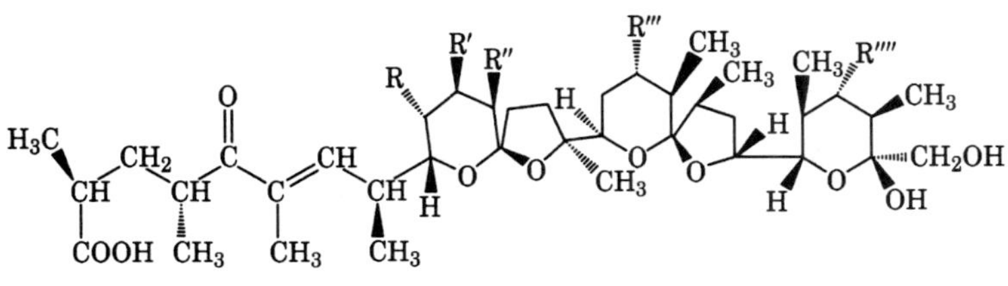

(**7**)

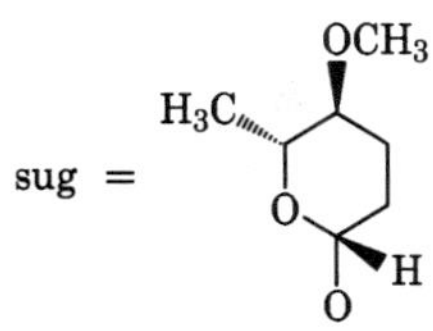

(**8**)

Two commercially important polyethers, narasin (**9**, R = CH_3) and salinomycin (**9**, R = H) have an unusual trispiroketal linkage.

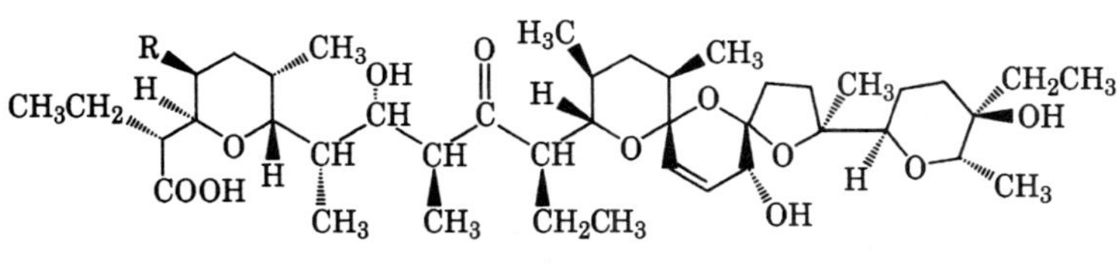

(**9**)

Table 1. Polyether Antibiotics

Name[a]	CAS Registry Number	Molecular formula	Producing organism	Mp, °C	Reference I[b]	Reference S[c]
22 to 23 Carbon backbone (22,23 C), 0-Spiroketal						
ferensimycin A (5057 A)	[*83852-59-9*]	$C_{34}H_{60}O_{10}$	*Streptomyces* sp. 5057 FERM BP-62	133–135[d]	10	10
ferensimycin B (5057 B)	[*83852-60-2*]	$C_{35}H_{62}O_{10}$	*Streptomyces* sp. 5057 FERM BP-62	143–145[d]	10	10
lysocellin (X14537 A)	[*55898-33-4*]	$C_{34}H_{60}O_{10}$	*Streptomyces cacaoi* var. *asoensis, Streptomyces longwoodenis* ATCC 29251)	158–160[d]	11,12	13
X14873 A	[*88263-37-0*]	$C_{35}H_{62}O_{11}$	*Streptomyces* sp. X 14873 ATCC 31679	154[d]	14	14
X14873 G	[*88263-35-8*]	$C_{34}H_{60}O_{8}$	*Streptomyces* sp. X 14873 ATCC 31679	152–153	14	14
X14873 H	[*88263-36-9*]	$C_{34}H_{62}O_{9}$	*Streptomyces* sp. X 14873 ATCC 31679	145–146	14	14
X14889 A	[*97671-96-0*]	$C_{33}H_{60}O_{8}$	*Streptomyces* sp. X 14889 NRRL 15517	149–150	15	15
X14889 C	[*97671-95-9*]	$C_{33}H_{58}O_{10}$	*Streptomyces* sp. X 14889 NRRL 15517	139–141[d]	15	15
X14889 D	[*97671-94-8*]	$C_{33}H_{58}O_{7}$	*Streptomyces* sp. X 14889 NRRL 15571	117	15	15
24 C, 0-Spiroketal						
inostamycin	[*129905-10-8*]	$C_{38}H_{68}O_{11}$	*Streptomyces* sp. MH816-AF15	181–181[d]	16	16
iso-lasalocid A	[*54156-67-1*]	$C_{34}H_{54}O_{8}$	*Streptomyces lasaliensis* NRRL 3382	183–185[d]	17	17
lasalocid A (X537 A)	[*25999-31-9*]	$C_{34}H_{54}O_{8}$	*Streptomyces lasaliensis* NRRL 3382	100–109	18	19

lasalocid B	[*55051-86-0*]	$C_{35}H_{56}O_8$	*Streptomyces lasaliensis* NRRL 3382	86–87	20	20
lasalocid C	[*55051-84-8*]	$C_{35}H_{56}O_8$	*Streptomyces lasaliensis* NRRL 3382	97–100	20	20
lasalocid D	[*55051-82-6*]	$C_{35}H_{56}O_8$	*Streptomyces lasaliensis* NRRL 3382	102–104	20	20
lasalocid E	[*55051-80-4*]	$C_{35}H_{56}O_8$	*Streptomyces lasaliensis* NRRL 3382	90	20	20
24 C, 1-Spiroketal						
CP54883	[*112396-63-1*]	$C_{41}H_{61}Cl_2O_{12}$	*Actinomadura routienii* Huang N365-41	330–340[d]	21	22
25 C, 0-Spiroketal						
CP78545	[*124150-86-3*]	$C_{33}H_{58}O_7$	*Streptomyces* sp. N731-45	94–98	23	23
zincophorin (Griseochelin, M144255)	[*91920-88-6*]	$C_{33}H_{60}O_7$	*Streptomyces griseus* NCIB 11504, ZIMET 43681	66–70	24,25	24,26
26 C, 1-Spiroketal						
CP47433	[*74758-62-6*]	$C_{47}H_{82}O_{14}$	*Actinomadura macra* ATCC 31286	89–99	27	28
CP47434	[*74758-61-5*]	$C_{46}H_{80}O_{14}$	*Actinomadura macra* ATCC 31286	230–238[d]	27	28
CP80585	[*128345-23-3*]	$C_{37}H_{64}O_{11}$	*Streptomyces* sp. ATCC 53862		29	29
CP82996	[*127288-12-4*]	$C_{50}H_{86}O_{16}$	*Actinomadura* sp. ATCC 53764	115–117	30	30
deoxylaidlomycin (A712)	[*102674-88-4*]	$C_{37}H_{62}O_{11}$	*Streptoverticillium eurocidicum* SANK 61484	194–196[d]	31	31
			Streptoverticillium olivoreticuli IMET 43861		32	

Table 1. (Continued)

Name[a]	CAS Registry Number	Molecular formula	Producing organism	Mp, °C	Reference	
					I[b]	S[c]
laidlomycin (AB78, TS822)	*[56283-74-0]*	$C_{37}H_{62}O_{12}$	*Streptoverticillium eurocidium* ATCC 31175, *Streptomyces olivoreticuli* A14939	151–153	33,34	34,35
monensin A	*[17090-79-8]*	$C_{36}H_{62}O_{11}$	*Streptomyces cinnamonensis* ATCC 15413	103–105	36	1
monensin B	*[30485-16-6]*	$C_{35}H_{60}O_{11}$	*Streptomyces cinnamonensis* ATCC 15413	227–228[d]	36	37
monensin C	*[31980-87-7]*	$C_{37}H_{64}O_{11}$	*Streptomyces cinnamonensis* ATCC 15413	212–214[d]	36	37
27 C, 1-Spiroketal						
6270 B	*[122548-05-4]*	$C_{44}H_{74}O_{13}$	*Nocardiopsis* sp. FERM BP-717		38	38
A82810 (CP84657)	*[127759-24-4]*	$C_{45}H_{78}O_{14}$	*Actinomadura fibrosa* NRRL 18348, *Actinomadura* sp. ATCC 53708	255–257[d]	39,40	39,40
cationomycin	*[82987-42-6]*	$C_{45}H_{70}O_{15}$	*Actinomadura azurea* FERM BP-83	108–112	41	41
kijimicin	*[129297-22-9]*	$C_{37}H_{64}O_{11}$	*Actinomadura* sp. MI215-NF3	217–218[d]	42	42
portmicin (A80190)	[103521-25-1]	$C_{44}H_{76}O_{14}$	*Nocardiopsis* sp. 6270 FERM BP-717, *Actinomadura oligospora* NRRL 15878	115–118	43,44	44,45
30 C, 0-Spiroketal						
UK41637	*[83532-94-9]*[d]	$C_{46}H_{78}O_{16}$	*Actinomadura cremea* ATCC 31676	196–198[d]	46	46

30 C, 1-Spiroketal						
6016	[*69421-39-2*]	$C_{46}H_{78}O_{16}$	*Streptomyces albus* 6016	192–195[d]	47	48
A204 A	[*43110-10-7*]	$C_{49}H_{84}O_{17}$	*Streptomyces albus* NRRL 3384	96–98	49	50
A204 B	[*65208-37-9*][d]		*Streptomyces albus* NRRL 3384		49	[e]
A28695 B	[*42617-35-6*]	$C_{48}H_{82}O_{17}$	*Streptomyces albus* NRRL 3883	122–124	51	52
A80438 (demethoxylonomycin A, SF2437)	[*108044-86-6*]	$C_{43}H_{74}O_{13}$	*Streptomyces pactum* NRRL 15970, *Streptomyces bobili* NRRL 15971, *Streptomyces* sp. FERM P-8469	160–162[d]	53,54	53,55
abierixin	[*100634-16-0*]	$C_{40}H_{68}O_{11}$	*Streptomyces albus* NRRL B-1865	83–85	56	56
BL580 beta	[*53414-72-5*]	$C_{47}H_{80}O_{16}$	*Streptomyces hygroscopicus* NRRL 5647		57	58
BL580 delta	[*66389-75-1*]	$C_{47}H_{80}O_{16}$	*Streptomyces hygroscopicus* NRRL 8180	157–161[d]	59	60
carriomycin (T42082)	[*65978-43-0*]	$C_{47}H_{80}O_{15}$	*Streptomyces hygroscopicus* ATCC 31080	120–122	61	62
CP70228	[*101621-31-2*]	$C_{53}H_{90}O_{18}$	*Actinomadura roseorufa* ATCC 39697	113–123	63	63
CP70828	[*101621-30-1*]	$C_{54}H_{92}O_{18}$	*Actinomadura roseorufa* ATCC 39697	133–136	63	63
2-epi-mutalomycin	[*124918-41-8*]	$C_{41}H_{70}O_{12}$	*Streptomyces mutabilis* NRRL 8086	166–177[d]	64	64
28-epi-mutalomycin	[*124986-36-3*]	$C_{41}H_{70}O_{12}$	*Streptomyces mutabilis* NRRL 8086	187–191[d]	64	64
epi-nigericin	[*108266-17-7*]	$C_{40}H_{68}O_{11}$	*Streptomyces hygroscopicus* NRRL B-1865	188.5[d]	65	65
etheromycin (CP38295, C20-12, T50417)	[*59149-05-2*]	$C_{48}H_{82}O_{16}$	*Streptomyces* C20-12 FERM P-2736, *Streptomyces hygroscopicus* ATCC 31050	135–138	66,67	67,68

Table 1. (*Continued*)

Name[a]	CAS Registry Number	Molecular formula	Producing organism	Mp, °C	Reference I[b]	Reference S[c]
grisorixin (K358)	[*31357-58-1*]	$C_{40}H_{68}O_{11}$	*Streptomyces griseus*	75–80	69	70
K41 A (A32887)	[*53026-37-2*]	$C_{48}H_{82}O_{18}$	*Streptomyces hygroscopicus* FERM P-1342, *Streptomyces albus* NRRL 11109	196–198[d]	90,91	92,93
K41 B	[*72017-85-7*]	$C_{54}H_{92}O_{20}$	*Streptomyces hygroscopicus* FERM P-1342	185–186[d]	94	95
LL-C23201 delta	[*107020-14-4*]	$C_{47}H_{80}O_{15}$	*Streptomyces olivaceogriseus* NRRL 15357	164–168[d]	71	71
lonomycin A (DE3936, emericid, A218)	[*58785-63-0*]	$C_{44}H_{76}O_{14}$	*Streptomyces ribosidificus* ATCC 31051, *Streptomyces hygroscopicus* FERM P-3159, *Streptomyces hygroscopicus* DS 24367	188–189[d]	72–74	75–77
lonomycin B	[*68567-60-2*]	$C_{44}H_{26}O_{14}$	*Streptomyces ribosidificus* ATCC 31051	181–182	78	79
lonomycin C	[*68537-50-8*]	$C_{44}H_{74}O_{14}$	*Streptomyces ribosidificus* ATCC 31051	186–187[d]	78	79
maduramicin alpha (LL-C23024, prinicin, X14868 A)	[*79356-08-4*]	$C_{47}H_{80}O_{17}$	*Actinomadura yumaense* NRRL 12515, *Nocardia* sp. X 14868 ATCC 31585	193–194[d]	80,81	59,81
maduramicin beta (X14868 C)	[*79331-53-6*]	$C_{46}H_{78}O_{17}$	*Actinomadura yumaense* NRRL 12515, *Nocardia* sp. X 14868 ATCC 31585	172–175[d]	80,81	59,81
27-methoxyseptamycin	[*125131-53-5*]	$C_{49}H_{84}O_{17}$	*Actinomadura* sp. ATCC 53676		82	82
mutalomycin (S11743 A)	[*62618-08-0*]	$C_{41}H_{70}O_{12}$	*Streptomyces mutabilis* NRRL 8088	163–171[d]	83	84

nigericin (K178, X464, polyetherin A, duamycin)	*[28380-24-7]*	$C_{40}H_{68}O_{11}$	*Streptomyces violaceniger* NRRL B1356, *Streptomyces hygroscopicus* E-749, 325-15	170–172	85–87	88,89
semduramicin (UK61689)	*[113378-31-7]*	$C_{46}H_{76}O_{16}$	*Actinomadura roseorufa* ATCC 53666	167[d]	96	96
septamycin (A28695 A, BL580 alpha)	*[54927-63-8]*	$C_{48}H_{82}O_{16}$	*Streptomyces hygroscopicus* NRRL 5678; NRRL 5647, *Streptomyces albus* NRRL 3883	164–166[d]	51,57,97	52,98,99
TM582	*[98791-39-0]*	$C_{40}H_{68}O_{11}$	*Streptomyces hygroscopicus* TM 582	80–82	100	100
UK58852	*[101621-29-8]*	$C_{52}H_{88}O_{18}$	*Actinomadura roseorufa* ATCC 39697	123–126	63	63
W341 C	*[110368-36-0]*	$C_{47}H_{80}O_{16}$	*Streptomyces* W341 C		101	101
X14868 D	*[79331-54-7]*	$C_{47}H_{80}O_{17}$	*Nocardia* sp. X 14868 ATCC 31585	194–195	81	81
30 C, 2-Spiroketals						
A130 B	*[73492-07-6]*[d]	$C_{54}H_{90}O_{16}$	*Streptomyces hygroscopicus* ATCC 21840		102	102
A130 C	*[73522-76-6]*	$C_{47}H_{78}O_{13}$	*Streptomyces hygroscopicus* ATCC 21840		102	102
CP53607 (X14931 A)	*[84680-56-8]*	$C_{40}H_{66}O_{11}$	*Streptomyces halstedii* ATCC 31812, *Streptomyces* sp. X 14931	199–204[d]	103,104	103,104
CP80219	*[123286-64-6]*	$C_{47}H_{78}O_{14}$	*Streptomyces hygroscopicus* ATCC 53626	176–180[d]	105	106
CP82483	*[121962-58-1]*	$C_{47}H_{78}O_{14}$			107	107
dianemycin	*[38565-33-9]*	$C_{47}H_{78}O_{14}$	*Streptomyces hygroscopicus* NRRL 3444	72–74	108	109
eudusamycin (CP63517)	*[100242-41-9]*	$C_{47}H_{78}O_{14}$	*Streptomyces endus* subsp. *aureus* ATCC 39574	95–105	110	110
19-epidianemycin (CP60993)	*[93218-58-7]*	$C_{47}H_{78}O_{14}$	*Streptomyces hygroscopicus* ATCC 39205	193–205[d]	111	111

Table 1. *(Continued)*

Name[a]	CAS Registry Number	Molecular formula	Producing organism	Mp, °C	Reference	
					I[b]	S[c]
lenoremycin (A130 A, Ro 21-6150)	[*51257-84-2*]	$C_{47}H_{78}O_{13}$	*Streptomyces hygroscopicus* X14563, ATCC 21840	87–92	112,113	114,115
leuseramycin (TM531 A)	[*73537-10-7*]	$C_{47}H_{78}O_{13}$	*Streptomyces hygroscopicus* ATCC 31590	88–91	116	116
moyukamycin	[*96827-80-4*]	$C_{47}H_{76}O_{13}$	*Streptomyces hygroscopicus* FERM BP-274		43,85, 86,89, 117,118[d]	117
TM531 B	[*80118-77-0*]	$C_{46}H_{76}O_{14}$	*Streptomyces hygroscopicus* ATCC 31590	252–253[d]	119	120
TM531 C	[*80118-78-1*]	$C_{47}H_{78}O_{15}$	*Streptomyces hygroscopicus* ATCC 31590	190–191	119	120
TM531 D	[*87579-00-8*]	$C_{46}H_{76}O_{14}$	*Streptomyces hygroscopicus* ATCC 31590		121	121
TM531 E	[*87579-01-9*]	$C_{46}H_{76}O_{14}$	*Streptomyces hygroscopicus* ATCC 31590		121	121
X14934A (CP47224)	[*96998-05-9*]	$C_{48}H_{80}O_{15}$	*Streptomyces* sp. X14934 NRRL 15518, *Streptomyces hygroscopicus* ATCC 31337	156–158[d]	122,123	122
30 C, Trispiroketal						
deoxy-epi-narasin	[*70051-99-9*]	$C_{43}H_{72}O_{10}$	*Streptomyces aureofaciens* NRRL 11181		124	124
deoxy-epi-salinomycin	[*64129-77-1*]	$C_{42}H_{70}O_{10}$	*Streptomyces albus* ATCC 21838	180	125	125
deoxynarasin	[*70022-35-4*]	$C_{43}H_{72}O_{10}$	*Streptomyces aureofaciens* NRRL 11181		124	124
deoxysalinomycin	[*64003-50-5*]	$C_{42}H_{70}O_{10}$	*Streptomyces albus* ATCC 21838		125	125

narasin A (A28086)	*[55134-13-9]*	$C_{43}H_{72}O_{11}$	*Streptomyces aureofaciens* NRRL 5758	98–100	126	127
narasin B	*[58439-94-4]*	$C_{43}H_{70}O_{11}$	*Streptomyces aureofaciens* NRRL 5758	150–153	128	128
narasin D	*[58334-78-4]*	$C_{44}H_{74}O_{11}$	*Streptomyces aureofaciens* NRRL 5758		128	128
salinomycin	*[53003-10-4]*	$C_{42}H_{70}O_{11}$	*Streptomyces albus* ATCC 21838	112–113	129	130
SY4 (5-hydroxysalinomycin)	*[110947-74-5]*	$C_{42}H_{70}O_{12}$	*Streptomyces albus* FERM P-419		131	131
SY8 (18,19-dihydrosalinomycin)	*[111057-09-1]*	$C_{42}H_{72}O_{11}$	*Streptomyces albus* FERM P-419		131	131
SY9 (20-oxosalinomycin)	*[83564-04-9]*	$C_{42}H_{68}O_{11}$	*Streptomyces albus* FERM P-419		132	132
31 C, 1-Spiroketal						
hidamycin (CP51532)	*[84331-31-7]*	$C_{42}H_{72}O_{13}$	*Actinomadura* sp. MI429-38F1; *Actinomadura verrucosospora* ATCC 31466	196–199[d]	133,134	133,135
31 C, Trispiroketal						
chloronoboritomycin A (X14766 A)	*[75217-55-9]*	$C_{43}H_{63}ClO_{14}$	*Streptomyces malachitofuscus downeyi* ATCC 31547	160	136	137
noboritomycin A	*[68508-45-2]*	$C_{43}H_{64}O_{14}$	*Streptomyces noboritoensis* NRRL 8123	235–237[d]	118	84
noboritomycin B	*[68508-46-3]*	$C_{44}H_{66}O_{14}$	*Streptomyces noboritoensis* NRRL 8123	220–222[d]	118	84
32 C, 0-Spiroketal						
ionomycin	*[56092-81-0]*	$C_{41}H_{72}O_9$	*Streptomyces conglobatus* ATCC 31005	205–206[f]	138	139

Table 1. (*Continued*)

Name[a]	CAS Registry Number	Molecular formula	Producing organism	Mp, °C	Reference	
					I[b]	S[c]
32 C, Trispiroketal						
CP44161	[*66771-51-5*]	$C_{43}H_{66}O_{10}$	*Dactylosporangium salmoneum* ATCC 31222	155–157	140	141
34 C, 0-Spiroketal						
CP73064	[*118674-05-8*]	$C_{44}H_{76}O_{13}$	*Streptomyces* sp. ATCC 53523	214–216[d]	142	142
37 C, 0-Spiroketal						
alborixin (CP38986, S14750)	[*57760-36-8*]	$C_{48}H_{84}O_{14}$	*Streptomyces albus* 3840, *Streptomyces flaveolus* ATCC 31100	100–105	143,144	145
X206	[*36505-48-3*]	$C_{47}H_{82}O_{14}$	*Streptomyces* sp. X 206	133–145	18	146
Acyl tetronic acids						
A80577 (SF2487)	[*123484-11-7*]	$C_{42}H_{64}O_{12}$	*Actinomadura verrucospora* NRRL 18236, *Actinomadura* sp. FERM P-9063	250–252[d]	147,148	147,148
tetronasin (ICI139603, M139603)	[*75139-06-9*]	$C_{35}H_{54}O_8$	*Streptomyces longisporoflavus* NCIB 11426	176–178[d]	149	150
tetronomycin	[*82206-10-8*]	$C_{34}H_{50}O_8$	*Streptomyces* sp. NRRL 11266	107–110[d]	151	151

Pyrrole ethers						
A23187 (calcimycin)	[*52665-69-7*]	$C_{29}H_{37}N_3O_6$	*Streptomyces chartreusis* NRRL 3882	181–182	152	153
A83094 A (deethylindanomycin)	[*117615-33-5*]	$C_{29}H_{39}NO_4$	*Streptomyces setonii*		154	154
AC7230	[*108266-12-2*]	$C_{28}H_{34}N_2O_7$	*Dactylosporangium* sp. AC7230	>300[d]	155	155
cafamycin	[*112303-17-0*]	$C_{30}H_{41}NO_4$	*Streptomyces* sp.	152–153	156	156
CP61405	[*99623-84-4*]	$C_{26}H_{30}N_2O_7$	*Streptomyces routienii* ATCC 39446	334–335	157	157
homoindanomycin	[*129258-44-2*]	$C_{32}H_{45}NO_4$	*Streptomyces galbus* NCAIM (P) 001036		158	158
indanomycin (X14547 A)	[*66513-28-8*]	$C_{31}H_{43}NO_4$	*Streptomyces antibioticus* NRRL 8167	138–141	159	160
X14885 A	[*83917-57-1*]	$C_{27}H_{32}N_2O_4$	*Streptomyces* sp. X 14885 NRRL 12350	264–266[d]	161	161
Others						
BL580 zeta	[*69522-23-2*][d]		*Streptomyces hygroscopicus* NRRL 1108		60	[e]
UK44579	[*83589-30-4*][d]		*Actinomadura cremea* ATCC 31676		46	[e]

[a]Alternative names are given in parenthesis.
[b]References corresponding to isolations.
[c]References corresponding to structure determinations.
[d]Value given is for the sodium salt.
[e]Structure not yet determined.
[f]Value given is for the calcium salt.

Ionomycin (**10**), a 32 C, 0-spiroketal, has a profound specificity for Ca^{2+} and is an important biochemical tool. The beta-carbonyl moiety provides a second charged ligation point, enabling this compound to form 1:1 complexes with divalent cations.

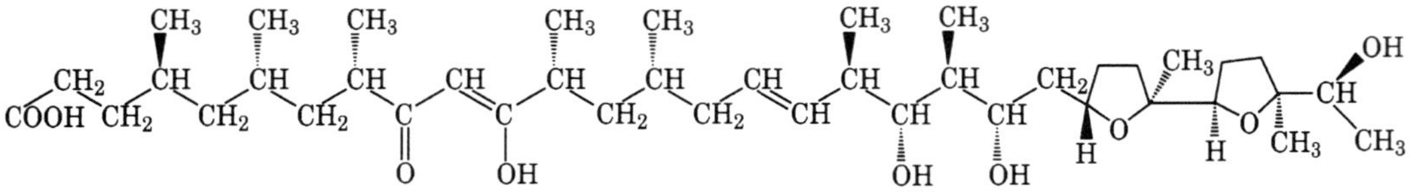

(**10**)

X206 (**11**, R = H) and alborixin (**11**, R = CH_3) are the only known representatives of the 37 C backbone subset.

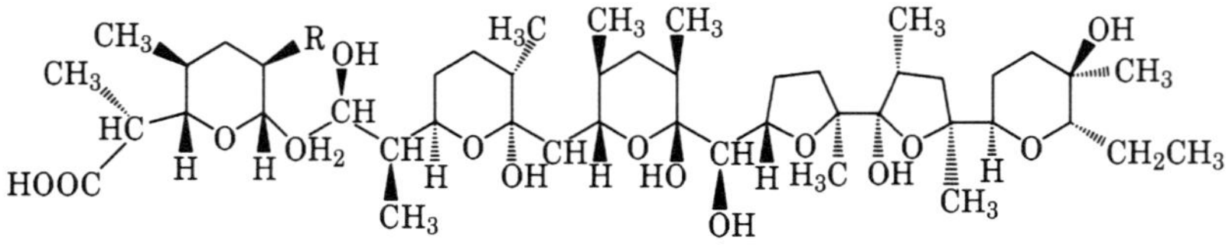

(**11**)

Tetronasin (**12**) is an example of the acyl tetronic acid class and A23187 (**13**, R = $NHCH_3$, R′ = R″ = CH_3) and related compounds, CP61405 (**13**, R = OH, R′ = R″ = H), X14885 A (**13**, R = OH, R′ = CH_3, R″ = H), and AC7230 (**13**, R = OH, R′ = R″ = CH_3) are examples of the pyrrole ether group.

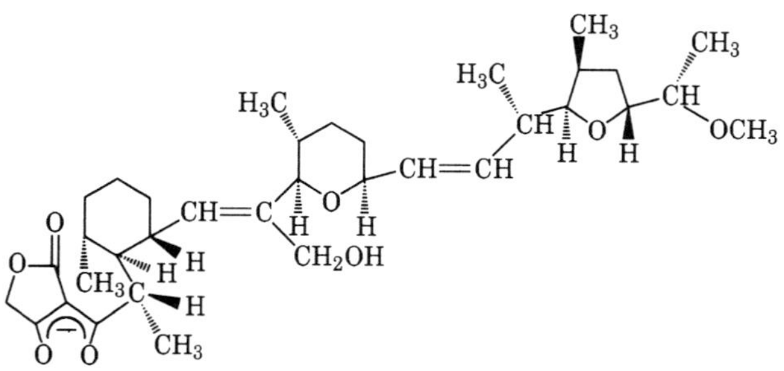

(**12**)

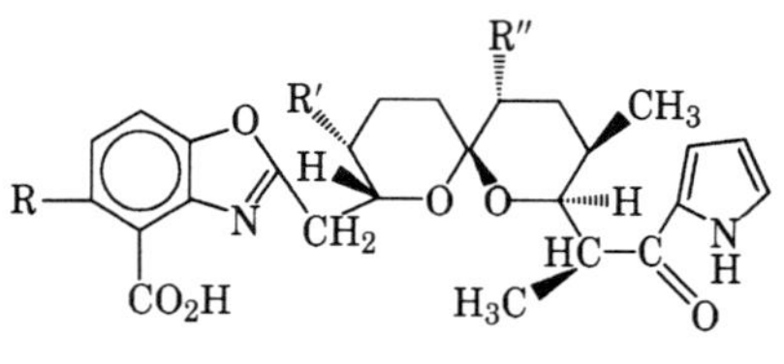

(**13**)

Purification and Production

Polyethers are usually found in both the filtrate and the mycelial fraction, but in high yielding fermentations they are mostly in the mycelium because of their low water-solubility (162). The high lipophilicity of both the free acid and the salt forms of the polyether antibiotics lends these compounds to efficient organic solvent extraction and chromatography (qv) on adsorbents such as silica gel and alumina. Many of the production procedures utilize the separation of the mycelium followed by extraction using solvents such as methanol or acetone. A number of the polyethers can be readily crystallized, either as the free acid or as the sodium or potassium salt, after only minimal purification.

Polyethers such as monensin, lasalocid, salinomycin, and narasin are sold in many countries in crystalline or highly purified forms for incorporation into feeds or sustained-release bolus devices (see CONTROLLED-RELEASE TECHNOLOGY). There are also mycelial or biomass products, especially in the United States. The mycelial products are generally prepared by separation of the mycelium and then drying by azeotropic evaporation, fluid-bed driers, continuous tray driers, flash driers, and other types of commercial driers (163). In countries allowing biomass products, crystalline polyethers may be added to increase the potency of the product.

Biological Activities

The polyether antibiotics exhibit a broad range of biological, antibacterial, antifungal, antiviral, anticoccidial, antiparasitic, and insecticidal activities. They improve feed efficiency and growth performance in ruminant and monogastric animals. Only the anticoccidial activity in poultry and cattle, and the effect on feed efficiency in ruminants such as cattle and sheep are of commercial interest.

Table 2 gives antimicrobial data for representative polyethers. In general, the polyethers are effective against gram-positive and anaerobic bacteria, but ineffective against gram-negatives (164). The antibacterial properties play an important, though poorly understood, role in the improvement of feed efficiency in ruminants (165). High toxicity has limited the use of these compounds as antibacterial and antifungal agents. The toxicity varies somewhat from species to species, and the commercial polyethers are well tolerated in poultry and cattle at the use level. Horses are subject to lethal toxicity at relatively low (1–2 mg/kg) dosages (166).

Treponema hyodysenteriae, a causative agent of swine dysentery, is sensitive to polyether antibiotics at low concentrations *in vitro.* In pigs, lasalocid was effective in controlling dysentery at levels of 0.005–0.05% in feeds (167). Several species of *Mycoplasma* are inhibited *in vitro* at a MIC range of 2.0–25 μ/mL of polyethers including narasin, carriomycin, and K41 (164).

Antiviral Activity. Tissue culture studies have demonstrated activity against a wide range of viruses including the veterinary pathogens, transmissible gastroenteritis virus, infectious canine heptatitis, Newcastle disease virus, infectious bovine rhinotracheitis, and pseudorabies (168). Monensin, narasin, septamycin, and nigericin were effective in treating an infection of transmissible gastroenteritis virus in piglets at doses of 0.1–100 mg/kg (168). The mechanism of inhibition has not been characterized, but it is probably related to the ionophoretic properties of these antibiotics. Monensin has been shown to inhibit the intracellular transport of viral membrane proteins of cells infected with Semliki Forest virus (169). The formation of syncytia, normally observed when T-lymphoblastoid cell line (CEM) cells are cocultivated with human immunodeficiency virus (HIV-1)-infected T-cell leukemia cell line (MOLT-3) cells, was significantly inhibited in the presence of monensin (170). This observation suggests that

Table 2. ***In Vitro*** **Antimicrobial Activity of the Polyether Antibiotics**

Polyether antibiotic	Minimum inhibitory concentration, μg/mL[a]								
	Staph. aureus ATCC No. 6538P	*Sarcina lutea* 9341	*Bacillus sp. E* 27859	*Bacillus subtilis* 558[b]	*Bacillus megaterium* 8011	*Bacillus sp. TA* 27860	*Mycobacterium phlei* 355	*Streptomyces cellulosae* 3313	*Paecilomyces varioti* 26820
monensin	3.1	12.5	0.4	1.6	12.5	1.6	12.5	6.3	[c]
nigericin	0.2	0.1	0.004	0.1	0.1	0.02	0.4	0.4	0.8
grisorixin	0.4	0.2	0.1	0.2	0.1	0.4	0.2	0.2	3.1
salinomycin	3.1	3.1	0.2	0.8	0.2	0.8	6.3	3.1	3.1
narasin	0.8	1.6	0.2	0.9	0.4	0.4	3.1	3.1	
lonomycin	1.6	12.5	1.6	3.1	0.8	1.6	12.5	6.3	50
X206	0.2	0.8	0.02	0.4	0.2	0.2	0.2	0.8	0.8
dianemycin	1.6	3.1	0.2	3.1	3.1	1.6	6.3	6.3	6.3
lenoremycin	0.2	0.2	0.02	0.4	0.1	0.4	0.2	1.6	0.8
septamycin	0.8	1.6	0.006	1.6	1.6	0.2	1.6	3.1	6.3
A204A	3.1	12.5	0.8	6.3	12.5	1.6	12.5	12.5	25
CP38952	12.5	6.3	1.6	6.3	3.1	3.1	3.1	12.5	12.5
lasalocid	1.6	3.1	0.2	1.6	3.1	1.6	12.5	6.3	
iso-lasalocid	12.6	6.3	1.6	1.6	1.6	3.1	6.3	25	12.5
lysocellin	0.8	0.4	0.03	0.2	0.4	0.4	3.1	1.6	1.6
A23187	0.2	0.04	0.2	0.1	0.8	0.04	1.6	6.3	
ionomycin		12.5	12.5	25	6.3				

[a]Lowest twofold dilution giving zone of inhibition in agar-well diffusion assay.
[b]NRRL collection number.
[c]Indicates no activity up to 50 μg/mL.

the viral glycoproteins in the treated cells were not transported to the cell surface from the Golgi membrane.

In Bulgaria, a preparation containing nigericin is used topically as an antiherpes drug (171).

Anticoccidial Activity. The 1968 report that claimed monensin has activity against *Eimeria* sp., particularly *E. tenella, E. maxima,* and *E. acervulina,* greatly altered the prevention and control of coccidiosis in poultry (172). It is estimated that the polyether ionophores presently constitute more than 80% of the total worldwide usage of anticoccidials (173). Lasalocid and monensin have been approved for use in control of coccidiosis in cattle.

To combat more tolerant, ie, resistant coccidia, particularly *E. acerulina* and *E. maxima,* a 1:1 combination of narasin and nicarbazin, eg, 0.45 kg premix containing 36 g of each compound, has been marketed as Maxiban by Elanco. These two compounds work synergistically against less susceptible field strains.

The ionophoretic properties of polyether antibiotics are important in their mode of action against coccidia (174). The primary coccidicidal effect of monensin involves an influx of sodium ions which results in swelling and vacuolization of the parasite because of osmotic pressure effects. A secondary effect is the stimulation of glycolysis on the part of the parasite resulting in a depletion of carbohydrate stores and eventual death. *Toxoplasma* sp., coccidia related to *Eimeria* sp., are infective in humans by two routes: fecal–oral ingestion of oocysts from the definitive host, the cat; and ingestion of tissue cysts from the intermediate host through eating undercooked meat such as lamb or pork. Monensin has been shown to be effective against toxoplasmosis in cats (175) and sheep (176). No human studies have been reported.

Enhancement of Feed Efficiency in Ruminants. Polyether antibiotics have been shown to improve feed efficiency in ruminant animals such as cattle, sheep, and goats and also to enhance weight gain. This enhancement in growth performance has been shown to correlate with increased production of propionic acid in the rumen. A decrease in the production of acetic acid, butyric acid, and methane often accompanies the increased propionic acid production. These volatile fatty acids (VFA) are produced in the rumen by the degradation of carbohydrates by microorganisms. The polyethers apparently selectively inhibit certain of the microorganisms to achieve greater propionate production. Propionate is thought to be more effectively utilized in energy metabolism in the host animal than acetate or butyrate. The effect on production efficiency and mode of action has been reviewed (177,178). (See also GROWTH REGULATORS).

Because feed comprises over 80% of the cost of producing and fattening cattle, the maximum utilization of ever increasingly expensive rations is of upmost importance (179). Monensin under the trade name of Rumensin (Elanco Products) was introduced in 1976 at a recommended level of 30 ppm in cattle feed. Lasalocid having the trade name Bovatec (Hoffmann-LaRoche, Inc.) was marketed some years later.

Salinomycin (180) and narasin (128) have been reported to be effective in improving feed efficiency but neither has been marketed for this use. Laidlomycin propionate (Syntex, Inc.) and tetronasin (Coopers Animal Health, Inc.) have been under investigation in cattle and sheep (181,182).

The polyethers have been reported to increase growth performance in monogastric animals such as horses, swine, and rabbits. No commercial use has been approved for swine diets. The sensitivity of horses to the toxic effects of these antibiotics precludes consideration of their use.

Biosynthesis

Bacteria belonging to the order Actinomycetales are the organisms reported to produce all of the polyethers. Most are secondary metabolites of *Streptomyces* sp. with the species *hygroscopicus* and *albus* accounting for about one-third of the antibiotics. Other genera represented are *Streptoverticillium, Dactylosporangium, Actinomadura, Nocardia,* and *Nocardiopsis*. The taxonomy of these producing organisms has been reviewed (183).

As is the case in other antibiotic families, the polyethers are often produced as complexes of closely related compounds. For example, lasalocid A is the principal component in a six-membered complex from *Streptomyces lasaliensis;* NRRL 3382 (20) and leuseramycin are co-produced also with four minor factors from *Streptomyces hygroscopicus* ATCC 31590 (119,121). The use of isotopically labeled substrate and subsequent analysis of the labeling patterns have established the polyketide route as the principal pathway for biosynthesis (184,185). The shortest polyether, lysocellin, is constructed from 11 short-chain fatty acid units, whereas the longest, X-206 and alborixin, are assembled from 18 subunits. Most, however, are assembled from 15 subunits giving the C-30 backbone. After cyclization of the ether rings the molecules can be modified by the addition of methyl groups or sugar moieties.

The Cane-Celmer-Westley unified stereochemical model is an elegant proposal relating the stereochemistry of thirty different polyethers (186). According to this hypothesis, the cyclic ether groups arise by a cascade of cyclization steps on a polyepoxide intermediate, which arises from a polyene precursor having double bonds in the (*E*)-configuration. However, there has been little experimental evidence to support this scheme. In studies on the biosynthesis of monensin, putative triene intermediates were synthesized but no incorporation of these compounds into monensin A or B were found (187). Other incorporation experiments using (*E*)- and (*Z*)-nerolidol suggest that the enzymes from *S. cinnamonensis* preferentially oxidize bonds of (*Z*)-configuration (188). As an extension of the work on oxidative cyclization in the biosynthesis of clavulanic acid (see ANTIBIOTICS, β-LACTAMS, β-LACTAMASE INHIBITORS), the possibility of an all (*Z*)-triene precursor for monensin has been proposed (189).

Structure Determination and Synthesis

Because of the complexity of the polyether antibiotics little progress has been made in structure determination by the chemical degradation route. X-ray methods were the techniques most successfully applied for the early structure elucidations. Monensin, X206, lasalocid, lysocellin, and salinomycin were included in nineteen distinct polyether x-ray analyses reported in 1983 (190). Use of mass spectrometry (191), and ^{1}H (192) and ^{13}C nmr (141) are also reviewed. More recently, innovative developments in these latter techniques have resulted in increased applications for structure determinations. For example, heteronuclear multiple bond connectivity (hmbc) and homonuclear Hartmann-Hahn spectroscopy were used to solve the structure of portimicin (**14**) (193). Fast atom bombardment mass spectrometry was used in solving the structures of maduramicin alpha and co-factors (58).

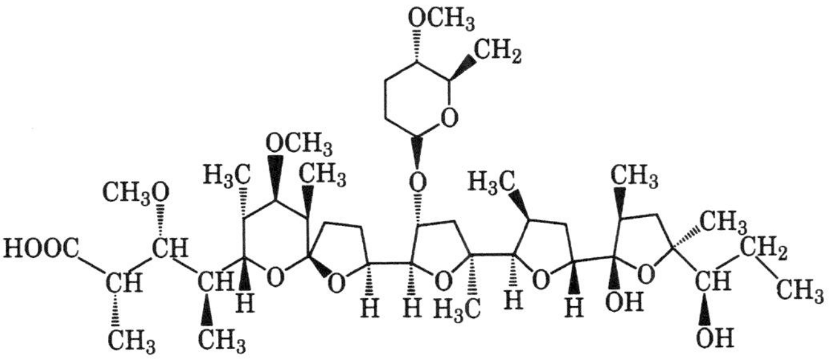

(**14**)

Total synthesis of polyether antibiotics presents a challenging task, requiring a high degree of stereo-, regio-, and chemoselectivity for each step. This subject was reviewed in 1983 (194), and since then many more compounds have been conquered. The asymmetric synthesis of a ferensimycin synthon (195) and total convergent syntheses of ionomycin (196) and X206 (197) have been reported. A method for the assembly of polypropionates with excellent diastereoselectivity has also been reported (198). Another total synthesis of ionomycin utilized a novel sulfur-assisted organocuprate displacement of a secondary tosylate with complete inversion of configuration (199). Other significant advances include the syntheses of monensin and lasalocid (200). A synthesis of zincophorin utilizes the hetero-Diels-Alder reaction in synthesis of both the polyol and pyran portions of the molecule (201,202). The stereospecific cyclization of a beta-hydroxy ketone into the spiroketal of A23187 under basic conditions has been carried out (203) and a total convergent synthesis of indanomycin uses the readily available levoglucosan as a starting material for the tetrahydropyranylacetic acid moiety (204).

Economic Aspects

The worldwide usage of polyether antibiotics for controlling coccidiosis is approximately \$190 million compared to a total market of \$210–220 million. Monensin and salinomycin represent about 65–70% of this market; lasalocid, narasin, and maduramicin make up the remainder. Other compounds for coccidiosis control include nicarbazine, halofunginone, amprolium, and robenidine. Worldwide usage is in excess of 3 million kg of product.

The total world market for the use of ionophores for feed efficiency improvement in ruminants is approximately \$80–\$90 million. The United States is the largest market. Lasalocid and monensin are the only members of this class cleared for use. Outside the United States, salinomycin is used in limited quantities. Worldwide usage is about 1.5 million kg.

The primary manufacturers are Hoffmann-LaRoche, Inc., Pfizer, Inc., Agri-Bio Corp. (a division of A. H. Robins), Elanco Products Co. (a division of Eli Lilly and Co.), Hoechst-Roussel Agri Vet Co., American Cyanamid Co., and Kaken Pharmaceutical Co.

Table 3 lists the polyether antiobitics used as poultry anticoccidial drugs in the United States. Recently, lasalocid and monensin have been approved for use in bovine coccidiosis at levels in feed of 11–33 g/t.

Table 3. Poultry Anticoccidial Agents used in the United States

Generic name	Trade name	Company	Year of first commercial use	Registered use level, g/t
maduramicin	Cygro	American Cyanamid Co.	1985	5.0–6.0
monensin	Coban	Elanco Products (division of Eli Lilly and Co.)	1971	99–120
lasalocid	Avatec	Hoffmann-LaRoche, Inc.	1977	75–125
narasin	Monteban	Elanco Products (division of Eli Lilly and Co.)	1981	60–79
salinomycin	Bio-Cox	Agri-Bio (division of A. H. Robins)	1982	44–66

BIBLIOGRAPHY

"Antibiotics, Polyethers" in *ECT* 3rd ed., Vol. 3, pp. 47–64, by J. W. Westley, Hoffmann-La Roche, Inc.

1. A. Agtarap, J. W. Chamberlin, M. Pinkerton, and L. Steinrauf, *J. Am. Chem. Soc.* **89,** 5737–5739 (1967).
2. H. A. Lardy, D. Johnson, and W. C. McMurray, *Arch. Biochem. Biophys.* **78,** 587–597 (1958).
3. B. C. Pressman, E. J. Harris, W. S. Jagger, and J. M. Johnson, *Proc. Natl. Acad. Sci. U.S.A.* **58,** 1949–1956 (1967).
4. D. Ammann, *Ion-Selective Microelectrodes: Principles, Design and Application,* Springer-Verlag, Berlin, 1986.
5. A. H. J. Maas, F. B. T. J. Boink, N. E. L. Saris, R. Sprokholt, and P. D. Wimberley, eds., *Methodology and Clinical Applications of Ion Selective Electrodes,* Vol. 7, IFCC Technical Secretariat, Copenhagen, 1986.
6. K. Tohda and co-workers, *Anal. Chem.* **62,** 936–942 (1990).
7. J. W. Westley, ed., *Polyether Antibiotics,* Vol. 1, Marcel Dekker, Inc., New York, 1982, pp. 1–20.
8. J. Berdy, *CRC Handbook of Antibiotic Compounds,* Vol. 5, CRC Press, Inc., Boca Raton, Fla., 1981, pp. 461–462.
9. J. W. Westley, *J. Antibiot.* **29,** 584–586 (1976).
10. Y. Kusakabe and co-workers, *J. Antibiot.* **35,** 1119–1129 (1982).
11. E. Ebata, H. Kasahara, K. Sekine, and Y. Inoue, *J. Antibiot.* **28,** 118–121 (1975).
12. B. L. T. Prosser and N. J. Palleroni, *Int. J. Syst. Bacteriol.* **26,** 319–322 (1976).

13. N. Otake, M. Koenuma, H. Kinashi, S. Sato, and Y. Saito, *J. Chem. Soc., Chem. Commun.,* 92–93 (1975).
14. J. W. Westley and co-workers, *J. Antibiot.* **39,** 1704–1711 (1986).
15. U.S. Pat. 4,537,956 (Aug. 27, 1985), C.-M. Liu and J. W. Westley (to Hoffmann-La Roche, Inc.).
16. M. Imoto and co-workers, *J. Nat. Prod.* **53,** 825–829 (1990).
17. U.S. Pat. 3,944,573 (Mar. 16, 1976), J. W. Westley (to Hoffmann-La Roche, Inc.).
18. J. Berger, A. I. Rachlin, W. E. Scott, L. H. Sternbach, and M. W. Goldberg, *J. Am. Chem. Soc.* **73,** 5295–5298 (1951).
19. S. M. Johnson, J. Herrin, S.-J. Liu, and I. C. Paul, *J. Am. Chem. Soc.* **92,** 4428–4434 (1970).
20. J. W. Westley and co-workers, *J. Antibiot.* **27,** 744–753 (1974).
21. W. P. Cullen and co-workers, *J. Antibiot.* **40,** 1490–1495 (1987).
22. J. Bordner, P. C. Watts, and E. B. Whipple, in Ref. 21, pp. 1496–1505.
23. J. P. Dirlam and co-workers, *J. Antibiot.* **42,** 1213–1220 (1989).
24. H. A. Brooks, D. Gardner, and J. P. Poyser, *J. Antibiot.* **37,** 1501–1504 (1984).
25. U. Grafe and co-workers in Ref. 24, pp. 836–846.
26. L. Radics, *J. Chem. Soc., Chem. Commun.,* 599–601 (1984).
27. U.S. Pat. 4,148,882 (Apr. 10, 1979), W. D. Celmer and co-workers (to Pfizer, Inc.).
28. J. Tone and co-workers, *Curr. Chemother. Infect. Dis., Proc. Int. Congr. Chemother., 11th* **1,** 469–470 (1980).
29. U.S. Pat. 4,920,050 (Apr. 24, 1990), W. P. Cullen, J. R. Hauske, G. J. Kostek, H. Maeda, and J. Tone (to Pfizer, Inc.).
30. J. P. Dirlam and co-workers, *J. Ind. Microb.* **6,** 135–142 (1990).
31. Y. Fukuzawa and co-workers, *Sankyo Kenyusho Nempo* **37,** 57–63 (1985).
32. U. Grafe, R. Schlegel, C. Stengel, W. Ihn, and L. Radics, *J. Basic Microbiol.* **29,** 149–155 (1989).
33. F. Kitame and co-workers in Ref. 20, pp. 884–888.
34. Jpn. Pat. Appl. Kokai 76-9790 (Jan. 26, 1976), A. Tamura, R. Furuta, and H. Kotani (to Dainippon Pharmaceutical Co., Ltd.).
35. F. Kitame and N. Ishida in Ref. 9, pp. 759–761.
36. M. E. Haney, Jr. and M. M. Hoehn, *Antimicrob. Agents Chemother.* **1967,** 349–352 (1968).
37. M. Gorman, J. W. Chamberlin, and R. L. Hamill, *Antimicrob. Agents Chemother.* **1967,** 363–368 (1968).
38. Eur. Pat. Appl. 293,787 (Dec. 7, 1988), Y. Kusakabe, H. Suzuki, and H. Kudo (to Kaken Pharmaceutical Co., Ltd.).
39. R. L. Hamill and co-workers, *29th Interscience Conference on Antimicrobial Agents and Chemotherapy* Houston, Tex., 1989, Abstr. 423.
40. J. P. Dirlam and co-workers, *J. Antibiot.* **43,** 668–679 (1990).
41. G. Nakamura, K. Kobayashi, T. Sakurai, and K. Isono, *J. Antibiot.* **34,** 1513–1514 (1981).
42. Y. Takahashi and co-workers in Ref. 40, pp. 441–443.
43. Y. Kusakabe, N. Takahashi, Y. Iwagawa, and A. Seino, in Ref. 21, pp. 237–238.
44. U.S. Pat. 4,582,822 (Apr. 15, 1986), R. L. Hamill and R. C. Yao (to Eli Lilly and Co.).
45. H. Seto and co-workers, *Tetrahedron Lett.* **28,** 3357–3360 (1987).
46. Brit. Pat. Appl. 2,090,243 (July 7, 1982), A. B. Penrose, J. C. Ruddock, R. Shibakawa, and J. Tone (to Pfizer, Inc.).
47. Y. Kusakabe, S. Mitsuoka, Y. Omuro, and A. Seino, *J. Antibiot.* **33,** 1437–1442 (1980).
48. H. Seto and co-workers, *J. Antibiot.* **32,** 244–246 (1979).
49. U.S. Pat. 3,705,238 (Dec. 5, 1972), R. L. Hamill and M. M. Hoehn (to Eli Lilly and Co.).

50. N. D. Jones, M. O. Chaney, J. W. Chamberlin, R. L. Hamill, and S. Chen, *J. Am. Chem. Soc.* **95**, 3399–3400 (1973).
51. U.S. Pat. 3,839,558 (Dec. 23, 1971), R. L. Hamill and M. M. Hoehn (to Eli Lilly and Co.).
52. D. E. Dorman and co-workers in Ref. 47, pp. 252–255.
53. U.S. Pat. 4,824,863 (Apr. 4, 1989), R. L. Hamill, R. Yao, and W. N. Nakatsukasa (to Eli Lilly and Co.).
54. Jpn. Pat. Appl. Kokai 60-130,394 (July 11, 1985), T. Sasaki (to Meiji Seika Kaisha, Ltd.).
55. Meiji Seika Kaisha, Mar. 31, 1988, personal communication.
56. L. David, H. L. Ayala, and J.-C. Tabet, *J. Antibiot.* **38**, 1655–1663 (1985).
57. U.S. Pat. 3,812,249 (May 21, 1974), J. H. E. J. Martin and S. Kantor (to American Cyanamid Co.).
58. M. M. Siegel, W. J. McGahren, K. B. Tomer, and T. T. Chang, *Biomed. Environ. Mass Spectrom.* **14**, 29–38 (1987).
59. U.S. Pat. 4,138,481 (Feb. 6, 1979), J. H. E. J. Martin and M. R. Hertz (to American Cyanamid Co.).
60. U.S. Pat. 4,132,779 (Jan. 2, 1979), M. R. Hertz and J. H. Martin (to American Cyanamid Co.).
61. A. Imada and co-workers, *J. Antibiot.* **31**, 7–14 (1978).
62. H. Nakayama, N. Otake, H. Miyamae, S. Sato, and Y. Saito, *J. Chem. Soc., Perkin Trans. 2*, 293–295 (1979).
63. U.S. Pat. 4,746,650 (May 24, 1988), W. P. Cullen, H. Maeda, J. C. Ruddock, and J. Tone (to Pfizer, Ltd.).
64. T. Fehr and co-workers in Ref. 23, pp. 897–902.
65. R. Barrada, G. Dauphin, and L. David, *J. Org. Chem.* **52**, 2388–2391 (1987).
66. Jpn. Pat. Appl. 76-91396 (Aug. 10, 1976), A. Seino, C. Nagatsu, H. Akutsu, and Y. Kusabe (to Kaken Chemical Co., Ltd.).
67. U.S. Pat. 4,129,578 (Dec. 12, 1978), W. D. Celmer and co-workers (to Pfizer, Inc.).
68. M. J. O. Anteunis and N. A. Rodius, *Bull. Soc. Chim. Belg.* **87**, 753–764 (1978).
69. P. Gachon, A. Kergomard, T. Staron, and C. Esteve in Ref. 11, pp. 345–350.
70. P. Gachon, A. Kergomard, H. Veschambre, C. Esteve, and T. Staron, *J. Chem. Soc., Chem. Commun.*, 1421–1423 (1970).
71. U.S. Pat. 4,628,046 (Dec. 9, 1986), D. P. Labeda, J. J. Goodman, and J. H. Martin (to American Cyanamid Co.).
72. S. Omura, S. Michinori, S. Machida, and J. Sawada in Ref. 9, pp. 15–20.
73. M. Oshoma and co-workers in Ref. 9, pp. 354–365.
74. L. Ninet and co-workers, *Experiencia* **32**, 319–321 (1976).
75. N. Otake and M. Koenuma, *Tetrahedron Lett.*, 4147–4150 (1975).
76. K. Yamazaki, K. Abe, and M. Sano in Ref. 9, pp. 9–12.
77. C. Riche and C. Pascard-Billy, *J. Chem. Soc., Chem. Commun.*, 951–952 (1975).
78. T. Mizutani and co-workers in Ref. 47, pp. 1224–1230.
79. H. Seto and co-workers in Ref. 61, pp. 929–932.
80. Belg. Pat. 894,762 (Apr. 21, 1983), D. P. Labeda and co-workers (to American Cyanamid Co.).
81. C.-M. Liu and co-workers, *J. Antibiot.* **36**, 343–350 (1983).
82. Eur. Pat. Appl. 317,231 (May 24, 1989), J. P. Dirlam, J. Tone, W. P. Cullen, and H. Maeda (to Pfizer, Inc.).
83. T. Fehr, H. D. King, and M. Kuhn, *J. Antibiot.* **30**, 903–907 (1977).
84. T. Fehr and co-workers in Ref. 48, pp. 535–536.
85. R. L. Harned, P. H. Hidy, C. J. Corum, and K. L. Jones, *Antibiot. Chemother.* **1**, 591–596 (1951).

86. J. Shoji and co-workers, *J. Antibiot.* **21,** 402–409 (1968).
87. Jpn. Pat. 70-26719 (Sept. 2, 1970), T. Misato and co-workers (to Kaken Chemical Co., Ltd.).
88. L. K. Steinrauf, M. Pinkerton, and J. W. Chamberlin, *Biochem. Biophys. Res. Commun.* **33,** 29–31 (1968).
89. T. Kubota, S. Matsutani, M. Shiro, and H. Koyama, *Chem. Commun.,* 1541–1543 (1968).
90. N. Tsuji and co-workers in Ref. 9, pp. 10–14.
91. U.S. Pat. 4,133,876 (Jan. 9, 1979), R. L. Hamill and M. M. Hoehn (to Eli Lilly and Co.).
92. M. Shiro, H. Nakai, K. Nagashima, and N. Tsuji, *J. Chem. Soc., Chem. Commun.,* 682–683 (1978).
93. J. L. Occolowitz, D. E. Dorman, and R. L. Hamill, *J. Chem. Soc., Chem. Commun.,* 683–684 (1978).
94. Jpn. Pat. Appl. Kokai 80-38352 (Mar. 17, 1980), N. Tsuji (to Shionogi and Co., Ltd.).
95. N. Tsuji, N. Nagashima, Y. Terui, and K. Tori in Ref. 48, pp. 169–172.
96. U.S. Pat. 4,912,130 (Mar. 27, 1990), A. C. Goudie and N. D. A. Walshe (to Pfizer, Inc.).
97. C. Keller-Juslen, H. D. King, Z. L. Kis, and A. von Wartburg, *J. Antibiot.* **27,** 854–859 (1975).
98. M. P. Kunstmann, 1976, personal communication.
99. T. J. Petcher and H.-P. Weber, *J. Chem. Soc., Chem. Commun.,* 697–698 (1974).
100. Jpn. Pat. Appl. Kokai 85-105679 (June 11, 1985), K. Mizogami, M. Yamagishi, S. Omura, N. Otake, and H. Seto (to Taisho Pharmaceutical Co., Ltd.).
101. R. S. Wehbie, C. Runsheng, and H. A. Lardy, *J. Antibiot.* **50,** 887–893 (1987).
102. N. Tsuji, Y. Terui, K. Nagashima, K. Tori, and L. F. Johnson in Ref. 47, pp. 94–97.
103. U.S. Pat. 4,361,649 (Nov. 30, 1982), W. D. Celmer, W. P. Cullen, R. Shibakawa, and J. Tone (to Pfizer, Inc.).
104. J. W. Westley and co-workers in Ref. 24, pp. 813–815.
105. PCT Intl. Appl. WO 88-10309 (Dec. 29, 1988), J. P. Dirlam, W. P. Cullen, H. Maeda, and J. Tone (to Pfizer, Inc.).
106. J. P. Dirlam, L. A. Presseau-Linabury, and D. A. Koss in Ref. 43, pp. 727–730.
107. J. R. Hauske and G. Kostek, *J. Org. Chem.* **54,** 3500–3504 (1989).
108. R. L. Hamill, M. M. Hoehn, G. E. Pittenger, J. Chamberlin, and M. Gorman, *J. Antibiot.* **22,** 161–164 (1969).
109. E. W. Czerwinski and L. K. Steinrauf, *Biochem. Biophys. Res. Commun.* **45,** 1284–1287 (1971).
110. J. R. Oscarson and co-workers in Ref. 23, pp. 37–48.
111. U.S. Pat. 4,707,493 (Nov. 17, 1987), W. D. Celmer, W. P. Cullen, H. Maeda, J. C. Ruddock, and J. Tone (to Pfizer, Inc.).
112. C.-M. Liu and co-workers in Ref. 9, pp. 21–28.
113. T. Kubota, H. Hinoh, M. Mayama, K. Motokawa, and Y. Yasuda in Ref. 11, pp. 931–934.
114. J. F. Blount, R. H. Evans, Jr., C.-M. Liu, T. Hermann, and J. W. Westley, *J. Chem. Soc., Chem. Commun.,* 853–855 (1975).
115. H. Koyama and K. Utsumi-Oda, *J. Chem. Soc, Perkin Trans. 2,* 1531–1536 (1977).
116. T. Mizutani and co-workers in Ref. 47, pp. 137–143.
117. H. Nakayama and co-workers in Ref. 56, pp. 1433–1436.
118. C. Keller-Juslen, H. D. King, M. Kuhn, H.-R. Loosli, and A. von Wartburg in Ref. 61, pp. 820–823.
119. U.S. Pat. 4,359,583 (Nov. 16, 1982), T. Mizutani and co-workers (to Taisho Pharmaceutical Co., Ltd.).
120. T. Mizutani and co-workers in Ref. 41, pp. 1369–1373.
121. Jpn. Pat. Appl. Kokai 83-121290 (July 19, 1983), T. Mizutani and co-workers (to Taisho Pharmaceutical Co., Ltd.).

122. U.S. Pat. 4,510,317 (Apr. 9, 1985), C.-M. Liu and J. W. Westley (to Hoffmann-LaRoche, Inc.).
123. U.S. Pat. 4,150,152 (Apr. 17, 1979), W. D. Celmer and co-workers (to Pfizer, Inc.).
124. U.S. Pat. 4,141,907 (Feb. 27, 1979), W. M. Nakatsukasa, G. G. Marconi, N. Neuss, and R. L. Hamill (to Eli Lilly Co.).
125. J. W. Westley, J. F. Blount, R. H. Evans, Jr., and C.-M. Liu in Ref. 83, pp. 610–612.
126. L. D. Boeck and co-workers, *Devel. Indus. Microbiol.* **18,** 471–485 (1977).
127. H. Seto, T. Yahagi, Y. Miyazaki, and N. Otake in Ref. 83, pp. 530–532.
128. U.S. Pat. 4,035,481 (July 12, 1977), D. H. Berg, R. L. Hamill, and M. M. Hoehn (to Eli Lilly and Co.).
129. Y. Miyazaki and co-workers in Ref. 20, pp. 814–821.
130. H. Kinashi, N. Otake, H. Yonehara, S. Sato, and Y. Saito, *Tetrahedron Lett.,* 4955–4958 (1973).
131. Jpn. Pat. Appl. Kokai 86-247398 (Nov. 4, 1986), Y. Miyazaki and co-workers (to Kaken Pharmaceutical Co., Ltd.).
132. Jpn. Pat. Appl. Kokai 82-85390 (Nov. 17, 1980), T. Nakamura and A. Shibata (to Kaken Chemical Co., Ltd.).
133. K. Dobashi and co-workers in Ref. 23, pp. 629–632.
134. U.S. Pat. 4,195,079 (Mar. 25, 1980), W. D. Celmer and co-workers (to Pfizer, Inc.).
135. J. Tone and co-workers, *21st Interscience Conference on Antimicrobial Agents and Chemotherapy,* Chicago, Ill., 1981, Abstr. 186.
136. C.-M. Liu and co-workers in Ref. 41, pp. 133–138.
137. J. W. Westley and co-workers in Ref. 41, pp. 139–147.
138. W. C. Liu and co-workers in Ref. 61, pp. 815–819.
139. B. K. Toeplitz, A. I. Cohen, P. T. Funke, W. L. Parker, and J. Z. Gougoutas, *J. Am. Chem. Soc.* **101,** 3344–3353 (1979).
140. U.S. Pat. 4,081,532 (Mar. 28, 1978), W. D. Celmer and co-workers (to Pfizer, Inc.).
141. H. Seto and N. Otake in Ref. 7, Vol. 2, 1983, p. 356.
142. U.S. Pat. 4,751,317 (June 14, 1988), W. P. Cullen, J. R. Oscarson, J. Tone, and H. Maeda (to Pfizer, Inc.).
143. C. Delhomme, A. Kergomard, G. Kergomard, and T. Staron, *J. Antibiot.* **28,** 682–695 (1976).
144. U.S. Pat. 4,022,885 (May 10, 1977), W. D. Celmer, W. P. Cullen, C. E. Moppett, J. R. Oscarson, and J. B. Routien (to Pfizer, Inc.).
145. H. Seto and co-workers in Ref. 48, pp. 970–971.
146. J. F. Blount and J. W. Westley, *J. Chem. Soc., Chem. Commun.,* 533 (1975).
147. Eur. Pat. Appl. 305,086 (Mar. 1, 1989), R. L. Hamill and R. C. F. Yao (to Eli Lilly and Co.).
148. M. Hatsu and co-workers in Ref. 40, pp. 259–266.
149. U.S. Pat. 4,279,894 (July 21, 1981), D. H. Davies and G. L. F. Norris (to Imperial Chemical Industries, Ltd.).
150. D. H. Davies, E. W. Snape, P. J. Suter, T. J. King, and C. P. Falshaw, *J. Chem. Soc., Chem. Commun.,* 1073–1074 (1981).
151. C. Keller-Juslen and co-workers in Ref. 10, pp. 142–150.
152. U.S. Pat. 3,923,823 (Dec. 2, 1975), R. M. Gale, C. E. Higgens, and M. M. Hoehn (to Eli Lilly and Co.).
153. M. O. Chaney, P. V. Demarco, N. D. Jones, and J. L. Occolowitz, *J. Amer. Chem. Soc.* **96,** 1932–1933 (1974).
154. S. H. Larsen, L. D. Boeck, F. P. Mertz, J. W. Paschal, and J. L. Occolowitz, *J. Antibiot.* **41,** 1170–1177 (1988).
155. S. Yaginuma, M. Awata, N. Muto, K. Kinoshita, and K. Mizuno in Ref. 21, pp. 239–241.
156. N. V. Murenets and co-workers, *Antibiot. Med. Biotekhnol.* **32,** 811–814 (1987).

157. U.S. Pat. 4,547,523 (Oct. 15, 1985), W. D. Celmer, W. P. Cullen, H. Maeda, and J. Tone (to Pfizer, Inc.).
158. Hung. Pat. HU T049-909 (Nov. 28, 1989), P. Toth and co-workers (to Gyogyszerkutato Interzet.).
159. C.-M. Liu and co-workers in Ref. 48, pp. 95–99.
160. J. W. Westley and co-workers in Ref. 48, pp. 100–107.
161. J. W. Westley and co-workers in Ref. 81, pp. 1275–1278.
162. R. L. Hamill and L. W. Crandall in M. J. Weinstein and G. H. Wagman, ed., *Antibiotics, Isolation, Separation and Purification,* Elsevier Scientific Publishing Co., New York, 1978, pp. 479–520.
163. Brit. Pat. 1,296,584 (Nov. 15, 1972), R. A. Kennedy and S. L. McCormick, Jr. (to Eli Lilly and Co.).
164. C.-M. Liu, in Ref. 7, pp. 43–102.
165. J. A. Russell and H. J. Strobel, *Appl. Environ. Microbiol.* **55,** 1–6 (1989).
166. T. Matsuoka, *J. Am. Vet. Med. Assoc.* **169,** 1098–1100 (1976).
167. U.S. Pat. 3,947,586 (Mar. 30, 1976), R. E. Messersmith (to Hoffmann-LaRoche, Inc.).
168. U.S. Pat. 3,995,027 (Nov. 30, 1976), C. Gale and L. R. McDougald (to Eli Lilly and Co.).
169. G. Griffiths, P. Quinn, and G. Warren, *J. Cell Biol.* **96,** 835–850 (1983).
170. R. Pal, R. C. Gallo, and M. G. Sarngadharan, *Proc. Natl. Acad. Sci., U.S.A.* **85,** 9283–9286 (1988).
171. P. Andonov and co-workers, *MBI, Med. Biol. Inf.* 6–9 (1989).
172. R. F. Shumard and M. E. Callender in Ref. 36, pp. 369–377.
173. T. K. Jeffers in P. Yvore, ed., *Coccidia and Intestinal Coccidiomorphs,* INRA Publishers, Versailles, France 1989, pp. 295–308.
174. C. K. Smith, II and R. B. Galloway, *J. Parasitol.* **69,** 666–670 (1983).
175. J. K. Frenkel and D. D. Smith, *J. Parasitol.* **68,** 851–855 (1982).
176. D. Buxton, D. A. Blewett, A. J. Trees, C. McColgan, and J. Finlayson, *J. Comp. Pathol.* **98,** 225–236 (1988).
177. W. G. Bergen and D. B. Bates, *J. Anim. Sci.* **58,** 1465–1483 (1984).
178. J. B. Russell, *J. Anim. Sci.* **64,** 1519–1525 (1987).
179. M. D. Ruff in Ref. 7, pp. 303–332.
180. N. Otake, *Drugs Pharm. Sci.* **22,** 721–742 (1984).
181. U.S. Pat. 4,431,665 (Feb. 14, 1984), A. F. Kluge and R. D. Clark (to Syntex, Inc.).
182. S. J. Bartle, R. L. Preston, and J. H. Bailie, *J. Anim. Sci.* **66,** 1502–1507 (1988).
183. B. L. T. Prosser and N. J. Palleroni, in Ref. 7, pp. 21–43.
184. J. W. Westley, *J. Nat. Prod.* **49,** 35–74 (1986).
185. D. O'Hagan, *Nat. Prod. Reports* **6,** 205–219 (1989).
186. D. E. Cane, W. D. Celmer, and J. W. Westley, *J. Am. Chem. Soc.* **105,** 3594–3600 (1983).
187. D. S. Holmes, J. A. Sherringham, U. C. Dyer, S. T. Rusell, and J. A. Robinson, *Helv. Chim. Acta* **73,** 239–259 (1990).
188. D. S. Holmes, D. M. Ashworth, and J. A. Robinson in Ref. 187, pp. 260–271.
189. C. A. Townsend and A. Basak, *Tetrahedron,* in press.
190. E. N. Duesler and I. C. Paul in Ref. 7, Vol. 2, 1983, pp. 87–195.
191. J. L. Occolowitz and R. L. Hamill in Ref. 7, Vol. 2, 1983, pp. 197–244.
192. M. J. O. Anteunis in Ref. 7, Vol. 2, 1983, pp. 245–334.
193. H. Seto and co-workers, *Tetrahedron Lett.* **28,** 3357–3360 (1987).
194. Y. Kishi, in Ref. 7, Vol. 2, 1983, pp. 1–50.
195. D. A. Evans and R. P. Polniaszek, *Tetrahedron Lett.* **27,** 5683–5686 (1986).
196. D. A. Evans, R. L. Dow, T. L. Shih, J. M. Takacs, and R. Zahler, *J. Am. Chem. Soc.* **112,** 5290–5313 (1990).
197. D. A. Evans, S. L. Bender, and J. Morris, *J. Am. Chem. Soc.* **110,** 2506–2536 (1988).

198. D. A. Evans, J. S. Clark, R. Metternich, V. J. Novack, and G. S. Sheppard, *J. Am. Chem. Soc.* **112,** 866–868 (1990).
199. S. Hanessian, N. G. Cooke, B. DeHoff, and Y. Sakito in Ref. 198, pp. 5276–5290.
200. W. C. Still and co-workers, *Lect. Heterocycl. Chem.* **9,** 33–42 (1987).
201. S. J. Danishefsky, H. G. Selnick, M. P. DeNinno, and R. P. Zelle, *J. Am. Chem. Soc.* **109,** 1572–1574 (1987).
202. S. J. Danishefsky, H. G. Selnick, R. E. Zelle, and M. P. DeNinno in Ref. 197, pp. 4368–4378.
203. D. P. Negri and Y. Kishi, *Tetrahedron Lett.* **28,** 1063–1066 (1987).
204. M. P. Edwards, S. V. Ley, S. G. Lister, B. D. Palmer, and D. J. Williams, *J. Org. Chem.* **49** (1984).

General Reference

J. W. Westley, ed., *Polyether Antibiotics,* Vols. 1 and 2, Marcel Dekker, Inc., New York, 1982–1983.

LOUISE W. CRANDALL
ROBERT L. HAMILL
Lilly Research Laboratories

TETRACYCLINES

The tetracyclines are a group of antibiotics having an identical 4-ring carbocyclic structure as a basic skeleton and differing from each other chemically only by substituent variation. Figure 1 shows the principal tetracycline derivatives now used commercially.

The first tetracycline discovered was produced by a soil organism, *Streptomyces aureofaciens,* and is now known as chlortetracycline [*57-62-5*] (**2**), $C_{22}H_{23}ClN_2O_8$ (1). This compound ushered in a new era in antibacterial chemotherapy because it was effective orally and against a broad range of gram-positive and gram-negative bacteria. Chlortetracyline was joined by the second member of the family, oxytetracycline [*79-57-2*] (**3**), $C_{22}H_{24}N_2O_9$, produced by another actinomycete, *Streptomyces rimosus,* in 1950 (2). Tetracycline [*60-54-8*] (**1**), $C_{22}H_{24}N_2O_8$, discovered in 1953, lacks the 7-chloro of chlortetracycline (**2**) and the 5-hydroxyl group of oxytetracycline (**3**). Tetracycline was produced either by reductive dechlorination of (**2**) (3,4) or by a direct fermentation (5). 6-Demethylchlortetracycline [*127-33-3*], $C_{21}H_{21}ClN_2O_8$, (**4**) was discovered (6) as a metabolite of a mutant strain of the original *Streptomyces aureofaciens* and was introduced in 1958.

The three tetracyclines most recently marketed were made by a semisynthetic pathway. The first of these were methacycline (6-methylene oxytetracycline) [*914-00-1*] (**5**), $C_{22}H_{22}N_2O_8$, (7), and its reduction product doxycycline [*564-25-2*] (**6**), $C_{22}H_{24}ClN_2O_8$ (7,8). The latter compound is a potent antibiotic which is well-absorbed and slowly excreted thus allowing small and infrequent (once or twice a day) dosage schedules. Finally, the most recent addition to the commercial tetracyclines is minocycline [*10110-90-8*] (**7**), $C_{23}H_{27}N_3O_7$ (9), which is also well-

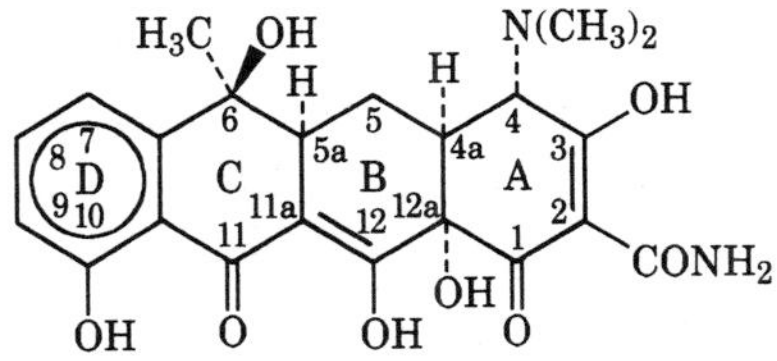

(1)

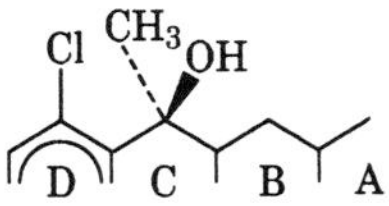

(2)

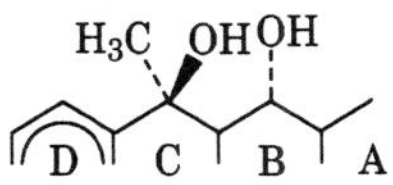

(3)

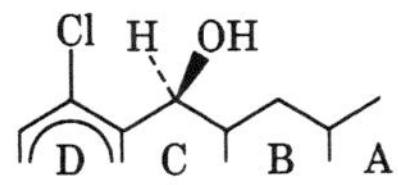

(4)

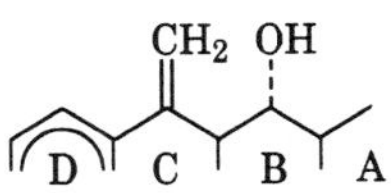

(5)

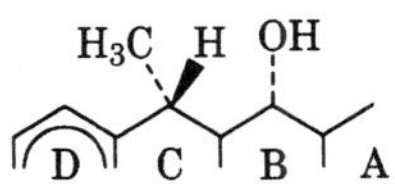

(6)

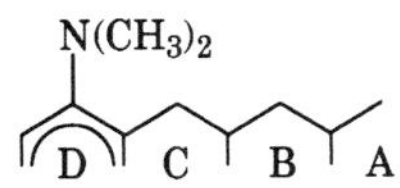

(7)

Fig. 1. Tetracycline (**1**) and its derivatives: chlortetracycline (7-chlorotetracycline) (**2**); oxytetracycline (5-hydroxytetracycline) (**3**); demeclocycline (6-demethyl-7-chlorotetracycline) (**4**); methacycline (6-demethyl-6-deoxy-5-hydroxy-6-methylenetetracycline) (**5**); doxycycline (6α-deoxy-5-hydroxytetracycline) (**6**); and minocycline (6-demethyl-6-deoxy-7-dimethylamino tetracycline) (**7**). Substituents at positions not specifically shown or mentioned remain as in (**1**). Courtesy of Blackwell Scientific Publications, Ltd.

absorbed and slowly excreted. Minocycline protects mice against staphylococcal infections which are resistant to most tetracyclines, penicillins, and many other antibiotics. In addition, it shows a marked superiority over other tetracyclines when tested against large numbers of randomly occurring, hospital-isolated gram-positive bacteria (10).

During the years from 1948 to 1952 considerable research was devoted to determining the structures of the tetracyclines. The gross structure (**3**) was determined by a combination of degradation sequences and spectral studies (11). Structure (**2**) was also published (12). Structure (**2**) was also determined by a different degradative pathway (13). The more subtle points of structure, such as the stereochemical configurations, were later determined by x-ray methods (14) and the absolute configuration by degradation. Structures of the tetracyclines developed since 1952 are easily arrived at by comparison to the structures of earlier compounds.

Physical Properties

In general, the tetracyclines are yellow crystalline compounds that have amphoteric properties (Fig. 2) (15). They are soluble in both aqueous acid and aqueous base. The acid salts tend to be soluble in organic solvents such as 1-butanol, dioxane, and 2-ethoxyethanol. In fact, 1-butanol is used to extract the salts from aqueous solution.

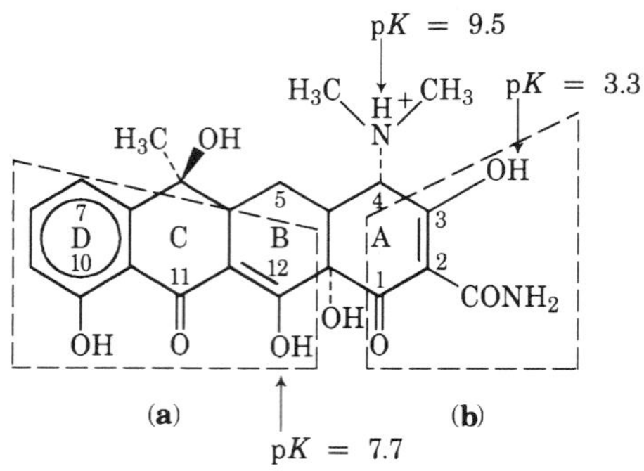

Fig. 2. Tetracycline (**1**) indicating the titratable hydrogens and showing (**a**) the BCD-chromophore and (**b**) the A-chromophore (15).

The tetracyclines are strong chelating agents. Both the A-ring and 11,12 β-diketone systems are active sites for chelation (16). This ability to chelate to metals, such as calcium, results in tooth discoloration when tetracycline is administered to children (17).

Each tetracycline possesses a characteristic uv-absorption spectrum and this property is used extensively in structure elucidation (12,13). This spectrum results from the contribution of two chromophores: the BCD ring system gives a λ_{max} at approximately 350 nm and the A-ring a λ_{max} at approximately 265 nm.

Traditionally, paper chromatography was used to monitor chemical modification reactions to determine the composition of the reaction mixture (13,18–20). The tetracyclines were usually detected by uv fluorescence on the paper strip or by bioautography where the areas on the paper strip that inhibited the growth of microorganisms were determined. The transference of paper chromatography solvent systems to column chromatography has been accomplished and since 1978, high pressure liquid chromatography (hplc) has been used with excellent success (21). Mass spectral analysis (22) and nmr (9) have become useful tools for tetracycline structure determinations.

Semisynthetic Modifications

The tetracycline molecule (**1**) presents a special challenge with regard to the study of structure–activity relationships. The difficulty has been to devise chemical pathways that preserve the BCD ring chromophore and its antibacterial properties. The lability of the 6-hydroxy group to acid and base degradation (12,13), plus the ease of epimerization (23) at position 4, contribute to chemical instability under many reaction conditions.

Under acidic conditions, dehydration to an anhydrotetracycline [*20154-34-1*] (**8**), $C_{22}H_{22}N_2O_7$, occurs; under basic ones, ring C opens to an isotetracycline [*3811-31-2*] (**9**), $C_{22}H_{24}N_2O_8$. The anhydrotetracyclines, such as (**8**), appear to exhibit a mode of antibacterial action, but it is unlike that of tetracycline (24). Epimerization (23,25,26) at C-4 occurs in a variety of solvents within the pH range 2–6, particularly in acetic acid (25). A number of anions (27) facilitate this reaction.

The reverse process, from 4-epitetracycline [*79-85-6*], $C_{22}H_{24}N_2O_8$, to tetracycline, is promoted by chelation with ions such as calcium and magnesium (28).

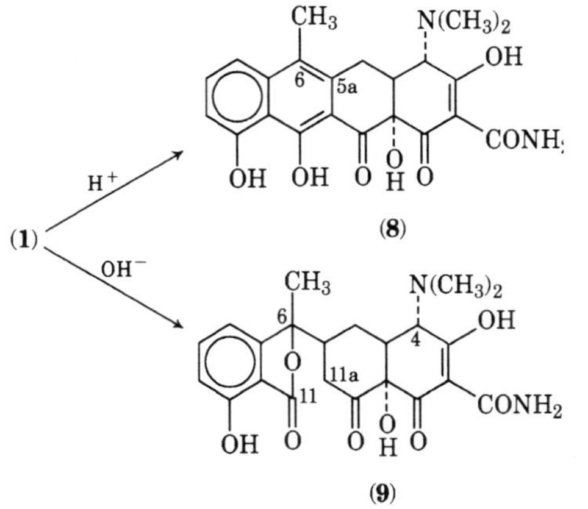

Conversion of the C-2 amide to a biologically inactive nitrile, which can be further taken via a Ritter reaction (29) to the corresponding alkylated amide, has been accomplished. When the 6-hydroxyl derivatives are used, dehydration occurs at this step to give the anhydro amide. Substituting an *N*-hydroxymethylimide for isobutylene in the Ritter reaction yields the acylaminomethyl derivative (30). Hydrolysis affords an aminomethyl compound. Numerous examples (31–35) have been reported of the conversion of a C-2 amide to active Mannich adducts which are extremely labile and easily undergo hydrolysis to the parent tetracycline. This reverse reaction probably accounts for the antibacterial activity of these tetracyclines.

Reactions at the C-4 carbon atom have been studied. With the exception of (**3**), reaction with methyl iodide (36,37) converts the 4-amino group to a quaternary amine with a concomitant loss of antibacterial activity. Treatment of this quaternary derivative with zinc in acetic acid results in a selective removal of the 4-dimethylamino grouping (36). This deamination can also be accomplished by photolysis (38). A transformation (18) involving the C-4 position has resulted in the synthesis of 4-hydroxy-6-methylpretetramid [*2011-31-6*], $C_{20}H_{15}NO_7$, an important precursor (39) in the biosynthesis of tetracycline. Another unusual reaction (40,41) yielded the 4,6-hemiketals: reactions of (**1**) or 6-demethyltetracycline [*987-02-0*] (**10**), $C_{21}H_{22}N_2O_8$, with positive halogens, or cupric or mercuric acetates yield the corresponding 4,6-hemiketals (**11**) and (**12**).

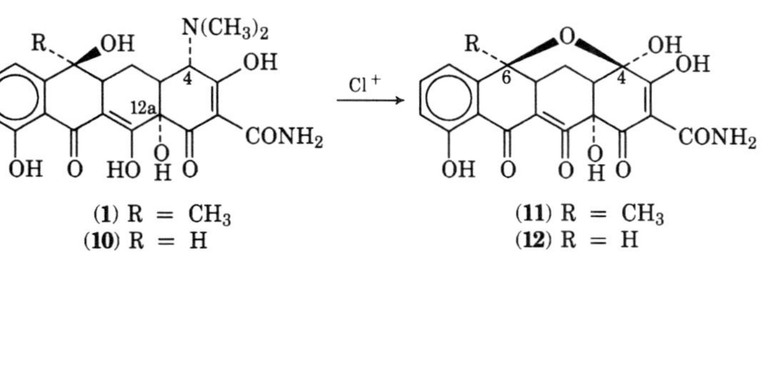

(**1**) R = CH_3 (**11**) R = CH_3
(**10**) R = H (**12**) R = H

The hemiketal products (**11**) and (**12**) have been converted to the corresponding oximes, hydrazones, and substituted amines (40,41). Although many of these derivatives exhibit substantial antibacterial activity, they are generally less active than the parent tetracyclines.

Reactions at the C-5 position of the tetracycline molecule have been limited to the introduction of an alkoxy group (42) and the acetylation of the hydroxyl group (43) in 5-hydroxytetracyline. Neither of these modifications improved the biological activity of the molecule.

The isolation of the 6-deoxytetracyclines (44) led to other chemical modifications of (**1**). 6β-Deoxytetracycline [*5614-03-9*] (**13**), $C_{22}H_{24}N_2O_7$, was prepared by catalytic hydrogenolysis of tetracycline (**1**), resulting in an inversion (45) of the configuration at the C-6 position, but retention of antibacterial activity. Catalytic reduction (7,8) of the 6-methylene derivative (**14**) yields both the 6α-methyl (**15**) and 6β-methyl compound (**13**). The 6α-isomer (**15**) is reported (7,45) to be more active than the 6β-isomer (**13**). The α-isomer, doxycycline (**6**), is an example of a semisynthetic tetracycline that has become commercially useful.

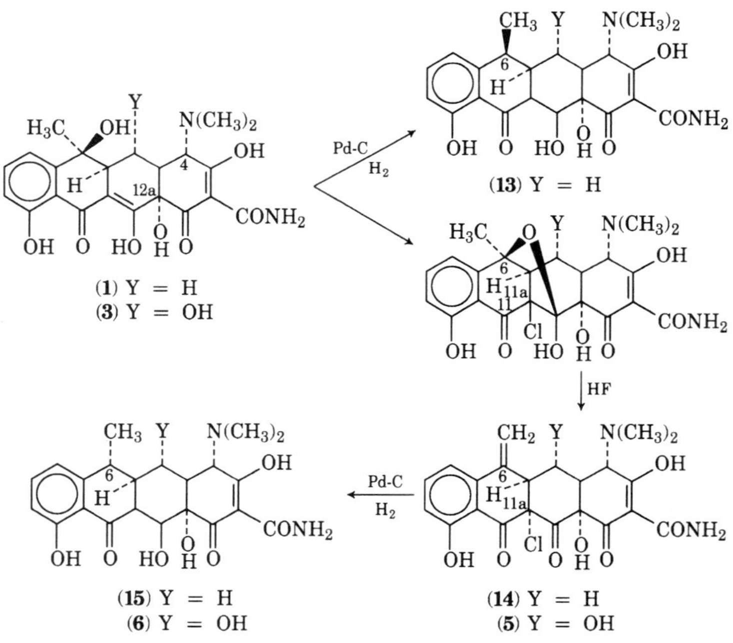

The 6-fluoro isomers, 6-deoxy-6-demethyl-6α-fluorotetracycline [*24333-20-8*], $C_{21}H_{21}FN_2O_7$, and 6-deoxy-6-demethyl-6β-fluorotetracycline [*24333-21-9*], $C_{21}H_{21}FN_2O_7$, have been prepared and showed relatively high *in vitro* and *in vivo* biological activities compared to the parent tetracyclines.

The increased chemical stability of the 6-deoxytetracylines allows chemical modification with retention of biological activity: electrophilic substitutions have been carried out at C-7 and C-9 under strongly acidic conditions (46–53).

Reactions of 6-deoxy-6-demethyltetracycline [*808-26-4*] (**16**), $C_{21}H_{22}N_2O_7$, with electrophiles, such as nitrate ion (49), bromomium ion (46,47) (from *N*-bromosuccinimide), or *N*-hydroxymethylphthalimide (53), yielded 7-substituted tetracyclines. In the case of the nitration reaction, both the 7- and 9-nitro isomers (**17**, X = NO_2, Y = H) and (**17**, X = H, Y = NO_2) were obtained.

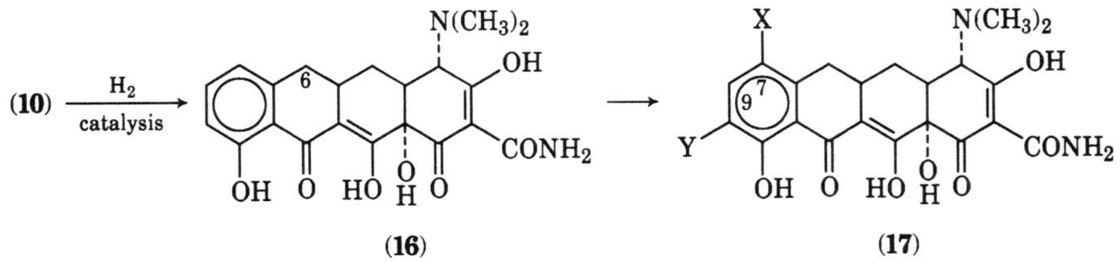

A similar series of reactions can be carried out for the 6-methyl series, (**13**) and (**15**), where catalytic reduction affords the amino compounds, ie (**17**, X = NH_2, Y = H). Oxidation (54) of tetracyclines using the Udenfriend reagent has yielded 9-hydroxytetracyclines and disubstituted (C-7 and C-9) products (48) can also be obtained. Substituent assignments are made from nmr spectral interpretations. The 7- and 9-methyl tetracyclines have been prepared and reported to retain biological activity (55).

Although many of the tetracycline derivatives showed useful *in vivo* activity against a wide spectrum of pathogenic organisms, few showed any significant improvement in overall activity when compared to tetracycline. The exception is minocycline (**7**) (9) which exhibits superior activity against tetracycline-sensitive organisms and many tetracycline-resistant strains of gram-positive bacteria.

Structure–Activity Correlations

There are a number of tetracycline structural features that are prerequisites for biological activity. The linear arrangement of the rings, coupled with the phenolic β-diketone system, is essential (56,57). Any structure variation at the 11a position results in loss of activity. The C-11 to C-12 β-diketone system has exceptional chelating qualities, and probably is involved in the binding of the tetracyclines to ribosomes (58), in the interactions with bacterial repressor proteins (58), and in transport of tetracyclines into the bacterial cell. The amide hydrogen can be replaced by a methyl group, but larger residues, if not rapidly cleaved in water, bring about a reduction in activity.

The configuration at the chiral centers C-4a, C-5a, and C-12a determine the conformation of the molecule. In order to retain optimum *in vitro* and *in vivo* activity, these centers must retain the natural configuration. The hydrophobic part of the molecule from C-5 to C-9 is open to modification in many ways without losing antibacterial activity. However, modification at C-9 may be critical because steric interactions or hydrogen bonding with the oxygen atom at C-10 may be detrimental to the activity.

X-ray crystallographic studies (59) have defined the conformations and hydrogen bonding of the tetracyclines under nonpolar and polar conditions. These are shown in Figure 3. It is believed that the equilibrium between the zwitterionic and nonionized forms is of importance for the broad-spectrum antibacterial activity, membrane permeation, and pharmacokinetic properties.

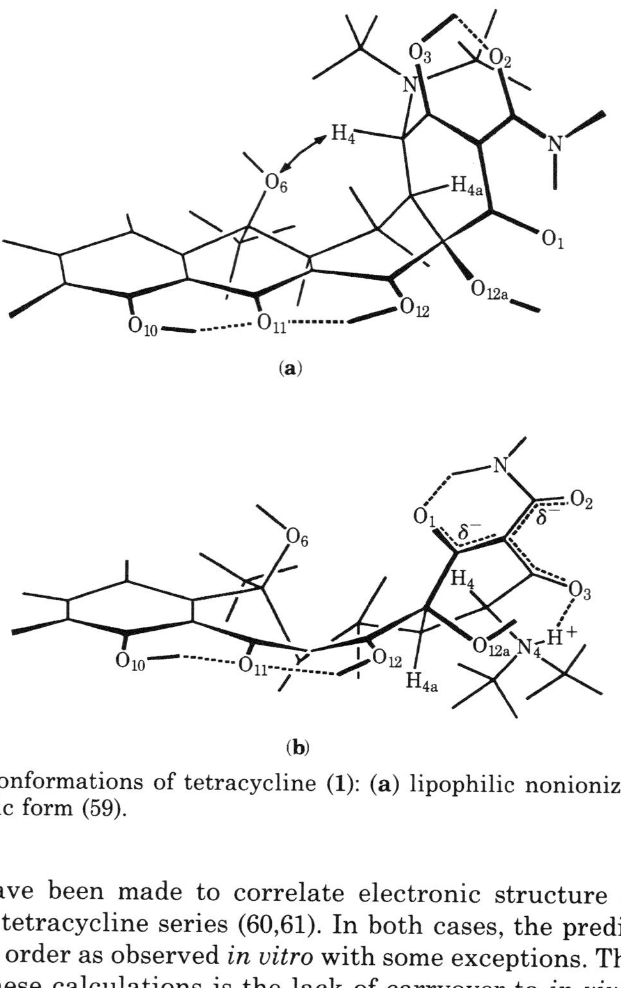

Fig. 3. Conformations of tetracycline (**1**): (**a**) lipophilic nonionized form, and (**b**) polar zwitterionic form (59).

Efforts have been made to correlate electronic structure and biological activity in the tetracycline series (60,61). In both cases, the predicted activities are of the same order as observed *in vitro* with some exceptions. The most serious drawback to these calculations is the lack of carryover to *in vivo* antibacterial activity. Attempts have also been made (62) to correlate partition coefficients and antibacterial activity. The stereochemical requirements are somewhat better defined. Thus 4-epitetracycline and 5a-epitetracycline [*65517-29-5*], $C_{22}H_{24}N_2O_8$, are inactive (63). The 6-epi compound [*19369-52-9*] is about one-half as active as the 6α (or natural) configuration.

The unexpected biological activities of tetracyclines, such as 5a-epi-6-epi-tetracycline [*19543-88-5*], $C_{22}H_{24}N_2O_8$, and 7-chloro-5a,11a-dehydro-6-epitetracycline [*22688-60-4*], $C_{22}H_{21}ClN_2O_8$, make predicting structure–activity relationships difficult (64). Aside from the C-2 amide Mannich-base derivatives, variation at other centers in the molecule, ie, C-4, 4a, 5a, 12a, decreases the biological activity.

Total Synthesis

The first synthesis of a tetracycline having full biological activity, 6-demethyl-6-deoxytetracycline (**16**) was reported in the early 1960s (76–79). This compound contained the important and difficult to obtain 4-dimethylamino group, but lacked the 6-hydroxyl. An alternative total synthesis of (**16**) was devised in 1965 (69) using methods that could lead to the synthesis of tetracyclines in large quantities and in good yields. The only total synthesis of a fermentation-produced tetracycline was carried out for oxytetracycline (**3**) which has six asymmetric centers (70). More recently the following nonnaturally tetracyclines have been synthesized: 5a-methyltetracyclines (59), 6-oxatetracyclines (59), and 6-thiatetracyclines (59). None of these compounds proved to be commercially important.

Manufacture

Most of the fermentation and isolation processes for manufacture of the tetracyclines are described in patents (71,72). Manufacture begins with the cultivated growth of selected strains of *Streptomyces* in a medium chosen to produce optimum growth and maximum antibiotic production. Some clinically useful tetracyclines (**2–4**) are produced directly in these fermentations; others (**5–7**) are produced by subjecting the fermentation products to one or more chemical alterations. The purified antibiotic produced by fermentation is used as the starting material for a series of chemical transformations (59).

The choice of the strain of microorganism is one of the important variables in the process. The strains to be used in manufacture are mutants of the original producer, which are chosen as the result of a planned program of mutant selection. Sometimes a spontaneous mutation occurs; usually, it is induced by mutagenic agents or irradiation of various sorts. The choice of the best strain depends on its ability to produce large amounts of the proper antibiotic in a reasonable time from ingredients that are economically feasible (73).

All steps of manufacture are monitored for antibiotic potency, purity, stability, and bioavailability by quality control methods. In addition, each batch of bulk powder and each lot of finished pharmaceutical dosage form must be certified by the FDA. During all stages of manufacture the FDA requires compliance with specified standards.

Economic Aspects

The total U.S. antibiotic market for 1990 was about $4.73 billion, $233 million of that was tetracyclines. The development of the semisynthetic β-lactam antibiotics (see ANTIBIOTICS, β-LACTAMS) and emergence of resistance to the tetracyclines has steadily diminished the clinical usefulness of tetracyclines.

In the United States, the manufacturers of fermentation-derived tetracyclines (**1**), (**2**), and (**3**) are the Lederle Laboratories, a division of American Cyanamid Co., Charles Pfizer Inc., Bristol Laboratories, and Rachelle Laboratories. There are also several manufacturers abroad. Tetracycline is now sold generically by many companies. Pfizer's doxycycline (**6**) and Lederle's minocycline (**7**), both semisynthetic tetracyclines, are the only members of the group that have increasing sales. Table 1 lists the commercial tetracyclines and the corresponding trade names.

Table 1. Tetracyclines Used for the Therapy of Infectious Diseases

Generic name	Trade name	Year of discovery
chlortetracyline	Aureomycin	1948
oxytetracycline	Terramycin	1948
tetracycline	Achromycin	1953
demeclocycline	Declomycin	1957
methacycline	Rondomycin	1965
doxycycline	Vibramycin	1967
minocycline	Minocin	1972

Biosynthesis

The overall biosynthetic pathway to the tetracyclines has been reviewed (74). Studies (75–78) utilizing ^{13}C labeled acetate and malonate and nmr analysis of the isolated oxytetracycline (**2**), have demonstrated the exclusive malonate origin of the tetracycline carbon skeleton, the carboxamide substituent, and the folding mode of the polyketide chain. Feeding experiments using [1-^{13}C, $^{18}O_2$] acetate and analysis of the nmr isotope shift effects, led to the location of the [$^{18}O_2$] derived oxygen substituents in oxytetracycline: carbons 1, 3, 10, 11, and 12 were labeled. The remaining oxygen substituents, those at carbons 5, 6, and 12 in oxytetracycline, originate from subsequent oxidation of the 4-hydroxy-6-methylpretetramid [*2011-31-6*].

Feeding studies using deuterated [1-^{13}C] acetate and subsequent location of the deuterated sites by the nmr isotope shift effects showed labeling at carbons 7 and 9. The absence of a detectable β-^{2}H isotope shift at carbon 4a indicates that only one of the carbon-5 hydrogens is acetate-derived and that this is stereospecifically eliminated by carbon-5α hydroxylation. 8-Methoxychlortetracyclines, with and without hydroxyl substitution at carbon-4a, have been isolated. These tetracyclines retain the original oxygen at C-8 from the polyketide chain (79–81).

Because of the continued commercial importance of the tetracyclines, a study of the genetics of the oxytetracycline (OTC) producing organism has been undertaken to improve the efficiency of antibiotic production. The biosynthetic genes of *Streptomyces rimosus* have been cloned by taking advantage of the fact that the OTC genes are clustered around at least one OTC resistance gene. Using this resistance gene as a selectable marker to detect hydridizing clones, a 30–40 kilobase piece of DNA surrounding the resistance gene, and containing some of the biosynthetic genes, was identified (82).

Anhydrotetracycline oxygenase from *Streptomyces aureofaciens,* which catalyzes the conversion of anhydrotetracycline to dehydrotetracycline, has been isolated and characterized as a flavin-dependent oxygenase (83). It consists of two subunits of mol wt $\approx$57,500 based on SDS/polyacrylamide–gel electrophoresis.

The cosynthetic factor 1 of *Streptomyces aureofaciens,* involved in the reduction of 5a,11a-dehydrochlortetracycline to chlortetracycline, has been identified as 7,8-didemethyl-8-hydroxy-5-deazariboflavin. This work was aided by comparison of spectral data with that of an authentic sample obtained from the hydrolysis of coenzyme F-420 (84).

Biological Aspects

It has been known for some time that tetracyclines are accumulated by bacteria and prevent bacterial protein synthesis (Fig. 4). Furthermore, inhibition of protein synthesis is responsible for the bacteriostatic effect (85). Inhibition of protein synthesis results primarily from disruption of codon–anticodon interaction between tRNA and mRNA so that binding of aminoacyl–tRNA to the ribosomal acceptor (A) site is prevented (85). The precise mechanism is not understood. However, inhibition is likely to result from interaction of the tetracyclines with the 30S ribosomal subunit because these antibiotics are known to bind strongly to a single site on the 30S subunit (85).

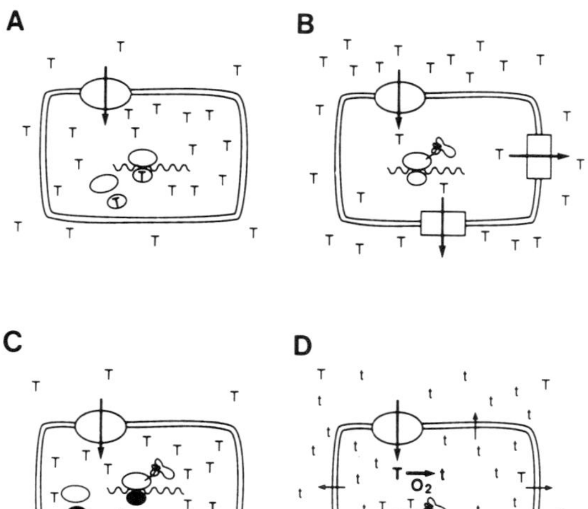

Fig. 4. Comparison of the three types of tetracycline resistance where T represents the tetracycline molecule; ⬭, a tetracycline transporter; and ∿∿, the ribosome; A shows the effect of tetracycline exposure on a sensitive cell; B, the efflux of resistance where a cytoplasmic membrane protein (□) pumps tetracycline out of the cell as fast as the tetracycline transporter takes it up; C, the ribosomal protection type of resistance where the ribosome is modified by (⬬) to block productive binding; and D, the tetracycline modification type of resistance where t is an inactive form of tetracycline. Reproduced with permission from reference 121. (See also refs. 85, 86, 88, 105–107, 109, and 111).

Studies designed to characterize the nature of the tetracycline binding domain have revealed that when bound to ribosomes, tetracycline protects base A892 in 16sRNA from reactivity toward dimethyl sulfate and enhances reactivity towards bases U1052 and C1054 (86). These results suggest that the 892–1054 region of 16sRNA contributes together with 30S ribosomal proteins (87) to the antibiotic binding domain.

Uses

The commercially available tetracyclines, listed in Table 1, have fairly similar antibacterial activities, apart from doxycycline (**6**) and minocycline (**7**), which show enhanced antibacterial activity compared to the others (59). Tetracyclines were the first prominent group of antimicrobial agents for which the term "broad spectrum" was used. That is, they exhibit activity against a wide range of gram-positive and gram-negative bacteria, including obligate anaerobes (88,89). *Mycoplasmas, rickettsiae, chlamydia,* and protozoan parasites are also, in general, susceptible to tetacyclines (88–90) (see also ANTIPARASITIC AGENTS).

Clinical Uses. The emergence of bacterial resistance to tetracyclines has limited the use of these agents as the drugs of first choice in the treatment of many infections for which they were previously effective. Nevertheless, they are still the treatment of first choice in the following cases: (*1*) for bacterial infections causing brucellosis, cholera, chancroid, granuloma inguinale, and Lyme disease (88,89,91); (*2*) for rickettsial infections, eg, typhus, scrubtyphus, and spotted fever (88,89); (*3*) for chlamydial infections, eg, psittacosis, lymphogranuloma venereum, trachoma, and inclusion conjunctivitis (88,89); (*4*) in the treatment of nonspecific urethritis because of *Chlamydia* or *Ureaplasmas* (88); and (*5*) in the treatment of acne vulgaris and rosacea (88,92).

Many sexually transmitted infections are polymicrobial in nature, ie, more than one type of microorganism is responsible for the infection (89,93,94). For example, pelvic inflammatory disease (PID) is usually caused by a concurrent infection with both *N. gonorrhoeae* and *C. trachomatis* (94). For these reasons many sexually transmitted diseases, including PID, are treated with mixtures of antibiotics to provide broad coverage for mixed, aerobic, and anaerobic infections (89). Tetracyclines play a role in such therapeutic regimens, eg, β-lactam antibiotics plus doxycycline are used in the therapy of PID (94).

It has been suggested that tetracyclines alone can be used for certain sexually transmitted syndromes, such as urethritis in which *N. gonorrhoeae, Chlamydia,* and *Ureaplasma* species are causative agents (89). Although β-lactam antibiotics remain the drugs of choice against *N. gonorrhoeae,* the tetracyclines are active against the associated pathogens. However, the increasing prevalence of tetracycline resistance in *N. gonorrhoeae* (95) is likely to seriously challenge this therapeutic strategy.

Tetracyclines are used as alternative drugs in a variety of circumstances when the patient is unable to take the drug of choice, eg, in patients allergic to penicillin (88,89). Tetracyclines are widely known to cause staining of teeth (and are therefore contra-indicated in children developing permanent teeth), photosensitivity, and, in the case of minocycline, vestibular toxicity. Details of these adverse effects and others associated with administration of tetracyclines have been comprehensively reviewed (96–101).

Veterinary Uses. Tetracyclines are widely used for veterinary therapy. The types of pathogens encountered are frequently different from those for which tetracyclines are used in humans (102). Tetracyclines are also used in animal husbandry as growth promoters (88,102) (see GROWTH REGULATORS, ANIMAL). Tetracyclines, eg, chlortetracycline, used in this way are usually administered to the animals orally, at subtherapeutic doses, either in the feed or drinking water (see FEEDS AND FEED ADDITIVES). The procedure frequently improves the efficiency of feed conversion, or the rate of weight gain in poultry and other animals reared under commercial conditions (88,102). The mechanisms responsible for growth promotion are still obscure but, in the case of chlortetracycline (**2**), probably result from suppression of deleterious organisms in the animal's intestine (102). Various antibiotics, including tetracyclines, have demonstrable biochemical effects on microorganisms at drug concentrations which are below those required to achieve complete inhibition of microbial growth (103). These effects include suppression of bacterial adhesion and interference with the secretion of extracellular toxins, both of which may relate to the growth-promoting activity of tetracyclines. Some countries, eg, the United Kingdom, have introduced legislation forbidding the addition of tetracyclines to animal feed for growth promotion purposes. See reference 102 for a discussion.

Resistance to Tetracyclines. The tetracyclines still provide inexpensive and effective treatment for several microbial infections, but the emergence of acquired resistance to this class of antibiotic has limited their clinical usefulness. Studies to define the molecular basis of resistance are underway so that derivatives having improved antibacterial spectra and less susceptibility to bacterial resistance may be developed. Tetracyclines are antibiotics of choice for relatively few human infections encountered in daily clinical practice (104), largely as a result of the emergence of acquired tetracycline-resistance among clinically important bacteria (88,105,106). Acquired resistance occurs when resistant strains emerge from previously sensitive bacterial populations by acquisition of resistance genes which usually reside in plasmids and/or transposons (88,106,107). Furthermore, resistance determinants contained in transposons spread to, and become established in, diverse bacterial species (106).

Mechanisms of Resistance. Three distinct biochemical mechanisms of resistance to tetracyclines have been identified and these are illustrated in Figure 4. The energy-dependent efflux of antibiotic mediated by resistance proteins located in the bacterial cytoplasmic membrane is illustrated in Figure 4B (85,88,105–109). The intracellular tetracycline concentration remains too low for effective binding to ribosomes. The complete sequence of the genes for several efflux proteins has been established and models for the organization of the proteins in the membrane have been proposed (109,110). Ribosomal protection, whereby tetracyclines no longer bind productively to the bacterial ribosome, is shown in Figure 4C (108,111). In a tetracycline-resistant cell, shown in Figure 4A, tetracycline accumulation within the cell is similar to that in the sensitive cell, but the ribosome is modified so that tetracycline no longer binds productively to the ribosome. Although the molecular basis of this resistance mechanism is not fully understood, it may involve modification of ribosomal RNA or protein which affects binding of antibiotic to the 30S ribosomal subunit (111). Figure 3D shows chemical alteration of the tetracycline molecule by a reaction in the cytoplasm that requires oxygen. This renders the drug inactive as an inhibitor of protein synthesis (108). The altered tetracycline then diffuses out of the cell. This mechanism of resistance is poorly characterized and may not be expressed in the natural habitat of most pathogens especially because foci of infection in the human body are usually poorly aerated (108).

Distribution of Resistance Determinants in Microorganisms. The majority of tetracycline resistance determinants are located on plasmids or transposons. These determinants have been grouped into classes defined by lack of cross-hybridization under stringent conditions (107,108). These classes, some of which are related, are given in Table 2. Classes A–D show DNA homologies of from 44–63% (108) and encode products having similar functions. Although tetracycline resistance determinants are widely distributed in microorganisms, certain classes appear to be restricted to particular groups of organisms. For instance, tetracycline efflux is the dominant mechanism of resistance in the *Enterobacteriaceae*, but this mechanism has not been reported in other organisms eg, *Neisseria* species.

Table 2. Classification and Distribution of Tetracycline Resistance Determinants in Microorganisms[a]

Representative family		
Class	Mechanism	Genus or species
A	efflux	*Enterobacteriaceae, Aeromonas, Vibrio, Pseudomonas*
B	efflux	*Enterobacteriaceae, Haemophilus, Vibrio, Yersinia*
C	efflux	*Enterobacteriaceae, Vibrio, Pseudomonas*
D	efflux	*Enterobacteriaceae, Aeromonas, Vibrio, Pasteurella*
E	efflux	*Escherichia aeromonas*
F	efflux	*Bacteroides fragilis*
G	unknown	*Vibrio anguillarum*
K	efflux	*Staphylococcus, Enterococus, Streptococcus,*[b] *Peptostreptococcus*[b]
L	efflux	*Bacillus, Staphylococcus, Streptococcus,*[b] *Enterococcus, Peptostreptococcus*[b]
M	ribosomal protection	*Neisseria, Mycoplasma, Ureaplasma, Haemophilus, Campylobacter, Clostridium, Enterococcus, Staphylococcus, Streptococcus,*[b] *Gardnerella, Kingella, Eikenella, Veillonella, Fusobacterium, Peptostreptococcus,*[b] *Clostridium difficile, Streptococcus pneumoniae*
N	unknown	*Strep agalactiae*
O	ribosomal protection	*Campylobacter, Lactobacillus, Streptococcus,*[b] *Enterococcus, Peptostreptococcus*[b]
P	unknown	*C. perfringens*
X	antibiotic modification	*Bacteroides* (cryptic)

[a]Refs. 85, 105, 107, 108, and 112.
[b]Certain isolates are reported to contain simultaneously K, L, M, and O determinants (112).

BIBLIOGRAPHY

"Tetracyclines" in *ECT* 1st ed., Vol. 13, pp. 771–810, by P. Regna, Chas. Pfizer & Co., Inc., C. H. Demos and L. M. Preuss, Lederle Laboratories, American Cyanamid Company; in *ECT* 2nd ed., Vol. 20, pp. 1–33, by R. K. Blackwood, Chas. Pfizer & Co., Inc.; in *ECT* 3rd ed., Vol. 3, pp. 64–78, by J. H. Boothe and J. J. Hlavka (Lederle Laboratories, a division of the American Cyanamid Company).

1. R. W. Broschard and co-workers, *Science* **109,** 199 (1949); B. M. Duggar, *Ann. N.Y. Acad. Sci.* **51,** 177 (1948).
2. A. C. Finlay and co-workers, *Science* **111,** 85 (1950); P. P. Regna and co-workers, *J. Am. Chem. Soc.* **73,** 4211 (1951).
3. J. H. Boothe and co-workers, *J. Am. Chem. Soc.* **75,** 4621 (1953).
4. L. H. Conover and co-workers, *J. Am. Chem. Soc.* **75,** 4622 (1953).
5. P. P. Minieri and co-workers, *Antibiot. Ann.* **1953–1954,** 81 (1953–1954).
6. J. R. D. McCormick and co-workers, *J. Am. Chem. Soc.* **79,** 4561 (1957).
7. R. K. Blackwood and co-workers, *J. Am. Chem. Soc.* **85,** 3943 (1963).
8. C. R. Stephens and co-workers, *J. Am. Chem. Soc.* **85,** 2643 (1963).
9. M. J. Martell, Jr. and J. H. Boothe, *J. Med. Chem.* **10,** 44 (1967); G. S. Redin in G. L. Hobby, ed., *Antimicrobial Agents and Chemotherapy,* Williams & Wilkins, Baltimore, Md., 1966, p. 371.
10. N. H. Steigbigel, C. W. Reid, and M. Finland, *Am. J. Med. Sci.* **255,** 179 (1968).
11. F. A. Hochstein and co-workers, *J. Am. Chem. Soc.* **75,** 5455 (1953).
12. C. R. Stephens and co-workers, *J. Am. Chem. Soc.* **76,** 3658 (1954); **74,** 4976 (1952).
13. C. W. Waller and co-workers, *J. Am. Chem. Soc.* **74,** 4981 (1952).
14. S. Hirokawa and co-workers, *Z. Kristallogr. Kristallgeom. Kristallphys. Kristallchem.* **112,** 439 (1959); J. Donohue and co-workers, *J. Am. Chem. Soc.* **85,** 851 (1963).
15. M. Schnarr, M. Matthies, and W. Lohmann, *Z. Naturforsch.* **34c,** 1156 (1979).
16. J. J. Hlavka and J. H. Boothe, eds., *Handbook of Experimental Pharmacology,* Vol. 78, Springer-Verlag, New York, 1985, p. 86.
17. H. Shwachman and A. Schuster, *Pediat. Clin. Am.* **3,** 295 (1956).
18. J. J. Hlavka and P. Bitha, *J. Am. Chem. Soc.* **87,** 1795 (1965).
19. J. J. Hlavka, P. Bitha, and J. H. Boothe, *J. Am. Chem. Soc.* **87,** 1795 (1965); **90,** 1034 (1968).
20. P. Bitha, J. J. Hlavka, and J. H. Boothe, *J. Med. Chem.* **13,** 89 (1970).
21. G. D. Mack and R. B. Ashworth, *J. Chromatog. Sci.* **16,** 93 (1978).
22. D. Hoffman, *J. Org. Chem.* **31,** 792 (1966).
23. J. R. D. McCormick and co-workers, *J. Am. Chem. Soc.* **79,** 2849 (1957).
24. K. Koschel and co-workers, *Biochem. Z.* **344,** 76 (1966); see Ref. 58.
25. C. R. Stephens and co-workers, *J. Am. Chem. Soc.* **78,** 1515 (1956).
26. A. P. Doerschuk, B. A. Bitler, and J. R. D. McCormick, *J. Am. Chem. Soc.* **77,** 4687 (1955).
27. E. G. Remers, G. Steiger, and A. Doerschuk, *J. Pharm. Sci.* **52,** 753 (1963).
28. U.S. Pat. 3,009,956 (Nov. 21, 1961), M. Noseworthy (to Chas. Pfizer & Co.).
29. C. R. Stephens and co-workers, *J. Am. Chem. Soc.* **85,** 2643 (1963); U.S. Pat. 3,028,409 (Apr. 3, 1962), C. R. Stephens (to Chas. Pfizer & Co.).
30. M. J. Martell, Jr., A. S. Ross, and J. H. Boothe, *J. Med. Chem.* **10,** 485 (1967).
31. G. C. Barrett, *J. Pharm. Sci.* **52,** 309 (1963).
32. A. Brunzell, *Acta. Chem. Scand.* **16,** 245 (1962).
33. R. Huttenrauch and J. Keiner, *Naturwissenschaften* **53,** 552 (1966).
34. W. J. Gottstein, W. F. Minor, and L. C. Cheney, *J. Am. Chem. Soc.* **81,** 1198 (1959).
35. W. Seidel, A. Soder, and F. Linder, *Munch. Med. Wschr.* **17,** 661 (1958); South Afr. Pat. Spec. 57 3169 (1957).

36. J. H. Boothe and co-workers, *J. Am. Chem. Soc.* **80,** 1654 (1958).
37. L. H. Conover, *Symposium Antibiotics of Mold Metabolites,* Spec. Publ. No. 5, The Chemical Society, London, UK, 1956.
38. J. J. Hlavka and P. Bitha, *Tetrahedron Lett.* **32,** 3843 (1966).
39. J. R. McCormick and co-workers, *J. Am. Chem. Soc.* **87,** 1793 (1965).
40. R. C. Esse and co-workers, *J. Am. Chem. Soc.* **86,** 3874, 3875 (1964); South Afr. Pat. Appl. 63 4791 (1963) (to American Cyanamid Co.); J. H. Van Den Hende, *J. Am. Chem. Soc.* **87,** 929 (1965).
41. R. K. Blackwood and C. R. Stephens, *J. Am. Chem. Soc.* **86,** 2736 (1964); *Can. J. Chem.* **43,** 1382 (1965).
42. M. Schach von Wittenau, F. Hochstein, and C. R. Stephens, *J. Org. Chem.* **28,** 2454 (1963).
43. M. Kolosov and co-workers, *Chemistry of Antibiotics,* Vol. 1, Academy of Science, Moscow, USSR, p. 180.
44. J. R. D. McCormick and co-workers, *J. Am. Chem. Soc.* **82,** 3331 (1960); South Afr. Pat. Appl. 58 513 (May 13, 1958) (to American Cyanamid Co.).
45. M. Schach von Wittenau and co-workers, *J. Am. Chem. Soc.* **84,** 2645 (1962).
46. J. J. Hlavka and co-workers, *J. Am. Chem. Soc.* **84,** 1426 (1962).
47. J. J. Hlavka and H. M. Krazinski, *J. Org. Chem.* **28,** 1422 (1963).
48. J. J. Hlavka, H. M. Krazinski, and J. H. Boothe, *J. Org. Chem.* **27,** 3674 (1962).
49. J. H. Boothe and co-workers, *J. Am. Chem. Soc.* **82,** 1253 (1960).
50. J. Petisi and co-workers, *J. Med. Pharm. Chem.* **5,** 538 (1962).
51. J. L. Spencer and co-workers, *J. Med. Chem.* **6,** 405 (1963).
52. J. J. Beereboom and co-workers, *J. Am. Chem. Soc.* **82,** 1003 (1960).
53. M. J. Martell, Jr., A. S. Ross, and J. H. Boothe, *J. Med. Chem.* **10,** 359 (1967).
54. P. Shu, *J. Am. Chem. Soc.,* 4529 (1966).
55. L. Bernardi and co-workers, *Il Farmaco Ed. Sci.* **30,** 1025 (1975).
56. M. Burstall, *Manuf. Chem.* **31,** 474 (1960).
57. J. Doluisio and A. Martin, *J. Med. Chem.* **6,** 16 (1963).
58. B. Rasmussen and co-workers, *Antimicrob. Agents Chemother.* **35,** 2306 (1991).
59. W. Rogalski in Ref. 16, p. 179.
60. A. Cammarata and co-workers, *Mol. Pharmacol.* **6,** 61 (1970).
61. F. Peradejordi, A. Martin, and A. Cammarata, *J. Pharm. Sci.* **60,** 577 (1971).
62. J. Colaizzi and P. Klink, *J. Pharm. Sci.* **58,** 1185 (1969); M. Schach von Wittenau, *Chemotherapy (Basel)* **13,** 41 (1968).
63. J. R. D. McCormick and co-workers, *J. Am. Chem. Soc.* **80,** 5572 (1958).
64. M. Martell, A. Ross, and J. Boothe, *J. Am. Chem. Soc.* **89,** 6780 (1967).
65. L. H. Conover and co-workers, *J. Am. Chem. Soc.* **84,** 3222 (1962).
66. R. B. Woodward, *Pure Appl. Chem.* **6,** 561 (1963).
67. W. Martin, H. Hartung, H. Urboch, and W. Durckheimer, *Tetrahedron Lett.,* 3513 (1973).
68. J. J. Korst and co-workers, *J. Am. Chem. Soc.* **90,** 439 (1968).
69. H. Muxfeldt and W. Rogalsky, *J. Am. Chem. Soc.* **87,** 933 (1965).
70. H. Muxfeldt and co-workers, *J. Am. Chem. Soc.* **90,** 6534 (1968).
71. R. C. Evans, *The Technology of the Tetracyclines,* Quadrangle Press, New York, 1968.
72. *Tetracycline Manufacturing Processes,* Noyes Development Corp., Park Ridge, N.J., 1969.
73. J. J. Goodman, in Ref. 16, p. 5.
74. J. A. Robinson, *Chem. Soc. Rev.* **17,** 383 (1988).
75. R. Thomas and D. J. Williams, *J. Chem. Soc., Chem. Commun.* 128, (1983).
76. *Ibid.,* 677, (1983).
77. *Ibid.,* 443, (1984).

78. *Ibid.,* 802, (1985).
79. M. Patel and co-workers, *J. Antibiot.* **40,** 1408 (1987).
80. *Ibid.,* 1414 (1987).
81. E. B. Smith and co-workers, *J. Antibiot.* **40,** 1419 (1987).
82. M. J. Butler and co-workers, *Mol. Gen. Genet.* **215,** 231 (1989).
83. I. Vancurova and co-workers, *Biochem. J.* **253,** 263 (1988).
84. J. R. D. McCormick and G. O. Morton, *J. Am. Chem. Soc.* **104,** 4014 (1982).
85. I. Chopra in Ref. 16, p. 317.
86. D. Moazed and H. F. Noller, *Nature* **327,** 389 (1987).
87. M. A. Buck and B. S. Cooperman, *Biochemistry* **29,** 5374 (1990).
88. I. Chopra and co-workers, *J. Antimicrob. Chemother.* **8,** 5 (1981).
89. B. A. Cunha in Ref. 16, p. 393.
90. T. D. Edlind, *Antimicrob. Agents Chemother.* **33,** 2144 (1989).
91. B. Cryan and D. J. M. Wright, *J. Antimicrob. Chemother.* **25,** 187 (1990).
92. J. J. Leyden, K. J. McGinley, and A. M. Kligman, *Arch. Dermatol.* **118,** 19 (1982).
93. K. E. Toomeg and R. C. Barnes, *Rev. Infect. Dis.* **12** (Suppl. 6), 645 (1990).
94. H. B. Peterson, E. I. Galaid, and J. M. Zenilman, *Rev. Infect. Dis.* **12** (Suppl. 6), 656 (1990).
95. *Morbidity Mortality Weekly Rep.* **39,** 284 (1990).
96. J. F. Noble, K. Kanegis, and D. W. Hallesy, *Toxicol. Appl. Pharmacol.* **11,** 128 (1967).
97. W. R. Wilson, *Mayo Clin. Proc.* **52,** 635 (1977).
98. E. M. Ory, in B. M. Kagan, ed., *Antimicrobial Therapy,* 3rd ed., W. B. Saunders, Philadelphia, Pa., 1980, p. 117.
99. J. Jeljasewicz in G. G. Grassi and L. D. Sabath, eds., *New Trends in Antibiotics: Research and Therapy,* Elsevier/North-Holland Biomedical Press, Amsterdam, The Netherlands, 1981, p. 45.
100. M. Jones and B. A. Cunha, *Ther. Drug Monit.* **4,** 137 (1982).
101. S. Savin and G. Houin, *Clin. Pharmacokinet.* **15,** 355 (1988).
102. R. H. Gustafson and J. S. Kiser in Ref. 16, p. 405.
103. I. Chopra and A. H. Linton, *Adv. Microb. Physiol.* **28,** 211 (1986).
104. M. Robbins, R. Marais, D. Felmingham, and G. L. Ridgway, *J. Antimicrob. Chemother.* **20,** 379 (1987).
105. S. B. Levy in L. E. Bryan, ed., *Antimicrobial Drug Resistance,* Academic Press, New York, 1984, p. 191.
106. A. D. Russell and I. Chopra, *Understanding Antibacterial Action and Resistance,* Ellis Horwood, New York, 1990.
107. S. B. Levy, *J. Antimicrob. Chemother.* **24,** 1 (1989).
108. A. A. Salyers, B. S. Speer, and N. B. Shoemaker, *Mol. Microbiol.* **4,** 151 (1990).
109. I. Chopra, *J. Antimicrob. Chemother.* **18,** 51 (1986).
110. B. Eckert and C. F. Beck, *J. Biol. Chem.* **264,** 11663 (1989).
111. E. K. Manavathu and co-workers, *Antimicrob. Agents Chemother.* **34,** 71 (1990).
112. M. C. Roberts and S. L. Hillier, *Antimicrob. Agents Chemother.* **34,** 261 (1990).

JOSEPH J. HLAVKA
GEORGE A. ELLESTAD
IAN CHOPRA
American Cyanamid Company

ANTIOBESITY DRUGS

Obesity is a complex and widespread disease with multiple etiology. As of 1985, an estimated 34 million adults in the United States were 20% or more above their ideal body weight and could benefit from weight loss therapy (1). Stress and environment are important contributing factors to the development of obesity, but there is also a significant genetic component as indicated in a number of adoption and twin studies (2–5). For example, it has recently been suggested that an abnormality in skeletal muscle metabolism among obese individuals contributes to their excessive weight gain (6). In addition to being a cosmetic problem, obesity is associated with an enhanced likelihood of developing chronic conditions including hypertension, coronary heart disease (7–9), diabetes (10), cancer, gall bladder disease, and arthritis (1).

In order to define the obese state in a clinical setting, it is necessary to have a means of estimating the amount of adipose (fat) tissue relative to lean body mass. Whereas highly accurate determinations of body composition require complex laboratory procedures, large clinical studies typically employ measures of skin-fold thickness (11) or more commonly, body mass index (BMI) as a quantitative measure of obesity.

$$\text{BMI} = \frac{\text{weight in kilograms}}{(\text{height in meters})^2}$$

Normal individuals have a BMI of 20–25 and those with a BMI in the range of 25–30 are generally not at sufficiently increased risk to justify medical intervention. People having a higher BMI and particularly those having a BMI of over 40 may be considered morbidly obese and are candidates for increasingly aggressive treatment (12). An important consideration is the location of the excess fat. Although obesity is a serious health risk factor for both sexes (13,14), the abdominal or android pattern of fat distribution typical of overweight males carries a higher risk of life threatening complications than the gluteal or gynoid fat distribution typical of overweight females (15,16).

Obesity is a difficult condition to treat. Dietary restriction of caloric intake is the first line therapy and is optimally combined with an exercise program to promote loss of fat relative to lean body mass (17). For the grossly obese (BMI > 40), invasive mechanical measures such as jaw wiring, gastric banding, and gastric by-pass have been attempted with at least limited success (18).

Appetite suppressants have been widely used as an adjunct to dietary restriction and sympathomimetic amines have traditionally been used for this purpose. These agents have not proven particularly useful and frequently cause unacceptable side effects, hence their popularity has been waning for several years. The most promising newer drugs work through a serotoninergic mechanism and hold considerable promise at least for certain obese patients.

The majority of formerly obese patients eventually regain excess weight lost and thus a truly successful program must include behavior modification. Whereas no drugs are available for long-term use in a maintenance program, some agents are being studied clinically, such as the thermogenic ones. Among individuals who have been overweight for long periods, metabolic set points, such as lipoprotein lipase activity, may be altered favoring reestablishment of the obese state (19). Lipoprotein lipase is responsible for the hydrolysis of circulating triglycerides into free fatty acids, facilitating fatty acid uptake into adipocytes prior to reesterification and storage as lipid droplets. An elevation of messenger RNA (mRNA) levels of this enzyme occurs in obese patients following weight reduction (19). This finding may provide a partial explanation at the molecular level for this set point alteration.

Sympathomimetic Agents

The sympathetic or adrenergic nervous system operates in juxtaposition to the parasympathetic nervous system to maintain homeostasis in response to physical activity and physical or psychological stress. Sympathomimetic neurotransmission is generally mediated by norepinephrine [*51-41-2*] (**1**), $C_8H_{11}NO_3$, released from presynaptic storage granules upon stimulation. A second endogenous sympathomimetic agent, epinephrine [*51-43-4*] (**2**), $C_9H_{13}NO_3$, is released systemically from the adrenal glands during emergencies as part of the "fight or flight" response. Based on the discovery of selective ligands, two broad subclasses of receptors termed α- and β-adrenergic receptors have been defined. Each receptor subclass has two or more subtypes. A large variety of peripheral organ functions are affected by the sympathetic nervous systems including heart rate, cardiac contractile force, blood pressure, bronchopulmonary tone, and metabolism. In the central nervous system, sympathetic nervous stimulation results in increases in wakefulness, psychomotor activity, and reduction of appetite (20).

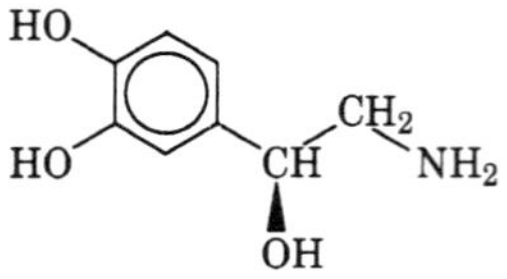

(1)

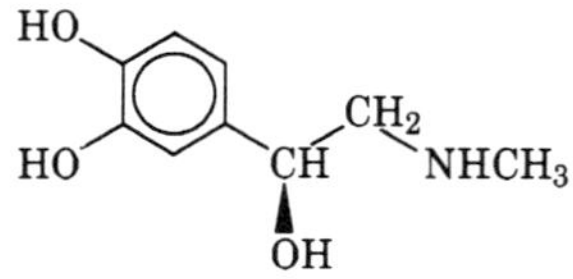

(2)

Compounds structurally related to the endogenous sympathomimetic amines epinephrine and norepinephrine have classically been employed as appetite suppressants. These agents, of which amphetamine [*300-62-9*], $C_9H_{13}N$, is the prototypical example, generally retain the phenethylamine, but lack the catechol moiety present in the natural substances. As a consequence, they are well absorbed after oral administration and readily distribute into the central nervous system where they exert their anorectic effects at hypothalamic appetite control centers. A component of their efficacy in promoting weight loss may be a result of a drug induced increase in metabolic rate through stimulation of thermogenesis or physical activity. These compounds act primarily by indirect mechanisms involving displacement of norepinephrine from presynaptic nerve storage vesicles or by prevention of its reuptake rather than by a direct effect at the receptor level (20). To a lesser extent, certain agents of this class affect dopaminergic or serotoninergic neurons. The overall pharmacological profile of members of the noncatechol sympathomimetic amine class depends on individual tissue distribution and mechanism of action.

Compounds available in the United States are listed in Table 1. Whereas they vary in degree, all of them share similar liabilities of cardiovascular side effects, the potential for central nervous system (CNS) stimulation, the development of tolerance, and abuse potential. All, with the exception of mazindol, are derivatives of phenethylamine. The introduction of an oxygen atom on the β-carbon of the side chain tends to reduce CNS stimulant properties without decreasing the anorectic activity. Following the Federal Controlled Drug Act of 1970, drugs were classified into one of five schedules according to medical utility and abuse potential.

Amphetamine and the other drugs in Schedule II have small sales and no legitimate role in antiobesity therapy, because of the pronounced potential for misuse. The compounds in Schedules III and IV, considered progressively safer, are prescribed more widely. They do not differ greatly in efficacy. However, the drug of choice for an individual depends on the patient's general health status with respect to subtle differences in side effect profiles. These agents are relatively effective over the course of at least several weeks and although tolerance is considered to be a problem, it does not always occur (12). Despite the prevalence of obesity in the United States, the use of prescription anorectic drugs has not increased in the past decade primarily as a result of concerns over abuse and general dissatisfaction with the long-term efficacy of sympathomimetics. As there are currently few alternatives for the drug therapy of obesity, the overall U.S. prescription market for antiobesity drugs has remained stagnant for several years and in 1989 amounted to only $75 million (21).

As indicated in Table 1, phenylpropanolamine [*14838-15-4*], $C_9H_{13}NO$, is the one agent listed which is available over the counter (OTC). It is present in a number of common diet aids and nasal decongestants having such names as Dexatrim, Contac, and Dimetapp (see also HISTAMINE AND HISTAMINE ANTAGONISTS). Because it has two asymmetric carbon atoms, phenylpropanolamine has the four possible stereoisomers (**3–6**) shown as Fischer projections. The compound sold as phenylpropanolamine in the United States is (±)-norephedrine and consists of equal amounts of D-(−) and L-(+)-norephedrine, (**3**) and (**4**), respectively, which have the erythro relationship between the amino and hydroxyl groups. The diastereomers with the opposite (threo) relationship between these groups are referred to as D-(−) and L-(+)-norpseudoephedrine, (**5**) and (**6**), respectively. In Europe and Australia the material sold as cathine and often referred to as phenylpropanolamine in the literature is (**6**). Although also used as an anorectic agent, (**6**) has greater CNS stimulant effects and a different toxicological profile from (±)-norephedrine (23). This situation has led to considerable confusion in the literature over side effects and efficacy of individual isomers (24).

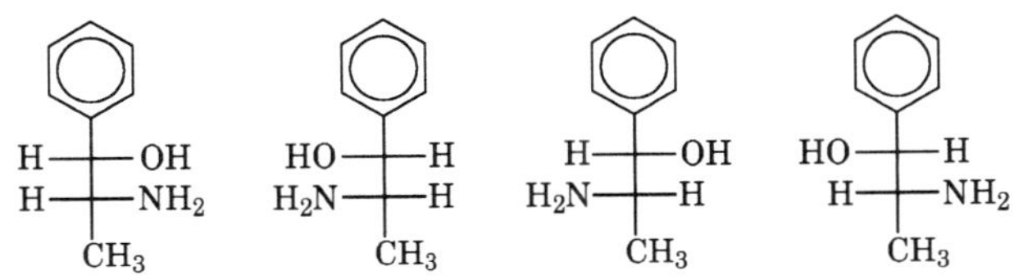

(**3**) (**4**) (**5**) (**6**)

(±)-Norephedrine is pharmacologically similar to other indirect acting sympathomimetic amines, but has little CNS stimulant activity. After high doses, it does affect peripheral α- and β-sympathetic receptors resulting in cardiovascular symptoms, but these are generally not troublesome at normal clinical doses. (Its efficacy as a nasal decongestant is related to vasoconstrictor effects on blood vessels in the nasopharynx.) (±)-Norephedrine is well absorbed after oral administration and is excreted largely unchanged in the urine over the course of 24 h (24). A number of clinical trials have focused on its efficacy in appetite suppression (25,26) indicating effectiveness in promoting weight loss by suppression of appetite. Because most of these trials were of short duration and rarely were run in comparison with other anorectic drugs, it is difficult to assess the relative utility of (±)-norephedrine with respect to the pharmacologically related prescription drugs in the longer term treatment of morbid obesity. Informal surveys indicate the product is largely used by nonmorbidly obese people to assist in short term weight loss programs following a period of unusual weight gain, as might occur for example during a vacation (25). U.S. drugstore and supermarket sales of (±)-norephedrine for the weight loss indication were estimated to be in the range of $100 million in 1989.

Serotoninergic Agents

Serotoninergic neurons utilize serotonin [*50-67-9*], $C_{10}H_{12}N_2O$, which is 5-hydroxytyrptamine (5-HT), as a neurotransmitter. Central serotoninergic neurons play an

Table 1. Sympathomimetic Agents in Use as Appetite Suppressants

Generic name[a]	CAS Registry Number[b]	Molecular formula	Chemical name	Trade name	Manufacturer[c]	Structure	Anorexic dose
				Schedule II			
amphetamine sulfate	[*60-13-9*]	$C_9H_{14}SO_4N$	(±)-α-methylbenzene-ethanamine sulfate	Benzedrine	Lannett	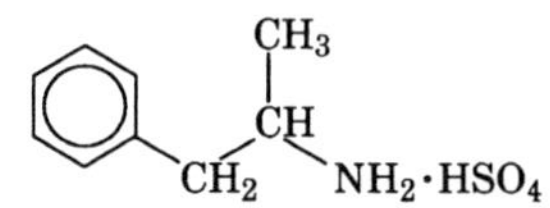	5–10 mg 3 × daily
dextroamphetamine sulfate	[*51-63-8*]	$C_9H_{14}NO_4S$	(*S*)-α-methylbenzene-ethanamine sulfate	Dexedrine	SKB[d]	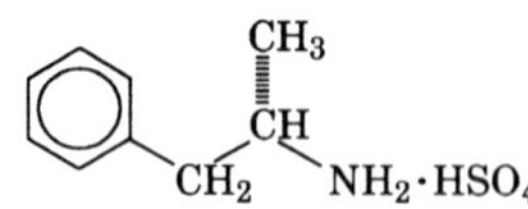	5–10 mg 3 × daily
methamphetamine hydrochloride	[*51-57-0*]	$C_{10}H_{16}ClN$	(*S*)-*N*,α-dimethyl-benzeneethanamine hydrochloride	Desoxyn	Abbott	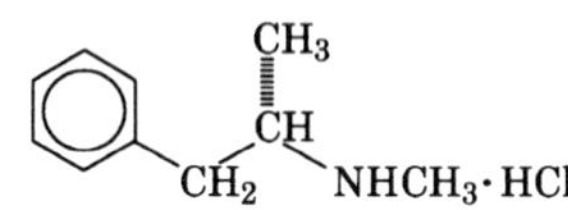	5–10 mg 1 × daily
phenmetrazine hydrochloride	[*1707-14-8*]	$C_{11}H_{16}ClNO$	2-phenyl-3-methyl-morpholine hydrochloride	Preludin	Boehringer Ingelheim	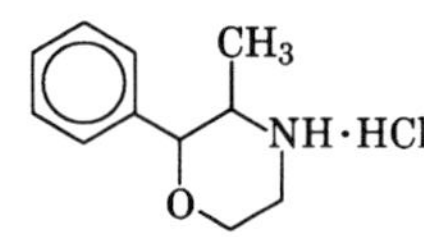	75 mg 1 × daily
				Schedule III			
phendimetrazine tartrate	[*50-58-8*]	$C_{16}H_{23}NO_7$	(2*S*, *trans*)-3,4-dimethyl-2-phenyl-morpholine tartrate	Plegine	Wyeth-Ayerst	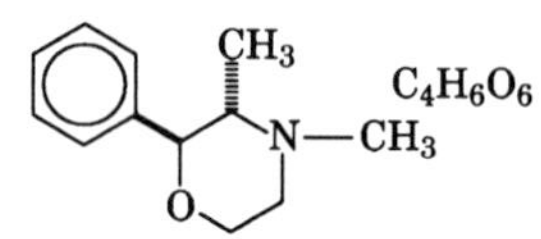	35 mg 2–3 × daily
benzphetamine hydrochloride	[*5411-22-3*]	$C_{17}H_{22}ClN$	(*S*)-*N*,α-dimethyl-*N*-(phenylmethyl)ben-zeneethanamine hydrochloride	Didrex	UpJohn	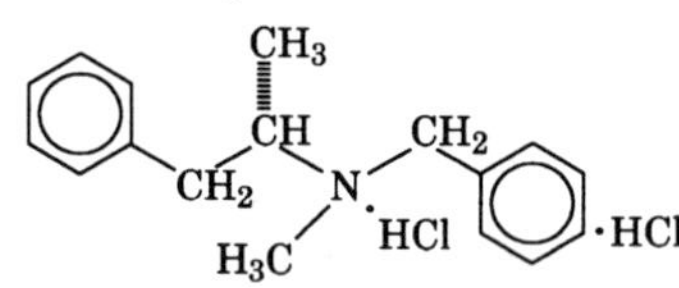	25–50 mg 1–3 × daily

				Schedule IV			
phenteramine hydrochloride	[*1197-21-3*]	$C_{10}H_{16}NCl$	α,α-dimethylbenzene-ethanamine hydrochloride	Fastin Adipex-P Ionamin	SKB[d] Lemmon Fisons	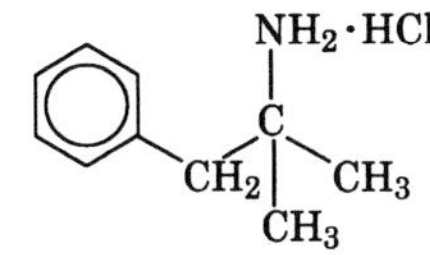	30 mg 1 × daily
diethylpropion hydrochloride	[*134-80-5*]	$C_{13}H_{20}ClNO$	2-(diethylamino)-1-phenyl-1-propanone hydrochloride	Tenuate Tepanil	Lakeside Rikker	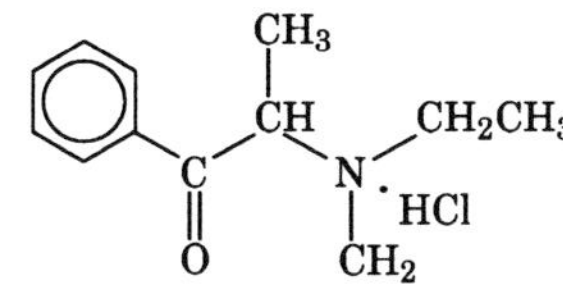	25 mg
mazindol	[*22232-71-9*]	$C_{16}H_{13}ClN_2O$	5-(4-chlorophenyl)-2,5-dihydro-3*H*-imidazo[2,1-a]iso-indol-5-ol	Sanorex Mazanor	Sandoz Wyeth-Ayerst	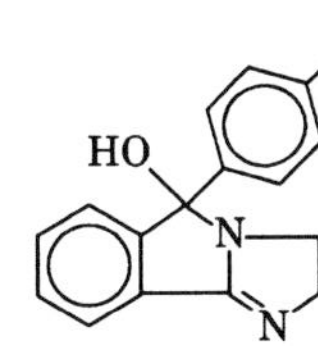	1 mg 1–3 × daily
				OTC[e]			
phenylpropanolamine hydrochloride[f]	[*154-41-6*]	$C_9H_{14}ClNO$	(±)-(*R*,*S*)-α-(1-amino-ethyl)benzene-methanol hydrochloride	Dexatrim Acutrim	Thompson Medical Company CIBA Consumer Pharmaceuticals	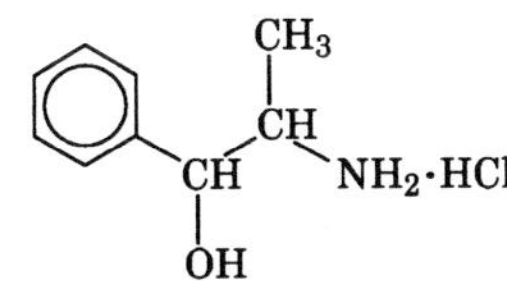	75 mg 1 × daily

[a]For physical properties and preparation, see Ref. 22.

[b]Most of these compounds have multiple registry numbers assigned to different salt forms and nomenclature variations. The registry number listed is that of the chemical name given.

[c]This listing is intended to be informative, not exhaustive, many of these compounds are distributed by several manufacturers.

[d]SmithKline Beecham, formerly SmithKline Beckman.

[e]OTC is over the counter.

[f]Also known as (±)-norephedrine hydrochloride.

important role in determining mood and food intake. Work with selective agonists and antagonists has led to the identification of a variety of receptor subtypes classified as 5-HT_{1a}, 5-HT_{1b}, 5-HT_{1c}, 5-HT_2, and 5-HT_3. Selective serotoninergic agents are receiving increased attention as antidepressants, antiemetics, and treatments for migraine. Stimulation of postsynaptic 5-HT_{1b} and 5-HT_{1c} receptors in the ventromedial hypothalamus has been implicated in the hypophagic response to serotonin agonists (27,28).

Fenfluramine, the structure of which is shown in Table 2, is a close analogue of amphetamine. However, fenfluramine acts as a serotoninergic agonist primarily by inhibiting serotonin reuptake and to a lesser extent by stimulation of serotonin release. This topic has been reviewed (29–31). Fenfluramine is a racemate. The (+)-enantiomer is twice as effective as the racemate as an appetite suppressant and has been shown in clinical trials to be relatively free of side effects (32,33). Dexfenfluramine [*3239-44-9*], $C_{12}H_{16}F_3N$, is currently in phase II clinical trials in the United States and is marketed in Europe by Servier. The U.S. license is Interneuran Pharmaceuticals. The (−)-enantiomer of fenfluramine, unlike the (+)-enantiomer, stimulates norepinephrine release and increases norepinephrine and dopamine metabolism (31). The catecholaminergic effects of the (−)-enantiomer may be largely responsible for the side effects encountered with fenfluramine. Rat studies indicate that whereas amphetamine tends to delay the onset of meals and decrease protein intake, fenfluramine decreases the rate of eating and promotes early meal termination with no effect on meal pattern or nutrient selection (34,35). Like many *N*-alkylphenethylamines, *N*-dealkylation is the primary metabolic pathway for both enantiomers of fenfluramine leading to norfenfluramine [*1886-26-6*] (**7**), $C_{10}H_{12}F_3N$. The (+)-metabolite is more potent as a stimulator of 5-HT release and less potent as an uptake inhibitor than its parent. It is also more persistent in the CNS and is thus in part responsible for the overall pharmacological profile of fenfluramine.

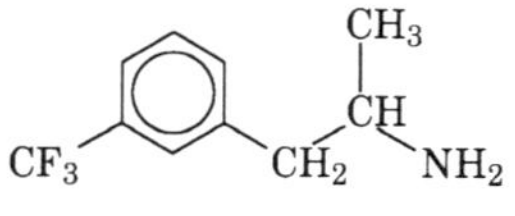

(**7**)

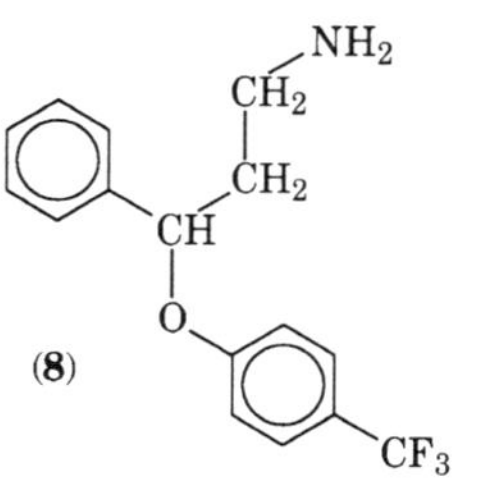

(**8**)

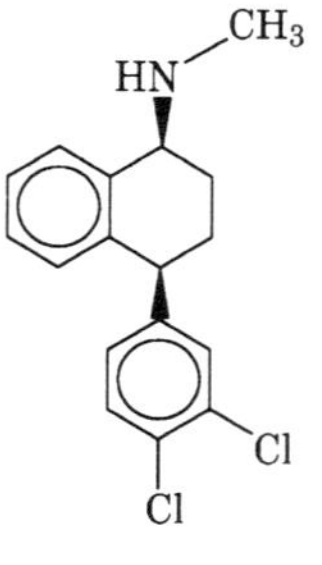

(**9**)

Fluoxetine, also shown in Table 2, is a serotonin reuptake inhibitor. It is more selective than fenfluramine and is being marketed by Eli Lilly as an antidepressant (36–38) (see PSYCHOPHARMACOLOGICAL AGENTS). Introduced in 1987, the 1990 sales were estimated at $700 million (39). Since it was observed in animal studies that fluoxetine decreased meal size and promoted weight loss, these findings have been extended to humans in a number of clinical trials. Fluoxetine has been consistently found to be effective in nondepressed humans and exhibits minimal side effects (40–43). The half-life of fluoxetine is 1–3 days and that of its

Table 2. Serotoninergic Agents as Appetite Suppressants

Generic name[a]	CAS Registry Number[b]	Molecular formula	Chemical name	Trade name	Manufacturer	Structure[c]	Anorexic dose
fenfluramine hydrochloride	*[404-82-2]*	$C_{12}H_{17}ClF_3N$	(±)-*N*-ethyl-α-methyl-3-trifluoromethylbenzene-ethanamine hydrochloride	Pondinim	Robbins	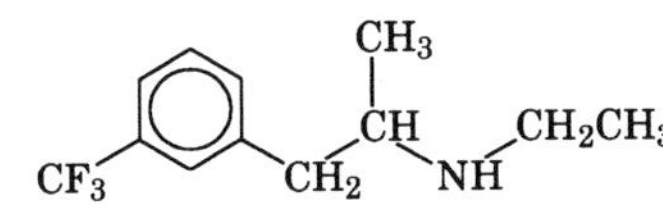	20 mg 3 × daily
fluoxetine hydrochloride	*[56296-78-7]*	$C_{17}H_{19}ClF_3NO$	(±)-*N*-methyl-γ-[4-(trifluoro-methyl)phenoxy]benzene-propanamine hydrochloride	Prozac	Lilly	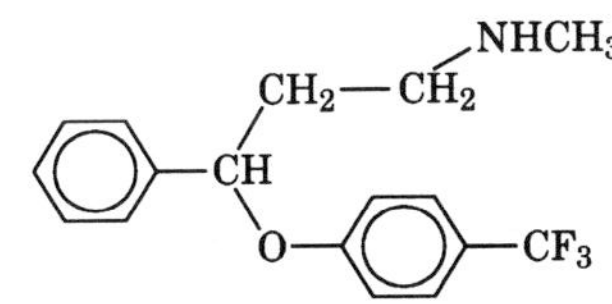	20 mg 3 × daily[d]

[a]For physical properties and preparation, see Ref. 22.
[b]These compounds have multiple registry numbers assigned to different salt forms and nomenclature variations.
[c]Structure does not reflect hydrochloride.
[d]Recommended antidepressant dose is 20 mg, 1–2 × daily.

principal metabolite, norfluoxetine [*56161-73-0*] (**8**), $C_{16}H_{16}F_3NO$, is 7–15 days (36). Fluoxetine is racemic and evidence has been presented indicating that both enantiomers are approximately equally active (44). An excellent synthesis relying on a catalytic chiral hydroboration reaction has been reported (45). It is likely that fluoxetine is being prescribed as an appetite suppressant although it is not yet approved for this indication. Other serotonin uptake inhibitors, such as sertraline [*79617-96-2*] (**9**), $C_{17}H_{17}Cl_2N$, which is being developed by Pfizer as an antidepressant, also has antiobesity activity and may eventually capture a part of this market (46,47).

Gut Hormones

Cholecystokinin (CCK) [*9011-97-6*] represents a family of gut and neurohormones of which the 33 amino acid peptide CCK-33 and the equipotent 8 amino acid peptide CCK-8 (**10**), are the most abundant circulating bioactive forms. Peripheral activities of CCK include stimulation of enzyme secretion from pancreatic acinar cells, inhibition of gastric emptying, and stimulation of vagal afferents terminating in the nucleus of the solitary tract and the area postrema. Exogenous administration of CCK to animals (48) or to humans (49,50) decreases meal size and in rats, elicits a sequence of behavior mimicking normal satiety. These effects have prompted the proposal that CCK released endogenously from the small intestine is a signaling hormone mediating postprandial satiety (51–53). Work with selective CCK antagonists (54) indicates that there are at least two distinct CCK receptors, a peripheral (CCK-A) type responsive to CCK-8 and a central (CCK-B) type responsive to the C-terminal tetrapeptide CCK-4. The precise role of CCK in terminating food intake remains to be defined, although animal studies suggest that both peripheral and central receptors are involved. Research is focused on developing selective CCK agonists which may be useful as adjuncts to dietary therapy by limiting meal size. Other peptides which have been suggested to play a role in regulating food intake include bombesin [*31362-50-2*], calcitonin, glucagon [*9007-92-5*], and corticotropin-releasing factor (55–57). However, the data suggesting a physiological role in regulating feeding behavior is not as strong for these compounds.

$$H_2N—Asp—Tyr(SO_3H)—Met—Gly—Trp—Met—Asp—Phe—NH_2$$

(10)

Lipid Adsorption Inhibitors

Dietary fat occurs largely in the form of triglycerides which must be hydrolyzed through the action of lipases, primarily pancreatic lipase, in the digestive tract prior to absorption. In a situation where maintenance or induction of weight loss is desired with a minimal impact on meal composition, an attractive approach to limiting caloric intake is to minimize the absorption of fat by inhibiting pancreatic lipase. Lipstatin is a pancreatic lipase inhibitor isolated from *Streptomyces toxytricini* (58–59). The corresponding tetrahydro derivative [*96829-58-2*] (**11**) has been shown to decrease fat absorption in mice, rats, and monkeys and to lower the rate of body weight gain in rats on a high calorie diet (60–62). This compound is undergoing clinical trials and potentially represents a new approach to antiobesity therapy.

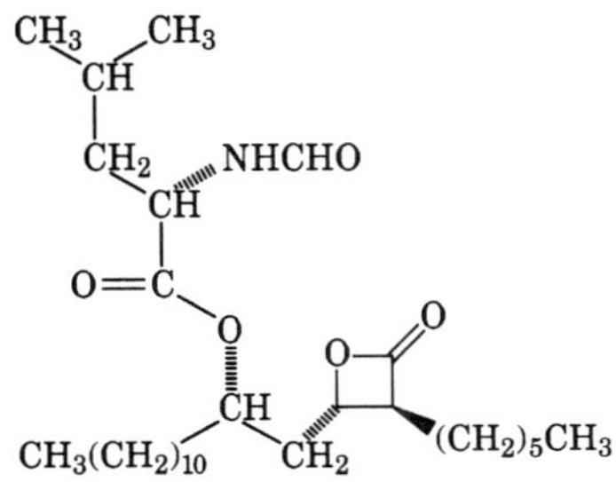

(**11**)

Thermogenic Agents

Drugs which stimulate or block peripheral adrenergic receptors have been in widespread use as, for example, antihypertensives, antianginal agents, and bronchodilators (see ANTIASTHMATIC AGENTS; CARDIOVASCULAR AGENTS). Detailed investigations with some of these has led to the proposal of a novel type of β-adrenergic receptor, termed the β_3-adrenergic receptor, which mediates catecholamine induced lipolysis in brown adipose tissue leading to increases in thermogenesis. The human gene coding this receptor has been identified and expressed in eukaryotic cells (63). Compounds such as ICI 201,651 [*107332-57-0*], $C_{19}H_{23}NO_6$, (**12**) (64) and BLR 37,344 [*90730-96-4*], $C_{19}H_{22}ClNO_4$, (**13**) (65) have been identified which selectively stimulate this receptor and lead to increases in metabolic rate and thermogenesis in rodents with little effect on food intake. In humans such compounds could be expected to decrease the efficiency of food utilization and could be particularly helpful to those patients whose metabolic set-points favor an inappropriate body mass.

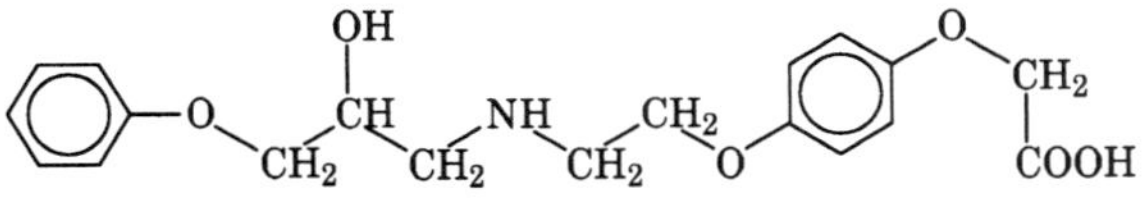

(**12**)

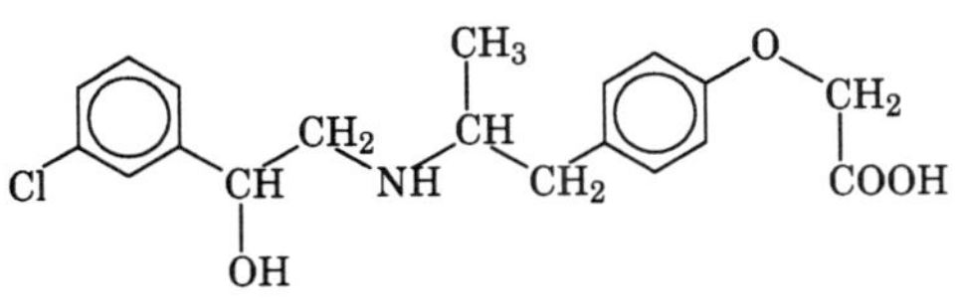

(**13**)

Economic Aspects

The most striking aspect of the antiobesity market is the relatively small fraction of dollars which are spent on drugs in comparison with diet foods, meal replacements, and other treatments for obesity such as diet centers. The overall market for the prescription adrenergic agents listed in Table 1 was about $75 million in 1989 and is not growing. The somewhat larger sales of phenylpropanolamine-based products reflect use by a wider audience than the morbidly obese as a consequence of their OTC availability. Several promising new agents are currently being developed. Foremost among these is the serotoninergic agent fluoxetine. Whereas it is not approved in the U.S. for use in obesity, some percentage of its approximately $700 million in 1990 sales were derived from this off-label use.

BIBLIOGRAPHY

"Appetite-Suppressing Agents" in *ECT* 3rd ed., Vol. 3, pp. 184–193, by S. T. Ross (Smith, Kline & French Laboratories).

1. *Ann. Int. Med.* **103,** 1073–1077 (1985).
2. C. Bouchard and co-workers, *N. Engl. J. Med.* **322,** 1477–1482 (1990).
3. A. J. Stunkard, J. R. Harris, N. L. Pedersen, and G. E. McClearn, *N. Engl. J. Med.* **322,** 1483–1487 (1990).
4. R. A. Price, R. J. Cadoret, A. J. Stunkard, and E. Troughton, *Am. J. Psychiatry* **144,** 1003–1008 (1987).
5. A. J. Stunkard and co-workers, *N. Engl. J. Med.* **314,** 193–198 (1986).
6. A. J. Wade, M. M. Marbut, and J. M. Round, *Lancet* **335,** 805–808 (1990).
7. R. J. Garrison, M. Feinleib, W. P. Castelli, and P. M. McNamara, *J. Am. Med. Assoc.* **249,** 2199–2203 (1983).
8. H. B. Hubert, P. M. Feinleib, P. M. McNamara, and W. P. Castelli, *Circulation* **67,** 968–977 (1983).
9. F. W. Ashley and W. B. Kannel, *J. Chronic Dis.* **27,** 103–114 (1974).
10. E. Barrett-Connor, *Epidemiol. Rev.* **11,** 172–181 (1989).
11. A. S. Jackson and M. L. Pollock, *Phys. Sportmed.* **13,** 76–90 (1985).
12. G. A. Bray, *Dis.-a-Mon.* **35,** 449–540 (1989).
13. J. E. Manson and co-workers, *N. Engl. J. Med.* **322,** 882–889 (1990).
14. A. Rissanen and co-workers, *Brit. Med. J.* **301,** 835–837 (1990).
15. A. H. Kissebah and co-workers, *J. Clin. Endocrinol. Metab.* **54,** 254–260 (1982).
16. M. Krotkiewski, P. Björntorp, L. Sjöström, and U. Smith, *J. Clin. Invest.* **72,** 1150–1162 (1983).
17. I. D. Caterson, *Drugs* **39**(Suppl. 3), 20–32 (1990).
18. J. F. Munro and co-workers, *Ann. N.Y. Acad. Sci.* **499,** 305–312 (1987).
19. P. A. Kern, J. M. Ong, B. Saffari, and J. Carty, *N. Engl. J. Med.* **322,** 1053–1059 (1990).
20. N. Weiner, in A. G. Gilman, L. S. Goodman, T. W. Rall, and F. Murad, eds., *The Pharmacological Basis of Therapeutics,* 7th ed., Macmillan, New York, 1985, pp. 145–169.
21. IMS Drug Store and Hospital Audit, IMS America, Plymouth Meeting, Pa., 1989.
22. *Merck Index,* 11th ed., Merck Corporation, Rahway, N.J., 1989.
23. J. P. Morgan, in J. P. Morgan, D. V. Kagan, and J. S. Brody, eds., *Phenylpropanolamine: Risks, Benefits, and Controversies,* Greenwood Press, Westport, Conn., 1985, pp. 180–194.
24. M. Weiner, in Ref. 23, pp. 25–36.
25. M. Weintraub, in Ref. 23, pp. 53–79.

26. H. Blumberg and J. P. Morgan, in Ref. 23, pp. 80–93.
27. G. A. Kennett, C. T. Dourish, and G. Curzon, *Eur. J. Pharmacol.* **141,** 429–435 (1987).
28. G. A. Kennett and G. Curzon, *Psychopharmacology* **96,** 93–100 (1988).
29. N. E. Rowland and J. Carlton, *Prog. Neurobiol.* **27,** 13–62 (1986).
30. C. Nathan and Y. Rolland, *Ann. N.Y. Acad. Sci.* **499,** 277–296 (1987).
31. S. Garattini and co-workers, *Appetite* **7,** 15–38 (1986).
32. N. Finer, D. Craddock, R. Lavielle, and H. Keen, *Curr. Ther. Res. Clin. Exp.* **38,** 847–854 (1985).
33. G. Enzi and co-workers, *Clin. Neuropharmacol.* **11**(Suppl. 1), S173–S178 (1988).
34. J. J. Wurtman and R. J. Wurtman, *Curr. Med. Res. Opin.* **6**(Suppl. 1), 28–33 (1979).
35. J. E. Blundell and co-workers, *Curr. Med. Res. Opin.* **6**(Suppl. 1), 34–52 (1979).
36. L. Lemberger, N. A. Farid, R. F. Bergstrom, and R. L. Wolen, *Int. J. Obesity* **11**(Suppl. 3), 157–161 (1987).
37. R. W. Fuller and D. T. Wong, *Drug Dev. Res.* **17,** 1–15 (1989).
38. R. W. Fuller, D. T. Wong, and D. W. Robertson, *Med. Res. Rev.* **11,** 17–34 (1991).
39. M. Waldholz, *Wall St. J.* B1 (July 18, 1990).
40. J. M. Ferguson and J. P. Feighner, *Int. J. Obesity* **11**(Suppl. 3), 163 (1987).
41. L. R. Levine, S. Rosenblatt, and J. Bosomworth, *Int. J. Obesity* **11**(Suppl. 3), 185 (1987).
42. R. L. Zerbe, *Int. J. Obesity* **11**(Suppl. 3), 191 (1987).
43. L. R. Levine and co-workers, *Int. J. Obesity* **13,** 635–645 (1989).
44. D. W. Robinson, J. H. Krushinski, R. W. Fuller, and J. D. Leander, *J. Med. Chem.* **31,** 1412–1417 (1988).
45. E. J. Corey and G. A. Reichard, *Tetrahedron Lett.* **30,** 5207–5210 (1989).
46. B. K. Koe, *J. Clin. Psychiatry* **51**(Suppl. B), 13–17 (1990).
47. J. Heym and B. K. Koe, *J. Clin. Psychiatry* **49**(Suppl. 40–45) (1988).
48. J. J. Antin and co-workers, *J. Comp. Physiol. Psychol.* **89,** 784–790 (1975).
49. H. R. Kissileff, F. X. Pi-Sunyer, J. Thornton, and G. P. Smith, *Am. J. Clin. Nutr.* **34,** 154–160 (1981).
50. N. Muurahainen, H. R. Kissileff, A. J. Derogatis, and F. X. Pi-Sunyer, *Physiol. Behav.* **44,** 645–649 (1988).
51. A. J. Silver and J. E. Morley, *Prog. Neurobiol.* **36,** 23–34 (1991).
52. G. P. Smith, in A. J. Stunkard and E. Stellar, eds., *Eating and Its Disorders,* Raven, New York, 1984, pp. 67–75.
53. C. A. Baile, C. L. McLaughlin, and M. A. Della-Fera, *Physiol. Rev.* **66,** 172–234 (1986).
54. R. F. Freidinger, *Med. Res. Rev.* **9,** 271 (1989).
55. S. F. Leibowitz, *Fed. Proc.* **45,** 1396–1403 (1986).
56. J. E. Morley and co-workers, *Peptides* **6**(Suppl. 2), 181–192 (1985).
57. J. E. Morley, *Endocr. Rev.* **8,** 256–287 (1987).
58. E. K. Weibel and co-workers, *J. Antibiot.* **40,** 1081–1085 (1987).
59. E. Hochuli and co-workers, *J. Antibiot.* **40,** 1085–1089 (1987).
60. S. Hogan and co-workers, *Int. J. Obesity* **11**(Suppl. 3), 35–42 (1987).
61. E. Fernandez and B. Borgström, *Biochim. Biophys. Acta* **1001,** 249–255 (1989).
62. P. Hadvary, H. Lengsfeld, and H. Wolfer, *Biochem. J.* **256,** 357–361 (1988).
63. L. J. Emorine and co-workers, *Science* **245,** 1118–1121 (1989).
64. B. R. Halloway, in H. Lardy and F. Stratman, eds., *Hormones, Thermgensis, and Obesity,* Elsevier, New York, 1989, pp. 477–484.
65. J. R. S. Arch and co-workers, *Nature (London)* **309,** 163–165 (1984).

JEFFERSON W. TILLEY
Hoffmann-LaRoche Corporation

ANTIPARASITIC AGENTS

ANTHELMINTICS

Human infections caused by worms (helminths) represent one of the most important public health problems in the world. Helminths form three main categories or phyla: Platyhelminths, flatworms; Aschelminthes, roundworms; and Nemathelminthes, thorny-headed worms. Platyhelminths consist of Trematoda and Cestoda. Members of the class Trematoda, or flukes (schistosomes), are slender leaf-shaped organisms that possess attachment organs in the form of cup-shaped hooks called suckers. Members of the class Cestoidea, or tapeworms, are flat and ribbonlike. These worms have serially repeated sections behind the neck and an attachment organ called the scolex. The Aschelminthes (class Nematoda) are roundworms, which have a basic cylindrical shape with major variations in proportions, size, and structure. The Nemathelminthes (Acanthocephala), or thorny-headed worms, are widely distributed among animals and generally do not involve a human host.

Over the last several years, substantial progress in the discovery and development of anthelmintic drugs has been made. Effective agents are available for most human gastrointestinal infections (Table 1); however, drugs that are effective in treating the extraintestinal complications of many helminthic infections are still needed.

Frequently, the treatment of helminthic diseases requires adjunct medication. Allergic reactions are commonly seen as a result of tissue invasion by worms or as a consequence of anthelmintic therapy. Antihistamines and corticosteroids may be necessary adjuncts to therapy. Anemia, indigestion, and secondary bacterial infections can also occur and may require concomitant therapy with hematopoietic drugs and appropriate antibiotics.

Treatment of Trematode Infections

Blood Flukes (Schistosomiasis). Three main species of blood flukes cause schistosomiasis in humans: *Schistosoma haematobium, S. mansoni,* and *S. japonicum.* The effectiveness of antischistosomal drugs depends on the reduction or arrest of egg production. Infection only occurs as a result of the penetration of the intact skin by free-swimming cercaria. After they develop to pre-adult forms in the skin and lung, the parasites migrate through host blood vessels to various tissues. Adult worms that are approximately 2 cm in length remain paired within the portal or intestinal vasculature of the host. The female, however, travels as far as possible toward the capillary beds to lay eggs, a portion of which embolize to the liver and induce granuloma formation. During the early stages of the disease, patients may experience fever, gastrointestinal distress, headache, and fatigue. In later stages of the disease, signs of hepatic fibrosis and ascites are seen. The number of worm pairs found in a patient may vary from a few to over 100. In untreated patients, adult worms are now generally assumed to live less than 5–10 years.

Praziquantel. This drug (**1**), $C_{19}H_{24}N_2O_2$, can be used to treat schistosomiasis caused by any one of the three major species. Praziquantel is an acylated pyrazino–isoquinoline derivative that has replaced the traditional (and more toxic) trivalent antimonial compounds.

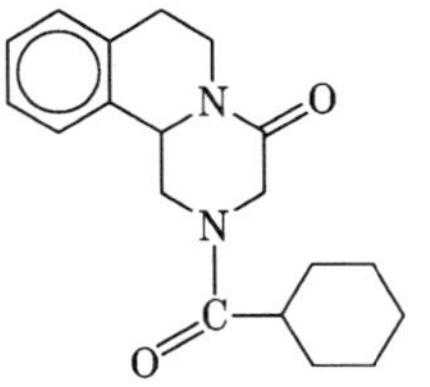

(**1**)

Table 1. Anthelmintics: Uses and Properties

Drug	Structure number	CAS Registry Number	Trade name	Physical properties	Solubility	Disease (organism)	Reference
praziquantel	**(1)**	*[55268-74-1]*	Biltricide solid, Cesol, Livera, Pontel, Prazi, Pyquiton	mp 136–138°C	practically insoluble in water; sparingly in alcohol, chloroform, and ether	flukes (*C. sinensis, F. buski, P. westermani, S. haematobium, S. mansoni, S. japonicum*), tapeworms (*H. nana, T. solium, T. saginata, D. latum*), and larval stage (cysticercosis)	1,2
oxamniquine	**(2)**	*[21738-42-1]*	Mansil, Vansil	pale yellow crystals	sparingly soluble in water; soluble in acetone, chloroform, and methyl alcohol	flukes (*S. mansoni*)	3
metrifonate	**(3)**	*[52-68-6]*	Bilarcil	white crystals, mp 83–84°C	1 g in 6.5 mL water, 33 mL chloroform, 5.9 mL ether	blood flukes (*Schistosoma haematobium*)	4
bithionol[a]	**(4)**	*[97-18-7]*	Bitin	crystals or gray-white powder, mp 188°C	insoluble in water, acetone, and 4% NaOH	lung flukes (*Paragonimu westermani*), sheep liver flukes (*Fasciola hepatica*)	5,6
tetrachloroethylene	**(5)**	*[127-18-4]*	Nema worm capsules, Perklone	colorless, nonflammable fluid, ethereal odor, deteriorates in warm climates, bp 121°C	soluble 1:10,000 in water; miscible with ethanol	intestinal fluke (*Fasicolopsis buski*)	7

niclosamide	**(6)**	[*50-65-7*]	Cestocida, Mansonil, Nasemo, Niclocide, Sulqui, Yomesan	pale yellow crystals, or yellow-white powder, mp 225–230°C	insoluble in water; sparingly soluble in ethanol	tapeworms (*T. saginata*, *T. solium*, and *D. latum*)	8
quinacrine	**(7)**	[*83-89-6*]	Atabrine hydrochloride[b]	yellow crystals,[b] mp ca 250°C	1 g in 36 mL water; slightly soluble in ethanol	pork tapeworm (*Taenia solium*)	9
piperazine citrate[c]	**(8)**	[*144-29-6*]	Antepar, Helmezine, Pinozan, Pinrou, Uvilon	white crystalline powder, mp ca 185°C (dec)	freely soluble in water; insoluble in ethanol	roundworms (*Ascaris lumbricoides*)	10
diethylcarbamazine citrate	**(9)**	[*1642-54-2*]	Banocide, Filarex, Hetrazan, Notezine	white crystalline powder, mp 141–143°C	soluble in water and hot ethanol; insoluble in acetone	lymphatic filariases (*Wuchereria bancrofti*, *Loa loa*, and other filaria)	11
pyrantel pamoate	**(10)**	[*22204-24-6*]	Antiminth, Cobantil, Combantin, Helmex, Pyrantin	white or yellow or tan crystalline powder, melts and decomposes, mp ca 250°C	insoluble in water and nearly so in ethanol	pinworms (*Enterobius vermiculus*), roundworms (*Ascaris lumbricoides*), hookworms (*Ancylostoma duodenale*, *Necator americanus*)	12
mebendazole	**(11)**	[*31431-39-7*]	Meben, Totamin, Vermox, Vorme	crystals or yellow amorphous powder, mp 289°C	very slightly soluble in water and most organic solvents	whipworms (*Trichuris trichiura*), hookworms (*Ancylostoma duodenale*, *Nector americanus*), filariasis (*Mansonella perstans*), roundworm (*Ascaris lumbricoides*), Trichinosis (*Trichinella spiralis*)	13

Table 1. (*Continued*)

Drug	Structure number	CAS Registry Number	Trade name	Physical properties	Solubility	Disease (organism)	Reference
thiabendazole	**(12)**	[*148-79-8*]	Foldan, Lombrstop, Mintezol, Minzolum, Nomoxiur, Thibenzole, Tiabenda, Triasox, TBZ	white crystals, mp ca 300°C	almost insoluble in water; slightly soluble in ethanol	hookworms (*Trichostrongylus sp.*, cutaneous larva migrans, *Angiostrongylus costaricensis*, and *Strongyloides stercoralis*)	14
ivermectin[a]	**(13)**	[d]	Eqvalan, Ivomec, Mectizan	off-white powder, mp 155–157°C	4 μg/mL water, insoluble in saturated hydrocarbons, very soluble in propylene glycol	filariasis (*Onchocerca volvulus*)	15

[a]Available from the CDC Drug Service.
[b]The dihydrochloride dihydrate [*69-05-6*].
[c]Formed from 3 mol piperazine (**8**) and 2 mol citric acid.
[d]See Figure 1.

Praziquantel is thought to alter the permeability of the worm's cell membrane to calcium, resulting in tetanic paralysis and the dislodging of schistosomes from their attachment sites. The drug also causes the disintegration of the schistosome tegument, which is followed by host phagocytosis (16). Adverse reactions include dizziness, headache, and gastrointestinal distress. Cure rates of approximately 85% have been achieved with a single treatment regimen. The principal producer of praziquantel in the United States is Miles Pharmaceutical of West Haven, Connecticut. Other manufactures include Bayer A.G. of Leverkusen, Germany and Chinoin of Budapest, Hungary.

Oxamniquine. This tetrahydroquinoline (**2**), $C_{14}H_{21}N_3O_3$, has been proven effective for the treatment of *S. mansoni* infections. It is clinically ineffective against *S. haematobium* and *S. japonicum* infections.

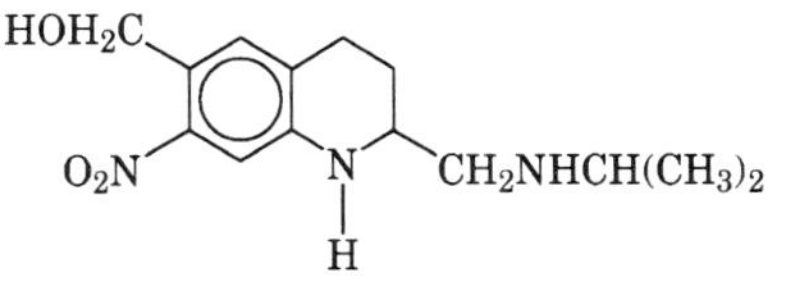

(**2**)

Its precise mode of action is not known, although it does exhibit anticholinergic properties (17). Oxamniquine is generally well tolerated, although dizziness, with or without drowsiness, does occur in at least one-third of the patients taking the drug. Allergic-type reactions, including fever, increased circulating immune complexes, pulmonary infiltrates, and pruritic skin rashes, may also occur. These allergic symptoms may be due to the death of the infecting parasites (18). A single dose is usually sufficient for strains of *S. mansoni* acquired in the Western Hemisphere; those acquired in East Africa may require longer periods of treatment. Oxamniquine is solely manufactured by Pfizer, Inc. in the United States, United Kingdom, and France.

Metrifonate. This organosphosphorus compound (**3**), $C_4H_8Cl_3PO_4$, was first used as an insecticide.

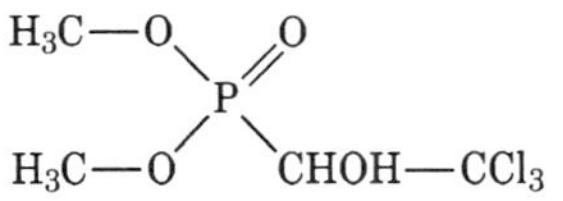

(**3**)

In the body, metrifonate converts to the active metabolite dichlorvos, (2,2-dichlorovinyl dimethyl phosphate), which is responsible for the inhibition of the enzyme acetylcholinesterase in the susceptible worm. This effect alone is unlikely to explain the antischistosomal properties of metrifonate (19). Clinically, metrifonate is effective only against infection caused by *S. haematobium*. Metrifonate is administered in three doses at 2-wk intervals (17). The drug is well tolerated. Side effects such as mild vertigo, nausea, and cramps are dose-related. This product is not available in the United States. The only manufacturer of metrifonate is Bayer A.G. of Leverkusen, Germany.

Lung and Liver Flukes. Paragonimus organisms (lung flukes) are about 1 cm long and normally live encapsulated within lung cysts. They primarily infect lung tissue, but may be found in other viscera or in the brain, producing tumorlike symptoms.

Fasciola organisms (liver flukes) are about 3 cm long. After extensive migration, young flukes invade and mature in the bile ducts of the liver, where heavy infections may result in fibrosis of the bile duct wall. Flukes migrating through the peritoneal cavity may become lodged in other locations, causing intense tissue reactions.

Bithionol. This compound (**4**), $C_{12}H_6Cl_4O_2S$, can be used to treat persons infected with either *Paragonimus* or *Fasciola* organisms. It is a phenolic compound similar in structure to hexachlorophene.

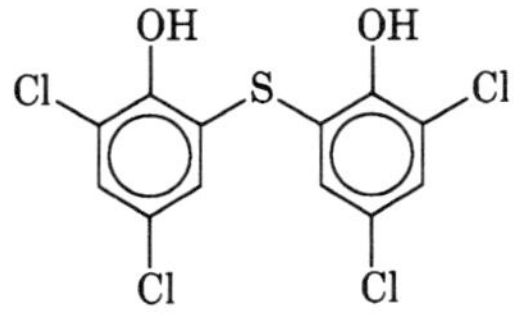

(**4**)

Bithionol and hexachlorophene have been used as anthelmintics in veterinary medicine. In the past, because of its antibacterial activity, bithionol was incorporated into at least 20 medicated skin cleansers manufactured in the United States; however, this agent proved to cause skin irritation, and its use as an antibacterial cleanser was discontinued.

Bithionol interferes with the neuromuscular physiology of helminths, impairs egg formation, and may cause defects in the protective cuticle covering the worm. At the biochemical level, the oxidative phosphorylation of the worm is inhibited.

Bithionol is an iron chelator and therefore may inactivate iron-containing enzyme systems (20). A 10-d oral regimen of bithionol may resolve the pathology ie, abnormalities of lung disease in about three months, but improvement in cerebral paragonimiasis and *Fasciola hepatica* (sheep liver fluke) infections are variable. Adverse effects include abdominal pain, diarrhea, vomiting, rashes, and photosensitivity. The drug is available in the United States through the CDC Drug Service of Atlanta, Georgia. The only manufacturer of bithionol is Tanabe Seiyaku of Osaka, Japan.

Praziquantel (**1**) (previously mentioned for the treatment of schistosomiasis) is also effective against all *paragonimus* species.

Chinese Liver Fluke. The adult worm of the Chinese liver fluke (*Clonorchis sinensis*) can grow to be 2 cm long. Worms infect the biliary tree where they cause local inflammation, diarrhea, and hepatomegaly in the acute infection. Progressive biliary obstruction and cirrhosis can occur in the more advanced disease state. The presence of 20–200 worms is common, but they may number over 20,000. Infection is the consequence of eating raw fish that contain viable parasites. Untreated worms can live for up to 30 years. Treatment is with praziquantil (**1**).

Intestinal Fluke. The intestinal fluke, *Fasciolopsis buski,* which can reach a length of 8 cm, lives attached to the wall of the small intestine. Rarely, untreated heavy infections may cause ulceration, toxemia, and death in children. Praziquantel (**1**) or tetrachloroethylene (**5**) are both effective in the treatment of *Fasciolopsis* infection.

Tetrachloroethylene. This chlorocarbon (**5**), $Cl_2C{=}CCl_2$, has been used as a solvent in dry cleaning and a metal degreaser as well as an anthelmintic. Tetrachloroethylene, C_2Cl_4, has replaced carbon tetrachloride, CCl_4, which is very hepatotoxic, in the treatment of intestinal flukes. Investigators do not agree on the mechanism of action of tetrachloroethylene or on its basic pharmacology. Some investigators believe that tetrachloroethylene dissolves the cell lipids in the helminthic neuromuscular system and interferes with neuromuscular function. In the treatment of flukes, a single oral dose is given on an empty stomach preceded by a day of low fat diet. A 24-h period of low fat intake is necessary because fat in the gastrointestinal tract increases drug absorption, thereby increasing its toxicity. This drug is primarily used in veterinary medicine (21).

Treatment of Cestode Infections

Fish tapeworm (*Diphyllobothrium latum*) grows to a length of 10 m with 4000 proglottides, or segments, each containing reproductive organs. The proglottides tend to remain attached to the intestinal wall and discharge eggs, which are passed into the feces. Besides ingesting host nutrients, this worm avidly sequesters vitamin B_{12} and folic acid. The closer to the stomach that the intestinal fluke is attached, the more vitamin the worm ingests. Depletion of host B_{12} supplies leads to the development of pernicious anemia, with associated neurological symptoms. There is usually only one worm in a patient and it can survive for 10 years.

Beef tapeworm (*Taenia saginata*) is found worldwide in people who eat rare beef and is most common in developing countries. Worms are often 5–10 m long with about 1000 proglottides. Usually a patient carries only one worm. Infected persons are frequently asymptomatic, but intestinal disturbances can occur. Detached sections of worm 0.5 cm × 2 cm may creep out of the end of the digestive tract.

Both the adult and the larval cysticerci (bladderworm) of *Taenia solium* (pork tapeworm) are able to live in humans; the parasite is found sporadically in uncooked pork. In the stomach, the larva is digested out of the pork flesh; it then grows and attaches to the wall of the small intestine. Maturity is reached in 5–12 weeks. The adult is 5 m long, and untreated adult worms may survive for 25 years.

Dwarf tapeworm (*Hymenolepis nana*) is the only human tapeworm that does not utilize an intermediate host. Infection is transmitted directly from person to person. *H. nana* is only 2–4 cm long, it is of universal distribution in mice and humans in temperate zones, where children, especially those in institutions, are most frequently infected. Although an infected person is often symptomless, this tapeworm can cause abdominal discomfort and diarrhea if infection is heavy.

Niclosamide. This drug (**6**) is a halogenated salicylanilide derivative, $C_{13}H_8Cl_2N_2O_4$.

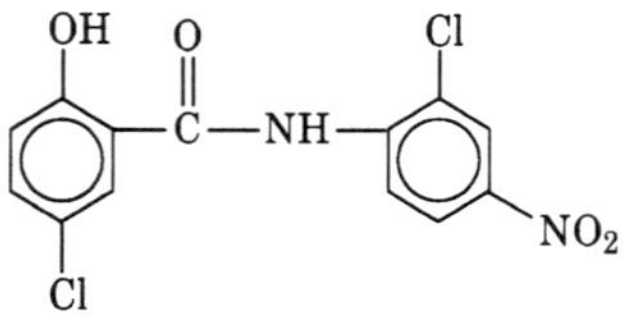

(**6**)

The modes of action for niclosamide are interference with respiration and blockade of glucose uptake. It uncouples oxidative phosphorylation in both mammalian and taenioid mitochondria (22,23), inhibiting the anaerobic incorporation of inorganic phosphate into adenosine triphosphate (ATP). Tapeworms are very sensitive to niclosamide because they depend on the anaerobic metabolism of carbohydrates as their major source of energy. Niclosamide has selective toxicity for the parasites as compared with the host because little niclosamide is absorbed from the gastrointestinal tract. Adverse effects are uncommon, except for occasional gastrointestinal upset.

Niclosamide causes the tapeworm head to disengage from the intestinal wall of the host and the body wall of the parasite to disintegrate. The first dose of niclosamide disposes of the worms in the intestinal cavity; however, any worms in developing stages in the tissues of the intestinal wall are unharmed. Therefore, dosage must be repeated in 5–7 days to destroy the worms that will finish their development and reenter the intestinal lumen.

Niclosamide given in a single oral dose is effective for the treatment of tapeworm infections. A large percentage of patients are cured, and adjunct therapy is not required when used against beef tapeworm (*T. saginata*), the dwarf tapeworm (*H. nana*), or the fish tapeworm of humans (*D. latum*) and most other human intestinal cestode infections. It is also effective in the treatment of pork tapeworm (*T. solium*), but not against infection with larval *T. solium* (cysticercosis) for which praziquantel (**1**) is preferred. Niclosamide is available in the United States and is manufactured by Miles Pharmaceutical of West Haven, Connecticut. Other manufacturers include Bayer A.G. of Leverkusen, Germany and Bellon Laboratories of Neuilly-sur-Seine, Cedex, France.

Quinacrine. Quinacrine (**7**), $C_{23}H_{30}ClN_3O$, is an acridine derivative. It is used in the form of the dihydrochloride dihydrate, $C_{23}H_{32}Cl_3N_3O{\cdot}2H_2O$, [*69-05-6*].

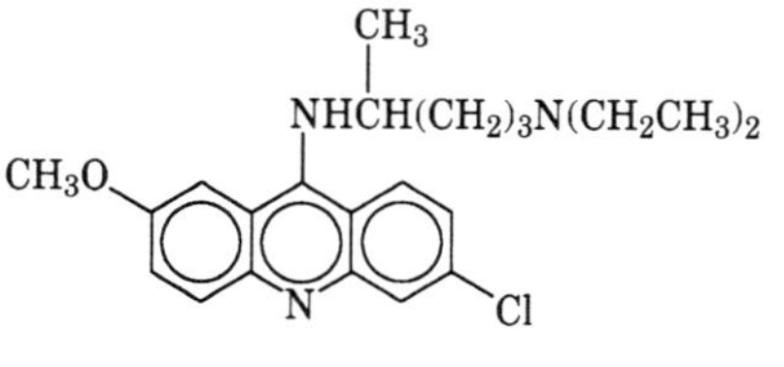

(**7**)

This drug was used prior to the availability of niclosamide and is considered less satisfactory for the treatment of tapeworms then niclosamide (24). It causes more side effects and produces severe nausea. Quinacrine, however, is preferred by some clinicians for the treatment of *Taenia solium* infection because, unlike niclosamide, it expels the worms intact, thus reducing the theoretical risk of cysticercosis (25).

Quinacrine concentrates in the scolex of the parasite and causes the muscles needed for holding onto the intestinal wall to relax. The worms are stained yellow and pass from the body, still alive. Quinacrine can intercalate with DNA and inhibit nucleic acid synthesis. It creates fluorescent bands in deoxyadenylate–deoxythmidylate-rich regions of DNA and has been used as a stain in the study of human genetics.

The drug is readily absorbed from the gastrointestinal tract and may persist in human tissues for two months after administration. Quinacrine is given as a single dose for the treatment of tapeworms; it is manufactured by Winthrop-Breon of New York, New York.

Paromomycin. Paromomycin [*7542-37-2*] is a broad-spectrum antibiotic that has been used in the treatment of *H. nana* and *T. saginata,* and amebiasis, and also as an antibacterial agent in cases of diarrhea and dysentery (see ANTIBIOTICS, OLIGOSACCHARIDES).

Treatment of Nematode Infections

Female roundworms of *Ascaris lumbricoides* are about 30 cm × 0.5 cm. The adults live and feed in the small intestine where they can cause abdominal discomfort or pain. Frequently, light infections are symptomless. The worm develops through four larval stages, with a molt between each successive stage. The eggs of the adult are deposited in the feces. Eggs do not hatch in the soil or enter the body through the cutaneous route. Infective eggs are swallowed and the larvae hatch in the stomach. Hatched larvae actively migrate into the lungs. Larvae continue to develop in the lungs, then migrate through the respiratory passages to the throat and are swallowed. In the stomach they molt and grow to the mature adult form. Untreated adult worms may live for a year and a half.

The pinworm (*Enterobius vermicularis*) is distributed worldwide in temperate zones. In cities, it is frequently a disease of households or institutions. Female worms are about 1 cm long. While growing to maturity, they live in the gut lumen, attaching by their mouths to the intestinal mucosa. Itching in the perianal region occurs when the mature female migrates out the anus to discharge her eggs. Usually an intestinal population is less than 100 worms; however, more than 5000 have been recovered from a single patient. The eggs are resistant to drying and may get into household dust and be inhaled or swallowed by household members, causing light infections with few worms in the adult relatives of infected children.

Whipworm (*Trichuris trichiura*) adult females are 5 cm long. These worms thread their entire body into the epithelium of the colon, where they feed on tissue juice and small amounts of blood. Infections of several hundred worms may cause irritation and inflammation of the mucosa, with abdominal pain, diarrhea, and gas. Eggs are discharged and passed into the feces. Infections result from the swallowing of eggs that are obtained directly from contaminated soil. Untreated adult worms live for years.

The two most common hookworms in humans are *Ancylostoma duodenale* and *Necator americanus. A. duodenale* adult worms are 8–13 mm in length. The filariform larva found in moist soils may be either ingested or penetrate the skin of its host. It is then carried through the circulatory system to the lungs and migrates up the respiratory tree into the digestive tract. The worms feed on intestinal tissue and blood. Some worms may persist in humans as long as nine years. Infestations cause cutaneous reactions, pulmonary lesions, intestinal ulcerations, and anemia.

American hookworm (*N. americanus*) adult worms, 7–11 mm in length, may live up to 15 years. The life cycle of the *N. americanus* is similar to *A. duodenale* except that *N. americanus* requires a pulmonary migration with development in the lungs before it can develop to the adult worm in the intestine.

Piperazine Salts. Piperazine citrate is highly effective against both *A. lumbricoides* (large roundworm) and *E. vermicularis* (pinworm). A large number of substituted piperazine derivatives exhibit anthelmintic activity; however, only diethylcarbamazine (**9**) is used clinically (26).

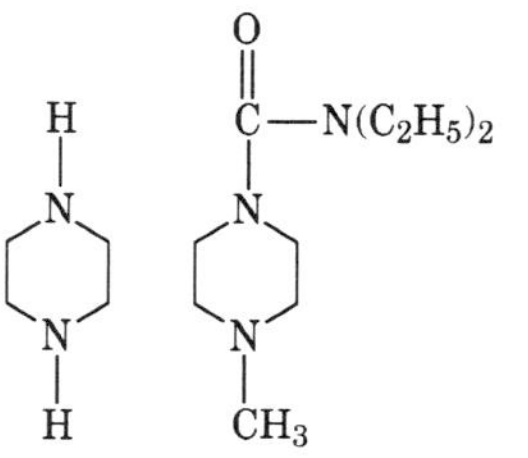

(**8**) (**9**)

Piperazine is taken orally by nonfasting subjects. For large roundworms, one dose a day on two consecutive days reportedly cures approximately 90% of cases treated (27). This drug can be used for pinworms with a 7-d course of treatment.

Piperazine causes flaccid paralysis of *Ascaris* by blocking the worms' ability to respond to acetylcholine, thus dislodging the worms from their position in the digestive tract. The worms are passed through the digestive tract still alive. The drug is well absorbed from the human intestine, and the reason for its selective toxicity is still uncertain.

Occasional side effects from piperazine include abdominal cramps, nausea, vomiting, and diarrhea. Rarely, patients may experience headache, vertigo, and tremors, or they may feel weak and have difficulty concentrating. Red patches can appear on the skin, sometimes with the flat, elevated, itching welts of urticaria. Piperazine is manufactured by Burroughs Wellcome Company in Research Triangle Park, North Carolina and Wallace Laboratories of Cranbury, New Jersey. Other manufacturers include Laboratories Clin Midy of Paris, France, Regent Laboratories Ltd. of London England, and Lennon Ltd. of Saxonwold, South Africa.

Pyrantel Pamoate. This drug (**10**), $C_{34}H_{30}N_2O_6S$, cures pinworm infections and is close to 100% effective against *Ascaris* when given in a single dose (28,29). Pyrantel pamoate is also a principal drug for the treatment of hookworm infections; thus it is useful in patients with mixed worm infections.

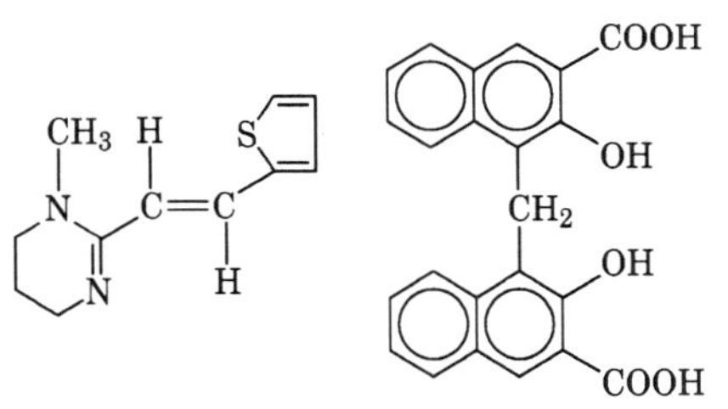

(10)

For hookworms, treatment is continued for three days. In roundworms, muscle tissue is stimulated continuously, resulting in spastic paralysis. The drug was introduced in veterinary medicine in 1966 and then applied to clinical medicine three years later.

Since about 85% of the administered dose is passed unchanged in the feces of the patient, selective toxicity of the drug can be attributed primarily to poor absorption. Side effects include abdominal pain, nausea, vomiting, diarrhea, loss of appetite, headaches, and vertigo or drowsiness. Skin rashes can also develop. Pyrantel pamoate is produced by Pfizer, Inc., New York, New York.

Mebendazole. In therapeutic dosages administered orally for three or four consecutive days, mebendazole (**11**) is highly effective against whipworm (*Trichuris trichiura*). In addition, a single dose usually cures pinworm infections. Mebendazole is also effective against trichinosis (*Trichinella spiralis*), hookworms (*Ancylostoma duodenale* and *Necator americanus*), and *Ascaris,* partially effective against threadworms (*Strongyloides*), and taeniid tapeworms. This benzimidazole derivative, $C_{16}H_{13}N_3O_2$, is a broad-spectrum anthelmintic (30) and is used in veterinary medicine as well as in clinical practice. It belongs to the same series of drugs as thiabendazole, $C_{10}H_7N_3S$ (**12**).

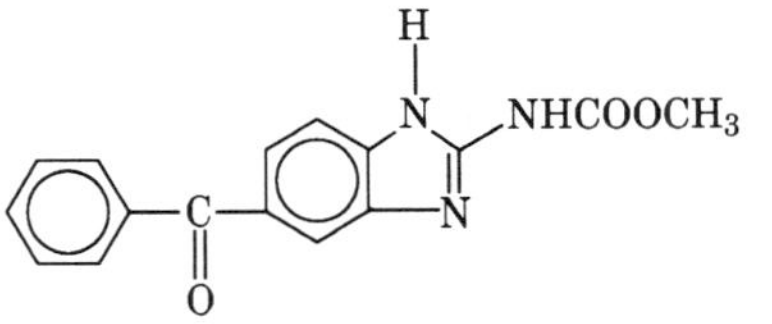

(11)

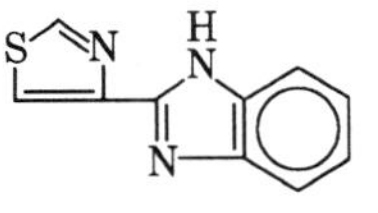

(12)

These drugs are closely related to fungicides such as benomyl [*17804-35-2*] (**13**), $C_{14}H_{18}N_4O_3$.

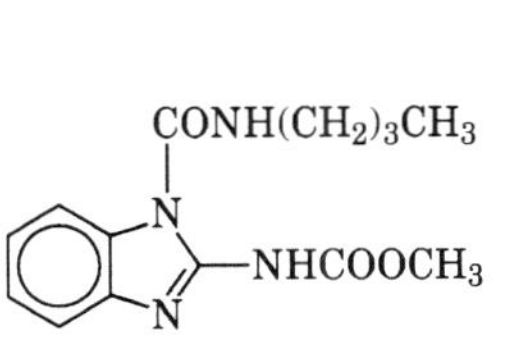

(13)

Mebendazole interferes with the glucose metabolism of helminths, irreversibly inhibiting glucose uptake. When the glycogen stores in the worms are depleted, ATP supplies fail. The worms are slowly immobilized, die, and are excreted over a period of three days. In developing *Trichuris* eggs, larvae fail to develop normally. Selective toxicity of mebendazole may be due to its antimicrotubule action on nematode intestinal cells with no effect on mammalian cells. Furthermore, relatively little drug is absorbed from the human gastrointestinal tract. Adverse effects in patients include abdominal pain and diarrhea. Mebendazole is manufactured by Janssen Pharmaceutica, Inc., Piscataway, New Jersey. Other manufacturers include Esteve Drug of Barcelona, Spain and Cadila Laboratories of Maninagar, Ahmedabad, India.

Treatment of Tissue Roundworm (Nematode) Infections

Many parasitic worms cause systemic infections outside the gastrointestinal tract. These include: *Strongyloides stercoralis* (threadworm), *Trichinella spiralis, Dracunculus medinensis,* and the several species of nematodes that cause filariasis (*Mansonella perstans* and *Onchocerca volvulus*).

S. stercoralis mature adult worms are between 2 and 3 mm long. Threadworms have two separate developmental stages. In the first stage, the worm has a complete free-living nonparasitic life cycle, and in the second stage, a filariform larvae penetrates the skin of its host and moves into cutaneous blood vessels. The larvae are then carried to the lungs. In the lungs, the migrating larvae break out of the pulmonary capillaries and migrate through the respiratory passages to the throat where they are swallowed. Once they are inside the intestine, they become imbedded in intestinal wall.

Trichinosis is the condition caused by the adult worm of *T. spiralis* which is between 1 and 2 mm long. The cysts are found in contaminated meat, primarily pork. When ingested, cysts release immature larvae, which invade the intestinal lining and develop into adult worms. Mature worms mate in the submucosa and deposit mobile larvae, which migrate into intestinal lymphatics. They are then disseminated throughout the body and become encapsulated within striated muscle fibers. Adult worms live about one month. Mebendazole (**10**) is highly effective against *T. spiralis* infections. Treatment with mebendazole for trichinosis is for 13 days.

Thiabendazole. This benzimidazole derivative (**12**) is an effective oral drug used in the treatment of intestinal roundworms and selected tissue parasites. Infections with *S. stercoralis* are commonly treated with thiabendazole for two days. Disseminated strongyloidiasis is treated for at least five days. Satisfactory results have been reported when used for *T. spiralis* infections; however, its effectiveness on encapsulated muscle larvae has not been clinically demonstrated. Extensively used in veterinary medicine, thiabendazole was the first broad-spectrum benzimidazole anthelmintic. It is widely used in the United States. The mechanism of action is not clearly understood, but it has been shown that thiabendazole inhibits the enzyme fumarate reductase, which is found specifically in the mitochondria of helminths (31).

One ppm of thiabendazole prevents *Ascaris* eggs from maturing. In therapeutic topical concentrations, it is used to prevent the development of dog and cat hookworm larvae that have invaded human skin, causing cutaneous larva migrans. In laboratory animals, thiabendazole has antipyretic, analgesic, and antiinflammatory properties, as well as anthelmintic properties. These properties are probably also exhibited in humans and may account for the symptomatic improvement in persons with early systemic trichinosis.

The adverse effects include digestive disturbances, neurological symptoms, and manifestations of allergic responses. As many as half of the patients taking it are incapacitated by some of these adverse reactions for several hours. Whether these symptoms are caused by hypersensitivity to the drug, the parasite, or by a manifestation of the disease is not known. Overall, effects are dose-related and transient.

Guinea Worm. Thiabendazole (**12**) and metronidazole (**14**) [*443-48-1*], $C_6H_9N_3O_3$, the broad-spectrum antiprotozoal agent

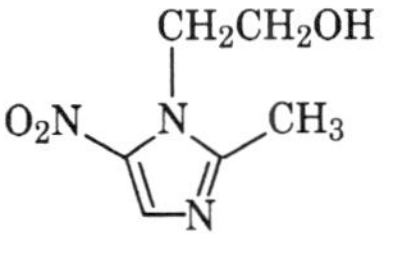

(**14**)

are used in the treatment of the guinea worm (*Dracunculus medinensis*) (see also ANTIPARASITIC AGENTS, ANTIPROTOZOALS). These stringlike worms, about 1 m long but less than 2 mm in diameter, can be seen and felt below the skin surface. The head of the worm is usually exposed in an ulcer on the foot. The far end of the worm is hooked, so that although it is possible to grasp the front end and pull, the worm does not slide out easily and is apt to break, causing infectious, allergic, and toxic consequences as the remainder of the worm under the skin dies and deteriorates. After drug therapy, the worm either exits or it can be pulled easily without danger of breakage. Although metronidazole and thiabendazole have no direct effect on guinea worms, their effectiveness against the guinea worm may be due to their antiinflammatory effects. If left untreated, within a month the worm may come out naturally, withdraw from the opening, and be resorbed or become calcified in the body.

Filarias. Diethylcarbamazine (**9**) (orally administered) has been successful for decades in the treatment of filariasis, eg, elephantiasis (32). It kills microfilariae in the blood of the filarial worms *Wuchereria bancrofti, Brugia malayi,* and *Loa loa.*

Diethylcarbamazine has limited antimicrofilarial activity against *Onchocerca volvulus.* Adults of *W. bancrofti,* the filarial worm causing elephantiasis, coil in the lymph system. Here females can attain a length of 10 cm. Over the years, tissue reactions result in obstruction to lymph return. Lymph nodes, lymph vessels, and the spleen become enlarged. The condition of elephantiasis is a late and unusual complication of filariasis, where the lower extremities of the body become edematous, enlarge, and over a period of time harden with a rough nodular skin.

Untreated adult worms live for 5–10 years. *Loa loa* females are about 6 cm long and migrate constantly through connective tissues. Calabar swelling occurs as a result of local responses to worms. The swelling lasts two or three days and then subsides. If a worm migrates into the anterior chamber of the eye, the patient may be able to see it. Sometimes one eye is swollen shut when a worm is in the vicinity. Untreated worms will live as long as 10 years.

Microfilariae in the tissues of the lung (ie, eosinophilic lung or tropical eosinophilia) produce symptoms of chest pain, cough, and asthma, which are often relieved by the administration of diethylcarbamazine.

Diethylcarbamazine. This derivative of piperazine (**9**), causes the microfilariae to become susceptible to phagocytosis by the macrophages in the reticuloendothelial system. Its pharmacological effects enhance the body's immunologic response. The mechanism of the filaricidal action on adult *Wuchereria* and *Loa* is unknown. The selective toxicity of diethylcarbamazine is as yet unexplained. It is well absorbed and widely distributed throughout the body and does not accumulate, even after repeated doses. The adverse effects of nausea, fever, headache, and dizziness may occur during diethylcarbamazine therapy. The simultaneous administration of corticosteroids and antihistamines is intended to reduce the occurrence of rash, itching, edema, and other severe allergic reactions to the disintegration of the worms. Diethylcarbamazine is used in veterinary and clinical medicine. It is manufactured by Lederle Laboratories of Pearl River, New York. Other manufacturers include the Wellcome Foundation Ltd. in Cheshire, England, Cidan of Castellon, Spain, and Specia S.A. in Paris, France.

Mansonella perstans adult worms live encysted in various peritoneal tissues (eg, pericardium, pleura, and peritoneum). *M. perstans* microfilariae are not affected by diethylcarbamazine but mebendazole has been successful in treating *M. perstans* filariasis (33).

Ivermectin. This drug (**15**) is a semisynthetic derivative of one of the avermectins, which are macrocyclic lactones produced by the actinomycete *Streptomyces avermitilis* (34) (Fig. 1). This antiparasitic agent has a broad spectrum of activity against nematode worms and insects in animals and is widely used in veterinary medicine (see ANTIPARASITIC AGENTS, AVERMECTINS). The drug acts on susceptible organisms by potentiating the release of γ-aminobutyric acid [*56-12-2*] (GABA) and binding the GABA at postsynaptic sites on the neu-

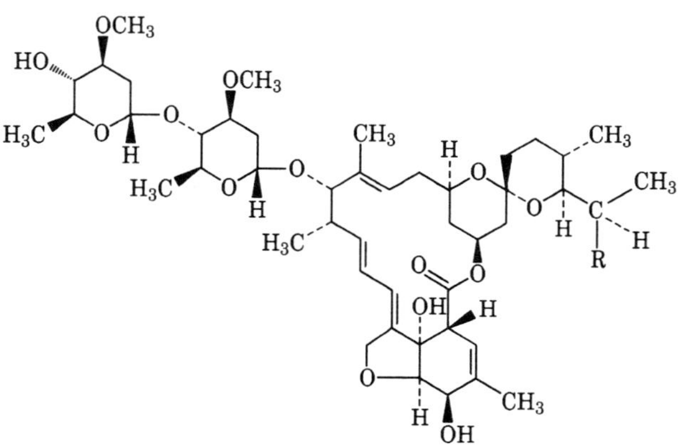

(**15**)

Fig. 1. Ivermectin [*70288-86-7*] is a mixture containing at least 80% 22,23-dihydroavermectin B_{1a} [*70161-11-4*], wherein R = C_2H_5, and not more than 20% B_{1b} [*70209-81-3*], wherein R = CH_3.

romuscular junction, thus paralyzing the nematode. Ivermectin acts slowly and has a long duration of action. Ivermectin has no effect on adult filariid nematodes, but it has proved to be superior to diethylcarbamazine as a microfilaricidal agent in the treatment of onchocerciasis. Neither drug completely cures the infection. More recent medical studies have also shown that ivermectin has a high activity against various intestinal nematodes (eg, *A. lumbricoides, T. trichiura,* and *E. vermicularis*) (35).

Systemic reactions are less severe than with diethylcarbamazine. The most commonly seen reactions are fever, rash, and lymph-node pain or swelling. Suppressive ivermectin therapy consists of a single oral dose every 6–18 months. The required duration of suppressive therapy is unknown, probably at least three years (36). Ivermectin is available from the CDC Drug Service on request. It is manufactured by Merck Sharp and Dohme in the United States and England.

BIBLIOGRAPHY

"Chemotherapeutics, Anthelmintic" in *ECT* 3rd ed., Vol. 5, pp. 451–468, by J. W. Ingalls, John Wiley & Sons, Inc.

1. U.S. Pat. 4,001,411 (Jan. 4, 1977), P. Andrews, J. Seubert, and H. Thomas (to Merck & Co.).
2. Ger. Pat. 2,362,539 (Dec. 17, 1973), P. Andrews, J. Seubert, and H. Thomas (to Merck & Co.).
3. U.S. Pat. 3,903,283 (Sept. 2, 1975), H. Richards (to Pfizer, Inc.).
4. U.S. Pat. 2,701,225 (Feb. 1, 1955), A. Lorenz (to Bayer AG).
5. Ger. Pat. 583,055 (Aug. 28, 1933), F. Muth (to I.G. Farbenind AG).
6. U.S. Pat. 2,849,494 (Aug. 26, 1958), R. H. Cooper and K. Goldberg (to Monsanto).
7. U.S. Pat. 3,040,109 (June 9, 1959), R. E. Feathers and R. H. Rogerson (to Pittsburgh Plate Glass).
8. U.S. Pat. 3,079,297 (Feb. 26, 1963), E. Schranfstraller and R. Gonnert (to Bayer AG).
9. U.S. Pat. 2,113,357 (Apr. 15, 1939), F. Mietzsch and H. Mauss (to Winthrop Chemical Co.).
10. U.S. Pat. 2,901,482 (Aug. 25, 1959), G. F. Mackenzie and K. L. Turbin (to The Dow Chemical Company).
11. U.S. Pat. 2,467,893; 2,467,894; 2,467,895 (Apr. 19, 1949), S. Kushner and L. Brancone (to American Cyanamid Co.).
12. South Afr. Pat. 6,800,516 (June 27, 1968), R. V. Kasubrick and J. W. McFarland (to Chas. Pfizer, Inc.).
13. U.S. Pat. 3,657,267 (Feb. 18, 1971), J. C. H. VanGelder (to Janssen).
14. U.S. Pat. 3,017,415 (June 16, 1962), L. Sarett (to Merck & Co.).
15. U.S. Pat. 4,199,569 (Aug. 4, 1979), J. Chabala and M. Fisher (to Merck & Co.).
16. P. Andrews and co-workers, *Med. Res. Rev.* **3,** 147 (1983).
17. *Control of Schistosomiasis,* Technical report Series, WHO No. 728, World Health Organization, Washington, D.C., 1985.
18. N. Katz, F. Zicker, and J. P. Pereira, *Rev. Inst. Med. Trop. Sao Paulo* **18,** 371–377 (1976).
19. A. Bloom, *Acta. Pharmacol. Toxicol.* **49,** 109–113 (1981).
20. *Informational Material for Physicians, Bithionol,* Centers for Disease Control, Atlanta, Ga., 1986, pp. 1–18.
21. M. Abramovicz, ed., *Med. Lett.* **30,** 18 (1988).
22. J. Putter and Z. Parasitenk, **34,** 23 (1970).

23. J. Putter and P. Andrews, *Conference of the German Society for Parasitology,* Wuppertal, Apr. 9–11, 1970.
24. F. Richards and P. M. Schantz, *Lancet* **1,** 1264 (1985).
25. Ref. 21, p. 15.
26. O. D. Standen in R. J. Schnitzer and F. Harding, eds., *Experimental Chemotherapy,* Vol. 1, Academic Press, Inc., New York, 1963, pp. 702–892.
27. R. H. R. White and J. W. Scopes, *Lancet* **1,** 256–258 (1960).
28. N. E. Pitts and J. R. Migliardi, *Clin. Pediatr.* **13,** 87–94 (1974).
29. T. S. Bumbalo, D. J. Fugazzotto, and J. V. Wyczalek, *Am. J. Trop. Med. Hyg.* **18,** 50–52 (1969).
30. H. VandenBossche, *Biochem. Pharmacol.* **29,** 1981–1990 (1980).
31. H. M. Robinson and C. S. Samorodin, *Arch. Dermatol.* **112,** 1757–1760 (1976).
32. F. Hawking, *Adv. Pharmacol. Chemother.* **16,** 129–194 (1979).
33. M. Wahlgren and J. Frolov, *Trans. R. Soc. Trop. Med. Hyg.* **77,** 422 (1983).
34. W. C. Campbell and co-workers, *Science* **212,** 823–828 (1983).
35. C. Naquira and co-workers, *Am. J. Trop. Med. Hyg.* **40,** 304–309 (1989).
36. *Informational Material for Physicians, Ivermectin,* Centers for Disease Control, Atlanta, Ga., 1986, pp. 1–23.

General References

P. C. Beaver, R. C. Jung, and E. W. Cupp, *Clinical Parasitology,* 9th ed., Lea and Febiger, Philadelphia, Pa., 1984, 825 pp.

P. E. C. Manson-Bahr and D. R. Bell, *Mansons's Tropical Diseases,* 19th ed., Bailliere Tindall, Philadelphia, Pa., 1987, 1557 pp.

A. S. Benson, ed., *Control of Communicable Diseases in Man,* 14th ed., American Public Health Association, Washington, D.C., 1985, 485 pp.

E. Braunwald and co-workers, eds., *Harrison's Principles of Internal Medicine,* 11th ed., McGraw-Hill, New York, 1987, pp. 805–829.

W. C. Campbell and R. S. Rew, eds., *Chemotherapy of Parasitic Diseases,* Plenum Press, New York, 1986, 655 pp.

T. Cheng, *General Parasitology,* Academic Press, New York, 1973, 355 pp.

L. L. Gustafsson, B. Beermann, and Y. A. Abdi, *Handbook of Drugs for Tropical Parasitic Infections,* Taylor and Francis, Philadelphia, Pa., 1987, 151 pp.

Health Information for International Travel 1989, U.S. Department of Health and Human Services, Publication No. (CDC) 89-8280, 1989.

K. F. Lampe and co-workers, *AMA Drug Evaluations,* W. B. Saunders, Philadelphia, Pa., pp. 1565–1613.

E. K. Markell and M. Voge, *Medical Parasitology,* 5th ed., W. B. Saunders, Philadelphia, Pa., 1984, 374 pp.

J. F. Reynolds and co-workers, *Martindale, The Extra Pharmacopeia,* 29th ed., The Pharmaceutical Press, London, 1989, pp. 47–69, 505–521.

G. D. Schmidt and L. Roberts, *Foundations of Parasitology,* 3rd. ed., C. V. Mosby, St. Louis, Mo., 1985.

E. A. Swinyard in A. R. Gennaro, ed., *Remington's Pharmaceutical Sciences,* 17th ed., Mack Publishing Company, Easton, Pa., 1985, pp. 1234–1240.

M. Verderame and J. Mackiewicz in M. Verderame, ed., *CRC Handbook of Chemotherapeutic Agents,* Vol. 2, CRC Press, Boca Raton, Fla., 1986, pp. 95–178.

L. T. Webster in L. S. Goodman, A. Gilman, T. W. Rall, and F. Murad, eds., *The Pharmacological Basis of Therapeutics,* 7th ed., Macmillan, New York, 1985, pp. 1004–1028.

JOHN BECHER
CDC Drug Service

ANTIMYCOTICS

Since the 1960s, there has been such a rapid evolution in the pharmacotherapy of mycoses that three phases of the development of antimycotic agents can be distinguished: (*1*) compounds that existed before griseofulvin (1939); (*2*) compounds presently available on the market, including itraconazole (1987); (*3*) new antimycotics that are being studied now and will be prescribed tomorrow. Both milestones (griseofulvin and itraconazole) are intended for oral administration. Topical treatment of superficial dermatomycoses (trichophytosis, favus, and microsporosis) has been complemented by systemic oral treatment.

The world market for antimycotics in 1980 was valued at a total of \$350 million. About 10% of this came from the sale of systemic (intravenous or oral) antifungals. In 1990, the antifungal world market was estimated at \$1560 million of which almost \$280 million was accounted for by systemic antifungals. This important increase in market value has induced more interest in research activities in the field of antifungals leading to more and more costly new treatment modalities. The five most prescribed antifungals today are miconazole, clotrimazole, ketoconazole, nystatin, and econazole. As this list indicates, among antifungal drugs the azoles are most significant; roughly 70% of all antifungals used today belong to this chemical category.

Though highly pathogenic, fungi represent only a minority; the importance of opportunistic infections is increasing from year to year. Human resistance against fungal infections is undermined by a number of factors and both exogenous and endogenous factors are capable of increasing the pathogenicity of certain fungi. Fungal growth may be stimulated when the host is treated with antibiotics, oral contraceptives, or cytostatics. Exogenous factors such as protheses, catheters, and valves may give rise to an increased incidence of mycoses. In immunocompromised patients and diabetics, saprophytic fungi may become pathogenic due to disturbances in the internal environment. These patients may present chronic mucocutaneous candidosis (CMC) due to *Candida albicans*. *Aspergillus fumigatus* and *A. niger* may become moderately to highly pathogenic in the presence of previously mentioned predisposing factors.

To mycologists, each fungal infection has something specific, either in its symptomatology or its etiology. However, this is less obvious to practitioners. The incidence and the severity of the pathology are sometimes underestimated. Mycoses may be classified as follows:

Superficial	Deep	
dermatophytosis	aspergillosis	histoplasmosis
trichophytosis	blastomycosis	maduromycosis
microsporosis	candidosis	paracoccidioidomycosis
epidermophytosis	chromomycosis	sporotrichosis
pityriasis	coccidioidomycosis	
candidosis	cryptococcosis	

The physician must decide whether to opt for systemic or topical treatment and which arguments will convince the patient to use one or both forms for the minimum period recommended. It is typical for mycoses that the period of treatment generally exceeds one week and may even last a few months. There is still a strong tendency to treat the organ or the part of the organ affected by the pathogenic microorganism. This is justifiable in the case of localized unifocal mycoses of the skin caused by dermatophytes.

Oral treatment offers the advantage of bringing all the lesions at all sites under control, in addition to the absence of unpleasant cosmetic effects. In certain cases, it may be preferable to use oral treatment for *C. albicans* vaginitis and for extensive and persistent pityriasis versicolor, a skin disorder caused by *Pityrosporum orbiculare*. In the case of onychomycosis, a combination treatment, topical plus systemic, is required. It is preferable to use oral treatment for deep and systemic mycoses, though intravenous or intrathecal treatment is sometimes required.

The final choice of a suitable antimycotic and the route of administration are determined by many factors: safety of the antimycotic, easy administration, broad-spectrum activity, and rapid clinical improvement associated with mycological cure.

Topical Antimycotics

Older Compounds. Some antimycotics have been used for a long time; Whitfield's ointment is a typical example (1907). The ointment usually contains 6% benzoic acid [*65-85-0*], $C_7H_6O_2$, and 3% salicylic acid [*69-72-7*], $C_7H_6O_3$. The action is attributed to the keratolytic effect of the salicylic acid and the direct effect of benzoic acid on the fungus.

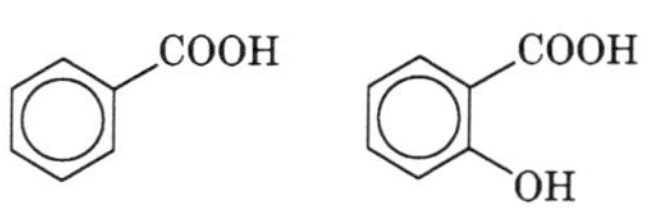

The main advantage of this ointment is its low price. This combination has now been replaced by more active modern antimycotics. Long ago, the antimycotic effect of aliphatic carboxylic acids with an increasing number of C-atoms was discovered. The optimum is 11 C-atoms and $CH_2{=}CH(CH_2)_8COOH$, undecylenic acid [*112-38-9*], $C_{11}H_{20}O_2$, is used in several ointments.

Other drugs are a combination of a keratolytic agent (mesulfen [*135-58-0*], $C_{14}H_{12}S_2$) and the well-known antimycotic zinc undecylenate [*557-08-4*], $C_{22}H_{38}O_4Zn$.

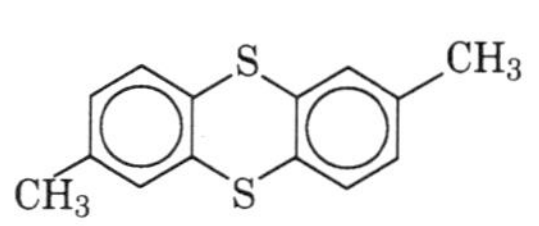

mesulfen

This cream also contains zinc naphthenate, terpineol, chlorcresol, and methyl salicylate [*119-36-8*]. The latter has a deeply penetrative effect. It combines antimycotic and antibacterial properties and is used to treat swimmer's eczema. The cream must not be used on open skin areas infected by bacteria. Some foot powders are a combination of zinc undecylenate, boric acid, hexachlorophene, and salicylic acid.

Tolnaftate [*2398-96-1*], $C_{19}H_{17}NOS$, which is active against dermatophytes, is the active component in another antifungal powder; it also contains cetylpyridinium chloride and talcum venetum.

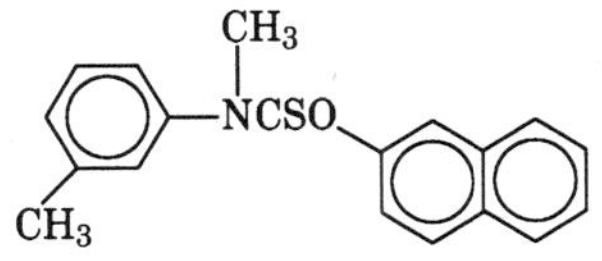

tolnaftate

In addition to tolnaftate and cetylpyridinium chloride, the ointment also contains poly(ethylene glycol). This preparation is not active against *C. albicans*.

ANTIMYCOTIC ANTIBIOTICS

Nystatin. This compound (**1**) belongs to the group of polyene antimycotic antibiotics (Table 1). Since the early 1960s, approximately 200 polyenes have been isolated from different *Streptomyces* strains.

Table 1. Polyene Antimycotic Antibiotics[a]

Structure number	Name	CAS Registry Number	Molecular formula	Trade name
(**1**)[b]	nystatin	[*1400-61-9*]	*b*	Fungicidin, Biotanal, Diastatin, Candex Mycostatin[c], Mioronal, Multilind, Nystan, Nystarescent
(**2**)	natamycin	[*7681-93-8*]	$C_{33}H_{47}NO_{13}$	Pimaricin, Mycophyt, Myprozine, Natacyn, Pimatucin, Synogil
(**3**)	amphotericin B	[*1397-89-3*]	$C_{47}H_{73}NO_{17}$	Amphozone, Fungizone[c], Fungilin, Ampho-Mioronal
(**4**)[c]	candicidin	[*1403-17-4*]	*d*	Levorin, Candeptin, Candimon, Vanobid
	filipin	[*11078-21-0*]	*e*	Filimarisin
	homycin	[*1403-71-0*]	*e*	Primamycin
(**5**)	etruscomycin[f]	[*13058-67-8*]	$C_{36}H_{53}NO_{13}$	Etruscomycin
	trichomycin[g]	[*1394-02-1*]	*e*	Cabimicina, Trichonat

[a]See Figure 1.
[b]Nystatin has three biologically active components: A_1, A_2, and A_3. Figure 1 depicts A_1.
[c]Squibb is the U.S. producer.
[d]A 4-component mixture; candicidin D shown in Figure 1 is the primary component.
[e]Multicomponent mixture.
[f]Also known as lucensomycin.
[g]Also known as hachimycin.

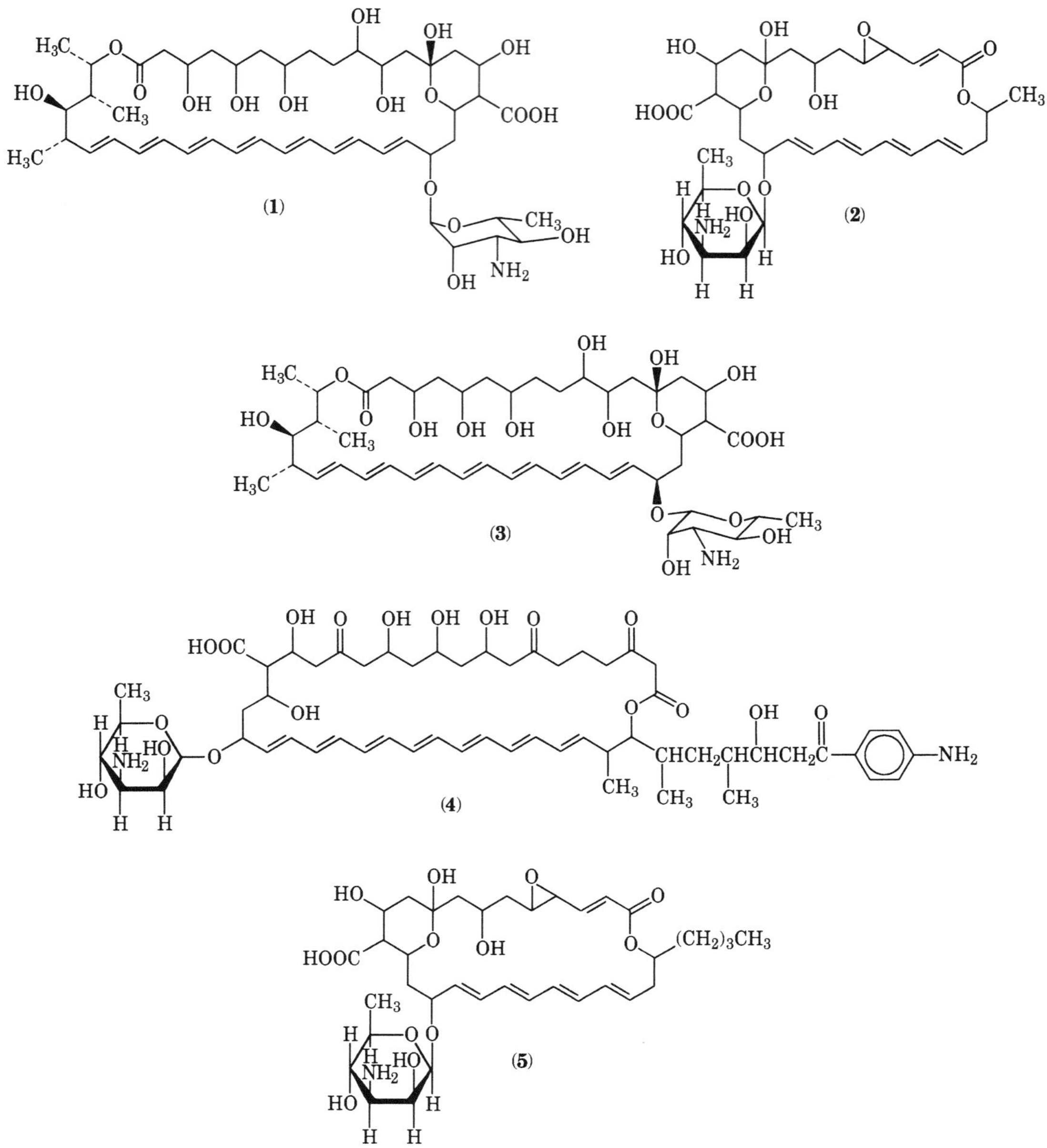

Fig. 1. Polyene antibiotics (see Table 1).

Nystatin is obtained from *Streptomyces noursei*. The polyenes are characterized by a macrolide ring and differ from one another by the number (12–37) of carbon atoms in the ring structure, the number (6–14) of hydroxyl groups, and the presence or absence of a carbohydrate (1). Polyenes alter the membrane permeability of sensitive cells by forming a complex with the sterol present in the membranes. Due to this binding to sterols, potassium ions are lost. Both nystatin and amphotericin B bind more strongly to ergosterol than to cholesterol.

Ergosterol is the principal sterol in the membrane of yeasts and fungi, whereas the cell membrane in mammals contains mainly cholesterol. This difference in binding capacity is probably responsible for the selectivity (1,2). Since bacteria, with the exception of *Mycoplasma* and *Acholeplasma,* contain no sterols; polyenes have no antibacterial activity.

Nystatin has a local fungicidal and fungistatic action against *C. albicans.* This polyene is not water-soluble. It is absorbed to a limited extent from the digestive tract, which limits the field of indications after oral administration to candidoses of the oral cavity, the esophagus, and the intestines. Oral administration of large doses may cause gastrointestinal disorders (nausea, vomiting, diarrhea). Nystatin is too toxic for parenteral administration. More information concerning chemical, mycological, and clinical properties of the polyenes is available (1,3,4).

Nystatin is mainly used to treat vaginal and oral infections and localized skin lesions, including *Candida* intertrigo and *Candida* nappy dermatitis. It may also be used as prophylaxis during treatment with antibiotics.

Nystatin (100,000 IU) is also available in combination with neomycin sulfate [*1405-10-3*] (35,000 IU), polymyxin B sulfate [*1405-20-5*] (35,000 IU), acetarsol [*97-44-9*] (150 mg), and dimethicone [*8050-81-5*] (2,500 mg). One or two ovules per day are inserted vaginally for at least 6–12 days. This combination has an antibacterial, antimycotic, and antitrichomonas action (see also ANTIBIOTICS, OLIGOSACCHARIDES; ANTIPARASITIC AGENTS, ANTIPROTOZOALS).

Natamycin. Natamycin or pimaricin (**2**) is a polyene with only four conjugated double bonds. This tetraene is obtained from *Streptomyces natalensis.* Like the other polyenes, pimaricin induces K^+ release from cells with membranes containing 20–40 mol % ergosterol. This is also the ergosterol concentration in the membrane of *Saccharomyces cerevisiae.*

Natamycin is not water-soluble or soluble in most organic solvents. It is not absorbed in the digestive tract. It is indicated for skin and nail infections with *C. albicans,* intertrigo and fissures at the corners of the mouth (perleche) caused by *C. albicans, Candida vulvitis,* and vaginitis. Natamycin plays an important role in the treatment of mycotic keratitis. Natamycin also appears to be active against the protozoan *Trichomonas vaginalis.* Side effects are nausea or diarrhea during oral treatment. Dosage is application of cream several times a day.

Combination creams or ointments contain 3.5 mg neomycin base and 10 mg hydrocortisone [*50-23-7*] per g, in addition to 10 mg natamycin. The combination as a lotion contains 1.75 mg neomycin and 5 mg hydrocortisone/g, in addition to 10 mg natamycin. This combination has an antiinflammatory, antibacterial, and antimycotic action. It is applied 2–4 times per day.

Amphotericin B. Amphotericin B (**3**), an important polyene antibiotic, is administered almost exclusively via the intravenous route and is therefore discussed in more detail under the systemic antimycotics. The vaginal tablets contain 50 mg amphotericin B, and 100 mg tetracycline base per tablet (see also ANTIBIOTICS, TETRACYCLINES). The tablets for oral use contain 50 mg amphotericin B, 250 mg tetracycline base, and 125 mg sodium hexametaphosphate. A combination ointment contains 1 mg fludrocortisone acetate, 2.5 mg neomycin, 0.25 mg gramicidin, and 1 g plastibase in addition to 30 mg amphotericin B (see also ANTIBIOTICS, PEPTIDES).

SYNTHETIC ANTIMYCOTICS

Because of their limited activity, small spectrum, and side effects, the older topical antimycotics have generally been surpassed by newer antimycotic chemotherapeutic agents. These newer antimycotics for topical use include the imidazole derivatives clotrimazole, miconazole, econazole, isoconazole, sulconazole, fenticonazole, oxiconazole, bifonazole, butoconazole, zinoconazole, tioconazole, and the triazole derivative, terconazole (Table 2) (5–7). The introduction of the azole derivatives represents a milestone in the treatment of mycoses.

Clotrimazole. The imidazole derivative clotrimazole (**6**) was introduced in 1969. Clotrimazole [*23593-75-1*] or 1-(*o*-chloro-α,α-diphenylbenzyl)imidazole is a water-insoluble antimycotic for topical application, with a broad-spectrum activity against mycoses of the skin and the vagina.

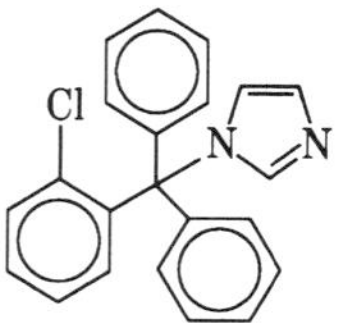

(**6**)

This antimycotic, which was introduced for topical use, shows *in vitro* activity against dermatophytes, species of *Candida, Aspergillus, Coccidioides immitis, Histoplasma capsulatum, Cryptococcus neoformans,* and *Madurella* species. *Petriellidium (Allescheria) boydii* and *Phialophora* species are less sensitive.

Clotrimazole and other azole derivatives have a different mode of action than the polyenes, eg, amphotericin B. The latter bind to the ergosterol present in the membranes of yeasts and fungi, but azole derivatives inhibit the cytochrome P-450 dependent biosynthesis of ergosterol (8–11). This inhibition not only results in a reduction of ergosterol, but also in an accumulation of C-14 methyl sterols. They disturb membrane permeability, inhibit cell replication, and are basically responsible, in combination with the reduction of ergosterol levels, for the antifungal action.

Clotrimazole is only partly absorbed in the digestive tract. It is a microsomal enzyme (cytochrome P-450) inductor and induces its own metabolism. This explains why it is not active *in vivo* against the etiologic agents of deep mycoses, including *H. capsulatum, C. immitis,* and *C. neoformans*. Enzyme induction also produces increased levels of other liver enzymes, such as transaminases. Because of this and possible side effects such as nausea, vomiting, and hallucinations (cerebral toxicity), the use of clotrimazole is limited to topical application. Topical application of clotrimazole is well tolerated; local allergic reactions and generalized skin irritations may occur occasionally.

Miconazole. Miconazole nitrate [*22832-87-7*] (Fig. 2), the 1-phenethylimidazole derivative first described in 1969, interferes at low doses with the cytochrome P-450 dependent ergosterol biosynthesis in yeasts and fungi. The result is accumulation of C-14 methylated sterols on the one hand and reduction of the ergosterol levels in the membranes on the other hand (12). In analogy to clotrimazole, this leads to a disturbance in the membranes; it results in inhibition

Table 2. Azole Antimycotics[a]

Name	CAS Registry Number	Structure number	Molecular formula	Trade name
clotrimazole	*[23593-75-1]*	**(6)**	$C_{22}H_{17}ClN_2$	Canesten, Canifug, Expecid, Lotrimin, Mycofug, Pedisafe, Rimazole, Tibatin, Trimysten
miconazole	*[22916-47-8]*	**(7a)**	$C_{18}H_{14}Cl_4N_2O$	Fungisdin, Aflorix[b], Albistat[b], Andergin[b], Brentan[b], Conoderm[b], Constite[b], Daktor[b], Deralbine[b], Dermonistat[b], Florid[b], Micatin[b], Monistat[b], Prilagin[b], Vodol[b]
econazole	*[27220-47-8]*	**(7b)**	$C_{18}H_{15}Cl_3N_2O$	Ecostatin[b], Ifenec[b], Micofugal[b], Micogin[b], Palavale[b], Pargin[b], Pevaryl[b], Spectazole[b]
isoconazole	*[27523-40-6]*	**(7c)**	$C_{18}H_{14}Cl_4N_2O$	Fazol[b], Travogen[b]
ketoconazole	*[65277-42-1]*	**(10)**	$C_{26}H_{28}Cl_2N_4O_4$	Fungarest, Fungoral, Ketoderm, Ketoisdin, Nizoral, Orifungal M, Panfungol
butoconazole	*[64872-76-0]*	**(11)**	$C_{19}H_{17}Cl_3N_2S$	Exelgyn[b], Femstat[b], Gynomyk[b]
oxiconazole	*[64211-46-7]*[b]	**(12)**	$C_{18}H_{13}Cl_4N_3O$	Myfungar[b], Oceral[b], Oxistat[b]
omoconazole	*[74512-12-2]*	**(13)**	$C_{20}H_{17}Cl_3N_2O_2$	Fangorex[b], Fongarex[b]
sulconazole	*[61318-90-9]*	**(14)**	$C_{18}H_{15}Cl_3N_2S$	Exelderm[b], Myk[b], Sulcosyn[b]
itraconazole	*[84625-61-6]*	**(18)**	$C_{35}H_{38}Cl_2N_8O_4$	Sporanox
fluconazole	*[86386-73-4]*	**(19)**	$C_{13}H_{12}F_2N_6O$	Diflucan, Triflucan
saperconazole	*[110558-57-3]*			
fenticonazole	*[72479-26-6]*		$C_{24}H_{20}Cl_2N_2OS$	Falvin[b], Lomexin[b]
bifonazole	*[60628-96-8]*		$C_{22}H_{18}N_2$	Amycor, Amolmen, Bedriol, Mycospor, Mycosporan
terconazole	*[67915-31-5]*		$C_{26}H_{31}Cl_2N_5O_3$	Fungistat, Panlomyc, Terazol, Tercospor
tioconazole	*[65899-73-2]*		$C_{16}H_{13}Cl_3N_2OS$	Fungibacid, Trosyd, Trosyl, Zoniden
zinoconazole	*[80168-44-1]*			

[a]U.S. producers include Ortho Pharmaceutical (eg, Monistat), Schering-Plough (eg, Lotrimin), and Syntex (Femstat).
[b]The nitrate.

of cell replication, mycelium development (in *C. albicans*), and finally, cell death. High concentrations of miconazole, which may be achieved with topical use, disturb the orientation of phospholipids in the membranes, which produces leaks (13).

Miconazole *[22916-47-8]* has a potent antifungal action against dermatophytes and *Candida* species. It is also active against gram-positive bacilli and

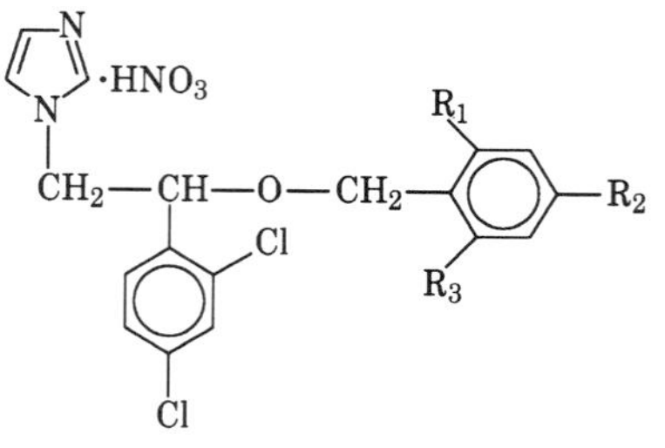

(7)

Fig. 2. Miconazole nitrate (**7a**) $R_1 = R_2 = Cl$, $R_3 = H$; econazole nitrate (**7b**) $R_1 = R_3 = H$, $R_2 = Cl$; isoconazole nitrate (**7c**) $R_1 = R_3 = Cl$, $R_2 = H$.

cocci. *In vitro,* it is also effective against *Trichomonas, Leishmania,* and *Plasmodium.* Miconazole is only moderately absorbed in the digestive tract. After intravenous administration, it is rapidly metabolized.

This drug is indicated for infections of skin and nails due to dermatophytes or *Candida* species and vulvovaginal infections due to *Candida* species. Miconazole is used orally for prophylactic purposes in patients who have been treated with cytostatics or immunosuppressants, in particular, to prevent candidosis of the mouth and digestive tract. Side effects include possible nausea during oral treatment. Miconazole tablets may increase the anticoagulant effect of coumarin derivatives. Topical treatment with cream or powder is well tolerated. Local irritation and allergic reactions of the skin and mucosa occur only rarely.

Daktacort is a combination product. It combines a broad-spectrum antimycotic with antibacterial action against gram-positive bacilli and cocci, miconazole and hydrocortisone. This corticosteroid was added for its anti-inflammatory and antipruriginous properties. Daktacort is indicated for skin infections due to dermatophytes or *Candida* species associated with inflammatory symptoms. For this reason, daktacort is used mainly during the initial phase of treatment. Afterwards, it is possible to switch to a cream containing only miconazole nitrate.

Econazole. Econazole and isoconazole are structurally very similar to miconazole; they differ only in the pattern of chlorine substitution on the terminal phenyl ring. Chemically, econazole (Fig. 2, **7b**) is closely related to miconazole. The clinical indications are almost identical. Econazole [*27220-47-9*] is available as a vaginal cream and ovule containing respectively 1% and 150 mg econazole nitrate. For mycoses of the skin, several forms are available including cream, spray-powder, spray-solution, and lotion. All these formulations contain 1% econazole nitrate [*68797-31-9*]. The dosages are practically identical to those of miconazole.

Isoconazole. Like econazole, both the chemical structure and the clinical indications of this imidazole derivative (Fig. 2, **7c**) closely resemble those of miconazole. It is available as a vaginal tablet (300 mg isoconazole nitrate per tablet). In some countries isoconazole [*27523-40-6*] is also available as a cream for dermatological use and as a cream and ovules for treatment of vaginal candidosis.

Naftifine. Naftifine (**8**) belongs to the allylamines, a new class of antimycotics (14). It is used to treat superficial mycoses and is particularly active against dermatophytes.

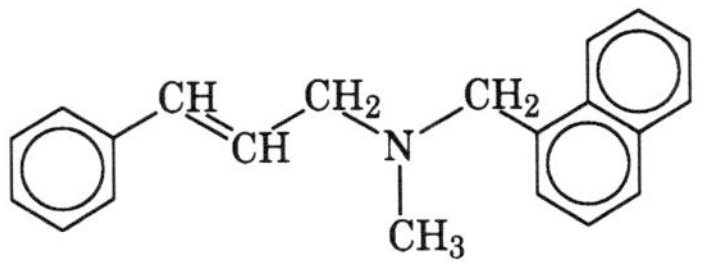

(**8**)

Like the azole derivatives, it inhibits the biosynthesis of ergosterol. However, naftifine [*65472-88-0*] does not inhibit the cytochrome P-450 dependent C-14-demethylase, but the epoxidation of squalene. Squalene epoxidase catalyzes the first step in the conversion of squalene via lanosterol to ergosterol in yeasts and fungi or to cholesterol in mammalian cells. The squalene epoxidase in *C. albicans* is 150 times more sensitive to naftifine, $C_{21}H_{21}N$, than the enzyme in rat liver (15). Naftifine is available as a 1% cream.

Ciclopiroxolamine. This ethanolamine salt, $C_{14}H_{24}N_2O_3$, of ciclopirox (**9**) is a topically active antimycotic, available as a cream and powder (16).

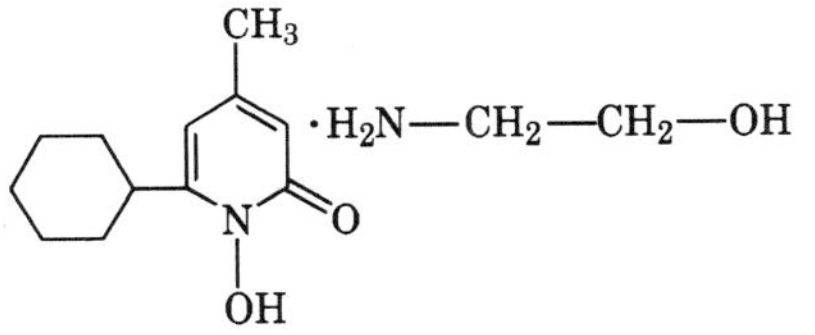

(**9**)

The biochemical basis for its action in fungi has not been totally elucidated. Ciclopiroxolamine [*41621-49-2*] is used mainly in the treatment of mycoses of the skin and nails.

Ketoconazole. Initial observations indicating that oral administration of ketoconazole (**10**) produced good results in seborrheic eczema and dandruff, led to the development of a 2% cream and a 2% shampoo (scalp gel) of this antimycotic (17,18). Naturally, these two topical forms of ketoconazole [*65277-42-1*] are highly active against superficial mycoses.

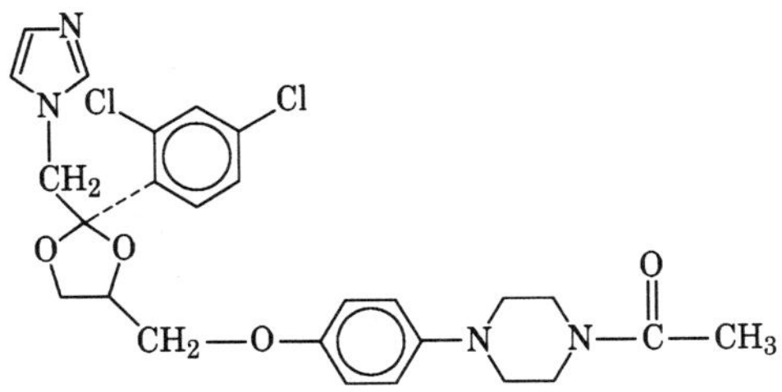

(**10**)

The main application of these two topical forms is the treatment of seborrheic eczema and dandruff. Ketoconazole's potent pityrosporicidal effect at concentrations of 100 ng/mL to 1 μg/mL forms the rationale of this activity.

Haloprogin. 3-Iodo-2-propynyl 2,4,5-trichlorophenyl ether, $C_9H_4Cl_3IO$, is an almost insoluble antifungal for topical use, available in the United States. Haloprogin [*777-11-7*] was developed in 1963 and it is active against dermatophytes. The drug's effect against *Candida* sp. is rather limited.

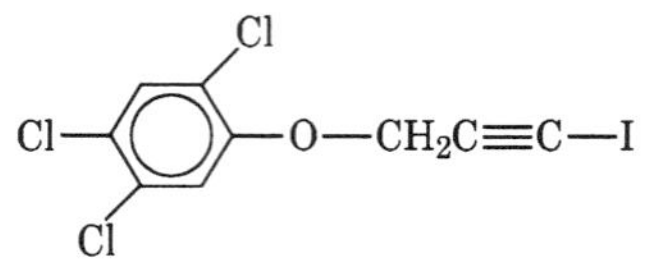

Developmental Topical Antimycotics. Most of the topical antimycotics still being developed are azole derivatives (mostly imidazole), including butoconazole, oxiconazole, omoconazole, zinoconazole, and sulconazole (Fig. 3). Generally speaking, these newer azoles offer no additional therapeutic advantages compared to existing substances. These drugs are already available in certain countries, but are not yet commercialized on a worldwide scale.

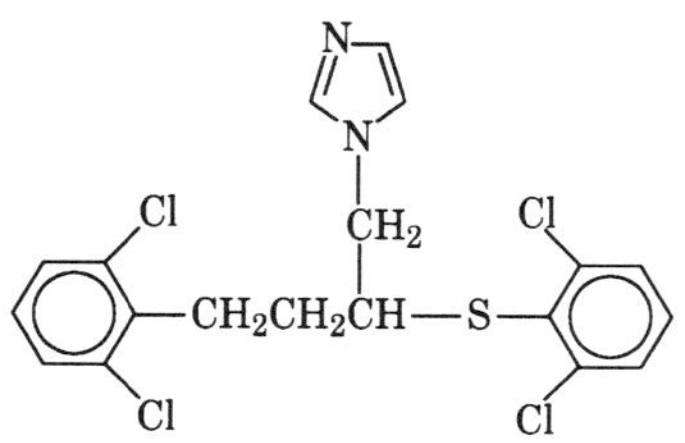

(11)

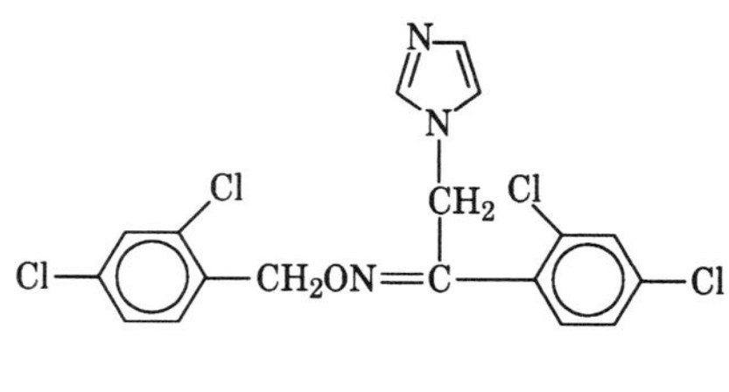

(12)

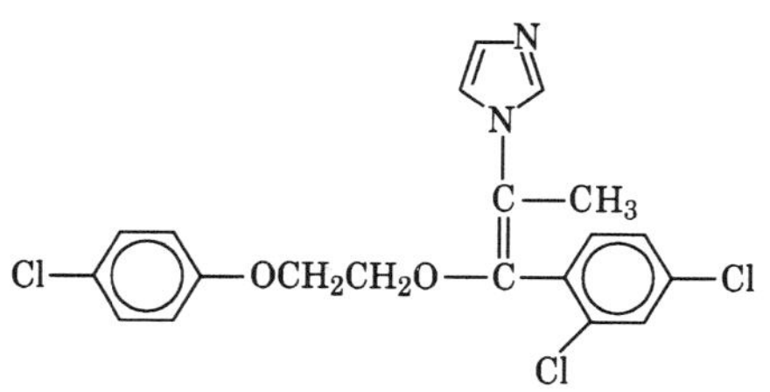

(13)

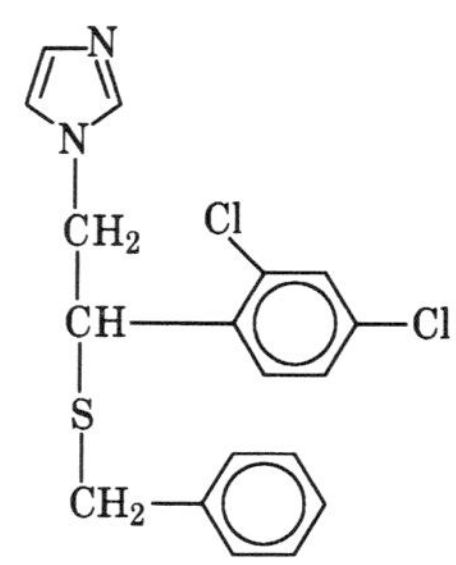

(14)

Fig. 3. Developmental topical antimycotics. Butoconazole **(11)**, oxiconazole **(12)**, omoconazole **(13)**, and sulconazole **(14)**.

Amorolfine [*78613-35-1*] is a molecule that is totally different from this group; it is a morpholine derivative **(15)** (19). This substance has a broad-spectrum activity and has a fungicide effect at fairly low concentrations. The development of a nail varnish for treatment of onychomycoses could be very interesting.

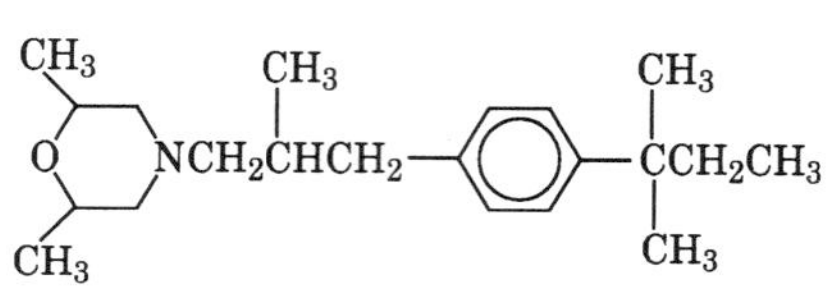

(15)

Systemic Antimycotics

Potassium Iodide. When potassium iodide [*7681-11-0*] is administered orally for several (6–8) weeks, a therapeutic effect may be obtained in the subcutaneous form of sporotrichosis. Amphotericin B is used intravenously to treat systemic sporotrichosis. The KI dosage is usually a saturated solution in water (1 g/mL). The usual oral dose is 30 mg/kg/d. Children should receive five droplets, three times a day (after meals); the dose may be increased to 15–20 droplets. Side effects include digestive disorders, swelling of the salivary glands, and lacrimation. Thyroid function tests may be disturbed.

Griseofulvin. Although this antibiotic was first isolated in 1939 and was described as a metabolite of *Penicillium griseofulvum* (20), its action against *Microsporum* and *Trichophyton* was only discovered in 1958. Several modes of action were suggested for (+) 7 chloro-4,6-dimethoxycoumaran-3-one-2-spiro-1′-(2′-methoxy-6′-methylcyclohex-2′-en-4′-one) ($C_{17}H_{17}ClO_6$) (**16**), this antibiotic derived from *P. griseofulvum.*

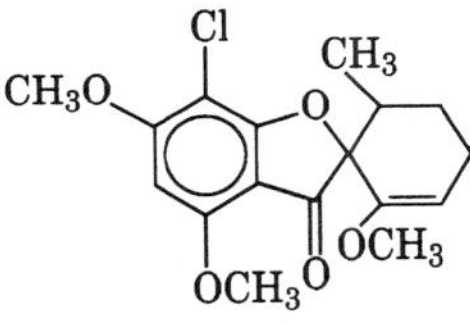

(**16**)

It interferes with the synthesis of the hyphal walls, the biosynthesis of nucleic acids, and the synthesis of chitin. The interaction with microtubules has also been described. The sensitivity of a cell seems to depend particularly on the ability to form griseofulvin–nucleic acid complexes. Further information concerning griseofulvin is available (21).

After oral administration, griseofulvin [*126-07-8*] is absorbed moderately. The absorption is increased when it is administered together with oil or fat. Micronization also increases the absorption greatly in the digestive tract; griseofulvin with a particle diameter of 2.7 μm reaches twice the plasma levels of griseofulvin with a particle diameter of 10 μm. Griseofulvin is found in the outer layers of the stratum corneum. Perspiration probably plays an important role in bringing it there.

Indications are mycoses of the skin, hair, and nails due to species of *Trichophyton, Epidermophyton floccosum,* and *Microsporum.* Yeasts and bacteria are not sensitive. Griseofulvin has a very weak effect against *Aspergillus.* Side effects may include headaches and gastrointestinal disorders, but they are usually only temporary. Occasionally, urticaria, erythema, or photosensitivity are observed. Griseofulvin is a strong inducer of the mixed function oxidase in the liver. It therefore has a stimulating effect on the conversion of various other drugs, including oral anticoagulants. It also affects the biosynthesis of porphyrin; this may result in accumulation of protoporphyrin in hepatocytes.

In the case of tinea corporis, treatment may last longer than six weeks. The treatment of onychomycosis may last up to twelve months, certainly if the toenails are involved. Caution is required in patients with impaired kidney or liver function or hematological disorders. Also, use is contraindicated during pregnancy.

Amphotericin B. This heptaene macrolide antibiotic (**3**) is produced by *Streptomyces nodosus*. It has fungistatic and fungicide properties and is effective against *Candida* species and the etiologic agents of systemic mycoses, including *Histoplasma capsulatum, Blastomyces dermatitidis, Coccidioides immitis,* and *Cryptococcus neoformans*. It is also active against *Sporothrix schenckii*. Like nystatin, this polyene has a greater affinity for ergosterol than cholesterol. Consequently, much lower levels of amphotericin B are required to induce potassium release in *C. albicans* than in mammalian cells. Potassium release is secondary to the increased flux of protons (1). In yeasts, the proton gradient plays an important role in the function of the plasma membrane. The transport of amino acids and other nutrients depends on this proton gradient (22).

Absorption is minimal after oral and topical administration. In order to obtain a systemic effect, the compound must be administered intravenously. Plasma levels are fairly stable during the first 24 hours; only a small fraction is eliminated via urine during this period. The polyene can be detected in urine up to three weeks after treatment. Diffusion of amphotericin B into cerebrospinal fluid is only moderate. More information concerning polyene antibiotics is available (1–4,23). Several side effects should be taken into account during intravenous administration. The main side effects are noted in the glomeruli of the renal parenchyma and include thickening of the basal membrane with hyalinisation, and in the tubuli, where focal or generalized degeneration and nephrocalcinosis may occur. Hypokalemia is an important consequence of nephrotoxicity. Headache, nausea, and vomiting are usually the first signs of toxicity. The dosage should be reduced until these side effects disappear. Vomiting, fever, and headache may be prevented by the administration of antihistamines and corticosteroids. Amphotericin B is administered over a 6-h period by a slow intravenous infusion. The recommended concentration is 0.1 mg/mL. The initial dose is 0.1–0.2 mg/kg/d. The dose is then increased to 1 mg/kg/d. The daily dose must never exceed 1.5 mg/kg. The duration of the treatment depends on the nature and the extent of the infection.

A marked improvement is generally noted after 4–8 weeks of treatment. Treatment is often continued until a total dose of 3 g is reached. In the case of coccidioidomycosis, for example, treatment with 0.4–0.8 mg/kg/d may last months. The polyene is administered intrathecally to treat *Coccidioides* meningitis. However, the results are only moderate. It is very important to check renal and hepatic function during treatment with amphotericin B.

Flucytosine. Flucytosine (**17**) or 5-fluorocytosine (4-amino-5-fluoro-2-pyrimidone, 5-FC), $C_4H_4FN_3O$, is a pyrimidine derivative, that is efficient against *Candida albicans, Cryptococcus neoformans,* and *Torulopsis glabrata*.

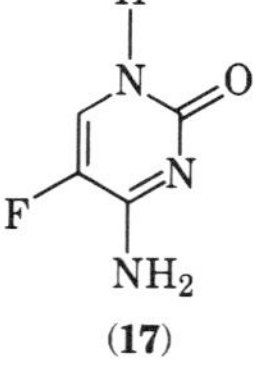

(**17**)

It is also active against *Aspergillus, Phialophora,* and *Cladosporium.* The protozoa *Acanthamoeba culbertsoni* and *Leishmania* are sensitive to flucytosine (1,24).

Flucytosine [*2022-85-7*] is well absorbed in the digestive tract, which is why oral administration is preferable. Plasma levels of 30–40 mg/L are obtained after a dose of 30 mg/kg body weight. Approximately 90% of the pyrimidine derivative is found unaltered in urine, indicating that it is highly suitable for the treatment of renal candidosis. High concentrations were also noted in cerebrospinal fluid; the average concentration is approximately 75% of the plasma concentration.

With the aid of cytosine permease, flucytosine reaches the fungal cell where it is converted by cytosine deaminase into 5-fluorouracil [*51-21-8*]. Cytosine deaminase is not present in the host, which explains the low toxicity of 5-FC. 5-Fluorouracil is then phosphorylated and incorporated into RNA and may also be converted into 5-fluorodeoxyuridine monophosphate, which is a potent and specific inhibitor of thymidylate synthetase. As a result, no more thymidine nucleotides are formed, which in turn leads to a disturbance of the DNA-synthesis. These effects produce an inhibition of the protein synthesis and cell replication (1,23,24). 5-Fluorouracil cannot be used as an antimycotic. It is poorly absorbed by the fungus to begin with and is also toxic for mammalian cells.

Flucytosine-resistant strains can develop very rapidly. These mutants may have a disturbed 5-FC-metabolism, or a compensatory mechanism for the disturbed nucleic acid functions. No cytosine permease was found in a resistant *Cryptococcus neoformans* strain, whereas cytosine deaminase was absent in resistant *C. albicans* strains. A deficiency of uridine monophosphate pyrophosphorylase occurs frequently in resistant *C. albicans* strains (1).

The dosage of flucytosine is 150–200 mg/kg orally in four portions every six hours. A 1% flucytosine solution has been developed for intravenous administration. In some countries, a 10% ointment is also available. In patients with normal renal function, flucytosine is seldom toxic, but occasionally severe toxicity may be observed (leukopenia and thrombocytopenia). Plasma levels should be determined and the dose in patients with impaired renal function should be checked. Liver function tests (transaminases and alkaline phosphatase) should be performed regularly. In some patients with high flucytosine plasma levels, hepatic disorders have been observed (24).

Miconazole. Miconazole (Fig. 2, **7a**) is also available as a sterile solution for intravenous infusion. Miconazole has a therapeutic effect on systemic mycoses due to *C. albicans, Aspergillus fumigatus, Cryptococcus neoformans, Blastomyces dermatitidis, Histoplasma capsulatum, Coccidioides immitis, Paracoccidioides brasiliensis,* and *Petriellidum boydii.*

The recommended dosage produces plasma levels that exceed the *in vitro* MIC-values for the previously mentioned fungi. Doses >9 mg/kg generally produce plasma levels of ≥1 mg/L. Large doses may cause loss of appetite, nausea, and vomiting. Very rapid infusion may lead to transitory cardiac arrhythmia and tachycardia. During prolonged and repeated intravenous administration, liver function tests are recommended. Cremophor, the solvent, may induce hypertriglyceridemia.

Ketoconazole. For treatment of systemic mycoses with amphotericin B or miconazole, the patient must be admitted to a hospital. This is not always possible, particularly in areas where systemic mycoses occur frequently, nor is it always desirable, because of the expense. For these reasons, it was desirable to find an antimycotic that combined safety and broad-spectrum activity with oral administration. Ketoconazole (**10**), which is orally active, met most of these requirements. This inhibitor of the ergosterol biosynthesis is an *N*-substituted imidazole, that differs from its precursors by the presence of a dioxolane ring (6,7). Ketoconazole is rapidly absorbed in the digestive system after oral administration. Sufficient gastric acid is required to dissolve the compound and for absorption. Therefore, medication that affects gastric acidity (for example, cimetidine and antacids) should not be combined with ketoconazole.

Plasma levels of 3–5 μg/mL are obtained two hours after administration of 200 mg ketoconazole. No accumulation in the bloodstream was noted after a 30-wk treatment with this dose. The half-life is approximately eight hours. When ketoconazole is taken with meals, higher plasma levels are obtained. Distribution studies using radioactive ketoconazole in rats show radioactivity mainly in the liver and the connective tissue. Radioactivity is also present in the subcutaneous tissue and the sebaceous glands. After one dose of 200 mg in humans, ketoconazole is found in urine, saliva, sebum, and cerumen. Like miconazole, the mode of action is based on inhibition of the cytochrome P-450 dependent biosynthesis of ergosterol. This results in disturbed membrane permeability and membrane-bound enzymes (8,10,23,25).

Ketoconazole is active against dermatophytes (*Microsporum, Trichophyton, Epidermophyton*), yeasts (*Candida* species, *Cryptococcus*), and the dimorphous fungi (*Coccidioides, Histoplasma,* and *Paracoccidioides*). Activity has also been demonstrated in patients with aspergillosis and in a limited number of patients with chromomycosis and sporotrichosis; promising results were obtained. Some *Leishmania* species are sensitive.

It is indicated for infections of the skin and nails due to dermatophytes and/or yeasts, yeast infection of the gastrointestinal tract, chronic recurring vaginal candidosis, and systemic fungal infections. Ketoconazole is well tolerated by most patients. Gastrointestinal symptoms or pruritus have been noted in 1–3% of the patients. Occasionally, liver enzymes may be increased temporarily. Normalization usually occurs during treatment. Of the estimated 4,000,000 patients treated, 300 developed hepatitis, ie, one per 15,000 patients (26). Treatment with ketoconazole should be discontinued when clinical and/or laboratory indications of hepatitis are present (7,27). In the case of prolonged treatment, it is advisable to check liver function twice a month during the first month of treatment, followed by monthly checks. With large doses (600 mg–1.2 g/d) a reversible inhibition of testosterone synthesis is observed (28). The period of treatment, with the exception of vaginal candidosis, depends on the clinical and mycological results. In the case of recurrent vaginal candidosis, 400 mg/d is prescribed for five days, but more recent studies with *Pityriasis versicolor* have demonstrated that shorter treatments also produce good results. Ketoconazole is contraindicated during pregnancy.

Itraconazole. Itraconazole (**18**) is a highly lipophilic compound with a triazole structure. Compared to ketoconazole, itraconazole has a broader spectrum (including *Aspergillus* spp.) (29,30) and an *in vitro* activity that is 10 times higher than ketoconazole for most species.

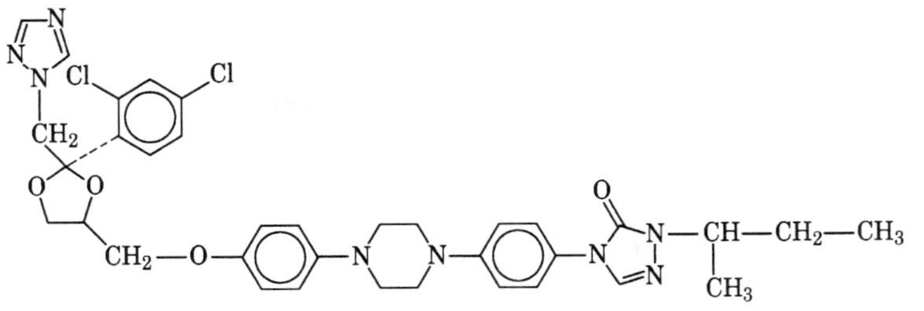

(**18**)

This *in vitro* superiority is also confirmed by animal experiments. Compared to ketoconazole, itraconazole [*84625-61-6*] is a much more selective inhibitor of the cytochrome P-450 dependent ergosterol biosynthesis in yeasts and fungi (29,31). Pharmacokinetically, itraconazole has a much longer bioavailability ($\pm$24-h) and a greater tissue affinity. In general, itraconazole tissue levels are significantly higher than the corresponding plasma levels. It has been established toxicologically that therapeutic doses of itraconazole do not interfere with steroid metabolism in humans. Itraconazole does not affect the liver as much as ketoconazole.

There are few side effects including headache, nausea, vertigo, intestinal disorders, etc. Itraconazole is used to treat vaginal candidosis, pityriasis versicolor, tinea corporis–tinea cruris, tinea pedis–tinea manus, and most of the systemic mycoses including aspergillosis cryptococcal meningitis. Itraconazole is contraindicated during pregnancy (32).

Fluconazole. This substance (**19**) is a water-soluble bis-triazole tertiary alcohol (33). Fluconazole [*86386-73-4*] has a broad spectrum, but has little activity *in vitro*. However, animal experiments reveal a broad spectrum and a potent effect.

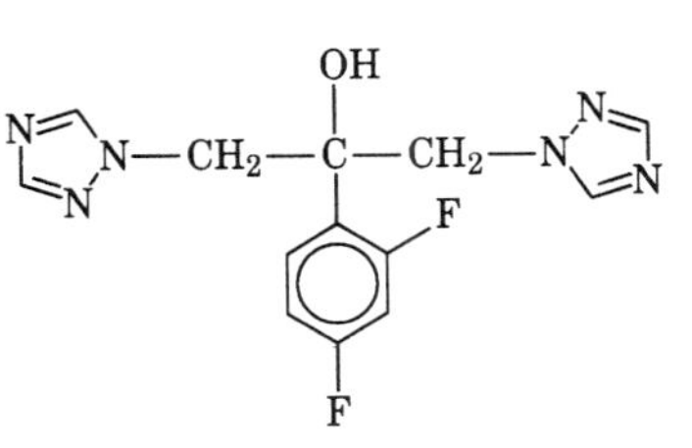

(**19**)

At the tested daily dose of 50 mg, fluconazole appears to be slightly less active against dermatophytes. The substance is used mainly to treat vaginal candidosis (a single capsule of 150 mg) and oral and esophageal candidosis (50 mg od for 14 d). In a number of countries, ie, England, the maximal period of treatment is 14 d.

Toxicological studies have demonstrated that there are no important problems with fluconazole. Therapeutic doses of fluconazole may cause enzyme induction in the liver. This suggests that interactions with other drugs cannot be excluded. The side effects are similar to those of itraconazole and include nausea, headache, and vertigo. Occasionally, increased liver enzymes may be noted. Like itraconazole, fluconazole is contraindicated during pregnancy.

Future Antimycotics for Systemic Treatment. Two new antimycotics for systemic use have now reached the stage of clinical development. The first is a triazole and fluoride analogue of itraconazole. This compound (saperconazole) is extremely active against *Aspergillus* spp. and slightly more soluble. Consequently, intravenous administration might be possible (34). The second molecule is terbinafine [*91161-71-6*], an allylamine, $C_{21}H_{25}N$, that appears to be particularly active against dermatophytes, just like topical naftifine (35).

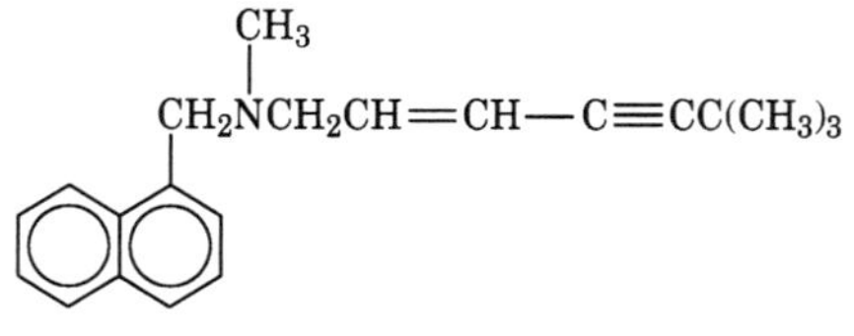

Both substances will probably become available for more widespread clinical use in the following three to four years.

BIBLIOGRAPHY

1. D. Kerridge and L. E. Whelan in A. P. J. Trinci and J. F. Ryley, eds., *Mode of Action of Antifungal Agents,* British Mycological Society Symposium 9, Cambridge University Press, Cambridge, UK, 1984, pp. 343–375.
2. E. F. Gale, E. Cundliffe, P. E. Reynolds, M. H. Richmond, and M. J. Waring, *The Molecular Basis of Antibiotic Action,* 2nd ed., John Wiley & Sons, Inc., 1981, pp. 201–219.
3. G. Medoff and G. A. Kobayashi in D. C. E. Speller, ed., *Antifungal Chemotherapy,* John Wiley & Sons, Inc., New York, 1980, pp. 3–33.
4. G. Medoff, J. Brajtburg, and G. S. Kobayashi, *Ann. Rev. Pharmacol. Toxicol.* **23,** 303–330 (1983).
5. R. J. Holt in Ref. 3, pp. 107–147.
6. H. B. Levine, *Ketoconazole in the Management of Fungal Disease,* Adis Press, Sydney, Australia, 1982.
7. H. E. Jones in R. Fleischmajer, ed., *Progress in Diseases of the Skin,* Vol. 2, Grune & Stratton, New York, 1984, pp. 217–249.
8. H. Vanden Bossche and co-workers, *Pestic. Sci.* **15,** 188–198 (1984).
9. H. Vanden Bossche, G. Willemsens, P. Marichal, W. Cools, and W. Lauwers in Ref. 1, pp. 21–41.
10. H. Vanden Bossche in M. McGinnes, ed., *Current Topics in Medical Mycology,* Vol. 1, Springer-Verlag, New York, 1985, pp. 313–351.
11. H. Vanden Bossche and co-workers, *Pestic. Sci.* **21,** 289–306 (1987).
12. H. Vanden Bossche, G. Willemsens, W. Cools, W. F. J. Lauwers, and L. Le Jeune, *Chem. Biol. Interact.* **21,** 59–78 (1978).
13. H. Vanden Bossche and co-workers, *Biochem. Pharmacol.* 2609–2617 (1982).
14. G. Petranyi, N. S. Ryder, and A. Stütz, *Science* **224,** 1239–1241 (1984).
15. N. S. Ryder in C. Nombela, ed., *Microbial Cell Wall Synthesis and Autolysis,* Elsevier Science Publishers, Amsterdam, The Netherlands, 1984, pp. 313–312.
16. B. Farkas, L. Kem;aaeny, and M. Csato in E. Tümbay, ed., *The Efficacy of Cy-*

clopyroxolamine in the Treatment of Mycosis and Its Influence on the Keratinocyte Killing Activity, The Turkish Microbiological Society No. 5, 1986, p. 35.
17. J. D. Stratigos and co-workers, *J. Am. Acad. Dermatol.* **19**(5) (part 1), 850–853 (1987).
18. G. Cauwenbergh, P. De Doncker, P. Schrooten, and H. Degreef, *Int. J. Dermatol.* **25**(8), 541 (1986).
19. A. Polak and D. M. Dixon in R. A. Fromtling, ed., *Recent Trends in the Discovery, Development and Evaluation of Antifungal Agents,* J. R. Prous Science Publishers, SA, 1987, pp. 555–573.
20. A. E. Oxford, H. Raistrick, and P. Simonart, *Biochem. J.* **33,** 240–248 (1939).
21. R. R. Davies in Ref. 3, pp. 49–82.
22. F. Foury and A. Goffeau, *J. Biol. Chem.* **250,** 2354–2362 (1974).
23. H. Vanden Bossche, G. Willemsens, and P. Marichal, *CRC Crit. Rev. Microbiol.* **15,** 57–72 (1987).
24. H. J. Scholer in Ref. 3, pp. 35–106.
25. H. Vanden Bossche and co-workers, *Antimicrob. Agents Chemother.* **17,** 922–928 (1980).
26. J. H. Lewis, H. J. Zimmerman, G. D. Benson, and K. G. Ishak, *Gastroenterology* **86**(3), 503–513 (1986).
27. P. A. J. Janssen and J. E. Symoens, *Am. J. Med.* 80–85 (1983).
28. H. Vanden Bossche, W. Lauwers, G. Willemsens, and W. Cools in F. Schroeder, ed., *European Organisation for Research on Treatment of Cancer,* Monograph 2, Alan E. Liss Inc., New York, 1985, pp. 187–196.
29. J. Van Cutsem, F. Van Gerven, M-A. Van De Ven, M. Borgers, and P. A. J. Janssen, *Antimicrob. Agents Chemother.* **26,** 527–534 (1984).
30. H. Vanden Bossche, D. W. R. Mackenzie, and G. Cauwenbergh, eds., *Aspergillus and Aspergillosis,* Plenum Press, New York, 1988.
31. H. Vanden Bossche in Ref. 19, pp. 207–221.
32. G. Cauwenbergh and P. De Doncker, *Drug Dev. Res.* **8,** 317–323 (1986).
33. P. F. Troke in Ref. 19, pp. 103–112.
34. J. Van Cutsem and P. A. J. Janssen, *28th Interscience Conference on Antimicrobial Agents and Chemotherapy,* Los Angeles, Calif., 1988, p. 141.
35. N. S. Ryder in Ref. 19, pp. 451–459.

GEERT CAUWENBERGH
Janssen Research Foundation

ANTIPROTOZOALS

Diseases caused by protozoa affect more people than those brought on by any other biological cause (1). There are over 60,000 species of protozoa, of which some 10,000 are parasitic. In humans, protozoa chiefly infect the gastrointestinal tract, vagina, urethra, blood, and blood-forming organs. Malaria is the most widespread of the protozoan diseases, and is responsible for the greatest number of deaths due to infection. Although protozoan diseases occur throughout the world, they impact most severely on people of tropical areas where there is widespread malnutrition, minimal health education, and poor sanitation. Fatal protozoal infections are occurring to an increased extent in the more developed countries among immunosuppressed individuals, especially those with AIDS. Domesticated and wild animals are also extensively affected and can harbor and propagate infectious protozoa. Protozoan diseases of poultry and ruminants deprive large numbers of people, especially in Africa, of much needed food.

The word protozoa is derived from the Greek *protos* and *zoon,* meaning first and animal, respectively. Protozoa are a highly varied group of one-celled animals that reproduce by both sexual and asexual means. Given suitable conditions, ie, host susceptibility and lack of immunity, their numbers can increase by the tens of thousands in the course of one day. They are eukaryotic (having a well-defined nucleus), are surrounded by a membrane through which nutrients can pass, and their cytoplasm contains organelles that enable them to generate energy, digest nutrients, respire, grow, and reproduce. Protozoa are capable of motility and propel themselves by means of cilia, flagella, or pseudopods. Accurate differentiation and identification of infective protozoa facilitates the choice of treatment, but is frequently a difficult task.

Progress in finding new chemotherapeutic agents to combat protozoal disease has been slow. The urgency of the problem has been dramatized by the realization that AIDS patients usually die from opportunistic diseases, many of which are protozoan in origin. There are no satisfactory treatments for many of the protozoal diseases; where there are proven drugs available, parasites in many parts of the world are developing resistance to them. This resistance has necessitated the continual design, synthesis, and testing of new classes of prophylactic and therapeutic agents. Despite the alarming view by a World Health Organization spokesman in 1990 that the war against tropical diseases is being lost, only a modest proportion of the world's research resources is being devoted to devising new treatments. Unfortunately, most pharmaceutical firms do not anticipate adequate financial reward from solving the health problems of developing countries.

This article discusses the main diseases caused by protozoa and the important chemotherapeutic antiprotozoal agents currently in use or in an advanced state of development. Not all alternative proprietary names are given for each drug. There are excellent sourcebooks on protozoology and parasitology (2–8).

Amebiasis

Amebiasis, a widespread disease of humans, causes an estimated 500 million cases (without the inclusion of China) annually. It is believed to be the third leading cause of death due to parasites. The disease is caused by pathogenic strains of the protozoan *Entamoeba histolytica,* which exists both in a stable infective cyst form and in a more fragile, potentially invasive, trophozoite form. Amebiasis is transmitted by drinking water contaminated with fecal matter from infected individuals or by eating food contaminated by infected humans or flies. Symptoms of the disease are generally centered around the gastrointestinal tract. They include amebic dysentery and sometimes appendicitis or fulminant colitis. The disease of the intestine may range from acute dysentery with chills, fever, and blood or mucoid diarrhea (amoebic dysentery), to mild abdominal discomfort with diarrhea, containing blood or mucus. Its infection is often asymptomatic. Blood-borne propagation can cause liver abscesses, and less commonly, infection of the brain or lungs. Skin ulceration may occur by extension of intestinal lesions. Recently, it was shown that *E. histolytica* has both pathogenic and nonpathogenic strains, with characteristic isoenzyme electrophoresis patterns, that occur in symptomatic and asymptomatic persons, respectively.

Acute intestinal amebic dysentery is most commonly treated with metronidazole (**1**, Flagyl) (Table 1). An alternative drug is iodoquinol (**2**, diiodohydroxyquin, diiodohydroxyquinoline [*83-73-8*]), which has also been used advantageously in combination with metronidazole or oxytetracycline (**3**, Terramycin [*79-57-2*]). For patients too sick to take iodoquinol orally, emetine (**4**) or dehydroemetine (**5**, 2,3-dehydroemetine) hydrochloride is administered either subcutaneously or intramuscularly. The latter sometimes is followed by chloroquine (**6**, Aralen) phosphate or iodoquinol. Because emetine and dehydroemetine cause cardiac arrhythmias, muscle weakness, and inflammation at the injection site, these toxic compounds are used primarily for patients whose lives are threatened by the disease. Antibiotics, such as tetracycline (**7**, Achromycin [*60-54-8*]) and paromomycin (**8**, Humatin), are also effective against moderate intestinal amebiasis.

Extraintestinal (eg, hepatic) amebiasis is treated with metronidazole and can be followed by iodoquinol or a combination of dehydroemetine or emetine hydrochloride with chloroquine phosphate. Iodoquinol is the drug of choice for asymptomatic amebiasis, whereas diloxanide furoate (**9**, Furamide) has been used successfully to treat symptomatic and asymptomatic intestinal amebic cyst carriers.

Other compounds that have shown promising properties *in vitro* against *E. histolytica* are niridazole (**10**), tinidazole (**11**, Fasigyn, Simplotan), gossypol (**12**), bithionol [(**13**), 2,2′-thiobis(4,6-dichlorophenol)], and pimaricin (**14**, Natamycin [*7681-93-8*]). The chemotherapy of amebiasis has been reviewed (9).

Related amoeba that cause infections of lower incidence but greater severity are the invasive organisms *Naegleria fowleri* and species of *Acanthamoeba* such as *A. castellanii* and *A. polyphaga*. These amoeba are free-living in the soil or warm fresh water, occur widely, and affect many animals but have no known animal reservoir. Entry of *Naegleria* into the body usually is by the nasal passages, whereas *Acanthamoeba* can enter via the eye, skin, or lung. They cause primary amebic meningoencephalitis (PAM), an infection of the central nervous system that is usually fatal. Symptoms in humans include severe headache, confusion, nausea, seizures, and coma. Frequently the olfactory tract is affected. These infections are considered opportunistic and are seen to a greater extent in immunosuppressed humans (due to AIDS, steroids, radiotherapy, or chemotherapy) than in those who are healthy.

Naegleria is treatable with intravenous amphotericin B (**15**, Fungizone), a toxic drug that must be used with caution.

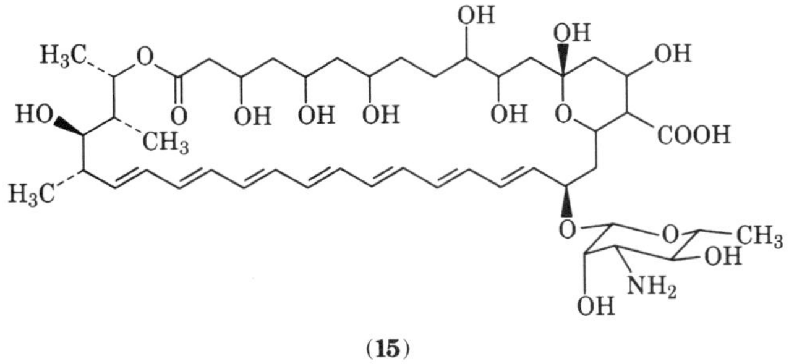

(**15**)

Table 1. Amebiasis Antiprotozoal Agents[a]

Structure number	Compound name	CAS Registry Number	Molecular formula	Structure
(1)	metronidazole[b,c,d,e,f]	*[443-48-1]*	$C_6H_9N_3O_3$	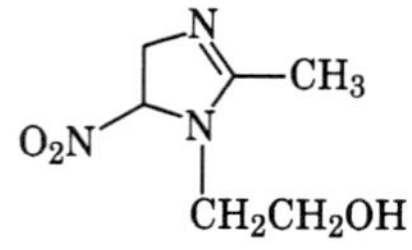
(2)	iodoquinol[b,d]	*[83-73-8]*	$C_9H_5I_2NO$	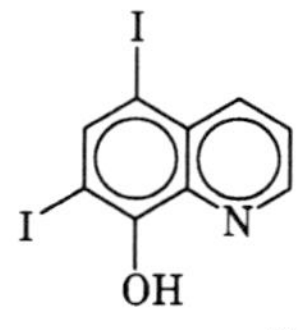
(3)	oxytetracycline[b,g,h]	*[79-57-2]*	$C_{22}H_{24}N_2O_9$	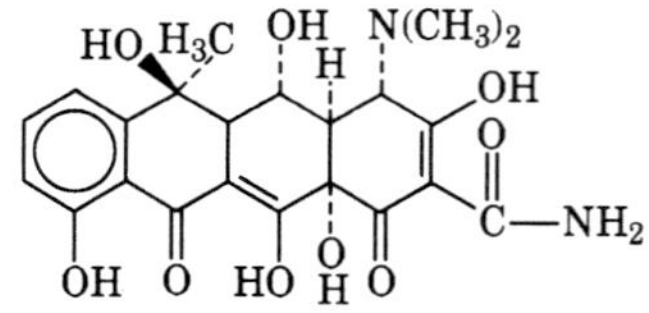
(4)	emetine (HCl)	*[316-42-7]*	$C_{29}H_{40}N_2O_4{\cdot}2ClH$	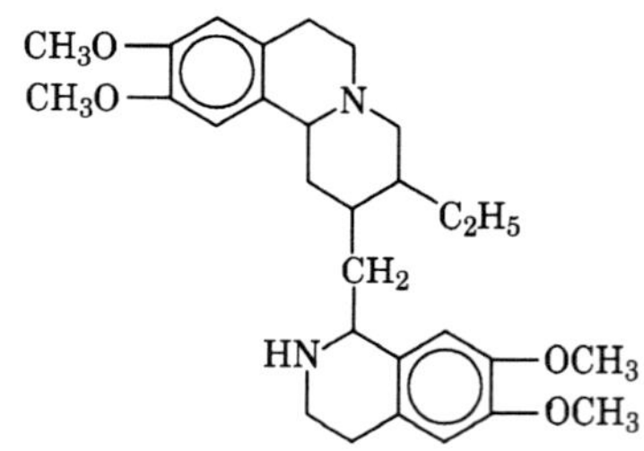
(5)	dehydroemetine	*[4914-30-1]*	$C_{29}H_{38}N_2O_4$	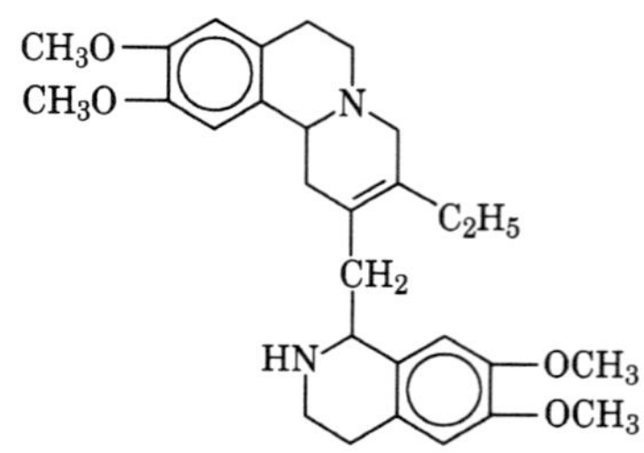
(6)	chloroquine[g,i,j] (diphosphate)	*[50-63-5]*	$C_8H_{26}ClN_3{\cdot}2H_3O_4P$	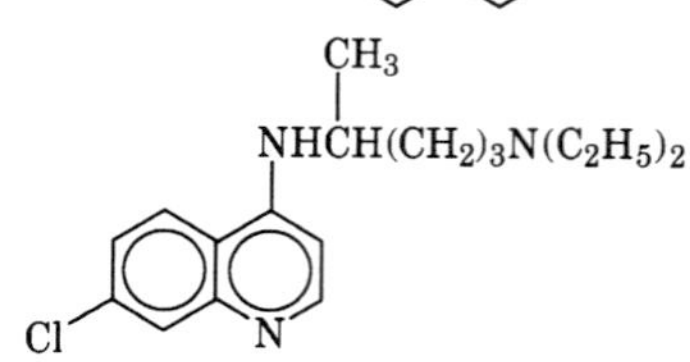
(7)	tetracycline[b,j]	*[60-54-8]*	$C_{22}H_{24}N_2O_8$	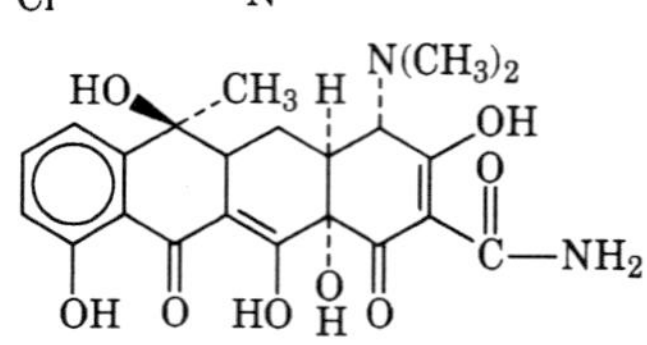

Table 1. ***(Continued)***

Structure number	Compound name	CAS Registry Number	Molecular formula	Structure
(8)	paromomycin[b,c,k]	*[59-04-1]*	$C_{23}H_{45}N_5O_{14}$	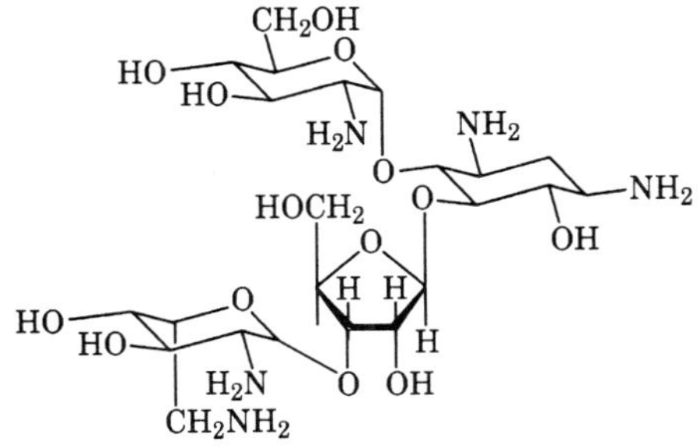
(9)	diloxanide furoate	*[3736-81-0]*	$C_{14}H_{11}Cl_2NO_4$	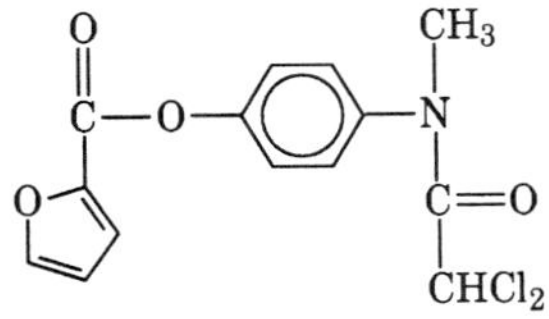
(10)	niridazole[f]	*[61-57-4]*	$C_6H_6N_4O_3S$	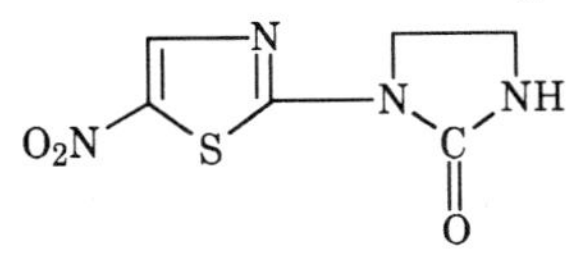
(11)	tinidazole[b,c,d,e,f]	*[19387-91-8]*	$C_8H_{13}N_3O_4S$	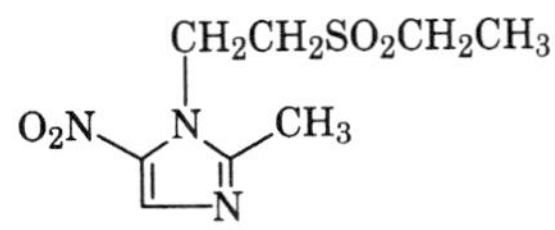
(12)	gossypol	*[303-45-7]*	$C_{30}H_{30}O_8$	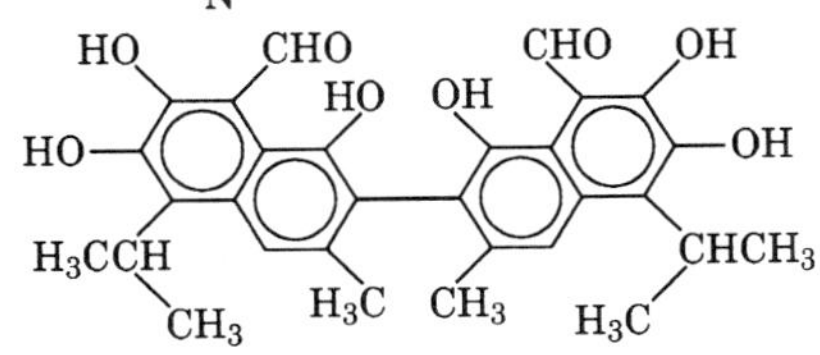
(13)	bithionol	*[97-18-7]*	$C_{12}H_6Cl_4O_2S$	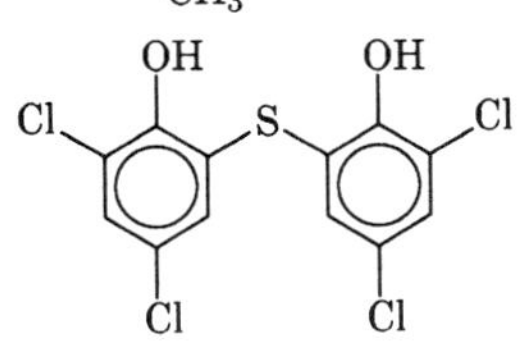
(14)	pimaricin	*[7681-93-8]*	$C_{33}H_{47}NO_{13}$	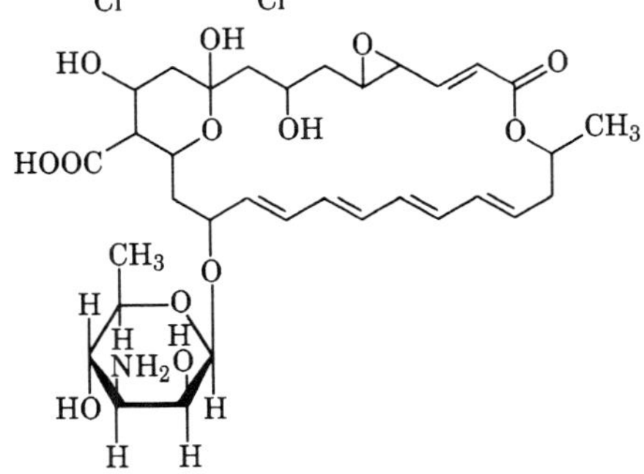

Table 1. ***(Continued)***

Structure number	Compound name	CAS Registry Number	Molecular formula	Structure
(15)	amphotericin B[k]	*[1397-89-3]*	$C_{47}H_{73}NO_{17}$	in text
(16)	miconazole	*[22916-47-8]*	$C_{18}H_{14}Cl_4N_2O$	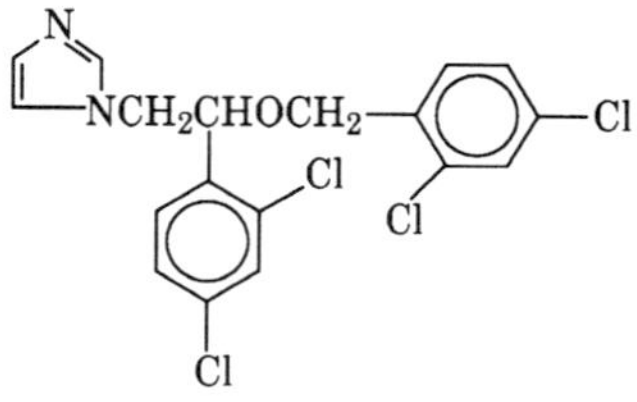
(17)	rifampin[j]	*[13292-46-1]*	$C_{43}H_{58}N_4O_{12}$	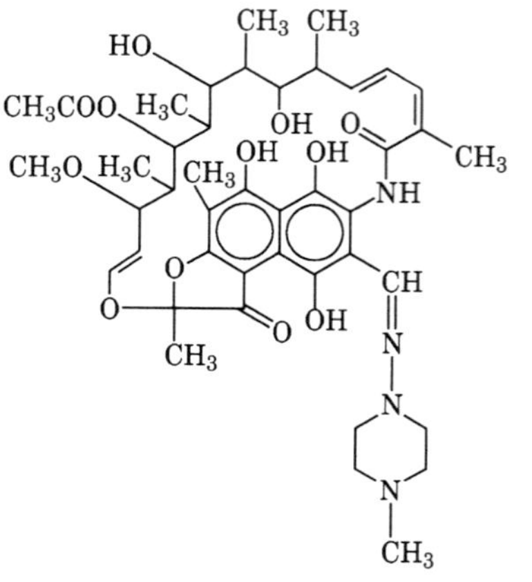
(18)	minocycline (HCl)	*[13614-98-7]*	$C_{28}H_{27}N_3O_7{\cdot}ClH$	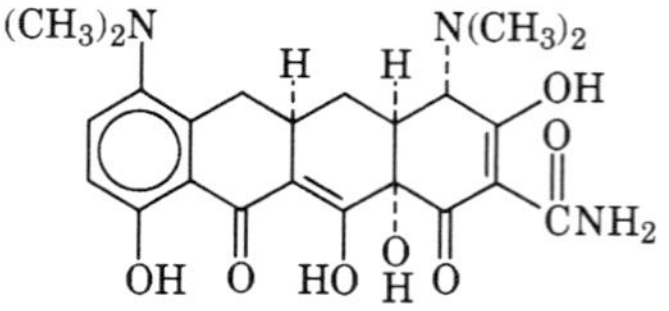

[a]Other applications of these agents are indicated in footnotes to the table.
[b]Balantidiasis.
[c]Giardiasis.
[d]Hexamitosis.
[e]Histomoniasis.
[f]Trichomoniasis.
[g]Anaplasmosis.
[h]Theileriasis.
[i]Babesiasis.
[j]Malaria.
[k]Leishmaniasis.

A combination of amphotericin B, miconazole (**16**), and rifampin (**17**) was used to successfully cure one patient. In addition, tetracycline (**7**) and minocycline (**18**) have been recommended although their clinical efficacy have not been established. No proven therapeutic agents exist for treating *Acathamoeba* infections; however, the phenothiazines, trifluoperazine *[117-89-5]* and chlorpromazine *[50-53-3]*, show promise *in vitro*.

Anaplasmosis

Anaplasmosis is a tick-borne disease of cattle and other ruminants, such as deer in the western United States and elk in the former USSR. It is caused by the protozoan *Anaplasma marginale,* so named because the organism appears to be devoid of cytoplasm and resides at the margin within an erythrocyte. At least 30 species of ticks have been shown to be capable of transmitting anaplasmosis under laboratory conditions. Horseflies are also vectors of the disease. Anaplasmosis is mild in young calves and increases in severity, being frequently fatal to cattle above the age of three. Although the disease may be asymptomatic, it is sometimes manifest by anorexia, weakness, fever, anemia, jaundice, reduced milk production, depression, dehydration, difficult respiration, and abortion. Infection can also cause infertility in bulls. The disease in its acute form occurs frequently in purebred animals and high producing milk cows, which may die in a few hours after the start of infection.

Imidocarb (**19**) dihydrochloride [*5318-76-3*], the most effective drug against anaplasmosis, is reported to be curative in a single dosage if given early in the infection during the rise in parasitemia. Multiple doses are required if the animals are treated at a later stage or if a relapse occurs (Table 2). Some antibiotics active against the disease include the tetracyclines chlortetracycline (**20**, Aureomycin [*57-62-5*]) and oxytetracycline (**3**), which are now given in long-acting pharmaceutical dosage forms because of the need for extended treatment. Many

Table 2. Anaplasmosis Antiprotozoal Agents[a]

Structure number	Compound name	CAS Registry Number	Molecular formula	Structure
(**19**)	imidocarb[b]	[*27885-92-3*]	$C_{19}H_{20}N_6O$	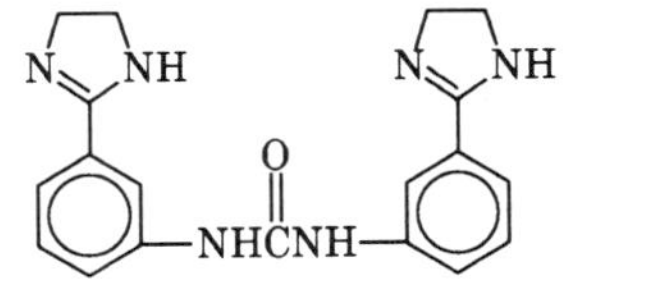
(**20**)	chlortetracycline[b,c,d]	[*57-62-5*]	$C_{22}H_{23}ClN_2O_8$	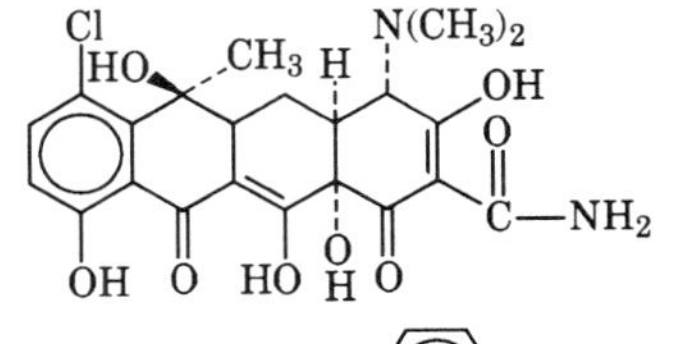
(**21**)	amodiaquine[e]	[*86-42-0*]	$C_{20}H_{22}ClN_3O$	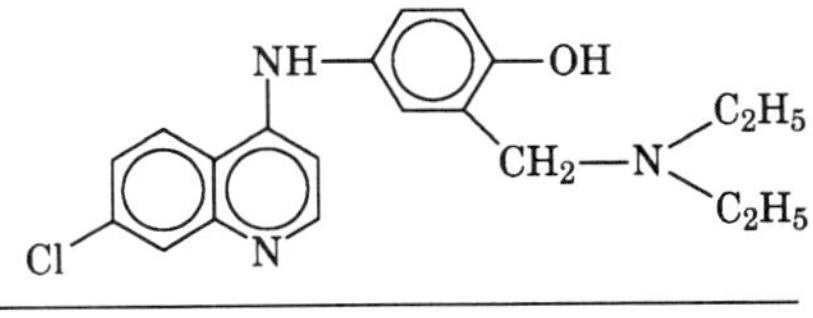

[a]Other applications of these agents are indicated in footnotes.
[b]Theileriasis.
[c]Balantidiasis.
[d]Hexamitosis.
[e]Malaria.

other drugs, including arsenicals, antimonials, antimalarials, eg, chloroquine (**6**) and amodiaquine (**21**, the dihydrochloride dihydrate Camoquin [*6398-98-7*]), and dyes have been employed in the past in an effort to cure or prevent anaplasmosis. α-Ethoxyethylglyoxal dithiosemicarbazone (Gloxazone) has a marked inhibitory effect on *A. marginale* but was found to be too toxic to lactating cattle.

Babesiasis

Babesiasis (babesiosis, piroplasmosis), primarily a disease of cattle, is caused by a protozoan related to the malaria parasite. The protozoan, *Babesia microti,* after ingestion by its vector, the ixodid hard-bodied tick, grows inside the gut of the insect and spreads throughout its body. Infection of the vertebrate host, which can be either a wild or domesticated animal, occurs when the tick obtains a blood meal from it. Humans do not appear to be a reservoir for the infection and are not commonly affected by the disease (10). However, they can acquire babesiasis by blood transfusion. The consequences of infection by *B. microti* in humans can range from being asymptomatic to manifestation of a severe and prolonged illness. Typically, there is a gradual onset of fever, chills, sweating, generalized myalgia, fatigue, anemia, and renal insufficiency and failure. Infections by *Babesia divergens* are associated with cattle and are characterized in humans by chills, high fever, nausea, vomiting, and severe hemolytic anemia. Individuals infected with *B. divergens* generally lack functioning spleens and the disease has been fatal in a large percentage of cases. Related infections include *Babesia bigemina,* the cause of Texas cattle fever, and *Babesia bovis,* responsible for hemoglobinuric fever in European cattle.

Babesiasis in humans is frequently self-limiting. Quinine [*130-95-0*] (**22**) plus clindamycin (**23**) is considered the treatment of choice and has also been used effectively to treat transfusion babesiasis (Table 3). Quinine, either alone or in combination with pyrimethamine (**24**), has failed to eliminate experimental *B. microti* in hamsters. Babesiasis contracted in the United States has been successfully treated with chloroquine (**6**). However, when tested in infected hamsters, chloroquine failed to clear the infection, as did the other antimalarials sulfadiazine (**25**) and pyrimethamine (**24**). There was only a limited response to pentamidine (**26**) and diminazene aceturate [**27**, 4,4′-(diazoamino)dibenzamidine aceturate, Berenil, also diminazene dilactate Babesin]. Pentamidine has been used successfully to treat the symptoms of the disease in humans and in cats infected with *Babesia felis.* Diminazene aceturate is effective against experimental *B. microti* in jirds, but not *in vitro* against *Babesia equi.* Imidocarb (**19**) dipropionate is a therapeutic and prophylactic treatment for babesiasis in cattle, gerbils, and other domestic animals. The urea, amicarbalide (**28**) diisethionate [*3671-72-5*] (3,3′-diamidinocarbanilide diisethionate, Diampron) is also used for babesiasis in cattle. *Babesia gibsoni* in dogs has been treated effectively with phenamidine isethionate (**29**). Another urea, Babesan, (**30**), 1,3-di-6-quinolinylurea as the bismethosulfate salt ie, quinuronium sulfate, $C_{23}H_{26}N_4O_9S_2$, is used for veterinary purposes.

Table 3. Babesiasis Antiprotozoal Agents[a]

Structure number	Compound name	CAS Registry Number	Molecular formula	Structure
(22)	quinine (sulfate)[b]	*[804-63-7]*	$C_{20}H_{24}N_2O_2{\cdot}{1/2}H_2O_4S$	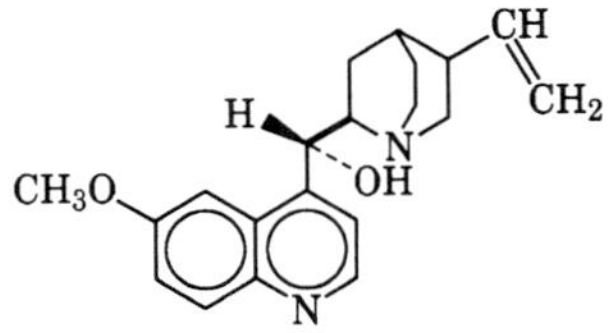
(23)	clindamycin[b,c,d]	*[18323-44-9]*	$C_{18}H_{33}ClN_2O_5S$	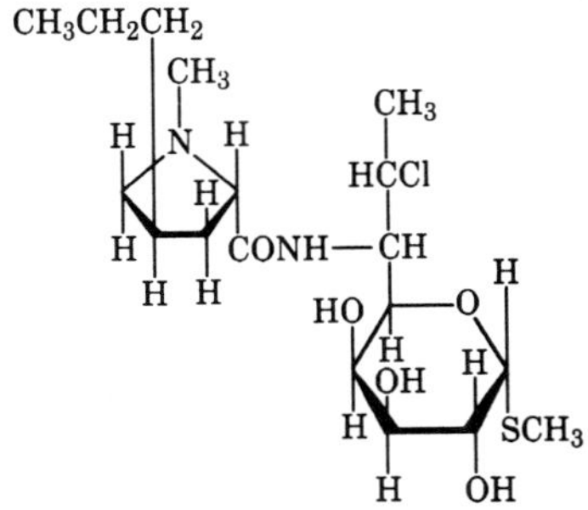
(24)	pyrimethamine[b,c,d,e]	*[58-14-0]*	$C_{12}H_{13}ClN_4$	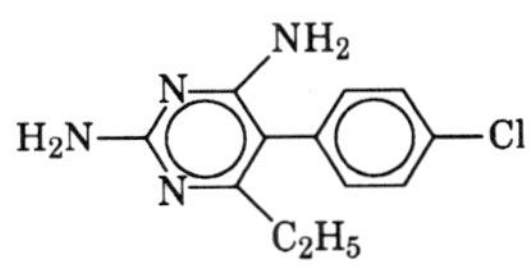
(25)	sulfadiazine[c,d,f]	*[68-35-9]*	$C_{10}H_{10}N_4O_2S$	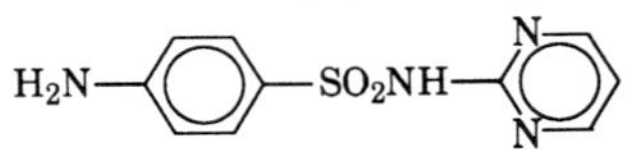
(26)	pentamidine[g]	*[100-33-4]*	$C_{19}H_{24}N_4O_2$	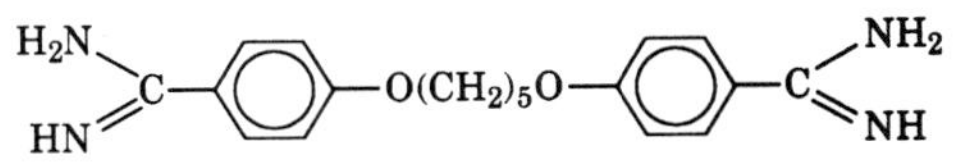
(27)	diminazene aceturate[c,h]	*[908-54-3]*	$C_{22}H_{29}N_9O_6$	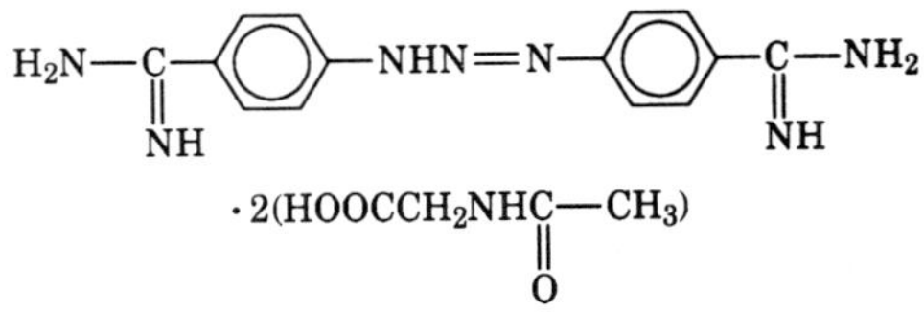
(28)	amicarbalide	*[3459-96-9]*	$C_{15}H_{16}N_6O$	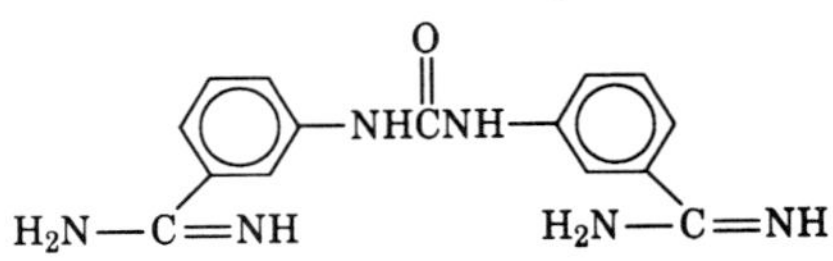
(29)	phenamidine isethionate	*[620-90-6]*	$C_{18}H_{26}N_4O_9S_2$	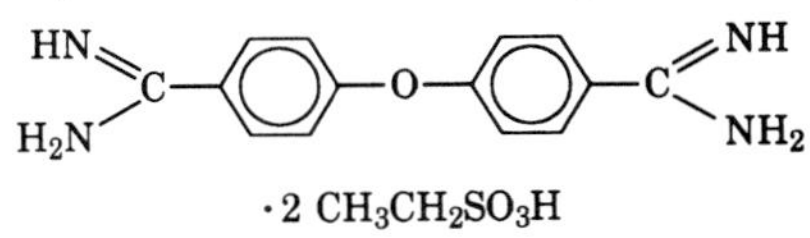
(30)	babesan	*[532-05-8]*	$C_{19}H_{14}N_4O$	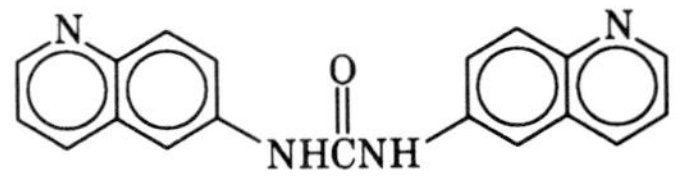

[a]Other applications are indicated by footnotes.
[b]Malaria.
[c]Pneumocystosis.
[d]Toxoplasmosis.
[e]Coccidiosis.
[f]Balantidiasis.
[g]Leishmaniasis.
[h]Trypanosomiasis (African trypanosomiasis).

Balantidiasis

Balantidiasis (balantidiosis, balantidial dysentery), an intestinal disease seen almost worldwide, is caused by the large ciliated protozoan, *Balantidium coli*. The organism is usually found in the lumen of the large intestine of humans and animals. Cysts formed in the lumen of the colon or in freshly evacuated feces of humans or domesticated and wild animals, can colonize the colon and terminal ileum of new hosts by the latter's ingestion of contaminated food or water. The hog has been found to be the most heavily parasitized host. Its association with the rat may be a means for maintaining a reservoir infection in the two animals.

Balantidiasis in humans is manifest by chronic episodes of intermittent diarrhea and constipation, symptoms similar to those of amebiasis. The patient may also have abdominal pain, tenderness over the colon, anorexia, nausea, severe weight loss, and weakness. The disease may be fatal and, before the availability of a treatment, was the cause of death in approximately 30% of infected individuals.

The infection can be cured most readily with tetracycline (**7**), oxytetracycline (**3**), or chlorotetracycline (**20**). Metronidazole (**1**) and iodoquinol (**2**) are also effective. Additional effective drugs include paromomycin (**8**), tinidazole (**11**), sulfadiazine (**25**), and carbarsone [*121-59-4*] (**31**, *p*-ureidobenzenearsonic acid, $C_7H_9AsN_2O_4$, Ameban).

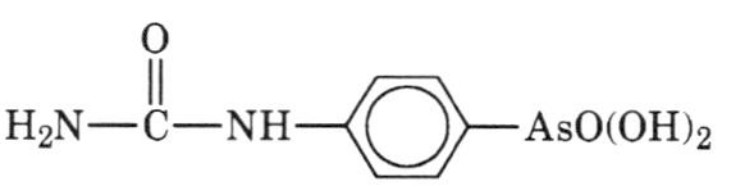

(**31**)

Coccidiosis

Coccidiosis is a widespread disease that occurs most often in fowl, such as chickens and turkeys, and other farm animals (cows, sheep, swine, horses, and rabbits) (11). In chickens the disease has caused severe economic losses. Coccidiosis also occurs in ox, water buffalo, zebu, bighorn sheep, wild goat, alpaca, lion, puma, fox, mink, parakeet, Canada goose, snow goose, and camel, among others. It is seen only rarely in humans, and dogs and cats are only occasionally infected.

Coccidiosis in chickens, as with cattle, is frequently caused by a combination of infective *Eimeria* protozoa. *Eimeria tenella* is the most important, but *Eimeria nectrix* is more common. *Eimeria maxima, Eimeria acervulina,* and *Eimeria mivati* are moderately pathogenic; *Eimeria brunetti* is highly pathogenic, but relatively uncommon. In chickens, young birds become infected by ingesting with their food oocysts shed in the feces of older birds. Diseased birds exhibit diarrhea, emaciation, and anorexia. The disease results in a marked decrease in the production of eggs, as well as a ca 10% mortality caused by the ingestion of a large number of oocysts.

Anticoccidial drugs are administered in the drinking water or feed of poultry and are primarily prophylactic, rather than curative, in nature. The most common ones in use are salinomycin (**32**, Coxistac [*53003-10-4*]) and monensin (**33**), a carboxylic acid which is used as its water-soluble sodium salt (Coban, Rumensin). These polyether ionophoric antibiotics are capable of complexing and interfacing with specific alkali metal cations to render them lipid-soluble and diffusible across parasite mitochondrial membranes. Related useful polyether antibiotics are narasin [*55134-13-9*] and lasalocid (**34**, Lasalocid A) (see ANTIBIOTICS, POLYETHERS). The potent ionophoric antibiotic maduramicin (**35**, Cygro [*84878-61-5*]) is currently being evaluated as an alternative for use against resistant strains of *Eimeria* inasmuch as resistance to the entire class of ionophoric compounds is presenting serious problems. Strains resistant to all coccidiostats arise rapidly because of the widespread use of these drugs. Sulfa drugs, such as sulfadimethoxine (**36**) and sulfaquinoxaline (**37**) are members of the first nontoxic class of anticoccidials (see ANTIBACTERIAL AGENTS, SYNTHETIC, SULFONAMIDES). Other types of compounds that have been considered treatments for the control of coccidiosis are the arsenicals, such as roxarsone (**38**, 4-hydroxy-3-nitrophenylarsonic acid [*121-19-7*]); the alkylidene diphenols; 3,3′-dinitrodiphenyl disulfide; the nitrofurans exemplified by nitrofurazone (**39**, 5-nitro-2-furaldehyde semicarbazone [*59-87-0*], Amifur, Furacin); the triazinones, diclazuril (**40**) and toltrazuril (**41**, Hycox, Maxicox); the carbanilide complexes such as nicarbazin [**42**, bis(4-nitrophenyl) urea 1:1 complex with 4,6-dimethyl-2(1*H*)-pyrimidone, Nicarb]; pyrimethamine (**24**); clopidol (**43**, Coyden); robenidine {**44**, 1,3-bis[(*p*-chlorobenzylidene)amino]guanidine}; and halofuginone (**45**) and its hydrobromide [*64924-67-0*], Stenorol (Table 4).

In cattle, infections by a single species of protozoan are unusual; combinations of the protozoa such as *Eimeria zuernii, Eimeria bovis,* and *Eimeria auburnensis* are more typical. Bovine coccidiosis occurs mainly in calves from three weeks to six months old. Adult animals are usually the carriers but tend to be asymptomatic. Infection in calves occurs by ingestion of the oocysts with their feed and/or water. As with poultry, the severity of the disease is a function of the number of oocysts ingested. Small numbers may cause no disease, whereas a large profusion ultimately may cause death. Severe disease is marked by anemia, weakness, emaciation, and anorexia. Monensin (**33**) is effective against coccidia in both cattle and sheep. Clopidol (**43**), halofuginone (**45**), and amprolium (**46**) are administered as feed additives. Other coccidiostats that have been employed are nitrofurazone (**39**), lincomycin (**47**), lasalocid (**34**), and salinomycin (**32**). Sulfonamides are only partially effective.

Coccidiosis is seen frequently in puppies and kittens and is responsible for diarrhea and even death if the animals are maintained in unsanitary conditions. For dogs and cats there exists no satisfactory treatment; however, the disease is self-limiting. Human infections can frequently be traced to these animals.

Cryptosporidiosis, an intestinal infection caused by protozoa of *Cryptosporidium* species is a taxonomically related disease (12). The disease affects animals, such as calves, lambs, and chickens, and infects humans worldwide, especially infants and children in developing countries. Symptoms range from mild self-limiting diarrhea and abdominal pain to a potentially fatal extreme diarrhea that results in weight loss and poor nutritional absorption. Increasingly, this opportunistic disease occurs in those with a suppressed immune system, especially those with AIDS. The only effective treatment for cryptosporidiosis in AIDS patients who do not respond readily to therapy is spiramycin (**48**).

Table 4. Coccidiosis Antiprotozoal Agents

Structure number	Compound name	CAS Registry Number	Molecular formula	Structure	Applications
(32)	salinomycin	[*53003-10-4*]	$C_{42}H_{70}O_{11}$	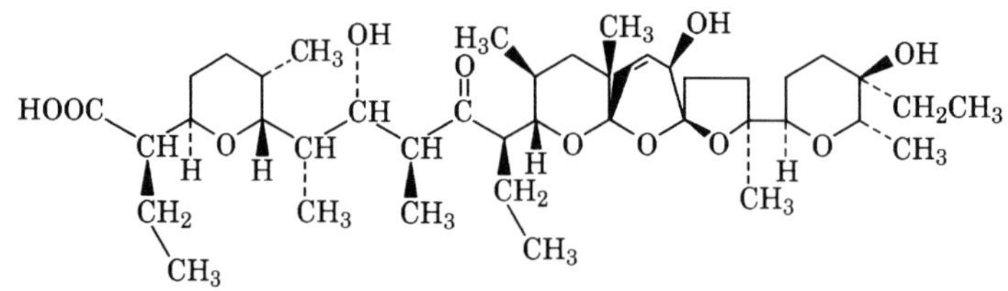	fowl, cattle, sheep
(33)	monensin[a]	[*17090-79-8*]	$C_{36}H_{62}O_{11}$	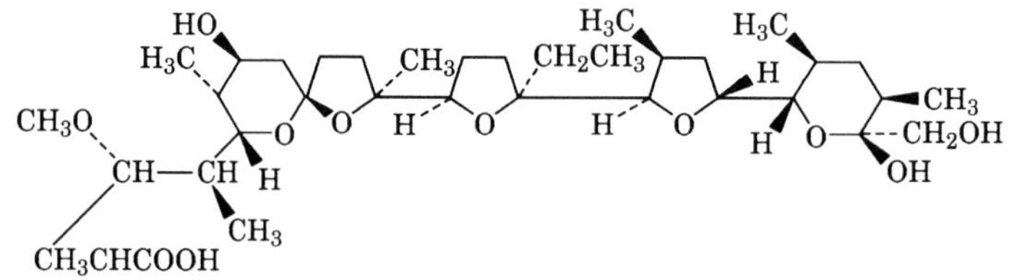	fowl, cattle, sheep
(34)	lasalocid A	[*25999-31-9*]	$C_{34}H_{54}O_8$	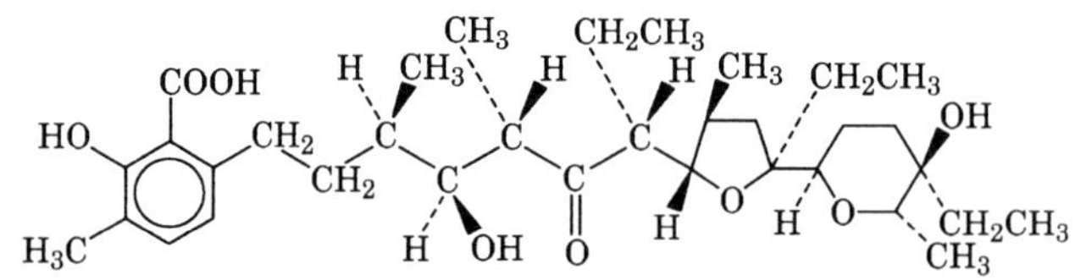	fowl, cattle, sheep
(35)	maduramicin	[*84878-61-5*]	$C_{47}H_{83}NO_{17}$	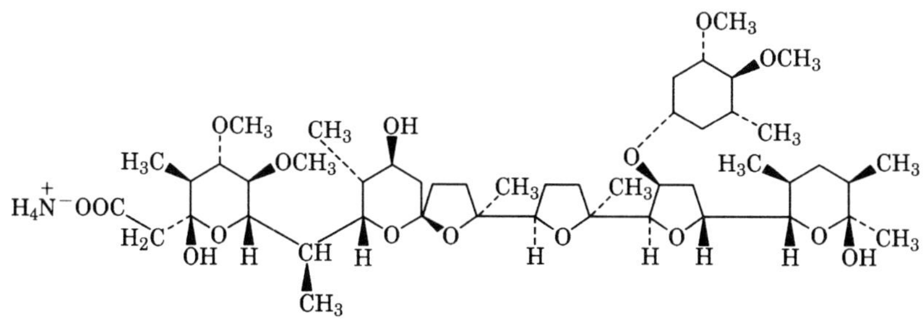	fowl

(36)	sulfadimethoxine	[*122-11-2*]	$C_{12}H_{14}N_4O_4S$	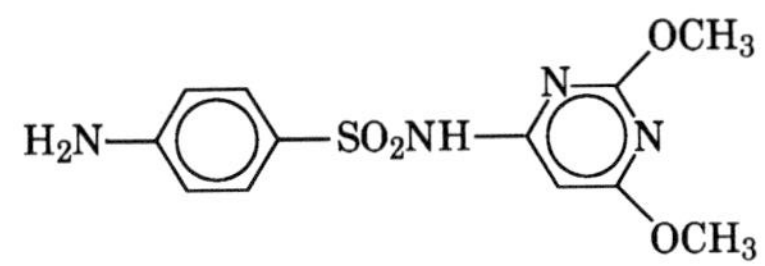	fowl
(37)	sulfaquinoxaline	[*59-40-5*]	$C_{14}H_{12}N_4O_2S$	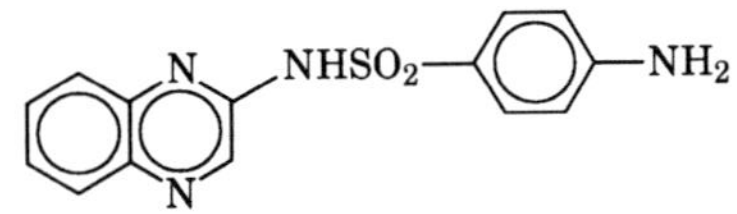	fowl
(38)	roxarsone	[*121-19-7*]	$C_6H_6AsNO_6$	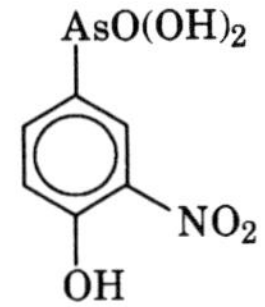	fowl
(39)	nitrofurazone	[*59-87-0*]	$C_6H_6N_4O_4$	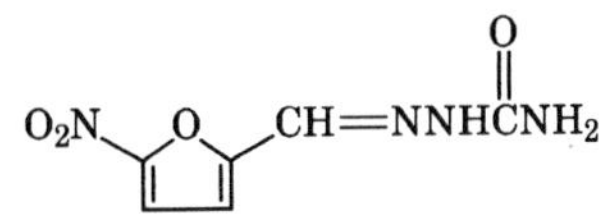	fowl, cattle, sheep
(40)	diclazuril	[*101831-37-2*]	$C_{17}H_9Cl_3N_4O_2$	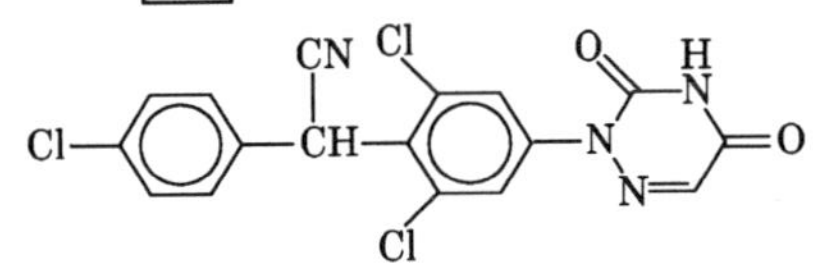	fowl
(41)	toltrazuril	[*69004-03-1*]	$C_{18}H_{14}F_3N_3O_4S$	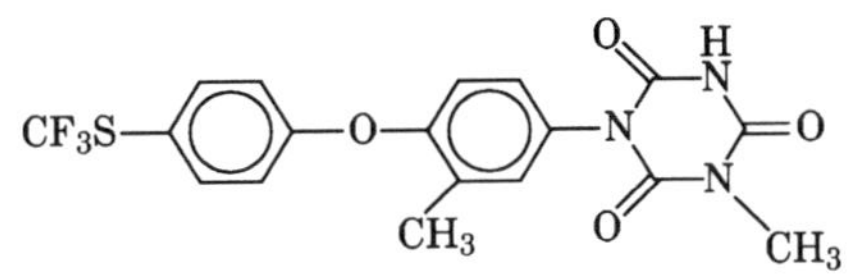	fowl
(42)	nicarbazin	[*330-95-0*]	$C_{13}H_{10}N_4O_5 \cdot C_6H_8N_2O$	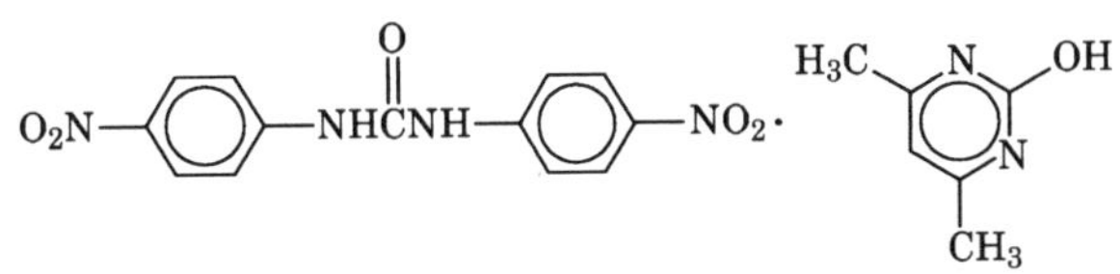	fowl

Table 4. (*Continued*)

Structure number	Compound name	CAS Registry Number	Molecular formula	Structure	Applications
(43)	clopidol	[*2971-90-6*]	$C_7H_7Cl_2NO$	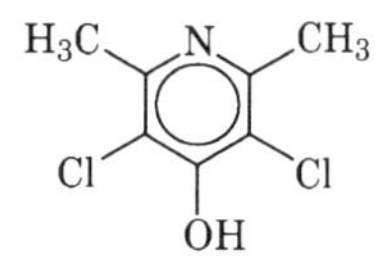	fowl, cattle, sheep
(44)	robenidine	[*25875-51-8*]	$C_{15}H_{13}Cl_2N_5$	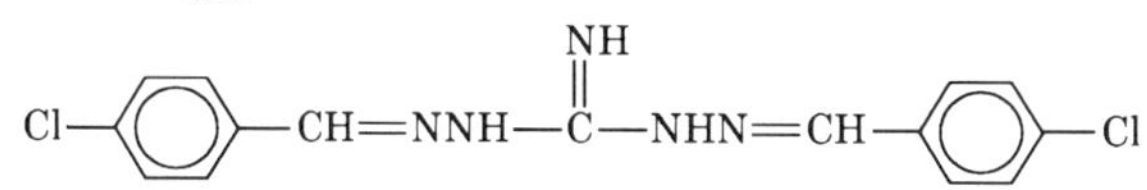	fowl
(45)	halofuginone[b]	[*55837-20-2*]	$C_{16}H_{17}BrClN_3O_3$	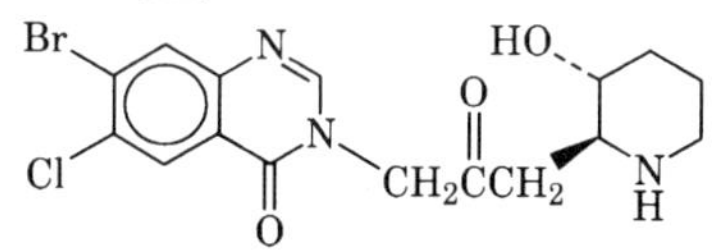	fowl, cattle, sheep
(46)	amprolium	[*121-25-5*]	$C_{14}H_{19}N_4Cl$	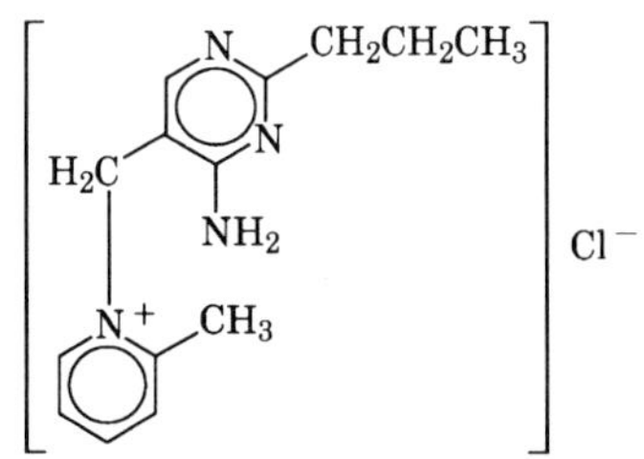	cattle, sheep

(47)	lincomycin	[*154-21-2*]	$C_{18}H_{34}N_2O_6S$	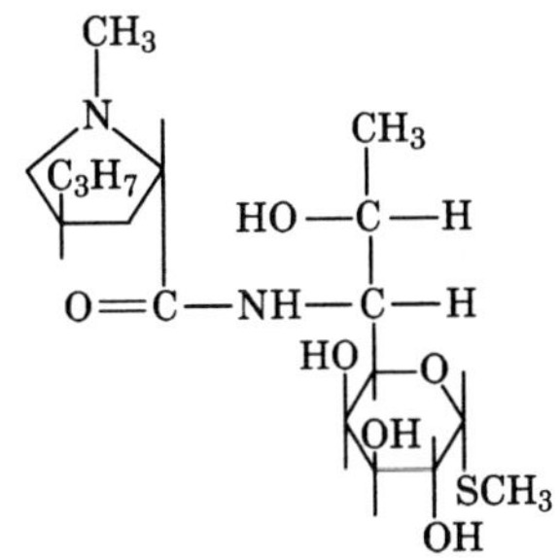	cattle, sheep
(48)	spiramycin[c]	[*8025-81-8*]	mixture	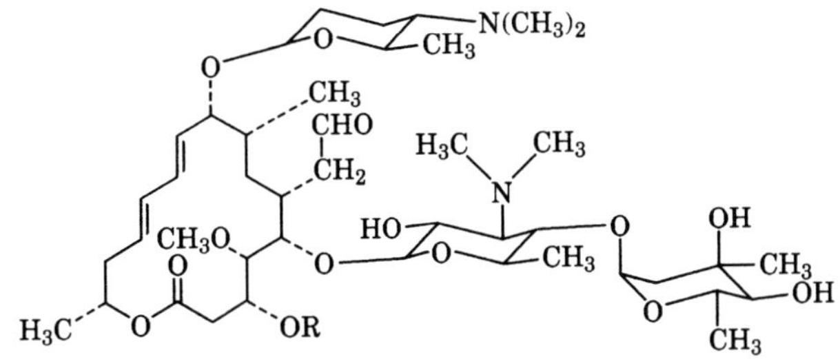	

[a]Also used to treat histomoniasis.
[b]Also used to treat theileriasis.
[c]Also used to treat toxoplasmosis.

Giardiasis

Giardiasis is a water-borne enteric disease of protozoan origin that occurs throughout the world. It is the most prevalent protozoal disease found in humans in the United States; the Rocky Mountain region is a particularly highly endemic area. It is also common in Central America, the former USSR, and India. The infection of the small intestine is caused by the flagellated protozoan, *Giardia lamblia*. The disease is transmitted by drinking water that has been fecally contaminated and poorly purified, or by person-to-person transmission of fecally contaminated food. Besides humans, the host for the protozoan includes the domesticated dog and certain wild animals, especially the beaver. The disease is more prevalent in children than in adults, and is especially common in those attending daycare centers. It is increasing in prevalence among male homosexuals.

Untreated giardiasis may be self-limiting and may produce no apparent symptoms, regardless of the duration of the infection. Alternatively, it may continue for numerous weeks or months and be extremely debilitating with symptoms that include severe and chronic diarrhea, abdominal cramps, weakness, anorexia, vomiting, fatigue, and weight loss. Because fats and fat-soluble nutrients are malabsorbed, the disease may be responsible for some of the mortality seen in underdeveloped countries. In severe giardiasis, there is inflammation and damage to the duodenal and jejunal mucosa.

Treatment of giardiasis is most successful with quinacrine (**49**, mepacrine) or its hydrochloride eg, Atabrine dihydrochloride. The drug may cause side effects such as dizziness, headache, nausea, vomiting, and reversible yellow staining of the skin. Alternatively, metronidazole (**1**) is recommended by the Centers for Disease Control, but because of its potential for mutagenesis or carcinogenesis should not be administered to children or pregnant women. Failures that occur with both drugs necessitate repetition of the therapy. Another drug reported to be effective is furazolidone (**50**), perhaps even more so in children than quinacrine (**49**) (Table 5). The aminoglycoside paromomycin (**8**) is recommended for pregnant women because the usual drugs are contraindicated. Tinidazole (**11**) has also been used successfully, but its activity could not be confirmed in a murine model. The disease has also been treated with acranil {**51**, 1-[(6-chloro-2-methoxy-9-acridyl)amino]-3-(diethylamino)-2-propanol dihydrochloride}. Further details pertaining to giardiasis are available (13).

Hexamitosis

Hexamitosis is a disease of chickens, turkeys, quail, and pheasants in which there is an infectious catarrhal enteritis in the duodenum and small intestine. The disease, caused by the protozoan *Hexamita meleagridis*, occurs in the United States, the United Kingdom, South America, and parts of Europe. It is primarily a disease of young birds under 10 weeks old in which mortality may be as high as 80%. The infected animals develop diarrhea, lose weight rapidly, appear listless and weak, and may eventually die. Hexamitosis is transmitted through contaminated food and water. Carrier birds are adults that have survived earlier in-

Table 5. Giardiasis Antiprotozoal Agents[a]

Structure number	Compound name	CAS Registry Number	Molecular formula	Structure
(**49**)	quinacrine[b]	[*83-89-6*]	$C_{23}H_{30}ClN_3O$	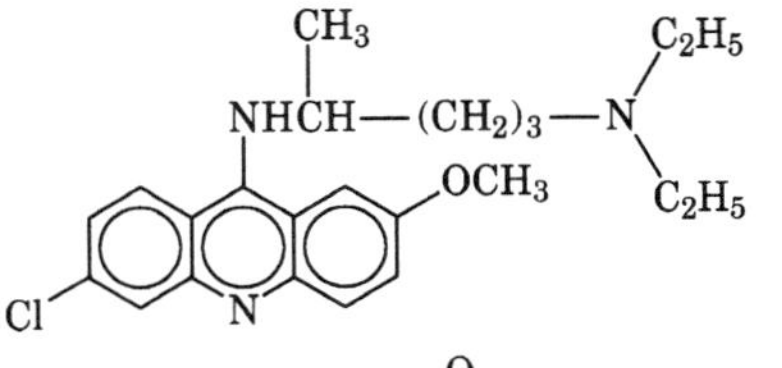
(**50**)	furazolidone[c]	[*67-45-8*]	$C_8H_7N_3O_5$	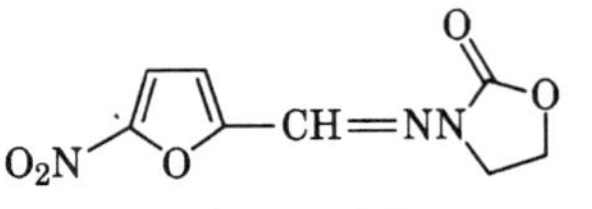
(**51**)	acranil[d]	[*1684-42-0*]	$C_{21}H_{28}Cl_3N_3O_2$	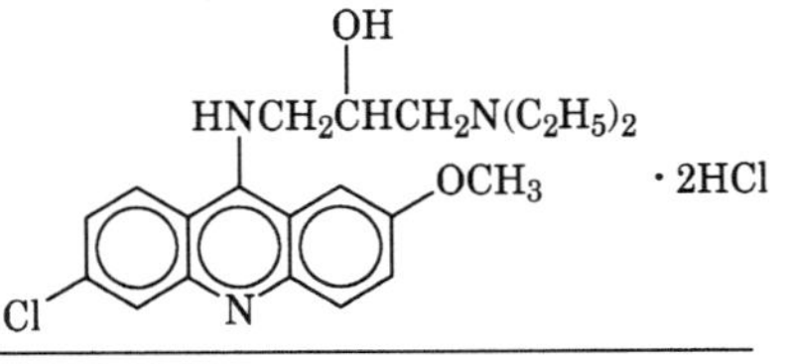

[a]Other applications are indicated in footnotes.
[b]Malaria.
[c]Histomoniasis.
[d]Hexamitosis.

fections. Hot weather and overcrowding may exacerbate the severity of the outbreak.

There is no effective treatment for hexamitosis. Penicillin (**52**), oxytetracycline (**3**), chlorotetracycline (**20**), Enheptin (**53**, 2-amino-5-nitrothiazole [*121-66-4*]), and streptomycin (**54**) sulfate [*3810-74-0*] were found to have limited value. Hexamitosis in carrier pigeons caused by *Hexamita columbae* was treated successfully with ronidazole (**55**, Dugro) (Table 6). An experimental infection of nude mice with *Hexamita muris* was treated with dimetridazole (**56**, 1,2-dimethyl-5-nitro-1*H*-imidazole [*557-92-8*]), metronidazole (**1**), tinidazole (**11**), and acranil (**51**). All compounds lowered or suppressed the fecal discharge of cysts, but the latter reappeared when the 1–3 week treatment terminated. In a related study, dimetridazole (**56**) controlled the clinical disease in mice but did not eliminate the infection.

Histomoniasis

Histomoniasis (histomonas, enterohepatitis, blackhead disease) is primarily an affliction of chickens and turkeys, but it also affects wild populations of peafowl, guinea fowl, pheasant, grouse, quail, and partridge. The disease occurs throughout the world and extensive outbreaks have been responsible for enormous losses of domesticated fowl; however, good sanitation reduces the number of affected animals. Because chickens are carriers of the disease, they must be separated from turkeys, which are more readily infected. Young turkeys from 3 to 12 weeks

Table 6. Hexamitosis and Histomoniasis Antiprotozoal Agents

Structure number	Compound name	CAS Registry Number	Molecular formula	Structure
(52)	penicillin	[*1406-05-9*]		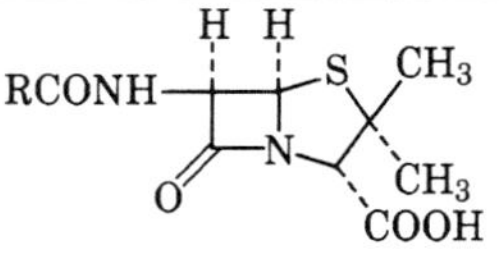
(53)	enheptin	[*121-66-4*]	$C_3H_3N_3O_2S$	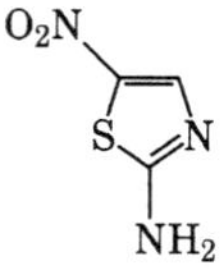
(54)	streptomycin	[*57-92-1*]	$C_{21}H_{39}N_7O_{12}$	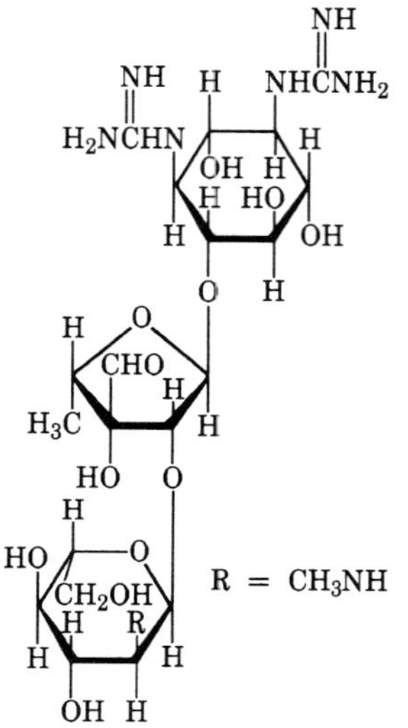
(55)	ronidazole	[*7681-76-7*]	$C_6H_8N_4O_4$	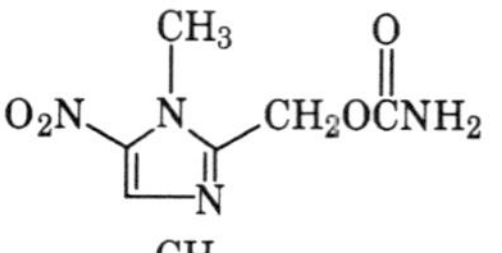
(56)	dimetridazole	[*551-92-8*]	$C_5H_7N_3O_2$	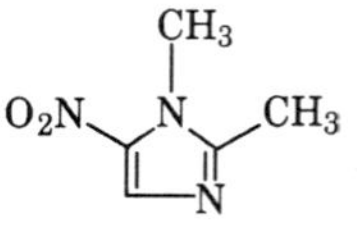
(57)	acetylenheptin	[*140-40-9*]	$C_5H_5N_3O_3S$	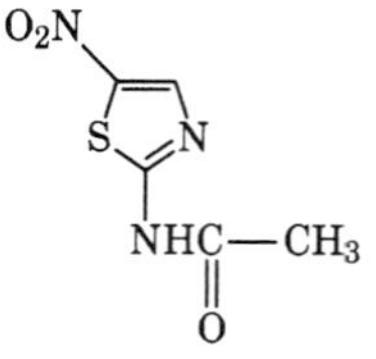
(58)	nithiazide	[*139-94-6*]	$C_6H_8N_4O_3S$	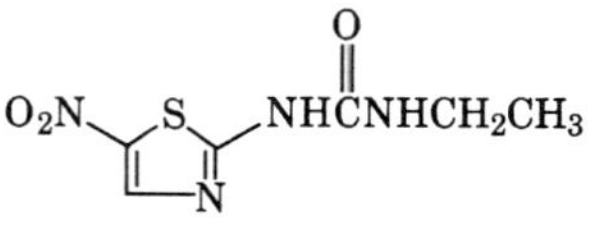
(59)	ipronidazole	[*14885-29-1*]	$C_7H_{11}N_3O_2$	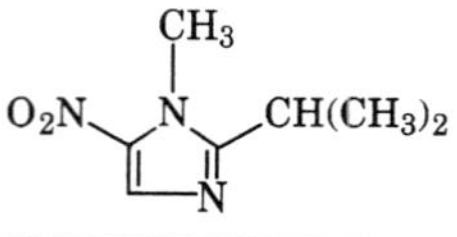

of age are particularly susceptible to the disease and may die within three days of the appearance of the first symptoms. Mortality decreases with age. Diseased birds look weak and droopy, may have diarrhea due to enterohepatitis, the head may become darkened, and death may ensue.

Histomoniasis is caused by the protozoan, *Histomonas meleagridis,* found in the liver and cecum of birds. The protozoans from the cecum are more infective than those from the liver. Transmission primarily occurs by ingestion of the eggs of the cecal worm, *Heterakis gallinarium,* which carries the protozoan and can survive several years in the soil.

A favored treatment for histomoniasis is dimetridazole (**56**) (Table 6). Other useful antiprotozoal agents include the thiazoles, Enheptin (**53**), acetylenheptin (**57**, 2-acetamido-5-nitrothiazole [*140-40-9*], aminitrozole, Enheptin-A), and nithiazide [**58**, 1-ethyl-3-(5-nitro-2-thiazolyl)urea [*139-94-6*], Hepzide], which are administered continuously into the feed for prevention and suppression. Furazolidone (**50**) and ipronidazole (**59**, Ipropran [*14885-29-1*]) have also been used. The latter has been evaluated, in addition, as a growth promotant. In an *in vitro* study, metronidazole (**1**), monensin (**33**), and tinidazole (**11**) showed good antihistomonal activity. Arsenicals that have shown activity are 4-nitrophenylarsonic acid and carbarsone (**31**). The drugs mentioned above, in order to prevent relapses, typically require continuous administration in the drinking water of the poultry until a few days prior to their slaughter, and are responsible for reduced egg production.

Leishmaniasis

Leishmaniasis affects some 12 million humans annually in an area where 350 million are at risk. It is a complex of at least two protozoan diseases, consisting primarily of cutaneous and visceral forms. A mucocutaneous form is considered by some to be another distinct variety. Clinical manifestations of the disease range from an asymptomatic infection to an infection in which there is considerable destruction of cutaneous tissue and mucous membranes. Leishmaniasis can often be fatal, especially in the visceral form. The seriousness of the disease depends on the state of the immunological system of the host and the species of parasite that inflicts the damage (14). The different species of *Leishmania* that are responsible for the disease can be difficult to distinguish from one another and methods to identify them have occupied the efforts of many researchers. The various species may be localized geographically or may overlap, and each is responsible for somewhat different symptoms. Although the human clinical manifestations may be categorized into different types, infections by more than a single species may be present at one time.

The disease is initiated by a bite from an infected female sand fly of the genus *Phlebotomus*. The injected parasites usually multiply near the site of the original infection, causing a lesion. The sand fly vector becomes infected when it acquires a blood meal from the skin or peripheral blood of a parasitized host and ingests the protozoans in the amastigote stage. The amastigotes develop into promastigotes in the intestine of the sand fly and eventually migrate to its proboscis. The saliva of the sand fly appears to enhance the survival of the injected promastigotes which, after phagocytosis, are transformed into the amastigote phase. The numerous reservoirs for the protozoans include humans, dogs, foxes, cats, rodents, and horses.

Cutaneous leishmaniasis is characterized by one or more slowly healing superficial ulcers that may be painful. These lesions are liable to further infection and may remain as open sores or become hard, wartlike nodules. The form of cutaneous leishmaniasis referred to as New World disease is caused by species of the *Leishmania (Viannia) braziliensis* complex *(L. braziliensis, L. guyanensis, L. panamanesis,* and *L. peruviana*) and species of the *Leishmania (Leishmania) mexicana* complex *(L. mexicana, L. amazonensis,* and *L. venezuelensis)* in the western hemisphere. Old World disease (oriental sore, Delhi or Baghdad boil) is caused by *Leishmania tropica* and *Leishmania major* in Asia, Africa, and southern Europe, and *Leishmania aethiopica,* restricted to eastern Africa. The incubation period for cutaneous leishmaniasis ranges from a few weeks to several months. Spontaneous cures can take place in a period ranging from one month to several years.

Visceral leishmaniasis is also known as kala azar, meaning black fever or black sickness in Hindi, because of the characteristic pigmentary changes that occur in the skin. The disease is endemic in much of the tropical and subtropical regions of southern Europe and the Mediterranean, the Middle East, India, Africa, and South and Central America. It is caused by species of the *Leishmania (Leishmania) donovani* complex [*L. donovani* (Old World), *L. infantum* (Old World), and *L. chagasi* (New World)]. In this systemic form of leishmaniasis, the parasites invade internal organs. It is characterized in patients by an enlarged liver and spleen, fever, weight loss, hemorrhage, leukopenia, and anemia. There is abdominal discomfort due to the spleen and liver involvement. Death due to severe diarrhea, pneumonia, and gastrointestinal bleeding may ensue if the disease is not treated. Following treatment, patients may relapse with a leishmanial form that solely affects the skin (post-kala azar dermal leishmaniasis). Rarer forms of leishmaniasis include chronic relapsing or recidivans forms of the cutaneous disease and diffuse cutaneous leishmaniasis.

Certain species of the *L. braziliensis* complex may cause mucocutaneous leishmaniasis, a form seen most frequently in northern and central South America. The patient first develops skin infections which later metastasize to produce highly disfiguring mucocutaneous lesions of the oronasopharynx.

Antimony compounds have been used to treat leishmaniasis ever since tartar emetic (antimony potassium tartrate) was discovered early in the 20th century to have efficacy against the mucocutaneous form of the disease. The cutaneous form has been treated with tartar emetic formulated in an ointment. Many side effects have been seen with this trivalent antimonial, some of which can be ascribed to the difficulty of obtaining pure antimony for its manufacture. These side effects include toxicity to the heart, liver, and kidneys. Other promising trivalent antimonials have been abandoned in favor of pentavalent antimonials with lower toxicity.

Pentavalent antimony preparations constitute the primary treatments for all forms of leishmaniasis, the most important of which are sodium stibogluconate (**60**, Pentostam [*16037-91-5*]) and glucantime (*N*-methylglucamine antimonate, $C_7H_{18}NO_8Sb$, meglumine antimonate) (Table 7). The actual chemical structure of

Table 7. Leishmaniasis Antiprotozoal Agents

Structure number	Compound name	CAS Registry Number	Molecular formula	Structure
(60)	sodium stibogluconate	[*16037-91-5*]	$C_{12}H_{20}O_{17}Sb_2{\cdot}9H_2O{\cdot}3Na$	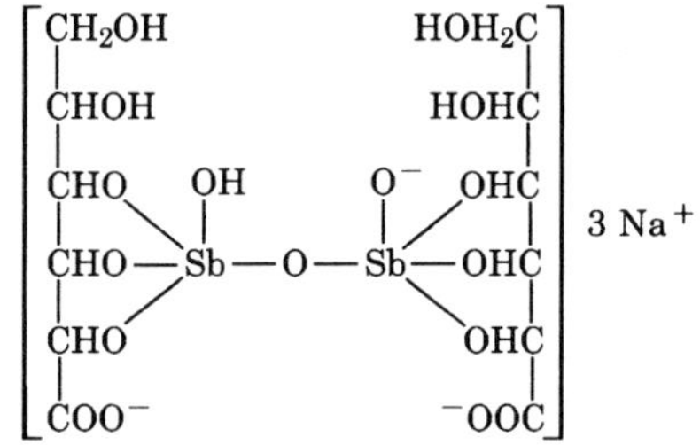
(61)	stilbamidine	[*122-06-5*]	$C_{16}H_{16}N_4$	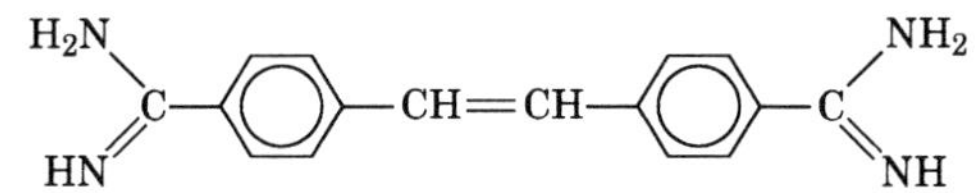
(62)	allopurinol	[*315-30-0*]	$C_5H_4N_4O$	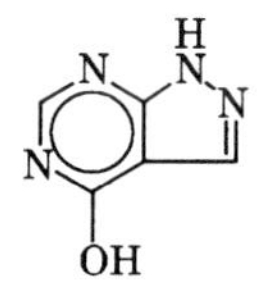
(63)	ketoconazole	[*65277-42-1*]	$C_{26}H_{28}Cl_2N_4O_4$	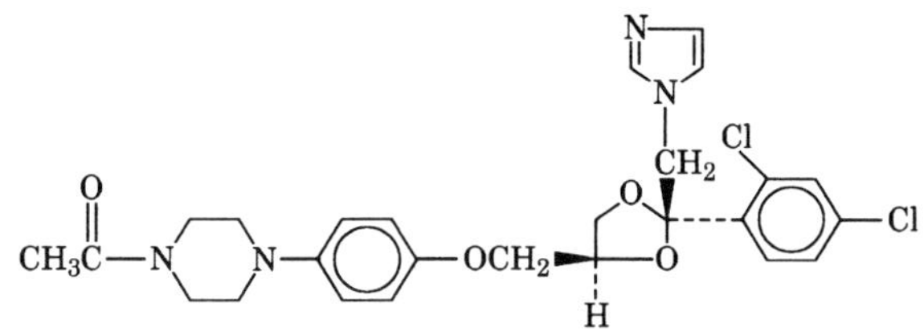

Sodium stibogluconate [*16037-91-5*
Stilbamidine [*122-06-5*], *3:000*
Allopurinol [*315-30-0*], *3:000*
Ketoconazole [*65277-42-1*], *3:000*

the antimonials is unknown. Studies indicate that they are mixtures of compounds with widely differing molecular weights. Use of glucantime is favored in Latin America and French-speaking countries. Cardio- and hepatotoxicity have been observed when the drug is administered over the typical 3–4 weeks of treatment. Liposome preparations of Pentostam and glucantime substantially increased the activity of these antimonials under experimental conditions. In 1990, an estimated 10,000 out of 200,000 cases of kala azar reported in India were unresponsive to treatment with antimonials; in Kenya, the strains were even more resistant to antimony. Although antimony-resistant strains of leishmania can be treated with diamidines, the latter compounds are more toxic. Pentamidine (**26**) as the isethionate, is the most widely used diamidine and has a high cure rate. Another diamidine, stilbamidine (**61**, 4,4′-diamidinostilbene [*122-06-5*]), is also effective.

Amphotericin B (**15**) is an antifungal macrolide antibiotic produced by *Streptomyces nodosus* that has been used as an alternative, albeit more toxic, drug to the antimonials. It acts as a leishmanicide against the visceral and mucocutaneous forms of the disease. To overcome its potentially severe nephrotoxicity, the drug must be administered over an extended period of time.

Because of the outbreak of antimony-resistant leishmaniasis and the need to develop an orally-administered therapy, the use of many other compounds has been considered. Those that appear to have clinical utility are allopurinol (**62**), ketoconazole (**63**), and both systemically and topically applied paromomycin (**8**) (see ANTIPARASITIC AGENTS, ANTIMYCOTICS).

Malaria

Malaria affects an estimated 270 million people and causes 2–3 million deaths annually, approximately one million of which occur in children under the age of five. While primarily an affliction of the tropics and subtropics, it has occurred as far north as the Arctic Circle. The disease essentially has been eradicated in most temperate-zone countries, but some 1100 cases of malaria in U.S. citizens returning from abroad were reported to the Centers for Disease Control during 1990. Malaria is seen today in Southeast Asia, Africa, and Central and South America. It is on the increase in Afghanistan, Brazil, China, India, Mexico, the Philippines, Sri Lanka, Thailand, and Vietnam. Escalation of the disease is because of the discontinued use of the insecticide DDT which effectively kills mosquito larvae, but has been found to be toxic to livestock and wildlife. Also, chloroquine (**6**), a reliable drug for the prophylaxis and treatment of falciparum malaria, is ineffective in many parts of the world because of the spread of drug-resistant strains.

Malaria is transmitted by the bite of an infected female *Anopheles* mosquito, one of the few species of the insect capable of carrying the human malaria parasite. The responsible protozoa are from the genus *Plasmodium,* of which only four of some 100 species can cause the disease in humans. The remaining species affect rodents, reptiles, monkeys, birds, and livestock. The species that infect humans are *Plasmodium falciparum, Plasmodium vivax, Plasmodium malariae,* and *Plasmodium ovale*. Note that concomitant multiple malaria infections are commonly seen in endemic areas, a phenomenon that further complicates choice of treatment.

An infection by *P. falciparum* has an incubation period (time from mosquito bite to clinical symptoms) of 6–25 days (average 12). Infection from this parasite can result in severe anemia, renal failure, pulmonary edema, shock, jaundice, cerebral malaria, and death. The life cycle of *P. falciparum* begins with a bite by an infected female mosquito in need of a blood meal for the reproductive stage of her life. Some parasites in the sporozoite stage are injected with her saliva, which contains material to prevent rapid clotting of the wound. The sporozoites enter the circulatory system and reach the liver in about one hour. The parasites grow and multiply by asexual division inside liver cells for 5–7 days. Then, as merozoites, having multiplied some 40,000-fold, they leave the liver to enter the bloodstream and invade red blood cells where they grow and further multiply over a period of 1–4 days. The merozoites emerge from the ruptured erythrocytes in a synchronous manner in about 48 hours. This results in the patient experiencing the clinical symptoms of the disease, namely, chills with rising temperatures, followed by fever and intense sweating. In addition, there can be severe headache, fatigue, dizziness, nausea, vomiting, anorexia, and diarrhea. Repetition of these symptoms occur on alternate days in this so-called tertian malaria. The falciparum parasites also cause the red cells to adhere to blood vessels, resulting in clumps that diminish the flow of blood to organs such as the brain, liver, and kidneys. Some released blood forms reinfect other red cells, where they again multiply leading to a further concerted rupturing and a return of the malaria symptoms. Some other merozoites develop into the sexual forms, ie, male or female gametocytes. Repetition of the malaria life cycle begins when gametocytes are acquired by an uninfected mosquito upon biting an infected individual. The gametocytes enter the mosquito stomach, undergo fertilization, and the zygotes thus formed migrate to the salivary glands of the insect where they are capable of being transmitted to another human by a bite. Chloroquine-resistant strains of *P. falciparum* have been found in Southeast Asia, Africa, and South America.

Plasmodium vivax, responsible for the most prevalent form of malaria (benign tertian), has an incubation period of 8–27 days (14 average). A variety seen in northern and northeastern Europe has an incubation period as long as 8–10 months. The disease can cause splenic rupture and anemia. Relapses (renewed manifestations of erythrocytic infection) can occur with this type of malaria. Overall, *P. vivax* is still susceptible to chloroquine; however, resistant strains have been reported from Papua New Guinea and parts of Indonesia. *Plasmodium malariae,* the cause of quartan malaria, has an incubation period of 15–30 days and its asexual cycle is 72 hours. This mildest form of malaria can cause nephritis in addition to the usual symptoms. It is a nonrelapsing type of malaria but the red blood cell infection can last for many years. No resistance to chloroquine by this plasmodium has been reported. *Plasmodium ovale,* responsible for ovale tertian malaria, has an incubation period of 9–17 days (15 average). Relapses can occur in people infected with this plasmodium. No chloroquine resistance has been reported for this parasite.

Antimalarials. Antimalarials can be categorized according to their mode of action (15).

Tissue Schizonticides. These eradicate the liver stages of the parasite and thereby prevent their entry into the blood. As a class, therefore, they are useful for prophylaxis. Some tissue schizonticides can act on the long-lived tissue forms (hypnozoites) of *P. vivax* and *P. ovale* and thus can cure the latter infections by preventing relapses.

Blood Schizonticides. These destroy the erythrocytic stages of the parasites and are useful for the clinical cure of falciparum malaria or suppression of relapsing infections.

Gametocytocides. These annihilate the sexual forms of the plasmodia (gametocytes) and also destroy the stages of the parasites in the *Anopheles* mosquito.

Sporontocides. These act against the sporozoites and oocysts in the mosquito and thereby prevent the transmission of the disease.

It should be noted that drugs may operate by more than one mechanism, and may possess a specific mode of action against one species of plasmodium but lack efficacy against others. In addition, antimalarial drugs may be classified according to their structural types.

Quinine (**22**), the first known antimalarial, is a 4-quinolinemethanol that has served as a model for the design of numerous antiplasmodial drugs. Its history began in mid-seventeenth century Peru when the Incas told the Jesuits of the medicinal properties of the bark of an evergreen mountain tree they called quinaquina (later called cinchona). The bark, when made into an aqueous infusion, was capable of curing malaria. The use of cinchona spread to Europe and the alkaloid from it, quinine, was isolated in 1820. Quinine, an extremely bitter substance, has been used by millions of malaria sufferers. In recent times, it has been employed successfully to treat chloroquine-resistant strains of *P. falciparum* but it frequently fails to provide a complete cure of the infection. Quinine acts on the asexual blood forms of the plasmodia in a manner slower than many synthetic drugs. Overdoses cause tinnitus and visual disturbances, side effects that disappear on withdrawal of the drug. It can also cause premature contractions in women in late stages of pregnancy. Although quinine is a favored antimalarial for parenteral administration, it is nevertheless hazardous by this route. Quinidine (**64**), has been shown to be even more effective in combatting the disease (Table 8). However, it has undesirable cardiac side effects that reduce its suitability as an antimalarial. Mixtures of cinchona alkaloids, known as totaquine, are easier to produce and have been employed in treatment. Totaquine has been standardized to contain a minimum of 15% quinine.

The success of quinine inspired the search for other antimalarials. The greatest impetus for the development of synthetic drugs came this century when the two World Wars interrupted the supply of cinchona bark to the combatants. A structurally related 4-quinolinemethanol is mefloquine (**65**, Lariam [*51773-92-3*]), which now serves as an effective alternative agent for chloroquine-resistant *P. falciparum.* This is a potent substance that requires less than one-tenth the dose of quinine to effect cures. There are some untoward side effects associated with this drug such as gastrointestinal upset and dizziness, but they tend to be transient. Mefloquine is not recommended for use by those using beta-blockers, those whose job requires fine coordination and spatial discrimination, or those with a history of epilepsy or psychiatric disorders. A combination of mefloquine with Fansidar (a

Table 8. Malaria Antiprotozoal Agents[a]

Structure number	Compound name	CAS Registry Number	Molecular formula	Structure
		4-Quinolinemethanols		
(64)	quinidine (sulfate)	[*50-54-4*]	$C_{20}H_{24}N_2O_2 \cdot \frac{1}{2}H_2O_4S$	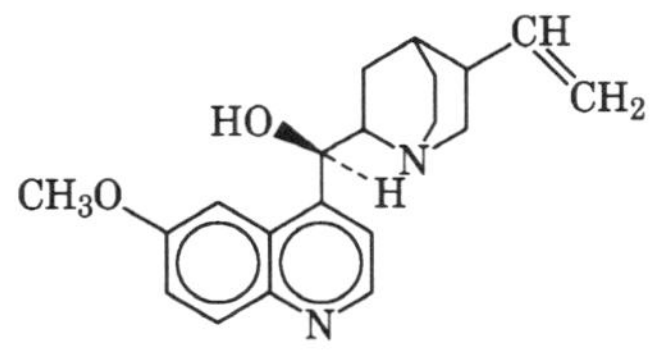
(65)	mefloquine	[*53230-10-7*]	$C_{17}H_{16}F_6N_2O$	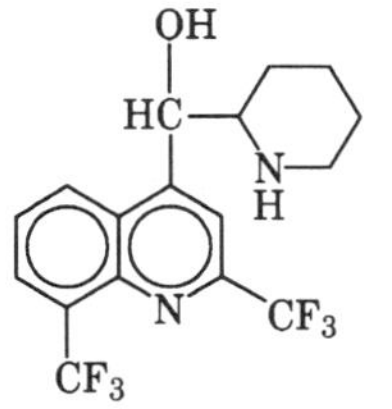
		Phenanthrene-methanol		
(66)	halofantrine	[*69756-53-2*]	$C_{26}H_{30}Cl_2F_3NO$	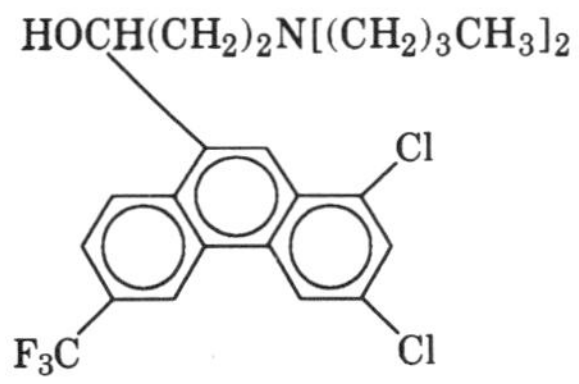
		4-Aminoquinoline		
(67)	hydroxy-chloroquine	[*118-42-3*]	$C_{18}H_{26}ClN_3O$	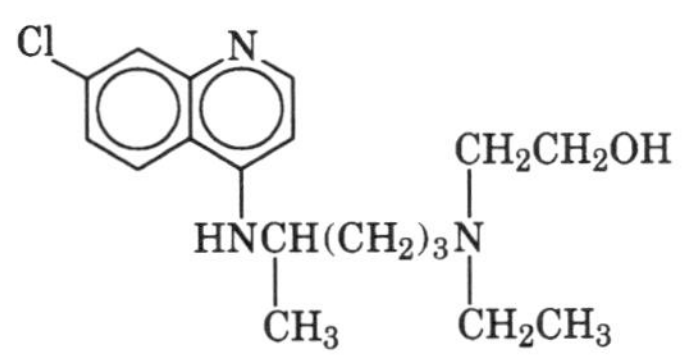
		8-Aminoquinolines		
(68)	pamaquine[b]	[*491-92-9*]	$C_{19}H_{29}N_3O$	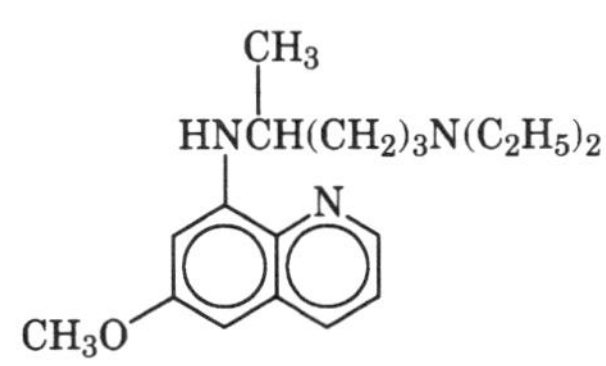
(69)	primaquine (phosphate)[c,d]	[*63-45-6*]	$C_{15}H_{21}N_3O \cdot 2H_3O_4P$	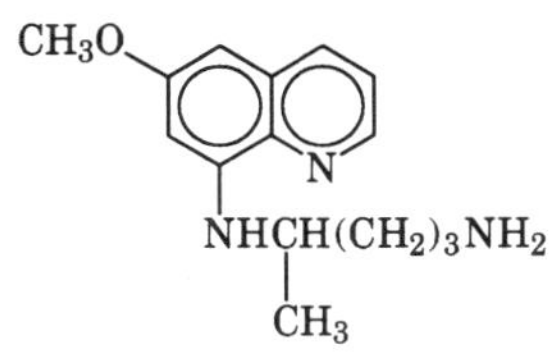
		Antifolates		
(70)	dapsone	[*80-08-0*]	$C_{12}H_{12}N_2O_2S$	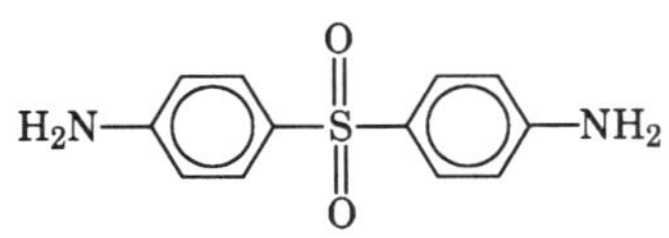

Table 8. (Continued)

Structure number	Compound name	CAS Registry Number	Molecular formula	Structure
(71)	sulfadoxine[c]	*[2447-57-6]*	$C_{12}H_{14}N_4O_4S$	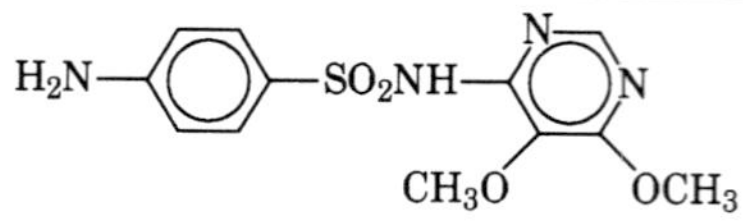
(72)	sulfalene[e]	*[152-47-6]*	$C_{11}H_{12}N_4O_3S$	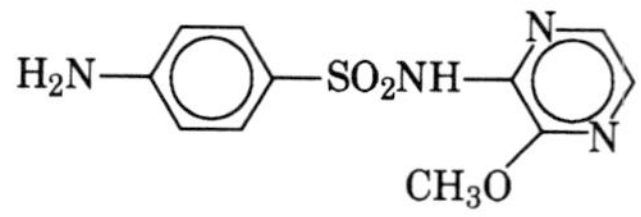
(73)	trimethoprim[c,e]	*[738-70-5]*	$C_{14}H_{18}N_4O_3$	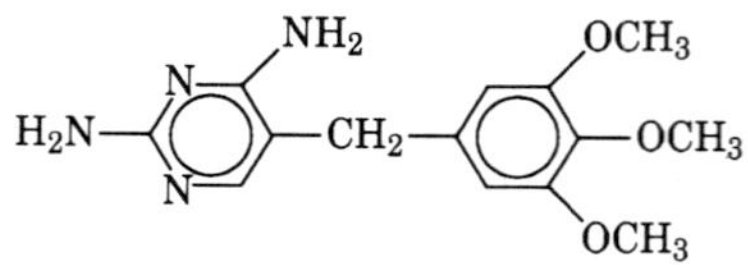
(74)	chlorguanide	*[500-92-5]*	$C_{11}H_{16}ClN_5$	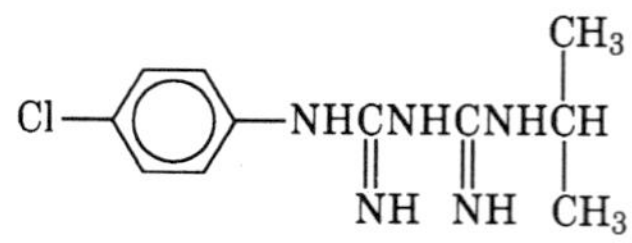
(75)	cycloguanil	*[516-21-2]*	$C_{11}H_{14}ClN_5$	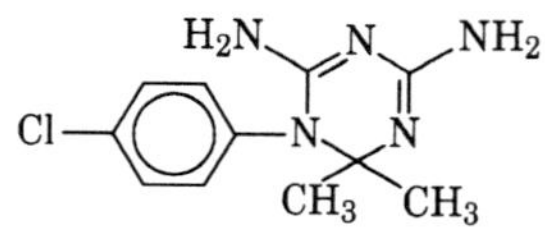
			Others	
(76)	menoctone[b]	*[14561-42-3]*	$C_{24}H_{32}O_3$	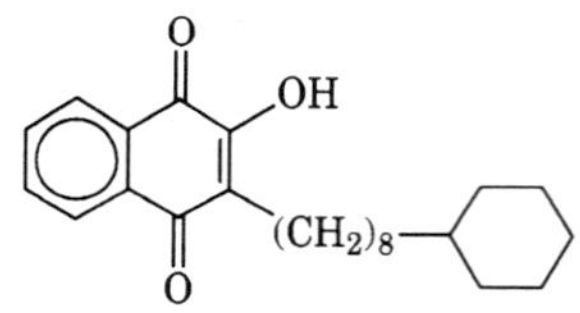
(77)	pyronaridine	*[74847-35-1]*	$C_{29}H_{32}ClN_5O_2$	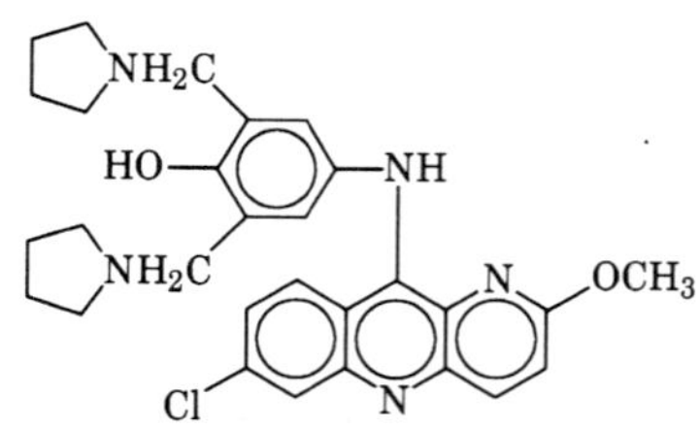
(78)	doxycycline	*[564-25-0]*	$C_{22}H_{24}N_2O_8$	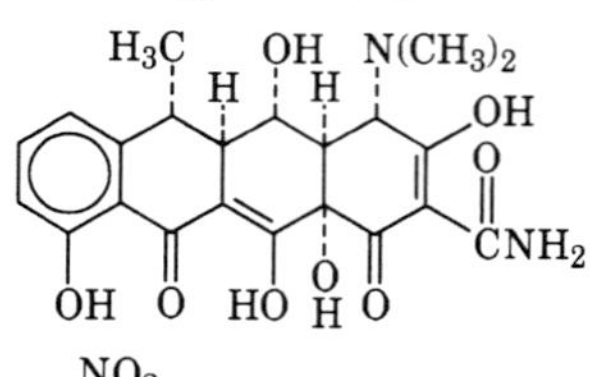
(79)	chloram-phenicol	*[56-75-7]*	$C_{11}H_{12}Cl_2N_2O_5$	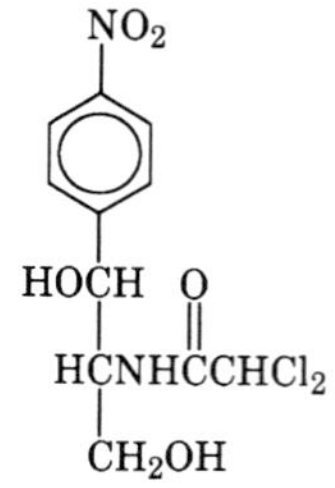

Table 8. (*Continued*)

Structure number	Compound name	CAS Registry Number	Molecular formula	Structure
(80)	febrifugine	[*24159-07-7*]	$C_{16}H_{19}N_3O_3$	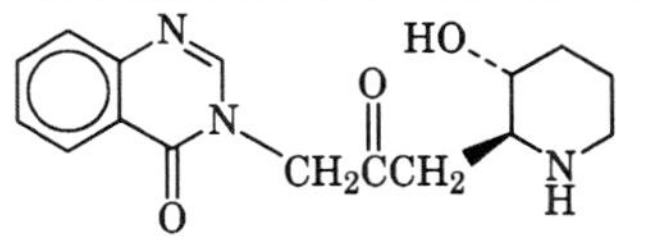
(81)	artemisinin	[*63968-64-9*]	$C_{15}H_{22}O_5$	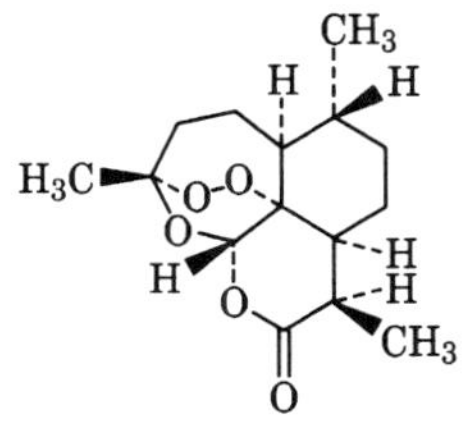
(82)	artemether	[*71963-77-4*]	$C_{16}H_{26}O_5$	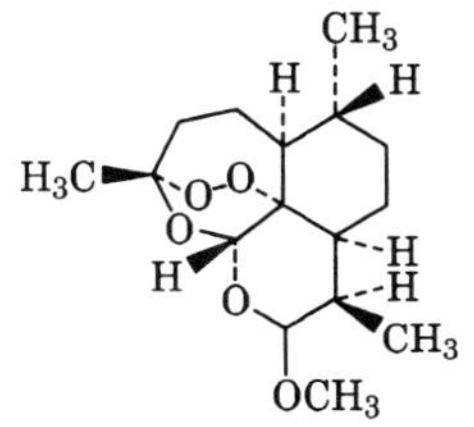
(83)	arteether	[*75887-54-6*]	$C_{17}H_{28}O_5$	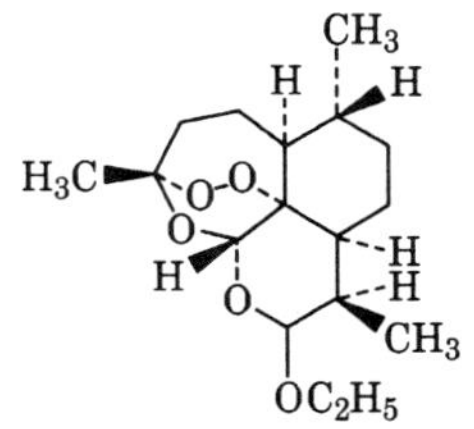
(84)	artesunic acid	[*88495-63-0*]	$C_{19}H_{28}O_8$	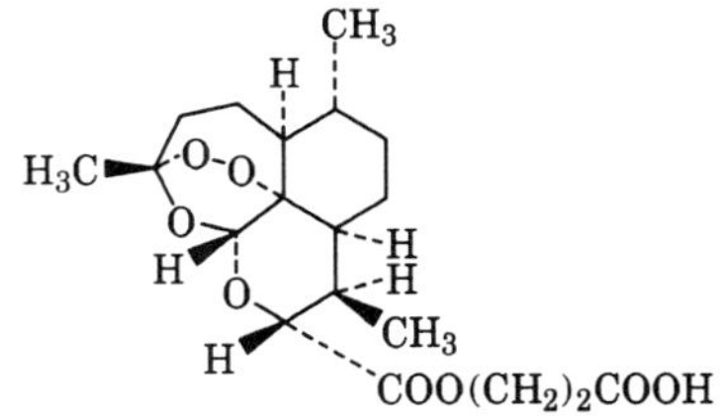
(85)	artelinic acid	[*109637-83-4*]	$C_{23}H_{30}O_7$	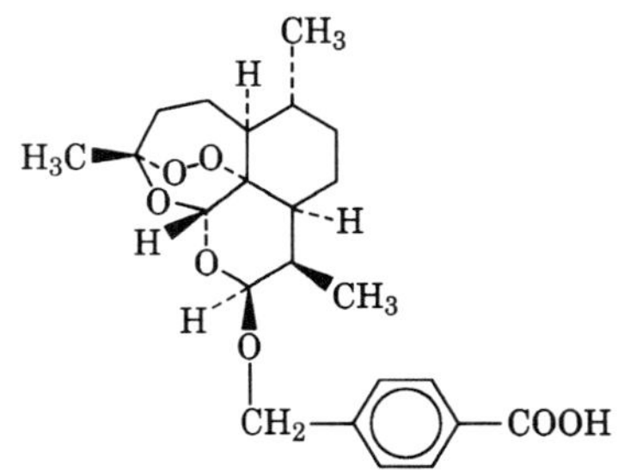

[a]Other applications are indicated in footnotes.
[b]Theileriasis.
[c]Pneumocystosis.
[d]American trypanosomiasis.
[e]Toxoplasmosis.

mixture of pyrimethamine and sulfadoxine) is known as Fansimef but its use is not recommended. Resistance to mefloquine has been reported even though the compound has not been in wide use.

The best example of the class of phenanthrene-methanols is halofantrine (**66**, Halfan [*36167-63-2*]), a drug that is effective against chloroquine-resistant malaria and is now being evaluated in Africa. It produces temporary gastrointestinal disturbances.

Chloroquine (**6**) is the most effective of the hundreds of synthesized 4-aminoquinolines and was the antimalarial of choice until resistance developed to it in Colombia and Southeast Asia in 1962. It is still one of the most important antimalarial agents in use for suppression (prophylaxis) in endemic areas where there remains sensitivity to the drug. It acts as a blood schizonticide and rarely produces serious side effects in individuals taking it prophylactically. Ironically, it was the widespread prophylactic use that most likely led to drug resistance. For treatment of the disease, the phosphate salt is administered orally, or the chloride, parenterally. The minor side effects include gastrointestinal upset, headache, dizziness, blurred vision, and pruritus (itch), but discontinuance of the drug is seldom necessary. Those who have used chloroquine continually for more than 6 years, either for prophylaxis or in the treatment of arthritis, are recommended to have ophthalmologic exams to check for possible retinal damage. It was demonstrated that chloroquine resistance can be reversed *in vitro* by treating *P. falciparum* parasitized erythrocytes with the calcium channel blocking drug verapamil (16). Subsequently, other calcium channel blockers and antihistamines have been found to be effective in overcoming chloroquine resistance *in vitro* and in rodents. Desipramine (Norpramin) and other antidepressants have been shown to reverse chloroquine resistance in monkeys. Alternatives to chloroquine for treatment are amodiaquine (**21**) and hydroxychloroquine (**67**), a chloroquine modification in which one of the ethyl groups on the tertiary nitrogen atom of the side chain is replaced with hydroxyethyl. These compounds are less effective, and amodiaquine is more toxic than chloroquine. The 4-aminoquinolines also act as gametocytocidal agents against *P. vivax, P. ovale, P. malariae* and immature gametocytes of *P. falciparum.*

The 8-aminoquinolines are effective against both the primary and secondary tissue forms and sexual blood forms of the parasites. At toxic levels they are also active against asexual blood forms in humans. Pamaquine (**68**, plasmochin, plasmoquine), the oldest useful member of the class of 8-aminoquinolines, was synthesized in Germany in the mid-1920s. Primaquine [*90-34-6*] (**69**), which is less toxic and more effective, is the most widely used of the 8-aminoquinolines. It is gametocytocidal against all species of human malaria parasites and is an anti-relapsing drug, but has the undesirable ability to cause severe hemolysis in glucose-6-phosphate dehydrogenase (G6PD) deficient individuals. This accounts for its recommended administration in limited doses over 1–2 weeks and its low utilization. During pregnancy, it is considered inadvisable for primaquine to be taken as it could be passed on to a G6PD-deficient fetus, thereby causing hemolytic anemia. Primaquine is generally used in conjunction with a blood schizonticide such as chloroquine (**6**), amodiaquine (**21**), or pyrimethamine (**24**) and is used for the prevention of relapses only against *P. vivax* and *P. ovale.*

The drugs known as antifolates act as effective blood schizonticides. Unfortunately, the parasites readily develop resistance to them. Most antifolates show poor oral tolerance, absorption, and host toxicity. They fall into two types depending on the mechanisms by which they operate.

The first are competitors of PABA (*p*-aminobenzoic acid) and thus interrupt host *de novo* formation of the tetrahydrofolic acid required for nucleic acid synthesis. Examples of drugs that fall into this group are the sulfones and sulfonamides. The most well-known of the sulfones is dapsone (**70**, 4,4′-diaminodiphenyl sulfone, DDS), whose toxicity has discouraged its use. Production of folic acid, which consists of PABA, a pteridine unit, and glutamate, is disturbed by the substitution of a sulfonamide (structurally similar to PABA). The antimalarial sulfonamides include sulfadoxine (**71**, Fanasil [*2447-57-6*]), sulfadiazine (**25**), and sulfalene (**72**, sulfamethoxypyrazine [*152-47-6*], Kelfizina). Compounds of this group are rapidly absorbed but are cleared slowly.

The second type of antifolates bind preferentially with, and thus selectively inhibit, the enzyme dihydrofolate reductase contained in the plasmodia. This interferes with the ability of the malaria parasites to convert dihydrofolate to tetrahydrofolic acid. In the erythrocyte host, however, dihydrofolate reductase is considerably less sensitive to these drugs and the blood cells are capable of utilizing exogenous folate. Dihydrofolate reductase inhibitors are potent blood schizonticides that act on the asexual blood stages. Members of this class include the pyrimidines, pyrimethamine (**24**), and trimethoprim (**73**). Pyrimethamine is a slow-acting drug that is not recommended for use in acute attacks and is potentiated by combination with sulfadoxine (**71**) (mixture is called Fansidar), dapsone (**70**) (mixture is called Maloprim), sulfalene (**72**) (mixture is called Kelfimeta or Metakelfin), or chloroquine (**6**) (mixture is called Darachlor). It is an effective suppressant against *P. falciparum*. The mode of action of trimethoprim (**73**) is similar to that of pyrimethamine (**24**). Because it is also slow-acting, it is generally used in combination with the fast-acting sulfonamide, sulfalene (**72**), to give an effective treatment against *P. falciparum*. Related structurally and functionally are the so-called open and cyclic triazines, exemplified by chlorguanide (**74**, proguanil [*500-92-5*], or the hydrochloride, Paludrine), a biguanide, and cycloguanil (**75**, Camolar [*609-78-9*]), respectively. The compounds have low toxicity and are also effective as causal prophylactics, ie, they destroy parasites before they enter the red blood cells.

Quinones and naphthoquinones were explored during the World War II Antimalarial Drug Program. Now that chloroquine resistance is a serious problem, compounds of this group such as menoctone (**76**) are being reinvestigated.

Quinacrine (**49**) is an acridine that was used extensively from the mid-1920s to the end of World War II. It acts much like chloroquine and is reasonably effective. Because it causes the skin to turn yellow and, in high doses, causes yellow vision, the drug is no longer in use as an antimalarial. Pyronaridine (**77**), a 1-azaacridine developed in China, appears to be effective against mefloquine-resistant, but not entirely against chloroquine-resistant, strains of *P. falciparum*.

Some antibiotics, such as the tetracyclines, tetracycline (**7**), doxycycline (**78**), and minocycline (**17**), chloramphenicol (**79**), and clindamycin (**23**) have modest antimalarial properties, but are slow-acting.

A Chinese traditional herbal treatment for malaria obtained from the roots of *Dichroa febrifuga* is called Ch'ang Shan and was investigated in the 1940s.

Febrifugine (**80**), the alkaloid responsible for its activity, was isolated and found to be considerably more active than quinine in experimental infections. Unfortunately, the drug caused nausea and vomiting in humans. Synthesized analogues were generally less effective than the parent.

The most promising group of new antimalarials is based on the compound artemisinin (**81**), an unusual sesquiterpene lactone endoperoxide first isolated in China in 1972 and then in the United States from the weed *Artemisia annua* (17). Artemisinin, also called qinghaosu, meaning in Chinese "extract of green herb," is a rapid-acting and relatively nontoxic therapeutic agent. Its low oil and water solubility has been increased by making certain structural modifications to the molecule. Oil-soluble artemether (**82**), the methyl ether of the lactol dihydroartemisinin, has been administered clinically in China by the intramuscular route. The closely related ethyl ether, arteether (**83**), is being developed by the World Health Organization as an alternative oil-soluble analogue. The sodium salts of artesunic acid (**84**) and artelinic acid (**85**) are the most interesting water-soluble modifications of artemisinin. If administered intravenously, these sodium salts could be advantageously applied in the treatment of cerebral malaria where rapid return to consciousness of the comatose patient and reduction of the parasitemia are the primary clinical goals. Although the water-soluble compound sodium artelinate is more stable in aqueous solution than sodium artesunate, only the latter has been tested clinically. Artelinic acid and related compounds effectively eliminated or prevented the establishment of parasitemia in *P. berghei*-infected mice when they were administered by the transdermal route. Artemisinin and its derivatives have been used successfully in China with several thousand *P. falciparum* and *P. vivax* patients; because this class of drugs is new, parasite resistance has not yet been encountered.

Pneumocystosis

The organism responsible for pneumocystosis in humans, *Pneumocystis carinii,* having the characteristics of both protozoan and fungi, has defied simple taxonomic classification. Recent evidence suggests that it falls into the fungi category. Acute pneumocystosis rarely strikes healthy individuals, although the organism is harbored by a wide variety of animals and most people without any apparent adverse effect. *P. carinii* becomes active only in those individuals who have a serious impairment of their immunologic systems (18). Typically, this opportunistic disease occurs in AIDS patients, 80% of whom ultimately contract *P. carinii* pneumonia, and is one of the main causes of mortality. The disease also occurs in those being administered immunosuppressive drugs for, eg, organ transplantation or treatment of malignant disease. In addition, pneumocystosis is seen in malnourished infants whose immunological systems are impaired. The disease is characterized by a severe pneumonia caused by a rapid multiplication of the organisms almost exclusively in lung tissue. If left to run its course, the acute disease is generally fatal.

The drug of choice is the antifolate mixture, trimethoprim (**73**)-sulfamethoxazole (**86**) (mixture is called co-trimoxazole, Bactrim, Septra) given orally or intravenously (Table 9). It is well tolerated if given in low doses. Also used is pyrimethamine (**24**) combined with sulfadiazine (**25**) or sulfadoxine (**71**). Pentamidine (**26**) isethionate is administered parenterally but is generally not prescribed for prophylactic use because there is a high incidence of hypotension, renal failure, and pain and tissue injury at the injection site. However, there have been successful prophylactic trials with the drug in the aerosolized form for inhalation. The polyamine-inhibitor eflornithine, (**87**, α-difluoromethylornithine, DFMO; [*67037-37-0*] the hydrochloride monohydrate is Ornidyl) is undergoing evaluation as a treatment for *P. carinii*. A combination of clindamycin (**23**) and primaquine (**69**) shows great promise against this protozoan in clinical trials when administered for prophylaxis or for therapy in mild to moderately severe pneumocystis pneumonia. Either drug alone is not effective.

Theileriasis

Theileriasis (theileriosis) is a tick-transmitted protozoan disease of cattle seen primarily in central and eastern Africa. The disease not only affects domesticated cattle, but also ox, zebu, water buffalo, and African buffalo. The most significant form of the disease is caused by the protozoan *Theileria parva (Piroplasma kochi, Piroplasma parvum, Theileria kochi, Theileria lawrencei)* and is known as African Coast fever, East Coast fever, bovine theileriasis, or Corridor disease. The most important vector of the disease is the tick, *Rhipicephalus appendiculatus,* which harbors the protozoan through one molt. About 90–100% adult cattle infected with *T. parva* die, but calves seem to be less susceptible. After receiving a bite from an infected tick, there is an incubation period of 8–25 days before the appearance of the disease. In the acute form, the animals have a fever, cease to eat, may have a nasal discharge and lachrymate, have general weakness, and diarrhea. Rapid and labored breathing precede death.

There are other milder forms of theileriasis that resemble the disease caused by *T. parva* infection. They also primarily affect cattle and are transmitted by ticks. Causative organisms include *Theileria annulata,* responsible for tropical thieleriasis, tropical piroplasmosis, Egyptian fever, and Mediterranean Coast fever. These affect ox, zebu, and water buffalo. Endemic areas include northern Africa, southeastern Europe, the southern part of the former USSR, and Asia. *Theileria mutans* is the cause of benign bovine thieleriasis, Marico calf disease, and mild gall sickness. It affects ox and zebu in Africa, Asia, southern Europe, England, the former USSR, Australia, and North America.

There are no fully effective therapeutic agents for the treatment of theileriasis. Chlorotetracycline (**20**) and oxytetracycline (**3**) have therapeutic activity during the incubation period. Pamaquine (**68**) was reported to have a specific effect on the erythrocyctic forms. Other drugs with limited efficacy are imidocarb (**19**) on *T. annulata,* halofuginone (**45**) on both *T. annulata* and *T. parva,* and the naphthoquinones menoctone (**76**), parvaquone (**88**), and buparvaquone (**89**) on *T. parva*. Methotrexate (**90**) has been found to be active *in vitro* (Table 9).

Table 9. Various Antiprotozoal Agents

Structure number	Compound name	CAS Registry Number	Molecular formula	Structure
			Pneumocystosis	
(86)	sulfamethoxazole[a]	[*723-46-6*]	$C_{10}H_{11}N_3O_3S$	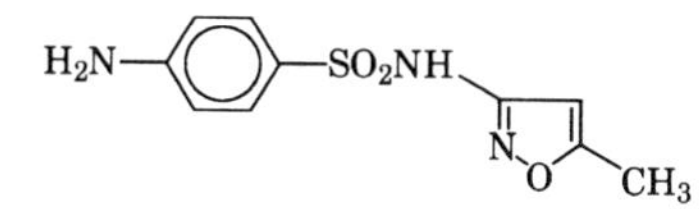
(87)	eflornithine[b]	[*67037-37-0*]	$C_6H_{12}F_2N_2O_2$	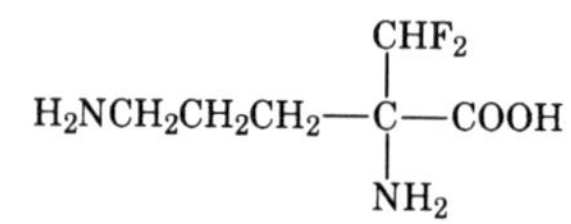
			Theileriasis	
(88)	parvaquone	[*4042-30-2*]	$C_{16}H_{16}O_3$	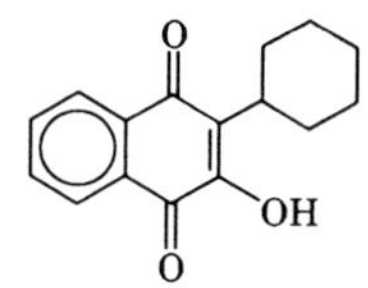
(89)	buparvaquone	[*88426-33-9*]	$C_{21}H_{26}O_3$	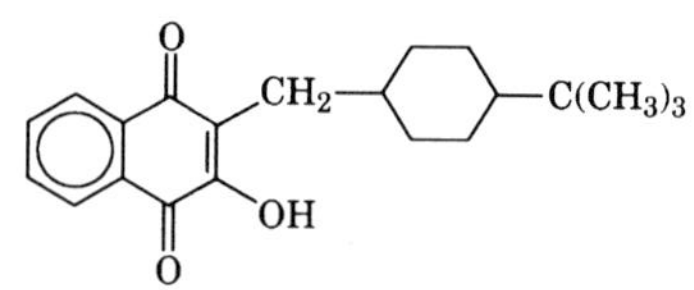
(90)	methotrexate	[*59-05-2*]	$C_{20}H_{22}N_8O_5$	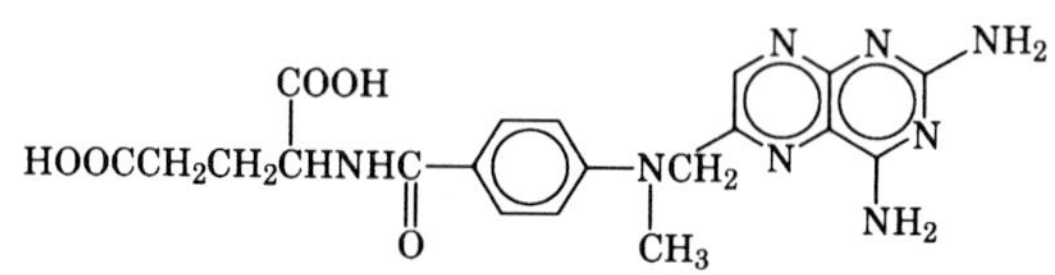
			Trichomoniasis	
(91)	ornidazole	[*16773-42-5*]	$C_7H_{10}ClN_3O_3$	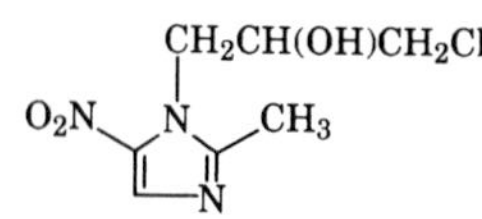

[a]Also used for toxoplasmosis.
[b]Also used for African trypanosomiasis.

Toxoplasmosis

Toxoplasmosis, a coccidial disease, affects both humans and animals throughout the world. Approximately one-third of the population has antibodies to it but is asymptomatic. Although rare in the past, toxoplasmosis is an increasingly important opportunistic disease that afflicts immunocompromised hosts such as post-transplant patients taking immunosuppressive drugs and AIDS patients. The disease is responsible for a central nervous system infection with symptoms in humans that include fever, headache, swollen lymph glands, cough, sore throat, nasal congestion, anorexia, and skin rash. Complications include neurological abnormalities, seizures, and fatal toxoplasmic encephalitis. In pregnant women, toxoplasmosis can cause miscarriage, premature birth, or blindness in the fetus and is a leading cause of birth defects. In the newborn child, the disease may manifest itself primarily in the eye as chorioretinitis (fetal inflammation of the retina), but it may also cause hydrocephalus, microcephalus, carditis, and hepatitis.

Toxoplasmosis is caused by the protozoan *Toxoplasma gondii*. In humans, the disease is generally acquired by eating inadequately cooked meat or by association with infected felines. Meat becomes infected when cows or sheep graze in pastures contaminated by fecal matter from infected felines. Cats, the most important reservoir of the disease, become carriers by eating infected rodents or by the ingestion of infected feline feces. Pregnant women are advised to avoid eating insufficiently cooked meat, avoid having close contact with cats, and, especially, avoid cleaning feline litter boxes.

The cyst of *T. gondii* is persistant, making long-term therapy a requirement in patients whose immune systems are compromised. Treatment of cerebral or ocular toxoplasmosis is generally accomplished with antifolate compounds consisting of a combination of pyrimethamine (**24**) and a long-acting sulfonamide such as sulfadiazine (**25**). Sometimes the latter is replaced by triple sulfonamides or sulfalene (**72**). There is evidence that the related mixture of trimethoprim (**73**) plus sulfamethoxazole (**86**) (mixture is called Bactrim) is nearly as effective but less toxic than the pyrimethamine combinations. Additional drugs for toxoplasmosis treatment are the antibiotics spiramycin (**48**), used alone or in combination with pyrimethamine or a sulfonamide, and clindamycin (23). Clindamycin has been used by itself (reportedly not with great success) and in combination with pyrimethamine to treat chorioretinitis and toxoplasmic encephalitis in AIDS patients. It is especially useful for patients who cannot tolerate extended regimens of sulfonamides.

Trichomoniasis

Trichomoniasis is a widespread disease of the urogenital tract caused by the protozoan *Trichomonas vaginalis*. This sexually transmitted disease affects humans only and has no known animal reservoirs (19). While it primarily affects women, *T. vaginalis* can infect men as well. The disease is marked in women by the development of purulent, frothy, vaginal discharge with a highly disagreeable odor. The role of *T. vaginalis* in producing inflammation of the male genital tract is not fully understood. The protozoan primarily invades the urethra, but may also enter the seminal vesicles or prostate. In a vast majority of infected males, the disease is asymptomatic, perhaps because they are protected by the high zinc content of prostatic secretions. Successful invasion of the prostate by the protozoan can lead to painful urination.

Treatment by the 5-nitroimidazole class of antiprotozoal agents can destroy most strains of *T. vaginalis;* however, some strains have been reported to be resistant to these drugs. The agent of choice is orally administered metronidazole (**1**), the first systemic antitrichomonal agent, introduced in 1960. Other members of this therapeutic group include niridazole (**10**), tinidazole (**11**), and ornidazole (**91**) but, in general, they offer no distinct advantage over metronidazole (Table 9). It is recommended that both female and male sexual partners be treated simultaneously to avoid reinfecting one another. Nitroimidazoles have been demonstrated to be mutagenic in bacteria and carcinogenic in high doses in laboratory animals. It is advised that pregnant women in the first trimester of pregnancy refrain from taking these drugs even though there is no clinical evidence of fetal abnormalities being caused by them. Other nitro-heterocycles have been synthesized in large number and many show promising *in vitro* activity.

Trypanosomiasis

The disease trypanosomiasis has two geographically delineated forms, the so-called African and American varieties. Because of their distinct differences they will be discussed separately (20).

African Trypanosomiasis. According to the World Health Organization, African trypanosomiasis affects some 25,000 people annually in 36 subSaharan countries where 50 million inhabitants are at risk. It is the cause of sleeping sickness, a disease transmitted by the bite of either the female or male tsetse fly. The tsetse fly is infected by ingesting blood from an animal or human containing trypanosomes. The elimination of these vectors is difficult because they thrive on both domestic cattle and wild animals. Trypanosomiasis in cattle, known as nagana, has seriously limited the production of this primary food source and has thwarted economic growth in endemic areas. African trypanosomiasis is caused by *Trypanosoma brucei* whose subspecies, *Trypanosoma brucei gambiense* and *Trypanosoma brucei rhodesiense,* have identical ultrastructures in humans. *T. b. gambiense* is responsible for an unevenly advancing sickness, whereas *T. b. rhodesiense* leads to a more acute form with rapid progression.

A human bitten by an infected tsetse fly develops a small nodule (primary or trypanosomal chancre) at the site of skin penetration from which the organisms may be isolated. These protozoa, whose life cycle is complex, change form, multiply, and invade the bloodstream via the lymphatic system in a matter of weeks or months. The trypanosomes are responsible for causing the clinical form of the disease, which includes severe headache, anemia, local edema, skin rashes, fever, and liver and/or spleen damage. Lymph nodes become enlarged and, in the advanced form of the disease, damage occurs to the central nervous system that causes the sleeping sickness phase. Lesions develop in the brain and, in some instances, in the myocardium. The disease has a high rate of mortality if left untreated. In the case of *T. b. rhodesiense,* death may ensue in several months.

The early stages of African trypanosomiasis, ie, before central nervous system involvement and the sleeping sickness take effect, can be best treated with intravenous suramin sodium (**92**), a compound with a dyelike structure (Table 10). Suramin is most effective against *T. b. rhodesiense,* but has also been used against *T. b. gambiense* infection. The compound causes side effects such as nausea, photophobia, and peripheral neuropathy which disappear shortly after conclusion of administration. Because the drug is unable to pass the blood–brain barrier, prompt treatment of patients is essential. Suramin in combination with tryparsamide (**93**, Tryparsone [*554-72-3*]) is an alternative that has been investigated. Pentamidine (**26**), as the isethionate or methanesulfonate salt, has been used successfully against *T. b. gambiense,* but the compound also does not enter the central nervous system and is not effective in advanced disease. It can cause severe side effects such as syncope, hypotension, hypoglycemia, vomiting, abdominal pain, and a peripheral neuropathy. Pentamidine (**26**) has been administered to large populations prophylactically against *T. b. gambiense.* This probably accounts for the appearance of resistant strains.

For late-stage disease, in which the central nervous system is implicated, the compound of choice until recently was melarsoprol (**94**, Mel B, Arsobal [*494-79-1*]) for *T. b. gambiense* or *T. b. rhodesiense.* The drug, administered intravenously, is a solution containing a combination of BAL (2,3-dimercaptopropanol) and the trivalent arsenic compound, melarsen oxide. Not only can the drug cause serious side effects such as intense dermal irritation, myocarditis, and renal and hepatic damage, but it is also responsible for death in 5% of patients. Resistance has developed to melarsoprol.

The first new treatment in 40 years that is showing great promise is eflornithine (**87**). It is an irreversible inhibitor of ornithine decarboxylase that blocks putrescine biosynthesis in *T. brucei in vitro* and inhibits the growth and multiplication of the parasite *in vivo.* The drug, administered orally or parenterally, is relatively nontoxic and is effective in eliminating the parasite *T. b. rhodesiense* and, to a greater extent, *T. b. gambiense.* Late-stage disease is rapidly reversed, causing comatose sleeping sickness patients to regain consciousness. Eflornithine was approved for human use by the FDA in late 1990. Side effects associated with oral administration are diarrhea, nausea, and vomiting. There is a significant relapse rate. A mixture of eflornithine (**87**) and nifurtimox (**95**, Lampit, Bayer 2502) shows good activity against *T. b. gambiense.* The former compound is also reported to potentiate the effect of melarsoprol in infected mice.

A combination of salicylhydroxamic acid (SHAM) and glycerol, although capable of destroying bloodstream trypanosomiasis, is less effective against infections involving the central nervous system.

American Trypanosomiasis. American trypanosomiasis, known as Chagas' Disease, is limited to South and Central America, where it affects 16–18 million people annually in an area where 90 million are at risk. Although only an estimated 1% of infected individuals contract the disease, poverty and poor housing exacerbate it. There is a particularly high incidence of the disease in children.

The protozoan causative agent, *Trypanosoma cruzi,* is harbored in domesticated animals, such as dogs, cats, and pigs, as well as wild animals, eg, rats, bats, foxes, opossums, and monkeys. These reservoirs can infect the vector, the night-feeding, blood-sucking triatomid bug. This insect, after biting a person usually on

Table 10. Trypanosomiasis Antiprotozoal Agents

Structure number	Compound name	CAS Registry Number	Molecular formula	Structure
		African trypanosomiasis		
(92)	suramin sodium	*[129-46-4]*	$C_{51}H_{34}N_6Na_6O_{23}S_6$	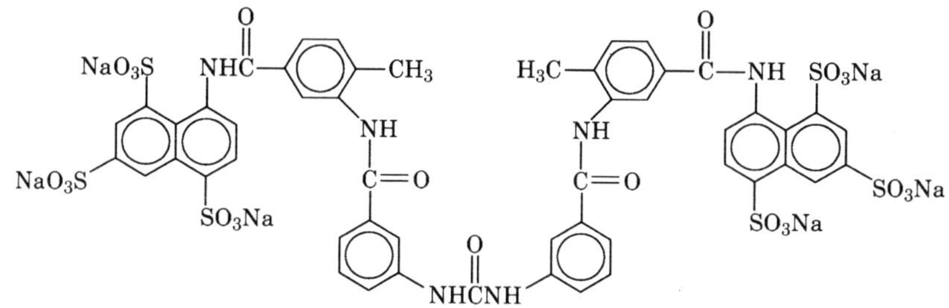
(93)	tryparsamide	*[554-72-3]*	$C_8H_{10}AsN_2NaO_4$	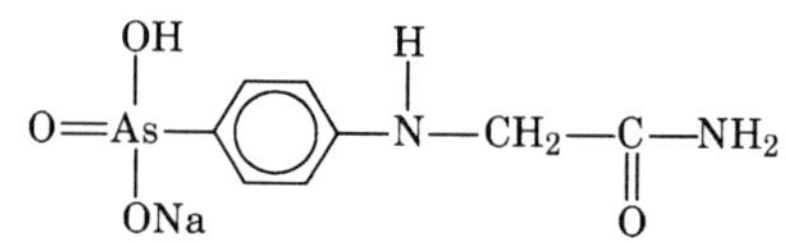
(94)	melarsoprol	*[494-79-1]*	$C_{12}H_{15}AsN_6OS_2$	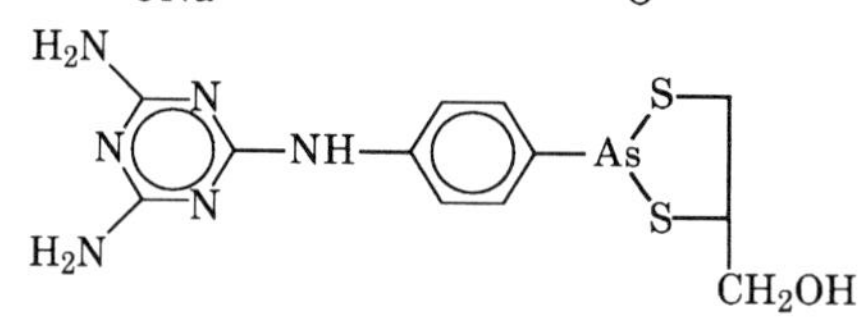
(95)	nifurtimox[a]	*[23256-30-6]*	$C_{10}H_{13}N_3O_5S$	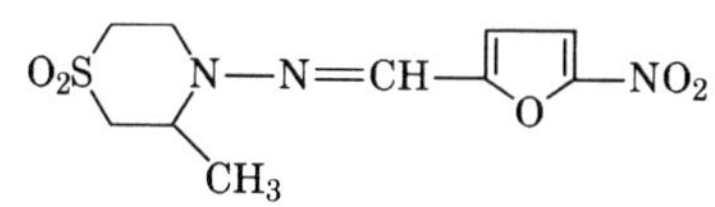
		American trypanosomiasis		
(96)	benznidazole	*[22994-85-0]*	$C_{12}H_{12}N_4O_3$	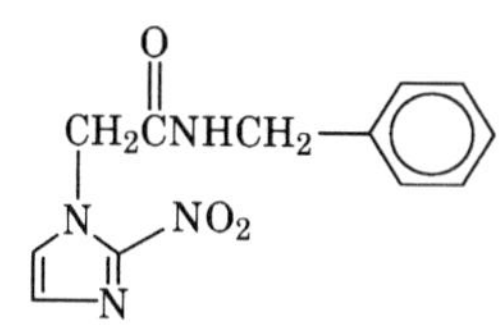

[a]Also used to treat American trypanosomiasis.

the face, defecates near the location of the wound. Parasites from the insect feces may either enter the human host at the point of the bite and/or by fecal contamination of mucus membranes, usually those of the mouth or eyes. This causes a skin lesion (called a primary chagoma) that consists of a small, reddish, slightly painful nodule. One of the eyelids may become swollen. The injected trypanosomes enter many kinds of cells of the host where they multiply and become amastigotes. They then differentiate into trypomastigotes that circulate in the bloodstream and are distributed to all parts of the body. They do not replicate until they enter another cell or are ingested by another blood-sucking insect vector.

The disease has two phases. The acute phase is of approximately two months' duration in which the parasite spreads. Although the disease may be asymptomatic, more typically it is marked by mild fever, colonic and esophageal dilation, liver and spleen enlargement, erythromatous rash, and acute myocarditis. In children, this phase of the disease can be serious and even fatal. During the chronic phase there is slow tissue destruction that can persist over many years. There may be cardiac and/or digestive involvement, with the former being the primary cause of death in Chagas' Disease. Despite its potential seriousness, the chronic form can also be asymptomatic.

The nitrofuran nifurtimox (**95**) is the most effective drug against *T. cruzi,* being cidal against the trypomastigote and amastigote forms. It is active against both the extra- and intracellular form of the parasite. In combination with corticosteroids it has prevented myocardial inflammation and destroyed the parasites within the heart. Side effects of the drug tend to be mild and include nausea, vomiting, insomnia, nervous excitation, vertigo, and skin rashes. Cardiac symptoms are treated symptomatically. Benznidazole (**96**, *N*-benzyl-2-nitro-1-imidazoleacetamide, RO 7-1051), an alternative drug, can prevent the spread of the parasites from one tissue to another although relapses are common. Primaquine (**69**) can destroy the extracellular trypanosomes in the blood, but is not effective against the intracellular forms of the parasite.

BIBLIOGRAPHY

"Chemotherapeutics, Antiprotozoal" in *ECT* 3rd ed., Vol. 5, pp. 513–542, by E. J. Martin and H. C. Zell, Food and Drug Administration, and B. T. Poon, Walter Reed Army Institute of Research.

1. G. T. Strickland, ed., *Hunter's Tropical Medicine,* 7th ed., W. B. Saunders, Philadelphia, Pa., 1991, p. 547.
2. K. S. Warren and A. A. F. Mahmoud, eds., *Tropical and Geographical Medicine,* 2nd ed., McGraw-Hill, New York, 1989.
3. M. Katz, D. D. Despommier, and R. W. Gwadz, *Parasitic Diseases,* 2nd ed., Springer-Verlag, New York, 1988.
4. J. H. Leech, M. A. Sande, and R. K. Root, eds., *Parasitic Infections,* Churchill Livingston, New York, 1988.
5. G. T. Strickland, ed., *Hunters's Tropical Medicine,* 7th ed., W. B. Saunders, Philadelphia, Pa., 1991, pp. 546–679.
6. W. C. Campbell and R. S. Rew, *Chemotherapy of Parasitic Diseases,* Plenum Press, New York, 1986.

7. G. C. Cook, *Parasitic Disease in Clinical Practice,* Springer-Verlag, London, 1990.
8. N. D. Levine, *Veterinary Protozoology,* Iowa State Press, Ames, 1985.
9. R. Knight, *J. Antimicrob. Chemother.* **6,** 577 (1980).
10. T. K. Ruebush, *Trans. Roy. Soc. Trop. Med. Hyg.* **74,** 149 (1980).
11. P. L. Long, ed., *Coccidiosis in Man and Domestic Animals,* CRC Press, Boca Raton, Fla., 1990.
12. S. Tzipori, *Adv. Parasitol.* **27,** 63 (1988).
13. S. L. Earlandsen and E. A. Meyer, eds., *Giardia and Giardiasis,* Plenum Press, New York, 1984.
14. J. D. Berman, *Rev. Infect. Dis.* **10,** 560 (1988); W. Peters and R. Killick-Kendrick, eds., *The Leishmaniases in Biology and Medicine,* Academic Press, New York, 1987.
15. W. Peters and W. H. G. Richards, eds., *Antimalarial Drugs,* Vols. 1 and 2, Springer-Verlag, New York, 1984; L. J. Bruce-Chwatt, ed., *Chemotherapy of Malaria,* WHO, Geneva, Switzerland, 1981; *Practical Chemotherapy of Malaria,* WHO Technical Report 805, 1990.
16. S. K. Martin, A. M. Oduola, and W. K. Milhous, *Science (Washington, D.C.),* **235,** 899 (1987).
17. D. L. Klayman, *Science (Washington, D.C.),* **228,** 1049 (1985).
18. W. T. Hughes, *Pneumocystis carinii,* Vols. 1 and 2, CRC Press, Boca Raton, Fla., 1987.
19. B. M. Honigberg, ed., *Trichomonads Parasitic in Humans,* Springer-Verlag, New York, 1989.
20. T. M. Leach and C. J. Roberts, *Pharmac. Therap.* **13,** 91 (1981); D. H. Molyneux and R. W. Ashford, *The Biology of Trypanosoma and Leishmania, Parasites of Man and Domestic Animals,* Taylor and Francis, London, 1983.

DANIEL L. KLAYMAN
Walter Reed Army Institute of Research

AVERMECTINS

In 1976 scientists at the Merck Corporation discovered a complex of eight closely related natural products, subsequently named avermectins A_{1a} through B_{2b}, in a culture of *Streptomyces avermitilis* MA-4680 (NRRL8165) originating from a soil sample collected at Kawana, Ito City, Shizuoka Prefecture, Japan and isolated by the Kitasato Institute. Their structures are shown in Figure 1 (1–6). They are among the most potent anthelmintic, insecticidal, and acaricidal compounds known.

The avermectins are closely related to another group of pesticidal natural products, the milbemycins. First described by Japanese workers, milbemycins were later found to be more abundant in nature than the avermectins (7–12). Both the avermectins and milbemycins are sixteen-membered lactones, with a spiroketal system containing two six-membered rings. The principal difference between them is that the avermectins have an α-L-oleandrosyl-α-L-oleandrosyl disaccharide attached at the 13-position whereas the milbemycins have no 13-substituent. Milbemycin structures are shown in Figure 2.

Two avermectins, abamectin (the avermectin B_1) [*71751-41-2*] and ivermectin [*70288-86-7*], which is saturated at C22–C23, have been commercialized to date. These two marketed avermectins have been described in considerable detail (Table 1) (13).

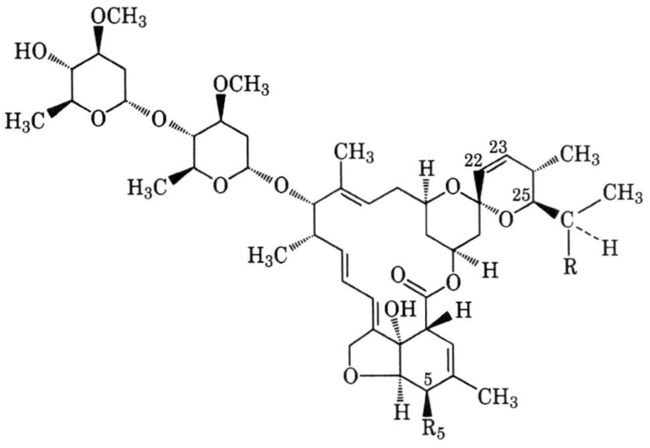

Fig. 1. The avermectins show variations at positions 5, 22–23, and 25. In the A series $R_5 = OCH_3$; for the B series $R_5 = OH$. The designation 1 corresponds to $\underset{22}{—CH}{=}\underset{23}{CH}—$ as shown. The designation 2 indicates $\underset{22}{—CH_2}—\underset{23}{CHOH}$. In ivermectin, C-22–C-23 is saturated. The avermectins are isolated as mixtures wherein the major component a has a *sec*-butyl group at position 25 ($R = C_2H_5$), and component b has an isopropyl group ($R = CH_3$) at C-25.

Avermectin	CAS Registry Number	Molecular formula
A_{1a}	[*65195-51-9*]	$C_{49}H_{74}O_{14}$
A_{1b}	[*65195-52-0*]	$C_{48}H_{72}O_{14}$
A_{2a}	[*65195-53-1*]	$C_{49}H_{76}O_{15}$
A_{2b}	[*65195-54-2*]	$C_{48}H_{74}O_{15}$
B_{1a}	[*65195-55-3*]	$C_{48}H_{72}O_{14}$
B_{1b}	[*65195-56-4*]	$C_{47}H_{70}O_{14}$
B_{2a}	[*65195-57-5*]	$C_{48}H_{74}O_{15}$
B_{2b}	[*65195-58-6*]	$C_{47}H_{72}O_{15}$

Table 1. Trade Name Registrations by the Merck Corporation[a]

Trade name	Delivery form	Recipient
Abamectin		
Avid	foliar spray	
Agri-Mek	foliar spray	
Vertimek	foliar spray	
Affirm	bait	fire ants
Avomec	injection	cattle
Duotin	injection	cattle
Ivermectin		
Ivomec	injection	cattle, sheep, goats, pigs, camels
	oral liquid	cattle, sheep
	paste, pour on	cattle
Ivomec F	injection	cattle
Oramec	oral liquid	sheep, goats
Eqvalan	paste	horses
	oral liquid	horses
Heartguard 30	tablet or chewable	dogs
Cardomec	tablet	dogs
Cardotek	tablet	dogs
Mectizan	tablet	humans

[a]Ref. 12.

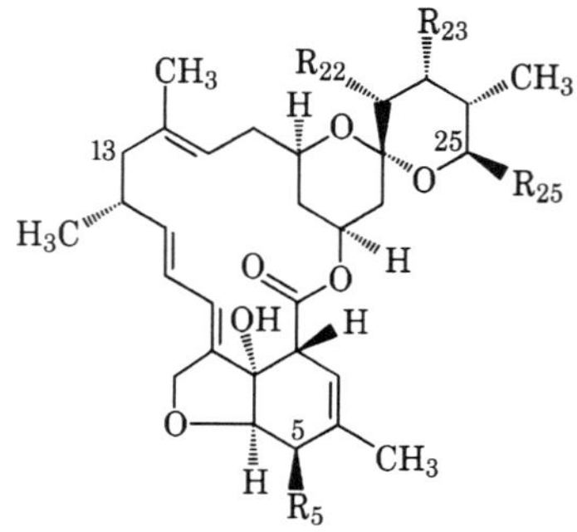

Fig. 2. Milbemycin structures in the milbemycin Alfa series. R_5 = OH, OCH_3; R_{22} = H, OH; R_{23} = H, $OCOCH(CH_3)(CH_2)_3CH_3$; R_{25} = CH_3, C_2H_5. In the anthelmintic F-28249, antibiotics S541 series R_5 = OH, OCH_3; R_{22} = H; R_{23} = OH; R_{25} =

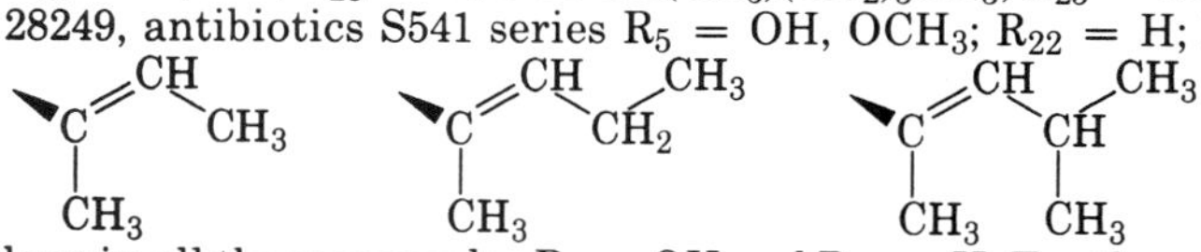

. Some specific examples follow; in all these examples R_5 = OH and R_{22} = H. For the milbemycins, R_{23} = H; R_{23} = OH for nemadectin.

Name	CAS Registry Number	R_{25}	Molecular formula
milbemycin Alfa$_1$ (A_3)	*[51596-10-2]*	CH_3	$C_{31}H_{44}O_7$
milbemycin Alfa$_3$ (A_4)	*[51596-11-3]*	CH_2CH_3	$C_{32}H_{46}O_7$
milbemycin D	*[77855-81-3]*	$CH(CH_3)_2$	$C_{33}H_{48}O_7$
anthelmintic F-28249-Alpha (nemadectin)		$C(CH_3)CHCH(CH_3)_2$	$C_{36}H_{52}O_8$

Abamectin

Avermectin B_1 is the most effective of the avermectin family of natural products against agriculturally important insects and mites (14). It has been commercialized for agricultural use under the nonproprietary name abamectin. This mixture of avermectins contains at least 80% of avermectin B_{1a} ($C_{48}H_{72}O_{14}$) and not more than 20% of avermectin B_{1b} ($C_{47}H_{70}O_{14}$). Abamectin for use in foliar spray applications is formulated as an emulsifiable concentrate. It is being developed for use in a number of horticultural and agronomic crops to control mites and insects. A summary of its biological activity is shown in Table 2 and a summary of its agricultural applications is shown in Table 3.

Abamectin is also used to control the imported red fire ant *Solenopsis invicta*. For this use abamectin is formulated as a bait together with soybean oil and corn grits. Worker ants transport the bait to the colony, the queen becomes sterile, and the colony is eliminated after 12 to 21 weeks. Similar effects on the fecundity of other female insects at nonlethal doses have been reported (16,17).

Ivermectin

Selective reduction of the 22,23-olefin of avermectin B_1 yields the 22,23-dihydro derivative assigned the nonproprietary name ivermectin (18). The structure

Table 2. Biological Activity of Abamectin (Avermectin B_1) Against Mites and Insects[a]

Common name	Species name	LC_{90}(ppm)
Mite species[b]		
citrus rust mite	*Phyllocoptruta oleivora*	0.02
two-spotted spider mite	*Tetranychus urticae*	0.03
strawberry mite	*Tetranychus turkestani*	0.08
European red mite	*Panonychus ulmi*	0.04
citrus red mite	*Panonychus citri*	0.24
broad mite	*Polyphagotarsonemus*	0.03
Insect species[c]		
Colorado potato beetle	*Leptinotarsa decemlineata*	0.03
tomato hornworm	*Manduca secta*	0.02
Mexican bean beetle	*Epilachna varivestes*	0.20
pea aphid	*Acyrthosiphon pisum*	0.40
cabbage looper	*Trichoplusia ni*	1.0
corn earworm	*Heliothis zea*	1.5
southern armyworm	*Spodoptera eridania*	6.0

[a]Ref. 15.
[b]Contact effect against adult mites.
[c]Foliar residue bioassay.

Table 3. Principal Agricultural Applications of Abamectin Foliar Spray

Common name	Species name	Application
armyworm	*Spodoptera exigua*	vegetables[a]
	Spodoptera eridania	
broad mite	*Polyphagotarsonemus latus*	citrus
citrus red mite	*Panonychus citri*	citrus
citrus rust mite	*Phyllocoptruta oleivora*	citrus
citrus thrip	*Scirtothrips citri*	citrus
cotton aphid	*Aphis gossypii*	cotton
deciduous tree nut		miscellaneous crops
European red mite	*Panonychus ulmi*	pears
leafminer	*Liriomyza trifolii*	ornamentals, vegetables[a]
	Liriomyza sativae	vegetables[a]
Lepidoptera		cotton
pear psylla	*Psylla pyricola*	pears
pear rust mite	*Epitrimerus pyri*	pears
strawberry		miscellaneous crops
spider mite	*Tetranychus urticae*	cotton
	Tetranychus turkestani	cotton
	Tetranychus pacificus	cotton
	Tetranychus cinnabarinus	cotton
thrips species		ornamentals
tomato pinworm	*Keiferia lycopersicella*	vegetables[a]
tomato russet mite	*Aculops lycopersici*	vegetables[a]
two-spotted spider mite	*Tetranychus urticae*	vegetables[a], pears, ornamentals

[a]Celery, tomato, pepper.

shown depicts the 25-*sec*-butyl derivative [*70161-11-4*], but it should be noted that both commercial products contain up to 20% of the 25-isopropyl analogue [*70209-81-3*].

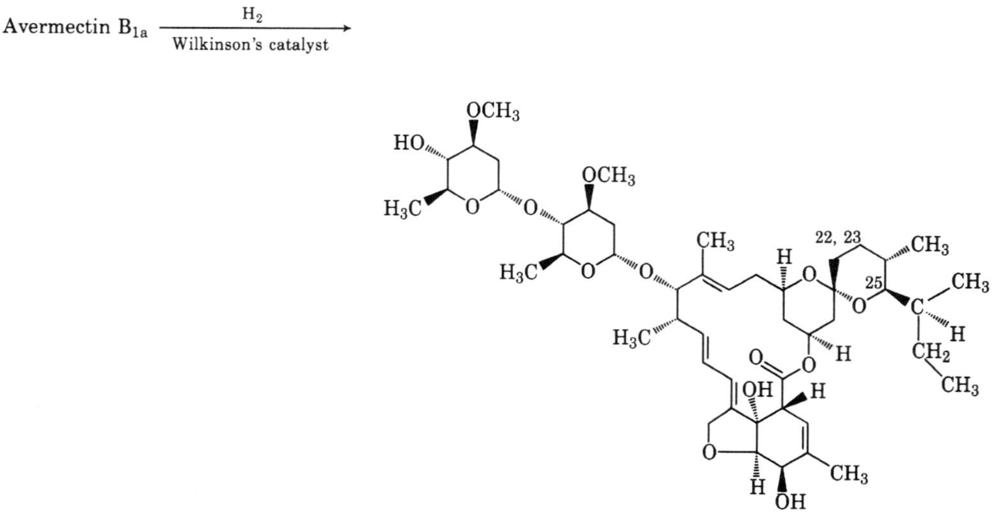

Ivermectin is active against two significant phyla of animal parasite: the Nemathelminthes or nematodes (roundworms) and the Arthropoda (insects, ticks, and mites). Ivermectin is inactive against platyhelminthes (flukes and tapeworms).

Ivermectin has an extremely broad spectrum of antinematodal activity in a variety of domestic animals. Indeed, among the many nematodes against which it has been tested, none has been found that is not affected by ivermectin during at least one stage of the life cycle. In all but a few instances the drug is highly active against both immature and mature worms (19).

Perhaps the most striking aspect of the antinematodal action of ivermectin is its potency. This varies widely from species to species and stage to stage, but in all cases the minimum effective dosage is much less than that of other anthelmintics. Among the most sensitive parasites are immature *Dirofilaria immitis* in dogs (0.001 mg/kg) and *Oesophagostomum radiatum* and *Dictyocaulus viviparus* in cattle (0.05 mg/kg). Among the least sensitive nematodes are *Nematodirus helvetianus* in cattle, and *Toxascaris leonina,* and perhaps immature *Trichuris vulpis,* in dogs. Two parasite life cycle forms against which activity has not been demonstrated even at high dosage (more than 1.0 mg/kg) are adult *Dirofilaria immitis* (heartworm) in dogs and immature *Trichinella spiralis* in the muscle of mice and pigs even though other life cycle stages are affected. The vast majority of nematodes, regardless of host species or stage of worm development, are highly susceptible to ivermectin when the drug is given as a single oral or parenteral dose at 0.1–0.2 mg/kg.

The activity of ivermectin against the filarial parasite *Dirofilaria immitis* in dogs suggested a possible role for the control of filarial parasites of humans (20). It has been extensively tested in human onchocerciasis and is now considered to be the drug of choice. In a single yearly oral dose, it suppresses microfilariae in the skin and eyes and, in most cases, prevents the progression of the disease to blindness. Table 4 shows the results of a 30-patient double-blind study recorded over one year.

Table 4. Ivermectin and Diethylcarbamazine Against *Onchocerca volvulus* Infections[a]

Study day	Placebo[b]	Ivermectin[c]	DEC[d]
−1	99.4	130.4	100.3
2	108.2	38.8	27.0
4	99.7	14.1	14.4
8	105.1	6.6	4.1
14	125.9	2.2	6.8
28	102.6	0.6	9.2
90	84.5	1.0	18.0
180	65.3	2.9	21.8
270	80.8	5.0	27.5
360	93.0	11.8	45.1

[a]Numbers represent the skin density of microfilariae (21).
[b]Ten patients received placebo.
[c]Ten patients received a single oral dose of ivermectin, 12 mg.
[d]Ten patients received diethylcarbamazine [*90-89-1*] (DEC) daily orally for eight days: total dose 1.3 g.

Two forms of lymphatic filariasis are found in India. The Bancroftian form is the most common and accounts for more than 90% of the disease whereas Brugian filariasis accounts for the rest. In a study carried out in India (6) in 40 patients with *Wuchereria Bancrofti* filariasis treated with single oral doses, all of the dose levels chosen (25, 50, 100, and 200 mg/kg) were efficacious in clearing microfilariae from the blood of all patients treated. However, after three months some microfilaria recurred in the blood of most patients (Table 5). Further studies are planned and some are underway using different doses and regimens. Ivermectin still appears to hold promise as a new treatment for lymphatic filariasis.

Table 5. Efficacy[a] of Ivermectin in the Treatment of *Wuchereria Bancrofti* Filariasis[b]

Single oral dose, mg/kg	Day						
	0	1.5	5	12	30	90	180
25	761	2.9	<1	<1	5.2	42.9	98
50	1154	3.3	<1	<1	3.5	103.6	92.3
100	610	3.0	<1	<1	<1	19.9	95.9
200	478	<1	<1	<1	1.5	43.7	70.8

[a]Geometric mean microfilariae/mL.
[b]Ref. 22.

Ivermectin is widely used as an endectocide for cattle as an injectable, oral, topical, or slow release bolus; for sheep as an injectable or oral formulation; for swine as an injectable; for horses as a paste or drench; and for goats as an injectable or oral formulation. Ivermectin has recently been introduced for heartworm prophylaxis in dogs and it is being studied for use with cats, many other mammals, birds, fish, and reptiles.

Ivermectin is used in cattle, sheep and horses at 0.2 mg/kg; swine at 0.3 mg/kg; dogs at 0.006 mg/kg; and man at 0.05–0.2 mg/kg. It is effective against parasitic nematodes, grubs, lice, mites, ticks, and bots. Ivermectin is not active against tapeworms, flatworms, bacteria, or fungi.

Biosynthesis

The proposed pathway for the biosynthesis of the avermectins (Fig. 3) has been described in a review (23). Some of the details are yet to be elucidated, although the steps, in general, are based on firm evidence from four types of studies: incorporation of labeled precursors, conversion of putative intermediates by producing strains and blocked mutants, *in vitro* measurement of biosynthetic enzymes, and studies with enzyme inhibitors. The biosynthesis of the oleandrose units was elucidated from studies using $^{3}H/^{14}C$ and ^{13}C labeled glucose, which indicated a direct conversion of glucose to oleandrose.

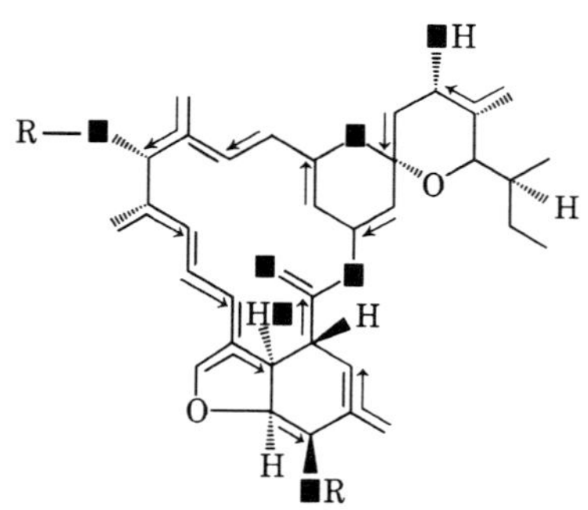

Fig. 3. Biosynthetic scheme of avermectins, ■ = ^{18}O, CH_3COOH, CH_3CH_2COOH.

Chemistry

Stability. Avermectins are highly lipophilic substances and dissolve in most organic solvents such as chloroform, methylene chloride, acetone, alcohols, toluene, cyclohexane, dimethylformamide, dimethyl sulfoxide, and tetrahydrofuran. Their solubility in water is correspondingly low, only 0.006–0.009 ppm (= mg/L).

Chemical stability studies are monitored by silica gel thin-layer chromatography (tlc) or by high performance liquid chromatography (hplc) using a reverse-phase C_{18} coated column (24). Hplc peaks or tlc spots are visualized by their uv absorption at 245 nm; the tlc spots can also be detected by ceric sulfate or phosphomolybdic acid staining.

Avermectins are acid-sensitive. Although stable in anhydrous glacial acetic acid solution at room temperature, aqueous CH_3COOH and stronger acids, such as dilute HCl, H_2SO_4, and *p*-toluenesulfonic acid in methanol solution, more rapidly cleave off the first sugar. Stronger concentrations of methanolic or aqueous acids are necessary to hydrolyze off the second sugar to give the aglycone.

Strong bases such as methanolic potassium hydroxide, sodium methoxide, or 1,8-diazabicyclo [5.4.0] undec-7-ene (DBU), cause epimerization at the C-2 carbon or shift the beta–gamma double bond into conjugation with the lactone carbonyl (Fig. 4) (25,26).

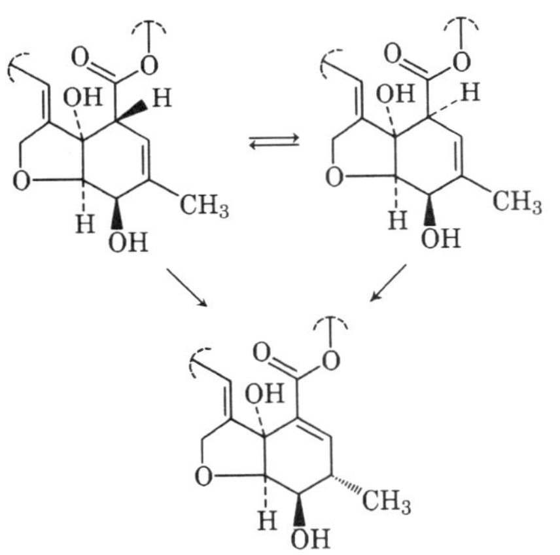

Fig. 4. Treatment of ivermectin with methanolic potassium hydroxide solution (25).

Uv light below 280 nm rapidly isomerizes the (*E*) (*trans*) 8,9- and 10,11-double bonds to the 8,9- and 10,11-(*Z*)-isomers (Fig. 5) (27).

Reactivity. Avermectin B_1 and ivermectin both contain two secondary and one tertiary hydroxy group. The two secondary alcohols are readily acetylated with acetic anhydride in pyridine, although at 0°C, a fair amount of the 4″-monoacetate, but not the 5-monoacetate, is obtained. With more reactive acid chlorides, however, it is not possible to obtain the 4″-*O*-monoacyl derivative, and it is necessary to protect the 5-OH as a 5-*O*-*tert*-butyldimethysilyl (TBDMS) derivative. This can be prepared selectively because of the hindered C-4″ position (28). Removal of the 5-protecting group is accomplished with *p*-toluenesulfonic acid monohydrate (p-TsOH·H_2O) in CH_3OH (1% at RT for 20–30 min; longer reaction times generate substantial amounts of monosaccharide) or with an

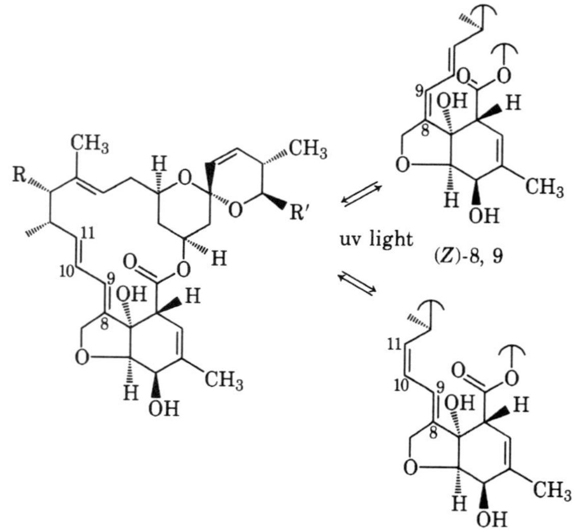

Fig. 5. Photoisomerization of avermectins. R = α-L-oleandrosyl-α-L-oleandrosyl; R′ = *sec*-C_4H_9 or *i*-C_3H_7 (see Fig. 1) (27).

HF–pyridine–THF mixture (29). It is sometimes possible to prepare 5-monoacyl derivatives directly from unprotected ivermectin by careful use of one equivalent of acylation reagent, particularly when the latter is bulky (eg, bis(trichloroethoxy)phosphoryl chloride). Many 4″-*O*-substituted derivatives have been prepared as they retain considerable bioactivities; however, 5-*O*-substitution products lose most anthelmintic activities.

Analytical Procedures

Since the avermectins exhibit unprecedented potency, they are used at unusually low doses of 6–300 μg/kg, which makes the detection and isolation of residues and metabolites from animal tissue a new challenge. For this reason a sensitive analytical assay requires a derivative suitable for detection at concentrations down to 1/10 or 1/100 of one ppm. Ivermectin and avermectin B_1 are therefore converted into an aromatic derivative which allows detection by fluorescence absorbance. To achieve this derivatization, avermectin B_1, ivermectin, or their derivatives are heated with acetic anhydride in pyridine at 100°C for 24 h (30). The reaction time can be reduced to 1 h by using *N*-methylimidazole as a catalyst (31). The resultant 4″-*O*-acetyl-2,5,6,7-bisdehydro analogue is detected after hplc separation and characterized by a distinct retention time. Using a C_2-extraction cartridge, with 22,23-dihydro avermectin B_{1a} monosaccharide as an internal standard, it is possible to detect as little as 1 ng/mL of ivermectin in plasma or milk (32). This procedure was adapted as a confirmatory assay by hydrolysis of the avermectin derivatives to aglycone, monosaccharide, and parent disaccharide, subsequent aromatization, and finally detection of three distinct hplc peaks corresponding to the acetylated bisdehydro aglycone, monosaccharide, and parent compound peaks via fluorescence detection. A procedure using chemical ionization mass spectrometry–mass spectrometry (ms/ms) was developed for a confirmatory assay of ivermectin tissue residues in cattle (33). It is also possible to determine avermectins directly after extraction, using a synthetically isomerized avermectin derivative as an internal standard for hplc with uv detection (25). A hplc-reverse isotope dilution assay is available for the determination of ivermectin residues in animal tissue (34). A similar fluorescent derivative containing a 5-phenol group obtained by dehydration of the 5-keto derivative of ivermectin with ammonium acetate was also reported (35). An extraction procedure allowing direct hplc analysis with uv detection of ivermectin in bovine serum at concentrations as low as 2 ppb is available (36). Monoclonal antibodies for immunoassay of avermectins have been prepared (37).

Monosaccharides and Aglycones

The macrocyclic lactone of all avermectins has an α-L-oleandrosyl-α-L-oleandrosyloxy substituent at carbon 13, which is a 2-deoxy sugar glycoside, relatively sensitive to acid hydrolysis or alcoholysis. A solution of ivermectin in methanol containing a strong acid such as 1% sulfuric acid readily gives a good yield of the aglycone after 16–24 h at RT. When 2-propanol is substituted for methanol in this reaction, a good yield of the monosaccharide is obtained, as the bulkier 2-propanol preferentially attacks the sterically less hindered 1″-position over the 1′-position, which is attached directly to the C-13 of the macrocycle, which in turn is flanked on either side by the C-12 and C-14 methyl groups, respectively (Fig. 6) (38). These procedures readily yield the monosaccharides of ivermectin and avermectin B_1 and the aglycone of ivermectin. The preparation of the aglycone of avermectin B_1, however, is complicated: under the reaction conditions, addition of methanol to the 22,23-double bond occurs, yielding mainly the two epimeric 22,23-dihydro-23-methoxy monosaccharides and/or aglycones. The aglycone of the avermectin B_1 therefore must be prepared with aqueous acid; it can be obtained as a 1:1 mixture of aglycone and monosaccharide using 10% sulfuric acid in aqueous THF solution (38).

Removal of the 13-hydroxy group from avermectin aglycones gives 13-deoxyavermectin aglycones which are closely related to certain milbemycins. For example, ivermectin was converted into its aglycones (22,23-dihydroavermectin B_{1a} and B_{1b} aglycones); the 5-hydroxy group was then protected as an *O-tert*-butyldimethylsilyl group. The 13-hydroxy group was substituted by chloro; the 13-chloro group was removed by reduction; C-5-OH was deprotected; and, finally, the C-25 B_{1a} and B_{1b} homologues were separated into 13-deoxy-22,23-dihydroavermectin B_{1a} and B_{1b} aglycones (39). Interestingly, the latter was subsequently also obtained from a milbemycin fermentation and given the name milbemycin B41-D (see Fig. 2 for structure B_{1b}).

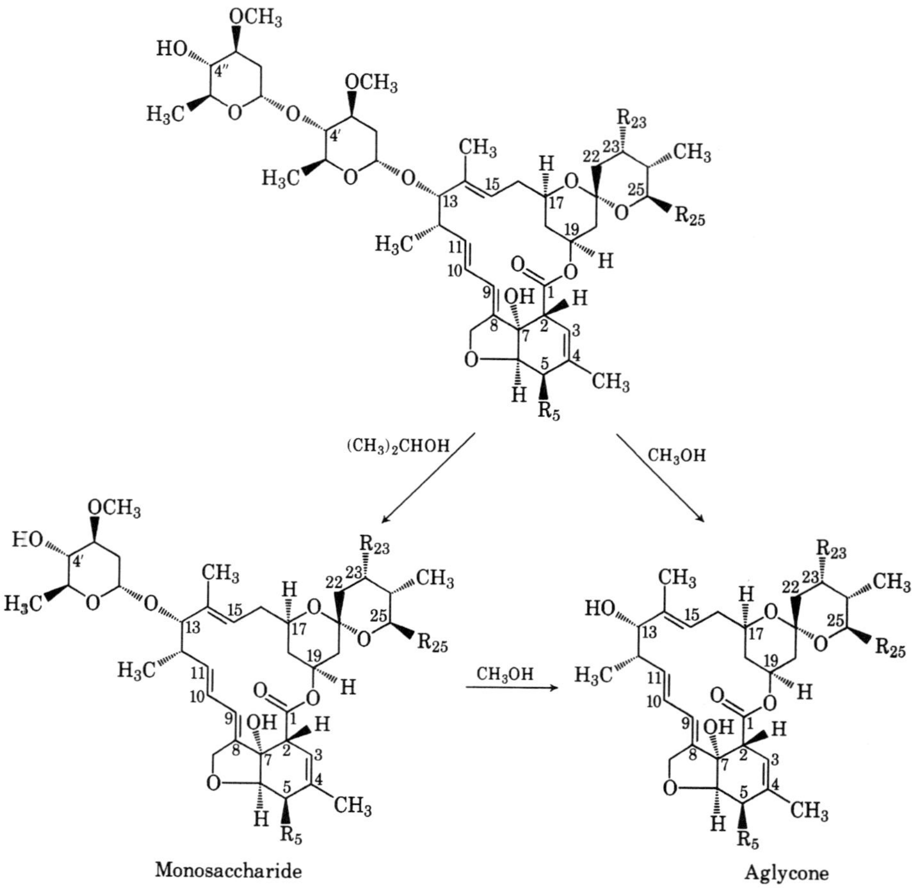

Fig. 6. Avermectin monosaccharides and aglycones. R_{25} = *sec*-C_4H_9; R_5 = OCH_3 or OH; R_{23} = OH or H. The reaction conditions are 18°C, 16 h, 1% H_2SO_4 in the solvent indicated (38).

Chemical Reactions

Oxidations. Avermectin B_1 and ivermectin have two secondary hydroxy groups susceptible to oxidation to the corresponding ketones. Because of the allylic nature of the C-5 hydroxy group its selective oxidation to 5-oxoavermectin is readily accomplished with manganese dioxide (40). Although tautomerization to the enol and elimination of the remaining tertiary C-7 hydroxy group leads to a phenol (35), the 5-oxo derivatives are nonetheless stable enough to allow further chemical reactions. Particularly important is the stereospecific reduction with $NaBH_4$ to the 5-beta alcohol with natural configuration, which is used to prepare tritiated avermectin derivatives. Reaction with trifluoroacetic anhydride, however, gives the 2,5,6,7-tetradehydro phenol analogue. Use of mild oxidation procedures, such as oxalyl chloride–DMF (Swern oxidation), gives 4″,5-dioxo avermectin derivatives or, after protection of the C-5 hydroxy group as a *tert*-butyldimethylsilyl derivative, 5-*O*-*tert*-butyldimethylsilyl-4″-oxoavermectins, which are valuable intermediates for further 4″-modified analogues. Sodium borohydride reduction of the 4″-keto function gives a mixture of 4″-epimeric hydroxy compounds, with the unnatural axial 4″-hydroxy group the major product and the natural epimer the minor one. Reductive amination with $NaCNBH_3$ and NH_4OOCCH_3 likewise yields an epimeric mixture of 4″-aminoderivatives with interesting biological properties (41).

The avermectins also possess a number of allylic positions that are susceptible to oxidative modification. In particular the 8a-methylene group, which is both allylic and alpha to an ether oxygen, is susceptible to radical oxidation. The primary product is the 8a-hydroperoxide, which has been isolated occasionally as an impurity of an avermectin B_1 reaction (such as the catalytic hydrogenation of avermectin B_1 with Wilkinson's rhodium chloride-triphenylphosphine catalyst to obtain ivermectin). An 8a-hydroxy derivative can also be detected occasionally as a metabolite (42) or as an impurity arising presumably by air oxidation. An 8a-oxo-derivative can be obtained by oxidizing 5-*O*-protected avermectins with pyridinium dichromate (43). This also can arise by treating the 8a-hydroperoxide with base.

The allylic 4a-methyl group can be oxidized with SeO_2-catalyzed *tert*-butylhydroperoxide to a 4a-hydroxy analogue (44).

The double bonds of avermectins react with *m*-chloroperbenzoic acid to give 3,4-, 8,9-, and 14,15-epoxides. The 8,9-epoxide is the primary product and can be isolated in good yield (45). The 8,9-epoxide was opened by aqueous acids to the 8,9-diol (46). The 3,4-diol can be obtained readily and regiospecifically by osmium tetroxide oxidation. Neither peracids nor OsO_4 will attack the 22,23-double bond.

Reductions Including Ivermectin Preparation. Reductions of avermectin B_1 containing five double bonds with Pd or Pt catalysts proceed at almost comparable rates at the two disubstituted 10,11- and 22,23-double bonds to give a mixture of 10,11- and 22,23-dihydro derivatives. Further hydrogenation leads to 10,11,22,23-tetrahydro- and 3,4,10,11,22,23-hexahydro-derivatives; exhaustive hydrogenation leads to 3,4,8,9,10,11,22,23-octahydro analogues. Under no circumstances is the 14,15-double bond hydrogenated. For the preparation of ivermectin (22,23-dihydro avermectin B_1), Wilkinson's homogeneous $[(C_6H_5)_3P]_3RhCl$ catalyst has been employed for the regiospecific hydrogenation of the avermectin B_1 22,23-double bond, which is the only disubstituted cis double bond (18). To prepare 10,11-dihydro avermectin B_1, the regioselective reaction with *N*-bromacetamide was used to make a 11,10-bromohydrin, which can then be dehalogenated and deoxygenated to the desired product (47).

Radiolabeled Derivatives. Since ivermectin (= 22,23-dihydroavermectin B_1) is obtained by catalytic reduction of avermectin B_1, the same procedure using tritium gas conveniently affords tritiated ivermectin (22,23-$^3[H]$-22,23-dihydroavermectin B_1). The preparation of a tritiated derivative containing a 22,23-double bond starts with the readily available 5-ketone, which is reduced with $^3[H]$-sodium borohydride stereospecifically to a 5-$^3[H]$-derivative (40). Carbon-14 labeled avermectins can be obtained by a biosynthetic process using sodium (1-^{14}C)propionate as labeled precursor (48).

Interconversions of Avermectins. Methylation of avermectins B_1 and B_2 leads to the corresponding derivatives of the A series (49). A procedure involving the oxidation of the 5-methoxy group with mercuric acetate and $NaBH_4$ reduction of the 5-keto-intermediate allows the conversion of the A to the B components (50). The 23-hydroxy group of the "2" components, after selective protection of the other secondary hydroxy groups, is converted to a thionocarbonate, which can be eliminated to give the 22,23-double bond of the "1" components; alternatively it can be reduced with tributyltin hydride to the 22,23-dihydro derivatives (= ivermectins) (51).

Syntheses

Avermectin aglycones, monosaccharides, and the naturally occurring disaccharides themselves have been further modified by attaching various sugars to the different hydroxy groups. Most of these methods have used 1-bromo sugars via the Konigs-Knorr procedure (52,53). More recently, the use of 1-phenylthio and 1-fluoro sugars has resulted in better yields of those glycosides (54). A mixture of the two anomeric methyl glycosides of oleandrose is obtained by acid methanolysis of avermectins, and these have been converted to L-oleandrose (55) and the 1-phenylthio glycoside (54). Total synthesis of avermectins including the glycosylation of avermectin aglycones by interesting, new methods have been reported recently by several research groups (56–59). Two excellent reviews of partial and total synthesis of avermectins and milbemycins have been published (60,61).

Economic Aspects

The worldwide acceptance of ivermectin in livestock production and health care of companion animals has made it the largest selling animal health drug. Abamectin is in commercial use as an agricultural pesticide and its applications are continuing to expand. Ivermectin under the trademark Mectizan is now considered the drug of choice for treatment of human onchocerciasis. There is no commercial return since it is provided free of charge to the World Health Organization. Ivermectin also shows considerable promise for the treatment of human lymphatic filariasis and strongyloidiasis.

BIBLIOGRAPHY

1. R. W. Burg and co-workers, *Antimicrob. Agents Chemother.* **15,** 361 (1979).
2. G. Albers-Schönberg and co-workers, *J. Am. Chem. Soc.* **103,** 4216 (1981).
3. J. P. Springer, B. H. Arison, J. M. Hirschfield, and K. Hoogsteen, *J. Am. Chem. Soc.* **103,** 4221 (1981).
4. M. H. Fisher and H. Mrozik, in S. Omura, ed., *Macrolide Antibiotics,* Academic Press, New York, 1984, pp. 553–606.
5. H. G. Davies and R. H. Green, *Natl. Prod. Rep.* **3,** 87 (1986).
6. M. H. Fisher, in N. F. Janes, ed., *Recent Advances in the Chemistry of Insect Control,* Royal Society of Chemistry, London, 1985, pp. 53–72; H. Mrozik, *BCPC Mono* No. 32, *Biotechnology and its Application to Agriculture,* 1985, pp. 133–143; H. Mrozik, *Topics in Medicinal Chemistry,* 1987, pp. 245–254; T. Blizzard, M. H. Fisher, H. Mrozik, and T. L. Shih, in G. Lukacs and M. Ohno, eds., *Recent Progress in the Chemical Synthesis of Antibiotics,* Springer-Verlag, Berlin, Heidelberg, Germany, 1990, pp. 65–102.

7. Y. Takiguchi and co-workers, *J. Antibiot.* **33,** 1120 (1980).
8. H. Mishima, J. Ide, S. Muramatsu, and M. Ono, *J. Antibiot.* **36,** 980 (1983).
9. Eur. Pat. Appl. 170,006 (1986), J. B. Wood and co-workers (to American Cyanamid Co.).
10. Brit. Pat. Appl. 2,166,436A (1986), J. B. Wood and co-workers (to Glaxo Corp.).
11. G. T. Carter, J. A. Nietsche, and D. B. Borders, *J. Chem. Soc. Chem. Commun.* 402 (1987).
12. M. J. V. Ramsay and co-workers, *Tetrahedron Lett.* **28,** 5353 (1987).
13. W. C. Campbell, ed., *Ivermectin and Abamectin,* Springer, New York, 1989, pp. 1–363.
14. J. Putter and co-workers, *Experientia* **37,** 963 (1981).
15. R. A. Dybas and A. St. J. Green, *Proc. 1984 Brit. Crop Protect. Conf.-Pests Dis.,* **31,** 947–954 (1984).
16. C. S. Lofgren and D. F. Williams, *J. Econ. Entolmol.* **75,** 798–803 (1982).
17. L. Strong and T. A. Brown, *Bull. Entomol. Res.* **77,** 357–389 (1987).
18. J. C. Chabala and co-workers, *J. Med. Chem.* **23,** 1134 (1980).
19. Ref. 13, pp. 215–259.
20. Ref. 13, pp. 311–323.
21. M. Lariviere and co-workers, *Lancet* **2,** 174 (1985).
22. V. Kumaraswami and co-workers, *J. Am. Med. Assoc.* **259,** 3150 (1988).
23. T. S. Chen, O. D. Hensens, and M. D. Schulman, in Ref. 13, pp. 55–72.
24. T. W. Miller and co-workers, *Antimicrob. Agents Chemother.* **15,** 368–371 (1979).
25. J. V. Pivnichny, J. S. K. Shim, and L. A. J. Zimmerman, *Pharm. Sci.* **72,** 1447 (1983).
26. J. V. Pivnichny and co-workers, *J. Agric. Food Chem.* **36,** 826 (1988).
27. H. Mrozik and co-workers, *J. Org. Chem.* **53,** 1820 (1988).
28. H. Mrozik and co-workers, *J. Med. Chem.* **25,** 658–663 (1982).
29. C. Bliard and co-workers, *J. Chem. Soc., Chem. Commun.,* 368 (1987).
30. J. W. Tolan and co-workers, *J. Chromatogr.* **190,** 367–376 (1980).
31. P. C. Tway, J. S. Wood, and G. V. Downing, *J. Agric. Food Chem.* **29,** 1049–1063 (1981).
32. R. Chiou, R. J. Stubbs, and W. F. Bayne, *J. Chromatogr.* **416,** 196–202 (1987).
33. P. C. Tway and co-workers, *Biomed. Mass Spectrom.* **11,** 172–176 (1984).
34. S. H. L. Chiu and co-workers, *J. Agric. Food Chem.* **33,** 99–102 (1985).
35. J. D. Stong, *Anal. Chem.* **59,** 266–270 (1987).
36. D. D. Oehler and J. A. Miller, *J. Assoc. Off. Anal. Chem.* **72,** 59 (1989).
37. D. J. Schmidt and co-workers, *J. Agric. Food Chem.* **38,** 1763 (1990).
38. H. Mrozik and co-workers, *J. Org. Chem.* **47,** 489–492 (1982).
39. H. Mrozik and co-workers, *Tetrahedron Lett.* **24,** 5333 (1983).
40. J. C. Chabala, A. Rosegay, and M. A. R. Walsh, *J. Agric. Food Chem.* **29,** 881 (1981).
41. H. Mrozik and co-workers, *Experientia* **45,** 315 (1989).
42. D. L. Bull and co-workers, *J. Agric. Food Chem.* **32,** 94–102 (1984).
43. U.S. Pat. 4,547,491 (1985), H. Mrozik and F. S. Waksmunski (Merck & Co., Inc.).
44. U.S. Pat. 4,457,920 (1984), H. Mrozik (Merck & Co., Inc.).
45. U.S. Pat. 4,530,921 (1985), H. Mrozik (Merck & Co., Inc.).
46. A. B. Smith, III and A. S. Thompson, *Tetrahedron Lett.* **26,** 4279 (1985).
47. T. L. Shih, H. Mrozik, J. Ruiz-Sanchez, and M. H. Fisher, *J. Org. Chem.* **54,** 1459 (1989).
48. C. C. Ku and S. C. Hwang, *J. Labelled Compd. Radiopharm.* **22,** 451–459 (1985).
49. U.S. Pat. 4,200,581 (1980), M. H. Fisher, A. Luisi, and R. L. Tolman (Merck & Co., Inc.).
50. H. Mrozik, P. Eskola, and M. H. Fisher, *J. Org. Chem.* **51,** 3058 (1986).
51. H. Mrozik, P. Eskola, and M. H. Fisher, *Tetrahedron Lett.* **23,** 2377–2378 (1982).
52. U.S. Pat. 4,134,973 (1979), M. H. Fisher and R. L. Tolman (Merck & Co., Inc.).
53. U.S. Pat. 4,203,976 (1980), M. H. Fisher and R. L. Tolman (Merck & Co., Inc.).
54. K. C. Nicolaou, R. E. Dolle, D. P. Papahatjis, and J. L. Randall, *J. Am. Chem. Soc.* **106,** 4189 (1984).
55. H. Els, W. D. Celmer, and K. Murai, *J. Am. Chem. Soc.* **80,** 3777–3782 (1958).
56. S. Hanessian, D. Dube, and P. J. Hodges, *J. Am. Chem. Soc.* **109,** 7063 (1987).

57. S. J. Danishefsky and co-workers, in Ref. 56, p. 8117.
58. S. J. Danishefsky and co-workers, in Ref. 56, p. 8119.
59. M. J. Ford, J. G. Knight, S. V. Ley, and S. Vile, *SYNLETT*. 331–332 1990).
60. M. T. Crimmins, W. G. Hollis, Jr., and R. O'Mahony, in Atta-ur-Rahman, ed., *Studies Directed Toward the Total Synthesis of the Milbemycins and Avermectins in Studies in Natural Products Chemistry,* Vol. 1, Part A, Elsevier, Amsterdam, The Netherlands, 1988, p. 435.
61. T. Blizzard, M. H. Fisher, H. Mrozik, and T. L. Shih, in G. Lukacs and M. Ohno, eds., *Recent Progress in the Chemical Synthesis of Antibiotics,* Springer-Verlag, Berlin, Heidelberg, Germany, 1990.

MICHAEL H. FISHER
HELMUT MROZIK
Merck, Sharp and Dohme Research Laboratories

ANTIVIRAL AGENTS

Viral infections are among the greatest causes of human morbidity, and it is estimated that in developed countries more than 60% of all the episodes of human illness result from viral infections (1). High virus infection rates also occur among pets, livestock, and plants. The high morbidity and the resulting economic loss caused by these infections have generated tremendous efforts in recent years to develop means to combat viral infections using antiviral agents.

Since viruses propagate only within living cells, the development of antiviral drugs which would disrupt the viral replication without affecting the metabolism of the host cell was initially believed to be difficult, if not impossible. However, dramatic progress in viral molecular biology has now made it possible to identify enzymatic processes which are unique to virus-infected cells. As a consequence, it is becoming feasible to design chemical compounds which identify infected cells, block a specific step in viral replication, and leave uninfected cells unharmed.

Viruses are one of the smallest biological entities (except viroids and prions) that carry all the information necessary for their own reproduction. They are unique, differing from procaryotes and eucaryotes in that they carry only one type of nucleic acid as genetic material, which can be transported by the virus from one cell to another. Viruses are composed of a shell of protein enclosing a core of nucleic acid, either ribonucleic acid (RNA) or deoxyribonucleic acid (DNA), that codes for viral reproduction. The outer shell serves as a protective coat to keep the nucleic acid intact and safe from enzymatic destruction. In addition to their protein coat, some viruses contain an outer covering known as an outer envelope. This outer envelope consists of a lipid or polysaccharide material.

The principal obstacle encountered by a virus is gaining entry into the cell. The cells are protected by a cell wall of a thickness comparable to the size of the virus. The virus must first become attached to the cell surface. Much of the specificity of a virus for a certain type of cell lies in its ability to attach to the surface of that specific cell. Durable contact is thus important if the virus is to infect the host cell. The ability of the virus and the cell surface to interact is a property of both the virus and the host cell. The fusion of viral and host-cell membranes, based on unique chemical recognition factors at the cellular surface, allows the intact viral particle or, in certain cases, only its infectious nucleic acid to enter the cell. This process also may involve a type of enzymatic digestion or breakdown of the host plasma membrane at the site of attachment. It is not only during initial infection of a cell that viruses have to cross cell membranes, however. Although the phenomenon of syncitia formation (for HIV infection) and lysing (for picornavirus infection) are observed, cell-to-cell transfer must also occur if the virus is to effectively multiply and cause a substantial infection.

Once inside the host cell, the virus must replicate its own nucleic acid. To do this, it often uses part of the normal synthesizing machinery of the host cell. If the virus is to continue its growth cycle, viral nucleic acid and viral protein must be properly transported within the cell, assembled into the infective virus particle, and ultimately released from the cell. All of these fundamental processes involve an intimate utilization of both cellular and viral enzymes. Certain enzymes that are involved in this process are specifically supplied by the invading virus. It is this type of specificity that can provide the best basis for antiviral chemotherapy. Thus an effective antiviral agent should specifically inhibit the viral-encoded or virus-induced enzymes without inhibition of the normal enzymes involved in the biochemical process of the host cell. Virus-associated enzymes have been reviewed (2,3) (Table 1).

Table 1. Virus-Associated Enzymes

Active enzymes	Viruses
neuraminidase	influenza virus, paramyxovirus
protein kinase	orthopoxvirus, herpes virus, oncovirus
nucleases	mastadenovirus, orthopoxvirus, oncovirus
nucleotide phosphohydrolases	orthopoxvirus
DNA-dependent RNA polymerase	orthopoxvirus
RNA-dependent RNA polymerase	influenza virus, paramyxovirus, vesiculovirus
RNA-dependent DNA polymerase (reverse transcriptase)	oncornavirus, human immunodeficiency virus (HIV)

Because viruses multiply only within the host cells and are totally dependent on these cells for nutrition and reproduction, effective antiviral agents must penetrate the cell to prevent viral multiplication. The ideal antiviral agent should not interfere with the cell's normal defense mechanisms against viral infection but should complement the normal cellular immunity and humoral antibody response to viral reproduction. During the early stages of infection, the activity of several enzymes already present in the host cell increases because of the need of the virus for increased synthesis of the nucleic acid and proteins required by its progeny. This increased activity may stimulate certain other host-controlled enzymic activity.

Viruses contain either RNA or DNA, and this nucleic acid composition forms the basis for their classification. Although viruses are known to infect bacteria, insects, plants, animals, and humans, this discussion is restricted to the important viruses of vertebrates. The relevant viruses are summarized in Table 2,

Table 2. Important[a] Viruses of Vertebrates

Family	Genus	Virus	Type of infection induced
		DNA group	
Adenoviridae	*Mastadenovirus*	numerous adeno types	respiratory and ophthalmic
Herpetoviridae	*Herpesvirus*	herpes types 1 and 2	ophthalmic, central nervous system, genital cutaneous, oral, upper respiratory
		cytomegalo	generalized of infant respiratory, glandular
		varicella (herpes zoster)	nervous, cutaneous
		pseudorabies	central nervous system of livestock
		infectious rhinotracheitis	respiratory of livestock, fowl
Poxviridae	*Orthopoxvirus*	variola (smallpox)	generalized
		vaccinia	cutaneous
Papovaviridae	*Polyomavirus*	polyoma, SV 40	tumors of rodents
		RNA group	
Orthomyxoviridae	*Influenzavirus*	influenza	respiratory
Paramyxoviridae	*Paramyxovirus*	parainfluenza	respiratory
		mumps	glandular
		Newcastle disease	respiratory and nervous of fowl
	Morbillivirus	measles	generalized
		canine distemper	generalized in dogs
	Pneumovirus	respiratory syncytial	respiratory
Picotnaviridae	*Enterovirus*	polio	nervous
		coxsackie	respiratory, nervous, cardiovascular
		echo	nervous, intestinal
	Rhinovirus	numerous rhino types	respiratory
		foot and mouth disease	generalized of livestock
	Calicivirus	feline calici	respiratory of cats
		vesicular exanthema	generalized of swine
Rhabdoviridae	*Vesiculovirus*	vesicular stomatitis	generalized of livestock
	Lyssavirus	rabies	nervous
Coronaviridae	*Coronavirus*	corona	respiratory
		infectious bronchitis	respiratory of fowl
Togaviridae	*Alphavirus*	eastern, western, Venezuelan equine encephalomyelitis	nervous, respiratory
		Semliki Forest	nervous
	Flavivirus	yellow fever	nervous
		numerous encephalitides, including Japanese B, St. Louis, Russian spring-summer	nervous
		dengue	joints, cutaneous
	Rubivirus	rubella	skin, generalized nervous

Table 2. *(Continued)*

Family	Genus	Virus	Type of infection induced
Bunyaviridae	*Bunyavirus*	numerous encephalitides, including bunyamwera and California	nervous
Arenaviridae	*Arenavirus*	lymphocytic choriomeningitis	nervous
		lassa	generalized
Retroviridae	Types B and C *Oncovirus*	numerous leukemias, including friend, Gross Moloney, Rauscher	oncogenic of lower animals
		rous sarcoma, HIV	oncogenic of fowl, aids

[a]Based on disease incidence.

using the nomenclature and taxonomy recommended by the International Committee on Taxonomy of Viruses (4,5).

Extensive efforts have been made over the years to develop drugs for the safe and effective treatment of viral infections. Numerous such compounds have been synthesized and evaluated against several viruses. A large number of these compounds have been found active in cell culture and/or experimental animals. Only a few of these compounds have shown efficacy in clinical trials. The majority of antiviral drugs which are under clinical development today generally interrupt viral nucleic acid synthesis. These compounds often do not affect host cell metabolism and possess considerable selectivity against virus-induced enzymes. This article discusses agents exhibiting significant antiviral activity against viral infections in animal model systems.

Agents Active Against DNA Viruses

One of the simplest molecules found to inhibit the replication of DNA viruses in animals is phosphonoformic acid [*4428-95-9*] (PFA, **1**) CH_3O_5P. Both PFA (as the trisodium salt CNa_3O_5P, foscarnet [*63585-09-1*]) and its homologue phosphonoacetic acid [*4408-78-0*] (PAA, **2**) $C_2H_5O_5P$, were developed by Astra Pharmaceuticals (6) and show selective inhibition of DNA polymerase in various herpes viruses.

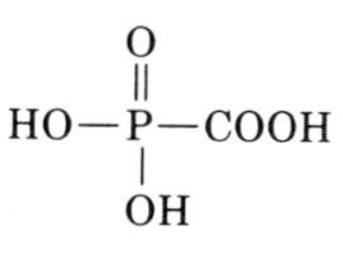

(1)

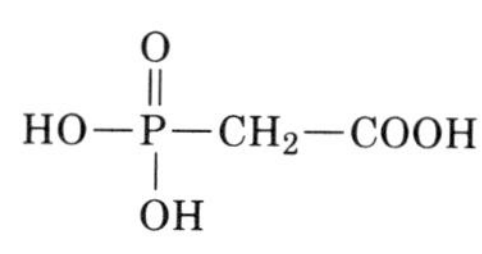

(2)

PFA has recently undergone clinical evaluation in humans for the treatment of recurrent genital herpes, hepatitis B viral infection, and acquired immunodeficiency syndrome (AIDS), as well as cytomegalovirus (CMV) infection of bone marrow and renal transplant patients (7). PFA was found to be effective against genital herpes when applied topically as a 3% cream, herpes labiales (cold sore of the lip), and CMV. However, the results have not been generally encouraging when the compound was tried against AIDS. The thio-analogues of PFA [*102805-86-7*], CH_3O_4PS and PAA [*102805-84-5*], $C_2H_5O_4PS$, (**3** and **4**, respectively) are also effective inhibitors of HSV-1 DNA polymerase (8). Carbonyl bisphosphonic acid [*14255-62-0*], $CH_4O_7P_2$, (**5**) has been shown (9) to inhibit the reverse transcriptase of human immunodeficiency virus (HIV).

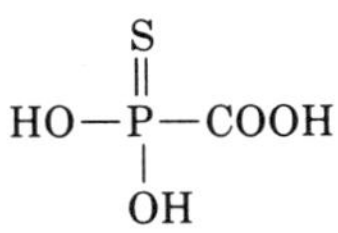

(3)

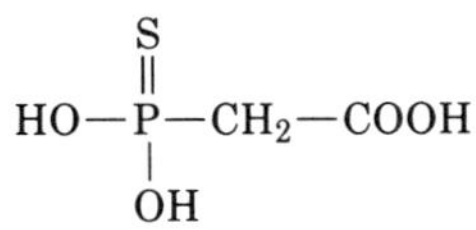

(4)

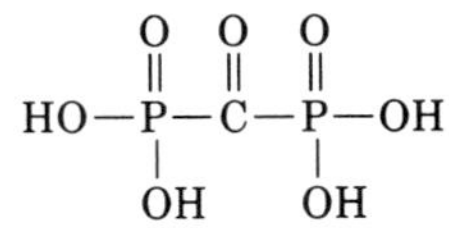

(5)

Thiosemicarbazones have long been used as antiviral agents, principally against pox viruses of the vaccinia family. One compound of this series, the isatin derivative (**6**) $C_9H_8N_4OS$, has been used prophylactically to prevent outbreaks of smallpox in humans (10) and to inhibit the protein synthesis in poxvirus-infected cells. The molecular mechanics relating to this property are still not known (11), though the binding of a metal ion may be a key factor (12). Methisazone [*1910-68-5*] $C_{10}H_{10}N_4OS$ (1-methyl-3-thiosemicarbazone of 2-oxoindole, **7**), one of the more active in the isatin-3-thiosemicarbazone [*487-16-1*] series, has been used in the treatment and prevention of smallpox and vaccinia infections that develop as complications of smallpox vaccination (13). Replacement of the sulfur atom in the thiosemicarbazone moiety by an oxygen atom leads to a loss of antiviral activity. Methisazone has no significant effect on vaccinia virus DNA synthesis (14) but seems to inhibit late protein synthesis by a mechanism that remains to be elucidated.

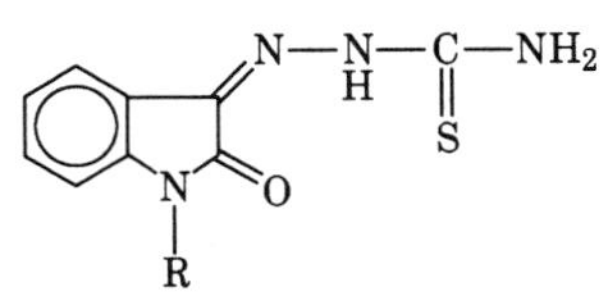

(**6**) R = H
(**7**) R = CH_3

Levamisole [*14769-73-9*] (6-phenyl-2,3,5,6-tetrahydroimidazo[2,1-*b*]thiazole, **8**) $C_{11}H_{12}N_2S$, was found to be effective against herpes virus infections in humans (15,16). It acts as a modulator of host resistance mechanisms, especially as an enhancer of cellular immunity.

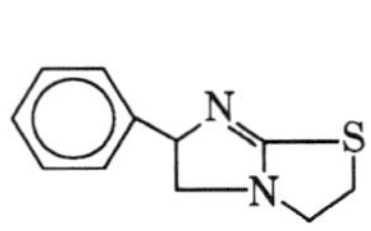

(8)

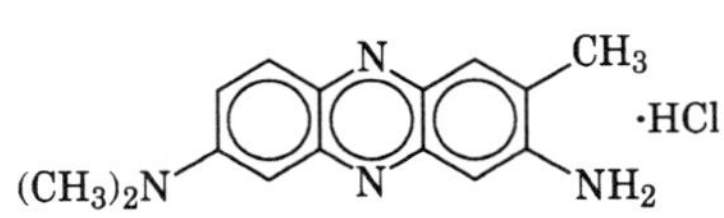

(9)

Certain heterocyclic dyes such as neutral red [*553-24-2*], $C_{15}H_{17}ClN_4$, (**9**), acridine orange [*65-61-2*], and proflavine [*92-62-6*], which act by binding to viral nucleic acid, absorbing visible light, and inactivating the virus by oxidation, have shown antiviral activity of clinical potential against cutaneous herpes virus infections (17). However, concern has been raised that photoinactivation of herpes viruses may be clinically dangerous because the inactivated virus may be capable of transforming normal cells to a malignant state (18). Moreover, these dyes have a strong affinity for the nuclei of normal cells and may damage their genetic makeup (19,20).

Several pyrimidine bases have been found to inhibit herpes virus-induced keratitis in rabbits (21,22), eg, 2-amino-4,6-dichloropyrimidine [*56-05-3*] $C_4H_3Cl_2N_3$, (**10**) and 1-allyl-6-chloro-3,5-diethyluracil [*20938-38-9*], $C_{11}H_{15}ClN_2O_2$, (**11**). Compound (**11**) exhibits efficacy in treating herpetic skin and mucus diseases in humans (22). An important aspect of these aglycones is their continued antiviral activity late in the viral replicative cycle. Mechanistic studies with these compounds suggest that they inhibit the assembly of viral proteins into procapsides, subsequently preventing the coating of available infectious viral nucleic acids which form mature virus particles (21).

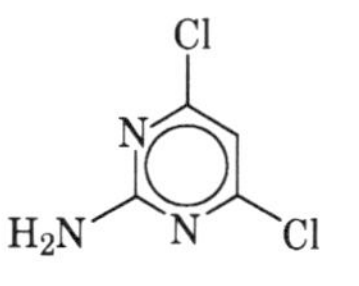

(**10**)

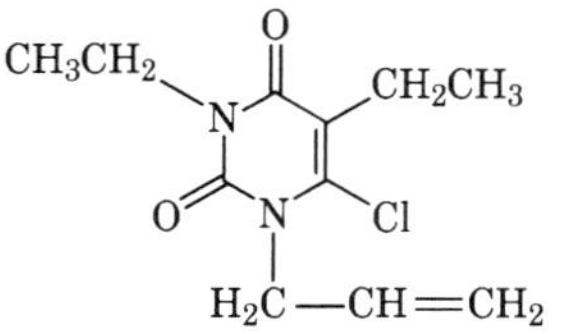

(**11**)

The most successful clinical antiviral agents belong to the nucleoside category. Nucleoside analogues with potent antiviral activity have been known since idoxuridine [*54-42-2*] and trifluridine [*70-00-8*] were shown to be efficacious against herpes keratitis more than 30 years ago. However, despite their antiviral efficacy the therapeutic usefulness of these early nucleosides was limited because of mutagenic, teratogenic, carcinogenic, cytostatic, or cytotoxic side effects. Recently the specificity of antiviral action of this class of compounds has been significantly improved; potent and highly selective nucleoside antiviral agents have now been developed. These are either antiviral nucleosides which are only activated by virus-infected cells, or antiviral agents which specifically inhibit virus-induced enzymes required for viral replication.

Idoxuridine. The first clinically effective nucleoside antiviral agent was 5-iodo-2′-deoxyuridine [*54-42-2*] (idoxuridine, IdU, **12**) $C_9H_{11}IN_2O_5$, synthesized (23) in 1959 (Fig. 1). It was used for the treatment of herpes keratitis in humans (24) and was the first successful use in humans of a synthetic antiviral drug against a herpes virus. It unequivocally demonstrated that IdU prevents the development of severe ocular infections and blindness. FDA approval of IdU for this use quickly followed; the drug was sold by SmithKline Beecham for 20 years for use by ophthalmologists. Recently it has been replaced by more effective and less toxic nucleoside analogues. Moreover, IdU's utility for treatment of systemic herpes infections has been limited by the drug's toxicity. In any event, the human use of IdU became a prototype for the search for other synthetic nucleoside antiviral agents. The other halogenated 2′-deoxyuridines have had mixed reports of activity in experimental systems and have not found clinical usefulness as antiviral agents (25,26).

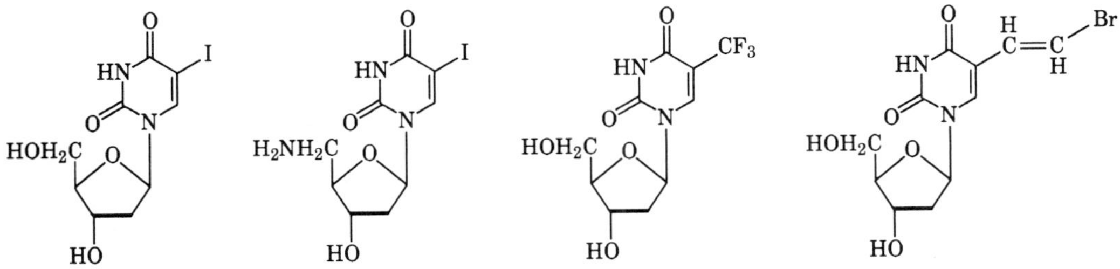

(12) (13) (14) (15)

Fig. 1. Halogenated 2′-deoxyuridine derivatives.

5′-Amino-5-iodo-2′,5′-dideoxyuridine [*56045-73-9*] (**13**) $C_9H_{12}IN_3O_4$, was synthesized in 1975 (27) and was found effective against herpes keratitis in rabbits (28). This compound is markedly less cytotoxic than IdU, indicating that it may have a safer and more specific mode of antiviral activity. A potential limitation of this group of nucleosides is their specificity, for they fail to inhibit all strains of herpes viruses. The specific antiviral activity of (**13**) is considered to be a result of the incorporation of the 5′-*N*-phosphate into both viral and host DNA in infected cells, but not into the DNA of normal cells. Phosphorylation of (**13**) occurs only in herpes virus-infected cells, brought about by a virus-induced thymidine kinase (29).

Trifluridine, $C_{10}H_{11}F_3N_2O_5$, (5-trifluoromethyl-2′-deoxyuridine [*70-00-8*], F_3TdU, **14**) was first prepared (30) in 1962. It is used for topical therapy of herpes virus-infected eyes. It is especially useful for treating infections that are resistant to IdU therapy. Like IdU, trifluridine is incorporated into DNA in place of thymidine in both infected and uninfected cells. But it is 10 times more potent than IdU against herpes keratitis in rabbits and 10 times more soluble in water. Trifluridine is also useful in treating human cytomegalovirus (HCMV), but its toxicity to bone marrow may limit its clinical use.

Another 2′-deoxyuridine derivative, (*E*)-5-(2-bromovinyl)-2′-deoxyuridine [*69304-47-8*], BVdU, (**15**) $C_{11}H_{13}BrN_2O_5$, was reported in 1979 (31). BVdU differs from IdU and F_3TdU by being specifically phosphorylated in the 5′-position by herpes simplex virus type-1 (HSV-1) induced thymidine kinase. This restricts its action to cells infected by HSV-1. It is less active against genital herpes (HSV-2). HSV-1-induced thymidine kinase converts BVdU to the corresponding 5′-mono- and diphosphate, but HSV-2-induced thymidine kinase stops at the stage of the 5′-phosphate of BVdU. Apparently, cellular kinases phosphorylate BVdU-5′-diphosphate to the corresponding 5′-triphosphate, which inhibits HSV-1 DNA polymerase to a greater extent than similar cellular DNA polymerases.

BVdU is incorporated into DNA, but since it is phosphorylated specifically in the virally infected cell, incorporation is mainly confined to such cells. Evidence suggests that BVdU also inhibits the biosynthesis of HSV-1 glycoproteins, which may contribute to the inhibition of the HSV-1 virus (32). BVdU is degraded by thymidine phosphorylase more rapidly than the natural substrate, thymidine. This rapid enzymic degradation may present a problem in its clinical use. Moreover, herpes viruses develop resistance to BVdU, apparently because of mutant viruses that have lower thymidine kinase activity. G. D. Searle has dropped further development of BVdU because of increased animal tumor incidence induced by prolonged dosing (1).

Cytosine Derivatives. 1-β-D-Arabinofuranosylcytosine [*147-94-4*] (ara-C, **16**), $C_9H_{13}N_3O_5$, reportedly has had significant therapeutic effects in patients with localized herpes zoster, herpes eye infections, and herpes encephalitis (33), although several negative results have also been reported (34) (Fig. 2). Ara-C, also known as cytarabine, is quite toxic and is only recommended for very severe viral infections. It is rapidly deaminated in humans to the relatively inactive ara-U. Ara-C is converted in the cell to the 5′-monophosphate by deoxycytidine kinase, followed by formation of the corresponding di- and triphosphate. The triphosphate has been shown to inhibit DNA polymerase.

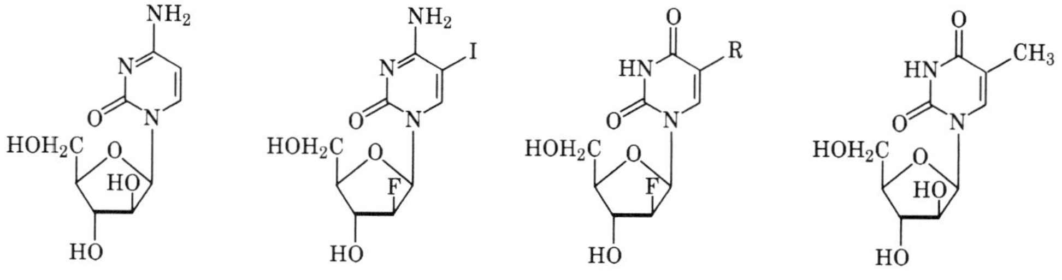

(**16**) (**17**) (**18**) R = CH_3 (**19**) R = CH_2CH_3 (**20**)

Fig. 2. 1-β-D-Arabinofuranosylcytosine (ara-C) and derivatives.

The 3′,5′-cyclic phosphate of ara-C has shown significant antiviral activity *in vivo* with an efficacy greater than that exhibited by ara-C itself (35). It was speculated that the 3′,5′-cyclic phosphate moiety may inhibit the deamination of ara-C, thus causing the increased *in vivo* potency. A number of derivatives of ara-C have been prepared in an effort to improve on antiviral activity and to reduce the toxicity. One such derivative is 2′-fluoro-5-iodo-1-β-D-arabinofuranosylcytosine [*69123-90-6*] (FIAC, **17**), synthesized (36) in 1979. It is active against certain DNA viruses. FIAC, $C_9H_{11}FIN_3O_4$, strongly inhibits HSV-1 and HSV-2 *in vivo*. This selective activity is due in large part to the HSV-encoded thymidine kinase. Clinical studies with FIAC against herpes infections in immunosuppressed cancer patients have shown evidence that this nucleoside can abort cutaneous varicella zoster. The remarkable metabolic stability of the *N*-glycosyl linkage of FIAC may account for the potent activity *in vivo*. When given to mice, FIAC is substantially deaminated to 2′-fluoro-5-iodo-1-β-D-arabinofuranosyluracil [*69123-98-4*], $C_9H_{10}FIN_2O_5$, which is also a significant antiviral agent active against herpes viruses.

FIAC also strongly inhibits HCMV and Epstein-Barr virus (EBV) *in vitro*, the two viruses known not to induce a specific viral thymidine kinase for their replication. However, HCMV may stimulate cellular kinases that can anabolize FIAC to its 5′-triphosphate, which specifically inhibits the HCMV-encoded DNA polymerase. This selective activity suggests that FIAC should be evaluated against HCMV infections. FIAC-triphosphate incorporated into DNA has shown strong *in vitro* activity against the DNA polymerases of human hepatitis B virus (HBV) and of woodchuck hepatitis virus (WHV) (37).

Several uracil nucleoside analogues of FIAC also have been prepared. Among these are 2′-fluoro-5-methylarabinosyluracil [*69256-17-3*] (FMAU, **18**), $C_{10}H_{13}FN_2O_5$, and the corresponding 5-ethyl derivative FEAU (**19**) [*83546-42-3*], $C_{11}H_{15}FN_2O_5$. Both are effective *in vivo* against HSV-1 and HSV-2. FMAU has shown dramatic results in the treatment of WHV in woodchucks and thus may prove effective against HBV infections in humans. However, FMAU is associated with severe toxicity. By contrast, FEAU induced a sustained, although less dramatic, decrease of viral replication without apparent toxic effect (38).

Yet another pyrimidine nucleoside is ara-T (1-β-D-arabinofuranosylthymine [*605-23-2*], $C_{10}H_{14}N_2O_6$, **20**) isolated from Caribbean sponges in 1950 and first synthesized in 1957 (39). Ara-T inhibits herpes DNA replication and is selectively toxic to cells infected by herpes virus but does not inhibit DNA synthesis in uninfected cells. *In vivo* studies with ara-T against experimental HSV-1-induced encephalitis in mice showed remarkable efficacy; ara-T is as potent as the monophosphate of ara-A (**23**) which will be discussed. Ara-T is being developed as an antiherpes agent by Yamasa Shoyu Co. in Japan, although some neurotoxicity has been noted in animal studies with this agent.

An interesting azapyrimidine nucleoside reported to have significant antiviral activity *in vivo* is 6-azauridine [*54-25-1*] (**21**). 6-Azauridine, $C_8H_{11}N_3O_6$, exhibits activity *in vivo* against vaccinia and herpes virus infections (40), and has been reported to be effective against herpes simplex and smallpox virus infections in humans, although side effects have curtailed the use of this drug (25). The basis for the antiviral activity of 6-azauridine has appeared to be conversion of this nucleoside to the 5′-monophosphate, which inhibits orotidylic acid decarboxylase (41). Another active azapyrimidine nucleoside with antiherpes virus activity *in vivo* is 5,6-dihydro-5-azathymidine [*57350-36-4*] (**22**), which was isolated from the fermentation broth of *Streptomyces platensis* var clarensis (42). The antiviral activity of (**22**), $C_9H_{15}N_3O_5$, may be due to inhibition of thymidine phosphorylation in the infected cell (42).

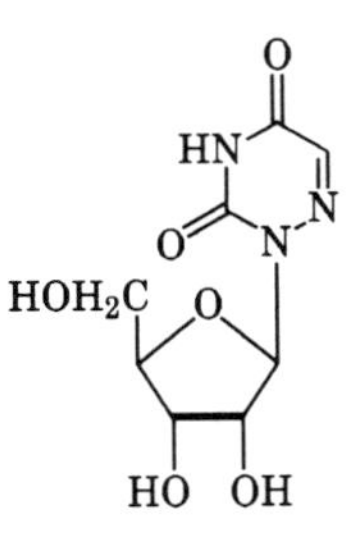

(**21**)

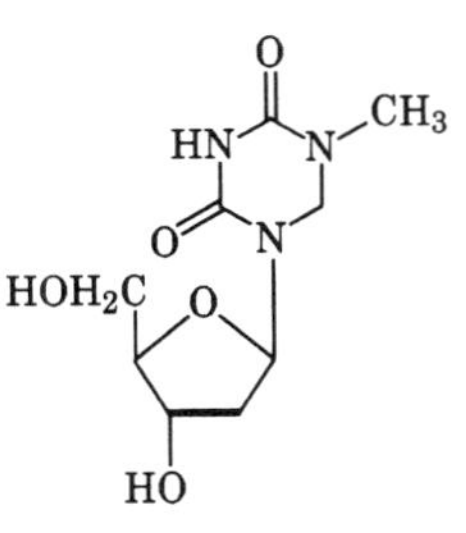

(**22**)

Purine Nucleoside Derivatives. A number of purine nucleoside analogues are also found to be active against several DNA viruses (Fig. 3). The clinically active antiviral drug ara-A (9-β-D-arabinofuranosyladenine [*5536-17-4*], vidarabine, **23**) is active against a number of DNA viruses *in vivo* and also inhibits certain RNA tumor viruses which replicate through a DNA intermediate (43). Ara-A, $C_{10}H_{13}N_5O_4$, was first synthesized in 1960 (44) and later discovered as a

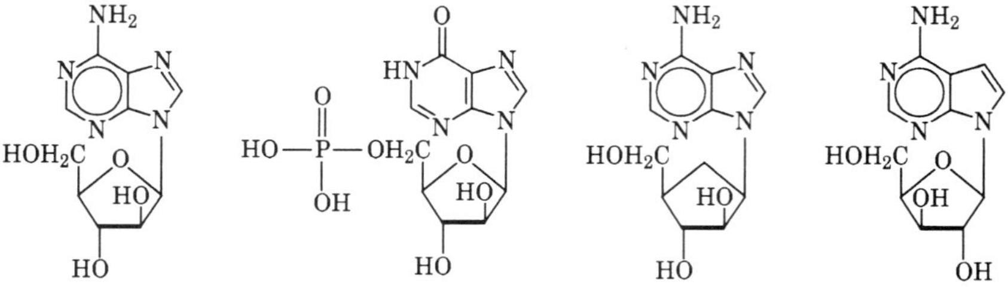

(23) (24) (25) (26)

Fig. 3. Purine nucleoside analogues found to be active against DNA viruses.

nucleoside antibiotic by Warner-Lambert researchers. The *in vivo* antiviral activity of ara-A against intracerebral vaccinia infection in mice and herpes simplex viral keratitis in hamsters, as well as intracerebral HSV-1 infections in mice, was reported by the Southern Research Institute. Ara-A is now marketed in the United States by Warner-Lambert for the systemic treatment of the life-threatening viral illness herpes encephalitis. It also is effective in treating neonatal herpes and varicella zoster (chicken pox), in addition to replacing idoxuridine in treating herpes keratitis (1).

Because it is relatively insoluble, ara-A must be administered by slow infusion, requiring large volumes of fluids. It is also rather toxic, causing nausea, vomiting, and diarrhea. In addition, chromosomal damage caused by ara-A has been observed in blood, and the compound is teratogenic, carcinogenic, and mutagenic in certain animals.

Ara-A is rapidly deaminated by adenosine deaminase in animals to give 9-β-D-arabinofuranosylhypoxanthine [*7013-16-3*] (ara-Hx), $C_{10}H_{12}N_4O_5$, which is approximately 10 times less active than ara-A in cell culture experiments. It has been shown that various adenosine deaminase inhibitors greatly increase the effectiveness of ara-A in culture. However, ara-A-2′,3′-di-*O*-acetate [*65174-96-1*], $C_{14}H_{17}N_5O_6$, is resistant to deamination and is hydrolyzed by tissue esterases to the active ara-A (45). The minimum inhibitory concentration for HSV-1 or HSV-2 plaque reduction by ara-A-2′,3′-di-*O*-acetate was 6 to 10 times higher than that of ara-A or ara-AMP (46). In addition, inhibition of virus-induced lesions was observed in the HSV-2 genital infection of the guinea pig after topical application of 10% ara-A-2′,3′-di-*O*-acetate. On the basis of these results ara-A-2′,3′-di-*O*-acetate is considered an antiviral agent suitable for topical therapy of primary genital infection with herpes virus (47).

Ara-A-5′-monophosphate [*29984-33-6*] (ara-AMP), $C_{10}H_{14}N_5O_7P$, is more water-soluble than ara-A, and therefore can be used in higher dosage during the first hours of treatment of viral infections. Ara-AMP has been shown to decrease virion-associated DNA polymerase concentrations in ground squirrels carrying ground squirrel hepatitis virus. The hypoxanthine derivative, ara-HxMP [*54656-49-4*] (**24**) is more water-soluble, appears to have a similar antiviral spectrum to ara-A, and is considerably less toxic (48).

Topical application of ara-HxMP, $C_{10}H_{13}N_4O_8P$, significantly inhibited the development of keratitis-induced HSV-1, HSV-2, or vaccinia virus in the eyes of rabbits. Ara-HxMP also significantly controlled the development of HSV-1 or vaccinia viral-induced encephalitis in mice and was also active in preventing equine abortion viral deaths in hamsters. Clinical trials with ara-HxMP have not yet been reported.

Ara-A is phosphorylated in mammalian cells to ara-AMP by adenosine kinase and deoxycytidine kinase. Further phosphorylation to the di- and triphosphates, ara-ADP and ara-ATP, also occurs. In HSV-1 infected cells, ara-A also is converted to ara-ATP. Levels of ara-ATP correlate directly with HSV replication. It has recently been suggested that ara-A also may exhibit an antiviral effect against adenovirus by inhibiting polyadenylation of viral messenger RNA (mRNA), which may then inhibit the proper transport of the viral mRNA from the cell nucleus.

It is likely that ara-HxMP similarly exerts its antiviral activity in the form of the triphosphate, ara-HxTP, since ara-HxTP inhibits HSV-1 DNA polymerase (49). Another possible explanation of the antiviral activity of ara-HxTP is that it is metabolically converted to ara-AMP. In fact, it has been shown at Wellcome Research Laboratories that ara-HxMP is a substrate for adenylosuccinate synthetase, and that the resulting arabinofuranosyladenylosuccinate is cleaved to ara-AMP by adenylosuccinate lyase (1). The selective action of ara-A against HSV appears to be a consequence of the preferential inhibition of ara-ATP against HSV-1 and HSV-2 polymerases. Ara-ATP also inhibits normal cellular DNA polymerases, which may be the reason for its cellular toxicity. Also, it has been observed that ara-A is incorporated uniformly throughout the HSV-1 genome, which may result in defective viral DNA (50).

Another interesting adenosine analogue active against DNA viruses is cyclaradine [*69979-46-0*] (**25**), $C_{11}H_{15}N_5O_3$, the carbocyclic analogue of ara-A (51). In this nucleoside the furanose ring oxygen of ara-A is replaced by a methylene group. Cyclaradine is resistant to adenosine deaminase and, in contrast to ara-A, is highly active against HSV-2. Topical (intravaginal) treatment of genital herpes infections in guinea pigs with the 5′-methoxyacetyl derivative of cyclaradine beginning three days after infection largely prevented the development of viral lesions (52).

Of considerable interest is the xylofuranosyl analogue (**26**) [*64526-29-0*], $C_{11}H_{14}N_4O_4$, of tubercidin [*69-33-0*] (53). When (**26**) was administered intraperitoneally at 50 mg/kg per day, a significant reduction of the mortality rate of mice infected with HSV-2 was achieved (54). Compound (**26**) applied topically in 1% DMSO to intracutaneous HSV-2 in hairless mice gave 100% survivors without observable toxicity. This nucleoside would appear to have considerable potential for topical treatment of HSV-2 infections.

Adenine Derivatives. A number of aliphatic adenosine analogues (Fig. 4) which exhibit broad-spectrum antiviral activity have been prepared in recent years. One of the first such nucleoside analogues shown to have antiviral activity is (*S*)-9-(3-hydroxy-2-phosphonylmethoxypropyl)adenine [*92999-29-6*], $C_9H_{14}N_5O_5P$ [(*S*)-HPMPA, (**27**)]. This nucleotide analogue has been prepared (55) and reported (56) to be active against a number of DNA viruses *in vitro*. When given orally, (*S*)-HPMPA was active against a cutaneous HSV-1 model infection in hairless mice and a lethal systemic HSV-1 infection in mice (56). It is also active in the herpetic eye infection model in rabbits (57). (*S*)-HPMPA is not enzymatically dephosphorylated and penetrates cells directly as the phosphonate. It appears to

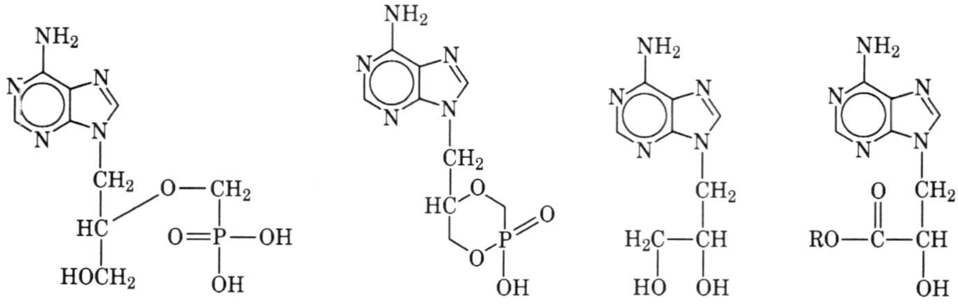

(27) (28) (29) (30)

Fig. 4. Aliphatic adenosine analogues with broad-spectrum antiviral activity.

be incorporated as a chain terminator. It is active against herpes thymidine kinase which lacks mutants resistant to acyclovir and BVdU.

The cyclic phosphonate of (*S*)-HPMPA, [(*S*)-cHPMPA [*106941-26-8*], $C_9H_{12}N_5O_4P$, **28**] also is active against HSV-1 and HSV-2 infection in mice. Both (**27**) and (**28**) are reported to be efficacious against adenovirus, vaccinia virus, as well as varicella-zoster virus infections (58).

The first synthesis of racemic [9-(2,3-dihydroxypropyl)adenine] DHPA, $C_8H_{11}N_5O_2$, was reported in 1965 (59), and the synthesis of the (*R*)- and (*S*)-enantiomers of DHPA was reported later (60). Subsequently (*S*)-DHPA [*54262-83-8*] (**29**) was shown to be active against a broad range of DNA and RNA viruses including vaccinia, HSV-1, HSV-2, measles, and vesicular stomatitis viruses. The racemic mixture, (*R,S*)-DHPA [*55904-02-4*] was almost as active as the (*S*)-enantiomer. However, the (*R*)-isomer of DHPA was inactive as an antiviral agent.

The antiviral activity of (*S*)-DHPA *in vivo* was assessed in mice inoculated intranasally with vesicular stomatitis virus; (*S*)-DHPA significantly increased survival from the infection. (*S*)-DHPA did not significantly reduce DNA, RNA, or protein synthesis and is not a substrate for adenosine deaminase of either bacterial or mammalian origin. However, (*S*)-DHPA strongly inhibits deamination of adenosine and ara-A by adenosine deaminase. Its mode of action may be inhibition of *S*-adenosyl-L-homocysteine hydrolase (61). Inhibition of SAH hydrolase results in the accumulation of SAH, which is a product inhibitor of *S*-adenosylmethionine-dependent methylation reactions. Such methylations are required for the maturation of vital mRNA, and hence inhibitors of SAH hydrolase may be expected to block virus replication by interference with viral mRNA methylation.

Adeninylhydroxypropanoic acid alkyl esters [(*R,S*)-AHPA esters, (**30**)] represent a new class of broad-spectrum antiviral agents, which are, like (*S*)-DHPA, targeted at SAH hydrolase (62). The free acid, (*R,S*)-AHPA, is only weakly active as an antiviral agent. Thus the alkyl moiety merely serves as a protecting group to facilitate uptake of AHPA by the cells. Within the cells, the AHPA alkyl esters would be hydrolyzed to release the free acid, which should then interact with SAH hydrolase.

Acyclic Purine Nucleosides. As a consequence of earlier studies (63–65) involving adenosine deaminase as a model enzyme for examining structural features required for antiviral activity, a number of acyclic purine nucleosides have emerged as antiviral agents (Fig. 5). The forerunner in this group of compounds is the guanosine analogue, acyclovir [*59277-89-3*] [9-(2-hydroxyethoxymethyl)guanine, **31**] (66). Acyclovir, $C_8H_{11}N_5O_3$, is proven to be active against herpes viruses, especially HSV-1 and HSV-2, and remains the drug of choice for control of HSV infections. It has been approved by FDA for general topical as well as oral and intravenous use in treating HSV-1 and HSV-2 infections. It has been marketed by Burroughs Wellcome Co. since 1982.

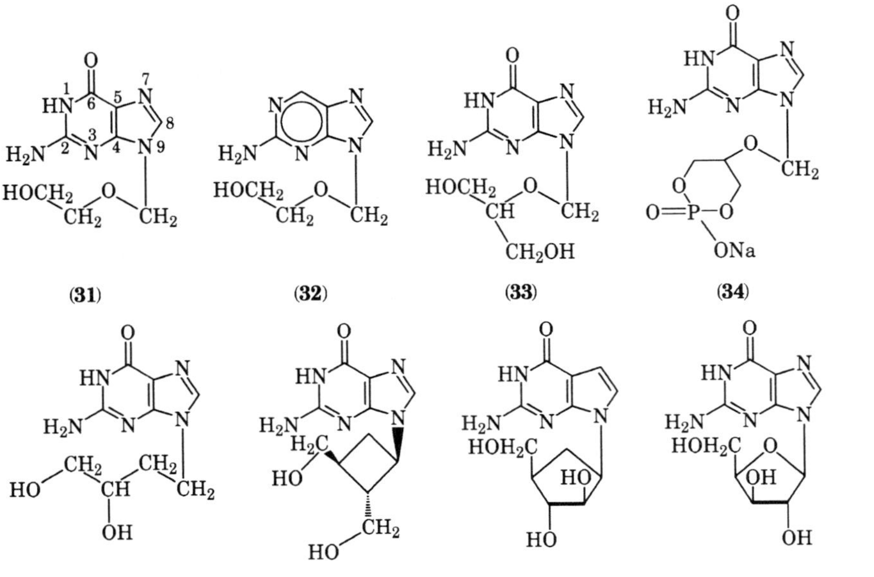

(31) (32) (33) (34) (35) (36) (37) (38)

Fig. 5. Acyclovir and its analogues.

Acyclovir is more effective the more serious the disease and the earlier it is given. It has been shown to be efficacious when used systemically in the prophylaxis of HSV infections in immunosuppressed patients, ie, bone marrow transplant recipients (67). Acyclovir therapy appears to be superior to ara-A in the treatment of herpes simplex encephalitis in humans (68).

Acyclovir has a low toxicity and is excreted unchanged in the urine (69). In patients with normal renal function, 10% of the acyclovir dose is recovered in the urine as 9-carboxymethoxymethylguanine, the only significant metabolite (70). It has a half-life of 2.5–3.0 h after intravenous administration (71). Renal toxicity occurred in early studies when the drug was given intravenously as a bolus, especially in poorly hydrated patients or in those with renal disease. Although its toxicity is low, acyclovir is quite insoluble in body fluids. A number of prodrugs of acyclovir have been studied, eg, 2′-*O*-glycylacyclovir and 6-deoxyacyclovir. These have greater solubility in aqueous medium and are better absorbed orally. 6-Deoxyacyclovir [*75128-58-4*] (**32**) $C_8H_{11}N_5O_2$, is metabolized rapidly by xanthine oxidase to acyclovir. When given orally, compound (**32**) gives levels of acyclovir similar to those reached with intravenous acyclovir.

The antiviral mechanism of action of acyclovir has been reviewed (72). Acyclovir is converted to the monophosphate in herpes virus-infected cells (but only to a limited extent in uninfected cells) by viral-induced thymidine kinase. It is then further phosphorylated by host cell guanosine monophosphate (GMP) kinase to acyclovir diphosphate [*66341-17-1*], which in turn is phosphorylated to the triphosphate by unidentified cellular enzymes. Acyclovir triphosphate [*66341-18-2*] inhibits HSV-1 viral DNA polymerase but not cellular DNA polymerase. As a result, acyclovir is 300 to 3000 times more toxic to herpes viruses in an HSV-infected cell than to the cell itself. Studies have shown that a once-daily dose of acyclovir is effective in prevention of recurrent HSV-2 genital herpes (1). HCMV, on the other hand, is relatively uninhibited by acyclovir.

The incorporation of acyclovir triphosphate into calf thymus DNA primer template has been shown to be much more rapid and extensive with HSV-1 DNA polymerase than with vero cell DNA polymerase α. This incorporation of acyclovir ceased after 15 min since the template is chain terminated by the acyclovir incorporation, as there is no 3′-hydroxyl group on which to continue elongation. The viral DNA polymerase is also inactivated by tight binding to the terminated template.

Several promising analogues of acyclovir have been discovered. Two clinically approved guanosine analogues are DHPG [9-(1,3-dihydroxy-2-propoxymethyl)guanine [*82410-32-0*], Ganciclovir, **33**] and DHPG cyclic phosphate (**34**). The synthesis and antiviral activity of DHPG has been reported simultaneously by several groups (73–77). DHPG, $C_9H_{13}N_5O_4$, is more soluble than acyclovir and has greater bioavailability. Oral administration of DHPG is fiftyfold more efficacious than acyclovir in the treatment of systemic or local HSV-1 or HSV-2 intravaginal infections in mice. Its mode of antiviral action seems to be generally similar to that of acyclovir, involving phosphorylation to the monophosphate by herpes viral thymidine kinase followed by phosphorylation by cellular kinases to the triphosphate, which is a potent inhibitor of herpes viral DNA polymerase and thus inhibits DNA viral replication. However, DHPG is reported to be a thirtyfold more efficient substrate for HSV-1 thymidine kinase than acyclovir and cellular GMP kinase phosphorylates DHPG monophosphate more readily than acyclovir monophosphate. These differences result in seven times more DHPG triphosphate in HSV-1 infected cells than acyclovir triphosphate 6 h after infection. DHPG is being developed by Syntex.

DHPG also is clinically effective against HCMV infections in immunosuppressed patients. It has been shown (78) that DHPG triphosphate inhibits cellular DNA polymerase at levels about 10 times higher than that required to inhibit HCMV-induced DNA polymerase. Treatment with DHPG of patients with AIDS and serious CMV infections resulted in suppression of CMV replication and improvement in CMV retinitis in seven out of eight patients (79). However, treatment of CMV viral pneumonia with DHPG has been disappointing (80).

The sodium salt of the cyclic phosphate of DHPG [*91516-85-7*], $C_9H_{11}N_5O_6P{\cdot}Na$ (**34**), (81), is much more soluble than DHPG and has the broadest spectrum of activity against DNA viruses of the acyclovir analogues (82,83). This cyclic phosphate inhibits HSV-1, HSV-2, HCMV, vaccinia virus, and adenovirus, but is inactive against RNA viruses. Its *in vitro* antiviral activity is about 10 times greater than that of DHPG. This cyclic nucleotide, which is not dependent on virus-specified thymidine kinase for antiviral activity, appears to inhibit DNA viruses by a mechanism different from those of acyclovir or DHPG. The synthesis of the derivatives of (**34**) has been reviewed (84).

3,4-Dihydroxybutylguanine (DHBG, **35**), [*86304-28-1*] $C_9H_{13}N_5O_3$, is another analogue of acyclovir, reported by the Astra Pharmaceutical group (85). DHBG is similar to acyclovir in its activity against HSV-1 keratitis in rabbits. With regard to HSV-2 infections in mice, DHBG appears somewhat superior to acyclovir; the *R* enantiomer (Buciclovir) is more active than the *S*. DHBG, however, is less effective than acyclovir against HSV-1-induced encephalitis in mice. Currently, DHBG is under clinical study as an antiherpetic agent by Astra Pharmaceuticals in Sweden.

(±)-(1α,2β,3α)-9-[2,3-Bis(hydroxymethyl)cyclobutyl]guanine [(±)-BHCG, **36**] is a recently synthesized (86) nucleoside analogue with potent and selective antiviral activity against members of the herpesvirus group, including human cytomegalovirus. (±)-BHCG [*126062-18-8*], $C_{11}H_{15}N_5O_3$, is readily phosphorylated by purified HSV-1 thymidine kinase, and BHCG triphosphate synthesized enzymatically is a selective inhibitor of HSV-1 DNA polymerase (86). (±)-BHCG did not inhibit host cell growth at concentrations at least one thousandfold higher than HSV-2 inhibitory concentrations. The activity against a thymidine kinase deficient HSV-2 mutant was twenty-fivefold poorer than against the parent virus, suggesting that phosphorylation is an important prerequisite for antiviral activity against HSV-2 (86). Subcutaneous administration of (±)-BHCG was protective against HSV-1 systemic infections in mice.

The ara-carbocyclic analogue of 7-deazaguanosine {(±)-2-amino-3,4-dihydro-7-[(1α,2α,3β,4α)-2,3-dihydroxy-4-(hydroxymethyl)-1-cyclopentyl]pyrrolo[2,3-*d*]-pyrimidin-4-one, **37**} [*97570-35-9*], $C_{12}H_{16}N_4O_4$, has been synthesized (87) and studies done against an HSV-1 encephalitis model at a total dose of 100 mg/kg given intraperitoneally over a 48 h period. This resulted in 4 of 15 survivors as compared to zero of 30 for the controls (88). In this model, in comparison, BVdU was only marginally active with 1 of 15 survivors. Compound (**37**) at 25 μg/mL inhibited viral DNA synthesis fourfold over cellular DNA synthesis (88).

Moderate *in vivo* antiherpes virus activity was demonstrated by 9-β-D-xylofuranosylguanine [*27462-39-1*] (xylo-G, **38**), $C_{10}H_{13}N_5O_5$, and the 5′-mono- and 3′,5′-cyclic phosphates of (**38**), although none was as active as ara-A (89). Generally, guanine base-modified analogues of acyclovir are less active than acyclovir because they are not readily phosphorylated by herpes thymidine kinase.

Viruses become resistant to these guanosine analogues by several different mechanisms. The most common resistant variant is deficient in viral thymidine kinase. Virus strains that are resistant to idoxuridine also have been found to be resistant to acyclovir because both drugs depend on viral thymidine kinase for activation. Acyclovir-resistant strains of HSV-1 have been isolated from certain patients receiving acyclovir therapy. With necessary repeated dosing against recurrent HSV-1, resistance could prove to be a primary clinical problem. Certain HSV-2 mutants with an altered viral DNA polymerase also are markedly resistant to DHPG. The problem of drug resistance and antiviral agents has been reviewed (90).

Agents Active Against RNA Viruses

A large number of α-hydroxybenzylbenzimidazole [*50-97-5*] (HBB, **39**), $C_{14}H_{12}N_2O$, derivatives has been prepared and extensively studied as selective inhibitors of the RNA-containing enteroviruses (91). Although none of these derivatives have shown any antiviral activity in animals, 1,2-bis(5-methoxy-2-benzimidazol-2-yl)-1,2-ethanediol [*16656-27-2*] (**40**), $C_{18}H_{18}N_4O_4$, was found to be active against an experimentally induced rhino virus infection in chimpanzees (92). However, the *in vivo* antiviral efficacy was accompanied by significant toxicity.

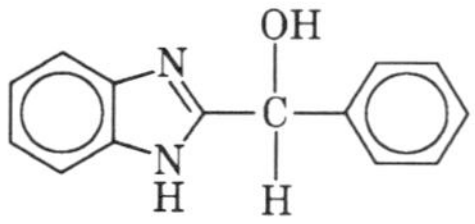

(**39**)

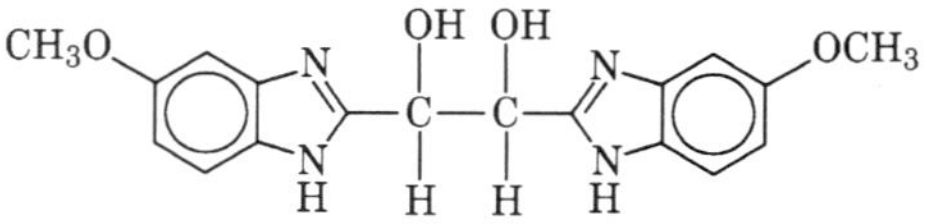

(**40**)

Amantadine hydrochloride [*665-66-7*] (1-adamantanamine hydrochloride, **41**), $C_{10}H_{17}N{\cdot}HCl$, (93) is a good example of a narrow-spectrum agent active only against influenza A virus. It became the first antiviral drug available for systemic use in the United States when it was approved by the FDA in 1966 for use against Asian influenza. In 1976, FDA approval was extended to the use of amantadine for the relief of symptoms of all influenza A strains. Amantadine is marketed by Du Pont de Nemours & Co., Inc. A structurally related drug, rimantadine hydrochloride [*1501-84-4*] $C_{12}H_{21}N{\cdot}HCl$, (α-methyl-1-adamantanemethylamine hydrochloride, **42**), is widely used in Russia to treat influenza A virus (94).

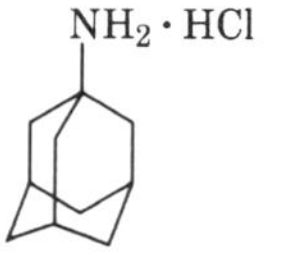

(**41**)

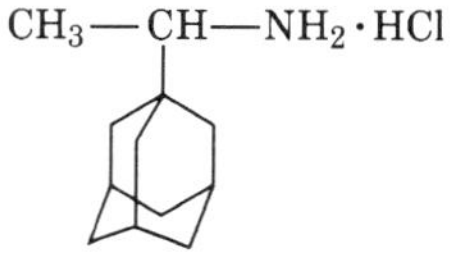

(**42**)

Both amantadine and rimantadine have been found to reduce the duration of influenza A-induced fever and malaise, and to lessen viral shedding. Prophylactic treatment has been recommended for high risk patients (95). It has been suggested that, in the presence of amantadine, the influenza virus attaches normally to cells, but once inside the cell the virus fails to initiate replication. Thus amantadine appears to inhibit the initiation of transcription at an early stage between uncoating and viral-specific RNA synthesis (96).

In most clinical trials, both (**41**) and (**42**) have provided protection against influenza A viruses in approximately 70% of the cases. Neither agent is effective against influenza B viral infections, and influenza A virus becomes resistant to both agents. This may account for the fact that both compounds have had difficulty being accepted in the United States as a valid treatment for influenza infections. In addition, amantadine causes mild transient side effects involving the central nervous system, such as insomnia, lightheadedness, and difficulty in concentration. The incidence of side effects is less with rimantadine. A number of derivatives of amantadine and rimantadine have been prepared and tested, some in clinical trials, but none has given overall results better than the original drugs (97).

Lipophilic β-Diketones. Arildone [*56219-57-9*] (**43**), $C_{20}H_{29}ClO_4$, and several other lipophilic β-diketones (Fig. 6) have exhibited significant *in vivo* activity against a number of RNA viruses. The most potent action is against poliovirus where arildone prevents the uncoating of the viral capsid (98). Arildone is moderately effective against the rhinoviruses. Extensive structure–activity studies with variation of the substituents on the phenyl ring and the β-diketone side chain of (**43**) led to WIN 51711 [*87495-31-6*] (**44**) with potent *in vitro* antirhinovirus and antipoliovirus activity (99,100). WIN 51711 (Disoxaril) is effective in a prophylactic regime against polio-induced death in mice at 100 mg/kg. However, at higher doses formation of crystalline urea was observed. Recently, a number of 2,6-disubstituted derivatives of WIN 51711, $C_{20}H_{26}N_2O_3$, have been prepared and evaluated against several rhinovirus serotypes (101). The 2,6-dichloro analogue was found to be highly effective *in vitro* against rhinoviruses with an MIC_{80} of 0.3 μM, and was also effective in preventing paralysis in mice infected with coxsackievirus A-9 (101).

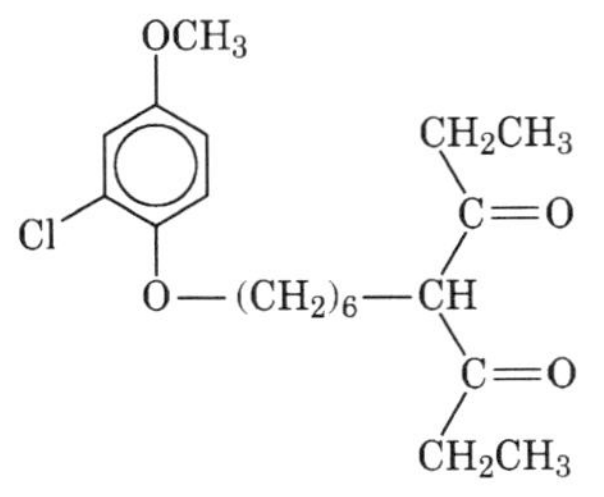

(**43**)

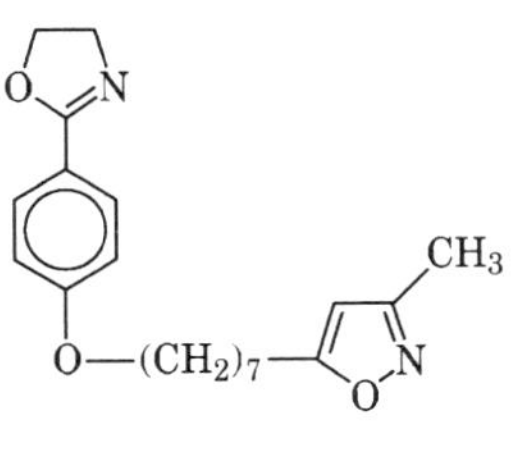

(**44**)

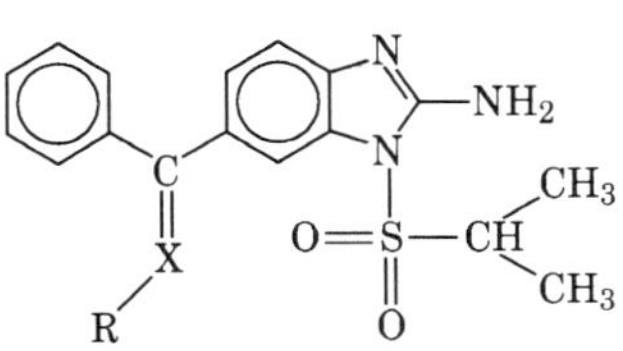

(**45**) X = N, R = OH
(**46**) X = CH, R = CH_3

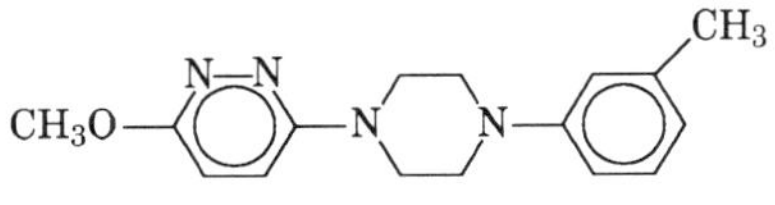

(**47**)

Fig. 6. Lipophilic β-diketones and derivatives active against RNA viruses, especially rhinovirus.

Enviroxime [*72301-79-2*] (**45**), $C_{17}H_{18}N_4O_3S$, a benzimidazole derivative, is found to be highly effective against several different strains of rhinovirus. When given intranasally to volunteers challenged with rhinovirus type-9, symptoms of rhinovirus infection and the virus titre of nasal wash specimens were reduced, and also there was a significant decrease in rhinorrhoea (102). Although enviroxime is active clinically, further studies are needed. The enviroxime surrogate, namely enviradene [*66334-86-9*] (**46**) $C_{19}H_{21}N_3O_2S$, has been shown to possess excellent *in vitro* antirhinovirus activity (103). A new antirhinoviral compound 3-methoxy-6-[4-(3-methylphenyl)-1-piperazinyl]pyridazine [*100241-46-1*] (R61837, **47**) $C_{16}H_{20}N_4O$, has been reported (104). R61837 is active against rhinoviruses and coxsackievirus A21, with MICs as low as 0.018 μM for some serotypes. R61838 has shown prophylaxis against HRV-9 using intranasal delivery (104).

Adenosine and Guanosine Analogues. Several adenosine analogues have been found to be active against both RNA and DNA viruses (Fig. 7). The prototype nucleoside in this group is 3-deazaadenosine [*6736-58-9*] (**48**), $C_{11}H_{14}N_4O_4$, first synthesized in 1966 (105). 3-Deazaadenosine inhibits the replication of RNA type C virus, HL-23, and Rous sarcoma virus (106). Avian sarcoma virus B77, grown in chicken embryo fibroblasts, is inhibited by 90% at 100 μM of (**48**). 3-Deazaadenosine is also active in cell culture against vesicular stomatitis (VSV), parainfluenza 3, and reovirus at less than 10 μg/mL, which is five to twentyfold lower than the host cell cytotoxic concentration. The broad-spectrum antiviral activity of this nucleoside is attributed (107) to its inhibition of *S*-adenosylhomocysteine hydrolase. This inhibition prevents the regeneration of the *S*-adenosylmethionine [*29908-03-0*] required to transfer a methyl group to viral mRNA. Methylation reactions, such as the methylation of the 5′-cap of viral mRNA, are essential to viral replication. 3-Deazaadenosine is not enzymatically deaminated or phosphorylated and can act both as an inhibitor of *S*-adenosylhomocysteine hydrolase and as a substrate of the *S*-adenosylhomocysteine hydrolase to yield 3-deazaadenosylhomocysteine, which has been shown to inhibit methylations *in vivo,* and which acts as a feedback inhibitor of the viral methyltransferases.

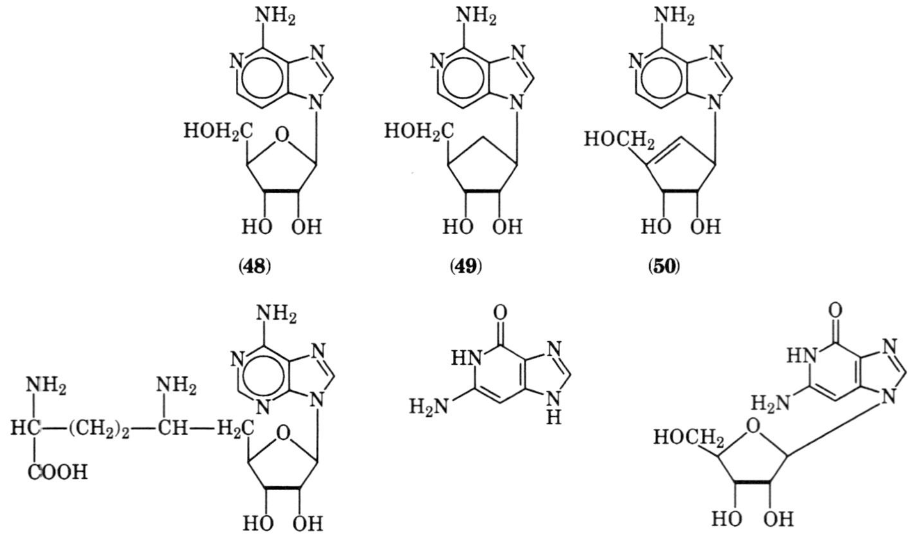

(**48**) (**49**) (**50**) (**51**) (**52**) (**53**)

Fig. 7. Adenosine and guanosine analogues with both RNA and DNA antiviral effectiveness.

Other adenosine analogues that inhibit both RNA and DNA viruses are (±)-3-deazaaristeromycin, [*81601-33-4*], neplanocin A [*72877-50-0*], and sinefungin [*58944-73-3*]. Their mode of antiviral action is believed to be similar to that of 3-deazaadenosine. The carbocyclic analogue of 3-deazaadenosine, [(±)-3-deazaaristeromycin, **49**] $C_{12}H_{16}N_4O_3$, was first synthesized (108) in 1982 and reported to be active against HSV-1, vaccinia virus, and HL-23 *in vitro*. Subsequently, it has been shown (109) that (**49**) protects newborn mice against a lethal infection of VSV. When given 1 h after infection, mortality was reduced from 90% in the controls to 48% in the treated mice. Nucleoside (**49**) is also active *in vitro* against the RNA viruses, Sindbis, measles, and parainfluenza 3 and reo-1 viruses; and the activity in general is superior to that of 3-deazaadenosine.

Neplanocin A (**50**), $C_{11}H_{13}N_5O_3$, is a naturally occurring nucleoside antibiotic (110) which was synthesized (111) in 1983. Neplanocin A inhibits vaccinia virus multiplication of mouse L-cells. The inhibition of vaccinia virus by (**50**) coincides with the inhibition of *S*-adenosylhomocysteine hydrolase in infected cells. Thus the resultant inhibition of macromolecular methylation reactions inhibits the production of new viral particles by preventing methylation of new viral mRNA. Although neplanocin A shows excellent antiviral activity against vaccinia, influenza, measles, and vesicular stomatitis viruses, it gives only marginal protection against a lethal infection of VSV in mice (112).

Sinefungin (**51**), $C_{15}H_{23}N_7O_5$, an antifungal antibiotic (113), is an analogue of *S*-adenosylhomocysteine. It is a potent inhibitor of both Newcastle disease virus (NDV) and vaccinia virus in mouse L-cells. Sinefungin is also active against cutaneous HSV-1 and vaginal HSV-2 in guinea pigs, and systemic herpes viral infections in mice (114). It inhibits methylation of NDV and vaccinia viral mRNA. Sinefungin is effective in preventing Rous sarcoma virus-induced transformation of chick embryo fibroblasts. Unfortunately, sinefungin is also an inhibitor of tRNA and protein methyltransferases in CEF cells, suggesting it may be too toxic for human use.

Both 3-deazaguanine [*41729-52-6*] (**52**) and 3-deazaguanosine [*56039-11-3*] (**53**), prepared by ICN Pharmaceuticals (115), have shown significant *in vitro* antiviral activity against the RNA viruses, parainfluenza 1 and 3, five strains of rhino viruses, vesicular stomatitis virus, and a number of DNA viruses including adenovirus, HSV-1, HSV-2, CMV, vaccinia virus, and pseudorabies virus (116). Although (**52**) and (**53**) show significant antiviral activity against parainfluenza and influenza A_2 infections in mice, the cumulative toxicity may limit their use for treatment of human viral infections. 3-Deazaguanine, $C_6H_6N_4O$, and 3-deazaguanosine, $C_{11}H_{14}N_4O_5$, are metabolized to the 5′-nucleotides (117) and are incorporated into DNA, which is believed to be responsible for the observed toxicity.

The naturally occurring nucleoside antibiotic pyrazofurin [*30868-30-5*] (**54**), $C_9H_{13}N_3O_6$, shows broad-spectrum antiviral activity and broad safety margin against both RNA and DNA viruses *in vitro* (118). However, in mice inoculated with Rift Valley Fever virus, 0.25 mg/kg of pyrazofurin administered daily for five days with treatment initiated a day before infection resulted in a survival rate of only 20%. Similar results were obtained for the treatment of Pichende virus infections in guinea pigs. Thus neither the potent efficacy nor the broad margin of safety noted *in vitro* are reflected in animal models (118). Attempts to improve on this antiviral activity by the synthesis of a number of pyrazofurin derivatives were unsuccessful (119).

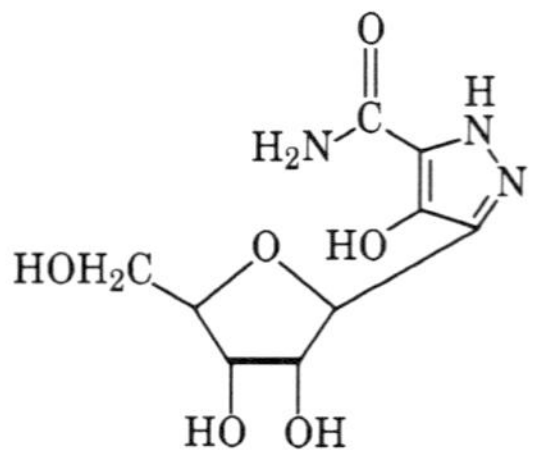

(54)

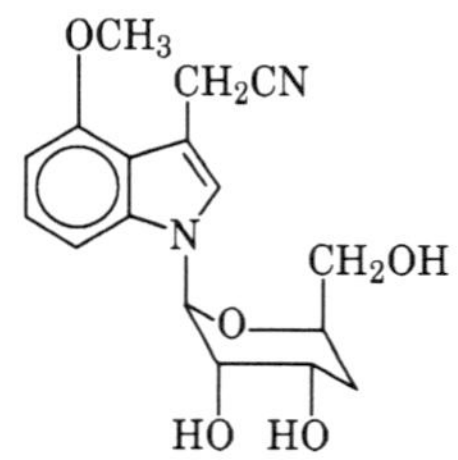

(55)

Another naturally occurring nucleoside antibiotic SF-2140 [*93207-27-3*], $C_{17}H_{20}N_2O_5$, a 3-cyanomethyl-4-methoxyindole nucleoside (**55**) is found to be as active as amantadine in protecting mice against an APR-8 strain of influenza A (120). This suggests that it may be possible to alter both the heterocyclic base and the carbohydrate moiety for greater specificity since no acute toxicity was observed by intraperitoneal injection of SF-2140 at 2 g/kg. These studies have not been confirmed by other laboratories.

Ribavirin and Structural Analogues. One of the broad-spectrum antiviral agents that emerged from ICN Pharmaceuticals is an azole ribonucleoside 1-β-D-ribofuranosyl-1,2,4-triazole-3-carboxamide, designated as ribavirin [*36791-04-5*] (**56**) (121) (Fig. 8). Ribavirin, $C_8H_{12}N_4O_5$, has been studied in more animals and against more viruses than any other antiviral agent known today (122). It is active in cell culture against approximately 85% of all viruses studied, although it is inactive against polio virus, certain coxsackie viruses, most corona viruses, pseudorabies virus, and HBV.

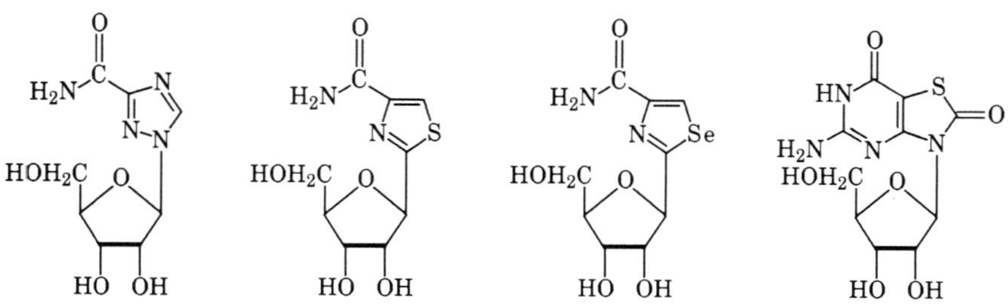

(56) **(57)** **(58)** **(59)**

Fig. 8. Ribavirin (**56**) and active structural analogues.

Ribavirin is not incorporated into the DNA or RNA of either mammalian or viral systems. It has been shown (123), however, that a high dosage of ribavirin given over a prolonged period to Rhesus monkeys results in anemia of red blood cells. This effect is dose related and reversible upon cessation of treatment. Guanosine partially reverses the antiviral effect of ribavirin against certain viruses.

The antiviral mode of action of ribavirin was originally thought to be based on the inhibition of inosine monophosphate (IMP) dehydrogenase by ribavirin

5′-monophosphate, which results in a depletion of cellular guanosine triphosphate (GTP) pools. This lowering of cellular GTP is a self-potentiating effect since GTP competes with ribavirin 5′-triphosphate (RTP), the most abundant form of ribavirin in the cell, in the inhibition of viral-specific RNA polymerase by acting as a GTP or ATP substrate analogue. With certain RNA and DNA viruses, RTP also inhibits the viral-specific mRNA capping enzymes, guanylyl transferase, and N^7-methyl transferase, so that either no viral protein is transcribed or, if it is synthesized, it is nonfunctional (Fig. 9).

This multisite inhibition of viral mRNA processing by RTP appears to be a result of biochemical differences between the viral and the natural cellular enzymes involved in capping mRNA. The viral enzymes are much more strongly inhibited, consequently, preventing viral protein synthesis or giving rise to short segments of viral protein that are nonfunctional. All capped viral mRNAs contain a purine exclusively in the 5′ penultimate position (B_1 in Fig. 9) and initiation of viral mRNA synthesis appears to be purine specific. In mammalian cell mRNA, B_1 may be a pyrimidine as well as a purine, and most often B_1 is cytosine and B_2 uracil.

The nucleotide form of ribavirin does not manifest its antiviral activity simply by lowering the GTP levels, but may indeed participate directly in binding to specific G proteins (124). Ribavirin has recently been studied as an inhibitor of vesicular stomatitis virus and La Crosse virus (125). Of the phosphorylated forms of the drug, ribavirin-5′-diphosphate was by far the most potent inhibitor of viral replication for these two viruses.

Clinical trials in Brazil in 1977 of the oral administration of ribavirin prompted the Brazilian government to approve its use for treating infectious hepatitis A (126). Because ribavirin is water-soluble, it can be dispersed readily in a fine-particle aerosol. Clinical trials (127) indicate that such aerosols reduce the fever, the systemic illness, and the presence of virus in respiratory secretions in patients infected with influenza A or B viruses. It has been found that aerosolized ribavirin used to treat both young adults and infants suffering from RSV infections significantly reduces the severity of the disease (128). The same type of treatment has been used successfully for infants with severe combined immunodeficiency disease, life-threatening RSV, and parainfluenza-3 infections. Ribavirin aerosol has been used successfully to treat adenoviral pneumonia in young children. In 1985, the FDA approved the use of ribavirin aerosol for treating severe infections of RSV, a disease often fatal to infants and children. Ribavirin is being marketed by ICN Pharmaceuticals.

In monkeys, ribavirin has potent antiviral activity against human toga, bunya, and arena viruses. When used intravenously to treat comatose patients infected with Lassa Fever virus in Sierra Leone, ribavirin reduced mortality from an expected 80% to only 10% (129). Ribavirin also is active against various viruses that cause hemorrhagic fevers in mice and monkeys. In model infections with arenaviruses in guinea pigs and monkeys, ribavirin has shown both prophylactic and therapeutic efficacy (130). In the People's Republic of China, ribavirin significantly reduced mortality in hemorrhagic fever patients with renal syndrome (130). Ribavirin is highly effective against sandfly fever virus in humans (131). Its only adverse effect in humans is manageable, reversible anemia. Intra-

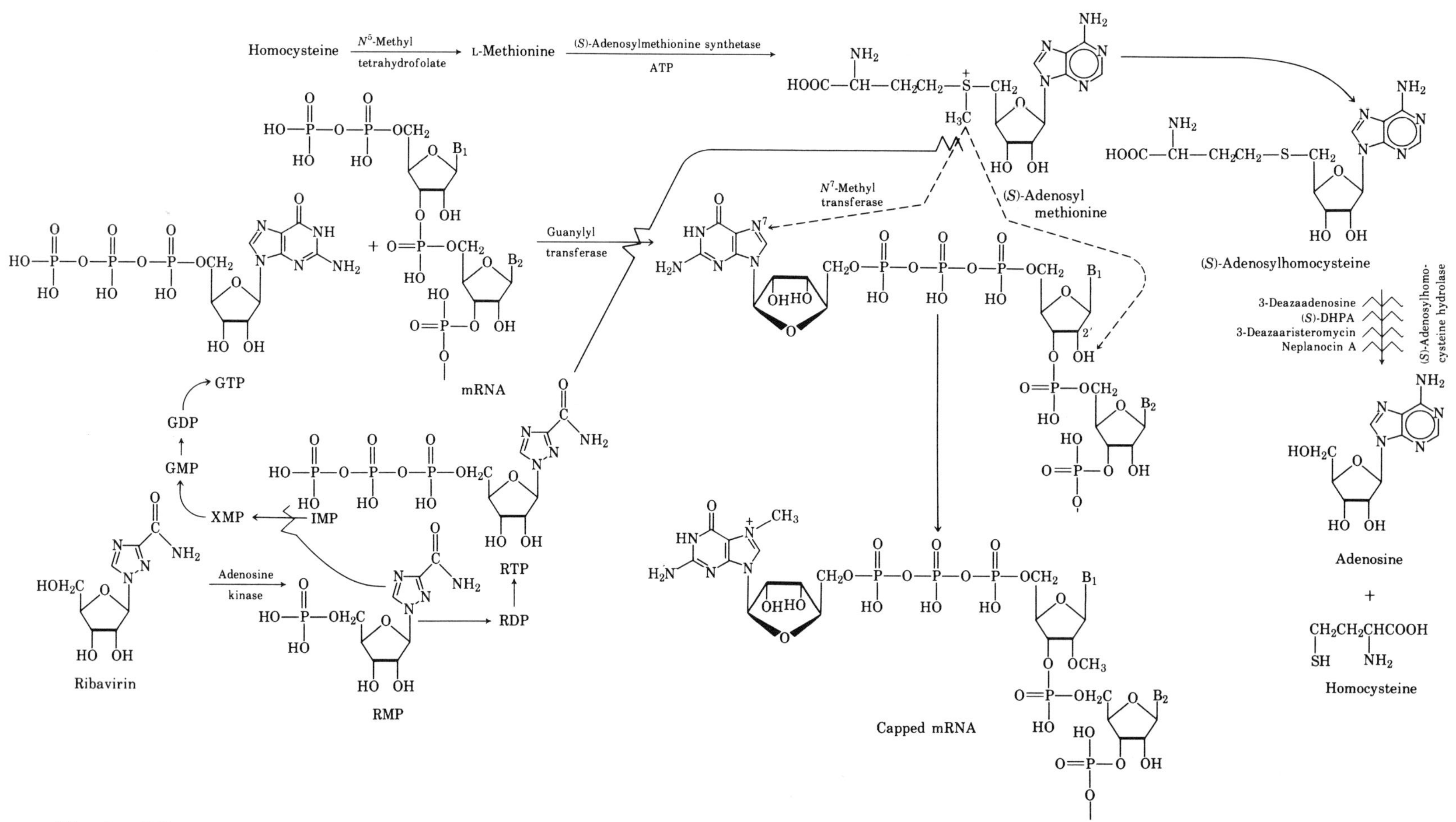

Fig. 9. Ribavirin 5′-monophosphate (RMP) strongly inhibits inosine monophosphate (IMP) dehydrogenase, preventing the conversion of IMP to xanthosine monophosphate (XMP). This, in turn, results in the lowering of guanosine triphosphate (GTP) levels within the cell. Ribavirin 5′-triphosphate (RTP) inhibits viral guanylyl transferase and viral N^7-methyl transferase, probably binding as a substrate analogue of GTP. The low levels of GTP accentuate the viral enzyme inhibition of RTP by competing poorly with RTP for the natural guanylate binding sites of viral-specific guanylyl transferase and N^7-methyl transferase. B_1 and B_2 are purines or pyrimidines as described in the text.

peritoneal treatment with ribavirin 2′,3′,5′-triacetate effected significant increases in both mean survival time and survival rate of dengue virus type-2 infected mice.

Unlike idoxuridine, BVdU, and acyclovir, viral strains susceptible to ribavirin have not been found to develop a resistance to the drug. The resistance against ribavirin is less likely because the drug exhibits multiple sites of antiviral action.

An azole nucleoside structurally related to ribavirin is tiazofurin [*60084-10-8*] (2-β-D-ribofuranosylthiazole-4-carboxamide, **57**) (Fig. 8). It was synthesized in 1977 (132). Tiazofurin, $C_9H_{12}N_2O_5S$, has shown good activity against tomato-spotted wilt virus and good *in vitro* activity against several RNA viruses (132,133), but the *in vivo* experiments against RNA viral-infected animals have been disappointing. However, tiazofurin is found to be efficacious in the treatment of yellow fever virus-infected mice. A combination of ribavirin and tiazofurin showed significant synergistic activity against a number of togaviruses, bunyaviruses, and arenaviruses *in vitro* (133).

Another azole nucleoside that was prepared in 1983 is selenazofurin [*83705-13-9*] (2-β-D-ribofuranosylselenazole-4-carboxamide, **58**) (134), which is structurally similar to ribavirin. *In vitro,* selenazofurin, $C_9H_{12}N_2O_5Se$, has a broad antiviral spectrum similar to ribavirin. It shows definite but marginal therapeutic effects against influenza A and B in mice (134). A combination of ribavirin and selenazofurin, used to treat guinea pigs infected with Pichinde virus, is superior to ribavirin alone and results in the survival of 80% of infected animals. Similarly, ribavirin in combination with selenazofurin has been shown to exhibit synergistic activity *in vitro* against mumps, measles, parainfluenza-3, and vaccinia viruses, as well as HSV-1 (135). It has been observed that selenazofurin has a positive effect against murine hepatitis infections in mice (136). Similarly, selenazofurin had significant effect when injected into rabies-infected mice by the intramuscular route (137).

Although the structures of ribavirin and selenazofurin are similar, they appear to exert their antiviral action at different enzyme sites along the same biochemical pathway. Selenazofurin forms the nicotinamide adenosine dinucleotide (NAD) analogue, which inhibits IMP dehydrogenase by binding in place of the NAD cofactor, and hence this potent reduction of guanylate pools is responsible for the antiviral effect of selenazofurin.

7-Thia-8-oxoguanosine {5-amino-3-β-D-ribofuranosylthiazolo[4,5-*d*]pyrimidine-2,7(3*H*,6*H*)-dione, **59**} [*122970-40-5*] synthesized by ICN Pharmaceuticals, (138) has shown broad-spectrum antiviral activity. 7-Thia-8-oxoguanosine, $C_{10}H_{12}N_4O_6S$, is highly active in mice and rats against Semliki Forest, San Angelo, benzi, rat corona, and encephalomyocarditis viruses when administered intraperitoneally before exposure to the virus (139). The compound was moderately effective in mice infected intraperitoneally with HSV-2 or intranasally with vesicular stomatitis virus. The mode of antiviral action of (**59**) *in vivo* may be due in part to the induction of interferon α (138).

The treatment of Semliki Forest, San Angelo, and benzi viruses in mice suggests 7-thia-8-oxoguanosine (**59**) may be useful in treating arthropod-borne diseases that are endemic to Asia, the Pacific, and Africa.

Agents Active Against Persistent Viral Infections

Persistent viral infection is a difficult challenge for antiviral chemotherapy. Retroviruses as a class are often found to be responsible for persistent viral infections. Retroviruses are unique RNA viruses characterized by the transcription of their single-stranded RNA into the double-stranded DNA of the host cell using the viral enzyme reverse transcriptase. AIDS is an example of such a persistent and latent human viral infection. Such viral infections may indeed be responsible for multiple sclerosis, Alzheimer's disease, and many other dementia of unknown etiology (140). The high incidence of AIDS dementia is due to direct brain infection by human immunodeficiency virus (HIV) (141). Controlling such persistent viral infections by antiviral chemotherapy would require penetration of the blood–brain barrier and prolonged treatment with a relatively nontoxic agent.

Following the identification of a retrovirus, HIV, as the etiological agent of AIDS (142–144), an intense effort has been made to identify drugs for the treatment or prevention of this debilitating, lethal disease (145–147). Several 2′,3′-dideoxyribonucleosides of purine and pyrimidine were discovered to be potent inhibitors of HIV replication *in vitro* (148–150). Considerable data has also accumulated on *in vitro* antiHIV testing of acyclic and carbocyclic nucleoside analogues (151). Although more than 100 nucleosides have been shown to have an *in vitro* antiHIV activity commensurate with development to clinical trials (150,152), only 3′-azido-3′-deoxythymidine [*30516-87-1*] (retrovir, zidovudine, AZT, **60**) $C_{10}H_{13}N_5O_4$, has become widely available for the treatment of AIDS (153,154), and approved by the FDA for treatment of advanced AIDS cases. Although clinically useful in many settings, AZT is associated with substantial toxicity, especially in myelosuppression, leading to serious hematological side effects (155). AZT was first prepared in 1964 (156), and the *in vitro* activity against HIV was initially described in 1985 (157). AZT-resistant HIV strains have been isolated from AIDS patients (158).

Basically, AZT is anabolically phosphorylated to AZT mono-, di-, and triphosphates by various enzymes (kinases) of a target cell (159). AZT-triphosphate competes with other phosphorylated pyrimidine nucleosides for incorporation into HIV DNA by the viral reverse transcriptase. Incorporation of the AZT-triphosphate into reverse transcriptase results in viral DNA chain termination. Reverse transcriptase is essential in the replicative cycle of HIV.

Optimism certainly exists that similar 2′,3′-dideoxynucleosides (Fig. 10) also will prove clinically effective in the treatment of HIV infection but complete clinical data are not yet available. It has been shown that at 10 μM 2′,3′-dideoxyadenosine [*4097-22-7*] (DDA, **61**), first prepared in 1964 (160), completely protected ATH8 cells exposed to HIV and enabled the cells to survive and grow. Although DDA, $C_{10}H_{13}N_5O_2$, is a potent DNA chain terminator of *E. coli,* resistance develops rapidly. Similar inhibition is also shown by 2′,3′-dideoxyguanosine [*85326-06-3*], $C_{10}H_{13}N_5O_3$, (161). 2′,3′-Dideoxycytidine [*7481-89-2*] (DDC, **63**), first synthesized in 1967 (162), reportedly (161) inhibits HIV at 0.5 μM and is presently undergoing clinical trials at NIH in AIDS patients. The principal problem with dideoxynucleosides is that they may be converted within certain cells to the

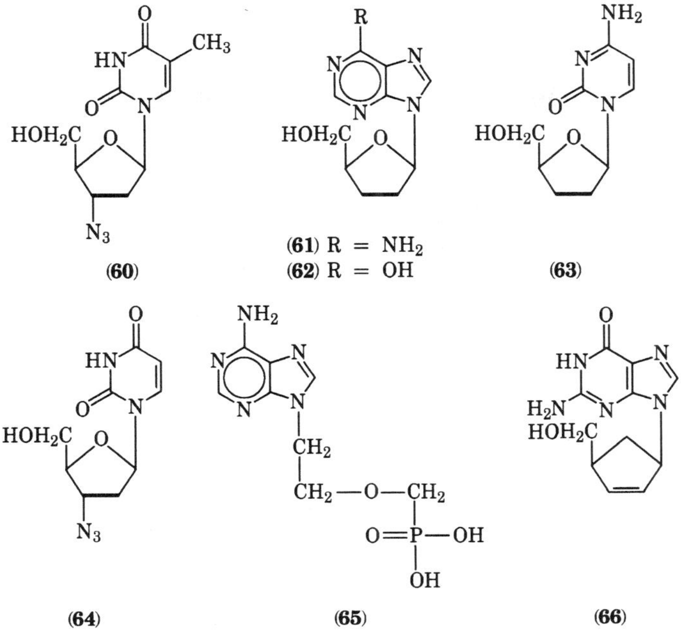

Fig. 10. 2′,3′-Dideoxynucleosides active against human immunodeficiency virus (HIV).

corresponding 5′-triphosphates, which could terminate host DNA synthesis or be incorporated into host DNA (163). The conversion of DDC, $C_9H_{13}N_3O_3$, to the 5′-triphosphate takes place in several mammalian cells and is not solely a property of HIV infected T-lymphocytes (164). Unlike AZT, DDC does not have significant hematologic toxicity and has not shown cross-resistance with AZT *in vitro*. However, DDC has been associated with a severe dose-related peripheral neuropathy. A combination therapy of AZT and DDC in patients with AIDS and advanced AIDS-related complex has been reported (165).

In a long-term clinical study (166) with 58 patients, the experimental drug 2′,3′-dideoxyinosine [*69655-05-6*] (DDI, **62**), was found to improve signs of immune functions in most patients with AIDS and AIDS-related complex without causing severe side effects, and it is at least as efficacious as AZT. In this study, 80% of AIDS patients were alive after 21 months of DDI treatment, compared with about 50% similarly treated with AZT. The common dosage level of DDI, $C_{10}H_{12}N_4O_3$, is 500–700 mg/d. Although none of the patients died from drug toxicity, DDI shows mild side effects such as insomnia, anxiety, irritability, and headaches. Other side effects associated with DDI include painful nerve damage to the feet, and occasionally pancreatitis, an inflammation of the pancreas. DDI, however, causes little or no bone marrow suppression. A combination of DDI and AZT to fight AIDS may be useful in achieving desired effects with lower doses of each drug and lessening overall side effects. DDI is being developed by Bristol-Myers Squibb Co. After an expedited review, the U.S. Food and Drug Administration (FDA) recently granted approval to Bristol-Myers Squibb Co. to market DDI (Videx) for the treatment of AIDS. The drug, approved initially for narrow use, can be used to treat both adult and pediatric AIDS patients who are intolerant to AZT.

Another dideoxypyrimidine nucleoside active against human immunodeficiency virus is 3′-azido-2′,3′-dideoxyuridine [*84472-85-5*] (AZDU or CS-87, **64**) $C_9H_{11}N_5O_4$. Since its synthesis, (167) CS-87 has been identified as a promising antiHIV agent (168) and is currently undergoing phase I clinical trials in patients with AIDS and AIDS-related complex. It appears to be less potent than AZT against HIV in a peripheral blood mononuclear (PBM) cell screening system and in MT-4 cell lines. This lower activity in PBM cells appears to be related to a lower affinity of CS-87 for the enzyme responsible for its initial phosphorylation (169). However, CS-87 has significantly lower toxicity on bone marrow cells than AZT (170) and penetration of the CNS as a 5′-dihydropyridine derivative.

All 2′,3′-dideoxynucleoside analogues are assumed to be intracellularly phosphorylated to their active form (5′-triphosphate), and then targeted at the virus-associated reverse transcriptase. The rate and extent of the 2′,3′-dideoxynucleosides phosphorylate to the 5′-triphosphates may be of equal or greater importance than the differences in the relative abilities of these 5′-triphosphates to inhibit the viral reverse transcriptase (171). At the level of viral reverse transcriptase, the 5′-triphosphate of AZT and other dideoxynucleosides may either serve as a competitive inhibitor with respect to the natural substrates or may act as an alternate substrate, thus leading to chain termination (172).

9-(2-Phosphonylmethoxyethyl)adenine [*106941-25-7*] (PMEA, **65**) (173), synthesized in 1987 (174), is foremost among the acyclic nucleoside analogues proven to be effective inhibitors of HIV-1 replication. The *in vitro* potency and selectivity of PMEA is comparable to the antiHIV-1 potency and selectivity of 2′,3′-dideoxyadenosine (175). Although less potent than AZT *in vitro,* PMEA, $C_8H_{12}N_5O_4P$, is markedly more potent than AZT as an *in vivo* inhibitor of retrovirus replication (176). In fact, PMEA has proven efficacious in the treatment of murine, feline, and simian retrovirus infections in mice, cats, and monkeys, respectively.

The mode of action of PMEA may be quite similar to the mechanism by which (*S*)-HPMPA accomplishes its selective inhibitory activity against herpes viruses. For PMEA to reach its active triphosphate form, it needs only two phosphorylation steps. The triphosphate derivative of PMEA has a much stronger affinity for HIV-1 reverse transcriptase than for cellular DNA polymerases (175). Whether it is actually incorporated into DNA and terminates the growing DNA chain is currently under investigation.

Carbocyclic 2′,3′-didehydro-2′,3′-dideoxyguanosine [*118353-05-2*] (carbovir, CBV, **66**), $C_{11}H_{13}N_5O_2$, synthesized in 1988 (177), is a promising candidate for the chemotherapy of AIDS. CBV inhibits HIV replication and HIV-induced cytopathic effects in a variety of human T-lymphoblastoid cell lines at concentrations approximately two hundred- to four hundredfold below its cytotoxic concentrations (177). CBV is as effective as AZT and DDC in reducing the expression of viral antigen in HIV-infected CEM cells (177). The antiviral potency and selectivity of carbovir is comparable to the anti-HIV-1 potency and selectivity of 2′,3′-dideoxyadenosine (178). The exact mode of antiviral action of carbovir has not yet been elucidated, but may be the modulating effect of intracellular nucleotides on 5′-nucleotidase activity (179).

The initial antiretroviral activity observed with ribavirin (180,181) prompted a study (182) of ribavirin against HIV *in vitro.* Viral suppression at 50 μg/mL and considerable lowering of reverse transcriptase activity in the presence of ribavirin was reported. Phase I oral dose finding clinical studies of ribavirin in 23 patients with lymphadenopathy syndrome (LAS) showed decreased viral replication and enhanced immune function (183) at several dose levels. Phase II clinical trials of ribavirin in the treatment of patients with LAS have shown that at 800 mg/d of ribavirin in a randomized, double-blind, placebo-controlled trial, 53 patients did not develop AIDS during a 24 week study, whereas 9 of 56 patients developed AIDS in the placebo group during the same period (184). Open clinical studies in five AIDS patients treated with 1200 mg, twice daily for two weeks, showed that in four of five AIDS patients, T4 cells doubled and then returned to the baseline level once the drug was discontinued (185). Also, six of nine patients initially positive for HIV in blood became negative during ribavirin treatment (186). A prolonged regimen of a three-day loading dose followed by 600 mg of ribavirin was given to patients for one year. Ribavirin may suppress the HIV virus and produce clinical benefits without the significant bone marrow suppressive effects caused by other drugs such as zidovudine (185).

The search for new antiviral agents is ongoing and extensive; not all viruses have been included here, and new viruses pathogenic to humans will continue to be identified. Novel nucleosides as well as nonnucleosidic compounds which possess greater potency and enzymic specificity for the future treatment of both acute and persistent human viral infection will continue to be discovered.

NOMENCLATURE

acyclovir	9-(2-hydroxyethoxymethyl)guanine
AHPA	adeninylhydroxypropanoic acid
AIDS	acquired immunodeficiency syndrome
ara-A	9-β-D-arabinofuranosyladenine
ara-AMP	5′-monophosphate of ara-A
ara-C	1-β-D-arabinofuranosylcytosine
ara-Hx	9-β-D-arabinofuranosylhypoxanthine
ara-HxMP	5′-monophosphate of ara-Hx
ara-T	1-β-D-arabinofuranosylthymine
ara-U	1-β-D-arabinofuranosyluracil
AZT	3′-azido-3′-deoxythymidine
BHCG	(1α,2β,3α)-9-[2,3-bis(hydroxymethyl)cyclobutyl]guanine
BVdU	(*E*)-5-(2-bromovinyl)-2′-deoxyuridine
carbovir	carbocyclic 2′,3′-didehydro-2′,3′-dideoxyguanosine
CS-87	3′-azido-2′,3′-dideoxyuridine
DDA	2′,3′-dideoxyadenosine
DDC	2′,3′-dideoxycytidine
DDI	2′,3′-dideoxyinosine
DHBG	9-(3,4-dihydroxybutyl)guanine
DHPA	9-(2,3-dihydroxypropyl)adenine
DHPG	9-(1,3-dihydroxy-2-propoxymethyl)guanine
DMSO	dimethyl sulfoxide
DNA	2′-deoxyribonucleic acid
EBV	Epstein-Barr virus

FEAU	2′-deoxy-2′-fluoro-5-ethylarabinosyluracil
FIAC	2′-deoxy-2′-fluoro-5-iodoarabinosylcytosine
FMAU	2′-deoxy-2′-fluoro-5-methylarabinosyluracil
FMDV	foot and mouth disease virus
F_3TdU	5-trifluoromethyl-2′-deoxyuridine
GMP	guanosine 5′-monophosphate
GTP	guanosine 5′-triphosphate
HBB	α-hydroxybenzylbenzimidazole
HBV	hepatitis B virus
HCMV	human cytomegalovirus
HIV	human immunodeficiency virus
HPMPA	9-(3-hydroxy-2-phosphonylmethoxypropyl)adenine
HSV-1	herpes simplex type-1 virus
HSV-2	herpes simplex type-2 virus
IdU	5-iodo-2′-deoxyuridine
IMP	inosine 5′-monophosphate
LAS	lymphadenopathy syndrome
NAD	nicotinamide adenosine dinucleotide
NDV	Newcastle disease virus
PAA	phosphonoacetic acid
PFA	phosphonoformic acid
PMEA	9-(2-phosphonylmethoxyethyl)adenine
ribavirin	1-β-D-ribofuranosyl-1,2,4-triazole-3-carboxamide
RNA	ribonucleic acid
mRNA	messenger RNA
RSV	respiratory syncytial virus
RTP	ribavirin 5′-triphosphate
selenazofurin	2-β-D-ribofuranosylselenazole-4-carboxamide
tiazofurin	2-β-D-ribofuranosylthiazole-4-carboxamide
VSV	vesicular stomatitis virus
WHV	woodchuck hepatitis virus
xylo-G	9-β-D-xylofuranosylguanine

BIBLIOGRAPHY

1. R. K. Robins, *Chem. Eng. News* **64,** 28 (1986).
2. W. M. Mitchell, in W. A. Carter, ed., *Selective Inhibitors of Viral Functions,* CRC Press, Cleveland, Ohio, 1973, p. 51.
3. B. R. McAuslan, *Life Sci.* **14,** 2085 (1974).
4. F. Fenner, *Intervirology* **7,** 4 (1976).
5. R. W. Sidwell and J. T. Witkowski in M. E. Wolff, ed., *Burger's Medicinal Chemistry,* 4th ed., John Wiley & Sons, Inc., New York, 1979, pp. 543–593.
6. E. Helgstrand and co-workers, *Science* **201,** 819 (1978).
7. B. Öberg, *Pharmac. Ther.* **40,** 213 (1989).
8. D. W. Hutchinson and S. Masson, *IRCS Med. Sci.* **14,** 176 (1986).
9. L. Vrang and B. Öberg, *Antimicrob. Agents Chemother.* **29,** 867 (1986).
10. C. J. Pfau, *Handbook Exp. Pharmacol.* **61,** 147 (1982).

11. J. A. Cooper, B. Moss, and E. Katz, *Virology* **96,** 381 (1979).
12. D. W. Hutchinson, *Antiviral Res.* **5,** 193 (1985).
13. D. M. McLean, *Ann. N.Y. Acad. Sci.* **284,** 118 (1977).
14. K. B. Easterbrook, *Virology* **17,** 245 (1962).
15. A. Kint and L. Verlinden, *New Engl. J. Med.* **291,** 308 (1974).
16. R. J. O'Reilly, A. Chibbaro, R. Wilmot, and C. Lopez, *Ann. N.Y. Acad. Sci.* **284,** 161 (1977).
17. J. L. Melnick and C. Wallis, *Ann. N.Y. Acad. Sci.* **284,** 171 (1977).
18. F. Rapp, *J. Am. Med. Assoc.* **225,** 456 (1973).
19. L. Weinstein and T. Chang, *New Engl. J. Med.* **289,** 725 (1973).
20. E. G. Friedrick, *J. Obstetr. Gynecol.* **43,** 304 (1974).
21. P. La Colla and co-workers, *Ann. N.Y. Acad. Sci.* **284,** 294 (1977).
22. K. K. Gauri and B. Rohde, *Klin.-Ther. Wochenschr.* **47,** 375 (1969).
23. W. H. Prusoff, *Biochim. Biophys. Acta.* **32,** 295 (1959).
24. H. E. Kaufman, *Proc. Soc. Exp. Biol. Med.* **109,** 251 (1962).
25. W. H. Prusoff and D. C. Ward, *Biochem. Pharmacol.* **25,** 1233 (1976).
26. J. Sugar and H. E. Kaufman in Ref. 2, p. 295.
27. Y-C. Cheng and co-workers, *J. Virol.* **15,** 1284 (1975).
28. D. M. Albert and co-workers, *J. Invest. Ophthalmol.* **15,** 470 (1976).
29. M. S. Chen, D. F. Ward, and W. H. Prusoff, *J. Biol. Chem.* **251,** 4833 (1976).
30. C. Heidelberger, D. G. Parsons, and D. C. Remy, *J. Am. Chem. Soc.* **84,** 3597 (1962).
31. A. S. Jones, G. Verhelst, and R. T. Walker, *Tetrahedron Lett.,* 4415 (1979).
32. S. Olofsson, M. Lundström, and R. Datema, *Virology* **147,** 201 (1985).
33. B. E. Juel-Jensen, *Brit. Med. J.* **1,** 406 (1973).
34. C. B. Lauter, C. J. Bailey, and A. M. Lerner, *Antimicrob. Agents Chemother.* **6,** 598 (1974).
35. R. W. Sidwell and co-workers, *Nature New Biol.* **242,** 204 (1973).
36. K. A. Watanabe and co-workers, *J. Med. Chem.* **22,** 21 (1979).
37. O. Hantz and co-workers, *Antiviral Res.* **4,** 187 (1984).
38. I. Fourel and co-workers, *Antimicrob. Agents Chemother.* **34,** 473 (1990).
39. J. J. Fox, N. Yung, and A. Bendich, *J. Am. Chem. Soc.* **79,** 2775 (1957).
40. G. A. Galegov, R. M. Bikbulatov, K. A. Vanag, and R. M. Shen, *Vop. Virusol.* **13,** 18 (1968).
41. R. E. Handschumaker and C. A. Pasternak, *Biochim. Biophys. Acta* **30,** 451 (1958).
42. H. E. Renis, *Antimicrob. Agents Chemother.* **13,** 613 (1978).
43. B. J. Sloan in D. Pavan-Langston, R. A. Buchanan, and C. A. Alford Jr., eds., *Adenine Arabinoside: An Antiviral Agent,* Raven Press, New York, 1975, p. 45.
44. W. W. Lee, A. Benetz, L. Goodman, and B. R. Baker, *J. Am. Chem. Soc.* **82,** 2648 (1960).
45. D. C. Baker, T. H. Haskell, S. R. Putt, and B. J. Sloan, *J. Med. Chem.* **22,** 273 (1979).
46. J. T. Richards, E. R. Kern, C. J. Overall Jr., and L. A. Glasgow, *Antiviral Res.* **2,** 27 (1982).
47. W. M. Shannon and co-workers, *Antimicrob. Agents Chemother.* **24,** 706 (1983).
48. G. R. Revankar and co-workers, *J. Med. Chem.* **18,** 721 (1975).
49. D. Derse, K. F. Bastow, and Y. C. Cheng, *J. Biol. Chem.* **257,** 10251 (1982).
50. J. C. Drach, in E. De Clercq and R. T. Walker, eds., *Targets for Design of Antiviral Agents,* Plenum Press, New York, 1984, p. 234.
51. R. Vince and S. Daluge, *J. Med. Chem.* **20,** 612 (1977).
52. R. Vince and co-workers, *Science* **221,** 1405 (1983).
53. E. De Clercq and co-workers, *J. Med. Chem.* **30,** 481 (1987).
54. E. De Clercq and M. J. Robins, *Antimicrob. Agents Chemother.* **30,** 719 (1986).
55. A. Holy, *Chemica Scripta* **26,** 83 (1986).
56. E. De Clercq and co-workers, *Nature* **323,** 464 (1986).

57. P. C. Maudgal, E. De Clercq, and P. Huyghe, *Invest. Ophthalmol. Vis. Sci.* **28,** 243 (1987).
58. E. De Clercq, in E. De Clercq and R. T. Walker, eds., *Antiviral Drug Development—A Multidisciplinary Approach,* Plenum Press, New York, 1987, pp. 97–122.
59. H. J. Schaeffer, D. Vogel, and R. Vince, *J. Med. Chem.* **8,** 502 (1965).
60. A. Holy, *Collect. Czech. Chem. Commun.* **40,** 187 (1975).
61. I. Votruba and A. Holy, *Collect. Czech. Chem. Commun.* **45,** 3039 (1980).
62. E. De Clercq and A. Holy, *J. Med. Chem.* **28,** 282 (1985).
63. L. M. Lerner and R. R. Rossi, *Biochemistry* **11,** 2272 (1972).
64. H. J. Schaeffer, S. Gurwara, R. Vince, and S. Bittner, *J. Med. Chem.* **14,** 367 (1971).
65. H. J. Schaeffer and D. F. Schwender, *J. Med. Chem.* **17,** 6 (1974).
66. H. J. Schaeffer and co-workers, *Nature* **272,** 583 (1978).
67. D. Gold and L. Corey, *Antimicrob. Agents Chemother.* **31,** 361 (1987).
68. R. J. Whitley and co-workers, *New Eng. J. Med.* **314,** 144 (1986).
69. O. L. Laskin and co-workers, *Antimicrob. Agents Chemother.* **21,** 393 (1982).
70. P. de Miranda and co-workers, *Clin. Pharmacol. Ther.* **26,** 718 (1979).
71. O. L. Laskin, *Clin. Pharmacokinet.* **8,** 187 (1983).
72. G. B. Elion, *Am. J. Med.* **73,** 7 (1982).
73. K. K. Ogilvie and co-workers, *Can. J. Chem.* **60,** 3005 (1982).
74. K. O. Smith, K. S. Galloway, K. K. Ogilvie, and U. O. Cherujan, *Antimicrob. Agents Chemother.* **22,** 1026 (1982).
75. J. C. Martin and co-workers, *J. Med. Chem.* **26,** 759 (1983).
76. D. F. Smee, J. C. Martin, J. P. H. Verheyden, and T. R. Matthews, *Antimicrob. Agents Chemother.* **23,** 676 (1983).
77. A. K. Field and co-workers, *Proc. Natl. Acad. Sci. USA* **80,** 4139 (1983).
78. K. B. Frank, J. F. Chiou, and Y. C. Cheng, *J. Biol. Chem.* **259,** 1556 (1984).
79. O. L. Laskin, C. M. Stahl-Bayliss, C. M. Kalman, and L. R. Rosecan, *J. Infect. Dis.* **155,** 323 (1987).
80. E. C. Reed, P. S. Dandliker, and J. D. Meyers, *Ann. Intern. Med.* **105,** 214 (1986).
81. R. L. Tolman and co-workers, *Biochem. Biophys. Res. Commun.* **128,** 1329 (1985).
82. J. Germershausen and co-workers, *Antimicrob. Agents Chemother.* **29,** 1025 (1986).
83. J. D. Karkas and co-workers, *Biochim. Biophys. Acta* **911,** 127 (1987).
84. M. MacCoss and co-workers, *Chemica Scripta* **26,** 113 (1986).
85. A. Larsson and co-workers, *Antimicrob. Agents Chemother.* **23,** 664 (1983).
86. A. K. Field and co-workers, *Antiviral Res.* **13,** 41 (1990).
87. M. Legraverend, R-M. N. Ngongo-Tekam, E. Bisagni, and A. Zerial, *J. Med. Chem.* **28,** 1477 (1985).
88. A. Zerial, M. Zerial, M. Legraverend, and E. Bisagni, *Ann. Inst. Pasteur/Virol.* **137E,** 317 (1986).
89. G. R. Revankar and co-workers, *J. Med. Chem.* **19,** 1026 (1976).
90. B. Loddo, *Pharmac. Ther.* **10,** 431 (1983).
91. I. Tamm and L. A. Caliguiri, in D. J. Bauer, ed., *International Encyclopedia of Pharmacology and Therapeutics,* Vol. 1, Pergamon Press, Oxford, UK, 1972, p. 115.
92. N. L. Shipkowitz and co-workers, *Appl. Microbiol.* **23,** 117 (1972).
93. K. Gerzon and co-workers, *J. Med. Chem.* **6,** 760 (1963).
94. A. W. Galbraith, *Brit. Med. Bull.* **41,** 381 (1985).
95. S. L. Barriere, *Pharm. Int.,* 139 (1984).
96. A. J. Hay, A. J. Wolstenholme, J. J. Skehel, and M. H. Smith, *EMBO J.* **4,** 3021 (1985).
97. D. L. Swallow and G. L. Kampfner, *Brit. Med. Bull.* **41,** 322 (1985).
98. J. J. McSharry, L. A. Caliguiri, and H. J. Eggers, *Virology* **97,** 307 (1979).
99. G. D. Diana and co-workers, *J. Med. Chem.* **28,** 748 (1985).
100. G. D. Diana, M. J. Otto, and M. A. McKinlay, *Pharmac. Ther.* **29,** 287 (1985).

101. G. D. Diana and co-workers, *J. Med. Chem.* **32,** 450 (1989).
102. R. J. Phillpotts and co-workers, *Lancet,* 1342 (1981).
103. S. M. Roberts, in Ref. 58, pp. 37–54.
104. K. Andries and co-workers, *Arch. Virol.* **101,** 155 (1988).
105. R. J. Rousseau, L. B. Townsend, and R. K. Robins, *Biochemistry* **5,** 756 (1966).
106. A. J. Bodner, G. L. Cantoni, and P. K. Chiang, *Biochem. Biophys. Res. Commun.* **98,** 476 (1981).
107. J. P. Bader, N. R. Brown, P. K. Chiang, and G. L. Cantoni, *Virology* **89,** 494 (1978).
108. J. A. Montgomery and co-workers, *J. Med. Chem.* **25,** 626 (1982).
109. E. De Clercq and J. A. Montgomery, *Antiviral Res.* **3,** 17 (1983).
110. M. Hayashi, S. Yaginuma, H. Yoshioka, and K. Nakatsu, *J. Antibiot.* **34,** 675 (1981).
111. M. Arita and co-workers, *J. Am. Chem. Soc.* **105,** 4049 (1983).
112. E. De Clercq, *Antimicrob. Agents Chemother.* **28,** 84 (1985).
113. R. L. Hamill and M. M. Holhn, *J. Antibiot.* **26,** 463 (1973).
114. U.S. Pat. 4,158,056 (July 13, 1978), R. Nagarajan.
115. P. D. Cook and co-workers, *J. Am. Chem. Soc.* **98,** 1492 (1976).
116. L. B. Allen and co-workers, *Antimicrob. Agents Chemother.* **12,** 114 (1977).
117. T. M. Page and co-workers, *Int. J. Biochem.* **18,** 957 (1986).
118. P. G. Canonico, P. B. Jahrling, and W. L. Pannier, *Antiviral Res.* **2,** 331 (1983).
119. C. R. Petrie and co-workers, *J. Med. Chem.* **29,** 268 (1986).
120. T. Ito and co-workers, *J. Antibiot.* **37,** 931 (1984).
121. J. T. Witkowski, R. K. Robins, R. W. Sidwell, and L. N. Simon, *J. Med. Chem.* **15,** 1150 (1972).
122. R. W. Sidwell, G. R. Revankar, and R. K. Robins, in D. Shugar, ed., *Intl. Enc. Pharmacol. Ther.* Sec. 116, Vol. 2, Pergamon Press, New York, 1985, pp. 49–108.
123. P. G. Canonico and co-workers, *Toxicol. Applied Pharmacol.* **74,** 155, 163 (1984).
124. R. K. Robins and co-workers, *Adv. Enzyme Reg.* **29,** 97 (1989).
125. P. Toltzis, K. O'Connell, and J. L. Patterson, *Antimicrob. Agents Chemother.* **32,** 492 (1988).
126. P. A. A. Galvao and I. O. Castro, *Ann. N.Y. Acad. Sci.* **284,** 278 (1977).
127. V. Knight and co-workers, *Lancet,* 945–949 (Oct. 31, 1981).
128. C. B. Hall and co-workers, *New Eng. J. Med.* **308,** 1443 (1983).
129. J. B. McCormick and co-workers, *New Engl. J. Med.* **314,** 20 (1986).
130. J. W. Huggins, *Rev. Inf. Dise.* **11**(Suppl. 4), S750 (1989).
131. J. W. Huggins, G. R. Kim, and K. T. McKee Jr., *VI International Congress of Virology,* Sendai, Japan, Sept. 1984.
132. P. C. Srivastava and co-workers, *J. Med. Chem.* **20,** 256 (1977).
133. J. W. Huggins, R. K. Robins, and P. G. Canonico, *Antimicrob. Agents Chemother.* **26,** 476 (1984).
134. P. C. Srivastava and R. K. Robins, *J. Med. Chem.* **26,** 445 (1983).
135. J. J. Kirsi and co-workers, *Antimicrob. Agents Chemother.* **26,** 466 (1984).
136. R. W. Sidwell and co-workers, *Antiviral Res.* **6,** 343 (1986).
137. F. Bussereau, M. Picard, J. Blancou, and P. Sureau, *Acta Virol.* **32,** 33 (1988).
138. K. Nagahara and co-workers, *J. Med. Chem.* **33,** 407 (1990).
139. D. F. Smee and co-workers, *Antimicrob. Agents Chemother.* **33,** 1487 (1989).
140. A. T. Haase, *J. Infect. Dis.* **153,** 441 (1986).
141. B. A. Navia, B. D. Jordan, and R. W. Price, *Ann. Neurol.* **17,** 517 (1986).
142. F. Barr'e-Sinoussi and co-workers, *Science* **220,** 868 (1983).
143. R. C. Gallo and co-workers, *Science* **224,** 500 (1984).
144. J. A. Levy and co-workers, *Science* **225,** 840 (1984).
145. S. Broder, H. Mitsuya, R. Yarchoan, and G. N. Pavlakis, *Annals Int. Med.* **113,** 604 (1990).

146. E. De Clercq, ed., in *Design of Anti-Aids Drugs,* Elsevier, New York, 1990, pp. 1–24.
147. H. Mitsuya, T. Shirasaka, and S. Broder, in Ref. 146, pp. 25–61.
148. P. Pauwels and co-workers, *Biochem. Pharmacol.* **37,** 1317 (1988).
149. P. Herdewijn and E. De Clercq, in Ref. 146, pp. 141–174.
150. M. Nasr, C. Litterst, and J. McGowan, *Antiviral Res.* **14,** 125 (1990).
151. J. Balzarini and E. De Clercq, in Ref. 146, pp. 175–194.
152. M. S. Hirsch and J. C. Kaplan, *Antimicrob. Agents Chemother.* **31,** 839 (1987).
153. H. D. Langtry and D. M. Campoli-Richards, *Drugs* **37,** 408 (1989).
154. S. Broder, *Med. Res. Reviews* **10,** 419 (1990).
155. D. D. Richman and co-workers, *N. Engl. J. Med.* **317,** 192 (1987).
156. J. P. Horwitz, J. Chua, and M. Noel, *J. Org. Chem.* **29,** 2076 (1964).
157. H. Mitsuya and co-workers, *Proc. Natl. Acad. Sci. USA* **82,** 7096 (1985).
158. B. A. Larder, G. Darby, and D. D. Richman, *Science* **243,** 1731 (1989).
159. R. Yarchoan and S. Broder, *N. Engl. J. Med.* **316,** 557 (1987).
160. M. J. Robins and R. K. Robins, *J. Am. Chem. Soc.* **86,** 3585 (1964).
161. H. Mitsuya and S. Broder, *Proc. Natl. Acad. Sci. USA* **83,** 1911 (1986).
162. J. P. Horwitz, J. Chua, M. Noel, and J. T. Donatti, *J. Org. Chem.* **32,** 817 (1967).
163. M. C. Starnes and J. C. Cheng, *J. Biol. Chem.* **262,** 988 (1987).
164. D. A. Cooney and co-workers, *Biochem. Pharmacol.* **35,** 2065 (1986).
165. T-C. Meng, M. A. Fischl, and D. D. Richman, *Am. J. Med.* **88**(5B), 27S (1990).
166. R. Yarchoan and co-workers, *Lancet* **336,** 526 (1990).
167. T-S. Lin and W. R. Mancini, *J. Med. Chem.* **26,** 544 (1983).
168. C. K. Chu and co-workers, *J. Med. Chem.* **32,** 612 (1989).
169. B. F. H. Eriksson, C. K. Chu, and R. F. Schinazi, *Antimicrob. Agents Chemother.* **32,** 1729 (1989).
170. C. K. Chu and co-workers, *J. Med. Chem.* **33,** 2188 (1990).
171. Z. Hao and co-workers, *Mol. Pharmacol.* **34,** 432 (1988).
172. M. H. St. Clair and co-workers, *Antimicrob. Agents Chemother.* **31,** 1972 (1987).
173. R. Pauwels and co-workers, *Antimicrob. Agents Chemother.* **32,** 1025 (1988).
174. A. Holy and I. Rosenberg, *Collect. Czech. Chem. Commun.* **52,** 2801 (1987).
175. E. De Clercq, in V. Claassen, ed., *Trends in Drug Research,* Elsevier Science Publishers, Amsterdam, The Netherlands, 1990, p. 133.
176. J. Balzarini and co-workers, *Proc. Natl. Acad. Sci. USA* **86,** 332 (1989).
177. R. Vince and co-workers, *Biochem. Biophys. Res. Commun.* **156,** 1046 (1988).
178. R. Vince and M. Hua, *J. Med. Chem.* **33,** 17 (1990).
179. L. L. Bondoc, Jr. and co-workers, *Biochemistry* **29,** 9839 (1990).
180. W. M. Shannon, *Ann. NY Acad. Sci.* **284,** 472 (1977).
181. R. W. Sidwell and D. F. Smee, *Antiviral Res.* **1,** 47 (1981).
182. J. B. McCormick, S. W. Mitchell, J. P. Getchell, and D. R. Hicks, *Lancet,* 1367 (Dec. 15, 1984).
183. R. B. Roberts and co-workers, *Clin. Res.* **34,** 678 (1986).
184. R. B. Roberts and co-workers, *J. AIDS* **3,** 884 (1990).
185. C. Crumpacker and co-workers, *Int. Conf. on AIDS,* Paris, France, June 1986, Abstr. No. 84, s14b.
186. C. Crumpacker and co-workers, *Ann. Int. Med.* **107,** 664 (1987).

GANAPATHI R. REVANKAR
Triplex Pharmaceutical Corporation

ROLAND K. ROBINS
ICN Nucleic Acid Research Institute

AUTOMATED INSTRUMENTATION

CLINICAL CHEMISTRY

Clinical chemistry, initiated in the 1940s, involves the biochemical testing of body fluids to provide objective information on which to base clinical diagnosis. Between 1946 and 1975 clinical testing had a period of rapid growth in the United States, when the volume of tests increased by about 15% a year. The ever-increasing demand for high quality, routine clinical testing stimulated the development of automated techniques, and as early as the 1960s automation in the clinical laboratory was the rule rather than the exception.

Automation made the tests easier and less costly to perform, resulting in still greater demand. The growth rate, about 10% per year throughout the 1980s and into the 1990s, resulted not from a proliferation of laboratories as much as from increased productivity. Whereas the volume of tests more than quadrupled, the total number of United States hospital and private laboratories was reduced by half from 1975 to 1985 (1). Although growth was initially driven by automation, in the 1990s the growth in U.S. laboratory testing may be attributed to several factors. Among them are the discovery of new diseases and the introduction of new therapies as well as better understanding of body chemistry, and the aging of the U.S. population, which increases the risk of contracting age-related illnesses requiring clinical chemistry analyses (2).

Assay Automation

Clinical chemistry analyzers are automated instruments used for measuring concentrations of the various chemical constituents of blood or other body fluids. For a discussion of the related category of instruments used for the measurement of blood cell parameters, see AUTOMATED INSTRUMENTS, HEMATOLOGY.

Before the advent of automation, the steps a clinical technologist had to follow in performing a single clinical chemistry assay included: sample preparation, identification, centrifugation, and filtering; reagent preparation; manual metering and addition of sample and reagents into a reaction vessel; mixing and incubation in the reaction vessel; optical measurement of the mixture in a separate cuvette; and calculation and recording of the results. With the exception of the sample and reagent preparation procedure, which is sometimes done separately, these steps are all performed by an automated analyzer such as that shown in Figure 1.

Tubes or cups containing the liquid samples to be analyzed are placed by the operator into an area of the analyzer known as the sampler. The sampler, which in this case is a circular tray, moves each specimen to a position where it can be accessed by the sample aspirating probes. A sampling probe aspirates an amount of sample and dispenses it into a reaction cuvette, which is most often made of clear glass or plastic. Precise metering and transfer of the sample is controlled by the sample aspirating–dispensing system. The chemical reagents necessary for performing an assay on that sample are aspirated from reagent containers and dispensed into the same cuvette by the reagent handling system. Immediately after dispensing the reagents into the cuvette, sample and reagents are thoroughly mixed, eg, by a motor-driven mixing paddle, in order to obtain a uniform reaction rate within each cuvette. The reaction cuvettes, most often arranged at the periphery of a circular reaction tray, are held for several minutes at a constant temperature of 37°C or 30°C in the incubator, or reaction chamber. During this time chemical reactions in the cuvettes are monitored by an optical detection system. At the end of the incubation period each cuvette is evacuated, washed, and dried at a washing station, to be ready for the next assay. The optical measurement results are converted into metabolite concentration units by the analytical processor computer. Collating the results of different assays into patient reports and storing the test results are usually done by a separate computer, known as results processor or data manager.

In some systems, known as continuous-flow analyzers, the reaction develops as the sample–reagent mixture flows through a conduit held at constant temperature. In such systems, the reaction cuvettes are replaced by optical reading stations called flow cells. In most analyzers, whether of discrete- or continuous-flow type, determination of electrolyte tests, eg, sodium and potassium levels, is

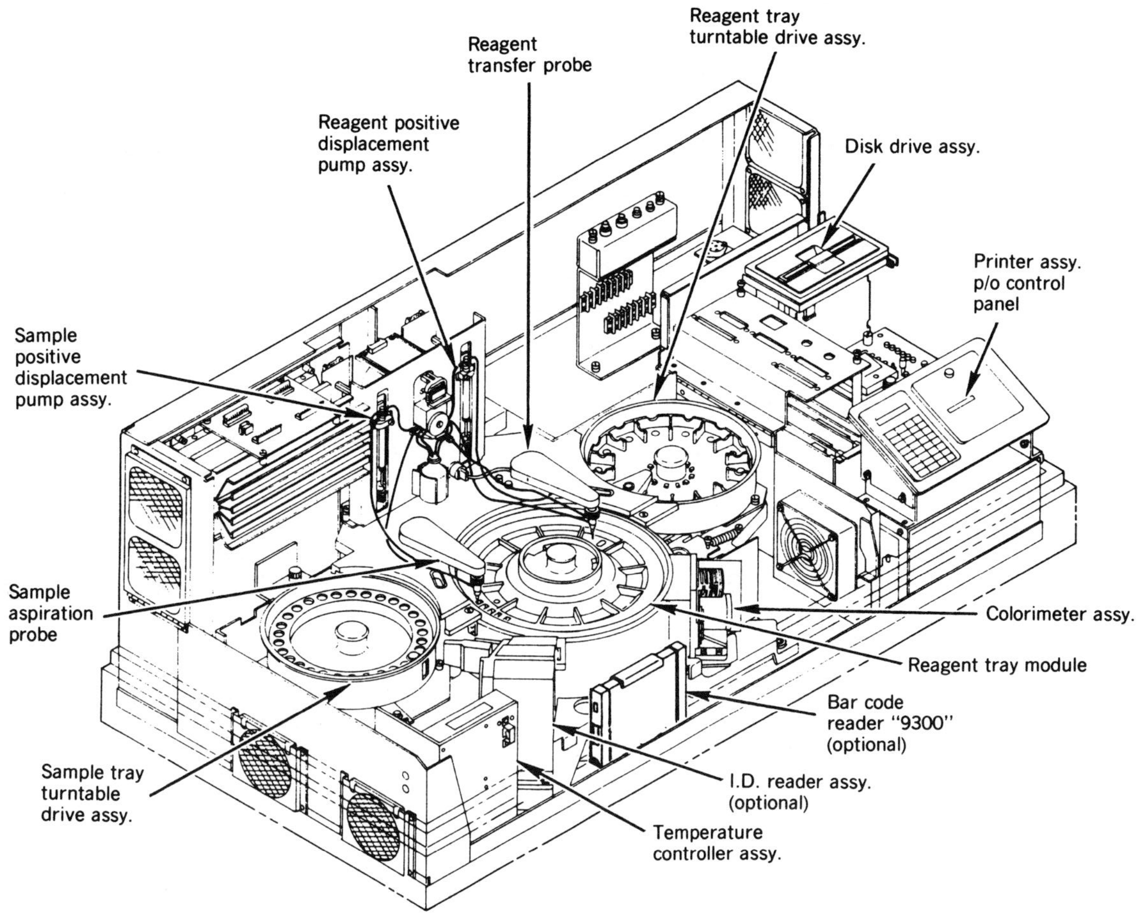

Fig. 1. Internal view of a clinical chemistry analyzer (Technicon RA-XT). Courtesy of Miles Inc.

done by a separate unit using the technique of ion-selective electrodes (ISE) rather than optical detection.

Automated methods are more reliable and much more precise than the average manual method: dependence on the technique of the individual technologist is eliminated. The relative precision, or repeatability, measured by the consistency of the results of repeated analyses performed on the same sample, ranges between 1% and 5% on automated analyzers. The accuracy of an assay, defined as the closeness of the result or of the mean of replicate measurements to the true or expected value (4), is also of importance in clinical medicine.

Technology

An automated system for clinical analysis consists of the instrument (hardware), the reagents, and the experimental conditions (time, temperature, etc) required for each determination. The reagents plus the experimental conditions are sometimes referred to as the chemistry of the system. The chemistry employed is generally similar to that used in manual assays because most automated assay methods have been adapted from the manual ones. However, automated analyzers rarely afford the flexibility of experimental procedure that is possible in manual analysis.

Table 1 lists several of the chemical determinations and the corresponding reactions utilized, which are available on automated clinical analyzers. With the exception of assays for various electrolytes, eg, Na^+, K^+, Cl^-, and CO_2, determination is normally done by photometric means at wavelengths in the ultraviolet and visible regions. Other means of assay include fluorescence, radioisotopic assay, electrochemistry, etc. However, such detection methods are normally required only for the more difficult assays, particularly those of serum or urine constituents at concentrations below $\mu g/L$. These latter assays are discussed more fully in the literature (3,4).

Essential features of an automated method are the specificity, ie, the assay should be free from interference by other serum or urine constituents, and the sensitivity, ie, the detector response for typical sample concentration of the species measured should be large enough compared to the noise level to ensure assay precision. Also important are the speed, ie, the reaction should occur within a convenient time interval (for fast analysis rates), and adequate range, the result for most samples should fall within the allowable range of the assay.

The assay methods listed in Table 1 for the various biochemical species can be classified according to reaction rate behavior, eg, end point vs kinetic methods, blanking schemes, or reaction principle and type of reagents employed.

End Point vs Kinetic Methods. Samples may be assayed for enzymes, ie, biocatalysts, and for other substances, all of which are referred to as substrates. The assay reactions for substrates and enzymes differ in that substrates themselves are converted into some detectable product, whereas enzymes are detected indirectly through their conversion of a starting reagent A into a product B. The corresponding reaction curves, or plots of detector response vs time, differ for these two reaction systems, as shown in Figure 2. Figure 2**a** illustrates a typical

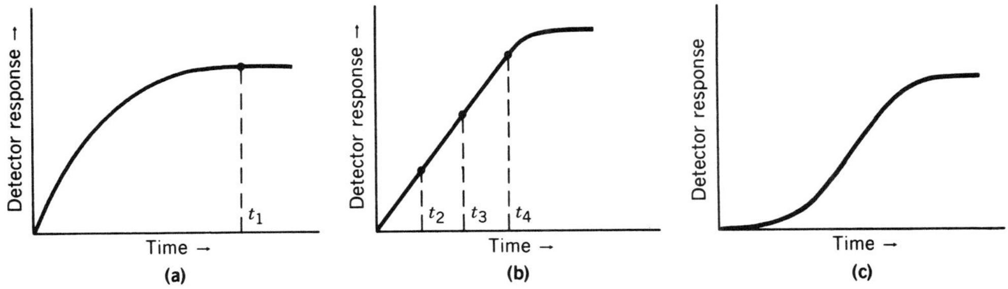

Fig. 2. Assay reaction curves for (**a**) substrates, (**b**) enzymes, and (**c**) enzymes exhibiting a lag phase or reduced reaction rate; where t_x is measurement time.

substrate reaction curve; Figure 2**b** shows a typical enzyme reaction curve (see ENZYME APPLICATIONS).

In substrate assays, the reaction product is measured at time t_1 (Fig. 2**a**) when the reaction is complete. The concentration of the species to be assayed is usually proportional to the net detector response at t_1, and generally there is no problem in relating this measurement to substrate concentration. These so-called end point assays correspond to a constant value and are less subject to errors occurring as a result of variations in experimental conditions (temperature, reagent concentrations, etc) than measurements made under rapidly changing conditions. In some cases, t_1 can be as long as 5–10 min, which would severely limit maximum analysis rates if such assays were performed individually. Automated analyzers address this problem in various ways. For example, most analyzers allow a series of samples to pass through the system one after the other. In this way each sample may spend up to 5–10 min within the system, but if n samples are in the system at any given time, the net analysis rate, expressed in samples per minute, is n divided by 5–10. Alternatively, all samples might be loaded into the system for parallel processing in the same 5–10 min. Continuous-flow analyzers are an example of the first scheme, whereas cuvette-based analyzers may work in the latter manner. A more important constraint in end point assays by automated analyzers is the need in some analytical reactions for multiple addition of different reagents at different times. Cuvette-based analyzers, for example, do not normally allow for more than two reagent additions.

For enzyme determinations, kinetic assay procedures are generally favored. These refer to the determination of the slope of the reaction curve. However, not all enzyme reaction curves are as simple as that shown in Figure 2**b**. Some reactions show a so-called lag phase, or reduced reaction rate, at the beginning of the reaction, as shown in Figure 2**c**. Other reactions level off quickly, much before the average time t_4, shown in Figure 2**b**. As a result, many clinical chemists prefer enzyme assays that are based on monitoring the reaction curve over a wide range in time so that the linear zero-order region can be readily identified and the slope calculated in this region. However, this method of monitoring represents a serious constraint on many analyzers; ie, if there is one detector, then the maximum analysis rate is $1/t_4$. Because t_4 is commonly 5–10 min, this would limit

Table 1. Clinical Tests Offered on Automated Analyzers[a]

Assay	Reaction
	Chemical
albumin	albumin + bromcresol green (BCG) $\longrightarrow$ albumin BCG[b]
calcium	calcium + cresolphthalein complexone $\longrightarrow$ complex[b]
chloride	$2Cl^- + Hg(SCN)_2 \longrightarrow HgCl_2 + 2\ SCN^-$ $3\ SCN^- + Fe^{3+} \longrightarrow Fe(SCN)_3$[b]
creatinine	creatinine + sodium picrate $\longrightarrow$ colored product[b]
iron	iron + Ferene-S $\longrightarrow$ colored complex
total bilirubin	bilirubin + sulfanilic acid + NO_2^- $\longrightarrow$ azobilirubin[b]
total protein	protein + $CuSO_4 \xrightarrow{OH^-}$ cuproproteinate complex
	Biochemical
alanine aminotransferase (ALT)	1-alanine + oxoglutarate $\xrightarrow{ALT}$ pyruvate + 1-glutamate pyruvate + NADH $\xrightarrow[\text{dehydrogenase}]{\text{lactate}}$ lactate + NAD
alkaline phosphatase	*p*-nitrophenyl monophosphate $\xrightarrow[\text{phosphatase}]{\text{alkaline}}$ *p*-nitrophenol[b] + PO_4^{3-}
aspartate transaminase (AST)	glutamate + *a*-ketoglutarate $\xrightarrow{\text{aspartate transaminase}}$ oxaloacetate oxaloacetate + NADH $\xrightarrow[\text{dehydrogenase}]{\text{malate}}$ malate + NAD
carbon dioxide	phosphoenolpyruvate + HCO_3^- $\xrightarrow[\text{phosphoenolpyruvate carboxylase}]{}$ oxalacetate + PO_3^{3-} oxalacetate + NADH $\xrightarrow[\text{dehydrogenase}]{\text{malate}}$ malate + NAD
cholesterol	cholesterol esters $\xrightarrow[\text{esterase}]{\text{cholesterol}}$ free cholesterol + fatty acids

	free cholesterol + $O_2 \xrightarrow[\text{oxidase}]{\text{cholesterol}}$ cholest-4-en-3-one + H_2O_2
	H_2O_2 + phenol + 4-aminoantipyrine $\xrightarrow{\text{peroxidase}}$ quinoneimine dye
creatine kinase (CK)	creatine phosphate + ADP $\xrightarrow[\text{kinase}]{\text{creatine}}$ creatine + ATP
	ATP + glucose $\xrightarrow{\text{hexokinase}}$ ADP + glucose-6-phosphate
	NAD + glucose-6-phosphate $\xrightarrow[\text{dehydrogenase}]{\text{glucose-6-phosphate}}$ NADH + 6-phosphogluconate
gamma glutamyl transferase (GGT)	glutamyl-*p*-nitroanalide + glycylglycine $\xrightarrow{\text{GGT}}$ glutamylglycylglycine + *p*-nitroaniline
glucose	ATP + glucose $\xrightarrow{\text{hexokinase}}$ ADP + glucose-6-phosphate
	glucose-6-phosphate + NAD $\xrightarrow[\text{dehydrogenase}]{\text{glucose-6-phosphate}}$ 6-phosphogluconate + NADH
lactate dehydrogenase (LD)	lactate + NAD $\xrightarrow[\text{dehydrogenase}]{\text{lactate}}$ pyruvate + NADH[b]
triglycerides	triglycerides $\xrightarrow{\text{lipase}}$ glycerol + fatty acids
	glycerol + ATP $\xrightarrow[\text{kinase}]{\text{glycerol}}$ glycerol-1-phosphate + ADP
	ADP + phosphoenolpyruvate $\xrightarrow[\text{kinase}]{\text{pyruvate}}$ ATP + pyruvate
	pyruvate + NADH[b] $\xrightarrow[\text{dehydrogenase}]{\text{lactate}}$ lactate + NAD
urea nitrogen	urea + $H_2O \xrightarrow{\text{urease}}$ 2 NH_3 + CO_2
	NH_3 + NADH + oxoglutarate $\xrightarrow[\text{dehydrogenase}]{\text{glutamate}}$ NAD + glutamate

[a]Coenzymes such as adenosine diphosphate (ADP), adenosine 5′-triphosphate (ATP), nicotinamide adenine dinucleotide (NAD), and nicotinamide adenine dinucleotide, reduced (NADH), are involved in some reactions (4).
[b]Chromophore, optical detection used.

maximum analysis rates to only 6–12 samples per hour, which would be much too slow for large clinical laboratories.

Early instruments ignored the assay time problem, made a single measurement at some time, as in end point analysis, and calculated an assumed slope dividing the net detector response by the time after initiation of the reaction. For many assays, this was satisfactory. However, frequent interferences were encountered with these single-point procedures, as use of so-called 340 nm chemistries increased for various enzymes. The 340 nm chemistry is based on the production of the coenzyme reduced nicotinamide adenine dinucleotide (NADH), which absorbs light at a wavelength of 340 nm. One solution is the use of parallel blank reactions, where one of the reaction reagents is missing. Subtraction of the resulting blank reaction slope from the total reaction slope gives the desired enzyme reaction curve. However, this approach greatly increases the cost of both equipment and reagents. More importantly, it does not allow for detection of reagent depletion, where high enzyme concentrations cause a bending of the reaction curve in a very short time, as shown in Figure 2**b** after t_4. Newer cuvette-based analyzers, such as that shown in Figure 1, use multiple detector readings or multiple detectors. In these analyzers, a batch of samples for the same assay is rotated through a common detector station so that multiple points along the reaction curve are collected for each sample. This multiplexing of the detector results in an overall analysis rate of n/t_4, where n is the number of samples in the batch. Thus these analyzers achieve high throughput rates, including a fairly complete definition of the total reaction curve, plus reliable determination of the zero-order slopes for all samples.

Blanking Schemes. Whereas most clinical assays rely on the specificity of the assay reaction to avoid interferences that would require a blank correction, certain ubiquitous interferences, eg, lipemic turbidity, hemoglobin in hemolyzed sera, bilirubin [*635-65-4*], $C_{33}H_{36}N_4O_6$, in icteric specimens, present special problems in many otherwise specific assays. Sample interferences can be classified as follows: (*1*) fixed absorbance in the absence of reagents, eg, turbidity; (*2*) fixed absorbances as a result of reaction with reagents, eg, free-glycerol interference in assays for triglycerides; and (*3*) changing absorbance as the result of slow reaction with reagents.

Each interference type may require a different elimination strategy, the simplest of which is to remove the potential interference before beginning the assay. Removal is commonly done in manual analysis, usually by deproteinization, eg, by addition of some denaturing agent such as trichloroacetic acid. This technique also tends to remove other large molecules such as bilirubin and hemoglobin, as well as eliminating sample turbidity. However, such procedures are not easily adapted to automated analysis, and in fact are generally not used. Modern assay methods generally employ mild reaction conditions which not only do not cause sample protein precipitation; these methods, because they frequently use highly specific enzymes as reagents, also have significantly reduced interference.

In cases where sample blank correction is required, a variety of techniques are employed. For interferences of type 1, a common technique is to use a separate determination in which an essential reagent for the primary reaction is eliminated; eg, the sodium nitrite, $NaNO_2$, in the usual reaction for bilirubin. The blank, or interference, response in the detector mimics that in the assay reaction, but the assay reaction itself does not occur. Subtraction of the blank from the assay result on a given sample then provides a correction for interferences of type 1. For interferences of type 2, a blank is run using the reagent that reacts with the interfering species. A modification of this technique uses two reagents. The sample is added to the first reagent, and the blank is measured. No assay reaction occurs. The second reagent, added later, contains the active component which produces the measurable assay response. A subtraction of the two measurements cancels out the unwanted contribution of the blank reaction to the desired response.

Separate sample blanking requires an additional analytical channel, and is therefore wasteful of both reagents and hardware. An alternative approach that is used on several automated systems, eg, Du Pont ACA, BM-Hitachi 704, Technicon RA-1000, is that of bichromatic analysis (5) where absorbance measurements are taken at two, rather than one, wavelength. When the spectral curves for the interference material and the chromogen of the species measured differ sufficiently, this can be an effective technique for reducing blank contributions to assay error. Bichromatic analysis is effective for blanks of both the first and second type.

Blanks of the third type are the most difficult to handle. Examples are found in the usual Jaffe reaction for creatinine, and in certain enzyme reactions. In the creatinine [*57-00-1*], $C_4H_9N_3O_2$, assay, if potential interferences are not removed by dialysis or some similar procedure, they exhibit both slower and faster reactions compared to those of creatinine. A two-point analysis can be utilized here with the measurement times selected to be later than the time required for fast-reacting interferences and before the time for significant reaction of slow-reacting interferences. This procedure is utilized in the creatinine assay on many automated chemistry analyzers.

Reactions and Reagents. The three general categories of reactions are chemical, biochemical, and immunological. Automated analyzers mainly utilize the assays listed in Table 1, which fall into one of the first two categories; trace-level assays are carried out by immunological methods (see IMMUNOASSAY). Chemical methods are based on reactions that do not normally take place in living systems, eg, the assay for calcium by the cresolphthalein complexone reaction. These assays are often relatively nonspecific, and may require removal of sample interferences or the use of blanking techniques. Chemical methods can be used by most types of automated clinical analyzers. Biochemical methods rely on reactions that play a role in the functioning of living systems; eg, the assay of glucose by the coupled enzyme system hexokinase plus glucose-6-phosphate dehydrogenase. Biochemical assays are often highly specific because the corresponding biochemical reactions in the body are specific. Automatic analyzers utilizing biochemical methods predominate in the marketplace because of greater specificity and reduction in cost through increased use.

Immunological methods suffer from many disadvantages, and automation has been attained only in the 1980s and 1990s. Immunological reactions utilize antibodies as reagents, and are often competitive binding procedures. Radioimmunoassay was the first technique employed for immunological methods. The use of nephelometric, turbidimetric, and fluorometric means of detection have gained ground; however, use of radioimmunoassay is decreasing in part because of the problems of radioactive waste disposal.

Measurement Methods

The majority of the various analyte measurements made in automated clinical chemistry analyzers involve optical techniques such as absorbance, reflectance, luminescence, and turbidimetric and nephelometric detection means. Some of these are illustrated in Figure 3. The measurement of electrolytes such as sodium and potassium have generally been accomplished by flame photometry or ion-selective electrode sensors (qv). However, the development of chromogenic ionophores permits these measurements to be done by absorbance photometry also.

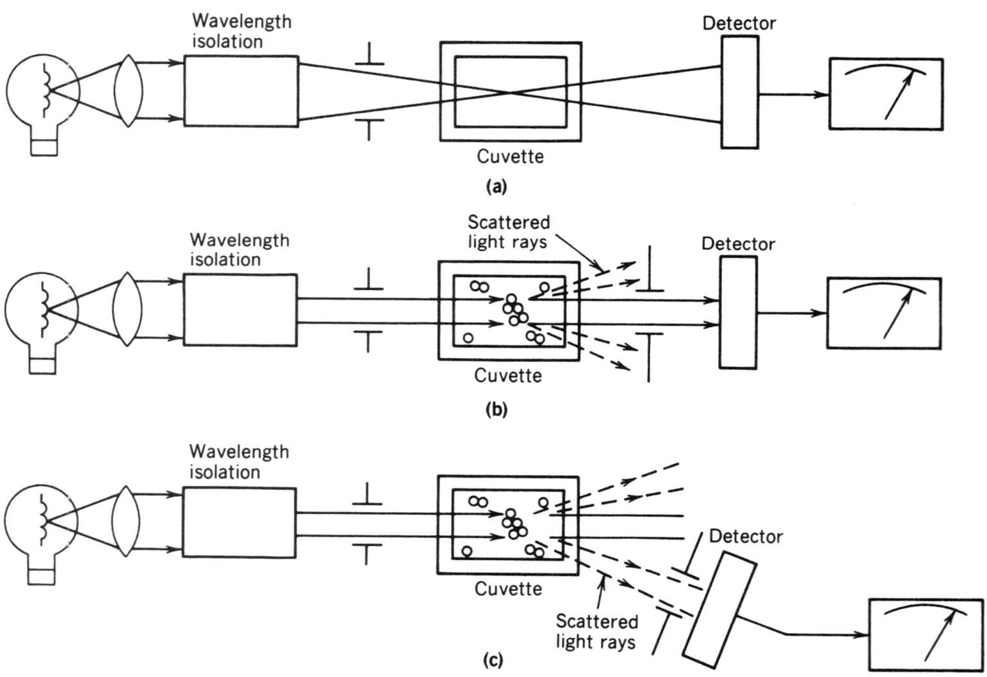

Fig. 3. Schematic of optical detection methods, (**a**) absorbance or colorimetric; (**b**) turbidimetric; and (**c**) nephelometric.

Absorbance. Analyte measurements in clinical analyzers using liquid reagents are most commonly performed by transmission of light, ie, by absorbance photometry or colorimetry (Fig. 3**a**). The liquid to be analyzed is either held in a cuvette or passed through a flowcell having transparent walls. A light beam traverses the liquid compartment and then falls on an optical detector (see PHOTO DETECTORS). Prior to striking the detector, a narrow wavelength band of radiant power centered at the wavelength of peak absorbance, the analytical wavelength, is isolated. Wavelength isolation, shown in Figure 3 as occurring before the light impinges upon the sample, may take place after the sample. In dilute solutions, analyte concentrations are proportional to the measured absorbance at the analytical wavelength in accordance with the Beer-Lambert law: $T = P/P_o = 10^{-\epsilon cl}$ where T is the relative transmittance, P_o is the beam power for a nul concentration sample, and P is the power for the sample being measured, c is concentration in moles per liter (M), l is the path length in centimeters of the absorber, and ϵ is the molar extinction coefficient having units $M^{-1}\text{cm}^{-1}$. The absorbance A is defined as $-\log_{10} T$, thus $A = \epsilon cl$ where l the path length, representes the length of the light path through the liquid chromophore in the flowcell or cuvette.

Reflectance. In dry chemistry analyzers, the diffuse reflectance of the chromophore is measured in the supporting solid matrix. The output response is approximated by the Kubelka-Munk function in which the diffuse reflectance of the sample depends only on the ratio of the absorption and scattering coefficients and not on their absolute values. The scattering coefficient is generally considered to be constant for a given matrix. Typically the liquid sample is deposited on one side of the reagent-impregnated strip or slide, and the optical measurement is made on the opposite side. In making diffuse reflectance measurements, it is important to minimize the specular component of reflection from erroneously affecting the reading.

Turbidimetry and Nephelometry. In contrast to classical absorbance methods, immunoassay reactions frequently involve agglutination in which the optical scatter signal of the agglutinated particles is measured by turbidimetric or nephelometric means. The principles of light scattering as it relates to analytical methods is discussed in reference 6.

Turbidimetry is considered a transmission type of measurement in which the detector is optically in line with the light source (Fig. 3**b**). Particles in the sample scatter the incoming beam away from the in-line detector, and consequently cause a loss in light power falling on the detector. This decrease in light intensity is measured and converted to optical density units relative to a solution without particles. In the absence of any absorbing species in the sample, the turbidity is proportional to the concentration of the particles, the scattering cross section, and the optical pathlength of the cuvette. The scattering cross section is a function of the relative particle size to wavelength ratio and to the refractive indexes of the particle and surrounding medium.

In nephelometry the detector is located at an angle to the light path and to the cuvette centerline. Essentially, only scattered light is collected by the detector (Fig. 3**c**). Although the detector signal level is considerably lower than in turbidimetric measurements, the signal is measured against an essentially zero background rather than a high signal level as in the turbidimetric technique. This results in an increase in the ratio of the maximum to minimum signals and generally is advantageous, providing the signal level is adequate, eg, where the electronic or shot noise is not significant, for the precision required.

In both turbidimetric and nephelometric measurements used for agglutination reactions, the relationship of the concentration of the analyte to the measured optical density or scatter signal is a complex one. Various optical and physical factors, such as the rate of formation of two- or three-particle agglutinates generated from the monomers, ie, unagglutinated particles, their orientation, and degree of multiple scattering, affect the signal to concentration relationship. Obviously the chemistry of the immunologic reaction is paramount to determining the kinetics and sensitivity of the reaction.

Fluorescence. The fluorescence detection technique is often used in clinical chemistry analyzers for analyte concentrations that are too low for the simpler absorbance method to be applied. Fluorescence measurements can be categorized into steady-state and dynamic techniques. Included in the former are the conventional simultaneous excitation–emission method and fluorescence polarization.

In time-resolved fluorescence, rare earths are frequently used as fluorescent labels. The fluorophores have large Stokes shifts, ie, shifts of the emitted light to a higher wavelength relative to the absorption wavelength, and comparatively long decay times, approximately 0.5 ms. This simplifies the optical detection and minimizes serum background effects, as the background fluorescence has essentially decayed to zero when the longer lifetime signal is being read out. The more research-oriented instruments, based on fluorescence lifetime measurements, utilize the more sophisticated impulse–time decay and the frequency and phase domain techniques.

The vast majority of detectors, used in absorbance and fluorescence measurements, operate in the analogue mode, in which the average of a large number of time-overlapped electron pulses constitute the measured current. In this technique the amplified electron pulses, generated initially by the photons striking the photocathode, are counted individually as one single count per photon, irrespective of their amplitudes. For very low radiant power measurements, such as may be found in some fluorescence applications, a photon counting method may be used advantageously.

Ion Selective Electrodes Technique. Ion selective (ISE) methods, based on a direct potentiometric technique (7) (see ELECTROANALYTICAL TECHNIQUES), are routinely used in clinical chemistry to measure pH, sodium, potassium, carbon dioxide, calcium, lithium, and chloride levels in biological fluids.

In potentiometry, the potential difference in volts is measured between two electrodes, one a reference electrode having a stable potential under controlled conditions, and the other a working or measuring electrode. In the working ISE, an ion-selective membrane is used to separate two solutions of different ionic strength. As the sample containing the analyte, eg, sodium ions, to be measured flows past the ISE, changes in electrical potential take place between the ionically constant inner surface, saturated KCl solution, and the ionically variable outer surface, the sample. These changes in electrical potential are automatically measured against those of the reference electrode. The potential difference (voltage) between the two electrodes is related to the activity of the ions in solutions, and the activity is proportional to the logarithm of the respective analyte concentration of the sample, according to the Nernst (8) equation. The analogue signal generated is digitized and then converted to reportable concentration units by the instrument's computer processing unit (CPU).

Classification of Clinical Chemistry Analyzers

Automated clinical analyzers can be divided into discrete and continuous-flow systems. In discrete systems, the sample–reagent mixtures from different specimens are kept in individual reaction cuvettes during incubation and optical reading. A notable exception in this category are the dry chemistry analyzers, where the cuvettes were replaced by individual slides coated with dry reagent films. In continuous-flow analyzers, the sample–reagent mixtures form liquid segments flowing through a tube. Adjacent segments are separated by air bubbles. Optical readings are taken at different locations along the tube, which is kept at the incubation temperature.

Clinical analyzers can also be classified according to their degree of flexibility. Most of the modern systems are random access analyzers, for which the tests on various specimens are performed in any order programmed by the operator. Some analyzers operate in batch or profile mode, ie, they perform the same test or group of tests on every sample until the system is reset for another test or group of tests.

The throughput range, number of assays performed per hour, of clinical analyzers reflects the diversity of the market. Systems that perform less than 400 tests per hour are usually referred to as small analyzers; medium analyzers cover the range from about 500 tests per hour to 1,500 tests per hour, while high throughput analyzers can process up to 10,000 tests per hour. Table 2 lists a number of representative automated systems and their characteristics. Whereas small analyzers, eg, the Ames CLINISTAT, can easily fit on a desktop, high throughput analyzers, eg, the Technicon DAX-96, can occupy a good portion of a laboratory.

Functional Elements of Clinical Chemistry Analyzers

Sample Handling System. Venous or capillary blood, urine, and cerebrospinal fluid are specimens routinely used in medical diagnostic testing. Of these biological fluids, the use of venous blood is by far the most prevalent. Collection devices such as syringes and partial vacuum test tubes, eg, Vacutainer, are used to draw ten milliliters or less of venous blood. At collection time, the test tubes are carefully labeled for later identification.

The specimen, as drawn, contains cells, platelets, fibrin, and particulates. For most chemistry tests, eg, determination of glucose, cholesterol, etc, it is necessary to first separate the cellular blood fraction which, if present during the assay, would interfere with the determination and adversely affect the accuracy of the measurement. The cellular components are separated by centrifugation. The centrifuge is generally not part of the automated analyzer. If anticoagulants are added to the sample, the resultant supernatant layer is called plasma; if not, the sample must be allowed to stand for approximately ten minutes before centrifugation to allow completion of the clotting process. In the latter case, the supernatant is referred to as serum (see FRACTIONATION, BLOOD).

After centrifugation, the specimens are placed in a holder specifically de-

Table 2. Features of Automated Analyzers

Manufacturer and system	Throughput per hour		Number of resident methods	Amount of sample needed per test, μL	Incubator temperature, °C	Optical system[a]	Distinctive features
	Samples	Tests					
Roche COBAS MIRA	variable	125	30	2–95	37	AMW,P	disposable cuvettes
Technicon RA-1000	variable	240	12	2–30	30, 37	AMW,P	no probe wash
Ames CLINISTAT	up to 80	up to 80		10	37	R	dry chemistry
Abbott SPECTRUM	variable	400–600	23 3	1.25–25	25, 30, 37	AMW,P,T	polychromatic sample blanking
Beckman SYNCHRON CX3	up to 75	600	8	122/8 tests	37	AMW,P	STAT analyzer[b] Beckman liquid calibrators and controls
Kodak EKTACHEM 700	300 max	600	26	10, 11	37	R	dry chemistry slides, routine/STAT[b]
Baxter PARAMAX	240	720	32	2–50	37	AMW,P	closed container sampling option; unit dose dry tablet reagent
BM Hitachi 737	300	1,200	23	3–20	30, 37	AMW,P	
Technicon CHEM 1	variable	1,800	35	1	30, 37	AMW,P	continuous-flow capsule chemistry
BM Hitachi 736-50	300	8,100	27	10	30, 37	AMW,P	
Technicon DAX-96	300	10,200	34	3–67	37	AMW,P	

[a]AMW = absorption, multiple wavelength per test; N = nephelometry; T = turbidimetry; P = potential; and R = reflectance.
[b]A specimen that has to be processed with priority vs other specimens is known as STAT sample.

signed to work with a particular instrument. Some clinical analyzers utilize sample carrier blocks; others use a rotary tray or a continuous flexible chain link. In some cases, an aliquot of the sample has to be poured into a secondary container, small cup, or tube which is then labeled and placed on the sample holder. Direct or primary tube sampling is preferable because it involves less sample handling and less risk of loss of patient identification. A trend in primary tube sampling is the introduction of closed tube sampling, which allows the proper amount of serum or plasma to be aspirated directly from the primary collection tube without having to first remove the stopper. One such technique is described schematically in Figure 4. The obvious advantage of such techniques is the significant reduction of risks associated with sample handling such as operator exposure to hepatitis B and HIV viruses.

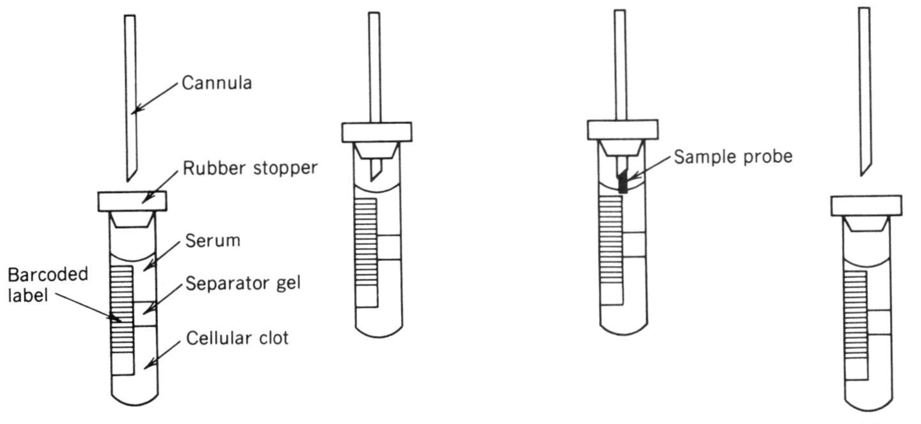

Fig. 4. Schematic of the Closed Container Sampling technique used in the Baxter PARAMAX analyzer showing (**a**) the collection tube with bar-coded label being brought into sampling position under the cannula; (**b**) the tube raised so that the cannula has penetrated the stopper; (**c**) the sample sensing probe coming through the cannula to aspirate the exact volume required for each assay; and (**d**) after sampling, where the tube is lowered away from the cannula. The stopper removes any sample from the cannula and reseals itself.

Once loaded, the sample holder is placed onto the carrier transport mechanism of the system, which is referred to as the sampler. The function of the sampler is to bring each specimen into a position where a sample of it can be taken by the aspiration probe, then to transport it further to the unloading area. Most analyzers allow continuous feeding, or loading, and unloading of samples. Priority sampling, which allows the sampler to temporarily interrupt sampling of routine specimens in order to aspirate from emergency, or STAT, specimens, is also a common feature. The analyzer has to keep track of sample identity. Some systems use optical detection devices to read a bar-coded label on each test tube or sample cup before aspirating from it. This approach is known as positive sample identification. In other systems, the sample holders, eg, carrier blocks, are assigned unique numbers that can be read by an optical device, then the specimens are identified by their position on the carrier block.

When a test result for a particular specimen is found to have an elevated, out of method range value, some analyzers, eg, Beckman CX3, can automatically repeat the sampling from the same specimen. For elevated concentrations, the precision of the optical system is reduced so an automatic dilution of the sample, eg, by aspiration of a reduced amount of sample, is provided during the second sampling.

Sample Aspiration and Dispensing. Clinical chemistry analyzers perform tests mostly on blood serum or plasma, and less often on other body fluids. Precisely metered amounts of the sample have to be aspirated rapidly and without allowing intersample contamination, known as sample carry-over. Typical sample volume aspirated per test ranges between 1 μL and 25 μL. Generally, pipetting is accomplished by motor-driven syringes connected through a fluid line to a thin aspiration probe (Fig. 1).

In most analyzers, sample carry-over is minimized by one or more of the following techniques: sensing the liquid level in the sample container, and limiting the probe immersion in the liquid to a few millimeters; or rinsing the aspiration probe inside and out after each immersion. Probe rinsing has been eliminated in a few analyzers by the use of an aspiration probe employing disposable tips. After aspiration of all the aliquots needed from a particular specimen, the tip is discarded, and a new one is loaded for the next sample. Yet another approach to carry-over prevention is where the aspirating probe is coated inside and out with a thin film of a chemically inert fluorocarbon oil. During immersion this coating prevents the sample from wetting the probe wall while allowing aspiration to take place.

Reagent Handling System. Many analyzers can be programmed to perform a wide variety of assays. However, reagents for only a limited number of tests, usually referred to as resident tests, are available on the instrument at any one time. The main reason for this limitation is that, for infrequently requested tests, the time period until reagent depletion may exceed the chemical stability time limit. Refrigerated compartments for the reagent containers, usually kept between about 4°C to 15°C, are generally provided on the medium to high throughput systems in order to extend the stability of the resident reagents. For most analyzers using liquid reagents, the reagent stability on the system ranges between 2 days to 30 days. In some systems, such as the Baxter Paramax, reagents are stored in tablet form and reconstituted as needed, thus extending on-board reagent stability to 60 days.

Precise metering of the amount of liquid reagent needed for a test is generally done using a motor-driven syringe. Other variants, such as peristaltic pumps or pneumatic syringes, are also employed. The amount of reagent used per test ranges between 0.1 mL and 0.7 mL. In some analyzers, a dedicated conduit links each reagent container to a dispensing nozzle above the cuvette. Other instruments use a single reagent probe to aspirate the reagents from the reagent containers and dispense them into the reaction cuvette. If the latter approach is used, reagent cross-contamination by the probe has to be avoided. This is usually achieved by a thorough rinsing of the probe after each reagent dispense. Reagent-to-reagent carry-over typically has to be kept to less than 1 part per 20,000.

Coating the reagent probe with an inert liquid eliminates the need for the rinsing operation.

Two notable technological developments related to the reagent handling system occurred in the last 1980s. One of these was the introduction of the dry slide reagent technology where the reagents needed for a particular assay are deposited in thin layers on a slide (Fig. 5). The reaction starts when a drop of sample, deposited on the upper slide surface, dissolves the reagents. A top coating layer, covering the dry reagents, helps the uniform spread of the sample drop and filters out interfering substances. The dye formed in the reaction is measured by reflectance spectrophotometry from the lower side of the slide, the light being reflected by the top coating layer. The dry slide technology eliminates the preparation and handling of liquid reagents, as well as the need to meter the amount of reagents for each test. A slide cartridge is provided for each resident test. After opening the cartridge, the reagents are stable for up to seven days.

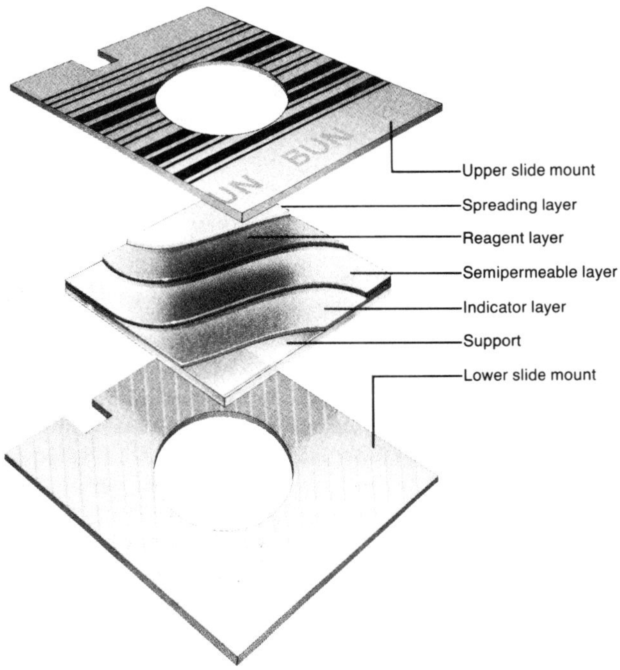

Fig. 5. Schematic of a dry reagent slide used in Kodak EKTACHEM analyzers. Courtesy of Kodak Instruments.

Another unique development, which significantly reduced the amount of reagent necessary for an assay, is an extension of the continuous-flow analyzer technology called capsule chemistry. The core of the system is a capillary Teflon tube coated on the inside with a thin, flowing film of fluorocarbon oil. Sample and liquid reagents, alternating with air bubbles, are aspirated into the tube forming a segmented flow. Each liquid segment surrounded by the film of immiscible fluorocarbon oil, forms a discrete capsule which does not contaminate the neighboring test capsules. Because of the small size of the capillary tube, the amount of liquid reagents per test is reduced to 14 μL, and the amount of sample to 1 μL per test. The small reagent volumes used allow on-board storage of reagent sufficient for up to 1200 tests for any of the resident methods. The reaction takes place while the liquid capsules are flowing through the capillary tube, and the photometric reading is taken through the tube wall, eliminating the need for reaction cuvettes (Fig. 6).

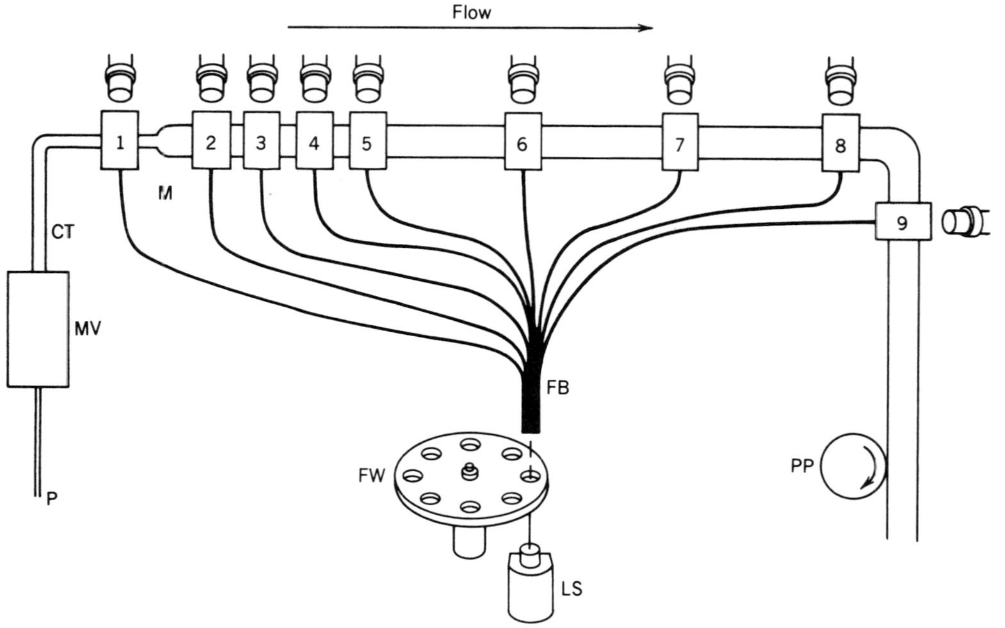

Fig. 6. Schematic of Technicon CHEM I Analyzer where P = aspiration; MV = metering valves; CT = capillary Teflon tube; M = reagent mixing zone; FW = filter wheel; FB = fiber optic bundle; LS = light source; PP = peristaltic pump. The colorimetric reading stations are marked 1 through 9.

Optical Detection System. The optical detection system includes dedicated or time-multiplexed arrangements of the various elements such as the light source, cuvette or flowcell, wavelength isolation elements, and detector. A common layout may have a single light source with light distribution via fiber optics to many cuvettes or flowcell read stations. On the output side of the reaction cuvette or flowcell, fiber optics pipe the light signals to an optical chopper, through dedicated interference filters, and to a common multiplexed photomultiplier. A variation of this arrangement uses a rotating mirror device to multiplex the light output of many cuvettes into the input of a single grating–array photodiode.

Light Sources. The majority of clinical chemistry methodologies result in analyte reactant products which are analyzed using either ultraviolet or visible light. Light sources include the tungsten halogen lamp, lasers, light-emitting diodes, and arc lamps such as deuterium, xenon, and xenon–mercury of the continuous or the flash lamp type. The tungsten lamp generally has greater short- and long-term stability than the arc lamps, and consequently is the preferred source for wavelengths above 340 nm. Lasers (qv) and light-emitting diodes, which operate at a single wavelength, with the exception of tunable dye lasers, are special-purpose sources used in single wavelength applications (see LIGHT GENERATION).

Reaction Cuvette/Flowcell. This critical element is often unique to the system, and differs to the greatest extent among various instruments. Wet chemical analyzers typically use a cuvette or flowcell to contain the reactant to be measured. The volume of this fluid may range from as little as 15 μL in flowcell analyzers to several hundred microliters in cuvette-based instruments. In absorbance measuring instruments, the light path lengths generally range from 1 to 10 mm. Because the path length of the cell affects the measured absorbance directly, the choice of this design parameter is carefully considered with regard to the system volume constraints.

Wavelength Isolation. There are several means of isolating the analytical wavelength needed for the measurement of the analyte. One of the most common methods is the use of a multilayer interference filter illustrated in Figure 7. When coupled with suitable blocking filters, this optical element has an extremely good rejection ratio of the out-of-band to in-band power, minimizing the stray light effects which can contribute to nonlinearity. The half-bandwidth of the filter, spectral bandwidth at half the peak transmission, is the fundamental filter specification, and this value should be small relative to the chromophore bandwidth. Bandwidths range from 2 to 20 nm. Where a large number of wavelengths are required for the system, a multiwavelength dispersive device such as a prism or grating is more suitable.

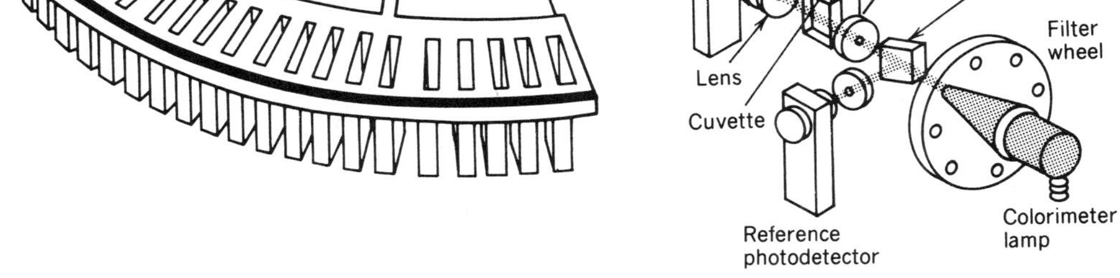

Fig. 7. Schematic of a reaction tray and colorimetric system.

Prisms and gratings have historically been used for spectral band isolation in spectrophotometers. These optical elements provide a wide range of wavelengths and have very narrow bandwidth characteristics when coupled with suitable entrance and exit slits. The mechanically scanned prisms or gratings provide sequential wavelength information, whereas the more recent grating/multielement array detector combinations produce almost simultaneous (in the millisecond range) information. Acousto-optical tunable filters are newer devices which are used to isolate narrow spectral bands. They can operate at scanning speeds at least 100 times faster than the mechanically scanned devices.

Detectors. The solid-state silicon photodiode has become ubiquitous in clinical instruments and is the detector of choice for the near ultraviolet–visible spectral range. Among the detector's attributes are small size, high quantum efficiency (especially in the visible–near infrared region), low noise equivalent power, wide linear range, excellent response times, and simplicity of interfacing to the data acquisition hardware. Silicon photodiodes are available in single-element packages and in multielement arrays.

Another class of detectors consists of the vacuum phototube devices. Included in this group is the photomultiplier detector, which has high gain but requires a high voltage power supply. It has an advantage over the silicon detector in that it has extremely low noise, and thus is useful in applications requiring extremely low level light detection and/or extremely fast response times. Additionally, the ultraviolet quantum efficiency of multialkali photocathodes is generally superior to that of the silicon photodiode.

The photoconductive detector is primarily used in the visible–infrared region rather than the ultraviolet–visible range.

Silicon charge coupled devices (CCDs), commonly used in solid-state video cameras and in research applications, are being applied to low light level spectroscopy applications. The main advantage of area array CCDs over linear photodiode detectors is the two-dimensional format, which provides simultaneous measurements of spatial and spectral data.

Computer System. The brain of the modern clinical chemistry analyzer is its computer system. The part of the computer system that controls the functional aspects of the analyzer is known as the process control computer or analytical processor (AP); the test results are handled by the data management computer, also known as the results processor (RP).

The analytical processor stores the sample identification information, and, checking against the worklist input by the operator, determines which tests should be performed on the current sample. The AP controls the aspiration of the right amount and type of reagents, dispenses the reagents together with the sample in the reaction cuvette, and collects the raw data from the detection system. Some tasks, eg, temperature control and motion profiles, are actually delegated to dedicated microprocessors reporting to the AP. These microprocessors are often referred to as embedded control processors. The analytical processor also monitors the status of various sensors in the analyzer and warns the operator of abnormal conditions by displaying messages on the computer monitor, or by triggering audible alarms. Other functions controlled by the AP include start-up and shut-down procedures, calibration procedures, and, in some analyzers, self-diagnosis of failure modes.

The results processor computes the test results from the raw data furnished by the AP and collates these results together with the demographic patient data into test reports. Test results falling outside normal limits are flagged on the report to speed up the diagnosis process. These data managers can also store thousands of patient reports in their current memory. Some of the more sophisticated systems also store the actual reaction curves used to determine the test results.

Whereas the analytical unit's form and function is somewhat determined by a particular technology, the results processor's approach may be influenced by a larger system strategy. The RP might be part of a single analyzer, a stand alone entity, or it could act as a single results processor for two or more analytical units. In another configuration, the RP could act as an interface between one or more analytical units or between one or more analytical units and a larger, more centralized computer system.

Economic Aspects

It is estimated that the worldwide clinical chemistry diagnostics market is about $3 billion. This amount includes an estimated $700 million in instrument sales, and $2.3 billion in sales of reagents and consumables. Some of the principal instrument manufacturers are Hitachi (Japan), Miles Laboratories/Technicon Instruments (United States), E. I. du Pont de Nemours (United States), Beckman Instruments (United States), Eastman Kodak (United States), Abbott Laboratories (United States), Olympus (Japan), Toshiba (Japan), Hoffmann-La Roche (Switzerland), and Ciba Corning Diagnostics (United States).

The proliferation of automated analyzers in the 1980s enticed an increased number of manufacturers to enter the market, increasing competition and lowering selling prices. In 1991 list prices for automated clinical chemistry analyzers ranged from $30,000 to $80,000 for small (120–400 tests per hour) analyzers, $100,000 to $175,000 for medium (500–1500 tests per hour) throughput, and $200,000 to $350,000 for very high throughput systems capable of processing up to 10,000 tests per hour.

Automated clinical analyzers became affordable, not only to small hospital laboratories, but to doctors' group practices and to individual doctor's offices as well.

BIBLIOGRAPHY

1. *Report #7076,* Biomedical Business International, 1987.
2. R. Gambino, *14th International Congress of Clinical Chemistry,* San Francisco, Calif., July 22–26, 1990.
3. N. W. Tietz, ed., *Textbook of Clinical Chemistry,* W. B. Saunders Co., Philadelphia, Pa., 1986.
4. L. J. Kaplan and A. J. Pesce, *Clinical Chemistry, Theory, Analysis, and Correlation,* The C. V. Mosby Co., 1989.
5. G. Kessler and L. R. Snyder, *Technicon International Congress,* Vol. 1, Mediad Press, Tarrytown, N.Y., 1977, pp. 28–35.
6. I. M. Kolthoff, P. J. Elving, and E. J. Meehan, *Optical Methods of Analysis,* Vol. 8, *Treatise on Analytical Chemistry,* Part I, John Wiley & Sons, Inc., New York, 1986.
7. G. Eisenman and co-workers, *Science* **126,** 831 (1957).
8. G. Eisenman, *Glass Electrodes for Hydrogen and Other Cations, Principles and Practices,* Marcel Dekker, Inc., New York, 1967, p. 2.

PAUL GHERSON
HOWARD LANZA
MILTON PELAVIN
DAN VLASTELICA
Miles Inc.

HEMATOLOGY

Hematology analyzers provide information about blood cells and their constituents. The three basic blood cell types are erythrocytes or red blood cells, first described in 1674 by Van Leeuwenhoek using a microscope of his own design; leukocytes or white blood cells, noted by Hewson in 1773; and thrombocytes or platelets, described in 1841 by Addison, who also noted the granules present in certain of the leukocytes. Hemoglobin, the principal nonaqueous component of red blood cells, was identified by Hoppe-Seyler in 1864 when it was shown that hemoglobin is a complex of four Fe^{2+} protoporphyrins, called hemes, and a protein, or globin, consisting of two alpha and two beta chains. The absorption spectrum of the oxidized form of hemoglobin showing that hemoglobin can absorb and release oxygen was also reported (1). Hemoglobin's physiological importance gives it the status of a primary hematologic constituent.

The first hematologic cell-counting chamber, the hemocytometer, was developed in 1873. A drop of a diluted suspension of blood cells was inserted by capillary action between a glass cover slip and the finely cross-hatched surface of a glass chamber. The blood cells were allowed to settle to a common focal plane. Then, using a microscope, the red blood cells (RBC) within the cross-hatched areas were counted. Eventually, technologists used the hemocytometer to count white blood cells (WBC) and platelets (PLT) as well, first destroying the red blood cells which predominated. Hemocytometers are still used on occasion.

In 1877 Ehrlich developed a triacid stain that distinguished the nucleus and cytoplasm of cells in a stained blood film and originated the white blood cell differential count (Diff). By applying the triacid stain to a blood film slide and examining the slide under a microscope, five white cell types were distinguishable by the shapes of the nuclei and by the varying affinity of cell granules for the stain (2). The five white blood cell types are lymphocytes (LYMPHS) which have round nuclei and no granules; monocytes (MONOS) which have indented nuclei and sparse granules; neutrophils (NEUTS) which have polymorphonuclear (PMN) or segmented nuclei and granules that have affinity for neutral dyes such as picric acid-rosaniline; eosinophils (EOS) which have PMNs and granules having affinity for acidophilic dyes, such as eosin; and basophils (BASOS) which have PMNs and granules having affinity for basophilic dyes, such as fuchsin. Most clinical laboratories still routinely use Ehrlich's method for white blood cell differential counts, determining the number of each cell type in a 100 cell count, albeit using other stains.

Technologists have measured a sample's hemoglobin content in various ways. In 1878 a method for measuring hemoglobin concentration in whole blood (HGB) using visual photometry was introduced. Hemoglobin concentration may be defined as the number of grams of hemoglobin per deciliter of whole blood. Samples of blood were diluted and the color compared to that of a solution of picrocarmine. Later samples were converted to carboxyhemoglobin and the color compared to that of a carboxyhemoglobin standard or converted to acid hematin through the addition of dilute hydrochloric acid and the brown color compared to that of a standard. In 1920 a method to convert hemoglobin to cyanmethemoglobin by lysing the red cells in distilled water and adding potassium ferricyanide and potassium cyanide was introduced. The ferricyanide oxidized the heme to the

Fe^{3+} form, yielding methemoglobin, and the cyanide converted methemoglobin to cyanmethemoglobin. This method was later modified by adding sodium bicarbonate, in dilute form, to reduce the turbidity of the lysed product. Hemoglobin concentration was determined by comparing cyanmethemoglobin's color to that of a reference material or by measuring its photometric absorption at 540 nm and converting to hemoglobin concentration. Drabkin's solution, the ferricyanide, cyanide, bicarbonate mixture, was used for many years for photometric hemoglobin measurements because it effectively converted all forms of cellular hemoglobin to a single, stable form for photometric measurement. In 1966 the International Committee for Standardization of Hematology (ICSH) adopted the cyanmethemoglobin photometric method, which is related to the use of Drabkin's solution. Technologists also use physical properties of hemoglobin, such as specific gravity and refractive index, to determine concentration.

A method for determining a sample's total hemoglobin content was developed in 1929. A tube of whole blood was centrifuged, packing the red cells at the bottom of the tube, and the percentage of total sample volume occupied by the RBC was determined by dividing the height of the packed cells by the total height of the sample. This percentage was called packed cell volume (PCV) or hematocrit (HCT). In combination with RBC measurements, the hematocrit provided a sample's mean red blood cell volume (MCV) (3).

$$MCV = (HCT/RBC) \times 10$$

Two additional parameters, the mean cellular hemoglobin concentration (MCHC) and the mean cellular hemoglobin (MCH), can be obtained.

$$MCHC = (HGB/HCT) \times 100$$
$$MCH = (HGB/RBC) \times 10$$

By 1935 a technologist could perform a complete blood count (CBC) consisting of: WBC in $10^3/\mu L$, RBC in $10^6/\mu L$, HGB in g/dL, HCT in %, MCV in fL (10^{-15} L), MCH in pg, MCHC in g/dL, and PLT in $10^3/\mu L$; and a complete white blood cell differential count, consisting of all white blood cell types, ie, the five normal types plus any other cells not normally found in peripheral blood.

The first commercially successful automated blood cell counter, the Model A Coulter Counter, was introduced in 1956. The Model A counted cells by using electrical impedance properties, and the Coulter Counter quickly became the instrument of choice for counting red and white blood cells. When it was later demonstrated that cell volume was roughly proportional to the electrical impedance signal amplitude, the Coulter Counter was modified to provide MCV as the mean of the individually measured red cell volumes. Impedance counters calculated HCT as (RBC × MCV)/10 and, by the early 1970s, these instruments also counted platelets. By the mid-1970s, impedance counters combined with photometric hemoglobinometers to produce CBCs.

In 1965 Technicon introduced the SMA 4/A Autocounter, which counted red and white cells by using light-scattering properties and determined a sample's HCT by measuring electrical conductivity. The instrument included a photometer for automated HGB measurements. The Autocounter was the basis for the first fully automated CBC analyzer including PLT, the Hemalog 8, introduced in 1970. The first commercial automated five-part Diff analyzer, the Hemalog D, introduced in 1971 by Technicon Instruments Corporation, produced five-part white blood cell differential counts by combining measurements of the light-scattering and cytochemical properties of white cells. By 1975 Geometric Data, Corning Medical and Scientific, Abbott Laboratories, and Perkin-Elmer each had instruments on the market. The latter instruments were automated slide readers equipped with computerized image analyzers (4). Commercial success was limited, however, because these instruments were only slightly faster than manual differential counting methods. Additionally, they were expensive to operate and suffered from questionable quality of the prepared blood film slides.

The first automated CBC/five-part Diff analyzer was introduced in 1981. This instrument used modified Hemalog D 90 technology as well as other light-scattering measurements and photometric hemoglobinometry. It was followed by an improved version, the H*1 (Technicon), that incorporated a novel method for measuring certain red blood cell parameters. Then in 1989 TOA introduced the NE-8000, which produces five-part differentials by combining impedance technology with cytochemistry; Unipath introduced the Cell-Dyn 3000, which uses light scattering only to produce five-part differentials; and Coulter introduced the Coulter STKS, which performs five-part differential counts by using impedance and light scattering.

Aperture Impedance Instruments

The aperture impedance principle of blood cell counting and sizing, also called the Coulter principle (5), exploits the high electrical resistivity of blood cell membranes. Red blood cells, white blood cells, and blood platelets can all be counted. In the aperture impedance method, blood cells are first diluted and suspended in an electrolytic medium, then drawn through a narrow orifice (aperture) separating two electrodes (Fig. 1). In the simplest form of the method, a d-c current flows between the electrodes, which are held at different electrical potentials. The resistive cells reduce the current as the cells pass through the aperture, and the current drop is sensed as a change in the aperture resistance.

Aperture impedance counters are flow cytometers, as are all current automated cell counters. Flow cytometers measure cells as the cells flow hydraulically, one at a time, through sensing zones. All flow cytometers, including aperture impedance counters, count fewer cells than actually pass through their sensing zones (6–11). When two or more cells are in a sensing zone simultaneously, by coincidence, the counter senses only one cell. A number of approaches may be taken to correct for count loss resulting from coincidence. The simplest is to dilute the sample to the point where the probability of coincidence is acceptably small (12,13). However, because this approach typically requires dilution factors of a million or more, it introduces dilution errors. Also, because dilution significantly reduces the number of cells counted, this approach also increases sampling error. An alternative approach is to model the probability of coinci-

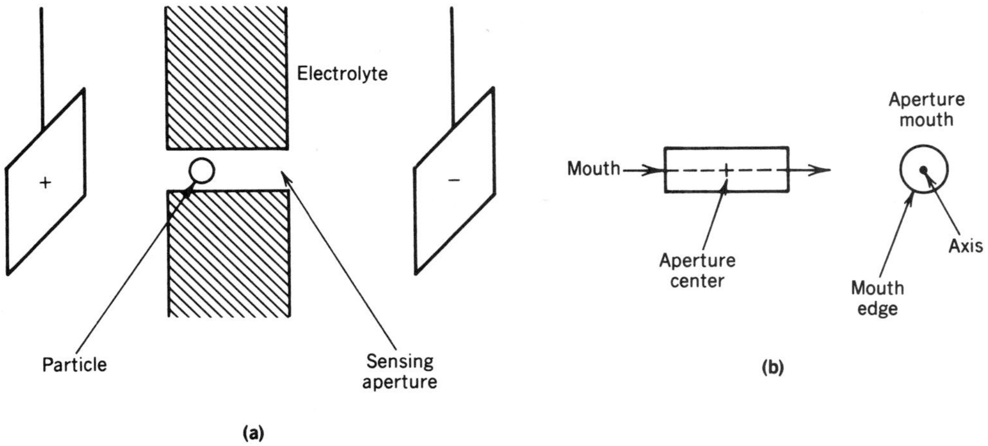

Fig. 1. (**a**) Schematic diagram of the electrical impedance method for counting and sizing blood cells. The electrodes are represented as charged rectangles. As particles flow through an aperture, the impedance and, hence, the flow of current are modified; (**b**) schematic diagram of the sensing aperture of a flow cytometer.

dence as a function of either sample concentration or the fraction of time the sensing zone is occupied by passing cells (dead time) (6–11). The model produces a coincidence-correction factor which is applied to the count. Still another approach is to reduce the size of the sensing zone, which reduces the probability of coincidence (7). In practice, commercial counters combine the latter two approaches.

Flow cytometer cell counts are much more precise and more accurate than hemocytometer counts. Hemocytometer cell counts are subject both to distributional (13) and sampling (14–16) errors. The distribution of cells across the surface of a hemocytometer is sensitive to the technique used to charge the hemocytometer, and nonuniform cell distribution causes counting errors. In contrast, flow cytometer counts are free of distributional errors. Statistically, count precision improves as the square root of the number of cells counted increases. Flow cytometer counts usually involve 100 times as many cells per sample as hemocytometer counts. Therefore, flow cytometry sampling imprecision is one-tenth that of hemocytometry.

Aperture impedance counters provide cell volume information as well as cell counts. The relationship between aperture resistance change and cell volume may be expressed as (17–20):

$$\frac{DR}{R} = f\frac{v}{V}$$

where DR = change in resistance resulting from the presence of a cell in the aperture, R = aperture resistance of the electrolyte alone, f = function of particle shape and orientation, v = cell volume, and V = effective aperture volume. Aperture impedance counters provide mean red cell volume (MCV) and mean platelet volume (MPV), both in femtoliters, as:

$$MCV = \frac{\Sigma \text{ red cell volumes}}{RBC}$$

$$MPV = \frac{\Sigma \text{ platelet volume}}{PLT}$$

Aperture impedance and most other automated counters measure MCV and RBC independently, in contrast to the manual methods where MCV and MCH accuracies depend on hemocytometer red cell count accuracy.

Automated counters can also provide cell volume distribution information because they measure volume on a cell-by-cell basis. Manual methods measure MCV indirectly. Automated counters provide red cell volume distribution width (RDW) and platelet volume distribution width (PDW) where RDW is the standard deviation of red cell volumes/MCV and PDW is the standard deviation of platelet volumes. RDW measured as a percentage and PDW in units of femtoliters have descriptive value. For example, a high RDW value suggests the presence of microcytes (small red cells) and/or macrocytes (large red cells). Microcytes may be associated with red blood cell iron deficiency and macrocytes with vitamin B-12 or folate deficiency. Automated counters also group cell volumes by frequency to form cell volume histograms. The histograms provide integrated views of cell volume distribution (Fig. 2).

Several factors affect aperture impedance cell volume measurements. DR/R depends on f, the shape and orientation of the cells passing through the aperture. For spherical objects $f = 1.5$, whereas for needlelike objects oriented along the cylindrical axis of the aperture, $f = 1$ (20–22). Additionally, as needle-shaped objects tumble, the f values change. Red cells are especially subject to shape and orientation effects in apertures because these cells are highly deformable and their shape in the aperture depends on the shear forces therein. Photographs show that as red cells pass through apertures at typical flow rates, they deform from their native biconcave disk shape to one approximating an oblate ellipsoid (cigar-shape) (22). If all red cells deformed to the same extent, deformability could be factored into f. However, red cell deformability depends on cellular hemoglobin concentration. High cellular hemoglobin concentration reduces red cell deformability thereby reducing elongation from shear forces. Less deformed red cells have higher f values than normally deformed cells. Therefore, aperture impedance counters often assign erroneously high cell volume values to red cells having high cellular hemoglobin concentration, and erroneously low cell volume values to cells where the cellular hemoglobin concentration is low. A fraction of the red cells may tumble about their long (deformation) axis as they pass through the aperture along the aperture wall (21–23), because of discontinuous changes in hydrodynamic forces near the wall. As the cells tumble, f increases and skews MCV upward.

Aperture impedance measurements of cell volume are affected by variations in the electric field strength that occur in the region of the aperture. The field is strongest at the edge of the aperture mouth and weakens as the aperture axis is

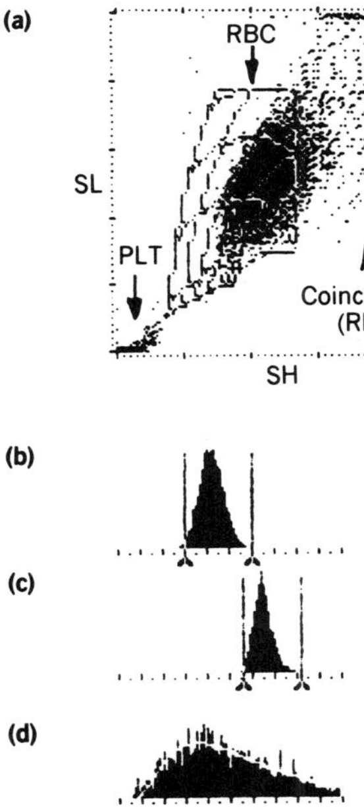

Fig. 2. Histograms representing (**a**), the frequency of occurrence of platelets (PLT) and red blood cells (RBC) and illustrating coincidence where SL is low angle scatter, 2°–3°, and SH is high angle scatter, 5°–15°; (**b**), the volume of RBC, from 0 to 200 fL; (**c**), the hemoglobin (HGB) concentration from 0 to 50 g/dL; and (**d**), the platelet volume from 0 to 20 fL.

approached (21,22) (Fig. 1**b**). The field also weakens from the aperture entrance to the center of the aperture. Therefore, DR/R and computed cell volumes are greater for cells passing along the aperture wall than for cells passing along the aperture axis.

A number of investigators (24–26) demonstrated that aperture impedance cell volume measurements could be improved by focusing or restricting cell flow to the aperture axis, thus reducing variations resulting from inhomogeneous electric and hydrodynamic fields. Counters that caused the cell suspensions and electrolyte solutions to flow concentrically through the apertures were designed. The electrolytes sheathed the cell suspensions, narrowing them and centering the cells on the aperture axis.

Aperture impedance measurements of cell volume must take into account the osmolality and pH of the medium. A hypotonic medium causes cells to swell; a hypertonic medium causes them to shrink. Some manufacturers of aperture impedance counters deliberately provide hypertonic electrolytic media for red blood cell measurements. The shrunken red cells not only become more nearly spherical and thus less affected by orientation, but also less deformable than cells in isotonic media and thus less affected by differences in hemoglobin content.

The basic aperture impedance method can produce three-part white cell differential counts. Impedance counters can distinguish three white cell types by size: the LYMPHS, mid-range cells including MONOS and BASOS, and granulocytes including NEUTS and EOS.

A refined version of the aperture impedance method which combines radio frequency (r-f) aperture currents and d-c currents, can produce four-part differentials. R-f aperture impedance is sensitive to the internal structure of white cells (27). R-f aperture impedance measurements distinguish BASOS from LYMPHS by virtue of BASOS lower nucleus-to-cytoplasm ratio. Therefore, the combined d-c/r-f aperture impedance counter distinguishes LYMPHS, MONOS, BASOS, and NEUTS + EOS. To distinguish EOS from NEUTS, and thereby produce five-part differentials, requires the use of light-scattering measurements.

In summation, aperture impedance counters provide information on WBC, RBC, HCT, MCV, PLT, MPV, RDW, PDW, and three-part Diff, for d-c current only, or four-part Diff using a d-c/r-f combination.

Light-Scattering Instruments

The light-scattering principle of cell counting is based on the observation that microscopic particles, such as blood cells, scatter into small (0–15°) angles, most of the visible light incident upon them (28). This principle is used to count red blood cells, white blood cells, and platelets. In the basic form of the light-scattering method, a dilute suspension of cells and a sheathing fluid having a matching refractive index flow concentrically through an optical flow cell onto which light is focused. The cells scatter the incident light as they pass through the flow cell, and the light scattered into a small-angle interval, typically 1–3 degrees, is sensed by an optical detector. The purpose of the sheath is to narrow the cell stream and reduce the cell passage to single file while centering the stream in the flow cell. The unscattered component of focused light is much larger than the scattered component and must be blocked from the detector so that it does not swamp the desired small-angle scatter signal (Fig. 3).

The basic single-angle interval light-scattering method cannot accurately measure individual red blood cell or platelet volumes, but it can provide MCV and MPV. Red cells are bi-concave disks, and platelets are rod to disk shaped. Scattering intensities depend on the orientation in the flow cell. Because the cells can interrupt the optical path in random orientations, individual scattering intensities are not proportional to cell volume. However, because thousands of cells of each type pass through the flow cell, the effects of orientation can be averaged. To a first approximation HCT and platelet crit (PCT), the percentage of blood sample volume occupied by platelets, is proportional to the sums of the scattering intensities of the red cells and platelets, respectively. MCV can be computed from HCT and RBC, whereas MPV can be computed from PCT and PLT. The accuracy of MCV determined by this method is tied to the RBC accuracy, as is the case for the manual MCV method. Ortho Instruments Corporation's ELT-8 uses these counting and sizing methods.

Light-scattering measurements of sphered cells are not subject to orientation effects, and a method for the rapid sphering and fixing of red cells for the purpose of measuring them in a light-scattering flow cytometer has been developed (29). The reagents used to lyse red blood cells, typically surfactants (qv), can also be used at lower concentrations to sphere them. In fact, the sphering process is part of the overall process of hemolysis (30). The red cell's sphered state is the

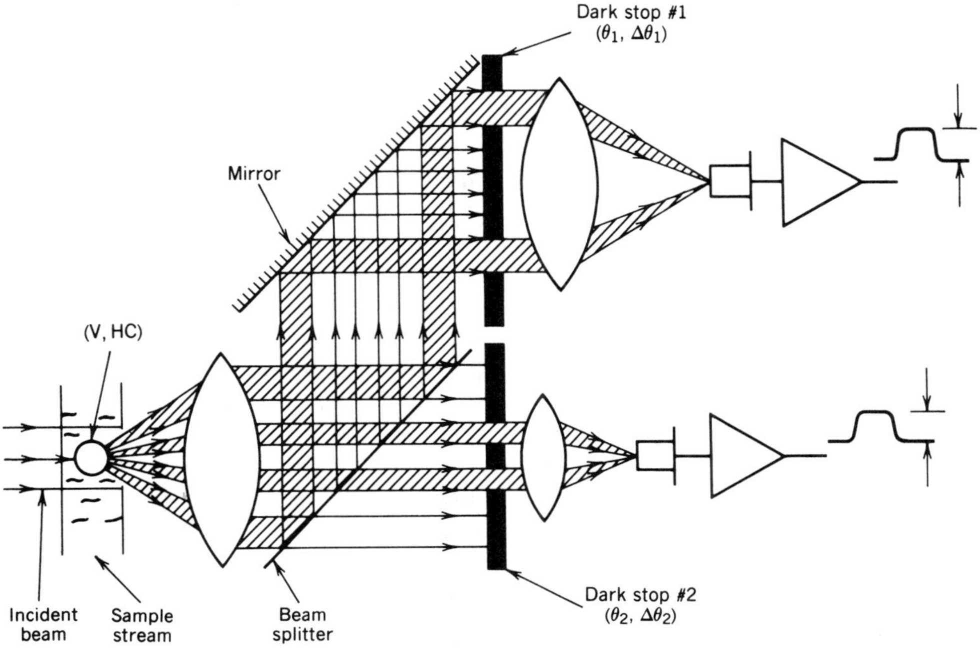

Fig. 3. Schematic diagram of a dark field system for measuring the light scattering by a spherical red blood cell where V = volume of sample and HC = hemoglobin concentration. Optical flow cell system having double-angular interval detection at angles θ_1 and θ_2.

last intact state before it is hemolyzed. Hemolysis is, however, undesirable because, depending on the extent of hemolysis, red cells are reduced in size or totally destroyed. The sphering agent concentration should not be too high, because it can hemolyze the red cells in samples having low red blood cell concentrations. However, the concentration of sphering agent must be high enough to sphere all the red cells in samples having high red blood cell concentrations. Fixing the cells after they are sphered prevents hemolysis in samples having low RBC values, yet permits concentrations of sphering agents that are high enough to sphere all red cells in concentrated red cell suspensions.

A single reagent capable of rapidly sphering and fixing all the red blood cells in a sample over a wide range of red blood cell concentrations has been developed. The reagent composition ensures that red cells are not fixed before they are sphered, and that the cells do not lyse before they are fixed. The sphering and fixing reagent contains sodium dodecyl sulfate (SDS) [*151-21-3*], a buffer, and gluteraldehyde [*111-30-8*], $C_5H_8O_2$. SDS is the sphering agent, the buffer provides an isotonic medium in which to isovolumetrically sphere red cells, and gluteraldehyde is the fixing agent. The Technicon H*6000 Hematology Analyzer incorporated this method in order to measure red cell volume on a cell-by-cell basis. In contrast to the light-scattering method for sphered and fixed red cells, the aperture impedance method for determining the MCV of unsphered red cells is subject to orientational effects.

Light-scattering measurements of red cell volume are accurate only if the measurements account for the effects of cell refractive index on scattering intensity. Basic single-angle interval light-scattering measurements of red cell volume, even for sphered and fixed red cells, are only accurate to within approximately 10%. Mie Scattering Theory predicts that the angular scattering intensity pattern for sphered red cells depends on cell refractive index as well as cell size (31,32). Red cell refractive index is linearly related to cellular hemoglobin concentration (33). Therefore, two normal red cells of the same volume but of different cellular hemoglobin concentration scatter light differently. A sphered cell's volume and the hemoglobin concentration can both be accurately determined from Mie Scattering Theory calculations by making simultaneous measurements of light-scatter intensity over two suitably chosen angle intervals (34). Two suitable angle intervals for red cells are 2–3 degrees, known as low angle scatter (SL), and 5–15 degrees, high angle scatter (SH) (Fig. 2). In contrast to the aperture impedance method, this method for determining red cell volume does not suffer from inaccuracies as a result of cellular hemoglobin concentration variability.

In principle, the two-angle interval method can produce all CBC parameters within a single measurement channel, uniquely providing cell-by-cell hemoglobin concentration. The mean of the concentrations provides an alternative (and direct) measurement of MCHC. The method also provides an alternative HGB measurement, because HGB may be set equal to (RBC × MCV × MCHC)/1000. This method, like the basic light-scattering method, uses the same flow cell to measure platelets and red cells with the result that the method is capable of providing the CBC parameters RBC, HGB, HCT, MCV, MCHC, MCH, and PLT. The method can also count a sample's white blood cells if the sample's red blood cells have been lysed.

The distribution of cellular hemoglobin concentrations within a blood sample, also known as hemoglobin distribution width (HDW), can be provided. HDW is the standard deviation of cellular hemoglobin concentrations measured in units of g/dL. HDW has descriptive value: a high HDW value suggests the presence of abnormally high numbers of hyperchromic and/or hypochromic red cells, ie, either high or low cellular hemoglobin concentration cells. Hypochromic cells are usually associated with anemia, whereas hyperchromic cells may be associated with hereditary spherocytosis or spherical blood cells, a condition in which the survival rate of the spherocytes is decreased. Instruments using the two-angle interval scatter method can display cellular hemoglobin concentration histograms which provide integrated views of cellular hemoglobin concentration distributions (Fig. 2).

In summation, the two-angle interval scattering method can provide CBC, RDW, PDW, PCT, and HDW. This two-angle interval method has been incorporated into the Technicon H*1 system which provides RBC, HCT, MCV, PLT, MPV, RDW, PDW, PCT, and HDW.

The single-angle interval light-scattering method counts white cells but cannot produce five-part Diffs. White cell light-scattering intensity depends on cell size, cell refractive index, and internal cell structure. White cell size ranges overlap and the cells differ in refractive index. Scattering measurements can distinguish among cell types based on differences in internal structure alone, but only if the cell types have the same overall size and refractive index.

Light-scattering measurements made over two suitably chosen angle intervals can combine with depolarized light-scattering measurements to provide five-part Diffs (35). Unipath's Cell-Dyn 3000 instrument measures white cell light scatter at 1–3 degrees and at 3–11 degrees in order to resolve LYMPHS, BASOS, MONOS, and NEUTS + EOS. The two-angle interval light-scattering method distinguishes the cell types based on size and refractive index and is similar to the Technicon H*1 RBC/PLT method. The Cell-Dyn 3000 instrument uses the scattering behavior of EOS granules to distinguish EOS from NEUTS. EOS granules are refractile, numerous, large, and closely packed. Light incident on the granules is scattered a number of times, from granule to granule, before reaching the detectors. If the incident light is polarized, multiple scattering depolarizes the light scattered onto the detectors. The other white cell types scatter polarized incident light once, but do not depolarize it. Depolarization as a result of multiple scattering is enhanced at high scattering angles. Thus the Cell-Dyn 3000 measures the depolarized and polarized components of light scattered at 90 degrees to incidence by white cells and plots depolarized scattering intensity vs polarized scattering intensity for each cell. A distinct EOS cluster forms on the plot.

Basic light-scattering measurements can be combined with other measurements to provide five-part Diffs. Coulter's STKS combines d-c/r-f aperture impedance and light-scattering measurements to provide five-part Diffs. The STKS lyses a sample's red cells, then shrinks the white cells by suspending them in a hypertonic medium. The system distinguishes among the shrunken cells based on cell size differences and internal structure differences related to r-f impedance and light-scattering characteristics. D-c/r-f measurements distinguish among LYMPHS, BASOS, MONOS, and NEUTS + EOS based on cell size and r-f impedance. Small-angle scattering measurements resolve EOS and NEUTS based on EOS granules scattering behavior.

Cytochemistry

Cytochemical techniques can be combined with light-scattering and absorption measurements to provide five-part Diffs. Cytochemistry concerns the chemical reactions of cell components. The reactions for automated white blood cell differential analysis include those that bind chromophores to the granules of white cell types, based on the presence of various substrates and enzymes in the granules. These reactions yield products suitable for light-scattering and optical absorption measurements. Other reactions exploit the differential resistance of white cell types to cytoplasmic stripping by the lysing action of surfactants. The reaction products are suitable for light-scattering and aperture impedance measurements. Optical absorption measurements made on intact cells in suspension are different from those made on dissolved hemoglobin. Measurements made on intact cells involve reduced overall scattering intensity and do not obey Beer's Law; measurements made on solutions involve dissolved chromophores and do obey Beer's law (see SPECTROSCOPY).

Although both the manual and light-scatter–cytochemistry methods of counting white cell differentials involve the staining of white cells, the staining and measurement techniques are different (36). Classic stains used in manual differential analyses, such as Wright's Stain and Giemsa Stain, consist mainly of azure dyes, which are basic, and eosin dyes, which are acidic. The DNA of white cell nuclei are azurophilic (basophilic) and are stained blue to magenta by the azure dyes, as are the granules of NEUTS and the DNA contained in the cytoplasm of LYMPHS and MONOS. The heparin in BASOS granules is also basophilic. The acidophilic (eosinophilic) proteins in EOS granules and nonazurophilic NEUT granules, however, are stained red by eosin. The white blood cell types are not distinguishable by azure–eosin staining specificity and intensity alone. In order to distinguish among the white blood cell types, a visual examination of differences in their size, shape, granular content, and nuclear structure is also necessary.

The Technicon Hemalog D was the first hematology analyzer to combine cytochemistry with light-scattering–optical absorption measurements to produce five-part Diffs. This instrument used cell-specific staining properties of the leukocytes as well as their light-scattering behavior to distinguish among the five normal leukocyte types (37,38). BASOS are unique in that the heparin in their granules reacts with alcian blue to precipitate a blue-green dye in the granules. MONOS alone contain lipase, an enzyme catalyzing the reactions of esters, and the substrates α-naphthyl butyrate and basic fuchsin combine, in the presence of lipase and a diazonium salt, to precipitate a red-brown azo dye. Only EOS are peroxidase-active below pH 3. Between pH 3 and pH 5, NEUTS are also peroxidase-active, though not as strongly as EOS, as are MONOS weakly peroxidase-active in this pH range. Therefore, EOS, NEUTS, and LYMPHS + BASOS + MONOS are distinguishable by peroxidase staining alone. By combining the three staining techniques, the Hemalog D could have distinguished among the five normal white cell types based on stain absorption alone. However, the Hemalog D achieved improved resolution of LYMPHS and MONOS by adjusting peroxidase staining conditions and measuring white cell light scattering as well as absorption in the peroxidase channel.

In order to use the leukocytyes' enzymatic properties, the Hemalog D carried out reactions under specific conditions. Because enzymes are easily damaged or released from cells, the cells were fixed to stop metabolism and to immobilize the enzymes without significantly changing their catalytic activity. Fixation was controlled so that plasma proteins did not precipitate and generate spurious particle counts, clogging the system. Red cells were lysed at elevated temperatures to destroy red cell and serum catalase and hemoglobin pseudoperoxidase activity, which would otherwise have completely consumed the peroxidase reaction substrates. Then plasma esterase inhibitor was inactivated by adding concentrated formaldehyde and ethylene glycol, in which the inhibitor is labile. The esterase inhibitor would otherwise have stopped the MONO reaction. In order to specifically bind alcian blue to heparin and not all nuclei, the BASO reaction was carried out at pH 2.2. The three staining reactions are

1. Peroxidase reaction for NEUTS, EOS, (MONOS)

$$\text{4-chloro-1-naphthol} + H_2O_2 \xrightarrow[\text{peroxidase}]{} \text{dark precipitate}$$

2. BASO reaction

$$\text{alcian blue} + \text{heparin (in BASO granules)} \longrightarrow \text{blue-green dye}$$

3. MONO reaction

$$\alpha\text{-naphthyl butyrate} + \text{basic fuchsin} \xrightarrow[\text{lipase}]{\text{MONOS}} \text{butyric acid} + \alpha\text{-naphthol}$$

$$\alpha\text{-naphthol} + \text{diazonium salt} \longrightarrow \text{red-brown azo dye}$$

The system measured small-angle light-scattering intensity and/or cell stain absorption in each of three reaction channels. In the MONOS channel, an absorption detector registered as MONOS those white cells that absorbed light at the MONOS stain-absorption wavelength as the cells passed through a flow cell. In the BASOS channel, white cells interrupted a broad wavelength band of focused light as they passed through a flow cell. Two scattering detectors, one with a green filter, the other with a red filter, registered scattering intensity. The analyzer counted as BASOS those cells that scattered less green light than red light. In the peroxidase channel, white cells scattered and absorbed light differentially, depending on their peroxidase activity, as they passed through a flow cell. The analyzer plotted small-angle light-scattering intensity versus absorption, for each cell. The plotted points formed clusters by cell type. Cluster-analysis algorithms distinguished among LYMPHS + BASOS, NEUTS, EOS, and, to a limited extent, MONOS. The Hemalog D also used the sum of the small-angle scatter signals in the peroxidase channel to provide WBC.

The Technicon H*6000 system used a different combination of cytochemistry and light-scattering–absorption measurements to produce five-part Diffs. The system retained the Hemalog D peroxidase channel but eliminated the MONO channel, thus reducing the number of system flow cells by one. The system relied on modified staining conditions and peroxidase channel cluster analysis to distinguish MONOS as well as LYMPHS + BASOS, NEUTS, and EOS (39). A series of reagents were developed that better fix monocytes, as well as the other white cell types, than did the Hemalog D reagents. The well-fixed cells are better able to retain peroxidase stain, stain more uniformly, and have narrower size distributions than the cells fixed by Hemalog D reagents. The resultant peroxidase channel clusters are tighter and thus easier to distinguish.

The Technicon H*1 and H*2 systems use a different method to distinguish BASOS, exploiting the BASOS resistance to cytoplasmic stripping. The BASOS reagent contains surfactant and phthalic acid plus hydrochloric acid. The surfactant lyses the red cells. Additionally, surfactant and phthalic acid, at low pH, strip the cytoplasm from all white cell types except BASOS. The reaction mixture then contains intact BASOS and the nuclei of the other white cell types. BASOS and nuclei pass through the same flow cell as the one used for RBC/PLT measurements. Intact BASOS and nuclei scatter light into the SL and SH detectors. BASOS scatter much more light into the SL detector than do nuclei because the intact BASOS are much larger. Thus the analyzer distinguishes BASOS by the high SL scattering intensity.

The H*1 and H*2 method of performing five-part Diffs is superior to the H*6000 method because a single stain is used instead of two; the same flow cell is used for BASOS measurements as for the RBC/PLT ones, thus reducing the number of flow cells from three to two; and the BASO channel provides more information about white cells because the BASO channel produces the lobularity index (LI), a measure of granulocyte maturity, in addition to producing a percentage of BASO and absolute BASO count.

The H*1 and H*2 systems plot SL intensity versus SH intensity for the nuclei. Two clusters form on the BASOS cytogram as shown in Figure 4. The left-hand cluster is comprised of LYMPHS and MONOS, which are both mononuclear (MN), or nonsegmented. The right-hand cluster is comprised of NEUTS and EOS, which are polymorphonuclear (PMN), or segmented. SH intensity is sensitive to refractive index and number of nuclear segments. The higher the refractive index or the greater the number of nuclear segments, the higher the SH intensity. Mature granulocytes have more nuclear segments than younger granulocytes. In addition, the more mature nuclei have higher refractive indexes than the less mature ones, because the more mature ones are denser. Therefore, the more mature segmented nuclei appear further to the right on the BASO cytogram than the less mature segmented nuclei. Technicon H*1 and H*2 systems define the lobularity index (LI) as the ratio of the PMN histogram mode channel to the MN histogram mode channel (Fig. 4).

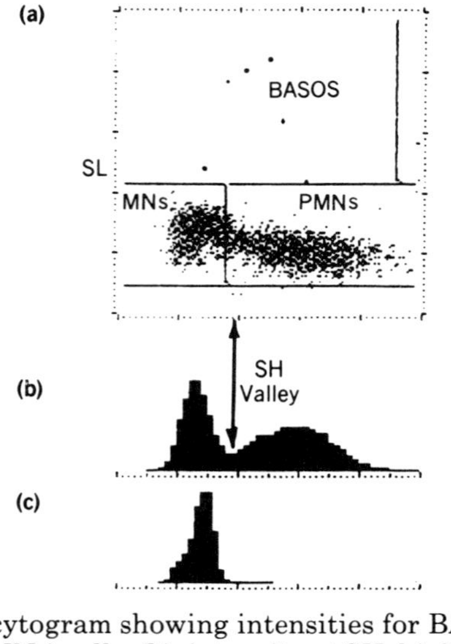

Fig. 4. (**a**), BASOS cytogram showing intensities for BASOS, mononuclear (MNs), and polymorphonuclear (PMNs) cells; (**b**), intensity at SH, high angle scatter; (**c**), intensity at SL, low angle scatter.

The TOA NE-8000 system combines d-c/r-f aperture impedance measurements with cytochemistry to produce five-part Diffs. The d-c/r-f impedance measurements are used to distinguish LYMPHS, mid-range cells, and granulocytes. The NE-8000 system lyses all white cell membranes except those of BASOS and identifies the BASOS by their higher d-c aperture impedance. Then, in a separate reaction, the NE-8000 lyses all white cell membranes except those of EOS and identifies EOS by their higher d-c aperture impedance.

Various method combinations can be used to produce five-part Diffs including d-c/r-f aperture impedance plus light scattering; two-angle interval light-scattering plus depolarized light scattering; light scattering plus cytochemistry;

and d-c/r-f aperture impedance plus cytochemistry. Flow cytometric methods produce five-part Diffs that are generally more precise and accurate than manual Diffs. However, manual methods can provide more information. White cell differential counts based on microscopic examination of blood smears are subject to sampling errors (14–16). Manual white cell differential counts are usually based on 100 cells. Therefore, for basophils which comprise 0–2% of the differential count, and for eosinophils (1–4%), and even for monocytes (2–8%), statistical sampling error is large. Flow cytometer differential counts, which are typically based on 5,000–10,000 cells, have much smaller sampling errors for these white cell types. Also, flow cytometers provide both direct absolute counts of each white cell type and differential counts, whereas manual methods provide only indirect absolute counts as the product of the differential count and WBC. Manual white cell differential counts are subject to distributional errors (40–43). Drawing a film of blood across the surface of a glass slide tends to force the larger, less dense blood cells to the edges of the slide. Uneven cell distribution across the surface of the slide can cause counting errors. Flow cytometric differential counts are not subject to distributional error. On the other hand, a technologist performing manual white blood cell differential counts can identify abnormal cell types that the automated analyzers cannot identify.

Photometric Hemoglobinometry

Most automated hemoglobinometry methods are derivatives of the manual ICSH reference method (44). The manual method follows three steps: (*1*) The blood sample is diluted in a reagent containing Triton X-100, a nonionic surfactant. The surfactant lyses the RBC membrane, releasing hemoglobin into solution. (*2*) Ferricyanide present in the diluent diffuses into the hemoglobin molecule and oxidizes heme from the Fe^{2+} to the Fe^{3+} form, yielding methemoglobin. (*3*) Cyanide diffuses into methemoglobin's interior and reacts with heme to yield cyanmethemoglobin (HiCN), which has a characteristic absorption spectrum that is measured spectrophotometrically.

The Technicon H*1 HGB automated method follows four steps: (*1*) The blood sample is diluted in a reagent containing lauryldimethylamine *N*-oxide (LO), an ionic surfactant and more powerful lytic agent than Triton X-100. The RBC membrane is lysed as is Step 1, above. (*2*) The combined action of surfactant and a pH 11.3 releases hemes from globin as a mixture of ferrous (Fe^{2+}) and ferric (Fe^{3+}) hemes. (*3*) Hemes are air-oxidized to ferric (Fe^{3+}) form. (*4*) The ferric hemes are ligated by cyanide and micellized by LO, forming reaction products with characteristic absorption spectra that are measured spectrophotometrically.

The automated method differs from the ICSH method chiefly in that oxidation and ligation of heme iron occur after the hemes have been released from globin. Therefore, ferricyanide and cyanide need not diffuse into the hemoglobin and methemoglobin, respectively. Because diffusion is rate-limiting in this reaction sequence, the overall reaction time is reduced from approximately three minutes for the manual method to 3–15 seconds for the automated method. Reaction sequences in the Coulter S+II and the Technicon H*1 and H*2 are similar. Moreover, similar reactions are used in the other Coulter systems and in the TOA and Unipath instruments.

Summary

All automated hematology analyzers are comprised of flow cytometers, which count and size red cells, white cells, and platelets, and photometric hemoglobinometers, which measure hemoglobin concentration. All analyzers produce CBCs and most also produce MPV, RDW, and PDW. Automated CBC analyzers use either the basic aperture impedance method or one of the light-scattering methods to count and size cells. Automated analyzers that also produce five-part Diffs use some combination of d-c/r-f aperture impedance, light scattering, and cytochemistry to do so. One disadvantage of automated systems is the inability to identify abnormal cell types.

BIBLIOGRAPHY

1. M. M. Wintrobe, *Blood Pure and Eloquent,* Vol. 11, McGraw-Hill Book Co., Inc., New York, 1980.
2. J. L. Binet, *Blood Cells* **6,** 507–511 (1980).
3. M. M. Wintrobe, *J. Lab. Clin. Med.,* (15), 287–289 (1929).
4. *Geometric Data: Hematrack Automated Differential System: Model-590, Operator's Manual,* Geometric Data, Wayne, Pa., 1982.
5. U.S. Pat. 2,656,508 (1953), W. H. Coulter.
6. J. L. Grant and co-workers, *Amer. J. Clin. Path.* **33,** 138–143 (1960).
7. M. Wales and co-workers, *Rev. Sci. Inst.* **32,** 1132–1136 (1961).
8. L. H. Princen and co-workers, *Rev. Sci. Inst.* **36,** 646–653 (1965).
9. W. B. Mercer, *Rev. Sci. Inst.* **37,** 1515–1520 (1966).
10. P. W. Helleman, *The Coulter Electronic Particle Counter: Aspects and Views in Counting and Sizing of Erythrocytes,* De Bilt Holland, the Netherlands, 1972, Chapt. 7–11.
11. U.S. Pat. 4,447,883 (1984), G. A. Farrell and co-workers.
12. A. R. Jones, in O. W. Van Assendelft and J. M. England, eds., *Advances in Hematological Methods: The Blood Count,* CRC Press, Boca Raton, Fla., 1982, Chapt. 5.
13. C. F. T. Mattern and co-workers, *J. Appl. Physiol.* **10,** 56–70 (1957).
14. C. L. Rumke, *Blood Cells* **11,** 137–140 (1985).
15. *Ibid.,* pp. 141–144.
16. D. W. Ross, *Blood Cells,* **11,** 145–157 (1985).
17. D. C. Golibersuch, *Biophys. J.* **13,** 265–280 (1973).
18. N. B. Grover and co-workers, *Biophys. J.* **9,** 1398–1414 (1969).
19. *Ibid.,* pp. 1415–1425.
20. N. B. Grover and co-workers, *Biophys. J.* **12,** 1099–1211 (1975).
21. C. S. Waterman and co-workers, *Clin. Chem.* **21**(9), 1201–1211 (1975).
22. B. B. Shank and co-workers, *J. Lab. Clin. Med.* **74**(4), 630–641 (1969).
23. J. Gutman, PhD dissertation, Max Planck Institute, Munich, Germany, 1965, p. 17.
24. V. Kachel, *J. Histochem. Cytochem.* **24**(1), 211–230 (1976).
25. J. T. Merrill and co-workers, *Rev. Sci. Inst.* **42**(8), 1157–1163 (1971).
26. R. Thom, in G. Izak, ed., *Modern Concepts in Hematology,* Academic Press, New York, 1972, pp. 191–200.
27. U.S. Pat. 3,502,974 (1970), W. H. Coulter.
28. H. C. van de Hulst, *Light Scattering by Small Particles,* John Wiley & Sons, Inc., New York, 1957, Chapt. 8.
29. U.S. Pat. 4,574,490 (1986), L. Ornstein.

30. E. Ponder, *Hemolysis and Related Phenomena,* Grune and Stratton, New York and London, 1971, pp. 10–49.
31. Ref. 28, Chapts. 9 and 12.
32. M. Kerker, *The Scattering of Light and Other Electromagnetic Radiation,* Academic Press, Inc., Orlando, Fla., 1969, Chapts. 3 and 4.
33. R. Barer, in A. W. Pollister, ed., *Physical Techniques in Biological Research* IIIa, Academic Press, New York, 1966, Chapt. 1.
34. U.S. Pat. 4,735,504 (1988), D. H. Tycko.
35. U.S. Pat. 5,017,497 (1991), G. de Grooth and co-workers.
36. H. Ansley and L. Ornstein, *Advances in Automated Analysis,* Vol. 1, Thurman Associates, Miami, Fla., 1971, pp. 437–446.
37. A. M. Saunders, *Clin. Chem.* **18**(8), 783–788 (1972).
38. H. P. Mansberg and co-workers, *J. Histochem and Cytochem* **22**(7), 711–724 (1974).
39. U.S. Pat. 4,801,549 (1989), J. Cremins and co-workers.
40. J. V. Dacie and co-workers, *Practical Hematology,* 5th ed., Churchill Livingstone, 1975, pp. 48–51.
41. A. Simmons, *Technical Hematology,* 2nd ed., J. B. Lippincott Co., 1976, pp. 100–103.
42. C. E. Seiverd, *Hematology for Medical Technologists,* 5th ed., Lea and Febiger, 1983, p. 255.
43. R. E. Miller, *Clinics in Laboratory Medicine,* Vol. 1, 1st ed., 1985, pp. 143–147.
44. M. J. Malin and co-workers, *J. Clin. Path.* **92**(3), 286–294 (1989).

DAVID ZELMANOVIC
ROQUE A. GARCIA
JOHN TURRELL
Technicon Instruments Corporation

BIOPOLYMERS, ANALYTICAL TECHNIQUES

Analytical techniques that utilize biopolymers, ie, natural macromolecules such as proteins, nucleic acids, and polysaccharides that compose living substances, represent a rapidly expanding field. The number of applications is large and thus uses herein are limited to chiral chromatography, immunology, and biosensors.

Biopolymers in Chiral Chromatography

Biopolymers have had a tremendous impact on the separation of nonsuperimposable, mirror-image isomers known as enantiomers. Enantiomers have identical physical and chemical properties in an achiral environment except that they rotate the plane of polarized light in opposite directions. Thus separation of enantiomers by chromatographic techniques presents special problems, whereas diastereomers have different chemical and physical properties and may be separated by more conventional methods (see CHROMATOGRAPHY). Direct chiral resolution by liquid chromatography (lc) involves diastereomeric interactions between the chiral solute and the chiral stationary phase. Because biopolymers are chiral molecules and can form diastereomeric interactions with chiral solutes, they are ideal for use as chiral stationary phases. This property has led to a rapid growth of chromatographic stationary phases utilizing biopolymers to separate chiral molecules.

Cyclodextrin Chromatographic Phases. *Properties and Structure.* Cyclodextrins (CDs) are natural macrocyclic polymers of glucose that contain from 6 to 12 D-(+)-glucopyranose units bonded through α-(1,4)-linkages. They are chiral, torodial-shaped molecules with all the glucose units in a C-1 (D) chair conformation (1,2). The structure of β-cyclodextrin is shown in Figure 1. The mouth of the molecule, facing outward, has the larger circumference of the openings, and contains the secondary hydroxyl groups on C-2 and C-3 of the glucose molecule. The primary hydroxyl groups, those attached to C-6 of the

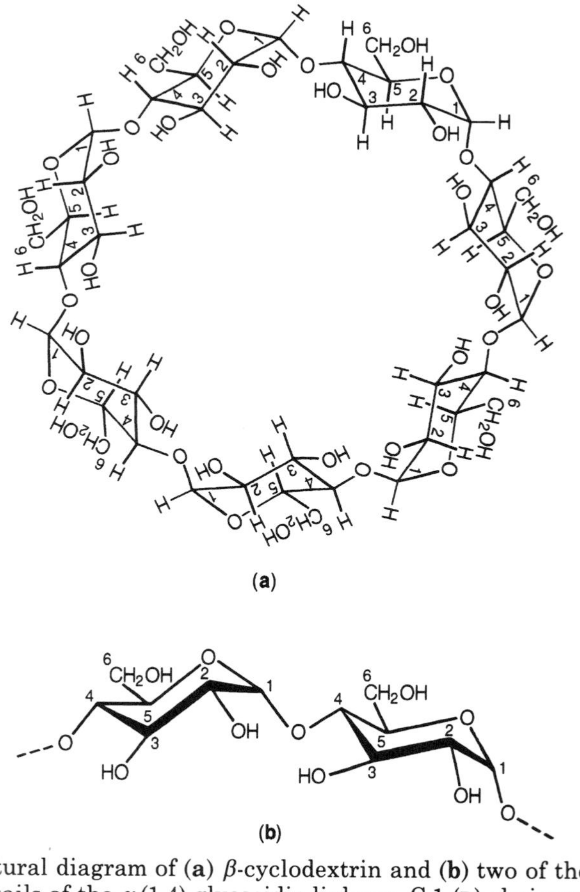

Fig. 1. Structural diagram of (**a**) β-cyclodextrin and (**b**) two of the glucopyranose units illustrating details of the α-(1,4) glycosidic linkage, C-1 (D) chair conformation, and the numbering system employed to describe the ring system.

glucose unit, are on the opposite end of the cyclodextrin, which forms the smaller opening. The cyclodextrin molecule is therefore shaped like a truncated cone. The primary hydroxyl groups on the truncated end can rotate to block the cavity partially, whereas the secondary hydroxyl groups at the mouth are held relatively rigid. The interior of the cavity consists of two rings of C–H groups with a ring of glucosidic oxygens in between. This makes the interior relatively hydrophobic in comparison to polar solvents such as water. The mouth of the cyclodextrin cavity, however, is hydrophilic. A Greek letter is used to denote the number of glucose units in the cyclodextrin molecule. For example, α for six, β for seven, γ for eight, and so on.

Some physical properties of the four most common cyclodextrins are listed in Table 1 (3). Other important properties are (*1*) cyclodextrins are nonreducing; (*2*) glucose is the only product of acid hydrolysis; (*3*) molecular weights are always integral numbers of 162.1, the value for glucose; (*4*) cyclodextrins are nontoxic; and (*5*) they do not appreciably absorb ultraviolet (uv) or visible light.

Table 1. Physical Properties of Cyclodextrins

Cyclodextrin	CAS Registry Number	Molecular formula	Number of glucose units	Cavity diameter, nm[a] External	Cavity diameter, nm[a] Internal	Aqueous solubility, *M*
α-CD	[*10016-20-3*]	$C_{36}H_{60}O_{30}$	6	1.37	0.57	0.114
β-CD	[*7585-39-9*]	$C_{42}H_{70}O_{35}$	7	1.53	0.78	0.016
γ-CD	[*17465-86-0*]	$C_{48}H_{80}O_{40}$	8	1.69	0.95	0.179
δ-CD	[*85220-53-7*]	$C_{54}H_{90}O_{45}$	9			v sol

[a]Depth of cavity is 0.78 nm for each cyclodextrin.

The enzyme responsible for producing cyclodextrins from starch, cyclodextrin transglycosylase (CTG), does not cleave a specific number of glucose units from starch. Homologues from 6 to 12 glucose units may be obtained as mixtures having small amounts of branched cyclic molecules and branched open-chain dextrins. A five-membered or smaller dextrin ring has never been obtained. This is most likely attributable to the considerable strain present in small rings.

Several procedures are used to control the ratios of cyclodextrins produced. One is addition of a substance to the reaction mixture that can greatly affect the formation of one specific cyclodextrin over another. For example, in the presence of 1-decanol and 1-nonanol, α-cyclodextrin is produced almost exclusively whereas hexane or toluene promote the production of β-cyclodextrin. Conversely both cyclodextrins are produced simultaneously in the presence of 1-heptanol (2,4).

Immobilization. The ability of cyclodextrins to form inclusion complexes selectively with a wide variety of guest molecules or ions is well known (1,2) (see INCLUSION COMPOUNDS). Cyclodextrins immobilized on appropriate supports are used in high performance liquid chromatography (hplc) to separate optical isomers. Immobilization of cyclodextrin on a solid support offers several advantages over use as a mobile-phase modifier. For example, as a mobile-phase additive, β-cyclodextrin has a relatively low solubility. The cost of γ- or α-cyclodextrin is high. Furthermore, when employed in thin-layer chromatography (tlc) and hplc, cyclodextrin mobile phases usually produce relatively poor efficiencies.

Cyclodextrin stationary phases utilize cyclodextrins bound to a solid support in such a way that the cyclodextrin is free to interact with solutes in solution. These bonded phases consist of cyclodextrin molecules linked to silica gel by specific nonhydrolytic silane linkages (5,6). This stable cyclodextrin bonded phase is sold commercially under the trade name Cyclobond (Advanced Separation Technologies, Whippany, New Jersey). The vast majority of all reported hplc separations on CD-bonded phases utilize this media which was also the first chiral stationary phase (csp) developed for use in the reversed-phase mode.

Applications. The first widely applicable lc separation of enantiomeric metallocene compounds was demonstrated on β-CD bonded-phase columns. Thirteen enantiomeric derivatives of ferrocene, ruthenocene, and osmocene were resolved (7). Retention data for several of these compounds are listed in Table 2, and Figure 2**a** shows the lc separation of three metallocene enantiomeric pairs. β-Cyclodextrin bonded phases were used to resolve several racemic and diastereomeric 2,2-binaphthyldiyl crown ethers (9). These compounds do not contain a chiral carbon but still exist as enantiomers because of the staggered position of adjacent naphthyl rings, and a high degree of chiral recognition was attained for most of these compounds (9).

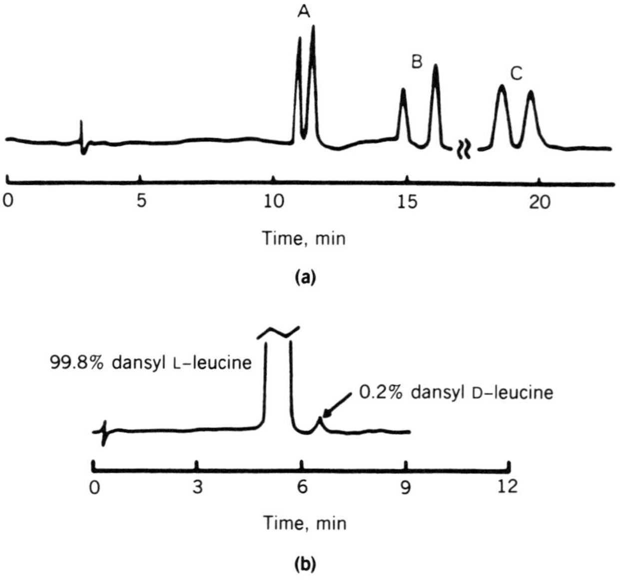

Fig. 2. Chromatogram showing (**a**) the lc separation of A, (±) (*S*)-(1-ferrocenylethyl)thioethanol; B, (±) 1-ferrocenyl-1-methoxyethane; and C, (±) 1-ruthenocenylethanol, on a 25-cm β-cyclodextrin column (see Table 2), and (**b**) the potential use of a β-cyclodextrin column to determine optical purity when one of the enantiomers is present at very low concentration (5,7). Courtesy of Preston Publications.

The β-CD column exhibits excellent selectivity for enantiomers of certain amino acid derivatives. Underivatized amino acids are apparently too small to bind tightly to the β-CD cavity and show no enantiomeric resolution. When a

Table 2. Retention Data for Racemic Compounds Separated on a β-Cyclodextrin Stationary Phase[a]

Compound	CAS Registry Number	Molecular formula	k′[b]	α[c]	R_s[d]
	Complexes				
1-ferrocenyl-1-methoxyethane		$C_{13}H_{17}FeO$	3.8	1.12	1.58
(*S*)-(1-ferrocenylethyl)thioethanol		$C_{14}H_{17}FeSO$	3.0	1.23	2.13
1-ruthenocenylethanol		$C_{12}H_{14}RuO$	4.6	1.11	1.56
alanine β-naphthylamide	[*74144-49-3*]	$C_{13}H_{14}N_2O{\cdot}HCl$	5.1	1.20	2.0
dansyl-leucine	[*102783-70-0*]	$C_{18}H_{24}N_2O_4S{\cdot}C_6H_{13}N$	3.0	1.40	2.4
dansylphenylalanine	[*42808-06-0*]	$C_{21}H_{22}N_2O_4S{\cdot}C_6H_{13}N$	3.1	1.23	1.1
	β-Adrenergic blocker				
propranolol hydrochloride	[*3506-09-0*]	$C_{16}H_{21}NO_2{\cdot}HCl$	2.78	1.04	1.40
	Antihistamine				
chlorpheniramine	[*132-22-19*]	$C_{16}H_{19}ClN_2$	5.86	1.07	1.51
	Sedative–anticonvulsants				
mephenytoin	[*50-12-4*]	$C_{12}H_{14}N_2O_3$	0.48	1.33	1.83
mephobarbital	[*115-38-8*]	$C_{13}H_{14}N_2O_3$	14.80	1.14	1.60
	Nonsteroidal antiinflammatory				
ketoprofen	[*22071-15-4*]	$C_{16}H_{14}O_3$	7.67	1.06	1.24

[a] Refs. 4, 5, and 8.
[b] k′ is the capacity factor for the first eluting isomer.
[c] α is the selectivity factor and is the ratio of the capacity factor of the last eluting isomer to the first eluting isomer.
[d] R_s is the resolution factor. $R_s = 2$ (distance between peaks)/(sum of the bandwidths of the two peaks).

substituent such as a dansyl group is present on the amino acid, strong inclusion complexes with β-CD are formed and baseline separation is achieved (see Table 2) (10). Either the amino or the carboxylate group of the amino acid can be derivatized to obtain chiral recognition. Derivatization of both groups, however, tends to reduce chiral recognition. It is possible to detect as little as 0.2% of one enantiomer in a racemic mixture as shown in Figure 2**b**, thus providing an extremely sensitive test of optical purity (5).

Table 2 gives chromatographic data for different classes of enantiomeric drugs resolved by β-CD bonded phases (8). Drugs for which resolution factors (R_s) greater than 1.0 were obtained include mephenytoin, ketoprofen, chlorpheniramine, and the barbiturates mephobarbital and hexobarbital. Cyclodextrin-bonded phases provide a rapid and specific technique for the pharmacological evaluation of racemic drugs.

Many diastereomers, geometric isomers, and epimers can be successfully resolved using cyclodextrin phases (5,10). For example, the four epimers of estriol [*50-27-1*], $C_{18}H_{24}O_3$, were separated using β-CD, and *cis*-benzo(a)pyrene and *trans*-benzo(a)pyrene were completely resolved on a γ-CD column. Diastereomeric drugs such as the cinchona alkaloids (qv) and antiestrogens have also been separated. Cyclodextrin columns are also of great utility in separating structural isomers such as the ortho-, meta-, and para- isomers of nitroaniline, xylene, cresols, nitrophenols, and substituted benzoic acids (11). Cyclodextrin bonded phases are used as nonconventional reversed phases for routine analyses as a result of the unusual selectivity of the cyclodextrin columns. Uses in routine analyses include the separation of a series of barbiturates, mycotoxins, polycyclic aromatic hydrocarbons, vitamins (qv) and selected dipeptides (12). Cyclodextrin stationary phases can be used in the normal-phase mode with hexane–isopropanol mobile phases. Solutes adsorb to the hydroxyl groups on the outside of the cyclodextrin, rather than forming inclusion complexes. Separations in the normal-phase mode tend to be analogous to those of diol columns.

Mechanism of Separation. There are several requirements for chiral recognition. (*1*) Formation of an inclusion complex between the solute and the cyclodextrin cavity is needed (4,10). This has been demonstrated by performing a normal-phase separation, eg, using hexane–isopropanol mobile phase, on a β-CD column. The enantiomeric solute is then restricted to the outside surface of the cyclodextrin cavity because the hydrophobic solvent occupies the interior of the cyclodextrin. (*2*) The inclusion complex formed should provide a relatively "tight fit" between the hydrophobic species and the cyclodextrin cavity. This is evident by the fact that β-CD exhibits better enantioselectivity for molecules the size of biphenyl or naphthalene than it does for smaller molecules. Smaller compounds are not as rigidly held and appear to be able to move in such a manner that they experience the same average environment. (*3*) The chiral center, or a substituent attached to the chiral center, must be near to and interact with the mouth of the cyclodextrin cavity. When these three requirements are fulfilled the possibility of chiral recognition is favorable.

The unidirectional 2- and 3-hydroxyl groups located at the mouth of the cyclodextrin cavity appear to be of particular importance in chiral recognition. This is seen in Figure 3, which shows computer-generated projections of the lowest free-energy inclusion complexes of (*R*)- and (*S*)-propranolol with β-CD (8). The (*R*)- and (*S*)-propranolol are placed identically inside the cyclodextrin cavity and the hydroxyl groups attached to the chiral carbon for the enantiomers are placed in the same position for ideal hydrogen bonding to the 3-hydroxyl group of the cyclodextrin. Important differences exist between the complexes with respect to the secondary amine group. In the (*R*) complex the respective bond distances between the nitrogen and the cyclodextrin 2- and 3-hydroxyl groups are 0.33 and 0.28 nm, respectively. This allows for two reasonable hydrogen bond interactions. The same amine group in the (*S*)-propranolol complex is positioned less favorably for hydrogen bonding: closest bond distances are 0.38 and 0.45 nm, respectively. These models suggest that (*R*)-propranolol can preferentially interact with β-cyclodextrin in a way that the (*S*)-isomer cannot, resulting in chiral recognition by the cyclodextrin molecule. These findings agree with the three-point attachment concept introduced in the 1950s (13).

α_1-Acid Glycoprotein Chromatographic Phases. *Properties and Structure.* α_1-Acid glycoprotein (α_1-AGP) has a molecular mass of about 41,000 and consists of a peptide chain having 181 amino acid residues and five carbohydrate units (14,15). Two cystine disulfide cross-linkages connect residues 5 and 147 and

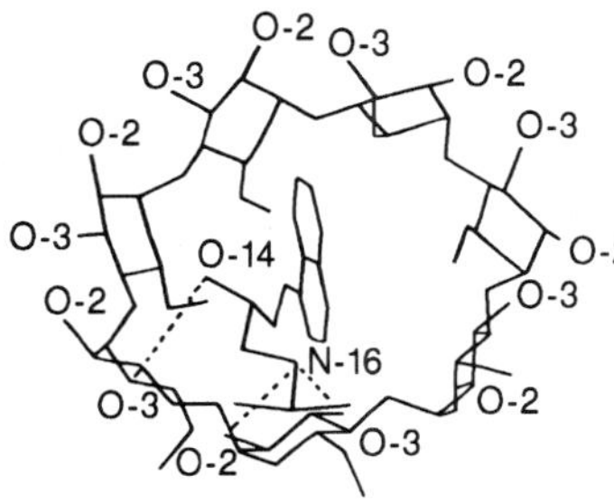

(a)

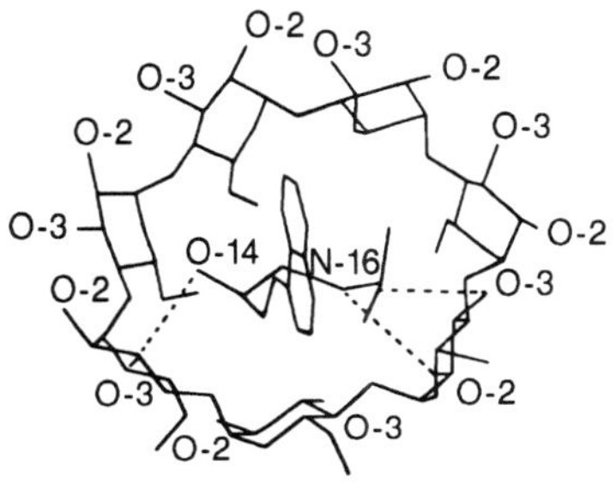

(b)

Fig. 3. Computer projections of β-cyclodextrin inclusion complexes of (**a**) (*R*)-propranolol and (**b**) (*S*)-propranolol from x-ray crystallographic data. Dotted lines represent potential hydrogen bonds (see text). The configurations shown represent the optimal orientation of each isomer on the basis of the highest degree of hydrogen bonding and complexation (8).

residues 72 and 164. The carbohydrate units comprise 45% of the molecule and contain sialic acid, hexosamine, and neutral hexoses. In phosphate buffer the isoelectric point of the protein is 2.7. AGP is a very stable protein and tolerates organic solvents as well as high temperatures. AGP columns may be used over a wide pH range without being denatured. Denaturation, as measured by changes in the optical rotation of the molecule, may be caused by boiling in distilled water or adding 10 *M* LiBr, 10 *M* urea, or 5 *M* HCl (14).

Immobilization. The solid-phase support used for bonding AGP is silica gel. The general preparation is based on bonding the protein by charge forces to diethylaminoethylsilica followed by a cross-linking procedure (16). A final reduction to secondary amines is performed using cyanoborohydride. The silica has a particle diameter of 10 μm, a large pore volume, and a large surface area. The commercially available EnantioPac AGP column (LKB Products, Bromma, Sweden) is manufactured using this procedure. A second generation AGP column, Chiral-AGP (ChromTech AB, Norsburg, Sweden), uses covalent linkage of the protein onto silica with cross-linking of adjacent protein molecules (17,18). This latter packing contains 5 μm spherical silica particles having a smaller pore volume and a smaller surface area giving a stationary phase that is more mechanically stable and resistant to hydrolytic attack. Bonding capacity of the chiral phase and loading capacity of solute are directly dependent on the amount of protein that is bound. The amount of AGP bound on the silica depends on the bonding technique and on the accessible surface area of the silica.

Applications. Retention and selectivity of solutes on AGP is regulated by the concentration and properties of modifiers in the aqueous-based mobile phase. Retention and enantioselectivity usually decrease with increasing concentration of uncharged organic modifiers such as methanol, ethanol, 2-propanol, and acetonitrile in the mobile phase. However, in some cases, enantioselectivity can be improved by adding uncharged modifiers to the mobile phase (19). For example, addition of 2-propanol as an organic modifier gave improved chiral resolution for mephenytoin [*50-12-4*], $C_{12}H_{14}N_2O_2$, and methylphenobarbital. On the other hand, the enantioselectivity for the tertiary amines mepivacaine and bupivacaine is unaffected by additions of up to 8% of propanol to the mobile phase, despite the fact that the retention of these solutes decreases markedly under these conditions (18) (see ANESTHETICS; HYPNOTICS, SEDATIVES, AND ANTICONVULSANTS).

Column retention is generally increased if the modifier has a charge opposite to that of the solute; a modifier having the same charge as the solute generally decreases retention. For some compounds the addition of charged modifiers is essential to achieve chiral recognition. The fenthiazin derivative, propiomazine [*362-29-8*], $C_{20}H_{24}N_2OS$, was not resolved on an AGP column when phosphate buffer, pH 7.55, was used as the mobile phase. After addition of 1 m*M* of the tertiary amine *N,N*-dimethyloctylamine (DMOA) to the mobile phase, however, the enantiomers of propiomazine were baseline resolved (20) and longer retention times were obtained. Increase in retention is believed to be caused by competition between the solute and DMOA in binding to the AGP. The changes in selectivity are believed to result from reversible changes in the conformation of AGP brought about by changes in pH. Similar effects on selectivity and retention have also been observed by addition of tetrapropylammonium bromide or tetrabutylammonium bromide (15,16,21,22).

Changes in the pH and temperature of the mobile phase can have a significant effect on both retention and selectivity. Changing the pH of the mobile phase has a profound effect on retention and selectivity for basic, acidic, and nonprotolytic compounds. For example, the separation factors for hexobarbital [*56-29-1*], $C_{12}H_{16}N_2O_3$ (weakly acidic) and metoprolol [*37350-58-6*], $C_{15}H_{25}NO_3$ (basic) (see CARDIOVASCULAR AGENTS), increase with increasing pH, but the enantioselectivity for stronger acids such as 2-phenoxypropionic acid ($pK_a = 4.6$) generally decreases (21). Column temperature also strongly influences retention and enantioselectivity. It has been generally reported that retention, resolution, and separation factors decrease with increasing temperature, whereas efficiency increases.

AGP columns have wide application for the direct separation of enantiomers of many different classes of drugs, amines, acids, and nonprotolytic compounds (18,23). Acidic drugs resolved include ibuprofen [*15687-27-1*], $C_{13}H_{18}O_2$, ketoprofen [*22071-15-4*], $C_{16}H_{14}O_3$, and naproxen [*22204-53-1*], $C_{14}H_{14}O_3$, and basic drugs such as disopyramide [*3737-09-5*], $C_{21}H_{29}N_3O$, tropicamide [*1508-75-4*], $C_{17}H_{20}N_2O_2$, atropine [*51-55-8*], $C_{17}H_{23}NO_3$, and homatropine [*87-00-3*], $C_{16}H_{21}NO_3$, have also been separated (21,24). Table 3 lists some racemic compounds that have been completely resolved using a Chiral-AGP column (18). The AGP columns are also commonly used in the determination of enantiomers present at low concentrations in biological fluids such as plasma and urine (19,25). Metoprolol was extracted from plasma and injected on a Chiral-AGP column for separation (25). It was possible to measure as little as 2 nmol/L plasma using fluorescence detection after separation.

Table 3. Baseline-Resolved Racemic Compounds Using Chiral-AGP

Compound	CAS Registry Number	Molecular formula	k′[a]	α[b]
alprenolol	[*13655-52-2*]	$C_{15}H_{23}NO_2$	17.9	1.19
atenolol	[*29122-68-7*]	$C_{14}H_{22}N_2O_3$	2.91	1.36
bupivacaine	[*2180-92-9*]	$C_{18}H_{28}N_2O$	6.66	1.29
cyamemazine	[*3546-03-0*]	$C_{19}H_{21}N_3S$	5.95	1.55
ephedrine	[*90-81-3*]	$C_{10}H_{15}NO$	3.86	1.34
ketamine	[*6740-88-1*]	$C_{13}H_{16}ClNO$	7.00	1.26
metoprolol	[*37350-58-6*]	$C_{15}H_{25}NO_3$	4.72	1.26
oxprenolol	[*6452-71-7*]	$C_{15}H_{23}NO_3$	16.1	1.29
pheniramine	[*86-21-5*]	$C_{16}H_{20}N_2$	11.3	1.33
pindolol	[*13523-86-9*]	$C_{14}H_{20}N_2O_2$	13.9	1.21
verapamil	[*52-53-9*]	$C_{27}H_{38}N_2O_4$	15.6	1.32
warfarin	[*81-81-2*]	$C_{19}H_{16}O_4$	8.18	1.39

[a]k′ is the capacity factor for the first eluting isomer.
[b]α is the selectivity factor. See Table 2.

Mechanism of Separation. Whereas the scientific basis of separation is well documented, ie, differences in binding between drug enantiomers and proteins, the mechanism for chiral recognition is not clearly understood. It is well known that the conformation of a native protein in solution can be altered by addition of organic modifiers and changes in pH. It is assumed that the AGP molecule has a high degree of flexibility even after bonding to the silica surface. Therefore adding modifiers to the mobile phase can alter the AGP molecule so that new chiral phases having different binding properties are induced (18,21).

Bovine Serum Albumin Chromatographic Phases. *Properties and Structure.* Bovine serum albumin (BSA) is a globular protein having a molecular mass of 66,210. It consists of 581 amino acids in a single chain, and 17 intrachain disulfide bridges form nine double loops (26). Having an isoelectric point of 4.7, BSA is a relatively acidic protein. It is highly soluble in water, but like most globular proteins it precipitates from solution at high salt concentrations. BSA exhibits hydrophobic character and numerous examples of organic compounds binding to albumins have been reported (27). Whereas hydrophobic interactions contribute greatly to the total affinity of organic ligands for BSA, there are other contributions to consider, mainly electrostatic interactions, hydrogen bonding, and charge-transfer processes.

The first observation of the enantioselective properties of an albumin was made in 1958 (28) when it was discovered that the affinity for L-tryptophan exceeded that of the D-enantiomer by a factor of approximately 100. This led to more studies in 1973 of the separation of DL-tryptophan [*54-12-6*], $C_{11}H_{12}N_2O_2$, on BSA immobilized to Sepharose (29). After extensive investigation of the chromatographic behavior of numerous racemic compounds under different mobile-phase conditions, a BSA-SILICA hplc column (Resolvosil-R-BSA, Macherey-Nagel GmvH, Duren, Germany) was introduced in 1983.

Retention and stereoselectivity on the BSA columns can be changed by the use of additives to the aqueous mobile phase (30). Hydrophobic compounds generally are highly retained on the BSA, and a mobile-phase modifier such as 1-propanol can be added to obtain reasonable retention times. The retention and optical resolution of charged solutes such as carboxylic acids or amines can be controlled by pH and ionic strength of the mobile phase.

Applications. Various *N*-derivatives of amino acids (qv) are resolvable on BSA columns. These *N*-amino acid derivatives include benzenesulfonyl-, phthalimido-, 5-dimethylamino-1-naphthalenesulfonyl- (DANSYL-), 2,4-dinitrophenyl- (DNP-), and 2,3,6-trinitrophenyl- (TNP-) derivatives (30). Amines such as Prilocain, (±)-2-(propylamino)-*o*-propionotoluidide, a local anesthetic (Astra Pharm. Co.), are also resolved on BSA. The aromatic amino acids DL-tryptophan, 5-hydroxy-DL-tryptophan, DL-kynurenine [*343-65-7*], $C_{10}H_{12}N_2O_3$, and 3-hydroxy-DL-kynurenine [*484-78-6*], and drugs such as warfarin, phenprocoumon, and benzodiazepine derivatives can be separated on BSA as well.

Cellulose Triacetate and Cellulose Derivatives. *Properties and Structure.* Cellulose [*9004-34-6*] (qv) and other polysaccharides have long been known to have chiral recognition properties. Cellulose is readily available, inexpensive, and has good chemical stability. Microcrystalline cellulose triacetate [*9012-90-3*] (MCA), which is commercially available, is the product of the heterogeneous acetylation of microcrystalline cellulose (31). Various cellulose ester derivatives supported on macroscopic silica gel are available as hplc columns from Diacel (Diacel Chemical Industries, Ltd., Tokyo, Japan). These columns have good mechanical stability and mobile phases such as hexane–2-propanol or alcohols are used. The mechanism of chiral recognition on cellulose-based phases is unknown, although hydrogen bonding and ligand inclusion play a part in resolution (32,33) (see CELLULOSE ESTERS).

Applications. MCA is used for the resolution of many classes of chiral drugs. Polar compounds such as amines, amides, imides, esters, and ketones can be resolved (34). A phenyl or a cycloalkyl group near the chiral center seems to improve chiral selectivity. Nonpolar racemates have also been resolved, but charged or dissociating compounds are not retained on MCA. Mobile phases used with MCA columns include ethanol and methanol.

The Diacel columns can be used for the separation of a wide variety of compounds, including aromatic hydrocarbons having hydroxyl groups, carbonyls and sulfoxides, barbiturates, and β-blockers (35,36). There are presently nine different cellulose derivative-based columns produced by Diacel Chemical Industries. The different columns each demonstrate unique selectivities so that a choice of stationary phases is available to accomplish a separation.

Miscellaneous Chiral Chromatographic Techniques. Cyclodextrins are often used as chiral mobile-phase additives in miscellaneous chromatographic techniques as well as for stationary phases for thin-layer chromatography (35,36). Gas chromatography (gc) columns consisting of derivatized cyclodextrins coated on the capillary column wall are also commercially available (Advanced Separation Technologies, Whippany, New Jersey). Chiral analytes that can be vaporized without degradation or racemization are suitable for analysis on the gc columns. Several reviews and books have been published on the various chiral chromatographic methods (35–37).

Biopolymers in Immunology

Biopolymers are employed in many immunological techniques, including the analysis of food, clinical samples, pesticides, and in other areas of analytical chemistry. Immunoassays (qv) are specific, sensitive, relatively easy to perform, and usually inexpensive. For repetitive analyses, immunoassays compare very favorably with many conventional methods in terms of both sensitivity and limits of detection.

Antigens. One condition that must be met for the application of an immunochemical method is that the analyte must be capable of stimulating an immune response leading to the formation of antibodies in the immunized animal. These antibodies can then be isolated and used as highly specific analytical reagents (immunoassays). Analytes that can combine with the corresponding antibodies are called antigens. There are physical and chemical restrictions on the types of analytes that may be used as immunoassay antigens. In general, large, rigid, chemically complex molecules make good antigens. For example, serum albumin, mol wt ca 60,000, is a good antigen because of its large size, complex structure, and good structural stability. Large repeating polymers such as lipids and carbohydrates, make poor antigens as a result of the simplicity of their structure, even though they may have high molecular weights and good structural stability. One molecule having mol wt 750, the smallest molecular mass that has demonstrated antigenic behavior, is *p*-azobenzene-arsonatetrityrosine (38).

A small analyte is not necessarily prevented from analysis by immunoassay, because a small molecule that is linked to a large antigenic molecule forms a new species having a modified surface structure. Small molecules made antigenic in this manner are called haptens. The larger molecule to which they are attached is called the carrier. The immune response to a hapten-carrier complex is actually triggered by two signals; one from the hapten and one from its carrier. This is necessary, otherwise the hapten would provoke an immune response on its own. Even with the dual signal response, the signal triggered by the complex is still hapten-specific because the immune response is in part triggered by the hapten.

Antibodies. Antibodies are proteins, found in many body fluids such as tears, saliva, and urine, that are present in highest concentrations in blood serum. Because antibodies are proteins (qv), they may be characterized by such physical properties as solubility, electrostatic charge, isoelectric point, and molecular weight. The particular proteins which exhibit antibody activity are the immunoglobulins (Ig). The principal immunoglobulin in blood serum is immunoglobulin G (IgG), the structure of which is similar to the other immunoglobins. IgG is a glycoprotein having mol wt 150,000. It has a Y shape (38) and its structure is shown in Figure 4. There are estimated to be approximately 10 million potential combinations of antigen-binding specificities resulting from

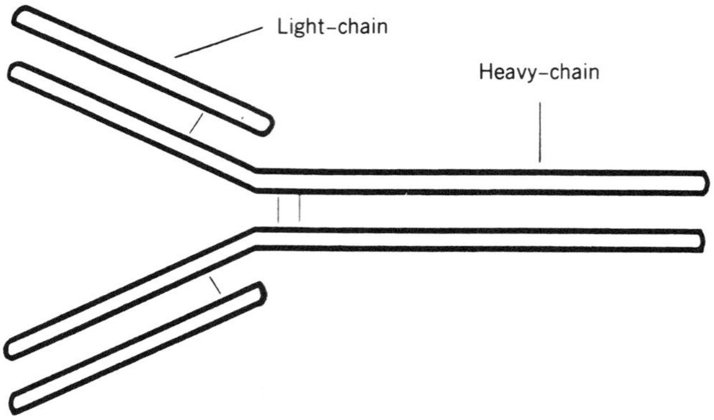

Fig. 4. Simple model of an IgG molecule showing light- and heavy-chain segments where a line (|) between the chains represents a disulfide bond.

light- and heavy-chain combinations in the immunoglobulins. The possibility of utilizing all of these combinations as reagents in immunochemical methods is highly interesting though improbable.

General Methodology. *Common Procedures.* The general analytical scheme for immunochemical methods is rather simple. The analyte of interest, the antigen (Ag), reacts with the analytical reagent, the corresponding antibody (Ab), forming an immunochemical antigen–antibody complex:

$$\text{Ag} + \text{Ab} \rightleftharpoons \text{Ag–Ab}$$

The immunochemical interaction between the antigen and antibody is very specific. By labeling either the antigen or antibody, the method's sensitivity is increased. The most frequently used labels to increase sensitivity are radionuclides (see RADIOISOTOPES) where the assay process is called radioimmunoassay (RIA), or enzymes where the assay is named enzyme immunoassay (EIA) (see ENZYME APPLICATIONS).

Labeling. Radioisotopic labeling, one of the first labeling methods used, is still prominent in assays where the use of nonradioisotopic labels has not been feasible (39). Labeling with enzyme to produce a spectrophotometrically detectable product is frequently used. Fluorescent labeling, used in conjunction with fluorescent polarization detection techniques, can be utilized for small molecules. A commercially available system uses this technique for monitoring drug levels in biological fluids (40). Rare-earth chelates such as those of europium [*7440-53-1*], Eu, are also successfully used as fluorescent labels (41). The rare-earth chelates have a large Stokes shift that helps to reduce interference from light scattering and from background fluorescence. In addition, these labels have fluorescent lifetimes lasting several hundred microseconds allowing delayed fluorescent sampling.

Chemiluminescent labels, in which the luminescence is generated by a chemical oxidation step, and bioluminescent labels, where the energy for light emission is produced by an enzyme–substrate reaction, are additional labeling types (39,42). Luminol [*521-31-3*], $C_8H_7N_3O_2$, and acridine [*260-94-6*], $C_{13}H_9N$, derivatives are often used as chemiluminescent labels.

Variations in Methods. The various immunochemical methods can differ in a number of ways. For example, the analytical reagent may be crude antiserum, monoclonal antibodies, isolated immunoglobulin fractions, etc. The conditions under which the method is run, detection of the antigen–antibody complex, and the techniques used to increase sensitivity or specificity of the reaction all may be varied.

Heterogeneous and Homogeneous Assays. The various immunochemical techniques may be roughly divided into two groups. The first involves homogeneous procedures in which the separation of the bound and free labeled analyte, ie, the radionuclide or the enzyme, is not necessary. In contrast, heterogeneous immunochemical techniques require separation of the bound and free labeled analyte. The most common of these techniques is the enzyme-linked immunosorbent assay (ELISA) (43). In this technique either the antibody or the antigen (analyte) is attached to a solid phase, such as the walls of a polystyrene tube or surface of a plastic bead. Both a competitive and a double antibody (sandwich) ELISA technique are available for measuring antigens.

Sandwich Assays. In the sandwich technique, the most widely used ELISA, the solid surface is first coated with an appropriate antibody as shown in Figure 5. The sample solution containing the antigen (analyte) is then added and allowed to react with the bound antibody on the solid surface. After the reaction any remaining unbound antigens are washed away. Then an enzyme-labeled antibody, specific for a different site on the antigen, is added in a known amount for reaction with the bound antibody–antigen complex. After the reaction any unbound enzyme-labeled antibodies are washed away. A substrate is added which, when acted upon by the bound enzyme, produces a color change the amount of which is a direct measurement of specific enzyme-conjugated bound antibody and therefore of antigen present.

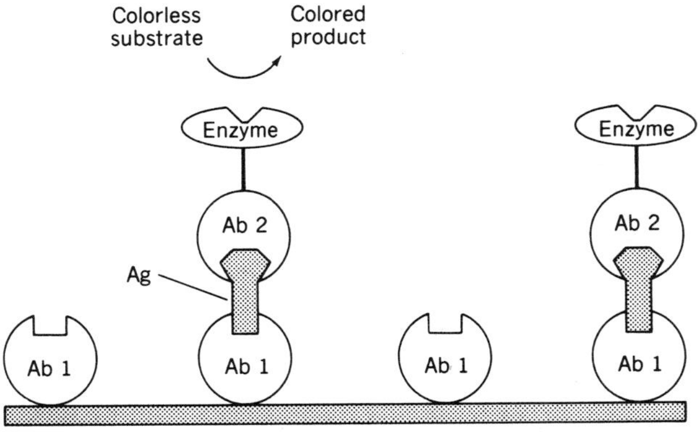

Fig. 5. Scheme of the last steps for the sandwich ELISA. Where Ab 1 represents the the surface-bound antibody, Ag, the antigen, and Ab 2, the enzyme-labeled antibody.

Competitive Assays. The next most widely used type of ELISA is the competitive assay (39,43). In this technique the analyte of interest is mixed with a known amount of enzyme-labeled antigen and both compete for a limited number of binding sites on an antibody that is adsorbed on a solid support. After binding the excess free enzyme-labeled antigen or excess test antigen is washed away. Then a substrate is added which is acted upon by the enzyme-labeled antigen yielding a colored product. The amount of color development is proportional to the amount of enzyme-labeled conjugate bound to the antibody. Little or no color change indicates that the unlabeled antigen of interest was present in the test solution and was bound to the antibody. A color change indicates that enzyme-labeled antigen was bound to the antibody, so that little or no unlabeled antigen was present in the original test solution.

Applications. Immunoassays are used in many different disciplines, having clinical, industrial, agricultural, and environmental applications. This technique has made possible rapid analysis of such varied analytes as viruses, toxins, hormones, foreign proteins, drugs, and insecticides.

As a result of regulations by the Food and Drug Administration, very sensitive and specific analysis techniques have been developed to test for food additives (qv) and possible contaminants in food, many of which are immunochemical methods. For example, botulism toxin and staphylococcal enterotoxins can be detected by ELISA. Animals for slaughter can be quickly screened for diseases such as brucellosis, tuberculosis, and cholera (43). The level of many regulatory enzymes in food can be monitored by immunoassays. Immunoassays are used in beer (qv) brewing to measure the levels of malt enzyme and amyloglucosidase, and in cheese production to measure the level of chymosin, microbial rennets, and beef pepsin [*9001-75-6*] (44). Immunoassays are also useful in the detection of foreign proteins added to foodstuffs. For example, the addition of cheaper meats, such as horse or kangaroo, into beef or pork products can be detected by immunoassays, as well as the inclusion of soybean protein in meat products. Immunochemical methods are often employed to control the production of gluten-free dietary products and hypoallergenic milk. Commercially produced ELISA immuno-kits for food analysis include tests for soya protein (Biokits Ltd., Clwyd, UK), meat species (Biokits Ltd., Clwyd, UK), papain [*9001-73-4*] (Labsystems Oy, Helsinki, Finland), *Salmonella* (Organon Teknika Corp., Durham, North Carolina), trichinosis (Agritech Systems, Portland, Maine; Idetek, San Bruno, California), aflatoxin B_1 [*1162-65-8*], $C_{17}H_{12}O_6$, (Neogen Corp., Lansing, Michigan; Agritech Systems, Portland, Maine; Immuno Systems, Biddeford, Maine; Biotech Research Lab, Rockville, Maryland; and Biokits Ltd., Clwyd, UK), aflatoxin M_1 [*6795-23-9*], $C_{17}H_{12}O_7$, (Neogen Corp., Lansing, Michigan), ochratoxin A (AFRC Food Research Institute, Norwich, UK; Biokits Ltd., Clwyd, UK), atrazine [*1912-24-9*] (Immuno Systems, Biddeford, Maine), antibiotic residues in milk (Angenics Inc., Cambridge, Massachusetts), zearalenone [*17924-92-4*], $C_{18}H_{22}O_5$, (Neogen Corp., Lansing, Michigan), and *Staphylococcus endotoxun* (Igen, Rockville, Maryland) (45).

Mycotoxins, toxic metabolites of some fungi, can be assayed by immunochemical techniques to determine concentration in animal feed and foodstuffs. Some of the analytes assayed in kits and the detection limits are listed in Table 4

(45). These assays are especially advantageous for routine analysis of large samples of foodstuffs (45,46).

Table 4. Detection Limits of Immunoassays Developed for Mycotoxins and Pesticides

Analyte	CAS Registry Number	Molecular formula	Method	Analyzed food	Detection limit, μg/kg[a]
			Mycotoxins		
aflatoxin B_1	[*1162-65-8*]	$C_{17}H_{12}O_6$	RIA, ELISA	peanuts, cereal	1–5
aflatoxin M_1			RIA, ELISA	milk	0.02–0.1[b]
ochratoxin A	[*303-47-9*]	$C_{20}H_{18}ClNO_6$	RIA, ELISA	cereals	0.2–2.0
zearalenone	[*17924-92-4*]	$C_{18}H_{22}O_5$	ELISA	cereals	20
			Pesticides		
parathion	[*56-38-2*]	$C_{10}H_{14}NO_5PS$	RIA	lettuce	10
diflubenzuron	[*35367-38-5*]	$C_{14}H_9ClF_2N_2O_2$	ELISA	milk	2
diclofop-methyl	[*51338-27-3*]	$C_{16}H_{14}Cl_2O_4$	EIA	sugarbeet, wheat, soybean	100–120
benomyl	[*17804-35-2*]	$C_{14}H_{18}N_4O_3$	RIA	fruit	500

[a]Detection limit is microgram of analyte per kilogram of sample unless otherwise noted.
[b]Units are microgram of analyte per liter of sample.

Immunochemical methods that utilize radioisotopic labeling can detect the use of anabolic sex hormones that increase the growth in meat animals. Stilbene [*588-59-0*], $C_{14}H_{12}$, trenbolone [*10161-33-8*], and zeranol [*55331-29-8*], $C_{18}H_{26}O_5$, can be successfully monitored by these immunoassay techniques (45). In order to prevent veterinary drugs from being transported to the human food chain, radioisotopic immunoassays were developed to monitor veterinary antibiotics such as penicillin and chloramphenicol [*56-75-7*], $C_{11}H_{12}Cl_2N_2O_5$, in meat, milk, and eggs (qv) (see ANTIBIOTICS; MEAT PRODUCTS; MILK AND MILK PRODUCTS).

Pesticide contamination can also be monitored by ELISA. Immunoassays for pesticides are advantageous in that time-consuming sample preparation and purification steps can be avoided or reduced and very high specificity can be obtained in many cases. Various sample kits are available to assay plant tissues, soil, water, and biological fluids. Commercial kits are also available for the determination of atrazine, chlordane [*57-74-9*], $C_{10}H_6Cl_8$, heptachlor [*76-44-8*], $C_{10}H_5Cl_7$, aldicarb [*116-06-3*], $C_7H_{14}N_2O_2S$, aldicarb sulfone, glyphosate [*1071-83-6*], $C_3H_8NO_5P$, and chlorpyrifos [*2921-88-2*], $C_9H_{11}Cl_3NO_3PS$ (47).

Another use of immunoassays is in clinical diagnosis. Biological fluids such as serum and bronchial secretions can be assayed for indications of various disorders, including leukemia, carcinomas, and cancers. Some of the enzymes assayed and the corresponding disorders are listed in Table 5 (44). These assays are based mainly on changes in enzyme concentrations that result from enzyme release from injured tissue or from the changed metabolism of an affected organ. The goal in clinical diagnosis is to develop highly sensitive methods so that symptoms can be detected at the earliest possible stage.

Table 5. Immunoassays of Enzymes in Clinical Diagnosis

Enzyme	Assayed medium	Detected disorder(s)
plasmin	serum	acute leukemia
pepsin, pepsinogen	serum	stomach ulcer, cancer, chronical gastritis
aspartate aminotransferase	serum	infarction, hepatitis
trypsin	serum	acute pancreatitis
pancreatic elastase	serum	pulmonar emphysema, acute pancreatic necrosis, arteriosclerosis
microbial elastase	bronchial secretions	chronical lung infections
acid phosphatase	serum	prostatic cancer
cystein proteinase	tumor medium	breast tumor
ribonuclease	serum	malignant diseases, renal failure
superoxide dismutase		lung cancer
glutamate-oxalacetate transferase	serum	differentiation of chronic liver disease
L-lactate dehydrogenase	serum	differentiation of heart cell damage

Biopolymers as Biosensors

Selectivity is an important consideration in analytical chemistry. Biologically derived polymers can be used as highly selective immobilized reagents in analytical applications. The first reported use of immobilized biopolymers as biosensors (qv) for the detection of an analyte was made in 1962 (48). Since that first reported use there has been a great deal of development and application of immobilized biopolymers in analytical chemistry.

Immobilized Enzymes. The immobilized enzyme electrode is the most common immobilized biopolymer sensor, consisting of a thin layer of enzyme immobilized on the surface of an electrochemical sensor as shown in Figure 6. The enzyme catalyzes a reaction that converts the target substrate into a product that is detected electrochemically. The advantages of immobilized enzyme electrodes include minimal pretreatment of the sample matrix, small sample volume, and the recovery of the enzyme for repeated use (49). Several reviews and books have been published on immobilized enzyme electrodes (50–52).

The enzyme can be immobilized on the electrode by several techniques (53). The simplest method, first used in 1962, is to trap an enzyme solution between the electrode surface and a semipermeable membrane. Another technique is to immobilize the enzyme in a polymer gel such as polyacrylamide which is coated on the electrode surface. Very thin-membrane films can be obtained by electropoly-

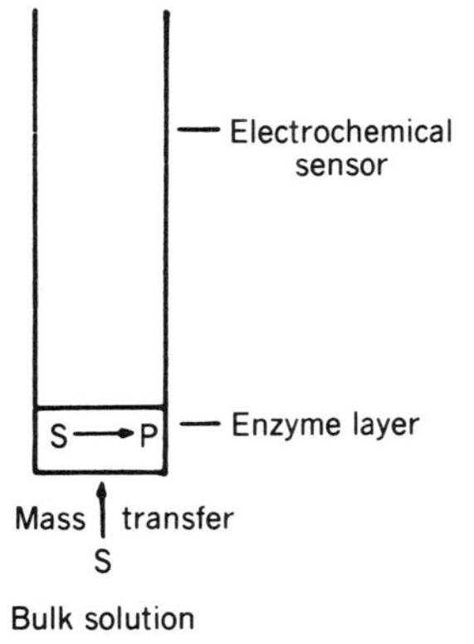

Fig. 6. Diagram of an immobilized enzyme electrode, where S is the substrate and P is the enzyme-bound substrate product.

merization techniques (49,54,55) using polypyrrole, polyindole, or polyphenylenediamine films, among others. These thin films (qv) offer the advantage of improved diffusion of substrate and product that improves the response time. The enzyme can also be covalently bound to the electrode surface either directly or by using a small linking molecule (56). Covalent immobilization has the advantage of improving the stability of the enzyme (57).

The response of the immobilized enzyme electrode can be made independent of the enzyme concentration by using a large excess of enzyme at the electrode surface. The electrode response is limited by the mass transport of the substrate. Using an excess of enzyme often results in longer electrode lifetimes, increased linear range, reduced susceptibility to pH, temperature, and interfering species (58,59). At low enzyme concentrations the electrode response is governed by the kinetics of the enzyme reaction.

Gas-Sensing Enzyme Electrodes. Potentiometry and amperometry are the most common electrochemical techniques to employ enzyme electrodes. Potentiometric gas-sensing and ion-selective electrodes have been converted into enzyme electrodes and used in various analytical applications (53). The gas-sensing electrodes for carbon dioxide and ammonia are most frequently converted to enzyme electrodes because of the lack of response to any dissolved ionic interferents. Decarboxylating or deaminating enzymes are immobilized to these gas-sensing electrodes so that the enzyme reaction product, CO_2 or NH_3, is detected. These potentiometric immobilized enzyme sensors are highly selective and are used for the detection of urea, creatinine, uric acid, amino acids (qv), and nucleotides, as well as other compounds (50,53). Amperometric electrodes are generally coupled with oxidase or dehydrogenase enzymes. Oxidase enzymes can be immobilized on a Clark oxygen electrode and used to detect the amount of oxygen consumed in the enzyme reaction. For example, in the determination of creatinine in blood serum, the enzymes creatinine amidohydrolase, creatine amidinohydrolase, and sarcosine oxidase are coimmobilized on the polypropylene membrane of a Clark oxygen electrode (60). The enzymes catalyze the decomposition of creatinine with the consumption of oxygen, which is monitored by the Clark electrode. The oxidase enzymes can also be trapped on a platinum electrode and used to measure the amount of hydrogen peroxide produced in the enzyme reaction. For example, glucose oxidase [*9001-37-0*] covalently attached to platinum wire via glutaraldehyde [*111-30-8*], $C_5H_8O_2$, was used to determine glucose [*50-99-7*], $C_6H_{12}O_6$, levels by monitoring the amount of hydrogen peroxide formed (56). Dehydrogenase enzymes immobilized on glassy carbon or platinum electrodes are used to measure nicotinamide NADH or electron-transfer mediators such as ferrocyanide (53,60). Amperometric immobilized enzyme electrodes are used in the determination of glucose and other sugars, amino acids, cholesterol [*57-88-5*], $C_{27}H_{46}O$, creatinine, uric acid, and alcohols (53,60).

Multienzyme Electrodes. Coupling the reactions of two or more immobilized enzymes increases the number of analytes that can be measured. An electroinactive component can be converted by an enzyme to a substrate that is subsequently converted by a second enzyme to form a detectable end product (57). For example, a maltose [*69-79-4*], $C_{12}H_{22}O_{11}$, sensor uses the enzymes glucoamylase and glucose oxidase, which convert maltose to glucose and then to gluconic acid [*526-95-4*], $C_6H_{12}O_7$, with the production of hydrogen peroxide, which is detected using a platinum anode (53).

Multienzyme electrodes can increase sensitivity from micromolar to nanomolar detection levels (53,57). In this case the substrate is converted to a detectable product by one enzyme, then that product is recycled into the initial substrate by another enzyme resulting in an amplification of the response signal. For example, using lactate oxidase and lactate dehydrogenase immobilized in poly(vinyl chloride), an amplification of 250 was obtained for the detection of lactate (61).

Miscellaneous Biopolymeric Electrodes. Whereas immobilized enzyme electrodes are most commonly used in electrochemical systems, other types of biopolymers are also employed. For example, a catalytic biosensor uses immobilized antibodies capable of catalyzing chemical reactions (62). In this case, acidic products formed by the catalytic antibody are then detected electrochemically. A micro-pH electrode is used and the catalytic antibody is immobilized in a porous membrane. Biopolymers are also used as semipermeable electrode coatings. This use of biopolymers rejects undesired interferents from the electrode-sensing surface while allowing the transport of the analyte. For example, cellulose acetate is used in this manner as a size-discrimination coating for electrodes (63,64). Lipid coatings prevent polar electroactive compounds from reaching the electrode-sensing surface while letting hydrophobic substances pass through. Cast films of palmitic or stearic acids (65,66) and phosphatidylcholine (67,68) are also used for this purpose. A mixed lipid layer consisting of phosphatidylcholine and cholesterol has improved mechanical stability as well as permeable selective properties (61).

Enzyme Immunosensors. Enzyme immunosensors are enzyme immunoassays coupled with electrochemical sensors. These sensors (qv) require multiple steps for analyte determination, and either sandwich assays or competitive binding assays may be used. Both of these assays use antibodies for the analyte of interest attached to a membrane on the surface of an electrochemical sensor. In the sandwich assay type, the membrane-bound antibody binds the sample antigen, which in turn binds another antibody that is enzyme-labeled. This immunosensor is then placed in a solution containing the substrate for the labeling enzyme and the rate of product formation is measured electrochemically. The rate of the reaction is proportional to the amount of bound enzyme and thus to the amount of the analyte antigen. The sandwich assay can be used only with antigens capable of binding two different antibodies simultaneously (53).

In the competitive binding assay, sample antigen competes with enzyme-labeled antigen for the antibody binding sites on the membrane. The immunosensor is then placed in a solution containing the substrate for the labeling enzyme and the rate of reaction is measured electrochemically. The reaction rate is inversely proportional to the concentration of the sample antigen. This approach is applicable to many analytes including small hapten molecules such as therapeutic drugs.

Enzyme immunosensors are employed for the determination of Hepatitis B surface antigen, IgG, alpha-fetoprotein, estradiol, theophylline, insulin [*9004-10-8*], and albumin (69,70). However, these immunosensors generally have slow response times and slow reversibility (57).

Enzyme immunosensors are used in flow injection systems and liquid chromatography to provide automated on-line analyses (71–73). These systems are capable of continuously executing the steps involved in the immunoassays, including the binding reactions, washing, and the enzyme reaction, in about 10 minutes.

Economic Aspects

Enantiomeric separations are expected to continue to have a considerable economic impact on the development of new drugs and therapy in the biomedical field. The Food and Drug Administration (FDA) has issued a set of guidelines on New Drug Applications in which the question of stereochemistry was approached directly for the manufacturing of drug substances (74). As of 1987, the FDA requires knowledge of the molecular structure of the drug substance. For chiral compounds this includes identification of all chiral centers. Whereas FDA guidelines do not discuss conditions under which a determination of absolute configuration is essential, it would be appropriate for supporting the manufacture of optically pure drugs (75). These requirements have led to a rapid growth in techniques that measure optical purity of drug substances. Listed in Table 6 are 1991 prices and manufacturers of chiral columns.

The importance of immunoassays for food monitoring and in the detection of diseases is expected to continue to grow as techniques and detection limits improve. In 1991, prices for most immunoassay kits ranged from \$250 to \$800 depending on the specific kit.

The development of biosensors is expected to benefit monitoring therapeutic drug levels, office testing, and implantable devices because of the advantages of cost-saving automation and data handling. A number of enzyme-based electrodes are commercially available and their manufacturers are given (69). Biosensors are expected to become the predominant sensor technology by the year 2000. The largest market for biosensors is expected to be in clinical laboratories and drug monitoring (76) (see AUTOMATED INSTRUMENTATION–CLINICAL CHEMISTRY).

Table 6. Chiral Stationary Phases, Manufacturers, and Prices

Stationary phase	Manufacturer	Price[a], $
CYCLOBOND I, β-cyclodextrin	Advanced Separation Technologies Whippany, N.J.	425
CHIRAL-AGP, α_1-acid glycoprotein	Chrom Tech AB Norsburg, Sweden	1100
Enantiopac AGP, α_1-acid glycoprotein	LKB Products Bromma, Sweden	
RESOLVOSIL-BSA, bovine serum albumin	Macherey Nagel Germany	925
CHIRACEL-O series, cellulose derivatives	Diacel Chemical Industries Tokyo, Japan	1080

[a]Prices are from 1991 catalogs.

BIBLIOGRAPHY

1. M. L. Bender and M. Komiyama, *Cyclodextrin Chemistry,* Springler-Verlag, Berlin, 1978.
2. J. Szejtli, *Cyclodextrins and Their Inclusion Complexes,* Akademiai Kiado, Budapest, 1982.
3. W. L. Hinze, *Sep. Purif. Methods* **10,** 159 (1981).
4. T. J. Ward and D. W. Armstrong, *J. Liq. Chromatogr.* **9**(2,3), 407 (1986).
5. D. W. Armstrong and W. DeMond, *J. Chromatogr. Sci.* **22,** 411 (1984).
6. U.S. Pat. 4,539,399 (1985), D. W. Armstrong.
7. D. W. Armstrong, W. DeMond, and B. P. Czech, *Anal. Chem.* **57,** 481 (1985).
8. D. W. Armstrong, T. J. Ward, R. D. Armstrong, and T. E. Beesley, *Science* **232,** 1132 (1986).
9. D. W. Armstrong, T. J. Ward, A. Czech, B. P. Czech, and R. A. Bartsch, *J. Org. Chem.* **50,** 5556 (1985).
10. D. W. Armstrong and co-workers, *Anal. Chem.* **57,** 234 (1985).
11. C. A. Chang, Q. Wu, and L. Tan, *J. Chromatogr.* **361,** 199 (1986).
12. D. W. Armstrong, W. DeMond, W. L. Hinze, and T. E. Riehl, *J. Liq. Chromatogr.* **8,** 261 (1985).
13. C. E. Dalgleish, *J. Chem. Soc.,* 3940 (1952).
14. K. Schmid, in F. W. Putman, ed., *The Plasma Proteins,* Academic Press, New York, 1975, p. 184.
15. G. Schill, I. W. Wainer, and S. A. Barkan, *J. Liq. Chromatogr.* **9,** 641 (1986).
16. J. Hermansson, *J. Chromatogr.* **298,** 67 (1984).
17. Chiral-AGP, *Application Note NO. 1,* Chrom Tech AB, Norsburg, Sweden, 1988.
18. J. Hermansson, *Trends Anal. Chem.* **8,** 251 (1989).
19. M. Enquist and J. Hermansson, *Chirality* **1**(3), 209 (1989).
20. J. Hermansson, *J. Chromatogr.* **316,** 537 (1984).
21. J. Hermansson and M. Eriksson, *J. Liq. Chromatogr.* **9,** 621 (1986).
22. G. Schill, I. W. Wainer, and S. A. Barkan, *J. Liq. Chromatogr.* **365,** 73 (1986).
23. J. Hermansson and G. Schill, in M. Zief and L. J. Crane, eds., *Chromatographic Chiral Separations,* Vol. 40, Marcel Dekker, New York, 1988, p. 245.
24. E. Arvidsson, S. O. Jansson, and G. Schill, *J. Chromatogr.* **506,** 579 (1990).

25. B. A. Persson, K. Balmer, P. O. Lagerstrom, and G. Schill, *J. Chromatogr.* **500,** 629 (1990).
26. Th. Peters, Jr., in Ref. 14, Vol. 1, p. 133.
27. M. C. Meyer and D. E. Guttman, *J. Pharm. Sci.* **57,** 895 (1968).
28. R. H. McMenamy and J. L. Oncley, *J. Biol. Chem.* **223,** 1436 (1958).
29. K. K. Stewart and R. F. Doherty, *Proc. Nat. Acad. Sci.* **70,** 2850 (1973).
30. S. Allenmark, *J. Liq. Chromatogr.* **9,** 425 (1986).
31. A. Ichida and T. Shibata, in Ref. 23, pp. 219–243.
32. Y. Okamoto, M. Kawashima, and K. Hatada, *Chem. Lett.,* 739 (1984).
33. Y. Okamoto, I. Okamoto, and K. Hatada, *J. Am. Chem. Soc.* **106,** 5357 (1984).
34. G. Blaschke, *J. Liq. Chromatogr.* **9,** 341 (1986).
35. M. Zief and L. J. Crane, eds., *Chromatographic Chiral Separations,* Vol. 40, Marcel Dekker, New York, 1988.
36. D. W. Armstrong and S. M. Han, *CRC Crit. Rev. Analyt. Chem.* **19**(3), 175 (1988).
37. S. Hara and J. Cazes, *J. Liq. Chromatogr.* **9**(2,3), (1986).
38. I. R. Tizard, *Immunology,* Saunders College Publishing, New York, 1984, p. 13.
39. E. Stinshoff, W. Stein, W. G. Wood, and P. Laska, *Anal. Chem.* **59,** 339R (1987).
40. R. C. Morton and E. P. Diamandis, *Anal. Chem.* **62,** 1841 (1990).
41. E. P. Diamandis and T. K. Christopoulos, *Anal. Chem.* **62,** 1149A (1990).
42. Y. Tatsu and S. Yoshikawa, *Anal. Chem.* **62,** 2103 (1990).
43. D. Monroe, *Anal. Chem.* **56,** 920A (1984).
44. J. Kas, L. Fukal, and P. Rauch, *Trends Anal. Chem.* **5,** 205 (1986).
45. L. Fukal and J. Kas, *Trends Anal. Chem.* **8,** 112 (1989).
46. B. Mattiasson, M. Nilsson, P. Berden, and H. Hakanson, *Trends Anal. Chem.* **9,** 317 (1990).
47. A. Klauser, *Bio-Technol.* **5,** 551 (1987).
48. L. C. Clark and C. Lyons, *Ann. N.Y. Acad. Sci.* **102,** 29 (1962).
49. C. Malitesta, F. Palmisano, L. Torsi, and G. Zambonin, *Anal. Chem.* **62,** 2735 (1990).
50. G. C. Guilbault, *Analytical Uses of Immobilized Enzymes,* Marcel Dekker, New York, 1984, p. 112.
51. R. K. Kobos, *Trends Anal. Chem.* **102,** 29 (1987).
52. G. G. Guilbault and J. M. Kauffmann, *Biotechnol. Appl. Biochem.* **9,** 95 (1987).
53. R. K. Kobos, *Trends Anal. Chem.* **6,** 6 (1987).
54. Y. Kajiya, H. Sugai, C. Iwakura, and H. Yoneyama, *Anal. Chem.* **63,** 49 (1991).
55. M. Umana and J. Waller, *Anal. Chem.* **58,** 2979 (1986).
56. S. K. Bey, G. J. Moody, and J. D. R. Thomas, *Analyst* **114,** 29 (1989).
57. M. Thompson and U. J. Krull, *Anal. Chem.* **63,** 393A (1991).
58. M. Aizawa, in T. Seyama, K. Fueki, J. Shiokawa, and S. Suzuki, eds., *Proceedings of the International Meeting of Chemical Sensors, Fukuoka,* Elsevier, Amsterdam, The Netherlands, 1983, p. 683.
59. B. Walters, *Anal. Chem.* **55,** 498A (1983).
60. V. K. Nguyen, C. Wolff, J. L. Seris, and J. Schwing, *Anal. Chem.* **63,** 611 (1991).
61. J. Wang and Z. Lu, *Anal. Chem.* **62,** 826 (1990).
62. G. F. Blackburn and co-workers, *Anal. Chem.* **62,** 2211 (1990).
63. G. Sittampalam and G. S. Wilson, *Anal. Chem.* **55,** 1608 (1983).
64. J. Wang and L. D. Hutchins, *Anal. Chem.* **57,** 1536 (1985).
65. K. Tanaka and R. Tamamushi, *J. Electroanal. Chem. Interfacial Electrochem.* **61,** 171 (1987).
66. I. Uchida, A. Ishino, T. Matsue, and K. Itaya, *J. Electroanal. Chem. Interfacial Electrochem.* **266,** 455 (1989).
67. O. J. Garcia, P. A. Quintela, and A. E. Kaifer, *Anal. Chem.* **61,** 979 (1989).

68. O. Chastel, J. M. Kauffmann, G. J. Patriarche, and G. D. Christian, *Anal. Chem.* **61,** 171 (1989).
69. F. Scheller and co-workers, *Analyst* **114,** 653 (1989).
70. F. Schubert, D. Kirstein, K. L. Schroder, and F. W. Scheller, *Anal. Chim. Acta* **169,** 391 (1985).
71. H. Liu, J. C. Yu, D. S. Bindra, R. S. Givens, and G. S. Wilson, *J. Anal. Chem.* **63,** 666 (1991).
72. P. C. Gunaratna and G. S. Wilson, *Anal. Chem.* **62,** 402 (1990).
73. J. Emneus and L. Gorton, *Anal. Chem.* **62,** 263 (1990).
74. *Guidelines for Submitting Supporting Documentation in Drug Applications for the Manufacture of Drug Substances,* Office of Drug Evaluation and Research (HFD-100), FDA, Rockville, Md., 1987, pp. 3, 4.
75. W. H. DeCamp, *Chirality* **1,** 2 (1989).
76. *Biosensors: Today's Technology, Tomorrow's Products,* Technical Insights Incorporated, Lee, N.J., 1987.

TIMOTHY WARD
Millsaps College

BIOSENSORS

The detection of trace and low levels of biologically active substances is among the most significant and challenging analytical technologies (see TRACE AND RESIDUE ANALYSIS). Devices that enable accurate determination of low concentrations of biologically important compounds have far reaching applications in the fields of medicine, ecology, food science, biotechnology, and military science. Many detection methods employ chemical systems derived directly from biological organisms. Any discrete sensing device that relies on a biologically derived component as an integral part of its detection mechanism is known as a biosensor (see BIOPOLYMERS, ANALYTICAL TECHNIQUES).

Basic Components

All biosensors are composed of two basic parts, the molecular recognition component and the transducer component. The molecular recognition component is typically a complex chemical system usually extracted or derived directly from a biological organism. The function of the molecular recognition component is to interact specifically with a target compound, ie, the chemical to be detected. The molecular recognition component must be capable of discerning the presence of a target in a solution containing possibly hundreds of other compounds, some of which may have molecular structures closely resembling that of the target. The term selectivity is used to quantify the degree to which a molecular recognition system responds specifically to a target while rejecting compounds having related molecular structures. There are three principal classes of molecular recognition components used in biosensors: enzymes or catalytic proteins (qv); immunoglobins, which are biological macromolecules that selectively bind to foreign substances invading an organism; and chemoreceptors, biomolecular units responsible for sensory reception in living organisms (see ENZYME APPLICATIONS). Each class has advantages and disadvantages, and each class comprises a vast number of highly selective possible molecular recognition components.

Once the molecular recognition component interacts with its target, a user-readable signal must be generated. This is the purpose of the transducer component. Ideally, the generated signal should not only report the presence of target but should also relay information concerning the amount or concentration of target compound in the test solution. As of this writing, there were four main types of analytical transducer components employed in biosensors: optical, electrochemical, field effect transistor, and piezoelectric.

The term sensitivity is used to quantify the ability of a biosensor to reliably report target level. There are two separate uses of the term. First, the detection limit of a biosensor is the lowest concentration of target for which the biosensor provides a reliable, reproducible, and unambiguous response. Increasing the sensitivity of a biosensor frequently refers to decreasing the biosensor's detection limit. Second, sensitivity can also refer to the smallest change in concentration from some nominal level of target compound that the biosensor can unambiguously report. Increasing sensitivity in these latter terms is usually a more difficult task than decreasing the detection limit. Herein sensitivity relates to determining differences in concentration. Detection limit is considered as a separate parameter.

Biosensor design involves consideration of many factors. Molecular recognition components are frequently large complex biomolecules or biological macroassemblies and usually perform efficiently within narrow ranges of pH, temperature, and ionic strength. Degradation and the limited lifetimes of biologically derived components can be important. Additionally, immobilization of enzymes sometimes results in decreased target activity. Factors influencing the selection of an appropriate transducer component include compatibility with the molecular recognition component and the environment in which the biosensor is to be used. Development and optimization of biosensor techniques is a rapidly growing research area; as of the early 1990s, publications on fundamental and applied aspects of biosensing were appearing at a rate approaching 400 papers per year.

Biosensors Based on Chemoreceptors

Chemoreceptors, complex biomolecular macroassemblies, are responsible in part for an organism's ability to sense chemicals in its environment. Most organisms rely heavily on trace level chemical detection to identify food, locate a mate, and sense danger. Chemoreceptors are quite common in biology and can be found in taste buds and the olfactory system. In a biological organism, arrays of highly selective chemoreceptors are employed in chemical sensing. A hound dog, for example, is not only capable of sensing trace amounts of chemicals from dead cells shed from a person to be tracked, but the dog can actually identify the individual from which the cells came. Chemoreceptors are also involved in neural communications and a variety of other nonsensory mechanisms. For example, hormones (qv) are chemical messengers that are sometimes used to trigger organ membranes to change permeability properties by binding to a chemoreceptor located in the membrane. Chemoreceptors have high selectivity, making them excellent candidates for use in biosensors.

Although the integration of complex chemoreceptor arrays, such as those of the olfactory system, into biosensors remains a challenge, research into the use of single chemoreceptors is well underway (1–8). Entire chemoreceptor-containing membranes dissected directly from certain crustaceans have been incorporated. The use of intact structures has the advantage of relatively easy preparation. These materials are readily available and are in an optimal state for molecular recognition. In addition, intact receptor assemblies have been evolved for precisely the purpose of detecting trace chemicals in solution.

In one biosensor design, the chemoreceptive nerve fibers of the antennules of the blue crab *Callinectes sapidus* are connected to a micropipet electrode. This assembly has been termed a receptrode (2). The receptrode created from *Callinectes sapidus* responds to the presence of amino acids (**1**) (qv) in concentrations as low as 10^{-9} *M*.

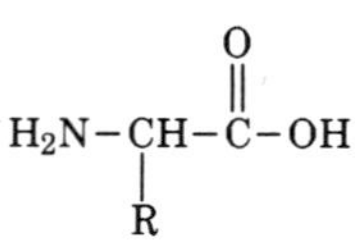

(**1**)

Moreover, the receptrode has an extremely rapid response time, requiring only two to three milliseconds to fully respond to a target concentration change. The response times of conventional chemical sensors (qv) are typically from several seconds to several minutes. The receptrode exhibits a response that follows the empirical relationship

$$R = \frac{R_{\max}}{\left[1 + \left(\frac{K}{C}\right)^n\right]}$$

where $R_{\max}$ is the maximum observed response, K is the equilibrium constant for the binding of the amino acid to the chemoreceptor, C is the molar concentration of target amino acid, and n is the cooperativity coefficient, a constant included to compensate for differences in sensitivity between different receptors.

This relationship is very similar to the Michaelis-Menton equation commonly used in enzyme kinetics. Interestingly, this receptrode is sensitive within a much wider range of target concentrations than that predicted by Michaelis-Menton kinetics possibly because the crab aesthestasc, hairlike structures in the antennule that contains the chemoreceptors, is actually a chemoreceptor array with up to 500 distinct receptors having different detection limits. The living blue crab is known to detect amino acid concentrations of 10^{-15} *M*, six orders of magnitude better than the detection limit of the biosensor created from the same antennule because the transducer component of the receptrode biosensor is a single-micropipet electrode and cannot take full advantage of the extremely low detection limits afforded by the receptor array. Similar biosensors that detect hormones, nucleotides, drugs, and toxins, using chemoreceptor-based molecular recognition components extracted from crayfish, lobster, and catfish, have also been created (5,6).

Chemoreceptors have also been isolated from sensory organs, purified, and used as molecular recognition components in biosensors. A biosensor is being developed to detect low levels of the neurotransmitter acetylcholine [*51-84-3*], $C_7H_{16}NO_2$, (**2**) (7). The biosensor comprises an acetylcholine receptor-containing lipid membrane coupled to an ion-sensitive field effect transistor (ISFET). An ISFET is a semiconductor device that has a channel below it and is insulated from a gate above it (see SEMICONDUCTORS). On either side of the channel are layers of *n*-type silicon, one called the source, the other the drain. Electrical conduction in the channel is related to changes in the electric field above the gate, which is in turn very sensitive to the ionic strength of a solution in contact with the gate. A schematic of a chemoreceptor-modified ISFET biosensor is shown in Figure 1. In this biosensor application, the receptor-impregnated membrane, extracted from the electric fish *Torpedo Californica,* is purified and coated onto the gate of the transistor. On binding acetylcholine, the local ionic strength of the solution in the vicinity of the chemoreceptor membrane increases. Consequently, the source-drain current in the ISFET increases. Changes in source-drain current are easily

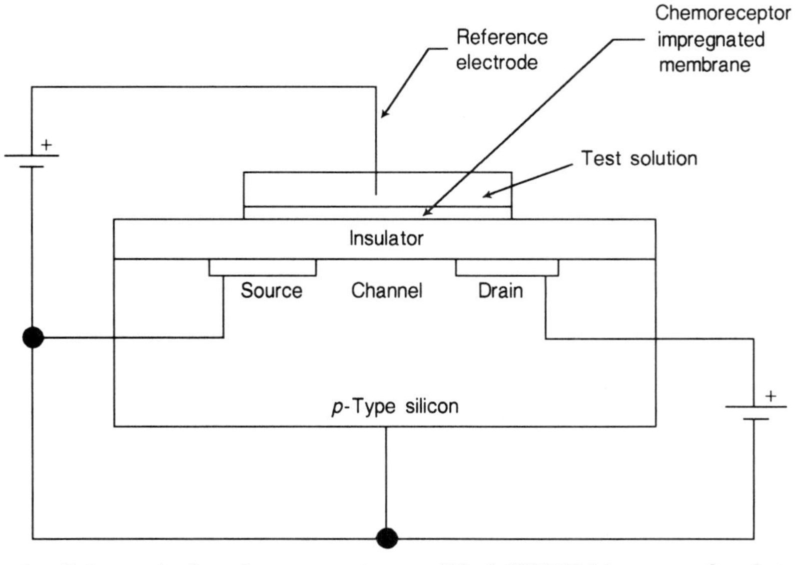

Fig. 1. Schematic for chemoreceptor-modified ISFET biosensor for detection of acetylcholine where the source and drain are both *n*-type silicon.

detected by a simple ammeter and can be related to the concentration of acetylcholine in the test solution. Micromolar quantities of acetylcholine are utilized.

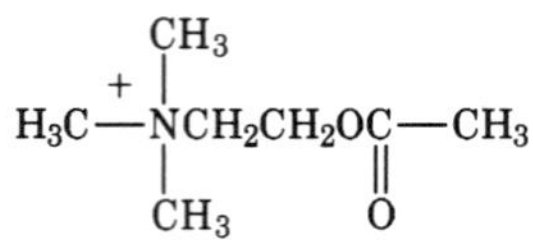

(2)

The low detection limit, high sensitivity, and fast response times of chemoreceptor-based biosensors result primarily from the extremely high binding constants of the receptor R for the target substrate S. The receptor–substrate binding may be described

$$R + S \rightleftharpoons R\text{–}S$$

where R–S is the bound receptor–substrate complex.

The binding constant is defined by the equilibrium constant for the reaction

$$K = \frac{[R\text{–}S]}{[R][S]}$$

where the square brackets denote molar concentrations.

Receptor–substrate-binding constants are typically between 10^9 and 10^{12} M^{-1}, at equimolar ratios, implying that when a receptor–substrate complex forms, only one ten-thousandth of one percent of substrate molecules remain unbound.

As of this writing chemoreceptor-based biosensors are not yet on the commercial market. Only a few chemoreceptors have been isolated and their substrates identified. Moreover, those chemoreceptors that have been isolated are fragile and have limited lifetimes.

Biocatalytic Biosensors

In a biocatalytic biosensor the molecular recognition component is an enzyme. Enzymes, macromolecular catalysts that are manufactured by plants and animals, affect the rates of biochemical reactions. Virtually all of the millions of chemical reactions involved in life processes have associated enzymes controlling the rates. Collectively, there are several thousand enzymes known and perhaps many thousand more yet to be discovered.

Enzymes are basically specialty proteins (qv) and consist of amino acids, the exact sequence of which determines the enzyme structure and function. Although enzyme molecules are typically very large, most of the chemistry involving the enzyme takes place in a relatively small region known as the active site. In an enzyme-catalyzed reaction, binding occurs at the active site to one of the molecules involved. This molecule is called the substrate. Enzymes are highly selective, although most enzymes can be fooled into binding molecules having structures closely resembling that of the substrate. Enzyme–substrate binding has often been thought of as being analogous to a lock and key. Where every lock has a specific key that opens it, every enzyme has a specific substrate to which it binds. Once the substrate has undergone conversion to the reaction product, the active site is freed and is ready to participate in another substrate binding.

The high selectivity of enzymes for their substrates is exploited in numerous ways for use in biosensors. In the simplest type of biocatalytic biosensor, an example of which is shown in Figure 2, enzyme molecules are immobilized onto the surface of an electrode. The electrochemical potential of the enzyme-modified electrode is monitored as the electrode is placed in a test solution. If the test solution contains the target compound, ie, the enzyme's substrate, the electrode potential changes. The magnitude of the change is related to the concentration of target in the test solution. Such enzyme-modified electrodes are only useful for enzymes that catalyze oxidation-reduction (redox) reactions or change solution pH. Table 1 lists some common redox enzymes, their substrates, and their potential for use in biosensing.

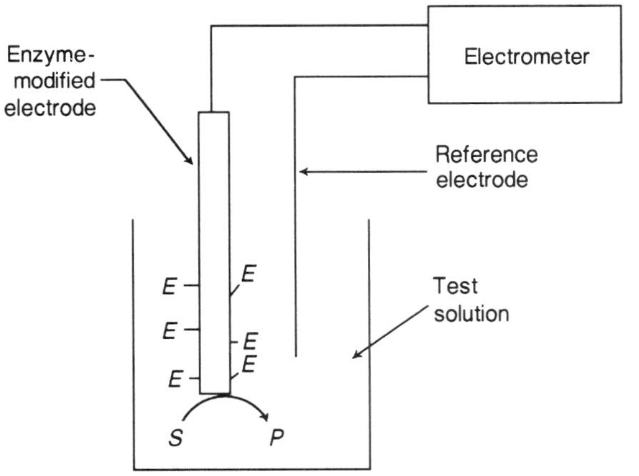

Fig. 2. Schematic of an electrochemical biocatalytic biosensor showing enzyme *E* immobilized onto the electrode where *S* is the substrate (target) and *P* is the product of the enzyme–substrate reaction.

Another potential use of redox enzymes in biosensing involves the conversion of enzymes from electrical insulators to electrical conductors (8–12). An electrically conductive redox enzyme is expected to have greatly enhanced efficacy in electrochemically transduced biosensors. In the native, ie, unmodified, enzyme the active site is generally surrounded by insulating polypeptide regions, representing a formidable barrier to electron transfer from the enzyme to the electrode. Chemical modification of these protein regions can, however, render the regions conductive, imparting facile electrical accessibility to the active site. A typical modification involves the covalent bonding of redox moieties to several sites on the enzyme, usually at the lysine residues. Glucose oxidase [*9001-37-0*], for example, has been modified using ferrocene [*101-54-5*], $C_{10}H_{10}Fe$, groups, which act as electron relays, mediating conduction to the electrode. Another approach has become known as electrical wiring of redox enzymes. In this method, the

Table 1. Redox Enzymes, Substrates, and Biosensing Potential

Enzyme	CAS Registry Number	Substrate	Potential usage
lactate dehydrogenase	[*9001-60-9*]	lactate	test for metabolic acidosis
D-β-hydroxybutyrate dehydrogenase	[*9001-55-2*]	β-hydroxybutyrate	diabetes testing
glyceraldehyde-3-phosphate dehydrogenase		glyceraldehyde-3-phosphate	test for glycolysis
alcohol dehydrogenase		ethanol	blood alcohol testing
glucose-6-phosphate dehydrogenase	[*9001-40-5*]	glucose-6-phosphate (link to glucose)	diabetes testing
malate dehydrogenase	[*9001-64-3*]	L-malate	test for citric acid cycle
glycerol-1-phosphate dehydrogenase		glycerol-1-phosphate (link to triglycerides)	triglyceride testing
L-glutamate dehydrogenase		L-glutamate	test for citric acid cycle
6-phosphogluconate dehydrogenase	[*9001-82-5*]	6-phosphogluconate	test for pentose phosphate
isocitrate dehydrogenase	[*9001-58-5*]	L-isocitrate	test for citric acid cycle
monoamine oxidase	[*9001-66-5*]	monoamine	drug testing, neurotransmitters, amphetamine, cocaine

redox enzyme is electrostatically complexed to a synthetic polymer that contains a high concentration of redox sites. The polymer provides a conduction pathway or molecular wire between the enzyme and the electrode.

The simple cases where one enzyme is employed afford a limited scope of potential targets. Usually two or more enzyme reactions are coupled, as exemplified by the development of a piezoelectrically-transduced biocatalytic biosensor that couples two enzyme reactions to detect glucose [*492-62-6*], $C_6H_{12}O_6$, (**3**) (13). In this biosensor a quartz radio crystal is functionalized with the enzyme glucose-6-phosphate dehydrogenase. As shown in Figure 3, a thin film of Prussian blue [*14038-43-8*], $C_{18}N_{18}Fe_7$, is then coated onto the crystal.

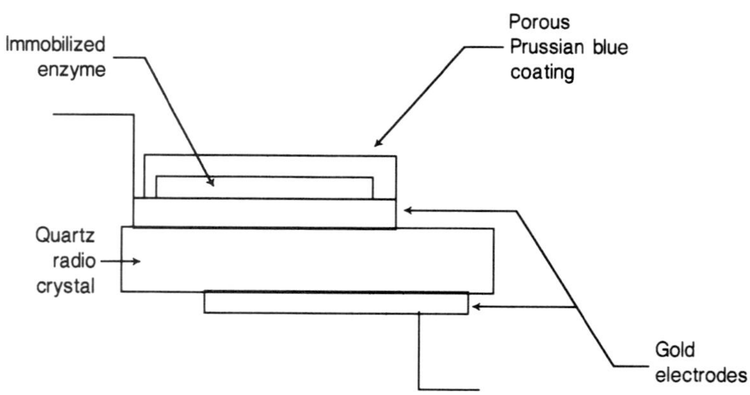

Fig. 3. Detail of enzyme-modified quartz radio crystal used in a piezoelectric biocatalytic biosensor.

The resonant frequency of the crystal is inversely proportional to the mass of the Prussian blue coating. When the immobilized enzyme acts on its substrate, glucose-6-phosphate [*54010-71-8*], $C_6H_{11}O_9P$, (**4**), electrons are transferred to the Prussian blue. In order to maintain electrical neutrality, cations such as K^+ or Na^+ present in the test solution are taken up by the Prussian blue film, resulting in an increased film mass that can easily be detected as a frequency change in the radio crystal (Fig. 4). In practice, a second enzyme, hexokinase [*9001-50-8*] is added to the test solution. Hexokinase forms glucose-6-phosphate from glucose. Hence, the target is glucose. This biosensor has a detection limit of approximately 10 μM glucose, and it is highly selective. Unfortunately, the response times are on the order of several minutes because of the slow diffusion rates of the cations into the Prussian blue film.

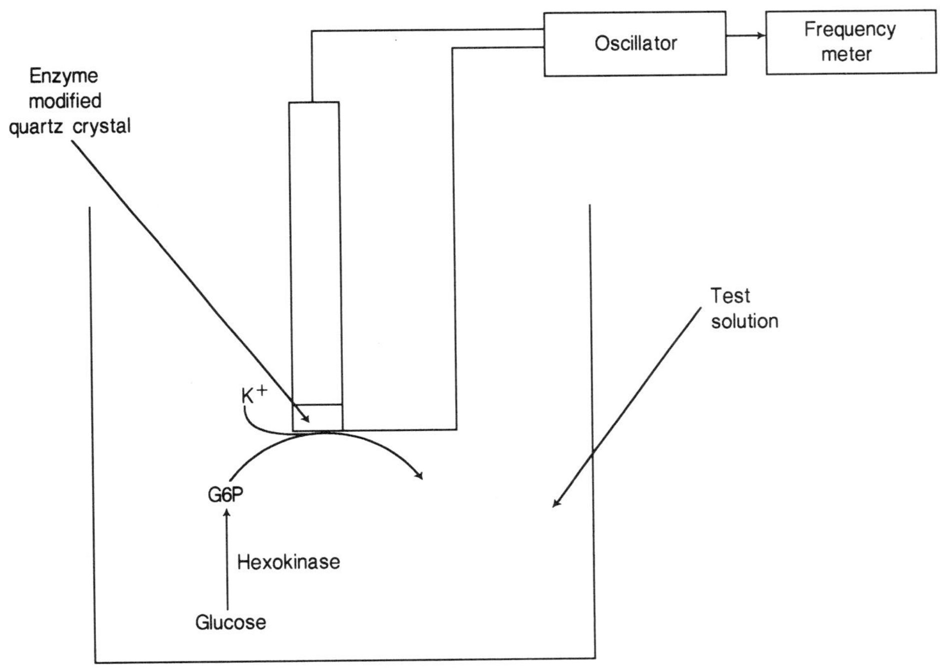

Fig. 4. Schematic of a multisequence biosensor in which the target glucose is first converted to glucose-6-phosphate, G6P, in the test solution by hexokinase. G6P then reacts selectively with glucose-6-phosphate dehydrogenase immobilized on the quartz crystal surface. Electrons released in the reaction then chemically reduce the Prussian blue film (see Fig. 3), forcing an uptake of potassium ions. The resulting mass increase is manifested as a frequency change in the crystal.

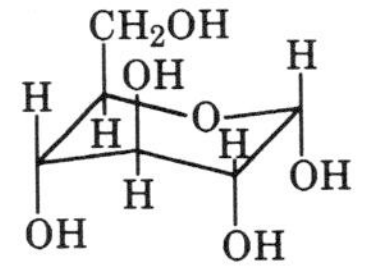

(**3**)

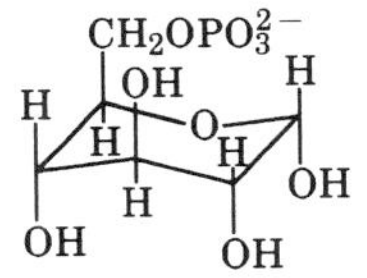

(**4**)

Several biocatalytic biosensors employing optical transducer components have also been developed (14). One ingenious design for the detection of glucose (15,16) involves immobilization of the enzyme glucose oxidase [*9001-37-0*] onto the end of an optical fiber, along with a fluorescent dye such as pyrenebutyric acid [*3443-45-6*], $C_{20}H_{16}O_2$, or perylenedibutyrate [*93838-72-3*], $C_{28}H_{24}O_4$. Glucose oxidase selectively catalyses the reaction of glucose to gluconic acid [*526-95-4*], $C_6H_{12}O_7$.

$$\text{glucose} + O_2 \longrightarrow \text{gluconic acid} + H_2O_2$$

The dye is excited by light supplied through the optical fiber (see FIBER OPTICS), and its fluorescence monitored, also via the optical fiber. Because molecular oxygen, O_2, quenches the fluorescence of the dyes employed, the intensity of the fluorescence is related to the concentration of O_2 at the surface of the optical fiber. Any glucose present in the test solution reduces the local O_2 concentration because of the immobilized enzyme resulting in an increase in fluorescence intensity. This biosensor has a detection limit for glucose of approximately 100 μM; response times are on the order of a minute.

Several other biosensors have been developed using this oxygen-quenched fluorescence approach. Target species include ethanol [*64-17-5*], C_2H_6O, hydrogen peroxide [*7722-84-1*], H_2O_2, lactate, and xanthine [*69-89-6*], $C_5H_4N_4O_2$, using alcohol oxidase, catalase [*9001-05-2*], lactate oxidase, and xanthine oxidase, respectively. An additional technique for biocatalytic biosensors involves the firefly chemiluminescent reaction (17):

$$\text{ATP} + \text{luciferin} + O_2 \longrightarrow \text{AMP} + \text{oxiluciferin} + CO_2 + \text{pyrophosphate} + \text{light}$$

Adenosine triphosphate [*56-65-5*] (ATP) and adenosine monophosphate [*16955-44-5*] (AMP) are relatively small biomolecules common to all organisms and are readily available. Luciferin is a biomolecule extracted from firefly lanterns. This reaction is catalyzed by the enzyme luciferase, also extracted from firefly lanterns. A biosensor has been developed in which luciferase is immobilized onto the tip of an optical fiber, which is immersed in test solution. Luciferin is then added to the test solution. The target is ATP. Any ATP in the test solution catalyzes the chemiluminescent reaction at the tip of the optical fiber. Through the use of photon counting techniques, ATP concentrations as low as 2.8×10^{-10} M can be detected. This scheme may prove very promising for the future of biosensors, because ATP is a cofactor in many biochemical processes. In addition, ATP is produced or consumed in numerous enzyme reactions, making this scheme generalizable to a large number of potential targets by enzyme coupling.

Immunological Biosensors

The molecular recognition components of immunological biosensors are antibodies. Antibodies, manufactured when a foreign organism invades an animal's blood, are proteins that constitute part of an animal's immune system. The antibodies bind to and incapacitate the invader and are highly specific. Specific antibodies can be made by injecting an animal, usually a rabbit, with target substance, allowing time for the animal to produce the target-specific antibodies, and then extracting the antibodies from the animal's blood. A substance that produces antibodies when injected in an animal is said to be immunogenic. Antibodies can be made toward bacteria, viruses, fragments of microorganisms, large biomolecules, and even some small organic molecules that have been appropriately derivitized to make them immunogenic. The immunogenic molecule, organism, or fragment that generated the antibody is called the antigen. The isolated antibodies bind the specific antigen toward which they were produced exhibiting extremely high selectivity and forming an antibody–antigen complex (see IMMUNOASSAY).

An electrochemically-transduced immunological biosensor is being developed (18) in which antibodies specific to the Wassermann antigen, produced during a syphilis infection, are immobilized in a cellulose membrane separating two solution compartments. One compartment is coupled to a reference electrode; the other contains test solution. Upon exposure to Wassermann antigen in the test solution, an electrochemical potential change occurs between the two solutions. This change in potential is measured by an electrometer.

Another very clever technique, this one employing monoclonal antibodies, forms the basis of a device being marketed by Molecular Devices Corp., Menlo Park, California. The device is capable of detecting down to attomole (10^{-18} mol) quantities of target protein or polynucleotide (19–23). One application of this technique is as a sensor for DNA. Pharmaceuticals (qv) manufactured using recombinant DNA techniques may consequently contain trace amounts of potentially harmful DNA contaminants. The ability to assay for such contaminants is of paramount importance in the production of safe drugs. The test solution is first incubated with a conjugate of biotin [*58-85-5*], $C_{10}H_{16}N_2O_3S$, (**5**) and single-stranded DNA-binding protein (SSB) extracted from *Escherichia coli*. Any DNA in the test solution is captured by the SSB forming a biotin–SSB–DNA complex. The solution is then incubated with streptavidin, a biomolecule present in *Streptomyces avidinii* that binds biotin with extreme avidity. Biotin, otherwise known as vitamin H, is a growth factor found in every living cell (see VITAMINS, BIOTIN). This latter incubation step forms a streptavidin–biotin–SSB–DNA complex that then reacts with monoclonal anti-DNA antibody complexed to the enzyme urease [*9002-13-5*]. Any DNA in the test solution therefore ultimately forms a streptavidin–biotin–SSB–DNA–anti-DNA–urease complex. The solution is filtered through a biotin-coated nitrocellulose membrane, which binds the streptavidin component of the complex.

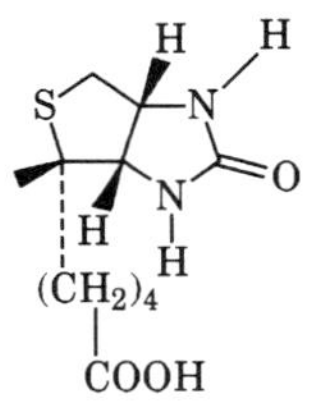

(**5**)

A technique known as light-addressable potentiometric sensor (laps) is then used to assay the presence of the complex by detection of the urease (22). Laps employs a semiconducting silicon wafer in contact with an electrolyte solution through a thin insulating layer. The electrochemical potential of the solution is poised with a controlling electrode. An alternating photocurrent is induced in the system by illumination of the wafer using an intensity modulated light-emitting diode (LED). The conductivity of the insulating barrier, and consequently the photocurrent, is highly sensitive to pH changes in the solution. The nitrocellulose membrane is compressed against the laps device and the complex containing urease, present in the membrane in proportion to the concentration of DNA in the test solution, produces the pH change. This laps DNA sensor has a reported detection limit of 20 pg total DNA. This technique can be generalized to other enzymes producing pH change. The technique is so sensitive that only 600,000 molecules of enzyme are needed to produce a detectable signal.

Under development are biosensors coupling the laps technology with living microorganisms. The metabolic processes of the microorganisms cause pH changes at the surface of the laps wafer. Such biosensors are called microphysiometers or cytosensors. One application of a microphysiometer is directed at replacing the well-known and somewhat controversial Draize test for ocular irritancy, in which compounds are placed in rabbit eyes to test the capacity to irritate (21) (see COSMETICS). A correlation has been found between a compound's ocular irritancy properties and its ability to inhibit metabolism in keratinocytes.

Factors Affecting Biosensor Detection Limits

Biosensor designers are striving for lower detection limits, higher sensitivity, greater selectivity, and lower occurrences of false-positive signals. The detection limit for a biosensor is related to both the nature of the molecular recognition component and the transducer component. In general, antibodies and chemoreceptors are well-suited for low detection limit biosensors because of the high binding constants for target molecules. In intact chemoreceptors the receptor–transducer interface is bridged by a nerve bundle that functions to convert receptor–substrate binding events to electrochemical signals. Whereas the direct output of the chemoreceptor may be quite subtle at trace target levels, it is nevertheless capable of triggering an action potential in the attached nerve. The action potential is an electrical signal, mediated by the motion of ions, that travels along the nerve axon. The action potential produces an electrochemical signal many times larger and thus more easily detectable than the initial trigger.

In general, low level detection is masked by the noise level inherent in any measuring device. Electrochemical methods are susceptible to electrical interference from external sources, variations in reference electrode parameters resulting from aging or contamination, and interference from redox contaminants in the test solution. Piezoelectric methods measure minute mass and microviscosity changes concurrent with molecular recognition events. These methods are highly susceptible to interference from substances in the test solution that can adsorb to the quartz crystal such as proteins. Optical methods that involve the onset of luminescence on molecular recognition are limited primarily by the detection limits of the biological component. However, those methods in which molecular

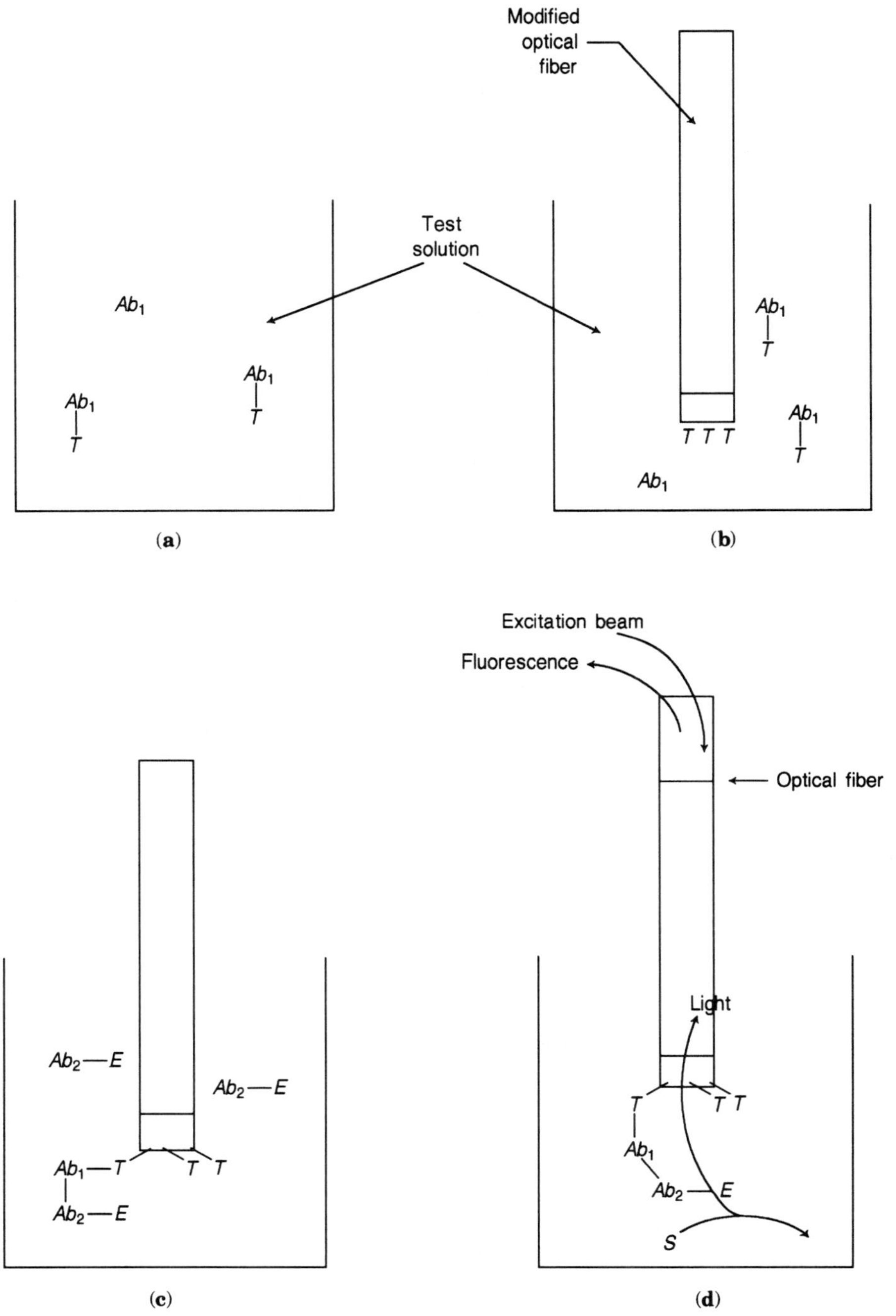

Fig. 5. (**a**) Antibody Ab_1 is added to test solution where some or all of Ab_1 becomes complexed with target, *T*. (**b**) A fiber-optic probe containing covalently attached target is inserted into the solution. Unbound Ab_1 binds to probe. (**c**) The probe is inserted into solution containing enzyme-modified detector antibody Ab_2-*E*, which binds to probe if any Ab_1 is attached. (**d**) The probe is inserted into a solution producing a fluorescent compound, which is then detected via the optical fiber.

recognition results in a decrease of luminescence from a nominal level are much less sensitive.

A clever method frequently employed to decrease detection limits in biosensors is enzyme amplification. An example is enzyme multiplied immunoassay technology (emit). In this method antibody is formed for the target compound, isolated, and complexed with the target chemically attached to an enzyme which catalyzes an indicator reaction producing a deeply colored substance. The antibody–antigen–enzyme complex (Ab–T–E) is immobilized in a gel layer on a plastic strip. On exposure to target, competitive binding for the antibody occurs displacing some enzyme-tagged target. The freed enzyme then catalyzes the indicator reaction deeply coloring an area on the strip. Because the enzyme is catalytic, only trace amounts need be released to produce a readable positive.

A variation of this methodology, illustrated in Figure 5, is known as enzyme-linked immunosorbant assay (elisa) (24). Target molecules T are covalently attached onto the tip of an optical fiber. Antibody Ab_1 specific to the target is then added to the test solution. Any target in the test solution binds to the antibody. The amount of free, ie, unbound, antibody is then assayed by inserting the fiber-optic probe in the test solution. Free antibody binds to the probe via antibody–antigen complexation with the attached target molecule. The probe is then inserted into another solution containing a second antibody Ab_2 called the detector antibody, which selectively binds the first antibody. The detector antibody is modified with an enzyme that produces a fluorescent compound and the fluorescence is monitored via the fiber-optic.

BIBLIOGRAPHY

1. S. F. Hallowell and G. A. Rechnitz, *Anal. Lett.* **20,** 1929 (1987).
2. S. L. Belli and G. A. Rechnitz, *Fresenius Z. Anal. Chem.* **331,** 439 (1987).
3. S. L. Belli and G. A. Rechnitz, *Anal. Lett.* **19,** 403 (1986).
4. S. L. Belli, R. M. Buch, and G. A. Rechnitz, *Anal. Lett.* **20,** 327 (1987).
5. G. A. Rechnitz, *Chem. Eng. News,* 24 (Sept. 5, 1988).
6. R. M. Buch and G. A. Rechnitz, *Anal. Chem.* **61,** 533A (1989).
7. I. Karube, *Biotechnol. Ser.* **5,** 209 (1985).
8. A. Heller, *Acc. Chem. Res.* **23,** 128 (1990).
9. Y. Degani and A. Heller, *J. Phys. Chem.* **91,** 1285 (1987).
10. Y. Degani and A. Heller, *J. Am. Chem. Soc.* **110,** 2615 (1988).
11. A. Heller and Y. Degani, in G. Dryhurst and K. Niki, eds., *Redox Chemistry and Interfacial Behavior of Biological Molecules,* Plenum Publishing Corp., New York, 1988, pp. 151–170.
12. Y. Degani and A. Heller, *J. Am. Chem. Soc.* **111,** 2357 (1989).
13. G. C. Komplin, *Dissertation Abstracts,* Ann Arbor, Mich., 1990.
14. O. S. Wolfbeis, ed., *Fiber Optic Chemical Sensors,* Vol. 2, Chemical Rubber Co. Press, Boca Raton, Fla., 1991.
15. J. I. Peterson, P. V. Fitzgerald, and D. K. Buckhold, *Anal. Chem.* **56,** 62, 1984.
16. D. W. Lubbers and N. Opitz, *Sens. Act.* **4,** 641 (1983).
17. L. J. Blum, S. M. Gautier, and P. R. Coulet, *Anal. Chem.* **21,** 717 (1988).
18. M. Aizawa, *ACS Symp. Ser.* **269,** 447 (1985).

19. H. M. McConnell, P. Rice, G. H. Wada, J. C. Owicki, and J. W. Parce, *Curr. Opin. Struct. Biol.* **1,** 647 (1991).
20. J. C. Owicki and J. W. Parce, *Nature* **344,** 271 (1990).
21. H. G. Wada and co-workers, *Alternatives to Animal Testing and Experimentation,* submitted for publication Nov. 1991.
22. D. G. Hafeman, J. W. Parce, and H. M. McConnell, *Science* **240,** 1182 (1988).
23. J. Briggs and P. R. Panifili, *Anal. Chem.* **63,** 850 (1991).
24. T. Vo-Dinh, G. D. Griffin, and K. R. Ambrose, *Appl. Spectrosc.* **40,** 696 (1988).

General References

Recent Advances in the Development and Analytical Application of Biosensing Probes, Vol. 20, Chemical Rubber Co. Press, Boca Raton, Fla., 1988.

A. P. F. Turner, I. Karube, and G. S. Wilson, eds., *Biosensors: Fundamentals and Applications,* Oxford University Press, New York, 1987.

WILLIAM PIETRO
York University

CARDIOVASCULAR AGENTS

The cardiovascular system consists of the heart and a complex network of conduit blood vessels (aorta and vena cavae), arteries, arterioles, capillaries, venules, and veins. The function of the system is to maintain all living cells of the body with an almost constant internal and external environment. To achieve this goal there is a continuous flow of nutrients and oxygen to the cells and a continuous removal of waste products and carbon dioxide from the cells. The heart functions as a pump distributing its output to the pulmonary and peripheral circulations via this network of blood vessels and capillaries. The heart and the blood vessels, under both neural, eg, central and autonomic nervous systems (see NEUROREGULATORS), and humoral, eg, bloodborne, factors control, are also affected by the physical properties of the blood such as volume, viscosity, and the concentration of electrolytes. Under certain conditions the heart and blood vessels may malfunction causing various disorders or diseases which can lead to irreversible damage and/or death. In the highly developed countries, mortality from cardiovascular disease is extremely high. In Japan, cerebrovascular disease predominates, whereas in Europe and the United States, the heart is the primary target. Cardiovascular agents are used to restore normal function to the cardiovascular system.

ANTIARRHYTHMIC AGENTS

Cardiac arrhythmias are an important cause of morbidity and mortality: approximately 400,000 people per year die from myocardial infarctions (MI) in the United States alone. Individuals with MI exhibit some form of dysrhythmia within 48 h. Post-mortem examinations of MI victims indicate that many die in spite of the fact that the mass of ventricular muscle deprived of its blood supply is often quite small. These data suggest that the cause of death is ventricular fibrillation and that the immediate availability of a safe and efficacious antiarrhythmic agent could have prolonged a number of lives. The goals of antiarrhythmic therapy are to reduce the incidence of sudden death and to alleviate the symptoms of arrhythmias, such as palpitations and syncope. Several excellent reviews of the mechanisms of arrhythmias and the pharmacology of antiarrhythmic agents have been published (1,2).

Properties of the Cardiovascular System

The Ionic Basis of Membrane Activity. Almost all living cells maintain specific internal chemical environments that are different from their external environments. In cardiac cells the principal ions involved in maintaining membrane activity are sodium, Na^+; potassium, K^+; chloride, Cl^-; and calcium, Ca^{2+}. The internal (i) and external (o) concentrations of these ions are: Na^+_o = 140 mM, Na^+_i = 30 mM; K^+_o = 4 mM, K^+_i = 140 mM; Cl^-_o = 104 mM, Cl^-_i = 30 mM; Ca^{2+}_o = 2 mM, Ca^{2+}_i = 100 nM. Ions diffuse through the cell membrane via protein structures, called channels, that span the membrane. The resting cell membrane is permeable to K^+ but not to Na^+ and the low concentration of Na^+_i is maintained by an energy-dependent Na^+/K^+ adenosine triphosphatase (ATPase) pump that exchanges three Na^+_i for two K^+_o. The resting transmembrane potential of −90 mV, ie, the interior of the cell membrane is negative with respect to the exterior of the membrane, is determined by the ionic gradients across the membrane and the selective permeability of the membrane to these ions. When the cell depolarizes there is an increase in the permeability of the cell membrane to Na^+_o, sodium channels are activated, and Na^+ influx rapidly increases. The interior of the membrane becomes positive with respect to the exterior. The membrane potential at this time is +10 mV. As the cell membrane potential moves from −90 mV, becoming less negative, there is an increased inward flux of Ca^{2+} at about −60 to −50 mV accompanied by the activation of calcium channels. At +10 mV Na^+ influx is terminated by activation of the Na^+/K^+ ATPase pump and Na^+ efflux begins. The cell begins to repolarize. The first phase of repolarization is thought to result from an outward K^+ conductance. The second phase is a result of the slow Ca^{2+} conductance; the third phase is a result of an outward K^+ current. When the cell is fully repolarized back to its resting membrane potential, phase 4 repolarization, the ionic gradients are re-established. Calcium is removed from the cell by a Na^+–Ca^{2+} exchange mechanism (3).

The electrical activity of individual cells can be studied using microelectrodes and recorded as transmembrane monophasic action potentials. The monophasic action potential of atrial and ventricular muscle cells and Purkinje fibers is characterized by a very rapid upstroke corresponding to phase 0, depolarization from Na^+ influx through fast sodium channels. In atrial and ventricular muscle this is followed by a slow repolarization phase, phase 2, from a calcium inward current, and a rapid repolarization phase, phase 3, from an outward K^+ current. Purkinje fibers have three phases of repolarization: phase 1, thought to result from an outward K^+-conductance, a rapid repolarization; and phases 2 and 3 as in atrial and ventricular muscle. Phase 4 repolarization is the resting membrane or diastolic potential during which atrial and ventricular muscle cells and Purkinje fibers are quiescent and fully repolarized. Cells of the sinoatrial (SA) and atrioventricular (AV) nodes exhibit action potentials in which during diastole or phase 4, the cells are slowly depolarizing (diastolic depolarization). This is followed by a relatively modest rapid upstroke, phase 0 depolarization, and then a relatively slow repolarization phase, phase 3. The spontaneously and slowly depolarizing and repolarizing phases of the SA node, shown in Figure 1**a** (3), are characteristic of pacemaker cells.

Properties of Excitable Tissues. *Automaticity.* Special cardiac cells, such as SA and AV nodal cells, His-bundle cells, and Purkinje fibers, spontaneously generate an impulse. This is the property of automaticity. Ectopic sites can act as

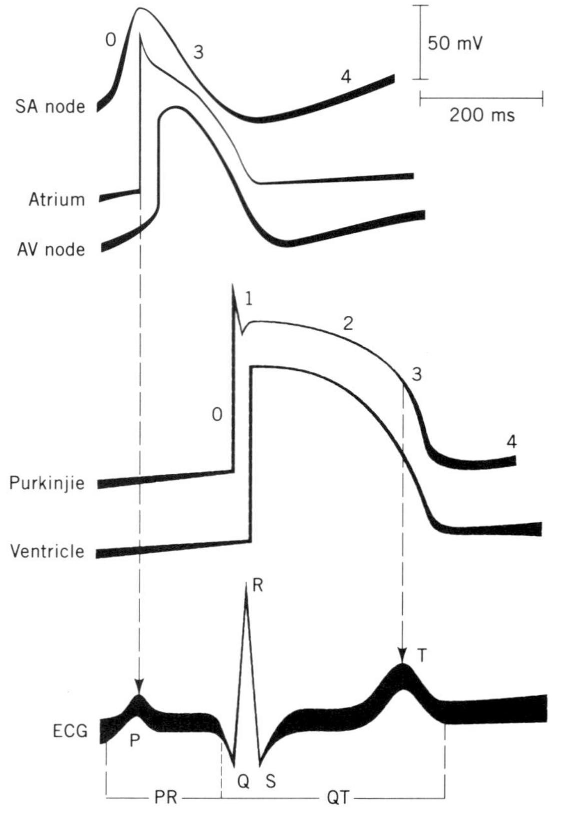

(a)

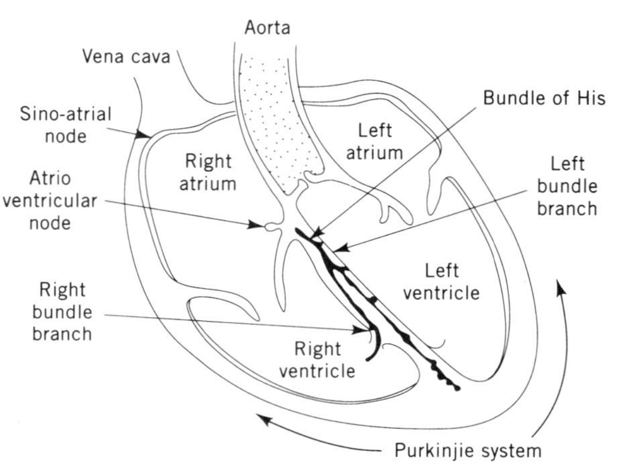

(b)

Fig. 1. (**a**) Myocardial cell transmembrane potentials, where the numbers and letters refer respectively to the phases and waves described in the text. ECG is electrocardiogram. (**b**) Conduction system of the heart.

pacemakers if the rate of phase 4 depolarization or resting membrane potential is increased, or the threshold for excitation is reduced.

Enhanced automaticity occurs in hypoxia, hypokalemia, hypercarbia, excessive sympathetic nervous system stimulation, or high concentrations of catecholamines. These conditions may lead to arrhythmias. Decreased automaticity may also lead to production of arrhythmias by enhancing ectopic activity in latent pacemakers (ectopic foci) or by altering conductivity and refractoriness in conduction pathways of myocardium.

Excitability. Excitability refers to electrical responsiveness of the heart to various stimuli by the generation of local excitatory currents, action potentials, or fibrillation.

Conductivity. Conductivity is an electrical property of excitable tissue which ensures that if one area of a membrane is excited to full activity, that area excites adjacent areas. Conduction of an impulse varies directly with the rate of development of phase 0 and the amplitude of the action potential. Phase 0 is faster, and amplitude of the action potential is greater, the more negative the transmembrane potential at the time of initiation of the impulse. Conduction velocity is faster when phase 0 is fast.

Refractoriness. The altered responsiveness of the cell membrane during recovery from previous activity is described by refractoriness. During repolarization phase 1, 2, and 3, the cell is either nonresponsive (absolute refractory period) or less responsive (effective and relative refractory periods) to the propagation of an impulse. During the absolute refractory period the cell cannot be excited regardless of the strength of the stimulating current applied, whereas during the effective refractory period the cell can be excited by applying greater stimulating currents, but the impulse is not propagated. During the relative refractory period the cell can be excited but the impulse is propagated at slower velocities, and finally during phase 4 when the cell is fully repolarized the cell can be stimulated with normal currents and the impulse is fully propagated (4).

The Cardiac Cycle. The heart (Fig. 1**b**) performs its function as a pump as a result of a rhythmical spread of a wave of excitation (depolarization) that excites the atrial and ventricular muscle masses to contract sequentially. Maximum pump efficiency occurs when the atrial or ventricular muscle masses contract synchronously (see Fig. 1). The wave of excitation begins with the generation of electrical impulses within the SA node and spreads through the atria. The SA node is referred to as the pacemaker of the heart and exhibits automaticity, ie, it depolarizes and repolarizes spontaneously. The wave then excites sequentially the AV node; the bundle of His, ie, the penetrating portion of the AV node; the bundle branches, ie, the branching portions of the AV node; the terminal Purkinje fibers; and finally the ventricular myocardium. After the wave of excitation depolarizes these various structures of the heart, repolarization occurs so that each of the structures is ready for the next wave of excitation. Until repolarization occurs the structures are said to be refractory to excitation. During repolarization of the atria and ventricles, the muscles relax, allowing the chambers of the heart to fill with blood that is to be expelled with the next wave of excitation and resultant contraction. This process repeats itself 60–100 times or beats per minute in normal humans in the resting state and represents the basal rate. Highly trained healthy athletes may exhibit heart rates of 30–40 beats/min (bradycardia), yet have similar or greater outputs to those of the average human; ie, exercise makes the heart a more efficient pump. With exercise or emotional excitement, heart rate temporarily increases above the basal rate (tachycardia), and as long as excitation begins in the SA node and follows the path described, the individual is said to have a normal sinus rhythm. An ectopic rhythm develops when excitation arises from sites other than the SA node (ectopic foci).

The electrocardiogram (ECG) is the body surface (skin) recording of the wave of electrical excitation of the heart. The ECG is the temporal summation of the electrical activity of the individual cells of the heart. The relationship of the activation of the single cells of the wave of excitation to the ECG is shown in Figure 1**a**. The P wave of the ECG represents the depolarization and the wave of electrical activity (conduction) over the atrial myocardium. The P–R interval represents the time necessary for conduction of impulses (propagation) through the AV node, the bundle of His, and the Purkinje fibers. The QRS complex represents the spread of the wave of excitation over the ventricular myocardium. During the QRS period the atria repolarize, and during the S–T segment, the entire ventricular mass is depolarized. The T wave represents the repolarization of the ventricular myocardial cells (4).

Cardiac arrhythmias or dysrhythmias are disturbances of the normal regular rhythm which may be caused by an abnormality in the site of impulse generation, its rate or regularity, or its propagation or conduction (1,2). The more commonly encountered cardiac arrhythmias are

Arrhythmias Originating in the Sinus Node: Sinus bradycardia; Sick sinus syndrome; Sinus tachycardia

Disorders of Impulse formation:

- Ectopic focus
  - premature atrial depolarization (contraction)
  - premature junctional (AV node) depolarization
  - premature ventricular depolarization
- Ectopic rhythm
  - atrial ectopic tachyarrhythmia
  - ventricular tachyarrhythmia
  - paroxysmal tachyarrhythmia
  - paroxysmal supraventricular tachyarrhythmia
  - paroxysmal atrial tachycardia
- Atrial flutter
- Atrial fibrillation
- Ventricular tachycardia
- Ventricular fibrillation

Disorders of Impulse Conduction:

- Reentry mechanism
- Intranodal (AV node) reentry
- Extranodal reentry
- Reentrant tachyarrhythmia

Conduction Blocks:

The conduction of an impulse can be slowed or stopped at any point along the conduction system of the heart. Slowing of conduction is called first-degree block, block of some impulses is called second-degree block, and a block of all impulses is called third-degree block.

Sino-atrial nodal disease
Atrio-ventricular block

Classification of Antiarrhythmic Agents. Similarities in the electrophysiologic properties of various antiarrhythmic agents led to a proposal of classification in 1970 (5). Several other classifications have evolved since then (6–8); the most recent is the most popular (9). Classifications may be unsatisfactory for several reasons. Classifications are usually based on one primary activity and may not acknowledge differences among agents in the same group or similarities among agents in different groups, so that the electrophysiological effects of each agent may not be fully appreciated. Additionally, the electrophysiological effects are often obtained from normal myocardial preparations and the abnormalities of cellular electrophysiology which cause arrhythmias in a diseased myocardium may be uniquely related to the disease process (10). Nevertheless, antiarrhythmic agent classifications have been useful as a mnemonic device.

Class I Antiarrhythmic Agents: The Sodium Channel Blockers

The Class I antiarrhythmic agents inactivate the fast sodium channel, thereby slowing the movement of Na^+ across the cell membrane (1,2). This is reflected as a decrease in the rate of development of phase 0 (upstroke) depolarization of the action potential (1,2). The Class I agents have potent local anesthetic effects. These compounds have been further subdivided into Classes IA, IB, and IC based on recovery time from blockade of sodium channels (11). Class IB agents have the shortest recovery times ($t_{1/2}<1$ s); Class IA compounds have moderate recovery times ($t_{1/2}$ usually < 9 s); and Class IC have the longest recovery times ($t_{1/2}$ usually > 9 s).

The Class I agents decrease excitability, slow conduction velocity, inhibit diastolic depolarization (decrease automaticity), and prolong the refractory period of cardiac tissues (1,2). These agents have anticholinergic effects that may contribute to the observed electrophysiologic effects. Heart rates may become faster by increasing phase 4 diastolic depolarization in SA and AV nodal cells. This results from inhibition of the action of vagally released acetylcholine [*51-84-3*], which allows sympathetically released norepinephrine [*51-41-2*] (NE) to act on these structures (1,2).

The Class I agents have many similar side effects and toxicities. The anticholinergic side effects include dry mouth, constipation, and urinary hesitancy and retention. Common gastrointestinal (GI) side effects include nausea, vomiting, diarrhea, and anorexia. Cardiovascular adverse effects are hypotension, tachycardia, arrhythmias, and myocardial depression, especially in patients with congestive heart failure. Common central nervous system (CNS) side effects are headache, dizziness, mental confusion, hallucinations, CNS stimulation, paraesthesias, and convulsions.

Class IA Antiarrhythmic Agents. Class IA antiarrhythmic agents decrease automaticity, ie, depress pacemaker rates, especially ectopic foci rates; produce moderate depression of phase 0 depolarization and thus slow conduction in atria, A-V node, His-Purkinje system, and ventricles; prolong repolarization, ie, lengthen action potential duration; increase refractoriness; and depress excitability. These electrophysiological effects are manifested in the ECG by increases in the PR, QRS, and QT intervals.

Quinidine. Quinidine, an alkaloid obtained from cinchona bark (*Cinchona* sp.), is the dextrorotatory stereoisomer of quinine [*130-95-0*] (see ALKALOIDS). The first use of quinidine for the treatment of atrial fibrillation was reported in 1918 (12). The sulfate, gluconate, and polygalacturonate salts are used in clinical practice. The drug is given mainly by the oral (po) route, rarely by the intravenous (iv) route of administration. It is the most frequently prescribed po antiarrhythmic agent in the United States. The clinical uses of quinidine include suppression of atrial and ventricular extrasystoles and serious ventricular arrhythmias (1–3).

After po administration, the drug is almost completely absorbed from the GI tract and undergoes first-pass metabolism in the liver. From 70 to 87% of the dose is bioavailable, depending on the salt used, and 10 to 20% of the absorbed dose is free in the circulation. The drug binds mainly to albumin, but also enters the red blood cell and binds to hemoglobin. At equilibrium, the concentrations in red blood cells and plasma may be equal. Peak plasma concentrations are achieved in 1–4 h depending on the salt used. Therapeutic plasma concentrations are 3–6 μg/mL; toxic reactions occur above 8 μg/mL. The metabolites of quinidine are 3-hydroxyquinidine, quinidine-*N*-oxide, 2′-oxyquinidine, and *O*-desmethylquinidine. Only 3-hydroxyquinidine has antiarrhythmic activity and achieves plasma levels about one-third of those achieved by the unchanged drug. The kidneys excrete 75–90% of the dose as metabolites, sulfonate or glucuronate conjugates, and the remainder as unchanged drug (1,2).

The cardiovascular adverse effects associated with quinidine therapy are hypotension and tachycardia, both of which are related to its α-adrenoceptor blocking actions. The tachycardia may be a reflex adjustment to the fall in blood pressure or may also be a direct action of the drug on sympathetic nerve terminals leading to an increased release of NE. Quinidine also produces ringing in the ears (cinchonism) (1,2).

Procainamide. Procainamide hydrochloride is a benzamide, synthesized to prolong the therapeutic effects of the local anesthetic procaine [*59-46-1*] (13) (see ANESTHETICS). The drug is effective in a wide range of supraventricular and ventricular arrhythmias (14).

Procainamide may be administered by iv, intramuscular (im), or po routes. After po dosing, 75–90% of the drug is absorbed from the GI tract. About 25% of the amount absorbed undergoes first-pass metabolism in the liver. The primary metabolite is *N*-acetylprocainamide (NAPA) which has almost the same antiarrhythmic activity as procainamide. This is significant because the plasma concentration of NAPA relative to that of procainamide is 0.5–2.5. In terms of drug metabolism there are two groups of patients: those that rapidly acetylate and those that slowly acetylate procainamide. About 15–20% of the drug is bound to plasma proteins. Peak plasma concentrations are achieved in 60–90 min. Therapeutic plasma concentrations are 4–10 μg/mL. Plasma half-lives of procainamide and NAPA, which are excreted mainly by the kidneys, are 2.5–4.5 and 6 h, respectively. About 50–60% is excreted as unchanged procainamide (1,2).

About 30% of the patients on chronic procainamide dosing develop a systemic lupuslike syndrome consisting of arthralgia, myalgia, skin rash, and fever. Patients who are slow acetylator phenotypes may be prone to this condition. Some may exhibit pleuropneumonic involvement and hepatomegaly (1,2).

Disopyramide. Disopyramide phosphate, a phenylacetamide analogue, is a racemic mixture. The drug can be administered po or iv and is useful in the treatment of ventricular and supraventricular arrhythmias (1,2). After po administration, absorption is rapid and nearly complete (83%). Binding to plasma protein is concentration-dependent (35–95%), but at therapeutic concentrations of 2–4 μg/mL, about 50% is protein-bound. Peak plasma concentrations are achieved in 0.5–3 h. The drug is metabolized in the liver to a mono-*N*-dealkylated product that has antiarrhythmic activity. The elimination half-life of the drug is 4–10 h. About 80% of the dose is excreted by the kidneys, 50% is unchanged and 50% as metabolites; 15% is excreted into the bile (1,2).

Class IB Antiarrhythmic Agents. Class IB antiarrhythmic agents produce less inhibition of the inward sodium current than Class IA agents. In normal myocardial tissue, phase 0 may be unaffected or minimally depressed. However, in ischemic or infarcted tissue, phase 0 is depressed. Myocardial tissue exposed to Class IB agents exhibits decreased automaticity, shortened action potential duration, ie, shortened repolarization, and shortened refractory period. Excitability of the myocardium is not affected and conduction velocity is increased or not modified. The refractory period is shortened less than its action potential duration, thus the ratio of refractory period to action potential duration is increased by these agents. The net effect is increased refractoriness. The PR and QT intervals of the ECG are shortened and the QRS interval is unchanged (1,2).

Lidocaine. Lidocaine hydrochloride, an anilide, was originally introduced as a local anesthetic in 1943 and found to be a potent antiarrhythmic in 1960. The compound is a reverse amide of procainamide. Lidocaine is generally considered to be the drug of choice in the treatment of ventricular arrhythmias and those originating from digitalis glycoside toxicity (1,2,15–17).

Lidocaine is given iv. Po dosing is inadequate because after rapid absorption, about 70% of the drug undergoes extensive first-pass metabolism in the liver to monoethylglycine xylide (MEGX) and glycine xylide (GX). MEGX has 80–100% and GX 10–25% of the antiarrhythmic potency of lidocaine, respectively. Plasma concentrations of MEGX may be 50–200% and GX 10–100% those of lidocaine. Lidocaine is 60–80% bound to plasma protein. Time to peak plasma concentration depends on the rate of IV dosing but is usually achieved in 45–90 seconds. Therapeutic plasma concentrations are 1.5–5.0 μg/mL, and concentrations above 5 μg/mL may be toxic. The elimination half-life after a bolus iv dose is 8 min; the elimination half-life after a 24 h iv infusion is about 100 min. The drug is eliminated by the kidneys. Ten percent is unchanged and the remainder is in the form of inactive metabolites (1,2).

Phenytoin. Phenytoin sodium is sodium diphenylhydantoin [*630-93-3*] which is structurally related to the barbiturates. It was originally introduced as an anticonvulsant (18) (see HYPNOTICS, SEDATIVES, AND ANTICONVULSANTS) and later found to have antiarrhythmic properties (19), although not approved by the FDA for any arrhythmic indications. Phenytoin is effective in the treatment of ventricular arrhythmias associated with acute MI and with digitalis toxicity (20). It is not very effective in treatment of supraventricular arrhythmias (20).

Phenytoin's absorption is slow and variable yet almost complete absorption eventually occurs after po dosing. More than 90% of the drug is bound to plasma protein. Peak plasma concentrations are achieved in 1.5–3 h. Therapeutic plasma concentrations are 10–20 μg/mL but using fixed po doses, steady-state levels are achieved in 7–10 days. Phenytoin is metabolized in the liver to inactive metabolites. The plasma half-life is approximately 22 h. Phenytoin is excreted primarily in the urine as inactive metabolites and <5% as unchanged drug. It is also eliminated in the feces and in breast milk (1,2). Prolonged po use of phenytoin may result in hirsutism, gingival hyperplasia, and hypersensitivity reactions evidenced by skin rashes, blood dyscrasias, etc (1,2).

Mexilitene. Mexilitene hydrochloride, a phenyl ether, is a po active congener of lidocaine. It is used clinically for suppression of ventricular arrhythmias (1,2).

Mexilitene is well absorbed from the GI tract and less than 10% undergoes first-pass hepatic metabolism. In plasma, 60–70% of the drug is protein bound and peak plasma concentrations are achieved in 2–3 h. Therapeutic plasma concentrations are 0.5–2.0 μg/mL. The plasma half-life of mexilitene is 10–12 h in patients having normal renal and hepatic function. Toxic effects are noted at plasma concentrations of 1.5–3.0 μg/mL, although side effects have been noted at therapeutic concentrations. The metabolite, *N*-methylmexilitene, has some antiarrhythmic activity. About 85% of the drug is metabolized to inactive metabolites. The kidneys excrete about 10% of the drug unchanged, the rest as metabolites. Excretion can also occur in the bile and in breast milk (1,2).

Moricizine. Moricizine, a phenothiazine derivative, was synthesized and developed in Russia, where it has been in general use since 1971. FDA approval of the new drug application (NDA) for use in the United States was granted in 1991. It is effective against atrial and ventricular arrhythmias (1,2,21).

Following po administration moricizine is completely absorbed from the GI tract. The drug undergoes considerable first-pass hepatic metabolism so that only 30–40% of the dose is bioavailable. Moricizine is extensively (95%) bound to plasma protein, mainly albumin and α_1-acid glycoprotein. The time to peak plasma concentrations is 0.42–3.90 h. Therapeutic concentrations are 0.06–3.00 μg/mL. Using radiolabeled moricizine, more than 30 metabolites have been noted but only 12 have been identified. Eight appear in urine. The sulfoxide metabolite is equipotent to the parent compound as an antiarrhythmic. Elimination half-life is 2–6 h for the unchanged drug and known metabolites, and 84 h for total radioactivity of the labeled drug (1,2).

Tocainide. Tocainide is a po active primary amine analogue of lidocaine. It consists of the (*S*)-(−) and the more active (*R*)-(+) enantiomers. The drug is effective in the treatment of ventricular arrhythmias, especially those following acute myocardial infarctions (1,2,22).

Tocainide is rapidly and well absorbed from the GI tract and undergoes very little hepatic first-pass metabolism. Unlike lidocaine which is ~30% bioavailable, tocainide's availability approaches 100% of the administered dose. Food delays absorption and decreases plasma levels but does not affect bioavailability. Less than 10% of the drug is bound to plasma proteins. Therapeutic plasma concentrations are 3–9 μg/mL. Toxic plasma levels are >10 μg/mL. Peak plasma concentrations are achieved in 0.5–2 h. About 30–40% of tocainide is metabolized in the liver by deamination and glucuronidation to inactive metabolites. The metabolism is stereoselective and the steady-state plasma concentration of the (*S*)-(−) enantiomer is about four times that of the (*R*)-(+) enantiomer. About 50% of the tocainide dose is eliminated by the kidneys unchanged, and the rest is eliminated as metabolites. The elimination half-life of tocainide is about 15 h, and is prolonged in patients with renal disease (1,2,23).

Some rare toxic effects that may result from tocainide therapy are a lupus-like syndrome, skin rash, pulmonary complications, and hematologic abnormalities, eg, agranulocytosis (1,2,24).

Class IC Antiarrhythmic Agents. Class IC antiarrhythmic agents have marked local anesthetic effects. They slow the rapid inward sodium current producing marked phase 0 depression and slow conduction. Action potential duration of ventricular muscle is increased, ie, prolonged repolarization, but decreased in the His-Purkinje system by these agents. The effects on the ECG are increased PR interval, marked prolongation of the QRS interval, and increased QT interval (1,2).

Encainide. Encainide hydrochloride, a benzamide derivative, is effective in the treatment of ventricular arrhythmias and refractory ventricular tachycardia (1,2).

Encainide is almost completely absorbed from the GI tract. Food may delay absorption without altering its bioavailability. The drug is rapidly metabolized in 90% of the patients to two principal metabolites, *O*-demethylencainide (ODE) and 3-methoxy-*O*-demethylencainide (MODE), while the other 10% metabolize encainide slowly with little or no ODE or MODE formed. Encainide, ODE, and MODE are extensively protein bound: 75–80% for encainide and ODE and 92% for MODE. Peak plasma concentrations are achieved in 30–90 min. Therapeutic plasma concentrations are very low; the concentrations of ODE and MODE are approximately five times those of encainide. The findings with the metabolites are significant because ODE is 2–10 times and MODE, 1–4 times more effective than encainide as antiarrhythmics. The half-lives for encainide in fast and slow metabolizers is 1–2 h and 6–12 h, respectively. The elimination half-life for ODE is 3–4 h and for MODE 6–12 h in fast metabolizers. Excretion occurs through the liver and kidneys (1,2).

Adverse effects of encainide therapy are dose-related. Diplopia and a bad taste in the mouth have been reported (1,2,25).

Flecainide. Flecainide acetate, a fluorobenzamide, is a derivative of procainamide, and has been reported to be efficacious in suppressing both supraventricular and ventricular arrhythmias (26–29). The drug is generally reserved for patients with serious and life-threatening ventricular arrhythmias. Flecainide depresses phase 0 depolarization of the action potential, slows conduction throughout the heart, and significantly prolongs repolarization (30). The latter effect indicates flecainide may possess some Class III antiarrhythmic-type properties (31).

Flecainide is well absorbed and 90% of the po dose is bioavailable. Binding to plasma protein is only 40% and peak plasma concentrations are attained in about 1–6 h. Three to five days may be required to attain steady-state plasma concentrations when multiple doses are used. Therapeutic plasma concentrations are 0.2–1.0 μg/mL. Flecainide has an elimination half-life of 12–27 h, allowing twice a day dosing. The plasma half-life is increased in patients with renal failure or low cardiac outputs. About 70% of the flecainide in plasma is metabolized by the liver to two principal metabolites. The antiarrhythmic potency of the meta-*O*-dealkylated metabolite and the meta-*O*-dealkylated lactam, relative to that of flecainide is 50 and 10%, respectively. The plasma concentrations of the two metabolites relative to that of flecainide are 3–25%. Flecainide is mainly excreted by the kidneys, 30% unchanged, the rest as metabolites or conjugates; about 5% is excreted in the feces (1,2).

Propafenone. Propafenone hydrochloride, an arylketone, is structurally similar to the β-adrenoceptor blocking agents. It has been in use in the former West Germany since 1977 and was introduced in the United States in 1990. Its effects may result from a combination of weak calcium channel blocking, weak nonselective β-adrenoceptor blocking, and sodium channel blocking activity. Propafenone is effective in treating supraventricular tachyarrhythmias, ventricular ectopic beats, and ventricular arrhythmias. It is the most frequently prescribed medication for ventricular arrhythmias in Europe (32).

About 97% of po dose is absorbed from the GI tract. The drug undergoes extensive first-pass hepatic metabolism and only 12% of the po dose is bioavailable. More than 95% is protein bound and peak plasma concentrations are achieved in 2–3 h. Therapeutic plasma concentrations are 0.064–1.044 μg/mL. The drug is metabolized in the liver to 5-hydroxypropafenone, which has some antiarrhythmic activity, and to inactive hydroxymethoxy propafenone, glucuronides, and sulfate conjugates. Less than 1% of the po dose is excreted by the kidney unchanged. The elimination half-life is 2–12 h (32).

Patients taking propafenone may experience a metallic taste after ingesting the drug (32).

Indecainide. Indecainide hydrochloride is a po active antiarrhythmic agent that received FDA approval in 1989, but it has not been marketed as of this writing. Chemically, it is 9-[3-(isopropylamino)propyl]fluorine-9-carboxamide [*74517-78-5*]. The drug has potent activity against premature ventricular complexes (PVCs) and ventricular tachycardias. Indecainide has no effect on sinus node function, atrial or ventricular effective refractory periods (32,33).

Therapeutic blood levels range from 0.3–0.9 μg/mL. Mean elimination half-life is 9 h (32).

Class I Antiarrhythmic Agents Under Development. *Lorcainide.* Lorcainide hydrochloride, introduced in the former West Germany in 1981, is investigational in the United States. The compound is an acetanilide derivative having local anesthetic properties. It is available in po and iv formulations and is an effective treatment for suppressing supraventricular extrasystoles, repetitive atrial tachycardias, and ventricular arrhythmias. Lorcainide is a long-acting agent allowing for twice a day dosing schedules (32).

Absorption is complete and bioavailability is about 100% at steady state during continuous po dosing. There is extensive hepatic first-pass metabolism to norlorcainide and hydroxylated metabolites. Norlorcainide is equipotent and equieffective to lorcainide in antiarrhythmic activity. Therapeutic plasma concentrations are 0.15–0.50 μg/mL. The elimination half-life of lorcainide is about 8 h and that of norlorcainide, 26 h (32). Lorcainide has low cardiovascular toxicity (32).

Pirmenol. Pirmenol hydrochloride, a pyridine methanol derivative, is a racemic mixture. It has Class 1A antiarrhythmic activity, ie, depression of fast inward sodium current, phase 0 slowing, and action potential prolongation. The prolongation of refractory period may be a Class III property. This compound has shown efficacy in converting atrial arrhythmias to normal sinus rhythm (34,35).

Class II Antiarrhythmic Agents: The β-Adrenoceptor Blocking Agents

β-Adrenoceptor blocking agents (β-adrenoceptor blockers) are often called β-blockers. However, the former term more accurately describes the way in which these drugs act. They bind selectively to β-adrenoceptors in a competitive and reversible way to antagonize the responses of neurally-released, humoral or exogenously-administered sympathomimetic amines. Some of these Class II agents have local anesthetic properties and are described as "quinidinelike" because their cardiac electrophysiological properties are similar to those of Class IA agents (1,2,11).

β-Adrenoceptors have been subdivided into β_1- and β_2-adrenoceptors. A third subset called nontypical β-adrenoceptors or β_3-adrenoceptors have been described but are still the subject of debate. In terms of the interactions with various subsets of β-adrenoceptors, some antagonists are nonselective in that they antagonize the effects of activation of both β_1- and β_2-adrenoceptors, whereas others are selective for either β_1- or β_2-adrenoceptors. β_1- and β_2-adrenoceptors coexist in almost all organs but generally, one type predominates. The focus herein is on the clinically relevant β_1-adrenoceptor-mediated effects on heart and on β_2-adrenoceptor-mediated effects on smooth muscles of blood vessels and bronchioles, the insulin-secreting tissue of the pancreas, and skeletal muscle glycogenolysis for side effects profile (36).

Agents that preferentially block β_1-adrenoceptor-mediated effects on the heart, resulting in reduced rate, force, and velocity of cardiac contractility, are classified as cardioselective β-adrenoceptor blockers. Thus these agents would be beneficial in patients having arrhythmias that are present with peripheral vascular disease, asthma, and insulin-dependent diabetes because the β-adrenoceptors in the organs involved are of the β_2-adrenoceptor subtype and blocking the β_2-adrenoceptors would worsen the disease states. However, cardioselectivity may be lost at high doses or concentrations of the β-adrenoceptor blocker. Some β-adrenoceptor blocking agents have local anesthetic properties resulting from membrane stabilizing and have effects on cardiac cells similar to those described for the Class 1 antiarrhythmic agents, ie, they reduce the rate of rise of the action potential without affecting action potential duration. This effect may be apparent at concentrations in excess of their therapeutic range (36).

Some β-adrenoceptor blockers have intrinsic sympathomimetic activity (ISA) or partial agonist activity (PAA). They activate β-adrenoceptors before blocking them. Theoretically, patients taking β-adrenoceptor blockers with ISA should not have cold extremities because the drug produces minimal decreases in peripheral blood flow (smaller increases in resistance). In addition, these agents should produce minimal depression of heart rate and cardiac output, either at rest or during exercise (36).

There have been a number of long-term trials with various β-adrenoceptor blockers in patients surviving acute MI (37–39) that demonstrated a reduction in mortality, sudden death, and nonfatal re-infarctions. The term cardioprotective has been used to describe this effect for the drugs studied. The mechanism of the cardioprotective effect is not understood and whether this property applies to all β-adrenoceptor blockers is unknown. Timolol [*26839-75-8*], $C_{13}H_{24}N_4O_3S$, metoprolol, atenolol, and propranolol have been shown to possess this cardioprotective property (37–39) (Table 1).

In addition to the properties described, β-adrenoceptor blockers may be lipophilic or hydrophilic in nature depending on the extent to which they partition between water and organic solvents. Lipophilic agents more readily cross the blood–brain barrier and are more likely to produce adverse CNS effects, eg, drowsiness, lassitude, impaired concentration, hallucinations, and nightmares. Lipophilic agents are eliminated by the liver and blood levels, duration of action, and toxicological potential may be increased in the presence of liver disease. Hydrophilic agents, on the other hand, are eliminated by the kidneys and blood levels, duration of action, and toxic potential may be increased in the presence of impaired kidney function. These factors have prompted the search for β-adrenoceptor blockers that have a balanced clearance profile, wherein the drugs are equally excreted by the liver and kidneys and elimination is not affected by dysfunction of either organ (36).

Prolonged exposure of β-adrenoceptor agonists down-regulates β-adrenoceptors, ie, their number decreases and they become less responsive. On the other hand, prolonged exposure to β-adrenoceptor antagonists (those without ISA) up-regulates β-adrenoceptors, ie, their numbers increase and they become more responsive. Therefore, patients on β-adrenoceptor blocker therapy should be withdrawn from this medication gradually (40).

The properties of β-adrenoceptor blockers that contribute to antiarrhythmic effects are antagonism of neural/humoral β-adrenergic activity, and antagonism of catecholamine-mediated electrophysiological properties, ie, increase refractory period and decrease in the rate of diastolic depolarization, ie, decrease automaticity and slow atrioventricular conduction (1,2).

Propranolol. Propranolol hydrochloride, considered the prototype of the β-adrenoceptor blocking agents, has been in use since 1964. It is a nonselective, highly lipid-soluble β-adrenoceptor blocker having no ISA. It is a mixture of (+) and (−) enantiomers, and the (−) enantiomer is the active moiety. The local anesthetic effects of propranolol are equipotent to those of lidocaine [*137-58-6*], $C_{14}H_{22}N_2O$, (see ANESTHETICS). Therapeutic effects include termination of catecholamine-induced arrhythmias, conversion of SA nodal tachycardias (including flutter and fibrillation) and AV nodal tachyarrhythmias to normal sinus rhythm, digitalis-induced arrhythmias, and ventricular arrhythmias (1,2). The drug also has cardioprotective properties (37,39).

Greater than 90% of a po dose of propranolol is absorbed from the gastrointestinal (GI) tract. Propranolol undergoes extensive first-pass hepatic metabolism and only 25% of the parent compound is bioavailable. Peak plasma concentrations are achieved in about 2 h. About 93% is bound to plasma protein. The drug is metabolized by *O*-dealkylation, side-chain oxidation, glucuronide conjugation, and ring oxidation. The ring hydroxylated metabolites, eg, 4-hydroxypropranolol, have β-adrenoceptor blocking activity. Only 1–4% of administered drug is excreted unchanged in the urine. Its elimination half-life is 3–4 h. The (+) enantiomer of propranolol is eliminated more slowly than the (−) enantiomer (1,2).

Toxic effects of propranolol are related to its blocking β-adrenoceptor blocking actions. They include cardiac failure, hypotension, hypoglycemia, and bronchospasm. Propranolol is lipophilic and crosses the blood–brain barrier. Complaints of fatigue, lethargy, mental depression, nightmares, hallucinations, and insomnia have been reported. GI side effects include nausea, vomiting, diarrhea, and constipation (1,2).

Acebutolol. Acebutolol hydrochloride is a hydrophilic, cardioselective β-adrenoceptor blocker that has about 1/25 the potency of propranolol in this regard. The drug has moderate ISA and weak membrane stabilizing activities. It is approved for the treatment of hypertension and ventricular arrhythmias, especially PVCs. Acebutolol should produce minimal depression of heart rate because of its ISA (32).

Acebutolol is well absorbed from the GI tract. It undergoes extensive hepatic first-pass metabolism. Bioavailability of the parent compound is about 40%. The principal metabolite, *N*-acetylacebutolol, has antiarrhythmic activity and appears to be more cardioselective. Binding to plasma proteins is only 26%. Peak plasma concentrations of acebutolol are achieved in 2.5 h, 3.5 h for *N*-acetylacebutolol. The elimination half-lives of acebutolol and its metabolite are 3–4 and 8–13 h, respectively. The compounds are excreted by the kidneys (30–40%) and by the liver into the bile (50–60%). About 40% of the amount excreted in the urine is unchanged acebutolol, the rest as metabolites (32).

In addition to the toxic effects indicated for propranolol, acebutolol may produce hair loss and elevate antinuclear antibody titers (32).

Esmolol. Esmolol hydrochloride, an ultrashort acting cardioselective β_1-adrenoceptor blocker having no ISA or membrane stabilizing activity, is minimally lipid-soluble. The β-adrenoceptor blocking activity is 0.07 that of propranolol. It is used primarily for short-term management of supraventricular tachyarrhythmias, including atrial fibrillation and flutter, perioperative supraventricular arrhythmias, and postoperative hypertension and acute myocardial ischemia (41).

Esmolol is iv administered. Maximal β-adrenoceptor blockade occurs in 1 min. Its elimination half-life is about 9 min. Full recovery from β-adrenoceptor blockade is within 30 min after stopping the infusion. The therapeutic plasma concentrations are 0.4–1.2 μg/mL. It is metabolized by hydrolysis in whole blood by red blood cell esterases resulting in the formation of a primary acid metabolite and free methanol. The metabolite is pharmacologically inactive. The resulting methanol levels are not toxic. Esmolol is 55% bound to plasma protein, the acid metabolite only 10%. Less than 2% of parent drug and the acid metabolite are

Table 1. Antiarrhythmic Agents

Generic name	CAS Registry Number	Molecular formula	Trade name	Structure
		Class IA agents (Sodium channel blockers)		
disopyramide phosphate	[*22059-60-5*]	$C_{21}H_{29}N_3O$	Norpace	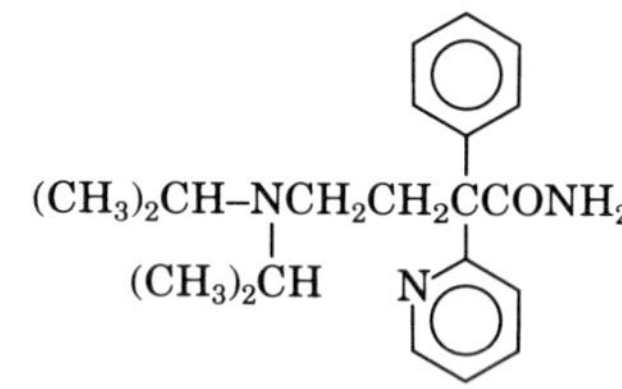
procainamide hydrochloride	[*614-39-1*]	$C_{13}H_{22}ClN_3O$	Procan, Pronestyl	
quinidine gluconate[a] quinidine sulfate[b]	[*7054-25-3*] [*6591-63-5*]	$C_{26}H_{36}N_2O_9$ $C_{40}H_{50}N_4O_8S$	Duraquin, Cin-Quin Quindex, Quinora	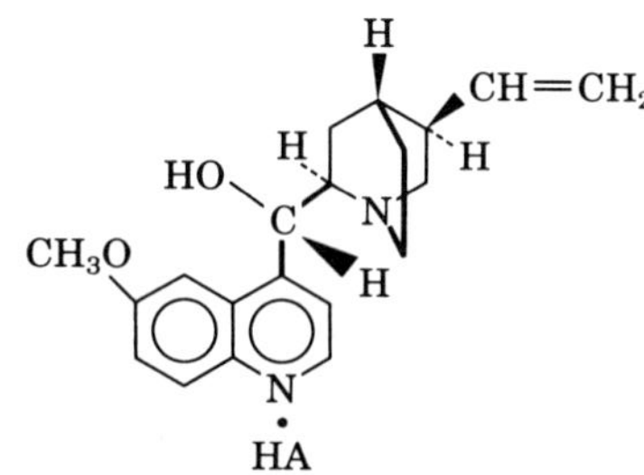
		Class IB agents (Sodium channel blockers)		
lidocaine hydrochloride	[*73-78-9*]	$C_{14}H_{23}ClN_2O{\cdot}H_2O$	Xylocaine, Dalcaine	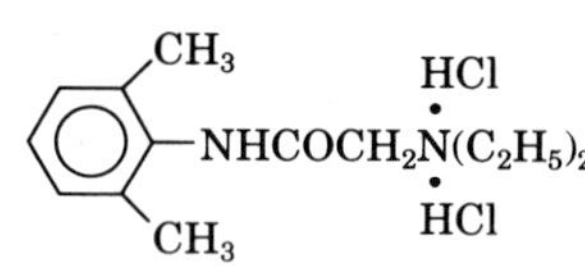

mexiletine hydrochloride	[*5370-01-4*]	$C_{11}H_{18}ClNO$	Mexitil	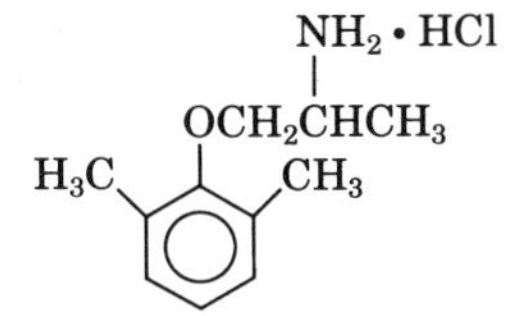
moricizine	[*31883-05-3*]	$C_{22}H_{25}N_3O_4S$	Ethmosin	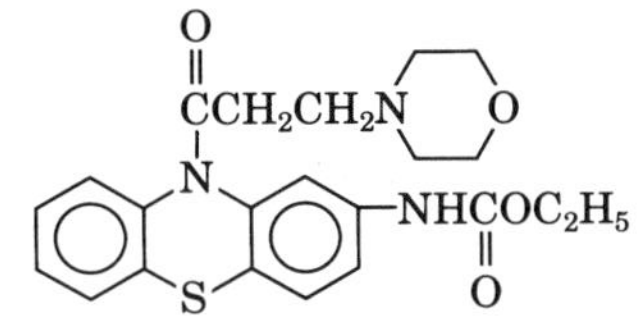
phenytoin sodium	[*630-93-3*]	$C_{15}H_{11}N_2NaO_2$	Dilantin	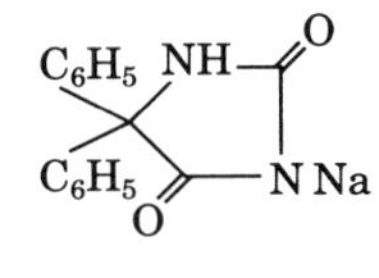
tocainide	[*41708-72-9*]	$C_{11}H_{16}N_2O$	Tonocard	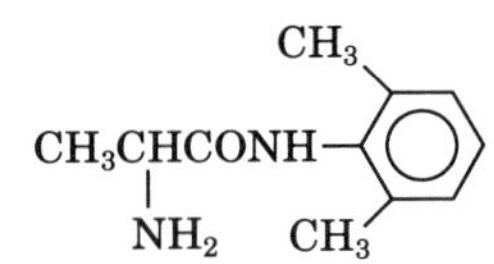
		Class IC agents (Sodium channel blockers)		
encainide hydrochloride	[*66794-74-9*]	$C_{22}H_{29}ClN_2O_2$	Encaid	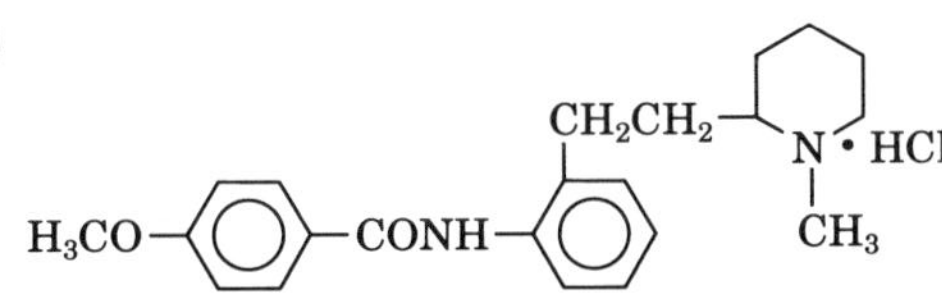

Table 1. (*Continued*)

Generic name	CAS Registry Number	Molecular formula	Trade name	Structure
flecainide acetate	[*54143-56-5*]	$C_{17}H_{20}F_6N_2O_3$	Tambocor	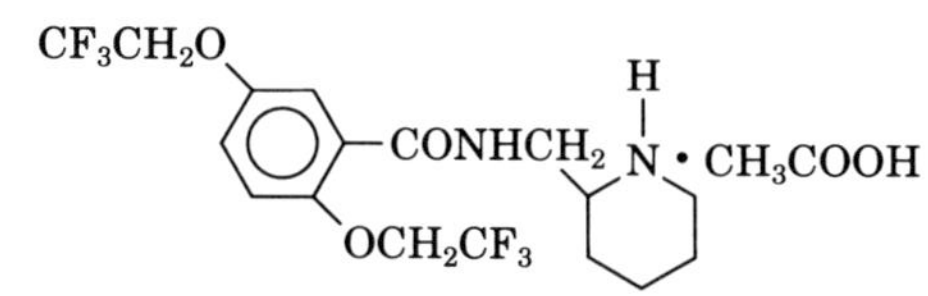
indecainide hydrochloride	[*73681-12-6*]	$C_{20}H_{24}N_2O$	Decabid	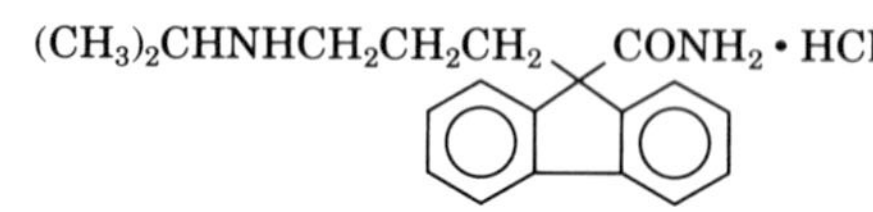
propafenone hydrochloride	[*34183-22-7*]	$C_{21}H_{28}ClNO_3$		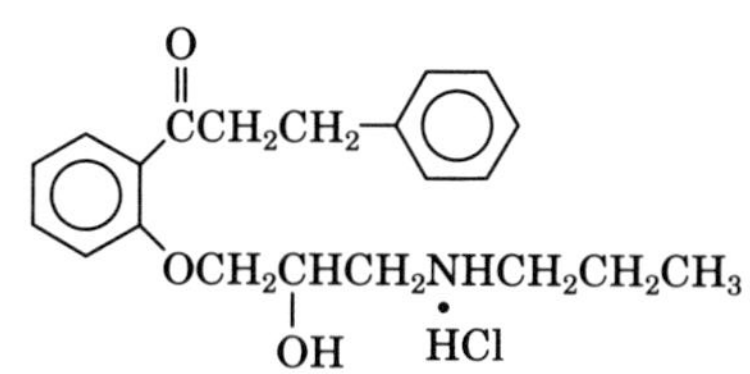
Other Class I agents				
lorcainide hydrochloride	[*58934-46-6*]	$C_{22}H_{28}Cl_2N_2O$		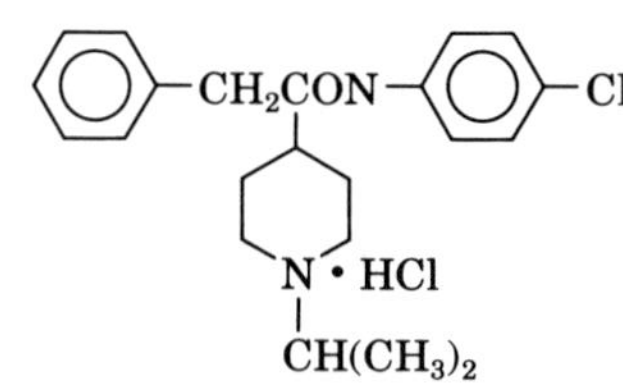

pirmenol hydrochloride	*[61477-94-9]*	$C_{22}H_{31}ClN_2O$		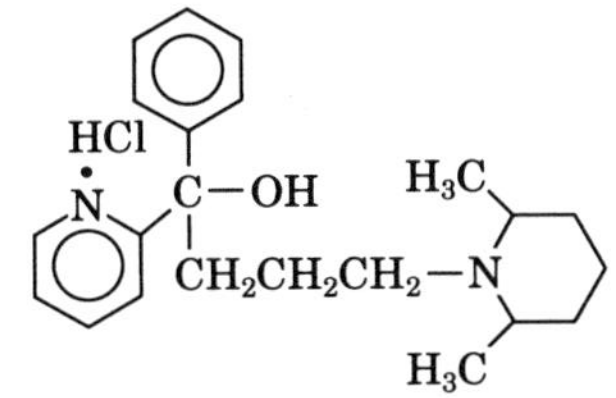

Class II agents (β-Adrenoceptor blockers)

acebutolol hydrochloride	*[37517-30-9]*	$C_{18}H_{29}ClN_2O_4$	Sectral	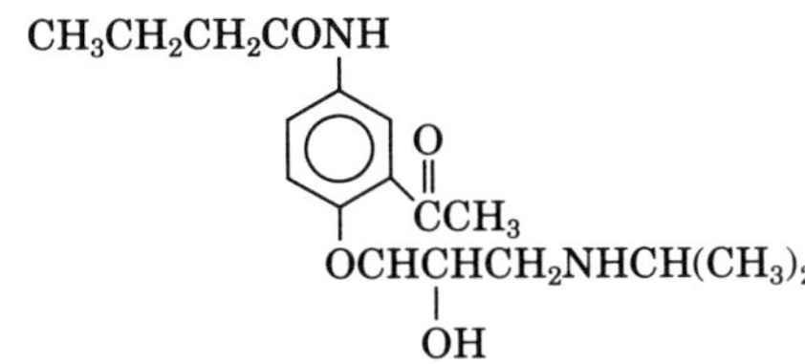
bopindolol	*[62658-63-3]*	$C_{23}H_{28}N_2O_3$		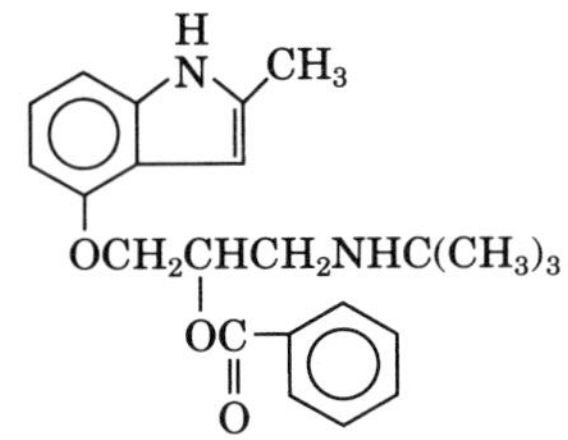
carteolol hydrochloride	*[51781-21-6]*	$C_{16}H_{25}ClN_2O_3$	Cartrol	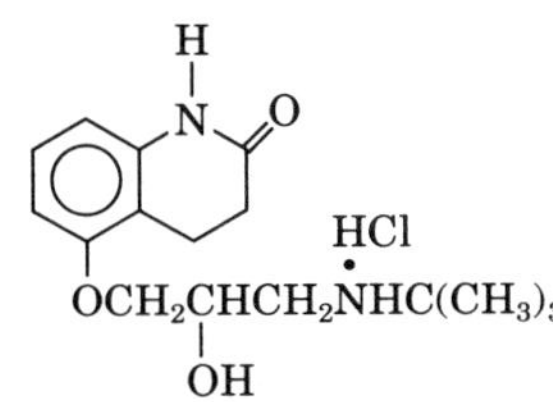

Table 1. (*Continued*)

Generic name	CAS Registry Number	Molecular formula	Trade name	Structure
esmolol hydrochloride	[*81161-17-3*]	$C_{16}H_{26}ClNO_4$	Brevibloc	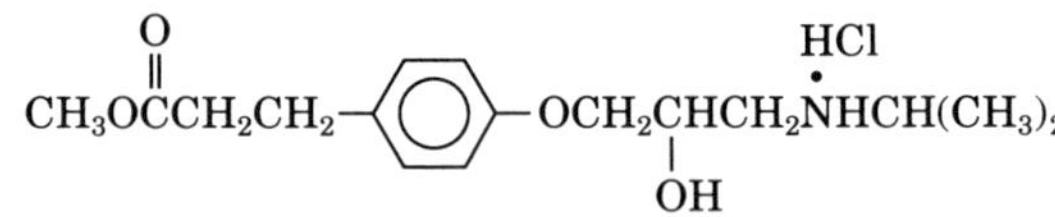
flestolol sulfate	[*88844-73-9*]	$C_{15}H_{22}FN_3O_4{\cdot}H_2SO_4$		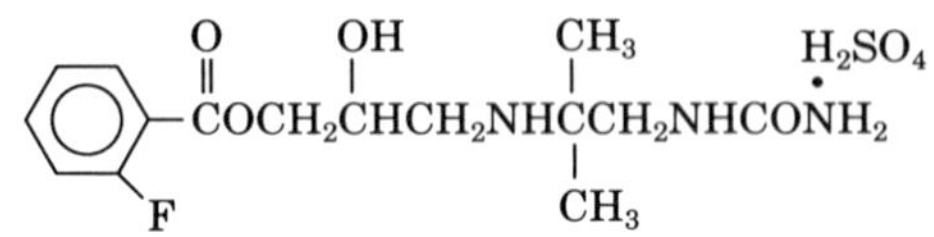
metoprolol tartrate	[*56392-17-7*]	$C_{15}H_{25}NO_3$	Lopressor	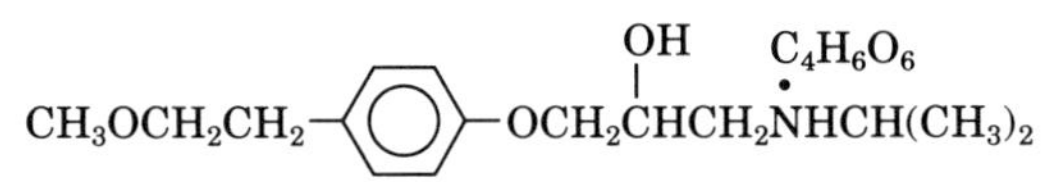
propranolol hydrochloride	[*319-98-9*]	$C_{16}H_{22}ClNO_2$	Inderal	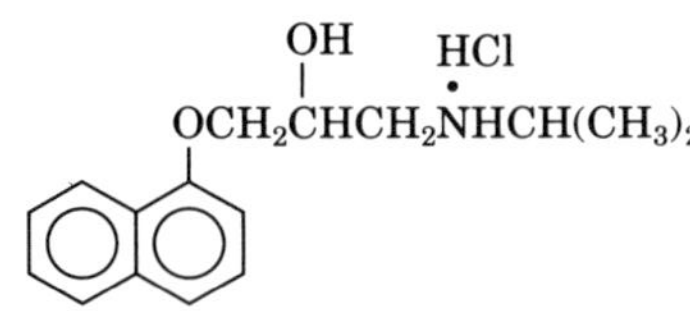
		Class III agents		
amiodarone	[*1951-25-3*]	$C_{25}H_{29}I_2NO_3$	Cordarone	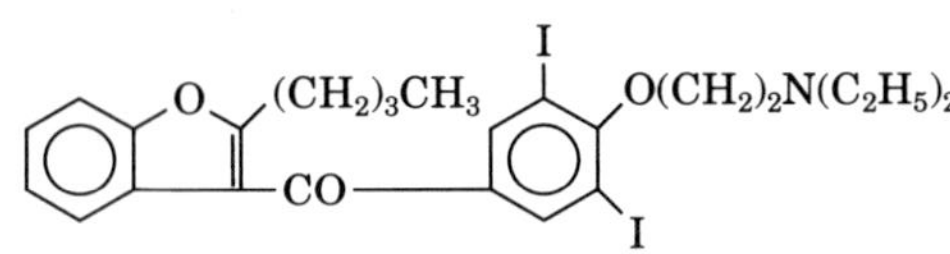

bretylium tosylate	[*61-75-6*]	$C_{18}H_{24}BrNO_3S$	Bretylol	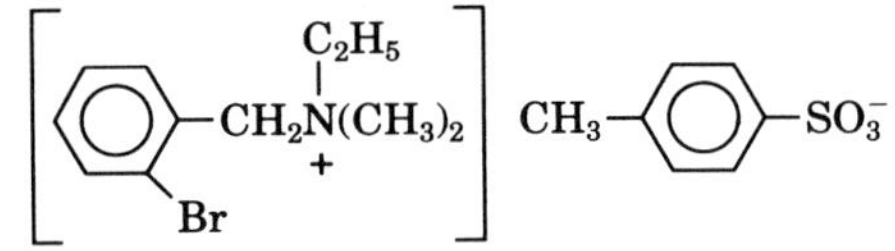
clofilium phosphate	[*60379-03-3*]	$C_{21}H_{39}ClNO_4P$		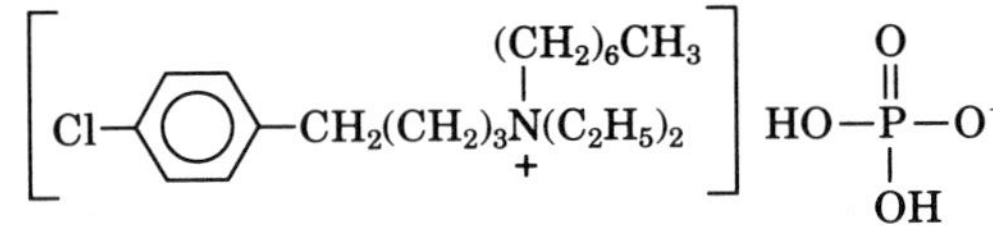
sotalol hydrochloride	[*959-24-0*]	$C_{12}H_{21}ClN_2O_3S$		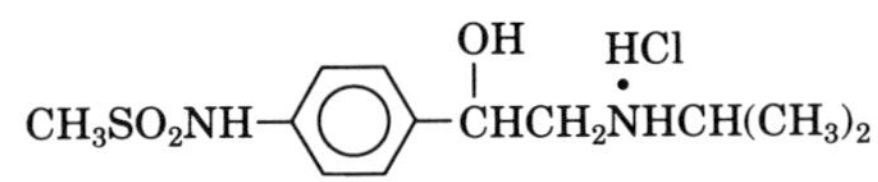
		Class IV agents		
verapamil hydrochloride	[*152-11-4*]	$C_{27}H_{39}ClN_2O_4$	Calan	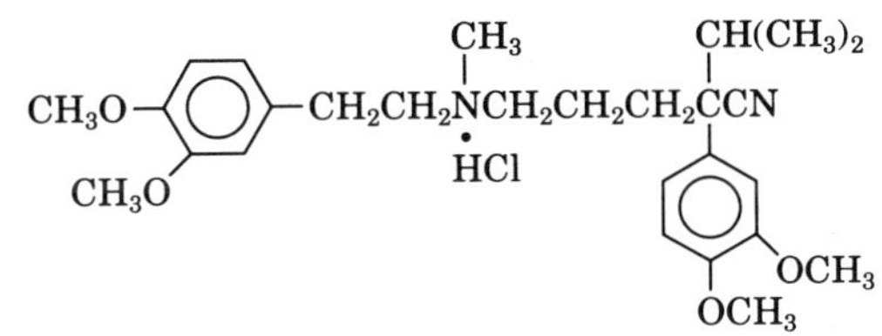
		Class V agents		
alinidine hydrobromide	[*33178-86-8*]	$C_{12}H_{14}BrCl_2N_3$		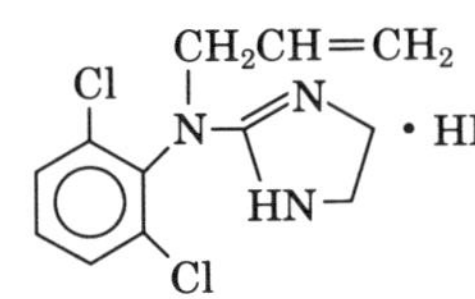

Table 1. (*Continued*)

Generic name	CAS Registry Number	Molecular formula	Trade name	Structure
		Autonomic agents		
atropine sulfate	[*5908-99-6*]	$C_{34}H_{48}N_2O_{10}S$		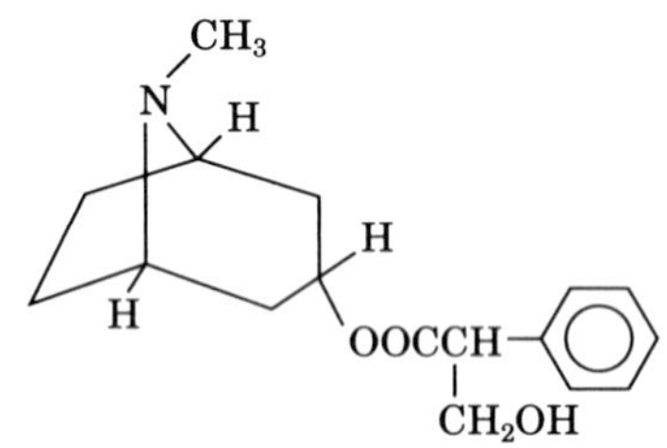
edrophonium chloride	[*116-38-1*]	$C_{10}H_{16}ClNO$	Tensilon	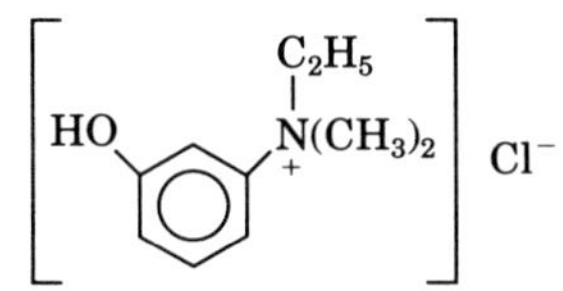
isoproterenol hydrochloride	[*51-30-9*]	$C_{11}H_{18}ClNO_3$	Isuprel, Norisodrine	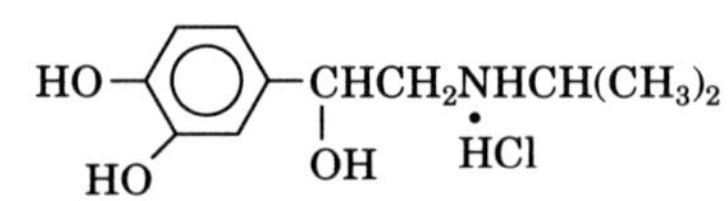
phenylephrine hydrochloride	[*61-76-7*]	$C_9H_{14}ClNO_2$		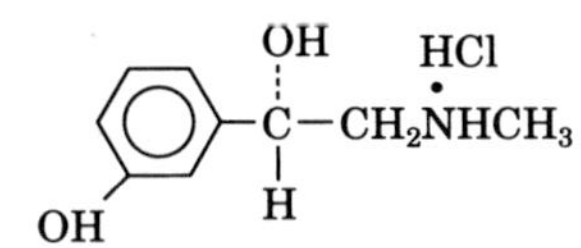

Digitalis glycosides

digitoxin[c]	*[71-63-6]*	$C_{41}H_{64}O_{13}$	Cardigin, Crystodigin, Digisidin, Linidigin	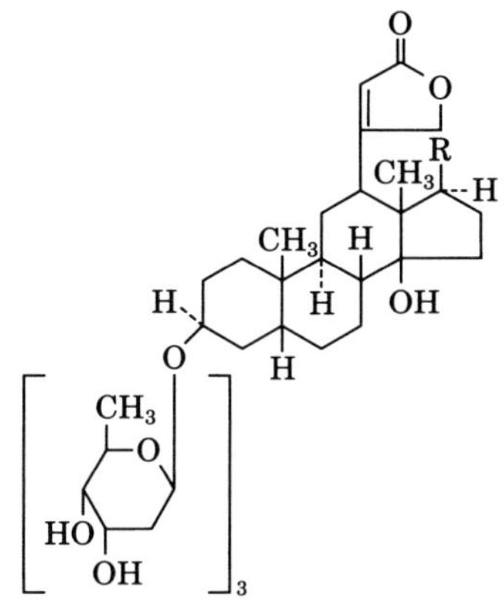
digoxin[d]	*[20830-75-5]*	$C_{41}H_{64}O_{14}$	Lanoxin	

Unclassified agents

acrihellin	*[67696-82-6]*	$C_{29}H_{38}O_7$		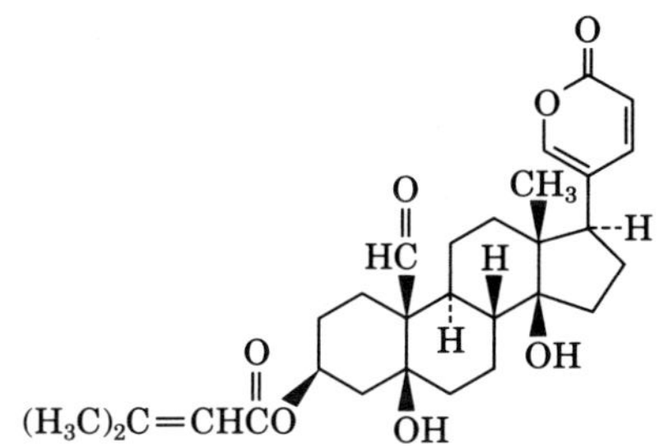
adenosine	*[58-61-7]*	$C_{10}H_{13}N_5O_4$	Adenocard	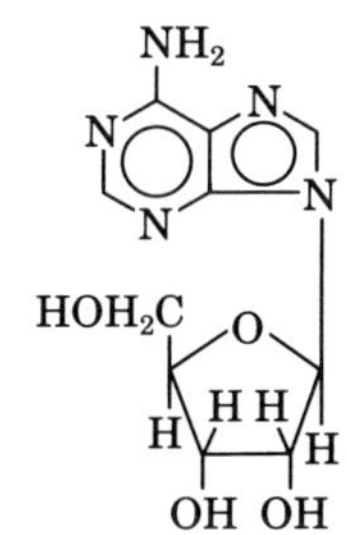

Table 1. (*Continued*)

Generic name	CAS Registry Number	Molecular formula	Trade name	Structure
asocainol	[*77400-65-8*]	$C_{27}H_{31}NO_3$		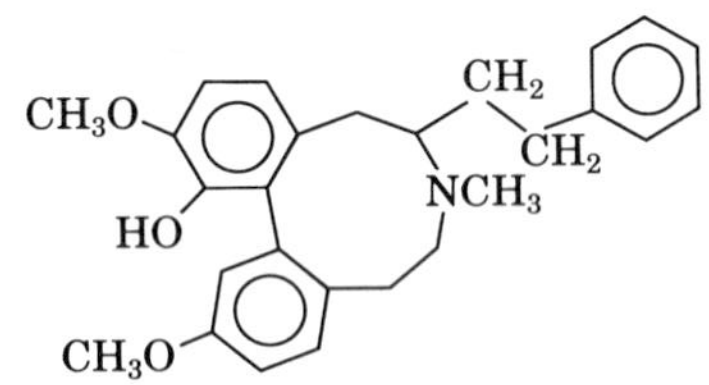
cifenline succinate	[*100678-32-8*]	$C_{22}H_{24}N_2O_4$	Cipralan	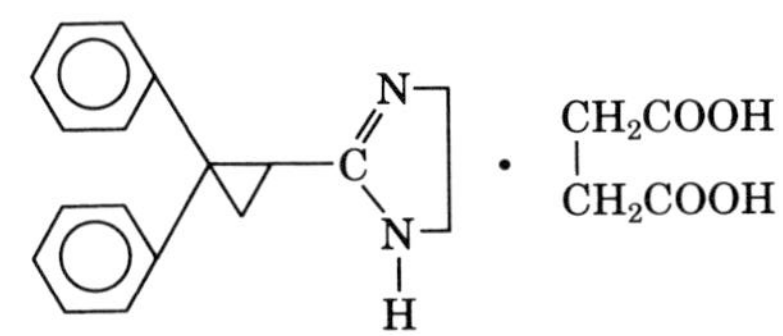

[a]HA = gluconic acid.
[b]HA = H_2SO_4.
[c]R = H.
[d]R = OH.

excreted by the kidneys. Plasma levels may be elevated and elimination half-lives prolonged in patients with renal disease (41).

Because of its brief half-life and minimal lipid solubility, the side effects of esmolol are transient and include hypotension, cold extremities, dyspnea (from bronchospasms), bradycardia, nausea, vomiting, and headaches (41).

Other β-Adrenoceptor Blocking Agents. Several other β-adrenoceptor blocking agents are in development as antiarrhythmic agents. These include carteolol, flestolol, and bopindolol (see Table 1).

Carteolol hydrochloride, a hydrophilic, long-acting, nonselective β-adrenoceptor blocker, has ISA and minimal membrane stabilizing activity. The drug has antihypertensive, antiarrhythmic, and antianginal effects (stable exertional angina) (42). About 78–83% of a po dose of carteolol is absorbed from the GI tract. The rest is excreted in the feces. Carteolol is 80–90% bioavailable. It undergoes hepatic first-pass metabolism. About 40–68% of the oral dose is excreted in the urine as unchanged carteolol, the rest as metabolites or conjugated compounds (3–18%). The primary metabolite in humans is 8-hydroxy-carteolol. Four other metabolites have been identified in animal studies (42). The side effects and toxicity of carteolol are similar to those associated with β-adrenoceptor blocking agent therapy (42).

Flestolol sulfate is a nonselective, ultrashort acting β-adrenoceptor blocker. It has no ISA and produces weak inhibition of the fast sodium channel. The drug is under clinical investigation for supraventricular tachyarrhythmias, unstable angina, and acute MI. In humans, flestolol has hemodynamics and electrophysiologic effects similar to those of other β-adrenoceptor blockers. The pharmacokinetics of flestolol are similar to those of esmolol. It is 50 times more potent than esmolol and the elimination half-life is 7.2 min. Recovery from β-adrenoceptor blockade is 30–45 min after stopping iv infusions. The drug is hydrolyzed by tissue esterases and no active metabolites of flestolol have been identified (41).

Bopindolol is a long-acting, nonselective β-adrenoceptor blocker. It has mild membrane stabilizing activity and ISA. *In vivo*, the compound is hydrolyzed to its active metabolite. Because of this prodrug feature the onset of action is slower than other available β-adrenoceptor blockers. Preliminary pharmacokinetic studies indicate that the compound is well absorbed, is 70% bioavailable, and peak plasma levels are achieved in about 2 h. Whereas its elimination half-life is 4–8 h, β-adrenoceptor blocking action (~40%) is still apparent after 48 h. The drug is being studied in hypertension, angina, and arrhythmias (43).

Class III Antiarrhythmic Agents

The Class III antiarrhythmic agents markedly prolong action potential duration and effective refractory period of cardiac tissue. The QT interval of the ECG is markedly prolonged.

Amiodarone. Amiodarone, an iodinated benzofuran derivative that has structural similarities to the thyroid hormones (qv), was originally introduced as a coronary vasodilator and antianginal agent in Europe in 1967 (44,45). It was subsequently found to have potent antiarrhythmic actions (46). The compound has a wide spectrum of antiarrhythmic activity against supraventricular and ventricular arrhythmias (47,48).

The electrophysiological effects of amiodarone may be a composite of several properties. In addition to prolonging action potential duration and refractory period in all tissues of the heart, the compound is an effective sodium channel blocker (49), calcium channel blocker (50), and a weak noncompetitive β-adrenoceptor blocking agent (51). Amiodarone slows the sinus rate, markedly prolongs the QT interval, and slightly prolongs the QRS duration (1,2).

Amiodarone dilates arteriolar vascular smooth muscle, especially coronary arteries, and thus exhibits antianginal effects. Its effects on the peripheral vasculature to decrease resistance leads to a decrease in left ventricular stroke work and a decrease in myocardial oxygen consumption. The drug rarely produces hypotension that requires discontinuation of the drug (1,2).

The GI absorption of the drug after po administration is slow and variable with estimates ranging from 20–55%. Once absorbed, 96% of the drug is bound to plasma proteins and other tissues on the body. Whereas peak plasma concentrations may be achieved in 3–7 h, the onset of antiarrhythmic action may occur in 2–3 days or more. This may result, in part, from distribution to and concentration of the drug in adipose tissue, liver, spleen, and lungs. Therapeutic plasma concentrations are 1–2 μg/mL, although there appears to be no correlation between plasma concentration and antiarrhythmic activity. The plasma half-life after discontinuation of the drug varies from 13–103 days. The drug is metabolized in the liver and the principal metabolite is desethylamiodarone. The primary route of elimination is through the bile. Less than 1% of the unchanged drug is excreted in the urine. The drug can also be eliminated in breast milk and through the skin (1,2).

Amiodarone exhibits many prominent adverse side effects. The drug is deposited as yellowish-brown microcrystals in almost every organ. About 10% of the patients observe halos or have blurred vision resulting from deposits in the cornea. About 10–25% of the patients exhibit photodermatitis because of deposits in the skin with 5% developing a grayish blue skin discoloration. Pulmonary inflammation and fibrosis with concomitant loss of pulmonary function occurs in 10–15% of patients. Because iodine accounts for about 40% of the molecular weight of amiodarone, thyroid function may be increased or decreased upon prolonged use. Neurological reactions include tremors, staggering gait and impaired ambulation, ataxia, paresthesias, and sleep disturbances (vivid dreams or nightmares). About 10% of the patients become constipated (1,2).

Bretylium. Bretylium tosylate, a bromobenzyl quaternary ammonium compound, is a relatively simple molecule having a complex pharmacological profile. The drug was originally introduced as an antihypertensive agent (52) and later found to possess antiarrhythmic/antifibrillatory properties (53,54). The electrophysiological effects of bretylium are prolongation of action potential duration (delayed repolarization), effective refractory period, and elevation of fibrillation threshold of the myocardium (1,2). Immediately upon iv injection the drug is taken up into postganglionic adrenergic nerves releasing norepinephrine [*51-41-2*] (NE), $C_8H_{11}NO_3$, resulting in an increase in heart rate and in hypertension because of an increase in vascular resistance (sympathomimetic actions). Subsequently the drug exhibits adrenergic neuronal blocking action and prevents the release of NE from the same postganglionic nerve terminals. The latter activity is the pharmacological basis of its antihypertensive effects. Bretylium is primarily indicated for life-threatening ventricular tachycardias and fibrillation (55–57).

Because bretylium is poorly absorbed from the GI tract (~10%), it is administered iv or im. Very little drug is protein bound in plasma. Bretylium is taken up by an active transport mechanism into and concentrated in postganglionic nerve terminals of adrenergically innervated organs. Peak plasma concentrations after im injections occur in about 30 min. Therapeutic plasma concentrations are 0.5–1.0 μg/mL. Bretylium is not metabolized and >90% of the dose is excreted by the kidneys as unchanged drug. The plasma half-life is 4–17 h (1,2).

Adverse effects with bretylium include hypotension, nausea, vomiting, parotid gland tenderness, and arrhythmogenesis (1,2).

Sotalol. Sotalol hydrochloride was introduced as a competitive, nonselective β-adrenoceptor blocker (58) and its potency has been estimated at 1/10–1/50 (59,60) to 1/100 (61) that of propranolol. Its electrophysiological effects consist of increased action potential duration (prolonged repolarization) with no change in the upstroke velocity of phase 0 depolarization (no effect on the sodium current) (62). The electrophysiological effects of sotalol are independent of its β-adrenoceptor blocking actions (63) because the (+)-isomer, which is devoid of the β-adrenoceptor blocking action, produces effects similar to those of the (−)-isomer and the (+),(−)-racemate on increasing refractoriness. Sotalol will antagonize catecholamine-induced effects on the myocardium. The compound produces an increase in myocardial contractile force that is not adrenergically mediated (no ISA) (64). The compound is effective in the treatment of supraventricular tachycardias (65), suppression of PVCs (66), nonsustained ventricular tachycardias (67), and inducible sustained ventricular tachycardias (68).

Sotalol is rapidly and almost completely (>90%) absorbed. Bioavailability of absorbed drug is 89–100%. Peak plasma levels are achieved in 2–4 h. Sotalol is 50% bound to plasma proteins. Plasma half-life of the compound is about 5.2 h. No metabolites of sotalol have been identified indicating little metabolism. The drug is excreted mainly by the kidneys (80–90%) and about 10% is eliminated in the feces. The plasma half-life is prolonged in patients having renal failure. Kinetics of the compound are not affected by changes in liver function (1,2). Sotalol has all the adverse effects of β-adrenoceptor blockers including myocardial depression, bradycardia, transient hypotension, and proarrhythmic effects (1,2).

Other Class III Antiarrhythmic Agents. Clofilium phosphate is a benzenebutanaminium derivative that has highly specific Class III antiarrhythmic activity. It is orally active, has a rapid onset of action, and a reasonably long duration of antiarrhythmic activity. In preliminary clinical studies, clofilium has shown efficacy against spontaneous ventricular tachycardias (69).

Class IV Antiarrhythmic Agents: The Calcium Antagonists

The calcium antagonists have also been known as calcium channel blockers, calcium entry blockers, slow channel blockers, or calcium channel antagonists. The term calcium antagonist is the accepted descriptive term for this class of drugs. The World Health Organization (WHO) has subdivided the calcium antagonists into six classes (70) represented by Roman numerals. The phenylalkylamines, eg, verapamil, constitute Class I; dihydropyridines, eg, nifedipine, Class II; benzothiazepines, eg, diltiazem, Class III; flunarizinelike compounds, Class IV; prenylaminelike compounds, Class V; and other agents in Class VI.

The general biochemical mechanism for contraction in cardiac, skeletal, or smooth muscle is dependent on the availability of a critical intracellular calcium concentration. Calcium enters cells by a variety of mechanisms. In cardiac cells, calcium enters through slow channels that are 100 times more selective for calcium than sodium. This occurs during phase 2 (plateau phase of repolarization) of the action potential, at which time there is a slow inward current of calcium through these channels. Phase 2 repolarization is depressed or shortened by calcium antagonists. In the SA and AV nodes, the calcium antagonists inhibit slow response action potentials, slow conduction, and prolong the PR interval of the ECG. The only calcium antagonist approved for antiarrhythmic indications is verapamil [*52-53-9*], $C_{27}H_{38}N_2O_4$, although diltiazem $C_{22}H_{26}N_2O_4S$, [*42399-41-7*], and others have similar electrophysiologic properties and may be useful antiarrhythmic agents (1,2).

Verapamil. Verapamil hydrochloride (see Table 1) is a synthetic papaverine [*58-74-2*], $C_{20}H_{21}NO_4$, derivative that was originally studied as a smooth muscle relaxant. It was later found to have properties of a new class of drugs that inhibited transmembrane calcium movements. It is a (+),(−) racemic mixture. The (+)-isomer has local anesthetic properties and may exert effects on the fast sodium channel and slow phase 0 depolarization of the action potential. The (−)-isomer affects the slow calcium channel. Verapamil is an effective antiarrhythmic agent for supraventricular AV nodal reentrant arrhythmias (V1-2) and for controlling the ventricular response to atrial fibrillation (1,2,71–73).

After po dosing, verapamil's absorption is rapid and almost complete (>90%). There is extensive first-pass hepatic metabolism and only 10–35% of the po dose is bioavailable. About 90% of the drug is bound to plasma proteins. Peak plasma concentrations are achieved in 1–2 h, although effects on AV nodal conduction may be apparent in 30 min (1–2 min after iv administration). Therapeutic plasma concentrations are 0.125–0.400 μg/mL. Verapamil is metabolized in the liver and 12 metabolites have been identified. The principal metabolite, norverapamil, has about 20% of the antiarrhythmic activity of verapamil (3). The plasma half-life after iv infusion is 2–5 h whereas after repeated po doses it is 4.5–12 h. In patients with liver disease the elimination half-life may be increased to 13 h. Approximately 50% of a po dose is excreted as metabolites in the urine in 24 h and 70% within five days. About 16% is excreted in the feces and about 3–4% is excreted as unchanged drug (1,2).

The side effects and toxic reactions to verapamil include upper GI upset, constipation, dizziness, headaches, flushing and burning, edema, hypotension, bradycardia, and various conduction disturbances. Verapamil has negative inotropic activity and may precipitate heart failure in patients having ventricular dysfunction (1,2).

Class V Antiarrhythmic Agents: Selective Bradycardic Agents

Class V agents have antiarrhythmic activity in sinus tachyarrhythmias. The bradycardia produced by these agents results from effects on the SA node to reduce the slope of phase 4 diastolic depolarization and prolonging SA node action potential duration. The effects are not mediated by a vagal sensitizing action, a muscarinic or cholinergic effect, a β-adrenoceptor blocking effect, or an effect on calcium or sodium channels. It appears that these agents may restrict anionic membrane currents (74). As of this writing only one compound has been studied that has this property.

Alinidine. Alinidine bromide (see Table 1) is the *N*-allyl analogue of clonidine, an α-adrenoceptor agonist. Alinidine produces greater effects on rate than on contractile force (75). This distinguishes alinidine from the calcium channel antagonists, the cholinergic agents, or Class I antiarrhythmic agents. The compound is being evaluated in patients having hyperkinetic heart syndrome (76,77), coronary artery disease (78,79), angina (80,81), sinus tachycardia, and acute MI (82).

Autonomic Drugs Used in the Treatment of Arrhythmias

Drugs that mimic or inhibit the actions of neurotransmitters released from parasympathetic or sympathetic nerves innervating the heart may also be used to treat supraventricular bradyarrhythmias, heart block, and supraventricular tachyarrhythmias. Those used in the treatment of arrhythmias may be found in Table 1.

Edrophonium. Edrophonium chloride is a rapid acting anticholinesterase agent. It inhibits the enzyme cholinesterase that inactivates acetylcholine and thus prolongs the actions of acetylcholine at sites of cholinergic innervation, eg, the vagus and the heart. Thus its pharmacologic and electrophysiologic effects are similar to those of acetylcholine. The heart rate is slowed by slowing diastolic depolarization and decreasing automaticity in SA and AV nodes. It slows conduction in the AV node. In atrial muscle, action potential duration is decreased (effective refractory period is decreased) and conduction velocity is increased. It can be used to terminate supraventricular tachycardias that do not respond to maneuvers to stimulate the vagus. The actions of edrophonium are transient (about 10 min duration). Adverse effects are salivation, sweating, miosis, and GI disturbances. Sinus arrest and complete heart block are also possible with the use of edrophonium (83,84).

Atropine. Atropine sulfate is an alkaloid obtained from the Belladonna plant (see ALKALOIDS). It is an anticholinergic agent and is a competitive inhibitor of acetylcholine, the neurotransmitter released from the vagus nerve which innervates the heart. Vagal stimulation and acetylcholine slows the heart rate by decreasing phase 4 diastolic depolarization in SA and AV nodes. It decreases the duration of the action potential of atrial muscle but not of ventricular muscle. Acetylcholine slows conduction velocity in AV node but increases conduction velocity in atrial muscle. This is reflected by a prolonged PR interval in the ECG. Atropine effects are opposite those of acetylcholine and it is useful for the treatment of bradyarrhythmias by blocking the actions of vagal stimulation. It increases conduction velocity in the AV node (85).

Atropine is well absorbed orally and can be administered parenterally. About 50% of an administered dose is metabolized in the liver, the rest is excreted in the urine unchanged. About 77–94% of an administered dose is excreted in the urine. The pharmacological half-life of atropine is about 2–3 h (85).

Adverse effects with atropine therapy include dry mouth, myosis, loss of visual accommodations, constipation, and urinary retention. The drug can also produce flushing, hyperthermia, delirium, tachycardia, and exacerbate glaucoma (85).

Isoproterenol. Isoproterenol hydrochloride is a nonselective β-adrenoceptor agonist that is chemically related to NE. It mimics the effects of stimulation of the sympathetic innervation to the heart which are mediated by NE. It increases heart rate by increasing automaticity of the SA and AV nodes by increasing the rate of phase 4 diastolic depolarization. It is used in the treatment of acute heart block and supraventricular bradyarrhythmias, although use of atropine is safer for bradyarrhythmias following MI (86).

Isoproterenol may also be used in cardiac arrest. It is often administered as an iv or intracardiac bolus along with sustained external cardiac massage to circulate the drug and stimulate the SA pacemaker to resume automaticity (86).

Isoproterenol is given sublingually or by iv. It is metabolized by monoamine oxidase and catechol-*O*-methyltransferase in brain, liver, and other adrenergically innervated organs. The pharmacological effects of isoproterenol are transient because of rapid inactivation and elimination. About 60% is excreted unchanged. Adverse effects using isoproterenol therapy include nervousness, hypotension, weakness, dizziness, headache, and tachycardia (86).

Phenylephrine. Phenylephrine hydrochloride is an α_1-adrenoceptor agonist. Phenylephrine produces powerful vasoconstrictor and hypertensive responses. This results in baroreceptor activation of a reflex bradycardia and thus is useful in the treatment of supraventricular tachyarrhythmias. Unlike epinephrine [*51-43-4*], the actions of which are relatively transient, phenylephrine responses are more sustained (20 min after iv dosing and 50 min after subcutaneous dosing) (86).

Digital Glycosides in the Treatment of Arrhythmias

The digitalis glycosides (see Table 1) play an important role in the treatment of supraventricular tachyarrhythmias, particularly atrial fibrillation. These agents have direct and indirect (vagomimetic or vagal sensitizing effects) actions on various parts of the heart. In general, the therapeutic effect is to decrease AV conduction by increasing refractoriness of the AV node and to slow the heart. Thus the ventricle does not respond to the tachycardic rates of the SA node or ectopic sites in the atrium (87).

Digoxin. Digoxin is one of the digitalis glycosides used in the treatment of supraventricular tachyarrhythmias. A po dose of tablet or elixir is 60–85% absorbed from the GI tract. Food delays absorption but the total amount of drug absorbed is unchanged. Onset of therapeutic effect is 0.5–2 h for po and 5–30 min for iv preparations. About 25% of an administered dose is protein-bound. Peak serum levels are observed in 45 min to 2 h. About 90% of the po dose is excreted unchanged and about 10% is excreted as metabolites. Excretion is primarily through the kidneys. About 60–90% of an administered dose is excreted in seven days. Patients having impaired kidney function have higher blood levels and longer elimination half-lives of digoxin (87).

Adverse reactions to digoxin include anorexia, vomiting, diarrhea, dizziness, headaches, visual disturbances, and cardiac arrhythmias. Allergic reaction such as urticaria, skin eruptions, fever, and edema have been reported (87).

Digitoxin. Digitoxin is a cardiac glycoside obtained from *Digitalis purpurea*. Digitoxin is indicated in the treatment of atrial flutter, atrial fibrillation, and supraventricular tachycardia. Its electrophysiologic and adverse effects are similar to those described for digoxin (87).

Because digitoxin is a nonpolar, lipophilic glycoside, absorption from the GI tract is complete. About 90% of the drug in plasma is tightly bound to protein. It is metabolized in the liver to many metabolites, including digoxin which is the only pharmacologically active metabolite. The drug is excreted via the bile into feces. The elimination half-life of digitoxin is seven to nine days (87).

Unclassified Approved Antiarrhythmic Agents

Adenosine, 6-amino-9-β-ribofuranosyl-9-*H*-purine (see Table 1), is an endogenous nucleoside found in all cells of the body. Its ubiquitousness suggests that adenosine functions as an autocoid and that its actions are mediated by specific receptors on the plasma membranes of all cells.

Adenosine is not active orally, but administered as an iv bolus drug adenosine rapidly eliminates supraventricular tachycardias within 1–2 min after dosing. The drug slows conduction through the AV node. Adenosine is rapidly removed from the circulation by uptake into red blood cells and vascular endothelial cells. Thus the plasma half-life is less than 10 s. Adenosine is rapidly metabolized to inosine or adenosine monophosphate and becomes part of the body pool for synthesis of adenosine triphosphate.

Adenosine in large doses produces vasodilation resulting in facial flushing, lightheadedness, dizziness, and hypotension. Shortness of breath and chest pressure is seen in about 10% of the patients.

Antiarrhythmic Agents Under Development. Three potential antiarrhythmic agents (Table 1) are in various stages of development.

Asocainol. Asocainol, a dibenzazonine derivative, has sodium channel (Class I) and calcium channel (Class IV) blocking activity that accounts for the antiarrhythmic activity. Preliminary studies indicate that the compound is effective against ventricular arrhythmias (88). Additional studies are needed to establish efficacy, toxicological potential, and pharmacokinetic profile.

Cifenline. Cifenline succinate, also known as cibenzoline, is an imidazoline derivative that is under development in the United States as an antiarrhythmic agent. It is a sodium channel blocker (Class IA), markedly prolongs action potential duration (Class III), and has calcium channel blocking activity (Class IV) (89). Cifenline has efficacy against ventricular arrhythmias (90–92). The pharmacokinetic profile and toxicological potential of cifenline is not well established at this writing (93).

Achrihellin. Acrihellin, a synthetic cardiac glycoside, is an ester of the corresponding aglycon of hellebrin [*13289-18-4*], a bufodienolide. The compound has

Table 2. Incidence of Worsening of Arrhythmias

Antiarrhythmic agent	Incidence, %
Class IA	
quinidine sulfate	16.1–17.0
procainamide hydrochloride	7.5–12.2
disopyramide phosphate	5.7–6.3
aprindine hydrochloride	13.2–13.3
Class IB	
mexilitene hydrochloride	10.3–10.9
tocainide hydrochloride	9.2–23.1
Class IC	
encainide hydrochloride	25.0
lorcainide	7.9
flecainide	14.3
propafenone	11.3
Class II	
propranolol hydrochloride	6.5–14.6
metoprolol	15.6–18.5

electrophysiological properties of the cardiac glycosides and antiadrenergic properties. These properties suggest utility in the treatment of supraventricular tachycardias (94).

Antiarrhythmic Drug Therapy Problems

Drugs used to treat and prevent cardiac arrhythmias have relatively small therapeutic dose to toxic dose ratios. Whereas antiarrhythmic drugs have been shown to be effective in the treatment of a wide variety of arrhythmias, worsening of ventricular arrhythmias (proarrhythmia) is a frequent and serious side effect of antiarrhythmic drugs and may lead to sudden death (11,94,95). Table 2 lists the incidence of worsening of arrhythmias by several antiarrhythmic agents. The effects on decreasing mortality have not been demonstrated. To determine whether these agents have effects on mortality, the National Heart, Lung, and Blood Institute in 1985 conducted a one-year preliminary trial in 10 medical centers called "The Cardiac Arrhythmia Pilot Study" (CAPS). The safety and efficacy of four antiarrhythmic agents were compared to a placebo in 502 post-acute myocardial infarction (MI) patients having premature ventricular complexes (PVCs) and nonsustained ventricular tachycardias. The results indicated that encainide suppressed PVCs 79%, flecainide 83%, moricizine 66%, imipramine 52%, and placebo 37%. Encainide, flecainide, and moricizine were considered safe and effective for suppressing PVCs in post-MI patients (96). The National Institutes of Health

(NIH) began a follow-up study in 1987 called "The Cardiac Arrhythmia Suppression Trial" (CAST), using 1500 patients, to determine whether encainide, flecainide, and moricizine would decrease the incidence of sudden cardiac death in post-MI patients when compared to a placebo. In 1989 an interim review of CAST revealed a significantly higher incidence (3.6-fold) of deaths in the encainide or flecainide groups than in the placebo group. Encainide and flecainide were removed from the study and the trial continued with moricizine and placebo treatments. The moricizine/placebo trial of CAST was also terminated in mid-1991 because there was a higher incidence of deaths in the moricizine-treated group than in the placebo group. The Food and Drug Administration (FDA) Cardiorenal Advisory Committee recommended that encainide, flecainide, and other Class IC agents were to be used for life-threatening or symptomatic arrhythmias and not for nonlife-threatening arrhythmias. It was additionally concluded that Class IA and IB drug monographs should have some type of warning and description of the CAST results including the statement: "No benefit with respect to mortality has been shown for any antiarrhythmic drug." Class IA and IB agents were not restricted to use for life-threatening arrhythmias (97).

ANTIANGINAL AGENTS

Angina pectoris is the principal symptoms of ischemic heart disease, which manifests itself as a dramatic, terrorizing, pressing poststernal pain. It occurs suddenly and can be severe. The location and character of the pain may vary but often radiates from the sternum to the left shoulder and over the flexor surface of the left arm to the tips of the medial fingers. It can be induced by exercise, anxiety, overeating, or stress and is often relieved quickly by rest. Other factors, such as decreased oxygen carrying capacity of the blood, or reduced aortic pressure may be involved. The attack may be transient and damage to the ischemic myocardium may be minimal or it may result in an acute MI and/or death. It is usually accompanied by ST segment changes in the ECG depending on the condition. The pain results from temporary ischemia of the myocardium when the oxygen supply is insufficient to meet the metabolic demands, either by a decrease in blood supply or an exceedingly large increase in oxygen requirements of the myocardium or both. For a drug to be efficacious in angina it should improve myocardial oxygen supply (increase blood flow) or reduce myocardial oxygen consumption or have both actions.

There are different classifications of angina. Stable angina is defined as angina in which no change in frequency, duration, or precipitating causes of the attacks have occurred in the past 30 days. Duration of pain is less than 15 min and occurs on effort or exertion. Unstable angina is defined as angina in which there is a progressive changing pattern, ie, an increase in frequency, severity, and/or duration of pain (crescendo angina), or there is a new onset of angina present for less than 30 days. Unstable angina may occur at rest or on exertion. Variant angina (Prinzmetal's angina) is angina resulting from coronary artery spasm (98).

Nitrate Therapy in Angina

Nitrates may produce relief of angina although differently in the various subsets of angina. For example, in stable exertional angina, the action of nitrates on decreasing myocardial oxygen demand is greater than that on relaxing coronary arteries. In unstable angina, nitrates reduce left ventricular wall tension and dilate eccentric stenotic coronary arteries, whereas in Prinzmetal's angina the primary effect of nitrates is to relax coronary vasospasm (98).

The precise mechanism of nitrate action is not clearly understood and may be a combination of many factors. The basic pharmacologic action of nitrates is a relaxation of most vascular smooth muscle, eg, vascular, bronchial, gastrointestinal, uretal, uterine, etc. Vascular smooth muscle relaxation is a direct effect that is not mediated by adrenergic receptors or endothelium. The nitrates react with sulfhydryl containing receptors in vascular smooth muscles to form *S*-nitrosothiol, which in turn activates guanylate cyclase to form cyclic-guanosine monophosphate [*7665-99-8*] (cGMP), $C_{10}H_{12}N_5O_7P$ (99).

Nitrates may release prostacyclin [*35121-78-9*], $C_{20}H_{32}O_5$, from the endothelial lining of the vasculature and thereby relax vascular smooth muscle. Prostacyclin is a potent endogenous vasodilator and inhibitor of platelet aggregation. Nitrates may be effective in relieving anginal pain by redistributing coronary blood flow to the ischemic myocardium following coronary artery dilation, or by improving the perfusion of the subendocardium. Reduction of myocardial oxygen demand has been proposed as part of the mechanism for relieving angina. In support of this hypothesis are the observations that nitrates also relax venous capacitance vessels, cause venous pooling, and decrease venous return to the heart. These actions on the venous side of the circulation reduce preload, left ventricular volume, and ventricular wall tension. Combined with the action on arteriolar resistance to reduce afterload and increase cardiac output, nitrates reduce the external work of the ventricles and, in turn, decrease oxygen requirements of the myocardium (98,99). Nitrate antianginal agents are shown in Table 3.

Nitroglycerin. Nitroglycerin, originally synthesized in 1847, was introduced as a drug for the treatment of angina pectoris in 1879. It is available in sublingual tablets, ointments, and transdermal nitroglycerin delivery systems. The transdermal nitroglycerin adhesive or band-aid patches overcome problems of handling controlled dosing, ie, eliminate the peaks and valleys of plasma levels associated with conventional therapy, to achieve a prolonged duration of action. These patches consist of an outer impermeable layer, a middle and/or inner drug reservoir layer, and are bordered by adhesive to attach the unit to the patient's skin. Nitro-Disc controls the rate of drug delivery by dispersing nitroglycerin in a solid polymer and uses a gel-like polymer matrix. Transderm-Nitro uses a semipermeable rate-controlling membrane between a reservoir of nitroglycerin in a viscous oil and the skin. The driving force for the nitroglycerin movement is the concentration gradient which is highest in the drug reservoir of the unit, lower through the layers of the unit, and lowest at the skin, providing a constant release of nitroglycerin into the systemic circulation. The stratum corneum of the skin is the rate determinant for the movement of the nitroglycerin. All systems are used once a day because nitroglycerin is delivered continuously for at least 24 h. The units should be applied to a hairless region of the body such as upper arm or chest and application sites should be alternated each day. The units may be worn while bathing, showering, or swimming (see CONTROLLED RELEASE TECHNOLOGY, PHARMACEUTICALS; DRUG DELIVERY SYSTEMS).

Nitroglycerin remains the drug of choice for treatment of angina pectoris. It has also been found useful for the treatment of congestive heart failure, myocardial infarction, peripheral vascular disease, such as Raynaud's disease, and mitral insufficiency, although the benefits of nitroglycerin in mitral insufficiency have been questioned.

Sublingual nitroglycerin preparations are rapidly absorbed from the oral mucosa. Peak blood levels and relief from angina occur within 1–4 min. Duration of action is 15–20 min although venous plasma nitroglycerin levels are not detectable after 10 min. Po nitroglycerin is more slowly absorbed. Maximal effects are observed in 30–60 min, and duration of action orally is 4–6 h. Nitroglycerin ointment is also slowly absorbed but its duration of action is 4–8 h. Site and area of application of the ointment appear to affect the plasma levels achieved. Using transdermal nitroglycerin systems, detectable blood levels are noted in about one hour and peak levels are noted at 4–8 h which are maintained for 24 h. Once nitroglycerin is absorbed, it is bound to plasma protein. Nitroglycerin has a large volume of distribution suggesting it is widely distributed in the body. Nitroglycerin is rapidly metabolized in the liver and other organs by glutathione organic nitrate reductase and in the blood of some species. The drug is excreted in the urine as the 1,2- and 1,3-glycerol dinitrates and some glycerol mononitrates which are pharmacologically inactive.

It was originally thought that po nitroglycerin therapy would be inadequate because absorption into portal blood, which passes directly into the liver, would allow hepatic denitration to occur on first-pass, clearing the blood of most of the drug before it reached the systemic circulation. However, glutathione organic nitrate reductase becomes saturated, slowing inactivation and allowing therapeutic nitroglycerin levels to be achieved. Drugs that increase or decrease liver microsomal enzymes may affect nitroglycerin metabolism, eg, phenobarbital [*50-06-6*], $C_{12}H_{12}N_2O_3$, (increase) and alcohol (decrease).

The principal side effects of nitroglycerin therapy are headache, nausea, flushing, dizziness, and weakness. Syncope and tachycardia can result from postural hypotension produced by nitroglycerin. The tachycardia can be attenuated by co-administration of β-adrenoceptor blocking agents. Methemoglobinemia following high doses can also occur because of oxidation of hemoglobin by nitrate ions. Whereas drug rashes resulting from ingestion of nitroglycerin are rare, there have been reports of contact dermatitis upon use of nitroglycerin ointment.

Decreased sensitivity or tolerance to the vascular effects of nitroglycerin following prolonged exposure to high doses of the drug have been reported (61). Dependence on nitrate has been observed in munitions workers exposed to glyceryl trinitrate [*55-63-0*], and prompt withdrawal of nitrate therapy in patients having coronary insufficiency has resulted in worsening of symptoms. Thus a gradual reduction from chronic high doses of nitroglycerin is recommended.

Isosorbide. Isosorbide dinitrate is a derivative of 1,4:3,6-dianhydro-D-glucitol [*652-67-5*], $C_6H_{10}O_4$. It is available in sublingual, chewable, regular, or time-

Table 3. Antianginal Agents

Generic name	CAS Registry Number	Molecular formula	Trade name	Structure
			Nitrates	
erythrityl tetranitrate	*[7297-25-8]*	$C_4H_6N_4O_{12}$	Cardilate	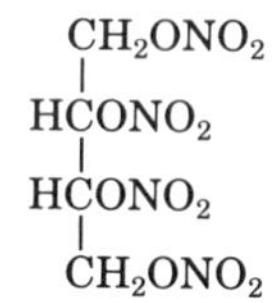
isosorbide dinitrate[a] isosorbide mononitrate[b]	*[87-33-2]* *[16051-77-7]*	$C_6H_8N_2O_8$ $C_6H_9NO_6$	Sorbitrate, Dilatrate	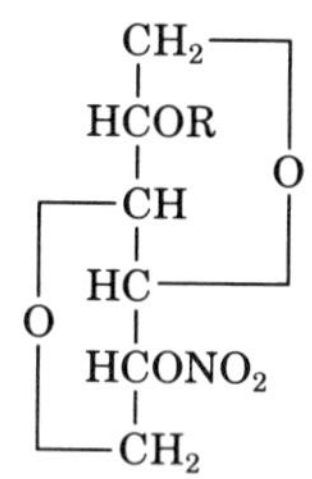
nitroglycerin	*[55-63-0]*	$C_3H_5N_3O_9$	Deponit, Tridil, Nitrogard, Nitro-Disc, Nitrol, Transderm-Nitro, Nitro-Dur, Nitronet, Sustarden	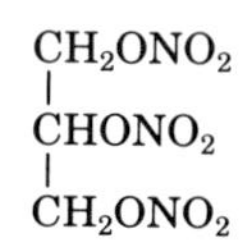
pentaerythritol tetranitrate	*[78-11-5]*	$C_5H_8N_4O_{12}$	Peritrate	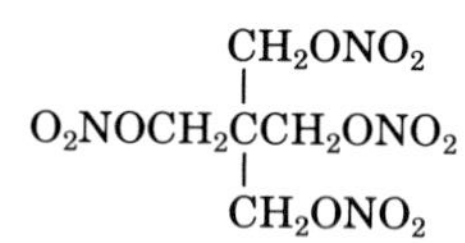

Calcium channel antagonists

Name	CAS Registry Number	Formula	Trade name	Structure
amlodipine besylate[c] amlodipine maleate[d]	[*111470-99-6*] [*88150-47-4*]	$C_{26}H_{31}ClN_2O_8$ $C_{24}H_{29}ClN_2O_9$		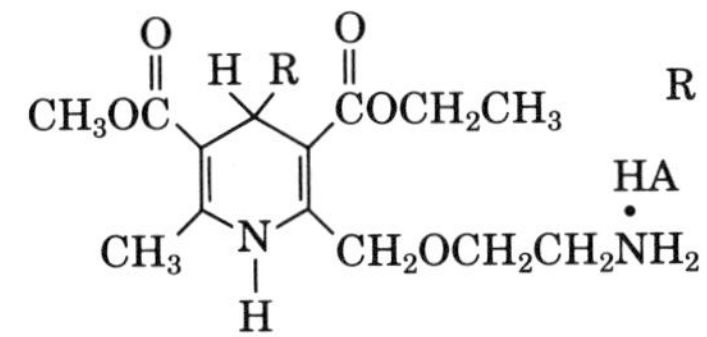 R = 2-chlorophenyl
bepridil hydrochloride	[*74764-40-2*]	$C_{24}H_{35}ClN_2O \cdot H_2O$		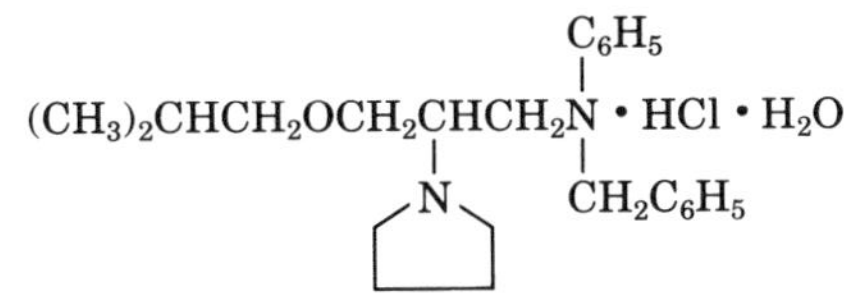
dilitiazem hydrochloride	[*33286-22-5*]	$C_{22}H_{27}ClN_2O_4S$	Cardizem	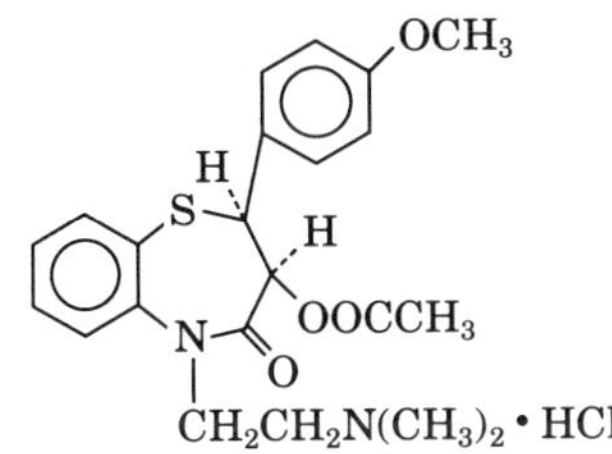
nicardipine hydrochloride	[*54527-84-3*]	$C_{26}H_{30}ClN_3O_6$	Cardene	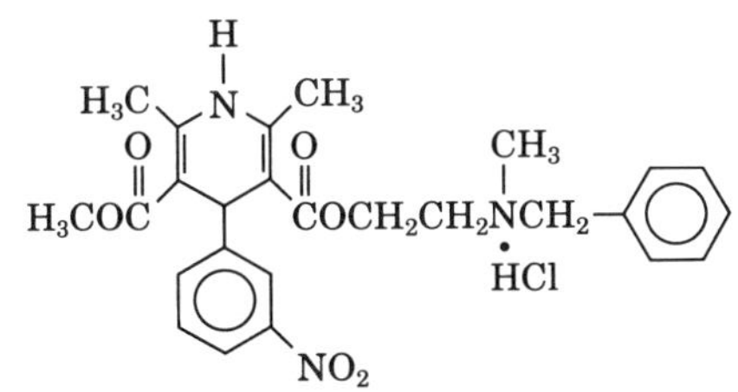

Table 3. (*Continued*)

Generic name	CAS Registry Number	Molecular formula	Trade name	Structure
nifedipine	[*21829-25-4*]	$C_{17}H_{18}N_2O_6$	Adalat, Procardia	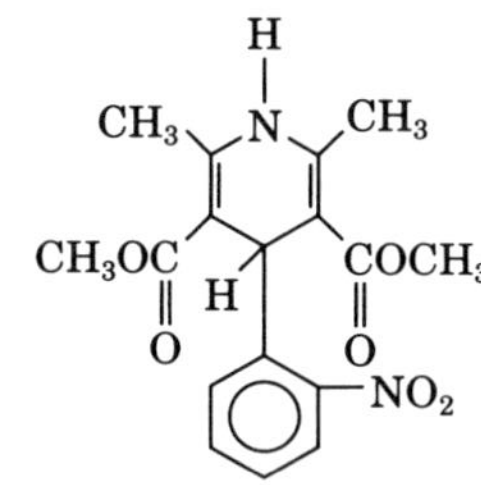
nisoldipine	[*63675-72-9*]	$C_{20}H_{24}N_2O_6$		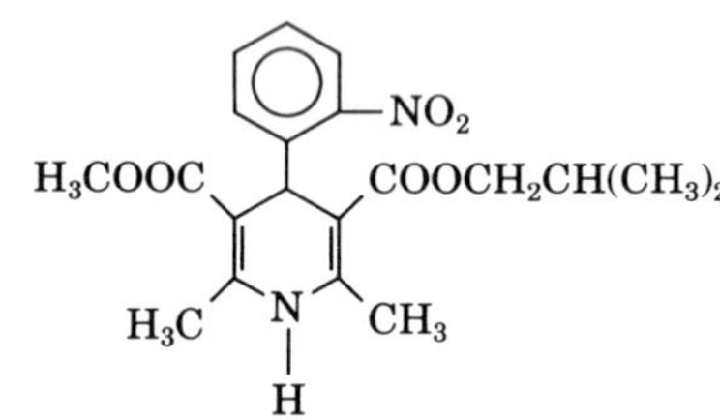
			β Adrenoceptor blockers	
atenolol	[*29122-68-7*]	$C_{14}H_{22}N_2O_3$	Tenormin	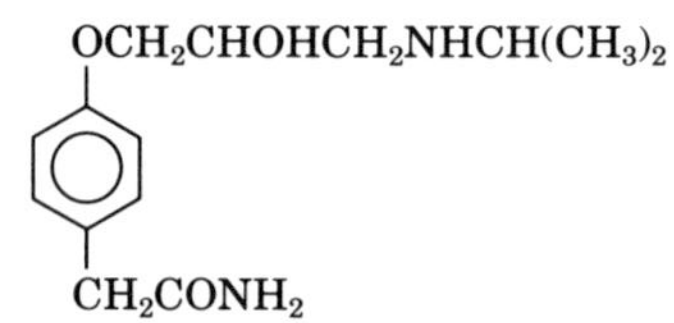
betaxolol hydrochloride	[*63659-19-8*]	$C_{18}H_{30}ClNO_3$		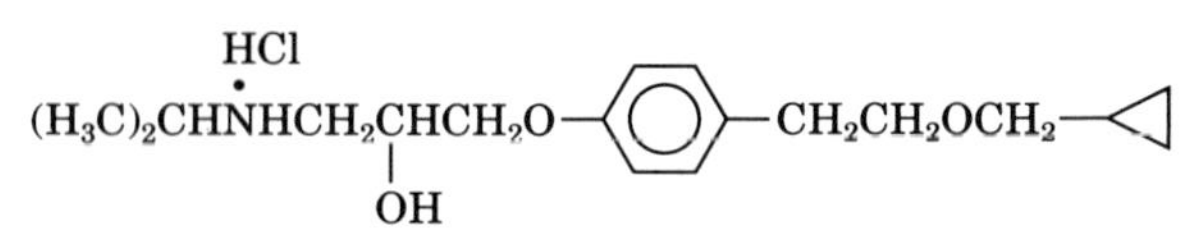

bevantolol hydrochloride	[*42864-78-8*]	$C_{20}H_{28}ClNO_4$	Vantol	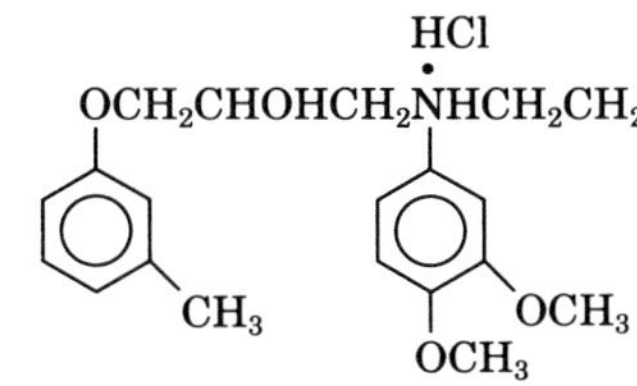
bisoprolol fumarate	[*104344-23-2*]	$C_{36}H_{64}N_2O_{12}$	Monocor	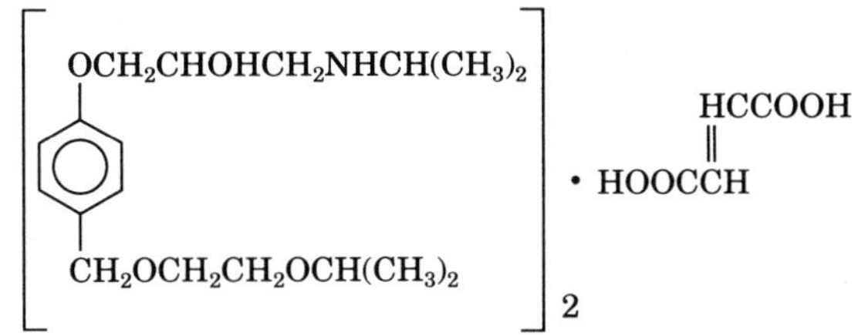
celiprolol hydrochloride	[*57470-78-7*]	$C_{20}H_{34}ClN_3O_4$		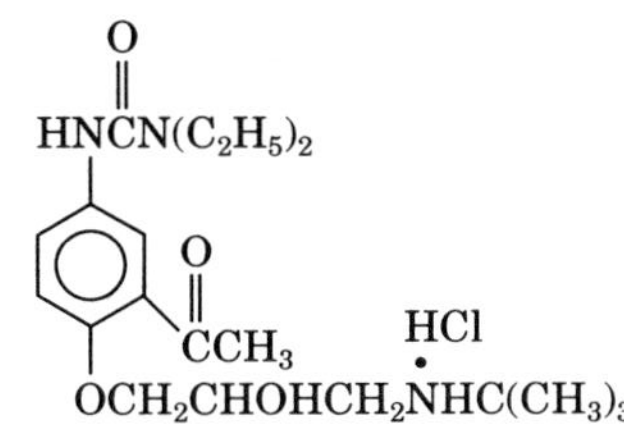
nadolol	[*42200-33-9*]	$C_{17}H_{27}NO_4$	Corgard	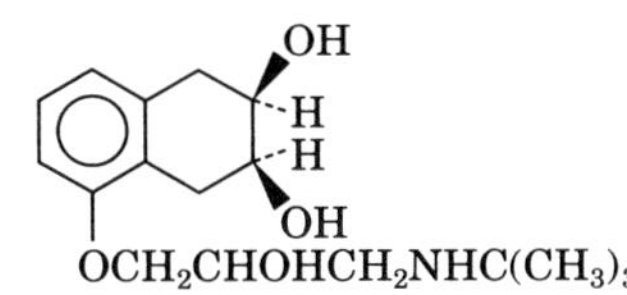

[a] R = NO_2.
[b] R = H.
[c] HA = benzene sulfonic acid.
[d] HA = maleic acid.

release tablets. The drug has also been used iv. The sublingual and chewable tablets are used just prior to anginal-provoking situations (exertion) because of their short duration of action (3–5 min). The regular tablet has a duration of action of up to 3 h, and the time-release tablet up to 8 h (99).

Isosorbide is rapidly absorbed and undergoes rapid first-pass metabolism by the liver. The bioavailability of the sublingual and chewable tablets is 59% and 22%, respectively, for the regular tablet. Isosorbide is metabolized to isosorbide-2-mononitrate and isosorbide-5-mononitrate, both of which have pharmacologic activity. The elimination half-lives of sublingual and po isosorbide dinitrate are 1 and 4 h, respectively. Those of the 2- and 5-mononitrate metabolites are 1.5–3.1 h and 4–5.6 h, respectively. The two metabolites prolong the elimination half-life of the dinitrate. Adverse effects with isosorbide are similar to those described for nitroglycerin (99).

Erythrityl. Erythrityl tetranitrate is a racemic mixture of 1,2,3,4-butanetetrol tetranitrate. It is slightly less effective than nitroglycerin in reducing anginal pain but has a more prolonged duration of action. The tablet can be used sublingually or swallowed. The peak effect of swallowed doses is less but of longer duration (99).

Erythrityl is readily absorbed from the GI tract. It undergoes extensive first-pass metabolism in the liver by glutathione organic nitrate reductase. Time to onset of effect is 5–10 min by sublingual administration and 20–30 min when swallowed. The duration of effects for the two routes are up to 3 and 6 h, respectively. Adverse effects are similar to those described for nitroglycerin (99).

Pentaerythritol. Pentaerythritol tetranitrate, the tetranitrate derivative of 2,2-bis(hydroxymethyl)-1-3-propanediol [*115-77-5*], $C_5H_{12}O_4$, is available in both regular and sustained release tablet preparations. Its efficacy and pharmacokinetic and toxic properties are similar to those described for erythrityl tetranitrate (99).

Calcium Antagonist Therapy in Angina

Calcium antagonists (Table 3) alter the availability of calcium by affecting the entry of this ion into its receptor site(s). Compounds, such as verapamil (see Table 1), nifedipine, and diltiazem, were utilized for a long time for the treatment of angina and certain peripheral vascular diseases in Europe and Asia before their introduction into the United States. Calcium antagonists are used mainly in the treatment of various types of angina, hypertension, and certain arrhythmias but may have potential therapeutic applications in heart failure, acute myocardial infarction, cardioprotection, cardiomyopathy, cerebral vasospasm, and other vasospastic syndromes as well.

Calcium plays a vital role in excitation–contraction coupling, and failure to maintain intracellular calcium homeostasis results in cell death. The availability of the calcium antagonists also provides a powerful tool for basic studies of excitation–contraction coupling, stimulus–excretion coupling, and other specific physiological functions.

Calcium and Vascular Smooth Muscle Contraction. Calcium acts on a number of sites associated with the control of the cytoplasmic calcium concentration. Vascular smooth muscle contraction can be initiated by the opening of the slow calcium channel allowing influx of extracellular calcium through the sarcolemmal membrane into the cytoplasmic compartment. The intracellular calcium concentration increases to $1 \times 10^{-5}\ M$, a threshold concentration necessary to initiate contraction.

In the presence of calcium, the primary contractile protein, myosin, is phosphorylated by the myosin light-chain kinase initiating the subsequent actin-activation of the myosin adenosine triphosphatase activity and resulting in muscle contraction. Removal of calcium inactivates the kinase and allows the myosin light chain to dephosphorylate myosin which results in muscle relaxation. Therefore the general biochemical mechanism for the muscle contractile process is dependent on the availability of a sufficient intracellular calcium concentration.

The calcium antagonists interfere with the entry of calcium through the membrane slow calcium channel and therefore prevent intracellular calcium from reaching the critical concentration necessary to initiate contraction. However, these drugs may be acting through other mechanisms to prevent the elevation of the free calcium ion concentration in the cell.

The calcium antagonists produce beneficial effects in angina by either reducing coronary artery spasm, slowing heart rate, or decreasing contractile force. These effects increase oxygen supply or decrease myocardial oxygen consumption. The clinical effects are increased duration of exercise, reduction in anginal episodes, or may be a reduction in the amount of nitrate therapy consumed to reduce the anginal episode.

The cardiac effects of the calcium antagonists, ie, slowed rate (negative chronotropy) and decreased contractile force (negative inotropy), are prominent in isolated cardiac preparations. However, in the intact circulation, these effects may be masked by reflex compensatory adjustments to the hypotension that these agents produce. The negative inotropic activity of the calcium antagonists may be a problem in patients having heart failure, where contractility is already depressed, or in patients on concomitant β-adrenoceptor blockers where reflex compensatory mechanisms are reduced.

The side effects or toxic effects that the calcium antagonists have in common are hypotension, facial flushing, headache, dizziness, weakness, sedation, skin rash, edema, constipation, and abdominal discomfort (nausea, vomiting, and epigastric pressure).

Verapamil. Verapamil hydrochloride is a phenylalkylamine and is considered the prototype of the Class I calcium channel blockers. Verapamil is also a potent inhibitor of coronary artery spasm and is useful in Prinzmetal's angina and in unstable angina at rest. Verapamil produces negative chronotropic and inotropic effects. These two actions reduce myocardial oxygen consumption and probably account for the effectiveness of verapamil in chronic stable effort angina (98,99). Moreover, verapamil is an effective antihypertensive agent.

Diltiazem. Diltiazem hydrochloride, a benzothiazepine derivative, is considered the prototype of the Class III calcium channel blockers. The (+)-cis form of the compound is the pharmacologically active isomer. Diltiazem dilates all vascular smooth muscle equally, ie, it lacks vascular selectivity. It has potent coronary artery vasodilator properties and relieves the vasospasm of Prinzmetal's or variant angina. Its efficacy in exertional angina probably results from the ability to reduce myocardial oxygen consumption by reducing heart rate and peripheral resistance (reduce the afterload of the heart). Diltiazem is also an effective antihypertensive agent, and in addition to a cardiac negative chronotropic effect it has a slight negative inotropic effect. The electrophysiological effects include prolongations of the P–R interval, AV conduction time, and the A–H interval (1,2,98,99).

The absorption of diltiazem from the GI tract after po dosing is nearly complete. It undergoes extensive first-pass metabolism in the liver and only 40% of the po dose is bioavailable. About 70–80% of the drug is bound to plasma protein. The onset of action is about 15 min and the peak effect occurs at 30 min. Diltiazem is metabolized by deacetylation, and *N*- and *O*-demethylation. The principal metabolite, desacetyldiltiazem has 25–50% of the pharmacological activity and 10–20% of the plasma levels of the parent compound. About 30–35% of the drug is excreted by the kidneys as metabolites and 2–4% as unchanged drug. Almost 60% is excreted in the feces as metabolites. The elimination half-life of diltiazem is 3–4.5 h. Elimination half-life and plasma levels are increased in patients having liver disease (1,98,99).

The side effects of diltiazem therapy are less than those of verapamil or nifedipine therapy. Side effects occur in about 4% of the patients (1,98,99).

Nifedipine. Nifedipine is a dihydropyridine and is considered the prototype of the Class II calcium channel antagonists. The dihydropyridines are photolabile compounds and rapidly degrade when exposed to ultraviolet or visible light. At therapeutic concentration, nifedipine dilates vascular smooth muscle without compromising myocardial or skeletal muscle function (tissue selective). Nifedipine has little effect on cardiac muscle contractility in patients, but in isolated cardiac tissue preparations, nifedipine has more potent negative inotropic effects than verapamil. Nifedipine shortens the P–R interval and at high doses increases conduction time. No effect on A–H or H–V interval has been noted, nor does it have a tendency to produce AV block. Nifedipine is indicated in vasospastic angina by relaxing coronary artery spasm and in chronic stable angina by reducing oxygen consumption through reduction of afterload (peripheral vasodilation) (1,98,99).

Po administered nifedipine is almost completely absorbed. The onset of action is 20 min and peak effects occur at 1–2 h. The principal route of elimination is through hepatic metabolism by oxidation to hydroxycarboxylic acid and the corresponding lactone. These metabolites are pharmacologically inactive. Almost 70–80% of drug is eliminated in the urine during the first 24 h. About 15% is excreted in the feces. The elimination half-life of nifedipine is about 1–2.5 h (1,98,99). Frequency of occurrence of side effects in patients is about 17% with about 5% requiring discontinuation of therapy (1,98,99).

Nicardipine. Nicardipine hydrochloride is a dihydropyridine that is less photolabile than others of the Class II calcium channel blockers. It is tissue selective, producing relaxation of coronary vasculature at dose levels that produce little or no negative inotropic effects. In isolated cardiac preparations it produces a moderate negative inotropic effect and prolongs AV conduction time. Nicardipine has been shown to increase exercise tolerance, and reduce nitroglycerin consumption and the frequency of anginal attacks. It is indicated for the management of patients with chronic stable angina (1,2,98,99).

Nicardipine is almost completely absorbed after po administration. Administration of food decreases absorption. It undergoes extensive first-pass metabolism in the liver. Systemic availability is dose-dependent because of saturation of hepatic metabolic pathways. A 30 mg dose is ~35% bioavailable. Nicardipine is highly protein bound (>95%). Peak plasma concentrations are achieved in 0.5–2.0 h. The principal path of elimination is by hepatic metabolism by hydrolysis and oxidation. The metabolites are relatively inactive and exert no pharmacological activity. The elimination half-life is 8.6 h. About 60% of the dose is excreted in the urine as metabolites (<1% as intact drug) and 35% as metabolites in the feces (1,2,98,99).

Other Calcium Antagonists. Nisoldipine is a lipophilic, light-sensitive, vascular-selective dihydropyridine analogue. Dog experiments have shown that this compound effectively increased coronary blood flow and cardiac output, and decreased total peripheral resistance and systemic blood pressure. It produces a reflex tachycardia that is blocked by propranolol. Po doses significantly prolonged exercise duration and time to onset of angina in patients, by 20% (100). The antianginal effects are similar to those of nifedipine (101–103). Nisoldipine may be beneficial in hypertension (104) but not in congestive heart failure (105).

Amlodipine besylate or maleate is a lipophilic dihydropyridine analogue that is moderately light-sensitive. The compound is potent, long-acting, and selective for vascular smooth muscle. It is being studied for the treatment of angina. Unlike nifedipine, amlodipine has a very long elimination time, half-life of >33 h after iv dosing, suggesting that the drug may be given once a day. The bioavailability with po doses is 52–88% (106).

Bepridil hydrochloride, a Class VI calcium antagonist that is equiselective for heart and vascular smooth muscle, was introduced in France as an antianginal agent. It has the electrophysiological properties of Class I, ie, inhibits the fast Na channel, and Class IV antiarrhythmic agents, ie, inhibits the slow calcium channel. Clinical trials in the United States for evaluation of the antiarrhythmic effects have been discontinued because of a number of deaths in patients receiving the drug. The drug is effective in angina, reducing both the frequency of attacks and prolonging exercise time. Efficacy in angina may result from a bradycardic effect, coronary and peripheral vasodilatation, and cardiodepressant effect. Its electrophysiological effects include prolongation of the effective refractory periods of AV node, atrial and ventricular muscle, slowing of sinus rate, and prolonged QT interval in the ECG (107).

Absorption after po dosing is fairly complete. It undergoes extensive first-pass metabolism in the liver and is 60% bioavailable. It is extensively bound (99%) to α_1-acid glycoproteins. Bepridil is almost completely metabolized in the liver. Seventeen metabolites have been identified but only the 4-hydroxy-*N*-phenylbepridil has some pharmacological activity. The elimination half-life is 33–42 h (107).

β-Adrenoceptor Blocker Therapy in Angina

The β-adrenoceptor blockers are effective antianginal agents because they reduce myocardial oxygen consumption and improve myocardial perfusion by several mechanisms. A decrease in heart rate increases diastolic filling time which allows for increased coronary perfusion. By decreasing blood pressure and myocardial contractility the work load of the heart and ventricular wall tension are reduced, thereby decreasing myocardial oxygen demand. The β-adrenoceptor blockers improve exercise tolerance and reduce chest pain in patients having stable exertional angina. Some of the β-adrenoceptor blockers have cardioprotective properties in patients post-MI, reducing the risk of sudden death. These agents are also useful for attenuating the reflex positive inotropic and chronotropic (tachycardia) effects in patients on nitroglycerin therapy, and in reducing the amount of nitroglycerin consumed (98,99).

Propranolol. Propranolol (Table 1), a Class II antiarrhythmic agent, is useful in the management of hypertrophic subaortic stenosis, especially for the treatment of exertional or other stress-induced angina by improving blood flow. The drug can increase exercise tolerance in patients suffering from angina. Propranolol has been shown to have cardioprotective action in post-MI patients (37–39,98,99,108).

Metoprolol. Metoprolol tartrate (Table 1), also a Class II antiarrhythmic agent, is a lipophilic, cardioselective β_1-adrenoceptor blocking agent having no ISA. The selectivity for β_1-adrenoceptors may be lost at high concentrations of the drug. Metoprolol has been shown to have cardioprotective effects in post-MI patients (37–39,98,99,108).

The absorption of metoprolol after po dosing is rapid and complete. The drug undergoes extensive first-pass metabolism in the liver and only 50% of the po dose in bioavailable. About 12% of the plasma concentration is bound to albumin. The elimination half-life is 3–7 h and less than 5% of the po dose is excreted unchanged in the urine. The excretion of the drug does not appear to be altered in patients having renal disease (98,99,108).

The adverse effects noted with metoprolol result from the β-adrenoceptor blocking actions and are similar to those noted for propranolol (98,99,108).

Nadolol. Nadolol (Table 3) is a hydrophilic, nonselective β-adrenoceptor blocking agent having no ISA and no membrane-stabilizing activity. It is useful for the treatment of hypertension and chronic stable exertional angina (98,99,108).

Absorption of nadolol after po dosing is variable, averaging about 30%. The presence of food does not affect absorption. There is no hepatic first-pass metabolism and peak plasma concentrations are achieved in 3–4 h after po doses. About 30% of the plasma concentration is protein bound. The elimination half-life of nadolol is 20–24 h, allowing once a day dosing. The drug is excreted unchanged by the kidneys and its excretion is delayed in patients having renal failure (98,99,108).

Atenolol. Atenolol is a lipophilic, cardioselective β-adrenoceptor blocking agent having no ISA. The cardioselectivity decreases with increasing doses or concentrations of the drug. Atenolol is indicated for the long-term management of patients having angina pectoris as a result of coronary atherosclerosis because the β-adrenoceptor blocking actions can reduce catecholamine-induced increase in blood pressure, heart rate, and myocardial contractility, together with myocardial oxygen consumption. Atenolol has also been shown to reduce the incidence of sudden death in post-MI patients (98,99,108).

After po doses, atenolol is rapidly but incompletely absorbed (~50%) from the GI tract, and 50% is excreted unchanged in the feces. Six to 16% of the plasma concentration is bound to protein. Atenolol undergoes little first-pass metabolism. Peak plasma concentrations occur in 2–4 h after po doses. The elimination half-life of atenolol is 6–7 h. Excretion of absorbed drug is mainly by the kidneys and elimination can be impaired in patients having renal failure. The adverse effects of atenolol are similar to those seen for propranolol therapy (98,99,108).

Other β-Adrenoceptor Blocking Agents. Carteolol hydrochloride (Table 1) is also a Class II antiarrhythmic agent. In three separate studies in patients having angina pectoris, carteolol was considered effective as evidenced by a reduction in the frequency and severity of anginal episodes, reduction in the amount of nitroglycerin consumed, improvement of ECG parameters, or an increase in the duration of treadmill exercise (42).

Betaxolol hydrochloride is a lipophilic, cardioselective β-adrenoceptor blocker having no ISA and little membrane-stabilizing activity. The drug is as equieffective and equipotent as atenolol. It is well absorbed from the GI tract, but does not undergo extensive first-pass metabolism in the liver. Its elimination half-life is 15–20 h. It is metabolized in the liver (~84%) to two principal inactive metabolites and one minor active metabolite. About 16% of the drug is excreted unchanged in the urine. Excretion of the drug is unchanged in patients having renal or liver impairment (43).

Bisoprolol fumarate is a long-acting, cardioselective β-adrenoceptor blocker, and is the most potent cardioselective β-adrenoceptor blocker available. Bisoprolol has no ISA. At high concentrations it has membrane-stabilizing activity. The drug has a "balanced clearance", ie, half is excreted by the kidneys and half is eliminated by the liver and its excretion is not affected by functional impairment of either organ. It is approved in Europe for hypertension and is being studied in angina (43).

Celiprolol hydrochloride is a hydrophilic, long-acting, cardioselective β-adrenoceptor blocker. In addition to ISA, the compound has weak α_2-adrenoceptor blocking action at high concentrations. No membrane-stabilizing activity has been noted with celiprolol. Food interferes with the absorption of celiprolol. Bioavailability is about 30 and 74% in fed and fasted subjects, respectively. The drug is not metabolized by the liver but is excreted through the bile into the feces and via the kidneys. Peak drug levels occur in 2–3 h and its elimination half-life is about 5 h. The drug is being developed as an antihypertensive and antianginal agent. It has shown efficacy in treating stable angina (43).

Bevantolol hydrochloride is a moderately lipophilic, long-acting, cardioselective β-adrenoceptor blocker. It has no ISA but has membrane-stabilizing activity. The drug is in use in Europe for the treatment of hypertension and angina. It is rapidly absorbed from the GI tract. Peak plasma levels occur in 1–2 h. It is metabolized extensively in the liver to a metabolite that has some ISA. It is excreted by the liver and the kidneys and excretion is delayed in patients having kidney failure.

AGENTS USED IN THE TREATMENT OF CONGESTIVE HEART FAILURE

Congestive heart failure (CHF) is an ever present, growing health problem in the United States. Over 2 million people have been diagnosed as having CHF, and more are expected to be diagnosed as the mean age of the population increases (109). CHF presents a special problem because of its high annual mortality rate of 20–50% (110). This high mortality rate is largely the lack of a pharmacologic agent that can improve the prognosis for cure above and beyond symptomatic relief. As of this writing there is a great deal of research centered on improved pharmacotherapeutic agents for the treatment of CHF.

Physiology and Biochemistry of Congestive Heart Failure

The heart, a four-chambered muscular pump has as its primary purpose the propelling of blood throughout the cardiovascular system. The left ventricle is the principal pumping chamber and is therefore the largest of the four chambers in terms of muscle mass. The efficiency of the heart as a pump can be assessed by measuring cardiac output, left ventricular pressure, and the amount of work required to accomplish any required amount of pumping.

When the heart begins to fail as a pump, a number of pathophysiologic processes occur. Perhaps the easiest to perceive is simply that some of the blood within the left ventricle is not completely expelled during muscle contraction or systole. This subsequently requires more work during future systoles as additional blood returns from the pulmonary vein. A sort of congestion begins to develop as the heart cannot develop enough pumping ability to provide adequate oxygenated blood to the arterial circulation or removal of metabolic waste through the venous circulation.

There can be a number of underlying causes of CHF. The most prevalent is the lack of oxygenated blood reaching the heart muscle itself because of coronary artery disease with myocardial infarction (111). Hypertension and valvular disease can contribute to CHF as well, but to a lesser extent in terms of principal causes for the disease.

Both high energy containing adenosine triphosphate (ATP) and calcium are integral components of cardiac muscle contraction. Hydrolysis of ATP is coupled to cross-linking of the actin–myosin contractile proteins, resulting in mechanical shortening of the functional unit of cardiac muscle, the sarcomere. Calcium is involved in binding of the contractile protein troponin, but the quantity of calcium entering the myocardial cell is finely regulated, concomitantly regulating the extent of cardiac muscle contraction. Imbalance in any aspects of the chemical process of contraction can result in reduced pumping ability, and contribute to CHF.

Therapy of Congestive Heart Failure

Many of the drugs used to combat congestive heart failure are inotropic agents, some of which are shown in Table 4. Inotrope is a derivation of the Greek *ino* (fiber) and *tropikos* (changing or turning). A positive inotropic agent is therefore one that increases cardiac muscle contractility associated with CHF.

The Cardiac Glycosides. As early as the 1700s the medicinal properties of the leaves of *Digitalis purpurea* (foxglove), which was used by physicians of that time for treatment of dropsy, was described (112). The leaves of *Digitalis purpurea* yield digitoxin (Table 1); whereas *Digitalis lanata* is used for production of digoxin (Table 1) through mild alkaline hydrolysis. The seeds of *Strophanthus gratus* is used to obtain ouabain, which is a rapidly acting glycoside. Digoxin is the most commonly prescribed cardiac glycoside in the United States, because of its intermediate duration of action, various routes of administration, and ease in which serum levels can be determined (113).

The cardiac glycosides are composed of a steroid aglycone (genin) in combination with from 1 to 4 moles of a sugar, some of which are acetylated. The genins, generally considered to poison through convulsion, have some activity for use in CHF, but the sugar moiety contributes both to potency and to transport. The glycosides remain an enigma to cardiologists, however, because of the combination of extreme potency and low margin of safety.

Mechanism of Action. Digitalis has particular efficacy in the treatment of congestive heart failure because of direct inotropic effects on the heart which serve to increase the force of myocardial contraction, resulting in increased cardiac output and alleviation of the congestion. The mechanism mediating this positive inotropic effect is through direct inhibition of the enzyme Na^+, K^+-ATPase, which in turn promotes elevated intracellular calcium. Digitalis also has direct effects on the electrical activity of the heart, affecting the normal physiologic function of the sinoatrial node, the atrioventricular node, and the Purkinje fibers. The ramifications of the direct effect of digitalis on cardiac electrical activity is a slowing of heart rate, but can also include digitalis-induced arrhythmias (114).

Digitalis is also beneficial in the treatment of CHF because of direct effects on the central nervous system. Efferent vagus nerve activity is promoted by digitalis administration, resulting in an indirect slowing of heart rate. Digitalis in relatively high concentrations can enhance efferent sympathetic nervous system activity, purportedly inhibit reuptake of norepinephrine, and ultimately promote arrhythmogenesis (114).

Moreover, digitalis has indirect effects on the circulation, which in normal hearts results in a small increase in arterial pressure, peripheral resistance, and cardiac output (114). The effects of digitalis on the circulation of an individual experiencing congestive heart failure are much more dramatic, however. The increased cardiac output, for example, increases renal blood flow which can relieve in part the edema of CHF associated with salt and water retention (114).

When administered orally, digoxin bioavailability ranges from 40 to 90% depending on the manufacturing process. Peak plasma levels occur within 3 h after oral administration; peak effects occur 2 h later. The drug is eliminated primarily through the kidney (114).

Ouabain, structurally related to digitalis, is a very potent analogue often used in experimental studies. In the 1980s, a number of critically controlled clinical studies were undertaken to better understand the benefits and risks of digoxin therapy (115–117). It was concluded from these studies that patients having mild CHF, ie, not severe enough to limit daily activities, did not benefit from digoxin,

Table 4. Inotropic Agents

Generic name	CAS Registry Number	Molecular formula	Trade name	Structure
		Glycosides		
ouabain octahydrate	[*630-60-4*]	$C_{29}H_{44}O_{12}$		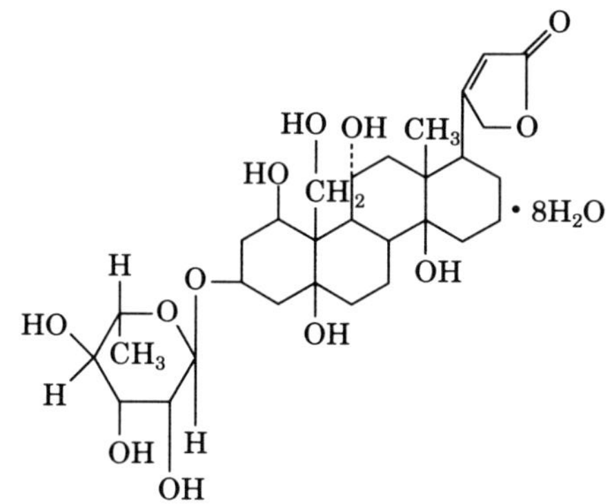
		Catecholamines		
dobutamine hydrochloride	[*49745-95-1*]	$C_{18}H_{24}ClNO_3$	Dobutrex	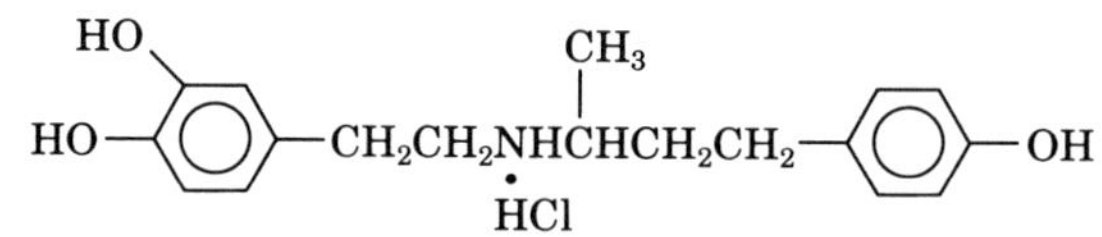
dopamine hydrochloride	[*62-31-7*]	$C_8H_{12}ClNO_2$	Dopastat, Intropin	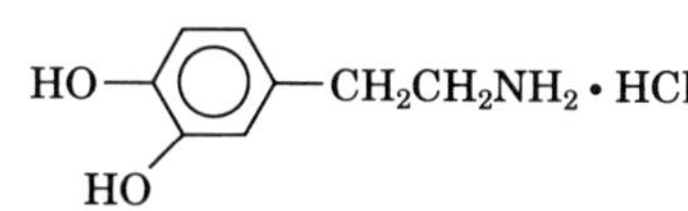

epinephrine bitartrate[a]	[*51-42-3*]	$C_{13}H_{19}NO_9$	Epitrate, Suprarenin	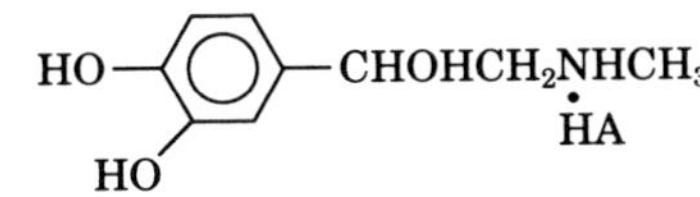
		Phosphodiesterase inhibitors		
amrinone	[*60719-84-8*]	$C_{10}H_9N_3O$	Inocor	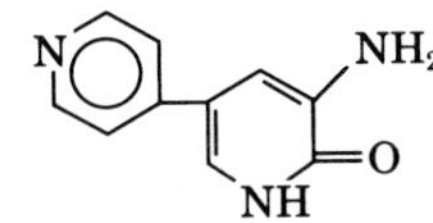
milrinone	[*78415-72-2*]	$C_{12}H_9N_3O$	Primacor	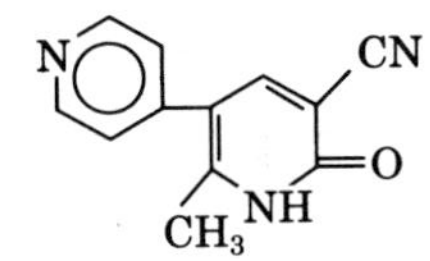
		Angioten converting enzyme inhibitors		
captopril	[*62571-86-2*]	$C_9H_{15}NO_3S$	Capoten	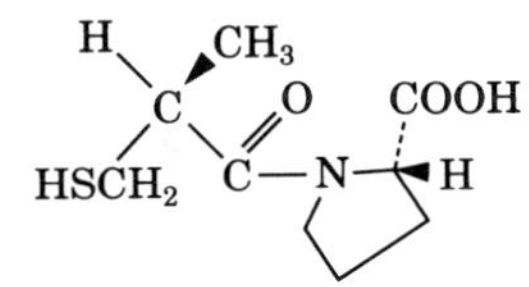
enalapril maleate[b]	[*76095-16-4*]	$C_{24}H_{32}N_2O_9$	Vasotec	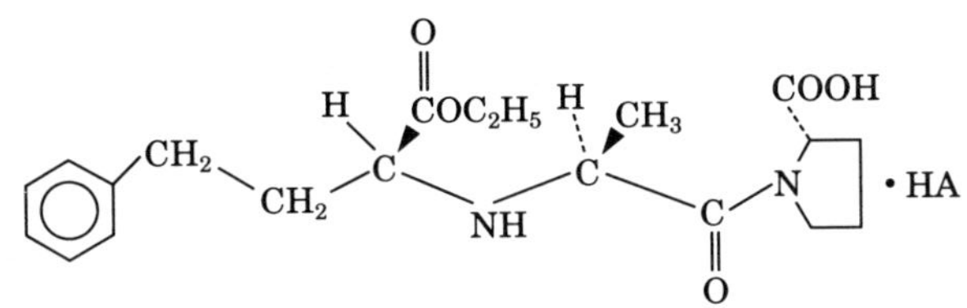

[a]HA = tartaric acid.
[b]HA = malonic acid.
[c]Registry number given is for anhydrous material.

even though the drug remains an appropriate therapy for patients having more extensive left ventricular dysfunction (118).

Cardiac Glycoside Toxicity. Digoxin remains a drug with considerable limitations because of toxicity, among which the most dangerous are the electrophysiologic effects. The drug can slow atrial rate and atrioventricular node conduction, leading to arrhythmia (119). In fact, because almost every variety of ventricular arrhythmia has been associated with digitalis toxicity (113), the drug can complicate, not eradicate, the treatment of CHF. Importantly, antiarrhythmic agents such as lidocaine and phenytoin can suppress digoxin-induced arrhythmia. More recently high affinity antidigoxin antibodies, specifically the *Fab* fragment, have become widely available to clinicians (120). The antibodies have greatly increased the safety of prescribing digitalis.

Nondigitalis Inotropic Agents. A number of inotropic agents that are not structurally related to the cardiac glycosides have been investigated for treatment of CHF. All are somehow mechanistically related to the complex series of events involved in normal cardiac muscle contraction, and in terms of pumping ability are capable of increasing left ventricular contractility. Many of these drugs do not perform in diseased cardiac tissue in a manner similar to that observed in normal, nondiseased muscle. Many of these agents act via the β-adrenergic receptors of the sympathetic nervous system (SNS).

Catecholamines. Simplistically, the SNS is composed of nerves capable of synthesizing and releasing various catecholamines that can bind appropriate molecular receptors post-synaptically. Dependent on their synthetic structure, they have the ability to bind, and therefore stimulate various post-synaptic events through α, β, and dopaminergic receptors.

Each receptor has a specific role, often vasodilation, which serves to facilitate the ability for a failing heart to pump against a reduced systemic pressure. Side effects of catecholamines include arrhythmias and tachyphylaxis. In addition, potent vasodilators such as isoproterenol [*7683-59-2*], $C_{11}H_{17}NO_3$, have a dual effect, namely increased contractility through direct cardiac muscle stimulation (121). If combined with significant reductions in blood pressure, isoproterenol can produce detrimental effects through a combination of reduced venous return yet increased oxygen requirements of the rapidly contracting cardiac muscle. The orally active catecholamine pirbuterol [*38677-81-5*], $C_{12}H_{20}N_2O_3$, has many of these classic side effects (121). These drugs have been selectively developed to affect specific receptor subtypes through radioligand binding studies using microsomal membrane preparations from animal tissue. This rapid method of screening for selective agonists and antagonists has produced considerable quantities of chemical structures that have potential as drugs. Each results in positive, yet relatively short-term efficacy in patients having CHF (122).

Dobutamine, a positive inotrope from the sympathomimetic amine family, was introduced in the early 1980s and possesses an asymmetric carbon atom resulting in two enantiomers. The commercial product is a 50:50 racemic mixture of the optical isomers, producing a combined β-adrenergic agonist and α-adrenergic antagonist effect (123) from the (+) enantiomer. The (−) enantiomer is a potent α-adrenoceptor agonist (124). The positive inotropic activity is a result of the two enantiomers, especially as relates to functional, simultaneous effects on cardiac and vascular smooth muscle. Because of this unique combination of activity, dobutamine is superior to isoproterenol, dopamine, epinephrine, and norepinephrine, primarily because of dobutamine's lesser effect on blood pressure reduction. At doses of 2.5–15 μg/(kg·min) dobutamine increases stroke volume, while reducing peripheral resistance. Even after drug infusion is stopped, the patient can continue to benefit (125). Because it is administered by infusion, the drug is not approved by the FDA for outpatient administration.

Drugs Acting on the β-Adrenoceptor. The increase in plasma norepinephrine observed in heart failure is considered to be responsible to a large part for a reduced β-adrenoceptor number in cardiac membranes (126). This down-regulation of β-adrenoceptors affects predominantly the β_1-subtype whereas the β_2-subtype remains normal. In 1982, experimental data from normal and failing myocardium, noting both a reduced binding capacity and post-receptor adenyl cyclase-mediated papillary muscle contraction in the failing heart, was published (127). These changes in the β-adrenocepter system of patients having CHF have therapeutic implications against treatment with β-antagonists.

The use of selective β-antagonists for treatment of CHF has included the β_1-blocker metoprolol (Table 1) and results of clinical trials suggest long-term beneficial effects. Selective β_1-antagonists have also been tested, an example of which is xamoterol [*81801-12-9*], $C_{16}H_{25}N_3O_5$, which is (±)-*N*-(2-hydroxy-3-(4-hydroxyphenoxy)propylamino)ethyl morphine-4-carboxamide. Xamoterol exhibits approximately 50% of the activity of isoproterenol, and serves to provide modest inotropic effects (128, 129).

Phosphodiesterase Inhibitors. Because of the complexity of the biochemical processes involved in cardiac muscle contraction, investigators have looked at these pathways for other means of drug intervention for CHF. One of the areas of investigation involves increased cyclic adenosine monophosphate [*60-92-4*] (cAMP) through inhibition of phosphodiesterase [*9025-82-5*] (PDE). This class of compounds includes amrinone, considered beneficial for CHF because of positive inotropic and vasodilator activity. The mechanism of inotropic action involves the inhibition of PDE, which in turn inhibits the intracellular hydrolysis of cAMP (130). In cascade fashion, cAMP-catalyzed phosphorylation of sarcolemmal calcium-channels follows, activating the calcium pump (131). A series of synthetic moieties including the bipyridines, amrinone and milrinone, piroximone and enoximone, [*77671-31-9*], $C_{12}H_{12}N_2O_2S$, all of which have been shown to improve cardiac contractility in short-term studies, were developed (132,133). These drugs initially had a wider therapeutic index than cardiac glycosides, and did not exhibit receptor down-regulation associated with β-adrenoceptor blockers. Questions have since surfaced concerning the detrimental effects of increased contractility upon accelerated deterioration of already diseased cardiac tissue, ie, these drugs can lead to shortened patient survival (134,135). An additional detrimental side effect is arrhythmogenesis (136).

Angiotensin Converting Enzyme Inhibitors. The angiotensin converting enzyme (ACE) inhibitors represent another nonglycoside method of treating CHF. A number of these drugs are in various stages of development. The first such was introduced as captopril [*62571-86-2*], $C_9H_{15}NO_3S$, in the 1980s. This drug produced reductions in both preload and afterload, and although originally marketed as an antihypertensive, this profile indicated potential efficacious therapy for CHF (137,138). The effect in acute heart failure can be seen in 30–360 min, depending on the specific ACE inhibitor employed (139). Controlled clinical studies have shown the ACE inhibitors are at least equal in effectiveness to cardiac glycosides in treatment of CHF (140), and may actually elicit a positive effect on the pathophysiologic mechanism of CHF by inhibiting angiotensin II formation.

ACE inhibitors can be administered with diuretics (qv), cardiac glycosides, β-adrenoceptor blockers, and calcium channel blockers. Clinical trials indicate they are generally free from serious side effects. The effectiveness of enalapril, another ACE inhibitor, in preventing patient mortality in severe (Class IV) heart failure was investigated. In combination with conventional drugs such as vasodilators and diuretics, a 40% reduction in mortality was observed after six months of treatment using 2.5–40 mg/d of enalapril (141). However, patients complain of cough, and occasionally rash and taste disturbances can occur.

Atrial Natriuretic Factor. Atrial natriuretic factor (ANF) has natriuretic, vasorelaxant, and diuretic properties (142). It was originally discovered in atrial tissue, from which extracts were made and subsequently injected systemically. The responses evoked from the atrial extracts established the heart as an endocrine organ.

A number of different types of ANF have been described, but the original molecule is a polypeptide of 28 amino acids (qv). The prepro-form is released from the atria by stretch, suggesting that increased blood volume in a poorly contracting atria compromised by CHF could be one pathophysiologic condition in which the hormone would prove efficacious. As of this writing, this peptide has not received FDA approval. However, a number of clinical studies have been performed. For example, a single 33 μg bolus of ANF was injected into normal volunteers and patients having volume retaining disorders (143), including CHF. Patients with end-stage renal failure and CHF showed blunted hemodynamic responses to ANF, whereas the healthy control patients also exhibited greater urinary excretion of sodium.

1 5 10 15 20 25
Ser-Leu-Arg-Arg-Ser-Ser-Cys-Phe-Gly-Gly-Arg-Met-Asp-Arg-Ile-Gly-Ala-Gln-Ser-Gly-Leu-Gly-Cys-Asn-Ser-Phe-Arg-Tyr

α-hANF

Although the potential of ANF has not been realized as a marketed drug for treatment of CHF, the fact that a small peptide can have such profound effects on hemodynamics and renal physiology has led to considerable excitement. This enthusiasm is based in part on the advent of molecular biological techniques for ease in production of large quantities of naturally occurring peptides, as well as the ability to use these techniques for understanding how the peptide is processed beginning at the level of genomic DNA (see PROTEIN ENGINEERING).

ATHEROSCLEROSIS AND ANTIATHEROSCLEROSIS DRUGS

Formation of atherosclerotic plaques (atheromas) remains one of the central processes in cardiovascular death, including coronary heart disease. Complications from atheromas in coronary arteries include myocardial infarction and stroke. This latter is a common manifestation of cerebral vascular plaque formation.

Atherosclerotic plaques begin as fatty streaks which progress to defined plaques containing cholesterol [*57-88-5*], $C_{27}H_{46}O$, eventually being overlaid with connective tissue (144). The atheroma can be initiated by vascular injury, and once formed can be the site for thrombus formation and blood platelet deposition, which eventually become calcified. The existence of such formations in arterial walls can produce serious reductions in blood flow to the organs being supplied by these diseased arteries, producing local ischemia. The atheromas therefore present an especially dangerous health problem when situated in arteries supplying blood to cardiac muscle and brain.

Reduction in serum lipids can contribute significantly to prevention of atherosclerosis. In 1985 a consensus report indicating that for every 1% reduction in serum cholesterol there is a 2% reduction in adverse effects of coronary heart disease was issued (145). Recommended serum cholesterol concentration was 200 mg/dL for individuals under 30 years of age, and individuals having concentration 240 mg/dL and LDL-cholesterol over 160 mg/dL should undertake dietary modification and possibly pharmacotherapy (146). Whereas the initial step in reducing serum cholesterol is through reduction of dietary cholesterol intake, a number of drugs are available that can affect serum lipid profile (see FAT SUBSTITUTES). The pathway to cholesterol synthesis is shown in Figure 2.

The primary transporter of cholesterol in the blood is low density lipoprotein (LDL). Once transported intracellularly, cholesterol homeostasis is controlled primarily by suppressing cholesterol synthesis through inhibition of β-hydroxy-β-methyl gluterate–coenzyme A (HMG–CoA) reductase, acyl CoA–acyl transferase (ACAT), and down-regulation of LDL receptors. An important drug in the regulation of cholesterol metabolism is lovastatin, also known as mevinolin, MK-803, and Mevacor, which is an HMG–CoA reductase inhibitor (Table 5).

A fermentation product of the fungus *Aspergillus terreus,* lovastatin was originally described as a cholesterol-lowering agent in 1980 (147). Through inhibition of the rate-limiting enzyme of cholesterol synthesis, HMG–CoA reductase, the drug interferes with mevalonic acid [*150-97-0*], $C_6H_{12}O_4$, synthesis. Lovastatin is actually a pro-drug, which is eventually hydrolyzed in the liver to its active, β-hydroxylated form (148). Oral lovastatin is prescribed in a once a day

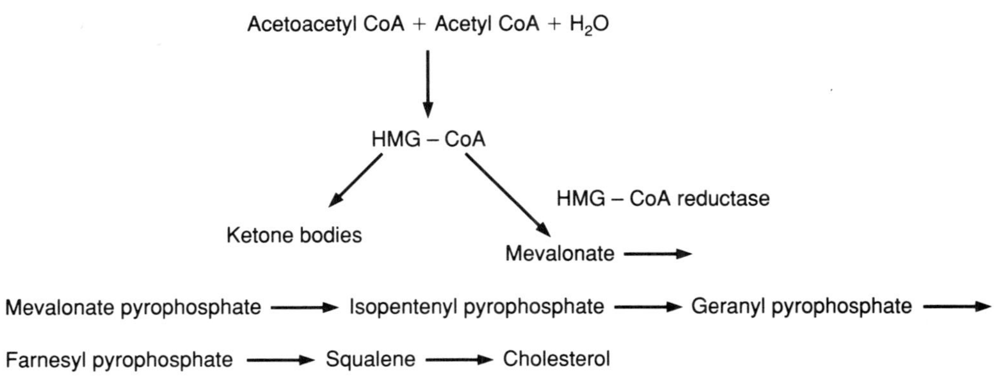

Fig. 2. Pathway of cholesterol synthesis.

Table 5. Antiatherosclerosis Drugs

Generic name	CAS Registry Number	Molecular formula	Trade name	Structure
		HMG-CoA reductase inhibitor		
lovastatin MK-803 mevinolin	*[75330-75-5]*	$C_{24}H_{36}O_5$	Mevacor	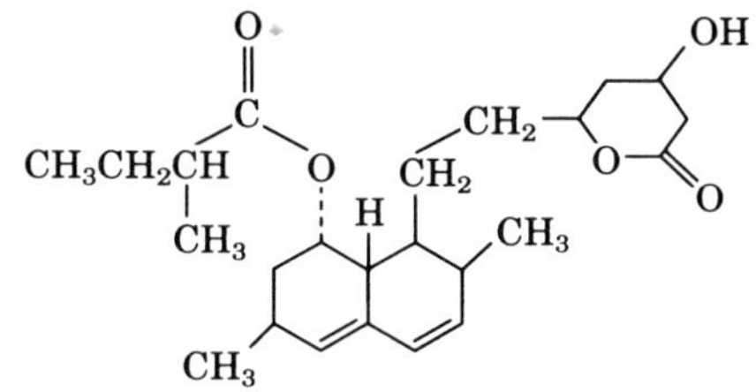
		Lipid lowering drugs		
bezafibrate	*[41859-67-0]*	$C_{19}H_{20}ClNO_4$		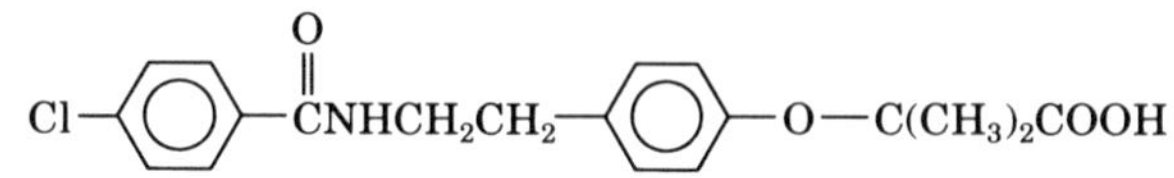
clofibrate	*[637-07-0]*	$C_{12}H_{15}ClO_3$	Atromid-S	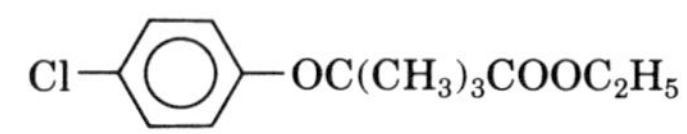
fenofibrate	*[49562-28-9]*	$C_{20}H_{21}ClO_4$		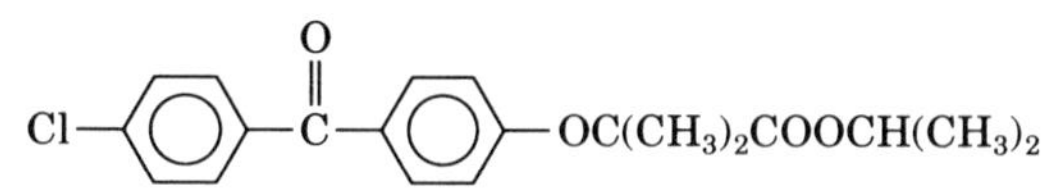
gemfibrozil	*[25812-30-0]*	$C_{15}H_{22}O_3$		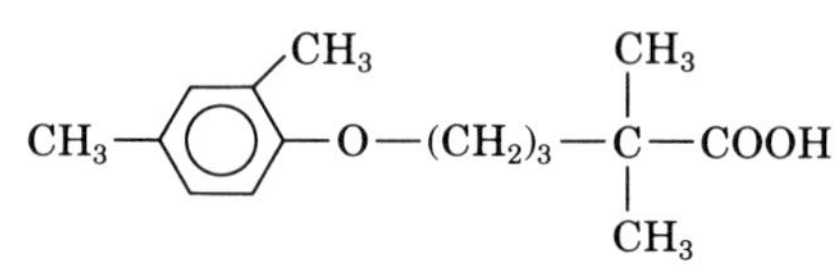
		Antioxidants: Cholesterol lowering agents		
probucol	*[23288-49-5]*	$C_{31}H_{48}O_2S_2$	Lorelco	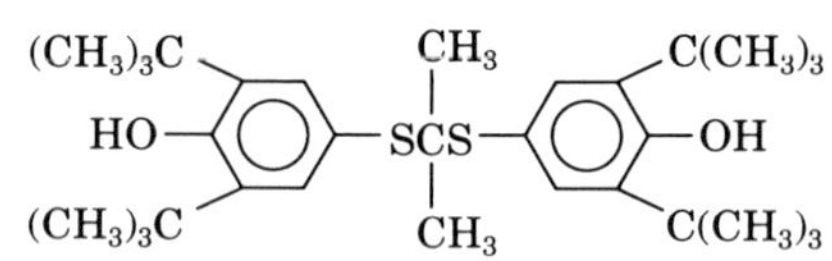

regimen of 20–80 mg/d. It exhibits significant plasma protein binding, and only about 5% of the oral dose reaches the general circulation.

In the first clinical studies with lovastatin, pre-drug serum cholesterol values of 150–300 mg/dL were shown to be decreased as much as 25% with a dosage of 15 mg twice daily for just over a week (149). Whereas the drug shows few adverse side effects, gastrointestinal disturbances, including diarrhea and abdominal pain, are the most common.

Bile Acid Sequestrants. The bile acid binding resins, colestipol [*26658-42-4*] and cholestyramine, are also effective in controlling serum cholesterol levels (150). Cholestyramine, a polymer having mol wt $> 10^6$, is an anion-exchange resin. It is not absorbed in the gastrointestinal tract, is not affected by digestive enzymes, and is taken orally after being suspended in water (151).

Colestipol, also an anion-exchange copolymer, is composed of diethylenetriamine [*111-40-0*], $C_4H_{13}N_3$, and 1-chloro-2,3-epoxypropane [*106-89-8*], C_3H_5ClO. Its metabolism and formulation is quite similar to cholestyramine. Because bile acids are synthesized from cholesterol, bile acid sequestrants such as cholestyramine and colestipol, through binding bile acids in the intestine, can affect cholesterol metabolism. The drug–bile acid complex is eventually excreted in the feces, effectively reducing the total cholesterol pool in the body. In fact, these drugs have been shown to be effective in decreasing both LDL-cholesterol and total serum cholesterol. Use, however, is limited to patients having hypercholesterolemia not associated with hypertriglyceridemia, also known as Type II hyperlipoproteinemia. The drug-induced reduction in cholesterol can be dramatic; as much as 50% reduction can be observed over long-term therapy (152). Thus control of serum cholesterol can contribute to the antiatherosclerotic attributes of this class of drug.

Because colestipol and cholestyramine are not absorbed, but simply pass through the body by the GI tract, few severe side effects occur. Patients often complain of distaste and constipation, however. More severe side effects such as GI bleeding are relatively uncommon (151).

Fibric Acid Derivatives. Fibric acid derivatives have been used since the 1960s for control of blood lipid levels. The four most well-known compounds in this class are fenofibrate, gemfibrozil, clofibrate, and bezafibrate, shown in Table 5. Fenofibrate has been the most popular fibric acid derivative in Europe, and has more recently been investigated extensively in clinical trials in the United States (153). Bezafibrate is structurally related to clofibrate, and has recently been reviewed by Western European and Australian clinical trials (154).

Clofibrate, shown to be an effective lipid lowering drug in the early 1960s, is the ethyl ester of *p*-chlorophenoxy isobutyric acid. Gemfibrozil is a structural congener of clofibrate.

The mechanism of action of the fibrates in controlling plasma lipids appears to be primarily through mediation of lipoprotein lipase activity. In addition, gemfibrozil stimulates apolipoprotein AI synthesis (155), presumably leading to increased high density lipoprotein (HDL) particles. In fact, the mechanistic effects of the fibrates in lowering serum cholesterol are complex, from indirect regulation of lipoprotein synthesis, to regulation of key enzymes and receptors involved in lipid metabolism. These drugs are well absorbed after oral administration. For example, in single-dose studies of 400 mg/d, almost complete absorption of benzafibrate was observed, followed by virtually 100% excretion within 48 h (156).

Adverse side effects of the fibric acid derivatives are similar to those of the bile acid sequestrants. Patients complain of GI effects, including nausea and vomiting. Generally speaking, these drugs are most often prescribed for Type IIa and IIb hyperlipidemia, and are effective in reducing plasma triglycerides and very low density lipoprotein (VLDL) cholesterol. Based on the increasing evidence that reduced cholesterol can contribute to prevention of atherosclerosis, these drugs can lead to reduction in cardiovascular disease.

Nicotinic acid [*59-67-6*], $C_6H_5NO_2$, has been known as a hypolipidemic since the mid-1950s. Dosages of 2–8 g daily reduce serum triglycerides primarily through decreasing the production of very low density lipoprotein. Side effects include pruritus and gastrointestinal disturbances. This drug is suggested for use in patients not responding to dietary interventions for lowering blood lipids.

Other Cardiovascular Agents Effecting Atherosclerosis. A large amount of clinical data is available concerning serum lipid profiles in patients subjected to drug therapy for other cardiovascular diseases. Atheroma, for example, may be the underlying cause of hypertension and myocardial infarction. There are on the order of 1.5 million heart attacks per year in the United States (155).

β-Adrenoceptor Antagonists or Blockers. Although having no overt effects on total plasma cholesterol, β-adrenoceptor blockers (Table 1) are somewhat effective in reducing acute cardiac insult resulting from atherosclerosis. By decreasing myocardial oxygen requirements, β-blockers can effectively decrease death from myocardial infarction, which is contributed to largely by atherosclerotic coronary artery disease. Because β-blockers can inhibit catecholamine-receptor agonist responses, including atheroma development, these drugs can slow the formation of plaques that could prove potentially dangerous (157). Mechanistically, this may occur by reducing endothelial cell permeability to lipoproteins, inhibiting ACAT, or simply reducing blood flow and heart rate (158).

Calcium Channel Blockers. Because accumulation of calcium is one of the facets of the more involved process leading to atherosclerosis, it would follow that the antihypertensive calcium channel blockers might be effective in preventing atheroma. Both verapamil (Table 1) and nifedipine (Table 3) have been shown to stimulate the low density lipoprotein (LDL) receptor (159). This specific receptor-mediated pathway could theroetically improve lipid metabolism in the arterial wall, and thereby prove antiatherogenic. These effects have been proven in animals.

Probucol. Probucol is an antioxidant that is effective in lowering LDL cholesterol. Whereas probucol was known to lower cholesterol after relatively simple clinical trials (160), its mechanism of action as an antioxidant in the treatment of atherosclerosis is quite novel. Probucol has been shown to have the ability to produce regression of atherosclerotic lesions in animal models (161). Probucol therefore represents a novel class of pharmaceutical agent for the treatment of atherosclerosis. This effect occurs mechanistically, in part, by preventing oxidation of LDL, a necessary step in foam cell formation. This antioxidant activity has been shown in laboratory experiments and its activity in lowering LDL cholesterol in human studies is well documented (162).

ANTIHYPERTENSIVE AGENTS

Hypertension is one of the two principal risk factors of many cardiovascular diseases, such as coronary heart disease (CHD), stroke, and CHF. Individuals are considered hypertensive if their systolic arterial blood pressure is over 140 mm Hg (18.7 Pa) or their diastolic arterial blood pressure is over 90 mm Hg (12 Pa). Over 60 million people, or one-third of the adult population in the United States are estimated to be hypertensive (163). About 90% of these patients are classified as primary or essential hypertensive because the etiology of their hypertension is unknown. It is generally agreed that there is a very strong genetic or hereditary component to this disease.

Arterial blood pressure at any given time is determined by the balance of the vasopressor and vasodepressor systems operating in various organs and tissues of the body. Some of the principal vasopressor systems identified are the adrenergic or sympathetic nervous system; the renin–angiotensin–aldosterone system (164); the thromboxane and the prostaglandin $F_2\alpha$ systems (165,166); vasopressor peptides; endothelin (167), endothelium-derived contracting factors (EDCF) (168,169); the kidney blood volume regulation system (170,171); and the vasopressin system (172). Some of the primary vasodepressor systems are the cholinergic or parasympathetic nervous system, the kallikrein–kinin system, prostaglandin E_2 and prostacyclin (epoprostenol, PGI_2) (166,173,174), atrial natriuretic factor (ANF) (175), endothelium-derived relaxing factors (EDRF) (168,169), and the renal neutral-lipids system. The central nervous system (CNS) can exert either a vasopressor or vasodepressor influence, depending on the level of homeostasis. Overactivity of the vasopressor systems with normal vasodepressor systems causes hypertension. On the other hand, normal vasopressor systems having underactivity of deficiency of opposing vasodepressor systems theoretically can also lead to hypertension.

It is well accepted that hypertension is a multifactorial disease. Only about 10% of the hypertensive patients have secondary hypertension for which causes, ie, partial coarctation of the renal artery, pheochromacytoma, aldosteronism, hormonal imbalances, etc, are known. The hallmark of hypertension is an abnormally elevated total peripheral resistance. In most patients hypertension produces no serious symptoms particularly in the early phase of the disease. This is why hypertension is called a silent killer. However, prolonged suffering of high arterial blood pressure leads to end organ damage, causing stroke, myocardial infarction, and heart failure, etc. Adequate treatment of hypertension has been proven to decrease the incidence of cardiovascuar morbidity and mortality and therefore prolong life (176–183).

Treatment of Hypertension

Treatment of essential or primary hypertension emphasizes not only the lowering of the elevated blood pressure, but also individualized therapy for each patient, providing each patient with minimized unnecessary side effects. The patient's cardiovascular morbidity and mortality should be decreased and end organ damage reversed or reduced (184,185).

Of the patients diagnosed to have essential hypertension, about 70% have mild hypertension where the diastolic blood pressure is from 90 to 104 mm Hg (12–13.9 Pa), 20% have moderate hypertension, ie, 105–114 mm Hg (14–15.2 Pa), and 10% have severe hypertension, >115 mm Hg (15.3 Pa). Once sustained hypertension is determined, nonpharmacological treatment as initial therapy is generally recommended for six months only for patients having diastolic blood pressure less than 100 mm Hg (13.3 Pa) when no evidence of target end organ damage is detected. Nonpharmacological measures include reductions in salt intake, weight, and alcohol consumption, and increases in potassium intake and physical exercise. Drug therapy is generally life-long for most hypertensive patients.

The Joint National Committee on the Detection, Evaluation, and Treatment of High Blood Pressure detailed a recommendation in 1988 (JNC-IV) that emphasized an individualized or patient-oriented (demographic) approach. Factors affecting each patient should be taken into consideration in choosing an antihypertensive drug. The first drug should be either an angiotensin converting enzyme (ACE) inhibitor, a β-adrenoceptor blocker, a calcium channel blocker, or a diuretic. One drug can normalize the blood pressure of about 50–60% of the hypertensive patients and the combination of two complementary drugs normalizes that of over 80% patients. White, young, hyperdynamic, hypertensive patients start with an ACE inhibitor or a β-adrenoceptor blocker; older patients respond well to a calcium channel blocker or a diuretic as do black hypertensive patients; patients having left ventricular hypertrophy respond well to an ACE inhibitor, a calcium channel blocker, or an α_2-adrenoceptor agonist; and patients having concomitant CHF, stroke, diabetes, or hypercholesterolemia can start with an ACE inhibitor.

Patients having high plasma renin activity (PRA) (>8 ng/(mL·h)) respond best to an ACE inhibitor or a β-adrenoceptor blocker; those having low PRA (<1 ng/(mL·h)) usually elderly and black, respond best to a calcium channel blocker or a diuretic (184). β-Adrenoceptor blockers should not be used in patients who have diabetes, asthma, bradycardia, or peripheral vascular diseases. The thiazide-type diuretics (qv) should be used with caution in patients having diabetes. Likewise, β-adrenoceptor blockers should not be combined with verapamil or diltiazem because these drugs slow the atrioventricular nodal conduction in the heart. Calcium channel blockers are preferred in patients having coronary insufficiency diseases because of the cardioprotective effects of these drugs.

For patients having a moderate level of PRA, an ACE inhibitor is often tried first. The alternative is a calcium channel blocker. If neither one can control the elevated blood pressure, a combination of the two is used to block both vasoconstrictor mechanisms. The alternative combination is an ACE inhibitor and a diuretic (184,186). Other classes of antihypertensive agents are used if proven to be superior to these first-line agents, or added to the first-line drugs to increase efficacy or to decrease side effects. The goal is to give each patient the fewest number of drugs, using the smallest effective amounts having the lowest frequency of dosing and minimum side effects (184,187). Ie, optimal therapy should reduce all risk factors of coronary heart diseases, reverse the hemodynamic abnormalities present by preserving cardiac output and tissue perfusion, and lower total peripheral resistance. The challenge is to choose the best antihypertensive drug for concomitant existing diseases while maintaining a good quality of life (188–199).

The various antihypertensive agents classified according to mechanism are as follows:

Agents Affecting the Renin–Angiotensin System

Angiotensin converting enzyme inhibitors
- captopril
- enalapril
- lisinopril
- perindopril, ramipril, spirapril, quinapril, alacepril, delapril, ceranapril, zofenopril, fosinopril, benazepril, libenzapril

Angiotensin II receptor antagonists
- DuP 753, DuP 532
- EXP-9993, EXP-3174
- L-158,809

Renin inhibitors
- Enalkiren (A-64662)
- CGP 38 560 A
- CP 71362
- KRI-1230, U-71039, ES6864, PD-134672

Agents Affecting the Adrenergic Nervous System

Ganglionic blockers
- chlorisondamine
- trimethaphan
- hexamethonium

Neuronal blockers
- guanethidine
- guanadrel, bretylium, debrisoquin, bethanidine

Neuronal norepinephrine depleting agents
- reserpine

α-Adrenoceptor blockers
- phentolamine
- prazosin
- terazosin
- doxazosin, bunazosin

β-Adrenoceptor blockers
- propranolol
- metoprolol
- atenolol
- bisoprolol
- nadolol, acebutolol, pindolol, betaxolol, celiprolol, carvedilol, bucindolol, brefanolol

α/β-Adrenoceptor blockers
- labetalol

Calcium Channel Blockers

- verapamil
- diltiazem
- nifedipine
- nicardipine
- felodipine
- amlodipine
- nitrendipine
- isradipine
- lacidipine, benidipine

Diuretics

Low ceiling diuretics
- hydrochlorothiazide, other thiazides
- chlorthalidone
- quinethazone
- metolazone

High ceiling diuretics
- furosemide
- bumetanide
- ethacrynic acid

Aldosterone antagonists
- spironolactone

Potassium sparing diuretics
- triamterene
- amiloride

Centrally Acting Agents

methyldopa
clonidine
guanabenz
guanfacine
rilmenidine

Vasodilators

hydralazine
minoxidil
nitroprusside
cadralazine

Potassium Channel Openers

pinacidil
cromakalim, lemakalim
nicorandil
diazoxide, minoxidil

Antihypertensive agents are shown in Table 6 (200–202).

Agents Affecting the Renin–Angiotensin System

The renin–angiotensin system (RAS) plays an important role in the regulation of arterial blood pressure in the body (203,204). The enzyme renin [*9015-94-5*], which is synthesized and released from the kidney, catalyzes the formation of angiotensin I [*9041-90-1*] a decapeptide, from the substrate angiotensinogen, an α_2-globulin, synthesized in the liver and released into the blood stream. ACE, a nonspecific carboxypeptidase, cleaves a dipeptide from the carboxyl terminal of angiotensin I to form angiotensin II [*11128-99-7*], an octapeptide. The blood vessels of the pulmonary circulation are the main site for this conversion. Angiotensin II is further cleaved to the heptapeptide, angiotensin III [*12687-51-3*] by the action of tissue aminopeptidase [*9031-94-1*]. Angiotensin II is the most potent product of this cascade. Angiotensin I is relatively inactive, whereas angiotensin III is less potent than angiotensin II as a vasoconstrictor but is as active as angiotensin II in stimulating the release of aldosterone [*52-39-1*], $C_{21}H_{28}O_5$.

The primary effects of angiotensin II are (*1*) producing vasoconstriction directly or indirectly through activation of the adrenergic nervous system; (*2*) eliciting cardiac stimulation; (*3*) stimulating aldosterone and vasopressin secretion; and (*4*) evoking dipsogenesis. All these actions, directly or indirectly, produce elevation of blood pressure (203). Theoretically, renin inhibitors, ACE inhibitors, and angiotensin II receptor antagoists can be potent antihypertensive agents.

Angiotensin Converting Enzyme (ACE) Inhibitors. ACE inhibitors are very efficacious in reducing high blood pressure. The antihypertensive activity is better correlated with the vascular tissue ACE inhibition than plasma ACE inhibition. Plasma ACE activity returns to predrug level while significant lowering of the blood pressure is still being observed. Teprotide [*35115-60-7*], a nonapeptide ACE inhibitor, was demonstrated to have potent antihypertensive effects after iv administration. The drawback of teprotide is lack of oral activity. Captopril (Table 4), a sulfhydryl containing small molecule, was the first practical nonpeptide ACE inhibitor in humans (205). Then came enalapril (Table 4) and lisinopril (206), the nonsulfhydryl containing ACE inhibitors (202).

ACE inhibitors are used in the first-line therapy of hypertension (207–213). Initially it was thought that ACE inhibitors were efficacious only in patients having high plasma renin activity (PRA) such as in renal or malignant hypertension. Captopril, however, was found to be efficacious in patients having essential hypertension, particularly those with high or moderately high PRA. Only about 20% of patients having essential hypertension have high to moderately high PRA, but ACE inhibitor used as monotherapy is effective in 50–60% of all patients. The reason may be that many other antihypertensive drugs (such as diuretics), calcium channel blockers, and a low salt diet activate the RAS and patients become renin dependent and are responsive to drugs affecting the RAS (214), such as ACE inhibitors, renin inhibitors, and angiotensin II receptor antagonists.

ACE inhibitors lower the elevated blood pressure in humans with a concomitant decrease in total peripheral resistance. Cardiac output is increased or unchanged; heart rate is unchanged; urinary sodium excretion is unchanged; and potassium excretion is decreased. ACE inhibitors promote reduction of left ventricular hypertrophy.

Other mechanisms of action, such as potentiation of the bradykinin [*58-82-2*], $C_{50}H_{73}N_{15}O_{11}$, effect, and increased production of prostaglandins have been reported to explain, in part, the antihypertensive effects of ACE inhibitors during chronic therapy (215,216). The inhibition of angiotensin II production is considered the primary, if not the only, mechanism of action of ACE inhibitors, even though the circulating angiotensin II levels may not accurately reflect the situation. Headache, dizziness, fatigue, skin rash, and cough are the most common side effects of ACE inhibitors. Cough in patients treated with an ACE inhibitor is considered to be caused by the build-up of bradykinin and prostaglandins. ACE is thought to be the same enzyme that catabolizes bradykinin.

Angiotensin II Receptor Antagonists. Many potent peptide angiotensin II receptor antagonists were known before the advent of the ACE inhibitors, but none were active after oral administration. One such peptide, saralasin [*34273-10-4*], $C_{42}H_{65}N_{13}O_{10}$, was approved for use in humans as a diagnostic tool. In addition to antagonistic effects, saralasin produces an initial agonist (hypertensive) action, even though it is much weaker than angiotensin II. The lack of oral efficacy and its intrinsic agonistic effects make saralasin an impractical drug.

The patent literature contains a series of nonpeptide imidazole derivatives that had angiotensin II receptor antagonistic effects (217). These compounds were found to be weak angiotensin II receptor antagonists (218,219), yet they were competitive antagonists, suggesting that a small nonpeptide molecule could compete with the octapeptide, angiotensin II, for the active site of the receptor. Several nonpeptide angiotensin II receptor antagonists have been synthesized, some of which are shown in Figure 3. DuP 753 is a potent orally active drug in clinical trials (220,221); L-158,809 has been reported to be 100–180 times more potent than DuP 753 as an angiotensin II receptor antagonist (222). The spectrum of antihypertensive activity should be similar to that of the ACE inhibitors, but with different side effects because the angiotensin II receptor antagonists do not potentiate the action of bradykinin.

Renin Inhibitors. Renin catalyzes the first and rate limiting step of the RAS and is very specific for the conversion of angiotensinogen to angiotensin I. Although renin inhibitors have been large molecules, particularly renin antibodies, and not orally active (203,223), several dipeptide and dipeptide transition-state compounds have been reported to inhibit renin activity and lower blood pressure

Table 6. Antihypertensive Agents

Generic name	CAS registry number	Molecular formula	Trade name	Structure
		Angiotensin converting enzyme inhibitors		
alacepril	[*74258-86-9*]	$C_{20}H_{26}N_2O_5S$		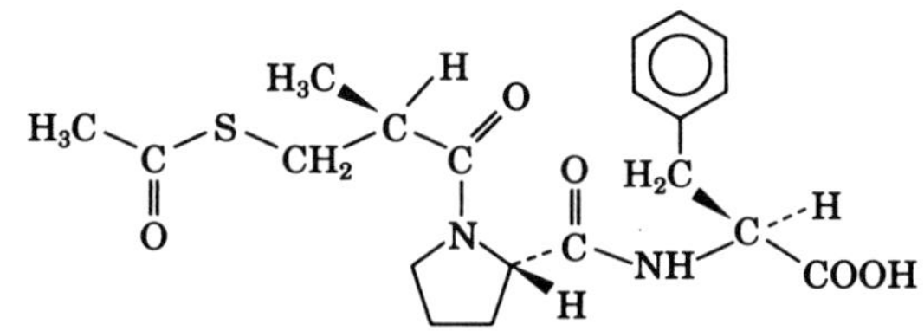
benazepril hydrochloride	[*86541-74-4*]	$C_{24}H_{28}N_2O_5 \cdot HCl$	Lotensin	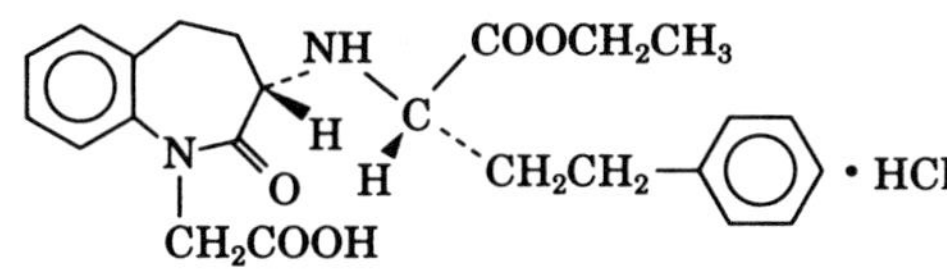
ceronapril	[*111223-26-8*]	$C_{21}H_{33}N_2O_6P$		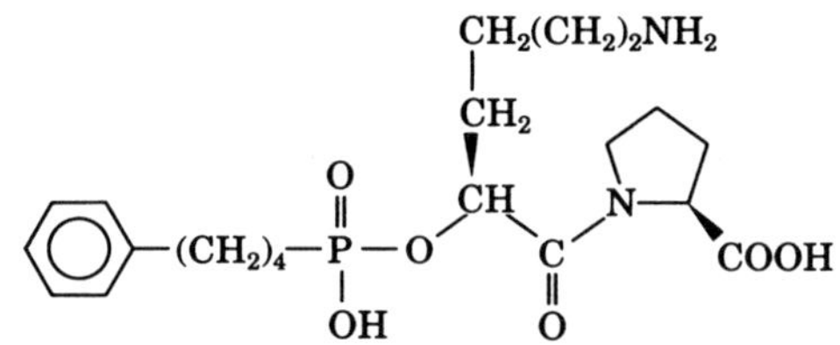
delapril hydrochloride	[*83435-67-0*]	$C_{26}H_{33}ClN_2O_5$		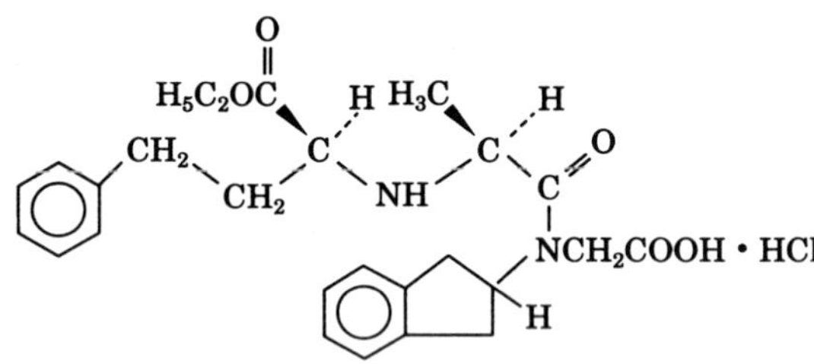

fosinopril sodium	[*88889-14-9*]	$C_{30}H_{45}NNaO_7P$	Monopril	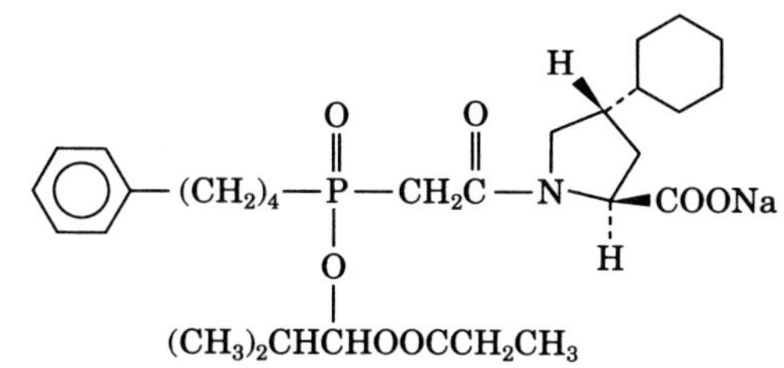
libenzapril	[*109214-55-3*]	$C_{18}H_{25}N_3O_5$		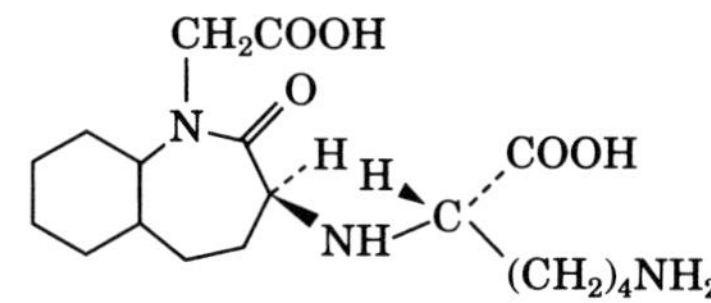
lisinopril (anhydrous)	[*83915-83-7*] [*76547-98-3*]	$C_{21}H_{31}N_3O_5 \cdot 2H_2O$	Prinivil, Zestril	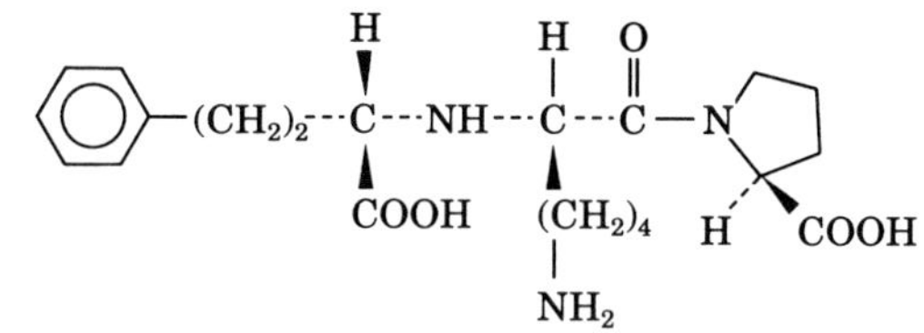
perindopril	[*82834-16-0*]	$C_{19}H_{32}N_2O_5$		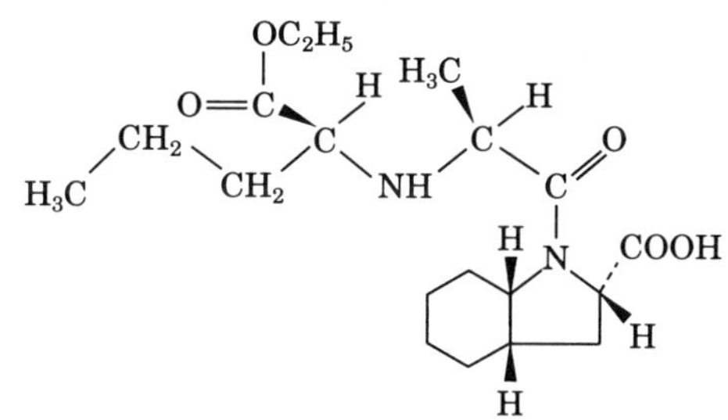

Table 6. (*Continued*)

Generic name	CAS Registry Number	Molecular formula	Trade name	Structure
quinapril hydrochloride monohydrate	[*90243-99-5*]	$C_{25}H_{30}N_2O_5 \cdot HCl \cdot H_2O$		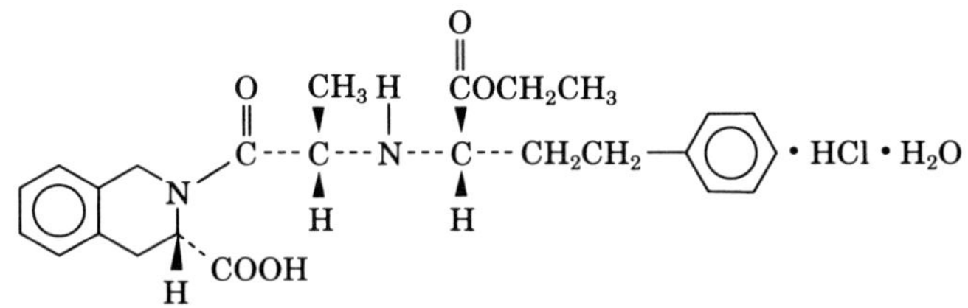
ramipril	[*87333-19-5*]	$C_{23}H_{32}N_2O_5$	Cardace	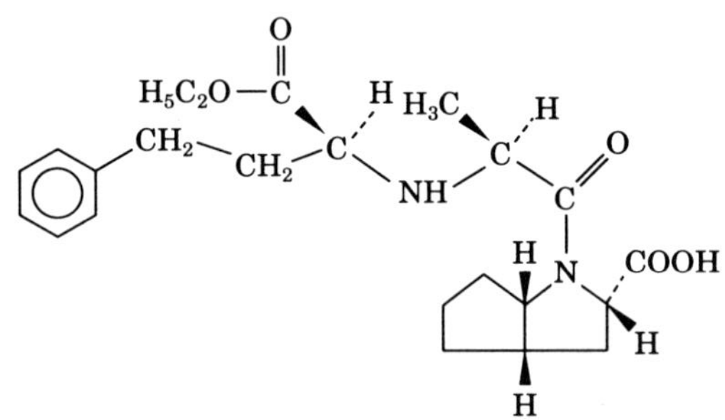
spirapril hydrochloride	[*94841-17-5*]	$C_{22}H_{30}N_2O_5S_2 \cdot HCl$		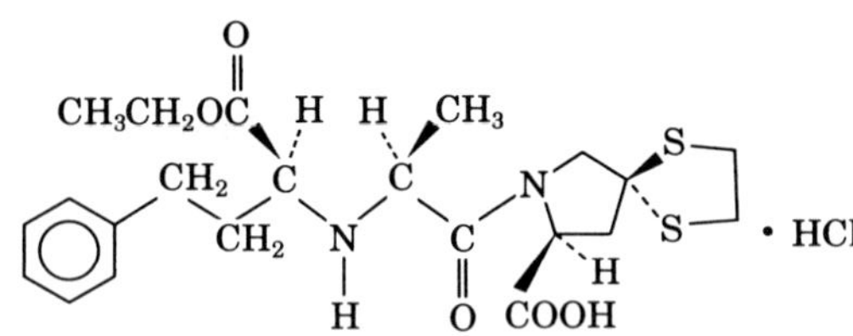

zofenopril calcium	[*81938-43-4*]	$C_{22}H_{23}NO_4S_2 \cdot \frac{1}{2}Ca$	Zoprace	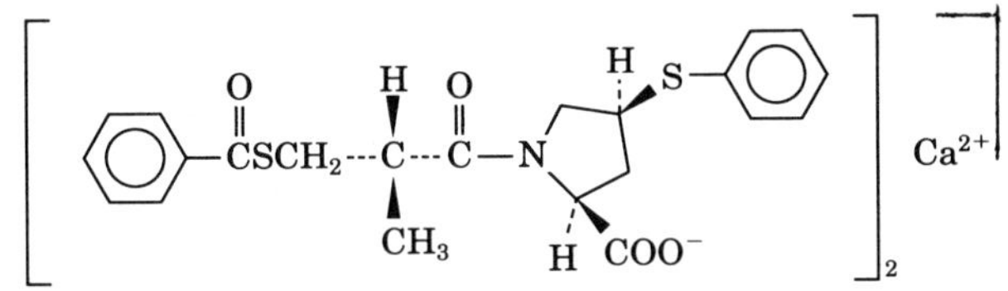
			Ganglionic blockers	
chlorisondamine chloride	[*69-27-2*]	$C_{14}H_{20}Cl_6N_2$	Ecolid	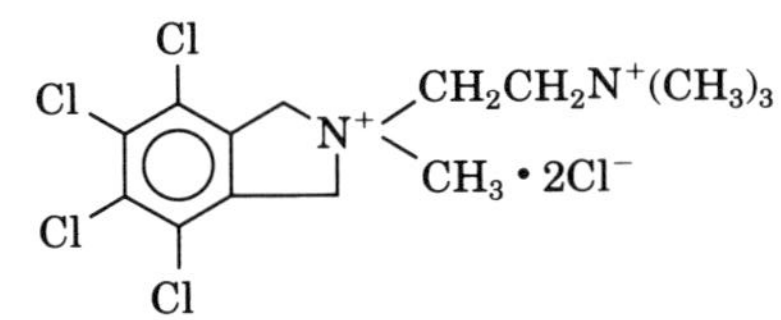
hexamethonium bromide	[*55-97-0*]	$C_{12}H_{30}Br_2N_2$	Bistrium	$(CH_3)_3N^+(CH_2)_6N^+(CH_3)_3 \cdot 2Br^-$
trimethaphan camsylate	[*68-91-7*]	$C_{32}H_{40}N_2O_5S_2$	Arfonad	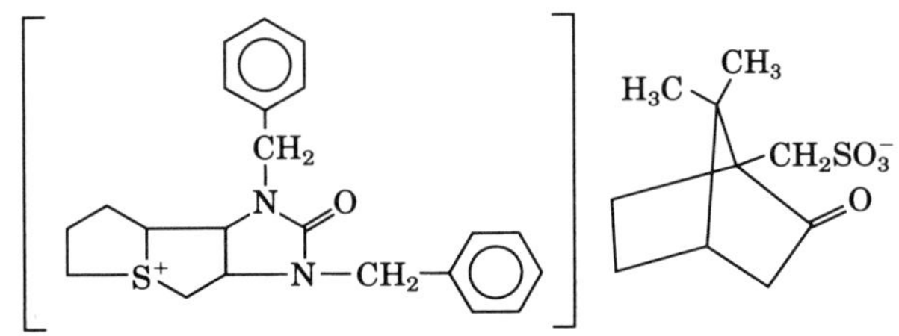

Table 6. (*Continued*)

Generic name	CAS Registry Number	Molecular formula	Trade name	Structure
		Neuronal blockers		
bethanidine sulfate	[*114-85-2*]	$C_{20}H_{32}N_6O_4S$	Meobentine	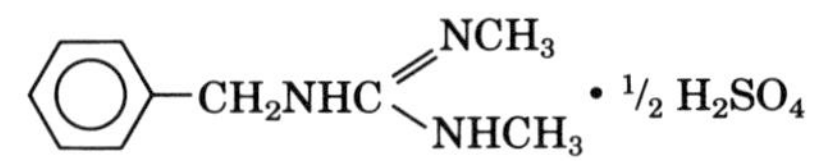
debrisoquin sulfate	[*1131-64-2*]	$C_{20}H_{28}N_6O_4S$	Declinax	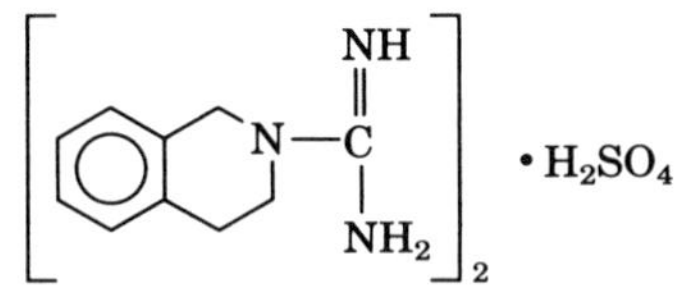
guanadrel sulfate	[*40580-59-4*]	$C_{20}H_{40}N_6O_8S$	Hylorel	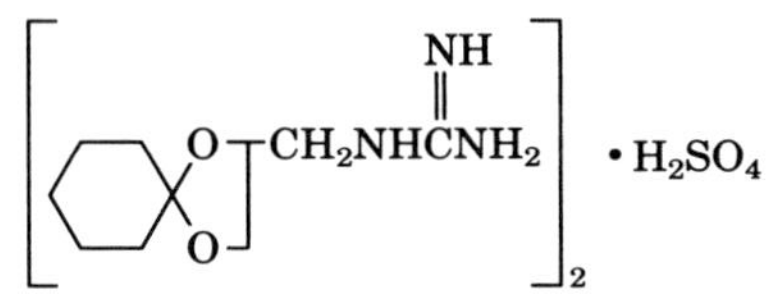
guanethidine monosulfate	[*645-43-2*]	$C_{10}H_{24}N_4O_4S$	Ismelin	
		Neuronal norepinephrin depleting agents		
reserpine	[*50-55-5*]	$C_{33}H_{40}N_2O_9$	Serpasil, Reserpoid	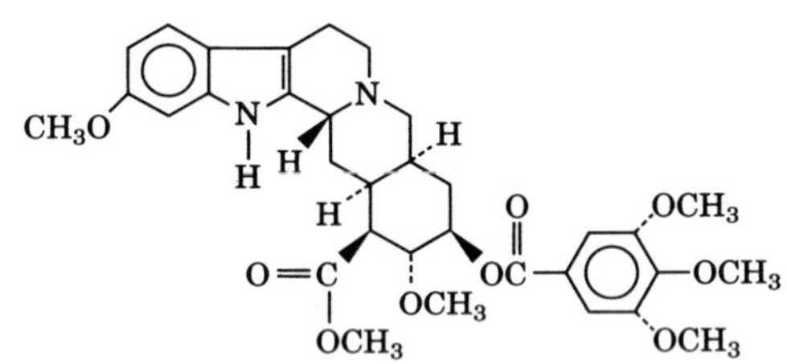

α-Adrenoceptor blockers

bunazosin	[*80755-51-7*]	$C_{19}H_{27}N_5O_3$		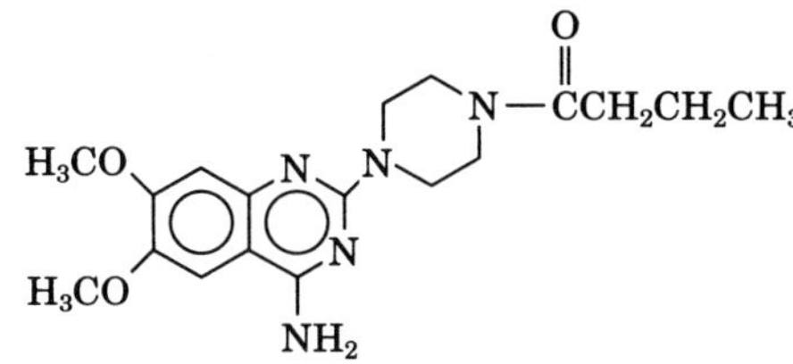
doxazosin mesylate	[*77883-43-3*]	$C_{24}H_{29}N_5O_8S$		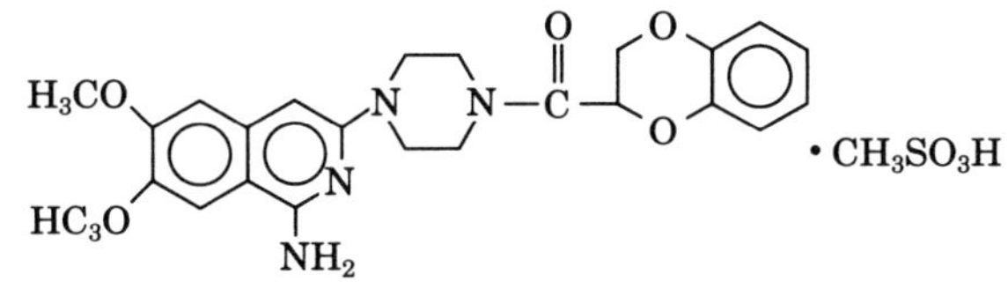
phentolamine mesylate	[*65-28-1*]	$C_{18}H_{23}N_3O_4S$	Regitine	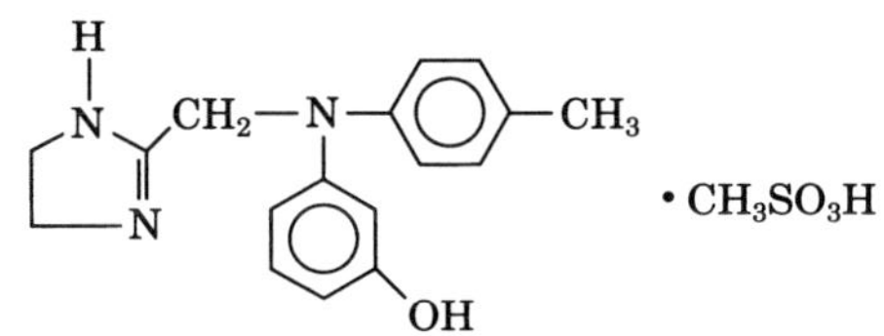
prazosin hydrochloride	[*19237-84-4*]	$C_{19}H_{22}ClN_5O_4$	Minipress	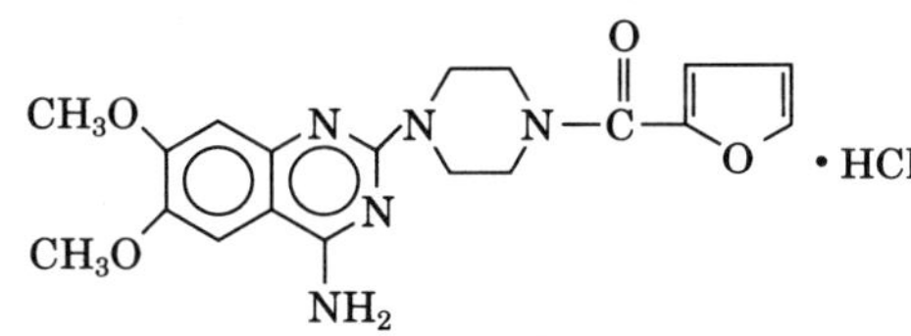
terazosin hydrochloride dihydrate	[*70024-40-7*]	$C_{19}H_{26}ClN_5O_4 \cdot 2H_2O$	Hytrin	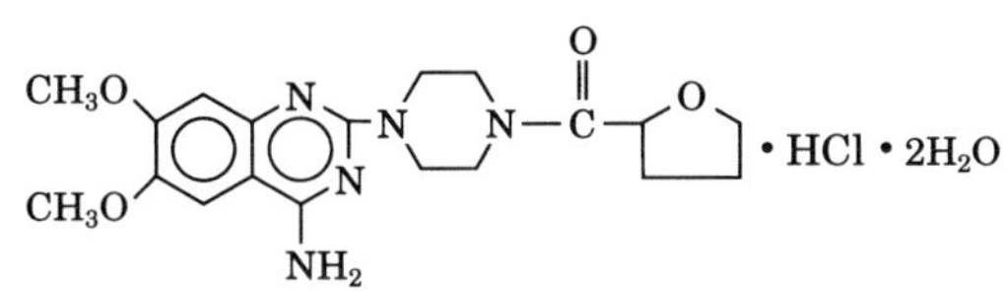

Table 6. (*Continued*)

Generic name	CAS Registry Number	Molecular formula	Trade name	Structure
		β-Adrenoceptor blockers		
brefanolol	[*104051-20-9*]	$C_{22}H_{28}N_2O_2$		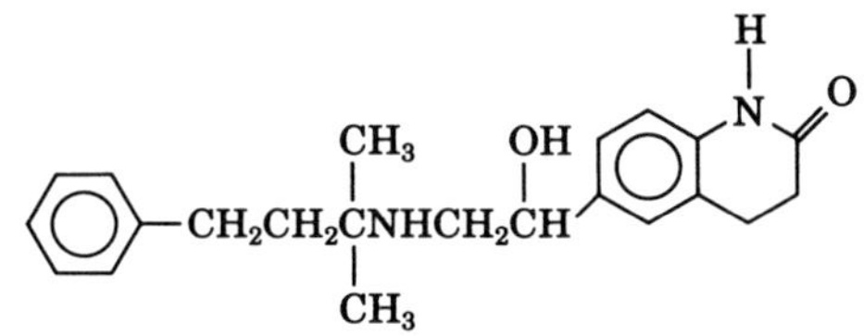
bucindolol hydrochloride	[*70369-47-0*]	$C_{22}H_{25}N_3O_2 \cdot HCl$		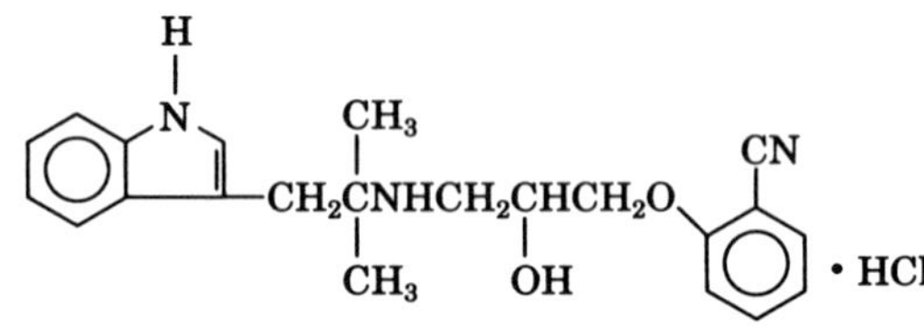
carvedilol	[*72956-09-3*]	$C_{24}H_{26}N_2O_4$		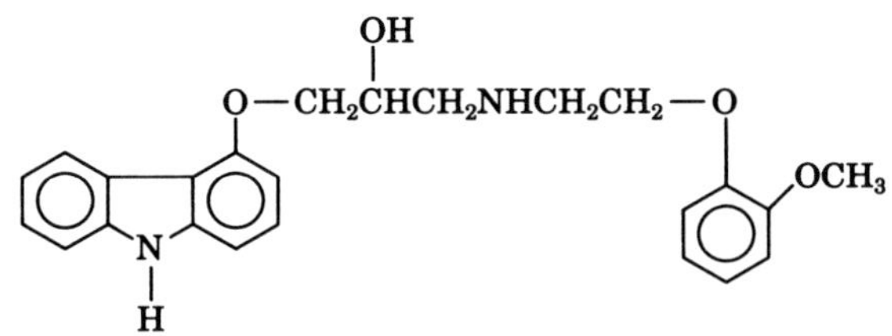
pindolol	[*13523-86-9*]	$C_{14}H_{20}N_2O_2$	Visken	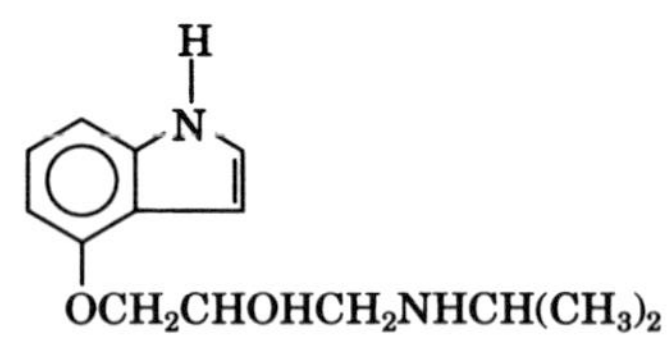

Calcium channel blockers

benidipine	[*105979-17-7*]	$C_{28}H_{31}N_3O_6$		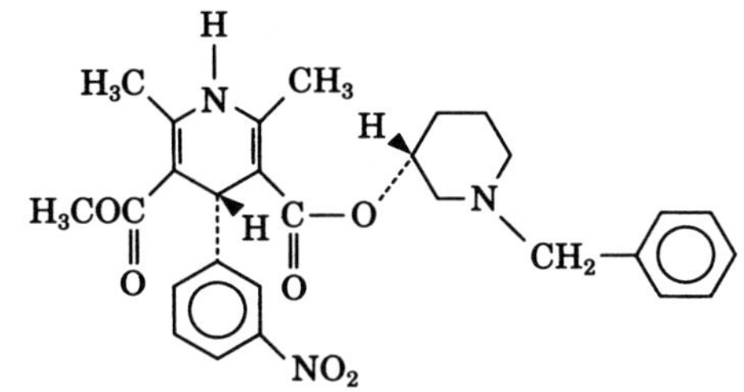
felodipine	[*72509-76-3*]	$C_{18}H_{19}Cl_2NO_4$		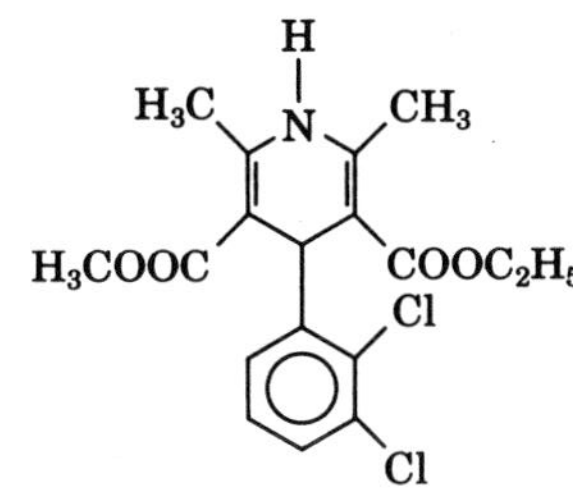
isradipine	[*75695-93-1*]	$C_{19}H_{21}N_3O_5$	DynaCirc	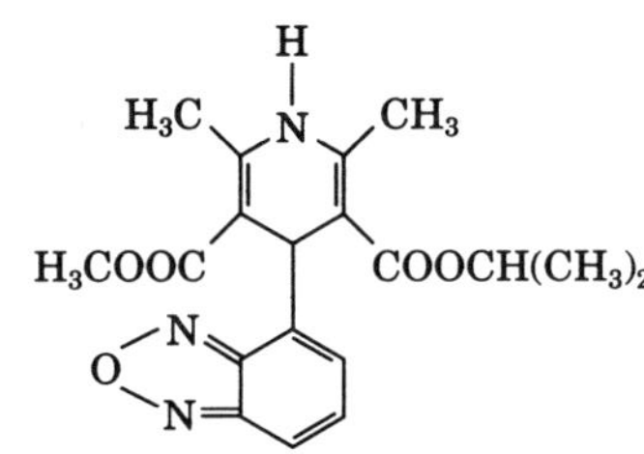

Table 6. (*Continued*)

Generic name	CAS Registry Number	Molecular formula	Trade name	Structure
lacidipine	[*103890-78-4*]	$C_{26}H_{33}NO_6$		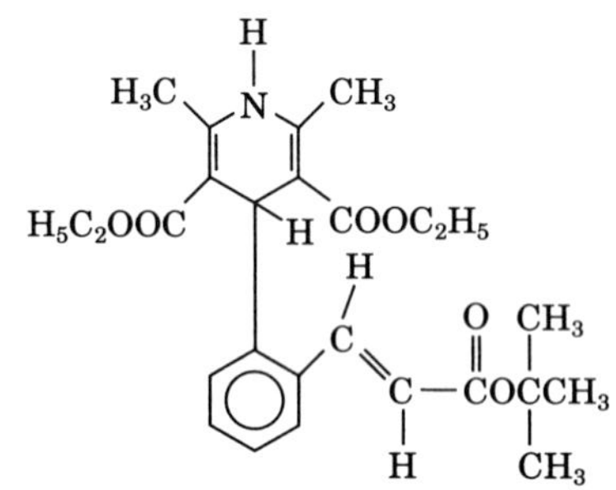
nitrendipine	[*39562-70-4*]	$C_{18}H_{20}N_2O_6$	Baypress	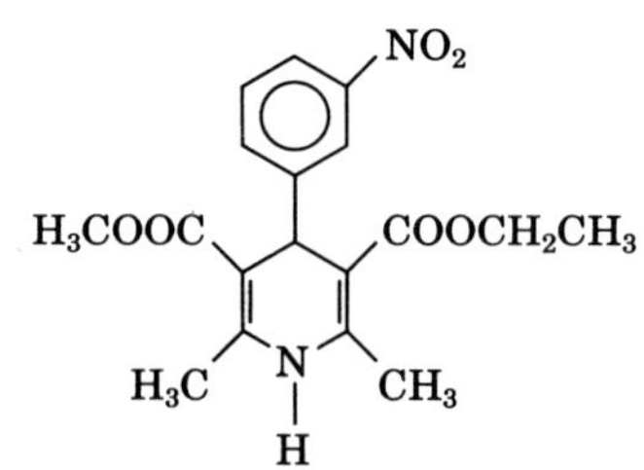
		Diuretics		
amiloride hydrochloride dihydrate	[*17440-83-4*]	$C_6H_9Cl_2N_7O{\cdot}2H_2O$	Midamor	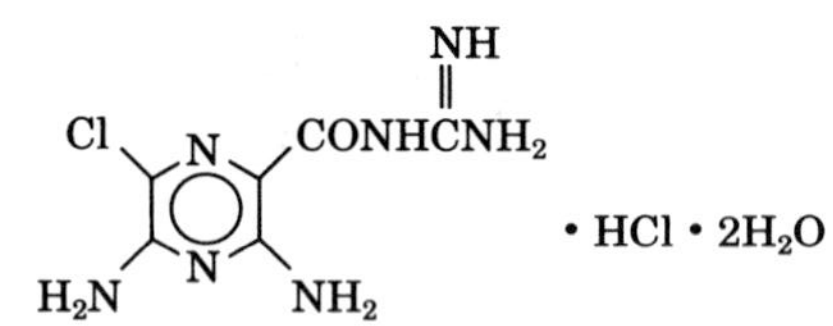

bumetanide	[*28395-03-1*]	$C_{17}H_{20}N_2O_5S$	Bumex	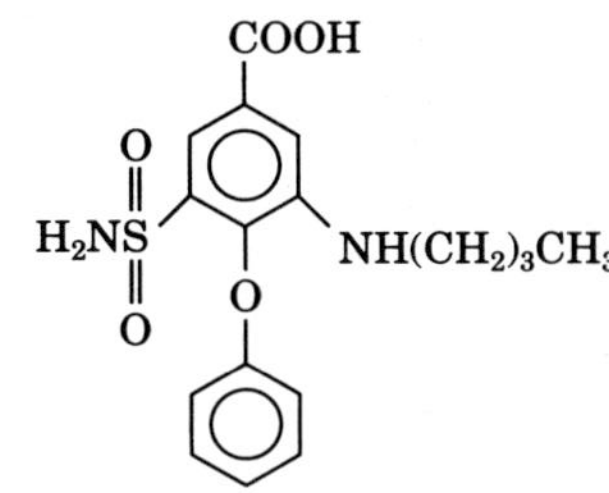
chlorthalidone	[*77-36-1*]	$C_{14}H_{11}ClN_2O_4S$	Hygroton	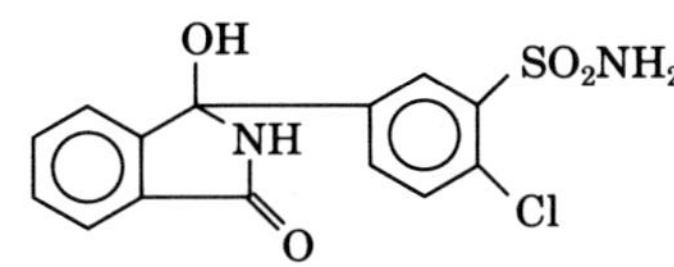
ethacrynic acid	[*58-54-8*]	$C_{13}H_{12}Cl_2O_4$	Edecrin	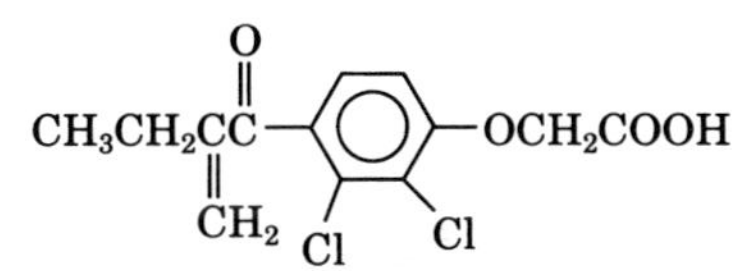
furosemide	[*54-31-9*]	$C_{12}H_{11}ClN_2O_5S$	Lasix, Disal	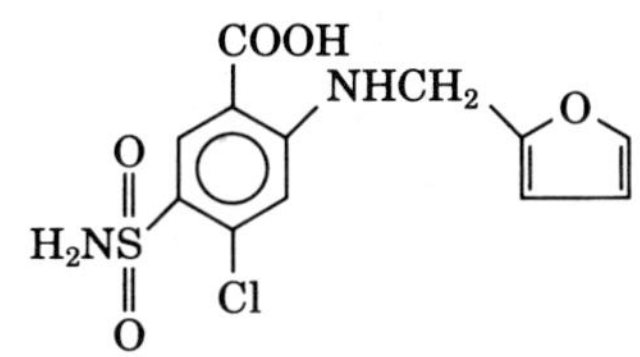
hydrochlorothiazide	[*58-93-5*]	$C_7H_8ClN_3O_4S_2$	Esidrix, HydroDiuril	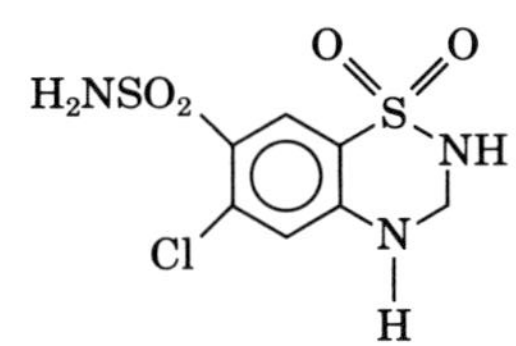

Table 6. (*Continued*)

Generic name	CAS Registry Number	Molecular formula	Trade name	Structure
metolazone	[*17560-51-9*]	$C_{16}H_{16}ClN_3O_3S$	Zaroxolyn, Mykrox, Diulo	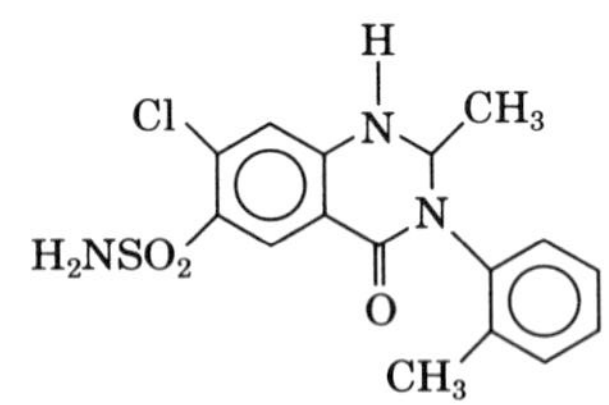
quinethazone	[*73-49-4*]	$C_{10}H_{12}ClN_3O_3S$	Hydromox	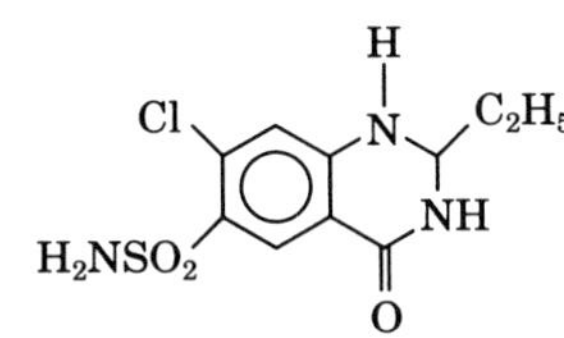
spironolactone	[*52-01-7*]	$C_{24}H_{32}O_4S$	Aldactone	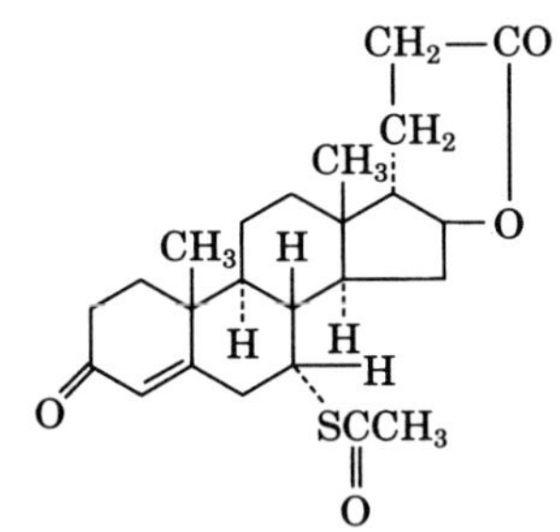

triamterene	[*396-01-0*]	$C_{12}H_{11}N_7$	Dyrenium	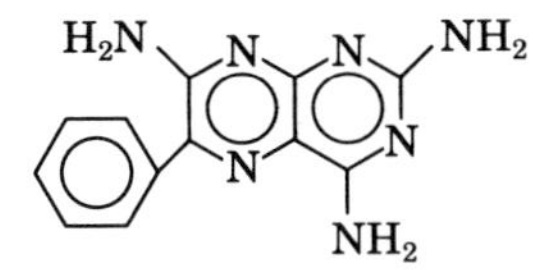
		Centrally acting antihypertensive agents		
clonidine hydrochloride	[*4205-91-8*]	$C_9H_{10}Cl_3N_3$	Catapress	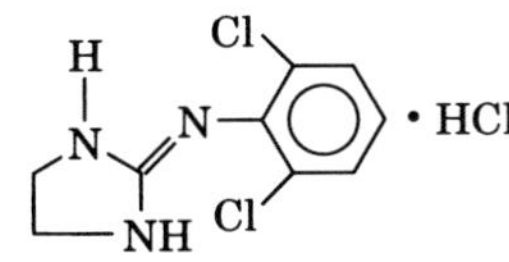
guanabenz acetate	[*23256-50-0*]	$C_{10}H_{12}Cl_2N_4O_2$	Wytensin	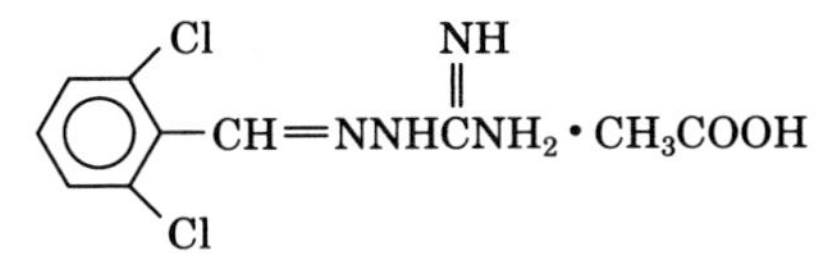
guanfacine hydrochloride	[*29110-47-2*]	$C_9H_{10}Cl_3N_3O$	Tenex	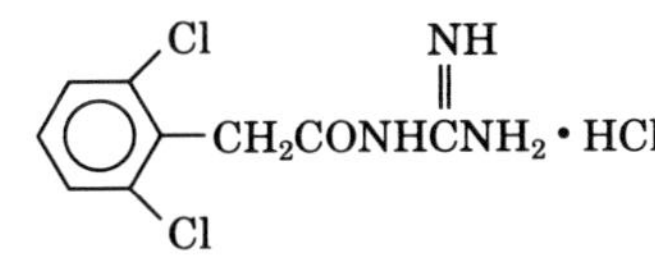
methyldopa	[*555-30-6*]	$C_{10}H_{13}NO_4$	Aldomet	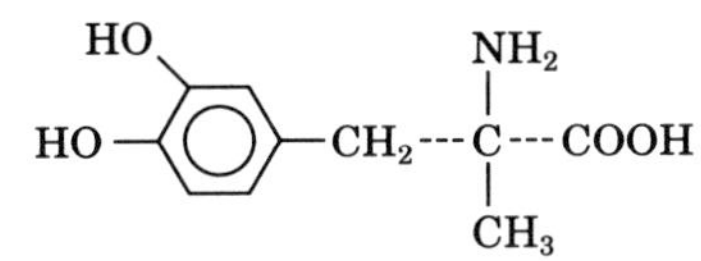
rilmenidene	[*54187-04-1*]	$C_{10}H_{16}N_2O$		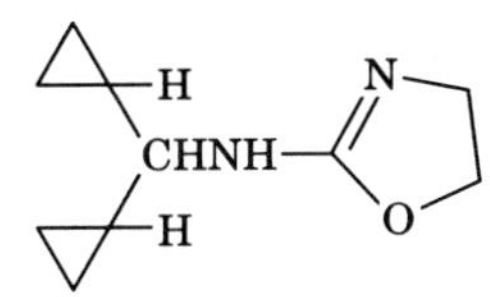

Table 6. (*Continued*)

Generic name	CAS Registry Number	Molecular formula	Trade name	Structure
		Vasodilators		
cadralazine	*[64241-34-5]*	$C_{12}H_{21}N_5O_3$		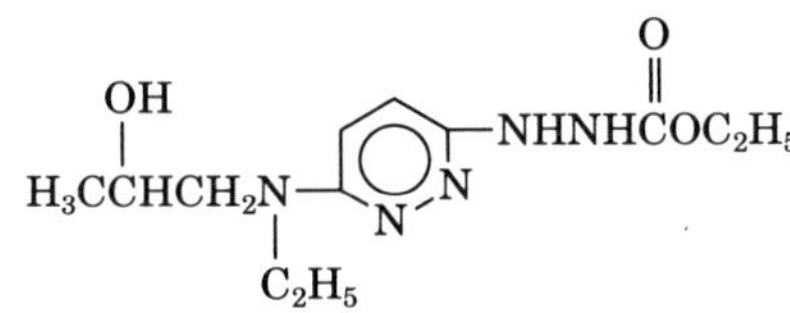
hydralazine hydrochloride	*[304-20-1]*	$C_8H_9ClN_4$	Apresoline	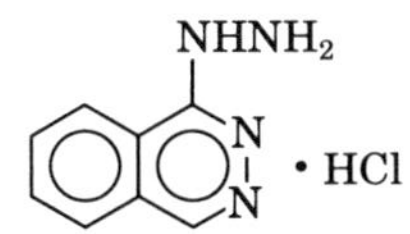
minoxidil	*[38304-91-5]*	$C_9H_{15}N_5O$	Loniten	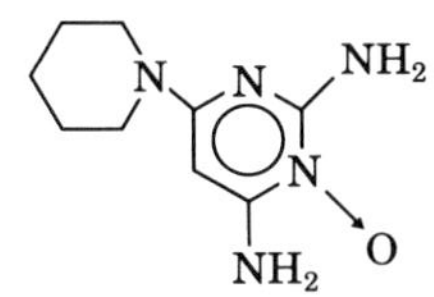
nitroprusside sodium dihydrate	*[13755-38-9]*	$C_5FeN_6O{\cdot}2H_2O{\cdot}2Na$	Nipride	$Na_2[Fe(CN)_5NO] \cdot 2H_2O$
		Potassium channel openers		
cromakalim	*[94470-67-4]*			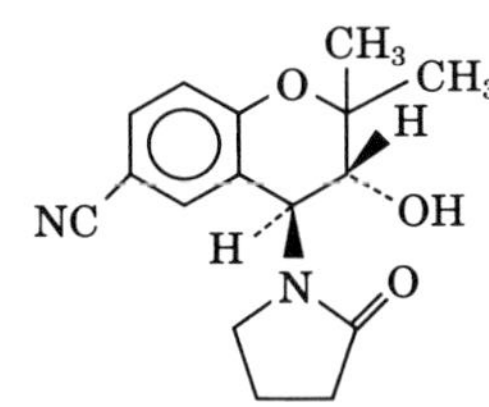

diazoxide	[*364-98-7*]	$C_8H_7ClN_2O_2S$	Hyperstat	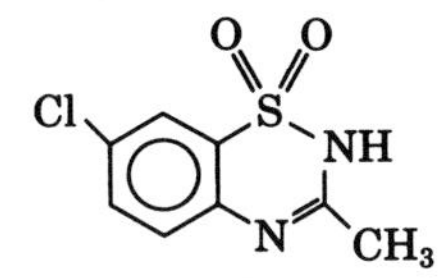
nicorandil	[*65141-46-0*]	$C_8H_9N_3O_4$		
pinacidil	[*85371-64-8*]	$C_{13}H_{19}N_5 \cdot H_2O$	Pindac	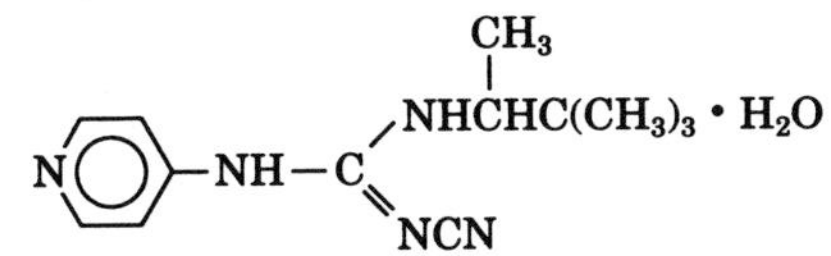

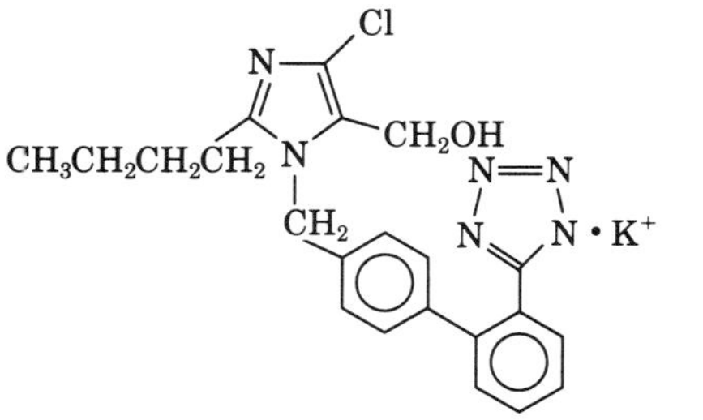

DuP 753

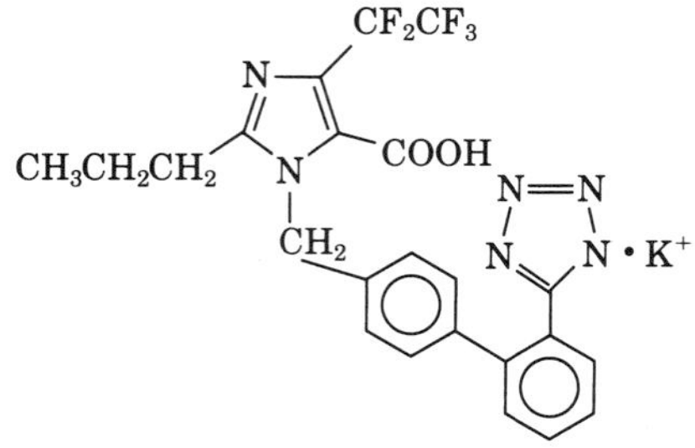

DuP 532

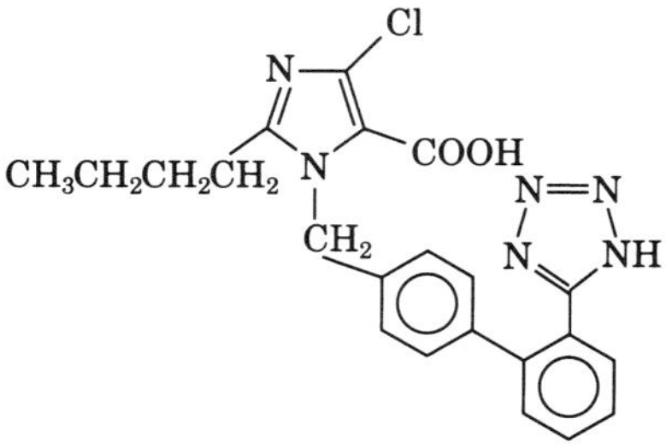

EXP-3174

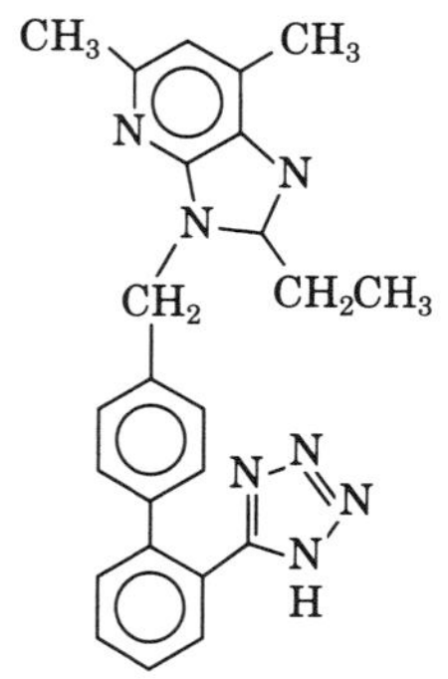

L-158,609

Fig. 3. The Angiotensin II receptor antagonists.

in humans following iv administration. However, there is no indication that any of these compounds are orally active at reasonably low doses (202,203,224–226). The transition-state inhibitors are stable structures resembling the transition state of the enzyme reaction and are bound more tightly to renin than renin is to the substrate.

In animal models, renin inhibitors after iv administration have been shown to be equally efficacious as compared to ACE inhibitors in lowering blood pressure, and to have similar effects on the cardiovascular and renal parameters. Some of the renin inhibitors tested are enalkiren (A-64662), CGP 38 560A, CP 71362, KRI-1230, U-71039, ES6864, and PD-134672. The structures of enalkiren and the formula for PD-134672 are shown in Figure 4.

Agents Affecting the Adrenergic Nervous System

Historically, the vasopressor systems, particularly the adrenergic nervous system, drew the most attention in hypertension research and many drugs affecting this system have been introduced. An agent achieves antagonism of this system either by decreasing the number of nerve impulses traveling down the nerves, by blocking ganglia, by blocking the release of the adrenergic neurotransmitter, norepinephrine, by depleting the norepinephrine stores in the neurons, or by blocking

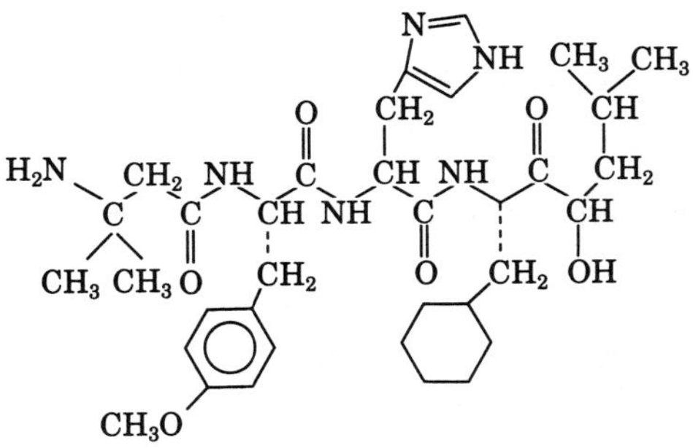

enalkiren

([1*S*-(1*R**, 2*S**, 3*R**)]-*N*-(4-morpholinyl-sulfonyl)-L-phenylalanyl-3-(2-amino-4-thiazolyl)-*N*-[1-cyclohexylmethyl-2,3-dihydroxy-5-methylhexyl]-L-alaninamide)

PD-134,672

Fig. 4. Renin inhibitors.

the adrenoceptors in the blood vessels. Vascular tone is maintained mainly by the adrenergic nervous system, and decreasing the tone by any of these mechanisms causes a lowering of blood pressure.

Ganglionic Blockers. Ganglionic blocking agents (see Table 6), such as hexamethonium, trimethaphan, and chlorisondamine, block the autonomic ganglia and thus decrease sympathetic outflow. Tolerance develops quickly to these agents. Because of such disturbing side effects as severe orthostatic hypotension, sexual dysfunction, dry mouth, and urinary retention, these drugs have become obsolete.

Adrenergic Neuronal Blockers. The adrenergic neuronal blocking agents, guanethidine, bretylium, debrisoquin, and guanadrel (Table 6), produce hypotension by blocking the release of norepinephrine from the nerve terminals of adrenergic neurons. These drugs are taken up by the neurons and decrease sympathetic tone, heart rate, cardiac output, and total peripheral resistance. Some deplete norepinephrine stores and thus produce an initial sympathomimetic response increasing these various responses. Orthostatic hypotension, severe sexual dysfunction and impairment, fluid retention, and diarrhea are the primary side effects of this class of agents which is obsolete and used only when no other agents work.

Neuronal Norepinephrine Depleting Agents. Reserpine (Table 6) is the most active alkaloid derived from Rauwolfia serpentina. The principal antihypertensive mechanism of action primarily results from depletion of norepinephrine from peripheral sympathetic nerves and the brain adrenergic neurons. The result is a drastic decrease in the amount of norepinephrine released from these neurons, leading to decrease in vascular tone and lowering of blood pressure. Reserpine also depletes other transmitters including epinephrine, serotonin [*50-67-9*], $C_{10}H_{12}N_2O$, and dopamine [*51-61-6*], $C_8H_{11}NO_2$ (see NEUROREGULATORS). Reserpine is efficacious in all forms of hypertension, particularly with the concomitant use of a thiazide diuretic. The most serious side effect is mental depression. Other side effects include nasal congestion, sedation, drowsiness, lethargy, decreased libido, impotence, and nightmares.

α-Adrenoceptor Blockers. Nonselective α-adrenoceptor blockers (Table 6), such as phentolamine, which block both α_1- and α_2-adrenoceptors, produce vasodilation by antagonizing the effects of endogenous norepinephrine. They also produce severe tachycardia and have been replaced by selective α_1-adrenoceptor blockers, such as prazosin, terazosin, and doxazosin, which do not usually cause severe tachycardia.

Prazosin, a selective α_1-adrenoceptor antagonist, exerts its antihypertensive effect by blocking the vasoconstrictor action of adrenergic neurotransmitter, norepinephrine, at α_1-adrenoceptors in the vasculature (200,227,228). Prazosin lowers blood pressure without producing a marked reflex tachycardia. It causes arteriolar and venular vasodilation, but a significant side effect is fluid retention. Prazosin increases HDL cholesterol, decreases LDL cholesterol, and does not cause glucose intolerance.

Doxazosin, a prazosin derivative, is a highly selective α_1-adrenoceptor blocker (229,230). It has no significant antagonistic effects on presynaptic or postsynaptic α_2-adrenoceptors. Blocking activation of α_1-adrenoceptors prevents the breakdown of phosphatidylinositol and the resulting vasoconstriction. Lowering of blood pressure by doxazosin is not accompanied by an increase in heart rate or cardiac output. Its duration of action is as long as 24 h. Doxazosin in chronic hypertension treatment reduces total cholesterol, LDL cholesterol, and triglyceride levels, and increases HDL cholesterol.

Terazosin is selective α_1-adrenoceptor blocker having hypotensive efficacy equal to that of prazosin. Terazosin has a longer duration of action and better gastrointestinal absorption profile than prazosin.

β-Adrenoceptor Blockers. There is no satisfactory mechanism to explain the antihypertensive activity of β-adrenoceptor blockers (see Table 1) in humans particularly after chronic treatment (228,231–233). Reductions in heart rate correlate well with decreases in blood pressure and this may be an important mechanism. Other proposed mechanisms include reduction in PRA, reduction in cardiac output, and a central action. However, pindolol produces an antihypertensive effect without lowering PRA. In long-term treatment, the cardiac output is restored despite the decrease in arterial blood pressure and total peripheral resistance. Atenolol (Table 1), which does not penetrate into the brain is an efficacious antihypertensive agent. In short-term treatment, the blood flow to most organs (except the brain) is reduced and the total peripheral resistance may increase.

Better antihypertensive effect of β-adrenoceptor blockers is found in patients having high PRA and most are not efficacious in patients having low PRA or in elderly patients. β-Adrenoceptor blockers usually lower arterial blood pressure about 10 mm Hg (1.3 kPa). Side effects include lethargy, dyspnea, nausea, dizziness, headache, impotency, cold hands and feet, vivid dreams and nightmares, bronchospasm, bradycardia, and sleep disturbances.

β-Adrenoceptor blockers for the treatment of hypertension include: (*1*) the cardioselective β_1-adrenoceptor blockers without intrinsic sympathomimetic activity (ISA), ie, atenolol (Table 3), bisoprolol (Table 3), and metoprolol (Table 1); (*2*) the cardioselective with ISA, ie, acebutolol (Table 1); (*3*) the noncardioselective without ISA, ie, propranolol (Table 1) and timolol [*26839-75-8*], $C_{13}H_{24}N_4O_3S$; and (*4*) the noncardioselective with ISA, ie, oxprenolol [*6452-71-7*], $C_{15}H_{23}NO_3$, and pindolol.

α- and β-Adrenoceptor Blocking Agents. Labetalol possesses both α- and β-adrenoceptor blocking effects, blocking both β_1- and β_2-adrenoceptors, but only

α_1-adrenoceptors (234,235). It also possesses β_2-adrenoceptor agonistic (vasodilatory) effects, and produces significant reduction in blood pressure and a slight decrease in heart rate. It is efficacious in mild, moderate, and severe hypertension. The cardiac output is maintained because of an increase in stroke volume. Exercise-induced blood pressure increases are blunted and the heart rate is reduced. In chronic treatment, PRA is reduced. Labetalol does not influence the blood cholesterol, triglyceride, or lipoproteins. The side effects are gastrointestinal discomfort and dizziness.

Calcium Channel Blockers

The extracellular concentration of calcium is approximately 2,500 to 25,000 times higher than that of intracellular calcium, 2.5 mM vs 0.1 to 1.0 μM. This gradient is maintained by regulating the calcium influx through receptor as well as voltage operated channels. Voltage operated channels are divided into three subtypes: L, N, and T. Calcium channel blockers, such as verapamil (Table 1), diltiazem (Table 3), or nifedipine (Table 3) only block the L-type channels. Contraction of vascular smooth muscle is governed by the intracellular free calcium, regulated by the voltage operated calcium channels. When the channels are open, the extracellular calcium moves in the cell across the membrane through a great electrochemical gradient. The process initiates the contraction.

Another mechanism in initiating the contraction is agonist-induced contraction. It results from the hydrolysis of membrane phosphatidylinositol and the formation of inositol triphosphate (IP_3). IP_3 in turn triggers the release of intracellular calcium from the sarcoplasmic reticulum and the influx of more extracellular calcium. The third mechanism in triggering the smooth muscle contraction is the increase of calcium influx through the receptor-operated channels. The increased cytosolic calcium enhances the binding to the protein, calmodulin [*73298-54-1*].

Calcium channel blockers reduce arterial blood pressure by decreasing calcium influx, resulting in a decrease in intracellular calcium (236,237). The arterial smooth muscle tone decreases, thereby decreasing total peripheral resistance. The increase in vascular resistance in hypertension is found to depend much on calcium influx. Calcium channel blockers reduce blood pressure at rest and during exercise. They decrease the transmembranous calcium influx or entry that lead to a net decrease of intracellular calcium and therefore the vascular tone falls, as does blood pressure.

In addition to the vascular effects, calcium channel blockers (Table 6) such as isradipine, nifedipine (Table 3), and nitrendipine, produce natriuretic effects by increasing fractional sodium excretion and free water clearance. They also reduce distal tubular sodium reabsorption.

Calcium channel blockers normalize the blood pressure in about 80% of hypertensive patients older than 60 years of age, 50% of those between 40 and 60 years of age, and only 20% of patients under 40 years of age. Thus calcium channel blockers are best for patients who are elderly and have low PRA and mostly ineffective in patients who have high PRA. This responsiveness profile is very similar to that of the diuretics.

Nifedipine, verapamil, and diltiazem are all efficacious in the treatment of mild and moderate hypertension, but nifedipine is more efficacious than diltiazem and verapamil in the control of severe hypertension. Nifedipine does not cause significant reflex tachycardia or orthostatic hypotension. Nifedipine benefits the older and black patients and patients with low PRA.

Calcium channel blockers cause more pronounced lowering of blood pressure in hypertensive patients than in normotensive individuals. Generally, all calcium channel blockers cause an immediate increase in PRA during acute treatment in patients having hypertension but PRA is normalized during chronic treatment despite the sustained decrease in blood pressure. These agents also do not generally produce sodium and water retention, unlike the conventional vasodilators. This is because they produce diuretic effects by direct actions on the kidney.

It has been suggested that in a large subset of hypertensive patients there is a slight but constant increase of traffic of free calcium influx and as a result there is an increase of intracellular free calcium leading to additional release of calcium from intracellular storage sites. The end result is a slight but chronic increase in vascular tone and peripheral resistance. Over time, constant adaptation of this slight increase of vascular tone inches the arterial blood pressure up to the hypertensive level. Calcium channel blockers block the inward movement of calcium into the cells of vascular smooth muscles, and are thus considered to be the drugs that correct the fundamental abnormalities of hypertension. Calcium channel blockers have been shown in animal experiments to prevent and cause regression of atherosclerosis (238,239).

Verapamil (Table 1), the first slow channel calcium blocker synthesized to selectively inhibit the transmembrane influx of calcium ions into cells, lowers blood pressure in hypertensive patients having good organ perfusion particularly with increased renal blood flow. Sustained-release verapamil for once a day dosing is available for the treatment of hypertension. Constipation is a prominent side effect. Headache, dizziness, and edema are frequent and verapamil can sometimes cause AV conduction disturbances and AV block. Verapamil should not be used in combination with β-adrenoceptor blockers because of the synergistic negative effects on heart rate and contractile force.

Nifedipine (Table 3) is a potent vasodilator that selectively dilates resistance vessels and has fewer effects on venous vessels. It does not cause reflex tachycardia during chronic therapy. Nifedipine is one of the first-line choices for black or elderly patients and patients having concomitant angina pectoris, diabetes, or peripheral vascular diseases. Nifedipine, sublingually, is also suitable for the treatment of hypertensive emergencies. Nifedipine does not impair sexual function or worsen blood lipid profile. The side effects are flushing, headache, and dizziness.

Diltiazem inhibits calcium influx via voltage-operated channels and therefore decreases intracellular calcium ion. This decreases smooth muscle tone. Diltiazem dilates both large and small arteries and also inhibits α-adrenoceptor activated calcium influx. It differs from verapamil and nifedipine by its use dependence. In order for the blockade to occur, the channels must be in the activated state. Diltiazem has no significant affinity for calmodulin. The side effects are headache, edema, and dizziness.

Diuretics

The principal mechanism of the hypotensive effect of diuretics (qv) is salt and fluid depletion, leading to reduction in blood volume (200,240). Acute effects lead to a decrease in cardiac output and an increase in total peripheral resistance. However, during chronic administration, cardiac output and blood volume return toward normal and total peripheral resistance decreases to below pretreatment values. As a result, the blood pressure falls. The usual reduction in blood volume is about 5%. A certain degree of sustained blood volume contraction has to occur before the blood pressure decreases. The usual decrease in blood pressure achieved using a diuretic is about 20/10 mm Hg (2.7/1.3 kPa) (systolic/diastolic pressures.

Intake of a large amount of sodium chloride negates the antihypertensive effects of diuretics. Other mechanisms, such as direct vasodilating action, decreased responsiveness to vasopressor agents, stimulation of prostacyclin [*35121-78-9*], $C_{20}H_{32}O_5$, production, and reduction in the intracellular calcium concentration, etc, only play a minor role in the overall antihypertensive effects of diuretics.

Diuretics, such as those of the thiazide type, have been the cornerstone of first-line antihypertensive treatments for decades. However, popularity and use have eroded as a result of increases in sudden death in patients on diuretic therapy, and unfavorable effects on blood lipids profile, ie, increasing cholesterol and triglyceride. These effects have been implicated as possible causes for the lack of decrease in the mortality rate resulting from acute MI in patients treated with a diuretic (187,240,241). However, diuretics do protect against stroke and CHF.

Diuretics are needed to return to normal the expanded extracellular volume that other antihypertensive agents produce, such as fluid retention and blood volume expansion, via compensatory mechanisms of the body. The loss of efficacy of antihypertensive agents can be restored if a diuretic is used concomitantly. In the treatment of hypertension, high ceiling or loop diuretics, such as furosemide, ethacrynic acid, and bumetanide, are no more efficacious than the thiazide-type of diuretics. In fact, these agents cause more side effects, such as dehydration, metabolic alkalosis, etc, and therefore, should not be used except in situations where rapid elimination of fluid volume is clearly indicated.

Diuretics can cause hypokalemia, hyperglycemia,and hyperuricemia. After long-term treatment they may increase serum triglyceride and cholesterol (242).

Aldosterone Antagonists. Spironolactone is a competitive antagonist of aldosterone and is effective in lowering the the hypertension not only of patients having primary aldosteronism, but also of many patients having essential hypertension whose aldosterone levels are not elevated. It is particularly efficacious in patients having low PRA. The mechanism of action is by antagonizing the sodium-retaining and potassium-wasting effects of aldosterone at the Na^+-K^+ exchange sites of the renal distal tubule. Spironolactone prevents or reverses hypokalemia in patients treated with the low and high ceiling diuretics.

Gynecomastia may be associated with the use of spironolactone and is reversible upon withdrawal of the drug.

Potassium Sparing Diuretics. Triamterene and amiloride, potassium sparing diuretics, by themselves produce only slight antihypertensive effects. The main use is to prevent or to treat the hypokalemia induced by thiazide-type and high ceiling loop diuretics, such as furosemide and bumetanide.

Centrally Acting Agents

Clonidine, methyldopa (α-methyldopa), guanabenz, and guanfacine lower arterial blood pressure by activating the α_2-adrenoceptors in the pontomedullary region of the brain (200). The central hypotensive activity is blocked by α-adrenoceptor blockers, such as phentolamine [*50-60-2*], $C_{17}H_{19}H_3O$, and yohimbine [*146-48-5*], $C_{21}H_{26}N_2O_3$. Side effects include drowsiness, sedation (except rilmenidine [S3341]), dry mouth, sexual dysfunction, and depression. Tricyclic antidepressants, such as imipramine [*50-49-7*], $C_{19}H_{24}N_2$, and desipramine [*50-47-5*], $C_{18}H_{22}N_2$, are known to counteract the central antihypertensive effects of clonidine (see PSYCHOPHARMACOLOGICAL AGENTS).

Methyldopa. Methyldopa reduces arterial blood pressure by decreasing adrenergic outflow and decreasing total peripheral resistance and heart rate having no change in cardiac output. Blood flow to the kidneys is not changed and that to the heart is increased. It causes regression of myocardial hypertrophy.

Methyldopa, through its metabolite, α-methylnorepinephrine formed in the brain, acts on the postsynaptic α_2-adrenoceptor in the central nervous system. It reduces the adrenergic outflow to the cardiovascular system, thereby decreasing arterial blood pressure. If the conversion of methyldopa to α-methylnorepinephrine in the brain is prevented by a dopamine β-hydroxylase inhibitor capable of penetrating into the brain, it loses its antihypertensive effects.

Methyldopa is effective in mild, moderate, and severe hypertension but a thiazide-type diuretic is needed to overcome the fluid retaining side effect. Methyldopa has been shown to prevent and induce regression of ventricular hypertrophy in hypertensive patients. The principal side effects are sedation, drowsiness, nasal congestion, fluid retention, and in rare occasions, hemolytic anemia.

Clonidine. Clonidine decreases blood pressure, heart rate, cardiac output, stroke volume, and total peripheral resistance. It activates central α_2-adrenoceptors in the brainstem vasomotor center and produces a prolonged hypotensive response. Clonidine, most efficaciously used concomitantly with a diuretic in long-term treatment, decreases renin and aldosterone secretion.

When clonidine is withdrawn abruptly, patients may experience a rebound hypertensive phenomenon, wherein blood pressure rises rapidly to a level higher than the predrug level. These patients may experience symptoms of headache, tachycardia, agitation, and nervousness. If rebound hypertension occurs, resumption of clonidine therapy or administration of phentolamine reduces the blood pressure. For clonidine withdrawal, the dose should be reduced gradually over a two-week period. The principal side effects are sedation, dry mouth, drowsiness, dizziness, and fatigue.

Guanabenz. Guanabenz has been shown to be efficacious in mild and moderate hypertension. Sympathetic tone is decreased with a resultant decrease in total peripheral resistance. In hypertensive patients, guanabenz tends to decrease heart rate and very seldom causes severe fluid retention. It does not suppress PRA but it has a cholesterol and triglyceride lowering effect in chronic therapy. Guanabenz does not cause glucose intolerance, so it is suitable to be used in diabetic patients. The principal side effects are sedation, dry mouth (1 in 4 patients), asthenia, and dizziness. Rebound hypertensive syndrome also occurs in patients experiencing abrupt discontinuation of guanabenz.

Guanfacine. Guanfacine, used in patients having mild to moderate hypertension, can lower blood pressure 50/25 mm Hg (systolic/diastolic) in hypertensive patients. Side effects such as sedation, dry mouth, and asthenia are less as compared to those of guanabenz and clonidine. Guanfacine reduces blood cholesterol and triglyceride and does not cause glucose intolerance.

Rilmenidine. Rilmenidine is a central α_2-adrenoceptor agonist and has been shown to be a potent centrally acting antihypertensive agent without the prominent side effect of sedation.

Vasodilators

Vasodilators dilate or relax the smooth muscles of the vasculatures directly or indirectly by releasing endogenous vasodepressors or antagonizing the endogenous vasopressors or vasopressor systems (200,243,244). Vasodilators may interfere with the entry, intracellular release, and utilization of calcium, the activation of the protein kinase C system, cGMP formation, and EDRF turnover.

Hydralazine. Hydralazine causes vasodilation in all primary vascular beds and has more pronounced effects on capacitance than on resistance blood vessels. Despite the hypotension it produces, hydralazine increases renal blood flow and cardiac output. PRA increases with its use. Tachycardia, headache, dizziness, and water and sodium retention are principal side effects of hydralazine therapy.

Diazoxide. Diazoxide, chemically related to the thiazide diuretics, is a vasodilator having no diuretic effects. In fact, diazoxide causes marked salt and water retention. It has been suggested that diazoxide has potassium channel opener effects. Because of rapid protein binding, diazoxide has to be given by rapid intravenous injection. Its use is limited to the treatment of hypertensive emergencies or crises. Diazoxide causes marked tachycardia, increases PRA, and causes hyperglycemia.

Minoxidil. Minoxidil is one of the most potent vasodilators available for the treatment of hypertension. Minoxidil is a potassium channel opener acting directly on the smooth muscle of the arterioles to decrease peripheral resistance and blood pressure. In long-term treatment, minoxidil causes severe fluid retention, tachycardia, and hair growth (hirsutism). It stimulates the adrenergic nervous system and the renin-angiotensin system. Because of side effects, minoxidil is reserved for use in hypertensive patients having renal failure or in patients having hypertension that is refractory to other antihypertensive drugs.

Nitroprusside. Nitroprusside is a potent, fast-acting vasodilator that has to be administered intravenously by infusion. It relaxes arterial and venous vascular smooth muscle. Its use is mainly in hypertensive crises. Its effects terminate as soon as infusion of the drug is stopped.

Potassium Channel Openers

The resting membrane potential of most excitable cells is around −60 to −80 mV. This gradient is maintained by the activity of various ion channels. When the potassium channels of the cell open, potassium efflux occurs and hyperpolarization results. This decreases calcium channel openings, which in turn prevents the influx of calcium into the cell leading to a decrease in intracellular calcium in the smooth muscles of the vasculature. The vascular smooth muscles then relax and the systemic blood pressure falls.

Vasodilators having potassium channel opener activity relax smooth muscle in low (5–20 m*M*) K^+ concentration but not in high (>80 m*M*) K^+ solution, and glibenclamide [*10238-21-8*] should be able to reverse the induced relaxation (242,245–249). They also increase potassium efflux.

There are at least 13 primary types of K^+ channels known. In addition, within each type there are several subtypes. The best known chemical classes of potassium channel openers are nicorandil, pinacidil, and cromakalim. They are all potent smooth muscle relaxants. Pharmacologically, they behave as classical vasodilators, lowering blood pressure and causing tachycardia and fluid retention.

Nicorandil. Nicorandil is a potassium channel opener that can lower blood pressure 21, 20, and 29 mm Hg after single oral doses of 10, 20, and 30 mg, respectively (250). There are no significant changes in heart rate. Headache is the primary side effect. Nicorandil has potent coronary vasodilator effects. It causes sustained vasodilation of arteriolar resistance and venous capacitance blood vessels, thus reducing cardiac preload and afterload.

Pinacidil. Pinacidil is a potent vasodilator, acting through potassium channel opening effects (242,251–253). Its antihypertensive effect is greater than that of hydralazine and prazosin in chronic treatment. Its fast onset of action also makes it suitable for use in hypertensive crisis.

Pinacidil has a short half-life and most human studies were carried out in slow-release formulations. The reduction in blood pressure produced by pinacidil is accompanied by tachycardia and fluid retention. Plasma catecholamines and renin activity are increased. Other side effects are headache, dizziness, and asthenia.

Cromakalim. Cromakalim has a long half-life (254). Cromakalim at an oral dose of 1.5 mg in humans significantly lowers blood pressure 19/12 mm Hg (systolic/diastolic pressure). It increases renal blood flow, PRA, and heart rate. Cromakalim has bronchodilating activity that is beneficial for hypertensive asthmatic patients. Because of some undesirable effects seen in cardiac papillary muscles of animals on long-term treatment, future clinical trials are to be carried out using the active enantiomer, lemakalim (BRL 38227).

THROMBOLYTIC AGENTS

It has been well documented that the primary cause of acute myocardial infarction (AMI) is coronary arterial thrombosis (255). Thrombosis formation occurs at sites of ulcerated or fissured atheromatous plaques in the coronary circulation (256,257). Reperfusion of the coronaries to the ischemic/infarcted area within 6 h after the onset of AMI can salvage the myocardium and limit infarct size (258–260). Furthermore, the better the preservation of the ventricular function, the better the survival rate. Progressive irreversible myocardiac damage occurs 30 min after ischemia starts. Therefore, the faster the coronary arterial thrombus is lyzed, the less irreversible damage done on the myocardium. It is imperative that once an AMI is diagnosed, iv thrombolytic agent should be administered as quickly as possible.

The success of thrombus lysis depends mainly on how large the thrombus is and whether any blood flow still remains. The outcome is better the larger the surface of the entire thrombus exposed to the thrombolytic agent. As the clot ages, the polymerization of fibrin cross-linking and other blood materials increases and it becomes more resistant to lysis. Therefore, the earlier the thrombolysis therapy starts, the higher the frequency of clot dissolution. Thrombolytic agents available are listed in Table 7 (261–276).

Strepokinase, prourokinase, and APSAC are indirect plasminogen activators, whereas t-PA and urokinase are direct plasminogen activators. All these activators convert the formation of the active proteolytic enzyme plasmin [*9001-90-5*] from the proenzyme, plasminogen [*9001-91-6*] by cleaving the Arg_{560}–Val_{561} bond. After iv administration, all five thrombolytic agents achieve approximately the same incidence of reperfusion rate (60–70%) if used optimally. t-PA and prourokinase require simultaneous heparin [*9005-49-6*] administration when used; the other thrombolytics do not. Reocclusion rate is higher with t-PA if heparin use is not maximized. The half-life is short (5–8 min) using t-PA and prourokinase, medium (16–23 min) using urokinase and streptokinase, and long (105 min) using APSAC. When used at efficacious doses, each has similar degrees of bleeding complications. The least expensive thrombolytic agent is streptokinase, which is about one-tenth of the others. No single agent is superior when all factors are taken into consideration. Whereas t-PA is clot selective, it has a very short half-life and simultaneous use of heparin is required. Bleeding is therefore as much of a problem with t-PA as the less clot selective agent, such as streptokinase. Streptokinase causes severe systemic fibrinogen degradation leading to extensive hypofibrinogenemia. Under this condition, reocclusion is less and the dissolution

Table 7. Thrombolytic Agents

Generic name	CAS Registry Number	Trade name
tissue plasminogen activator (alteplase t-PA)	[*105857-23-6*]	Activase
streptokinase	[*9002-01-1*]	Streptase, Kabikinase
urokinase	[*9039-53-6*]	Abbokinase, Breokinase, Win-kinase
anisoylated plasminogen-streptokinase activator (APSAC, anistreplase)	[*81669-57-0*]	Eminase
prourokinase (scu-PA)	[*82657-92-9*]	Sandolase

of the thrombus is more complete. This may explain why streptokinase has been shown, in most studies, to be as efficacious as t-PA. The most recent comparison study reporting the equal efficacy of t-PA, APSAC, and streptokinase tends to confirm this point. Most large trials show these agents reduce patient mortality from 16 to 30%.

The dosage of each thrombolytic agent should be high enough to produce a lysis state long enough for complete clot dissolution to occur. Aspirin, a platelet aggregation inhibitor, has been demonstrated to decrease mortality of AMI by itself and to further decrease that of the thrombolytic agents if used concomitantly (277–279). The use of combinations of aspirin, heparin, vasodilator, and the thrombolytic agent decreases the incidence of rethrombosis.

Tissue-Type Plasminogen Activator (t-PA). t-PA is a physiologic activator of plasminogen in the body. It is a protein containing 527 amino acids having mol wt ~64,000, and is mainly produced in the endothelial cells lining blood vessels. It is also made in the liver, spleen, lungs, muscle, and other tissues. The commercial preparation derives from biotechnological cloning and production techniques, ie, it is recombinant t-PA (see GENETIC ENGINEERING; PROTEIN ENGINEERING). t-PA can occur in either a one- or two-chain form. Both forms are active but the two-chain form is longer acting (261,264,266,280–283).

t-PA has a much higher affinity for fibrin and fibrin-bound plasminogen as compared to the circulating plasminogen. For this reason, under normal conditions generalized (systemic) fibrinogenlysis in the circulation is thought not to occur. However, in the treatment of AMI, the large amount of t-PA used and infused for a prolonged period causes generalized fibrinogenolysis because of the activation of circulating plasminogen. In the treatment of AMI, a bolus intravenous injection of 10 mg is followed by an iv infusion of 50 mg/h for one hour and 20 mg/h for two hours. The half-life of t-PA in the circulation is about 4–5 min. It is cleared by the liver. The activation of plasminogen by t-PA is accelerated by heparin. The primary side effect is bleeding or hemorrhage.

Streptokinase. Streptokinase is a single-chain protein containing 415 amino acids, mol wt of 45,000 to 50,000 (261,284–286). It is produced by β-hemolytic streptococci and is not an enzyme *per se*. Only after streptokinase combines with plasminogen on a 1:1 basis to form a streptokinase–plasminogen complex, to expose the active site of the plasminogen portion, does it become an active species. This complex then becomes a potent enzyme activator and converts plasminogen to plasmin. The complex has a half-life of about 23 min in the circulation. The usual dose used is 1.5 million units to ensure enough streptokinase is left after being neutralized by streptokinase antibodies and inhibitors in the blood stream. Thrombolytic activity lasts for 3 to 4 h. Streptokinase administration also causes fibrinolysis and fibrinogenolysis, through activation of plasma plasminogen, in the circulation and depletes plasma fibrinogen.

The mortality is usually reduced from 12% in the control group to 9–10% in the streptokinase group. Side effects are bleeding and hemorrhage, fever, and in rare occasions, anaphylaxis.

Urokinase. Urokinase is a direct plasminogen activator that is produced in the kidney and endothelial cells (259,261,263,264,266). It was first isolated from human urine, but now it is produced by cultured human fetal kidney cells. It is a double chain of 411 amino acids, mol wt 54,000–57,000, having a half-life of 16 min. It is metabolized in the liver. Urokinase is given at 4500 units per kilogram in a loading dose and then at 4500 units per kilogram per hour by a continuous iv infusion. Urokinase converts plasminogen to plasmin with greater affinity for the fibrin-bound plasminogen. However, during therapy of AMI, the circulating plasminogen is also activated which catalyzes the breakdown of fibrin and also fibrinogen, causing a generalized (systemic) lytic state. Hemorrhage is the most serious side effect.

Acylated (Anisolyated) Plasminogen–Streptokinase Activator Complex (APSAC). After intravenous administration streptokinase becomes active only after forming a complex with plasminogen. This activated form is rapidly inactivated in the plasma and also causes generalized (systemic) fibrinogenolysis. Therefore, most of the active complex cannot reach the arterial thrombus in the coronary circulation. To overcome this, the active site (catalytic center) of the preformed streptokinase–plasminogen complex is protected or blocked by an acyl group and APSAC is formed (259,261,263,264,266,287–289). However, the fibrin-binding sites are not changed, therefore APSAC binds rapidly to the fibrin clot. So the local dissolution of the clot is preferentially obtained. Deacylation occurs slowly in plasma.

In acylated form, the APSAC offers a prolonged half-life (105 min), greater lysis potency, and less systemic lytic state. APSAC is used at 30 U (30 mg) by iv injection.

Prourokinase. The single-chain urokinase-type plasminogen activator (scu-PA) urokinase does not have high fibrin specificity and therefore causes systemic fibrinogenolysis. In the early 1980s, a single-chain precursor of urokinase (prourokinase) was isolated and purified in human urine and cultured cell lines. It is produced using recombinant DNA biotechnology techniques. It was found that prourokinase was much more fibrin-specific than urokinase and is not effective in activating plasma plasminogen. Its mechanism of action was distinctly different from that of t-PA.

Prourokinase is a single-chain protein containing 411 amino acids (261,265,274,275). In clinical uses scu-PA does not bind to fibrin only and its use causes a decreased plasma fibrinogen of 80%. Its half-life in the circulation is 5 min. It is cleared by the liver. It is used at 40–70 mg over 1 h and heparin is needed simultaneously. Fibrin specificity and thrombolytic efficacy are similar to that of t-PA.

ECONOMIC ASPECTS

The cardiovascular agents are the fastest growing segment of the world pharmaceutical market. In 1990, the world ethical pharmaceutical market was $77.5 billion. The overall growth rate between 1978–1988 was 18% per annum but was expected to slow 10% per annum between 1988–1993. Despite the projected slowing of the total market, cardiovascular agents are expected to continue to make substantial progress in all world markets. Estimated worldwide sales of cardiovascular agents for 1993 are greater than $38 billion. U.S. sales alone are expected to be in excess of $10 billion. The estimated 1993 worldwide sales of cardiovas-

Table 8. 1993 Worldwide Sales of Cardiovascular (CV) Agents[a]

Therapeutic area	Sales, $\$ \times 10^9$	CV market share, %
hypertension	~12.6	34.6
angina	~5.2	14.1
congestive heart failure	~6.2	17.0
other cardiovascular diseases, eg, arrhythmias	~12.1	33.0

[a]Estimated.

cular agents for the treatment of specific cardiovascular diseases are summarized in Table 8 (290,291).

The projected 1993 U.S. sales of various classes of cardiovascular agents are summarized in Table 9 (290,291). Estimates for worldwide sales of antihypertensives alone in 1992 are approximately $10 billion. Thus the number of cardiovascular agents in development is numerous, eg, >45 antiarrhythmics; however, the probability of many becoming therapeutic agents is extremely small.

Table 9. 1993 U.S. Sales of Cardiovascular (CV) Agents of Various Classes of Cardiovascular Agents[a]

Class of CV agent	Sales, $\$ \times 10^{9a}$
antihypertensives	~6.2
antiatherosclerotics	~1.8
thrombolytics	~0.6
nitrates, antiplatelet agents	~0.7
antiarrhythmics	~0.4
other agents	~0.5

[a]Estimated.

BIBLIOGRAPHY

"Cardiovascular Agents" in *ECT* 1st ed., Vol. 3, pp. 211–224, by W. Modell, Cornell University Medical College; in *ECT* 2nd ed., Vol. 4, pp. 510–524, by W. Modell, Cornell University Medical College; in *ECT* 3rd ed., Vol. 4, pp. 872–930, by L. Goldman, Lederle Laboratories, American Cyanamid Co.; in *ECT* 3rd ed., Suppl. Vol. pp. 172–190, by P. Cervoni and F. Lai, Lederle Laboratories, American Cyanamid Co.

1. B. R. Lucchesi, in M. Antonacchio, ed., *Cardiovascular Pharmacology*, 3rd ed., Raven Press, N.Y., 1990, p. 369.
2. J. T. Bigger and B. F. Hoffman, in A. G. Gilman and co-eds., *Goodman and Gilman's, The Pharmacologic Basis of Therapeutics*, 8th ed., Pergamon Press, N.Y., 1990, p. 840.
3. D. T. Mason and co-workers in *Cardiovascular Drugs: Antiarrhythmic, Antihypertensive and Lipid Lowering Drugs*, Vol. 1, ADIS Press, Sydney, Australia, 1977, p. 75.
4. J. Ross, Jr. and co-workers, in J. B. West, ed., *Best and Taylor's Physiological Basis of Medical Practice: Cardiovascular System, Section 2*, 12th ed., Williams and Wilkins, Baltimore, Md., 1990, p. 109.

5. E. M. Vaughn Williams, in E. Sandoe and co-eds., *Symposium on Cardiac Arrhythmias*, Elsinore, Denmark, 1979; A. B. Astra, Södertälje, Sweden, 1979, p. 449.
6. A. L. Bassett and B. F. Hoffman, *Ann. Rev. Pharmacol.* **11,** 143 (1971).
7. B. N. Singh and O. Hauswirth, *Am. Heart J.* **87,** 367 (1974).
8. D. C. Harrison and co-workers, in D. C. Harrison and G. K. Hall, eds., *Cardiac Arrhythmias: A Decade of Progress*, 1981, p. 315.
9. E. M. Vaughn Williams, *J. Clin. Pharmacol.* **24,** 129 (1984).
10. L. S. Ghettes, *Mod. Concepts Cardiovasc. Dis.* **XLVIII,** 13 (1979).
11. K. R. Courtney, *Cardiovasc. Drugs Ther.* **1,** 117 (1987).
12. W. Frey, *Wien Klin. Wochenschr.* **55,** 849 (1918).
13. L. M. Mark and co-workers, *J. Pharmacol. Exp. Ther.* **102,** 5 (1951).
14. J. T. Bigger and R. H. Heissenbuttel, *Progr. Cardiovasc. Dis.* **11,** 515 (1969).
15. J. Frieden, *Am. Heart J.* **70,** 713 (1965).
16. B. Lown and M. Wolf, *Circulation* **44,** 130 (1971).
17. B. Bergdahl and co-workers, *Acta Med. Scand.* **204,** 311 (1978).
18. T. J. Putnam and H. H. Merritt, *Science* **85,** 525 (1937).
19. A. S. Harris and R. H. Kohernat, *Am. J. Physiol.* **136,** 505 (1950).
20. R. H. Helfant and co-workers, *Am. Heart J.* **77,** 315 (1969).
21. J. Morganroth and co-workers, *Am. Heart J.* **98,** 621 (1979).
22. A. Pottage, *Am. J. Cardiol.* **52,** 24C (1983).
23. R. A. Allan-Narker and co-workers, *Am. J. Cardiol.* **18,** 725 (1984).
24. B. Holmes and co-workers, *Drugs* **26,** 93 (1983).
25. B. Chesnie and co-workers, *Am. J. Cardiol.* **52,** 495 (1983).
26. J. L. Anderson and co-workers, *N. Eng. J. Med.* **305,** 473 (1981).
27. H. J. Duff and co-workers, *Am. J. Cardiol.* **48,** 1133 (1981).
28. S. Soon and co-workers, *Am. J. Cardiol.* **48,** 80 (1981).
29. C. R. Webb and co-workers, *J. Am. Coll. Cardiol.* **8,** 214 (1986).
30. B. Olsson and N. Evardson, *Am. Heart J.* **102,** 864 (1981).
31. J. E. Poole and co-workers, *Am. J. Cardiol.* **52,** 240 (1983).
32. E. L. Michelson and L. S. Dreifus, in W. H. Frishman and E. L. Michaelson, eds., *Medical Clinics of N.A.: Cardiovascular Pharmacotherapies II*, W.B. Saunders, Co. Philadelphia, Pa., 1988, p. 276.
33. J. Morganroth and co-workers, *J. Clin. Pharmacol.* **24,** 414 (1984).
34. B. Reichardt and co-workers, *Naunyn-Schmiedeberg Arch. Pharmacol.* **341,** 462 (1990).
35. L. K. Tolvenen and co-workers, *Brit. Heart J.* **55,** 176 (1985).
36. J. Tamargo and E. Delpón, *J. Cardiovasc. Pharmacol.* **16**(Suppl. 5), S10 (1990).
37. Norwegian Multicenter Study Group, *N. Engl. J. Med.* **304,** 801 (1981).
38. β-Blocker Heart Attack Trial Research Group, *JAMA* **247,** 1704 (1981).
39. V. Hansteen and co-workers, *Brit. Med. J.* **284,** 155 (1982).
40. G. Haeusler, *J. Cardiovasc. Pharmacol.* **16**(Suppl. 5), S1 (1990).
41. W. H. Frishman and co-workers, in Ref. 32, p. 359.
42. G. E. Sander and T. E. Giles, in A. Scriabine, ed., *New Cardiovascular Drugs, 1986*, Raven Press, New York, 1986, p. 19.
43. J. V. Okopski, *J. Clin. Pharm. Ther.* **12**, 369 (1988).
44. R. Charlier and co-workers, *Arch. Int. Pharmacodyn.* **139,** 255 (1962).
45. G. Deltour and co-workers, *Arch. Int. Pharmacodyn.* **139,** 247 (1962).
46. B. N. Singh and E. M. Vaughn Williams, *Brit. J. Pharmacol.* **39,** 657 (1970).
47. M. B. Rosenbaum and co-workers, *Am. Heart J.* **106,** 57 (1983).
48. G. V. Naccarrelli and co-workers, *Pharmacotherapy* **5,** 298 (1985).
49. J. W. Mason and co-workers, *Circ. Res.* **55,** 277 (1984).
50. S. Nattiel and co-workers, *Circulation* **76,** 442 (1987).

51. P. Polster and J. Broekhuysen, *Biochem. Pharmacol.* **25,** 131 (1976).
52. A. L. A. Boura and co-workers, *Lancet ii*, 17 (1959).
53. P. E. Leveque, *Nature (London)* **207,** 203 (1965).
54. M. B. Bacaner, *Am. J. Cardiol.* **17,** 528 (1966).
55. *Ibid.*, **21,** 530 (1968).
56. R. W. Dhurandhar and co-workers, *Can. Med. Assoc. J.* **105,** 161 (1971).
57. R. W. Dhurandhar and co-workers, *Heart Lung* **9,** 265 (1980).
58. P. N. Lish and co-workers, *J. Pharmacol. Exp. Ther.* **149,** 161 (1965).
59. J. R. Blinks, *Ann. N.Y. Acad. Sci.* **139,** 673 (1967).
60. M. J. Lewis and co-workers, *Eur. J. Pharmacol.* **86,** 171 (1983).
61. G. Aberg, and co-workers, *Life Sci.* **8,** 353 (1969).
62. H. C. Strauss and co-workers, *Circ. Res.* **26,** 661 (1970).
63. R. Kato and co-workers, *J. Am. Coll. Cardiol.* **7,** 116 (1986).
64. A. Kaumann and J. R. Blinks, *Naunyn-Schmiedeberg. Arch. Pharmcol.* **311,** 305 (1980).
65. J. Rizos and co-workers, *Am. J. Cardiol.* **53,** 1022 (1989).
66. J. L. Anderson and co-workers, *J. Am. Coll. Cardiol.* **8,** 752 (1986).
67. K. Nademanee and co-workers, *Circulation* **72,** 555 (1985).
68. J. Senges and co-workers, *Circulation* **69,** 577 (1984).
69. M. Steinberg and co-workers in A. Scriabine, ed., *New Drugs Annual: Cardiovascular Drugs*, Vol. 2, Raven Press, N.Y., 1984, p. 103.
70. P. M. Vanhoutte, *Am. J. Cardiol.* **59,** 3A (1987).
71. L. Schmaroth and co-workers, *Pflügers Arch.* **396,** 79 (1983).
72. D. Feigel and M. Ravid, *J. Electrocardiogr.* **12,** 129 (1979).
73. R. A. Kates and co-workers, *Clin. Pharmacol. Ther.* **30,** 44 (1981).
74. J. S. Millar and E. M. Vaughn Williams, *Lancet i*, 291 (1981).
75. W. Kobinger and co-workers, *Naunyn-Schmiedeberg Arch. Pharmacol.* **306,** 255 (1979).
76. W. Reiterer and B. Staneck, *Acta Med. Aust.* **8,** 182 (1980).
77. B. Staneck and co-workers, *Eur. J. Clin. Pharmacol.* **24,** 31 (1983).
78. H. Löllgen and co-workers, *Z. Kardiol.* **70,** 425 (1981).
79. W. Reiterer, *Acta Med. Aust.* **8,** 182 (1981).
80. J. Meinertz and co-workers, *Z. Kardiol.* **69,** 720 (1980).
81. J. Meinertz and co-workers, *Circulation* **64**(Suppl. IV), 294 (1980).
82. M. L. Simoons and P. G. Hugenholtz, *Am. J. Cardiol.* **49,** 980 (1982).
83. P. Taylor, in Ref. 2, p. 122.
84. P. Taylor, in Ref. 2, p. 131.
85. J. Heller Brown, in Ref. 2, p. 150.
86. B. B. Hoffman and R. J. Lefkowitz, in Ref. 2, p. 150.
87. B. F. Hoffman and T. J. Bigger, Jr., in Ref. 2, p. 814.
88. F. Späh, in Ref. 42, p. 129.
89. J. S. Millar and E. M. Vaughn Williams, *Brit. J. Pharmacol.* **75,** 469 (1982).
90. K. F. Browne and co-workers, *J. Am. Coll. Cardiol.* **3,** 857 (1984).
91. J. B. Kostos and co-workers, *J. Am. Coll. Cardiol.* **4,** 372 (1984).
92. D. S. Miura and co-workers, *Am. Heart J.* **109,** 827 (1985).
93. K. H. Dangman and D. S. Miura, in Ref. 42, p. 107.
94. B. G. Montgomery, *JAMA* **241,** 1771 (1979).
95. J. D. Minardo and co-workers, *N. Engl. J. Med.* **319,** 257 (1988).
96. The Cardiac Arrhythmia Pilot Study (CAPS) Investigators, *Am. J. Cardiol.* **61,** 501 (1988).
97. The Cardiac Arrhythmia Suppression Trial (CAST) Investigators, *N. Engl. J. Med.* **321,** 406 (1989).

98. J. K. Michelson and co-workers, in Ref. 1, p. 293.
99. F. Murad, in Ref. 2, p. 764.
100. J. Lam and co-workers, *J. Am. Coll. Cardiol.* **6,** 447 (1985).
101. P. A. Crean and co-workers, *Am. Heart J.* **113,** 261 (1987).
102. U. Brügman and co-workers, *Abstract, Eur. Heart J.* **1**(Suppl. 5), 274 (1984).
103. L. M. Lopez and co-workers, *Am. Heart J.* **110,** 991 (1985).
104. F. Pasanisi and co-workers, *Eur. J. Clin. Pharmacol.* **29,** 21 (1985).
105. M. Packer and co-workers, *Circulation,* **75**(Suppl.), 556 (1987).
106. J. L. Faulkner, *Brit. J. Clin. Pharmacol.* **22,** 21 (1986).
107. Bepridil, *Lancet, ii,* 278 (1988).
108. *Physician's Desk Reference,* 45th ed, Medical Economics Co., Oradell, N.J., 1991.
109. R. Gorlin, *Cardiovasc. Rev. Rep.* **4,** 767 (1983).
110. J. N. Cohn and co-workers, *N. Engl. J. Med.* **314,** 1547 (1986).
111. M. R. Sanders, J. B. Kostis, and W. H. Frishman, in W. H. Frishman and S. Charlap, eds., *Medical Clinics of N.A.: Cardiovascular Pharmacotherapy III*, W. B. Saunders, Philadelphia, Pa., 1989, p. 293.
112. W. Withering, "An Account of the Foxglove and Some of Its Medicinal Uses," C. G. J. Robinson and F. Robinson, in F. A. Williams and T. E. Keys, eds., *Classics of Cardiology*, Shuman, New York, 1941, p. 231.
113. R. A. Kelly, *Am. J. Cardiol.* **65,** 10E (1990).
114. B. F. Hoffman and J. T. Bigger, in, A. G. Gilman, L. S. Goodman, T. W. Rall, and F. Murad, eds., *The Pharmacologic Basis of Therapeutics*, 7th ed., Macmillan, New York, 1985, pp. 716–747.
115. A. J. Taggart and co-workers, *J. Cardiovasc. Pharmacol.* **5,** 229 (1983).
116. L. Fleg and co-workers, *Am. J. Med.* **73,** 244 (1982).
117. D. C. Lee and co-workers, *N. Engl. J. Med.* **306,** 699 (1982).
118. T. Smith, *N. Engl. J. Med.* **318,** 358 (1988).
119. M. R. Rosen and co-workers, *Am. Heart J.* **89,** 391 (1975).
120. T. W. Smith, E. B. Braunwald, and R. A. Kelly, in E. B. Braunwald, ed., *Heart Disease: A Textbook of Cardiovascular Medicine*, W.B. Saunders, Philadelphia, Pa., 1988, p. 485.
121. E. H. Sonnenblick and co-workers, *N. Engl. J. Med.* **300,** 17 (1979).
122. M. R. Bristow and co-workers, *N. Engl. J. Med.* **307,** 205 (1982).
123. T. C. Majerus and co-workers, *Pharmacotherapy* **9,** 245 (1989).
124. R. R. Ruffolo and co-workers, *J. Pharmacol. Exp. Ther.* **224,** 46 (1983).
125. C. V. Leirer and co-workers, *Circulation* **56,** 468 (1972).
126. M. C. Bogaert and N. Fraeyman, *Am. J. Med.* **90**(Suppl. 5B), 5B-105 (May 1991).
127. M. R. Bristow and co-workers, *N. Engl. J. Med.* **307,** 205 (1982).
128. R. W. F. Campbell, *Brit. J. Clin. Pharmacol.* **28,** 595 (1989).
129. H. F. Marlow, *Brit. J. Clin. Pharmacol.* **28,** 235 (1989).
130. P. Honerjager, *Eur. Heart J.* **10**(Suppl. C), 25 (1989).
131. A. A. Alousi and co-workers, *Circulation* **73,** 10 (1982).
132. R. J. Axelrod and co-workers, *J. Am. Coll. Cardiol.* **9,** 1124 (1987).
133. H. C. Herrmann and co-workers, *J. Am. Coll. Cardiol.* **9,** 1117 (1987).
134. C. S. Maskin and co-workers, *Am. J. Med.* **73,** 113 (1982).
135. M. Packer and co-workers, *Circulation* **70,** 1038 (1984).
136. A. M. Katz, *Circulation* **73,** 184 (1986).
137. J. Klugeret and co-workers, *Am. J. Cardiol.* **49,** 1667 (1982).
138. M. Packer and co-workers, *Circulation* **68,** 803 (1983).
139. P. Amadio and co-workers, *Post. Med.* **87,** 223 (1990).
140. B. Magnani and co-workers, *Post. Med.* **62**(Suppl. I), 153 (1986).
141. The Consensus Trial Study Group, *N. Engl. J. Med.* **316,** 1429 (1987).

142. I. G. Grozier and co-workers, *Clin. Exp. Pharmacol. Physiol.* **15,** 173 (1988).
143. J. M. Heim and co-workers, *Klin. Wochenschr.* **68,** 709 (1990).
144. E. B. Smith and co-workers, *J. Atheroscl. Res.* **7,** 171 (1967).
145. NIH Consensus Conference, *JAMA* **253,** 2080 (1985).
146. NIH Expert Panel, *Arch. Intern. Med.* **148,** 36 (1988).
147. A. W. Albertset and co-workers, *Proc. Natl. Acad. Sci.* **77,** 3957 (1980).
148. W. H. Frishman and R. C. Rapier, in Ref. 114, p. 440.
149. J. A. Tobert and co-workers, *Atherosclerosis* **41,** 61 (1982).
150. Lipid Research Clinics Program, *JAMA* **251,** 351 (1984).
151. M. Ast and W. H. Frishman, *J. Clin. Pharmacol.* **30,** 99 (1990).
152. P. T. Kuo and co-workers, *Circulation* **59,** 199 (1979).
153. R. H. Knopp, *Cardiology* **76**(Suppl. 1), 14 (1989).
154. J. P. Monk and P. A. Todd, *Drugs,* **33,** 539 (1987).
155. J. M. Cruickshank and J. C. Smith, *Pharmac. Ther.* **42,** 385 (1989).
156. H. Lederman and B. J. Kaufmann, *Int. Med. Res.* **9,** 516 (1981).
157. J. R. Kaplan, *Eur. Heart J.* **8,** 928 (1987).
158. J. M. Cruickshank, *Circulation* **82**(Suppl. II), 60 (1990).
159. R. Paoletti and co-workers, *Ann. NY Acad. Sci.* **522,** 390 (1988).
160. R. E. Tedeschi and co-workers, *Noure Presse Med.* **30,** 3021 (1980).
161. R. W. Wissler and D. Vesselinovitch, *Appl. Pathol.* **1,** 89 (1983).
162. C. J. Schwartz, *Am. J. Cardiol.* **62,** 1B (1988).
163. *1990 Heart and Stroke Facts*, American Heart Association, 1990, p. 2.
164. J. H. Laragh, in J. Genest and co-eds., *Hypertension: Physiopathology and Treatment*, 2nd ed., McGraw-Hill Book Co., New York, 1983, p. 615.
165. P. S. Chan and P. Cervoni, in M. M. Cohen, ed., *Biological Protection with Prostaglandin*, Vol. 1, CRC Press, Inc., Boca Raton, Fla., 1985, p. 197.
166. P. S. Chan and P. Cervoni, *Drug Dev. Res.* **7,** 341 (1986).
167. T. Masaki and M. Yamagisawa, *Cardiovasc. Drug Rev.* **8,** 373 (1990).
168. R. F. Furchgott and P. M. Vanhoutte, *FASEB J.* **3,** 2007 (1989).
169. T. G. Luscher and co-workers, in J. H. Laragh and B. M. Brenner, eds., *Hypertension: Pathophysiology, Diagnosis and Management*, Raven Press, New York, 1990, p. 637.
170. A. C. Guyton and co-workers, in Ref. 169, p. 1029.
171. P. S. Chan, in S. B. Day, ed., *Integrated Medicine, A Companion to the Life Sciences*, Vol. II, Van Nostrand Reinhold Co., New York, 1981, p. 184.
172. K. H. Berecek and B. H. Swords, *Hypertension* **16,** 213 (1990).
173. P. S. Chan and co-workers, *J. Pharmacol. Exp. Ther.* **226,** 726 (1983).
174. P. S. Chan and co-workers, *Drug Dev. Res.* **18,** 75 (1989).
175. S. A. Atlas and J. H. Laragh, in Ref. 169, p. 861.
176. Veterans Administration Cooperative Study Group, *J. Am. Med. Assoc.* **202,** 116 (1967).
177. Veterans Administration Cooperative Study Group, *Circulation* **45,** 991 (1972).
178. S. W. MacMahon and co-workers, *Prog. Cardiovasc. Dis.* **29**(Suppl. 1), 99 (1986).
179. R. H. Grimm, Jr., *Drugs* **31**(Suppl. 1), 13 (1986).
180. L. Hansson, *Drugs* **36**(Suppl. 3), 110 (1988).
181. N. M. Kaplan, *Am. J. Med.* **90**(2A), 19S (1991).
182. R. Gorlin, *Am. Heart J.* **121,** 658 (1991).
183. N. K. Hollenberg, *Am. J. Med.* **90**(2A), 2S (1991).
184. J. H. Laragh, *Hypertension* **13**(Suppl. 1), 1–103 (1989).
185. F. O. Simpson, *Aust. N. Z. J. Med.* **20,** 726 (1990).
186. R. R. Townsend and O. B. Holland, *Arch. Intern. Med.* **150,** 1175 (1990).
187. M. C. Houston, *Am. Heart J.* **117,** 911 (1989).
188. K. H. Rahn, *Am. J. Cardiol.* **65,** 82G (1990).

189. M. Burnier and co-workers, *Drugs* **39**(Suppl. 1), 32 (1990).
190. L. Hansson, *J. Hypertens.* (Suppl. 8), S13 (1990).
191. F. H. Messerli, *Postgrad. Med.* **88,** 127 (1990).
192. D. Ballantyne, *J. Hum. Hypertens.* **4**(Suppl. 2), 35 (1990).
193. P. Leren, *Clin. Exp. Hypertens.* [*A*] **12,** 761 (1990).
194. M. H. Weinberger, *Cardiovasc. Drugs Ther.* **4**(Suppl. 2), 379 (1990).
195. R. B. Taylor, *Am. Fam. Physician.* **42**(Suppl. 5), 29S (1990).
196. L. Hansson, *Am. J. Hypertens.* **4,** 84S (1991).
197. A. Breckenridge, *Am. J. Hypertens* **4,** 79S (1991).
198. R. Gorlin, *Am. Heart J.* **121,** 670 (1991).
199. H. R. Black, *Am. Heart J.* **121**(2(Pt. 2)), 707 (1991).
200. P. S. Chan and P. Cervoni, in P. B. Goldberg and J. Roberts, eds., *CRC Handbook on Pharmacology of Aging*, CRC Press, Inc., Boca Raton, Fla., 1983, p. 51.
201. J. G. Gerber and A. S. Nies, in Ref. 2, p. 784.
202. *Scrip, Hypertension, Therapy, Research and Market Opportunities*, PJB Publications Ltd., Surrey, UK, 1989, p. 1.
203. M. J. Antonaccio and J. J. Wright, in Ref. 1, p. 201.
204. H. Gavras, *Circulation* **81,** 381 (1990).
205. D. B. Case, E. H. Sonnenblick, and J. H. Laragh, eds., *Captopril and Hypertension*, Plenum Publishing Corp., New York, 1980, pp. 3–230.
206. A. A. Patchett and co-workers, *Nature* **288,** 280 (1980).
207. J. H. Laragh, *Am. J. Hypertens* **3,** 257S (1990).
208. A. E. Fletcher and C. J. Bulpitt, *J. Hum. Hypertens.* **4** (Suppl. 4), 45 (1990).
209. L. E. Rarnsay and W. W. Yeo, *J. Hum. Hypertens.* **4** (Suppl. 4), 1 (1990).
210. J. H. Bauer, *Am. J. Hypertens.* **3,** 331 (1990).
211. H. R. Brunner, *Hosp. Pract.* **25,** 71 (1990).
212. D. McAreavey and J. I. S. Robertson, *Drugs* **40,** 326 (1990).
213. B. Waeber, J. Nussberger, and H. R. Brunner, in Ref. 169, p. 2209.
214. P. S. Chan, M. A. Ronsberg, and P. Cervoni, *Fed. Proc.* **43,** 1346 (1984).
215. P. M. Vanhoutte and co-workers, *Brit. J. Clin. Pharmacol.* **28,** 95S (1989).
216. M. R. Ujhelyi, R. K. Ferguson, and P. H. Vlasses, *Pharmacotherapy* **9,** 351 (1989).
217. U.S. Pat. 4,355,040 (1982), Y. Furukawa, S. Kishimoto, and K. Nishikawa (to Takeda Chem. Industries).
218. P. C. Wong and co-workers, *J. Pharmacol. Exp. Ther.* **247,** 1 (1988).
219. A. T. Chiu and co-workers, *Eur. J. Pharmacol.* **157,** 13 (1988).
220. J. V. Duncia and co-workers, *J. Med. Chem.* **33,** 1312 (1990).
221. D. J. Carini and co-workers, *J. Med. Chem.* **33,** 1330 (1990).
222. R. S. L. Chang and co-workers, *FASEB J.* **5,** A1575, A1576 (1991).
223. E. Haber, *J. Hypertens.* **7**(Suppl. 2), S81 (1989).
224. R. S. Boger and co-workers *Hypertension* **15,** 835 (1990).
225. H. D. Kleinert and co-workers, *Cardiovasc. Drug Rev.* **8,** 203 (1990).
226. M. J. Ryan and co-workers, *FASEB J.* **5,** A1575 (1991).
227. P. A. van Zwieten, *Cardiovasc. Drug Ther.* **3,** 121 (1989).
228. P. A. van Zwieten, in Ref. 1, p. 37.
229. S. H. Taylor, in F. Messerli, ed., *Cardiovasc. Drug Therapy*, W. B. Saunders, Philadelphia, 1990, p. 714.
230. A. H. Maslowski, *Am. Heart J.* **121** (1(Pt 2)), 323 (1991).
231. L. Hansson, in Ref. 229, p. 408.
232. P. Bolli, P. G. Fernandez, and F. R. Buhler, in Ref. 169, p. 2181.
233. J. Conway and A. Bilski, in D. Ganten and P. J. Mulrow, eds., *Pharmacology of Antihypertensive Therapeutics (Handbook Exp. Pharmacology)*, Vol. 93, Springer-Verlag, Berlin, Germany, 1990, p. 65.

234. G. Sponer, W. Bartsch, and R. G. Hooper, in Ref. 237, p. 131.
235. W. Flamenbaum and A. Dubrow, in Ref. 235, p. 573.
236. T. D. Giles, *Cardiovasc. Drug Rev.* **8,** 138 (1990).
237. K. E. Kim, *Am. Fam. Physician* **43,** 583 (1991).
238. C. V. Ram, *Am. J. Cardiol.* **66,** 291 (1990).
239. W. W. Parmley, *Am. J. Cardiol.* **66,** 161 (1990).
240. R. W. Gifford, in Ref. 235, p. 298.
241. W. M. Kirkendall, *Arch. Intern. Med.* **151,** 398 (1991).
242. H. G. Langford and co-workers, *J. Hum. Hyperten.* **4,** 491 (1990).
243. G. B. Weiss, R. J. Winquist, and P. J. Silver, in Ref. 1, p. 75.
244. E. P. MacCarthy, *Cardiovasc. Drug Rev.* **8,** 155 (1990).
245. N. S. Cook and U. Quast, in N. S. Cook, ed., *Potassium Channels*, Ellis Horwood, Chichester, UK, 1990, p. 181.
246. G. Edwards and A. H. Weston, *Trends Pharmacol. Sci.* **11,** 417 (1990).
247. C. Richer and co-workers, *Life Sci.* **47,** 1693 (1990).
248. A. H. Weston and co-workers, *Blood Vessels,* 306 (1990).
249. U. Quast and N. S. Cook, *Trends Pharmacol. Sci.* **10,** 431 (1989).
250. M. Kinoshita and K. Sakai, *Cardiovasc. Drug Ther.* **4,** 1075 (1990).
251. Symposiun on Pinacidil, *Drugs* **36**(Suppl. 7), 1 (1988).
252. H. A. Friedel and R. N. Brogden, *Drugs* **39,** 929 (1990).
253. M. R. Goldberg, *J. Cardiovasc. Pharmacol.* **12**(Suppl. 2), 541 (1988).
254. A. H. Weston, in Ref. 237, p. 643.
255. M. A. DeWood and co-workers, *N. Engl. J. Med.* **303,** 897 (1980).
256. V. Fuster and co-workers, *Circulation* **77,** 1213 (1988).
257. B. Stein and V. Fuster, *Cardiovasc. Drug Ther.* **3,** 797 (1989).
258. R. M. Gunnar and co-workers, *J. Am. Coll. Cardiol.* **16,** 249 (1990).
259. D. C. Collen and H. K. Gold, *Thrombosis Res. Suppl. X*, 105 (1990).
260. C. J. Pepine, *Am. J. Cardiol.* **64,** 2B (1989).
261. S. Sherry, *Cardiovasc. Drug Rev.* **6,** 1 (1988).
262. R. C. Becker and R. Harrington, *Arn. Heart J.* **121** (1(Pt. 1)), 220 (1991), and **121** (2(Pt. 1)), 627 (1991).
263. J. D. Rutherford and E. Braunwald, *Chest.* **97**(Suppl.), 1375 (1990).
264. D. Collen, *J. Cell Biochem.* **33,** 77 (1987).
265. R. W. Holden, *Radiology* **174,** 993 (1990).
266. R. Fears, *Pharmacol. Rev.* **42,** 201 (1990).
267. P. Sleight, *Eur. Heart J.* **11**(Suppl. F), 1 (1990).
268. K. G. Ro and H. V. Anderson, *Postgrad. Med. J.* **88,** 79 (1990).
269. T. Yasuda and H. K. Gold, *Thrombosis Res. Suppl. X* (1990).
270. A. J. Tiefenbrunn and B. E. Sobel, *Fibrinolysis* **3**, 1 (1989).
271. P. Oldershaw, *Postgrad. Med. J.* **64,** 915 (1988).
272. C. L. Grines and A. N. DeMaria, *J. Am. Coll. Cardiol.* **16,** 223 (1990).
273. J. H. Chesebro, L. Badimon, and V. Fuster, *Am. J. Cardiol.* **65,** 12C (1990).
274. J. Loscalzo, *Chest.* **97**(Suppl.), 1175 (1990).
275. G. Di Minno and co-workers, *Pharmacol. Res.* **21,** 153 (1989).
276. M. Samama, *Eur. Heart J.* **11**(Suppl. F), 15 (1990).
277. C. H. Hennekens, *Chest.* **97**(Suppl.), 151S (1990).
278. C. H. Hennekens and co-workers, *Circulation* **80,** 749 (1989).
279. M. W. Webster, J. H. Chesebro, and V. Fuster, *Hematol. Oncol. Clin. N. Am.* **4,** 265 (1990).
280. G. Agnelli, *Chest.* **97**(Suppl.), 161S (1990).
281. S. D. Rogers, L. B. Riemersma, and S. D. Clements, *Pharmacotherapy* **7,** 111 (1987).
282. R. C. Becker and co-workers, *Am. Heart J.* **121,** 220 (1991).

283. M. H. Prins and J. Hirsh, *Am. J. Cardiol.* **67,** 3A (1991).
284. S. Sherry and S. H. Taylor, in Ref. 235, p. 1479.
285. S. Sherry and V. J. Marder, *Ann. Intern. Med.* **114,** 417 (1991).
286. K. L. Goa and co-workers, *Drugs* **39,** 693 (1990).
287. R. Fears, *Sem. Throm. Hemostasis* **15,** 129 (1989).
288. S. J. Crabbe, A. M. Grimm, and L. E. Hopkins, *Pharmacotherapy* **10,** 115 (1990).
289. M. A. Mungr and E. A. Forrence, *Clin. Pharmacy* **2,** 530 (1990).
290. *Prospects*, Healthcare Forecasting, Inc., Issue No. 28, 8-18-2, 1990.
291. *Scrip-Hypertension Therapy, Research and Market Opportunities*, 2nd ed., PJB Publications Ltd., Pharmabooks, Ltd., New York, 1991.

PETER CERVONI
DAVID L. CRANDALL
PETER S. CHAN
Lederle Laboratories, American Cyanamid Company

CELL CULTURE TECHNOLOGY

Cell culture processes, the *in vitro* growth of animal, insect, or plant cells on a large scale to manufacture biochemicals of commercial importance, have been used for some time for the manufacture of viral vaccines (see VACCINE TECHNOLOGY). Significant growth in this technology, primarily because of the advent of recombinant DNA methods (see GENETIC ENGINEERING) for the production of therapeutic proteins (qv) and hybridoma technology for production of monoclonal antibodies (see IMMUNOASSAY) occurred in the late 1980s and early 1990s. The need for cell culture technology stems mainly from the fact that bacteria do not have the capability to perform many of the post-translational modifications that most large proteins require for *in vivo* biological activity. These modifications include intracellular processing steps such as protein folding, disulfide linkages, glycosylation, and carboxylation.

Historically, large-scale cell culture technology traces its roots to the production of viral vaccines and other therapeutically important secreted products of human and primate cells such as those listed in Table 1. Most of these products were made by cells grown either in suspension or attached to microcarriers. Such cells were cultured in stirred tank reactors adapted from bacterial fermentation (qv) processes. The primary challenges in adapting microbial fermenters to the rigors of cell culture technology were reducing the risk of microbial contamination, modifying the agitation system to provide low shear agitation to the shear-sensitive mammalian cells, and providing adequate oxygenation under the low shear conditions mandated by the cells (see AERATION, BIOTECHNOLOGY).

Although the number of cell culture-derived products on the market as of this writing is relatively small, there are several products in clinical trials and many more in development. Hybridomas, which are obtained by fusing a myeloma cell and lymphoid cells from mice immunized with a specific antigen, are used to produce monoclonal antibodies. Monoclonals have already found many uses in the diagnostic industry by virtue of extreme specificity of binding to an antigen of interest (see BIOSENSORS; MEDICAL DIAGNOSTIC REAGENTS). They have also found use for immunopurification and as therapeutics (see BIOPOLYMERS, ANALYTICAL TECHNIQUES; IMMUNOTHERAPEUTIC AGENTS). The quantities of monoclonals needed for diagnostics are relatively small and thus the scale of production is also small. Examples of uses of monoclonals in diagnostics are for

Table 1. Historical Products of Cell Culture Technology

Product	Cell line	Process[a]	Year introduced	Reference
FMD[b] vaccine	baby hamster kidney	SC	1962	1
rabies vaccine	dog kidney	SM	1978	2
interferon	human namalwa	SC	1979	3
polio vaccine	monkey kidney	SM	1980	4

[a]SC = suspension culture; SM = stirred microcarrier.
[b]FMD = foot-and-mouth disease.

detection of polypeptide hormones (qv), eg, human chorionic gonadatropin and human growth hormone; detection of tumor markers, such as alpha-fetoprotein; infectious diseases, such as chlamydia, herpes, and varicella; and cell surface antigens, eg, T-cells and leukemia typing.

Recombinant DNA technology has already provided several products of therapeutic interest from mammalian cells. Table 2 gives examples of products from mammalian cells, the use, and the technology used for production. Technology development for these products has centered around the differences in characteristics of mammalian versus microbial cells, notably, the shear sensitivity and susceptibility to contamination of the mammalian lines.

Although the focus of this article is mainly on mammalian cells, the technologies described herein also apply in principle to insect and plant cells.

Characteristics of Mammalian Cells

Environmental Conditions. Mammalian cells *in vivo* are maintained in a carefully balanced homeostatic environment and thus have evolved to require fairly stringent environmental conditions. These cells differ significantly from bacterial cells in that they lack a rigid cell wall and are hence much more shear sensitive. Many animal cells are also attachment dependent, needing a surface to grow on. Many of the cell culture technologies provide the low shear, high surface area environment needed for the mammalian cells. Another approach, however, is to adapt cells to suspension culture and select cells that are less shear sensitive permitting the use of fermentation technology for the culture of animal cells. The optimum environmental parameters depend on cell type and are specific to cell type. Typical ranges for some of these parameters are listed in Table 3.

Nutritional Requirements. The nutrient requirements of mammalian cells are many, varied, and complex. In addition to typical metabolic requirements such as sugars, amino acids (qv), vitamins (qv), and minerals, cells also need growth factors and other proteins. Some of the proteins are not consumed, but play a catalytic role in the cell growth process. Historically, fetal calf serum of 1–20 vol % of the medium has been used as a rich source of all these complex protein requirements. However, the composition of serum varies from lot to lot, introducing significant variability in manufacture of products from the mammalian cells.

Table 2. Therapeutic Products from Mammalian Cells

Product	Company	Cell[a]	Process[b]	Therapeutic use
tissue plasminogen activator (tPa)	Genentech	CHO	SC	heart attacks
erythropoietin[c]	Amgen	CHO	ARB	anemia
factor VIII	Cutter/Miles	CHO	SC	hemophilia
OKT-3	Ortho	hybridoma	MA	prevention of transplant rejection
centoxin	Centocor	hybridoma	SC	septic shock

[a]CHO = Chinese hamster ovary.
[b]SC = suspension culture; ARB = automated roller bottles; MA = mouse ascites.
[c]CAS Registry Number is [*11096-26-7*].

Table 3. Environmental Parameters for Mammalian Cell Cultivation

Parameter	Range	Typical value
pH	6.6–7.6	7
temperature, °C	33–39	37
dissolved oxygen, Pa[a]	0.7–40	10
osmolarity, mOsm/kg[b]	280–360	300
dissolved CO_2, Pa[a]	0.9–20	7
tolerable shear rate, s^{-1}	0–3000	1500

[a]To convert Pa to mm Hg, multiply by 7.5.
[b]mOsm = milliosmolar or milliosmole where an osmole equals a mole of solute divided by the number of ions formed per molecule of the soluble, ie, one mole of sodium chloride is equivalent to two osmoles of sodium chloride and 1 *M* NaCl = 2 Osm NaCl.

Serum is expensive, the 1992 price was $400/L, and supply depends on cattle supply. Use of this serum also poses significant difficulties in validating processes for absence of viral contamination. Hence, a goal in cell culture technology is to develop serum-free media for cell culture. Much work has gone into developing serum-free media and a sizable portion of cell culture research is devoted to this project. A detailed discussion of serum-free media development is available (5).

Several generic media formulations have been developed for growth and cultivation of mammalian cells and are commercially available. Each contains amino acids, inorganic salts for providing the right osmolarity, essential minerals, and buffering capacity; vitamins, and energy sources such as glucose. Whereas mixtures of the different formulations are often used to optimize growth and productivity of cell lines, many of the basal media need to be supplemented with serum or other appropriate proteins for promoting cell growth. Some of the more commonly used media formulations include: minimum essential medium (MEM) for a broad spectrum of mammalian cells; basal media eagle (BME) for diploid or primary mammalian cells; Dulbecco's modified eagle media (DMEM) for a broad spectrum of cells; CMRL media for Earl's "L" cells and monkey kidney cells; Fischer's media for murine leukemic cells; Iscove's modified Dulbecco's media (IMDM) for rapidly proliferating high density cell cultures; McCoy's media for human lymphocytes; Ham's F10 and F12 for Chinese hamster ovary cells and other mammalian cells; and RPMI 1630/1640 for suspension cells and human leukemic cells. These media are available from Gibco Laboratories (Grand Island, New York), Irvine Scientific (Santa Ana, California), Sigma Chemical Co. (St. Louis, Missouri), Hyclone Laboratories (Logan, Utah), and JRH (Lenexa, Kansas) among others.

Another essential nutrient not supplied with the media is oxygen. The oxygen consumption rate of mammalian cell cultures is much lower than that of bacterial ones because cell densities are much lower than those achieved in bacterial cultures. The oxygen consumption rate varies from cell line to cell line, but the range has been reported to be as wide as 0.05–0.5 mmol/(10^9 cells·h) (6). Hence, designing oxygenation systems for mammalian cells is a function of the cells being used. Use of the worst case scenario may lead to costly overdesign. This is especially so if silicone tube oxygenators are being considered. Direct sparging in the reactor to accomplish oxygen transfer often leads to cell damage unless protective agents, such as pluronic polyols, are used (7). However, as long as the pluronic is nontoxic to the cells and is compatible with downstream processing steps, this use is probably the most efficient route to oxygenating cell culture systems. Most cell lines also require a small amount of dissolved carbon dioxide for growth, especially at low cell densities.

Kinetics of Cell Growth and Product Formation. Mammalian cells grow at a much slower rate than bacterial and yeast cells (see YEASTS). The maximum specific growth rates for mammalian cells range from 0.01 to 0.05 h^{-1}, corresponding to cell doubling times of 14 to 70 hours depending on the cell line and environmental conditions. Most primary cell lines are anchorage dependent and need a surface to grow on. They are also contact inhibited, ie, they stop growing once the surface is confluent. Alternatively, most of the cell lines used industrially for recombinant products and monoclonal antibodies are not attachment dependent. For example Chinese hamster ovary (CHO) cells are commonly used as host cells for recombinant products. These cells are transformed by tumor viruses and do not require a surface to grow on. They do, however, prefer to grow on surfaces and requirement for serum factors diminishes significantly when they are attached to surfaces. Thus CHO cells can be cultured for several weeks in protein-free media if grown on microcarriers, whereas they require serum proteins such as fetuin to grow in suspension. Most hybridomas used for making monoclonal antibodies are attachment independent, grow in suspension, and have minimal requirement for serum proteins. It is necessary to adapt these cells to serum-free media for several days before they start growing well in these media.

Cell growth kinetics of mammalian cells can be described by the typical lag, exponential growth, then stationary and death phases. The exponential phase may be described adequately by a Monod type of kinetic model when the growth rate is much larger than the death rate. At low growth rates, it is necessary to include cell death kinetics to account for the lower viability and to predict the cell viability. Toward the end of the exponential culture, cells are also subject to growth inhibition from metabolic by-products such as lactate and ammonia. Hence, for continuous processes, a comprehensive model should contain terms for cell growth, based on the limiting substrate concentration, cell death, and inhibition kinetics.

Product formation kinetics in mammalian cells has been studied extensively for hybridomas. Most monoclonal antibodies are produced at an enhanced rate during the G_0 phase of the cell cycle (8–10). A model for antibody production based on this cell cycle dependence and traditional Monod kinetics for cell growth has been proposed (11). However, it is not clear if this cell cycle dependence carries over to recombinant CHO cells. In fact it has been reported that dihydrofolate reductase, the gene for which is co-amplified with the gene for the recombinant protein in CHO cells, synthesis is associated with the *S* phase of the cell cycle (12). Hence it is possible that the product formation kinetics in recombinant CHO cells is different from that of hybridomas.

Cell Culture Processes

A wide variety of mammalian cells are used in industrial practice and to accommodate the diversity, a variety of cell culture technologies have evolved depending on the cell line and product characteristics.

Batch Suspension Culture. The batch suspension culture is perhaps the simplest technology available. It is adapted from traditional bacterial fermentation (qv) technology by changing the impellers to low shear marine propeller type, thus reducing the shear forces to which cells are subjected. Oxygenation is achieved either by direct sparging, if the cells are not subject to damage by bursting bubbles or are protected by surface-active agents such as pluronic polyols, or by membrane oxygenation where gas-permeable tubing is inserted into the fermentor.

Most hybridomas can be grown in batch suspension culture. Recombinant CHO cells can also be adapted for growth in suspension. However, CHO cells often require serum or expensive serum proteins to grow in this manner. The applicability of this process hinges on whether an inexpensive serum-free medium can be developed for the cells in question and whether the proteins used for serum-free media development, eg, bovine serum albumin [*9048-46-8*] can be effectively separated from the product during downstream processing. Viral vaccines are often produced in suspension culture reactors because the product is isolated from the cells and presence of serum is not a hindrance to purification. For example, FMD vaccine is produced in 3000-L batch suspension culture reactors (13). Genentech produces tPA in 10,000-L batch fermenters using recombinant CHO cells. It is believed that these cells are grown to high densities in serum-containing media, then shifted to a low protein production medium for harvesting the secreted product. Batch suspension culture is also used for large-scale production of monoclonal antibodies. Celltech has scaled up airlift suspension culture reactors to 2000-L scale for the production of monoclonals (14).

A batch suspension culture reactor consists of a stirred tank typically having a height to diameter ratio of 1:1 to 3:1, fitted with a low shear impeller. The tank has a hemispherical bottom to avoid stagnant zones, because agitation level is very low, and the agitator shaft is either magnetically coupled to the motor or is coupled with a double mechanical seal to protect the culture from contaminating microorganisms. The vessel is pressure rated for in-place steam sterilization. Medium is filtered into the fermentor through a 0.1 micrometer absolute filter. Inoculum is added via a steam sterilized connection from the inoculum fermentor. The seed fermentor is typically one-tenth the size of the production fermentor. Hence, for large-scale production purposes, a long inoculum train is required. Oxygenation may be effected by either direct sparging into the fementor or by using a coil of gas permeable tubing, either silicone or microporous Teflon tubing. The vessel is fitted with temperature, dissolved oxygen, and pH probes. pH is controlled by addition of carbon dioxide (qv) to lower the pH and a suitable base, eg, dilute NaOH or $NaHCO_3$, to raise it. Dissolved oxygen is controlled by addition of air and/or pure oxygen. Insulated jackets are used for controlling the temperature. The vessel is manufactured from 316 L stainless steel and polished to a high degree, typically 240 grit followed by electropolishing, for ease of cleaning. Other materials used for seals, etc, are restricted to medical-grade silicone, Teflon, Viton, and borosilicate glass to ensure that toxic materials do not leach into the culture and affect the cells.

Figure 1 shows a schematic of a typical cell culture manufacturing facility using batch suspension culture. The downstream processing of harvest from the fermentor usually consists of a clarification step, either a centrifuge or a microfilter, followed by a concentration/diafiltration step using a tangential flow ultrafiltration membrane of an appropriate molecular weight cutoff. The concentrated

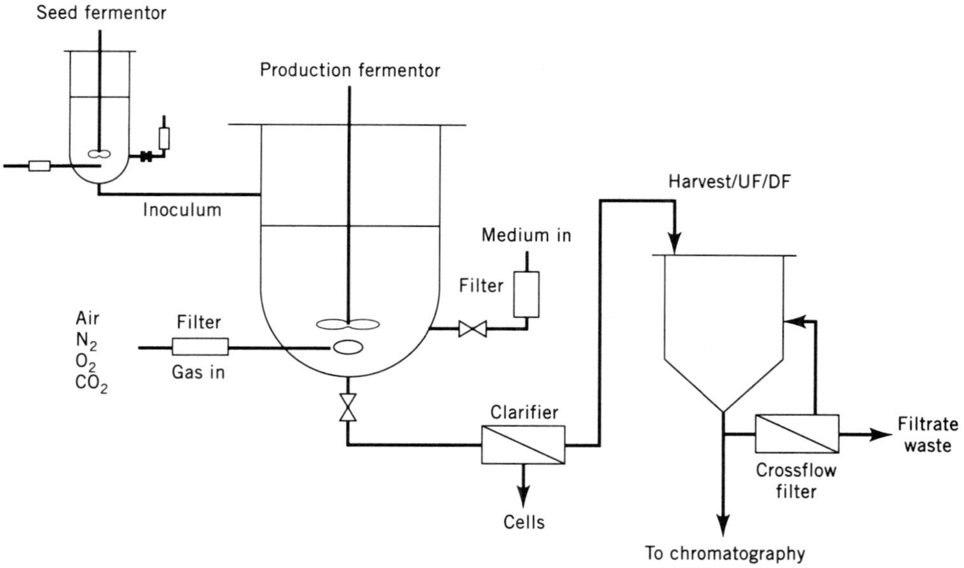

Fig. 1. Schematic of a process for batch suspension culture of mammalian cells, where UF is ultrafiltration and DF is diafiltration.

protein is further purified by a series of chromatography (qv) steps. The downstream processing steps are usually similar regardless of the cell culture technology being used.

Batch suspension culture has many inherent advantages. It is relatively simple to operate and scale up and less susceptible to microbial contamination. The homogeneous nature of the process makes process control (qv) and optimization easier, and from a regulatory point of view, it is the easiest to define and validate. This is therefore the process of choice if the cells can be grown in suspension in a relatively inexpensive medium. A disadvantage of the process is that labile products may degrade during the long batch periods, especially because of the presence of proteases released by lysed cells toward the end of the process. For such products, a continuous process having low reactor residence times may be more suitable.

Batch Microcarrier Process. For attachment dependent cells, the batch microcarrier process is the equivalent of the batch suspension process except for the fact that cells are attached to microcarriers. Cells attached on the surface of microcarriers are far more sensitive to shear forces than suspension cells because microcarriers are much larger than suspended cells. The damage to cells is theorized to be caused primarily by turbulent eddies when the eddy size becomes smaller than the particle size of microcarriers. Another mechanism for cell damage, especially at high bead concentration, is the bead to bead collision frequency. These mechanisms of cell damage in microcarrier cultures are discussed in the literature (15,16). Because of this limitation on agitation power input, agitator design is of great importance in microcarrier reactors. High efficiency impellers, which maximize flow and minimize shear, are utilized. The shear effects mentioned here are applicable to solid microcarriers where the cells are attached to the surface. Most recently, some macroporous carriers have been introduced that prevent the shear damage by providing attachment surfaces on the internal pores of the carriers. However, the attachment rate on such microcarriers is slower than that on solid microcarriers. This makes the porous microcarriers more suitable for long-term perfusion cultures.

Many microcarriers are now available commercially for mammalian cell culture. The choice of microcarrier depends to some extent on the cell line being used and whether a batch or continuous process is being contemplated. Table 4 lists some of the microcarriers commercially available. The Cytodex family of beads is probably the most widely used. Cytodex 2 is recommended for cells having fibroblastlike morphology; Cytodex 3 is recommended for cells having an epithelial-like morphology. In long-term serum-free cultures, Cytodex 2 cells tend to retain cells longer than Cytodex 3, whereas the latter is useful when available inoculum density is low. In some cases, productivity is affected by the surface characteristics. For example, some cells have higher productivity on negatively charged polystyrene, eg, Biosilon, than the positively charged dextran. In designing a microcarrier process, it is recommended that a quick screening experiment be conducted to assess the suitability of the microcarriers available. A more extensive review of various types of commercial and noncommercial microcarriers and their applications is available (17).

In addition to attachment dependent cells, a batch microcarrier process may also be used for other cells that can grow in the attached mode because it allows the use of totally protein-free media. Many cells can survive for long periods of time in completely protein-free medium if attached to a surface. However, microcarrier reactors face several practical difficulties. First, if the cells are attachment dependent, generating sufficient inoculum for large-scale reactors involves trypsinization of microcarriers from a smaller reactor and reattachment in the larger reactor. The exposure to trypsin [*9002-07-7*] has to be controlled carefully in order to minimize cell damage. Second, the switch to a protein-free medium entails draining the reactor and refilling with fresh medium, requiring settling of the beads for long periods of time, especially in large-scale reactors. During this time, the cells may be deprived of oxygen and other nutrients and may be affected adversely. Hence batch microcarrier processes are suitable for vaccine manufac-

Table 4. Microcarriers for Stirred Tank Reactors

Trade name	Manufacturer	Diameter, μm	Characteristics
Cytodex 1	Pharmacia, Sweden	147–248	dextran, high positive charge
Cytodex 2	Pharmacia, Sweden	135–200	dextran, positive charge
Cytodex 3	Pharmacia, Sweden	141–211	collagen-coated dextran
Biosilon	Nunc, Denmark	160–300	polystyrene, negative charge
Cultispher	Hyclone, United States	170–270	gelatin, macroporous
Bioglas	Solo Hill, United States	90–150	glass-coated plastic
Bioplas	Solo Hill, United States	90–150	cross-linked polystyrene

ture where presence of serum may not be a hindrance. Finally, the low shear requirements make oxygenation much more difficult than in suspension culture. Gas sparging tends to carry all the microcarriers into the foam layer because the beads tend to adhere to the bubbles as they rise through the liquid. One approach to solving this problem is to aerate the liquid in a rotating or vibrating cage, separating the microcarriers from the bubbles. The movement of the cage prevents the cage from getting clogged, especially in a perfusion system, with cells and microcarriers. Alternatively, gas permeable tubing may be used. Many fermentor manufacturers now offer caged aeration and perfusion systems as an option for cell culture fermentors. These include Sulzer/MBR (Switzerland), Applikon (The Netherlands), New Brunswick Scientific (United States), Biolafitte (France), B. Braun (Germany/United States), and Chemap (Switzerland). B. Braun and Bioengineering (Switzerland) also offer silicone membrane oxygenation as an option. Membranteknik (Switzerland) offers silicone membrane oxygenating systems to retrofit existing fermentors for enhancing oxygenating capabilities. A good overview of microcarrier culture technology and an in-depth discussion of design issues and applications is available (18).

Microcarrier Perfusion Systems. Microcarriers may also be used in a continuous perfusion mode. In the perfusion mode, the reactor is constantly fed with fresh, sterile medium and product is harvested from the reactor at an equal rate. Perfusion systems have the advantage that the same cells can be maintained in a productive mode for several days, thus reducing costly downtime and the cost of expensive growth medium. Once the cells attach to the carriers and become confluent, they can be maintained in a productive mode in protein-free medium for several days or, depending on the cell line, for several months. High density perfusion systems are also useful for labile products, because low residence time in the reactor minimizes the exposure of the product to degradative enzymes and conditions. Continuous systems allow steady-state operation, which makes process control and process optimization easier. Continuous processes suffer two principal drawbacks. First, the risk of contamination is higher because of the increased complexity of the process. Additionally, contamination is much more costly for a continuous process than a batch process because the magnitude of product and labor loss is much higher. Second, if the cell line is subject to genetic instability, the reactor may lose productivity over a long period because of slow overgrowth of nonproducing cells. Further, it is harder to define a batch for regulatory purposes when the product is being made by a slowly changing population of cells, making validation of the process that much more difficult.

Figure 2 shows a typical microcarrier perfusion system utilizing a rotating screen to separate the microcarriers from the harvest liquid. Oxygenation using silicone tubing in the form of a helical coil is shown. Oxygenation can also be accomplished by sparging gas within this cage. Either method eliminates the possibility of damage to the cells from gas bubbles. The medium and harvest tanks are usually sized to hold at least three days worth of medium. Peristaltic or steam sterilizable diaphragm metering pumps are commonly used for pumping the medium and harvest. Steam sterilizable connections are used to maintain sterility during medium feed and harvest operations. Although the screen is shown as mounted on the agitation shaft, it is possible to attach the screen to a separate shaft driven by a second motor. The latter system has the advantage that the rotational speed of the screen can be optimized for microcarrier rejection without affecting the mixing and shear characteristics of the agitator. Other methods for

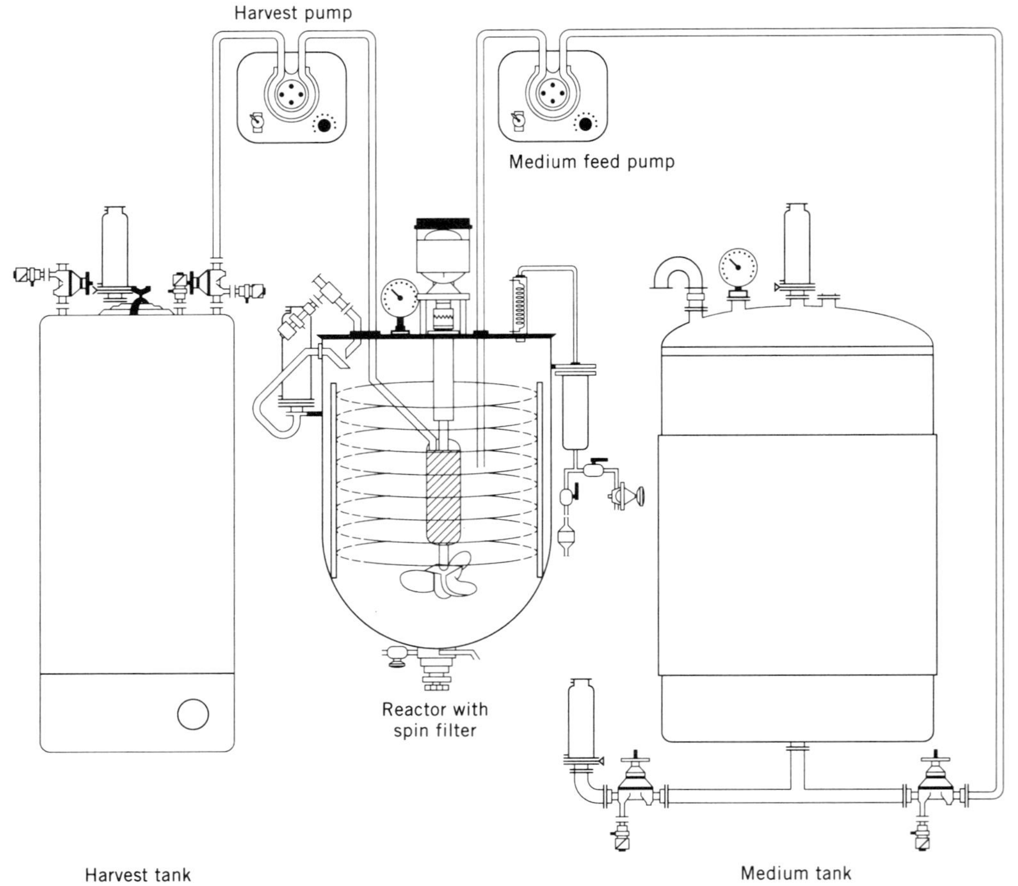

Fig. 2. Schematic of a process for microcarrier perfusion culture using a spin filter device.

retaining microcarriers within the reactor include settling towers, based on gravity settling of microcarriers, and "self-cleaning" static screens (19). In long-term continuous operation, cells tend to clump up and detach from nonporous microcarriers. For this reason, the macroporous carriers may be more suitable for long-term perfusion cultures.

Process design for microcarrier processes involves determination of the surface area of carriers required to accomplish the production in a given time. This can be translated from small-scale T-flask culture experiments. There is a practical limit to the surface area per unit volume that can be accommodated in a reactor. For example, for carriers having an average diameter of 150 μm, 30,000 cm^2/L is a reasonable limit. Beyond this limit the collision severity becomes a factor in cell damage. The reactor volume is determined based on these numbers and provision is made for the necessary oxygen transfer depending on the specific oxygen transfer rate of the cells. The procedures used for designing and scaling up microcarrier reactors have been described (20).

Fluidized-Bed Systems. Another approach to obviating the shear damage problem and to provide sufficient oxygenation to a high density cell culture system is to employ a fluidized-bed system utilizing macroporous microcarriers. Verax

(Lebanon, New Hampshire) has developed a system to culture cells at high densities in a fluidized-bed reactor where the carriers are fluidized by the recirculating medium which is continuously oxygenated by a membrane gas exchanger outside the reactor. The cells are immobilized in macroporous carriers made of collagen. Fresh medium is continuously added to this loop and equivalent amount of product harvested at the same time. The porous carriers are weighted to allow high enough recirculation rates through the bed so that sufficient oxygen can be provided to all the cells. pH, dissolved oxygen, and temperature sensors are also located in the recirculation loop and these parameters are controlled via a computer-based control system. A schematic of this process is shown in Figure 3; examples of usage are available (21).

Advantages of this system include: both anchorage dependent and independent cells can be cultured in this system without the danger of adverse shear effects; most of the control and instrumentation is located in the external loop and it is therefore possible to build in redundant loops and increase the reliability of the process; high (up to 3×10^8 cells/mL of matrix) cell densities allow short (2–5 h) residence times, which is helpful for labile proteins that are subject to degradation. This system also has the advantages of high cell density cultures while providing a relatively homogeneous environment for the cells. On the other hand the system has some limitations. Because of the limit on recirculation flow rate, set by the specific gravity of the matrix, the depth of the fluidized bed is limited to about 1.5 m. Scale-up beyond this is possible only by increasing bed diameter. Moreover, an oxygen-transfer gradient is imposed on the bed because of the rapid rate of oxygen consumption prohibiting the precise optimization of dissolved oxygen. Finally, as with any continuous system, the complexity of the system makes it harder to set up, although it is easier to operate, and the costs associated with contamination can be very high.

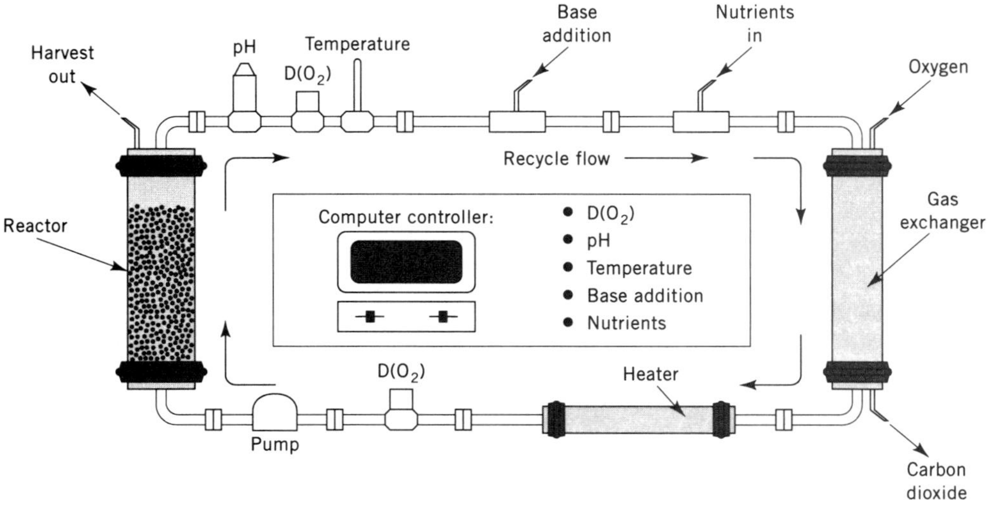

Fig. 3. Schematic of a fluidized-bed process for mammalian cell culture where $D(O_2)$ = dissolved oxygen. Courtesy of Verax Corp.

Another version of a fluidized-bed reactor has been introduced by Vogelbusch (Austria). This reactor has an internal recirculation loop set up by means of a high flow impeller. The system utilizes porous glass beads for immobilizing the cells. However, glass beads may not work with all types of cells.

Fixed-Bed Reactors. A fixed-bed reactor has been developed by Charles River Biotechnical Services (Wilmington, Massachusetts) for the culture of both anchorage dependent and independent cells. A ceramic matrix having multiple internal channels through which the recirculating medium flows is used. A smooth matrix is employed to immobilize attachment dependent cells whereas another matrix having surface pores serves as the immobilization support for other cells. The process concept is similar to that of the recirculation loop for the fluidized-bed system. Figure 4 shows a schematic of this system. Because the bed is fixed, recirculating flux rates can be higher than in fluidized systems.

Scale-up in fixed-bed reactors is limited by the maximum size of the matrix that can be manufactured as a monolith. Hence, this system is applicable for small- to medium-scale production of antibodies and other proteins, usually for

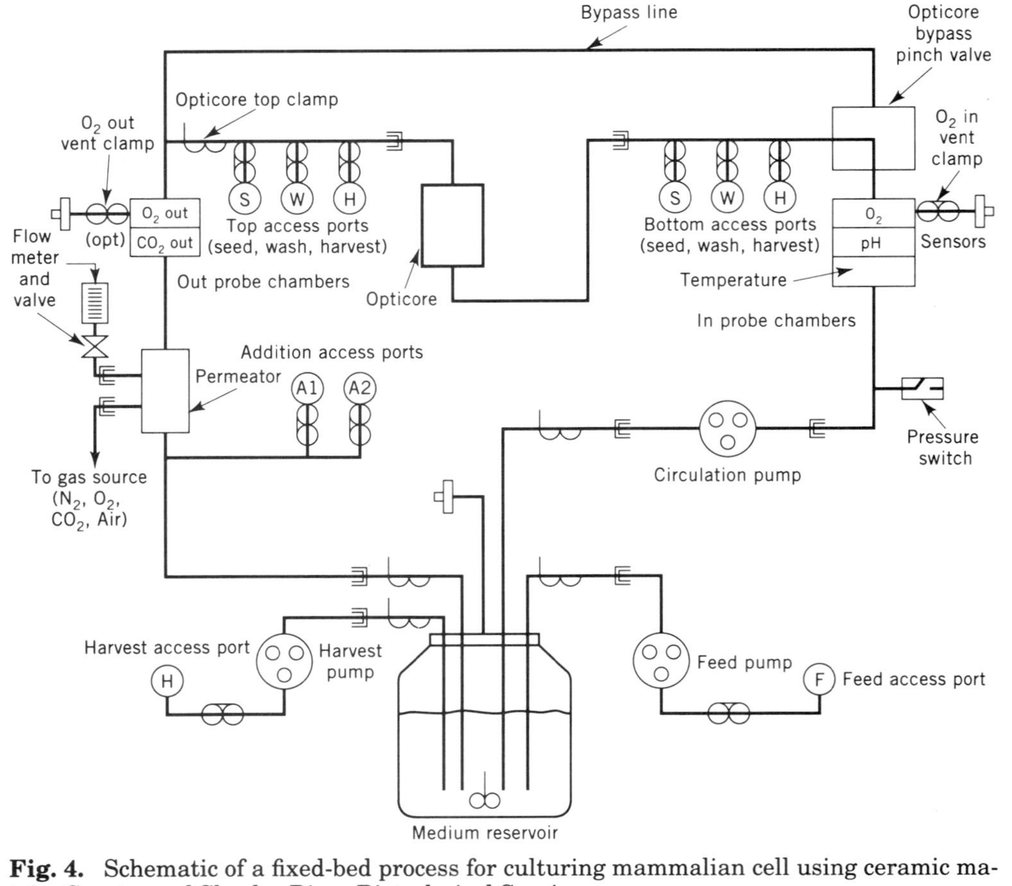

Fig. 4. Schematic of a fixed-bed process for culturing mammalian cell using ceramic matrix. Courtesy of Charles River Biotechnical Services.

the diagnostic market. This system has been described in greater detail in the literature (22).

Another type of fixed-bed system is the hollow-fiber reactor. A hollow-fiber device consists of a bundle of hollow fibers, usually made of hollow anisotropic plastic fibers that allow diffusion of molecules smaller than a specified molecular weight cut-off, potted at both ends of a plastic shell (see HOLLOW-FIBER MEMBRANES). The cells are immobilized in the extra capillary space (ecs) of the hollow fibers and the medium recirculates through the lumen of the fibers entering and exiting via headers at either end of the reactor. The cells are held in a static mode and grow to high tissuelike density in the ecs. The medium access to the cells is via diffusion and Starling flow. Thus shear forces on the cells are minimal. This system is especially suitable for cells that are extremely sensitive to shear induced injury. Another advantage is that the fibers can be specified such that the product, often a high molecular weight protein, can be retained within the ecs, while medium components and metabolic by-products can diffuse through. This arrangement allows for *in situ* concentration of the product to very high levels. Similarly, serum usage can be minimized by entrapping the high molecular weight protein components of serum in the ecs and using protein-free medium for perfusion.

Scale-up of this system is limited by the size limitation on hollow-fiber manufacturing. This system is also more suitable for the diagnostic markets where quantities of protein required are relatively low. A disadvantage of the system is that nutrient gradients are set up in the ecs leading to a nonhomogeneous environment making process control difficult. Endotronics has devised a way of solving this problem by pressure induced flow through the ecs (23). However, this increases the complexity of the system. Hollow-fiber devices are available through many vendors including Endotronics (Coon Rapids, Minnesota), Microgon (Laguna Hills, California), Unisyn Fibertec (San Diego, California), and Amicon (Beverly, Massachusetts). Some offer complete cell culture systems (Endotronics, Unisyn Fibertec, Amicon); others sell only the hollow-fiber device and leave the system design to the user.

Cell Recycle Systems. A variation of suspension culture reactors is a system where the cell concentration is increased by perfusing medium through the reactor continuously while retaining the cells in the reactor by means of a tangential flow filter. The tangential flow filter can be either a plate and frame filter, eg, the Prostack manufactured by Millipore, or a hollow-fiber device. Figure 5 gives an example of one such system. Cells are continuously recycled from the fermentor through a tangential flow filter. The filtrate removed from the filter is replaced by fresh medium from the medium tank. A small fraction of cells are either continuously or periodically purged from the reactor to maintain cell viability at a high level. This type of reactor provides the benefits of a high cell density system and maintains a homogeneous environment for the cells. A shortcoming of the system is that its applicability is limited to anchorage independent cells. It is also fairly complex to set up because of the external recirculation loop involved. It is possible to internalize the cell recycle by using an appropriately sized spin filter device like the one shown in Figure 2. However, such systems are prone to clogging by the cells after a few days of operation. These systems can be effectively

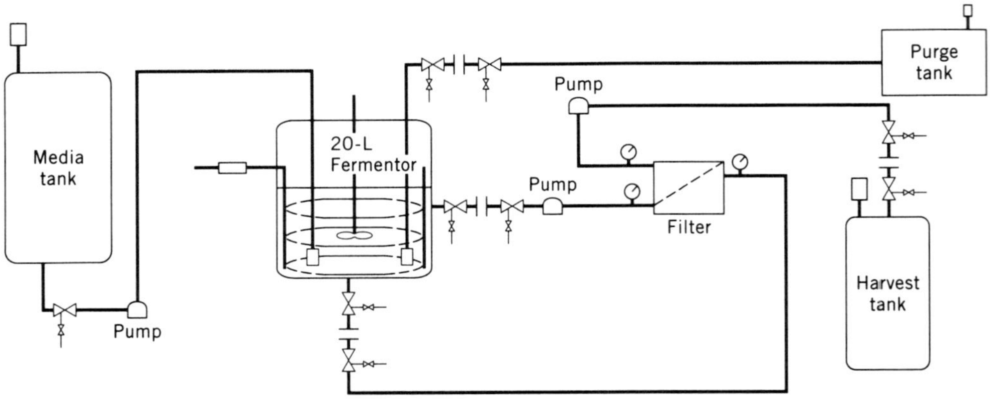

Fig. 5. A suspension perfusion culture process having cell recycle. System employed at Amgen using CHO cells for production of a recombinant protein.

utilized when there is a need to increase the capacity of existing stirred tank fermentors.

Other Systems. There are several other technologies utilized for specific purposes that do not fall into any of the above categories. For example, erythropoietin is produced in an automated roller bottle plant where cells are grown on the surface of roller bottles in a growth medium. Cells are then shifted to a serum-free production medium for harvesting. All operations are carried out aseptically in a clean room environment by an automated machine providing a high degree of reliability and consistency. Although roller bottles are often considered to be obsolete because of the labor and space intensive nature of the process, automation makes them a viable process for products where the volume requirements are not very high. The Technology Partnership (Cambridge, England) offers robotic systems for the production of cell culture-derived products using roller bottles and T-flasks (see ROBOTS, PROCESSING). Other systems attempt to scale up existing T-flask processes linearly by increasing the available surface area in a compact space and providing methods for media addition and harvest removal. Examples of such systems include the cell factory marketed by Nunc (Denmark) and the cell cube marketed by Costar (Cambridge, Massachusetts). Both systems are suitable for small-scale manufacturing, using attachment dependent cells. They have the advantage of assuring the fidelity of the process as it is scaled up from T-flasks.

Economic Aspects

As of early 1992, the market for cell culture-derived products approached $1 billion per year. The market is expected to grow substantially throughout the 1990s. Cell culture products include erythropoietin, 1991 sales of approximately $400 million, for the treatment of anemia associated with kidney dialysis, and tissue plasminogen activator, 1991 sales approximately $200 million, for treating heart attack victims with blocked arteries (see CARDIOVASCULAR AGENTS).

Process Economics. Relative economics of various cell culture processes depend heavily on the performance of the cell line in a system and on the cost of raw materials, particularly the medium. Models are usually developed for the various processes using productivity data obtained from small-scale experiments (see PILOT PLANTS AND MICROPLANTS). Often, for high value products, the process which ensures the shortest time to market may be the process of choice because of other economic criteria. This is especially true for pharmaceuticals (qv). Reliability concerns also often outweigh economic considerations in choosing a process for a high value product.

Continuous processes have lower labor costs but have higher failure risk. Batch processes can be started back up in a shorter period of time than can a complex continuous process. Batch processes are easier to take through the regulatory process than are continuous processes. Thus batch processes are often chosen for mammalian cell culture systems, even though continuous processes can offer significant cost advantages. Cell culture costs constitute only a small (10–30%) fraction of the overall cost of making a product.

Regulations and Standards

Most of the products derived from cell culture technologies are for therapeutic or diagnostic applictions and manufacture is regulated by the federal government through the Food and Drug Administration (FDA). The FDA requires that all drugs be manufactured in compliance with current Good Manufacturing Practices (cGMPs). Guidelines for cGMPs are provided through the *Code of Federal Regulations* (CFR) Title 21. Essentially, cGMPs require that all process steps and products be defined in a quantitative manner by the manufacturer, ie, specifications for all important processes must be developed and methods for testing and validating those steps must be identified. The National Institutes of Health have issued guidelines for safe experimentation using recombinant and mammalian cell lines (24).

The biotechnology (qv) industry has no formal standards for equipment manufacture and quality control as of this writing. The American Society of Mechanical Engineers (ASME) set up a task force in 1989, for developing relevant standards for biotechnology equipment.

Safety Considerations

The fact that cell culture-derived products are often injected into humans as therapeutic agents makes it imperative that there be no component in the final product that can pose a potential health risk to the patient. Health risks can be introduced into a product from many sources including: the cells themselves; raw materials, such as serum, media components, etc; materials used in purification, eg, antibodies; and external contamination. For a therapeutic product such risk factors are identified at the outset and ways of reducing them to acceptable levels are designed into the process. Before a product is released by the FDA the manufacturer has to demonstrate this risk reduction by rigorous validation of the process.

Some of the cells used in manufacturing are continuous or "immortal." Many of these have been shown to be tumorigenic in immunosuppressed animals. The cells also contain endogenous materials such as retroviruses and nucleic acids (qv), both of which can induce tumorigenesis, and immunogenic foreign proteins. Serum used in media can also introduce adventitious agents such as viruses and mycoplasma into the product. Other process chemicals, including cleaning agents, are low molecular-weight compounds that may be hazardous as well. Purification chemicals, such as monoclonals used for affinity purification, can be immunogenic to humans. Some of the potential health risks in mammalian cell culture processes and the methods used for risk reduction include:

cells	microfiltration
retroviruses	irradiation, sonication, heat, solvents, etc
nucleic acids	chromatography
cellular proteins	chromatography, ultrafiltration
bacterial contamination	microfiltration
process chemicals	diafiltration with appropriate buffers
serum proteins	affinity/ion-exchange chromatography

Most of these methods are commonly employed in the downstream processing of the desired cell culture technology product. Hence, most of the time it is only necessary to demonstrate that the designed process is reducing the putative risk factors to acceptable levels. Validation methods employed for risk reduction are discussed in the literature (25).

BIBLIOGRAPHY

1. P. B. Capstick, R. C. Telling, W. G. Chapman, and D. L. Stewart, *Nature* **195,** 1163 (1962).
2. A. L. van Wezel and G. Steenis, *Dev. Biol. Standard.* **40,** 69 (1978).
3. M. D. Johnston, G. Christofinis, G. D. Ball, K. H. Fantes, and N. B. Finter, *Dev. Biol. Standard.* **42,** 189 (1979).
4. A. L. van Wezel, C. A. M. van der Velden-de-Groot, and J. A. M. van Herwaarden, *Dev. Biol. Standard.* **46,** 151 (1980).
5. J. P. Mather, ed., *Mammalian Cell Culture: The Use of Serum-Free Hormone-Supplemented Media*, Plenum Press, New York, 1984.
6. J. N. Thomas, in A. S. Lubiniecki, ed., *Large Scale Mammalian Cell Culture Technology*, Marcel Dekker, New York, 1990, pp. 93–145.
7. B. Maiorella, D. Inlow, A. Shauger, and D. Harano, *Bio/Technol.* **6,** 1406 (1988).
8. R. A. Richieri, L. S. Williams, and P. C. Chau, *Cytotechnology* **5,** 243 (1991).
9. S. J. Kromenaker and F. Srienc, *Biotechnol. Bioeng.* **38,** 665 (1991).
10. N. G. Ray, S. B. Karkare, and P. W. Runstadler, *Biotechnol. Bioeng.* **33,** 724 (1989).
11. E. Suzuki and D. F. Ollis, *Biotechnol. Bioeng.* **34,** 1398 (1989).
12. B. D. Mariani, D. L. Slate, and R. T. Schimke, *Proc. Natl. Acad. Sci.* **78,** 4985 (1981).
13. P. J. Radlett, T. W. F. Pay, and A. H. M. Graland, *Dev. Biol. Standard.* **60,** 163 (1985).

14. M. Rhodes and J. Birch, *Bio/Technol.* **6,** 518 (1988).
15. M. S. Croughan, J.-F. Hamel, and D. I. C. Wang, *Biotechnol. Bioeng.* **29,** 130 (1987).
16. S. Lakhotia and E. T. Papoutsakis, *Biotechnol. Bioeng.* **39,** 95 (1992).
17. S. Reuveny, in Ref. 6, pp. 271–341.
18. R. Fleischaker, in B. K. Lydersen, ed., *Large Scale Cell Culture Technology*, Hanser, New York, 1987, pp. 59–79.
19. S. B. Karkare, S. T. Cole, R. Sachdev, V. Satyagal, and J. C. Fieschko, "Production of Rat Stem Cell Factor from BRL Cells by Microcarrier Perfusion Culture," paper presented at the *Engineering Foundation Conference on Biochemical Engineering VII*, Santa Barbara, Calif., 1991; *Annals of New York Academy of Sciences*, in press.
20. K. L. Nelson, *Biopharm.* **34** (Mar. 1988).
21. U.S. Pat. 4,978,616 (Dec. 18, 1990), R. C. Dean, Jr., P. V. Grela, S. B. Karkare, and P. W. Runstadler, Jr. (to Verax Corp.).
22. B. K. Lydersen, in Ref. 18, pp. 168–192.
23. M. D. Hirschel and M. L. Gruenberg, in Ref. 18, pp. 112–144.
24. *Fed. Reg.* **56**(138) (July 18, 1991).
25. A. S. Lubiniecki, M. E. Wiebe, and S. E. Builder, in Ref. 6, pp. 515–541.

SUBHASH B. KARKARE
Amgen, Inc.

CHEMOTHERAPEUTICS, ANTICANCER

Cancer is second only to cardiovascular disease as the principal cause of human mortality (see CARDIOVASCULAR AGENTS). As the median age of populations have risen, total deaths from cancer have increased. Treatment of cancer includes surgery, radiation, and chemotherapy. In those instances where a tumor is locally confined, has not metastasized, and is resectable, surgery alone or combined with local radiation may be curative. For the majority of patients, however, physicians are forced to rely on chemotherapy either as a means to attack residual disease following surgery and/or radiation, or as the primary treatment method. Chemotherapy encompasses the use of both cytotoxic agents and relatively nontoxic hormonal agents for the control of tumor growth (1).

Approaches to cytotoxic chemotherapy include special emphasis on drug targeting and toxicity alleviation. The directions in which new drug discovery strategies are moving and the criteria used for advancing compounds into clinical trials (2) are discussed herein, as are all of the drugs approved by the United States Food and Drug Administration (FDA) for the treatment of cancer as of this writing and those compounds in clinical trials.

Chemotherapeutic Agents

The majority of cytotoxic chemotherapeutic agents in clinical use were discovered empirically using cell cytotoxicity assays followed by assessment in an animal tumor model, usually an *in vivo* leukemia model such as L1210 or P388 (3). On the basis of the activity in those models and other murine solid tumor models, a large number of compounds advanced to human clinical trials where a low percentage were found to have sufficient activity in humans to justify approval as cancer chemotherapeutic agents. It is increasingly clear that these selection criteria were flawed (4). Concurrently, an increased knowledge of the biological complexity of cancer and the mechanism of action of drugs has permitted the design of more rational screens for the discovery of new lead structures and the design and synthesis of compounds targeted to specific receptors or enzymes involved in the pathogenesis of cancer (5).

At the preclinical level of discovery in the 1990s there is a greater reliance on the use of human tumor cell lines and clonogenic assays derived from primary explants (6). As the development of resistance to cytotoxic agents by tumor cells is better understood, it has become possible to assemble panels of cell lines consisting of a parent cell line and one or more resistant cell lines wherein the mechanism of resistance has been identified and quantitated, ie, expression of multidrug resistant (MDR) genes, increased glutathione [*70-18-8*] levels, and increased DNA repair. The use of mechanism-based screens such as specific oncogene tyrosine kinase assays, topoisomerase I and/or II assays, and tubulin assays has also become possible (5).

At the *in vivo* assay level, the classic ip-ip (interaperitoneal) *in vivo* model has been replaced as a selection criteria for advancement of new drug candidates to clinical trial. More stringent alternative models include subcutaneous or subrenal capsule implantation of tumor followed by intravenous drug dosing (7) and the human tumor xenograft models in nude mice (8).

In addition to a greater emphasis on *in vivo* evaluation as a means of assessing activity, there is also increased awareness of metabolism and pharmacokinetics in candidate selection (9). The use of prodrug strategies to improve water solubility, oral absorption, and other pharmaceutical properties has increased (10). Finally, the ability to assess potential toxic liabilities of a new drug candidate has improved as more predictive models for myelosuppression, ie, neutropenia and thrombocytopenia, emesis, and renal, cardiac, and hepatic toxicities, have been developed (11,12).

The modern drug discovery team consists of medicinal chemists, biologists, metabolism and pharmacokinetics specialists, and toxicologists working together to improve drug selection.

Drugs used in cancer chemotherapy or clinical trials are classified according to primary underlying mechanisms of action. However, many drugs operate through multiple mechanisms. Mechanisms include those of antimetabolites, DNA alkylating and/or cross-linking agents, DNA binding/cleaving agents, DNA topoisomerase interactive compounds, agents which act on tubulin structure, and hormones (qv) (Fig. 1). In addition to those drugs which have already been approved by the FDA, a number of investigational drugs are undergoing clinical evaluation and many others are in the pipeline.

Antimetabolites. Antimetabolites, which represent one of the earliest groups of anticancer agents, are listed in Table 1. Structures are shown in Fig. 2.

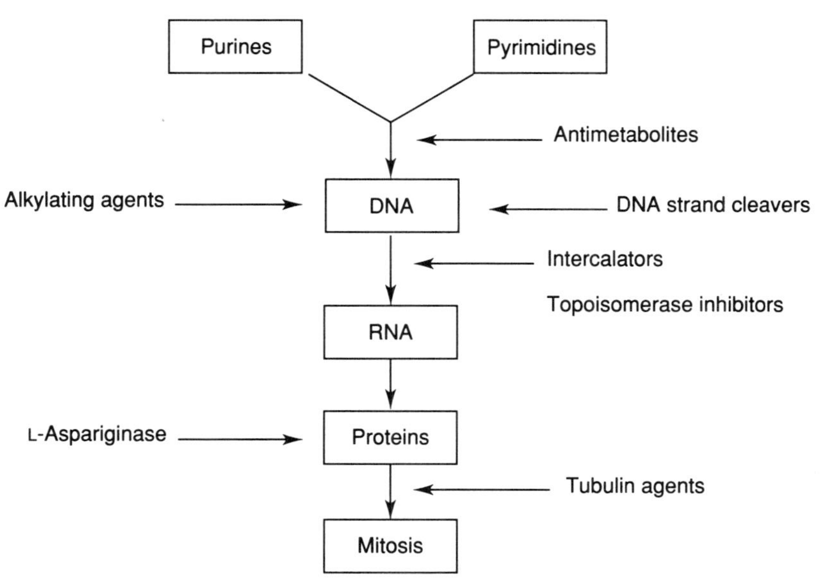

Fig. 1. Schematic of nucleic acid and protein synthesis and the steps leading to mitosis showing the common mechanisms of action and various classes of chemotherapeutic agents.

Table 1. Antimetabolites[a]

Drug (trade name)	CAS Registry Number	Molecular formula	Molecular weight	Structure number	Disease	Toxic effects
5-azacitidine[b] (Mylosar)	*[320-67-2]*	$C_8H_{12}N_4O_5$	244.21	(**1**)	acute myelogenous leukemia	nausea, vomiting; hepatic dysfunction; myelosuppression
cytarabine USP[b] (Cytosar)	*[147-94-4]*	$C_9H_{13}N_3O_5$	243.22	(**2**)	acute granulocytic leukemia (adults); acute lymphocytic leukemia (children); Hodgkin's disease	bone-marrow depression; hepatic toxicity; megaloblastosis; nausea; vomiting; diarrhea
gemcitabine[c]	*[95058-81-4]*	$C_9H_{11}F_2N_3O_4$	263.20	(**3**)	investigational drug; responses seen in Phase I trials in colon and nonsmall cell lung cancer	myelosuppression observed as dose limiting toxicity
floxuridine USP[d] (FUDR)	*[50-91-9]*	$C_9H_{11}FN_2O_5$	246.21	(**4**)	palliative treatment of gastrointestinal adenocarcinoma with liver metastases	severe hematological toxicity; gastrointestinal hemorrhage; nausea; vomiting; diarrhea; enteritis; stomatitis; erythema
fluorouracil USP[d] (Fluorouracil)	*[51-21-8]*	$C_4H_3FN_2O_2$	130.08	(**5**)	palliative treatment of carcinoma of colon, rectum, breast, stomach, and pancreas	bone-marrow depression; dermatitis; alopecia; nausea; vomiting; diarrhea; stomatitis; anorexia; GI ulcers; skin pigmentation

mercaptopurine USP[e] (Purinethol)	[*6112-76-1*]	$C_5H_4N_4S$	152.19	(**6**)	acute leukemia (more effective in children than in adults); chronic granulocytic leukemia	bone-marrow depression; hepatic toxicity; anemia; gastrointestinal (GI) ulceration; nausea; vomiting
thioguanine USP[e] (Tabloid)	[*154-42-7*]	$C_5H_5N_5S{\cdot}XH_2O$	167.19	(**7**)	acute leukemia; chronic granulocytic leukemia	bone-marrow depression; stomatitis; anorexia; nausea; vomiting
methotrexate[f] USP (Methotrexate)	[*59-05-2*]	$C_{20}H_{22}N_8O_5$	454.46	(**8**)	acute lymphocytic leukemia; meningeal leukemia; choriocarcinoma; chorioadenoma destruens; lymphosarcoma; osteogenic sarcoma; cancer of lung, neck, head, cervix; mycosis fungoides; hydatidiform mole high dose MTX followed by leucovorin rescue in nonmetastatic osteosarcoma	bone-marrow depression; renal and hepatic toxicity; enteritis; stomatitis; alopecia; abdominal distress; erythematous rash; oral and GI ulceration; diarrhea; nausea; vomiting
leucovorin[f] (calcium USP)	[*58-05-9*]	$C_{20}H_{21}CaN_7O_7$	511.51	(**9**)	high dose methotrexate rescue therapy in osteosarcoma	allergic sensitization
DDATHF[c] (Ly237147) (Lometrexol sodium)	[*120408-07-3*]	$C_{21}H_{23}N_5Na_2O_6$	487.42	(**10**)	investigational drug	

Table 1. (*Continued*)

Drug (trade name)	CAS Registry Number	Molecular formula	Molecular weight	Structure number	Disease	Toxic effects
trimetrexate[g]	[*52128-35-5*]	$C_{19}H_{23}N_5O_3$	369.42	(**11**)	investigational drug; partial remissions in soft tissue sarcomas observed	myelosuppression; mucositis; nausea; vomiting; skin rash
hydroxyurea USP[h] (Hydrea)	[*127-07-1*]	$CH_4N_2O_2$	76.05	(**12**)	chronic granulocytic leukemia; melanoma; cancer of ovary, head, neck	vomiting; anorexia; fever; bone-marrow depression; nausea; diarrhea

[a]See Figure 2.
[b]Upjohn.
[c]Lilly.
[d]Hoffmann-La Roche.
[e]Burroughs Wellcome.
[f]Lederle.
[g]Parke-Davis.
[h]Bristol-Myers Squibb.

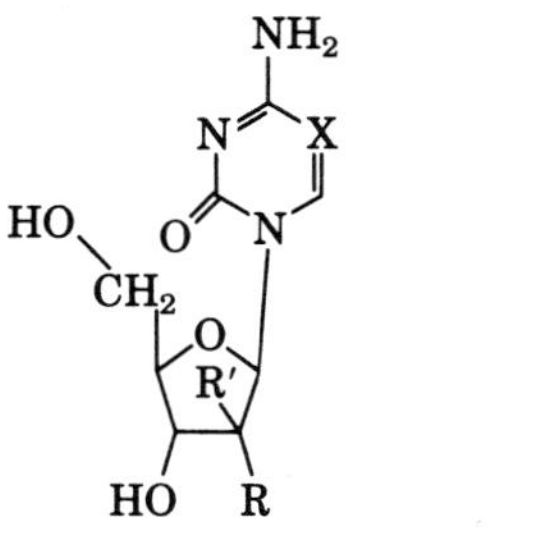

(1) X = N R′ = H R = OH

(2) X = CH R′ = OH R = H

(3) X = CH R′ = R = F

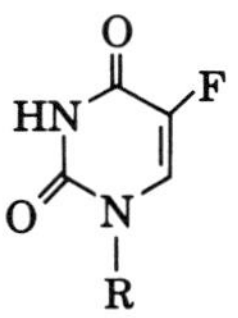

(4) R =

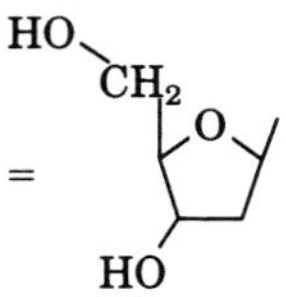

(5) R = H

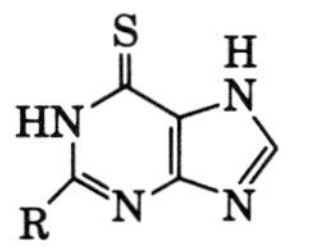

(6) R = H

(7) R = NH_2

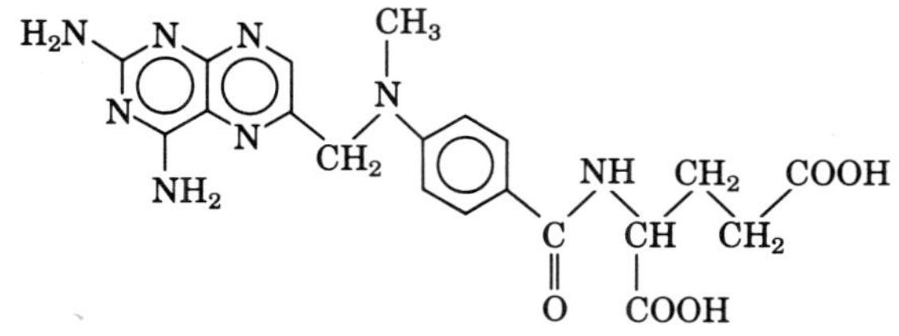

(8)

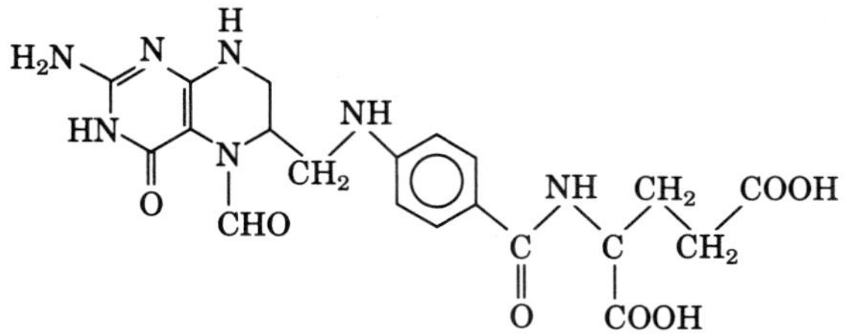

(9)

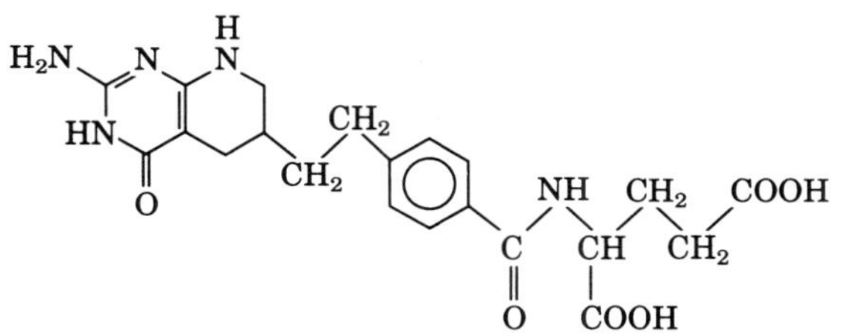

(10)

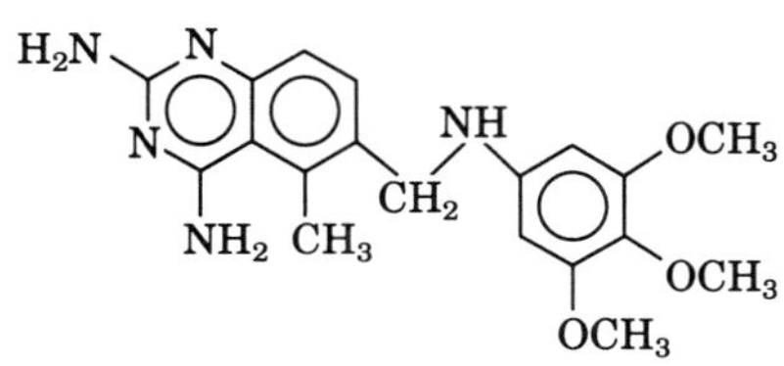

(11)

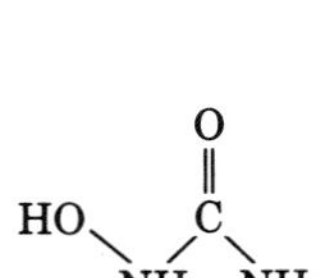

(12)

Fig. 2. Structures of antimetabolites listed in Table 1.

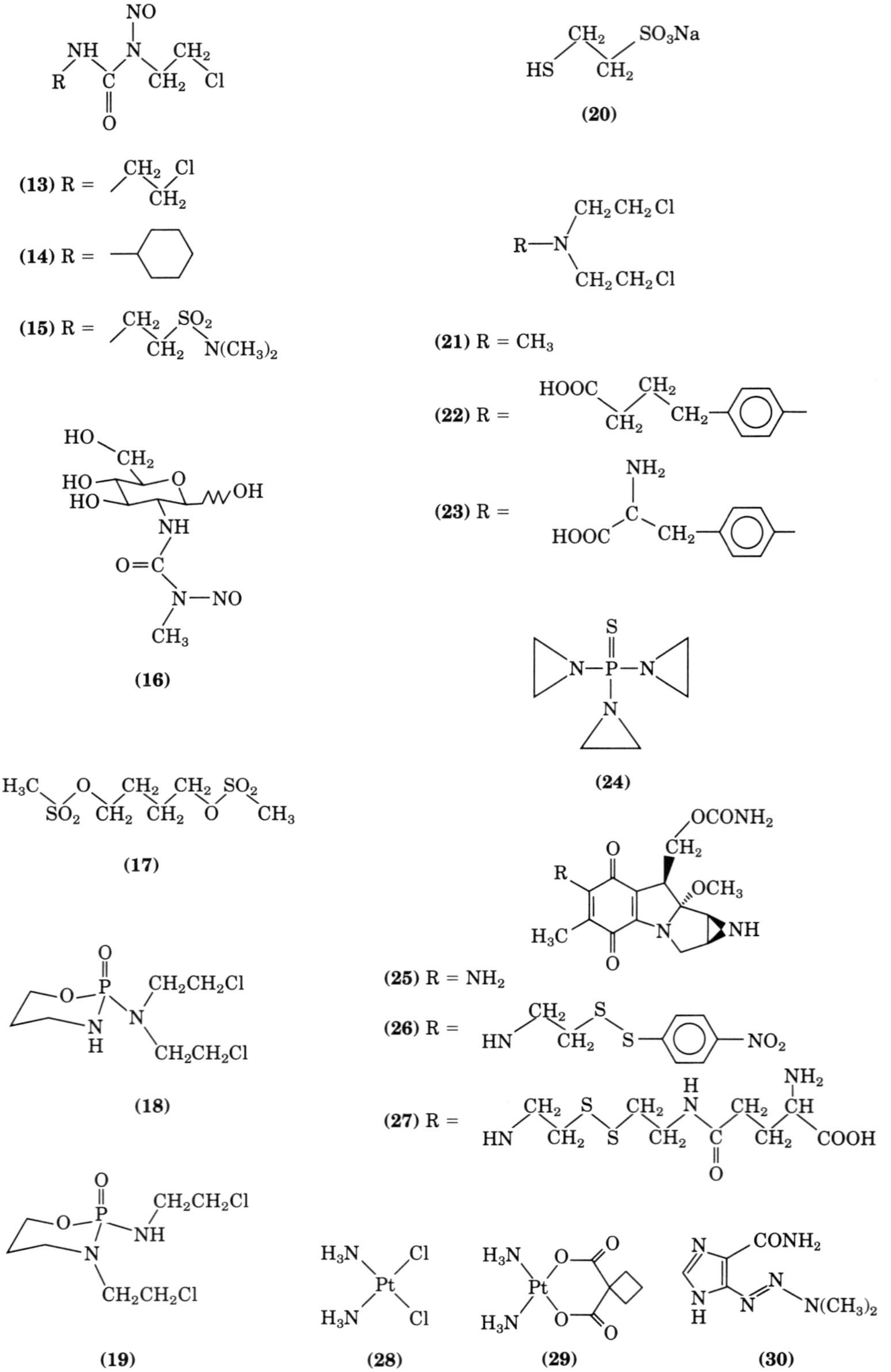

Fig. 3. Structures of DNA alkylating/cross-linking agents given in Table 2.

The classification of these drugs as antimetabolites stems from the mode of action as antagonists to the natural metabolic processes leading to either DNA, RNA, or protein synthesis (13) (see NUCLEIC ACIDS; PROTEINS). They either inhibit function of a key enzyme involved in protein synthesis or are recruited into the cell division process as DNA synthesis terminators. For example, methotrexate (8) is a folic acid [*59-30-3*], $C_{19}H_{19}N_7O_6$, antagonist and has a 100,000-fold greater affinity for the enzyme, dihydrofolate reductase [*9002-03-3*] (DHFR), than does the enzyme's natural substrate, folic acid. Inhibition of DHFR function, ie, conversion of folic acid to folinic acid [*58-05-9*] (leucovorin) (9), results in the arrest of purine and pyrimidine synthesis which culminates in cell death. The use of high dose methotrexate treatment followed by leucovorin rescue, ie, toxicity amelioration, has resulted in significant improvements in the treatment of nonmetastatic osteosarcoma. In the early 1990s, the use of high dose regimens followed by rescue with colony stimulating factors (CSFs) (14) has been extended to bone marrow transplantation.

Antimetabolites may be further classified as inhibitors of pyrimidine, purine, or glutamine metabolism. The compounds are cell cycle dependent.

DNA Alkylating/Cross-Linking Agents. This category includes compounds of diverse chemical classes (Table 2) eg, nitrosoureas (**13–16**) nitrogen mustards (**18–24**), mitomycins (**25–27**), and platinums (**28,29**), that have the ability to react covalently with DNA bases and to form inter- and intrastrand DNA cross-links (15–17). These compounds may also be responsible for the alkylation of proteins and protein–DNA linkages. The resulting lesions produced in the DNA result in disruption of cell growth and function, ultimately leading to cell death. The onset of action of this class of agents can be rapid, resulting in dramatic tumor shrinkage. However, because of the effects on normal cells, all compounds of this class are myelosuppressive and potentially teratogenic, mutagenic, and carcinogenic. In some cases the resulting myelosuppression has limited the ability of these drugs to be effective clinically. In other cases the rapid onset of resistance can occur; this limits utility as well as that of related analogues, because of cross resistance (18) (Fig. 3).

DNA Binding/Cleaving Agents. DNA binding and/or cleaving agents that have anticancer activity are listed in Table 3. Structures are shown in Fig. 4. All of the natural products (**31,32,35,36,38,39**) and analogues of natural products (**33,34,37**) in this category (Table 3) are able to bind DNA either as intercalators (**31–34**) or as minor groove binders (**35–37**), hence inhibiting DNA dependant RNA synthesis (**15–17**). Both bleomycin (**35**) and esperamicin A_1 (**36**) cleave DNA by forming free radicals in the immediate vicinity of the sugar–phosphate backbone. Activity as antitumor agents is related to the ability to induce irreparable lesions in DNA (15,19). Bleomycin generates oxygen free-radical species whereas esperamicin A_1 and a number of related natural products that include neocarzinostatin, dynemicin, and the calicheamicins, generate aryl diradical species, which abstract hydrogen atoms directly from the deoxyribose backbone (19). An analogue of the natural product CC1065 (**37**) has the unique property of being a DNA alkylating agent which recognizes poly-AT regions of the minor groove of DNA. It remains to be seen if the high potency and unique modes of action ascribed to these novel classes of agents can translate into clinically useful drugs (Figure 4) (20).

Table 2. DNA Alkylating/Cross-Linking Agents[a]

Drug (trade name)	CAS Registry Number	Molecular formula	Molecular weight	Structure number	Disease	Toxic effects
carmustine USP[b] (BiCNU)	[*154-93-8*]	$C_5H_9Cl_2N_3O_2$	214.05	(**13**)	Hodgkin's disease; non-Hodgkin's lymphomas; meningeal leukemia; brain tumor; multiple myeloma	bone-marrow depression; hepatic toxicity; nausea; vomiting
lomustine USP[b] (CeeNU)	[*13010-47-4*]	$C_9H_{16}ClN_3O_2$	233.70	(**14**)	malignant brain tumors; Hodgkin's disease	bone-marrow depression; hepatic toxicity
tauromustine[c]	[*85977-49-7*]	$C_7H_{15}ClN_4O_4S$	286.73	(**15**)	investigational drug responses observed in malignant melanoma	gastrointestinal; thrombocytopenia
streptozocin USP[d] (Zanosar)	[*18883-66-4*]	$C_8H_{15}N_3O_7$	265.22	(**16**)	metastatic islet cell carcinoma of the pancreas	bone-marrow depression; renal and hepatic toxicity; nausea; vomiting
busulfan USP[e] (Myleran)	[*55-98-1*]	$C_6H_{14}O_6S_2$	246.29	(**17**)	chronic granulocytic leukemia; other myeloproliferative disorders	bone-marrow depression; hyperuricemia; gynecomastia; amenorrhea; skin hyperpigmentation

cyclophosphamide USP[b] (Cytoxan)	[*6055-19-2*]	$C_7H_{15}Cl_2N_2O_2P$	279.10	(**18**)	acute and chronic lymphocytic leukemia; lung cancer; rhabdomyosarcoma; neuroblastoma; ovarian and mammary carcinoma; multiple myeloma; lymphosarcoma; Burkitt's lymphoma; Hodgkin's disease; retinoblastoma; mycosis fungoides	bone-marrow depression; hepatic toxicity; cystitis; alopecia; nausea; vomiting
ifosphamide USP[b] (Ifex)	[*3778-73-2*]	$C_7H_{15}Cl_2N_2O_2P$	261.09	(**19**)	germ cell testicular cancer; used in combination with mesna	myelosuppression; urotoxicity; alopecia; nausea; vomiting; CNS toxicities
mesna USP[b] (Mesnex)	[*19767-45-4*]	$C_2H_5NaO_3S_2$	164.17	(**20**)	prophylactic prevention of hemorrhagic cystitis	
mechlorethamine hydrochloride USP[f] (Mustargen)	[*55-86-7*]	$C_5H_{11}Cl_2N{\cdot}HCl$	192.52	(**21**)	Hodgkin's disease; non-Hodgkin's lymphomas; lymphosarcoma; cancer of breast, ovary, lung; neoplastic effusion	bone-marrow depression; nausea; vomiting; anorexia; diarrhea; local irritation
chlorambucil USP[e] (Leukeran)	[*305-03-3*]	$C_{14}H_{19}Cl_2NO_2$	304.22	(**22**)	chronic lymphocytic leukemia; cancer of ovary, breast, testis; Hodgkin's disease; non-Hodgkin's lymphomas	bone-marrow depression; nausea; vomiting

Table 2. (*Continued*)

Drug (trade name)	CAS Registry Number	Molecular formula	Molecular weight	Structure number	Disease	Toxic effects
melphalan USP[e] (Alkeran)	[*148-82-3*]	$C_{13}H_{18}Cl_2N_2O_2$	305.20	(**23**)	multiple myeloma; plasmacytic myeloma; cancer of breast and ovary	bone-marrow depression; nausea; vomiting; anorexia
thiotepa USP[g] (Thiotepa)	[*52-24-4*]	$C_6H_{12}N_3PS$	189.21	(**24**)	cancer of breast, ovary, lung, bladder; Hodgkin's disease; nonHodgkin's lymphomas; neoplastic effusion	bone-marrow depression; amenorrhea; anorexia; nausea; vomiting
mitomycin C USP[b] (Mutamycin)	[*50-07-7*]	$C_{15}H_{18}N_4O_5$	334.33	(**25**)	chronic myelogenous leukemia; reticulum cell sarcoma; Hodgkin's disease; non-Hodgkin's lymphomas; cancer of stomach, pancreas, lung; epithelial tumors	bone-marrow depression; renal toxicity; alopecia; stomatitis; anorexia; nausea; vomiting

BMY-25067[b]	[*95056-36-3*]	$C_{23}H_{25}N_5O_7S_2$	547.60	(**26**)	investigational drug	
KW2149[h]	[*118359-59-4*]	$C_{24}H_{34}N_6O_8S_2$	598.7	(**27**)	investigational drug	
cisplatin USP[b] (Platinol)	[*15663-27-1*]	$Cl_2H_6N_2Pt$	300.06	(**28**)	metastatic testicular tumors; metastatic ovarian tumors; advanced bladder cancer	nephrotoxicity; ototoxicity; myelosuppression; nausea; vomiting; allergic reaction
carboplatin USP[b] (Paraplatin)	[*41575-94-4*]	$C_6H_{12}N_2O_4Pt$	371.25	(**29**)	recurrent ovarian carcinoma	bone-marrow suppression; emesis; allergic reactions
dacarbazine USP[i] (DTIC)	[*4342-03-4*]	$C_6H_{10}N_6O$	182.18	(**30**)	malignant melanoma; Hodgkin's disease; soft tissue sarcomas	bone-marrow depression; flulike syndrome; alopecia; nausea; vomiting; anorexia

[a]See Figure 3.
[b]Bristol-Myers Squibb.
[c]Pharmacia.
[d]Upjohn.
[e]Burroughs Wellcome.
[f]Merck Sharp & Dohme.
[g]Lederle.
[h]Kyowa-Hakko.
[i]Dome.

Table 3. DNA Interactive Agents[a]

Drug (trade name)	CAS Registry Number	Molecular formula	Molecular weight	Structure number	Disease	Toxic effects
daunorubicin hydrochloride USP[b] (Cerubidine)	*[20830-81-3]*	$C_{27}H_{29}NO_{10}{\cdot}HCl$	563.99	(**31**)	acute lymphocytic and granulocytic leukemia; lymphomas	bone-marrow depression; cardiac toxicity; alopecia; stomatitis; GI disturbance
doxorubicin USP[c] (Adriamycin)	*[23214-92-8]*	$C_{27}H_{29}NO_{11}$	543.53	(**32**)	soft-tissue and osteogenic sarcomas; Hodgkin's disease; non-Hodgkin's lymphomas; acute leukemia; cancer of thyroid, breast, lung, genitourinary (GU) tract; Wilm's tumor; neuroblastoma	bone-marrow depression; cardiac toxicity; alopecia; stomatitis; GI disturbance
idarubicin hydrochloride[c] (Idamycin)	*[57852-57-0]*	$C_{26}H_{27}NO_9{\cdot}HCl$	533.96	(**33**)	acute myeloid leukemia in adults	bone-marrow suppression; cardiotoxicity; nausea; vomiting; alopecia
mitoxanthrone hydrochloride USP[d] (Novantrone)	*[70476-82-3]*	$C_{22}H_{28}N_4O_6{\cdot}2HCl$	517.41	(**34**)	acute nonlymphocytic leukemia including myelogenous promyelocytic, monocytic and erythroid acute leukemias	nausea; vomiting; alopecia; mucositis; stomatitus; myelosuppression; cardiotoxicity; allergic reaction; phlebitis
bleomycin sulfate USP[e] (Blenoxane)	*[11056-06-7]*	mixture of bleomycin A_2, B_2 as primary components		(**35**)	squamous cell carcinoma of head, neck, esophagus, skin, GU tract; testicular tumor; Hodgkin's lymphomas	pulmonary fibrosis; skin reactions; alopecia; nausea; vomiting; anorexia; fever; stomatitis

esperamicin A_1[e]	[*99674-26-7*]	$C_{59}H_{80}N_4O_{22}S_4$	1324.41	(**36**)	investigational drug	
Adozelesin[f] (U73,975)	[*110314-48-2*]	$C_{30}H_{22}N_4O_4$	502.30	(**37**)	investigational drug	
dactinomycin USP[g] (Cosmegen)	[*50-76-0*]	$C_{62}H_{86}N_{12}O_{16}$	1255.43	(**38**)	Wilm's tumor; Ewing's tumor; choriocarcinoma; testicular carcinoma; rhabdomyosarcoma; neuroblastoma; melanoma; soft-tissue and osteogenic sarcomas	bone-marrow depression; renal and hepatic toxicity; alopecia; mental depression; stomatitis; nausea; vomiting; diarrhea; anorexia; local irritation
plicamycin USP[b] (Mithracin)	[*18378-89-7*]	$C_{52}H_{76}O_{24}$	1085.16	(**39**)	testicular tumors; hypercalcemia and hypercalciuria associated with advanced malignancies	bone-marrow depression; hepatic and renal toxicity; hypocalcemia; hemorrhage; stomatitis; nausea; vomiting; anorexia; diarrhea
procarbazine hydrochloride USP[i] (Matulane)	[*366-70-1*]	$C_{12}H_{19}N_3O{\cdot}HCl$	257.76	(**40**)	Hodgkin's disease; non-Hodgkin's lymphomas; lung cancer	bone-marrow depression; neurological and dermatological toxicity; nausea; vomiting

[a]See Figure 4.
[b]Wyeth-Ayerst.
[c]Adria.
[d]Lederle.
[e]Bristol-Myers Squibb.
[f]UpJohn.
[g]Merck Sharp & Dohme.
[h]Miles.
[i]Hoffmann-La Roche.

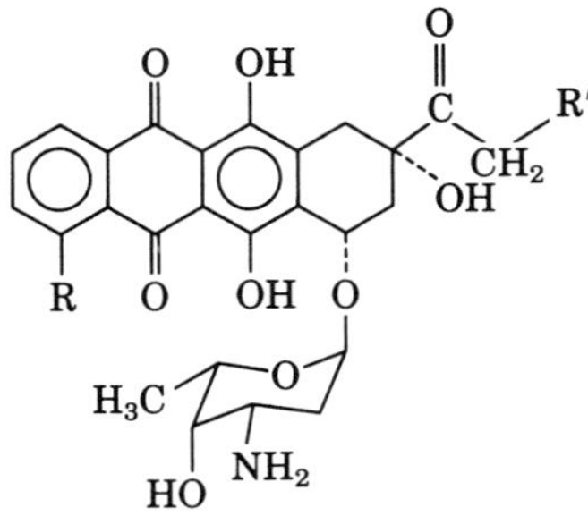

(31) R = OCH_3 R′ = H
(32) R = OCH_3 R′ = OH
(33) R = H R′ = H

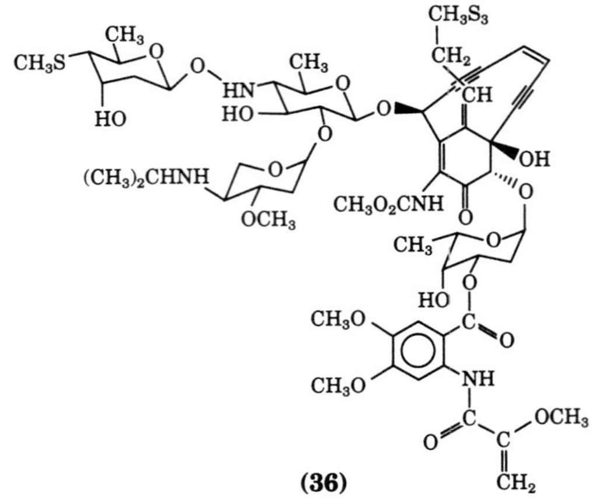

(36)

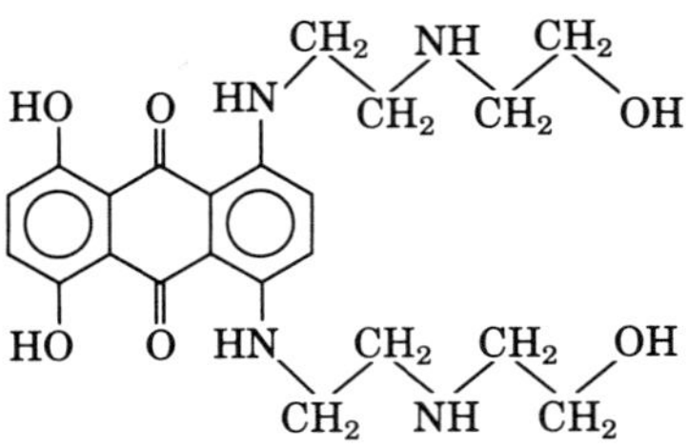

(34)

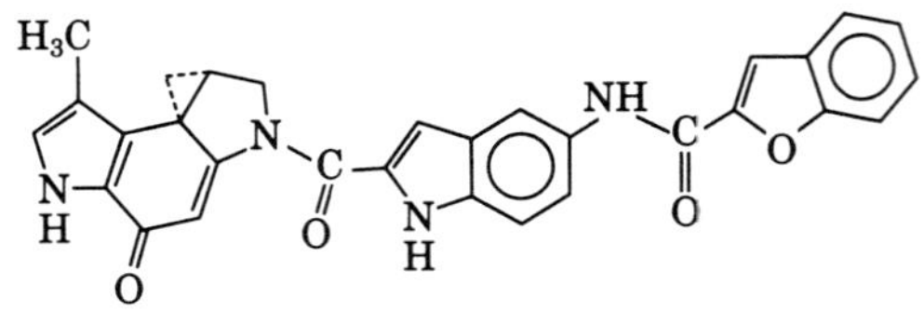

(37)

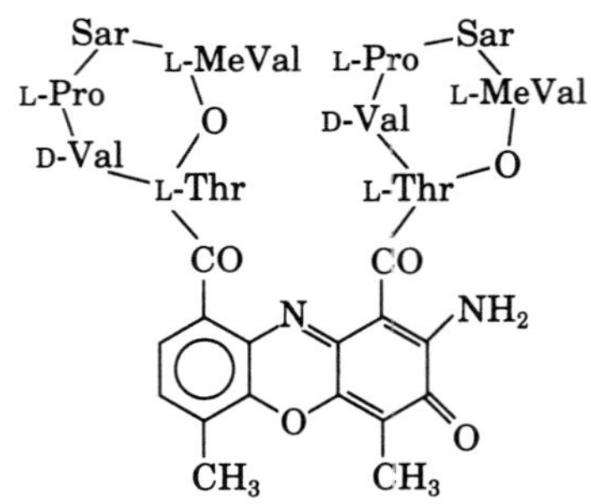

(38)

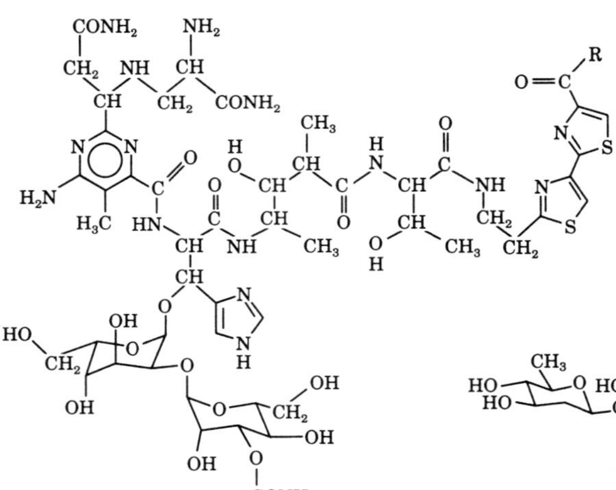

(35) R = $NH(CH_2)_3\overset{+}{S}(CH_3)_2$, Bleomycin A_2
R = $NH(CH_2)_4NHC(NH)NH_2$, Bleomycin B_2

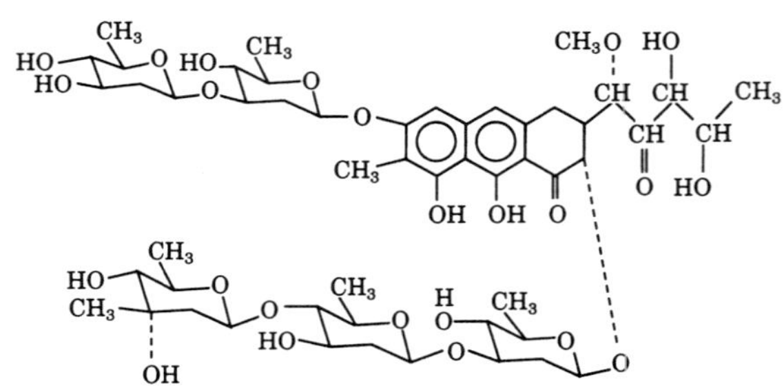

(39)

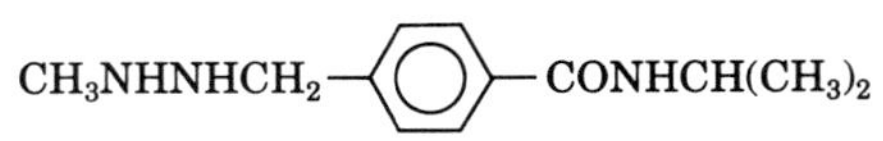

(40)

Fig. 4. Structures of DNA interactive agents are given in Table 3. In structure (**38**) L-MeVal is 1-*N*-methyl valine.

Topoisomerase Interactive Drugs. Topoisomerases I and II have emerged as interesting targets for the design of new anticancer agents (15,21) (Table 4). Etoposide (**41**) and related epipodophyllotoxin analogues (**42,43**) produce DNA strand scission via the mediation of topoisomerase II. Additional work with other clinically active agents, such as adriamycin (**32**) and amsacrine (**47**), indicate that these compounds interact with topoisomerase also. A number of drugs are known to work on topoisomerase II only. However, camptothecin [*7689-03-4*] and its analogues, CPT 11 (**44**) and topotecan (**45**), have been shown to target topoisomerase I (22). Encouraging solid tumor activity of these agents in Phase I trials has sparked an intense effort to identify further examples of topoisomerase I inhibitors (Fig. 5).

Tubulin Active Drugs. Vinblastin (**48**) and vincristin (**49**), two very useful natural products derived from Vinca rosea (periwinkle plant), were discovered in the early days of cancer chemotherapy (Table 5), (Fig. 6). (23). Vinblastin and vincristin are highly cell cycle dependent. They disrupt the mitotic spindle by promoting the disassembly of the microtubules essential for cell division. This results in cell death during replication. Vinblastin and vincristin differ minimally from one another structurally, but have different clinical utilities. A search for additional members of the Vinca alkaloid family has resulted in two new investigational drugs, vindesine (**50**) and navelbine (**51**) (24) (see ALKALOIDS), currently undergoing clinical trial. Initial results using navelbine indicate that this compound may find use in nonsmall cell lung cancer (25).

Taxol (**52**), a compound isolated from the pacific yew tree *Taxus brevifolia* (26), has shown high response rates in a variety of solid tumors including refractory ovarian cancer, refractory breast cancer, head and neck cancer, melanoma, and lung cancer. The mechanism of action for taxol is also linked with tubulin. However, unlike the Vinca alkaloids, taxol accelerates tubule formation and stabilizes it to depolymerization (27). The clinical development of taxol is dependent on identifying a renewable source. Once the FDA approves this drug for therapeutic use, large (kilogram) quantities are expected to be needed for the treatment of large patient populations. Approaches being investigated to address the problem of renewable source include cultivation, tissue culture, semisynthesis, and total synthesis (28). Resolution of supply problems also are needed to permit exploration of the full potential of this drug.

In addition to the Vinca alkaloids and taxol, a limited number of other cytotoxic agents have been demonstrated to act at the level of tubulin. This list includes colchicine [*64-86-8*] and the various podophyllotoxins and further underscores the importance of tubulin and microtubule formation as targets for anticancer drug action.

Hormones. Although not strictly cytotoxic, hormones (qv) have been used to control the environment of hormone dependent tumors such as those of the prostate, breast, and endometrium (29), ie, androgens are used to control the growth of estrogen dependent breast tumors, whereas estrogens control androgen dependent tumors of the prostate. Hormones that have anticancer activities are listed in Table 6. Structures are shown in Figure 7. In addition to the steroidal hormones (**57, 59–62**) several nonsteroidal hormones have been introduced (**54–56**) as have leutenizing hormone-releasing hormone (LHRH) agonists. These are often used in combinations. For example, LHRH analogues (**63,64**), antiestro-

Table 4. Topoisomerase Interactive Drugs[a]

Drug (trade name)	CAS Registry Number	Molecular formula	Molecular weight	Structure number	Disease	Toxic effects
etoposide USP[b] (Vepesid)	*[33419-42-0]*	$C_{29}H_{32}O_{13}$	588.56	(**41**)	refractory testicular tumors; small cell lung cancer	myelosuppression; mild to moderate nausea and vomiting; transient hypotension; allergic reactions; alopecia
etoposide phosphate[b]	*[117091-64-2]*	$C_{29}H_{33}O_{16}P$	712.51	(**42**)	investigational drug; prodrug of etoposide	prodrug of etoposide
teniposide[b]	*[29767-20-2]*	$C_{32}H_{32}O_{13}S$	656.67	(**43**)	refractory acute lymphocytic leukemia in children	myelosuppression; mild to moderate nausea and vomiting; transient hypotension; allergic reactions; alopecia
CPT-11[c]	*[100286-90-6]*	$C_{33}H_{38}N_4O_6{\cdot}HCl$	622.78	(**44**)	investigational drug; topoisomerase I inhibitor	
topotecan hydrochloride[d]	*[119413-54-6]*	$C_{23}H_{23}N_3O_5{\cdot}HCl$	457.91	(**45**)	investigational drug; topoisomerase I inhibitor	
elsamitrucin tartrate[b]	*[123303-98-7]*	$C_{33}H_{35}NO_{13}{\cdot}C_4H_6O_4$	771.37	(**46**)	investigational drug	
amsacrine[e] (Amsidyl)	*[51264-14-3]*	$C_{21}H_{19}N_3O_3S$	393.46	(**47**)	investigational drug	

[a]See Figure 5.
[b]Bristol-Myers Squibb.
[c]Yakult Honsha.
[d]Smith Kline Beecham.
[e]Parke-Davis.

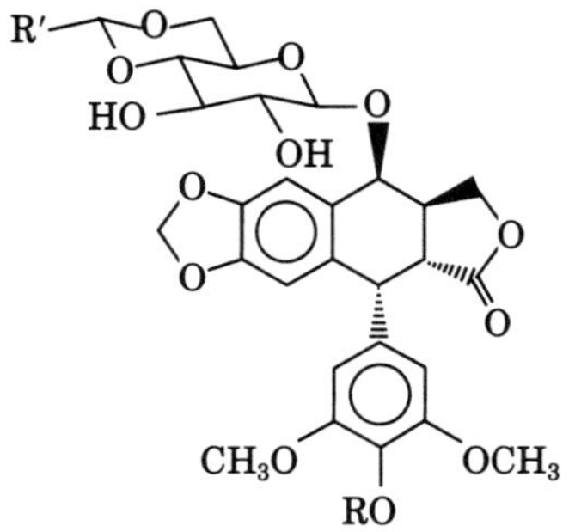

(41) R = H R′ = CH_3

(42) R = PO_3H_2 R′ = CH_3

(43) R = H R′ = 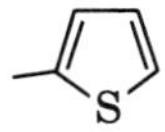

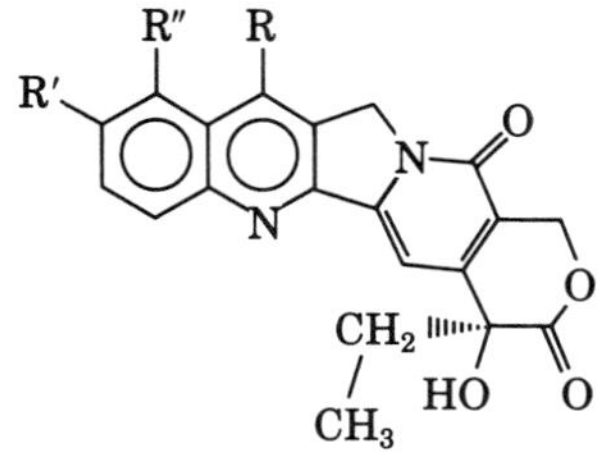

Camptothecin R = R′ = R″ = H

(44) R = C_2H_5 R′ = 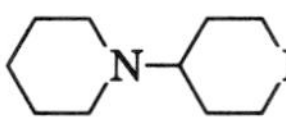 N—COO R″ = H

(45) R = H R′ = OH R″ = $CH_2N(CH_3)_2$

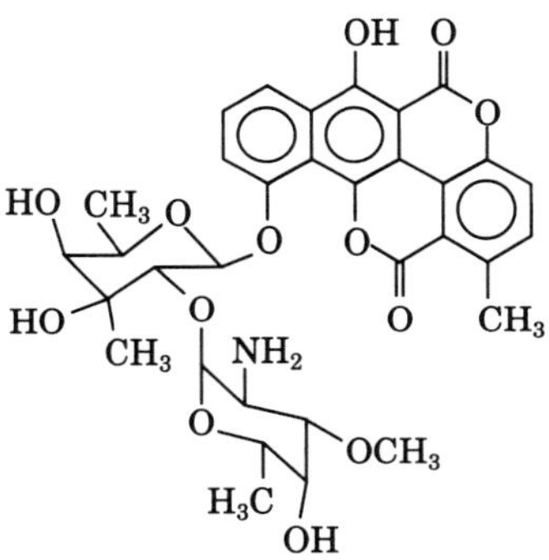

(46)

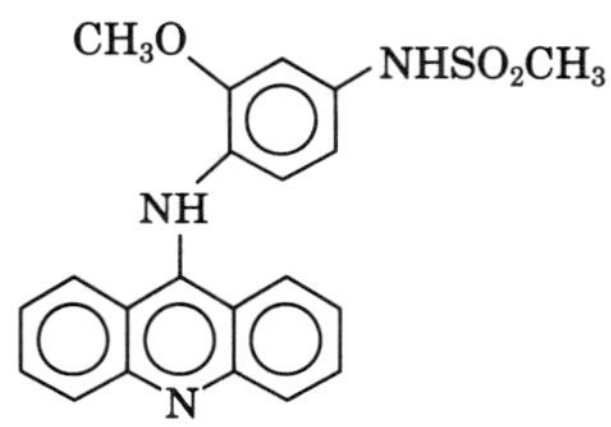

(47)

Fig. 5. Structures of topoisomerase interactive drugs given in Table 4.

gens (**54**), aromatase inhibitors (**62**), and progestins all are used in the control of breast cancer (30). LHRH analogues suppress estrogen production in the ovaries and are used in premenopausal breast cancer patients.

Aromatase inhibitors, which suppress estrogen production outside the ovaries, are useful for the treatment of tumors in postmenopausal women. Tamoxifen (**54**) is very well tolerated and is considered the drug of choice in post menopausal patients. The progestational agents medroxyprogesterone (**59**) and megestrol acetate (**60**) are used in the treatment of endometrial tumors to block overstimulation of the ovaries by estrogen. Megestrol acetate has been reported to have utility in reversing cachexia associated with cancer (31).

Miscellaneous Agents. Those chemotherapeutic agents, which do not fit into any of the classifications discussed, are listed in Table 7. Mitotane (**67**), a structural isomer of DDT, is used to induce chemical adrenalectomies in patients having adrenal cancer by reducing host levels of adrenocorticosteroids.

Table 5. Tubulin Active Drugs[a]

Drug (trade name)	CAS Registry Number	Molecular formula	Molecular weight	Structure number	Disease	Toxic effects
vinblastin sulfate USP[b] (Velban)	*[143-67-9]*	$C_{46}H_{58}N_4O_9 \cdot H_2SO_4$	909.06	(**48**)	Hodgkin's disease; lymphosarcoma; reticulum-cell sarcoma; neuroblastoma; choriocarcinoma; carcinoma of breast, lung, oral cavity, testis, bladder; acute and chronic leukemia; histiocytosis; mycosis fungoides	leukopenia; neurological toxicity (paresthesias, mental depression, loss of deep tendon reflexes, etc); dysfunction of autonomic nervous system (ileus, constipation, urinary retention, etc); alopecia; stomatitis; nausea; vomiting; local irritation
vincristin sulfate USP[b] (Oncovin)	*[2068-78-2]*	$C_{46}H_{56}N_4O_{10} \cdot H_2SO_4$	923.04	(**49**)	acute leukemia in children; lymphocytic leukemia; Hodgkin's disease; non-Hodgkin's lymphomas; Wilm's tumor; neuroblastoma; rhabdomyosarcoma.	neurological toxicity (paresthesias, foot drop, double vision, etc); constipation; ileus; alopecia; leukopenia (occasional);
vindesine sulfate[b] (Eldisine)	*[59917-39-4]*	$C_{43}H_{55}N_5O_7 \cdot H_2SO_4$	852.01	(**50**)	investigational drug	

navelbine[c] (Vinorelbine)	*[71486-22-1]*	$C_{45}H_{54}N_4O_8$	778.45	(**51**)	investigational drug; nonsmall cell lung cancer	
taxol[d]	*[33069-62-4]*	$C_{47}H_{51}NO_{14}$	853.92	(**52**)	investigational drug; refractory ovarian cancer; refractory breast cancer; melanoma; lung cancer; head and neck cancer	alopecia; neutropenia; hypersensitivity; mucositis; neuropathy
taxotere[e] (Docetaxol)	*[114977-28-5]*	$C_{43}H_{53}NO_{14}$	807.43	(**53**)	investigational drug	

[a] See Figure 6.
[b] Lilly.
[c] Pierre Fabre.
[d] Bristol-Myers Squibb.
[e] Rhône-Poulenc.

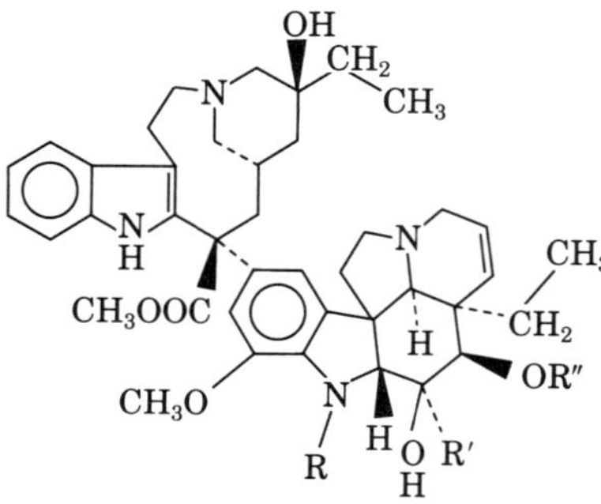

(48) R = CH_3 R′ = $COOCH_3$ R″ = CH_3CO^-
(49) R = CHO R′ = $COOCH_3$ R″ = CH_3CO^-
(50) R = CH_3 R′ = $CONH_2$ R″ = H

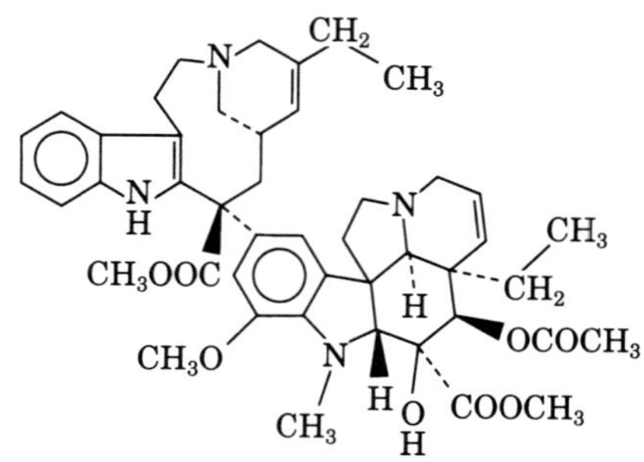

(51)

(52) R = C_6H_5 R′ = CH_3CO
(53) R = $OC(CH_3)_3$ R′ = H

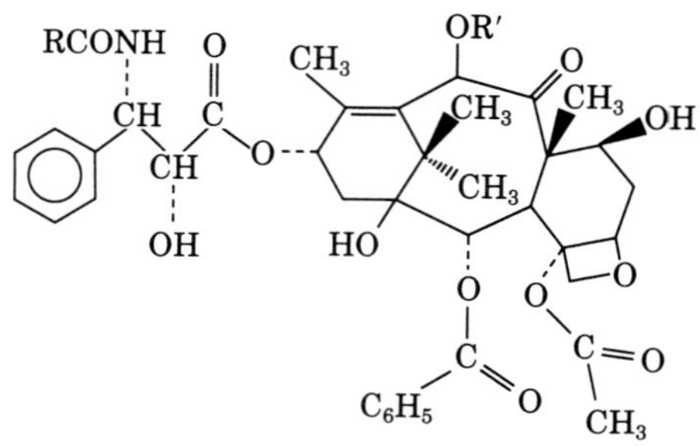

Fig. 6. Structures of tubulin active drugs listed in Table 5.

The retinoid isotretinoin (**68**) has been found to reduce the incidence of secondary malignancies in patients treated for head and neck cancer. In addition, the use of *trans*-retinoic acid in patients having M3 leukemia has been reported to induce complete, although temporary, remissions (32).

The novel agent sulofenur (**69**) has entered clinical trials based on its broad spectrum antitumor activity in tumor models, its unusual mechanism of action, and its lack of cross-resistance to other agents (33). In Phase I clinical trials, the drug was well tolerated and some clinical responses were noted.

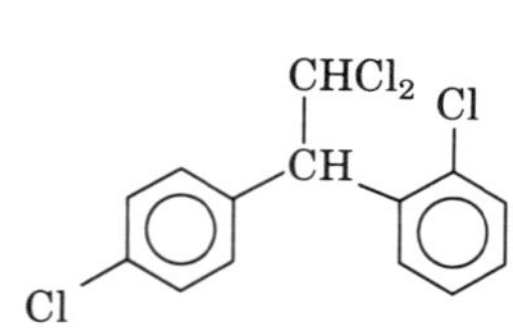

(67)

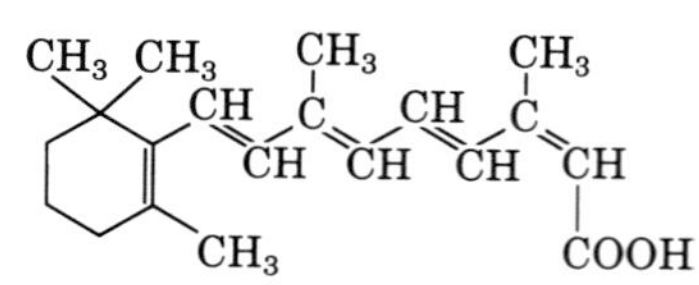

(68)

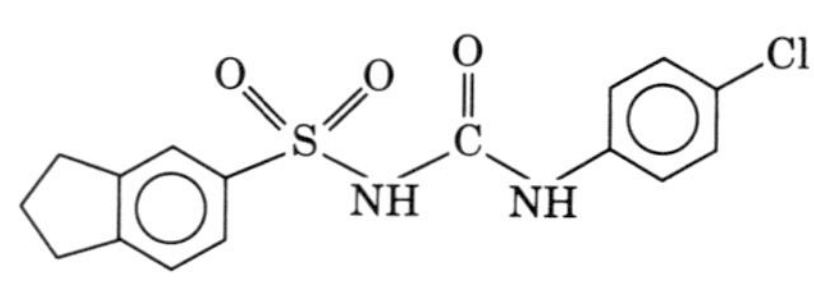

(69)

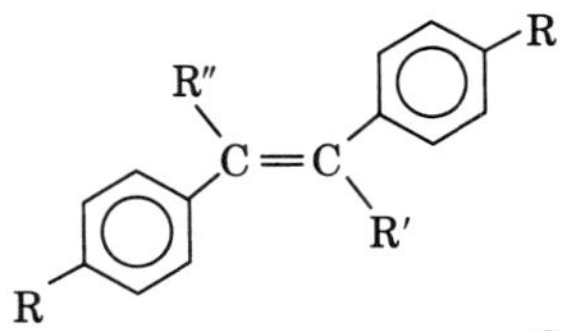

(54) R = H R′ = C_2H_5 R″ = 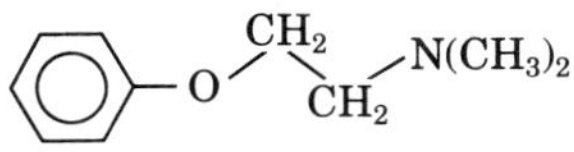

(55) R = O—P(=O)(OH)(OH) R′ = R″ = C_2H_5

(56) R = OCH_3 R′ = Cl R″ = 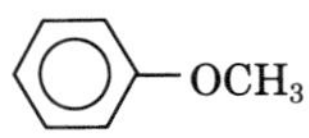

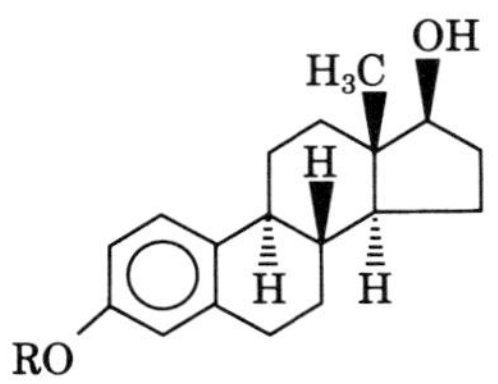

(57) R = H
(58) R = $CON(CH_2CH_2Cl)_2$

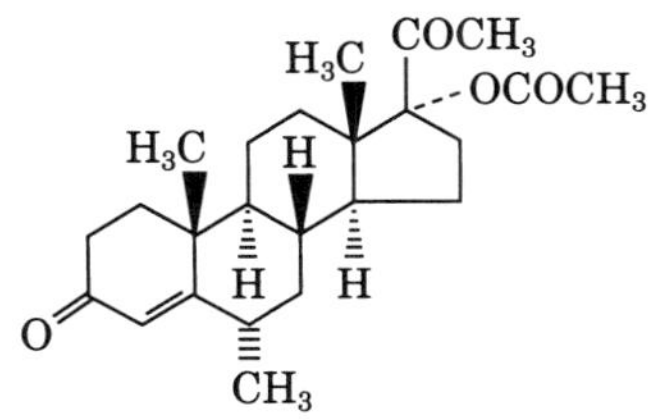

(59)

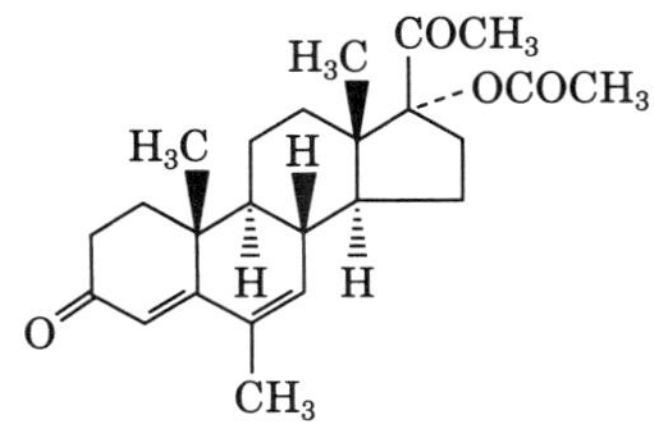

(60)

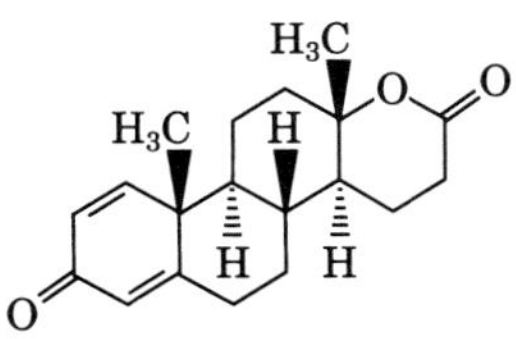

(61)

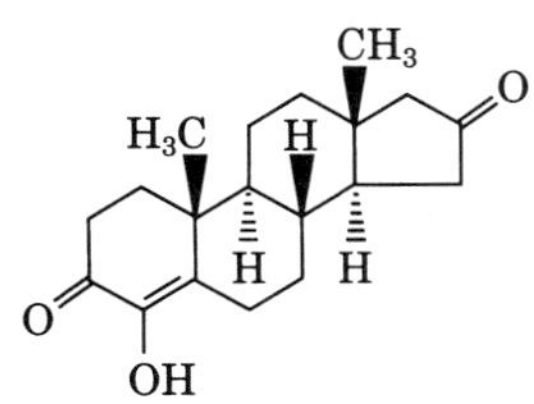

(62)

(63) H-5-OxoPro-His-Trp-Ser-Tyr-D-Ser(*t*-Bu)-Leu-Arg-Pro-$NHNHCONH_2$
(64) H-5-OxoPro-His-Trp-Ser-Tyr-D-Leu-Leu-Arg-ProNHCH_2CH_3•CH_3COOH
(65) H-D-Phe-Cys-Phe-D-Trp-Lys-Thr-Cys-Thr-OH (Cys–Cys disulfide bridge: S———S)

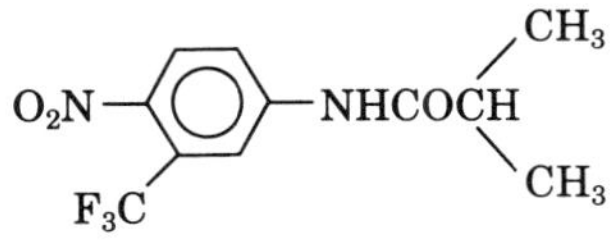

(66)

Fig. 7. Structures of hormone chemotherapeutic agents given in Table 6.

Table 6. Hormonal Therapy[a]

Drug (trade name)	CAS Registry Number	Molecular formula	Molecular weight	Structure number	Disease	Toxic effects
tamoxifen citrate USP[b] (Nolvadex)	*[54965-24-1]*	$C_{26}H_{29}NO \cdot C_6H_8O_7$	563.65	(**54**)	breast cancer	visual disturbances
diethylstilbestrol diphosphate USP[c] (Stilphostrol)	*[522-40-7]*	$C_{18}H_{22}O_8P_2$	428.31	(**55**)	prostatic carcinoma	fluid retention; hypercalcemia; common side effects of steroids
chlorotrianisene USP[c] (TACE)	*[569-57-3]*	$C_{23}H_{21}ClO_3$	380.87	(**56**)	androgen dependent carcinoma of the prostate	fluid retention; hypercalcemia; common side effects of steroids
estradiol USP[d] (Estrace)	*[50-28-2]*	$C_{18}H_{24}O_2$	272.39	(**57**)	breast cancer; prostatic carcinoma	fluid retention; hypercalcemia; common side effects of steroids
estramustine phosphate sodium USP[e] (Emcyt)	*[52205-73-9]*	$C_{23}H_{30}Cl_2N \cdot Na_2O_6P$	564.35	(**58**)	prostatic carcinoma	side effects because of estradiol; increased dyspnea; nausea; vomiting
medroxyprogesterone acetate[f] (USP Depo-provera)	*[71-58-9]*	$C_{24}H_{34}O_4$	386.53	(**59**)	metastatic endometrial carcinoma; renal carcinoma	fluid retention; hypercalcemia; common side effects of steroids
megestrol acetate USP[d] (Megace)	*[595-33-5]*	$C_{24}H_{32}O_4$	384.51	(**60**)	carcinoma of the breast or endrometrium	fluid retention; hypercalcemia; common side effects of steroids
testolactone USP[d] (Teslac)	*[968-93-4]*	$C_{19}H_{24}O_3$	300.40	(**61**)	breast cancer	fluid retention; hypercalcemia; common side effects of steroids

formestane[g] (Lentaron)	*[566-48-3]*	$C_{19}H_{26}O_3$	302.41	(**62**)	investigational drug; postmenopausal breast cancer	
goserelin USP[b] (Zoladex)	*[65807-02-5]*	$C_{59}H_{84}N_{18}O_{14}$	1269.43	(**63**)	prostatic carcinoma	bone pain; common hormonal side effects
leuprolide acetate USP[h] (Leupron)	*[74381-53-6]*	$C_{59}H_{84}N_{16}O_{12}\cdot C_2H_4O_2$	1269.47	(**64**)	prostatic carcinoma	common hormonal side effects
octreotide acetate USP[i] (Sandostatin)	*[79517-01-4]*	$C_{49}H_{66}N_{10}O_{10}S_2\cdot xC_2H_4O_2$	1019.24	(**65**)	mestastatic carcinoid tumors; vasoactive intestinal peptide-secretory tumors	nausea; diarrhea; loose stools; vomiting; abdominal pain; pain on injection
flutamide[j] (Eulexin)	*[13311-84-7]*	$C_{11}H_{11}F_3N_2O_3$	276.21	(**66**)	metastatic prostatic carcinoma in combination with LHRH agonist	diarrhea

[a]See Figure 7.
[b]ICI.
[c]Marion-Merrill Dow.
[d]Bristol-Myers Squibb.
[e]Pharmacia.
[f]Upjohn.
[g]CIBA-GEIGY.
[h]TAP.
[i]Sandoz.
[j]Schering.

Table 7. Chemotherapeutic Agents

Drug (trade name)	CAS Registry Number	Molecular formula	Molecular weight	Structure number	Disease	Toxic effects
mitotane USP[a] (Lysodren)	[*53-19-0*]	$C_{14}H_{10}Cl_4$	320.05	(**67**)	palliative treatment of inoperable adrenal cortical carcinoma	skin toxicity; vertigo; lethargy; somnolence; anorexia; nausea; vomiting; diarrhea
isotretinoin USP[b] (Accutane)	[*4759-48-2*]	$C_{20}H_{28}O_2$	300.44	(**68**)	investigational drug	
sulofenur[c]	[*110311-27-8*]	$C_{16}H_{15}ClN_2O_3S$	350.82	(**69**)	investigational drug; refractory ovarian carcinoma response seen in Phase I	anemia; methemoglobinemia
asparaginase[d] (Elspar)	[*9015-68-3*]				acute lymphocytic leukemia	hepatic, renal, and pancreatic toxicity; neurological effects; hypersensitivity reactions; clotting abnormalities; nausea

[a]Bristol-Myers Squibb.
[b]Hoffmann-La Roche.
[c]Lilly.
[d]Merck Sharp & Dohme.

The enzyme L-asparaginase is the only biological agent included in this review. Most normal cells are capable of synthesizing their own L-asparagine; leukemic cells are deficient in this regard. Use of the enzyme causes depletion of the exogenous L-asparagine necessary to the growth of the leukemia. Some remissions have been seen using this drug, however, they are usually transitory in nature because of the rapid emergence of resistance.

Toxicity

As a result of the life threatening nature of cancer and the general lack of therapeutically effective drugs for most cancers, doses of chemotherapeutic drugs in Phase I clinical trials are escalated until the emergence of a dose-limiting toxicity. The efficacy of these compounds in one or more tumor types is then established in Phase II/III clinical trials. The commonly observed dose-limiting toxicities include myelosuppression, gastrointestinal upset, and renal, hepatic, and cardiotoxicities.

The most common dose-limiting toxicity in cytotoxic chemotherapy is acute or cumulative suppression of bone marrow, ie, myelosuppression. To circumvent this, intermittent treatment schedules using suboptimal drug doses are often employed. Another approach involves either autologus, ie, derived from the same individual, or allogeneic, ie, derived from a different individual, bone marrow transplant accompanied by high dose therapy (34). Although therapeutic advantages have been achieved, patients are highly vulnerable to serious infections during the recovery period following reinfusion of bone marrow. When administered after infusion of the bone marrow, recombinant colony stimulating factors (CSFs) (14), granulocyte-macrophage-CSF (GM-CSF), and macrophage-CSF (M-CSF) stimulate the growth of white blood cells, particularly granulocytes, macrophages, and monocytes, hence shortening the recovery period so patients suffer fewer infections.

Cumulative organ toxicity also presents a significant obstacle for effective chemotherapy. In many cases, the severity of the toxicity impedes the broader use of an agent. Other specific toxicities are associated with specific agents, for example cardiotoxicity with adriamycin (**32**), renal toxicity with *cis*-platinum (**28**), and neurotoxicity with vincristine (**49**).

Toxicity Amelioration. Cancer researchers traditionally have not focused their attention on the question of toxicity amelioration. This is partly attributed to the lack of predictive animal models for human toxicities. For example, the preclinical rat model, used as a predictor of myelosuppression, has failed to predict myelosuppression in humans in clinical trials. In addition, reduction of one toxicity may result in the emergence of another, more serious problem. Research efforts to address the problem of toxicity amelioration has progressed in several directions. The three most prominent areas are analogue synthesis, chemoprotection, and drug targeting.

Analogue Synthesis. Two notable examples, in which analogues have greater therapeutic indexes than the parent drugs, have been identified in Phase I trials. These are carboplatin (**29**) and adozelesin (**37**) (35). Carboplatin's approval was based on its comparable efficacy to cis-platinum (**28**) and its more favorable toxicity profile, ie, reduced and delayed episodes of emesis, reduced ototoxicity, etc. On the other hand, adozelesin, a totally synthetic analogue of natural product CC1065, has demonstrated a similar potency and antitumor activity profile as its natural prototype but is devoid of the delayed death liability associated with the parent drug in animals (36).

Chemoprotection. The success of chemoprotection depends on a detailed biochemical knowledge of the drug target. The pharmacological and metabolic fate of the parent drug, its relationship to the observed organ toxicity, and how the observed toxicity can be prevented/lessened by chemical intervention without affecting the antitumor activity of the parent drug are all important factors. Administering an antidote to the parent drug to minimize the specific drug toxicity to organs and the hematopoetic, ie, blood forming, system, is being studied in various cancer therapies. For example, leucovorin (**9**) rescue therapy has been used successfully in the clinic as an antidote to the toxic effects of high dose methotrexate (**8**). The antidote renders methotrexate ineffective against the newly recruited stem cells after its initial cytotoxic insult. Other combinations under active clinical study to demonstrate effectiveness of chemoprotection are ifosphamide (**19**) and mesna (**20**) for hemorrhagic cystitis (37); cis-platinum (**28**) and WR-2721 [*20537-88-6*] for nephrotoxicity and neurotoxicity (38); and adriamycin (**32**) and ICRF-187 [*24584-09-6*] for cardiotoxicity (39).

Drug Targeting. Site-specific drug delivery involving a variety of approaches has been pursued to address the issue of drug toxicity in chemotherapy (40). Prodrug synthesis and liposome encapsulation of cytotoxic drugs, eg, adriamycin (**32**), have preceded more modern drug targeting approaches. The advent of hybridoma technology to produce large quantities of monoclonal antibodies directed toward antigenic epitopes on human cancer cells has made possible several novel approaches to selective delivery of cytotoxic agents to human tumors. The three most actively pursued research areas in this vein are the use of monoclonal antibody-drug conjugates (41), a combination approach involving the use of a targeted antibody–enzyme complex followed by administration of a suitable prodrug of an anticancer agent (42); and antibody-targeted radionuclides with α and β-particle emitting properties (43). *In vivo* results for antibody targeted cytotoxics against human tumor xenografts in nude mice have been very encouraging and several institutions are planning clinical trials to establish the proof of principle for such approaches. The antibody targeted radionuclides approach, which has been widely used in cancer diagnosis, is also showing promise for therapeutic utility (44).

Drug Resistance

Despite dramatic advances in the treatment of several human malignancies including Hodgkin's lymphomas and leukemias, drug resistance remains a pressing issue in cancer chemotherapy. Acquired or induced drug resistance afflicts practically all classes of cancer agents. It usually is manifested clinically subsequent to responsive therapy and cancer relapse following therapy often is fatal. The most recognized and studied mechanisms of drug resistance are attributed to multidrug resistance (MDR), gene amplification, DNA repair, topoisomerase II activity, and glutathione and metallothionein levels. Even with advances in understanding the biology and mechanism of drug resistance among different classes of antitumor agents, no real breakthrough appears imminent (45).

Research and clinical experience on drug resistance suggests that tumor cells are particularly adept at genetic selections leading to alterations in the structure, function, or synthesis of proteins involved in the antitumor drug action and detoxification. Multiple mechanisms of resistance have been shown to account for the resistance seen in the clinic (46).

Multidrug Resistance. MDR is characterized by reduced drug accumulation in tumor cells correlated with an enhanced drug efflux (47). This effect is common to a wide variety of cancer agents including the Vinca alkaloids, intercalators, and topoisomerase inhibitors. The drug efflux is attributed to the over expression of the MDR-1 gene, encoding a 170 kD glycoprotein (GP170) responsible for the energy (adenosine triphosphate) dependent nonspecific efflux of the drug from the MDR cells. In recent years, reversal of MDR by several classes of noncytotoxic agents, eg, verapamil [*52-53-9*], dipyridamole [*58-32-2*], progesterone [*57-33-0*], and dihydropyridine analogues, has been observed and studied at the preclinical level (48). Limited clinical trials are in progress to establish the validity of this approach. The exact mechanism of MDR reversal is not well understood but evidence suggests that these reversal agents either disrupt the efflux pump function or competitively inhibit drug binding to GP170 (49). The extent that MDR is responsible for the observed clinical resistance is being carefully monitored and evaluated.

Gene Amplification. This mechanism of drug resistance appears to be operative in drugs which act on *in vivo* enzymatic targets. For example, in the case of resistance to methotrexate (**8**), overexpression of the dihydrofolate reductase (DHFR) gene has been implicated. However, other mechanisms involving defects in drug uptake and polyglutamation also have been proposed and studied (50). Specifically, the multifactorial nature of methotrexate resistance represents a large therapeutic challenge. It is one example of how the heterogeneity of clinical resistance may be a serious problem.

DNA Repair. Resistance to alkylating agents, which constitute the bulk of chemotherapeutic agents, is ascribed to DNA repair and/or detoxification processes (51). Resistance usually manifests itself as an overproduction of repair enzymes. Overproduction of *O*-6-alkylguanosine transferase, an enzyme responsible for repairing lesions in DNA, is observed in resistance to nitrosoureas, mitomycin C (**25**), and melphlan (**23**). In the case of resistance to cis-platinum (**28**), tumors have demonstrated elevated levels of glutathione and the enzyme glutathione *S*-transferase. Glutathione is involved in both DNA repair and drug detoxification processes. Reversal of cis-platinum resistance has been observed in cells pretreated with glutathione depleting agents such as buthionine sulfoximine [*5072-26-4*].

Investigational Approaches

Chemotherapy has been impacted significantly by a small number of selected antitumor agents. In the 1970s it was adriamycin (**32**) and in the 1980s it was cis-platinum (**28**) and etoposide (**41**). Based on its unique mechanism of action and its clinical performance to date, taxol (**52**) may be the most important chemotherapeutic agent of the 1990s (26). However, the quest for novel chemotherapeutics still continues, guided by both rational mechanism-based screening and more relevant animal tumor models for preclinical evaluations. Approaches under investigation include the search for novel noncytotoxic agents which induce differentiation of cancer cells (52), inhibit tumor metastasis (53), and control growth factors (54). The dramatic advances made in recombinant DNA and hybridoma technology has caused a resurgence in the science of immunotherapy (see IMMUNOTHERAPEUTIC AGENTS) (55); the use of cytokines such as the interferons, interleukin-2 [*85898-30-2*] (IL-2), and tumor necrosis factor [*138415-31-3*] (TNF) as activators of the human immune system is under active study. The use of immunodulators, such as levamisole [*14769-73-4*], which stimulate and fortify the host's immunoreactivity toward tumors, also are being evaluated, but an unequivocal clinical demonstration of the effects of these biological response modifiers has not been made. In addition to use as selective drug delivery carriers, monoclonal antibodies are being employed as cytotoxic agents by the virtue of their natural effector mechanism (56). The clinical validity of all the above approaches are under careful scrutiny.

Resources are being marshalled to exploit and explore the possibility of cancer intervention at either the transcriptional or translational levels of gene activity. Specifically, as of 1992, antisense (RNA target) and antigene (DNA target) approaches (57,58) are being pursued at a number of institutions and several investigative therapies are pending FDA approval for initiation of human clinical trials.

Economics of Cancer Chemotherapy

The increased use of chemotherapy as a modality in the treatment of cancer has caused a corresponding increase in the market for anticancer agents. Table 8 lists the estimated 1990 worldwide sales for the most commonly utilized chemotherapeutic agents. Data for Japan are not included because of differences in medical and prescribing practices. Only a few anticancer drugs have achieved sales in excess of $100 million per year and many of the drugs discussed herein sell less than $10 million per year. Many of the approved drugs were introduced early on in cancer chemotherapy for the treatment of lymphomas and leukemias. Because of lack of patent coverage and the relatively small market size, these drugs are of relatively little commercial importance. Those agents which have achieved greater economic importance are newer, frequently used in combination chemotherapy, and of use in the treatment of solid tumors which comprise the bulk of reported cancer incidence. A number of the agents in clinical trials, such as taxol (**52**) and camptothecin analogues, are expected to have considerable economic impact based on activity in the treatment of the more common human malignancies, eg, lung, breast, colon, and ovary.

In addition to the market for cancer chemotherapeutic drugs, there is also a growing market for biologicals such as the interferons, having combined 1990

Table 8. Antitumor Drug Sales, 1990[a]

Chemotherapeutic drug	$\$ \times 10^6$
Antimetabolites	
doxifluridine	112
fluorouracil	18
methotrexate	110
hydroxyurea	26
cytarabine	30
other	12
Total	*308*
Alkylating agents	
cis-platinum	192
carboplatin	191
mitomycin C	74
cyclophosphamide	54
ifosphamide	15
chlorambucil	13
thiotepa	13
melphalan	10
Total	*562*
DNA interactive drugs	
doxorubicin	150
epirubicin	110
mitoxanthrone	75
bleomycin	60
daunorubicin	18
Total	*413*
Topoisomerase inhibitors	
etoposide	260
Tubulin active drugs	
vincristine	30
vindesine	25
Total	*55*
Hormones[b]	
tamoxifen	540
leuprolide	170
flutamide	150
goserelin	112
medroxyprogesterone	92
estramustine	76
megestrol acetate	70
Total	*1210*

[a]Estimated worldwide sales excluding Japan. Based on audited sales adjusted for under reporting. Sales of less than 10 million per annum not included.

[b]Most hormonal agents are used for other indications. It is not possible to estimate usage for antitumor purposes.

sales of ~\$560 million, and the colony stimulating factors, G-CSF, GM-CSF, etc. It is reasonable to expect an overall doubling of this market in the 1990s.

BIBLIOGRAPHY

"Cancer Chemotherapy" in *ECT* 2nd ed., Suppl. Vol., pp. 81–90, by C. J. Masur, Lederle Laboratories; "Chemotherapeutics, Antimitotic" in *ECT* 3rd ed., Vol. 5, pp. 469–489, by C. J. Masur and W. Pearl, Lederle Laboratories.

1. S. K. Carter, M. T. Bakowski, and K. Hellman, *Chemotherapy of Cancer*, John Wiley & Sons, Inc., New York, 1987.
2. J. A. Double and M. C. Bibby, *Drug Res. Dev.* **3**(5), 276 (1990).
3. T. H. Corbett, F. A. Valeriote, and L. H. Baker, *Invest. New Drugs* **5,** 20 (1987).
4. G. B. Grindey, *Cancer Cells* **2,** 163 (1990).
5. R. K. Johnson and R. P. Hertzberg, in *Annual Reports in Medicinal Chemistry*, Vol. 25, Academic Press, Inc., New York, 1990, p. 129.
6. R. H. Shoemaker and co-workers, in *Prediction of Response to Cancer Therapy*, Alan B. Liss, New York, 1988, pp. 265–286.
7. A. E. Bogden and co-workers, *Cancer*, **48,** 10 (1981).
8. B. C. Giovanella and co-workers, *Cancer*, **52,** 1146 (1983).
9. J. A. Double and M. C. Bibby, *J. Natl. Cancer Inst.* **81,** 988 (1989).
10. C. R. Gardner and J. Alexander, in *Drug Targeting*, Elsevier Science Publishers B.V. (Biomedical Division), New York, 1985, p. 145.
11. J. E. Schurig, A. P. Florczyk, and W. T. Bradner, *Cancer Chemother. Pharmacol.* **16,** 243 (1986).
12. J. E. Schurig and W. T. Bradner, in *Fundamentals of Cancer Chemotherapy*, McGraw-Hill Book Co., New York, 1987, p. 248–261.
13. K. T. Douglas, *Chem. Ind.*, 693 (1984).
14. D. Metcalf, *Science* **254,** 529 (1991).
15. J. D. Fisher and P. A. Aristoff, in *Progress in Drug Research*, Vol. 32, Birkhauser Verlag, Basel, Switzerland, 1988, 411–484.
16. K. T. Douglas, *Chem. Ind.*, 738 (1984).
17. K. T. Douglas, *Chem. Ind.*, 766 (1984).
18. M. M. Gottesman and I. Pastan, *Trends Pharmacol. Sci.* **9,** 54 (1988).
19. B. H. Long and co-workers, *Proc. Natl. Acad. Sci. USA* **86,** 2 (1989).
20. W. Wierenga "DNA-Minor Groove Binding Anticancer Agents" in *Cytotoxic Anticancer Drugs: Models and Concepts for Drug Discovery and Development*, Kluwer Academic Publishers, Boston, Mass., 1992, p. 105.
21. C. Jaxel and co-workers, *J. Biol. Chem.* **266,** 20418 (1991).
22. B. A. Chabner, *J. Clin. Oncology* **10,** 3 (1992).
23. K. Gerzon, in *Anticancer Agents Based on Natural Product Models*, Academic Press, New York, 1980, p. 271.
24. A. Krikorian and F. Breillout, *Onkologie* **14,** 7 (1991).
25. A. Depierre and co-workers, *Am. J. Clin. Oncol.* **14,** 115 (1991).
26. E. K. Rowinsky and R. C. Donehower, *J. Natl. Cancer Inst.* **83,** 1778 (1991).
27. P. B. Schiff and S. B. Horwitz, *Proc. Natl. Acad. Sci. USA* **77,** 1561 (1980).
28. D. G. Kingston, *Pharmac. Ther.* **52,** 1 (1991).
29. A. Borkowski, J. J. Body, and G. Leclercq, *Eur. J. Cancer Clin. Oncol.* **24,** 509 (1988).
30. L. P. Schacter and co-workers, *Semin. Oncol.* **17** (6 Suppl. 9), 38 (1990).
31. N. S. Tchekmedyian and co-workers, *Cancer*, **69,** 1268 (1992).
32. L. Degos, *Intl. J. Cell Cloning* **10,** 63 (1992).
33. J. J. Howbert and co-workers, *J. Med. Chem.* **33,** 2393 (1990).

34. R. J. O'Reilly, *N. Engl. J. Med.* **315,** 186 (1986).
35. L. L. Robert and co-workers, *Invest. New Drugs* **9,** 137 (1991).
36. J. P. McGovern and co-workers, *J. Antibiot.* **37,** 63 (1984).
37. J. Brock and J. Pohl, *Cancer Treat. Rev.* **10,** 33 (1983).
38. D. Glover and co-workers, *Int. J. Radiation Oncology Biol. Phys.* **16,** 1201 (1989).
39. J. L. Speyer and co-workers, *N. Engl. J. Med.* **319,** (1988).
40. D. R. Friend and S. Pangburn, *Med. Res. Rev.* **7,** 53 (1987).
41. G. A. Koppel, *Bioconjugate Chem.* **1,** 13 (1990).
42. P. D. Senter and co-workers, *Proc. Natl. Acad. Sci. USA* **85,** 4842 (1988).
43. D. Parker, *Chem. Soc. Rev.* **19,** 271 (1990).
44. S. T. Rosen and co-workers, *J. Clin. Oncol.* **5,** 562 (1987).
45. P. V. Woolley III and K. D. Tew, eds., *Mechanism of Drug Resistance in Neoplastic Cells*, Academic Press, Inc., New York, 1988.
46. A. L. Harris, *Dev. Oncol.* **58,** 107 (1990).
47. N. Kartner and V. Ling, *Sci. Am.* 44–51 (Mar. 1989).
48. A. Kiue and co-workers, *Cancer Res.* **50,** 3453 (1990).
49. I. Nogae and co-workers, *Biochem. Pharmacol.* **38,** 519 (1989).
50. C. B. Knight, P. C. Elwood, and B. A. Chabner, *Adv. Enzyme Regul.* **29,** 3 (1989).
51. H. Masuda and co-workers, *Cancer Res.* **48,** 5713 (1988).
52. R. Lotan and co-workers, *Cancer Res.* **50,** 3453 (1990).
53. Y. Iwamoto and co-workers, *Science* **238,** 1132 (1987).
54. C. C. Bascom and co-workers, *J. Cell Biochem.* **39,** 25 (1989).
55. K. A. Foon, *Cancer Res.* **49,** 1621 (1989).
56. I. Hellstrom, P. L. Beaumier and K. E. Hellstrom, *Proc. Natl. Acad. Sci. USA* **83,** 7059 (1986).
57. A. R. Vander Krol and co-workers, *BioTechniques* **6,** 958 (1988).
58. M. D. Matteucci and N. Bischofberger, *Annu. Rep. Med. Chem.* **26,** 287 (1991).

TERRENCE W. DOYLE
DOLATRAI M. VYAS
Bristol-Myers Squibb Company

INDEX

A

B

B

C

D

E

F

G

H

I

J

K

L

M

N

Q

R

S

T

U

V

W

X

Y

Z